Thermo-Haemo-Dynamics in Medical Engineering

Fabio Gori Ammannati

Editor

Thermo-Haemo-Dynamics in Medical Engineering

 Springer

Editor
Fabio Gori Ammannati
University of Rome Tor Vergata
Rome, Italy

ISBN 978-3-031-97213-3   ISBN 978-3-031-97214-0   (eBook)
https://doi.org/10.1007/978-3-031-97214-0

*Questo libro è dedicato
a Emilio Gori e Cesarina Ammannati,
miei genitori,
per la loro lezione di vita,
insegnamento umano e morale.
This book is dedicated
to Emilio Gori and Cesarina Ammannati,
my parents,
for their Life Lesson,
human and moral teaching.
Liber hic dedicatus est
Emilio Gori et Cesarina Ammannati,
parentibus meis,
pro eorum vita, doctrina,
humano et morali doctrina.
Mi Fabio clan cleva cen ziχ
apa Emilio Gori ati Cesarina Ammannati.*

Foreword by Franco Maceri

I am pleased to accept the invitation from my colleague and friend Prof. Fabio Gori Ammannati to introduce this book that collects his contributions and those of some of his Medical Engineering students, trained over a long period of time, at the University of Rome "Tor Vergata".

The Medical Engineering Degree began in the Academic Year 1998/1999, on my proposal, as Dean *pro tempore* of the Faculty of Engineering, shared by the Dean of the Faculty of Medicine, Prof. Alessandro Finazzi Agrò. We were united by the deep conviction that it was necessary to promote the development of new applications and technologies related to the impressive developments of life sciences in recent decades, to strengthen the cooperation between scholars of physiology and medicine, and specialists in the application of physical and chemical principles to the solution of practical problems, through mathematics.

It was therefore necessary and appropriate, through the tool of University education, to create a new preferential area of modern Engineering, capable of combining the intellectual instrumentation of knowledge, biological and medical, with effective analysis skills and concrete project capabilities, figures, therefore, that could deal with both frontier problems of biological and medical research and with problems related to patient care, tackling them with the wealth of methodologies and knowledge, particularly analytical-quantitative, that Engineering has to apply science for practical purposes.

The structure of the training project for the new professional figure, therefore, envisaged the acquisition by the students of the scientific laws that govern the behaviour of organised matter, both living and inert, as well as the intellectual and practical tools of mathematics; to all this was added the training in applications, based on both analysis and project synthesis profiles, with mathematical rigour.

This approach, followed for over 20 years, even after the introduction of the reform of university studies in Engineering, which introduced into the engineer's training process the two levels of Degree, has allowed, at Tor Vergata, the fully trained Medical Engineer, to have all the potential necessary to face a long professional life, such as:

- Versatility, which allows him to operate in different fields.
- Concreteness, which allows him to put his own creativity to good use through effective projects and realisations.
- Adaptability, which allows to evolve over time and which derives from profound basic training.

This volume gives concrete evidence of this: it indeed shows how the principles and application approaches of one of the founding disciplines of Engineering, Technical Physics, can open up to medical applications, with important results both in terms of understanding and of medically significant outcomes.

From this, one can draw—and I do so with the reader—the observation, both of the correctness of the training structure, and of the effectiveness of the teaching, which has given the young co-authors, and, certainly, other students who will follow, a full ability to realise their potential.

University of Roma Tor Vergata   Franco Maceri
Rome, Italy
†19 December 2021

Foreword by Gaetano Marrocco

Technical Physics is a transitional discipline, which constitutes the link between the understanding of natural phenomena, through mathematical laws and basic Physics, and the sciences of Engineering, which, starting from such physical laws, develop methods and models to dominate matter and create objects and devices with controlled functions. Indeed, the Technical Physics course is located in the first semester of the first year of the Master's Degree in Medical Engineering, precisely to mark the continuity between the foundational culture of the Bachelor's Degree with the engineering one that will be developed in the second two-year period. With Technical Physics, one enters the domain of Engineering, where, given the physical principle as a given, the objective shifts to the search for the optimal solution to problems, on the sizing of devices compatible with the boundary conditions, to the functional requirements and, finally, on their computerised control. But, before this, it is necessary to learn how natural phenomena can be dominated, with the strength of mathematics and numerical calculation, bearing in mind real applications, even very complex ones.

Professor Fabio Gori Ammannati has taught Technical Physics in our Medical Engineering Degree Course since its foundation, now over 27 years ago, and has made the significant effort to adapt and extend the traditional culture of this subject, well established in the disciplines of Thermodynamics, Thermo-Fluid-Dynamics, and Heat Transfer, to themes specific to the human body, where phenomena become more elusive and seem apparently to escape the constraint of equations and numbers.

This volume encapsulates the teaching experience and, above all, research of Prof. Gori Ammannati over many years of activity at our University. The chosen format is original because it consists in retracing the advancement of knowledge through the people who have been its creators, together with Prof. Gori Ammannati, and that are his students. Each chapter is indeed originated from the works of thesis of various degrees and doctorate (PhD) and is completed by a brief biography of the authors. It is therefore also a nice way to tell the stories of many of our graduates, who today are established professionals in industry, in research centres, who have followed unconventional paths, or who have, with passion, shared a piece of the scientific journey of Prof. Gori Ammannati.

The chapters reveal the complexity of Medical and Biomedical themes but also the power of modelling and simulation tools, which have allowed to replicate the functioning of complex biophysical processes, and of medical devices, to understand their operation and identify the limit cases of applicability.

The next step is that of the *Digital Twins*, that is the virtual representations of biological processes, through the equations and methods presented in the book, which can allow to simulate the dynamic effects of a medical device on the human body before to build it, to optimise its shape, size, and materials to best fulfil a specific function, to even customise the device for the specific patient, like a quality tailor with his esteemed client. This path leads to *Precision Medicine*, and that is to a patient-centric vision of Medicine, where the modelling of biological systems, and their interaction with matter and energy, will play a dominant role in the care of people in the coming years.

This book can therefore be considered a useful complement to the didactic narration, offering a vertical and in-depth view on some topics (cryosurgery, hyperthermia, thermoregulation, thermo-fluid-dynamics of stents, magneto-fluid-dynamics, noise, cardiac fluid dynamics acoustics) that will allow the reader, whether a student, a PhD student, or a young researcher, to appreciate the strength and potential of the proposed methods in application to real problems, on which to build the medical technologies of the future.

The structure of the book finally says another important thing: the visual presence of the co-authors of the research conveys a message of positivity. The physical-mathematical complexity should not scare: it has been mastered by young women and men in the final phase of their university training, well guided by an enlightened teacher. These methods are therefore within the reach of all students who follow our Degree Course in Medical Engineering. With effort and sacrifice, at the end of the journey, they will be well trained with the necessary skills to face challenging and exciting challenges towards the continuous improvement of human health.

Professor of Electromagnetic Fields　　　　　　　　　　　Gaetano Marrocco
Coordinator of the Degree Course in Medical Engineering
University of Rome Tor Vergata
Rome, Italy

Foreword by Salvatore Mangiafico

The book by Prof. Fabio Gori Ammannati, titled in the Italian version "Problematiche di Fisica Tecnica in Ingegneria Medica" and in the English version "Thermo-Haemo-Dynamics in Medical Engineering", which is a collection of works of his students during many years of teaching and research, is, in the eyes of a medical doctor, which I am, an interesting book.

The reading of the text, although it is necessarily superficial for me, due to lack of adequate cultural means, still allows me to advance some considerations that I feel like sharing with the future readers of the book: young university students in search of a direction in the studies of Medical Engineering.

In the work of Prof. Gori Ammannati, in addition to some chapters of high medical relevance, such as those on Haemo-Dynamics and on Magneto-Hydro-Dynamics, I am particularly interested in seeing the method with which biomedical problems are tackled (I recognised in the students of Prof. Gori Ammannati his imprint of the Master's setting).

It is interesting to understand how a clinical aspect is first analysed in biological mechanisms, and so far we are on the same road, to then be transformed, through physical analysis, into a mathematical model, on which to advance speculative models increasingly closer to the variability of biological phenomena, to finally arrive, with engineering method, at the development of a prototype, or, as happens in the most recent studies on Haemo-Dynamics, at the development of a predictive model of the pathology, and here the paths of the doctor and the medical-bio scientist meet again, just as they meet when, with the studies of Magneto-Hydro-Dynamics within the cardiovascular system, the possibility of increasing the concentration of drugs in specific tumour areas is hypothesised.

The meeting between medical-bio engineers and doctors is truly unique: they meet when they together scrutinise the future, researchers with cultures and languages deeply different from each other (strangers to each other one might say) who understand each other when they go beyond the horizon of the method, the scientific specifications and go further for the realisation of a dream.

It's the tension of the common project that leads to dialogue and the awareness that everyone has of themselves and the necessary presence and proximity of the other. Strange meeting theirs!

The engineer will increasingly be called to work together with doctors and doctors to work together with engineers. This is the work that awaits us!

Neurovascular Interventionist   Salvatore Mangiafico
Specialist in Neurology, Radiodiagnostics
and Neurosurgery
Florence, Italy

Contents

Introduction

Fabio Gori Ammannati

The introduction to the book takes its cue from my vision of *Fantastic Voyage*, a 1966 film directed by Richard Fleischer, whose plot is as follows [1].

"The United States and the Soviet Union have both developed technology that can miniaturize matter by shrinking individual atoms, but only for one hour. A scientist, Dr. Jan Benes, working behind the Iron Curtain, has figured out how to make the process work indefinitely. With the help of American intelligence agents, including agent Charles Grant, he escapes to the West and arrives in New York City, but an attempted assassination leaves him comatose with a blood clot in his brain that no surgery can remove from the outside. To save his life, Grant, Navy pilot Captain Bill Owens, medical chief and circulatory specialist Dr. Michaels, surgeon Dr. Peter Duval, and his assistant Cora Peterson are placed aboard a Navy ichthyology submarine at the Combined Miniature Deterrent Forces facilities. The submarine, named *Proteus*, is then miniaturized to 'about the size of a microbe', and injected into Benes' body. The team has 60 minutes to get to and remove the clot; after this, *Proteus* and its crew will begin reverting to their normal size, become vulnerable to Benes's immune system, and kill Benes. The crew faces many obstacles during the mission. An undetected arteriovenous fistula forces them to detour through the heart, where cardiac arrest must be induced to, at best, reduce turbulence that would be strong enough to destroy *Proteus*. As the crew faces an unexplained loss of oxygen and must replenish their supply in the lungs, Grant finds the surgical laser needed to destroy the clot was damaged from the turbulence in the heart, as it was not fastened down as it had been before: this and his safety line snapping loose while the crew was refilling their air supply lead Grant to suspect a saboteur is on the mission. The crew must cannibalise their wireless radio to repair the laser, cutting off all communication and guidance from the outside, although, because the submarine is nuclear-powered, surgeons and technicians outside Benes's body are still able to track their

F. Gori Ammannati (✉)
University of Rome Tor Vergata, Rome, Italy
e-mail: gori@uniroma2.it

F. Gori Ammannati (ed.), *Thermo-Haemo-Dynamics in Medical Engineering*,
https://doi.org/10.1007/978-3-031-97214-0_1

movements via a radioactive tracer, allowing General Alan Carter and Colonel Donald Reid, the officers in charge of CMDF, to figure out the crew's strategies as they make their way through the body. The crew is then forced to pass through the inner ear, requiring all outside personnel to make no noise to prevent destructive shocks, but while the crew is removing reticular fibres clogging the submarine's vents and making the engines overheat, a fallen surgical tool causes the crew to be thrown about and Peterson is nearly killed by antibodies, but they are able to reboard the submarine in time. By the time they finally reach the clot, the crew has only six minutes remaining to operate and then exit the body. Before the mission, Grant had been briefed that Duval was the prime suspect as a potential surgical assassin, but as the mission progresses, he instead begins to suspect Michaels. During the surgery, Dr. Michaels knocks out Owens and takes control of *Proteus* while the rest of the crew is outside for the operation. As Duval finishes removing the clot with the laser, Michaels tries to crash the submarine into the same area of Benes' brain to kill him. Grant fires the laser at the ship, causing it to veer away and crash, and Michaels to get trapped in the wreckage with the controls pinning him to the seat, which attracts the attention of white blood cells. While Grant saves Owens from the *Proteus*, Michaels is killed when a white blood cell consumes the ship. The remaining crew quickly swim to one of Benes' eyes and escape through a tear duct seconds before returning to normal size."

In that year, 1966, at 19 years old, I began the first two years of the Engineering degree, at the Faculty of Mathematical, Physical and Natural Sciences of the University of Florence, where I had grown up and lived. Despite the official start date of the lessons, which at the time was 5 November, being marred by the tragic flood of Florence on 4 November 1966 (fortunately, the only one of the 1900s), which required us young people to go through *the "edifying" experience of shovelling mud from the streets and cellars, carrying demijohns of water to the upper floors of apartments inhabited by elderly people and helping in every way everybody who were in difficulty,* at the beginning of 1967, after having cleaned several cellars full of mud, I managed to start attending the lectures.

Everything went well in the first two years of the Engineering Degree, and afterwards, I had to choose which Engineering direction to enrol in for the following three years of the Engineering Degree. At that time, Florence had not yet established the three-year Engineering Degree, and Florentine students had to go to other university locations, generally to Pisa or Bologna, which are the closest. Unfortunately, there were no orientation courses on the contents of the various Engineering Degrees, and my information was rather scarce. Finally, after evaluating various hypotheses, I chose to enrol in Chemical Engineering at the Faculty of Engineering in Bologna. The main motivation was that, in those years, it seemed a degree with many employment opportunities, given the high level reached by Chemical Engineering in Italy at that time.

This book also arises from this problem, which I felt then and which I believe is still present today among young people who have to choose the Engineering Degree, and the intention was to report in this book my experience of 50 years as a Technical Physics researcher who has dealt with issues related to Medical Engineering. The book aims to show to young people, intending to enrol in Engineering, but who have not yet chosen among its various courses, which problems can face and solve a student who enrols in Medical Engineering at the University of Rome "Tor Vergata", and who wants to pursue studies and research in Technical Physics. The book also aims to appeal to PhD students with a didactic approach, providing full listings of programmes written in some programming languages and a fairly detailed description of the procedures to use some of the numerical software employed.

As far as the terminology used in the book is concerned, it must be noted that the title of the book contains the term *Thermo-Haemo-Dynamics* (also *THD* in the rest of the book), which must be distinguished from the more common Fluid Dynamics for the following reasons. The common Fluid Dynamics approach is relative to the solution of the four scalar equations, known as Navier–Stokes equations, which include the mass conservation without chemical reactions (one scalar equation) and the momentum conservations (one vectorial equation or three scalar equations) of a fluid, which is generally considered to behave according to Newton's law and is usually called Newtonian.

The term *Thermo* of the title *Thermo-Haemo-Dynamics (THD) is* distinguishable from Fluid Dynamics because, besides the four equations of Navier–Stokes, it intends to solve one of the following equations: the scalar equation of conservation of energy (derived from the First Principle of Thermodynamics), which usually considers fluids following Fourier's law; the scalar equation of entropy conservation (derived from the Second Principle of Thermodynamics) and the scalar equation of mass diffusion of a chemical component, which usually follows Fick's law.

The term *Haemo* of the title *Thermo-Haemo-Dynamics (THD)* is distinguishable from Fluid Dynamics because it intends to solve the above-mentioned equations for a fluid such as blood, which does not follow Newton's law and is called non-Newtonian.

The book reports, in chronological order, seven chapters, two of which are divided into two parts and one into three. Each chapter reports studies and research that may have originated from homework from the course of "Thermo-Fluid-Dynamics of Biological Systems"; three-year degree theses; five-year degree theses; Specialist degree theses; master's degree theses; theses of second-level university master's in Thermo-fluid-dynamics (one year after the five-year, specialist or master's degree) and *PhD* theses. The author (Fabio Gori Ammannati) was the supervisor of all the papers or theses from which the studies reported in the book originated.

1   Chapter 2: Cryosurgery. Prediction of Freezing Front Penetration and of the Lesion Produced in the Tissue

Chapter "Cryosurgery. Prediction of Freezing Front Penetration and of the Lesion Produced in the Tissue" presents the study on *cryosurgery,* which started with the thesis of the second-level university master's, defended by the research doctor (*PhD*), Dr. Eng. Andrea Boghi in the academic year (A.Y.) 2007–2008, after the Specialist Degree in Medical Engineering and before the PhD.

Chapter "Cryosurgery. Prediction of Freezing Front Penetration and of the Lesion Produced in the Tissue", after describing some of the medical aspects, illustrates the monitoring techniques of the intervention and presents the main cryoprobes, the physics of cryosurgery, its mathematical setting with the relative analytical and numerical solutions, the prediction of the lesion produced in the tissue and the relative results. The numerical method derives from that used in the five-year degree thesis by Dr. Eng. Guido Veltri in Electronic Engineering (Biomedical specialisation), defended in Florence in 1976, on cryosurgery. The language used was *FORTRAN77*, which is still used today, even if in version *90* or subsequent, in the most modern approaches of Direct Numerical Simulation (*DNS*). The analytical solutions of the problems of *Stefan* and *Neumann* are still the subject of study in the course of Technical Physics for Medical Engineering of the author of the book. The tools with which the graphs were made in 1976 were curvilinear and ink writing, by hand but also with stencils, on opaque transparent sheets, overlaid on the graph paper on which it was drawn using pencils. The bibliographic tools were the university libraries of Florence and also those of Bologna and Pisa and the hospital library "Agostino Gemelli" in Rome.

The logo of chapter "Cryosurgery. Prediction of Freezing Front Penetration and of the Lesion Produced in the Tissue" can be Fig. 1, a mirrored image of Leonardo da Vinci's cold machine, as can be read on the right of the central figure, "Oncold" (*"Sulfredo"*). Based on this drawing by Leonardo da Vinci from 1492, the first cold machine, considered the ancestor of the refrigerator, was reconstructed in a project by Alessandro Vezzosi, founder and president of the Ideal Museum of Leonardo da Vinci. The work, carried out by Studio Emme in Florence, consists of a circular bellow with three air chambers made of "leather" (*"corame"*), 18 nozzles that focus the air on a central container and a small two-crank winch to be operated by hand. In this way, a jet of air is created, which is capable of maintaining or lowering the temperature of food or drinks in the container [2]. The autograph statement of the scientist explains the general principles underlying the origin of the cold due to heat deprivation or air movement. The explanation of the machine's operating principle is based on the analogy method or the similarity between the "supreme heat"(*"somo chaldo"*) produced by the "rockets"(*"razzi"*) of a concave mirror that focus on a point and the "supreme cold"(*"somo freed"*) produced by the blows of many bellows or the many "nozzles"(*"boccuccio"*) of a large bellow, focused towards a single point, a container to be refrigerated, as done by the author of the book and coworkers in jets [3–19].

Another logo of chapter "Cryosurgery. Prediction of Freezing Front Penetration and of the Lesion Produced in the Tissue" can also be Fig. 2 [20], which has the analytical solution of the *Neumann* problem and the numerical solution, drawn by the author with the aforementioned graphic tools. The choice of this figure as a logo is dictated by the procedure that was required at that time for the publication of the results of the numerical analysis, which had to, first of all, be independent of the size of the calculation grid used and then compared with the analytical solutions existing in the literature.

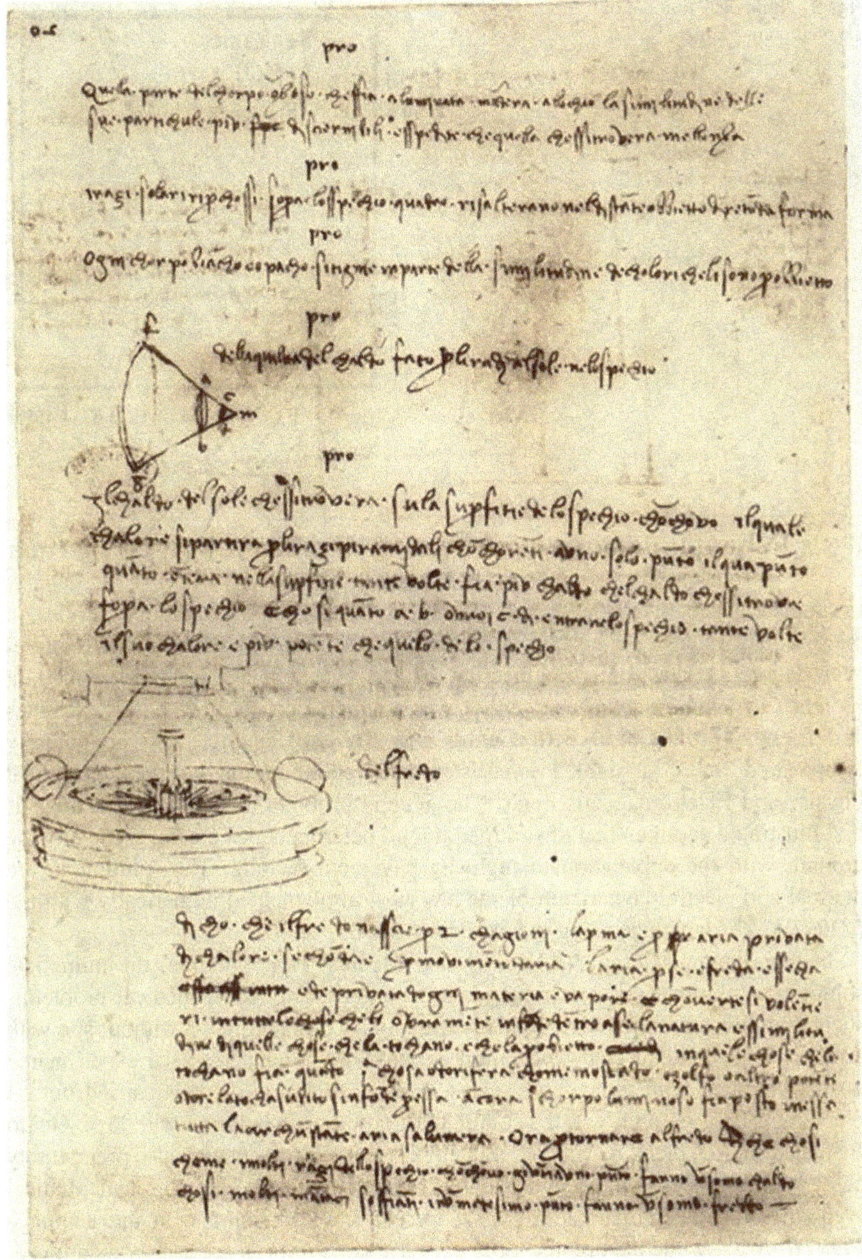

Fig. 1 Mirrored image of Leonardo da Vinci's cold machine

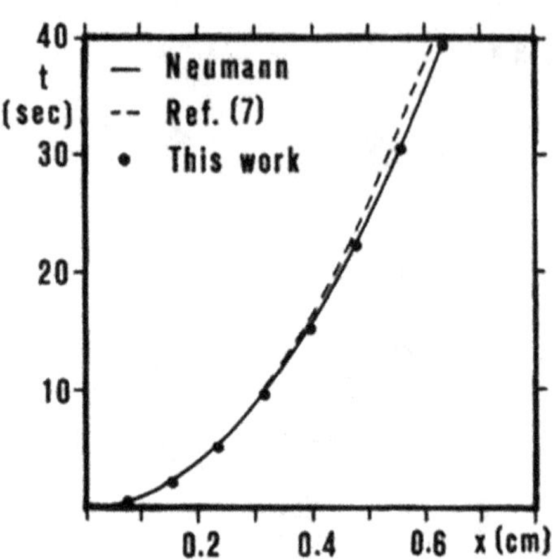

Fig. 2 Numerical solution of the Neumann problem

2   Chapter 3: Microwave Heating and Medical Hyperthermia

In chapter "Microwave Heating and Medical Hyperthermia", the author describes, in the first part, microwave heating, a research that started with the thesis of the five-year degree in Electronic Engineering of Dr. Eng. Luca Martini, defended in Florence in 1984, of which the author was a supervisor, together with colleagues and friends, Prof. Guido Biffi Gentili and Roberto Tiberio, who prematurely disappeared, in a far-sighted interdisciplinary collaboration between Technical Physics and Electromagnetic fields. The general problem of microwave heating is first illustrated and then that of stratified humid heterogeneous media, especially the ground, with the application to highway pavement heating. The solution to the thermal and electromagnetic problems has been implemented numerically with the *FORTRAN 77* language using the explicit method.

In the chapter Microwave Heating and Medical Hyperthermia, the numerical solutions of thermal and electromagnetic equations, applied to medical problems, such as applicators for medical hyperthermia, like a radiant dipole, without and with a reflector and without and with liquid cooling, are compared to the experimental results obtained with temperature measurements of liquid crystals carried out on phantoms. This research was the continuation of the collaboration with Prof. Guido Biffi Gentili, to which Dr. Marco Leoncini had also joined, but who also prematurely passed away. As for the second part of chapter Microwave Heating and Medical Hyperthermia, the numerical solutions were always obtained with the language *FORTRAN 77*, while the graphic tools were the same ones as previously mentioned. The bibliographic tools were the university libraries of Florence and also those of Bologna and Pisa.

The logo of chapter Microwave Heating and Medical Hyperthermia is Fig. 3, in which the temperature, the absorbed power and the degree of water saturation on a motorway floor, up to a depth of *25 cm*, hit by a microwave source are reported [21].

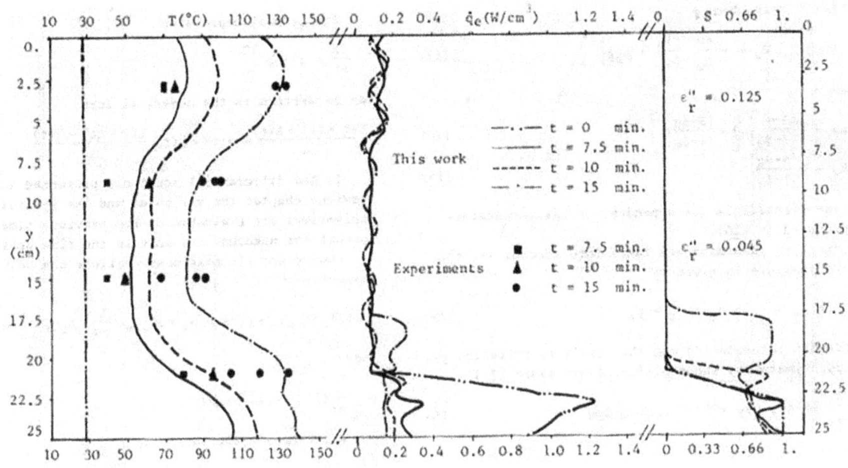

Fig. 3 Temperature, absorbed power and saturation degree on a motorway floor

The temperature has a wave-like trend (with maximums and minimums), which resembles those present in the Earth's atmosphere, as reported in Fig. 4 [22] (in Italian because more pictographic, [23]).

A qualitatively similar variation of the temperature in the atmosphere of Titan has been registered during the *Cassini–Huygens* mission [24], with the measurements carried on by the TEM thermometer of the Huygens Atmospheric Structure Instrument (HASI), designed and built (in laboratory version) in the Technical Physics and Energy Engineering Laboratory of "Tor Vergata" [25]. The temperature (K) profile of the atmosphere of Titan is reported on the abscissa of both figures in Fig. 5. In the top figure, the ordinates report the pressure (hPa) on the left and the altitude (km) on the right, up to 1400 km. In the bottom figure, the ordinates report the altitude (km) on the left only in the lower atmosphere, up to 150 km.

3 Chapter 4: Human Body Thermoregulation

Chapter "Human Body Thermoregulation" presents the study on the thermoregulation of the human body, originated from the thesis carried out by the research doctor (*PhD*), Dr. Eng. Ivan Di Venuta, in the homework course of "Thermo-Fluid-Dynamics of Biological Systems", exam of the Master's Degree Course in Medical Engineering given by the author.

The interest in the thermo-hygrometric comfort of the human body began during the lectures given by the author in the late 70 s within the course of Hygiene for medical doctors at the University of Florence and continued with the collaboration with the Santa Maria Annunziata Hospital in Florence for the numerical simulation of the problem of malignant hyperthermia during anaesthesia. The study of the

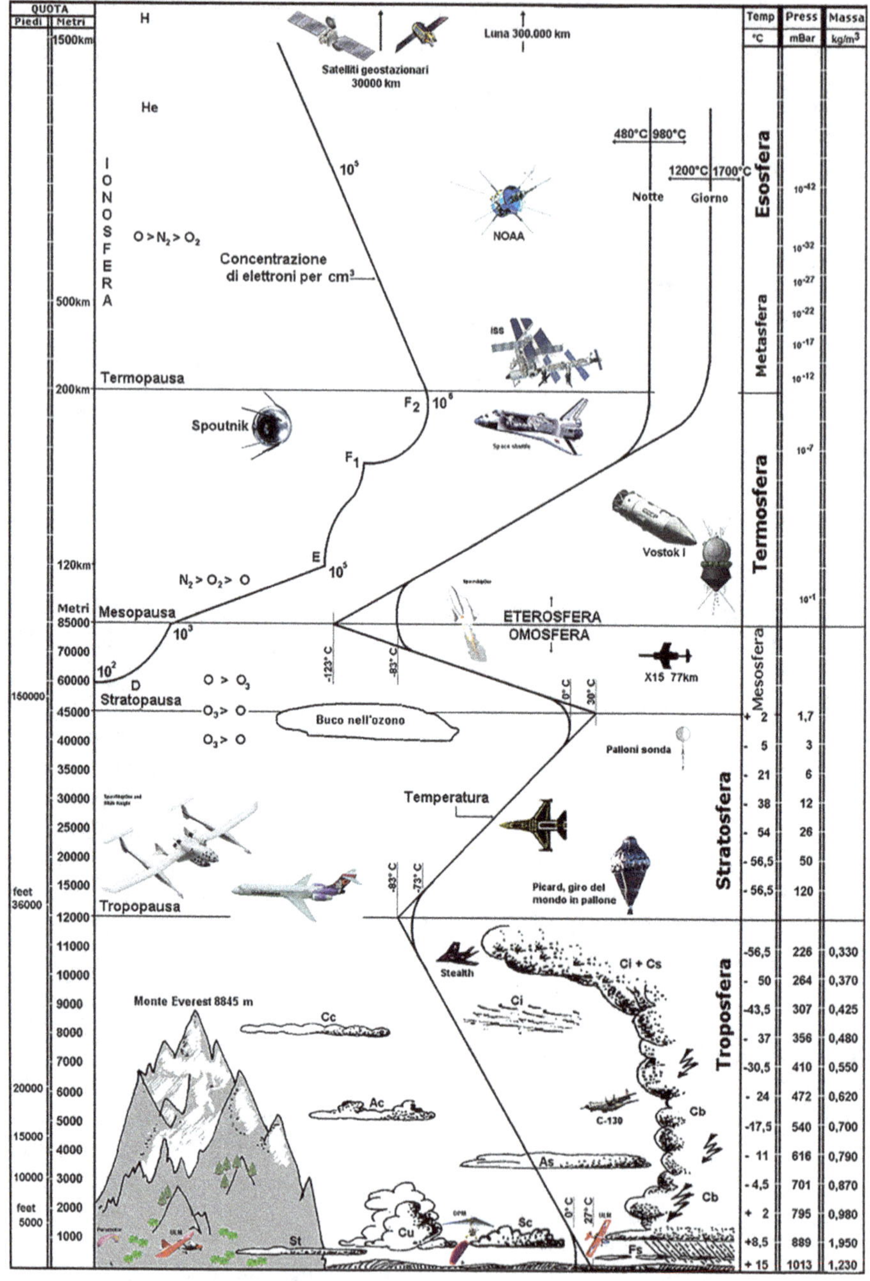

Fig. 4 Temperature profile in the Earth's atmosphere [22]

Fig. 5 Temperature (K) profile, on the abscissa, in the atmosphere of Titan. (figs. 2 and 5 of [24])

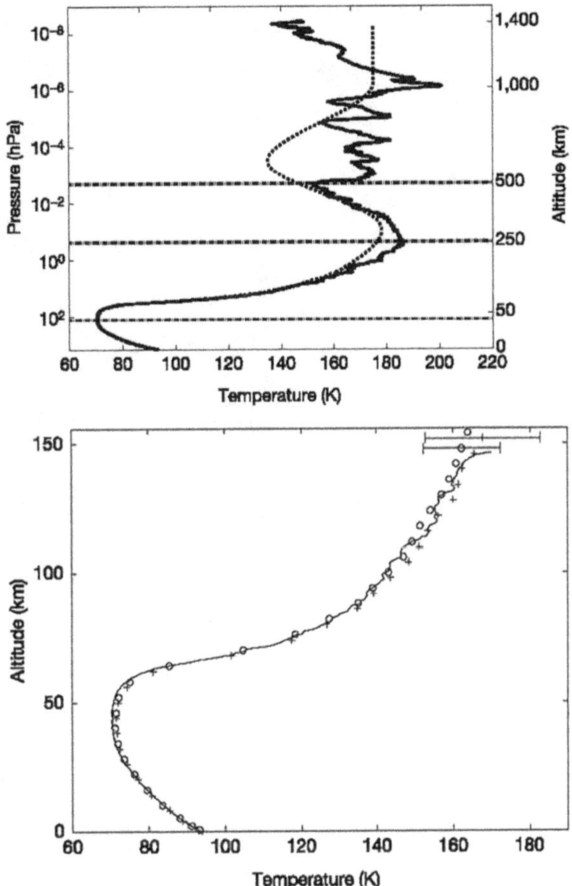

thermoregulation of the human body was then presented in the lectures of Technical Physics, carried out during the author's teaching at the Technical University (Polytechnic) of Milan, regarding the thermo-hygrometric comfort of the human body in the built environment [26, 27]. The numerical solutions of simple models of the human body are illustrated but with an increasing degree of approximation. The positive comparisons with the trend of the measurements obtained on patients under anaesthesia confirmed the validity of the simplified numerical approach. The numerical solutions were obtained with the language *FORTRAN77* and the program *MATLAB ODE45*, while the graphical tools have always been the same as mentioned earlier. The bibliographic tools have always been the university libraries of Florence, Bologna and Pisa and the Technical University of Milan (Milano Polytechnic).

The logo of chapter "Human Body Thermoregulation" can be Fig. 6 [28], in which Leonardo da Vinci shows, in the upper part, the abdominal organs and explains thermogenesis and, in the lower part, the heart and the papillary muscles and describes the mechanism of action of the heart.

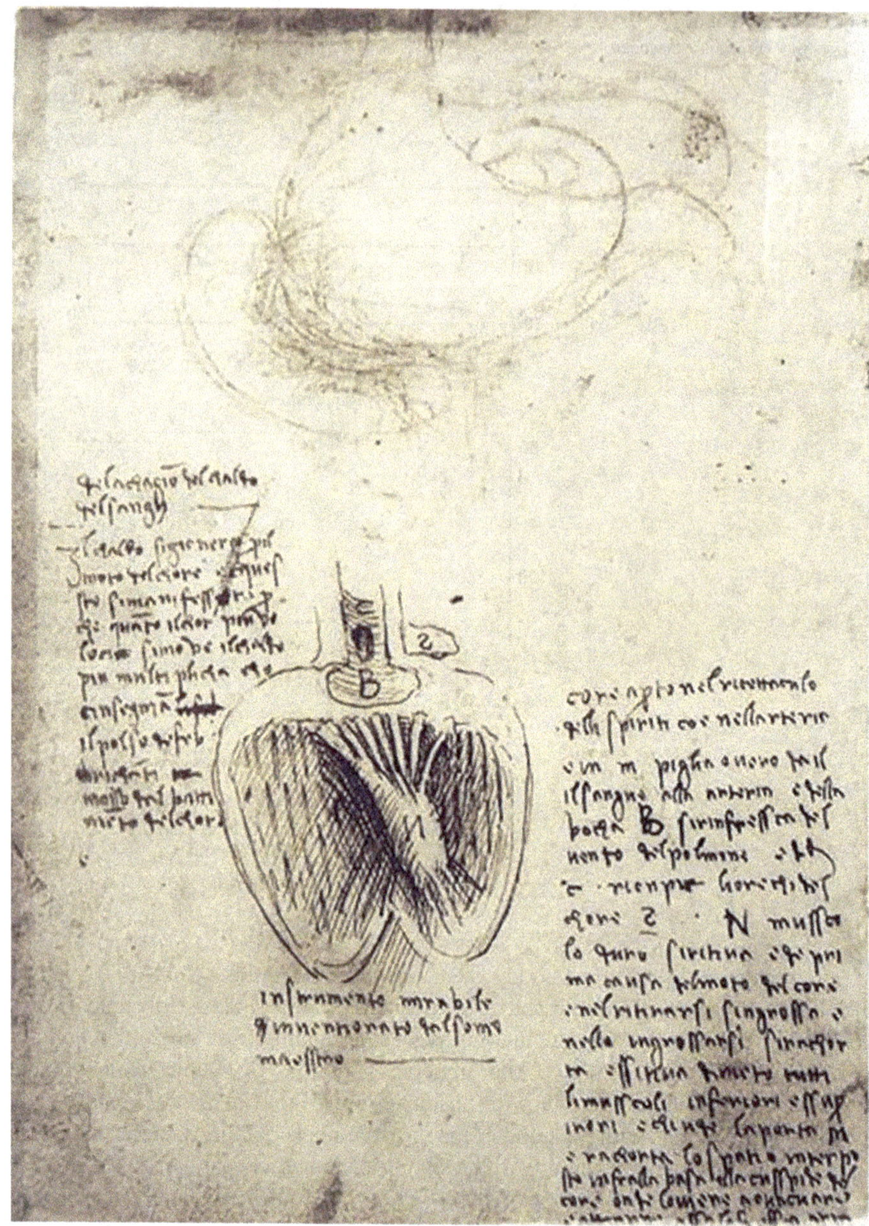

Fig. 6 Leonardo da Vinci: heart and papillary muscles. (Mirrored image of Sheet B12 r, [28])

As for thermogenesis and the cause of the heating of the blood, Leonardo states the following:

- About the reason of the heat of the blood. The heat is generated by the motion of the heart, and this is manifested because the faster the heart moves, the more the heat multiplies as the pulse of the feverish moved by the beating of the heart.
- *"dilachagio' del chaldo • dil sangh – • Il chaldo sigienera pil • moto del chore e cques • sto simanifess ta p • che quâto ilchor più ve • locie simove ilchaldo • più multi plicha cho • cinsegnjâ. (lifeb) • il polso defeb • brichâti (n) • mosso dal batti • mëto delchore."*

As for the mechanism of action of the heart, Leonardo states the following:

- Mirable instrument invented by the sum Master. Heart open in the receptacle of the spirits, that is, in the artery, and in M takes or gives the blood to the artery and from the mouth, B, is refreshed by the wind of the lung, and from C fills the ears of the heart, S. N, hard muscle, withdraws, and is the first cause of the motion of the heart, and, in withdrawing, thickens, and, in thickening, shortens, and pulls behind all the lower and upper muscles, and closes the door M, and shortens the space interposed between the base and the cusp of the heart, whence it comes to evacuate, and attract to itself the fresh air.
- *"instrumento mirabile • di innintionato dalsomo • maesstro– • core apto nel ricettaculo • di li spiriti co i nellarteria • e in m piglia o uero da il • ilsangue alla arteria e dilla • bocha B sirinfressca dil • uento del polmone eddi • .c . rienpie liorechi dil • chore S . N mussco • lo duro siritira e de pri • ma causa delmoto del core • e nel ritirarsi singrossa e • nello ingrossarsi sirachor • ta essitira dirieto tutti • li musscoli inferiori essup • irori e chiude la porta M • e rachorta lo spatio interpo • sto infralla basa ella cusspide del • core onde loujene a euacuare • e attrarre asse lafressca aria."*

Another logo of chapter "Human Body Thermoregulation" can also be Fig. 7, in which the temperatures, measured during a gynaecological procedure with anaesthesia, are compared with those obtained numerically, both with a complete model of the human body and with a model without a heat exchanger, which leads to an increase in tympanic temperature, as occurs in malignant hyperthermia [29].

4   Thermo-Haemo-Dynamics (THD) in Coronary Stents

Chapters "Thermo-Haemo-Dynamics (THD) in Coronary Stents: Numerical Newtonian Thermo-Haemo-Dynamics (THD) in Coronary Stents" and "Thermo-Haemo-Dynamics (THD) in coronary stents: Numerical Non-Newtonian Thermo-Haemo-Dynamics (THD) in Coronary Stents" deals with the Thermo- Haemo-Dynamics (THD) in coronary stents.

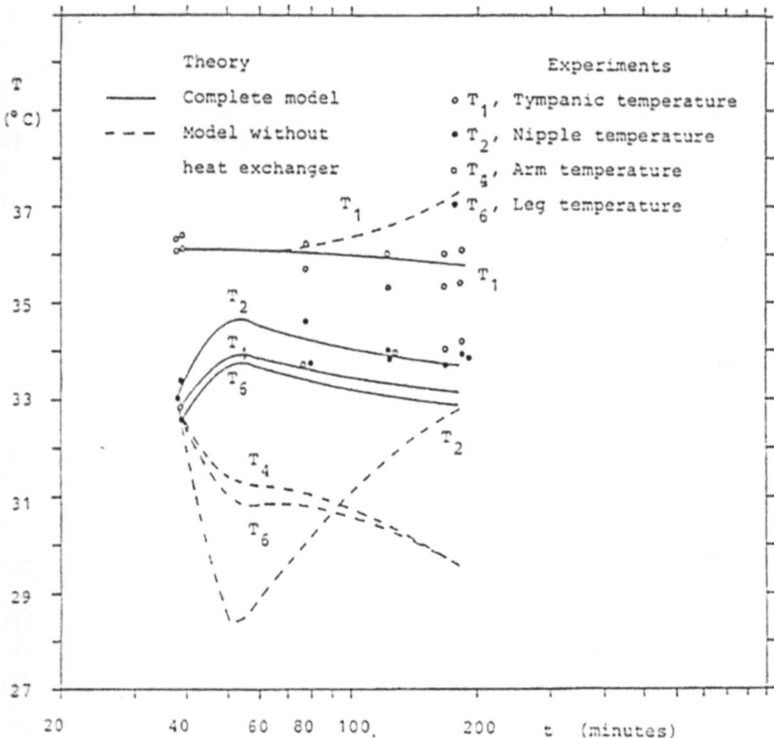

Fig. 7 Temperature variation in the human body during anaesthetic treatment

4.1 Chapter 5: Thermo-Haemo-Dynamics (THD) in Coronary Stents: Numerical Newtonian Thermo-Haemo-Dynamics (THD) in Coronary Stents

The chapter "Thermo-Haemo-Dynamics (THD) in Coronary Stents: Numerical Newtonian Thermo-Haemo-Dynamics (THD) in Coronary Stents" presents the study initiated with the Master's thesis in Medical Engineering in the A.Y. 2006–2007 by Dr. Ing. Mattia Amitrano, of which Andrea Boghi was a co-supervisor. After a careful examination of the literature on the subject, Mattia Amitrano introduces the problem of atherosclerosis and coronary stents, illustrating the pathophysiology of atherosclerosis, the anatomy and the cellular biology of the arterial wall, the treatments of re-vascularisation of ischaemic heart diseases, the angioplasty procedure and restenosis. He describes the coronary stents, the materials, the manufacturing techniques and the geometries. The fluid dynamics analysis of two *Cypher* coronary stents is then illustrated, with the description of the numerical analysis, the methods of construction and realisation of the geometry and the calculation grid. The results, obtained in steady and unsteady states, are obtained under the assumption that the blood is a Newtonian fluid. The evaluation parameter of endothelial permeability, *optical density*, is also introduced. In this first part, the comparison between peak-to-peak and peak-to-valley stents for the carotid artery carried out by Andrea Boghi and the author is also reported.

4.2 Chapter 6: Thermo-Haemo-Dynamics (THD) in coronary stents: Numerical Non-Newtonian Thermo-Haemo-Dynamics (THD) in Coronary Stents

In chapter "Thermo-Haemo-Dynamics (THD) in coronary stents: Numerical Non-Newtonian Thermo-Haemo-Dynamics (THD) in Coronary Stents", Andrea Boghi, Ivan Di Venuta and the author present the numerical solutions, in steady and unsteady state, of the non-Newtonian Haemo-Dynamics in drug-eluting stents used in coronary arteries, which undergo restenosis up to 90%.

The numerical tools of Chapter "Thermo-Haemo-Dynamics (THD) in coronary stents: Numerical Non-Newtonian Thermo-Haemo-Dynamics (THD) in Coronary Stents" are the programs *GAMBIT, SOLIDWORKS, FLUENT* and *MATLAB*. The graphic tools are *TECPLOT* and *PARAVIEW*. The bibliographic tools are *Scopus*, and *Web of Science*.

The logo of Chapter IV can be Fig. 8 [28], in which Leonardo illustrates the vessels of the abdominal viscera, liver, spleen and kidneys. At the beginning of the sheet, Leonardo describes the alteration of the intima of the vessels in the old man:

- One wonders why the veins in the old acquire great length, and yet they become tortuous; those, which used to be straight, become so thick with skin that they close and prohibit the movement of the blood, and from this comes death to the old, without disease. I judge that this thing increases more because it is closer to its nourishment; and, for this, being such a vein sheath of the blood, which nourishes the body, nourishes the veins much more, as they are closer to the blood.

- *"Dimâdasi pche leuene nevechi • acquisstino gran lùgeza (el) • e ancora si fan fressuo se • quelle che soleâ essere diritc essì • grossà tanto di pelle chella siri • chiude e proibissce ilmoto del • sanghue e di quj nasce la morte • alli vechi sanza malatia-gudico che quella cosa piu saumé • ti che piu vicina al suo nvtriméto • epquesto essendo tal uene guaina • del sangue che nutrissce ilcorpo no trissca tanto piu le vene quanto • esse son piu vicine alsâgue."*

Leonardo then describes the arteries of the abdomen and the cause of death in old people:

- The veins a b narrow so much in old people that the blood through them loses its motion, so it usually rots, but new blood can no longer penetrate them, which excavates it, which comes from the door of the stomach, so that the good blood

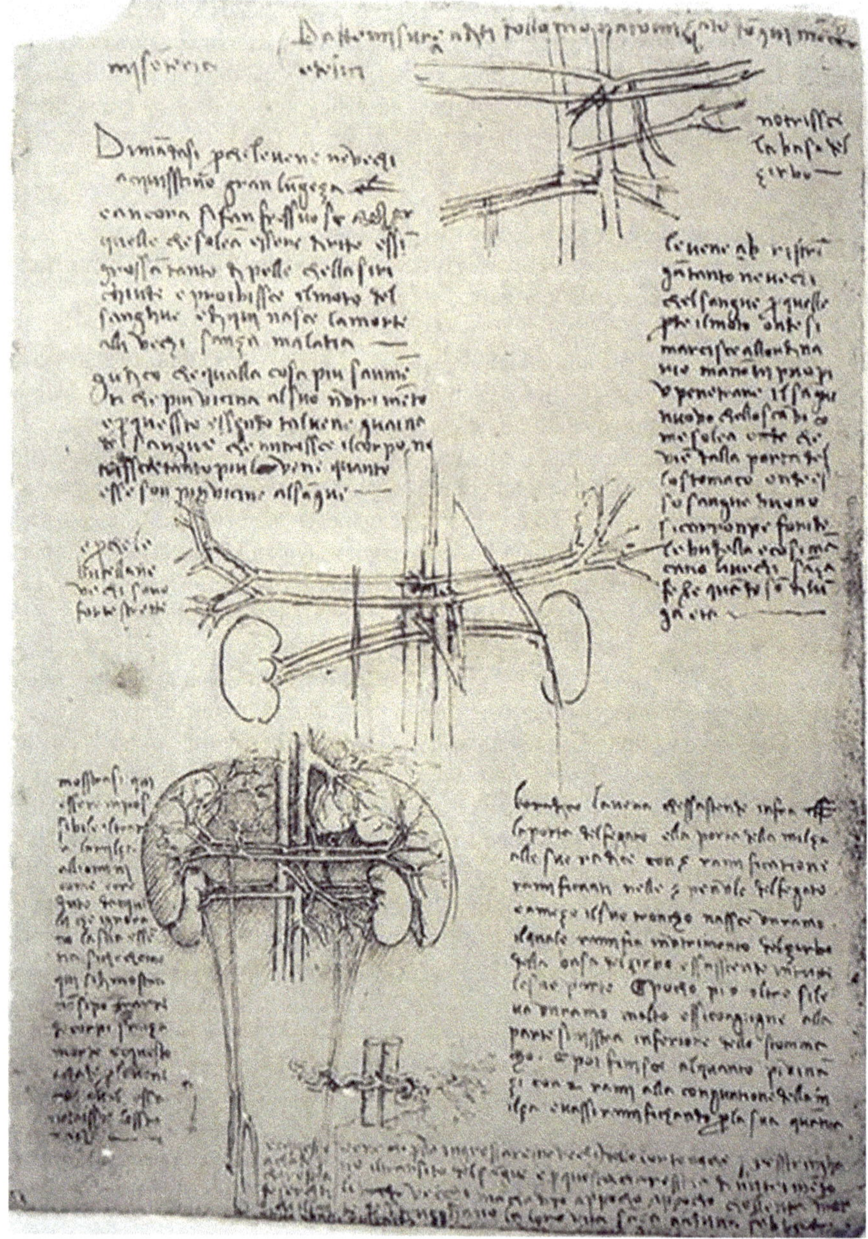

Fig. 8 Leonardo: vessels of the abdominal viscera, liver, spleen, kidneys and veins. (Mirrored image of sheet B 11 v, [28])

becomes corrupted in the bowels, and thus old people, without fever, when they are of long age, fail, and because the intestines are strong.

- *"le uene a b rjstrì • gâ tanto ne uechi • chel sangue pquelle • pde ilmoto onde si • marcisce allordina • rio manò uj puo pi • v penetrare il sâgue • nuovo chelloscavi co • me solea (onde) che • vié dalla porta del • lostomaco onde es • so sangue buono • sicorronpe fori de • le budella ecosi mà • cano liuechi sâza febe quàdo sò di lù • ga età e pche le • budella ne • vechi sono • forte."*

Leonardo, referring to the last figure at the bottom, describes the cause of death in the elderly:

- Veins, which, due to the thickening (in the elderly) of their tunics, restrict the transit of the blood, and, due to this scarcity of nourishment, the elderly, failing little by little, with slow death, destroy their life, without any fever. And this happens due to a lack of exercise because the blood does not warm up.
- *"vene che pllo ingrossare (ne vechi) delle lor tonjche (l) resstringha • no iltransito del sàgue e p questa charesstia di nutriméto • li (vechi) vecchi màchando appocho appocho chollenta mor • te desstrugghano la loro vita saza alchuna febbre e cquessto • achade p • charestia • deserciti • o che ilsan ghue nonsi risscalda."*

5   Chapter 7: Design of a Hydraulic Circuit for the Study of Vascular Problems

Chapter "Design of a Hydraulic Circuit for the Study of Vascular Problems" presents the study of the design of a hydraulic circuit for vascular problems, which started with the three-year/Bachelor's degree thesis in Medical Engineering by Vincenzo Galgano in the A.Y. 2008–2009, of which Ivano Petracci and Andrea Boghi were co-supervisors. In the first part of Chapter "Design of a Hydraulic Circuit for the Study of Vascular Problems", the medical aspects of the problem are described, illustrating the structure of the arteries, the arterial pressure and the electrical stimulus of the contraction of the blood, the coronary arteries, the carotid and the pathological conditions of the arteries, such as atherosclerosis and stenosis, and their pathogenesis. The state of the art is then illustrated, with the description of a technique to study the fluid dynamics, the *Particle Image Velocimetry (PIV)*, available in the Technical Physics and Energy Engineering Laboratory of "Tor Vergata," currently directed by Ivano Petracci. The experimental circuits that simulate the circulatory system present in the literature are then described. Finally, the designed circuit and the theoretical/numerical simulation of its operation are described, which allows a good reproducibility of the trend of the various cardiac quantities over time, in agreement with the data from the literature.

The logo of chapter "Design of a Hydraulic Circuit for the Study of Vascular Problems can be Fig. 9 [30] because Leonardo states that the "machine science"

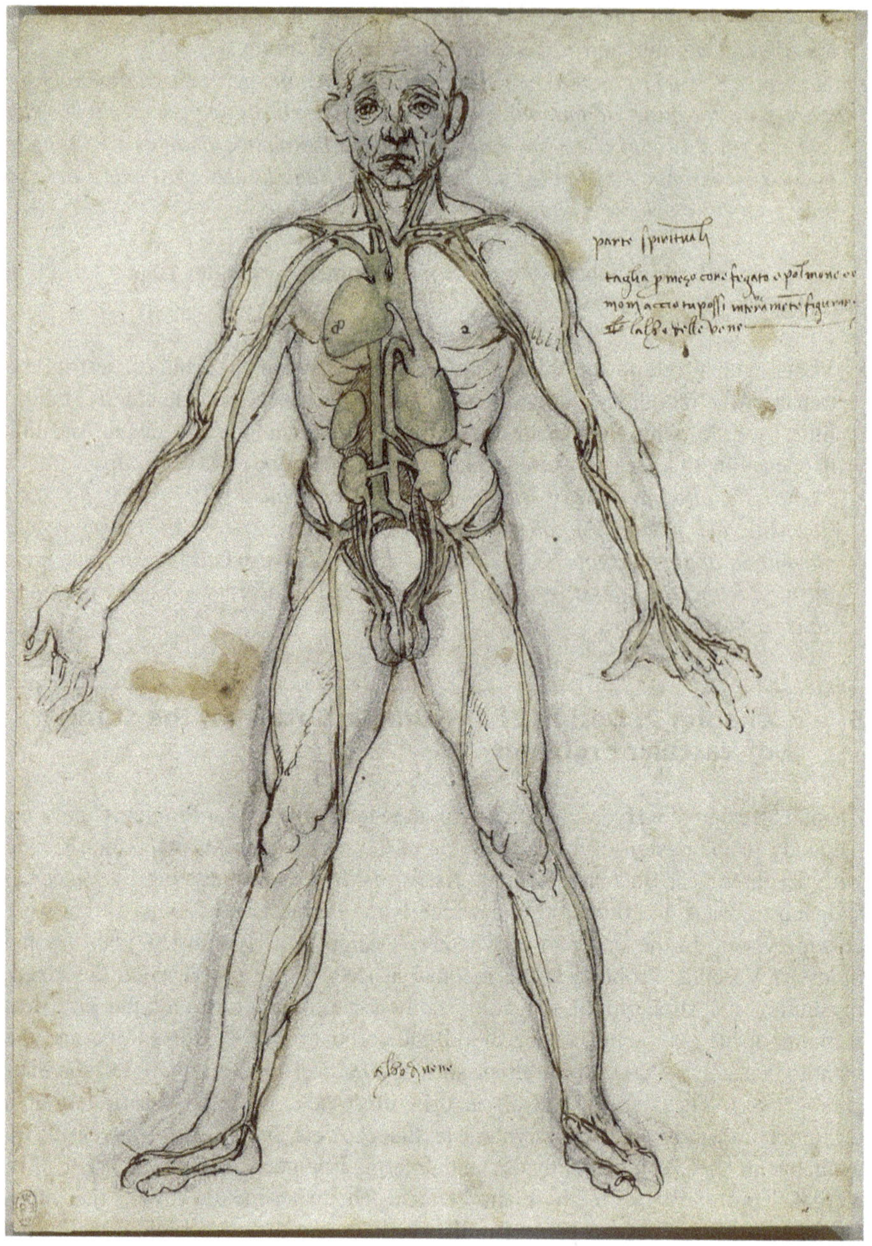

Fig. 9 Leonardo da Vinci: "vein tree." (Mirrored image of RL 12597r, recto, [30])

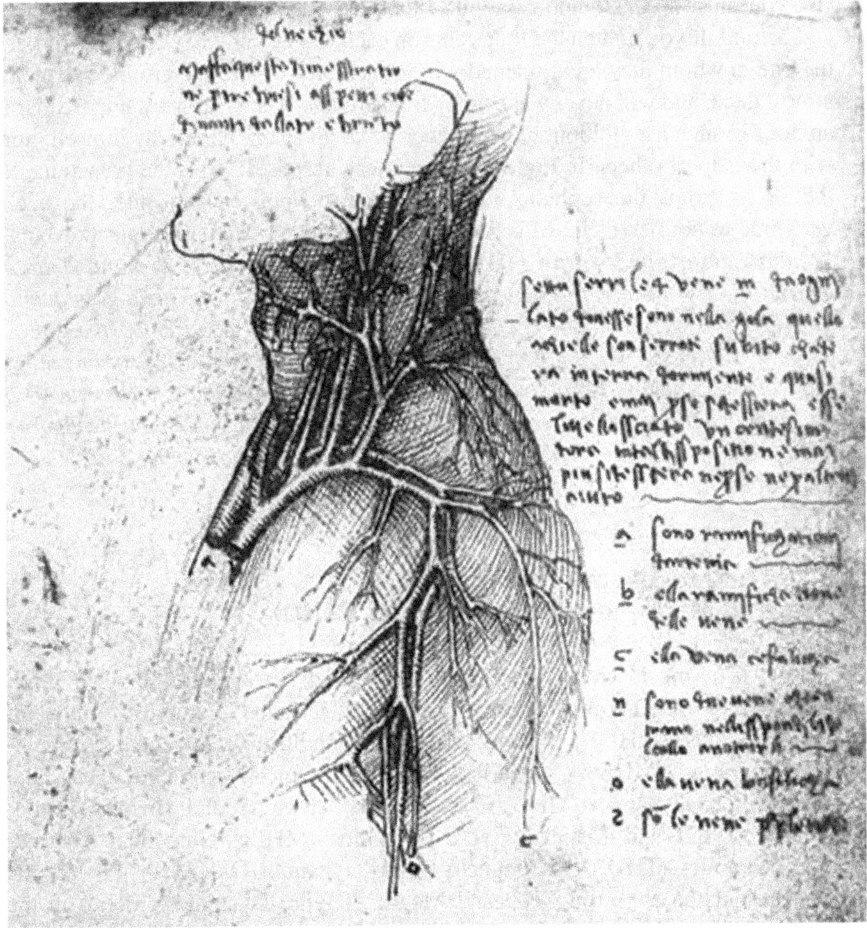

Fig. 10 Leonardo da Vinci: study of the arteries and veins of the head. (Mirrored image of Sheet B32v, [28])

("scientia macchinale"), which, "above all others is most useful" *("supra tutte le altre utilissima"),* is that by which "all living bodies... perform their operations" *("tutti li corpi animati… fanno le loro operazioni").* Figure 9 shows between the legs *"vein tree"* (*"albero di uene"*); top right: *"spiritual parts"* (*"parti spiritualj"*); below: "cut through the heart, liver, lungs, kidneys so you can fully figure the white of the veins" (*"taglia p mezo cor i fegato i polmoni i reni / accio tu possi interaminti figurare / lalbo dilli vene)."*

The logo can also be Fig. 10 [28], where Leonardo shows the supra-aortic vessels and titles it "of the old man" (*"del uechio"*). Leonardo then gives precepts for the demonstration of the vessels of the neck and of their importance for life:

- But you must do this demonstration in three aspects, i.e. in front, from the side and behind. If you tighten the four veins on each side, where they are in the throat, the one in whom they are tightened will immediately fall to the ground sleeping, almost dead, and will never wake up by himself; and if he is left for a hundredth of an hour in such a condition, he will never wake up again, neither by himself, nor with the help of others. In Fig. 10 (a) are artery branches; (b) is the branching of the veins; (c) is the cephalic vein; (n) are two veins, which enter the neck vertebrae to nourish them; o) is the basilic vein and (S) are the apoplectic veins.
- *"Maffa questa dimosstratio • ne p tre diuesi asspetti cioe • di nanti dallato e dirieto" • "settu serri le 4 vene m da ognj • lato douesse sono nella gola quello • achielle son serrate su bito chade • ra in terra dormjente e quasi / morto • e maj , pse sidesstera esse • luj e llassciato vn centesimo • dora intal djsspositione maj • più sidesstera ne ,pse ne ,paltruj • aiuto – • a sono ramjfichationj • darteria – • b ella ramjfichatione • delle uene – • c ella vena cefalicha • n sono due uene che en • trano nellisspondili de • l collo anotrir li – • o ella uena basilicha • S sò le uene popletiche."*

---

6   Thermo-Haemo-Dynamics (THD) and Magneto-Hydro-Dynamics (MHD)

Chapters "Thermo-Haemo-Dynamics (THD) and Magneto-Hydro-Dynamics (MHD): Bases of Thermo-Haemo-Dynamics (THD) and Numerical Magneto-Hydro-Dynamics (MHD)", "Thermo-Haemo-Dynamics (THD) and Magneto-Hydro-Dynamics (MHD): Numerical Thermo-Haemo-Dynamics (THD) and Magneto-Hydro-Dynamics (MHD) in Respiratory Airways", and Thermo-Haemo-Dynamics (THD) and Magneto-Hydro-Dynamics (MHD): Numerical Thermo-Haemo-Dynamics (THD) and Magneto-Hydro-Dynamics (MHD) in the Hepatic Artery present the study on THD and Magneto-Hydro-Dynamics (MHD) in the respiratory system and in the hepatic artery, which started with the *PhD* thesis of Dr. Eng. Flavia Russo, who carried out her *PhD* research at "Tor Vergata," with Andrea Boghi as co-supervisor.

6.1   Chapter 8: Thermo-Haemo-Dynamics (THD) and Magneto-Hydro-Dynamics (MHD): Bases of Thermo-Haemo-Dynamics (THD) and Numerical Magneto-Hydro-Dynamics (MHD)

In chapter "Thermo-Haemo-Dynamics (THD) and Magneto-Hydro-Dynamics (MHD): Bases of Thermo-Haemo-Dynamics (THD) and Numerical Magneto-Hydro-Dynamics (MHD)", the state of the art of Magneto-Hydro-Dynamics is illustrated, detailing the conservation equations of THD and MHD. The basics of numerical THD and Magneto-Hydro-Dynamics (MHD) are shown, and then, the relative solvers used by the software *OpenFOAM*, which is used to obtain the numerical results of the main basic cases of THD-MHD, are illustrated. One logo of chapter "Thermo-Haemo-Dynamics (THD) and Magneto-Hydro-Dynamics (MHD): Bases of Thermo-Haemo-Dynamics (THD) and Numerical Magneto-Hydro-Dynamics (MHD)" could be the frame from the film *"Fantastic Voyage"* [1], which shows a small spaceship moving in the blood along with the red blood cells, even though, nowadays, there is no longer a need to miniaturise men to perform some surgical procedures inside the human body.

6.2   Chapter 9: Thermo-Haemo-Dynamics (THD) and Magneto-Hydro-Dynamics (MHD): Numerical Thermo-Haemo-Dynamics (THD) and Magneto-Hydro-Dynamics (MHD) in Respiratory Airways

In chapter "Thermo-Haemo-Dynamics (THD) and Magneto-Hydro-Dynamics (MHD): Numerical Thermo-Haemo-Dynamics (THD) and Magneto-Hydro-Dynamics (MHD) in Respiratory Airways", the authors provide some medical information on the respiratory system, with the trachea, bronchi and the anatomy of the lungs. This is followed by the description of the three-dimensional (3D) modelling of anatomical structures, which starts from the acquisition of the image through Computed Tomography and ends with the reconstruction of three-dimensional surfaces. Finally, the second part concludes with the THD-MHD analysis in the respiratory tract. A logo of chapter "Thermo-Haemo-Dynamics (THD) and Magneto-Hydro-Dynamics (MHD): Numerical Thermo-Haemo-Dynamics (THD) and Magneto-Hydro-Dynamics (MHD) in Respiratory Airways" is Fig. 11 [28], where Leonardo shows the lung and explains the mechanism of respiration. Leonardo explains:

- When the lung has expelled the wind, and it decreases in quantity, by as much as the wind, which from it (exhaled) came out, then it must be examined where the space of the chest of the diminished lung attracts to itself the air, which fill its growth, with the fact that in nature there is no vacuum.
- *"(domandasi) quando ilpolmone a mandato fora iluento • e chelluj dimj nuissce di quantita p tanto quantera iluento. • che diluj [esalo] ussci allora sidebbe esamjnare donde losspatio • della chassa del dimj nuito pol mone attraggha asse laria che • rienpia ilsuo accresscimento con co sia che inatura nò • si da vachuo"*
- And after: it is still asked where, in the growth of the lung, it drives out the air from its receptacle, by what way this air escapes, and once it has escaped, where it is received.
- *"E ancora sidimanda donde nella cresscimento del polmone (s) • scaccj fori laria del suo ricettachulo p qual uja essa aria si • fuggha effuggita (doue) chee doue essa ericettata."*
- The lung is always filled with a quantity of air, even if it has driven out that air which is required for its exhalation, and when it is refreshed with new air, it rests on the ribs of the breast, and those somewhat dilates and pushes outwards, as is

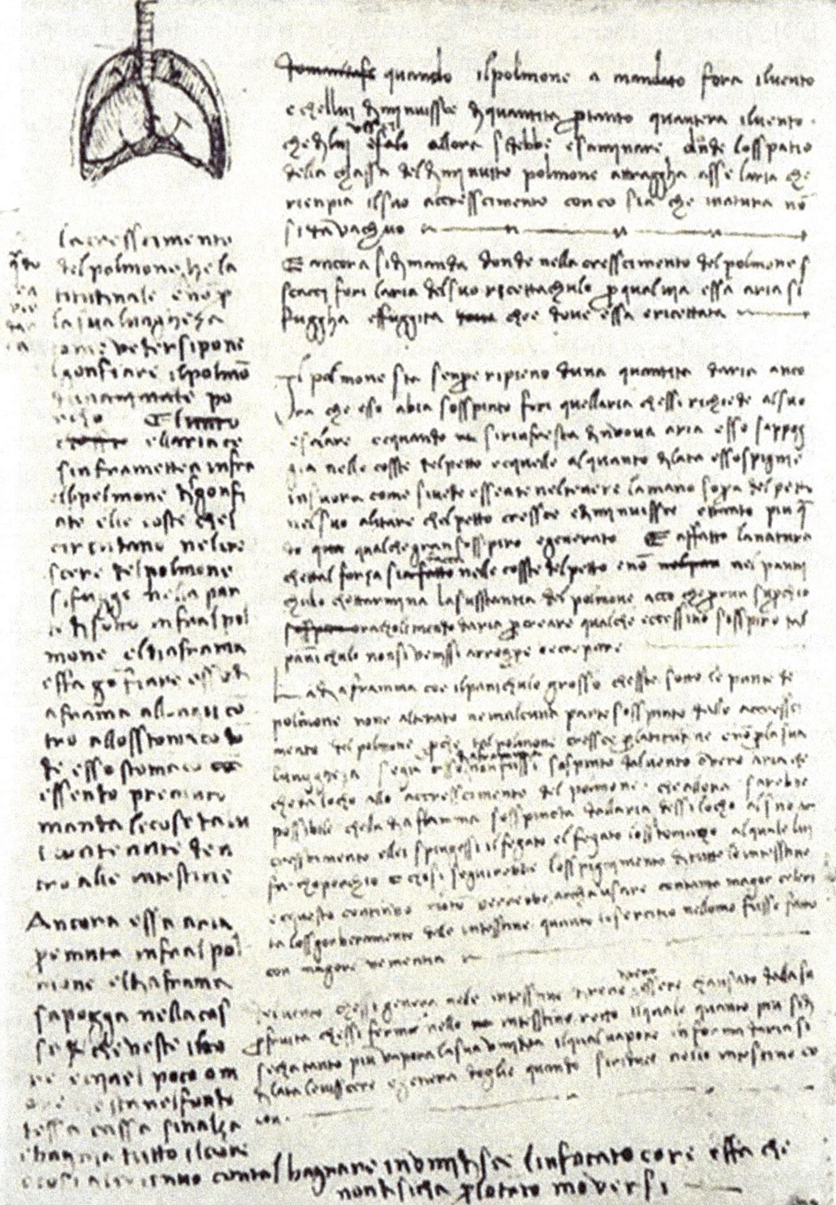

Fig. 11 Leonardo da Vinci: lung and mechanism of respiration. (Mirrored image of Sheet B17r, [28])

seen and felt, in holding the hand above the breast in its breathing, that the breast increases and decreases, and all the more so, when some great sigh is generated. And it has been nature that this force is made in the ribs of the breast, and not in the *panniculus*, which terminates the substance of the lung, so that by a sigh or recollection of air, to create some excessive sigh, such a *panniculus* would not come to break and crack.

- *"Il pol mone sta sen,pe ripieno duna quantita daria anco • ra che esso abia sosspinto fori quellaria chessi richiede alsuo • esalare e cquando (ri) sirinfresca di nvoua aria esso sappog • gia nelle cosste del petto e cquelle al quanto dilata essospignje • infuora come siuede essente neltenere la mano so,pa del petto • nel suo alitare chel petto cressce e dimjnuissce ettanto piu q • do (qua) qualche gran sosspiro egenerato E affatto la natura • chettal forza si(a fatto) facca nelle cosste del petto e nò (nel pan) nel pannj • chulo chettarmjna lasusstantia del polmone acco che ,prun su,pchio • (sosspiro o) racholimento daria ,p creare qualche eccessiuo • sosspiro tal • pànichulo nonsi venjssi arron,pe (o) ecrepare."*

- The growth of the lung, when it fills with air, is latitudinal and not along its length, as can be seen when inflating the lung of a (animal) pig, and the air that comes between the deflated lung and the ribs that surround it, as the lung grows, escapes to the lower part, between the lung and the diaphragm. And it causes the diaphragm to swell downwards, against the stomach, whence the stomach, being pressed, sends the things it contains into the intestines.

- *"lacresscimento • del polmone • qdo • sen • pie • dar • ia • he la • titudinale e nô , p • la sua lungheza • come veder sipo ne • l gonfiare ilpolmò • dun (anjmale) po • rcho . (E l unto • chessi) ellaria ce • sin fra mettea infra • ell polmone disgonfi • ato elle coste chel • cir cudano nel cre • scere del polmone • sifugge nella par • te di sotto in fral pol • mone el diaframa • effa gòfiare essodi • afràma allongu cò • tro allosstomaco do • de esso stomàco (cò) • essendo premuto • manda lecose da lu • i con tenute den • tro alle intestine."*

- Again, this air, pressed between the lung and the diaphragm, rests in the chest, which dresses the heart, and that little humour that is in the bottom of the chest rises and bathes the whole heart, and so continuously, with such wetting, moistens the fiery heart, and causes it not to dry up for so much movement.

- *"Ancora essa aria • ,pemuta infral pol • mone eldiaframa • sapogga nella cas • se (d) che veste ilco • re e cquel poco om • ore che sta nelfondo • dessa cassa sinalza • e bagnja tutto ilcore • e cosi alcotinuo contal bagnare invmjdisce linfocato core effa che • non disicha, plotàto moversi."*

Another logo of chapter "Thermo-Haemo-Dynamics (THD) and Magneto-Hydro-Dynamics (MHD): Numerical Thermo-Haemo-Dynamics (THD) and Magneto-Hydro-Dynamics (MHD) in Respiratory Airways" is Fig. 12 [28], where Leonardo shows the thoracic and abdominal organs of the pig. In the left part of Fig. 12, he indicates the thoracic and abdominal organs: *a b c d e the* lung, liver, spleen, stomach, diaphragm and spine. Leonardo explains: Let do this demonstration, a b c d e lung, liver, spleen, stomach, diphragm and spine *"faraj quessta dimòsstratione a b c d e polmone feghato milza stommaco diaflamma spina*

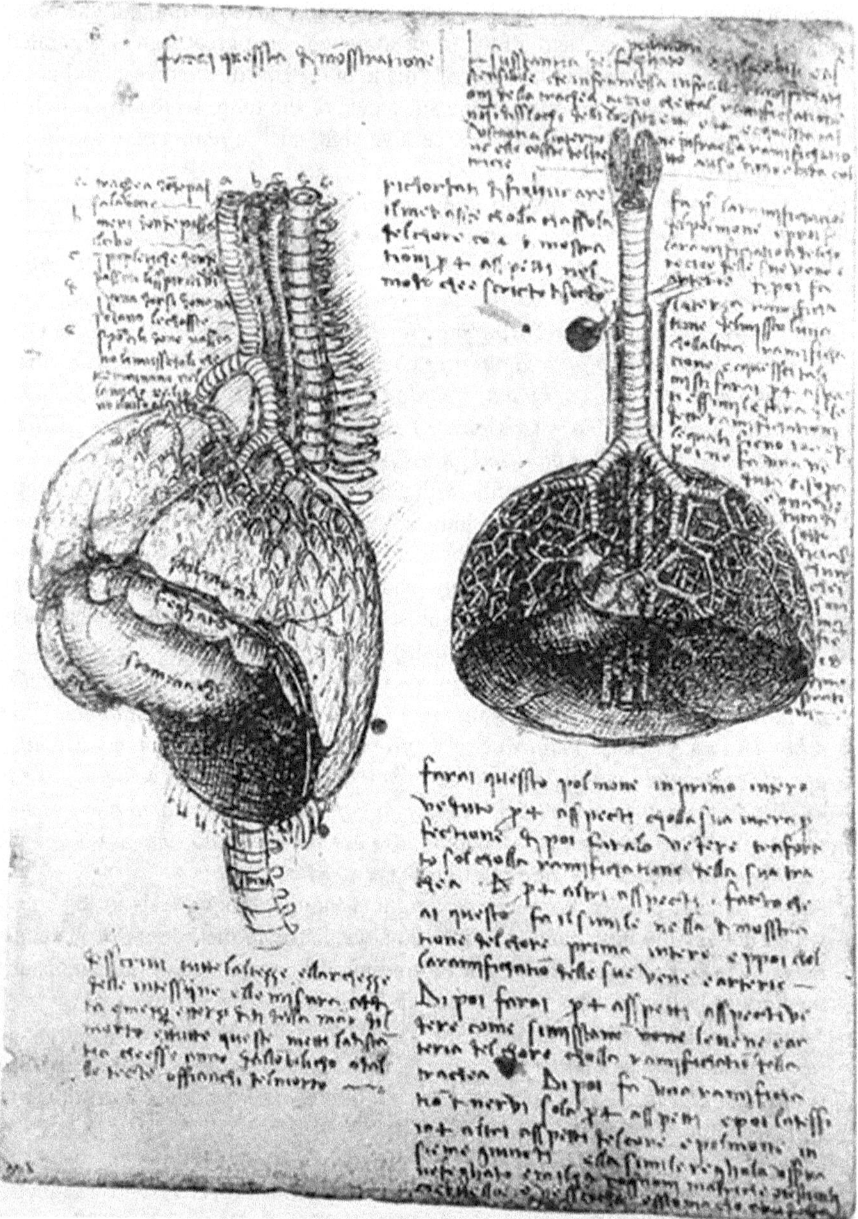

Fig. 12 Leonardo: lungs, heart and abdominal organs of the pig. (Mirrored image Gof Sheet B37v, [28])

(a) Trachea, where the voice passes
  - *"a trachea dòdepas • sa la vocie-"*.
(b) Oesophagus or *meri*, where the food passes
  - *"b meri donde passa • ilcibo-"*.
(c) Nerves or hypoplectics, where the vital spirits pass
  - *"c jpopletiche donde • passan lisspiriti vitalj"*
(d) Spinal column or spines of the back, where the ribs originate
  - *"d spina dorsi doue na • schano le chosste–"*
(e) Spinous processes or spondyls, where the muscles originate, which end in the nape and raise the face to the sky
  - *"e spòdili doue nassca • no li musscholi che • termjnano nel • la nucha ealza • no iluiso alcielo"*

Leonardo also describes the relationship of the lungs with the bronchi:
- The substance of the (liver) lung is expandable and extensible and is interspersed among the branches of the trachea so that such branching does not dislocate from their sites, and this substance interposes between this branching and the ribs of the chest like a soft quilt.
- *"L susstantia del (feghato) polmone e dilatabile e as • stensibile ede inframessa infralle ramjfichatj • onj della trachea accio che ttal ramjficha tione • nòsi disslochi delli lorsiti (e tt e de) e cquessta tal • sustantia sin terpo ne jnfraessa ramjfichatio • ne elle cosste del pe tto auso dimorbida col • tricje-"*.

6.3   Chapter 10: Thermo-Haemo-Dynamics (THD) and Magneto-Hydro-Dynamics (MHD): Numerical Thermo-Haemo-Dynamics (THD) and Magneto-Hydro-Dynamics (MHD) in the Hepatic Artery

In chapter "Thermo-Haemo-Dynamics (THD) and Magneto-Hydro-Dynamics (MHD): Numerical Thermo-Haemo-Dynamics (THD) and Magneto-Hydro-Dynamics (MHD) in the Hepatic Artery", the THD-MHD analysis in the hepatic artery is illustrated. The final part describes the method of particle treatment, used for the distribution of magnetic medicines, which, thanks to the use of a magnetic field, allows their absorption in specific areas of the human body, for example, tumour sites. The numerical tools of chapter "Thermo-Haemo-Dynamics (THD) and Magneto-Hydro-Dynamics (MHD): Numerical Thermo-Haemo-Dynamics (THD) and Magneto-Hydro-Dynamics (MHD) in the Hepatic Artery", VI are the programs *OpenFOAM, VMTK and MATLAB;* the graphic tools are *MATLAB* and *PARAVIEW* and the bibliographic tools are *Scopus* and *Web of Science*. The logo of Chapter VI can be Fig. 13, where Leonardo shows the vascular and biliary systems of the liver [28]. Leonardo describes the origin of all veins from the gibbous part of the heart in the top left:

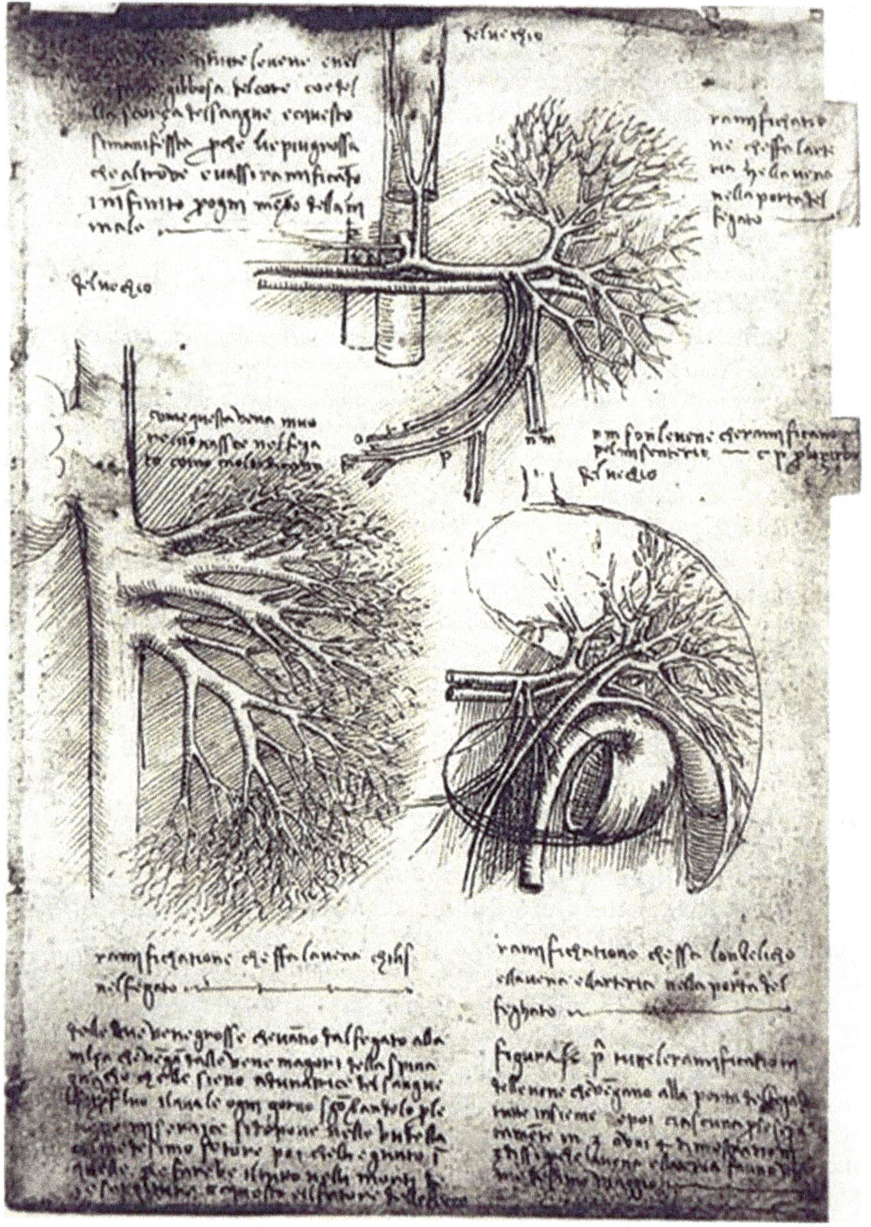

Fig. 13 Leonardo: vascular and biliary systems of the liver. (Mirrored image of Sheet B34v, [28])

- The root of all veins is in the gibbous part of the heart, that is of the bark of the blood; and this is manifested, because there it is larger than elsewhere, and it branches infinitely, for every member of the animal.
- *"Laradice ditutte leuene e nel • la parte gibbosa del core coe del • lla scorza delsangue e cquesto • simanifessta ,pche lie piu grossa • che altrove e uassi ramjficâdo • inìfinito ,pognj mëzzo dellanj • male-"*

In the top right, he shows the celiac trunk and its branches:

- About the old. The branching that the artery and the vein make in the liver's portal
- *"del uechio • ramjficatio • ne cheffa larte • ria hella uena • nella porta del fegato-"*

Leonardo also shows the mesenteric veins, where:

- n and m are the veins that branch for the mesentery and c and p for the spleen
- *"n m son leuene cheramj ficano • pel mjsenterio – c p ,plozirbo • del uehio"*

On the left side, he shows the origin of the portal:

- How this vein dies and does not originate in the liver, as many say, and the *chyli* vein, branching that the *chyli* vein makes in the liver
- *"come questa vena muo • re enò nassce nel fega • to come molti di-cano • ramj fichatione che ffa lauena chilis • nelfegato-"*

He indicates the veins from the liver to the spleen and their function:

- Of the two large veins, which go from the liver to the spleen, which come from the major veins of the spine, I judge that they are collectors of superfluous blood, which, every day clearing it through the mesenteric veins, is deposit in the intestines with the same stench, after it has arrived in those, which would do everything in the dead of the burials, and this is the stench of the faeces of the old.
- *"delle due vene grosse cheuâno dal fegato alla • mlza che végâ dalle vene magori della spina • gudicho che lle sieno adunatrice del sangue • su,pfluo il quale ognj gorno sgòbrandolo ,ple • uene mjserajce si depone nelle budella • col medesimo fetore poi chelli e gunto ì • quelle che farebe il tutto nelli mortj de • le sepolture e cquesto eil fetore delle fecce del uechio"*

Finally, he shows the following:

- The branching that the navel, the vein and the artery make in the liver's portal
- *"ramj fichatione che ssa lonbelicho • ella uena ellarteria nella porta del feghato-"*.

7   Cardiac Acoustics

7.1   Chapter 11: Cardiac Acoustics: Basics of Acoustics for the Evaluation of Environmental and Fluid Dynamic Noise

Chapter "Cardiac Acoustics: Basics of Acoustics for the Evaluation of Environmental and Fluid Dynamic Noise" discusses the basics of acoustics that are used for the study of environmental and aerodynamic noise due to wind turbines and air traffic, which started with the Master's thesis in Medical Engineering of Dr. Eng. Giulio La Bella in the A.Y. 2015–2016, of which Ivan Di Venuta was a co-supervisor. This chapter includes the problems of noise related to wind turbines and air traffic and their effects on health, such as, for example, on the cardiovascular system, sleep disturbances, damage to the auditory system, psycho-social effects and on cognitive performance. Subsequently, some basic information is given on the system of human hearing. The theoretical–analytical foundations of acoustics are then illustrated, with the description of the relative wave equations, divided into plane and spherical waves, of the energy, intensity and quantity of motion of sound waves, of acoustic monopoles, dipoles and quadrupoles. The unit of measurement of sound, the equivalent continuous noise level, the level of a single event and the equivalent sound pressure level are described. The last part accurately describes the equations of aero-acoustics, which deals with aerodynamic noise produced by turbulent flows or by aerodynamic forces acting on surfaces. Especially, the analogy of *Lighthill* and the subsequent formulation of *Ffwocs William and Hawkings*. The logo of chapter "Cardiac Acoustics: Basics of Acoustics for the Evaluation of Environmental and Fluid Dynamic Noise" can be the figure shown in [31], which reads "Wind turbine blades with owl wings, for quieter energy."

7.2   Chapter 12: Cardiac Acoustics: Acoustics in Cardiac Thermo-Haemo-Dynamics

Chapter "Cardiac Acoustics: Acoustics in Cardiac Thermo-Haemo-Dynamics" presents the study related to acoustic noise in cardiac Thermo-Haemo-Dynamics (THD), which started with the Master's thesis in Medical Engineering by Dr. Eng. Alex Colucci in the A.Y. 2016–2017, of which Ivan Di Venuta was co-supervisor. In the first paragraphs, the physiology and pathophysiology of the cardiac system is illustrated, and then, the physiology of tones, the pathophysiology of heart murmurs and the flow through a partial obstruction of the aortic valve apparatus are described. The fluid dynamics of valvular stenosis is discussed, with the model of one-dimensional current and the description of the stenotic valve. This is followed by the description of the hyperbolic wave equation, with the analytical and numerical solutions, with the explicit scheme of the finite differences. Some cases of analytical and numerical solutions are then reported, with their comparisons. Computational Thermo-Haemo-Dynamics is finally applied to a severely stenotic aortic valve, showing the two-dimensional geometry and the results in steady and unsteady states.

The final paragraphs show the application of computational acoustics to the two-dimensional geometry of a severely stenotic valve, comparing the results with systolic and diastolic acoustic signals. In chapter "Cardiac Acoustics: Acoustics in Cardiac Thermo-Haemo-Dynamics", the program is reported, written in *FORTRAN90*, for the numerical solution of the homogeneous hyperbolic wave equation, and the function, *User Defined Function (USF)*, is used to model the blood as a non-Newtonian fluid and set the pressure profile over time. The numerical tools of Chapter "Cardiac Acoustics: Acoustics in Cardiac Thermo-Haemo-Dynamics" . VIII are, in addition to the *FORTRAN90* language, the programs *MATLAB*, the graphic tools *MATLAB* and *PARAVIEW* and the bibliographic tools *Scopus, Web of Science and Google Scholar*. The logo of chapter "Cardiac Acoustics: Acoustics in Cardiac Thermo-Haemo-Dynamics" can be Fig. 14 [28], in which Leonardo illustrates the main vessels of the chest.

Leonardo illustrates the heart and vessels:

- The heart is the core, which generates the tree of veins; the veins that have their roots in the manure, that is, the miserable veins, go to deposit the acquired blood in the liver, from where the superior veins of the liver then nourish themselves.
- *"il core einocciolo che gienera lalbo delle vene • (e) lequalj vene an le radici nelletame cioe le • vene mjseraicie che van adipore loacqujssta • to sanghue nefeghato donde poi le uene supiori • del feghato sinutrichano" – • fa p' laramjfichatione delle • uene pse e poi lossa pse e • poi gu(gli) gnj lossa e uene ì • sieme"*.

Leonardo then makes a comparison with the roots and branches of plants:

- Never does the plant come from the branch because first it is the plant that branches, and first it is the heart that the veins
- *"mai la pianta nasscie dal • llarumjfichatione pche • prima e lla pianta che • essa ramj fichatione • e prima e ilchore chelle • vene"*.

The logo of chapter "Cardiac Acoustics: Acoustics in Cardiac Thermo-Haemo-Dynamics" can also be Fig. 15, which shows the heart and coronary arteries of the ox [30], with the tricuspid valve highlighted. Leonardo describes cardiac physiology through observation of the pig, describing its contraction:

- Passing with a pin "when the heart is stretched," (*"quando il core è allungato"*) (diastole); the same pin is pushed up when "the heart in its expulsion of blood gathers," (*"il core nella sua espulsione del sangue si raccorta"*) (systole) [30].

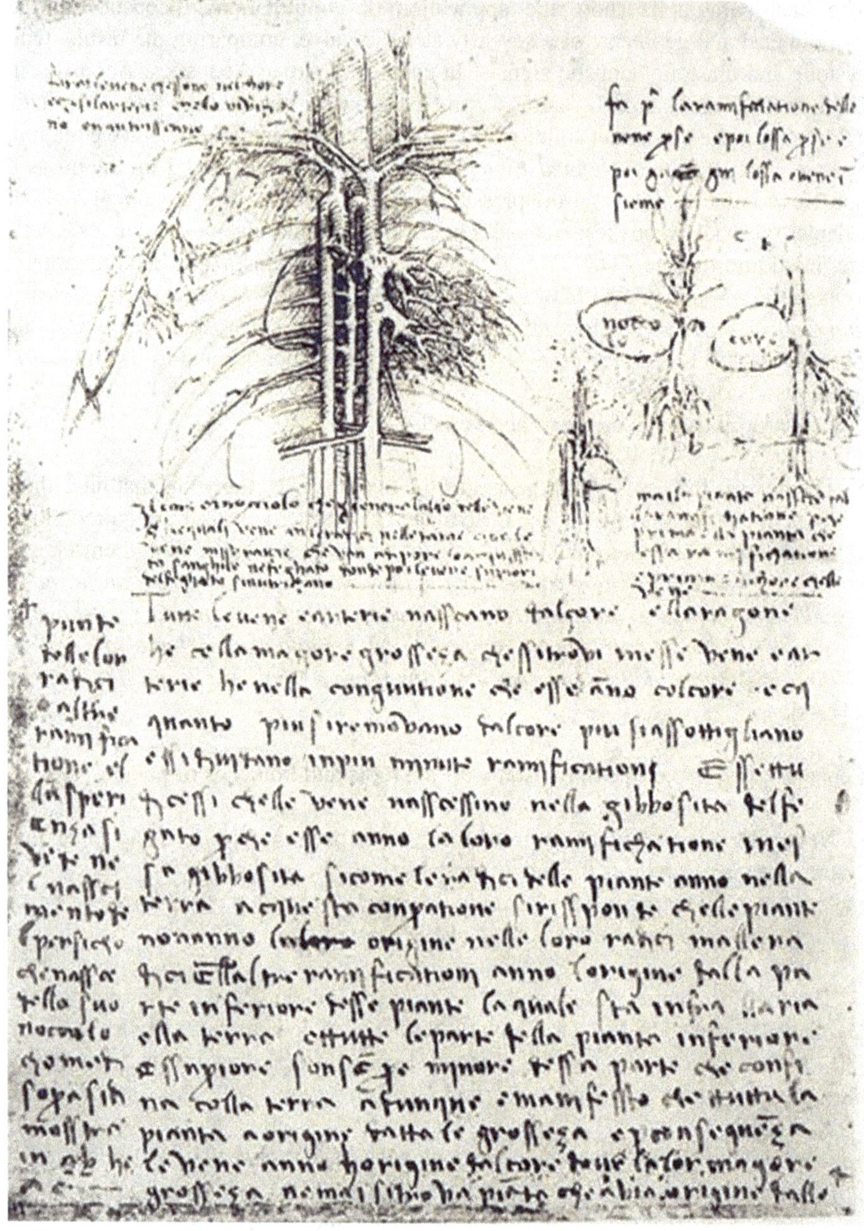

Fig. 14 Leonardo: main vessels of the thorax. (Mirrored of Sheet B11r, [28])

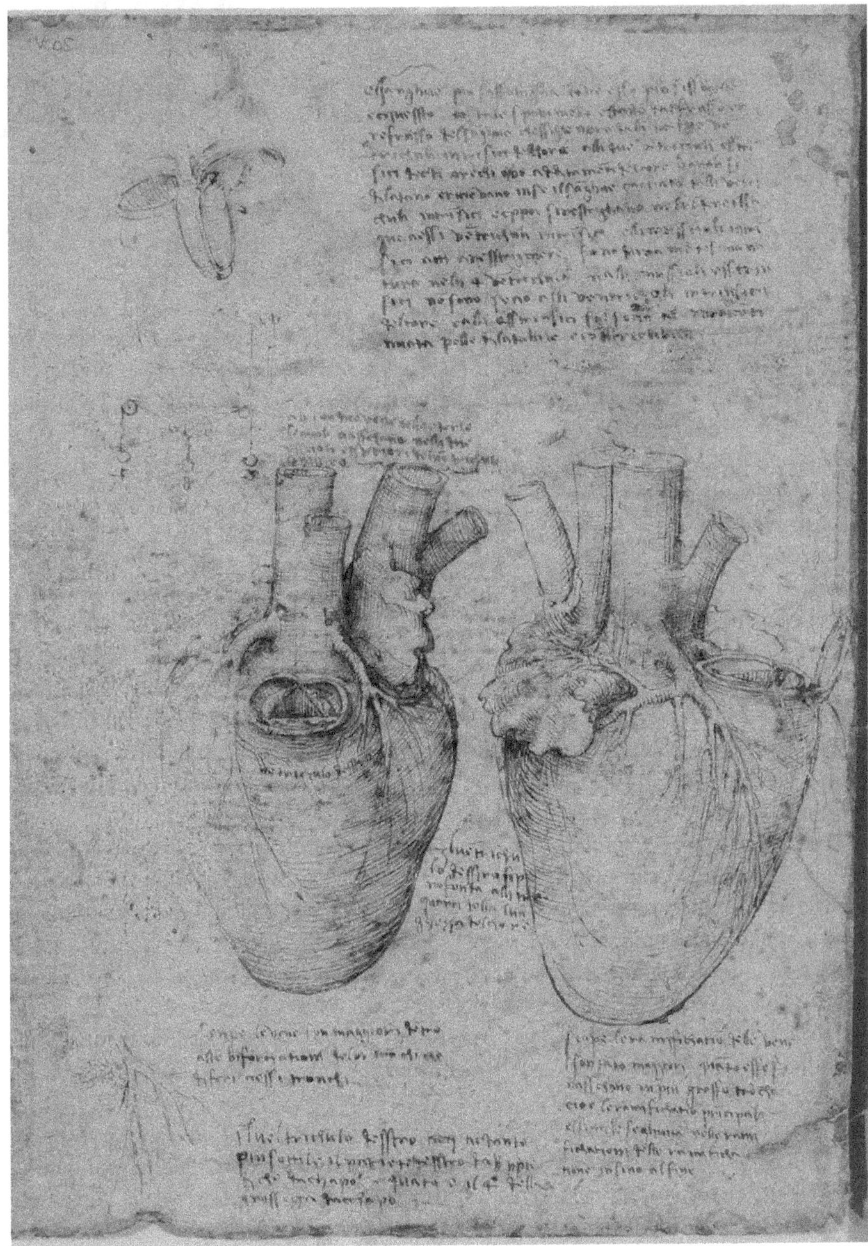

Fig. 15 Leonardo: the heart and the coronary arteries of the ox. (Mirrored image of sheet RL 19073v, [30])

Leonardo was the first to describe the four cardiac cavities:

- Distinguishing the ventricles from the "atria," ("*auricole*"), he recognises the muscular nature of the heart and the coronary arteries "the heart is a vessel made of dense muscle, enlivened and nourished by the artery and vein, like the other muscles" ("*il core è un vaso fatto di denso muscolo, vivificato e nutrito dall'arteria e vena, come son gli altri muscoli*"), "the coronary arteries arise from two external outlets of the left ventricle" ("*le coronarie nascono da due usciuoli esteriori del ventricolo sinistro*")

The importance of the heart and its autonomous contraction are described by Leonardo, who defines it a "wonderful instrument, intended by the Supreme Master" ("*instrumento mirabile, intenzionato dal Sommo Maestro*") and says that "it is very powerful above other muscles" ("*è potentissimo sopra gli altri muscoli*") and that "it moves by itself and does not stop, except eternally" ("*si muove da sé e non si ferma, se non eternamente*").

Leonardo has noted the link between the contraction of the left ventricle and its synchrony with the arterial pulse. He also describes the cardiac cycle: "heart motions" ("*moti del core*"), observing that the contraction of the atria coincides with ventricular diastole, while the ventricular systole coincides with the dilation of the atria. That the blood was pumped by the heart was clear to Leonardo, who indeed wrote "when water comes out of the earth, always aim at low places" ("*quando l'acqua esce dalla terra, mira sempre luoghi bassi*"); instead, the blood in living beings moves from the heart towards the periphery: "when a vein in the nose bursts, all the blood passes from below to the top" ("*quando scoppia una vena del naso, tutto il sangue passa da sotto in su*").

The logo of the chapter "Cardiac Acoustics: Acoustics in Cardiac Thermo-Haemo-Dynamics" can also be Fig. 16 [30], which shows the sketch of the left ventricular outflow tract in the top left-hand section, where he gives instructions on how to make a glass model of the same. To the right, there is an excised aortic valve apparatus to be fitted into the model and a drawing as to how one may construct an aortic valve. In the left-hand section below, there are drawings depicting the mitral valve with its cusps likened to a sail and the left ventricular outflow tract with eddies of blood passing, closing the aortic valve from the side, not from above [32].

Acknowledgements At the conclusion of this introduction, I would like to thank the University of Rome "Tor Vergata" for hosting me since 1992 and for continuing to offer me the opportunity to teach as a contract professor, also in the Academic Year 2024–2025. In this regard, I can note that with the A.Y. 2023–2024, I have completed 50 academic years of teaching, without even a single absence, which includes the years from 1974 to 1991 in Firenze, the Fall Course of 1988 in Stony Brook (USA), the years from 1990 to 1998 in Siena, from 1986 to 1990 in Reggio Calabria, from 1990 to 1994 at the Technical University (Politechnic) of Milan and from 1992 to the present at "Tor Vergata" University.

Fig. 16 Leonardo: aortic and mitral valves. (Mirrored image of sheet RL 19082r, [30])

A heartfelt thanks to colleagues and friends, particularly to the late Franco Maceri, who sadly left us on 19 December 2021, for his presentation; to Gaetano Marrocco, coordinator of the degree sourse in Medical Engineering and to M. D.

Salvatore Mangiafico, for their preface. Their considerations have captured all the aspects that I intended to highlight in this book.

I like to thank Prof. M. D. Orazio Schillaci, former rector of the University of Rome "Tor Vergata," for giving me the opportunity to organise and present, on the occasion *of "TorVergata40, the Symposium,* THERMO-HAEMO-DYNAMICS (THD) IN MEDICAL ENGINEERING AND IN THE MEDICINE OF THE FUTURE" [33], which was held on 25 October 2022 and saw the participation of Dr. Eng. Andrea Boghi and medical doctors Arturo Consoli, Valerio Da Ros, Salvatore Mangiafico, Enrica Mariano and Giuseppe Umana.

Many thanks to Prof. Gaetano Marrocco, coordinator of the Degree Course in Medical Engineering, for the organisation of the event "25 years of the Degree Course in Medical Engineering at 'Tor Vergata'," held on 9 May 2024. In that event, I was asked to make a presentation entitled "The teaching experience from the Engineering point of view." During that presentation, I described some events at the beginning of the Degree Course in Medical Engineering. In order to prepare some programmes of the Degree Course in Medical Engineering, I was asked by Prof. Franco Maceri to have some information about the Harvad-MIT Program in Health Sciences and Technology (HST), established in Boston (USA). During one of my frequent trips to the USA, I paid a visit, with a meeting on 6 August 1999, to Professors Paul Gray and Roger G. Mark of MIT, Boston (USA), asking for information about their Harvad-MIT Program in Health Sciences and Technology (HST). At that time, their institution was the only one that had started the educational–scientific collaboration between Medicine and Engineering. During the interview, the MIT colleagues outlined the research-teaching line that they indicated to the students of Medical Engineering at the MIT, which consisted of the knowledge of the four circulatory systems of the human body: arterio-venous, uro-genital, airways and cerebro-spinal. I also thank these colleagues from the MIT because some of the themes I have proposed to the students as a thesis topic were inspired by that visit–interview.

A sincere thanks to all the students who have worked with me on Bachelor's, Master's, Specialist's or Master's degrees; second-level Master's theses and PhD research. In a special way, I want to thank those who participated in the drafting of this work (in the order of academic seniority): Ivano Petracci, Andrea Boghi, Mattia Amitrano, Flavia Russo, Ivan Di Venuta, Giulio La Bella and Alex Colucci.

Regarding the students who are trained at the University of Rome "Tor Vergata" and also in the other Italian Universities where I have taught, I can affirm that they receive one of the best scientific preparation in the world. I can make this statement with knowledge of the facts, having attended, as a student, researcher and professor, the Italian Universities of Bologna, where I graduated, after following the three-year course in Chemical Engineering, and where I carried out the scholarship for young graduates; Firenze, where I followed the two-year course in Engineering and was an assistant and associate professor; Reggio Calabria, where I was a professor of Technical Physics; the Technical University of Milano (Milano Polytecnic), where I was a full professor and the University of Siena, where I was a contract professor. Furthermore, I know several foreign universities because I have collaborated with professors and students of *Master's* and *PhD* of *Imperial College* of London (e.g. *Prof. D. B. Spalding); Tokyo Institute of Technology, Tokyo, Japan (e.g. Prof. R. Echigo); Department of Physics, University of Rajasthan, Jaipur, India (e.g. Prof. D.R. Chaudhary); Institute of Fluid Science, Sendai, Japan (e.g. Prof. T. Aihara); Cornell University, New York, USA (e.g. Prof. R. Miller); University of Minnesota, Minneapolis, USA (e.g. Prof. R.J. Goldstein); University of Illinois* at *Chicago, Chicago, USA (e.g. Prof. J.P. Hartnett, Prof. L. Kennedy, Prof. W. Minkowycz and Prof. W.M. Worek) and University of New York* at *Stony Brook, USA (e.g. Prof. T.F. Irvine Jr)*, where I was teaching when I was appointed as professor during the autumn semester of 1988. It was the training and scientific preparation of our students that convinced me not to move to the USA, when I was appointed *professor* at the *University of New York* at *Stony Brook, USA*. Such scientific preparation allowed some of my students, after their degree, whether five-year, Specialist's or Master's, to be admitted directly to the *PhD* in some Universities of the *USA*, even recognising a third of the credits they necessarily had to follow to obtain the *PhD*, thus equating the Italian five-year degree with a Master's degree plus one year of *PhD*. I consider this a great recognition!

Regarding the preparation and training of our students, I like to recall an anecdote that occurred in the 90 s at the *University of Illinois* at *Chicago, USA*, with which I had established an agreement, according to which, after attending some of their courses, they could obtain the title of *Master of Science* from that University. The head of the department of that time, who hosted them, told me that, due to the mainly basic-deductive type of preparation of our students, when they arrived in his laboratory, they did not know anything about their instruments, unlike their colleagues, who had received an applied-inductive type of preparation and knew all the laboratory instruments, even the models and types of instruments. Well, when the head of the department returned to the laboratory after some weeks, he could see that our students had designed, proposed or implemented a new type of instrument for the research he had suggested!

TOR VERGATA 40

The 40th anniversary of the foundation of the University of Rome "Tor Vergata," entitled "TorVergata40 – FUTURE SIGHT," was organised during the week from 24 to 28 October 2022 (https://torvergata40.uniroma2.it). As part of this 40th anniversary, I proposed and moderated the meeting, titled *"Thermo-Haemo-Dynamics (THD) in Medical Engineering and in the Medicine of the Future."* The discussion took place at the Studios of the University of Rome "Tor Vergata" on 25 October 2022 from 12:30 to 13:30 and can be viewed at the following link [33]: https://torvergata40.uniroma2.it/eventi/la-termo-emo-dinamica-ted-nellingegneria-medica/

The meeting saw the participation, as speakers, of Dr. Eng. Andrea Boghi and medical doctors Arturo Consoli, Valerio Da Ros, Salvatore Mangiafico, Enrica Mariano and Giuseppe Umana, besides the author/moderator. The discussion was divided into two parts:

- Thermo-Haemo-Dynamics (THD) in Medical Engineering and in the Medicine *Today*
- Thermo-Haemo-Dynamics (THD) in Medical Engineering and in the Medicine of *the Future*.

Thermo-Haemo-Dynamics (THD) in Medical Engineering and in Medicine Today

In the first part, the speakers intervened in alphabetical order and presented their current activity, in relation to the theme of the meeting. Their speeches are reported in more detail in the following sections but can be heard (in Italian) in https://torvergata40.uniroma2.it/eventi/la-termo-emo-dinamica-ted-nellingegneria-medica/ [33]. Dr. Eng. Andrea Boghi made the presentation titled "Computational Thermo-Haemo-Dynamics (THD) in coronary stents and in the Magneto-Fluid-Dynamics (MHD) of respiratory and cardiovascular systems." M.D. Arturo Consoli made the presentation titled "The predictive role of computational THD and mass transport in the evolution of intracranial stenoses." M.D. Valerio Da Ros made the presentation titled "The predictive role of computational THD in the degree of occlusion of cerebral aneurysms treated with intra-saccular devices." M.D. Salvatore Mangiafico made the presentation titled "Cerebral Arterio Venous Malformations (AVM)." M.D. Enrica Mariano made the presentation titled "The applications of the principle of Torricelli and the Gorlin equation to cardiac THD." M.D. Giuseppe Umana made the presentation titled "Fluid-dynamics in the treatment of cerebral aneurysm residues after clipping vs coiling." Prof. Dr. Eng. Fabio Gori Ammannati made the presentation titled "Computational THD in the prediction of aneurysmatic and atherosclerotic events in the Circle of Willis."

Speech of Andrea Boghi

Boghi The first of my slides in https://torvergata40.uniroma2.it/eventi/la-termo-emo-dinamica-ted-nellingegneria-medica/ [33] shows the fluid dynamics in coronary stents, topic that we have dealt with over the years with some Medical Engineering degree theses. Through three-dimensional (3D) fluid dynamic simulation, the performance of various stent geometries and the influence of the degree of restenosis on fluid dynamics and stresses on the endothelium are evaluated. The shear stress on the vascular wall is very important for estimating the probability of vessel rupture and, therefore, the probability of thrombus promotion. We can see in

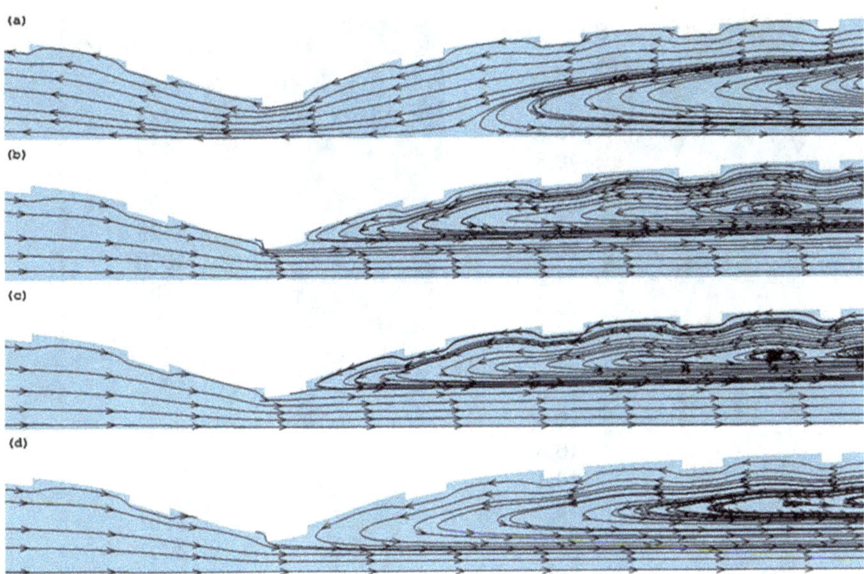

Fig. 17 Streamlines in a geometry with 90% degree of restenosis during the cardiac cycle after (**a**) 0.3 s (diastole), (**b**) 0.4 s, (**c**) 0.5 s and (**d**) 0.6 s (systole). (Fig. 7 of [34])

Fig. 17 [34] how the fluid dynamics changes depending on the pulsed blood flow and how the recirculation favours the deposition of lipid material and, therefore, the formation of restenosis in a geometry with 30% and 60% degree of restenosis. Figure 17 [34] shows the streamlines in a geometry with 90% degree of restenosis during the cardiac cycle at four instants of time: (a) 0.3 s (diastole), (b) 0.4 s, (c) 0.5 s, (d) 0.6 s (systole), remarking that the fluid dynamics changes depending on the pulsed blood flow and how the various recirculation favour the deposition of lipid material and, therefore, the formation of restenosis in a geometry with 90% degree of restenosis, compared to the similar results with 30% and 60% shown in figs. 5–6 of [34].

In Fig. 18 [35], you can observe the wall shear stress (WSS) in a geometry with 30% (left) and 90% (right) degree of restenosis during the cardiac cycle after 0.6 s (systole). The presence of redder areas indicates regions with greater wall shear stress (WSS), corresponding to the presence of the stenosis restriction, while the blue areas are those where the wall shear stress (WSS) is smaller.

The second of my slides in https://torvergata40.uniroma2.it/eventi/la-termo-emo-dinamica-ted-nellingegneria-medica/ [33] concerns the study of Magneto-Hydro-Dynamics (MHD) in the respiratory and in the Thermo-Haemo-Dynamics (THD) of the hepatic artery field, carried out during the Research Doctorate thesis of Dr. Eng. Flavia Russo [36]. In this interesting problem, magnetic particles of nano-metric dimensions are inhaled into the respiratory system (Fig. 19) or injected into the bloodstream in the iliac artery (Fig. 20).

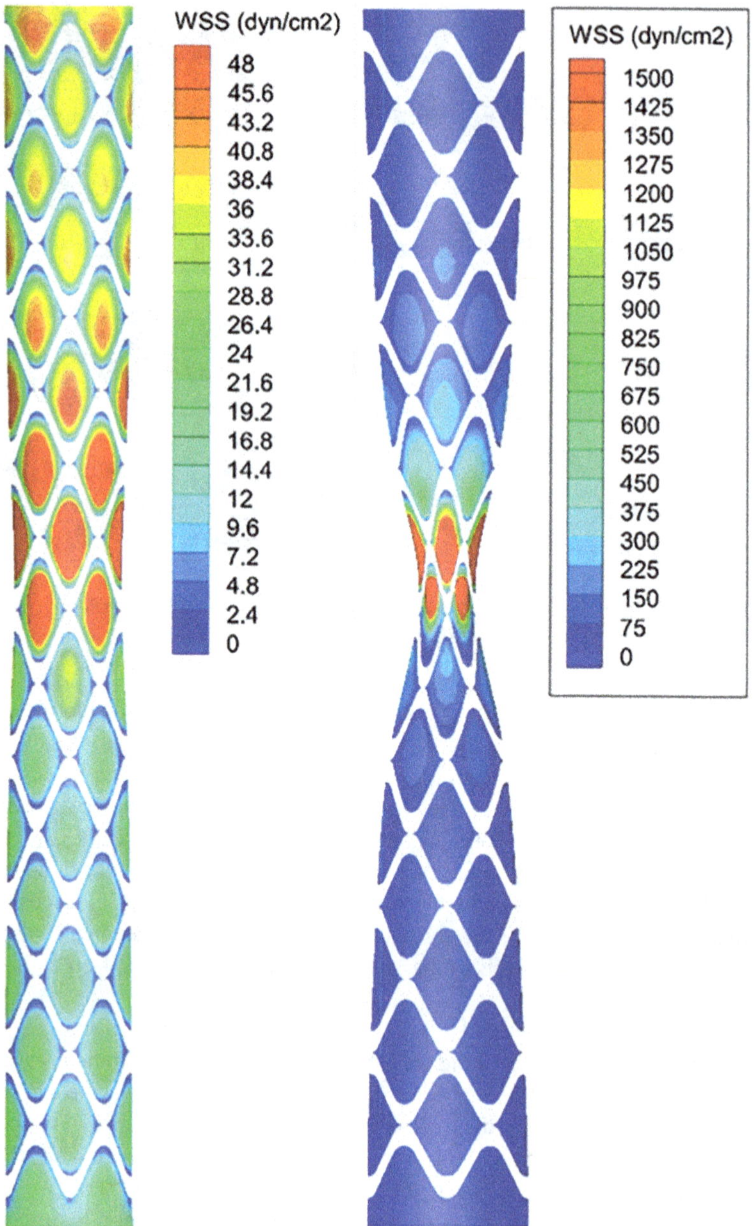

Fig. 18 Wall shear stress (WSS) in a geometry with 30% (left) and 90% (right) degree of restenosis during the cardiac cycle after 0.6 s (systole) [35]

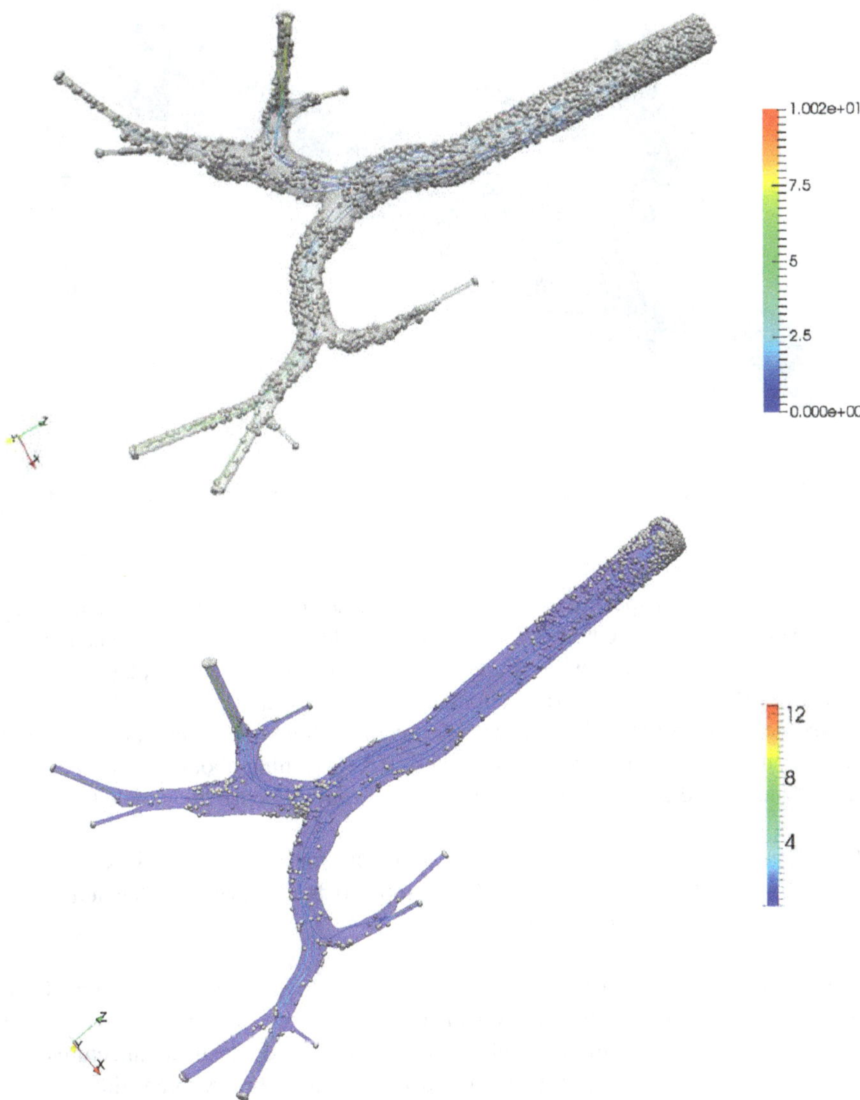

Fig. 19 Particle speed (m/s) in the lower respiratory tract after 2 s (scale on the right): (top) without external magnetic field and (bottom) with external magnetic field [36]

These particles are coated with drugs that can be used to treat various diseases, such as cancer. Thanks to the introduction of magnetic fields, we can guide these particles to the desired positions, thus limiting adverse effects. We have reconstructed the geometry of the blood vessels and the trachea. Subsequently, the typical equations of THD-MHD are written into calculation algorithms, which are then solved numerically. The magnetic field is generated by a single rectangular

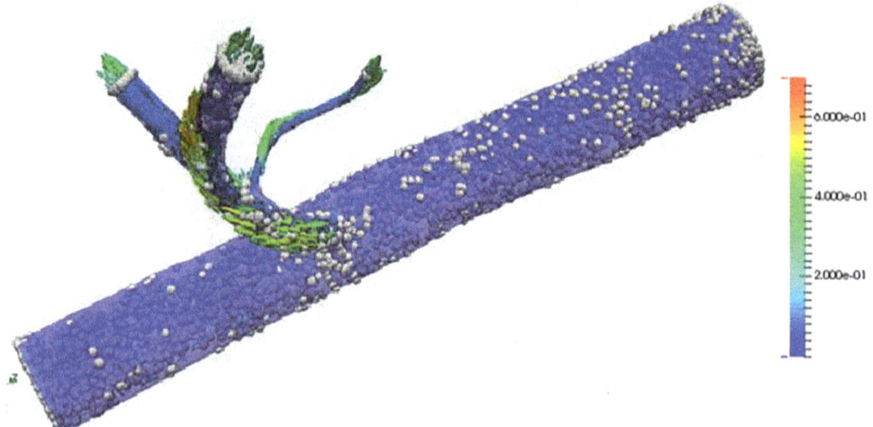

Fig. 20 Velocity field (m/s) in the Haemo-Dynamics of the iliac artery with nanoparticles after $t = 2$ s [36]

spiral. The patient's specific geometry is reconstructed from DICOM images using the VMTK software. A new solver, which couples the Lagrangian dynamics of magnetic nanoparticles with Eulerian fluid dynamics, is implemented in OpenFOAM to obtain numerical simulations. The resistive pressure, the periodic pulsed velocity profile at the inlet and the magnetic field of a rectangular spiral are implemented as boundary conditions. The interesting thing about this study is that the magnetic field does not seem to have a significant influence on particle transport.

Gori The question that I can ask you is important for engineers. Tell us about the computing capabilities of current PCs, compared to the needs of the numerical simulations we are talking about.

Boghi Compared to when we started, about 20 years ago, capabilities have increased enormously. I would say that the direction we are heading in is to use *Cloud* and *Server* systems remotely, which can have different architectures but usually have many processors that allow parallelisation. A very interesting thing is the advent of quantum computers because they can further increase the speed of calculation. However, to be able to use the quantum computer, you have to write the equations in a way that adapts to quantum calculation algorithms, and it is not something that will come immediately, but this is the direction we are moving in.

Speech of Arturo Consoli

Consoli First of all, I wanted to thank Gori for organising this meeting and for my invitation, which represents a fundamental moment of encounter for current Medicine and which can no longer do without collaboration with Medical Engineering. I present an activity that is actually the result of a chat we had with Gori several months ago, in which we tackled together the concept of THD and mass transport and how to apply them to clinical activity. The idea is to focus on a disease that has a chronic aspect like atherosclerosis, which we are all used to seeing at the level of the carotid arteries but which can also have an influence on the arteries of the intracranial circulation, as shown in the first of my slides in https://torvergata40.uniroma2.it/eventi/la-termo-emo-dinamica-ted-nellingegneria-medica/ [33], and which shows a medium cerebral arterial stenosis (mCAS).

Intracranical atherosclerotic disease (ICAD) is a chronic phenomenon determining the development of an atherosclerotic plaque in the arterial wall of cerebral arteries. This type of degenerative process is observed in about 5–6% of the Caucasian population, whereas it is significantly higher in Afro-American and Asian populations (up to 30–40%). The clinical presentation of this type of disease is often characterised by the onset of clinical deficits corresponding to the territory of the involved vessel (aphasia or dysarthria, hemiparesis/hemiplegia, neglect, sensitive deficits), with a typical fluctuating behaviour. Currently, a consensus about the therapeutic strategy has not been reached yet. The introduction of a dual anti-platelet therapy is considered the first-line approach with the possibility to treat the plaque by an endovascular approach in case of failure of the medical treatment.

The aim is to assess the predictive value of mass transport of LDL, glucose and oxygen in the arterial segments surrounding the plaque in order to investigate the accuracy and the effectiveness of this model to predict the behaviour of the plaque, in terms of stability or evolution, which could lead to the occlusion of the artery. Being a chronic disease, it can also have acute implications, and the study of mass transport could somehow predict what the evolution of this type of stenosis and plaques that form on the arterial walls of the intracranial circulation can be. This is a potential application of mass transport that could actually be useful for this type of patients. Also, because, to date, the medical literature regarding the treatment of these intracranial stenoses is not uniform, and there is no consensus in this sense.

The second of my slides in https://torvergata40.uniroma2.it/eventi/la-termo-emo-dinamica-ted-nellingegneria-medica/ [33] shows us the use of new imaging techniques, *"Cone Beam CT,"* more precise in terms of resolution compared to angiographic imaging, which allow us to acquire volumes with extremely thin layers. I think it can actually be suitable for this new type of analysis, correlated with laboratory clinical data, to be able to imagine a predictive model through THD and mass transport in this type of lesions, which, during the acute phase, can have extremely serious consequences. Three-dimensional acquisition and *Xper-CT-*acquired volumes are analysed, and mass transport of LDL, glucose and oxygen can be assessed. For this reason, the use of this type of acquisition, which we call in interventional neuroradiology *"Cone Beam CT,"* allows us to have an extremely defined image detail. Since the thicknesses of the *"slices"* of these volumes can be extremely thin *(< 1 mm),* these would be particularly suitable for this type of analysis. The idea is to be able to establish a predictive value capability using this type of technique that can give us information in the medium and long term regarding the evolution of this type of pathology and, therefore, the potential choices regarding treatment options.

Umana I would like to ask what is the clinical impact that this type of analysis could have.

Consoli The clinical impact is determined by the possibility, once a predictive model has been identified, that it can be used for a type of patient, to then personalise the therapeutic strategy. Therefore, a therapeutic approach can be chosen that can be more preventive and interventional or more conservative, depending on the patient profile that we will be able to define and that concerns his pathological condition at the very moment we see him, precisely because there is currently no consensus in the scientific community and in the literature regarding the best approach for this type of pathological conditions.

Speech of Valerio Da Ros

Da Ros Thank you for the opportunity to share our study regarding the THD applied to the treatment of cerebral aneurysms with intrasaccular devices, as shown in the first of my slides in https://torvergata40.uniroma2.it/eventi/la-termo-emo-dinamica-ted-nellingegneria-medica/ [33]. The intrasaccular devices, shown in the first of my slides [33], are systems introduced to the market about a decade ago for the treatment of bifurcation aneurysms of arteries with wide neck and complex anatomy. These are devices that, once inserted inside the aneurysmal sac, allow an interruption of the flow at the neck of the aneurysm, on one hand and, on the other, provide an optimal substrate for the formation of the neo-endothelium, necessary conditions for achieving a lasting occlusion of the aneurysm. However, the literature indicates that these devices, in the follow-up, can undergo changes in their morphology with consequent compaction and consequent risk of recanalisation of the aneurysmal sac. The mechanisms underlying the compaction of the system, however, are not entirely known.

The second of my slides in https://torvergata40.uniroma2.it/eventi/la-termo-emo-dinamica-ted-nellingegneria-medica/ [33] shows how, using rotational acquisitions with subsequent reconstruction of three-dimensional images, it is possible, together with the application of THD, to assess whether the inflow inside the aneurysmal sac can determine the compaction of the device and, therefore, a recanalisation of the aneurysm during the follow-up. Computational Thermo-Haemo-Dynamics (THD) study is performed using preoperative 3D rotational angiography. Data from 22 consecutive patients treated with the intrasaccular device for unruptured aneurysms located either at the middle cerebral artery or basilar tip are obtained. Mean follow-up period was 17 months. From digital subtraction, follow-up angiographies' patients are dichotomised into two groups: one with WEB compression and one without.

In the third of my slides in https://torvergata40.uniroma2.it/eventi/la-termo-emo-dinamica-ted-nellingegneria-medica/ [33], the ratio between flow at the parent vessel leading to the aneurysmal sac and the flow at the aneurysmal neck is calculated, obtaining numerical indices that demonstrate a correlation between the value of inflow at the aneurysmal sac and the probability of device compaction during patient follow-up. White arrows in the slide illustrate the cross sections where the flow in the parental artery and through the aneurysmal neck is calculated to obtain the THD inflow ratio. On the left, the aneurysm is exposed to high flow: all the parent artery flow is entering inside the aneurysm through the neck with associated outflow vortex re-entering inside the aneurysm (THD inflow ratio is 1.15). On the right side, there is an aneurysm less exposed to flow, with only less than one-fourth of the parent artery flow penetrating inside the sac (THD inflow ratio is 0.70). Colour scales are different between right and left. Despite it being clear that the recanalisation of the aneurysm is a process determined by multifactorial phenomena, thanks to the use of THD, we have obtained preliminary data capable of recognising the importance of the circulating flow inside the aneurysmal sac, which turns out to be one of the factors that conditions the anatomical outcome of the aneurysm post-treatment and, therefore, the stability of the aneurysm occlusion. Eleven devices presented "compression" during follow-up. Device compression is statistically correlated to the THD inflow ratio, although not to aneurysm volume, aspect ratio or neck size. The mechanisms underlying the worsening of aneurysms occlusion in patients treated with intrasaccular device, due to device compression, are most likely complex as well as multifactorial. However, it is apparent from our pilot study that a high arterial inflow is, at least, partially involved. Further studies are needed to increase our understanding of this phenomenon.

Mangiafico Valerio, thank you. Meanwhile, beautiful images of the cases. I wanted to ask you: if there is such an important relationship with the flow and, therefore, if with the fluid dynamics of the aneurysm, is it possible to predict in which treatments to use this device compared to others and in which type of aneurysms?

Da Ros Let's say that the purpose of the study was not to change or to identify a better strategy in the treatment of an aneurysm or of an endovascular approach rather than another but the information that THD gives us in pre-treatment planning and that of the choice of the most suitable system dimensions. In aneurysms where the inflow is particularly high, the tendency will be to choose a slightly oversized device, compared to the size of the aneurysm, in order to obtain an adhesion of the device to the wall and, therefore, overcome the compression forces induced by the flow in the face of a greater resistance of the device itself, given by its adhesion to the wall of the vessel.

Speech of Salvatore Mangiafico

Mangiafico Thank you, Professor Gori; thank you for the honour of being here at this meeting so important for the university, for the Italian Academy and for the entire scientific world. Truly, a great opportunity, and how to honour this occasion if not by showing in a very general way what are the methods we have available when we go to study the cerebral circulation.

The first of my slides in https://torvergata40.uniroma2.it/eventi/la-termo-emo-dinamica-ted-nellingegneria-medica/ [33] shows the comparison between images in cerebral Haemo-Dynamics. You see in the centre, in grey, a traditional angiographic image, subtracted geography, and then, on the left and right, coloured images. What is the meaning of these images, which is also the meaning of the medical research that we are carrying forward, and that is to give purely morphological images a content of information regarding the speed of blood in the cerebral circulation.

The second of my slides in https://torvergata40.uniroma2.it/eventi/la-termo-emo-dinamica-ted-nellingegneria-medica/ [33] shows micro-angiographic images of the micro-circulation and cerebral perfusion. So we are moving from a concept of pure morphological analysis, with a global haemo-dynamic ultra-structural analysis, and then with micro-catheters, also to study in detail the cerebral circulation. You see here, there is a selective image with a micro-catheter, which has reached the cerebral cortex, which injects the contrast medium, and then, we have images of venous return. So we really have a detail that we could say microscopic. If before, we were with a macroscopic investigation, this we could define it for the general public as super-super-selective. These information are accompanied by another important concept, which is that of Haemo-Dynamics, no longer global but local. Understanding what is the patency of the micro-circulation in many pathologies is fundamental.

Finally, in the third of my slides in https://torvergata40.uniroma2.it/eventi/la-termo-emo-dinamica-ted-nellingegneria-medica/ [33], the multiparametric evaluation of flow in a complex cerebral aneurysm and arterio-venous malformation (AVM) to predict the result of endovascular therapy is shown on the left. On the right, you see an arterio-venous malformation and a complete subversion of the normal cerebral architecture because the capillary filter is missing, and so we are in a condition of hyperflow with venous returns that are substantial. What we have understood in these years of work on this field of clinical applications and research is that the venous side of the malformation is the one that regulates cerebral Haemo-Dynamics, that is, the resistances that occur on the venous side and also inside the malformed nucleus downstream are those that determine the natural history and the haemorrhagic risk due to this malformation. So you understand how this study of Haemo-Dynamics associated with morphology and with super-selective study is the keystone to best treat patients and to offer an increasingly more customised approach. The malformations are different from one case to another; obviously, they are not superimposable, and therefore, it is necessary to have morphological and fluid dynamic information.

Future direction is micro-endovascular surgery based on perfusional magnetic or thermo-induced gradients applied by multiple super-selective micro-catheterisation of the target lesion. The meeting between engineers and doctors is just this. Going into a more complete research of what our experience is.

Mariano Thanks to Dr. Mangiafico for sharing these beautiful images. I wanted some more technical details on the treatment of cerebral arterio-venous malformations, spirals, coils and, specifically, of the cerebral circle. What is the influence on the fluid dynamics of the cerebral circle?

Mangiafico Certainly, the technique is that of super-selective micro-catheterisation, that is, with small-calibre micro-catheters, we reach the nest of the malformation, that is, the heart of the malformation, and inside this, normally, adhesive substances are injected, kinds of glues that close this malformed nest. The impact on Haemo-Dynamics is precisely the concept that I was saying before, that is, the regulation and control of vascular flow within the nest is linked to venous malformations, therefore to the alteration of venous Haemo-Dynamics. It is also intuitive; it is the exit route that tells us what the Haemo-Dynamics of the system is, never the entrance route. So it is interesting to study and understand this, and it is an absolutely new field.

Speech of Enrica Mariano

Mariano Many thanks from me to Professor Gori for this welcome invitation. I have been dealing with interventional cardiology for many years, especially the treatment of ischaemic heart disease by implantation of coronary stents. In recent years, I have become passionate about deepening the field of structural cardiology and, in particular, trans-catheter aortic valve implantation (TAVI). In the literature, you can see the example of some types of prostheses implanted during the TAVI procedure, the cruciality of pre-procedural planning and the images related to the intra-procedural steps of the procedure itself. Also, regarding possible areas of common research with THD, Prof. Gori, you can appreciate how the deflection of the "tips" of the valve prosthetic used during the TAVI procedure, and therefore, the potential calcification of the same is related to the duration and, therefore, to the same prognosis of the patient. Also, there is a difference between intra-annular and supra-annular aortic prostheses, also as regards the thrombogenicity of the valve itself. The amount of stent tip deflection in bioprosthetic heart valves is directly associated with the level of stress in the leaflets. Elevated leaflet stresses have been described in literature as being associated with calcification potential. High leaflet stresses are known to also lead to mechanical failure of bioprosthetic heart valve protheses. The study with a computational model of the THD of these prostheses could be very interesting to guide a strategy of treatment even more tailored and accurate for the patients themselves.

The first of my slides in https://torvergata40.uniroma2.it/eventi/la-termo-emo-dinamica-ted-nellingegneria-medica/ [33] shows how it is necessary to collaborate between doctors and engineers because, already in the study of the aortic valve area or mitral, there are a series of equations, starting from Torricelli's principle, which then need a suitable simplification. Without getting into technicalities that are not the subject of this meeting, it is shown how it was necessary to move from Gorlin's formula to the continuity equation formula.

The second of my slides in https://torvergata40.uniroma2.it/eventi/la-termo-emo-dinamica-ted-nellingegneria-medica/ [33] indicates that we need to move to a simplification of the Gorlin formula that involves the study of cardiac output; the period of diastolic, systolic ejection and heart rate. We have moved to a much less *time-consuming* and easier-to-process formula than the arch formula, for which the area of a valvular stenosis is calculated from the ratio between cardiac output and the square root of the pressure gradient.

The third of my slides in https://torvergata40.uniroma2.it/eventi/la-termo-emo-dinamica-ted-nellingegneria-medica/ [33] shows another topic dear to me, which is that of coronary reserve. Myocardial ischaemia results from an imbalance between myocardial oxygen supply and demand. Coronary blood flow provides the needed oxygen supply for any given myocardial oxygen demand and normally increases automatically from a resting level to a minimum level in response to increases in myocardial oxygen demand from exercise and neuro-humoral or pharmacological hyperemia stimuli. This increase from baseline to maximal flow has been termed coronary flow reserve.

We, interventional cardiologists, have understood for many years that we certainly need to integrate our diagnosis of ischaemic heart disease through mere "vessel luminology" by coronary angiographic study, not only with intravascular imaging methods, such as IVUS and OCT, but also with a study of THD. The coronary reserve is defined by the ratio between the pressure downstream of a stenosis and the aortic pressure. It is normally evaluated through a hyperemic simulation, for example, after venous or intracoronary infusion of adenosine. In addition to the coronary reserve, a series of other indices are found in the literature. The conclusion is that in these cases, the computational analysis on a twin of the conductance arteries is fundamental to better study these patients and to understand what is the index or the multiparametric indices to better treat the patient.

Da Ros I wanted to ask you, in light of this technological evolution that has been discussed and the evolution of the guidelines, how, in clinical practice, the indications for TAVI solutions have changed.

Mariano There are new guidelines from 2021, from the European Society of Cardiology, which have very clearly highlighted the fact that patients over 75 years old and who have a risk score, which is the Euroscore 2, and which is greater than 8%, should be candidates for trans-catheter aortic valve implantation and certainly with a class 1°, therefore a very high class. Another important factor that has been highlighted in these guidelines, and this is important for me, is the role of the team. That is, the fact that apparently different figures such as the clinician, the cardiac surgeon, the imaging expert and also the Thermo-Haemo-Dynamic (THD) expert should study each case, meeting together, perhaps every day, in a round table. Another important thing is to involve the patient in the therapeutic strategy. The patient, unless they are elderly, but also the younger one, must be involved in the therapeutic choice and must be able to decide whether or not to have a classic prosthesis—what are the risks, what is the durability and so on. The patient must be involved in the process, and this has been clearly highlighted in the new guidelines.

Speech of Giuseppe Umana

Umana Good morning, everyone. Thank you for the opportunity to share some research insights on a very common disease, which is intracranial haemorrhage and rupture of cerebral aneurysm. Cerebral aneurysms can be surgically treated by a craniotomy, i.e. by opening the skull and placing one or more clips at the base of the aneurysm or where the aneurysm originates from the anatomical vessel or from the non-pathological vessel. Alternatively, aneurysms can be treated endovascularly, as we have partly seen in the previous presentations, through the inside of the artery, reaching the aneurysm sac with devices that have the task of closing or filling from the inside the aneurysm sac and therefore of excluding it from the blood circulation. However, both these techniques carry on the risk that the aneurysm sac may not be completely excluded from the blood circulation. Therefore, a part of the aneurysm, which is not excluded from the blood circulation, regardless of the choice of the type of technique that is used, is called aneurysm residue.

Our research was born from some questions about the current management of patients suffering from cerebral aneurysm, which can be summarised in the first of my slides in https://torvergata40.uniroma2.it/eventi/la-termo-emo-dinamica-ted-nellingegneria-medica/ [33]. What is the difference, from the computational haemo-dynamic point of view, between the post-clipping aneurysm residue compared to the post-coiling, and, if there is any difference, what is the risk of rebleeding? And also, can we predict which treatment modality is safer, even in the case of an aneurysm residue?

There are some studies on post-coiling Computational Fluid Dynamics, as shown in the second of my slides in https://torvergata40.uniroma2.it/eventi/la-termo-emo-dinamica-ted-nellingegneria-medica/ [33], but there are no post-clipping studies, and as a result, there is no study in the literature that compares, from the THD point of view, the residues of these two techniques. Some of the parameters that can be studied from the fluid dynamics point of view include the average flow within the aneurysm, the concentration of the inflow jet and the average speed within the aneurysm.

The third of my slides in https://torvergata40.uniroma2.it/eventi/la-termo-emo-dinamica-ted-nellingegneria-medica/ [33] shows us that studying the fluid dynamics models in aneurysms and residues, both post-clipping and post-coiling, could help to better understand the risks related to the residues of these two methods. This could change the current approach to judging these two techniques and their respective results. Some parameters that can be studied are the mean aneurysm inflow, the concentration of the inflow jet and the mean aneurysm velocity.

We could potentially also modify the current therapeutic indications. We hypothesise, for example, that there may be a difference, from the fluid dynamics point of view, in the residual aneurysm post-clipping and post-coiling, as the forces that are exerted at the base of the aneurysm to exclude it from circulation are opposite. That is, in clipping, the forces are from the outside towards the inside and therefore, strangle the neck, also modifying, in part at times, the vessels that originate near the neck of the aneurysm. This may be one of the reasons why, sometimes, the aneurysm is not completely excluded with a clipping because, otherwise, it could cause the closure of a vessel that is born there nearby, compared to coiling, in which we fill the aneurysm sac from the inside, and therefore, it is dilated from the inside out. Our hypothesis would be precisely to try to better understand, from the fluid dynamics point of view, what happens at the neck level with these two methods, in order to better evaluate the results and therefore be able to give a more appropriate and safer indication to the patient.

Boghi I wanted to ask how you think the fluid dynamics study can influence the treatment of cerebral aneurysms in clinical practice.

Umana The idea would be to predict the success rate of one of the two techniques and the risk associated with the possible partial achievement of the objective. So I think this will be the line that connects all the presentations that have been made so far today, and that is to personalise the treatment based on the specific characteristics of the patient, even in the case of a partial achievement of the result, in such a way that it is a result, partial but effective, safe and lasting over time.

Speech of Fabio Gori Ammannati

Gori Ammannati I will make the final presentation of the research entitled "Numerical prediction of aneurysmal and atherosclerotic events in the Circle of Willis" in https://torvergata40.uniroma2.it/eventi/la-termo-emo-dinamica-ted-nellingegneria-medica/ [33], which also aims to stimulate the discussion for the second part of this event/conference. What is the circle or polygon of Willis? Something that doctors obviously know but that normal people do not know or do not know that they have it in their head. It is a hydraulic circuit that was discovered by Willis in the 1600s. What was done during the Master's thesis by Dr. Eng. Alessio Pignani? Images were taken, which are free in literature, of a circle of Willis of a healthy patient. In this circle of Willis, which for us is simply a hydraulic circuit where blood flows, we put blood that behaves like a non-Newtonian Casson-type fluid, and then we pushed it with physiological pressures, and we studied the THD and mass transport of the complete circle of Willis, meaning by mass transport, the transport of lipids, LDL, oxygen and glucose in the blood.

The first of my slides in https://torvergata40.uniroma2.it/eventi/la-termo-emo-dinamica-ted-nellingegneria-medica/ [33] shows, on the left, the positions where the literature reports it is most likely that aneurysms can occur in the circle of Willis, also depicted on the top of Fig. 21, Fig. 2 of [37]. As you can see, these are points where the blood flow bifurcates, crosses and so on. The bottom of Fig. 21 shows the results of the numerical simulation for the determination of the wall shear stress [38]. In yellow and red are indicated the areas where the values of the shear stress are higher than the physiological limits, and therefore, it is reasonable to expect that it tends to exert pressure on the arterial wall of the vessel over time and therefore to cause the possibility of an aneurysmatic event at that point.

The second of my slides in https://torvergata40.uniroma2.it/eventi/la-termo-emo-dinamica-ted-nellingegneria-medica/ [33] shows, on the left, the positions where the literature reports it is most likely that atherosclerotic events are more frequent in the circle of Willis also shown on the top of Fig. 22, fig. 3 of [37]. Subsequently, we studied the numerical prediction of atherosclerosis events. The bottom of Fig. 22 [38] shows the results of the numerical simulation for the determination of the

Fig. 21 Top: literature locations of intracranical saccular aneurysms, fig. 2 of [37]. Bottom: numerical wall shear stress [38]

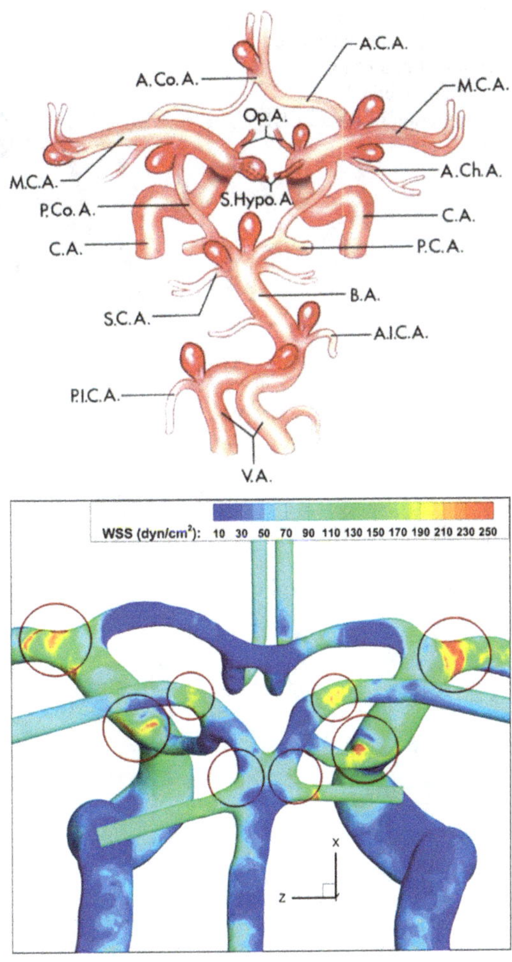

concentration of lipids, LDL [38]. I can say that the other numerical results related to glucose and oxygen are similar, so I have reported only this one.

Consoli I wanted to thank Gori for this extremely important reminder, for this type of activity, but I put myself in the shoes of the students and those who are approaching this field, who can see that the possibilities of collaboration with the world of modern medicine are multiplying more and more. I wonder what skills are required for a student, for example, of Medical Engineering, to be able to face this path of computational THD.

Gori Ammannati First of all, a clarification. When Professor Maceri proposed this degree course, it was meant to be a course in Medical Engineering, during which the students learn to know the behaviour of the healthy body, then if they know some

Fig. 22 Top: literature locations of atherosclerotic plaques. (fig. 3 of [37]). Bottom: numerical concentration of lipids, LDL [38]

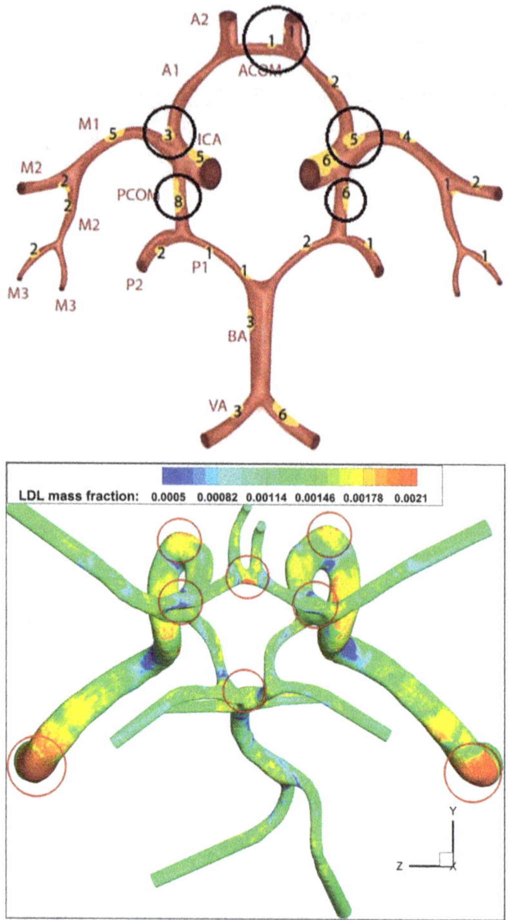

pathology better, but anyway they must know the healthy body. Something that does not happen in other degree courses in Biomedical Engineering, where machines are made by engineers for medical purposes, but the students do not know the human body. In conclusion, a Medical Engineer needs to know the human body and understand its way of operating. For example, this problem of the circle of Willis, was brought to my attention by Dr. Eng. Alessio Pignani, whom you know Arturo, because I, obviously, as a chemical engineer, did not know the circle of Willis, or, rather, I did not know I had it in my head.

Subsequently, it is necessary to know all the programming languages we need in numerical simulations, starting from the oldest, with which I started in the 70 s, which is FORTRAN, still valid, especially for the most recent simulations, the Direct Numerical Simulation (DNS), then moving on to C or C++, ending with MATLAB and, then, to know and learn the operation of numerical free software, more widespread as OpenFOAM. Hence, this further specialisation is required.

Thermo-Haemo-Dynamics (THD) in Medical Engineering and in Medicine of the Future

In the second part, Thermo-Haemo-Dynamics (THD) in Medical Engineering and in the Medicine of the Future, the speakers intervened on the three topics proposed by the moderator:

1. Is it possible to consider *Predictive Medicine*, thanks to the application of THD, to the arterio-venous system?
2. Is it possible to consider *Personalised Medicine*, thanks to the application of THD, to the images of the arterio-venous system of the individual patient?
3. How can *interdisciplinary collaboration between Medicine and Engineering* be encouraged and strengthened for these purposes?

The conclusions of the discussion, which followed the presentations of the first and second parts of the event and which can be listened to in the already-mentioned video [33], can be summarised here.

Regarding the first point, namely *Predictive Medicine*, it is considered desirable, even though it requires great efforts in research and substantial funding, as well as a more stringent and effective collaboration between Medicine and Engineering. The convenience that the national health system would have in being able to predict some diseases of the arterio-venous circulatory system or follow their evolution using computational THD, avoiding invasive imaging techniques, is also highlighted.

Regarding the second point, namely *Personalised Medicine*, computational THD is considered indispensable in all the issues presented, especially in the preoperative phase, that is, in the design phase of one type of intervention rather than another, for the specific patient.

Regarding the third point, namely the *interdisciplinary collaboration between Medicine and Engineering*, all the speakers consider it essential in order to competently and in coordination carry forward the first two points. This goal can be achieved with the introduction, in the two degree courses, of exams closer to the other degree course, such as clinical and pre-clinical exams in Medical Engineering, but, above all, with a deeper involvement of the two professional figures on cases of interest.

At the conclusion of this event, the moderator hopes that joint thesis proposals will be formulated between Medicine and Engineering that involve students and co-supervisors from the two disciplinary sectors, Medicine and Engineering.

Acknowledgements for the Figures The author thanks the *editor, ANTHELIOS (anthelios@anthelios.it),* for the kind concession to the reproduction of images *B11r, B11v, B12r, B17r, B32v, B34v and B37v,* from the volumes of the Tables of the Royal Library of Windsor, edited by Teodoro Sabachnikoff. The images are mirrored to be read more easily.

The author thanks the *Royal Collection Enterprises Limited 2024, Royal Collection Trust,* for the kind concession to the reproduction of images *RL 12597r, RL 19073v and RL 19082r.* The images are mirrored to be read more easily.

The author thanks the *publisher, Springer-Nature,* for the kind permission to reproduce images *2 and 5 of* the reference [24].

The author thanks the *publisher, Taylor and Francis,* for the kind permission to reproduce image *7 of* the reference [24].

The author thanks the *publisher, Elsevier,* for the kind permission to reproduce the *full paper of* the reference [37].

References

1. https://www.en.wikipedia.org/wiki/Fantastic_Voyage
2. https://www.comunicaffe.it/gelato-fu-quel-genio-di-leonardo-da-vinci-che-invento-la-macchina-del-freddo. August 27, 2024
3. Gori F et al (1998) Circumferential variation of heat transfer on three circular cylinders cooled by a slot jet of air. Int J Heat Technol 16(2):63–69
4. Gori F et al (2000) On the cooling effect of an air jet along the surface of a cylinder. Int Commun Heat Mass Transf 27(5):667–676
5. Gori F et al (2002) Cooling of finned cylinders by a jet flow of air. American Society of Mechanical Engineers, Process Industries Division (Publication), PID, 7, pp 117–122
6. Gori F et al (2003) Optimal slot height in the jet cooling of a circular cylinder. Appl Therm Eng 23(7):859–870. ISSN: 1359-4311
7. Gori F, Petracci I (2003) Heat Transfer Measurements and Numerical Simulations in the Cooling of a Circular Cylinder by a Slot Jet of Air. American Society of Mechanical Engineers, Heat Transfer Division (Publication) HTD, vol 374(1), pp 3–9, 2003 ASME International Mechanical Engineering Congress; Washington, DC; 15 November 2003 through 21 November 2003; Code 62617
8. Gori F et al (2003) Cooling of a finned cylinder by a jet flow of air. American Society of Mechanical Engineers, Process Industries Division (Publication) PID, vol 8, pp 147–152, 2003 ASME International Mechanical Engineering Congress; Washington, DC; 15 November 2003 through 21 November 2003; Code 62699
9. Gori F et al (2003) Pulsating jets cooling circular cylinders. American Society of Mechanical Engineers, Process Industries Division (Publication) PID, vol 8, pp 153–159, 2003 ASME International Mechanical Engineering Congress; Washington, DC; 15 November 2003 through 21 November 2003; Code 62699
10. Gori F et al (2005) Cooling of a finned cylinder by a jet flow of air. ASME J Heat Transf 127(12):1416–1421
11. Gori F et al (2007) Cooling of two smooth cylinders in row by a slot jet of air with low turbulence. Appl Therm Eng 27(14–15):2415–2425. ISSN: 1359-4311
12. Gori F et al (2011) Air cooling of a finned cylinder with slot jets of different height. Int J Therm Sci 50:1583–1593
13. Gori F et al (2012) Influence of turbulence on heat transfer upon a cylinder impinged by a slot jet of air. Appl Therm Eng 49:106–117
14. Gori F et al (2012) Experimental and numerical heat transfer on a cylinder cooled by two rectangular jets of different heights. In: ASME International Mechanical Engineering Congress and Exposition, Proceedings (IMECE), vol 7, Issue PARTS A, B, C, D, pp 2679–2688, ASME

2012 International Mechanical Engineering Congress and Exposition, IMECE 2012; Houston, TX; United States; 9 November 2012 through 15 November 2012; Code 100737

15. Gori F et al (2013) On the effect of the slot height in the cooling of a circular cylinder with a rectangular jet. Int Commun Heat Mass Transf 48:8–14

16. Gori F et al (2015) Influence of screen solidity ratio on heat transfer upon a cylinder impinged by a rectangular jet. Int J Heat Mass Transf 81:19–27

17. Petracci I et al (2016) Numerical simulation of the optimal spacing for a radial finned tube cooled by a rectangular jet. I-Average thermal results. Int J Therm Sci 104:54–67

18. Petracci I et al (2022) Forced convective heat transfer in a metallic foam cylinder cooled by a slot jet flow and comparison with a smooth cylinder and a full flow. Int J Heat Mass Transf 183: 122118

19. Di Venuta I et al (2023) Heat transfer on a flat wall due to a rectangular turbulent jet. Int Commun Heat Mass Transf 144:106769

20. Gori F (1976) Cryosurgical Probe: a numerical method for predicting the lesion produced in the tissue. University of Florence, Internal report of the Faculty of Engineering

21. Gori F et al (1984) Heat and mass transfer in soils during a microwave heating process. In: 22nd ASME-AIChE National Heat Transfer Conference, Niagara Falls (USA), ASME Paper 84-HT-78

22. https://it.wikipedia.org/wiki/Atmosfera_terrestre#/media/File:Atmosfeer_Atmosfera_-_Italiano.png

23. Gori F et al (2012) Lezioni di Termofluidodinamica (Lectures in Thermofluiddynamics). TEXmat

24. Fulchignoni M et al (2005) In situ measurements of the physical characteristics of Titan's environment, vol 438l8 December. https://doi.org/10.1038/nature04314

25. Ruffino G et al (1996) The temperature sensor on the Huygens probe for the Cassini mission: design, manufacture, calibration and tests of the laboratory prototype. Planet Space Sci 44(10): 1149–1162

26. Gori F (1996) Elementi di Termodinamica, Fisica Tecnica e Impianti (Elements of Thermodynamics, Technical Physics and Systems). Città Studi, Milan

27. Gori F (2000) Lezioni di Termodinamica (Lectures in thermodynamics). Città Studi Edizioni, Torino

28. Sabachnikoff T (1901) I Manoscritti di Leonardo da Vinci della Reale Biblioteca di Windsor, Dell'anatomia - Fogli B, Torino, Roux e Viarengo Editori. (The Manuscripts of Leonardo da Vinci of the Royal Library of Windsor, Anatomy - Sheets B, Turin, 1901, Roux and Viarengo Publishers)

29. Gori F (1991) Temperature variation of the human body during anesthetic treatments. Macroscopic and Microscopic Heat and Mass Transfer in Biomedical Engineering, K Diller, A Shitzer, eitors, 1992, International Symposium on Heat and Mass Transfer in Biomedical Engineering, ICHMT, Athens, September 119–133

30. Royal Collection Enterprises Limited 2024, Royal Collection Trust

31. https://www.focus.it/scienza/energia/le-pale-eoliche-ad-ali-di-gufo-per-energia-piu-silenziosa

32. Robicsek F (1991) Leonardo da Vinci and the Sinuses of Valsalva. Ann Thoracic Surg 52:328–335

33. https://torvergata40.uniroma2.it/eventi/la-termo-emo-dinamica-ted-nellingegneria-medica/

34. Di Venuta I et al (2017) Three-dimensional numerical simulation of a failed coronary stent implant at different degree of residual stenosis. Part I: fluid dynamics and shear stress on the vascular wall. Numer Heat Transf A Appl 71(6):638–652. https://urldefense.com/v3/__https:// www.tandfonline.com/__;!!O5Bi4QcV!EzQbvIn1n_9cyu2-9cTCYWaZTPFursHJ8FZYLftwrkzM2PC_EGNfiZKgdUQI5tWnPc_81NU3 DfghL6egunoQAKRoWVEr7w$

35. Di Venuta I (2014–2015) Influence of the degree of stenosis on the non-Newtonian blood flow in a stented coronary artery, Tesi di Laurea Magistrale in Ingegneria Medica. University of Rome Tor Vergata, A.A.

36. Russo F (2014/2015) Numerical simulation of magneto-hydro-dynamics inside cardiovascular and respiratory systems, PhD Thesis. University of Rome Tor Vergata, A.A

37. Pignani A et al (2021) Mass transfer and blood flow in a patient-specific three-dimensional Willis circle. Int Commun Heat Mass Transf 126:105369

38. Pignani A (2017–2018) Analisi numerica dell'emodinamica e del trasporto di massa nel circolo di Willis. Tesi di Laurea Magistrale in Termofluidodinamica dei Sistemi Biologici, A. A.

Prof. Dr. Eng. Fabio (name) Gori (family name of the father given at the birth) Ammannati (family name of the mother added on 2021)   was born in Montale (Pistoia) on 5 August 1947 to Emilio Gori and Cesarina Ammannati. On 16 July 2021, he added his mother's surname, Ammannati. He graduated in Chemical Engineering on 11 November 1971 with honours. In December 1971, he won a scholarship for young graduates at the Faculty of Engineering in Bologna, which he undertook from January 1972. In 1974, he became assistant professor and, in 1982, associate professor of Technical Physics at the Faculty of Engineering of the University of Florence, where he taught, also as a contract professor, until 1990. In 1978, he won a scholarship from the *British Council,* which he carried out at the *Imperial College* in London, where he collaborated with *Prof. D. Brian Spalding* on the topic of numerical analysis in the turbulent flow of liquid metals. During his time at the University of Florence, he established international scientific collaborations with *Prof. R. Echigo, Tokyo Institute of Technology, Tokyo, Japan; Prof. D.R. Chaudhary, Department of Physics, University of Rajasthan, Jaipur, India;* International Centre for Theoretical Physics *(ICTP)* in Trieste; *Centralny Osrodek Techniki Medycznej, Warsaw, Poland and Prof. T. Aihara, Institute of Fluid Science, Sendai, Japan.* In 1986, he won a National Research Council (Consiglio Nazionale delle Ricerche, CNR) scholarship, which he carried out at *Cornell University, Ithaca, New York,* where he collaborated with *Prof. R. Miller* on the topic of ground freezing. In the same year, he became professor in the Faculty of Engineering at the University of Reggio Calabria. In 1988, he was appointed as *professor* at the *University of New York at Stony Brook,* teaching the course, *Introduction to Fluid Dynamics,* in the autumn semester of the same year. During his stay, he collaborated with *Prof. T.F. Irvine Jr.* on the method of measuring thermal conductivity with a thermal probe and on the measurement of the isobaric thermal expansion coefficient of non-Newtonian fluids. In 1990, he moved to the Milano Polytechnic, where he taught Technical Physics and Systems until 1993. From 1991 until 1998, he was an adjunct professor at the Faculty of Engineering of the University of Siena. From 1992 to 2017, he was a professor of Technical Physics in the Faculty of Engineering at the University of Rome Tor Vergata, and from 2017, he is a contract professor. In 1994, he proposed the Research Doctorate (Italian *Ph.D.*) in Energy-Environment Engineering, of which he was the coordinator until 2011. In 1995, he proposed and directed the second-level Master's in Thermo-fluid-dynamics until 2016. From 1998 to 2001, he was the director of the Department of Mechanical Engineering. In 2001, he proposed and coordinated the programme, *Master of Science, Energy Engineering and Thermal and Fluid Dynamics,* in collaboration with the *Department of Mechanical Engineering, College of Engineering, University of Illinois at Chicago, USA,* which takes place at the University of Rome Tor Vergata but awards the title of *Master of Science* from the same American University. Since the late 90s, he has been a coordinator of the joint *Ph.D.* research programme with *Prof. R.J. Goldstein, Department of Mechanical Engineering, University of Minnesota, Minneapolis, USA,* and with *Prof. J.P. Hartnett, Prof. L. Kennedy, Prof. W. Minkowycz and Prof. W.M. Worek, Department of Mechanical Engineering, College of Engineering, University of Illinois at Chicago, Chicago, USA.* Since the mid-90s, he has proposed and directed the *Socrates–Erasmus* programme with *Prof. Mayinger, Technical University of Munich, Germany; Prof. van Steenhoven, Eindhoven Technical University, the Netherlands and Prof. C. Caro, Imperial College of Science, Technology and Medicine.* During his stay at Tor Vergata, he established scientific collaborations with the *Department of Mechanical Engineering, University of Minnesota, Minneapolis, USA,* collaborating with *Prof. R.J. Goldstein* on Thermo-fluid-dynamics and mass transport in gas turbine blades; the *Energy Resources Center, University of Illinois at Chicago,* collaborating with *Prof. J.P. Hartnett, Prof. W. Minkowycz* and *Prof. W.M. Worek*; the *Department of Mechanical Engineering, College of Engineering, University of Illinois at Chicago, collaborating with Prof. L. Kennedy and Prof.*

W.M. Worek; *Duke University, collaborating with Prof. A. Bejan* and *S. Mary's University, Halifax, Nova Scotia, Canada, collaborating with Prof. W. R. Tarnawski*. The bibliographic review of documents and citations, titled *PlosBiology Career*, places him within the top 2% of researchers in the field of *Mechanical Engineering and Transports*. Since 1992, he has been a tutor for over 30 PhD theses, supervisor for over 40 second-level Master's theses, supervisor for over 70 degree theses (five-year, Specialist and Master's), supervisor for over 10 *Master's degree in Mechanical Engineering* from the *University of Illinois at Chicago* and supervisor for over 20 theses *Erasmus-Socrates*.

Since the early 2000s, he has been a reviewer of international research projects on behalf of *Portuguese Science and Technology Foundation (FCT); Czech Science Foundation (GACR); European Science Foundation, Strasbourg, France,* on behalf of *FCT* and the *Shota Rustaveli National Science Foundation, Georgia.* Since 2017, the year of his retirement due to age limits, he has been a contract professor at the University of Rome Tor Vergata, where in the academic year 2024–2025, he has taught Technical Physics for the Master's Degree Course in Medical Engineering. M.D. Salvatore Mangiafico (Florence), scientific head of the NeuroVascular Base Camp, 2021, invited him to give a scientific presentation titled *"The engineering approach to the study of the Circle of Willis"* on 23 September 2021 at the Congress Centre, University of Rome La Sapienza, thus opening up a possible future collaboration.

Cryosurgery

Cryosurgery: Prediction of Freezing Front Penetration and of the Lesion Produced in the Tissue

Andrea Boghi and Fabio Gori Ammannati

1 Introduction and Medical Aspects

1.1 History of Cryosurgery

Already, in 2500 B.C., the Egyptians used cold to treat wounds and inflammations. Lorrey, a surgeon in Napoleon's service, had discovered that it improved amputation operations. Only in 1845 did cryosurgery become a treatment capable of destroying unwanted tissues, when James Arnott exposed a skin tumor to a partially frozen saline solution at a temperature of $-22\ ^\circ$C.

The use of cryosurgery in the treatment of diseases affecting internal organs occurred in 1961, when Cooper and Lee [1] presented the first cryostat for medical applications. A cryostat is defined as a cooling probe in which a cooling bath is generated within the probe, known today as a "*cryoprobe*." The first cryoprobe was used in 1967 by Lee [2] to treat brain tumors and Parkinson's disease. Over the years, cryosurgery has been tested on a wide variety of tumors and other tissues. But the concept of minimally invasive cryosurgery was introduced in the second half of the 80 s when many small probes, the so-called multiple probes, replaced a large cryoprobe. This improvement occurred as a result of the development of modern imaging techniques.

In 1993, the treatment of prostate cancer through cryosurgery became the first minimally invasive procedure, transitioning from an experimental state to a routine operation. The desire to make the surgical procedure minimally invasive creates a new level of difficulty, as it is necessary to have complete knowledge of the tissue

A. Boghi
Computational Science ltd, Southampton, UK
e-mail: a.boghi@computationalscience.co.uk

F. G. Ammannati (✉)
University of Rome Tor Vergata, Rome, Italy
e-mail: gori@uniroma2.it

F. Gori Ammannati (ed.), *Thermo-Haemo-Dynamics in Medical Engineering*,
https://doi.org/10.1007/978-3-031-97214-0_2

shape in three dimensions so that adjacent tissues can be preserved. Thanks to recent advances in *Joule–Thomson* cooling technology, the diameter of the cryoprobe is significantly reduced. To gain better control over the surgical procedure, the number of cryoprobes has further increased from 6 to 12 or more applied simultaneously. Although *Joule–Thomson* cooling has a lower cooling power compared to the heat transfer with liquid nitrogen, this feature is compensated by the increase in the number of cryoprobes. This gives rise to a new challenge: how to regulate the heat subtracted from the tissue by each probe to obtain the most efficient shape of the frozen region, i.e., the shape that allows the freezing of oncotic tissue without damaging the healthy tissue. A good example of a device that prevents freezing to preserve desired tissues during prostate cryosurgery is the urethral warmer, which is nothing more than a counter-current water exchanger inserted into a standard catheter [3, 4].

1.2   Cellular Response to Cryosurgery

The characteristics of tissue damage due to freezing have been known for several years and generally derive from two mechanisms. The first consists of direct damage to cells, caused by the formation of ice crystals, and the second leads to the blockage of microcirculation that occurs during the thawing period. The relative importance of these two effects is being discussed [5]. Some researchers investigated this aspect of the problem by evidencing that the speed of rewarming of the tissue can also influence the cell's destruction [6–10].

The formation of ice crystals and the consequent removal of water from the biological system produce a large number of harmful effects. The formation of intracellular ice is lethal for cells. In fact, cells, which are tightly packed in surrounding tissues, must withstand the stresses that the ice crystal exerts on the cell wall, i.e., normal stress and shear stress. There are also vascular effects such as, for example, the destruction of microcirculation. The loss of circulation prevents the cell from receiving nutrients and eliminating waste products.

The most important discovery at the molecular level, regarding cryosurgery, is that cells do not die from necrosis (death due to acute stress or cellular trauma) but from apoptosis (programmed cell death, finely regulated in its phases by complex biochemical processes). Apoptosis is a cellular mechanism that occurs in every tissue in order to ensure cell turnover. This mode of cell death is also present in various pathological conditions, such as cancer, cytotoxic chemotherapy, hormone ablation, and irradiation. Even in cryosurgery therapy, apoptotic cells are observed; after freezing, apoptotic cells are found in the periphery of the cryogenic lesion, as, in this area, the temperature is not low enough to kill all cells. On the periphery, some cells survive, and some die. Others remain for a few days in a state of uncertainty and then die, showing signs of apoptosis.

Kinetic experiments have shown that cells are susceptible to entering the state of apoptosis up to 8 h after thawing. The response to freezing varies from inflammation to destruction, depending on the extent of the freezing. Mild freezing only causes an inflammatory response, which can also be used therapeutically; more severe freezing causes the destruction of cells and tissues, which is the first requirement in the treatment of tumors, producing the state of coagulation necrosis in the frozen tissue in the days following thawing. There is some difference in cold sensitivity among different types of cells, and this is the basis of the so-called selective cryotherapy, whose aim is to destroy certain cells and leave others intact. The cryogenic lesion is characterized by a central zone, consisting of necrotic tissue of coagulation, while, in the periphery, only part of the cells will undergo death. This change in tissue develops a few days after freezing. Immediately after thawing, the tissue appears congested and hyperemic and quickly becomes edematous. The progression of necrosis becomes evident after 2 days. In the central region surrounding the cryo-probe, cell death is uniform, but at the periphery, in the region where the temperature is between 0 and -20 °C, some cells survive, others die, and some others are suspended between life and death. In this area, apoptotic cells are observed. The wound repair process begins in the periphery of the area in contact with living tissues. Inflammatory cells infiltrate the interstices, and new blood vessels grow in the damaged tissue. In a few weeks (or months), the dead tissue is slowly replaced by fibroblasts that deposit new collagen. The result is the formation of a contracted healed area [11].

1.3 Clinical Aspects

There has been growing enthusiasm when cryosurgery has been applied to the treatment of tumors. This became possible when new devices, based on the use of liquid nitrogen, pressurized argon, or nitric oxide, were introduced, as shown in Fig. 1 [12]. Liquid nitrogen is the coldest of the cryogenic liquids ($T_v = -195.8$ °C), but it cannot be used in probes that have a diameter smaller than 3 mm. Argon ($T_v = -185.7$ °C at a pressure of 0.1 MPa), pressurized for use in a device operating

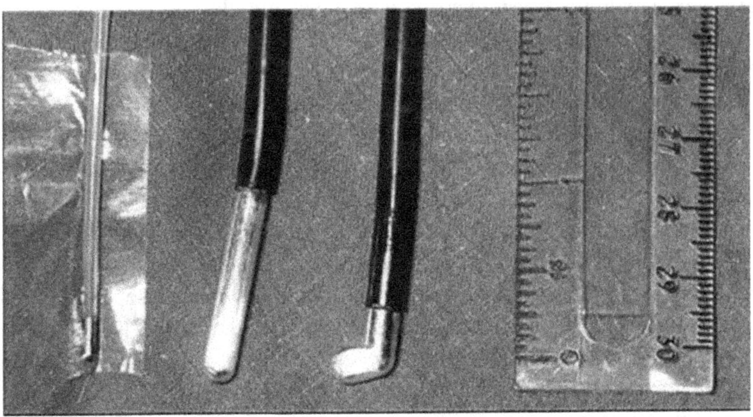

Fig. 1 Liquid nitrogen cryoprobes, Fig. 6 of [12]

according to the *Joule–Thomson* principle, requires the use of needles and probes with a diameter not smaller than 1.4 mm, which allows it to be used in the skin and in minimally invasive applications. Therefore, argon, along with nitric oxide, which has very similar characteristics, is very popular in cryosurgical applications. However, a smaller probe has a smaller freezing power. Applications using argon and nitric oxide require the presence of multiple probes to be implanted, thanks to guides. This procedure is widely used in the treatment of prostate cancer, where care must be taken not to touch the nearby regions to avoid causing infertility problems. In that case, it is good to have many small needles that make the temperature field more uniform and a lower ablation temperature in the target tissue. In addition to the tumor itself, a certain thickness of cells, immediately adjacent to the tumor, must be frozen, and in the case of the prostate, this thickness must be precisely regulated to avoid damaging the surrounding tissues. The effectiveness of cryosurgery is not only linked to the technique of freezing the tissue but also to the freeze–thaw cycle. Various studies have shown that the best results are obtained by freezing the tumor cells very quickly, reaching the lethal temperature as quickly as possible, up to the safety margin, which prevents damage to healthy tissues, and then causing the thawing of the tumor tissue very slowly. Furthermore, the freeze–thaw cycle must be repeated several times.

Every phase of the procedure can be harmful to the patient: the duration of freezing, or thawing, the temperature reached in the tissue, etc.; moreover, it must be taken into account that the tissue near the probe freezes quickly, but the tissue far from it freezes slowly, and the temperature profile is very irregular. Because of this, it will be appropriate to analyze the various phases in a more in-depth manner. The freezing phase must be as rapid as possible because only a rapid freezing allows the formation of intracellular ice, which is lethal for the cell itself. The intracellular ice forms in a wide range of temperatures, provided that the freezing speed is at least 20–50 °C/min. In more tightly packed cells, even lower speeds induce freezing. Moreover, the cryoprobe must have the lowest possible initial temperature before being introduced into the tissue. However, if the distance between the cable and the needle is very small, the freezing speed is reduced, and therefore, the freezing speed cannot be considered a primary factor in the success of the technique. For cryosurgery to be effective, the lethal temperature must be reached in all the tumor cells. The temperature at which cell death occurs is between −5 and −50 °C; however, to have the certainty of the effectiveness of the technique, it is ensured that the temperature is at the lower limit. The temperature that is actually used in clinic is between −40 and −50 °C and must be reached even in a certain thickness, called safety, outside of it. This temperature range ensures complete destruction. The cells must remain in the frozen state for at least 5 min to ensure the formation of ice crystals and the effects of recrystallization. However, the best duration of freezing has not yet been found. The most important factor in the success of the cryosurgical technique is identified as the thawing speed. For the technique to be successful, thawing must occur very slowly, and it is better if it happens without the support of heating elements. The long duration of thawing favors a series of events that destroys the cell: from damage due to the dissolution of molecules to structural changes of ice crystals during the melting and oxidative stress. During thawing, indeed, the ice crystals modify their structure and cause shear stresses that destroy the cell.

Over the years, it has been discovered that the effectiveness of the technique increases if the freezing–thawing cycle is repeated several times. The repetition of the cycle produces faster and more extensive freezing in the tissue, and the region of dead cells expands to the outer edge of the frozen region so that all cells in the frozen volume are dead. The first freezing cycle has the effect of increasing the thermal conductivity of the tissue, caused by the cell destruction, and the second cycle forces the tissue to extend those physicochemical changes by passing it for the second time through the same freezing conditions. The second cycle increases the necrosis of the tissue by 80% in the previously frozen volume. The repetition of the cycle has given excellent results in the treatment of prostate cancer. Also, the interval of repetition of the two cycles plays an important role in the development of the technique. A further advantage of a long thawing time is that there is complete vasoconstriction of the vessels in the frozen region, leaving the tissue in a hypothermic state. The presence of an interval between the two cycles leaves time for the microcirculation to stop completely so as not to bring more nourishment to the tissue. Moreover, an interval allows the progression of oxidative stress initiated in the thawing phase.

The process of freezing the tissue must be monitored with one or more techniques. Mounted on the cryoprobe needles are sensors, for example, thermocouples, that have proven to be quite accurate in monitoring therapy. Their role is to prevent the critical temperature from being reached in dangerous areas. In addition to this, the most common techniques of *imaging* are used in monitoring therapy, such as ultrasound (US), computed tomography (*CT*), and magnetic resonance imaging (*MRI*) (Fig. 2, [13]). The bottom right of Fig. 2 highlights the formation of ice crystals with irregular geometry and a rough profile around each cryoprobe. Precisely because it is so difficult to obtain an isothermal surface in the peripheral area, the one in which apoptosis prevails, it is useful to combine the surgical technique with a pharmacological therapy capable of promoting apoptosis. Such therapy could be carried out using cytotoxic chemotherapeutic drugs, irradiation, apoptotic promoters, or immune enhancers, which complete the destruction of cells in the cryogenic lesion. Experiments suggest that, when chemotherapy is given before freezing, even at not-too-low temperatures (-5 °C, -25 °C), the result is the complete apoptosis of the tissue. The choice of cytotoxic drug depends on the type of cancer found in the patient, and the possibility of applying the drugs in situ should not be excluded. The difficulty remains in the dose to be given to the patient. A similar problem occurs with irradiation. Although in vitro experiments have shown that frozen cells have a greater sensitivity to irradiation, nothing is known about the optimal exposure time.

1.4   Skin/Cutaneous Applications

Cryosurgery applied to the destruction of tumor cells in internal tissues represents a fairly recent application and, anyway, not the main one. The most frequent use is in

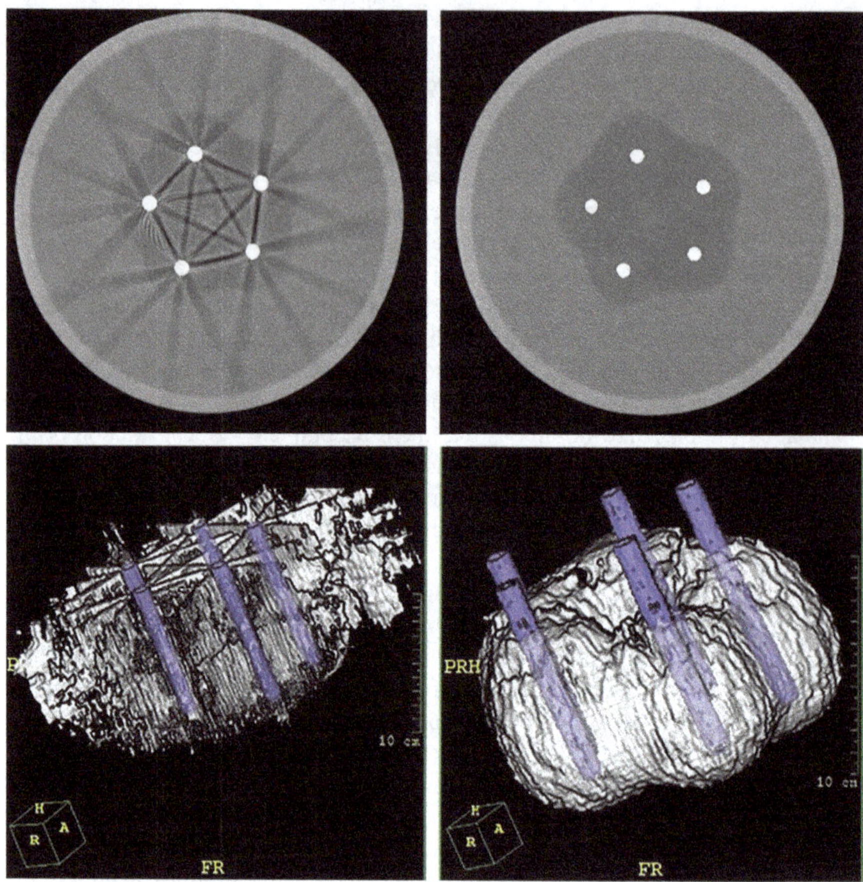

Fig. 2 Comparison of CT images of cryoprobes and ice balls in a homogeneous gelatin phantom before and after correction. (Top left) Image before correction, (Top right) image after correction, (Bottom left) 3D rendering before correction, and (Bottom right) 3D rendering after correction, Fig. 2 of [13]

skin applications, i.e., in the destruction of tumor cells but not only those that affect the skin. Therefore, it is used to treat basal cell and squamous cell skin carcinoma, taking advantage of the fact that the treatment is simple and does not leave keloids; multiple lesions can be treated in a single visit; and there is a huge speed of healing, which makes the treatment even more appreciated than radiation therapy.

Many cryogenic liquids are used in dermatological treatments, like carbon dioxide, nitric oxide, various *Freon* gases, and liquid nitrogen, which boils at $-196\,^{\circ}\mathrm{C}$ and is the most effective in cancer treatment. With liquid nitrogen, temperatures of $-25\,^{\circ}\mathrm{C} \div -30\,^{\circ}\mathrm{C}$ can be reached in 30–40 s if a sufficient amount of it is applied by using a spray or probes. The formation of ice crystals causes the rupture and death of cells, but greater damage is caused during thawing. Pigmented cells are the most sensitive to destruction, and keratinocytes are less so. Connective tissue is very resistant to damage, while capillaries and arterioles develop *microemboli* that contribute to tissue destruction. The large vessels remain inert, ensuring, in this way, a rapid revascularization and healing. Initially, there is a wet exudative tissue, followed by a dry scar tissue after about a month, and finally, the new pigmented skin.

The use of a cotton tip, soaked in liquid nitrogen, has been popular in the treatment of benign lesions; however, this method has been supplanted by the application of liquid nitrogen via spray. The spray is easy to use, and with it, benign, premalignant, and malignant lesions can be treated. Continuous freezing at a site for more than 30 s, when a sphere of ice of adequate size has formed around the tissue, can cause destruction of the collagen matrix. Light freezing causes dermo-epithelial separation, which is very useful in the treatment of benign lesions. The most sensitive cells of the epidermis are destroyed, while those in the dermis remain intact. Complications in treatment can be due to hypopigmentation, but studies and clinical experience show that re-pigmentation occurs after a few months for undamaged melanocytes within hair follicles or due to the migration of melanocytes away from the center of the lesion. The methods in which liquid nitrogen is sprayed on the lesion include the *timed spot freeze* and the *direct spray techniques* with a rotary or spiral path and the *paintbrush method. Timed spot freeze* allows the standard release of liquid nitrogen. The use of this technique maximizes the ability to destroy a lesion with the lowest morbidity rate. The freezing time is regulated with variables such as skin thickness, vascularization, and type of lesion. The technique is used with a spray, which can contain 300–500 ml of liquid nitrogen. If more than one freeze–thaw cycle is required, complete thawing must be ensured before starting the next cycle, in approximately 2–3 min. The technique achieves temperatures adequate for tissue destruction in lesions smaller than 2 cm in diameter, while, for larger lesions, several treatments must be overlapped. Variations of *timed spot freeze* include the rotary or spiral application and the *paintbrush method.* These techniques can be used for the treatment of larger benign lesions, which are not yet standardized and, therefore, cannot ensure a correct temperature for tissue destruction. While the spray technique can be used for most lesions, easily accessible from the outside, a cryoprobe attached to the liquid nitrogen spray can provide greater versatility, depending on the location and type of lesion. Various types and sizes of cryoprobes are available. Generally, a gelatinous interface is used between the skin and the cryoprobe. Cryoprobes are generally used in the treatment of small facial lesions (such as in the eyes), where the spread of the spray must be avoided. They are also useful in the treatment of vascular lesions, where the pressure of the cryoprobe can be used to remove blood from the tissue.

2   Monitoring of the Intervention

2.1   Imaging Techniques

When performing the intervention, the surgeon must know exactly which cells have to be destroyed, as well as all tumor cells and healthy tissues that have to be preserved; in short, he/she must know the position of the ice front. As long as the technique is used in the treatment of skin tumors, there are no problems because the frozen area is visible to the naked eye, but when the technique is used to treat brain or prostate cancer, we need more sophisticated tools to determine with certainty the position of the ice front. There is always a margin of uncertainty in the application of this technique, as different cells show different sensitivities to temperature lowering. In order to obtain information about the growth of the ice front, the simplest means is to observe its growth indirectly through the techniques of *imaging*.

2.2   Ultrasonography

The technique most commonly used in monitoring the procedure is one that is based on ultrasound (*US, ECHO,* etc.), which has the advantage of not releasing harmful radiation to the patient and of giving real-time information. Figure 3 [14] shows the ultrasonographic image of the formation of the ice sphere inside the goat's mammary glands using a transducer array at a frequency of 7 MHz. The frozen region appears as a dark area in the ultrasound image. The cryoprobe is perpendicular to the ultrasound transducer, which is applied from above, and therefore, the shadow of the cryoprobe is projected downward, Fig. 3a. A measurement of the ice ball is illustrated in Fig. 3d and appears to be 2 cm thick after 3 min of application. The diameter of the cryoprobe is 1.15 mm.

However, this technique has some flaws: It is two-dimensional because it gives information only on planes parallel to a given axis. This is the major disadvantage because, in heterogeneous and anisotropic tissues, like biological ones, one cannot expect the isothermal surfaces to present any symmetry, and therefore, having a two-dimensional view, one fails to fully grasp the advancement of the ice front and understand whether a certain region has been reached or not. This is very important because one might not have noticed a group of oncotic cells that escaped freezing and that could return to attack the tissue. Another disadvantage of ultrasound lies in its nature as mechanical waves. Indeed, sound waves are reflected in all directions by the ice, thus creating a "shadow" in the *US* image. In this way, all the structures that are beyond the frozen region cannot be seen. Consequently, the exact size of the ice front cannot be known. Moreover, the technique does not give information about the temperature field, so it is necessary to extract such information from external thermal sensors. For this reason, over time, alternative solutions have been sought that overcome the difficulties introduced by the use of ultrasound.

2.3   Magnetic Resonance

Magnetic resonance imaging (*MRI*) offers many advantages that make it suitable for use in cryosurgical intervention. The *MRI* provides high spatial and temporal resolution, and unlike the *US*, it is sensitive to temperature, which provides excellent contrast in discriminating between frozen and unfrozen tissues. As a result, it is possible to determine very accurately the size of the ice sphere in real time. However,

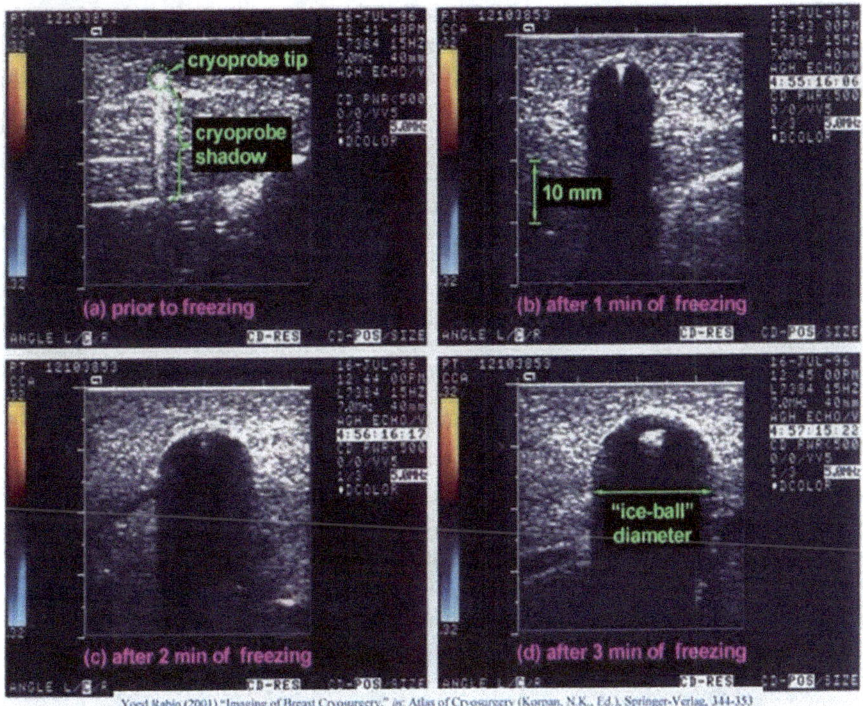

Fig. 3 Ultrasonography of the formation of the ice sphere inside the goat's mammary glands, Fig. 16.1.1 of [14]

the detection of the lethal lesion, caused in the target region, remains uncertain. Indeed, since the relaxation time decreases with temperature, the signal of *MRI* can no longer be measured when the temperature drops below 0 °C, at least considering the conventional return signal times (*echo times*). Consequently, the thermal history of the tissue can hardly be determined directly based on magnetic resonance. Therefore, frozen tissues require more refined methods to allow direct mapping of the ice sphere within the cryogenic domain.

A study, [15], also shows, in Fig. 2 of [15], the raw and processed images in gelatin, (a), (c), and (e), and ex vivo of the pig liver, (b), (d), and (f). The images (a), (b), (c), and (d) are acquired at two different times of the freezing cycle and an interval of 10 min. The images (e) and (f) present the results of images (2) and (3). In the raw images, the cryoprobe appears vertical, while the thermal sensors, indicated by dashed arrows, are seen as dark spots in the gelatin because they are inserted vertically into the sample, while in the pig liver, they appear horizontal. The clearest and most direct map of the evolution of the temperature field within the ice front is obtained by implanting thermal sensors at specific points near the target tissue. The obvious advantage of this procedure is the direct measurement of the temperature. The most obvious disadvantage is the very limited spatial resolution since the temperature field can only be measured at a limited number of points. For this reason, it is common to resort to numerical techniques that allow obtaining an indirect solution. The advantage of this technique is its accuracy and the possibility of calculating very complex three-dimensional temperature fields. However, numerical calculations are not real-time; therefore, the use of this method to deduce the temperature field remains a challenge. Research proceeds in this sense, and efforts are being made to speed up the calculation process to make the simulations effectively usable in diagnostics.

2.4   Computed Tomography (CT)

The *CT* is a valid alternative to the use of ultrasound, with the significant advantage of providing three-dimensional images of the ice sphere. Figure 2 [13] compares *CT* images of the cryoprobes and ice spheres in a gelatin dummy before and after correction: (a) image before correction, (b) image after correction, (c) three-dimensional rendering before correction, and (d) three-dimensional rendering after correction. Another advantage of the *CT* is that the temperature in the region where the density of the water–ice interface varies can be calibrated to allow the deployment of the isotherm contour, almost in real time. Unfortunately, these great advantages are compromised by the presence of artifacts, caused by the presence of metallic cryoprobes. One of the major causes of artifacts, due to the presence of metallic elements, is the correction that the software in use in the *CT* operates on hard X-rays. The theory of image reconstruction in *CT* assumes that the incident X beam is monoenergetic, but this can never be achieved, and practically, the energy spectrum is between 20 and 140 keV. To solve this problem, typically the software *CT* performs a digital filtering of the images to approximate the ideal attenuation that would be obtained with the monoenergetic beam. This correction is calibrated for biological tissues, and when trying to apply it to metallic objects, artifacts are created. Therefore, in the reconstruction made by the *CT*, what is noticed is a series of lines that depart from the cryoprobe throughout the image. Another source of artifacts is the poor signal-to-noise ratio, caused by the weakness of the photons of the incident beam passed through the metallic cryoprobe, which greatly attenuates its energy. To conclude the discussion on the use of *CT* in monitoring cryosurgery, it should be remembered that one disadvantage in using it is the fact that the patient absorbs a high dose of radiation as a result of continuous monitoring during the operation.

2.5   Optical Coherence Tomography (OCT)

In addition to more traditional techniques, such as *MRI*, *US*, and *CT*, it has been proposed to obtain images, at high resolution and in real time, of the ice front using an optical technique, namely the *optical coherence tomography (OCT)*. One of the advantages of this technique is that the *OCT* probe can be incorporated into a wide variety of instruments, such as catheters, endoscopes, and laparoscopes. The *OCT* is a technique similar to *US*, but it is based on the use of light. Using a super-luminescent diode (*SLD*) that emits broadband light through absorption and scattering phenomena, it propagates through the tissues. Since there are currently no sensors fast enough to distinguish between backscattered photons from tissues at different depths, a detection system based on coherent interferometry is required to detect the amount of backscattered light from various depths. By performing a lateral scan from the cryoprobe, the two-dimensional section is reconstructed.

The *OCT* also has advantages and disadvantages. First of all, within the cell, whose composition is watery, this technique can be used to detect freezing only after the surface has frozen. The advancement of the ice front that starts from the probe through the tissue is not evident, although some small local increase in the signal can be detected due to the increased reflectance of the water–ice interface. Instead, when the ice front crosses the lipid membrane, a strong reflective surface is created with a progressive decay of the signal as one moves away from the probe. The ability to discriminate between various tissues depends on the variation of the refractive index during freezing. Therefore, considering the greater ability to discriminate the signals obtained between ice and lipids, this technique is very suitable in all those applications where the tissues are rich in lipids, i.e., in skin applications. To monitor the ice front, very small interstitial probes are used around the region of interest. For the discrimination to be effective, there must be many probes, and it is preferred to insert them into the needles.

2.6   Current Impedance Technique

Another technique is the one based on the measurement of the impedance of biological tissues, *current impedance technique (CIT)*. This technique is noninvasive and provides information about the electrical characteristics of biological tissues within the body. It is based on the application of currents at low intensities and the measurement of the potentials that develop on the surface. A current, with a relatively low frequency, is applied, at about 5 mA, in such a way as to use the quasi-static approximation and ensure the safety of the tissue but high enough to be discriminated in frequency compared to biological signals. The most common procedure is based on the application of surface electrodes. It should be noted that the skull being very insulating, this technique cannot be applied as it is in the case of brain tumors without applying some modifications. The impedance of frozen tissues is much greater than that of unfrozen tissues; therefore, it is easy to detect the impedance jump, which must be studied in frequency behavior to verify the changes in inductive and capacitive components.

In this regard, there are two important indices that are used in the application of this technique, namely the *dynamic low-frequency impedance (DLFI)* and the *impedance changing rate (ICR)* [16, 17]. Thermocouples can be implanted in tissues but suffer from the disadvantage of being position-sensitive. Moreover, they can detect the temperature only at a particular point of the tissue, and there is the risk that the thermocouple moves the cells from the region where they need to be treated by cryosurgery. Magnetic resonance can theoretically monitor the temperature field, but it is not credible enough to be widely used for this purpose without the aid of thermocouples. The presence of the ice sphere within the conductive volume and its slow growth during the procedure alter the current lines that should bypass the sphere, electrically isolated. As a result, the developed potential also changes, and these changes are at the basis of impedance monitoring. The technique has been studied through two different approaches: The first one consists of the direct injection of current and the second one consists of induction through a magnetic field [18]. The second method also provides greater advantages in relation to the hardware to be used.

2.7 Thermography

A valid alternative could be to use infrared thermography to correlate the temperature reached with the extension of the necrotic area. Figure 4 [18] shows the distribution of potential (V) in the space, y (m) along the ordinate, and x (m) along the abscissa, for a coil at the coordinates 200, 0, and 1.7 (mm) (top) and the normalized current density (bottom) in the space y (m) along the ordinate and x (m) along the abscissa. In the images of Fig. 4, the isotherms appear asymmetrical due to the so-called *heat sink effect* because of the presence of the surrounding tissues. The *heat sink* effect is a phenomenon whereby the freezing effect decreases during the propagation of the ice front. This happens because ice is a good thermal insulator and, therefore, prevents a wide diffusion of the freezing effect during the treatment. Irregularities in the isotherms increase as the temperature rises. When the cryoprobe is removed after the first cycle, it takes 2.5 min for the temperature to rise above 0 °C. After each cycle, the frozen area grows like the thawing time, which, by the third cycle, reaches almost 4 min. The results show good agreement between the data recorded by the thermocouples and those of thermography. The thermocouples have a resolution of 2 °C and thermography of 0.5 °C, which makes the discrepancy between the measured temperatures up to a maximum of 3 °C. Histological sections also show that there is good agreement between the area representing the critical isotherm and the area of necrotic tissue.

Figure 5 [19] shows the isotherm surfaces, with a 1-cm thickness around a cryoprobe, applied to the shaved abdomen of a rat for 1 min. The isotherm at −20 °C (blue area) occupies 53% of the area of the visible ice sphere (red area). A disadvantage of infrared thermography is that it can only monitor surface temperature and cannot measure the penetration of the freezing effect. However, studies conducted on meat have given an idea of the shape and size of the ice front in the subepithelial region once the size of the ice surface is known. In the living animal, it can be assumed that this effect is lower due to the presence of blood vessels, which tend to heat the surrounding tissue.

Infrared thermography may appear as an accurate means to monitor the temperature reached in the therapy with a liquid nitrogen probe on a flat surface. Fig. 6 [19]

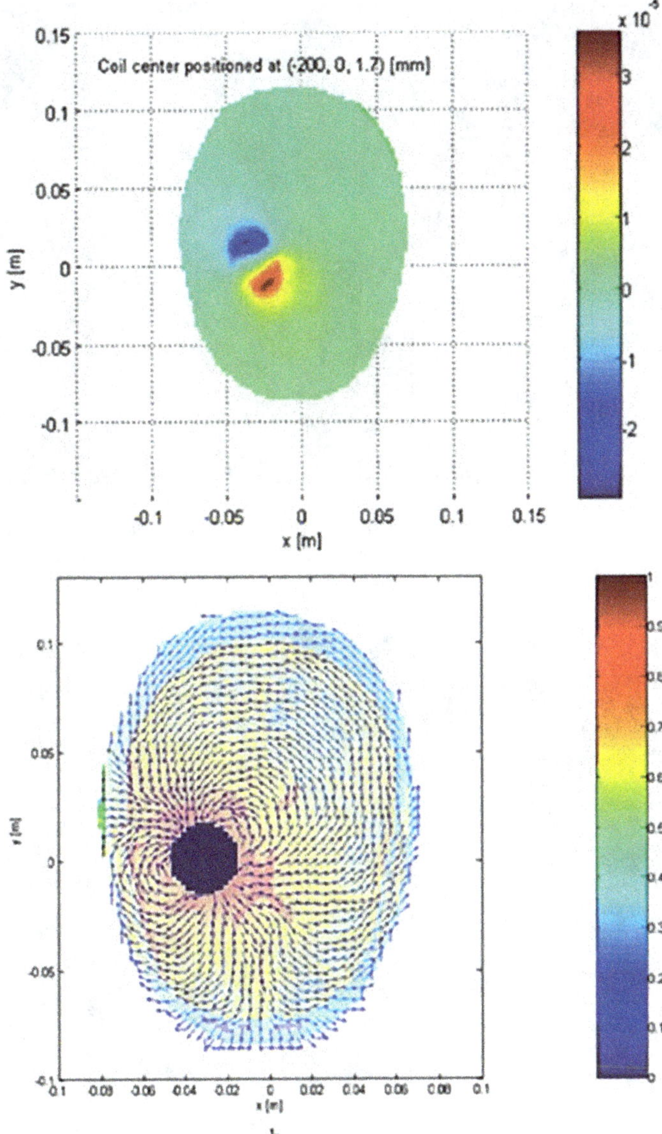

Fig. 4 (Top) Potential distribution (V) in the space y (m), along the ordinate, and x (m), along the abscissa. (Bottom) Normalized current density in the space y (m), along the ordinate, and x (m), along the abscissa, Fig. 5 of [18]

shows the isotherm surfaces 40 s after the probe is removed. The coldest area is at about $-16\ °C$. However, the equipment is expensive and delicate, requires maintenance, and cannot be considered a technique for use in clinical practice. In addition, the infrared camera is not accessible everywhere, and many surfaces are not flat, and therefore, the surface of the camera, not being orthogonal to that of the investigation,

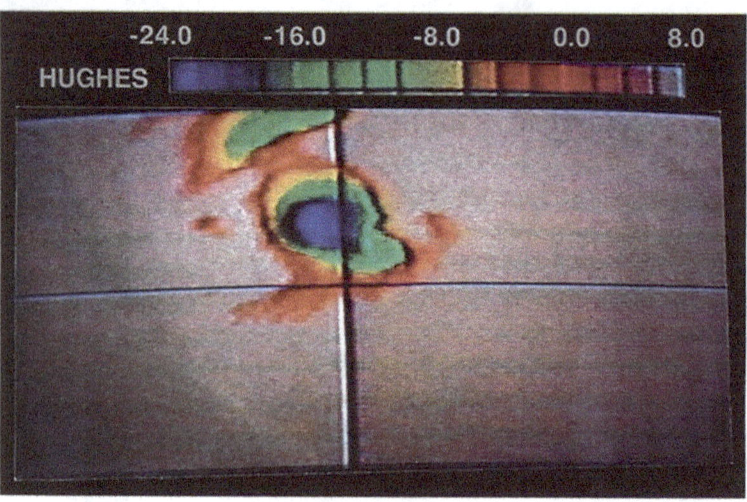

Fig. 5 Isotherm pattern of 1-cm, round cryoprobe applied to a shaved rat abdomen for 1 min. The −20 °C isotherm (the first blue isotherm) occupies 53% of the area of the visible ice ball (−2 °C red isotherm), Fig. 1 of [19]

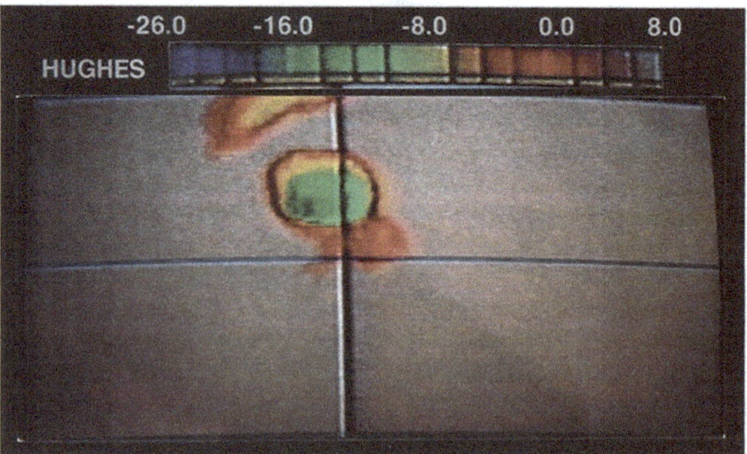

Fig. 6 Isotherm pattern 40 s after the removal of the probe. The coldest area is now at approximately −16 °C, Fig. 2 of [19]

can produce artifacts. The camera can also only view the surface of the lesion, and therefore, any information about the depth can be lost.

Since normal imaging techniques do not give us three-dimensionality, it is necessary to overcome this drawback by making a prediction of the temperature field using numerical analysis techniques. This becomes crucial, for example, in liver cancer, where the critical isotherm is −40 °C, but the *MRI* cannot penetrate at such low temperatures. Fig. 7 [20] shows the images taken at the end of the freezing cycle:

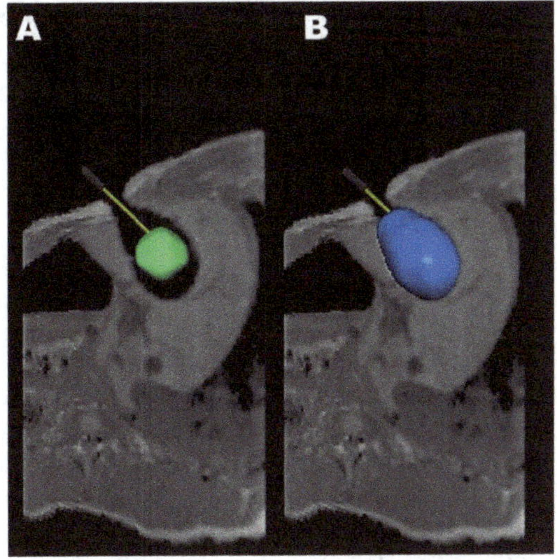

Fig. 7 Images taken at the end of the freezing cycle, Fig. 5 of [20]

In (A), the tumor (colored in green) is engulfed by an empty signal as a result of freezing, and in (B), the frozen region is colored blue (based on the segmentation of the images of the frozen region).

The estimation of the temperature field through the implementation of a mathematical model can help us understand the position of the critical isotherm when we are not able to see it. To do this, it is necessary that, attached to the cryoprobe, there is an instrument, generally of an optical type, that continuously measures the position of the probe with respect to a spatial reference. Therefore, the *software* provides a virtual environment in which the 3D scenario is composed from the medical image and the coordinates provided by the tracking system, enlarged, thanks to navigation systems. The scene is presented on a display unit, from which the clinician can observe it. The software must perform the operations of *image processing*, i.e., calculation of the temperature and generation of polygons for isothermal surfaces. Therefore, it needs the segmentation of the frozen tissue from the *MRI* images as input data. This can be done in a semi-automatic way by *thresholding* and refining the contours manually or using a median filter. Based on the segmented frozen region, the position of the cryoprobe, and its temperature, the 3D profile of the thermal field is calculated. Normally, the calculation is performed by connecting to a *server* system.

One of the limitations of this procedure is that the liver is considered rigid during registration, i.e., it does not take into account the mobility that the images can have due to respiration. However, it must be said that during the operation, the patient is under anesthesia and does not move too much anyway. We can assume, therefore, that the cryoprobe and the frozen tissue are rigidly coupled, and therefore, the position of the ice front can be inferred from that of the cryoprobe. The accuracy of this method is high. However, it must be ensured that the position of the tumor with respect to the probe is calculated faithfully. Another problem is to the long acquisition time, which for the *MRI* is about 6 s. This can cause significant delays in navigation. In addition, the total time required to calculate the thermal field is 5 min. The delay, however, can be acceptable since the procedure can last even 20 min.

3 Cryoprobes

3.1 Operating Principles

There are many types of cryoprobes, depending on the principles of operation and the applications for which they are designed. The first cryoprobe was introduced in [1] for clinical practice, and the freezing effect is given by the phase change of liquid nitrogen, which passes to the gaseous state. A more modern device is based on the *Joule–Thomson* effect, in which the freezing effect is obtained by passing a high-pressure gas through a small orifice. The expansion of the gas, which can be nitric oxide, carbon dioxide, or argon, causes the probe to cool down due to the *Joule–Thomson* effect. The advantage of this system is a quick response time, but its cooling capacity is modest. Another device can be obtained using the *Peltier effect*, the so-called thermoelectric cooling. The thermoelectric method cools down the probe due to the passage of a direct current in a junction made up of two different welded materials. Depending on the direction of the current, there is an addition or subtraction of heat transfer. However, the thermal efficiency of this type of device is very modest. Liquid nitrogen is used for cancer treatment because it has the highest cooling power, while liquids with a lower power can be used in the treatment of various inflammatory diseases.

This paragraph briefly illustrates the operating principle of a liquid nitrogen cryoprobe, an old model much larger than those in circulation and, therefore, very invasive. The cryoprobe in question works using a *Linde*-type freezing cycle, as shown in Fig. 8 [16]. The probe consists of three concentric stainless steel tubes. Nitrogen flows from a pressurized chamber in the central duct toward the hemispherical needle of the probe. At the end of the inner tube, circular holes of very small diameter (smaller than 0.3 mm) are made to allow the liquid nitrogen to reach the tip of the probe, covered by a copper sleeve, which aims to increase the heat transfer. The ring, formed between the innermost tube and the central one, allows the nitrogen, turned into vapor because of the latent heat exchange, to return to the

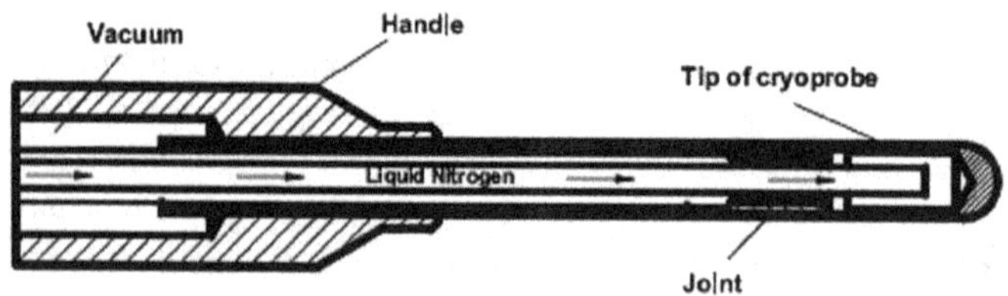

Fig. 8 Liquid nitrogen cryoprobe, Fig. 2 of [16]

external environment. The outermost tube is evacuated by means of a vacuum pump that aims to provide thermal insulation through the walls of the probe. The diameter of a probe of this type is around 5 mm, which is about the diameter of a pencil. From this, we can get an idea of how invasive an object of this type can be if we think of inserting a pencil, for example, into the abdomen. The fluid temperature is regulated based on the flow rate, so the temperature regulation is carried out by adjusting the fluid flow rate. In addition to the probe, we must consider the thermocouples inserted inside the patient to monitor the temperature. The ice sphere that you want to obtain from a probe that has an area of $\pi \times 2.5$ mm $\times 2.5$ mm is about 30 mm $\times$ 25 mm, which is about 75 times larger.

A new type of liquid nitrogen cryoprobe has been developed to combat trigeminal neuralgia and can be reasonably used in all those applications where you want to control pain. This cryoprobe, shown in Fig. 9 [21] has a diameter of 2.7 mm and is designed to avoid tissue dissections. Fig. 9 shows the construction of the tip of the cryoprobe. The external insulation, obtained by vacuum (chromium, nickel, and strontium), is indicated by (1) The internal insulation is indicated by (2) the return tube of the nitrogen liquid–gas mixture by (3) the delivery tube for the liquid nitrogen by (4) the copper tip by (5) the cold-welded joint by (6) the laser-melted

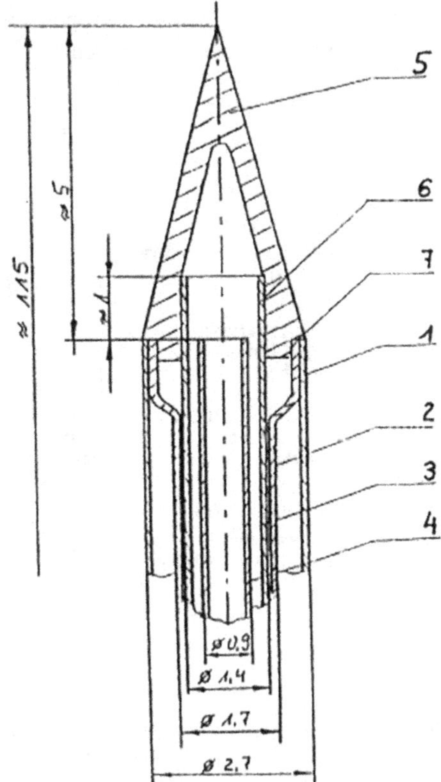

Fig. 9 Liquid nitrogen cryoprobe, Fig. 2 of [21]. Tip of cryoprobe: (1) outer vacuum insulation (chromium, nickel, and strontium), (2) inner vacuum insulation, (3) tube for returning nitrogen liquid–gas mixture, (4) feeding tube for liquid nitrogen, (5) copper tip, (6) junction hard-soldered, and (7) junction welded with laser

joint by (7) the system uses liquid nitrogen through a spray at a pressure of 0.4 mPa. A control system allows the adjustment of the time of cooling. The following characteristics of the cryoprobe are measured with thermal sensors: the tip temperature, the freezing speed, the freezing time interval, and the maximum diameter of the ice sphere surrounding the probe. The following cryobiological and cryo-physical data are recorded: temperatures below $-158\,°C$ at the tip in water; the temperature range between $-40\,°C$ and $-120\,°C$, reached at the nadir after 60 s; and the growth speed of 100 K/min. In water, the ice sphere reaches a diameter of 1 cm, and in gelatin of 4–5 mm. This procedure has given good results; the pain was reduced in 5 days and disappeared completely between 10 and 14 days. Moreover, the procedure is simple and short and does not cause stress to the patient. A loss of sensitivity is observed for 3–4 months, and the pain can return after 6 months. This, however, depends on the temperature reached during the operation.

Significant improvements in cryosurgical techniques have come with the use of a technique that employs subcooled liquid nitrogen at a temperature of $-209\,°C$ in a new device. Fig. 10 [22] shows the flow diagram of the main components of the *CMS AccuProbe System*. The subcooled heat exchange system provides liquid nitrogen to the disposable probes. The liquid nitrogen returns after diffusion. The subcooled state is obtained by passing pressurized liquid nitrogen through a heat exchanger immersed in a liquid nitrogen chamber at $-209\,°C$, maintained under vacuum. The subcooled liquid nitrogen circulates inside the probe. Thanks to a new probe design, the liquid nitrogen remains in the liquid state, does not evaporate, and

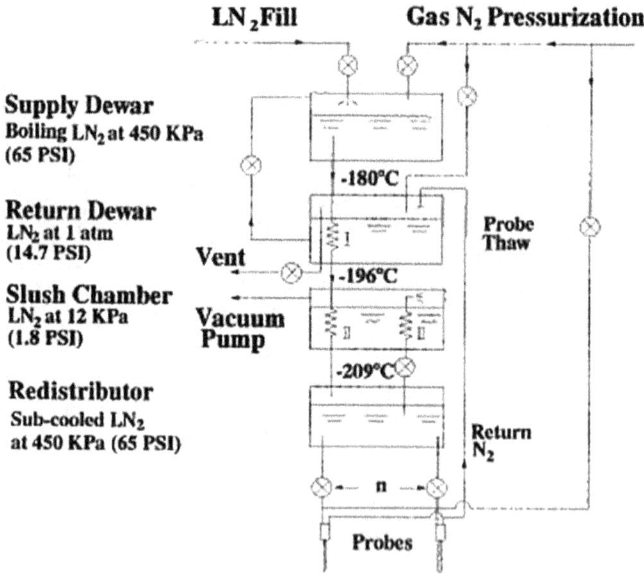

Fig. 10 Main components of the flow diagram of the CMS AccuProbe System, Fig. 1 of [22]. The subcooling heat exchange system provides liquid nitrogen flow to disposable cryoprobes. Liquid nitrogen is retrieved after delivery

is largely recovered on its return. The surface temperature of the probe hovers around $-165\ ^\circ C$ and $-195\ ^\circ C$, depending on the diameter of the probe, in the clinical use of the device. This means that the temperature gradient between the probe and the tissue is greater than that found in any other device. The steeper the temperature gradient, the greater the freezing capacity of the probe. With this device, a volume of 180 cc can be frozen, corresponding to a diameter of 7 cm, in 20 min of application.

This device also allows the cooling of *five* probes simultaneously. The use of multiple probes allows the overlapping of frozen areas in the treatment of large tumors and allows the modeling of the frozen tissue to the exact shape and size required for the treatment of the pathology. The new device allows the use of disposable probes, a choice in design that ensures high probe performance. On the other hand, reusable probes are subject to slow deterioration in their freezing capacity, and over time, their performance becomes poor. Probes are available in various diameters, lengths, and shapes, depending on the region in which they are to be used. Generally, the larger the probe, the greater its ability to freeze a given tissue in a certain time interval. It should be noted, however, that the performance of a probe of 3 mm and one of 8 mm is practically the same if used at the same temperature and with the same thermal load. The new probe that is 2 mm in diameter also has excellent freezing capabilities. In clinical practice, the way this system is used is to insert one or more probes into the target tissue; cool them in sequence, while they are all in position; and finally apply the maximum cooling power so that the freezing of the tissue is as fast as possible. Freezing continues until the entire volume of damaged tissue is included within the sphere of ice. In cancer treatment, a larger volume is included to increase the chances of survival.

A cryoprobe with a small diameter, in which liquid nitrogen circulates inside the boiling chamber on the tip, employing the *Joule–Thomson* effect as a principle of operation, has the advantage of not having moving parts, except in the compressor, which is separated by a reasonable distance from the final part of the cycle, Fig. 11, [22].

This simplicity allows for miniaturization, which is extremely important in the cryosurgical procedure, where the size of the probes is directly correlated with the invasiveness of the surgical procedure, Fig. 12, [16].

Recent advantages in *Joule–Thomson* systems have been achieved using a refrigerant mixture of liquid and gas, rather than pure gas. This has allowed the cryoprobe to have a cooling effect comparable to that of pure gas but with much lower operating pressures. In this way, the system can be constituted by a closed system, rather than an open one, and can be controlled by a small volumetric compressor. The advantage of using a closed system includes unlimited duration in the cryosurgical procedure, elimination of in situ exhalations, and reduction of operating costs [23].

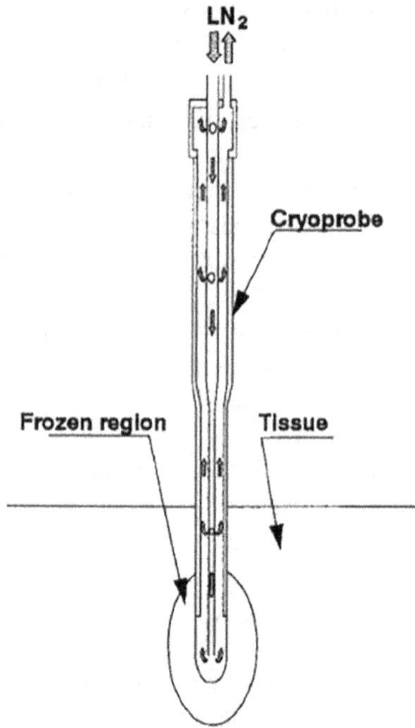

Fig. 11 Flow diagram of a small diameter cryoprobe, Fig. 2 of [22]

Fig. 12 Ice ball formed with the cryoprobe, Fig. 3 of [16]

3.2   Cryoheater

We have seen how important it is to control the advancement of the ice front in the cryosurgical technique. To do this, one must monitor the technique with sophisticated *imaging techniques* to prevent healthy, biologically important tissues from being involved. While all control mechanisms are based on monitoring the performance of the cryoprobe, a new technique deserves to be mentioned for the importance it could have in the future.

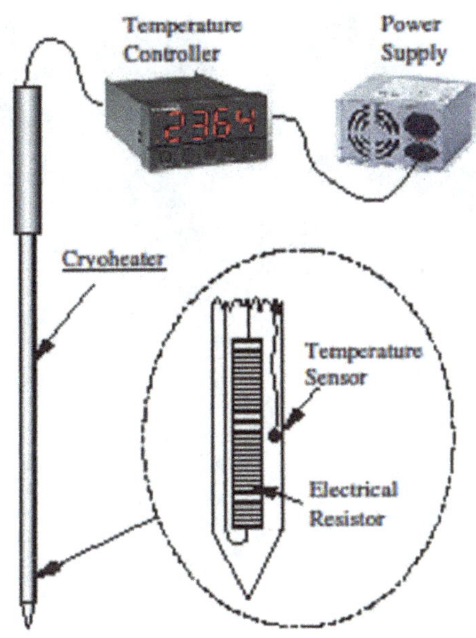

Fig. 13 Cryoheater, Fig. 1 of [3]. The cryoheater has the same external configuration as a standard minimally invasive cryoprobe

This technique controls the advancement of the ice front by heating the surrounding tissue and imposing the shape of the ice front through heating. There are two types of heaters in particular that are used in the technique: the *urethral warmer* and the *cryoheater*. Fig. 13 [3] shows the *cryoheater,* which has the same standard configuration as the *urethral warmer*. The *urethral warmer* is a water heat exchanger, which aims to maintain the temperature of the surrounding tissue at average body temperature, namely 37 °C. The *cryoheater* has the same purpose, but in this case, instead of using a heat exchanger, it uses an electrical resistance, and the heating occurs by the *Joule effect*. The most common configuration for the *urethral warmer* is that of the coaxial cylinder heat exchanger, with the internal duct constituting the delivery that receives water from an external tank and the external one constituting the return. Everything is inserted inside a catheter. The heating element is an electrical resistance, located at the tip of a long hypodermic needle. A temperature sensor is connected to the inner wall of the hypodermic needle in the middle of the ends of the heating element. Externally, the *cryoheater* looks exactly like a cryoprobe. The *cryoheater* is connected to a temperature controller through two cables: one for the temperature sensor, which is used as feedback for the control system, and the other for the heating element. The thermostat is connected to a power supply that provides electrical energy [3]. The *cryoheater* consists of a resistive cable wrapped around an insulating core, with a total resistance of 46 Ω and an outer diameter of 3 mm. The active length of the *cryoheater* is 20 mm. The *cryoheater* is activated before the start of freezing with a set-point temperature of 37 °C.

The effects of heating can clearly be seen in Fig. 14 [3] where a well-defined unfrozen region, with a thickness of 0.8 mm, is highlighted around the *cryoheater*. In the treatment, this region is intended to protect tissues from the destructive effects of

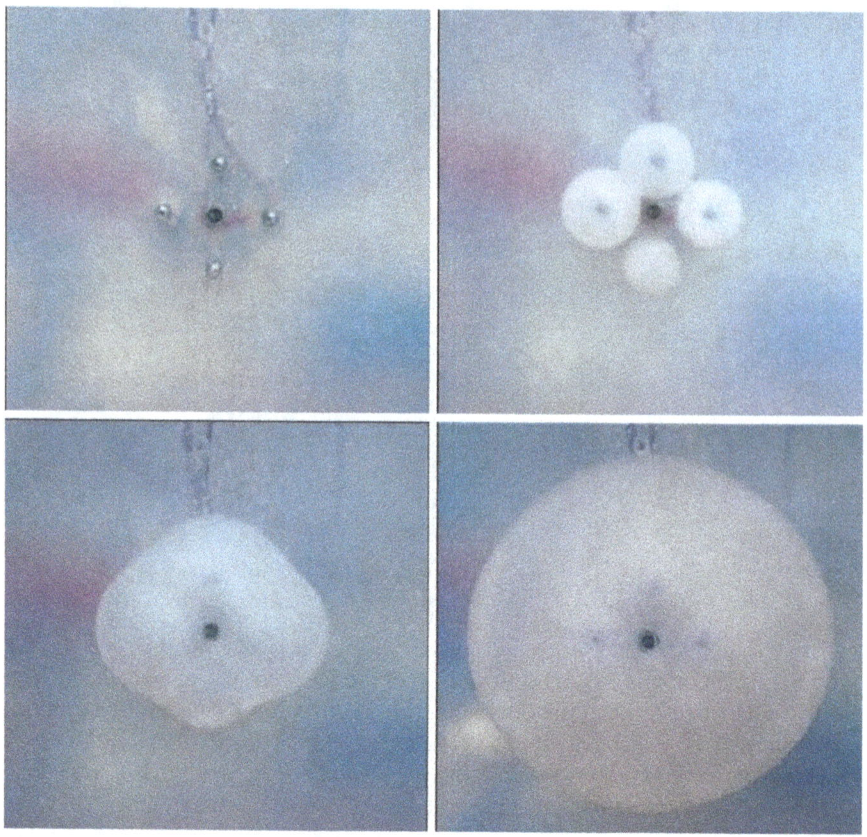

Fig. 14 Cross section with the growth of the ice sphere inside a gelatin solution, Fig. 3 of [3]

freezing. Figure 14 shows the cross section with the growth of the ice sphere inside a gelatin solution: (top left) before the start of the experiment, (top right) after 1 min, (bottom left) after 4 min, and (bottom right) after 10 min from the start of the experiment.

3.3　Mikus "Cryoprobe Endocare" Patent

To achieve apoptosis of the largest number of tumor cells, it is necessary to perform cycles of freezing–thawing. To do this, once the freezing phase is over, the fluid at the tip of the cryoprobe must heat up quickly, and so the cryoprobe heats up in order to achieve rapid thawing. Since the probe cannot be removed from the frozen tissue because it adheres firmly to it when it is frozen, there are methods to heat the cryoprobe through the flow of gas, condensation, electric heating, etc. The cryoprobe patented by Mikus et al., *Endocare* [24], has a mechanism for heating the cryoprobe of the *Joule–Thomson* type much faster than that due to the simple interruption of the gas. The system has a vent for the cooling gas delivery line. To stop the cooling effect, the delivery of the cooling gas is interrupted, and the vent valve is opened. The high-pressure gas tank in the delivery line is opened so that it can vent in the operating room and, therefore, does not dissipate through the *Joule–Thomson* nozzle. In this way, cooling is immediately interrupted, and heating begins as soon as the system adjusts to the heating mode.

3.4   LePivert Patent

The LePivert patent [25] concerns a cryoprobe that combines the effect of direct freezing (spray) and indirect freezing (contact with the surface). The choice of the two methodologies, direct and indirect, depends on many parameters, including the site of the lesion and the predictability of the effects of freezing. The closed-tip cryoprobe can use low-boiling-point refrigerants, such as liquid nitrogen, or *Joule–Thomson* expansion for fluids, such as nitrogen oxide, carbon dioxide, or gaseous argon.

Other techniques include the use of the latent heat of vaporization of gases, such as Freon. The frost produced by the tip is confined to a region near the area of the probe tip, where an icy sphere of frozen tissue forms. The main disadvantage of this technique is that the probe tip never reaches the lowest temperature allowed by the refrigerant due to the low effective thermal conductivity of the tip material, forcing the operator to use low-melting-point refrigerants. Using these refrigerants, it is necessary to change the position of the probe within the tissue and/or use multiple probes simultaneously to complete the freezing of the tissue if the target has a volume greater than a few cubic centimeters. In addition, the miniaturization required by endoscopic applications or catheter-based ones requires very small and flexible probes. Instead, these probes are large and cannot be easily moved.

The spray is able to cover a larger area than the closed-tip probe and has greater thermal efficiency. Generally, the spray is applied externally, and the operator estimates the effects of cryosurgery based on the tissue that appears to be frozen. Sometimes, if the lesion is deep, the operator must use invasive techniques to monitor the thermal effects, such as thermocouples inserted into the tissue, which is not desirable, nor practicable in all tissues. These measures are potentially risky, as they can spread tumor cells over a wide region of the tissue. Where possible, monitoring should be done using techniques such as *MRI, X-rays,* and *ultrasound.* However, sprays are not suitable for use inside the body due to the difficulty in controlling the release of cryogenic fluid and the effects of overspray on the surrounding tissue. In addition, the interstitial expansion of a gas or a refrigerant liquid can be harmful to tissues, as, by forcing gas bubbles inside the tissue under pressure, it risks swelling it or embolizing an open vessel.

The LePivert patent [25] allows us to overcome the difficulties associated with the use of sprays inside cavities using a two-stage freezing method. The apparatus also allows us to regulate the flow of the refrigerant in the spraying and monitoring of the ice sphere. This invention overcomes three major problems of cryosurgery: the need for a powerful refrigerant to treat malignant tumors, the temperature required to treat various types of tumors, and the need to use powerful refrigerants in very small probes, smaller than 3 mm in outer diameter, to have an efficient use of power. Powerful refrigerants are considered necessary, and in order to have greater effectiveness, it is necessary to achieve rapid freezing and a large ice sphere, conditions that are created with a high thermal gradient. However, since tissue can be destroyed even at temperatures of -2 °C, the conventional temperatures of -50 °C are not necessary if the freezing process is conducted in a way that destroys the vascular network. Therefore, it is also possible to use medium-power refrigerants, such as nitrogen dioxide, and monitoring should be focused on the modifications of the vascular bed.

4   Physics of the Change of Phase-Solidification

4.1   Solidification

Consider a pure substance. At the melting temperature T_f, the Gibbs free enthalpy, defined as

$$dg = dp/\rho - sdT \tag{1}$$

remains constant because the phase transition is an isothermal and isobaric transformation. Therefore, it must also have, in finite quantities,

$$\Delta G_f = \Delta H_f - T_f \Delta S_f = 0 \tag{2}$$

or

$$\Delta H_f = T_f \Delta S_f \tag{3}$$

where ΔH_f is the enthalpy variation, representing the latent heat, being ΔS_f positive for fusion. For other temperatures, beyond T_f, it must be

$$\Delta G = \Delta H - T\Delta S \approx \Delta H_f - T\Delta S_f = \Delta S_f(T_f - T) = \Delta S_f \Delta T_{\text{undercooling}} \tag{4}$$

The phenomenon of *supercooling* manifests in the liquid after the melting point, being positive the variation of free enthalpy and entropy. The driving force is therefore proportional to undercooling and takes into account the fact that enthalpy and entropy do not vary much with the temperature near the melting point [26]. The homogeneous nucleation of the solid phase only occurs at large ΔT, as, for example, in experiments without enclosures where a pure liquid is isolated from its environment. In general, the solidification consists of a heterogeneous nucleation, that of impurities made up of particles, or where the liquid comes into contact with the surface of the container. The speed v of the front transformation is related to the difference in the rates of liquid–solid and solid–liquid transition, where a barrier to the interface motion is present. Therefore,

$$v \propto \exp\left(-\frac{Q}{kT}\right)_{\text{solid}\,\to\,\text{liquid}} - \exp\left(-\frac{Q+\Delta G}{kT}\right)_{\text{liquid}\,\to\,\text{solid}} \tag{5}$$

from which

$$v \propto \exp\left(-\frac{Q}{kT}\right)\left(1 - \exp\left(-\frac{\Delta G}{kT}\right)\right) \tag{6}$$

and, for small ΔG,

$$v \propto \frac{\Delta G}{kT}\exp\left(-\frac{Q}{kT}\right) \tag{7}$$

or $v \propto \Delta T_{\text{undercooling}}$.

We can now see how the solute is distributed during freezing between the solid and liquid phases. By convention, we call the composition of the solid phase, in equilibrium with the liquid one, C^{SL} and, similarly, the composition of the liquid phase, in equilibrium with the solid one, C^{LS}. C_{M} represents the average composition of the substance. The partition coefficient k is defined as

$$k = \frac{C^{\text{SL}}}{C^{\text{LS}}} \tag{8}$$

and is often smaller than 1. In equilibrium conditions, the composition of the solid and liquid phases at all times of solidification is given by the phase diagram, and the proportions of the phases at each temperature are given by the lever rule. In practice, equilibrium will only be maintained at the interface, where the composition agrees with the phase diagram. Steady-state solidification occurs when the temperature is equal to the melting point; the composition of the solid phase, in equilibrium with the liquid one, is equal to the average composition of the substance; and there is no other solute distributed in the remaining liquid.

During the solidification process, at the steady state, the solid–liquid interface moves at a constant speed. An observer, positioned at the interface ($x = 0$), would not see changes in the solute distribution in front of the interface. It follows that any change due to diffusion is exactly compensated by the fact that the interface advances according to the equation

$$D\frac{d^2 C}{dx^2} - v\frac{dC}{dx} = 0 \tag{9}$$

where v is the speed of the solidification front, and D is the mass diffusivity.

A second equation can be deduced from the fact that the solute flow due to diffusion away from the interface must equal the speed at which the solute is distributed.

$$D\frac{dC}{dx}\bigg|_{x=0} = \left(C^{LS} - C^{SL}\right)v \tag{10}$$

Finally, using the boundary conditions, it is obtained

$$C = C_M + C_M \frac{(1-k)}{k} \exp\left(-\frac{v}{D}x\right) \tag{11}$$

The solute is distributed within the liquid, in front of the solidification front, causing a corresponding variation in the temperature of the liquid, under which freezing begins. However, there is a positive temperature gradient within the liquid, which increases the supercooled zone of liquid in front of the interface. This process is called constitutional supercooling because it is caused by composition changes. A small perturbation at the interface will expand within the supercooled liquid. This causes the dendrites to grow. It follows that a supercooled zone only forms when the temperature gradient of the liquid at the interface (T_L) is greater than the temperature gradient in the frozen phase:

$$\frac{\partial T_S}{\partial z}\bigg|_{z=h^-} < \frac{\partial T_L}{\partial z}\bigg|_{z=h^+} \tag{12}$$

and

$$\frac{\partial T_S}{\partial z}\bigg|_{z=h^-} < \Gamma\frac{\partial C_L}{\partial z}\bigg|_{z=h^+} \tag{13}$$

where

$$\Gamma = \frac{\partial T_L}{\partial C_L} \tag{14}$$

is the modulus of the slope of the liquid phase contour in the phase diagram. We can note that

$$\frac{dC_L}{dx}\bigg|_{x=0} = -\frac{\left(C^{LS} - C^{SL}\right)}{D}v \tag{15}$$

from which it follows that the minimum thermal gradient necessary for the stabilization of the front is given by

$$\frac{dT}{dx} < \frac{mC_M(1-k)}{kD}v \tag{16}$$

It is very difficult to avoid supercooling, in practice, because the required speed is really very small. Directional solidification with a flat front is only possible at low growth speeds. In most cases, the interface is unstable, and we have the formation of cells or dendrites. Convection can cause mixing, thus reducing the solute gradient in the liquid. A limiting case is when the liquid is uniform in composition during solidification. For a liquid of composition C_L, the composition of the solid phase at the interface is given by

$$C^S = kC^L \tag{17}$$

If f_S is the solidified fraction, it is deduced,

$$\left(C^L - kC^L\right)df = (1 - f)dC^L \tag{18}$$

from which it is obtained

$$C^S = kC_M(1 - f_s)^{k-1} \tag{19}$$

known as the *Scheil equation*.

The redistribution of solute increases when there is mixing in the liquid. There is a small or almost no segregation in front of the edge of the dendrites because the main partition is given by the solute, trapped between the lateral arms of the dendrites. The scale of this segregation can be reduced by increasing the speed of solidification because the space between the arms of the dendrites decreases. An approximate explanation of this phenomenon lies in the fact that a large undercooling allows the creation of a greater surface and, therefore, a smaller space between the dendrites.

4.2 Analytical Solutions

Water is initially at the uniform temperature of 37 °C, higher than the solidification temperature ($T_0 = 0$ °C), while the temperature of the probe is set at $T_{\text{probe}} = -100$ °C. The liquid is confined in a semi-infinite space, $x > 0$. At time $t = 0$, the temperature on the wall $x = 0$ is set at the temperature T_{probe}, which is smaller than the solidification temperature T_0 and is maintained at this temperature for all instants of time $t > 0$. The result is that the process of solidification begins on the face $x = 0$, and the solid–liquid interface moves along $x > 0$. The temperatures are unknown both in the liquid and solid phases.

In the following analysis, the temperature fields, for both liquid and solid phases, are determined, as well as the position of the ice front. The solution to this problem is known as the *Neumann solution*. We consider the problem with plane and cylindrical geometries. The problem data are as follows: $\rho_l = 998.2$ kg/m^3, $c_{pl} = 4182$ J/kg $\times$ K, $k_l = 0.6$ W/m $\times$ K, $\alpha_l = 1.43 \times 10^{-7}$ m^2/s, $\rho_s = 916.2$ kg/m^3, $c_{ps} = 2000$ J/(kg $\times$ K), $k_s = 2.23$ W/(m $\times$ K), $\alpha_s = 1.22 \times 10^{-6}$ m^2/s, and $L_h = 335$ kJ/kg.

4.2.1 Plane Geometry

The energy balance equation and the boundary condition for the solid phase are

$$\alpha_s \frac{\partial^2 T_s}{\partial x^2} = \frac{\partial T_s}{\partial t} \quad \{0 \le x \le s(t); t > 0\} \tag{20}$$

$$T_l(x, t) = T_{\text{probe}}; \quad \{x > 0, t = 0\} \tag{21}$$

and for the liquid phase are

$$\alpha_l \frac{\partial^2 T_l}{\partial x^2} = \frac{\partial T_l}{\partial t} \quad \{s(t) \le x < +\infty; t > 0\} \tag{22}$$

$$\lim_{x \to \infty} T_l(x, t) = T_{\text{blood}}; \quad t > 0 \tag{23}$$

The boundary conditions at the interface, $x = s(t)$ are

$$T_s(x, t) = T_l(x, t) = T_m; \quad \{x = s(t), t > 0\} \tag{24}$$

$$k_s \frac{\partial T_s}{\partial x}\bigg|_{x=s(t)} - k_l \frac{\partial T_l}{\partial x}\bigg|_{x=s(t)} = \rho_s L_h \frac{ds}{dt}; \quad t > 0 \tag{25}$$

The problem concerns a semi-infinite body, and the solutions are linear combinations of the constants and the *error function*. The Green function for the plane problem is the following

$$G_P(x, t | x', \tau) = \frac{1}{2\sqrt{\pi \alpha(t - \tau)}} \exp\left(-\frac{(x - x')^2}{4\alpha(t - \tau)}\right) \tag{26}$$

The *error function* is given by

$$\text{erf}\left(\frac{x}{2\sqrt{\alpha t}}\right) = \frac{1}{\sqrt{\pi}} \int_{-\infty}^{\frac{x}{2\sqrt{\alpha t}}} \exp\left(-\varsigma^2\right) d\varsigma \tag{27}$$

We suppose that the solutions are of the type

$$T_s(x, t) = T_{\text{probe}} + A \,\text{erf}\left(\frac{x}{2\sqrt{\alpha_s t}}\right) \tag{28}$$

for the solid phase and

$$T_l(x, t) = T_{\text{blood}} + B\left(\text{erf}\left(\frac{x}{2\sqrt{\alpha_l t}}\right) - 1\right) \tag{29}$$

for the liquid phase. The boundary and initial conditions are satisfied, and coefficients A and B remain to be determined. Given

$$\lambda = \frac{s(t)}{2\sqrt{\alpha_s t}} \tag{30}$$

we impose the conditions at the interface

$$T_{\text{probe}} + A\,\text{erf}(\lambda) = T_{\text{blood}} + B\left(\text{erf}\left(\lambda\sqrt{\frac{\alpha_s}{\alpha_l}}\right) - 1\right) \tag{31}$$

Note that, by definition, the coefficients of the general integral of the differential equation must be constants, and since the solidification temperature is constant, this implies that the coefficient λ is also a constant and does not vary over time. This, in turn, implies that

$$s(t) = 2\lambda\sqrt{\alpha_s t} \tag{32}$$

The ice front advances proportionally to the square root of time, with a proportionality constant that depends on the characteristics of the material. As for the coefficients A and B, we get

$$A = \left[T_m - T_{\text{probe}}\right]/\text{erf}(\lambda) \tag{33}$$

$$B = \left[T_m - T_{\text{blood}}\right]/\left[\text{erf}\left(\sqrt{\frac{\alpha_s}{\alpha_l}}\lambda\right) - 1\right] \tag{34}$$

By exploiting the condition of the thermal balance at the interface, we can find the unknown λ

$$
\begin{aligned}
&k_s\left[T_m - T_{\text{probe}}\right]\exp\left(-\lambda^2\right)/\text{erf}(\lambda) + k_l\sqrt{\frac{\alpha_s}{\alpha_l}} \\
&\times \left[T_m - T_{\text{blood}}\right]\exp\left(-\frac{\alpha_s}{\alpha_l}\lambda^2\right)/\left[\text{erf}\left(\sqrt{\frac{\alpha_s}{\alpha_l}}\lambda\right) - 1\right] = 2\sqrt{\pi}\rho_s L_h \alpha_s \lambda
\end{aligned}
\tag{35}
$$

The value of λ remains unknown, however, because the equation that is obtained is strongly nonlinear. Therefore, in order to obtain the solution, an iterative method is implemented, as, for example, the Gauss–Newton method. We set

$$
\begin{aligned}
g(\lambda) = {}&2\sqrt{\pi}\rho_s L_h \alpha_s \lambda - k_s \frac{\left[T_m - T_{\text{probe}}\right]}{\text{erf}(\lambda)}\exp\left(-\lambda^2\right) \\
&- k_l\sqrt{\frac{\alpha_s}{\alpha_l}}\frac{\left[T_m - T_{\text{blood}}\right]}{\left[\text{erf}\left(\sqrt{\frac{\alpha_s}{\alpha_l}}\lambda\right) - 1\right]}\exp\left(-\frac{\alpha_s}{\alpha_l}\lambda^2\right) = 0
\end{aligned}
\tag{36}
$$

Now, since

$$g(\lambda_{i+1}) = g(\lambda_i) + \left.\frac{\partial g}{\partial \lambda}\right|_{\lambda_i} (\lambda_{i+1} - \lambda_i) + o(\|\lambda_{i+1} - \lambda_i\|) \tag{37}$$

and imposing

$$g(\lambda_{i+1}) = 0; o(\|\lambda_{i+1} - \lambda_i\|) = 0 \tag{38}$$

we obtain

$$\lambda_{i+1} = \lambda_i - \left(\left.\frac{\partial g}{\partial \lambda}\right|_{\lambda_i}\right)^{-1} g(\lambda_i) \tag{39}$$

and

$$\begin{aligned}
\left.\frac{\partial g}{\partial \lambda}\right|_{\lambda_i} &= 2\sqrt{\pi}\rho_s L_h \alpha_s - k_s \frac{[T_m - T_{\text{probe}}]}{(\text{erf}(\lambda_i))^2} \frac{\exp(-2\lambda_i^2)}{\sqrt{\pi}} \\[2mm]
&\quad - k_l \frac{\alpha_s}{\alpha_l} \frac{[T_m - T_{\text{blood}}]}{\left[\text{erf}\left(\sqrt{\frac{\alpha_s}{\alpha_l}}\lambda_i\right) - 1\right]^2} \frac{\exp\left(-2\frac{\alpha_s}{\alpha_l}\lambda_i^2\right)}{\sqrt{\pi}} \\[2mm]
&\quad + 2\lambda_i k_s \frac{[T_m - T_{\text{probe}}]}{\text{erf}(\lambda_i)} \exp(-\lambda_i^2) \\[2mm]
&\quad + 2\lambda_i k_l \sqrt{\frac{\alpha_s}{\alpha_l}} \frac{[T_m - T_{\text{blood}}]}{\left[\text{erf}\left(\sqrt{\frac{\alpha_s}{\alpha_l}}\lambda_i\right) - 1\right]} \exp\left(-\frac{\alpha_s}{\alpha_l}\lambda_i^2\right)
\end{aligned} \tag{40}$$

From the iterative method, we derive the value for λ: $\lambda = 0.4355$ g $= 1.1369 \times 10^{-13}$ [27]. It is now possible to calculate the growth of the ice front with time, which is shown in Fig. 15, and the temperature distribution versus the distance from the probe, which is shown in Fig. 16.

Figure 15 presents the analytical solution of the thickness of the ice front, in *millimeters*, versus the time elapsed, in *seconds*, up to 30 s. After 10 s, the ice thickness is 3 mm; after 20 s, it is around 4.25 mm; and after 30 s, it is around 5.25 mm.

Figure 16 presents the temperature distribution, T $(°C)$, versus x (m) at six instants of time. T_1 is after 5 s, T_2 is after 10 s, T_3 is after 15 s, T_4 is after 20 s, T_5 is after 25 s, and T_6 is after 30 s. The curve in the frozen part presents a quasi-linear distribution, while in the unfrozen part, the curve has a marked exponential trend. The step change of the slope at the freezing point must be noted. Further on, it can be observed that the temperature of the body, 37 °C, is unchanged after 10 mm for all the times reported.

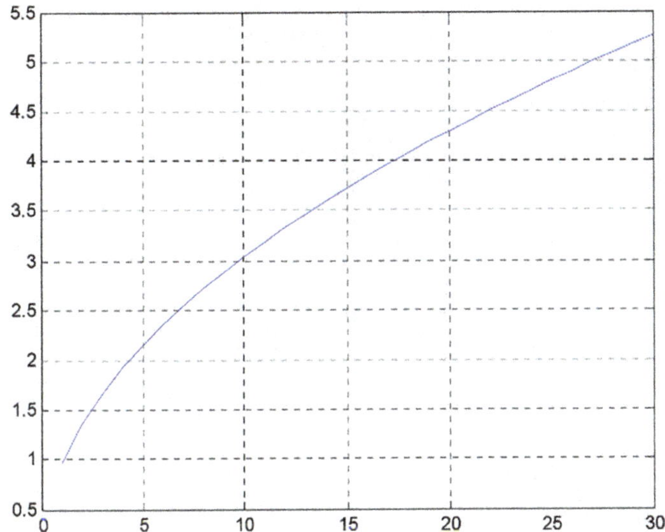

Fig. 15 Growth of the ice thickness (mm) versus time (s) in plane geometry [28]

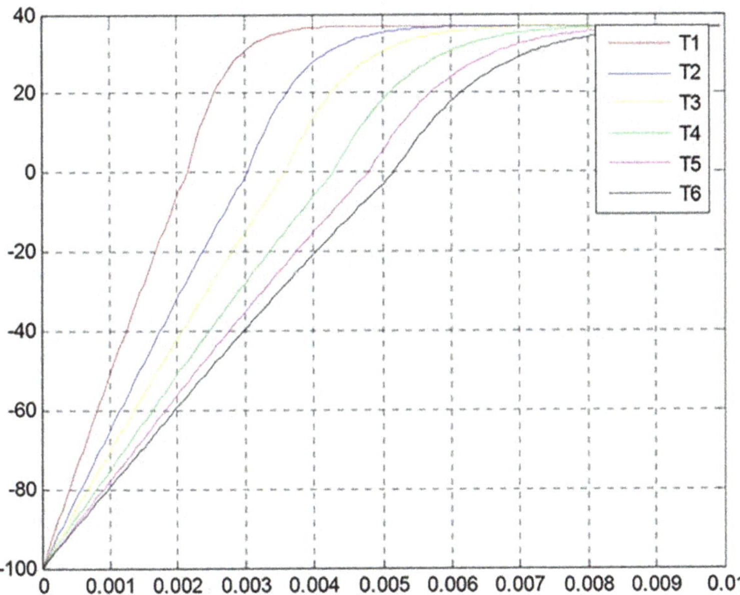

Fig. 16 Temperature distribution, T (°C), versus x (m) in plane geometry [28]

4.2.2 Cylindrical Geometry

A probe of the type *line heat source* is positioned on the abscissa $x = 0$ of a cylindrical coordinate reference system. The power delivered per unit length is Q. The energy balance for the solid phase leads to

$$\alpha_s \frac{1}{r} \frac{\partial}{\partial r}\left(r\frac{\partial T_s}{\partial r}\right) = \frac{\partial T_s}{\partial t}\{0 \leq r \leq s(t); t > 0\} \tag{41}$$

with the boundary condition

$$\lim_{r \to 0} 2\pi r k_s \frac{\partial T_s}{\partial r} = Q \tag{42}$$

and, for the liquid phase,

$$\alpha_l \frac{1}{r} \frac{\partial}{\partial r}\left(r\frac{\partial T_l}{\partial r}\right) = \frac{\partial T_l}{\partial t}\{s(t) \leq r \leq +\infty; t > 0\} \tag{43}$$

and

$$\lim_{r \to +\infty} T_l = T_{\text{blood}}; t > 0 \quad T_l = T_{\text{blood}}; t = 0 \tag{44}$$

The boundary conditions at the interface are

$$T_s(r,t) = T_l(r,t) = T_m; \{r = s(t), t > 0\} \tag{45}$$

$$k_s \frac{\partial T_s}{\partial r}\bigg|_{r=s(t)} - k_l \frac{\partial T_l}{\partial r}\bigg|_{r=s(t)} = \rho_s L_h \frac{ds}{dt}; t > 0 \tag{46}$$

The integral exponential is given by

$$\text{Ei}\left(-\frac{r^2}{4\alpha t}\right) = -\int_{\frac{r^2}{4\alpha t}}^{+\infty} \frac{\exp(-u)}{u} du \tag{47}$$

It is assumed that the solutions are of the type

$$T_s(r,t) = A - B\text{Ei}\left(-\frac{r^2}{4\alpha_s t}\right) \tag{48}$$

for the solid phase and

$$T_l(x,t) = T_{\text{blood}} - C\text{Ei}\left(-\frac{r^2}{4\alpha_l t}\right) \tag{49}$$

for the liquid phase. The following equalities hold

$$\frac{\partial T_s(r,t)}{\partial r} = -\frac{2B}{r}\exp\left(-\frac{r^2}{4\alpha_s t}\right) \tag{50}$$

$$\frac{\partial T_l(r,t)}{\partial r} = -\frac{2C}{r}\exp\left(-\frac{r^2}{4\alpha_l t}\right) \tag{51}$$

from which we obtain

$$B = -\frac{Q}{4\pi k_s} \tag{52}$$

The boundary condition at the interface leads to

$$T_m = A + \frac{Q}{4\pi k_s}\mathrm{Ei}\left(-\frac{s(t)^2}{4\alpha_s t}\right) = T_{\text{blood}} - C\mathrm{Ei}\left(-\frac{s(t)^2}{4\alpha_l t}\right) \tag{53}$$

As observed previously

$$\lambda = \frac{s(t)}{2\sqrt{\alpha_s t}} \tag{54}$$

with

$$s(t) = 2\lambda\sqrt{\alpha_s t} \tag{55}$$

and we obtain

$$A = T_m - \frac{Q}{4\pi k_s}\mathrm{Ei}\left(-\lambda^2\right) \tag{56}$$

$$C = [T_{\text{blood}} - T_m]/\mathrm{Ei}\left(-\lambda^2\frac{\alpha_s}{\alpha_l}\right) \tag{57}$$

Therefore, the values of the sought functions can be derived from

$$T_s(r, t) = T_m - \frac{Q}{4\pi k_s}\left(\mathrm{Ei}\left(-\lambda^2\right) - \mathrm{Ei}\left(-\frac{r^2}{4\alpha_s t}\right)\right) \tag{58}$$

$$T_l(x, t) = T_{\text{blood}} - [T_{\text{blood}} - T_m]\mathrm{Ei}\left(-\frac{r^2}{4\alpha_l t}\right)/\mathrm{Ei}\left(-\lambda^2\frac{\alpha_s}{\alpha_l}\right) \tag{59}$$

where the value of λ remains unknown. Then, the solution is obtained with an iterative method, as, for example, the Gauss–Newton method. We set

$$g(\lambda) = \rho_s L_h \alpha_s \lambda^2 - \frac{Q}{4\pi}\exp\left(-\lambda^2\right) - k_l[T_{\text{blood}} - T_m]\exp\left(-\lambda^2\frac{\alpha_s}{\alpha_l}\right)/\mathrm{Ei}\left(-\lambda^2\frac{\alpha_s}{\alpha_l}\right) = 0 \tag{60}$$

Now since

$$g(\lambda_{i+1}) = g(\lambda_i) + \frac{\partial g}{\partial \lambda}\bigg|_{\lambda_i} (\lambda_{i+1} - \lambda_i) + o(\|\lambda_{i+1} - \lambda_i\|) \tag{61}$$

and imposing

$$g(\lambda_{i+1}) = 0;\, o(\|\lambda_{i+1} - \lambda_i\|) = 0 \tag{62}$$

we obtain

$$\lambda_{i+1} = \lambda_i - \left(\frac{\partial g}{\partial \lambda}\bigg|_{\lambda_i}\right)^{-1} g(\lambda_i) \tag{63}$$

and

$$\frac{\partial g}{\partial \lambda}\bigg|_{\lambda_i} = 2\lambda \left(\rho_s L_h \alpha_s + \frac{Q\exp(-\lambda^2)}{4\pi}\right)$$

$$+ 2k_l(T_{\text{blood}} - T_m)\frac{\exp\left(-\lambda^2 \frac{\alpha_s}{\alpha_l}\right)}{\text{Ei}\left(-\lambda^2 \frac{\alpha_s}{\alpha_l}\right)}\left(\lambda\frac{\alpha_s}{\alpha_l} - \frac{1}{\lambda}\frac{\exp\left(-\lambda^2 \frac{\alpha_s}{\alpha_l}\right)}{\text{Ei}\left(-\lambda^2 \frac{\alpha_s}{\alpha_l}\right)}\right) \tag{64}$$

from which we derive $\lambda = -\text{j}\, 0.16843$, where j is the imaginary unit. It is now possible to calculate the growth of the ice front with time, which is shown in Fig. 17, and the temperature distribution versus the distance from the probe, which is shown in Fig. 18.

Figure 17 presents the analytical solution of the thickness of the ice front, in *millimeters*, versus the time elapsed, in *seconds*, up to 30 s. After 10 s, the ice thickness is around 1.2 mm; after 20 s, it is around 1.7 mm; and after 30 s, it is slightly above 2 mm. In light of this, it can be stated that, in cylindrical geometry, the ice front propagates more slowly than in plane geometry, and the speed is about half.

Figure 18 presents the temperature distribution, T $(°C)$, versus x (m) at six instants of time. T_1 is after 5 s, T_2 after 10 s, T_3 after 15 s, T_4 after 20 s, T_5 after 25 s, and T_6 after 30 s. Now, the curves present a marked exponential distribution in the frozen as well as in the unfrozen parts, and the step change of the slope at the freezing point is more evident. Further on, it can be observed that the temperature of the body, 37 °C, is unchanged after 7 mm for all the times shown.

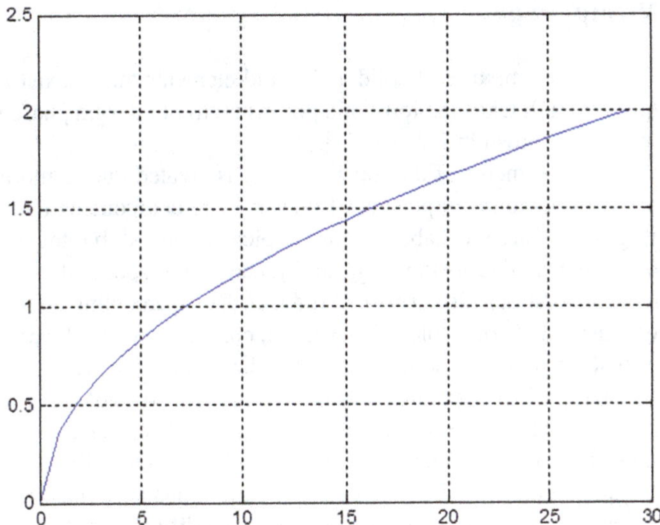

Fig. 17 Growth of the ice thickness (mm) versus time (s) in cylindrical geometry [28]

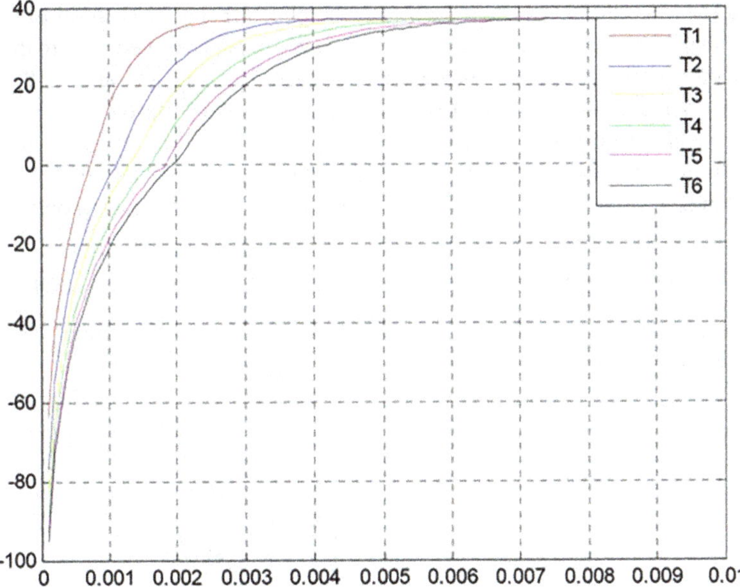

Fig. 18 Temperature distribution, T (°C), versus x (m) in cylindrical geometry [28]

4.3   "Mushy" Region

The *mushy region* is a mixture of solid and liquid elements that coexist in thermodynamic equilibrium. It takes the form of a porous matrix of a solid phase immersed in a liquid phase, as shown in Fig. 19 [29].

The convoluted geometry of the mushy region is created due to morphological instabilities that promote the expulsion of one or more components from the solid phase as it grows. These instabilities are typically caused by the constitutive supercooling, which is due to the large difference in the speed of heat diffusion between the various phases [30]. This fact reduces the supercooling and leads to the formation of a new mode of stable solidification, characterized by the presence of a mushy layer that separates the solid phase from the melted one. The heat and mass transfers in the *mushy layer* have major effects on the properties of the solid material produced, which is also responsible for the marked interest in the study of the solidification of binary fused substances, accompanied by a transition layer.

The dissolution is driven by a thermodynamic imbalance that results from composition variations, and its speed is limited by the diffusion of the constituents. The melting proceeds more quickly at the speed dictated by the heat transfer. Within the *mushy layer,* the internal dissolution tends to maintain the interstitial concentration of the local liquid by transporting the solute toward the microscale of the internal morphology. On the macroscale, however, the phase change from solid to liquid is still controlled by the speed of the heat transfer. We must, however, refer to the phase change on the microscopic scales, such as melting or freezing, although, at this level, the process is simply constituted by a diffusion of solute. Based on the different conditions, the melting of the *mushy region* can create a region completely melted or simply reduce the fraction of the solidified region. A significant challenge is the extension of the region that opposes complete melting. Before moving on to writing

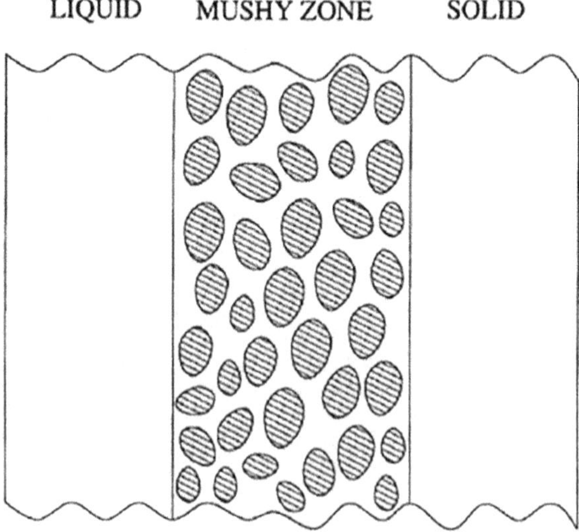

Fig. 19 Illustration of the "mushy region" in a slab of multicomponent material undergoing phase change, Fig. 1 of [29]

the balance equations, let us recall the equations and solutions of the problem in the one-dimensional case, with constant physical parameters, imposed wall temperature and neglecting the transport of solute in the solid phase

$$\rho_s c_{P,s}\frac{\partial T}{\partial t}=k_s\frac{\partial^2 T}{\partial z^2}, 0\leq z\leq h(t) \tag{65}$$

$$\rho_l c_{P,l}\frac{\partial T}{\partial t}=k_l\frac{\partial^2 T}{\partial z^2}, z\geq h(t) \tag{66}$$

$$\frac{\partial C}{\partial t}=D\frac{\partial^2 C}{\partial z^2}, z\geq h(t) \tag{67}$$

where $h(t)$ is the position of the solidification front. The boundary conditions are

$$T=T_B, z=0 \tag{68}$$

$$C\rightarrow c_0, z\rightarrow\infty \tag{69}$$

$$T\rightarrow T_\infty, z\rightarrow\infty \tag{70}$$

Now, we make the assumption that the front $z = h(t)$ is near equilibrium; therefore, it must be

$$T(h,t)=-\Gamma C(h^+,t), z=h(t) \tag{71}$$

The heat flow and the solute must be conserved on the front; therefore,

$$\rho_s L\frac{dh}{dt}=k_s\frac{\partial T}{\partial z}\bigg|_{z=h^-}-k_l\frac{\partial T}{\partial z}\bigg|_{z=h^+} \tag{72}$$

$$C(h^+,t)\frac{dh}{dt}+D\frac{\partial C}{\partial z}\bigg|_{z=h^+}=0 \tag{73}$$

where Γ is the slope of the liquid. The model admits similar solutions through the variables η and h in this form

$$\eta=\lambda z/h(t) \tag{74}$$

$$h(t)=\lambda\sqrt{4Dt+s} \tag{75}$$

where the parabolic growth, λ, is determined iteratively, and s is defined as

$$s=(h(0)/\lambda)^2 \tag{76}$$

The obtained solutions are the following:

$$T(\eta) = T_B + (T_h - T_B)\mathrm{erf}(\varepsilon_s\eta)/\mathrm{erf}(\varepsilon_s\lambda), \eta < \lambda \tag{77}$$

$$T(\eta) = T_\infty + (T_h - T_\infty)\mathrm{erfc}(\varepsilon_l\eta)/\mathrm{erfc}(\varepsilon_l\lambda), \eta > \lambda \tag{78}$$

$$C(\eta) = c_0 + (C_h - c_0)\mathrm{erfc}(\eta)/\mathrm{erfc}(\lambda), \eta > \lambda \tag{79}$$

where

$$\varepsilon_s = \sqrt{Dc_{P,s}\rho_s/k_s} \tag{80}$$

$$\varepsilon_l = \sqrt{Dc_{P,l}\rho_l/k_l} \tag{81}$$

and T_h and C_h are, respectively, the temperature and concentration calculated on the abscissa $z = h$. It should be noted that the temperature in front of the solidification front can become lower than the local temperature of the liquid. This phenomenon is called constitutional supercooling. The second phenomenon comes if the concentration gradient exceeds that of the temperature on the front, i.e., if the following condition applies:

$$\left.\frac{\partial T}{\partial z}\right|_{z=h^+} < \Gamma\left.\frac{\partial C}{\partial z}\right|_{z=h^+} \tag{82}$$

Now, substituting Eqs. 78 and 79 into Eq. 80, we get

$$(C_h - c_0) < (T_h - T_\infty)\frac{\varepsilon_l}{\Gamma}\frac{\mathrm{erfc}(\lambda)}{\mathrm{erfc}(\varepsilon_l\lambda)}\frac{\exp\left(-(\varepsilon_l\lambda)^2\right)}{\exp\left(-(\lambda)^2\right)} \tag{83}$$

If the right side of the equation equals the left, the last condition determines the onset of supercooling and, consequently, the critical values of the thermophysical variables that describe the validity of the solutions above for the flat front. Therefore, if the inequality remains true, we must use a model that contemplates the *mushy* layer in the solution of the equations. Let us now analyze the solidification of a binary molten phase with a *mushy layer,* in which the heterogeneous inclusions of the new phase grow so that this layer is completely supercooled. In this case, the *mushy layer* can be treated as independent from the precise morphology of the growing phase. The *mushy region* is also treated as a continuum, and its physical properties are considered functions of the local volume fraction of the liquid φ.

The heat and mass transfers are described by the heat conduction and the diffusion of mass in the *mushy layer,* where $a(t)$ and $b(t)$ are the *solid–mushy* and *mushy–liquid* boundaries, respectively.

$$(\varphi\rho_l c_{P,l} + (1-\varphi)\rho_s c_{P,s})\frac{\partial T}{\partial t} = \frac{\partial}{\partial z}\left((\varphi k_l + \varphi(1-\varphi)k_s)\frac{\partial T}{\partial z}\right) - \rho_l L\frac{\partial \varphi}{\partial t}, a(t) \leq z \leq b(t) \tag{84}$$

$$T = -\Gamma C, a(t) < z < b(t), \tag{85}$$

$$\varphi \frac{\partial C}{\partial t} = \frac{\partial}{\partial z}\left(\varphi D \frac{\partial C}{\partial z}\right) - C \frac{\partial \varphi}{\partial t}, a(t) \le z \le b(t) \tag{86}$$

We can note that there is no reason for φ being continuous in the model under consideration. The conditions of heat and mass flows imposed at the two interfaces are as follows:

$$\rho_l L \varphi_a \frac{da}{dt} = k_s \frac{\partial T}{\partial z}\bigg|_{z=a^-} - (\varphi_a k_l + (1 - \varphi_a)k_s)\frac{\partial T}{\partial z}\bigg|_{z=a^+} \tag{87}$$

$$C_a \varphi_a \frac{da}{dt} + D\varphi_a \frac{\partial C}{\partial z}\bigg|_{z=a^+} = 0 \tag{88}$$

$$\rho_l L \varphi_b \frac{db}{dt} = (\varphi_b k_l + (1 - \varphi_b)k_s)\frac{\partial T}{\partial z}\bigg|_{z=b^-} - k_l \frac{\partial T}{\partial z}\bigg|_{z=b^+} \tag{89}$$

$$C_b \varphi_b \frac{db}{dt} = D\varphi_b \frac{\partial C}{\partial z}\bigg|_{z=b^-} - D\frac{\partial C}{\partial z}\bigg|_{z=b^+} \tag{90}$$

We also have a further boundary condition for the equilibrium at the edge of the liquid:

$$\frac{\partial T}{\partial z}\bigg|_{z=b^+} = \Gamma \frac{\partial C}{\partial z}\bigg|_{z=b^+} \tag{91}$$

The last boundary condition shows that none of the liquid regions are supersaturated and that they are marginal, in the sense that Eq. 79 provides the smallest temperature gradient consistent with complete equilibrium.

---

5 Numerical Procedure for the Prediction of Temperature Distribution and Freezing Front Penetration

5.1 Numerical Method

In the previous paragraphs, we mentioned that, as it is difficult to monitor the progress of the freezing front produced in the tissue in real time by using imaging techniques, information about the temperature reached in the tissue can be extracted by using indirect methods. To do this, we need calculation algorithms that numerically solve the equations of heat transfer. Many authors have proposed calculation algorithms over the years to predict the progress of the ice front.

In this paragraph, we will present the method used in [31–34]. It should be emphasized that we are referring to probes that use liquid nitrogen cooling, while, today, for the reasons mentioned above, probes that use the *Joule–Thomson* effect is predominantly used [35], but this does not affect the discussion of heat transfer between the tissue and the probe, if anything between probe and tissue. The aim is to present a numerical method developed for predicting the growth of the freezing front penetration produced in the tissue, highlighting, in particular, an aspect hitherto neglected by previous works: the variation with temperature of the thermal conductivity of the tissue. In addition, the importance of blood perfusion on the freezing front penetration produced is discussed. The boundary conditions that can be adopted in the study of this problem are essentially of three types: constant heat flux at the wall, constant probe-wall temperature, and constant cryofluid temperature.

The first boundary condition of constant heat flux at the wall has been assumed in [36], where the blood perfusions and the heat generation in the tissue are neglected, and the solutions are found for a spherical geometry. The same boundary condition has been assumed in [37] for a plane geometry by assuming constant physical properties and taking into account the blood perfusion and metabolic heat. The second boundary condition of constant temperature of the wall of the probe is assumed in [38–40], which presented approximate numerical and analytical solutions in cylindrical and spherical geometries by assuming that the thermal capacities are negligible compared to the latent heat. The third boundary condition is assumed in [41], where the temperature of the probe surface is variable until it reaches that of the cryofluid. They present an integral approximate solution in the Cartesian coordinate, neglecting the heat capacities compared to the latent heat. The third condition seems to best describe the physical reality, but it has the drawback of requiring knowledge of the probe's characteristics to be applied, i.e., the material it is made of and its thickness.

Since the parameters characterizing the problem are numerous, it is deemed appropriate to simplify the problem by imposing the boundary condition of constant probe temperature, thus neglecting the temperature gradient between the probe surface and the cryofluid. In conclusion, the study refers to a geometry with constant surface temperature. The other boundary condition is that the tissue temperature at a sufficiently large distance from the tissue is equal to the body average, i.e., 37 °C. This assumption is plausible when examining a portion of tissue much larger than that affected by freezing. However, the work shows that this assumption does not significantly influence the results; i.e., the numerical method adopted is independent of this boundary condition.

A controversial parameter in the development of a predictive mathematical model is the blood perfusion, toward which we can adopt two hypotheses: linear trend between a null value at the freezing temperature and maximum value at the average body temperature, following the effects that freezing has on blood viscosity, or constant perfusion during cooling and null at the freezing temperature. The first hypothesis corresponds to a situation where the blood in the larger arteries does not clot when they are heated after freezing, and blood flow is thus restored. The second hypothesis corresponds to a situation where, following rapid cooling, the blood vessels do not have time to contract, and consequently, the blood continues to flow until, at the melting temperature, it freezes along with the tissue. Both of these hypotheses have been experimentally detected by several authors. In this work, we are considering the effects that both of these hypotheses have on the solutions obtained. The importance of such hypotheses, in comparison to that of variable thermal conductivity, is highlighted. The term for metabolic heat generation within the tissue is neglected, as it does not greatly influence the temperature field; however, its introduction does not pose difficulties in the numerical method. The tissue density and specific heat are assumed to be constant, for simplicity and because no data are found in the literature regarding such variation. Another advantage of the method used is that it is possible to implement any law for the variation of thermal conductivity with temperature. Most authors consider the thermal conductivity of biological tissues to be a piecewise constant, but since such tissues are mostly composed of water and the thermal conductivity of water varies with temperature, like that of ice, it seems correct to assume that the thermal conductivity of the tissue also varies with temperature.

In this study, a portion of tissue surrounding the probe, significantly larger than it, is considered. This tissue is ideally divided into infinitesimal parts that, by symmetry, have the same geometry as the probe. It is hypothesized that the probe is extensive enough to be able to neglect the edge effects it produces on the temperature distribution. It is also assumed that the surfaces, symmetrical to the probe, are isothermal, i.e., that the material is homogeneous and isotropic. Initially, the tissue is at body temperature (37 °C), the probe is immersed in the tissue, and there is heat transfer by conduction in the tissue. Based on the simplifying assumptions already discussed, the energy balance equations can be written as follows. For the unfrozen tissue zone, from now on indicated as liquid, i.e., with the subscript l, r *being* the generic one-dimensional space variable, we have

$$\rho_l c_{p,l} \frac{\partial T}{\partial t} = \frac{1}{r^{ige0}} \frac{\partial}{\partial r}\left(k_l(T) r^{ige0} \frac{\partial T}{\partial r}\right) + \dot{q}_g \tag{92}$$

Note that the geometry can be plane, cylindrical, or spherical, depending on whether the parameter $ige0$, respectively, takes the values 0, 1, or 2. Let us make explicit the derivatives in the previous equation

$$\rho_l c_{p,l} \frac{\partial T}{\partial t} = k_l(T)\left(\frac{\partial^2 T}{\partial r^2} + \frac{ige0}{r} \frac{\partial T}{\partial r}\right) + \frac{\partial k_l(T)}{\partial T}\left(\frac{\partial T}{\partial r}\right)^2 + \dot{q}_g \tag{93}$$

where ρ and c_p are constants. For the thermal conductivity, k, a relationship of the following type is assumed:

$$k_l(T) = k_{f,l}\left[1 + B_l(T - T_f)\right] \tag{94}$$

where $k_{f,l}$ is the value of the thermal conductivity at the reference temperature T_f, which can be any. In the program, the melting temperature is chosen.

The term of energy generation, $\dot{q}_g$, due to blood perfusion, is considered to be the effect of heat convection from the blood to the tissue. It is assumed that the blood enters the tissue at the average body temperature and exits at the temperature of the tissue in that particular position. Therefore, it is given by

$$\dot{q}_g = P\rho_b c_b (T_b - T) \tag{95}$$

where P is the blood perfusion, measured in s^{-1}, which can be assumed linearly variable with the temperature from the body value to the freezing one, where it becomes null, according to [42]:

$$P = G_y \frac{(T - T_f)}{(T_b - T_f)} \tag{96}$$

In this way, the term $\dot{q}_g$ is expressed by

$$\dot{q}_g = G_v \rho_b c_b \frac{(T_b - T)(T - T_f)}{(T_b - T_f)} \tag{97}$$

The other assumption is that the term P is a constant, according to the experimental observation of [43], because the blood vessels do not have enough time to contract due to the high speed of cooling, and the blood continues to flow until the freezing temperature is reached, when it freezes with the tissue. This assumption has been assumed in [31, 44, 45].

For the zone of frozen tissue, from now on indicated as solid, i.e., with the subscript s, the energy balance equation becomes

$$\rho_s c_{p,s} \frac{\partial T}{\partial t} = k_s(T) \left(\frac{\partial^2 T}{\partial r^2} + \frac{ige0}{r} \frac{\partial T}{\partial r} \right) + \frac{\partial k_s(T)}{\partial T} \left(\frac{\partial T}{\partial r} \right)^2 \tag{98}$$

having assumed for the thermal conductivity of the frozen tissue an equation of the type of Eq. 94, the blood perfusion null, and being ρ_s, $c_{p,s}$, the values that compete to the frozen tissue. For simplicity, we will tackle a one-dimensional study of the problem while still maintaining the dependence on time.

In conclusion, the equations of the thermal balance in unfrozen and frozen zones are, respectively,

$$\rho_l c_{p,l} \frac{\partial T}{\partial t} = k_l \frac{\partial^2 T}{\partial r^2} + \frac{dk_l}{dT} \left(\frac{\partial T}{\partial r} \right)^2 + \dot{q}_g \tag{99}$$

$$\rho_s c_{p,s} \frac{\partial T}{\partial t} = k_s \frac{\partial^2 T}{\partial r^2} + \frac{dk_s}{dT} \left(\frac{\partial T}{\partial r} \right)^2 \tag{100}$$

where r is the coordinate normal to the probe surface. The boundary conditions are

$$T = T_p, \text{ for } r = 0; \text{ and } T = T_b, \text{ for } r = +\infty \tag{101}$$

T_p being the temperature of the probe and T_b being the temperature of the body. On the separation surface between the frozen and unfrozen zones the energy balance equation applies:

$$k_s \left.\frac{\partial T}{\partial r}\right|_s - k_l \left.\frac{\partial T}{\partial r}\right|_l = L\rho_l \frac{dR}{dt} \tag{102}$$

where k_s, k_l, L, and ρ_l are, respectively, the thermal conductivity of the frozen tissue, the thermal conductivity of the unfrozen tissue, the latent heat of fusion, and the density of the liquid.

From a physical point of view, the previous equation can be derived from an energy balance on an arbitrary control volume, including the separation interface between solid and liquid, which, in integral form, is expressed as

$$\dot{Q} = \oint_{\partial\Omega} \vec{q} \cdot \hat{n} dA = L\frac{dM}{dt} = L\rho \frac{d}{dt} \int_\Omega dV = L\rho \oint_{\partial\Omega} \vec{v} \cdot \hat{n} dA \tag{103}$$

with $\vec{v} = \frac{d\vec{r}}{dt}$ being the advancement speed of the ice front.

The equations are made dimensionless by assuming

$$\Theta = \frac{(T - T_b)}{(T_p - T_b)} ; \tau = \frac{t}{t_r} ; R = \frac{r}{r_t} \tag{104}$$

where t_r is a reference time and r_t is the total length of the body and also by setting

$$C = \frac{k_l(T_b) - k_{l,f}}{k_{l,f}\Theta_f} ; M = \frac{k_{l,f}t_r}{\rho_l c_{p,l} r_t^2} ; N = \frac{G_v t_r \rho_b c_b}{\rho_l c_{p,l}} ; CC = \frac{k_s(T_s) - k_{s,f}}{k_{s,f}(\Theta_s - \Theta_f)} ; \tag{105}$$

$$M1 = \frac{k_{s,f}t_r}{\rho_s c_{p,s} r_t^2} ; H = \frac{k_{s,f}}{k_{l,f}} ; W = \frac{L\rho_l r_t^2}{t_r k_{l,f}(T_s - T_b)}$$

The Eqs. 99 and 100 become, for the unfrozen and frozen zones, respectively,

$$\frac{\partial\Theta}{\partial\tau} = M\left\{ (1 + C(\Theta_f - \Theta))\left(\frac{\partial^2\Theta}{\partial R^2} + \frac{ige0}{R}\frac{\partial\Theta}{\partial R}\right) - C\left(\frac{\partial\Theta}{\partial R}\right)^2 \right\} + N\Theta\left\{\frac{\Theta}{\Theta_f} - 1\right\} \tag{106}$$

$$\frac{\partial\Theta}{\partial\tau} = M1\left\{ (1 - CC(\Theta_f - \Theta))\left(\frac{\partial^2\Theta}{\partial R^2} + \frac{ige0}{R}\frac{\partial\Theta}{\partial R}\right) + CC\left(\frac{\partial\Theta}{\partial R}\right)^2 \right\} \tag{107}$$

while the boundary conditions are

$$\Theta = 1, R = 0; \Theta = 0, R = +\infty \tag{108}$$

and the boundary condition at the frozen–unfrozen interface is

$$H\frac{\partial\Theta}{\partial R}\bigg|_s - \frac{\partial\Theta}{\partial R}\bigg|_l = W\frac{\partial\Xi}{\partial\tau} \tag{109}$$

The tissue under examination is divided into a certain number of infinitesimal parts, distant from each other, ΔR. The finite difference method, employed for the numerical solution, uses the *forward difference* for the time derivative, while for the first and second spatial derivatives, it uses the *central difference*. Based on this scheme, Eqs. 106 and 107 become, for the unfrozen and frozen zones, respectively,

$$\Theta_m^{n+1} = \begin{aligned}&\Theta_m^n + \Delta\tau M\left\{1 + C\left(\Theta_f - \Theta_m^n\right)\right\}\left(\frac{\Theta_{m+1}^n - 2\Theta_m^n + \Theta_{m-1}^n}{\Delta R^2}\right.\\ &\left. + \left(\frac{ige0}{R}\right)\frac{\Theta_{m+1}^n - \Theta_{m-1}^n}{2\Delta R}\right) - \Delta\tau MC\left(\frac{\Theta_{m+1}^n - \Theta_{m-1}^n}{2\Delta R}\right)^2\\ &+ \Delta\tau\left\{N\Theta_m^n\left(\frac{\Theta_m^n}{\Theta_f} - 1\right)\right\}\end{aligned} \tag{110}$$

$$\begin{aligned}\Theta_m^{n+1} = &\,\Theta_m^n + \Delta\tau M1\left(1 + CC\{\Theta_m^n - \Theta_f\}\right)\left(\frac{\Theta_{m+1}^n - 2\Theta_m^n + \Theta_{m-1}^n}{\Delta R^2}\right) +\\ &\Delta\tau M1\left\{\left(1 + CC\{\Theta_m^n - \Theta_f\}\right)\left(\frac{ige0}{R}\right)\left(\frac{\Theta_{m+1}^n - \Theta_{m-1}^n}{2\Delta R}\right) + \left\{CC\left(\frac{\Theta_{m+1}^n - \Theta_{m-1}^n}{2\Delta R}\right)^2\right\}\right\}\end{aligned} \tag{111}$$

As far as the discretization of the two equations near the separation surface between frozen and unfrozen tissues is concerned, the *Crank* method is used [46, 47], which extends the discretization procedure used by the finite difference theory to a moving boundary.

This is equivalent to performing a discretization on function values that are not necessarily equally spaced. We then use the Lagrange formula:

$$f(x) = \sum_{j=0}^{n} l_j(x)f(a_j), \tag{112}$$

where

$$l_j(x) = \frac{p_n(x)}{(x - a_j)p_n'(a_j)} \tag{113}$$

$$P_n(x) = \prod_{j=0}^{n} (x - a_j) \tag{114}$$

If we limit ourselves to the three-point formula, $(n = 2)$, we get

$$\frac{1}{2}\frac{d^2 f(x)}{dx^2} = \frac{f(a_0)}{(a_0 - a_1)(a_0 - a_2)} + \frac{f(a_1)}{(a_1 - a_0)(a_1 - a_2)} + \frac{f(a_2)}{(a_2 - a_0)(a_2 - a_1)} \tag{115}$$

and

$$\frac{df(x)}{dx} = \left(\frac{(x - a_1) + (x - a_2)}{(a_0 - a_1)(a_0 - a_2)}\right) f(a_0) + \left(\frac{(x - a_2) + (x - a_0)}{(a_1 - a_2)(a_1 - a_0)}\right) f(a_1)$$
$$+ \left(\frac{(x - a_0) + (x - a_1)}{(a_2 - a_0)(a_2 - a_1)}\right) f(a_1) \tag{116}$$

We apply this formula near the moving boundary. Consider the domain divided into N elements, each of thickness ΔR, and admit that the boundary at the instant T_I is somewhere in the $(r + 1)$ element at the position $X_I = (r - 1 + p)\,\Delta R$, where p is a rational number such that $1 < p < 2$. Therefore, near the frozen contour, the discretization gives rise to the following approximations:

$$\left.\frac{\partial \Theta}{\partial R}\right|_g = \frac{1}{\Delta R}\left(\frac{p}{p + 1}\Theta_{m-2} - \frac{p + 1}{p}\Theta_{m-1} + \frac{2p + 1}{p(p + 1)}\Theta_f\right) \tag{117}$$

$$\left.\frac{\partial \Theta}{\partial R}\right|_l = \frac{1}{\Delta R}\left(\frac{2p - 5}{(p - 2)(p - 3)}\Theta_f + \frac{p - 3}{p - 2}\Theta_{m+1} + \frac{p - 2}{3 - p}\Theta_{m+2}\right) \tag{118}$$

$$\left.\frac{\partial \Theta}{\partial R}\right|_{m-1} = \frac{1}{\Delta R}\left(\frac{-p}{p + 1}\Theta_{m-2} + \frac{p - 1}{p}\Theta_{m-1} + \frac{1}{p(p + 1)}\Theta_f\right) \tag{119}$$

$$\left.\frac{\partial^2 \Theta}{\partial R^2}\right|_{m-1} = \frac{2}{\Delta R^2}\left(\frac{1}{p + 1}\Theta_{m-2} - \frac{1}{p}\Theta_{m-1} + \frac{1}{p(p + 1)}\Theta_f\right) \tag{120}$$

$$\left.\frac{\partial \Theta}{\partial R}\right|_{m+1} = \frac{1}{\Delta R}\left(-\frac{1}{(p - 2)(p - 3)}\Theta_f + \frac{1 - p}{p - 2}\Theta_{m+1} + \frac{2 - p}{3 - p}\Theta_{m+2}\right) \tag{121}$$

$$\left.\frac{\partial^2 \Theta}{\partial R^2}\right|_{m+1} = \frac{2}{\Delta R^2}\left(\frac{1}{(p - 2)(p - 3)}\Theta_f + \frac{1}{p - 2}\Theta_{m+1} + \frac{1}{3 - p}\Theta_{m+2}\right) \tag{122}$$

Equations 110 and 111 become, in the vicinity of the frozen tissue,

$$\Theta_{m+1}^{n+1} = \Theta_{m+1}^{n}$$
$$+ \Delta\tau \left\{ M \left[1 + C\left(\Theta_f - \Theta_m^n\right)\right] \left[\frac{2}{\Delta R^2}\left(\frac{\Theta_f}{(p-2)(p-3)}\right) + \frac{\Theta_{m+1}^n}{p-2} + \frac{\Theta_{m+2}^n}{3-p}\right]\right\}$$
$$+ \Delta\tau \left\{ M \frac{C}{\Delta R^2}\left[\frac{\Theta_f}{(2-p)(p-3)} + \frac{1-p}{p-2}\,\Theta_{m+1}^n + \frac{2-p}{3-p}\,\Theta_{m+2}^n\right]^2\right\}$$
$$+ \Delta\tau \left[N\,\Theta_{m+1}^n\left(\frac{\Theta_{m+1}^n}{\Theta_f} - 1\right)\right]$$

$$(123)$$

$$\Theta_{m-1}^{n+1} = \Theta_{m-1}^{n} + \Delta\tau\,M_1\,\frac{2}{\Delta R^2}\left[1 + CC\left(\Theta_{m-1}^n - \Theta_f\right)\right]\left[\frac{\Theta_{m-2}^n}{p+1} - \frac{\Theta_{m-1}^n}{p} + \frac{\Theta_f}{p\,(p+1)}\right]$$
$$+ \Delta\tau\,M_1\,\frac{CC}{\Delta R^2}\left[-\frac{p}{p-1}\,\Theta_{m-2}^n + \frac{p-1}{p}\,\Theta_{m-1}^n + \frac{\Theta_f}{p\,(p-1)}\right]^2$$

$$(124)$$

Equation 109, expressed in terms of the increase of the frozen tissue, becomes

$$p' = p + \frac{\Delta\tau}{W\Delta R^2}\left\{ H\left(\frac{p}{p+1}\,\Theta_{m-1}^n - \frac{p+1}{p}\,\Theta_m^n + \frac{1+2p}{p(p+1)}\,\Theta_f\right)\right.$$
$$\left. - \left(\frac{2p-5}{(p-2)(p-3)}\,\Theta_f + \frac{p-3}{p-2}\,\Theta_{m+1}^n + \frac{p-2}{3-p}\,\Theta_{m-2}^n\right)\right\}$$

$$(125)$$

At the beginning of cooling and until the freezing temperature is reached, Eq. 110 is used to determine the temperature distribution in the unfrozen tissue, while, when the temperature of the first node of the mesh reaches the solidification temperature, freezing begins. From this moment, it is assumed that the tissue begins to freeze on the surface of the probe, and the advancement of the frozen tissue is calculated, based on Eq. 125. During the time necessary for the frozen tissue to reach a thickness equal to that of the first element of the mesh, it is assumed that the temperature of this remains constant and equal to the melting temperature. The temperatures of the other nodes are always calculated with Eq. 110, and even in this period, they, following the cooling, continue to decrease. When the tissue has reached the first node of the mesh, the method of *Crank* can be fully applied since, from this moment, the tissue is divided into two distinct zones, one frozen and the other unfrozen. Assuming that the frozen front is between the nodes m and $m + 1$, the temperatures of the elements outside of $m + 2$ are determined using Eq. 110, the temperature of the element $m + 1$ with Eq. 123, the temperatures of the elements between m–2 and the probe with Eq. 111, the temperature of the element m–1 with Eq. 124, and the increase of the frozen tissue with Eq. 125. The temperature of the element m is determined by an interpolation between the temperatures of the elements $m-1$ and $m-2$ and that of the frozen front, f, assumed at the melting temperature, with the following equation:

$$\Theta_m^{n+1} = \frac{1-p}{p+1}\Theta_{m-2}^n + \frac{2p-2}{p}\Theta_{m-1}^n + \frac{2\Theta_f}{p(p+1)} \tag{126}$$

In this way, we determine the temperature distribution in the frozen and the unfrozen tissues and the increase in the frozen tissue. The convergence and stability of the numerical method are ensured by the condition

$$\frac{\Delta\tau}{\Delta R^2} \leq 0.5 \tag{127}$$

5.2   Software 2008_New_Cryo_Surgery©

From the work previously carried out in [31–34], we have developed a new version of the software, called *2008_Nuovo_Crio_Surgery©*, which has the following architecture and is reported after the following description.

The main program *BIST1.f90©* is the *main,* in which the various subroutines are contained. In the *main,* the number of nodes or elements is assigned, and the type of geometry is chosen: planar, cylindrical, or spherical. Within the *main,* there is a control block with conditional logic that skips the calculation of the initial tissue temperature if it is already at the melting temperature. If the tissue is already at the melting temperature, it is as if we have reached the desired conditions; however, this control is present there. In the *main,* the adopted variables are also made dimensional and are written to a file. The first subroutine we encounter is *CONDIN©,* where the values of the constants for frozen tissue, unfrozen tissue, and blood are assigned. There are also commands to write data to the screen that allow the operator to monitor what is happening. The subroutine *MAGLIA©* writes the adopted mesh in dimensionless form. The geometry of the tissue is considered one-dimensional, so in one case, the coordinate is perpendicular to the plane; in another, it is radial to the cylinder; and in another, it is radial to the sphere. To solve a differential equation numerically, the equations and the domain must be discretized. This implies dealing with discrete values instead of continuous ones, and therefore, the temperature value reached by the second node will always differ by a finite amount, however small it may be, from that of the first node. This means that the domain behaves as if it is unfrozen everywhere at the beginning for a finite number of iterations, which will be smaller, the smaller the increment between adjacent nodes. For this reason, at the beginning of freezing, the subroutine *TEMPEL©* is activated. This subroutine calculates the temperature field of the domain, as if it is unfrozen, and when the second node has reached the melting temperature, it stops and reports the dimensionless value of the instant at which the temperature has reached the melting point. At this point, the logical control discussed earlier is inserted; if the tissue temperature is already equal to the melting temperature, the *subroutine TEMPEL©* is skipped. The *subroutine TEMGH©* is a kind of *subroutine* startup, in the sense that it activates to describe what happens in the very short interval of time (infinitesimal from the analytical point of view) that elapses between the reaching of the melting temperature in the second node and the formation of the ice thickness in the same cell. The presence of this *subroutine* depends on the algorithm with which the ice thickness is calculated because the equations that describe the propagation of the ice thickness are different between the beginning and the steady state. The *subroutine TEMGHIA©*, finally, is the one that calculates the steady-state temperature field and involves three equations: the conservation of energy in the liquid, that in the solid, and the propagation of the ice front. There are also equations written for the solid–liquid interface, which, however, do not take into account the presence of the *mushy region,* but the algorithm requires them for the calculation of the ice thickness. In this *subroutine,* the *file* that describes the temperature trend as a function of time is also saved, and it does not have a predictable length in advance. At this point, the program terminates.

The structure is very similar to that of [31–34]. In addition, the process is automatized so that the change of a variable at one point in the program is immediately recognized by all other parts; it optimizes the calculation time, writes the variables in the units of measure of the international system, and stores the data in *files* for the visualization of the results. The program describes the formation of ice within a living tissue in which blood circulation is active and in which thermal conductivity depends on temperature. Several simulations are carried out to take into account the influence of various phenomena. The tissue is initially at the uniform temperature of 37 °C, higher than the solidification temperature ($T_0 = 0$ °C), while the temperature of the probe is set at $T_{\text{probe}} = -100$ °C. The probe has a radius of 0.5 mm. The tissue is confined in a semi-infinite space, $x > 0.5$ mm. At time $t = 0$, the temperature on the wall, $x = 0.5$ mm, is set at the temperature T_{probe}, which is smaller than the solidification temperature, T_0, and is maintained at this temperature for all instants of time, $t > 0$. The result is that the process of solidification begins on the face $x = 0.5$ mm, and the solid–liquid interface moves along $x > 0.5$ mm. The physical parameters of the tissue are taken equal to those of water as follows: $\rho_l = 998.2$ kg/m^3, $c_{pl} = 4182$ J/kg $\times$ K, $k_l = 0.558$ W/m $\times$ K, $\alpha_l = 1.43 \times 10^{-7}$ m^2/s, $\rho_s = 916.2$ kg/m^3, $c_{ps} = 2000$ J/(kg $\times$ K), $k_s = 2.23$ W/(m $\times$ K), $\alpha_s = 1.22 \times 10^{-6}$ m^2/s, and $L_h = 335$ kJ/kg.

5.2.1 List of the Code Cryosurgery (in FORTRAN)

```fortran
PROGRAM BIST1
INTEGER, PARAMETER::M=201
INTEGER, PARAMETER::IGE0=0
REAL, DIMENSION(M,2)::T
REAL, DIMENSION(M,1)::T_1
REAL, DIMENSION(M,1)::T_2
REAL, DIMENSION(M,1)::T_3
REAL, DIMENSION(M,1)::T_4
REAL, DIMENSION(M,1)::T_5
REAL, DIMENSION(M,1)::T_6
REAL, DIMENSION(M)::R
 COMMON/CTEMPO/DT,TEM,TFIN
 COMMON/CDATA/RS,TS,DR,RT,TFF,TF1,PERF,TE,TSD,TFU,TF
 COMMON/COST/AM,C1,AN,W,H,DR2,AM1,CC
 COMMON/CCOGH/COGHIA
!***** Menu' *****
WRITE(*,*)'*****************************************************************************'
WRITE(*,*)'***                                        ***'
WRITE(*,*)'***                                        ***'
WRITE(*,*)'***              CRYOSURGERY                      ***'
WRITE(*,*)'***                                        ***'
WRITE(*,*)'***                                        ***'
WRITE(*,*)'***                                        ***'
WRITE(*,*)'*****************************************************************************'
WRITE(*,*)
CALL CONDIN(M,T)
!THIS SUBROUTINE EVALUATES PARAMETERS OF FROZEN°TISSUE
!TEMPERATURE FIELD WITHOUT DIFFERENCE BETWEEN FROZEN AND NONFROZEN
WRITE(*,*)'ENTER'
READ*, B
CALL MAGLIA(IGE0,R,M)
!THIS SUBROUTINE DEFINES ONLY THE GEOMETRY PLANE°CYLINDRICAL°SPHERICAL,
!CORRESPONDING TO DIFFERENT°VALUE°OF IGE0. DEFINES THE VECTOR R OF COORDIN
ATES
WRITE(*,*)'ENTER'
READ*, B
IF(TFU.EQ.TE) GOTO 10 !DO°NOT°EVALUATE INITIAL FIELD°IF°FUSION TEMPERATURE IS
°INITIAL TEMPERATURE OF TISSUE
CALL TEMPEL(IGE0,R,M,T)
!THIS SUBROUTINE EVALUATE THE°TEMPERATURE FIELD AT°BEGINNENING OF°FREEZIN
G WHEN TISSUE°IS HOMOGENEOUS
!AND°ICE°IS°NOT°YET°PRESENT
WRITE(*,*)'ENTER'
READ*, B
```

```fortran
      GOTO 20
10    CONTINUE
      TEM = 0
      TF = TFF
20    CONTINUE
      CALL TEMGH(IGE0,R,M,T)
!THIS SUBROUTINE EVALUATE THE°TEMPERATURE FIELD UNTIL°ICE°HAS°REACHED°TH
E°FIRST°GRID
!AND°ICE°IS°NOT°YET°PRESENT
      WRITE(*,*)'ENTER?
      READ*, B
      CALL TEMGHIA(IGE0,R,M,T,T_1,T_2,T_3,T_4,T_5,T_6)
!THIS SUBROUTINE EVALUATE THE°PENETRATION°OF°THE°FREEZING°FRONT°AT°STEA
DY°STATE
      WRITE(*,*)'ENTER?
      READ*, B
      DO I=1,M
      T(I,2)= TE+(TSD-TE)*T(I,2)
      END DO
      DO I=1,M
      T_1(I,1)= TE+(TSD-TE)*T_1(I,1)
      END DO
      DO I=1,M
      T_2(I,1)= TE+(TSD-TE)*T_2(I,1)
      END DO
      DO I=1,M
      T_3(I,1)= TE+(TSD-TE)*T_3(I,1)
      END DO
      DO I=1,M
      T_4(I,1)= TE+(TSD-TE)*T_4(I,1)
      END DO
      DO I=1,M
      T_5(I,1)= TE+(TSD-TE)*T_5(I,1)
      END DO
      DO I=1,M
      T_6(I,1)= TE+(TSD-TE)*T_6(I,1)
      END DO
      OPEN(11,file='T_1.dat')
      DO 51 I=1,M
      WRITE(11,*) T_1(I,1)
51    CONTINUE
      CLOSE(11)
      OPEN(12,file='T_2.dat')
      DO 52 I=1,M
      WRITE(12,*) T_2(I,1)
52    CONTINUE
      CLOSE(12)
```

```fortran
      OPEN(13,file='T_3.dat')
      DO 53 I=1,M
      WRITE(13,*) T_3(I,1)
53    CONTINUE
      CLOSE(13)
      OPEN(14,file='T_4.dat')
      DO 54 I=1,M
      WRITE(14,*) T_4(I,1)
54    CONTINUE
      CLOSE(14)
      OPEN(15,file='T_5.dat')
      DO 55 I=1,M
      WRITE(15,*) T_5(I,1)
55    CONTINUE
      CLOSE(15)
      OPEN(16,file='T_6.dat')
      DO 56 I=1,M
      WRITE(16,*) T_6(I,1)
56    CONTINUE
      CLOSE(16)
      OPEN(17,file='R.dat')
      DO 19 I=1,M
      WRITE(17,*) RT*R(I)
19    CONTINUE
      CLOSE(17)
      OPEN(18,file='T.dat')
      DO 60 I=1,M
      WRITE(18,*) T(I,2)
60    CONTINUE
      CLOSE(18)
      STOP
      END
      SUBROUTINE CONDIN(M,T)
      REAL KLF
      REAL KLB
      REAL KSS
      REAL KSF
      INTEGER, INTENT(IN)::M
      REAL, DIMENSION(M,2)::T
      INTENT(INOUT)::T
      COMMON/CTEMPO/DT,TEM,TFIN
      COMMON/CDATA/RS,TS,DR,RT,TFF,TF1,PERF,TE,TSD,TFU,TF
      COMMON/COST/AM,C1,AN,W,H,DR2,AM1,CC
      COMMON/CCOGH/COGHIA
      ! DIMENSIONAL°TEMPERATURE
      TSD=-100. !PROBE°TEMPERAURE IN CELSIUS
      TE=37.  !!INITIAL°TEMPERATURE OF TISSUE
```

```
TFU=-0.000  !FUSION°TEMPERATURE IN CELSIUS
! DIMENSIONLESS°TEMPERATURE
TEA=0.  !DIMENSIONLESS°TEMPERATURE°OF°TISSUE
TS=1.   !DIMENSIONLESS°TEMPERATURE°OF°PROBE
TFF=(TFU-TE)/(TSD-TE) ! DIMENSIONLESS°TEMPERATURE°OF°FUSION°NOT ZERO
! CHANGE°OF°TFF
TF1= TFF +0.01
! TEMPORAL°PARAMETERS
TR= 1.  !DIMENSIONLESS°TIME
TFIN = 30.
!DT= 0.005 !<0.5*DR2
! UNFROZEN°TISSUE
! KL=KFL*(1+B*(T-TF))
CPL= 4182. !SPECIFIC°HEAT
RHOL=998.2 !DENSITY'
KLF=0.558
KLB=0.487
ALPHALB=KLB/(CPL*RHOL)
ALPHALF=KLF/(CPL*RHOL)
AL0= 335000.
PERAC=1.
AL=AL0*PERAC !LATENT°HEAT
! FROZEN°TISSUE
! KS=KFS*(1+B'*(TF-T))
CPS= 2000.
RHOS=916.2
KSF=2.15
KSS=3.44
ALPHASS=KSS/(CPS*RHOS)
ALPHASF=KSF/(CPS*RHOS)
! BLOOD
RHOB=1800.
CPB= 3000.
GV= 0.03 ! 0.0067  !s**(-1)
PERF=1.
! SONDA
RS0=0.0005
RT=0.014 !DIMENSIONAL COORDINATE
RTD=RT/RT
RS=RS0/RT
DR= (RT-RS0)/(M*RT) !
! PARAMETERS FOR EQUATIONS
AM= ALPHALF*TR/RT**2 ! 0.021875 !
CC= (KSS/KSF-1.)/(TFF-1) ! 0.22938 !
C1= (KLB/KLF-1.)/TFF   ! -0.98099 !
AN= GV*TR*((RHOB*CPB)/(RHOL*CPL)) ! 0.005762 !
W= RT**2*RHOL*AL/(TR*KLF*(TSD-TE)) ! -24.387 !
```

```
IF(TFU.EQ.TE) W=W*KLF/KSF
H= 3.52 !KSF/KLF
IF(TFU.EQ.TE) H=1.
DR2=DR**2 !0.04 !
AM1=ALPHASF*TR/RT**2 !0.17866 !
DT= 0.45*DR2 ! 0.0001 ! !STABILITY°CRITERION
! INITIAL°TEMPERATURES
DO 1 J=1,2
DO 1 I=2,M
T(I,J) =0.
1 CONTINUE
T(1,1)=TS
T(1,2)=TS
WRITE(6,100) (T(I,1),I=1,M)
100 FORMAT(///3X,20HINITIAL°TEMPERATURES(//2X,10E13.5))
WRITE(6,200) TSD,TFU,TE,TS,TFF,TEA,TR
200 FORMAT(///3X,22HPROBE°TEMPERATURE [C]=,E12.5,/3X,24HFSION°TEMPERATUR
E [C]=,E12.5,/3X,33HINITIAL°TEMPERATURE TISSUE [C]=,E12.5,/3X,25HDIMENSIONLESS
°TEMPERATURE,/3X,9HDI PROBE=,E12.5,/3X,11HDI FUSION=,E12.5,/3X,11HDI TISSUE=,
E12.5,/3X,25HREFERENCE°TIME [s]=,E12.5)
WRITE(6,201) RHOL,CPL,KLB,KLF,ALPHALB,ALPHALF
201 FORMAT(///3X,31HPHYSICAL°CHARACTERISTICS°TISSUE,///3X,21HUNFROZEN°TIS
SUE///3X,17HDENSITY [kg/m^3]=,E12.5,/3X,26HSPECIFIC°HEAT [J/kg*K]=,E12.5,/3X,38
HTHERMAL°CONDUCTIVITY°AT TE [W/m*K]=,E12.5,/3X,38HTHERMAL°CONDUCTIVITY°
AT T0 [W/m*K]=,E12.5,/3X,36HTHERMAL°DIFFUSIVITY°AT TE [m^2/s]=,E12.5,/3X,36HT
HERMAL°DIFFUSIVITY°AT T0 [m^2/s]=,E12.5)
WRITE(6,202) RHOS,CPS,KSS,KSF,ALPHASS,ALPHASF,AL
202 FORMAT(///3X,17HFROZEN°TISSUE,///3X,17HDENSITY [kg/m^3]=,E12.5,/3X,26HS
PECIFIC°HEAT [J/kg*K]=,E12.5,/3X,38HTHERMAL°CONDUCTIVITY°AT T7 [W/m*K]=,E12.
5,/3X,38HTHERMAL°CONDUCTIVITY°AT T0 [W/m*K]=,E12.5,/3X,36HTHERMAL°DIFFUSI
VITY°AT T7 [m^2/s]=,E12.5,/3X,36HTHERMAL°DIFFUSIVITY°AT T0 [m^2/s]=,E12.5,/3X,
22HLATENT°HEAT [J/kg]=,E12.5)
WRITE(6,203) RHOB,CPB,PERF,GV,100*PERAC
203 FORMAT(///3X,22HBLOOD°CHARACTERISTICS,//3X,24HBLOOD°DENSITY [kg/m^3]
=,E12.5,/3X,26HSPECIFIC°HEAT [J/kg*K]=,E12.5,/3X,21HBLOOD°PERFUSION=,E12.5,/3X,
40HBLOOD°FLOW°RATE PER UNIT OF MASS [1/s]=,E12.5,/3X,36HPERCENT OF WATER IN
 TISSUE [%]=,E12.5)
WRITE(6,204) RS0,RS,RT,RTD,RT*DR,DR
204 FORMAT(///3X,21HPROBE°CHARACTERISTICS,//3X,17HPROBE°RADIUS [m]=,E12.5,
/3X,16HE DIMENSIONLESS=,E12.5,/3X,19HTISSUE°RADIUS [m]=,E12.5,/3X,16HE DIMENS
IONLESS=,E12.5,/3X,20HDR DIMENSIONAL [m]=,E12.5,/3X,16HE DIMENSIONLESS=,E12.5
)
WRITE(6,205) AM,C1,AN,W,H,DR2,AM1,CC,DT,DR
205 FORMAT(///3X,3HAM=,E12.5,/3X,3HC1=,E12.5,/3X,3HAN=,E12.5,/3X,2HW=,E12.5,/
3X,2HH=,E12.5,/3X,4HDR2=,E12.5,/3X,4HAM1=,E12.5,/3X,3HCC=,E12.5,/3X,3HDT=,E12.
5,/3X,3HDR=,E12.5)
RETURN
```

```fortran
      END
      SUBROUTINE MAGLIA(IGE0,R,M)
      INTEGER, INTENT(IN)::M
      INTEGER, INTENT(IN)::IGE0
      REAL, DIMENSION(M)::R
      INTENT(INOUT)::R
      COMMON/CTEMPO/DT,TEM,TFIN
      COMMON/CDATA/RS,TS,DR,RT,TFF,TF1,PERF,TE,TSD,TFU,TF
      COMMON/COST/AM,C1,AN,W,H,DR2,AM1,CC
      COMMON/CCOGH/COGHIA
      R(1) = RS
      DO 1 I=2,M
      R(I) = R(I-1) +DR
    1 CONTINUE
! GEOMETRY
      IGE1 = IGE0-1
      IF(IGE1) 10,20,30
   10 CONTINUE
      WRITE (6,300) IGE0
  300 FORMAT (///3X,15HPLANE°GEOMETRY,10X,5HIGE0=,I2)
      GOTO 40
   20 CONTINUE
      WRITE (6,400) IGE0
  400 FORMAT (///3X,20HCYLINDRICAL°GEOMETRY,10X,5HIGE0=,I2)
      GOTO 40
   30 CONTINUE
      WRITE (6,500) IGE0
  500 FORMAT (///3X,17HSPHERICAL°GEOMETRY,10X,5HIGE0=,I2)
   40 CONTINUE
      WRITE (6,200) (R(I),I=1,M)
  200 FORMAT (///3X,15HDIMENSIONS°OF°GRID (//2X,10E13.5))
      RETURN
      END
      SUBROUTINE TEMGH(IGE0,R,M,T)
      INTEGER, INTENT(IN)::M
      INTEGER, INTENT(IN)::IGE0
      REAL, DIMENSION(M,2)::T
      INTENT(INOUT)::T
      REAL, DIMENSION(M)::R
      INTENT(INOUT)::R
      COMMON/CTEMPO/DT,TEM,TFIN
      COMMON/CDATA/RS,TS,DR,RT,TFF,TF1,PERF,TE,TSD,TFU,TF
      COMMON/COST/AM,C1,AN,W,H,DR2,AM1,CC
      COMMON/CCOGH/COGHIA
      INTEGER POSGHIA
      WRITE(6,901)
  901 FORMAT(3X,18HBEGINNING°OF°FREEZING,(//3X,10E13.5))
```

```
POSGHIA=1
! ICE°THICKNESS
P=1.
62 CONTINUE
I=POSGHIA
I=I+1
T(I,2)=TF
10 CONTINUE
I=I+1
IF(TFU.EQ.TE) GOTO 11
T(I,2)= T(I,1)+DT*(AM*((1.+C1*(TFF-T(I,1)))*(((T(I+1,1)-2.*T(I,1)+T(I-
1,1))/DR2)+(IGE0/R(I))*(T(I+1,1)-T(I-1,1))/(2*DR))-C1*((T(I+1,1)-T(I-
1,1))/(2*DR))**2)+AN*T(I,1)*((T(I,1)/TFF)-1.))
11 CONTINUE
IF(I.EQ.M-1) GOTO 20
GOTO 10
20 CONTINUE
J=POSGHIA
P=P+DT*(1./(W*DR2))*(H*(T(J,1)*(P/(P+1.))-
T(J,1)*((P+1.)/P)+TFF*((2.*P+1.)/(P*(P+1.))))-(TFF*((2.*P-5.)/((P-2.)*(P-
3.)))+T(J+1,1)*((P-3.)/(P-2.))+T(J+2,1)*((P-2.)/(3.-P))))
TEM=TEM+DT
M1=M-1
DO 1 I=2,M1
T(I,1)=T(I,2)
1 CONTINUE
900 FORMAT(//3X,18HTIME°ELAPSED t=,E13.5,//3X,19HICE°THICKNESS =,E13.5,//3X,2
1HGRID°TEMPERATURE,(//3X,10E13.5))
ICSE = DT
IF(TEM.EQ.ICSE) WRITE(6,909) TEM,P,(T(I,1),I=1,M)
ICSE = 5*DT
IF(TEM.EQ.ICSE) WRITE(6,909) TEM,P,(T(I,1),I=1,M)
ICSE = 10*DT
IF(TEM.EQ.ICSE) WRITE(6,909) TEM,P,(T(I,1),I=1,M)
ICSE =15*DT
IF(TEM.EQ.ICSE) WRITE(6,909) TEM,P,(T(I,1),I=1,M)
ICSE =20*DT
IF(TEM.EQ.ICSE) WRITE(6,909) TEM,P,(T(I,1),I=1,M)
ICSE = 25*DT
IF(TEM.EQ.ICSE) WRITE(6,909) TEM,P,(T(I,1),I=1,M)
ICSE = 30*DT
IF(TEM.EQ.ICSE) WRITE(6,909) TEM,P,(T(I,1),I=1,M)
ICSE = 35*DT
IF(TEM.EQ.ICSE) WRITE(6,909) TEM,P,(T(I,1),I=1,M)
ICSE = 40*DT
IF(TEM.EQ.ICSE) WRITE(6,909) TEM,P,(T(I,1),I=1,M)
ICSE = 45*DT
```

```fortran
      IF(TEM.EQ.ICSE) WRITE(6,909) TEM,P,(T(I,1),I=1,M)
 909  FORMAT(///3X,15HTIME°ELAPSED,E13.5,//3X,17HICE°THICKNESS,E13.5,//3X,21H
      GRID°TEMPERATURE,(//3X,10E13.5))
      IF(P.GE.2.) GOTO 40
      IF(POSGHIA.EQ.1)  GOTO 40
 40   CONTINUE
      P=P-1
      COGHIA=P
      WRITE(6,400) P
 400  FORMAT(3X,40HICE ON°FIRST°GRID,//3X,17HICE°THICKNESS,E13.5)
      WRITE(6,103) TEM, (T(I,1),I=1,M)
 103  FORMAT(//3X,15HTIME°ELAPSED,E13.5,///3X,21HGRID°TEMPERATURE,(//3X,10E
      13.5))
      RETURN
      END
      SUBROUTINE TEMGHIA(IGE0,R,M,T,T_1,T_2,T_3,T_4,T_5,T_6)
      INTEGER, INTENT(IN)::M
      INTEGER, INTENT(IN)::IGE0
      REAL, DIMENSION(M,2)::T
      INTENT(INOUT)::T
      REAL, DIMENSION(M,1)::T_1
      INTENT(OUT)::T_1
      REAL, DIMENSION(M,1)::T_2
      INTENT(OUT)::T_2
      REAL, DIMENSION(M,1)::T_3
      INTENT(OUT)::T_3
      REAL, DIMENSION(M,1)::T_4
      INTENT(OUT)::T_4
      REAL, DIMENSION(M,1)::T_5
      INTENT(OUT)::T_5
      REAL, DIMENSION(M,1)::T_6
      INTENT(OUT)::T_6
      REAL, DIMENSION(M)::R
      INTENT(INOUT)::R
      COMMON/CTEMPO/DT,TEM,TFIN
      COMMON/CDATA/RS,TS,DR,RT,TFF,TF1,PERF,TE,TSD,TFU,TF
      COMMON/COST/AM,C1,AN,W,H,DR2,AM1,CC
      COMMON/CCOGH/COGHIA
      INTEGER POSGHIA
      REAL P
      TEM0=TEM
      K=(TFIN-TEM0)/DT
      POSGHIA=2
      P=COGHIA
      OPEN(9,file='POSGHIA.dat', STATUS='UNKNOWN')
      OPEN(10,file='TEMPO.dat', STATUS='UNKNOWN')
      ! SPESSORE GHIACCIO
```

```
      WRITE(6,700) P,TEM,(T(I,1),I=1,M)
  700 FORMAT(3X,29HINITIAL°THICKNESS°OF°ICE,E12.5,//3X,32HINITIAL°TIME°OF°FREE
     ZING,3X,E13.5,//3X,28HTEMPERATURE°DISTRIBUTION,(//3X,10E13.5))
      GOTO 62
   80 CONTINUE
      K=K-1
      GOTO 71
   70 CONTINUE
      K=POSGHIA
   71 CONTINUE
      I=K
      T(I-2,2)=T(I-2,1)+DT*AM1/DR2*((1.+CC*(T(I-2,1)-TFF))*((T(I-1,1)-2.*T(I-2,1)+T(I-
     3,1))+(IGE0*DR/R(I-2))*(T(I-1,1)-T(I-3,1))/2)+CC/4.*(T(I-1,1)-T(I-3,1))**2)
      IF(K-2.EQ.2) GOTO 60
      GOTO 80
   60 CONTINUE
      I=POSGHIA
      T(I-1,2)=T(I-1,1)+DT*AM1/DR2*((1.+CC*(T(I-1,1)-TFF))*(2.*((1./(P+1.))*T(I-2,1)-
     (1./P)*T(I-1,1)+(1./(P*(P+1.)))*TFF)+(IGE0*DR/R(I-1))*(-(P/(P+1.))*T(I-2,1)+((P-
     1.)/P)*T(I-1,1)+(1./(P*(P+1.)))*TFF))+CC*(-(P/(P+1.))*T(I-2,1)+((P-1.)/P)*T(I-
     1,1)+(1./(P*(P+1.)))*TFF)**2)
      GOTO 61
   30 CONTINUE
   31 CONTINUE
   61 CONTINUE
      T(I,2)=((1.-P)/(P+1.))*T(I-2,2)+(2.*(P-1.)/P)*T(I-1,2)+(2./(P*(P+1.)))*TFF
   62 CONTINUE
      I=POSGHIA
      IF(POSGHIA.EQ.2) T(I,2)=((1.-P)/(P+1.) +2.*(P-1.)/P)*TS+(2./(P*(P+1.)))*TFF
      IF(POSGHIA.EQ.(M-1)) GOTO 20
      I=I+1
      IF(TFU.EQ.TE) GOTO 10
      T(I,2)=T(I,1)+DT*(AM*((1.+C1*(TFF-T(I,1)))*((2./DR2)*((1./((2.-P)*(3.-P)))*TFF+(1./(P-
     2.))*T(I,1)+(1./(3.-P))*T(I+1,1))+(IGE0/(DR*R(I)))*((1./((P-2.)*(3.-P)))*TFF+((1.-P)/(P-
     2.))*T(I,1)+((2.-P)/(3.-P))*T(I+1,1)))-C1*(1./DR2)*((1./((P-2.)*(3.-P)))*TFF+((1.-P)/(P-
     2.))*T(I,1)+((2.-P)/(3.-P))*T(I+1,1))**2)+AN*T(I,1)*((T(I,1)/TFF)-1.))
   10 CONTINUE
      IF(POSGHIA.EQ.(M-2)) GOTO 20
      I=I+1
      IF(TFU.EQ.TF) GOTO 19
      T(I,2)=T(I,1)+DT*(AM*((1.+C1*(TFF-T(I,1)))*(((T(I+1,1)-2.*T(I,1)+T(I-
     1,1))/DR2)+(IGE0/R(I))*(T(I+1,1)-T(I-1,1))/(2*DR))-C1*((T(I+1,1)-T(I-
     1,1))/(2*DR))**2)+AN*T(I,1)*((T(I,1)/TFF)-1.))
   19 CONTINUE
      IF(I.EQ.(M-1)) GOTO 20
      GOTO 10
   20 CONTINUE
```

```
J=POSGHIA
IF(POSGHIA.EQ.2) GOTO 2
P=P+DT*(1./(DR2*W))*(H*((P/(P+1.))*T(J-2,1)-((P+1.)/P)*T(J-
1,1)+((2.*P+1.)/(P*(P+1.)))*TFF)-(((2.*P-5.)/((P-2.)*(P-3.)))*TFF+((P-3.)/(P-
2.))*T(J+1,1)+((P-2.)/(3.-P))*T(J+2,1)))
GOTO 3
2 CONTINUE
P=P+DT*(1./(DR2*W))*(H*((P/(P+1.))*T(J-1,1)-((P+1.)/P)*T(J-
1,1)+((2.*P+1.)/(P*(P+1.)))*TFF)-(((2.*P-5.)/((P-2.)*(P-3.)))*TFF+((P-3.)/(P-
2.))*T(J+1,1)+((P-2.)/(3.-P))*T(J+2,1)))
3 CONTINUE
TEM=TEM+DT
!write(*,*) 'TIME°ELAPSED°IS= ',TEM
M1=M-1
DO 1 I=2,M1
T(I,1)=T(I,2)
1 CONTINUE
IF(TEM.GE.TFIN)   GOTO 100
IF(P.GE.2.) GOTO 40
IF(POSGHIA.EQ.2) GOTO 62
IF(POSGHIA.EQ.3) GOTO 60
IF(POSGHIA.GT.3) GOTO 70
40 CONTINUE
IF(TEM.LE.TFIN/6) GOTO 45
IF(TEM.LE.TFIN/3)  GOTO 46
IF(TEM.LE.TFIN/2)  GOTO 47
IF(TEM.LE.2*TFIN/3)  GOTO 48
IF(TEM.LE.5*TFIN/6)  GOTO 49
IF(TEM.LE.TFIN)  GOTO 50
GOTO 101
45 CONTINUE
DO 51 I=1,M
T_1(I,1)=T(I,1)
51 CONTINUE
GOTO 101
46 CONTINUE
DO 52 I=1,M
T_2(I,1)=T(I,1)
52 CONTINUE
GOTO 101
47 CONTINUE
DO 53 I=1,M
T_3(I,1)=T(I,1)
53 CONTINUE
GOTO 101
48 CONTINUE
DO 54 I=1,M
```

```fortran
      T_4(I,1)=T(I,1)
54    CONTINUE
      GOTO 101
49    CONTINUE
      DO 55 I=1,M
      T_5(I,1)=T(I,1)
55    CONTINUE
      GOTO 101
50    CONTINUE
      DO 56 I=1,M
      T_6(I,1)=T(I,1)
56    CONTINUE
      GOTO 101
101   CONTINUE
      WRITE(6,600) POSGHIA
600   FORMAT(3X,34HIL ICE°IS°AT°GRID°,5X,I4)
      WRITE(6,103) TEM, (T(I,1),I=1,M)
103   FORMAT(//3X,15HTIME°ELAPSED,E13.5,//3X,21HGRID°TEMPERATURE,(//3X,10E1
     3.5))
      WRITE(9,*) (RS+POSGHIA*DR)*RT
      WRITE(10,*) TEM
      P=P-1
      POSGHIA=POSGHIA+1
      IF(POSGHIA.GT.3) GOTO 70
      GOTO 60
100   CONTINUE
      WRITE(6,800)
800   FORMAT(3X,21HFROZEN°THICKNESS)
      WRITE(6,103) TEM, (T(I,1),I=1,M)
      CLOSE(9)
      CLOSE(10)
      RETURN
      END
      SUBROUTINE TEMPEL(IGE0,R,M,T)
      INTEGER, INTENT(IN)::M
      INTEGER, INTENT(IN)::IGE0
      REAL, DIMENSION(M,2)::T
      INTENT(INOUT)::T
      REAL, DIMENSION(M)::R
      INTENT(INOUT)::R
      COMMON/CTEMPO/DT,TEM,TFIN
      COMMON/CDATA/RS,TS,DR,RT,TFF,TF1,PERF,TE,TSD,TFU,TF
      COMMON/COST/AM,C1,AN,W,H,DR2,AM1,CC
      COMMON/CCOGH/COGHIA
      TEM= 0.
      WRITE(6,200) TEM
200   FORMAT(//3X,21HBEGINOF°COOLING,//3X,15HINITIAL°TIME =,E13.5)
```

```
      SOLID=0.
60    CONTINUE
      I=1
10    CONTINUE
      I=I+1
40    CONTINUE
      T(I,2)= T(I,1)+DT*(AM*((1.+C1*(TFF-T(I,1)))*(((T(I+1,1)-2.*T(I,1)+T(I-
     1,1))/DR2)+(IGE0/R(I))*(T(I+1,1)-T(I-1,1))/(2*DR))-C1*((T(I+1,1)-T(I-
     1,1))/(2*DR))**2)+AN*T(I,1)*((T(I,1)/TFF)-1.))
      IF(T(I,2).GT.TFF) GOTO 20  !DIMENSIONLESS°FUSION°TEMPERATURE
      IF(I.EQ.(M-1)) GOTO 50
      GOTO 10
20    CONTINUE
      IF(T(I,2).GT.TF1) GOTO 90
      GOTO 30
90    CONTINUE
      WRITE(6,103) TEM,(T(I,1),I=1,M)
      WRITE(6,103) TEM,(T(I,2),I=1,M)
      STOP
50    CONTINUE
      M1=M-1
      DO 1 I=2,M1
      T(I,1)=T(I,2)
1     CONTINUE
      TEM=TEM+DT
      IF(SOLID.EQ.1) GOTO 70
      GOTO 60
30    CONTINUE
      TF=T(I,2)
      WRITE(6,101) TF
101   FORMAT(/3X,30HFREEZING°TEMPERATURE,3X,E15.8)
      SOLID=1
      GOTO 10
70    CONTINUE
      WRITE(6,102) TEM
102   FORMAT(/3X,56HTEMPERATURE°AT°BEGINNING°OF °FREEZING°AT°TIME t=,E12.5)
      WRITE(6,103) TEM, (T(I,1),I=1,M)
103   FORMAT(//3X,15HTIME°ELAPSED,E13.5,/3X,20HGRID°TEMPEATURE,(//2X,10E13.5
     ))
      RETURN
      END
```

6   Numerical Solutions in Water or Tissue without Blood Perfusion

The numerical solutions for the change of phase-solidification of the tissue without blood perfusion employ the values of water as the constants involved in the problem: $\rho_l = 998.2$ kg/m^3, $c_{pl} = 4182$ J/kg*K, $k_l = 0.558$ W/m*K, $\alpha_l = 1.33*10^{-7}$ m^2/s, $\rho_s = 916.2$ kg/m^3, $c_{ps} = 2000$ J/(kg*K), $k_s = 3.44$ W/(m*K), $\alpha_s = 1.88*10^{-6}$ m^2/s, and $L = 335{,}000$ J/kg. The tissue is initially at the uniform temperature of $37\,°C$, while the temperature of the probe is set at $T_{\text{probe}} = -100\,°C$ on the probe radius equal to 0.5 mm. Two cases are investigated: Case 1: constant thermal conductivity and Case 2: variable thermal conductivity.

6.1   Case 1: Constant Thermal Conductivity

The following Figs. 20, 21, 22, 23, 24, and 25, present the growth of the ice thickness as a function of time and the temperature distribution along the distance from the probe for the three geometries with constant thermal conductivity of the frozen tissue equal to $k_s = 3.44$ W/(m K). The spatial integration step for the various geometries and for the various types of problem is not the same; the optimal one is chosen depending on the problem. This means that it is not possible to compare different problems in the same graph, or at least, it is difficult to do so.

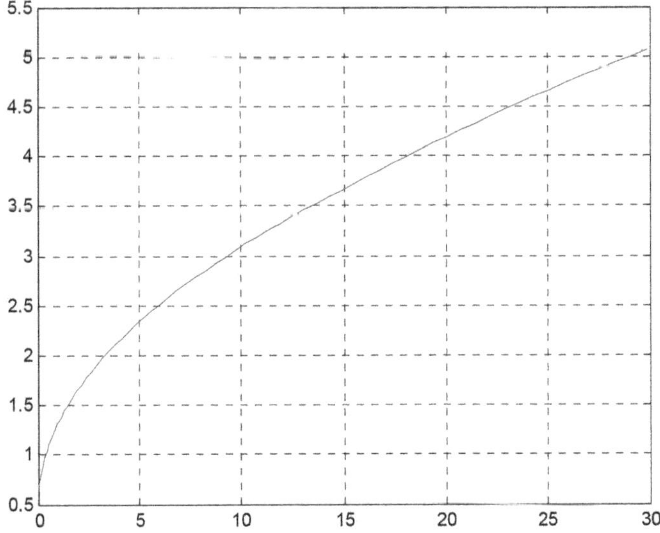

Fig. 20 Growth of the ice thickness (mm) with time (s) in plane geometry [28]

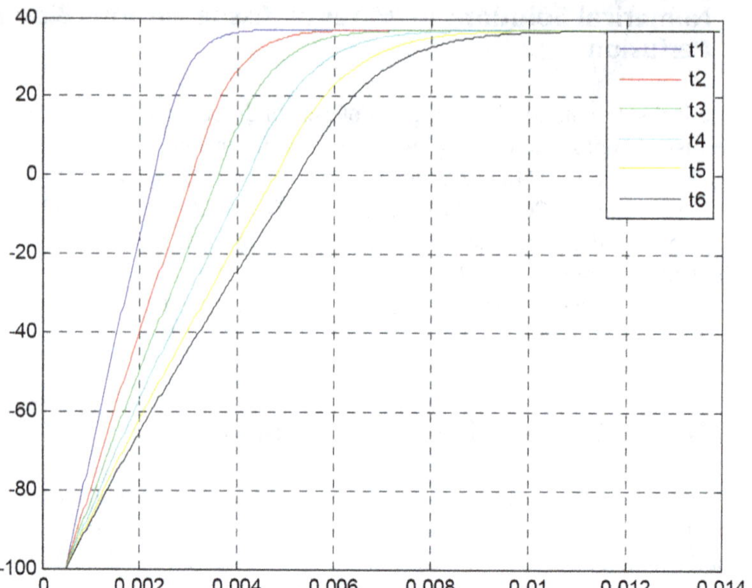

Fig. 21 Temperature distribution, T (°C), versus x (m) in plane geometry [28]

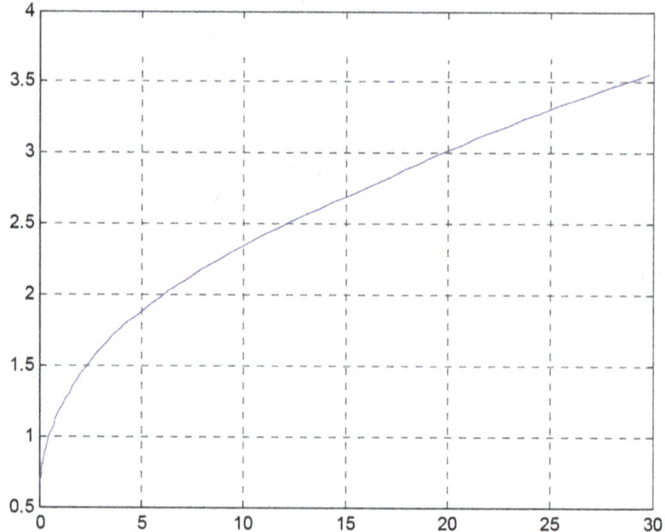

Fig. 22 Growth of the ice thickness (mm) with time (s) in cylindrical geometry [28]

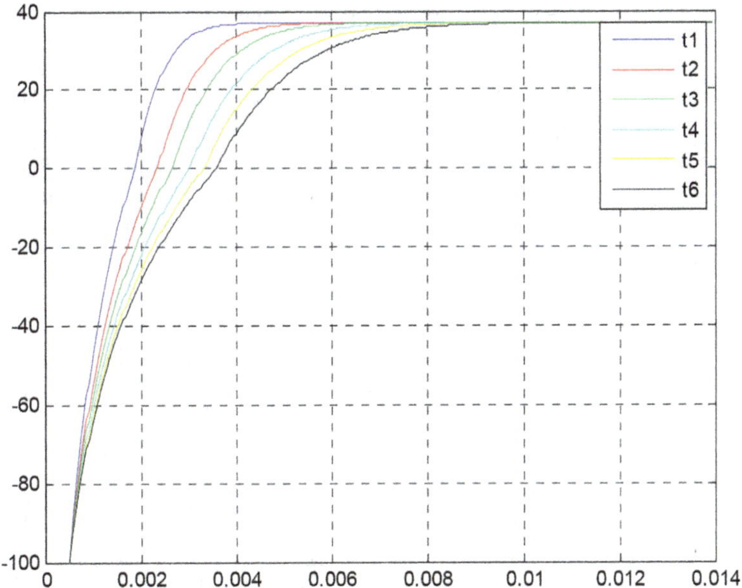

Fig. 23 Temperature distribution, T (°C), versus x (m) in cylindrical geometry [28]

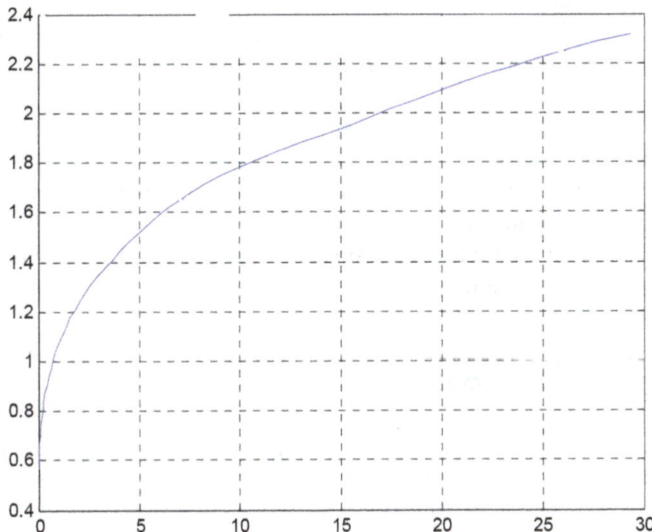

Fig. 24 Growth of the ice thickness (mm) with time (s) in spherical geometry [28]

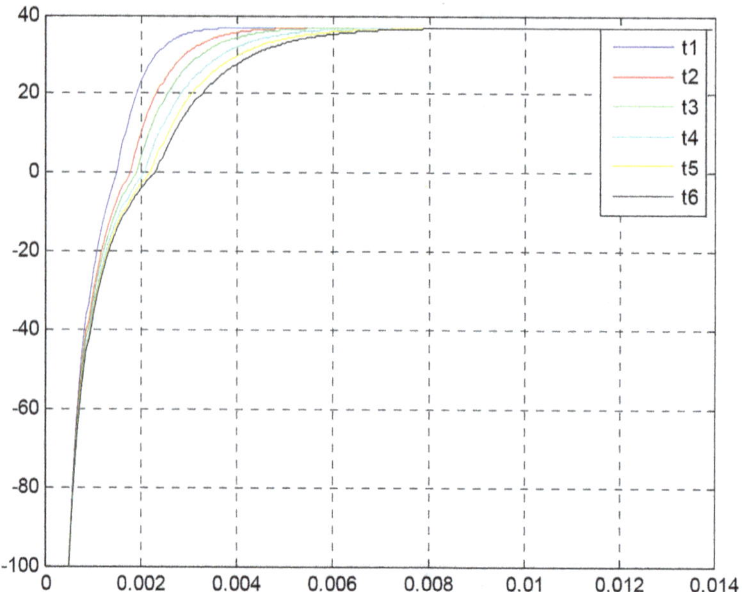

Fig. 25 Temperature distribution, T (°C), versus x (m) in spherical geometry [28]

6.1.1 Plane Geometry

Figure 20 reports the ice thickness growth for plane geometry. It is equal to 2.3 mm after 5 s, 3.1 mm after 10 s, 3.7 mm after 15 s, 4.2 mm after 20 s, 4.7 mm after 25 s, and 5.2 mm after 30 s.

Figure 21 presents the temperature distribution, T $(°C)$, versus $x\,(m)$ at six instants of time: t_1 is after 5 s, t_2 is after 10 s, t_3 is after 15 s, t_4 is after 20 s, t_5 is after 25 s, and t_6 is after 30 s. The temperature distribution in the frozen part presents a quasi-linear distribution, while in the unfrozen part, the curve has a marked exponential trend. The step change of the slope at the freezing point must be noted. Further on, it can be observed that the temperature of the body, 37 °C, is unchanged after 10 mm for all the times reported.

6.1.2 Cylindrical Geometry

Figure 22 reports the ice thickness growth for cylindrical geometry. It is equal to 1.8 mm after 5 s, 2.3 mm after 10 s, 2.7 mm after 15 s, 3.0 mm after 20 s, 3.3 mm after 25 s, and 3.6 mm after 30 s. As already observed in the analytical solution, the ice thickness is smaller in cylindrical geometry than in plane geometry.

Figure 23 presents the temperature distribution, T $(°C)$, versus $x\,(m)$ at the same six instants of time mentioned in plane geometry. The temperature distribution curve in the frozen part presents an exponential distribution, while in the unfrozen part, the curve has a marked exponential trend. The step change of the slope at the freezing point must be noted. Further on, it can be observed that the temperature of the body, 37 °C, is unchanged after 9 mm for all the times reported.

6.1.3 Spherical Geometry

Figure 24 reports the ice thickness growth for spherical geometry. It is equal to 1.5 mm after 5 s, about 1.8 mm after 10 s, 1.9 mm after 15 s, 2.1 mm after 20 s, 2.3 mm after 25 s, and 3.6 mm after 30 s. The growth of the ice thickness is smaller in spherical geometry than in cylindrical geometry.

Figure 25 presents the temperature distribution, T (°C), versus x (m) at the same six instants of time mentioned in cylindrical geometry. The temperature distribution curve in the frozen zone presents a more marked exponential trend than in cylindrical geometry, as well as in the unfrozen zone. The more marked step change of the slope at the freezing point must be noted. Further on, it can be observed that the temperature of the body, 37 °C, is unchanged after 8 mm for all the times reported.

6.2 Case 2: Variable Thermal Conductivity

Figures 26, 27, 28, 29, 30, and 31 [28] present the growth of the ice thickness as a function of time and the temperature distribution along the distance from the probe for the three geometries when the thermal conductivity is linearly variable in the frozen zone from $k_s = 3.44$ W/(m*K), at $T = -100$ °C, to $k_s = 2.15$ W/(m*K), at $T = 0$ °C. This means that the thermal conductivity of the frozen zone has a smaller effect on the freezing front penetration. The thermal conductivity of the unfrozen zone is assumed constant because its variation is negligible in the numerical results. The other variables are the same as in the constant thermal conductivity solution, but the thermal diffusivity varies accordingly.

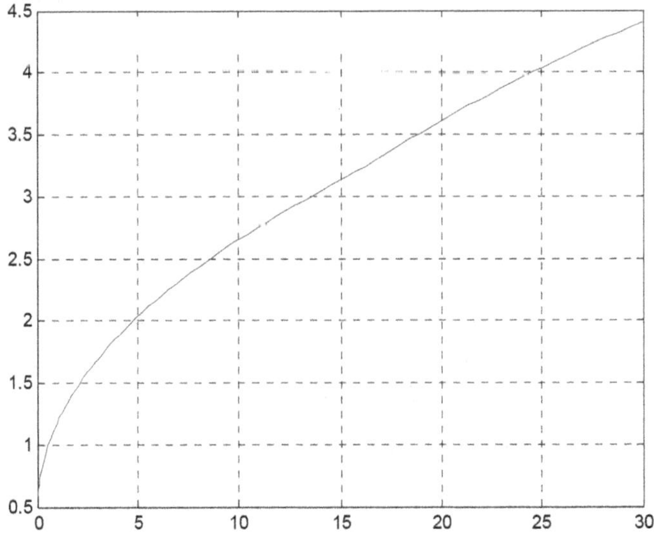

Fig. 26 Growth of the ice thickness (mm) with time (s) in plane geometry [28]

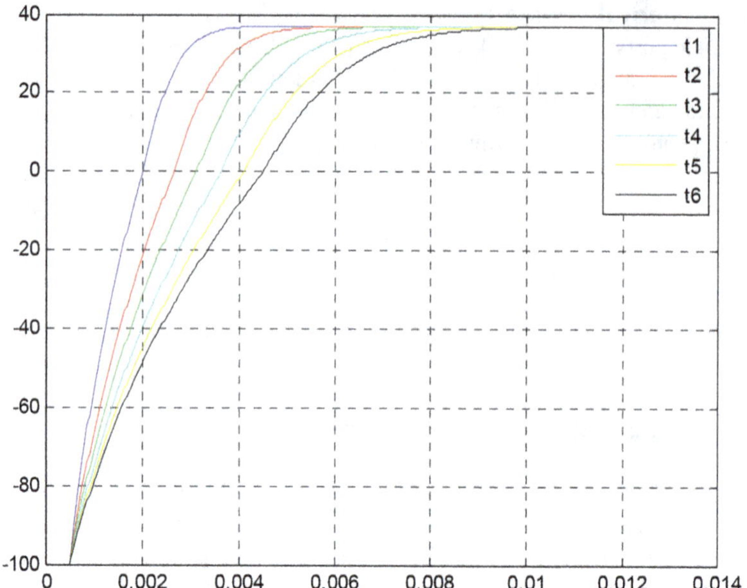

Fig. 27 Temperature distribution, T (°C), versus x (m) in plane geometry [28]

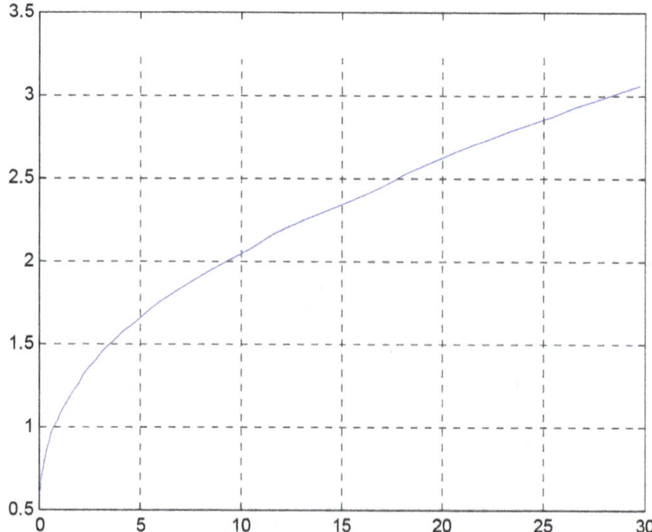

Fig. 28 Growth of the ice thickness (mm) with time (s) in cylindrical geometry [28]

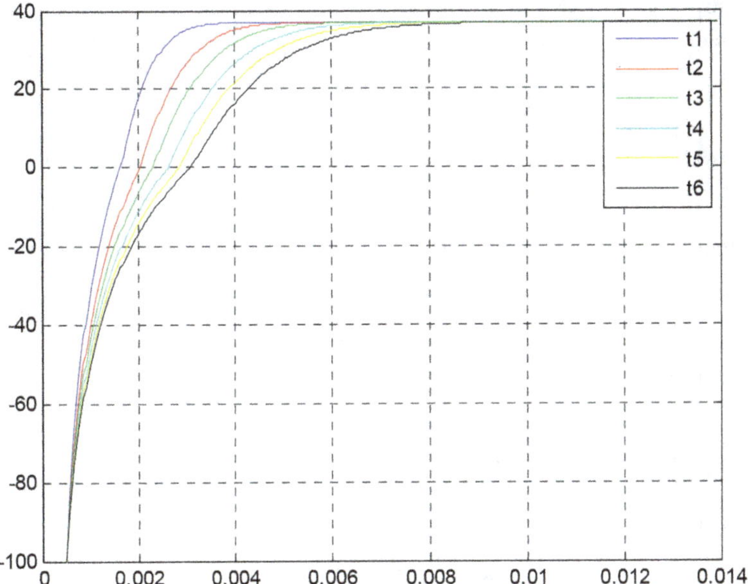

Fig. 29 Temperature distribution, T (°C), versus x (m) in cylindrical geometry [28]

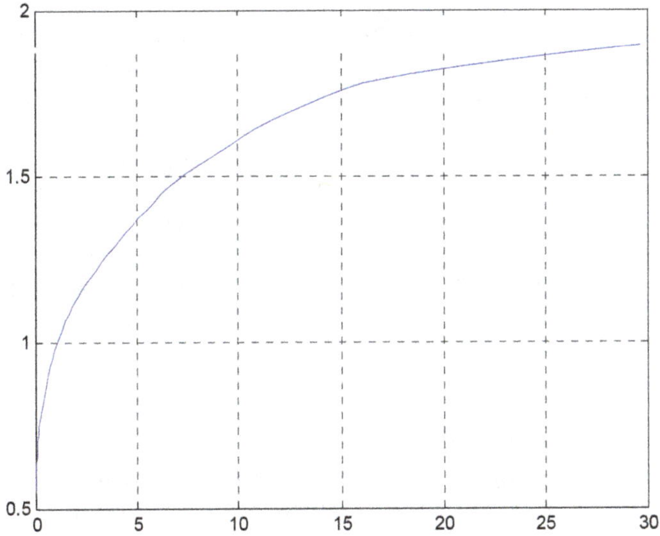

Fig. 30 Growth of the ice thickness (mm) with time (s) in spherical geometry [28]

6.2.1 Plane Geometry

Figure 26 reports the ice thickness growth for plane geometry. It is about 2.0 mm after 5 s, 2.7 mm after 10 s, 3.2 mm after 15 s, 3.6 mm after 20 s, 4.0 mm after 25 s, and 4.4 mm after 30 s. The ice front penetration is smaller with the variable thermal

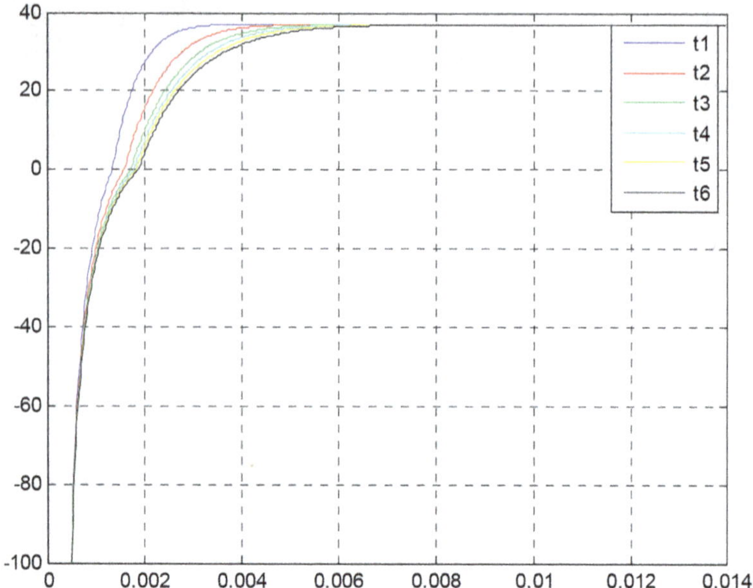

Fig. 31 Temperature distribution, T (°C), versus x (m) in spherical geometry [28]

conductivity in Case 2 than with the constant thermal conductivity in Case 1 because variable thermal conductivity decreases with the increase of the temperature.

Figure 27 presents the temperature distribution, T (°C), versus x (m) at six instants of time: t_1 is after 5 s, t_2 after 10 s, t_3 after 15 s, t_4 after 20 s, t_5 after 25 s, and t_6 after 30 s. The temperature distribution curve in the frozen part presents a more exponential distribution than in the case of constant thermal conductivity in Case 1, while in the unfrozen part, the curve has a marked exponential trend. The step change of the slope at the freezing point must be noted. Further on, it can be observed that the temperature of the body, 37 °C, is unchanged after 10 mm for all the times reported. The effect of the variable thermal conductivity in Case 2 on reducing the penetration of the freezing front is more evident here, as can be seen by observing the temperature distribution after 30 s.

6.2.2   Cylindrical Geometry

Figure 28 reports the ice thickness growth for cylindrical geometry. It is equal to 1.7 mm after 5 s, 2.1 mm after 10 s, 2.3 mm after 15 s, 2.6 mm after 20 s, 2.8 mm after 25 s, and 3.1 mm after 30 s. As already observed in the analytical solution, the ice penetration thickness is smaller in cylindrical geometry than in plane geometry. Further on, the ice front penetration is smaller with the variable thermal conductivity in Case 2 than with the constant thermal conductivity in Case 1 because it decreases with the temperature.

Figure 29 presents the temperature distribution, T (°C), versus x (m) at the same six instants of time mentioned in plane geometry. The temperature distribution curve in the frozen part presents a more pronounced exponential distribution than in the linear case. The step change of the slope at the freezing point must be noted. Further on, it can be observed that the temperature of the body, 37 °C, is unchanged after 9 mm for all the times reported. The effect of the variable thermal conductivity in Case 2 on reducing the penetration of the freezing front is more evident here, as can be seen by observing the temperature distribution after 30 s.

6.2.3 Spherical Geometry

Figure 30 reports the ice thickness growth for spherical geometry. It is equal to 1.3 mm after 5 s, about 1.6 mm after 10 s, 1.7 mm after 15 s, 1.8 mm after 20 s, 1.85 mm after 25 s, and 1.9 mm after 30 s. The growth of the ice thickness is smaller in spherical geometry than in cylindrical geometry. Further on, the ice front penetration is smaller with the variable thermal conductivity in Case 2 than with the constant thermal conductivity in Case1 because it decreases with the temperature.

Figure 31 presents the temperature distribution, T *(°C)*, versus *x (m)* at the same six instants of time mentioned in cylindrical geometry. The temperature distribution curve in the frozen zone presents a more marked exponential trend than in cylindrical geometry, as well as in the unfrozen zone. The more marked step change of the slope at the freezing point must be noted. Further on, it can be observed that the temperature of the body, 37 °C, is unchanged after 6 mm for all the times reported. The effect of the variable thermal conductivity in Case 2 on reducing the penetration of the freezing front is more evident here, as can be seen by observing the temperature distribution after 30 s.

7 Numerical Solutions for the Freezing Front Penetration in a Tissue with Blood Perfusion

The numerical solutions for the change of phase-solidification of the tissue in the presence of blood perfusion have employed the values of water as the constants involved in the problem: $\rho_l = 998.2$ kg/m^3, $c_{pl} = 4182$ J/kg × K, $k_l = 0.558$ W/m × K, $\alpha_l = 1.33 \times 10^{-7}$ m^2/s, $\rho_s = 916.2$ kg/m^3, $c_{ps} = 2000$ J/(kg × K), $k_s = 3.44$ W/(m × K), $\alpha_s = 1.88 \times 10^{-6}$ m^2/s, and $L = 335,000$ J/kg. The tissue is initially at the uniform temperature of 37 °C, while the temperature of the probe is set at $T_{probe} = -100$ °C on the probe radius equal to 0.5 mm. Two cases are investigated: Case 3: constant thermal conductivity in the presence of perfusion and Case 4: variable thermal conductivity in the presence of perfusion. The blood perfusion is equal to 0.003 s^{-1}. The presence of the blood perfusion reduces the freezing front penetration by a small amount.

7.1 Case 3: Constant Thermal Conductivity with Blood Perfusion

The following Figs. 32, 33, 34, 35, 36, and 37 [28], present the growth of the ice thickness as a function of time and the temperature distribution along the distance

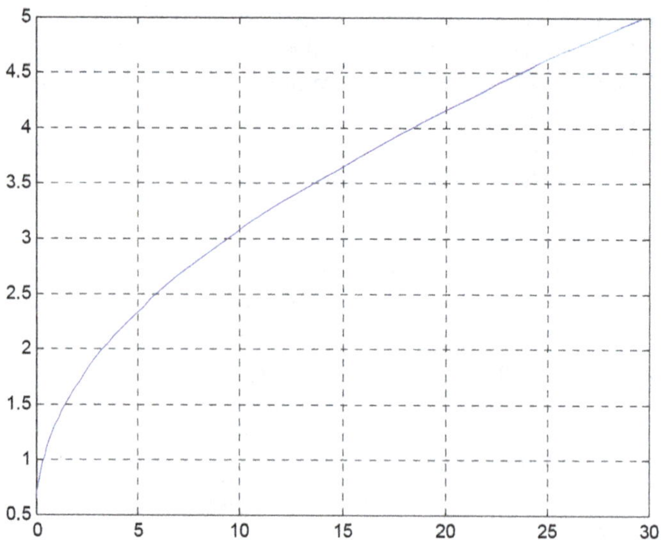

Fig. 32 Growth of the ice thickness (mm) with time (s) in plane geometry [28]

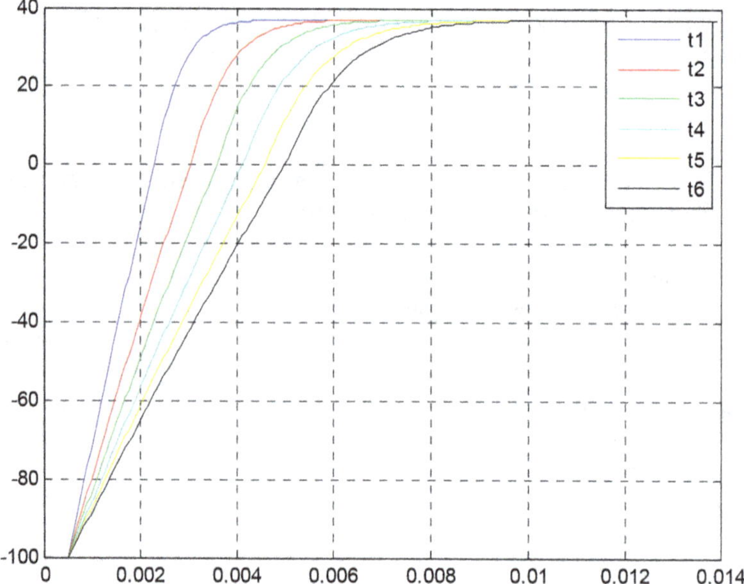

Fig. 33 Temperature distribution, T (°C), versus x (m) in plane geometry [28]

from the probe for the three geometries with constant thermal conductivity of the frozen tissue equal to $k_s = 3.44$ W/(m×K). The spatial integration step for the various geometries and for the various types of problem is not the same; the optimal one is chosen depending on the problem. This means that it is not possible to compare different problems in the same graph, or at least, it is difficult to do so.

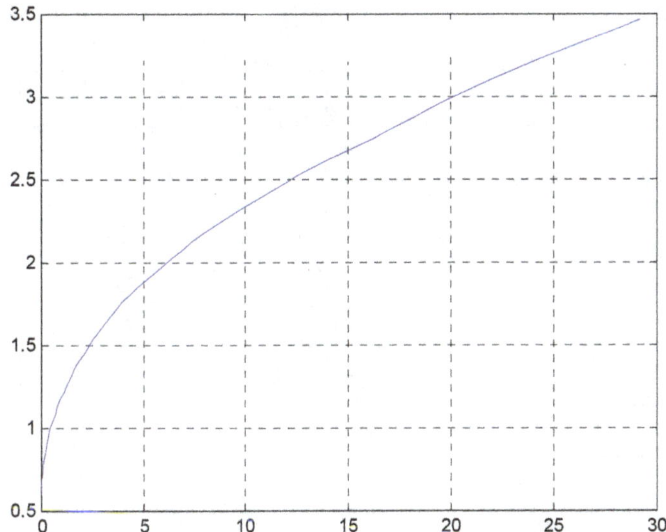

Fig. 34 Growth of the ice thickness (mm) with time (s) in cylindrical geometry, [28]

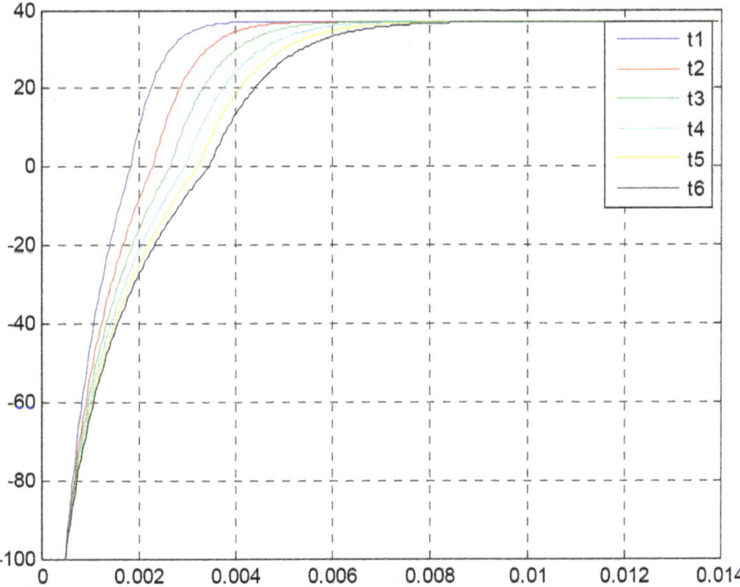

Fig. 35 Temperature distribution, T (°C), versus x (m) in cylindrical geometry [28]

7.1.1 Plane Geometry

Figure 32 reports the ice thickness growth for plane geometry. It is equal to 2.3 mm after 5 s, 3.1 mm after 10 s, 3.7 mm after 15 s, 4.2 mm after 20 s, 4.7 mm after 25 s, and 5.2 mm after 30 s. The ice front penetration is not changing very much as

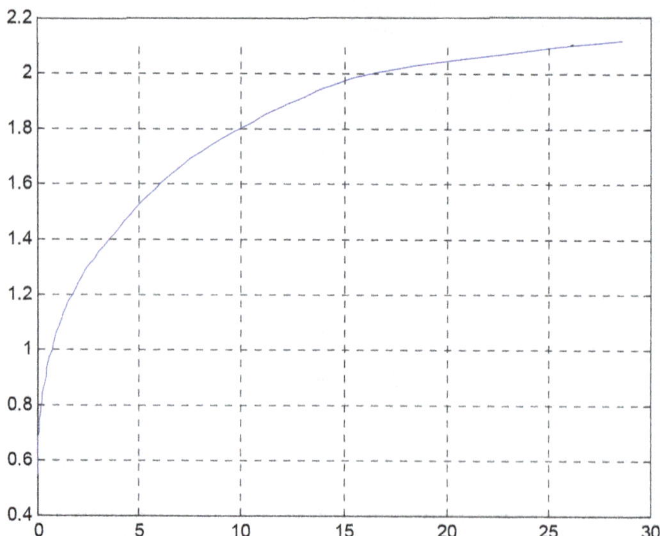

Fig. 36 Growth of the ice thickness (mm) with time (*s*) in spherical geometry, [28]

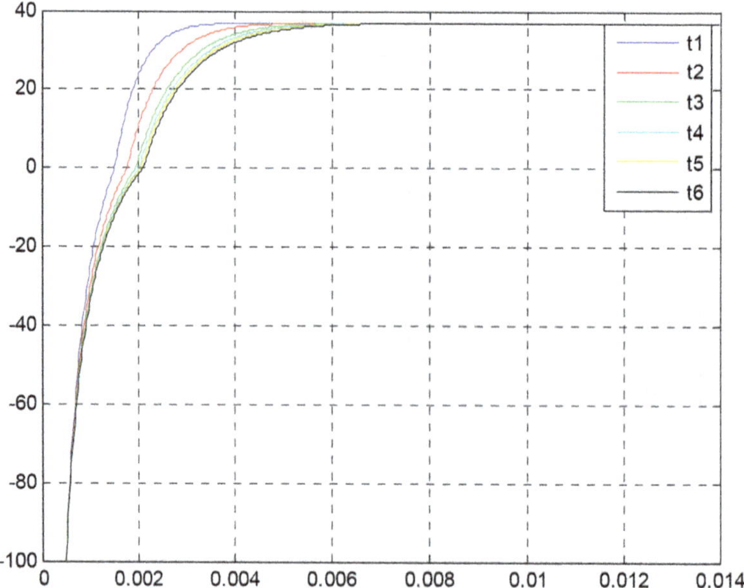

Fig. 37 Temperature distribution, T (°C), versus x (m) in spherical geometry [28]

compared to the case without blood perfusion, as can be viewed from the comparison, confirming the small influence of blood perfusion.

Figure 33 presents the temperature distribution, T (*°C*), versus x *(m)* at six instants of time: t_1 is after 5 s, t_2 is after 10 s, t_3 is after 15 s, t_4 is after 20 s, t_5 after is 25 s, and

t_6 is after 30 s. The temperature distribution curve in the frozen part presents a quasi-linear distribution, while in the unfrozen part, the curve has a marked exponential trend. The step change of the slope at the freezing point must be noted. Further on, it can be observed that the temperature of the body, 37 °C, is unchanged after 10 mm for all the times reported. The effect of blood perfusion on reducing the penetration of the freezing front is more evident here, as can be seen by observing the temperature distribution after 30 s.

7.1.2　Cylindrical Geometry

Figure 34 reports the ice thickness growth for cylindrical geometry. It is equal to 1.8 mm after 5s, 2.3 mm after 10s, 2.7 mm after 15s, 3.0 mm after 20s, 3.3 mm after 25s, and 3.6 mm after 30 s. As already observed in the analytical solution, the freezing front penetration is smaller in cylindrical geometry than in plane geometry. The ice front penetration is not changing very much as compared to the case without blood perfusion, as can be viewed from the comparison, confirming the small influence of blood perfusion.

Figure 35 presents the temperature distribution, T $(^\circ C)$, versus x (m) at the same six instants of time mentioned in plane geometry. The temperature distribution curve in the frozen part presents an exponential distribution, while in the unfrozen part, the curve has a marked exponential trend. The step change of the slope at the freezing point must be noted. Further on, it can be observed that the temperature of the body, 37°C, is unchanged after 9 mm for all the times reported. The effect of blood perfusion on reducing the penetration of the freezing front is more evident here, as can be seen by observing the temperature distribution after 30s.

7.1.3　Spherical Geometry

Figure 36 reports the ice thickness growth for spherical geometry. It is equal to 1.5 mm after 5s, about 1.8 mm after 10s, 1.95 mm after 15s, 2.05 mm after 20s, 2.1 mm after 25s, and less than 2.2 mm after 30s. The growth of the ice thickness is smaller in spherical geometry than in cylindrical geometry. The ice front penetration is not changing very much as compared to the case without blood perfusion, as can be viewed from the comparison, confirming the small influence of blood perfusion.

Figure 37 presents the temperature distribution, T $(^\circ C)$, versus x (m) at the same six instants of time mentioned in cylindrical geometry. The temperature distribution curve in the frozen zone presents a more marked exponential trend than in cylindrical geometry, as well as in the unfrozen zone. The more marked step change of the slope at the freezing point must be noted. Further on, it can be observed that the temperature of the body, 37 °C, is unchanged after 8 mm for all the times reported. The effect of blood perfusion on reducing the penetration of the freezing front is more evident here, as can be seen by observing the temperature distribution after 30 s.

7.2   Case 4: Variable Thermal Conductivity with Blood Perfusion

Figures 38, 39, 40, 41, 42, and 43 [28] present the growth of the ice thickness as a function of time and the temperature distribution along the distance from the probe for the three geometries when the thermal conductivity is linearly variable in the

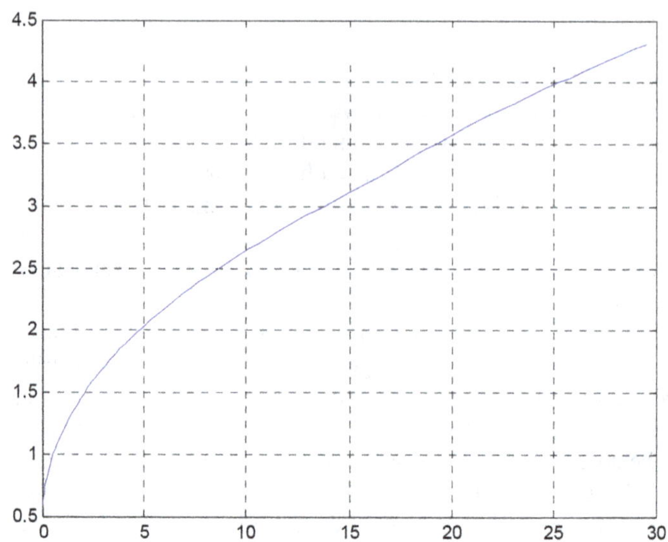

Fig. 38 Growth of the ice thickness (mm) with time (s) in plane geometry [28]

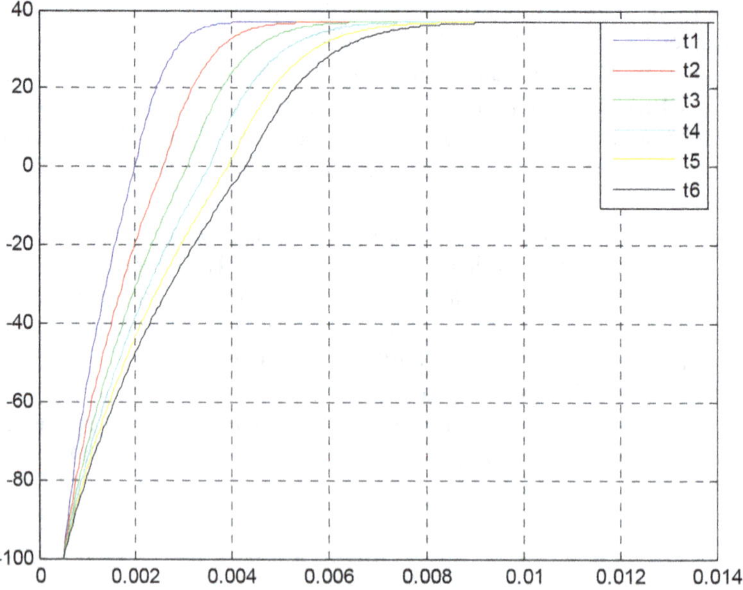

Fig. 39 Temperature distribution, T (°C), versus x (m) in plane geometry, [28]

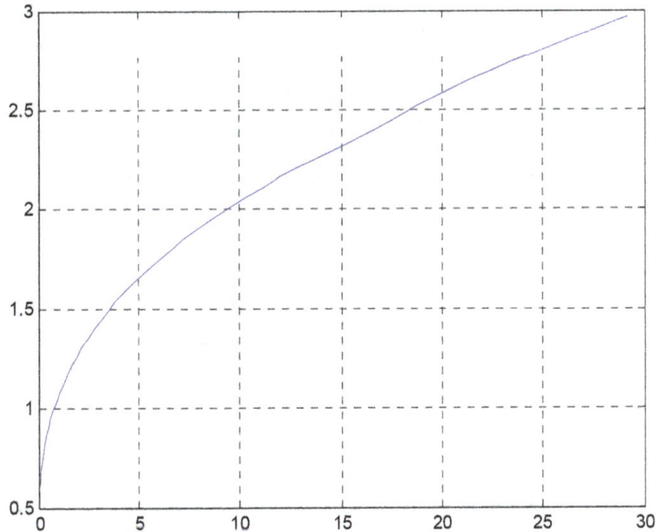

Fig. 40 Growth of the ice thickness (mm) with time (s) in cylindrical geometry [28]

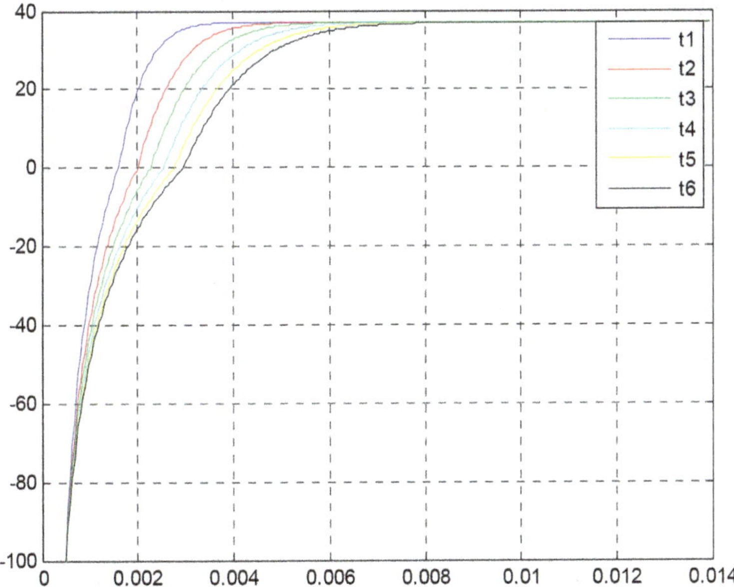

Fig. 41 Temperature distribution, T (°C), versus x (m) in cylindrical geometry, [28]

frozen zone from $k_s = 3.44$ W/(m×K), at $T = -100$ °C, to $k_s = 2.15$ W/(m×K), at $T = 0$ °C. This means that the thermal conductivity of the frozen zone has a smaller effect on the freezing front penetration. The thermal conductivity of the unfrozen water is assumed to be a constant because its variation is negligible in the numerical

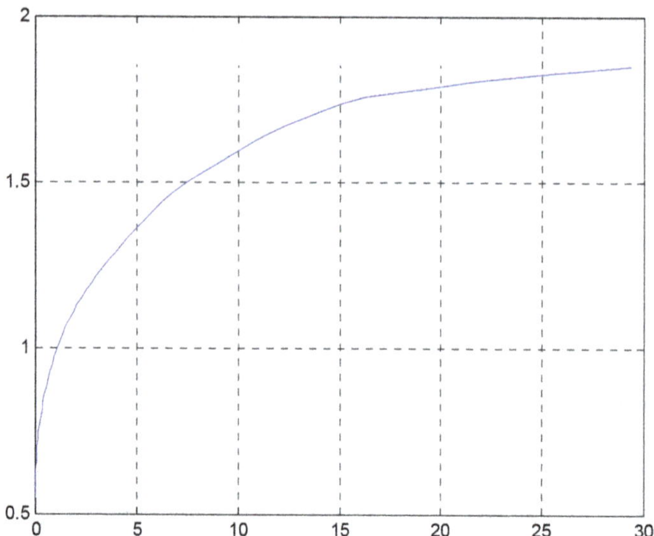

Fig. 42 Growth of the ice thickness (mm) with time (s) in spherical geometry [28]

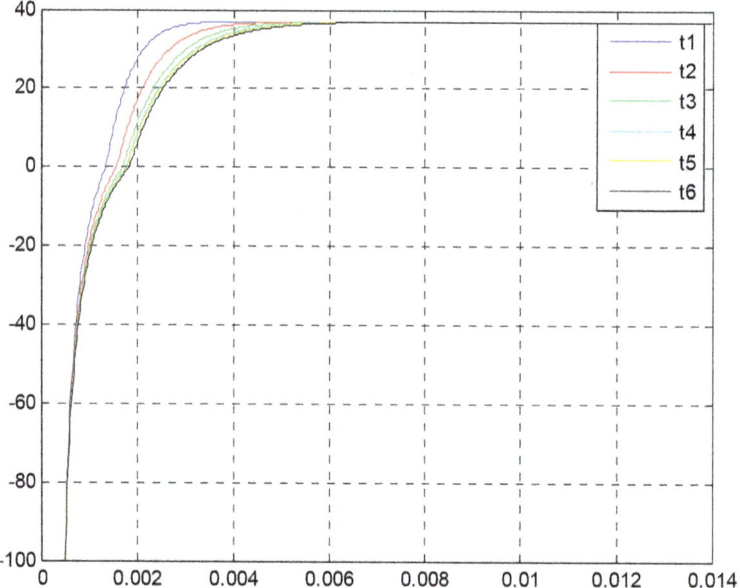

Fig. 43 Temperature distribution, T (°C), versus x (m) in cylindrical geometry [28]

results. The other variables are the same as in the constant thermal conductivity solution, but the thermal diffusivity varies accordingly.

The tissue is initially at the uniform temperature of *37 °C*, while the temperature of the probe is set at $T_{\text{probe}} = -100\,°C$ at 0.5 mm. The other variables are the same as in the constant thermal conductivity solution, but the thermal diffusivity varies accordingly.

7.2.1 Plane Geometry

Figure 38 reports the ice thickness growth for plane geometry. It is about 2.0 mm after 5 s, 2.7 mm after 10 s, 3.2 mm after 15 s, 3.6 mm after 20 s, 4.0 mm after 25 s, and 4.4 mm after 30 s. The ice front penetration is the same with variable thermal conductivity as with constant thermal conductivity in the presence of blood perfusion.

Figure 39 presents the temperature distribution, T $(°C)$, versus x (m) at six instants of time: t_1 is after 5 s, t_2 is after 10 s, t_3 is after 15 s, t_4 is after 20 s, t_5 is after 25 s, and t_6 is after 30 s. The temperature distribution curve in the frozen part presents a more pronounced exponential distribution than in the linear case. The step change of the slope at the freezing point must be noted. Further on, it can be observed that the temperature of the body, 37 °C, is unchanged after 10 mm for all the times reported. The effect of variable thermal conductivity on reducing the penetration of the freezing front is more evident here, as can be seen by observing the temperature distribution after 30 s.

7.2.2 Cylindrical Geometry

Figure 40 reports the ice thickness growth for cylindrical geometry. It is equal to 1.7 mm after 5 s, 2.05 mm after 10 s, 2.3 mm after 15 s, 2.6 mm after 20 s, 2.8 mm after 25 s, and 3.05 mm after 30 s. As already observed in the analytical solution, the ice thickness is smaller in cylindrical geometry than in plane geometry. The ice front penetration is the same with variable thermal conductivity as with constant thermal conductivity in the presence of blood perfusion.

Figure 41 presents the temperature distribution, T $(°C)$, versus x (m) at the same six instants of time mentioned in plane geometry. The temperature distribution curve in the frozen part presents a more pronounced exponential distribution than in the linear case. The step change of the slope at the freezing point must be noted. Further on, it can be observed that the temperature of the body, 37 °C, is unchanged after 9 mm for all the times reported. The effect of variable thermal conductivity on reducing the penetration of the freezing front is more evident here, as can be seen by observing the temperature distribution after 30 s.

7.2.3 Spherical Geometry

Figure 42 reports the ice thickness growth for spherical geometry. It is equal to 1.3 mm after 5 s, about 1.6 mm after 10 s, 1.7 mm after 15 s, 1.8 mm after 20 s, 1.85 mm after 25 s, and 1.9 mm after 30 s. The growth of the ice thickness is smaller in spherical geometry than in cylindrical geometry. The ice front penetration is the same with variable thermal conductivity as with constant thermal conductivity in the presence of blood perfusion.

Figure 43 presents the temperature distribution, T $(°C)$, versus x (m) at the same six instants of time mentioned in cylindrical geometry. The temperature distribution curve in the frozen part presents a more pronounced exponential distribution than in the linear case. The more marked step change of the slope at the freezing point must be noted. Further on, it can be observed that the temperature of the body, 37 °C, is unchanged after 6 mm for all the times reported. The effect of variable thermal conductivity on reducing the penetration of the freezing front is more evident here, as can be seen by observing the temperature distribution after 30 s.

8 Comparison of the Numerical Solutions in a Tissue with Blood Perfusion with the Experimental Freezing Front Penetration in a Living Tissue

The examination of the literature showed that the experiments performed by Teeter [48] are the most interesting, as far as the transient freezing of the biological tissue in an in vivo rabbit is concerned. The biological physical parameters assumed for the liver of the investigated animal are the following: $\rho_s\, c_{ps} = 0.96$ (cal/cm^3/°K) $= 4.17$ (J/m^{3}K), $\rho_l\, c_{pl} = 0.48$ (cal/cm^3/°K) $= 1.147$ (J/m^{3}K), $G_v = 100$ cm^3 (blood)/min/ 100 g(tissue) $= 1.67$ cm^3 (blood)/s/100 g(tissue) $=) = 1.67\ 10^{-6}$ m^3 (blood)/s/100 g (tissue). The body temperature is 37 °C, the probe temperature is -70.6 °C, the temperature at the beginning of freezing is 0 °C, and the temperature at the completion of freezing is -2 °C. The thermal conductivity of the unfrozen tissue is $k_l = 0.0012$ cal/(gr s °K) $= 0.50$ W/m/K and is variable as that of water. The thermal conductivity of the frozen tissue is $k_p = 0.0037$ cal/(gr s °K) $= 1.54$ W/m/K at the probe temperature, and it is $k_g = 0.0025$ cal/(gr s °K) $= 1.046$ W/m/K at the temperature of -2 °C. The thermal conductivity of the frozen tissue can be assumed a constant, as in Fig. 44; a continuous line; or linearly variable between the value at -70.6 °C and the value at -2 °C, as reported with the dotted line in Fig. 44.

Figure 45 presents the experimental results of [48] for the freezing front penetration (in *cm*) in the liver tissue of an in vivo rabbit after 10 s, 20 s, 30 s, 60 s, and 120 s from the beginning of freezing. The dashed line is the curve of the minimum squares that interpolates the experimental data. The numerical results are reported as continuous lines under different thermal conditions. The curves indicated as (1) and (2) are

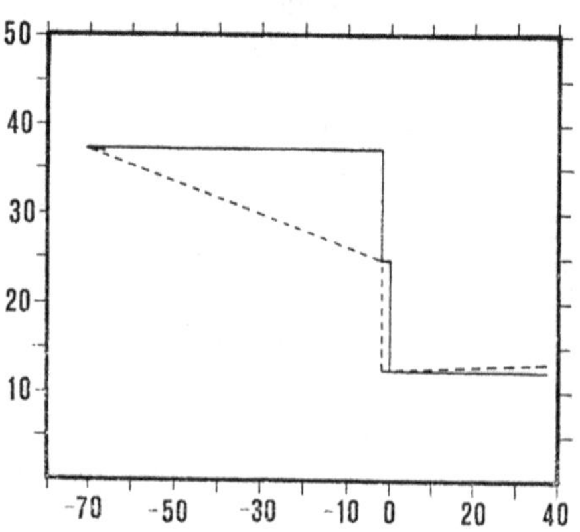

Fig. 44 Thermal conductivities of the tissue, cal/(cm s K), versus temperature (°C) [31]

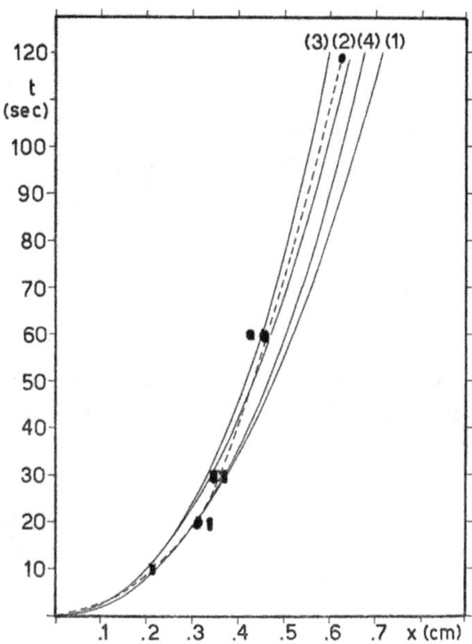

Fig. 45 Experimental results of [48] and numerical predictions of [31] for freezing front penetrations (cm) versus the time (seconds) elapsed from the beginning of freezing

obtained with the hypothesis of blood perfusion variable linearly from $G_v = 100$ cm^3 (blood)/min/100 g(tissue) $= 1.67$ cm^3 (blood)/s/100 g(tissue) $= 1.67 \cdot 10^{-6}$ m^3 (blood)/s/100 g(tissue) at the body temperature of $37\,°C$ down to 0 at $0\,°C$. Curve (1) assumes the thermal conductivity constant and is equal to $k_p = 0.0037$ cal/(gr s °K) $= 1.54$ W/m/K, while the curve indicated as (2) is obtained with the thermal conductivity variable linearly, as in Fig. 44, i.e., decreasing from the highest value at the probe temperature down to the value at $-2\,°C$. The results of curve (1) are in good agreement with the experiments up to 30 s, but they overpredict the freezing after 60 s and 120 s. On the contrary, curve (2) is in very good agreement with the experiments up to 120 s.

The curves indicated as (3) and (4) are obtained with the hypothesis of blood perfusion constant and are equal to $G_v = 100$ cm^3 (blood)/min/100 g(tissue) $= 1.67$ cm^3 (blood)/s/100 g(tissue) $- 1.67 \cdot 10^{-6}$ m^3 (blood)/s/100 g(tissue) at the body temperature of 37 °C, which becomes null at the freezing temperature of 0 °C. Curve (4) assumes the thermal conductivity constant and is equal to $k_p = 0.0037$ cal/(gr s °K) $= 1.54$ W/m/K, while the curve indicated as (3) is obtained with the thermal conductivity variable linearly, i.e., decreasing from the highest value at the probe temperature down to the value at $-2\,°C$. The results of curve (3) are in good agreement with the experiments up to 60 s, but they underpredict the freezing after 120 s. On the contrary, curve (4) is in good agreement with the experiments only up to 30 s but overpredicts the freezing after 60 s and 120 s.

The conclusion is that curves (2) and (3) are predicting the experimental results better than the other two. This means that the hypothesis of variable thermal conductivity is the most important one.

Fig. 46 Experimental temperatures (°C) [48] and numerical predictions [31] along the tissue (cm), at several instants of time (seconds) elapsed from the beginning of freezing

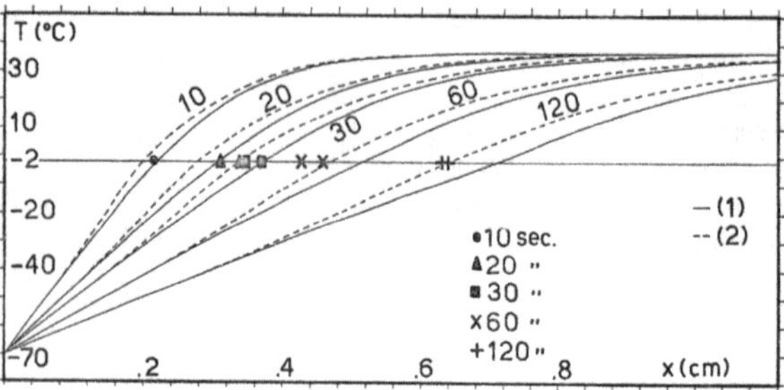

Figure 46 presents the experimental results of [48] for the freezing front penetration (cm) in the liver tissue of an in vivo rabbit after 10 s, 20 s, 30 s, 60 s, and 120 s from the beginning of freezing, together with the temperature distributions in the tissue, as predicted by curves (1) and (2) of Fig. 45. The numerical predictions of curve (1) are reported by the continuous line, while those of curve (2) are reported by the dashed line. It is evident that the results obtained with curve (2) are in better agreement with the experiments of [48].

9 Conclusions

The *software 2008_Nuovo_Crio_Surgery*© essentially confirms the results found in [31–34]. By comparing the results for the different geometries within a specific configuration, the first thing to notice is that freezing, in the case of plane geometry, propagates faster than in cylindrical geometry and the latter faster than in spherical geometry. The reason is simple: Virtually, in plane geometry, we have a medium with an infinite surface area that is at the temperature of the cryogenic fluid and, therefore, a greater heat flow. In the case of cylindrical geometry, the base area is fixed, but the height can be virtually infinite. This also implies an infinite exchange surface and, therefore, a large heat flow but not as much as in plane geometry, where we have two degrees of freedom, while here, there is only one degree of freedom. Finally, in the case of spherical geometry, the presence of a finite radius implies a finite probe surface and, therefore, a lower heat flow as compared to the other cases.

Acknowledgments for Figures The author thanks the *publisher, Sage Publications,* for the kind permission to reproduce image *6 of the reference* [12].

The author thanks the *publisher, IOP Publishing,* for the kind permission to reproduce image *2 of the reference* [13] and *images 1 and 3 of the reference* [3].

The author thanks the *publisher, Springer Nature,* for the kind permission to reproduce image *16.1.1 of the reference* [14], image *5 of the reference* [18], and images *2 and 3 of the reference* [16].

The author thanks the *publisher, Elsevier,* for the kind permission to reproduce images *1 and 2 of the reference* [19], image *5 of the reference* [20], image *2 of the reference* [21], images *1 and 2 of the reference* [22], and image *1 of the reference* [29].

References

1. Cooper IS, Lee ASJ (1961) Cryostatic congelation: a system for producing a limited, controlled region of cooling or freezing of biologic tissues. J Nerv Ment Dis 133:259–263
2. Lee ASJ (1967) Freezing probe for the treatment of tissue, especially in neurosurgery, US Patent, 3, 298, 371
3. Rabin Y, Stahovich TF (2003) Cyoheather as a means of cryosurgery control. Phys Med Biol 48(5):619–632
4. Graham GF (2001) Cryosurgery in the management of cutaneous malignancies. Clin Dermatol 19(3):321–327
5. Cooper IS (1969) Cryosurgery. Minerva Chir 10:1129
6. Meryman HT (1956) Mechanics of freezing in living cells and tissues. Science 124:515–521
7. Ling GR, Tien CL (1970) An analysis of cell freezing and deydratation. J Heat Transf ASME Ser C 92(3):393–397
8. Farrant J (1971) Cryobiology. The basis to cryosurgery. In: Cryogenics in surgery. Medical Examination Pub. Co. Inc.
9. Leibo SP, Mazur P (1971) The role of cooling rates in low temperature preservation. Cryobiology 8:447–452
10. Toscano WM, Cravalho EG, Silvares OM, Huggins CE (1975) The thermpdynamics of intracellular ice nucleation in the freezing of erythrocytes. J Heat Transf ASME Ser C 97: 326–332
11. Baust JB, Gage AA (2005) The molecular basis of cryosurgery. BJU Int 95(9):1187–1191
12. Maiwand MO, Asimakopoulos G (2004) Cryosurgery for lung cancer: clinical results and technical aspects. Technol Cancer Res Treat 3(2):143–150
13. Wei J, Sandison GA, Chen L, Liang Y, Xu LX (2002) X-ray CT high-density artefact suppression in cryosurgery. Phys Med Biol 47:319–326
14. Rabin Y (2001) In: Korpan NK (ed) Imaging of breast cryosurgery, atlas of cryosurgery. Springer-Verlag, Vienna, pp 344–353
15. Fournial R, Traoré AS, Laurendeau D, Moisan C (2004) An analytic method to predict the thermal map of cryosurgery iceballs in MR images. IEEE Trans Med Imag 23:122
16. Yu TH, Zhou YX, Liu J (2003) Dynamic low-frequency electrical impedance of biological materials subject to freezing and its implementation in cryosurgical monitoring. Int J Thermophys 24:513
17. Zlochiver S, Radai MM, Rosenfeld M, Abboud S (2002) Induced current impedance technique for monitoring brain cryosurgery in a two-dimensional model of the head. Ann Biomed Eng 30: 1172–1180
18. Zlochiver S, Rosenfeld M, Abboud S (2005) Contactless bioimpedance monitoring technique for brain cryosurgery in a 3D head model. Ann Biomed Eng 33(5):616–625
19. Pogrel MA, Yen CK, Taylor R (1996) A study of infrared thermographic assessment of liquid nitrogen cryotherapy. Oral Surg Oral Med Oral 81:396–401
20. Samset E, Mala T, Aurdal L, Balasingham I (2005) Intra-operative visualisation of 3D temperature maps and 3D navigation during tissue cryoablation. Comput Med Imaging Graph 29:499–505
21. Pradel W, Hlawitschka M, Eckelt U, Herzog R, Koch K (2002) Cryosurgical treatment of genuine trigeminal neuralgia. Br J Oral Maxillofac Surg 40(3):244–247
22. Baust J, Gage AA, Ma H, Zhang C-M (1997) Minimally invasive cryosurgery-technological advances. Cryobiology 34:373–384. Art. CY972017
23. Fredrickson K, Nellis G, Klein S (2006) A design method for mixed gas Joule–Thomson refrigeration cryosurgical probes. Int J Refrig 29:700–715
24. Mikus PW, Eum JJ, United States Patent No.: US 6,251,105 B1 June 26, 2001
25. LePivert, United States Patent No.: US 6,551,309 B1 Apr. 22, 2003
26. Bhadeshia HKDH. Materials science & metallurgy courses, metals and alloys. Cambridge University

27. Ozisik MN (1993) Heat conduction. Wiley-Interscience Publication
28. Boghi A (2007–2008) Cryosurgery, Software for predicting the lesion produced in the tissue, Second Level Master's thesis, University of Rome Tor Vergata, A.A
29. Bhattacharya M, Basak T, Ayappa KG (2002) A fixed-grid finite element based enthalpy formulation for generalized phase change problems: role of superficial mushy region. Int J Heat Mass Transf 45(24):4881–4898
30. Feltham DL, Worster MG (2000) Similarity solutions describing the melting of a mushy layer. J Cryst Growth 208:746–756
31. Gori F (1976) Cryosurgical probe: a numerical method for predicting the lesion produced in the tissue. University of Florence, Internal report of the Faculty of Engineering
32. Gori F (1977) Cryosurgical probe: a theoretical prediction of the freezing front penetration. In: First mediterranean conference on medical and biological engineering, Sorrento (Italy), I, 3–1/3–4
33. Gori F (1979) Numerical solutions of energy equation in cryosurgery. In: XV international congress of refrigeration, Venice (Italy), III, pp 325–329
34. Gori F (1981) On heat transfer in non-living materials around a cryosurgical probe. In: Numerical methods in thermal problems, II. Pineridge Press, Swansea, pp 230–239
35. Gori F (2007) Lezioni di Termodinamica (Lectures in thermodynamics). TeXmat
36. Barron RF (1968) Heat transfer problems in cryosurgery. J Cryosurgery 1:316–325
37. Rubinsly B, Shitzer A (1976) Analysis of a Stefan-like problem in a biological tissue around a cryosurgical probe. J Heat Transf 98(3):514–519
38. Cooper TE, Trezek GJ (1970) Analytical prediction of the temperature field emanating from a cryogenic surgical cannula. Cryobiology 2–3(7):79–93
39. Cooper TE, Trezek GJ (1971) Rate of lesion growth around spherical and cylindrical cryoprobes. Crybiology 4–6(7):183–190
40. Cooper TE, Trezek GJ (1971) Mathematical predictions of cryogenic lesion, cryogenics in surgery. Medical Examination Pub. Co. Inc
41. Warren RP, Bingham PE, Carpenter JD (1974) Heat flow in living tissue during cryosurgery, ASME Paper N. 74-WA/Bio-6
42. Hegnauer AM, Shuber WJ, Haterius HO (1950) Cardiovascular response of the dog to immersion hypothermia. Am J Phys 161:455–465
43. Rothenborg HW (1970) Cutaneous circulation in rabbits and humans before, during and after cryosurgical procedures measured by Xenon-133 clearance. Cryobiology 6(6):507–511
44. Chato JC (1968) A method for the measurement of the thermal properties of living tissue, thermal problems in biotechnology, ASME Symposium Series, 16–25
45. Walder HAD (1971) Experimental cryosurgery, cryogenics in surgery. Medical Examination Pub. Co. Inc
46. Crank L (1957) Two methods for the numerical solution of moving-boundary problems in diffusion and heat flow. Quart Jl Mech Appl Math X 2:220–231
47. Crank J (1975) Finite difference methods, moving boundary problems in heat flow and diffusion. Clarendon Press, Oxford
48. Teeter CL (1970) Development of a Freon cooled cryoprobe and an analysis of the associated temperature fields. Ph. D. thesis, University of Wisconsin

Andrea Boghi The Doctor of Research *(Ph. D.)*, Dr. Eng. Andrea Boghi, is the director of *Computational Science Ltd.*, a consultancy company in the field of software development and data analysis, which counts among its clients the British Air Traffic Agency *(NATS)* and the Bank of England, starting from 2018. He is the author of more than 30 international scientific publications indexed on the major scientific databases. For more than a decade, he has worked in academia in various European institutions, dealing with the modelling and simulation of transport phenomena, both for basic research and for industrial and biological applications. His academic journey began at the University of Rome "Tor Vergata," where, alongside research on the modelling and simulation of turbulent flows with variable laminar diffusivity and blood flow in arteries and *stents*, he also served as a contract lecturer for the courses of Technical Physics, Thermo-Fluid-Dynamics of Biological Systems and Numerical Calculation of Thermo-Fluid-Dynamic Systems (2006–2010). In the two-year period from 2010 to 2012, he was a contract researcher at the Department of Energy and Technology of the University of *Southampton* in the United Kingdom, where he dealt with modelling and simulation of bi-phase flows, characterised by spinodal decomposition. In 2012, he went to France to the Institute of Fluid Mechanics of Toulouse as a contract researcher for the OTE (Waves, Turbulence and Environment) research group, where he dealt with modelling and simulation of river and atmospheric fluid dynamics. He terminated his contract in 2013 to go to the University of *Cranfield* in the United Kingdom, where he was appointed *Senior Research Fellow*, becoming part of the academic staff. Here, for the following 5 years (2013–2018), he dealt with modelling and simulation of transport phenomena for various departments: Energy, Hydraulics, Environment and Agriculture, where he respectively dealt with transport of crude oil in pipelines, disinfection of tanks by chlorination and transport and diffusion of organic and inorganic material in soils, with associated absorption by plants. At the University of *Cranfield,* he also handled the course on modelling of atmospheric emissions in the two-year period from 2016 to 2017. Before leaving academia, he went again to the University of *Southampton* in 2018, where he dealt with the three-dimensional modelling of the transport of nitrates and carbonates at the air–water interface in soils. Throughout his period abroad, he maintained his collaboration with Fabio Gori Ammannati, supervising some students and collaborating on the production of scientific articles. In 2020, he obtained the national scientific qualification for the functions of associate university professor in the competitive sector of Technical Physics and Nuclear Engineering.

Fabio Gori Ammannati was born in Montale (Pistoia) on 5 August 1947 to Emilio Gori and Cesarina Ammannati. On 16 July 2021, he added his mother's surname, Ammannati. He graduated in Chemical Engineering on 11 November 1971 with honours. In December 1971, he won a scholarship for young graduates at the Faculty of Engineering in Bologna, which he undertook from January 1972. In 1974, he became assistant professor and, in 1982, associate professor of Technical Physics at the Faculty of Engineering of the University of Florence, where he taught, also as a contract professor, until 1990. In 1978, he won a scholarship from the *British Council,* which he carried out at the *Imperial College* in London, where he collaborated with *Prof. D. Brian Spalding* on the topic of numerical analysis in the turbulent flow of liquid metals. During his time at the University of Florence, he established international scientific collaborations with *Prof. R. Echigo, Tokyo Institute of Technology, Tokyo, Japan; Prof. D.R. Chaudhary, Department of Physics, University of Rajasthan, Jaipur, India;* International Centre for Theoretical Physics *(ICTP)* in Trieste; *Centralny Osrodek Techniki Medycznej, Warsaw, Poland and Prof. T. Aihara, Institute of Fluid Science, Sendai, Japan.* In 1986, he won a CNR scholarship, which he carried out at *Cornell University, Ithaca, New York,* where he collaborated with *Prof. R. Miller* on the topic of ground freezing. In the same year, he became professor in the Faculty of Engineering at the University of Reggio Calabria. In 1988, he was appointed as *professor* at the *University of New York at Stony Brook,* teaching the course, *Introduction to Fluid Dynamics,* in the autumn semester of the same year. During his stay, he collaborated with *Prof. T.F. Irvine Jr.* on the method of measuring thermal conductivity with a thermal probe and on the measurement of the isobaric thermal expansion coefficient of non-Newtonian fluids. In 1990, he moved to the Milano Polytechnic, where he taught Technical Physics and Systems until 1993. From 1991 until 1998, he was an adjunct professor at the Faculty of Engineering of the University of Siena. From 1992 to 2017, he was a professor of Technical Physics in the Faculty of Engineering at the University of Rome Tor Vergata, and from 2017, he is a contract professor. In 1994, he proposed the Research Doctorate (Italian *Ph.D.*) in Energy-Environment Engineering, of which he was the coordinator until 2011. In 1995, he proposed and directed the second-level Master's in Thermo-fluid-dynamics until 2016. From 1998 to 2001, he was the director of the Department of Mechanical Engineering. In 2001, he proposed and coordinated the programme, *Master of Science, Energy Engineering and Thermal and Fluid Dynamics,* in collaboration with the *Department of Mechanical Engineering, College of*

Engineering, University of Illinois at Chicago, USA, which takes place at the University of Rome Tor Vergata but awards the title of *Master of Science* from the same American University. Since the late 90s, he has been a coordinator of the joint *Ph.D.* research programme with *Prof. R.J. Goldstein, Department of Mechanical Engineering, University of Minnesota, Minneapolis, USA,* and with *Prof. J.P. Hartnett, Prof. L. Kennedy, Prof. W. Minkowycz and Prof. W.M. Worek, Department of Mechanical Engineering, College of Engineering, University of Illinois at Chicago, Chicago, USA.* Since the mid-90s, he has proposed and directed the *Socrates–Erasmus* programme with *Prof. Mayinger, Technical University of Munich, Germany; Prof. van Steenhoven, Eindhoven Technical University, the Netherlands and Prof. C. Caro, Imperial College of Science, Technology and Medicine.* During his stay at Tor Vergata, he established scientific collaborations with the *Department of Mechanical Engineering, University of Minnesota, Minneapolis, USA,* collaborating with *Prof. R.J. Goldstein* on Thermo-fluid-dynamics and mass transport in gas turbine blades; the *Energy Resources Center, University of Illinois at Chicago,* collaborating with *Prof. J.P. Hartnett, Prof. W. Minkowycz* and *Prof. W.M. Worek;* the *Department of Mechanical Engineering, College of Engineering, University of Illinois at Chicago,* collaborating with *Prof. L. Kennedy* and *Prof. W.M. Worek; Duke University, collaborating with Prof. A. Bejan* and *S. Mary's University, Halifax, Nova Scotia, Canada, collaborating with Prof. W. R. Tarnawski.* The bibliographic review of documents and citations, titled *PlosBiology Career,* places him within the top 2% of researchers in the field of *Mechanical Engineering and Transports.* Since 1992, he has been a tutor for over 30 PhD theses, supervisor for over 40 second-level university Master's theses, supervisor for over 70 degree theses (five-year, Specialist and Master's), supervisor for over ten *Master's degree in Mechanical Engineering* from the *University of Illinois at Chicago* and supervisor for over 20 theses *Erasmus-Socrates.*

Since the early 2000s, he has been a reviewer of international research projects on behalf of *Portuguese Science and Technology Foundation (FCT); Czech Science Foundation (GACR); European Science Foundation, Strasbourg, France,* on behalf of *FCT* and the *Shota Rustaveli National Science Foundation, Georgia.* Since 2017, the year of his retirement due to age limits, he has been a contract professor at the University of Rome Tor Vergata, where in the academic year 2024–2025, he has taught Technical Physics for the Master's Degree Course in Medical Engineering. M.D. Salvatore Mangiafico (Florence), scientific head of the NeuroVascular Base Camp, 2021, invited him to give a scientific presentation titled *"The engineering approach to the study of the Circle of Willis"* on 23 September 2021 at the Congress Centre, University of Rome La Sapienza, thus opening up a possible future collaboration.

Microwave Heating and Medical Hyperthermia

Microwave Heating and Medical Hyperthermia

Fabio Gori Ammannati

1 Microwave Heating

1.1 Introduction

This chapter describes microwave heating, which involves the transfer of electromagnetic energy from a microwave generator to various types of materials used for industrial, scientific, medical, domestic and commercial uses. They can be used to heat and dry food, timber and building materials, such as sand, cement, asphalt and textile fibres. The speed of this type of heating reduces the time compared to traditional heating systems, and some properties of the materials remain unchanged. For example, in cold drying, the colour, taste, shape and vitamin content of the food remain almost unchanged. For cooking food, a combined contemporary, conventional and microwave system is considered more economical than the other two separate ones [1]. For the drying of timber, or to evaporate water, the amount of energy required is reduced by a third compared to traditional systems. The fibre is only granulated if the internal steam pressure is too high and not due to rapid contraction of the dried fibres, as happens in normal heating. The hardening of the fibres can also be avoided by simply controlling the power output. The seasoning of the timber, up to about 10% moisture, requires about 100 min, while convection ovens require more than 10 days. Moreover, the timber is 10% stronger. Microwave defrosting is done in a few min, while it takes hours with water convection.

F. G. Ammannati (✉)
University of Rome Tor Vergata, Rome, Italy
e-mail: gori@uniroma2.it

F. Gori Ammannati (ed.), *Thermo-Haemo-Dynamics in Medical Engineering*,
https://doi.org/10.1007/978-3-031-97214-0_3

1.2   Physical Mechanism

The physical mechanism of microwave heating is described in detail in [2, 3]. Some dielectrics, like water, have molecules with a permanent electric moment, i.e. they are present even in the absence of an external electric field. Therefore, the molecule tends to elastically align its own electric field with the external one at the moment of its application. If this electric field is removed or reversed, the moment of the molecules decays exponentially. Since this phenomenon is mainly due to thermal agitation, the higher the temperature of the dielectric, the shorter the relaxation time of the molecules. The effect of thermal agitation can also be likened to the regime of viscous friction. At relatively low frequencies, the viscous forces are small, and the tendency of the dipoles to follow the polarity changes with the applied field, i.e. with a high dielectric constant, is high. When the period of the field is very small, the viscous forces prevent alignment, and there is a strong dependence of the dielectric's polarisation on the frequency of the external electric field. The dielectric constant is therefore relatively small. Between these two states, there is a state in which energy dissipation is maximum, as are electromagnetic losses. The state of maximum frequency dispersion for the relative electric constant, which for water is between 10^9 Hz and 10^{11} Hz, occurs with a maximum for losses around 18 GHz at room temperature.

Water is a very important substance for microwave heating because every material to be treated has a percentage between 5% and 50%, while the human body reaches values above 70% in certain organs. Water has highly polar molecules, i.e. very large polar moments, compared to those of the vast majority of solid dielectrics of industrial interest. Moreover, the temperature coefficient, typically negative for the imaginary part of the water's dielectric constant, has a thermostating action because it reduces the rate of temperature increase with the temperature itself. Another cause of energy absorption is the presence of free charges due to salts or other impurities present in the dielectric. Free electrons and ions flow more freely through the material, causing ohmic losses from to be more properly considered conduction imperfections rather than dielectric hysteresis. When the number of electrons of the rotating dipoles and/or the number of free electrons, possibly present in a material subject to a pulsating electro magnetic *(e.m.)* field, is large enough to induce a resulting field that equals the exciting one, the total field is null, and energy can no longer propagate within the body. This skin effect varies with the type of material, temperature and frequency. The penetration depth is the distance from the surface of the material exposed to radiation, in which the electric field has decayed to 37% of the surface value. For substances treated with microwaves, this depth is large, but for some ionic solutions of an organic type, it can be small, like meat (5 cm) or fat (1–2 cm).

1.3   Thermal Efficiency

Microwave heating occurs through thermal radiation, and the material and, largely, the water contained in it, absorb energy throughout the volume of the material. The thermal power that reaches the material is given by

$$\dot{Q} = \cos t f\, E^2 \frac{\varepsilon''}{\varepsilon_0} \tag{1}$$

where f is the frequency, E is the modulus of the electric field and ε'' is the imaginary part of the dielectric constant, while ε_0 is the value relative to the vacuum. The standard frequency values of microwave heating are 915, 2450, 5800 and 22,550 MHz. The maximum absorption frequency of water at room temperature is 18 GHz.

2   Microwave Heating of Heterogeneous Moist Media

The application discussed in [3] concerns the microwave heating of the first layers of the road pavement, used for the purpose of bringing the bituminous conglomerate from the solid state to the pasty one. The objective is to selectively heat the road pavement up to 25 cm in depth for the re-compaction, with a roller compressor, of the deteriorated asphalt. The temperatures required for this intervention are between 130 °C and 170 °C. For lower temperatures, the asphalt is too viscous, while, above 170 °C, the surface bark of the material carbonises with degeneration and production of harmful combustion gases. The problems are the development of steam, a considerable reflected and dispersed power, a significant development of temperatures in the pavement beyond 15 cm and, finally, high exposure times. The heating is carried out with a magnetron delivering the maximum power of 5 kW, with a horn reflector and absorbed load, for the inevitable power reflections, operating at the frequency of 2450 MHz.

2.1   Physical Properties and Dielectric Constant

The physical properties affected by electromagnetic and thermal phenomena are density, specific heat, porosity, permeability and the dielectric constant, which depends on the composition of the medium, humidity, temperature, microwave frequency, time and space. Various empirical models for the evaluation of the dielectric constant are summarised in [4, 5]. Based on the comparisons made in [4], it is agreed to use the following relationship [6, 7].

$$\sqrt{\varepsilon} = (1 - S)\emptyset\sqrt{\varepsilon_a} + (1 - \emptyset)\sqrt{\varepsilon_r} + [S\emptyset + (1 - S)\emptyset \rho_s/\rho_w]\sqrt{\varepsilon_w} \tag{2}$$

where ε is the dielectric constant, or permittivity, of the composite ground; ε_a that of the air; ε_r that of the solid and ε_w that of the water, while S is the water saturation of the ground, $\varnothing$ is the porosity of the ground, ρ_s is the density of the vapour and ρ_w is the density of the water.

3   Microwave Heating of Stratified Porous Media

The heating with plane waves that hit the surface of the ground perpendicularly is studied, not considering the edge effects. The material occupies the semi-infinite space indicated in Fig. 1, which is divided into N elementary layers, i.

3.1   Electromagnetic Model

The active power that transits in the material tends to be absorbed mainly where the dielectric losses are higher. Therefore, the electromagnetic field is exponentially attenuated, and the absorbed power varies with the square of the field. The electromagnetic power, per unit of volume, dissipated in each layer i of the ground, $\dot{q}_e$ (W/cm^3), is given by

$$\dot{q}_e = 0.278\,|E_i|^2 f_0 \varepsilon''_i 10^{-6} \tag{3}$$

where $|E_i|$ is the peak value of the electric field, f_0 is the working frequency and ε''_i is the imaginary part of the dielectric constant of the ground in each layer.

The electromagnetic model used is that of transmission lines, where a plane wave incident, normally on the surface of the material, is considered, which, at each separation surface between adjacent homogeneous layers, undergoes a reflection and a transmission towards the next layer [8]. Figure 1 shows that the ground in question is divided into N-1 layers, while the Nth layer extends to infinity and is

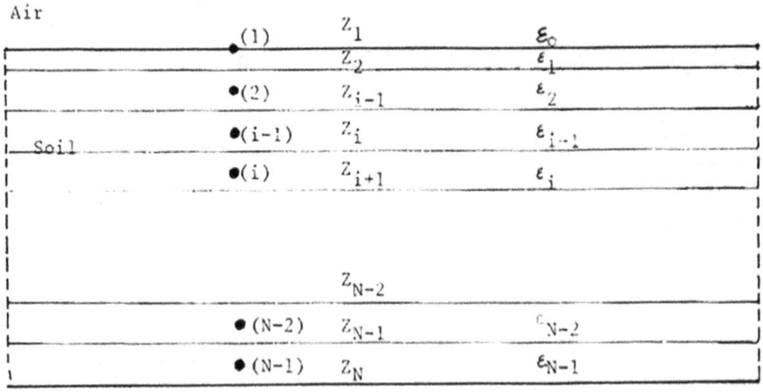

Fig. 1 Semi-infinite ground

considered adapted. According to the theory of transmission lines, the complex propagation constant of each layer i, γ_i, is given by

$$\gamma_i = \sqrt{\mu_0 \omega^2 (j\varepsilon''_i - \varepsilon'_i)} \tag{4}$$

where μ_0 is the magnetic permeability, ω is the angular frequency, ε'_i is the real part and ε''_i is the imaginary part of the dielectric constant of the ith layer, while the characteristic impedance, K_i, is given by

$$K_i = 1/\sqrt{(\varepsilon'_i - j\varepsilon''_i)/\mu_0} \tag{5}$$

According to the theory of lines, the characteristic impedance of each layer i, Z_i, is given by

$$Z_i = K_i \frac{Z_{i+1} + K_i \tan h(\gamma_i y_{hi})}{K_i + Z_{i+1} \tan h(\gamma_i y_{hi})} \tag{6}$$

The surface reflection coefficient of each layer i, Γ_i, is given by

$$\Gamma_i = \frac{K_i - Z_{i+1}}{K_i + Z_{i+1}} \tag{7}$$

and the transmission coefficient of each layer i, τ_i, is given by

$$\tau_i = \frac{K_i}{K_{i+1}} \frac{K_{i+1} + Z_{i+1}}{K_i + Z_{i+1}} \tag{8}$$

3.2 Thermo-Fluid-Dynamic (TFD) Model and Mass Transport

The phenomenon is considered to be in an unsteady state. Initially, the ground is in equilibrium with the pores filled with liquid water, water vapour and air, while the saturation conditions range from completely dry to saturated soil. The boundary condition for the soil layer is natural convection heat exchange, $\dot{Q}_c$, on the upper surface in contact with the ambient air, i.e.

$$\dot{Q}_c = \alpha A (T_1 - T_{ext}) \tag{9}$$

where α is the convection coefficient, A is the exchange surface, T_1 is the surface temperature and T_{ext} is the external air temperature, while the last layer is adiabatic. The boundary condition for mass transport on the last layer is the continuity of mass flow. The boundary conditions on pressure follow from the previous boundary conditions. In non-stationary mass transport, only convective transport is taken into account, and molecular diffusion is neglected, in accordance with [9].

The conservation of the mass of air is given by

$$\frac{\partial[\varrho_a(1-S)]}{\partial t} + \frac{\partial[(1-S)\varrho_a V_g]}{\partial y} = 0 \tag{10}$$

where ϱ_a is the air density, and V_g is the gas velocity.

The conservation of the mass of water is given by

$$\varrho_w\left\{\frac{\partial S}{\partial t} + \frac{\partial[(S-S_R)V_W]}{\partial y}\right\} + \frac{\partial[\varrho_a(1-S)]}{\partial t} + \frac{\partial[(1-S)\varrho_a V_g]}{\partial y} = 0 \tag{11}$$

where S_R is the residual saturation, which is the minimum value of connected liquid water, for which the water velocity is not zero [10].

The conservation of energy is written, considering the effective thermal conductivity of the ground, λ [11–13], and the energy generated in the ground due to microwaves. The conservation equations of momentum for the liquid and gaseous phases are used to simplify the energy conservation equation. For a ground with constant porosity and constant water density, the energy conservation equation, after some algebraic steps, is written as

$$\frac{1}{\varnothing}\left(\dot{q}_e + \lambda\frac{\partial^2 T}{\partial y^2}\right) = \frac{\partial}{\partial t}\left\{(1-S)(\varrho_S u_S + \varrho_a u_a) + S\,\varrho_w\,u_w + \frac{1-\phi}{\phi}\varrho_r\,u_r\right\} +$$

$$\frac{\partial}{\partial y}\left\{V_g(1-S)(\varrho_S h_S + \varrho_a h_a) + V_w(S-S_R)\varrho_w\,h_w\right\} +$$

$$- (1-S)V_g\frac{\partial p_g}{\partial y} - V_w(S-S_R)\frac{\partial p_w}{\partial y}$$

$$\tag{12}$$

The fluid velocities are calculated as follows, with Darcy's law, for the gas phase

$$(1-S)\varnothing V_g = -\frac{k_g}{\mu_g}\left(\frac{\partial p_g}{\partial y} + \varrho_g g\right) \tag{13}$$

and for the liquid phase

$$(S-S_R)\varnothing V_w = -\frac{k_w}{\mu_w}\left(\frac{\partial p_w}{\partial y} + \varrho_w g\right) \tag{14}$$

where

$$k_g = k\left(\frac{1-S}{1-S_R}\right)^2\left[1-\left(\frac{S-S_R}{1-S_R}\right)^2\right] \tag{15}$$

$$k_w = k \left(\frac{S - S_R}{1 - S_R} \right)^4 \tag{16}$$

and the viscosities are temperature-dependent [14]. Capillary phenomena are taken into account, and the pressure of the liquid water is given by

$$p_w = p_g - p_c = p_s + p_a - p_c \tag{17}$$

Hysteresis is neglected, and the capillary pressure, p_c, is approximated by empirical expressions [15]

$$\left(\frac{p_d}{p_c} \right)^2 (1 - S_R) + S_R = S \tag{18}$$

for $S \leq S_{in}$,

$$p_d \left(\frac{1 - S}{1 - S_{in}} \right) = p_C \tag{19}$$

for $S \geq S_{in}$.

Air is considered a perfect gas, while water vapour is considered superheated if the vapour pressure is lower than the corresponding saturation pressure. The saturation of the soil, S, is deduced from the desorption isotherm [9]. The following state equations are assumed

$$p_s = R_w \, Z_c \, T \, \varrho_s \tag{20}$$

$$du_s = c_{vs} \, dT \tag{21}$$

If the vapour pressure is greater than p_{sat}, the water vapour is assumed saturated, and the *Clapeyron* equation is used

$$\frac{dp_s}{dT} - \varrho_s \frac{h_s - h_w}{T} \tag{22}$$

Finally, the following empirical equations are derived from thermodynamic tables:

$$d\varrho_s = D_s \, dT \tag{23}$$

$$d(\varrho_s u_s) = R_s \, dT \tag{24}$$

3.3   Numerical Solution and Results

The differential equations are solved numerically with the finite difference scheme called *thanks and tubes* [16]. The electromagnetic and thermal variables are evaluated at the node *i*, while the fluid velocities are assumed between nodes *i*-1 and *i*. The general expression

$$\frac{\partial X}{\partial T} = a + b\,\frac{\partial T}{\partial T} \tag{25}$$

can be written in the form

$$\frac{X(t - \Delta t) - X(T)}{\Delta t} = a(t) + b(t)\frac{T(t + \Delta t) - T(t)}{\Delta t} \tag{26}$$

The variables and spatial derivatives are evaluated at the previous time so that the unknowns are only in the time derivatives. The energy conservation equation is written as

$$\frac{\partial}{\partial t}\left\{ (1 - S)(\varrho_s\, u_S + \varrho_a\, u_a) + S\,\varrho_w\, u_w + \frac{1 - \phi}{\phi}\varrho_r\, u_r \right\} = G(t) \tag{27}$$

The air mass conservation equation is

$$\frac{\partial[\varrho_a(1 - S)]}{\partial t} = M(t) \tag{28}$$

The water mass conservation equation is

$$\frac{\partial[\varrho_w(1 - S)]}{\partial t} = W(t) - \varrho_w\,\frac{\partial S}{\partial t} \tag{29}$$

which, if $\varrho_w \gg \varrho_s$, can be simplified to

$$\frac{\partial\varrho_s}{\partial t} = \frac{W(t)}{1 - S} - \frac{\varrho_w}{1 - S}\,\frac{\partial S}{\partial t} \tag{30}$$

The numerical solution of the water mass conservation equation is different if the water vapour is superheated or saturated. In the case of superheated vapour, it is obtained from the desorption isotherm and from Eq. 20

$$\frac{\partial S}{\partial t} = e\,\frac{\partial\varrho_s}{\partial t} + d\,\frac{\partial T}{\partial t} \tag{31}$$

which, with the help of Eq. 30, gives an equation similar to Eq. 25 for the variable S. The water mass conservation equation finally becomes

$$\frac{\partial\left[\varrho_S(1-S)\right]}{\partial t}=1+m\ \frac{\partial T}{\partial t} \tag{32}$$

which is similar to Eq. 25 for the variable $\varrho_S(1-S)$. For saturated water vapour, Eq. 30 is used, which gives, with Eq. 23,

$$\frac{\partial S}{\partial t}=n+r\ \frac{\partial T}{\partial t} \tag{33}$$

The numerical solution of the mass conservation equation for air is straightforward for the left-hand term of Eq. 28, while for the energy one, it differs for the case of saturated or superheated steam. The left-hand term of Eq. 27 uses Eqs. 32 and 21 for superheated steam, while it uses Eqs. 33 and 24 for saturated steam. In both cases, Eq. 27 reduces to the following equation

$$C(t)\ T^2(t+\Delta t)+D(t)\ T(t+\Delta t)+F(t)=0 \tag{34}$$

Once the temperature is calculated at the new time, all other variables are re-calculated. The results of the numerical simulations, obtained at a frequency of 2.45 GHz and with incident power equal to 125 kW/m^2, are reported in Figs. 2, 3, 4, 5 and 6.

Figure 2 shows the absorbed Electro magnetic *(EM)* power, W/cm^3, in the upper part, and the temperature, °C, in the lower part, in dry soil, with constant $\varepsilon'' = 0.045$.

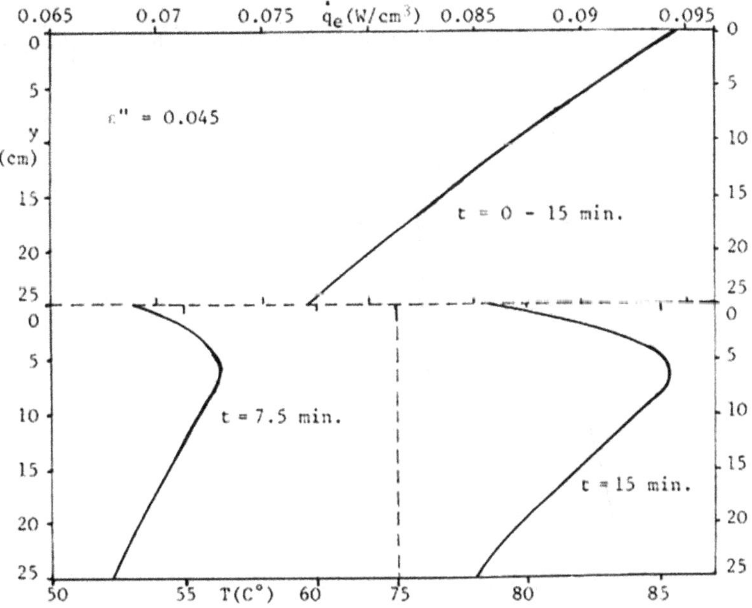

Fig. 2 Absorbed power (W/cm^3), top, and temperature in the soil (°C), bottom, versus y (cm)

On the ordinate, there is the distance from the surface (cm). The absorbed *EM* power does not depend on time and decreases almost linearly with the distance from the surface. The temperature increases over time, and after 15 min, a maximum is evidenced at a depth slightly greater than 5 cm.

Figure 3 presents the numerical results of the analogous case, with ε'' increasing with temperature, according to the relationship reported in the same figure, Fig. 3. On the ordinate, there is the distance from the surface, while, on the abscissa, the *EM* power absorbed in the upper part and the temperature in the lower part are given. The

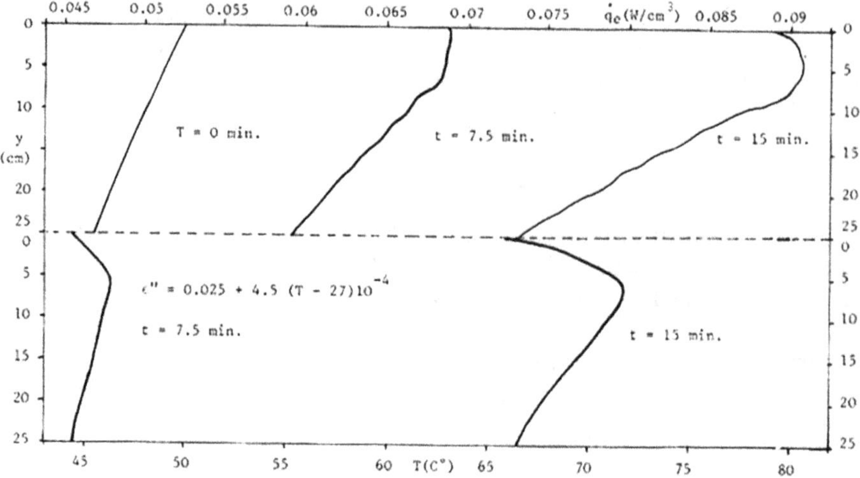

Fig. 3 Absorbed power (W/cm³), top, and temperature in the soil (°C), bottom, versus y (cm)

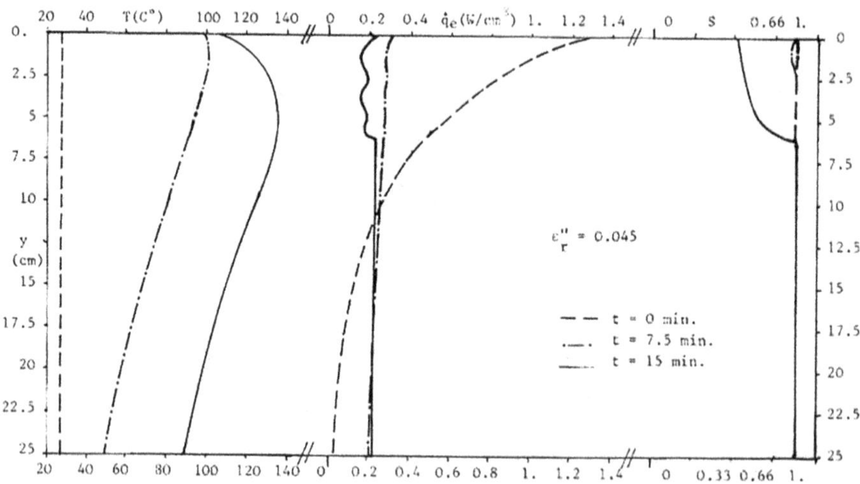

Fig. 4 Temperature (°C), absorbed power (W/cm³) and degree of saturation in the ground versus y (cm)

absorbed power increases over time, and after 15 min, it has a maximum at around 5 cm deep, as does the temperature.

Figure 4 reports the numerical results of the heating of a completely saturated ground, $S = 1$, at the beginning of the microwave treatment. The dielectric constant is constant and equal to $\varepsilon'' = 0.045$ in the ground. On the ordinate, there is the distance from the surface, while on the abscissa are reported, from left, the temperature, in °C,; the *EM* absorbed power, in W/cm^3 and the degree of saturation, S. Three curves are shown, at the initial instant of time, $t = 0$ min; after 7.5 min and after 15 min of heating. The ground temperature increases over time, and after 7 and 15 min, the partial vaporisation of the liquid in the upper part of the ground is recorded, with a maximum temperature around a depth of 6 cm. At the same time, there is a marked decrease in the absorbed power in the upper part, where the ground saturation has decreased.

Figure 5 reports the numerical results of the microwave heating at the initial instant of time and after 7.5, 10 and 15 min. The asphalt floor has a porosity equal to $\phi = 0.3$, residual saturation $S_R = 0.2$ and hydraulic permeability $k = 10^{-4}\ m^2$; effective thermal conductivity is a function of temperature according to the relation $\lambda = 0.000605 + 4.633\ T(°C)\ 10^{-6}\left[\frac{kW}{m\ K}\right]$. The electromagnetic model divides the ground into two layers of height 12.5 cm, with dielectric constant $\varepsilon'' = 0.125$ in the upper part and $\varepsilon'' = 0.045$ in the lower part. The initial floor saturation profile is assumed, as shown in Fig. 5, in accordance with the conditions of the experiments [17]. The experimental temperature results at the three moments of time are also reported [17]. At the beginning of the microwave heating, $t = 0$, the power is absorbed only in the final layer of the ground, where water is present, or where the saturation is $S = 1$. The resulting increase in temperature, up to partial vaporisation, produces a pressure gradient that moves the fluid upwards. The oscillatory trend of

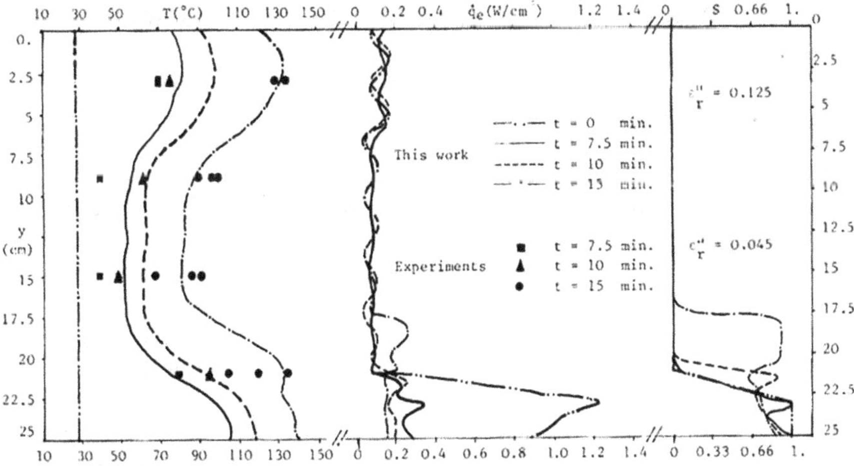

Fig. 5 Temperature (°C), absorbed power (W/cm^3) and degree of saturation in the motorway pavement versus y (cm)

the temperature with depth depends on the high value assumed for the dielectric constant in the upper part for dry ground, which shows a maximum at around 3 cm deep. The subsequent maximum temperature is recorded in the lower part of the ground, or at the maximum depth, where the saturation is the highest, or where the presence of water is the greatest. Figure 5 also reports the experimental results of the temperature, measured in [17], which are in reasonable agreement with the numerical ones.

The temperature has a wave-like trend (with maximums and minimums), which resembles that present in the stratosphere and the troposphere of the Earth, as reported in Fig. 6 [18] (in Italian because more pictographic [19]).

The results shown in Fig. 5 of the temperature profile also have some similarities with those measured by the thermometer, TEM, designed and prototyped in the Technical Physics and Energy Engineering Laboratory of the University of Rome *"Tor Vergata"* [20]. The TEM was inserted in the instrument *"Huygens Atmospheric Structure Instrument (HASI)"* of the Cassini Mission in order to measure the temperature of the atmosphere of Titan. The temperature measurements *(K)* in the atmosphere of Titan are on the abscissas of Fig. 7 [21]. The top of Fig. 7 reports, on the ordinates, the pressure *(hPa)*, on the left, and the altitude (km), on the right, in the first 1400 km of Titan's atmosphere. The bottom of Fig. 7 reports, on the ordinates, the altitude (km) in the first 160 km of Titan.

4   Applicators for Medical Hyperthermia

The possible applications for medical hyperthermia concern the treatment of neoplasms and other benign conditions. Some types of non-invasive applicators, also shaped as arrays, are proposed and clinically tested to produce therapeutic temperatures in deep tumours, preserving, as far as possible, the surrounding healthy tissues. The interest in endocavitary and interstitial applicators, which are more suitable for selective heating of deep organs or those located near natural body cavities, is more recent. The optimisation of such applicators, as far as the absorbed electromagnetic power and the temperature distribution in the tissue of interest are concerned, requires adequate theoretical tools, which allow us to identify and quantify the most significant design parameters. Moreover, the difficulty of monitoring the temperature of deep tissues, poorly accessible even with invasive techniques, requires a high accuracy of the theoretical models. A theoretical and numerical modelling of the electromagnetic and thermal behaviours of isolated dipole radiating elements, immersed in a biological tissue, is implemented.

The endocavitary and interstitial applicators with isolated radiating dipoles are schematically illustrated in Fig. 8. The interstitial ones can be with or without a cooling system. The endocavitary ones can be with or without a cooling system and with or without a reflector. The cooling system, in turn, can use air or liquid. The elements of the applicator that most influence the temperature distribution are the cooling system and the reflector because they limit the increase in temperature near the applicator wall and direct the *e.m.* energy mainly towards the tissues to be treated, preserving the healthy ones.

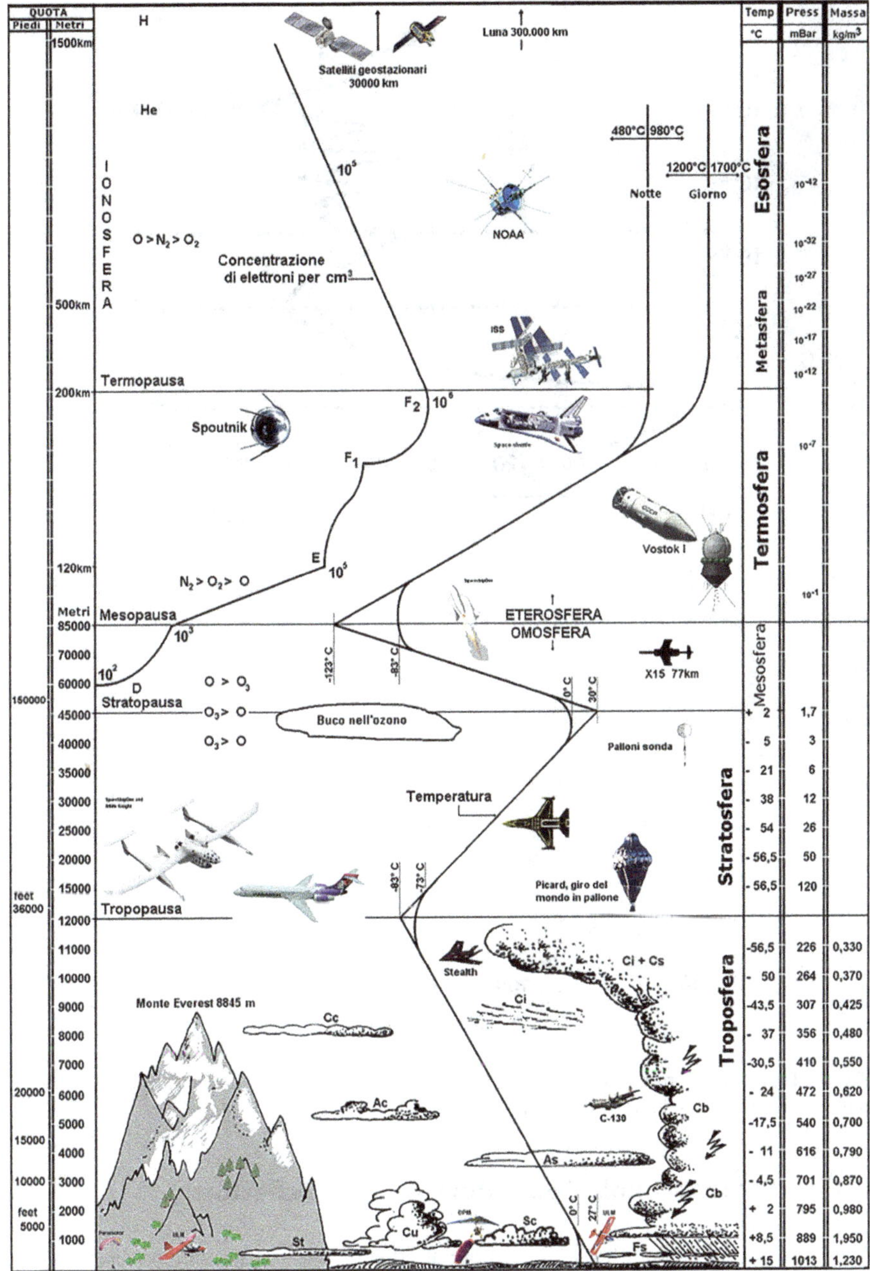

Fig. 6 Temperature profile in the Earth's atmosphere [18]

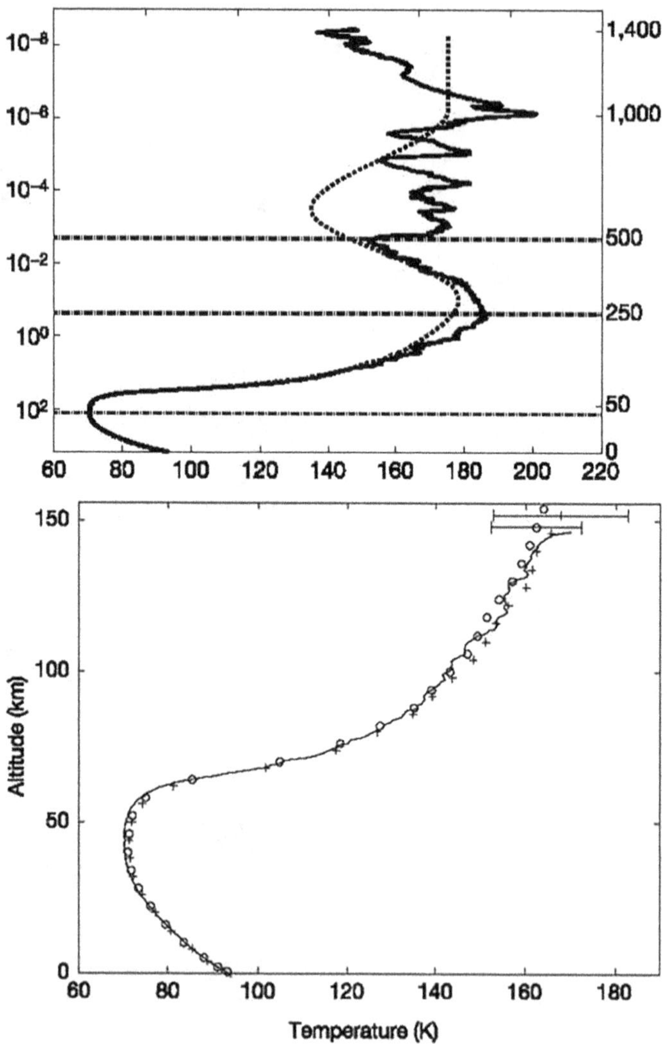

Fig. 7 Temperature profile (K), on the abscissas, of the atmosphere of Titan, measured by the TEM temperature sensor of the HASI instrument, figs. 2 and 5 of [21]

4.1   Radiant Dipole Applicators Without Reflectors

We consider an isolated radiating dipole applicator without a reflector inserted in a homogeneous medium, Fig. 9 [22]. The applicator used as an example has the following dimensions, Fig. 9: radiating element, $a = 1.8$ mm; insulator, $r_c = 2.2$ mm; inner diameter of the outer container, $r_1 = 5.15$ mm, or 5.55 mm or 5.75 mm; outer container, $r_2 = 6$ mm. The radiating element is an ideal conductor with electrical conductivity, $\sigma = \infty$. The insulator is made of plastic (Teflon) with electrical

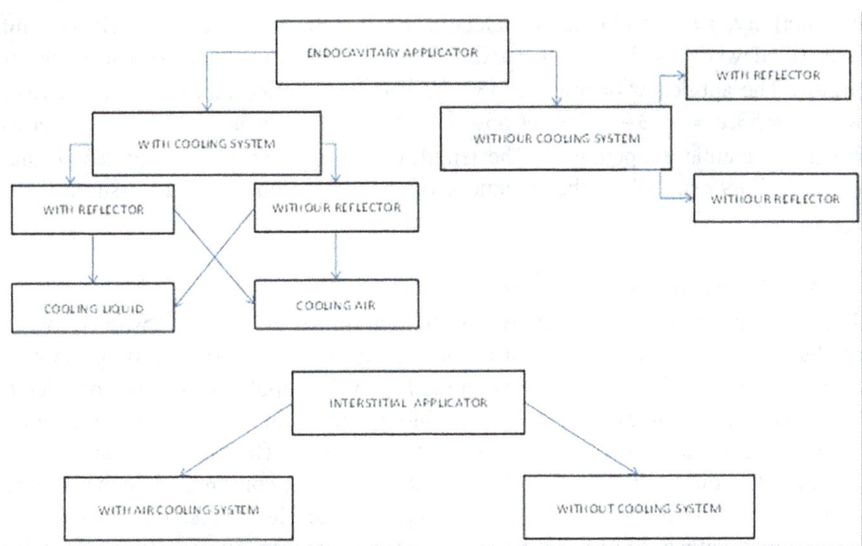

Fig. 8 Applicators for medical hyperthermia

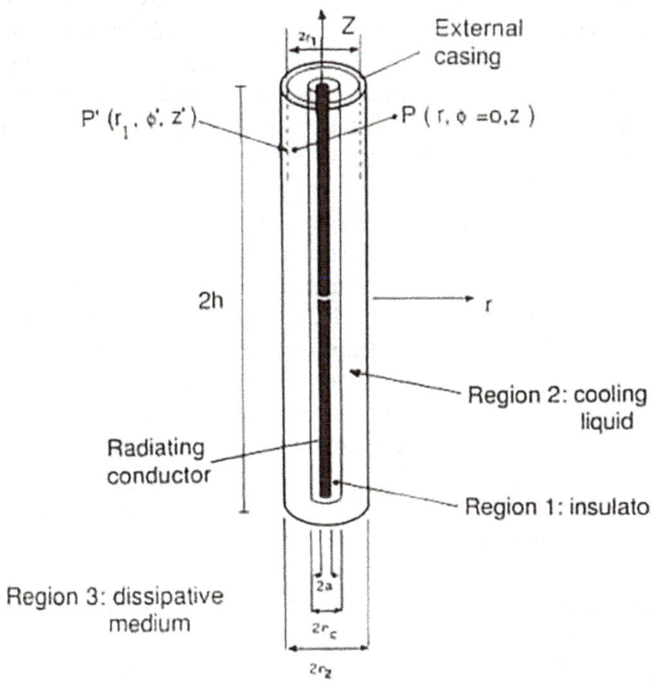

Fig. 9 Insulated radiating dipole applicator, fig. 1 of [22]

conductivity, $\sigma = 0$, and relative dielectric constant to vacuum, $\epsilon' = 2.1$. The coolant is distilled water, with $\epsilon' = 80, \sigma = 0.0864\frac{S}{m}$. The outer container is also made of plastic. The applicator operates at 432 *MHz* and is inserted into the muscle tissue, with $\epsilon' = 53, \sigma = 1.43\frac{S}{m}$. The net power feeding the applicator is 23 W in order to reach therapeutic temperatures. The length of the applicator is determined so that $\frac{\lambda}{2} = \frac{\pi}{\beta_L} = 7.89\ cm$, while the thickness of the outer container can assume three values.

4.1.1 Electromagnetic Model

The *e.m.* model developed allows us to determine the power absorbed being produced by the radiating element in the biological tissue, also including the *e.m.* effects due to the circulation of the cooling liquid. The applicator can be considered as an *e.m.* linear dipole, radiating in a biological tissue or equivalent phantom, homogeneous and dissipative, as represented in Fig. 9. The radiating conductor is coated with a thin insulating sheath, with a low dielectric constant, while the cooling liquid circulates in the annular region between the sheath and the casing. The insulating sheath is introduced to avoid contact between the radiating conductor and the cooling liquid and influences the wave number of the radiating element.

The theoretical model considers the applicator as a conductor with multi-layer insulation, radiating in an external dissipative medium, consisting of three dielectric layers, namely an insulating sheath, the cooling liquid with low losses at 432 MHz and the external containment casing. The theoretical analysis is developed on the basis of the theory of [23], regarding the current distribution on the radiating element, while the distribution of the *e.m.* field produced is calculated with a numerical procedure based on the method of [24]. To directly apply the theory of [23], it is necessary to introduce some assumptions: the presence of the outer sheath is negligible, since its thickness is much smaller than the wavelength; the effects due to the reflected wave, produced at the interface with the external dissipative medium, are negligible because the dielectric constant of the water liquid and of the biological tissue/equivalent phantom are comparable. The assumption on the external casing allows us to reduce the insulating system from three to two layers, while that on the reflected wall allows us to determine the effective wave number of the radiating element as a function of the dielectric characteristics of the cooling liquid. For the details of the procedure for the calculation of the current distribution on the central conductor, one can refer to [22, 25–29].

The *e.m.* field produced by the dipole in the external dissipative medium is calculated numerically in the following way:

- The electric field in the cooling liquid is calculated with the procedure of [24], once the current is known.
- The magnetic field $H_{2\phi}$ is obtained numerically with the finite difference method, [30], from Maxwell's equation.

$$\frac{\partial E_{2r}}{\partial z} - \frac{\partial E_{2z}}{\partial r} = -j\,\omega\,\mu_0\,H_{2\phi} \tag{35}$$

- The equivalent interface sources are calculated with the boundary conditions for $r = r_1 = r_2$.

$$H_{3\phi}(r_1, z) = H_{2\phi}(r_1, z) \tag{36}$$

$$E_{3r}(r_1, z) = \left(\frac{\epsilon''_2}{\epsilon''_3}\right) E_{2r}(r_1, z) \tag{37}$$

$$E_{3z}(r_1, z) = E_{2z}(r_1, z) \tag{38}$$

- The electric field in the medium is calculated numerically [24] by solving the following equations.

$$\begin{aligned}
E_{3z}(r, z) = \frac{1}{4\pi} \int_{-h}^{h} dz' \int_{0}^{\pi} r_1\, d\phi' \Big\{ -2j\,\omega\,\mu_0 H_{3\phi}(r_1, z')\Psi \\
+ 2\,E_{3z}(r_1, z')\frac{1}{R}\frac{\partial\Psi}{\partial R}(r_1 - r\cos\phi') - 2\,E_{3r}(r_1, z')\frac{1}{R}\frac{\partial\Psi}{\partial R}(z - z')\Big\}
\end{aligned} \tag{39}$$

$$\begin{aligned}
E_{3r}(r, z) = \frac{1}{4\pi} \int_{-h}^{h} dz' \int_{0}^{\pi} r_1\, d\phi' \Big\{ 2\,E_{3z}(r_1, z')\frac{1}{R}\frac{\partial\Psi}{\partial R}(z - z')\cos\phi' \\
+ 2\,E_{3r}(r_1, z')\frac{1}{R}\frac{\partial\Psi}{\partial R}(r_1\cos\phi' - r)\Big\}
\end{aligned} \tag{40}$$

with

$$\Psi = \frac{e^{-i K_3 R}}{R},\; K_3 = \beta_3 - j\alpha_3 = \omega(\mu_0\epsilon''_3)^{1/2},\; R = \left[(z - z')^2 + (r - r_1)^2 + 4\,r\,r_1\sin^2\frac{\phi'}{2}\right]^{1/2}$$

Numerical Results

The *e.m.*-deposited power, $G(r,z)$, mW/cm^3, is reported in Fig. 10, *with* and without coolant circulation. The continuous line is related to an applicator with coolant circulation, while the dashed line is related to the absence of a coolant. In the presence of cooling, the specific power remains very low in the coolant to increase in the tissue, with much higher electrical conductivity. In the absence of cooling, the absorbed power decreases monotonically, there being no discontinuity along r. The power deposited at the interface with the muscle tissue is higher than 20% in the presence of a coolant because water has a lower electrical conductivity than tissue. In

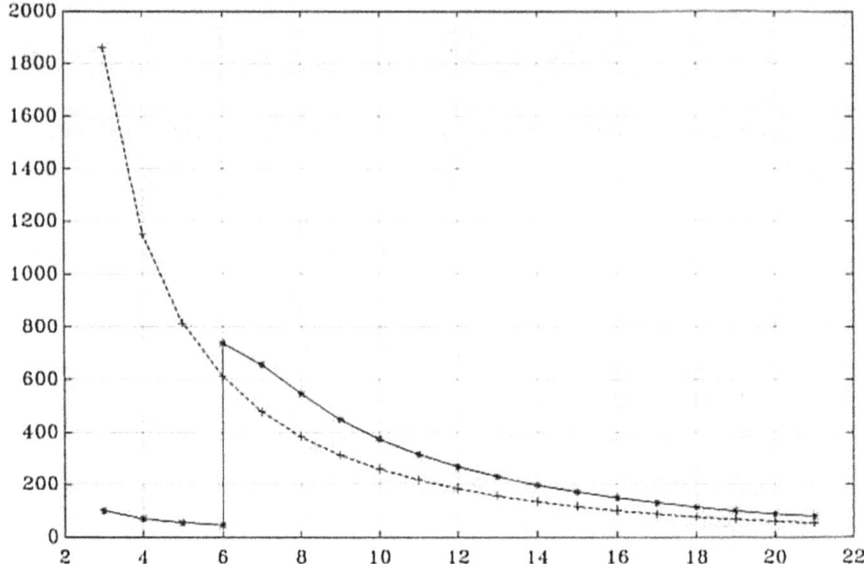

Fig. 10 Absorbed power, $G(r,z)$, (mW/cm³), on the ordinate, versus r(mm) at $z = 0$, fig. 3 of [22]

conclusion, the cooling system has a double positive effect on the distribution of absorbed power in the muscle tissue, as it allows the reduction of the gradient of absorbed power at the interface with the tissue and a greater and more uniform penetration into the tissue.

Figure 11 shows the absorbed power as a function of z at $r = 14$ mm in the presence of cooling. The maximum is for $z = 0$.

Figure 12 shows the three-dimensional view of the absorbed power, function of z and r, in the presence of a cooling system *(a)* and without cooling *(b)*. The maximum absorbed power with cooling is for $z = 0$. In the absence of a cooling system, there is a high gradient of absorbed power near the surface of the outer casing, in direct contact with the muscle tissue, and a rapid decay along r.

Figure 13 shows the curves of equal absorbed power in the muscle tissue in the presence of cooling.

4.1.2 Thermo-Fluid-Dynamic (TFD) Model in Living Tissue

The temperature distribution in a living muscle tissue can be calculated by solving the energy conservation equation, where the *e.m.* energy absorbed by the applicator, $G(r,z)$, the metabolic heat generated in the tissue, W_m, and the blood perfusion [31] are present. The reference thermo-fluid-dynamic scheme is shown in Fig. 14, where it is assumed that the cooling liquid flows in the region $-\infty \leq z \leq +\infty$ in axial symmetry.

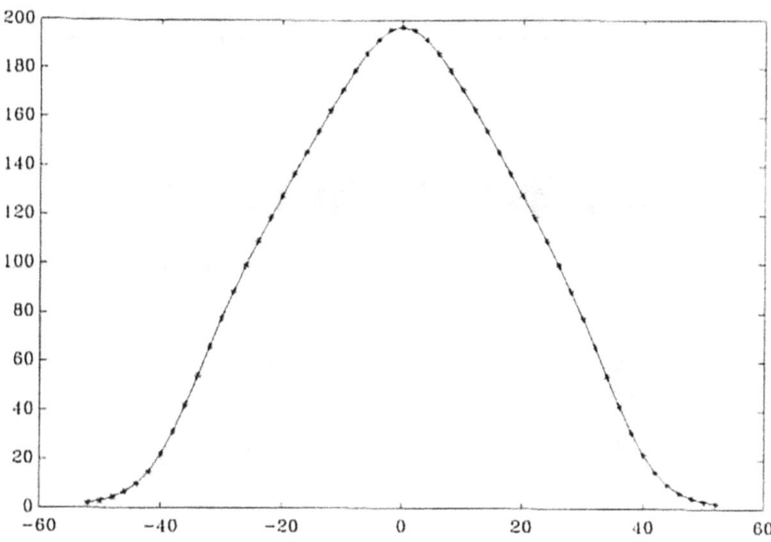

Fig. 11 Absorbed power, $G(r,z)$, (mW/cm^3), on the ordinate, versus z(mm), with cooling, at $r = 14$ mm, fig. 4 of [22]

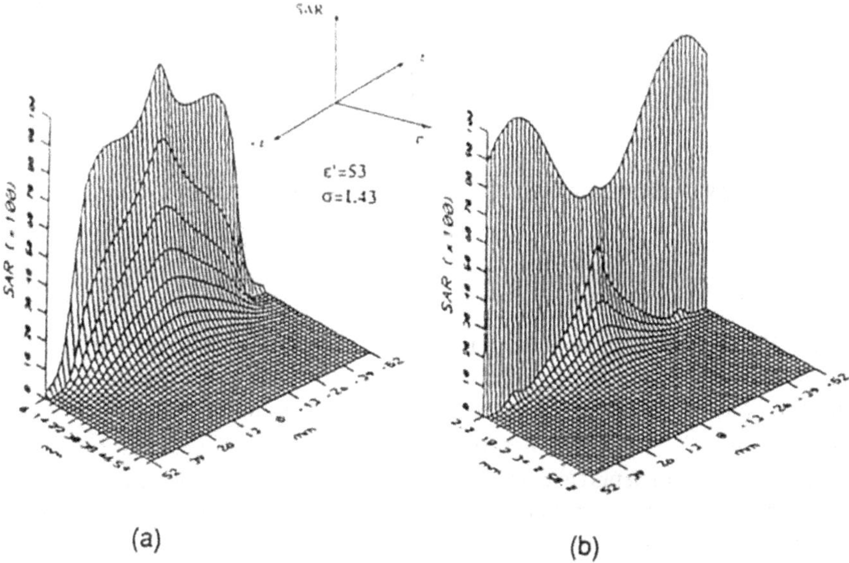

Fig. 12 Specific absorbed power, specific absorption rate (SAR), (mW/cm^3): (**a**) with cooling, (**b**) without cooling

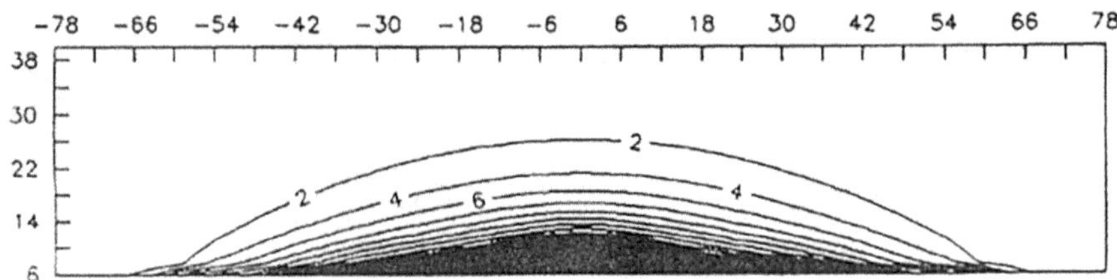

Fig. 13 Specific absorbed power, SAR (mW/cm^3) in the muscle tissue in the presence of cooling, fig. 7 of [26]

Fig. 14 Thermo-fluid-dynamic model, fig. 2 of [22]

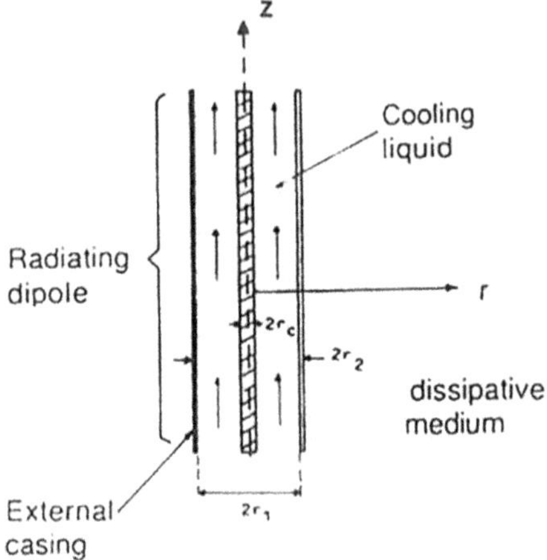

The energy conservation equation becomes, in cylindrical coordinates,

$$\frac{\partial^2 T}{\partial r^2}+\frac{1}{r}\frac{\partial T}{\partial r}+\frac{\partial^2 T}{\partial z^2}+\frac{1}{k_t}G(r,z)+\frac{1}{k_t}W_m-C(T-T_b)=\frac{1}{\alpha}\frac{\partial T}{\partial t} \tag{41}$$

where T is the tissue temperature, r and z are the coordinates, t is the time, k_t is the tissue thermal conductivity, ρ_t is the tissue density, c_t is the tissue-specific heat, α is the tissue thermal diffusivity, ρ_b is the blood density, c_b is the blood-specific heat, m_b is the volumetric blood flow rate per unit of mass of the tissue, $C=\frac{c_b\,\rho_b\,m_b\rho_t}{k_t}$ is the perfusion coefficient, $G(r,z)$ is the *e.m.* power per unit volume of tissue, W_m is the metabolic power generated per unit volume of tissue and T_b is the temperature of the blood entering the tissue.

The thermal boundary conditions are

$$-k_t\frac{\partial T}{\partial r}+U\,T=U\,T_{cl}, \tag{42}$$

for $r=r_2$ and $t\geq0$,

$$T = T_b \tag{43}$$

for $t = 0$,

$$T = T_b \tag{44}$$

where T_{cl} is the coolant temperature, $n = 10$ times the radius of the applicator, U the overall heat transfer coefficient, given by

$$U = \frac{1}{\left(\frac{r_2}{h^* r_1}\right) + \left(\frac{r_2}{k_{is}}\right) \ln\left(\frac{r_2}{r_1}\right)} \tag{45}$$

where h^* is the forced convection coefficient of water, and k_{is} is the thermal conductivity of the outer shell. The forced convection coefficient of water, h^*, is determined by the Nusselt number.

$$h^* = k \frac{Nu}{D_{eq}} \tag{46}$$

where $D_{eq} = 2(r_1 - r_c)$, and k is the thermal conductivity of water. In the case under consideration, the Nusselt number is assumed to be 4.98 for a fully developed flow in an annular duct, with $\frac{r_c}{r_1} \cong 0.4$ and constant heat flow rate at the wall between the tissue and the outer shell [32]. The energy conservation equation is solved with the implicit alternating directions method, known as U and $G(r, z)$, with a grid of 2 mm in the r and z directions, up to $r = 58$ mm and $z = 52$ mm.

Numerical Results

The numerical results of the temperature distribution at steady state in a biological tissue for an input power of 23 W are presented in Figs. 15, 16, 17, 18, 19, 20 and 21.

Figure 15 shows the temperature distributions, on the plane $z = 0$, with several different thicknesses of the outer shell, three thicknesses, d, for a coolant temperature, $T_{cl} = 16\ °C$, and a blood perfusion, $m_b = 4.5\ 10^{-6}$ (m³/kg s). The solid line (lower) represents the temperature obtained for a thickness $d = 0.25$ cm, for which $U = 235\ W/m^2\ K$, while the dashed line (intermediate) is obtained for $d = 0.45$ mm and $U = 179\ W/m^2\ K$, and the dotted line (upper) for $d = 0.85$ mm and $U = 119\ W/m^2\ K$. It can be observed that the decrease in thickness significantly reduces the temperature at the interface with the tissue, while the therapeutic values of the temperature are essentially unchanged. Conclusions to be taken into account in the design of the applicator.

Figure 16 shows the temperature distributions in the tissue, on the plane $z = 0$, with a blood perfusion, $m_b = 4.5\ 10^{-6}$ m³/kg s, for a thickness of the outer shell equal to $d = 0.45$ mm, with variable temperature of the coolant, T_{cl}. The solid line (lower) is obtained for a temperature $T_{cl} = 4\ °C$, for which $U = 176\ W/m^2\ K$, while the dashed line (intermediate) is obtained for $T_{cl} = 10\ °C$, for which $U = 178\ W/m^2\ K$, and the dotted line (upper) is obtained for $T_{cl} = 16\ °C$, for which $U = 180\ W/$

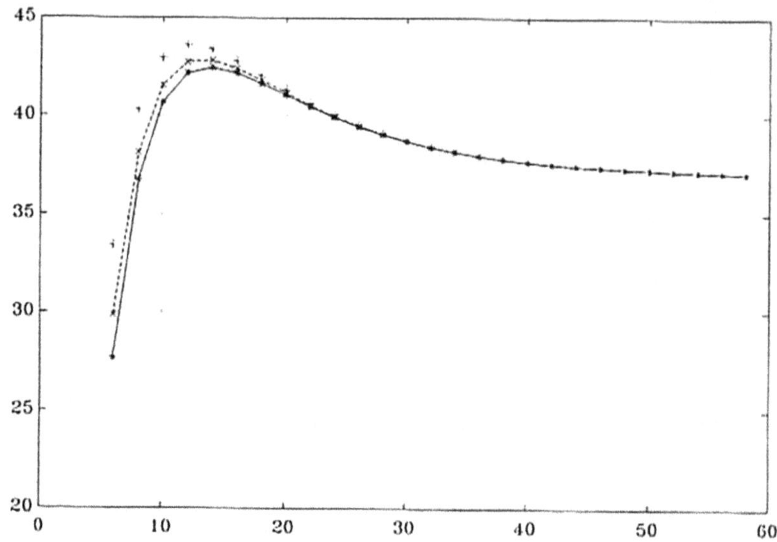

Fig. 15 Temperature distribution (°C), on the ordinate, versus distance (mm), varying with the thickness of the outer shell, fig. 6 of [22]

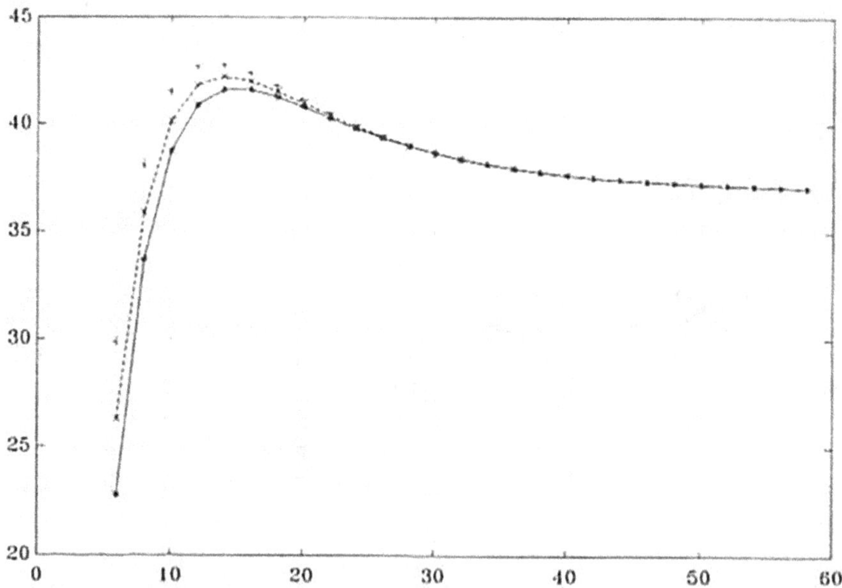

Fig. 16 Temperature distribution (°C), on the ordinate, versus distance (mm), varying with the temperature of the coolant, fig. 5 of [22]

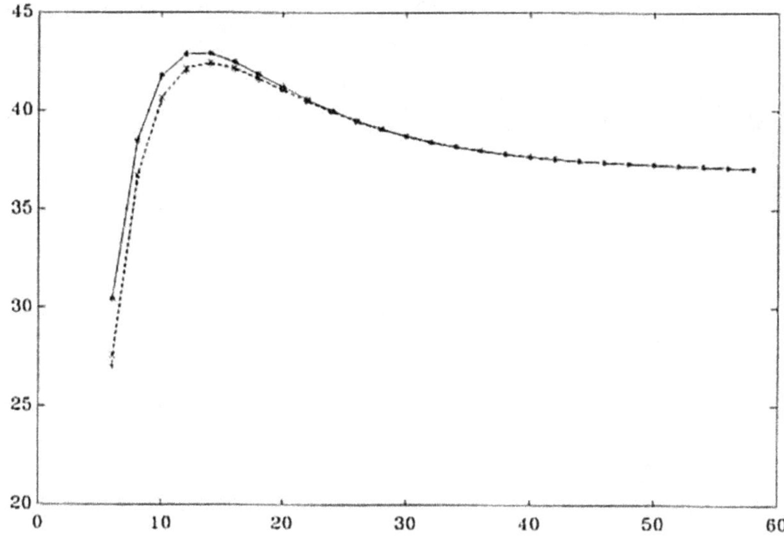

Fig. 17 Temperature distribution (°C), on the ordinate, versus distance (mm), varying with the flow rate of the coolant, fig. 7 of [22]

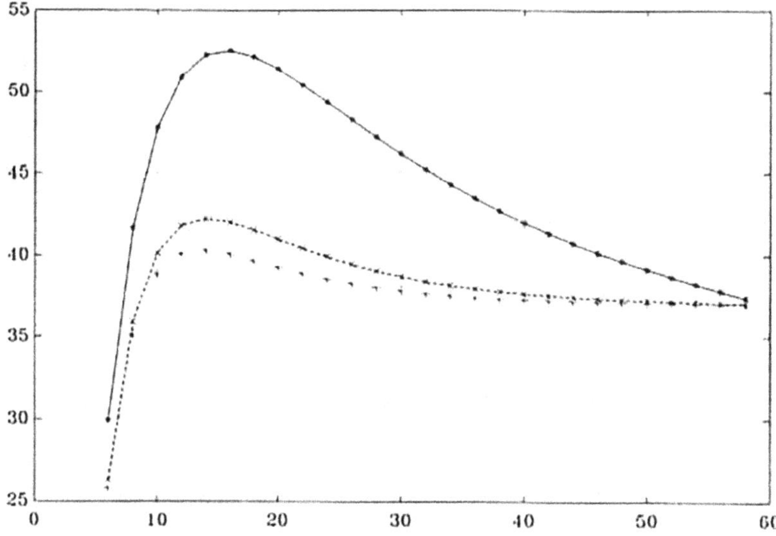

Fig. 18 Temperature distribution (°C), on the ordinate, versus distance (mm), varying with blood perfusion, fig. 8 of [22]

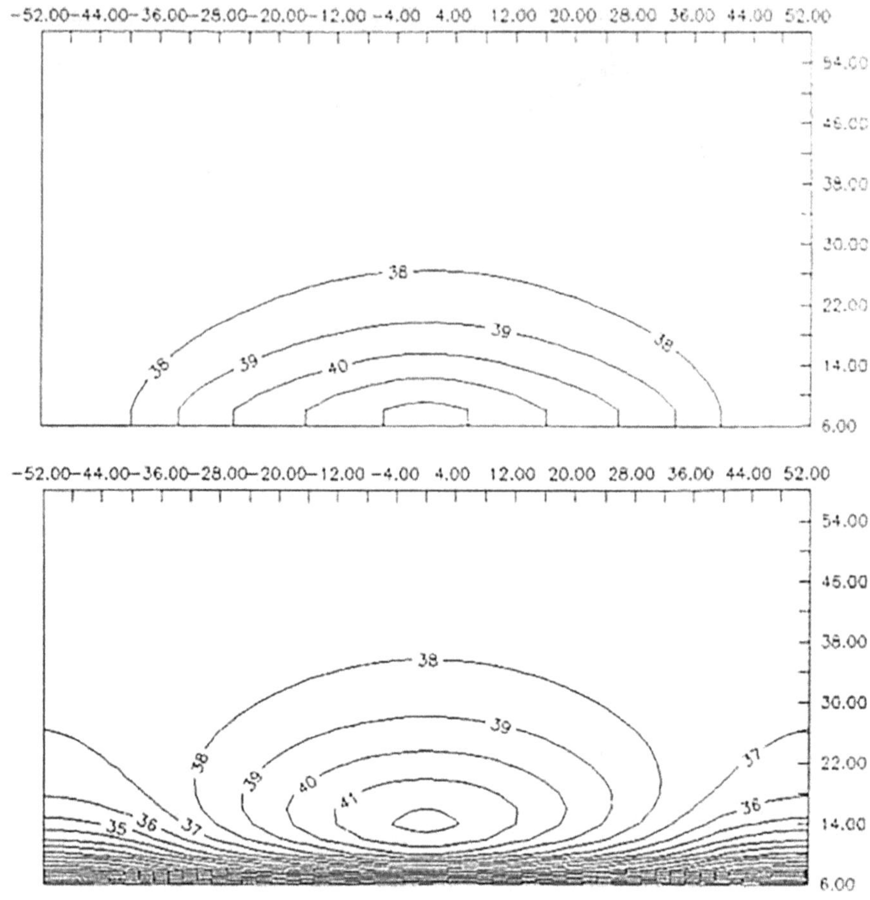

Fig. 19 Isotherms (°C) near the applicator (mm) in the absence (top) and presence (bottom) of cooling, fig. 9 of [22]

m^2 K. It can be observed that the decrease in the coolant temperature produces significant reductions in the temperature at the tissue interface, while the maximum therapeutic temperature values are substantially unchanged, with a slight decrease in the position of the maximum. Such variations are substantially similar to those reported in Fig. 15.

Figure 17 shows the temperature distributions on the plane $z = 0$, for a blood perfusion $m_b = 4.5 \ 10^{-6}$ m^3/kg s, a thickness $d = 0.85$ mm, a coolant temperature $T_{cl} = 10$ °C and a variable flow rate of the coolant. The solid line (upper) is obtained with a coolant flow rate equal to 1.9 L/min, which corresponds to laminar flow, with $Re = 2100$ and $Nu = 4.98$, and for which $U = 118$ W/m^2 K. The dashed line (intermediate) is obtained with a coolant flow rate equal to 3.3 L/min, which corresponds to turbulent flow, with $Re = 3600$ and $Nu = 38$, and for which $U = 156$ W/m^2 K. A decrease in the temperature at the tissue interface of about

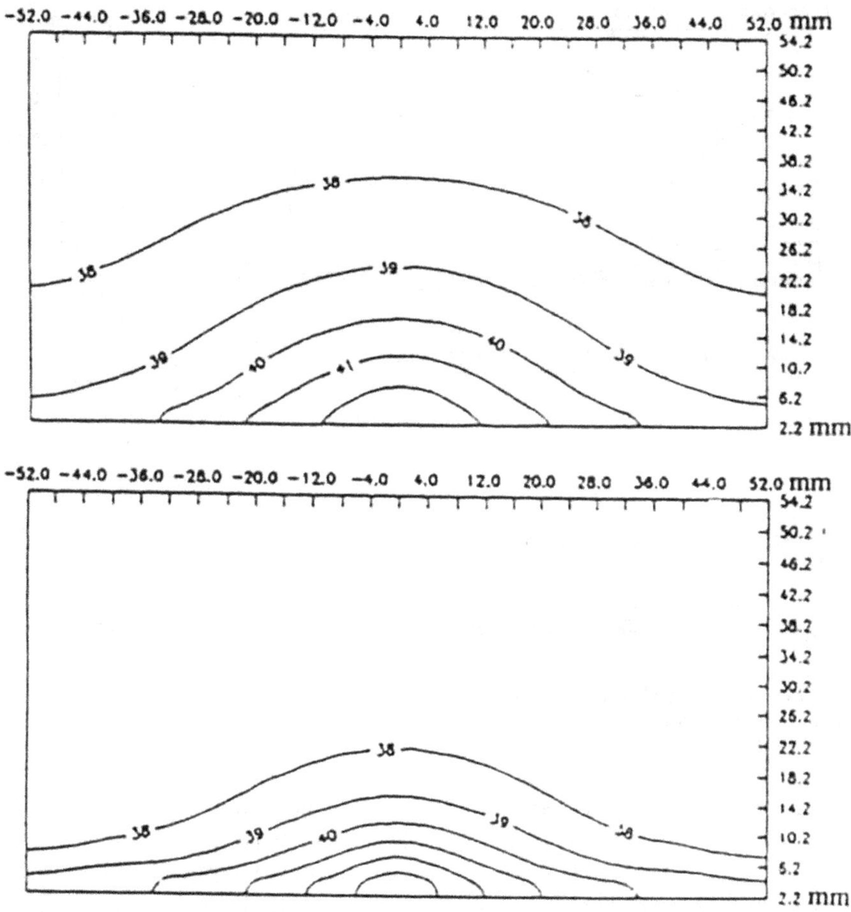

Fig. 20 Isotherms (°C) near the applicator (mm), varying with blood perfusion and without cooling

4 °C and of 1 °C in the maximum temperature value can be observed. A further increase in the value of the forced convection coefficient, h, up to the infinite value, for which $U = 164$ W/m^2 K, does not produce substantial variations in the temperature.

Figure 18 shows the temperature distributions on the plane $z = 0$, for a coolant temperature of $T_{cl} = 10$ °C, a thickness of $d = 0.85$ mm and variable blood perfusion. The solid line (upper) is obtained for a blood perfusion corresponding to a resting muscle, $m_b = 0.45\ 10^{-6}$ m^3/kg s. The dashed line (middle) is obtained with a blood perfusion corresponding to an active muscle, $m_b = 4.5\ 10^{-6}$ m^3/kg s. The dotted dashed line (lower) is obtained for a blood perfusion present in the brain, $m_b = 9\ 10^{-6}$ m^3/kg s. Figure 18 shows, with strong evidence, the dependence of the temperature distribution on blood perfusion.

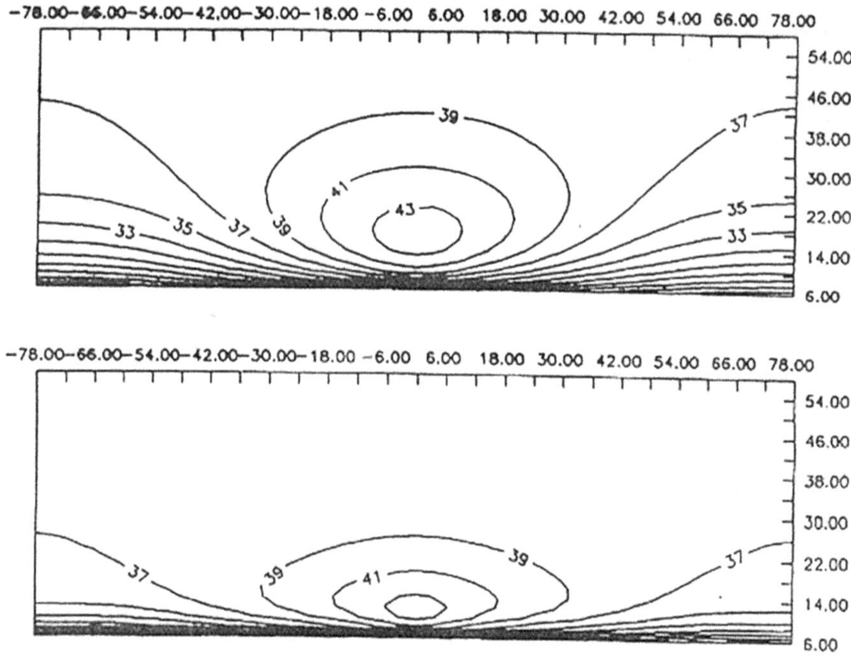

Fig. 21 Isotherms (°C) near the applicator (mm) at $z = 0$, varying with blood perfusion and cooling, fig. 8 of [26]

Figure 19 shows the isotherms obtained in the tissue, starting from the surface of the applicator, with a blood perfusion corresponding to an active muscle, $m_b = 4.5$ 10^{-6} m³/kg s, a thickness of $d = 0.45$ mm, in the absence or presence of a coolant circulation. The assumed values are $\epsilon' = 53$ and $\sigma = 1, 43$. The upper part of Fig. 19 shows the isotherms obtained with a heating power of 9 W, which produces the maximum temperature of about 42.2 °C but without fluid circulation or with stagnant fluid. The lower part of Fig. 19 shows the isotherms obtained with a heating power of 23 W, which produces the same maximum temperature of 42.2 °C, with a cooling fluid of 10 °C and laminar flow. In the presence of coolant circulation, the maximum temperature in the tissue is located about 8 mm from the applicator wall, while, in the absence of circulation, the maximum is on the applicator–tissue interface. The volume of heated tissue, at temperatures between 40 °C and 42.2 °C, increases by about 70%. Furthermore, it can be observed that the temperature in contact with the tissue decreases substantially in the presence of cooling, resulting in less discomfort for the patient involved.

Figure 20 shows the isotherms obtained in the tissue, with $\epsilon' = 53$ and $\sigma = 1.43$, starting from the surface of the applicator, which is equipped with only an insulating sheath and without a cooling system and, therefore, has a thickness of 2.2 mm, with variable blood perfusion. The upper part of Fig. 20 shows the isotherms obtained with a heating power of 2 W, which produces the maximum temperature of about

42–43 °C, for a blood perfusion corresponding to a resting muscle, $m_b = 0.45\ 10^{-6}$ m^3/kg s. The lower part of Fig. 20 shows the isotherms obtained with a power heating of 6 W, which produces a maximum temperature of about 42–43 °C, for a blood perfusion corresponding to an active muscle, $m_b = 4.5\ 10^{-6}$ m^3/kg s. In both cases, the maximum temperature is on the applicator–tissue interface. The volume of tissue heated to temperatures above about 40 °C decreases when the tissue moves from the resting condition to that of an active muscle.

Figure 21 shows the isotherms obtained in the tissue, with $\epsilon' = 53$ and $\sigma = 1.43$, starting from the surface of the applicator, with a thickness of $d = 0.45$ mm, in the presence of a circulating cooling fluid at 10 °C, after 20 min of application, in order to highlight the effects caused by blood perfusion. The upper part of Fig. 21 shows the isotherms obtained with a heating power of 22.5 W, which produces the maximum temperature of about 43.5 °C, for a blood perfusion corresponding to a resting muscle, $m_b = 0.45\ 10^{-6}$ m^3/kg s. The lower part of Fig. 21 shows the isotherms obtained with a heating power of 55.6 W, which produces the same maximum temperature of 43.5 °C, for a blood perfusion corresponding to a brain tissue, $m_b = 9\ 10^{-6}$ m^3/kg s. From the comparison of the two figures, it is clear that the increase in blood perfusion causes, along the entire length of the applicator, an increase in the temperature gradient in the region close to the surface of the outer casing of the applicator. The most significant effect is still the shift in the position of the maximum, with its distance from the surface of the applicator halving from the case with resting muscle to that with brain tissue.

4.1.3 Experimental Apparatus in Muscle-Equivalent Phantom

The experimental apparatus used for the evaluation of the thermal response of radiant applicators in a resting muscle tissue is reported in Fig. 22. The power generation and control system consists of a 432-*MHz* generator and a bidirectional power meter (*BIRD* 4431). The thermostat used for the cooling liquid, type *HAAKE*, includes the circulation pump of the cooling system. The applicator is immersed in the polyacrylamide phantom, inside which two liquid crystal films are calibrated for temperatures between 24 °C and 30 °C. They are positioned to show the isotherms, both on the plane passing through the axis of the dipole and on the one normal to it passing through the feeding gap [27]. The walls of the phantom's parallelepiped container are made of transparent material (plexiglass) to allow visual photographic capture (usually every 60 *s*). In Fig. 22, the colours and corresponding temperatures are also indicated.

Experimental Results

The experimental temperature distributions, obtained in the polyacrylamide phantom with the liquid crystal technique and displayed in the r, z plane of the applicator, are reported in Figs. 23, 24, 25 and 26, at different experimental conditions.

Figure 23 shows the image of the liquid crystals obtained in the case of initial phantom temperature, $T_{in} = 23$ °C, with a cooling liquid temperature, $T_{cl} = 4$ °C; an incident power, $P_{inc} = 29$ W and reference, $P_{rif} = 1$ W, after 5 min from the start of the power application. An ellipsoidal region, orange in colour, of about 26 °C, is evident, which encloses a green zone, of about 27 °C.

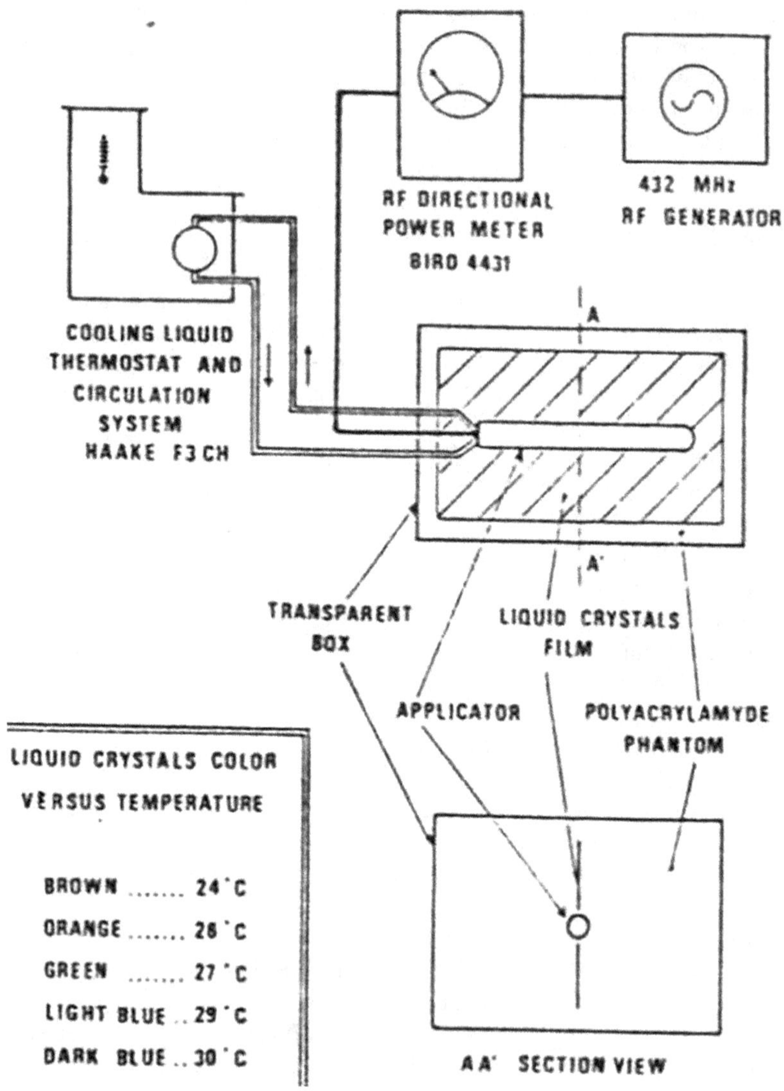

Fig. 22 Experimental apparatus, fig. 2 of [26]

Figure 24 shows the image obtained with the liquid crystals in case of an initial temperature of the phantom, $T_{in} = 23$ °C, with a coolant temperature, $T_{cl} = 4$ °C; an incident power, $P_{inc} = 29$ W and reference, $P_{rif} = 1$ W, after 20 min from the start of power application. There is evidence of an ellipsoidal region of orange colour of about 26 °C, which has expanded, compared to the image after 5 min, Fig. 23, and which encloses a green zone of about 27 °C and a dark blue zone of about 30 °C.

Fig. 23 Liquid crystal temperatures after 5 min for $T_{cl} = 4\,°C$

Fig. 24 Liquid crystal temperatures after 20 min for $T_{cl} = 4\,°C$

Fig. 25 Liquid crystal temperatures after 20 min for $T_{cl} = 20\,°C$

Figure 25 shows the image obtained with the liquid crystals in case of an initial temperature of the phantom, $T_{in} = 24\,°C$, with a coolant temperature, $T_{cl} = 20\,°C$; an incident power, $P_{inc} = 32$ W and reference, $P_{rif} = 1$ W, after 20 min from the start of the power application. The increase in the coolant temperature from $4\,°C$ to $20\,°C$ causes a less ellipsoidal shape, which closes in the extreme sections of the applicator, with an orange zone of about $26\,°C$, a green zone of about $27\,°C$ and a dark blue zone of about $30\,°C$.

Fig. 26 Liquid crystal temperatures after 20 min for $T_{cl} = 30\ °C$

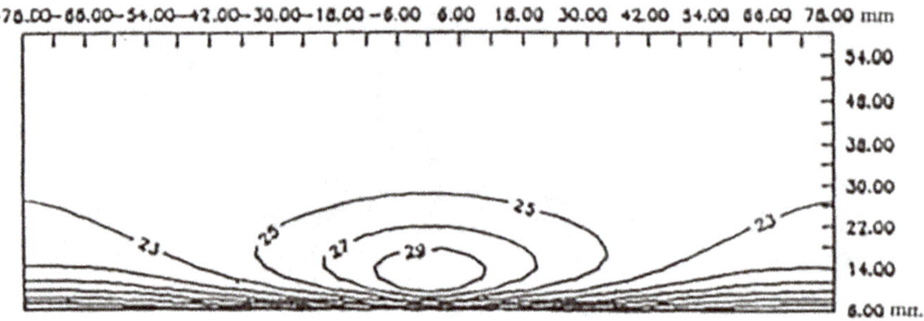

Fig. 27 Numerical temperatures (°C) after 5 min for $T_{cl} = 4\ °C$

Figure 26 shows the image obtained with the liquid crystals in case of an initial temperature of the phantom, $T_{in} = 24\ °C$, with a coolant temperature, $T_{cl} = 30\ °C$; an incident power, $P_{inc} = 32$ W and reference, $P_{ref} = 2$ W, after 20 min from the start of the power application. The increase in the coolant temperature to 30 °C causes a less ellipsoidal shape of the isotherms, with the lower isotherms not closing in the extreme sections of the applicator and an orange zone of about 26 °C, a green zone of about 27 °C and a dark blue zone of about 30 °C.

Numerical Results

The numerical temperature distributions in the r, z plane of the applicator are reported in Figs. 27, 28, 29 and 30, for different experimental conditions.

Figure 27 shows the isotherms obtained under conditions corresponding to the case already described for the experimental isotherms in Fig. 23. The numerical isotherms, between 25 °C and 29 °C, are well-highlighted and in good agreement with the experimental results obtained with the liquid crystals, Fig. 23.

Figure 28 shows the isotherms obtained under conditions corresponding to the case already described for the experimental isotherms in Fig. 24. The numerical isotherms, at 25°C and 27°C, have an ellipsoidal trend in good agreement with the experimental results obtained with the liquid crystals, Fig. 24.

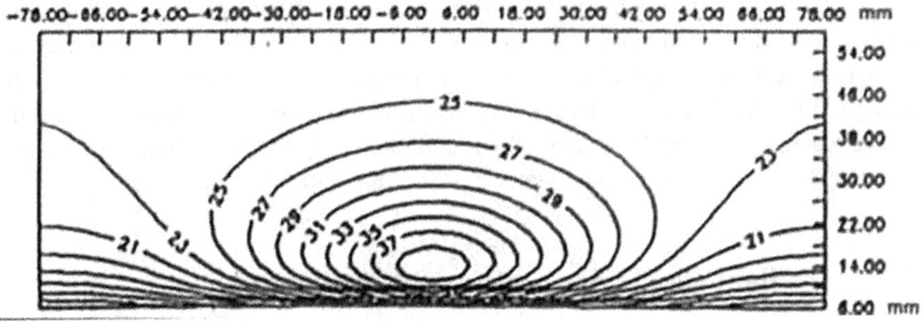

Fig. 28 Numerical temperatures (°C) after 20 min for $T_{cl} = 4\,°C$

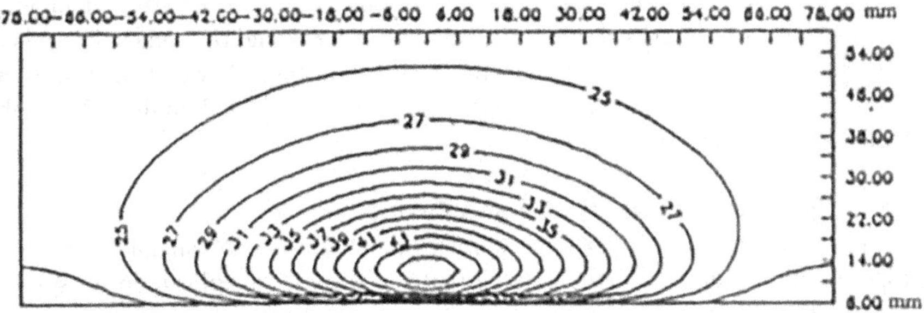

Fig. 29 Numerical temperatures (°C) after 20 min for $T_{cl} = 20\,°C$

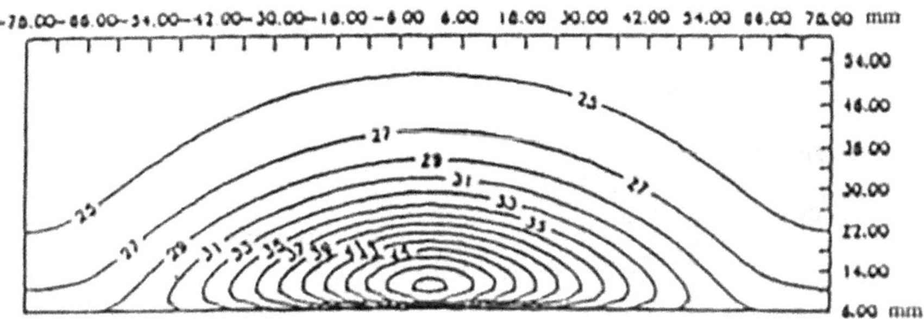

Fig. 30 Numerical temperatures (°C) after 20 min for $T_{cl} = 30\,°C$

Figure 29 shows the isotherms obtained under conditions corresponding to the case already described for the experimental isotherms in Fig. 25. The numerical isotherms, at 25 °C, 27 °C and 29 °C, have a trend in good agreement with the experimental results obtained with the liquid crystals experiments, Fig. 25, even if the curves do not close on themselves but end at the extreme sections of the applicator.

Figure 30 shows the isotherms obtained under the conditions corresponding to the case already described for the experimental isotherms in Fig. 26. The numerical isotherms are in good agreement with the experimental results obtained with liquid crystals, Fig. 26. The most obvious difference is due to the difference between the theoretical model, which assumes an indefinitely extended cooling, and the real applicator, in which the cooling water closes at the end of the applicator.

4.2 Radiant Dipole Applicators with Reflectors

In some endocavitary hyperthermic applications, such as in the treatment of prostate conditions, it is necessary to have an applicator that has the ability to direct the heating and concentrate the power deposition on the pathological tissues, in order to limit the heating of healthy tissues. The simplest, reliable and practical solution is to equip the dipole with a suitable reflector, as shown in Fig. 31, which shows, in the right part, the section perpendicular to the axis of the dipole and, in the left one, the longitudinal section.

4.2.1 Electromagnetic Model and Results

The most suitable *e.m.* model to estimate the directivity obtainable is the one proposed in [33], applied to the thin wire model of the dipole plus reflector system, shown in Fig. 32, where the continuous reflector is replaced by several wires parallel to the axis of the dipole. The main limitation is the difficulty in dealing with the

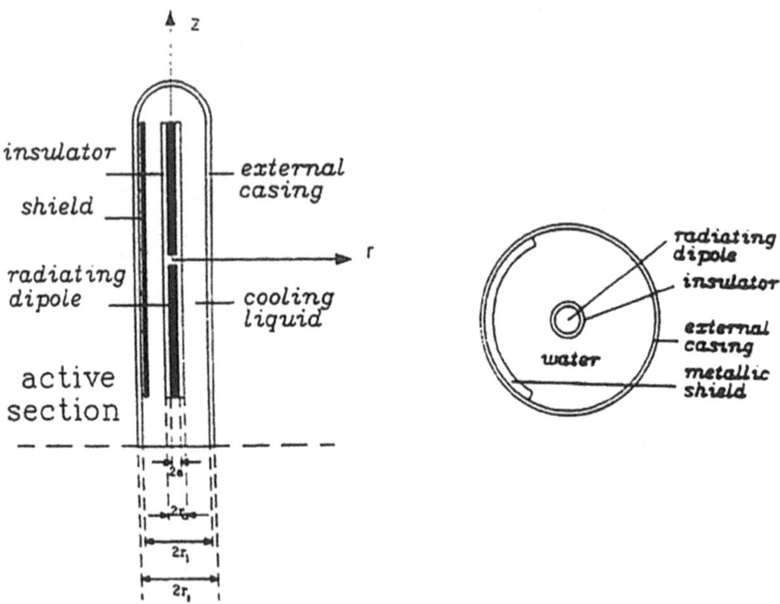

Fig. 31 Radiant dipole applicator with reflector, fig. 1 of [26]

Fig. 32 E.m. model of the applicator with reflector

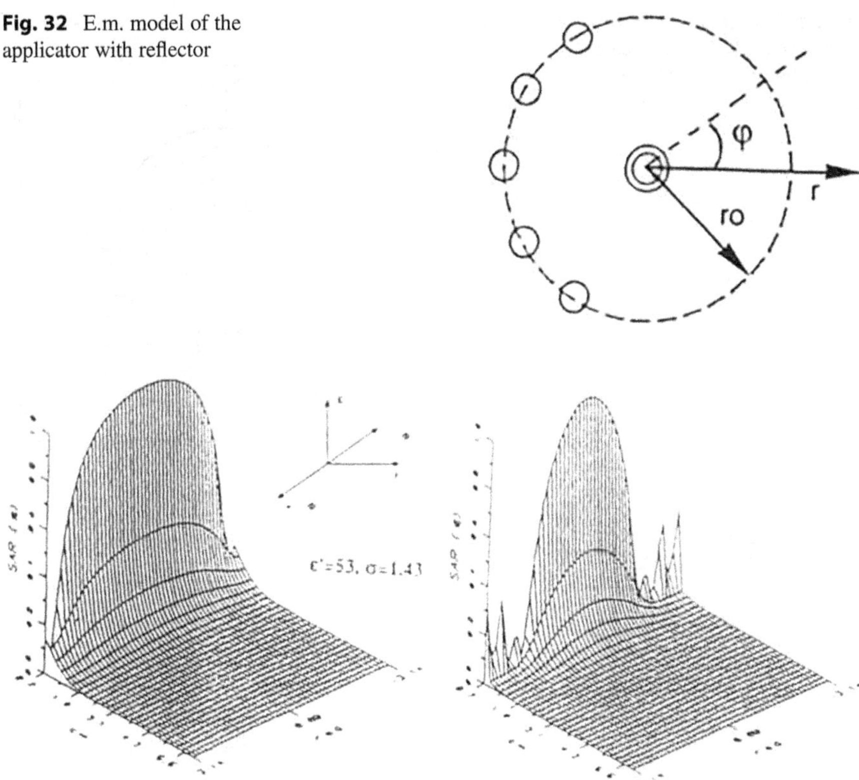

Fig. 33 Normalised absorbed power of the applicator (SAR) with reflector

interface between the cooling liquid and the tissue. The alternative method used is to consider the dipole-reflector system as immersed in a homogeneous tissue, and multiply the input power of the dipole by a corrective factor (also theoretically estimable) that takes into account the power lost in the volume normally occupied by the cooling water.

The left side of Fig. 33 shows, on the plane normal to the dipole axis, $z = 0$, the trends of the normalised absorbed power for an applicator equipped with a reflector consisting of a single thin conductive wire parallel to the dipole itself. The right side of Fig. 33, instead, shows the power for an applicator equipped with a reflector consisting of *five* wires arranged to cover an angle of 120°. In both cases, the filamentary elements that constitute the reflector are located at a distance of 6 mm from the applicator axis, which is immersed in the homogeneous medium, with dielectric properties equivalent to those of a muscle tissue, in the absence of blood perfusion. The reflector, however shaped, causes a reduction of the power deposited in the region of the tissue adjacent to it. Moreover, the larger the angle it covers, the greater is the directivity. The increase in absorbed power is modest in correspondence with the wires.

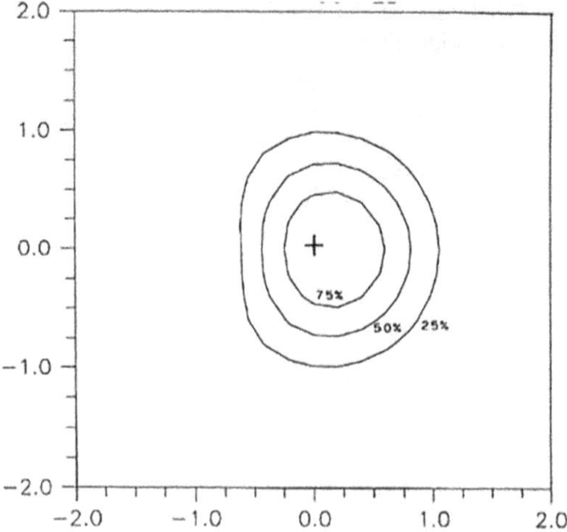

Fig. 34 Numerical absorbed power (IsoSAR) for a reflector consisting of a single wire as a function of distance (cm)

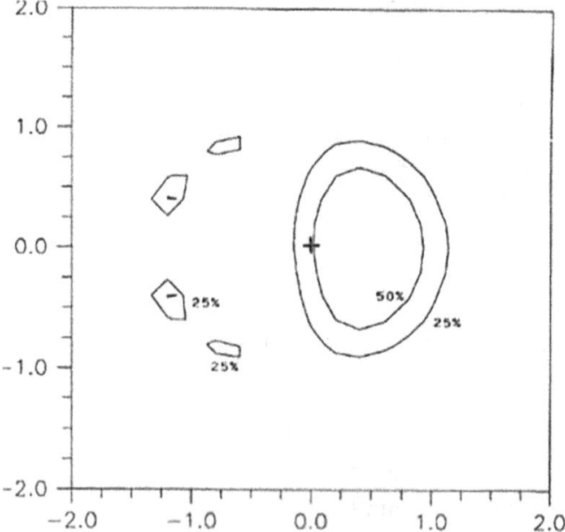

Fig. 35 Numerical absorbed power (IsoSAR) for a reflector consisting of five wires as a function of distance (cm)

The numerical distribution of the absorbed power, in the plane $z = 0$, is shown in Fig. 34 for the two reflectors consisting of a single wire and in Fig. 35 for those consisting of five wires. The values are normalised with respect to the maximum value at a 2-cm distance from the wall of the radiating dipole. The power distribution for the single wire case shows minimum values in the direction opposite to where the reflective wire is placed, showing a certain symmetry. In the case of the five-wire reflectors, there is a more marked directionality and a relative increase in absorbed power in correspondence with the wires. In conclusion, Figs. 34 and 35 show two main *e.m.* effects. A directional distribution due to the use of wire reflectors and a directionality that can be varied with the change in the angular extension of the reflectors.

4.2.2 Experimental Results

The temperature distributions, obtained experimentally in the polyacrylamide phantom and displayed with the liquid crystal technique, in the plane $z = 0$ of the applicator, are reported in Figs. 36, 37 and 38, for various experimental conditions.

Figure 36 shows the image of the liquid crystals in the case of an initial temperature of the phantom, $T_{in} = 21.5$ °C, with a coolant temperature, $T_{cl} = 4$ °C; an incident power, $P_{inc} = 31$ W and reference, $P_{rif} = 0.5$ W, after 5 min from the start of the power application. There is the presence of an orange zone, of about 26 °C, which encloses a small green zone, of about 27 °C. Both are closed curves separated from the applicator and indicate a pronounced directional effect.

Figure 37 shows the image obtained with the liquid crystals in the case of an initial temperature of the phantom, $T_{in} = 21.5$ °C, with a coolant temperature, $T_{cl} = 4$ °C; an incident power, $P_{inc} = 31$ W and reference, $P_{rif} = 0.5$ W, after 20 min from the start of the power application. There is a presence of an orange zone, of about 26 °C, which encloses a thin green zone, of about 27 °C, and a more extensive light and dark blue zone, of about 29 °C. All zones are separated from the applicator, indicating a pronounced directional effect. The low temperature of the coolant keeps the surface of the applicator at much lower temperatures than the maximum values but acceptable from a medical–biological point of view.

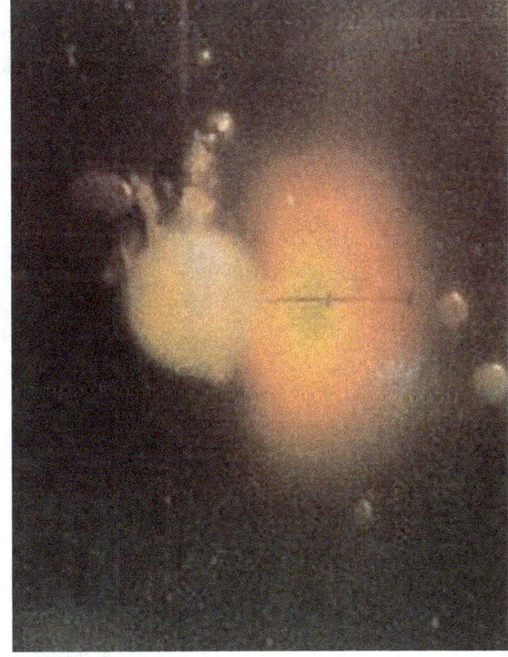

Fig. 36 Liquid crystal temperatures in the polyacrylamide phantom after 5 min for $T_{cl} = 4$ °C

Fig. 37 Liquid crystal temperatures in the polyacrylamide phantom after 20 min for $T_{cl} = 4\,°C$

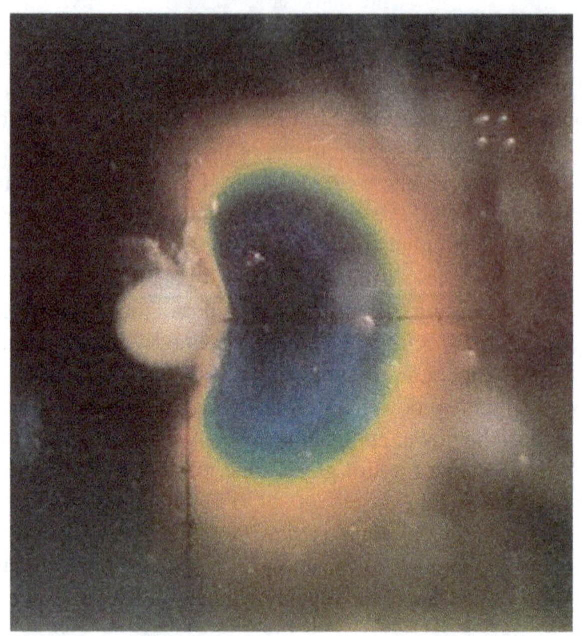

Fig. 38 Liquid crystal temperatures in the polyacrylamide phantom after 20 min for $T_{cl} = 20\,°C$

Fig. 39 Liquid crystal temperatures in the polyacrylamide phantom after 20 min for $T_{cl} = 30\,°C$

Figure 38 shows the image obtained with liquid crystals in the case of an initial temperature of the phantom, $T_{in} = 20.1\,°C$, with a cooling liquid temperature, $T_{cl} = 20\,°C$; an incident power, $P_{inc} = 27.6$ W and reference, $P_{rif} = 0.6$ W, after 20 min from the start of the power application. There is an orange zone of about 26 °C, a green zone of about 27 °C and a light and dark blue zone of about 29–30 °C. The increase in the cooling liquid temperature from 4 °C to 20 °C causes the isotherms to no longer be closed and separated from the applicator surface.

Figure 39 shows the image obtained with liquid crystals in the case of an initial temperature of the phantom, $T_{in} = 19.8\,°C$, with a cooling liquid temperature, $T_{cl} = 30\,°C$; an incident power, $P_{inc} = 31.5$ W and reference, $P_{rif} = 1$ W, after 20 min from the start of the power application. The increase in the cooling liquid temperature up to 30 °C causes the presence of orange zones of about 26 °C and green zones of about 27 °C, even in the region adjacent to the reflector. The curves have an almost circular shape, and the applicator is located inside them.

Figure 40 shows the image obtained with liquid crystals in the case of a single wire reflector. The initial temperature of the phantom is $T_{in} = 20\,°C$, with the cooling liquid temperature, $T_{cl} = 20\,°C$, and the image is taken after 20 min from the start of the power application. The temperature profile shows the presence of orange zones of about 26 °C, green zones of about 27 °C and blue zones of about 30 °C, only in the upper part, confirming the directional character of the *e.m.* power distribution.

Fig. 40 Liquid crystal temperatures in the polyacrylamide phantom after 20 min, for $T_{cl} = 20\,°C$ in an applicator with a single reflective wire

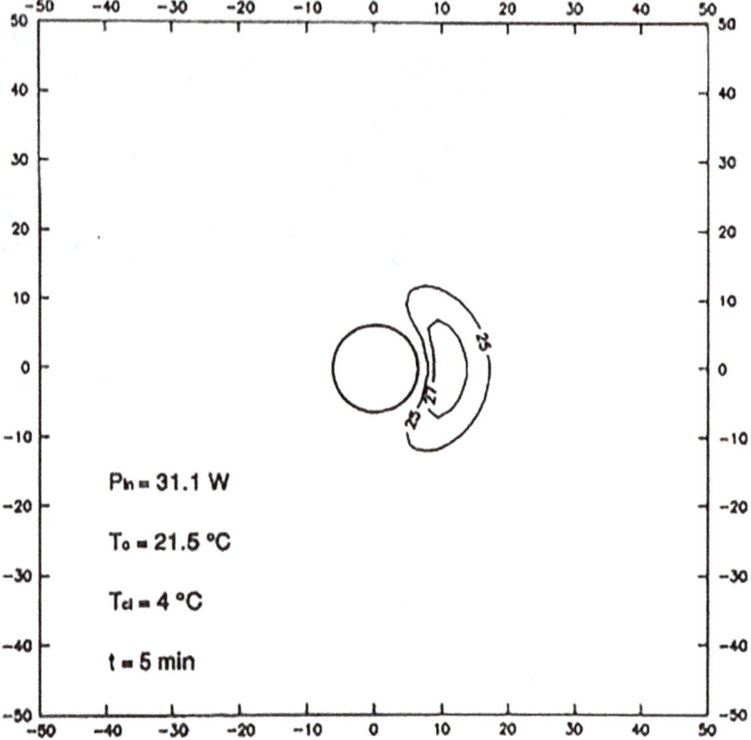

Fig. 41 Numerical temperatures (°C) after 5 min for $T_{cl} = 4\,°C$

4.2.3 Numerical Results

The temperature distributions obtained numerically, in the r, z plane of the applicator, are reported in Figs. 41, 42, 43 and 44 for different experimental conditions.

Figure 41 reports the numerical isotherms obtained in the conditions corresponding to the case already described for the experimental isotherms in

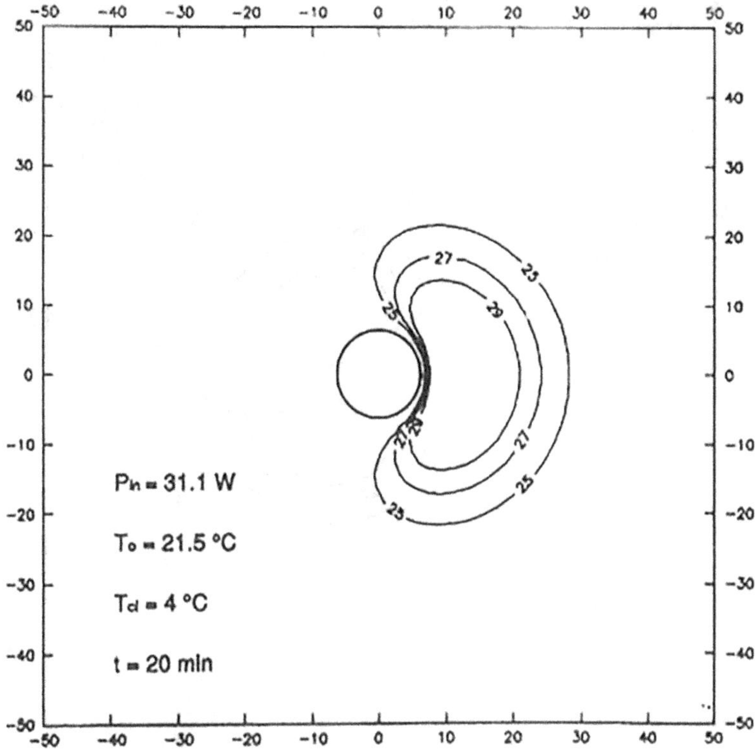

Fig. 42 Numerical temperatures (°C) after 20 min for $T_{cl} = 4\,°C$

Fig. 36. The numerical isotherms, between 25 °C and 29 °C, are well-highlighted and in good agreement with the experimental results obtained with the liquid crystals, Fig. 36.

Figure 42 reports the numerical isotherms obtained in the conditions corresponding to the case already described for the experimental isotherms in Fig. 37. The numerical isotherms, between 25 °C and 27 °C, have an ellipsoidal trend in good agreement with the experimental results obtained with the liquid crystals, Fig. 37.

Figure 43 reports the numerical isotherms obtained in the conditions corresponding to the case already described for the experimental isotherms in Fig. 38. The numerical isotherms at 25 °C, 27 °C and 29 °C have a trend in good agreement with the experimental results obtained with the liquid crystals, Fig. 38, even if the curves do not close on themselves but end in correspondence to the extreme sections of the applicator.

Figure 44 shows the numerical isotherms obtained under conditions corresponding to the case already described for the experimental isotherms in Fig. 39. The numerical isotherms are in good agreement with the experimental

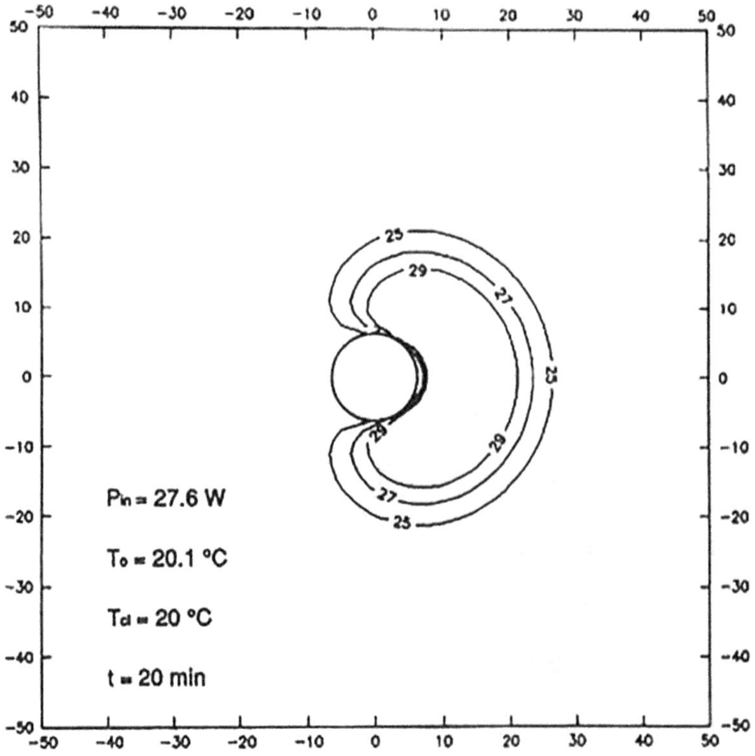

Fig. 43 Numerical temperatures (°C) after 20 min for $T_{cl} = 20\,°C$

results obtained with liquid crystals, Fig. 39. The most evident difference is due to the difference existing between the theoretical model, which assumes an indefinitely extended cooling, and the real applicator, in which the cooling water closes at the end of the applicator.

5 Conclusions

The conclusions on microwave heating concern the advantages that can be obtained, which are the reduction of processing times, resulting from a higher thermal efficiency compared to traditional systems; the reduction of raw material costs, for the same final quality of the product; the reduction of deformation processes and of the volume of the finished product; the reduction of the price/quality ratio of the treated product, once the plant costs have been amortised; the possibility to instantly control the energy levels involved in the heating process; the flexibility of insertion in a composite heating system and the absence of combustion products.

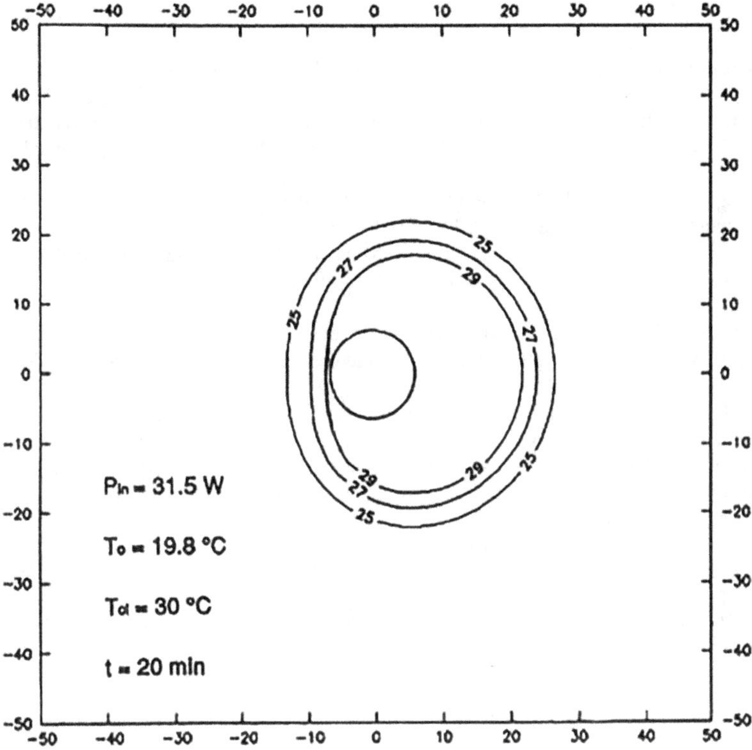

Fig. 44 Numerical temperatures (°C) after 20 min for $T_{cl} = 30\,°C$

The conclusions on endocavitary and interstitial applicators concern the interest in the hyperthermic treatment of pathologies of organs or tissues not easily accessible from the outside. The depth of the tissues involved exacerbates the problem of temperature monitoring, which is not yet solved, and increases the risk that the treatment is carried out in non-optimal conditions, at temperatures too low or too high. The thermal and electromagnetic models developed, in addition to being useful tools for analysis and optimisation available to the designer, can be used for adaptive temperature control for the estimation of some unknown parameters varying from subject to subject or within the same hyperthermic application, such as blood perfusion.

Acknowledgements for Figures The author thanks the **publisher Springer-Nature** for the kind permission to reproduce images 2 and 5 of the reference [21].

The author thanks the **publisher IEEE** for the kind permission to reproduce the images 1, 2, 3, 4, 5, 6, 7, 8 and 9 of the reference [22] and the images 1, 2, 7 and 8 of the reference [26].

References

1. Okress EC (1968) Microwave power engineering. Academic, New York
2. Hippel V (1961) Dielectric materials and applications. MIT Press, Cambridge, MA
3. Martini L (1983–84) Thermodynamic model of microwave heating for layered flat media. Application to the restructuring of motorway pavements, Degree Thesis in Electronic Engineering, University of Florence, A. A
4. Wang JR, Smugge TJ (1980) An empirical model for the complex dielectric permittivity of soils as a function of water content. IEEE Trans Geosci Remote Sensing 18:288
5. Shurko AM, Reutov EM (1982) Mixture formulas applied in estimation of dielectric and radiative characteristics of soils and grounds at microwave frequencies. IEEE Trans Geosci Remote Sensing 18(4):29–32
6. Gori F et al (1984) Heat and mass transfer in soils during a microwave heating process. In: 22nd ASME-AIChE National Heat Transfer Conference, Niagara Falls (USA), ASME paper 84-HT-78
7. Gori F et al (1987) Microwave heating of porous media. J Heat Transfer ASME 109(2):522–525
8. Wait JR (1953) Propagation of radio-waves over a stratified ground. Pergamon Press
9. Luikov AV (1966) Heat and mass transfer in capillary-porous bodies. Pergamon Press
10. Corey AR (1977) Mechanics of heterogeneous fluids in porous media. Water Resources Publications, Fort Collins
11. Gori F (1983) Prediction of effective thermal conductivity of two-phase media (including saturated soils). In: XVI International Congress of Refrigeration, Paris, B1-26, pp 525–531
12. Gori F (1983) A theoretical model for predicting the effective thermal conductivity of unsaturated frozen soils. In: The 4th international conference on permafrost, Fairbanks, AK, U.S.A, pp 363–368
13. Gori F et al (2002) Theoretical prediction of the thermal conductivity of soils at moderately high temperatures. J Heat Transfer, ASME 124(6):1001–1008
14. Schubert G, Strauss JM (1979) Steam-water counter-flow in porous media. J Geophys Res 84 (B4):1621–1628
15. McWorther D, Suwada DK (1977) Ground-water hydrology and hydraulics. Water Resources Publications, Fort Collins
16. Patankar SV (1980) Numerical heat transfer and fluid flow. McGraw Hill, New York
17. Botto M, Neri F (1981) Experimental research for the use of microwaves in the regeneration of bituminous conglomerates. In: Seminar on recycling of pavement materials, Road Research Program of the Organization for Economic Cooperation and Development (OECD), Rome, March 1981
18. https://it.wikipedia.org/wiki/Atmosfera_terrestre#/media/File:Atmosfeer_Atmosfera_-_Italiano.png
19. Gori F et al (2012) Lezioni di Termofluidodinamica, (Lectures in Thermofluidodynamics). TEXmat
20. Ruffino G et al (1996) The temperature sensor on the Huygens probe for the Cassini mission: design, manufacture, calibration and tests of the laboratory prototype. Planet Space Sci 44(10): 1149–1162
21. Fulchignoni M et al (2005) In situ measurements of the physical characteristics of Titan's environment. Nature 438. https://doi.org/10.1038/nature04314
22. Biffi Gentili G et al (1991) Electromagnetic and thermal models of a water cooled dipole radiating in a biological tissue. IEEE Trans Biomed Eng 38(1):98–103
23. King RW, Trembly BS, Strohbehn JW (1983) The electromagnetic field of an insulated antenna in a conducting or dielectric medium. IEEE Trans Microwave Theory Technol 31:574–583
24. Casey JP, Bansal R (1986) The near field of an insulated dipole in a dissipative dielectric medium. IEEE Trans Microwave Theory Technol 34:459–463

25. Biffi Gentili G et al (1991) Analisi del comportamento elettromagnetico e termico di applicatori interstiziali ed endocavitari (Analysis of the electromagnetic and thermal behaviour of interstitial and endocavitary applicators). Microonde – SMA 62–96:10
26. Biffi Gentili G et al (1991) A water-cooled EM applicator radiating in a phantom equivalent tissue-experiments and numerical analysis. IEEE Trans Biomed Eng 38(9):924–928
27. Biffi Gentili G et al (1990) Liquid crystals experiments and numerical predictions of heat conduction around a liquid cooled electromagnetic dipole. In: Ninth international heat transfer conference, Jerusalem, Israel, vol 5, pp 485–490
28. Biffi Gentili G et al (1992) Thermal model of an intracavitary hyperthermic microwaves applicator. ASME HTD 231:27–34. ASME – WAM, Advances in Biological Heat and Mass Transfer, Editor: J.J. McGrath, Book N. G00755 – 1992
29. Biffi Gentili G et al (1994) A dipole-type Intracavitary Hyperthermic applicator with a metallic reflector: experiments and theoretical analysis. Int J Hyperthermia 10(2):175–187. ISSN: 0265-6736
30. Ozisik N (1975) Heat conduction. McGraw-Hill, New York
31. Pennes HH (1948) Analysis of tissue and arterial blood temperatures in the resting human forearm. J Appl Physiol 1:93–122
32. Lundberg RE, McCuen PA, Reynolds WC (1963) Heat transfer in annular passages: hydrodynamically developed laminar flow with arbitrarily prescribed wall temperature or heat transfer. Int J Heat Mass Transfer 6:495–529
33. Richmond JH (1969) Computer analysis of three-dimensional wire antennas, Report 2708-4. Ohio State University Electroscience Laboratory, Department of Electrical Engineering

Fabio Gori Ammannati was born in Montale (Pistoia) on 5 August 1947 to Emilio Gori and Cesarina Ammannati. On 16 July 2021, he added his mother's surname, Ammannati. He graduated in Chemical Engineering on 11 November 1971 with honours. In December 1971, he won a scholarship for young graduates at the Faculty of Engineering in Bologna, which he undertook from January 1972. In 1974, he became assistant professor and, in 1982, associate professor of Technical Physics at the Faculty of Engineering of the University of Florence, where he taught, also as a contract professor, until 1990. In 1978, he won a scholarship from the *British Council*, which he carried out at the *Imperial College* in London, where he collaborated with *Prof. D. Brian Spalding* on the topic of numerical analysis in the turbulent flow of liquid metals. During his time at the University of Florence, he established international scientific collaborations with *Prof. R. Echigo, Tokyo Institute of Technology, Tokyo, Japan; Prof. D.R. Chaudhary, Department of Physics, University of Rajasthan, Jaipur, India;* International Centre for Theoretical Physics *(ICTP)* in Trieste; *Centralny Osrodek Techniki Medycznej, Warsaw, Poland and Prof. T. Aihara, Institute of Fluid Science, Sendai, Japan.* In 1986, he won a CNR scholarship, which he carried out at *Cornell University, Ithaca, New York,* where he collaborated with *Prof. R. Miller* on the topic of ground freezing. In the same year, he became professor in the Faculty of Engineering at the University of Reggio Calabria. In 1988, he was appointed as *professor* at the *University of New York at Stony Brook,* teaching the course, *Introduction to Fluid Dynamics,* in the autumn semester of the same year. During his stay, he collaborated with *Prof. T.F. Irvine Jr.* on the method of measuring thermal conductivity with a thermal probe and on the measurement of the isobaric thermal expansion coefficient of non-Newtonian fluids. In 1990, he moved to the Milano Polytechnic, where he taught Technical Physics and Systems until 1993. From 1991 until 1998, he was an adjunct professor at the Faculty of Engineering of the University of Siena. From 1992 to 2017, he was a professor of Technical Physics in the Faculty of Engineering at the University of Rome Tor Vergata, and from 2017, he is a contract professor. In 1994, he proposed the Research Doctorate (Italian *Ph.D.*) in Energy-Environment Engineering, of which he was the coordinator until 2011. In 1995, he proposed and directed the second-level Master's in Thermo-fluid-dynamics until 2016. From 1998 to 2001, he was the director of the Department of Mechanical Engineering. In 2001, he proposed and coordinated the programme, *Master of Science, Energy Engineering and Thermal* *and Fluid Dynamics,* in collaboration with the *Department of Mechanical Engineering, College of Engineering, University of Illinois at Chicago, USA,* which takes place at the University of Rome Tor Vergata but awards the title of *Master of Science* from the same American University. Since the late 90s, he has been a coordinator of the joint *Ph.D.* research programme with *Prof. R.J. Goldstein, Department of Mechanical Engineering, University of Minnesota, Minneapolis, USA,* and with *Prof. J.P. Hartnett, Prof. L. Kennedy, Prof. W. Minkowycz and Prof. W.M. Worek, Department of Mechanical Engineering, College of Engineering, University of Illinois at Chicago, Chicago, USA.* Since the mid-90s, he has proposed and directed the *Socrates–Erasmus* programme with *Prof. Mayinger, Technical University of Munich, Germany; Prof. van Steenhoven, Eindhoven Technical University, the Netherlands and Prof. C. Caro, Imperial College of Science, Technology and Medicine.* During his stay at Tor Vergata, he established scientific collaborations with the *Department of Mechanical Engineering, University of Minnesota, Minneapolis, USA,* collaborating with *Prof. R.J. Goldstein* on Thermo-fluid-dynamics and mass transport in gas turbine blades; the *Energy Resources Center, University of Illinois at Chicago,* collaborating with *Prof. J.P. Hartnett, Prof. W. Minkowycz* and *Prof. W.M. Worek*; the *Department of Mechanical Engineering, College of Engineering, University of Illinois at Chicago, collaborating with Prof. L. Kennedy and Prof. W.M. Worek; Duke University, collaborating with Prof. A. Bejan* and *S. Mary's University, Halifax, Nova Scotia, Canada, collaborating with Prof. W. R. Tarnawski.* The bibliographic review of documents and citations, titled *PlosBiology Career,* places him within the top 2% of researchers in the field of *Mechanical Engineering and Transports.* Since 1992, he has been a tutor for over 30 PhD theses, supervisor for over 40 second-level university Master's theses, supervisor for over 70 degree theses (five-year, Specialist and Master's), supervisor for over ten *Master's degree in Mechanical Engineering* from the *University of Illinois at Chicago* and supervisor for over 20 theses *Erasmus-Socrates.*

Since the early 2000s, he has been a reviewer of international research projects on behalf of *Portuguese Science and Technology Foundation (FCT); Czech Science Foundation (GACR); European Science Foundation, Strasbourg, France,* on behalf of *FCT* and the *Shota Rustaveli National Science Foundation, Georgia.* Since 2017, the year of his retirement due to age limits, he has been a contract professor at the University of Rome Tor Vergata, where in the academic year 2024–2025, he has taught Technical Physics for the Master's Degree Course in Medical Engineering. M.D. Salvatore Mangiafico (Florence), scientific head of the NeuroVascular Base Camp, 2021, invited him to give a scientific presentation titled *"The engineering approach to the study of the Circle of Willis"* on 23 September 2021 at the Congress Centre, University of Rome La Sapienza, thus opening up a possible future collaboration.

Human Body Thermoregulation

Human Body Thermoregulation

Ivan Di Venuta and Fabio Gori Ammannati

1 Thermoregulation Under Physiological Conditions

Thermoregulation is a physiological mechanism by which the body reacts to external thermal events in order to maintain its internal temperature within an appropriate range. The activation of the system can be triggered by various environmental factors (changes of temperature, humidity, heat sources, etc.), but also by exposure to electromagnetic radiation. The ability of the body to adapt to thermal events is present in the majority of living beings but reaches a high degree of perfection only in homoeo-thermic animals because they are equipped with a system of involuntary regulation based on the balance of two thermal phenomena: the production and dispersion of heat [1, 2].

In human beings, most vital organs operate, under normal conditions, at a temperature of around 37 °C, and in any case, within the range between 35.5 °C and 40 °C, no damage to the organism is detected. Temperature variations can occur due to physical exercise, age, emotional stress, digestion, alterations in heart rate, ambient temperature or be cyclical in women. Outside the limits of the considered range, the body is in conditions of extreme vulnerability (at low temperatures, there is a risk of freezing and at high temperatures, the coagulation of some proteins), so the thermoregulatory system is equipped with various control systems. A subject in normal conditions produces energy and dissipates it, which implies that, if temperature variations occur during a day, something must intervene to keep it constant.

I. Di Venuta
Rome, Italy

F. G. Ammannati (✉)
University of Rome Tor Vergata, Rome, Italy
e-mail: gori@uniroma2.it

© The Author(s), under exclusive license to Springer Nature Switzerland AG 2025
F. Gori Ammannati (ed.), *Thermo-Haemo-Dynamics in Medical Engineering*,
https://doi.org/10.1007/978-3-031-97214-0_4

The equilibrium can be expressed, in mathematical terms, by the energy balance equation:

$$M \pm W = \pm R \pm C \pm E \pm S \qquad (1)$$

The major producer of heat within the subject is the metabolic process (M): cells, to perform their functions, must oxidise food (indeed the first effect that is detected when a hypoglycaemic crisis occurs is a sensation of cold, and, having no more sugars to burn, it results in the inability to produce sufficient thermal energy). The energy provided by metabolism is partly converted into work and partly into internal energy (with greater blood flow to the muscles). In reality, the human body is a thermal machine whose maximum efficiency is around 20% (for well-trained athletes). Consequently, for 10 W/ m^2 of work produced at least 40% is given to the outside, or dispersed by sweating, to avoid an increase in internal temperature. The work can have a positive or negative sign. It is negative, for example, when braking downhill or when lowering weights from a shelf (transformation of potential energy into internal energy, which must be disposed of towards the outside, in addition, in this case, to metabolism). Due to the difficulty of evaluating this term, it is generally incorporated into metabolism and indicated with $M \pm W$.

On the right side of Eq. (1), there is a series of terms related to the processes that serve to compensate for the body's energy production. One way to exchange heat is thermal radiation (R), when the human body radiates outwards in infrared frequencies. A second method is the convective one (C), when the air, in contact with the body, which is normally at a higher temperature, follows convective motions directed upwards, which cool it. The term (E) is the latent heat given off by the evaporation of the skin's water or, more precisely, of the cells and subcutaneous capillaries. There are two mechanisms by which the water contained within the organism evaporates from the skin, namely perspiration, E_d (diffusion of vapour through the skin), and the evaporation of sweat, E_{sw} (sweat). It is also necessary to take into account the heat lost through respiration and due to the difference in temperature and humidity between the inhaled and exhaled air. The last term, S, relates to any energy storage.

Most of the systems involved in thermoregulation are connected to the surface and/or volume of the subject under examination; for this reason, the quantities involved in Eq. (1) are expressed in *Watts* (if the whole subject is considered) or in W/m^2 (if the quantities per unit of surface are considered). Moreover, these quantities are to be understood as temporal averages during the observation interval and at the thermal equilibrium. The production of thermal energy by a human being, at physical and mental rest, after at least 12 h from meals, and in a thermally neutral environment (33 °C), is far from being negligible and comes from the actual combustion process. It is interesting to note how each cell can be thought of as a micro–combustion chamber, and in total, for a man of the standard weight of 70 Kg, the oxygen consumption is about 250 ml/min, and the basal metabolic rate (*BMR*) is 1.2 W/Kg. However, this value is indicative because the ability to produce energy is a function of the biological state of the person but can be influenced by weight, diet, and very often, by endocrine and hormonal functions. Most of the metabolic heat is produced by the trunk, viscera, and the brain and transferred through the blood to the rest of the body. This is valid in the situation of immobility; indeed, even the mere movement of the limbs can contribute to increasing the body temperature. The average value of metabolic heat production is about 84 W, but this value can oscillate between 40 and 800 W/m^2 (or from 1 to 21 W/Kg), depending on the age, sex, size, and level of physical activity [3].

Physiological responses are triggered by the hypothalamus, which is the regulatory centre of the human body and receives afferents from all over the body through peripheral receptors located on the skin and through receptors located in the hypothalamus itself and in the large vessels. These thermo-receptors allow the mediation of information on the thermal state of each part of the body. The hypothalamus, therefore, processes the information received and is able, under physiological conditions, to coordinate the most appropriate response. As the ambient temperature varies, there can be three possibilities for an area below the cold critical temperature, one above the hot critical temperature, and an intermediate, or neutral, one. In the cold critical zone, the metabolic aspect prevails; in the intermediate zone, a vasomotor control prevails; and in the hot critical zone, there is, at first, a constant increase in the sweating and blood flow, and then, with increasing temperature, there begins an increase in metabolic production due to the fatigue of the thermoregulatory system.

A change in external temperature, when perceived by the central nervous system, induces an alteration of the muscular system, which manifests itself in a dilation or constriction of the blood vessels. This phenomenon occurs because the blood collects heat in the innermost parts of the body and transports it to the peripheral areas so that, in case of cooling, peripheral vaso-constriction reduces the influx of blood to the peripheral parts to avoid, as much as possible, the dispersion of heat. Conversely, if subjected to heating, vaso-dilation allows greater heat dispersion. The vasomotor control system is of great importance because it allows us to detect localised temperature variations, even far from the sensors, and transmit them to the whole body. During physical effort, the state changes significantly, and the blood at the outer periphery of the body can increase up to *ten* times to achieve greater dissipation of heat produced by the muscles due to the increase in local metabolic activity, and therefore, the thermoregulatory system must intensify its work.

In the human body, the temperature is not the same in all its parts. At points where there is the possibility of heat transfer, the temperature is lower than that of the innermost parts. The body is divided into an inner and an outer shell. In the inner shell, or *central core*, all the vital organs are located, which must be maintained at around 37–37.5 °C. The thickness of the outer shell is not regulated within narrow limits: it depends on the environment and the body's need to maintain temperature. The central core increases in size during exposure to heat due to peripheral vaso-dilation, as blood is an excellent heat carrier. If, instead, we expose the body to the cold, the central core significantly reduces its size, limiting it to the abdominal part only due to peripheral vaso-constriction.

2   Thermoregulation During Anaesthesia

General anaesthesia, both inhalation and intravenous, causes an alteration of the hypothalamic thermoregulatory centre, which results in the lack of thermoregulatory response of the body, up to significantly higher or lower temperatures than the physiological reference value. At the induction of general anaesthesia, the greatest drop in the temperature of the central core is noted: the reduction of the physiological threshold value for vaso-constriction results in a decrease of the tonic ortho-sympathetic discharge on the arteriolar system, with redistribution of heat from the hotter centre to the colder periphery. In practice, the central temperature drops, and the peripheral temperature rises, following the thermal gradient preset by the hypo-thalamus. After this initial phase, the central temperature continues to decrease due to the phenomena of environmental heat dispersion (conduction, convection, evapo-ration, and radiation) until the third or fourth hour of anaesthesia. At this point, the central temperature stabilises (phase of *plateau*): i.e. a dynamic equilibrium is created between heat production and heat loss. In general, we can say that after 3 or 4 h of anaesthesia, a decrease of about 3 °C is reached.

A mild degree of controlled hypothermia can be helpful in some surgical specialties, such as neurosurgery, cardiac surgery, and vascular surgery, or when one wants to protect tissues from ischaemic damage. However, accidental hypother-mia is always harmful, causing complications, both intra- and post-operative. During the intraoperative period, hypothermia is accompanied by a decrease in the *clear-ance* of anaesthetic drugs and curare, as well as an alteration of coagulation and blood count. In the post-operative period, hypothermia contributes to increasing the incidence of desaturation and myocardial ischaemic accidents. Hypothermic patients may also have a higher incidence of surgical wound infections, attributable to the increase in plasma cortisol concentration and the reduced production of inflamma-tory factors. Finally, the reduction in central temperature leads to the prolongation of awakening times from general anaesthesia and discharge from the *recovery room*, in addition to higher degrees of *discomfort* and post-operative pain. A room tempera-ture between 24 °C and 27 °C, with a relative humidity greater than 40%, is recommended. Insulation with reflective sheets can reduce heat loss (by thermal radiation), especially if at least 50% of the body surface is covered (an unusual situation in children). Given the importance of heat loss through evaporation, during anaesthesia, the use of humidifying filters is recommended, as well as rotating circuits with soda lime. Fluid heaters are recommended in case of the administration of large quantities.

The problem of hypothermia during anaesthesia remains a problem that is not yet fully solved. In some subjects, the problem diametrically opposite to hypothermia can occur, that is, malignant hyperthermia, which is a pharmacogenetic syndrome of skeletal muscles (with cases from 1/5000 to 1/50,000–100,000 anaesthetics), which is rare but potentially lethal. It can occur in genetically predisposed individuals, generally during surgical procedures, as a severe reaction to the administration of certain classes of drugs used for general anaesthesia, such as halogenated anaesthetic gases (halothane, isoflurane, sevoflurane, desflurane, and enflurane) and depolarising neuromuscular blockers, such as succinylcholine. Malignant hyperthermia crises are triggered by an uncontrolled increase in calcium levels in the fibres of the skeletal muscle. This initial phenomenon, in turn, causes an increase in oxidative metabolism, a massive and generalised contraction of the body musculature, and an uncontrolled and progressive increase in temperature. Affected individuals also experience an increase in carbon dioxide (hypercapnia) and potassium (hyperkalaemia) in the blood, a decrease in oxygen (hypoxaemia), and the destruction of many muscle fibres (rhabdomyolysis). If not promptly treated, it can cause cardiac arrest and, therefore, prove fatal or at least give rise to significant clinical consequences (renal failure, cerebral ischaemia, etc.).

An acute episode of malignant hyperthermia is successfully treated, thanks to diagnostic promptness, the immediate suspension of the administration of the triggering agent (the halogenated anaesthetic or the muscle relaxant), and the treatment of metabolic abnormalities (hyperventilation with oxygen at 100%). Therefore, continuous monitoring of temperature is requested during the anaesthetic treatment for a quick diagnosis of malignant hyperthermia and in order to measure heat loss in newborns or in older people during prolonged anaesthetic treatment or, again, for the control of induced hypothermia.

In analysing the trend of body temperature during anaesthesia, we proceed by considering the following increasingly complex models:

- The body is assimilated to a homogeneous sphere: this is the simplest model, in which the human body is represented as a sphere with homogeneous and constant physical properties at every point.
- The body is divided into two concentric spheres: the body is made of two concentric spheres, of which the inner one, representing the central core, is studied in the presence or absence of blood circulation (i.e. like a heat exchanger) to the external sphere.
- The body is divided into three parts: the body consists of a central part and two peripheral parts. The central part includes the trunk, the head, and the central part of the cardiovascular system. The two peripheral parts refer to the arms (with hands) and legs (with feet). Each part is composed of an external portion, which includes fat and skin, and an internal portion, which includes bones, muscles, etc.

2.1 Body Assimilated to a Homogeneous Sphere

The cooling of a sphere with constant physical properties is studied in the presence or absence of metabolic heat generation, and the numerical program in *MATLAB* is reported below [2].

2.1.1   List of the MATLAB Code

```matlab
clearall
closeall
clc
%Define the variables:
Ti=36.55; %Average body temperature (in °C)
Te=25; %External temperature, in the operating room (in °C)
h_e=5; %Convective heat transfer coefficient (in W*m^(-2)*°C^(-1))
rho=1000; %Density of the human body (in Kg/m^3)
M=70; %Mass of the individual (consider an average man of 70Kg)
Cv=3177; %Specific heat of the human body(J/(KG*K))
%Define the time axis:
t=0:100:300000; %time (in seconds)
%t=0:5:15000;
%Temperature trend
figure()
r=0.09;
T_1=Te+(Ti-Te)*exp(-3*h_e.*t/(rho*Cv)*(1/r));
plot(t,T_1,'r','linewidth',2)
hold on
r=0.15;
T_1=Te+(Ti-Te)*exp(-3*h_e.*t/(rho*Cv)*(1/r));
plot(t,T_1,'b','linewidth',2)
hold on
r=0.24;
T_1=Te+(Ti-Te)*exp(-3*h_e.*t/(rho*Cv)*(1/r));
plot(t,T_1,'g','linewidth',2)
hold on
grid on
legend('\fontsize{10} r= 9 cm','\fontsize{10} r= 15 cm','\fontsize{10} r= 24 cm')
%title('\fontsize{16} Temperature trend in absence of metabolic heat')
xlabel('t [s]','fontsize',16)
ylabel('T [°C]','fontsize',16)
text(10,36.55,'\leftarrow Ti = 36.55°C')
text(250000,25.5,'\downarrowTe=25°C')
set(gca,'fontsize',14)
set(gcf,'color','w')
%If I also consider the metabolic heat:
Qm=1.07*M; %Heat generated by metabolism (in Watt)
V=M/rho; %Body volume, obtained by dividing the body weight by the
density (in m^3)
%The temperature trend (in the presence of metabolic heat) is:
figure()
r=0.09;
```

```
a=Qm*rho*r/(3*h_e);
T_2=Te+a+(Ti-Te-a)*exp(-3*h_e.*t/(rho*Cv)*(1/r));
plot(t,T_2,'r','linewidth',2)
hold on
r=0.15;
a=Qm*rho*r/(3*h_e);
T_2=Te+a+(Ti-Te-a)*exp(-3*h_e.*t/(rho*Cv*r));
plot(t,T_2,'b','linewidth',2)
hold on
r=0.24;
a=Qm*rho*r/(3*h_e);
T_2=Te+a+(Ti-Te-a)*exp(-3*h_e.*t/(rho*Cv*r));
plot(t,T_2,'g','linewidth',2)
hold on
grid on
legend('\fontsize{10} r= 9 cm','\fontsize{10} r= 15 cm','\fontsize
{10} r= 24 cm')
%title('\fontsize{16} Temperature trend in presence of metabolic
heat')
xlabel('t [s]','fontsize',16)
ylabel('T [°C]','fontsize',16)
set(gca,'fontsize',14)
set(gcf,'color','w')
figure()
subplot(2,2,1)
Qm=0.02*M;
r=0.09;
a=Qm*rho*r/(3*h_e);
T_2=Te+a+(Ti-Te-a)*exp(-3*h_e.*t/(rho*Cv)*(1/r));
plot(t,T_2,'r','linewidth',2)
hold on
r=0.15;
a=Qm*rho*r/(3*h_e);
T_2=Te+a+(Ti-Te-a)*exp(-3*h_e.*t/(rho*Cv*r));
plot(t,T_2,'b','linewidth ',2)
hold on
r=0.24;
a=Qm*rho*r/(3*h_e);
T_2=Te+a+(Ti-Te-a)*exp(-3*h_e.*t/(rho*Cv*r));
plot(t,T_2,'g','linewidth',2)
hold on
grid on
legend('\fontsize{10} r= 9 cm','\fontsize{10} r= 15 cm','\fontsize
{10} r= 24 cm')
%title('\fontsize{16} Temperature trend in the presence of metabolic
```

```matlab
heat')
xlabel('t [s]','fontsize',16)
ylabel('T [°C]','fontsize',16)
set(gca,'fontsize',14)
set(gcf,'color','w')
grid on
subplot(2,2,2)
Qm=0.01*M
r=0.09;
a=Qm*rho*r/(3*h_e);
T_2=Te+a+(Ti-Te-a)*exp(-3*h_e.*t/(rho*Cv)*(1/r));
plot(t,T_2,'r','linewidth',2)
hold on
r=0.15;
a=Qm*rho*r/(3*h_e);
T_2=Te+a+(Ti-Te-a)*exp(-3*h_e.*t/(rho*Cv*r));
plot(t,T_2,'b','linewidth',2)
hold on
r=0.24;
a=Qm*rho*r/(3*h_e);
T_2=Te+a+(Ti-Te-a)*exp(-3*h_e.*t/(rho*Cv*r));
plot(t,T_2,'g','linewidth',2)
hold on
grid on
legend('\fontsize{10} r= 9 cm','\fontsize{10} r= 15 cm','\fontsize
{10} r= 24 cm')
%title('\fontsize{16} Temperature trend in presence of metabolic
heat')
xlabel('t [s]','fontsize',16)
ylabel('T [°C]','fontsize',16)
set(gca,'fontsize',14)
set(gcf,'color','w')
grid on
subplot(2,2,3)
t=0:5:15000;
Qm=0.02*M
r=0.09;
a=Qm*rho*r/(3*h_e);
T_2=Te+a+(Ti-Te-a)*exp(-3*h_e.*t/(rho*Cv)*(1/r));
plot(t,T_2,'r','linewidth',2)
hold on
r=0.15;
a=Qm*rho*r/(3*h_e);
T_2=Te+a+(Ti-Te-a)*exp(-3*h_e.*t/(rho*Cv*r));
plot(t,T_2,'b','linewidth',2)
```

```matlab
hold on
r=0.24;
a=Qm*rho*r/(3*h_e);
T_2=Te+a+(Ti-Te-a)*exp(-3*h_e.*t/(rho*Cv*r));
plot(t,T_2,'g','linewidth',2)
hold on
grid on
legend('\fontsize{10} r= 9 cm','\fontsize{10} r= 15 cm','\fontsize
{10} r= 24 cm')
%title('\fontsize{16} Temperature trend in presence of metabolic
heat')
xlabel('t [s]','fontsize',16)
ylabel('T [°C]','fontsize',16)
set(gca,'fontsize',14)
set(gcf,'color','w')
grid on
subplot(2,2,4)
t=0:5:15000;
Qm=0.01*M
r=0.09;
a=Qm*rho*r/(3*h_e);
T_2=Te+a+(Ti-Te-a)*exp(-3*h_e.*t/(rho*Cv)*(1/r));
plot(t,T_2,'r','linewidth',2)
hold on
r=0.15;
a=Qm*rho*r/(3*h_e);
T_2=Te+a+(Ti-Te-a)*exp(-3*h_e.*t/(rho*Cv*r));
plot(t,T_2,'b','linewidth',2)
hold on
r=0.24;
a=Qm*rho*r/(3*h_e);
T_2=Te+a+(Ti-Te-a)*exp(-3*h_e.*t/(rho*Cv*r));
plot(t,T_2,'g','linewidth',2)
hold on
grid on
legend('\fontsize{10} r= 9 cm','\fontsize{10} r= 15 cm','\fontsize
{10} r= 24 cm')
%title('\fontsize{16} Temperature trend in the presence of metabolic
heat')
xlabel('t [s]','fontsize',16)
ylabel('T [°C]','fontsize',16)
set(gca,'fontsize',14)
set(gcf,'color','w')
grid on
```

2.1.2 Homogeneous Sphere Without Metabolic Heat

Assuming the sphere with concentrated parameters, the temperature is not a function of spatial coordinates but only of time, that is, $T(x,y,z,t) \sim T(t)$. The energy balance equation (where it is assumed that the exchanged work and the variations of kinetic and potential energy are negligible) is reduced to the equality between the variation of internal energy of the sphere and the heat transfer by convection.

$$M \, c_v \, dT = h \, S \, (T_e - T) dt \tag{2}$$

where M is the body mass, c_v is the specific heat of the human body, h is the convective heat transfer coefficient, S is the body surface, and T_e is the external temperature. Integrating with respect to time Eq. 2 we obtain

$$\theta = \frac{T - T_e}{T_i - T_e} = e^{\frac{-h\,S\,t}{M\,c_v}} = e^{\frac{-3\,h\,t}{\rho\,R\,c_v}} \tag{3}$$

with T_i being the initial temperature of the sphere/body.

Figure 1 presents the temperature variation (°C) on the ordinate, as a function of the dimensionless time $\tau = \frac{\rho\,R\,c_v}{3\,h\,t}$, on the abscissa, both in a linear scale, in the cooling of three spheres of different diameters, 9 cm, 15 cm, and 24 cm, and in the absence of metabolic heat, starting from the average body temperature of 36.55 °C, down to the ambient temperature of 25 °C, assuming the following data for the spheres: $\rho = 1000 \, \frac{\text{kg}}{\text{m}^3}$, $c_v = 3204 \, \frac{\text{J}}{\text{kg K}}$, and $h = 4 \, \frac{\text{W}}{\text{m}^2\text{K}}$. It can be observed in Fig. 1 that as the radius increases, the body cools more slowly since, in the time constant of the exponential trend, $\tau = \frac{\rho\,R\,c_v}{3\,h\,t}$, the radius appears in the numerator.

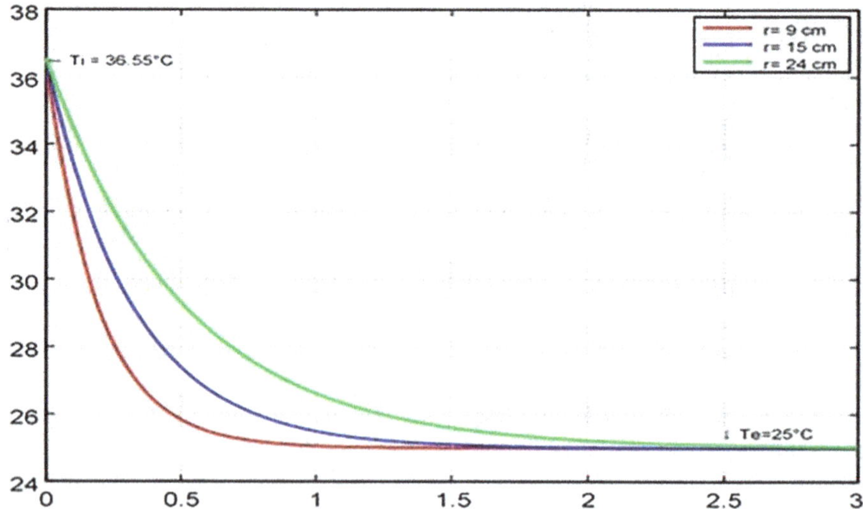

Fig. 1 Temperature (°C) versus time (s) in the cooling of three spheres/bodies of different diameters in the absence of metabolic heat [2]

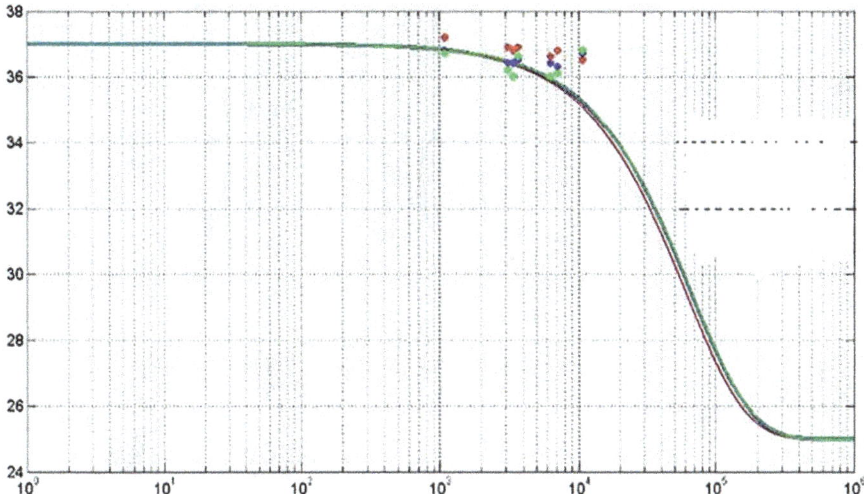

Fig. 2 Temperature (°C) versus time (s) of three spheres in the absence of metabolic heat and comparison with experimental data [2]

Figure 2 shows the temperature variations (in °C and linear scale) versus time (in seconds and logarithmic scale), assuming the same data for the spheres: $\rho = 1000$ $\frac{kg}{m^3}$, $c_v = 3204$ $\frac{J}{kg\,K}$, and $h = 4$ $\frac{W}{m^2K}$, except $T_e = 25\,°\,C$ *and* $T_i = 37.0\,°\,C$. The radius has the values of 0.25 m in order to have a sphere weight of 66 Kg, and the results are reported with a blue line. If the radius is 0.229 m in order to have a sphere weight of 50 Kg) the results are reported with a red line.; If the radius is 0.251 m in order to have a sphere weight of 66.5 Kg) the results are reported with a green line. These numerical results are compared with the experimental data obtained in the bodies of three women: the first, of height 1.6 m and weight 66 Kg, from now on referred to as D_1, and whose results are reported with blue dots; the second, of height 1.5 m and weight 50 Kg, from now on referred to as D_2, and whose results are reported with red dots; and the third, of height 1.68 m and weight 66.5 Kg, henceforth indicated by D_3, and whose results are reported with green points [1].

The results of this model, approximate but very simple, show that the temperature tends to the external temperature, 25 °C, there being no metabolic heat generated. The three trends are hardly distinguishable since the only parameter that can be varied between the three spheres is the radius.

2.1.3   Homogeneous Sphere with Metabolic Heat

If we consider the presence of metabolic heat in the sphere/body, the energy balance equation becomes

$$M\,c_v\,dT = h\,S\,(T_e - T)\,dt + dQ_m = h\,S\,(T_e - T)\,dt + q_m\,\rho\,V\,dt \qquad (4)$$

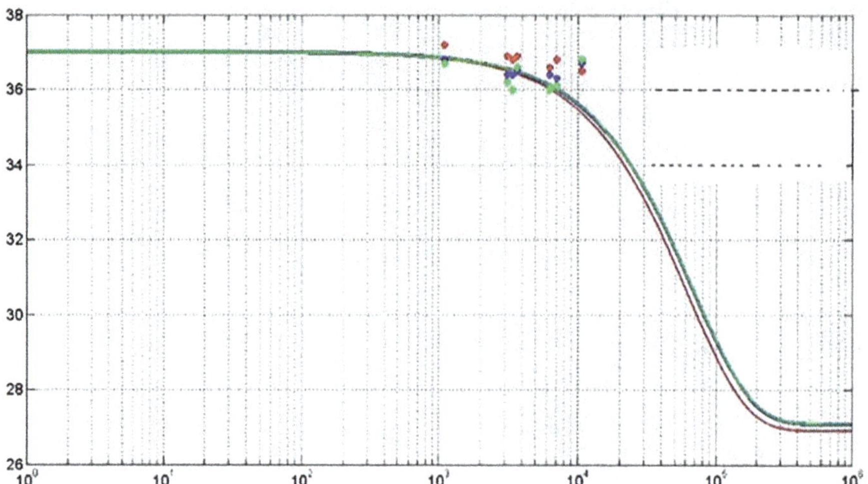

Fig. 3 Temperature (°C) versus time (s) in the cooling of three spheres, with a metabolic heat, $q_m = 0.1$ W/kg, and comparison with experimental data [2]

where q_m is the heat generated metabolically inside the sphere per unit of time and mass. Its integration leads to the following result:

$$\theta = \frac{T - T_e}{T_i - T_e} = e^{\frac{-h\,S\,t}{M\,c_v}} - \frac{\frac{q_m\,\rho\,V}{h\,S}}{T_i - T_e}\,e^{\frac{-h\,S\,t}{M\,c_v}} + \frac{\frac{q_m\rho\,V}{h\,S}}{T_i - T_e} \tag{5}$$

Figure 3 shows the temperature evolutions, calculated with the following data: $\rho = 1000\ \frac{kg}{m^3}$, $c_v = 3204\ \frac{J}{kg\,K}$, $h = 4\ \frac{W}{m^2K}$, $T_e = 25\ °C$, $T_i = 370\ °C$, *and* $q_m = 0.1$ W/ kg, and with radius taking the values 0.25 m, 0.229 m, and 0.251 m, as done previously in Fig. 2 . These trends are compared with the experimental data, obtained for the bodies of the three women, D_1, D_2, and D_3. The metabolic heat generated is low and incorrect, so the temperature has a decreasing trend for all the three radii and tends to a temperature around 27 °C. The spheres/bodies with the larger radii cool down more slowly and reach higher steady-state temperatures, albeit slightly.

Figure 4 shows the variations of the temperature with the following data: $\rho = 1000\ \frac{kg}{m^3}$, $c_v = 3204\ \frac{J}{kg\,K}$, $h = 4\ \frac{W}{m^2K}$, $T_e = 25\ °C$, $T_i = 37.0\ °C$, *and* $q_m = 1.07$ W/kg, and with the radius taking the values 0.25 m, 0.229 m, and 0.251 m. The results are then compared with the experimental data obtained for the three women, D_1, D_2, and D_3, as done previously. The temperature has an increasing trend with this experimental and correct value of the metabolic heat, reaching the steady-state values of about 45 °C for the red line and above 47 °C for the other two. This temperature trend could be seen as corresponding to the case of malignant hyperthermia, which is discussed later.

Figure 5 shows the temperature trends with the following input data: $\rho = 1000\ \frac{kg}{m^3}$, $c_v = 3204\ \frac{J}{kg\,K}$, $h = 4\ \frac{W}{m^2K}$, $T_e = 25\ °C$, $T_i = 37.0\ °C$, *and* $q_m = 2.0$ W/kg, and with

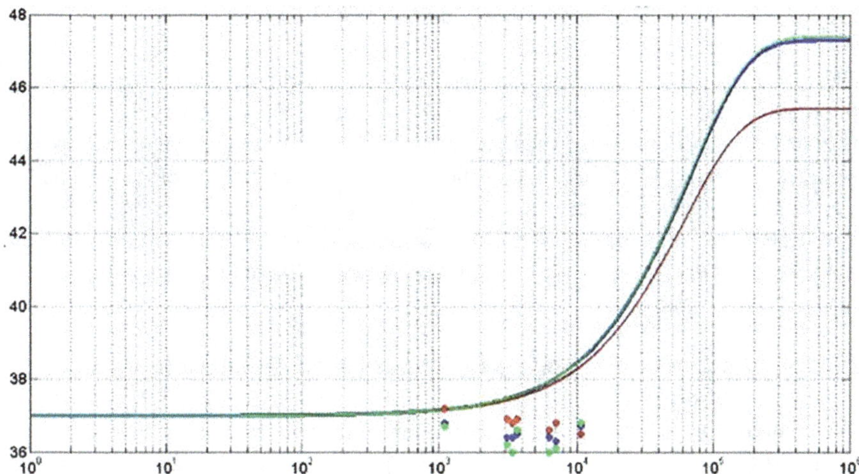

Fig. 4 Temperature (°C) versus time (s) in the cooling of three spheres, with a metabolic heat, $q_m = 1.07$ W/kg, and comparison with experimental data [2]

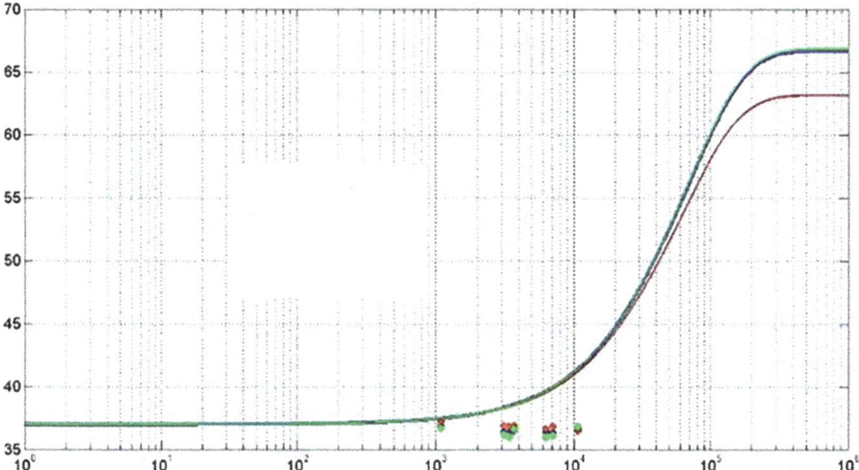

Fig. 5 Temperature (°C) versus time (s) in the cooling of three spheres, with metabolic heat, $q_m = 2.0$ W/kg, and comparison with experimental data [2]

radius equal to 0.25 m, 0.229 m, and 0.251 m. These trends are then compared with the experimental data obtained in the bodies of the three women, D_1, D_2, and D_3. The temperature has an increasing trend for this value of metabolic heat, with steady-state values of about 63°C for the red line and above 67 °C for the other two. This temperature trend could also be seen as corresponding to the case of malignant hyperthermia, which is discussed later.

2.2   Body Divided into Two Concentric Spheres

The human body is then approximated by two concentric spheres, of which the innermost represents the part including the skeleton, viscera, and muscles of the head, trunk, and limbs, while the outermost represents the outer layers, such as skin and fat. The study is carried out in the presence or absence of blood circulation (i.e. with or without the presence of a hypothetical heat exchanger) and in the case where the initial temperatures of the two concentric spheres are the same (a situation that is not very realistic, precisely because of the definition of inner and outer core) or different.

2.2.1   Two Concentric Spheres Without Blood Circulation or Heat Exchanger

The equations to be studied are now two and represent the energy balance of the inner part (indicated with 1) and of the outer part (indicated with 2). Therefore, the system of the two ordinary differential equations is

$$M_1 \, c_{v1} \frac{dT_1}{dt} = q_{m1} \, \rho_1 \, V_1 - U_1 \, S_1 (T_1 - T_2) \tag{6}$$

$$M_2 \, c_{v2} \frac{dT_2}{dt} = q_{m2} \, \rho_2 \, V_2 + U_1 \, S_1 (T_1 - T_2) - h_e \, S_e \, (T_2 - T_e) \tag{7}$$

If the coefficients are grouped as

$$k_1 = \frac{U_1 \, S_1}{M_1 \, c_{v1}}, q^*{}_{m1} = \frac{q_{m1} \, \rho_1 \, V_1}{M_1 c_{v1}}, k_2 = \frac{U_1 S_1}{M_2 c_{v2}}, k_3 = \frac{h_e S_e}{M_2 c_{v2}}, q^*{}_{m2} = \frac{q_{m2} \rho_2 \, V_2}{M_2 c_{v2}},$$

the solution of the two-equation system is

$$T_1 = A_1 \, e^{\lambda_1 t} + B_1 \, e^{\lambda_2 t} + C_1 \tag{8}$$

$$T_2 = T_1 - \frac{q^*{}_{m1}}{k_1} + \frac{dT_1}{dt} \frac{1}{k_1} \tag{9}$$

with

$$\lambda_1 = \frac{-(k_1 + k_2 + k_3) - \sqrt{(k_1 + k_2 + k_3)^2 - 4k_1 k_3}}{2}, \lambda_2$$

$$= \frac{-(k_1 + k_2 + k_3) + \sqrt{(k_1 + k_2 + k_3)^2 - 4k_1 k_3}}{2},$$

$$C_1 = T_e + \frac{q^*{}_{m2}}{k_3} + \frac{(k_2 + k_3)}{k_1 k_3} q^*{}_{m1}, \quad B_1 = \frac{q^*{}_{m1} - k_1(T_{10} - T_{20}) + C_1 \lambda_1 - \lambda_1 T_{10}}{\lambda_2 - \lambda_1}, A_1 = T_{10} - B_1 - C_1.$$

List of the MATLAB Code

The *MATLAB* calculation code for the solution of two concentric spheres without blood circulation is reported below [2].

```
clear all
close all
clc
%Define the variables:
h_e=5; %Convective heat transfer coefficient (in W*m^(-2)*K^(-1))
rho=1000; %Density of the human body (in Kg/m^3)
M=70; %Mass of the individual (considering an average man of 70Kg)
Cv=3177; %Specific heat of the human body (J/(KG*K))
%Considering metabolic heat:
Qm=0.005*M; %Heat generated by metabolism (in Watt)
Qm_in=0.887*Qm; %[W]
Qm_out=0.1146*Qm; %[W]
V=M/rho; %Volume of the body, obtained by dividing the body weight by the
density (in m^3)
%Defining the average specific heats of the internal part (skeleton and
blood)
%and of the external part (skin, fat and muscles):
Cv_in=3204; %[J/(Kg*K)]
Cv_out=3150; %[J/(Kg*K)]
%Defining the volumes, masses and densities :
rho_in=1000; %[Kg/m^3]
rho_out=1000; %[Kg/m^3]
V_in=0.6048*M/1000; %[m^3]
M_in=rho*V_in; %[Kg]
V_out=0.3952*M/1000; %[m^3]
M_out=rho*V_out; %[Kg]
Te=298.15;
%Defining the height for an average man of 70 Kg:
H=1.75; %[m]
%Defining the surfaces (total, internal and external) in [m^2]:
S=0.203*(H^0.725)*M^0.425;
S_in=0.8143*S;
S_out=S-S_in;
%Assuming natural convection for h_in and h_out:
h_out=5; %[W*m^(-2)*K(-1)]
h_in=7; %[W*m^(-2)*K^(-1)]
%To define the constants that appear in the equations of the temperatures
T_in
%and T_out:
k_1=h_in*S_in/(M_out*Cv_out);
k_2=h_in*S_in/(M_in*Cv_in);
```

```
k_3=h_out*S_out/(M_out*Cv_out);
k_4=Qm_out*rho_out*V_out/(M_out*Cv_out);
k_5=Qm_in*rho_in*V_in/(M_in*Cv_in);
k_t=k_5*(k_1+k_3)+k_2*k_4;
%To define the time axis:
t=100:80:100000000;
%To define the constants c1 and c2 (roots of the associated homogeneous):
c1=(-(k_1+k_2+k_3)+sqrt((k_1+k_2+k_3)^2-4*k_2*k_3))/2
c2=(-(k_1+k_2+k_3)-sqrt((k_1+k_2+k_3)^2-4*k_2*k_3))/2
%To define the initial temperature:
T_in0=309.65;
%To define the trends of the two temperatures:
T_in=(k_5/c1-c2/(c1-c2)*(T_in0-k_5/c1-Te-k_t/(k_2*k_3))).*exp(c1.
*t)...
+(c1/(c1-c2)*(T_in0-k_5/c1-Te-k_t/(k_2*k_3))).*exp(c2.*t)+Te+k_t/
(k_2*k_3);
T_out=(k_5/c1-c2/(c1-c2)*(T_in0-k_5/c1-Te-k_t/(k_2*k_3)))*
(c1/k_2+1).*exp(c1.*t)...
+(c2/k_2+1)*(c1/(c1-c2)*(T_in0-k_5/c1-Te-k_t/(k_2*k_3))).*exp(c2.
*t)+Te+k_t/(k_2*k_3)-k_5/k_2;
%To plot the trends of the two temperatures:
figure()
semilogx(t,T_in,'r',t,T_out,'b','linewidth',2)
hold on
grid on
legend('T_{in}','T_{out}')
xlabel('t [s]','fontsize',16)
ylabel('T [K]','fontsize',16)
set(gcf,'color','w')
```

The parameters chosen for the simulation in this case are $\rho = 1000\,\frac{\text{kg}}{\text{m}^3}$, $c_{v1} = 3204\,\frac{\text{J}}{\text{kg K}}$, $c_{v2} = 3150\,\frac{\text{J}}{\text{kg K}}$, $h_e = 5$ W/m^{2}K, $q_{m1} = 0.88\,q_m$, $q_{m2} = 0.12\,q_m$, $T_e = 25$ °C, and $T_i = 36.8$ °C. Figure 6 shows the analytical solution of the temperatures (in °C and linear scale) as a function of time (in seconds and logarithmic scale) of the two concentric spheres in the case where the initial temperatures of the central and peripheral parts coincide. The blue line reports the temperature of the central part, while the red line reports the temperature of the peripheral part. The experimental data obtained for case D_1 [1] are also reported, where the blue points are related to the central part, while the red points are related to the peripheral part.

Figure 7 shows the analytical solution of the temperatures (in °C and linear scale) as a function of time (in seconds and logarithmic scale) of the two concentric spheres in the case where the initial temperatures of the central and peripheral parts at the

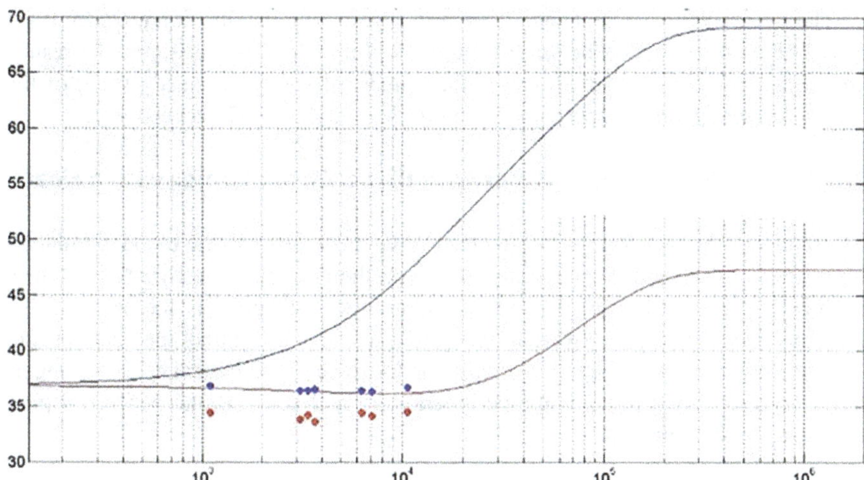

Fig. 6 Variation of the central (blue line) and peripheral (red line) temperatures (°C) versus time (s), with $T_{10} = T_{20} = 36.8\,°C$, with metabolic heat, $q_m = 1.07$ W/kg, but without heat exchanger, and comparison with case D_1 [2]

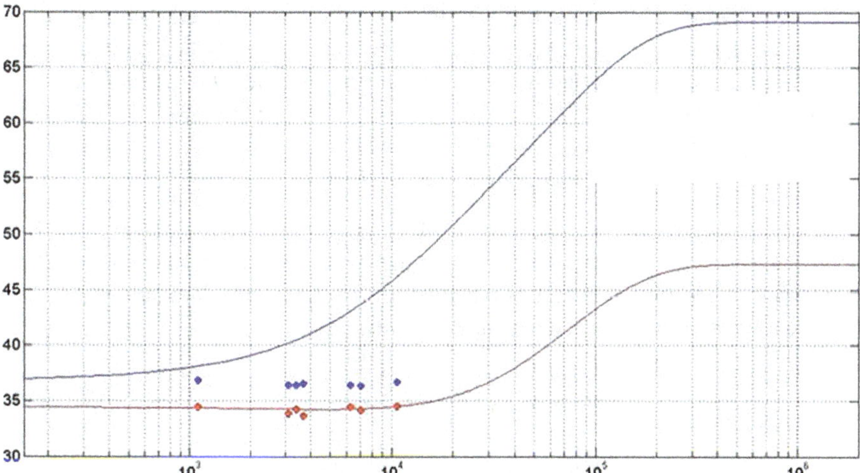

Fig. 7 Variation of the central (blue line) and peripheral (red line) temperatures (°C) versus time (s), with $T_{10} = 36.8°C$ and $T_{20} = 34.4°C$, with metabolic heat, $q_m = 1.07$ W/kg, but without heat exchanger, and comparison with case D_1 [2]

beginning are $T_{10} = 36.8°C$ and $T_{20} = 34.4°C$. The blue line shows the temperature of the central part, while the red line shows the temperature of the peripheral part. The experimental data obtained for the case D_1 are also reported, where the blue points are related to the central part, while the red points are related to the peripheral part.

The two equations, Eqs. 6 and 7, are also solved with the function *ODE45* of the *MATLAB software* under the same conditions of the previous cases, and the results are almost coincident. In both cases, the numerical values, at steady state, of the central and peripheral temperature are 69.08 °C and 47.29 °C, respectively.

2.2.2   Two Concentric Spheres with Blood Circulation or with a Heat Exchanger

The model of the two concentric spheres can take into account the blood circulation between the inner and outer spheres (i.e. with the presence of a hypothetical heat exchanger). The system of differential equations to integrate is the same as the previous one, with the only variant being that we have to subtract the heat per unit of time carried away by the blood circulation (or by the hypothetical heat exchanger) from the central part towards the peripheral part and add it to the peripheral part, according to the equation

$$Q_1 = G_b\, c_b\, (T_{\text{in}} - T_u) \tag{10}$$

where G_b is the blood mass flow rate; c_b is the specific heat of the blood; T_{in} is the temperature at the entrance of the heat exchanger, assumed to be equal to the temperature of the internal part, T_1; and T_u is the exit temperature from the heat exchanger, assumed to be equal to the temperature of the peripheral part, T_2.

The energy balance on a stretch dx of the heat exchanger, with p perimeter, and assuming T_1 variable along the heat exchanger but with T_2 constant, as in a thermostat, we obtain

$$G_b\, c_b dT_1 = U\, p\, dx\, (T_1 - T_2) \tag{11}$$

where U is the overall heat transfer coefficient in the heat exchanger (which is approximated with the convective heat transfer coefficient of the blood in the capillaries, neglecting the heat conduction in the very thin capillary wall). The total heat transfer in the heat exchanger becomes

$$Q_1 = G_b\, c_b\, (T_1 - T_2) \left[1 - e^{\frac{-U\,S}{G_b\,c_b}} \right] \tag{12}$$

where S is the total heat transfer surface in the heat exchanger.

The system of two differential equations to be solved are

$$M_1\, c_{v1}\, \frac{dT_1}{dt} = q_{m1}\, \rho_1\, V_1 - U_1\, S_1(T_1 - T_2) - G_b\, c_b\, (T_1 - T_2) \left[1 - e^{\frac{-U\,S}{G_b\,c_b}} \right] \tag{13}$$

$$M_2\, c_{v2}\, \frac{dT_2}{dt} = q_{m2}\, \rho_2\, V_2 + U_1\, S_1(T_1 - T_2) - h_e S_e(T_2 - T_e) + G_b c_b(T_1 - T_2) \left[1 - e^{\frac{-U\,S}{G_b\,c_b}} \right] \tag{14}$$

The variables are grouped as follows:

$$k_1 = \frac{U_1\, S_1}{M_1\, c_{v1}},\, q^*_{m1} = \frac{q_{m1}\, \rho_1\, V_1}{M_1\, c_{v1}},\, k_2 = \frac{U_1\, S_1}{M_2\, c_{v2}},\, k_3 = \frac{h_e\, S_e}{M_2\, c_{v2}},\, q^*_{m2} = \frac{q_{m2}\, \rho_2\, V_2}{M_2\, c_{v2}},\, g_1$$

$$= \frac{G_b\, c_b}{M_1\, c_{v1}},\, g_2\, \frac{G_b c_b}{M_2\, c_{v2}},$$

$$r = \left[1 - e^{\frac{-U\, S}{G_b c_b}}\right],\, f_1 = k_1 + g_1\, r,\, f_2 = k_2 + g_2\, r.$$

The system of the two equations is solved analytically by deriving one of the two equations with respect to time and then solving the second-order equation with the method of substituting the first derivative in time. The solution of the system leads to the following results:

$$T_1 = A_1 e^{\lambda_1\, t} + B_1\, e^{\lambda_2 t} + C_1 \tag{15}$$

$$T_2 = T_1 - \frac{q^*_{m1}}{f_1} + \frac{dT_1}{dt}\, \frac{1}{f_1} \tag{16}$$

where

$$\lambda_1 = \frac{-(f_1 + f_2 + k_3) - \sqrt{(f_1 + f_2 + k_3)^2 - 4 f_1 k_3}}{2},\, \lambda_2$$

$$= \frac{-(f_1 + f_2 + k_3) + \sqrt{(f_1 + f_2 + k_3)^2 - 4 f_1 k_3}}{2},$$

$$C_1 = T_e + \frac{q^*_{m2}}{k_3} + \frac{(f_2 + k_3)}{f_1 k_3}\, q^*_{m1},\, B_1 = \frac{q^*_{m1} - f_1(T_{10} - T_{20}) + C_1 \lambda_1 - \lambda_1 T_{10}}{\lambda_2 - \lambda_1},\, A_1 = T_{10} - B_1 - C_1.$$

List of the MATLAB Code
The *MATLAB* calculation code for the solution of two concentric spheres with blood circulation is reported below [2].

```
functiondT=myfun3(t,T)
h2=7;
he=5;
H=1.75;
M=70;
S_tot=0.203*(H^0.725)*M^0.425;
S2=0.8143*S_tot;
S1=S_tot-S2;
%Gs=0.00588;
Gs=0.0955; %[Kg/s]
Cb=3744;
rho=1000;
```

```
V=M/rho;
Nc=10^9; %total number of capillaries in humans
D=5*10^(-6); %capillary diameter in [m]
Lc=1.5*10^(-3); %average length of a single capillary in [m]
r_c=D/2; %capillary radius in [m]
k=0.60; %heat transfer coefficient of water
U=1/he+1/h2+r_c/k;
u_c=1/U; %global heat transfer coefficient
S_c=Nc*pi*D*Lc; %exchange surface of the capillaries in [m^2]
Qm=0.005*M;
rho_2=1000; %[Kg/m^3]
rho_1=1000; %[Kg/m^3]
V_2=0.6048*M/1000; %[m^3]
M2=rho*V_2; %[Kg]
V_1=0.3952*M/1000; %[m^3]
M1=rho*V_1; %[Kg]
Qm_2=0.887*Qm; %[W]
Qm_1=0.1146*Qm; %[W]
Cv_2=3204; %[J/(Kg*K)]
Cv_1=3150; %[J/(Kg*K)]
Te=298.15;
A=h2*S2+Gs*Cb*(1-exp(-u_c*S_c/(Gs*Cb)));
c1=Qm_2*rho_2*V_2/(M2*Cv_2);
c2=A/(M2*Cv_2);
c3=Qm_1*rho_1*V_1/(M1*Cv_1);
c4=A/(M1*Cv_1)
dT=[c3+c4*(T(2)-T(1))-((he*S1)/(M1*Cv_1))*(T(1)-Te);c1-c2*(T(2)-T
(1))];
Here is the main:
clearall
closeall
clc
%Define the time axis:
t=10:100:100000;
%Solution with the Matlab function ode45:
[t T]=ode45('myfun3',t,[304.55 309.85]);
figure()
semilogx(t,T(:,1),'r',t,T(:,2),'b','linewidth',2)
grid on
hold on
legend('T1 external','T2 internal')
set(gcf,'color','w')
xlim([10 5000])
%%%%%
% Comparison case 2 and case 3
```

```
%%%%%
%Solution with the Matlab function ode45:
[t TT]=ode45('myfun2',t,[304.55 309.85]);
figure()
semilogx(t,TT(:,1),'--r',t,TT(:,2),'--b','linewidth',2)
grid on
hold on
semilogx(t,T(:,1),'r',t,T(:,2),'b','linewidth',2)
legend('Test no exchange','Tint no exchange','Test with
exchange','Tint with exchange')
set(gcf,'color','w')
xlim([10 20000])
```

The solution of the equations, Eqs. 15 and 16, are obtained with the following parameters:

$$\rho = 1000\,\tfrac{kg}{m^3}\;,c_{v1} = 3204\;\tfrac{J}{kg\,K}\;,\quad c_{v2} = 3150\;\tfrac{J}{kg\,K}\;,\quad h_e = 4\,\tfrac{W}{m^2K}\;,\quad U_1 = 10\,\tfrac{W}{m^2K}\;,$$

$R_1 = 0.15$ m, $R_2 = 0.25$ m (total weight 66 Kg, case D_1), $q_m = 1.07\,\tfrac{W}{Kg}$, $q_{m1} = 0.88\;q_m$, $q_{m2} = 0.12\,q_m$, $T_e = 25\;°C$, $T_{10} = 36.8\;°C$, $T_{20} = 34.4\;°C$, $c_b = 3744\;\tfrac{J}{kg\,K}$, $G_b = 0.00098\,\tfrac{Kg}{s}$.

Figure 8 shows the analytical solution of the temperatures (in °C and linear scale) versus time (in seconds and logarithmic scale) of the two concentric spheres in the case where the initial temperatures of the central and peripheral parts at the beginning of the treatment are $T_{10} = 36.8\;°C$ and $T_{20} = 34.4\;°C$. The blue line reports

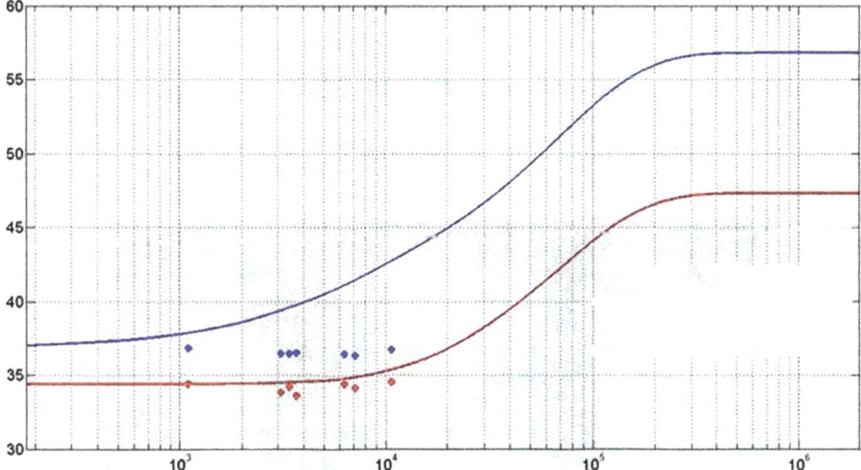

Fig. 8 Variation of central (blue line) and peripheral (red line) temperatures (°C) versus time (s), with $T_{10} = 36.8°C$ and $T_{20} = 34.4°C$, metabolic heat, $q_m = 1.07$ W/kg, with heat exchanger, and comparison with case D_1 [2]

the temperature of the central part, while the red one reports the temperature of the peripheral part. The temperatures obtained at steady state are about 57 °C for the internal part and about 47.5 °C for the peripheral part. A comparison can be made with the results obtained in the case of absence of blood perfusion, i.e. without a heat exchanger, which are 69 °C for the internal part and 47.5 °C for the peripheral part, respectively. The experimental data obtained for the case D_1, where the blue points are related to the central part, while the red points are related to the peripheral part, are also reported. In conclusion, as expected, blood perfusion redistributes heat and cools the internal part, thanks to the hypothetical heat exchanger, while the peripheral part is not too affected.

The two equations, Eqs. 13 and 14, are also solved under the same conditions of the previous cases with the *ODE45* program of the *MATLAB software,* and the two solutions, analytical and numerical, are almost coincident.

2.3   Body Divided into Three Parts

2.3.1   Equations

The body is assumed to be made up of three parts, a central part and two peripheral parts, each one modelled as a cylinder. The central part includes the trunk, head, and internal cardiovascular system (including the heart). The two peripheral parts refer to the arms (with the hands) and the legs (with the feet). Each part is composed of an external portion, which includes fat and skin, and an internal portion (bones, muscles, etc.). Thanks to the blood circulation (which works as a hypothetical heat exchanger), the internal portion of the central part is connected with the external portion of the central part and with the remaining four portions of the peripheral parts, as shown in Fig. 9.

The portions are indicated as follows:

- P1 is the internal portion of the central part where the heart is present, distributing the blood to the other portions.
- P2 is the external portion of the central part.

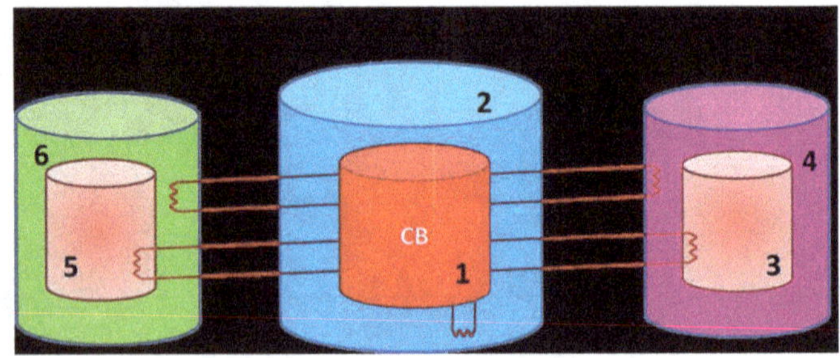

Fig. 9 Body divided into three parts and six portions: P1, P2, P3, P4, P5, and P6

- P3 is the internal portion of the peripheral part, legs, and feet.
- P4 is the external portion of the peripheral part, legs, and feet.
- P5 is the internal portion of the peripheral part, arms, and hands.
- P6 is the external portion of the peripheral part, arms, and hands.

The equations that represent the thermal balance for the six portions are as follows.

For the internal portion of the central part, P1, with temperature T_1,

$$\rho_1 \, c_{v1} \, V_1 \, \frac{dT_1}{dt} = Q_{g1} - R_s - Q_2 - Q_3 - Q_4 - Q_5 - Q_6 - U_1 \, S_1 \, (T_1 - T_2) \quad (17)$$

For the external portion of the central part, P2, with temperature T_2,

$$\rho_2 \, c_{v2} \, V_2 \, \frac{dT_2}{dt} = Q_{g2} + Q_2 - E_{t2} + U_1 \, S_1 \, (T_1 - T_2) - h_e \, S_2 \, (T_2 - T_e) \quad (18)$$

For the internal portion of the peripheral part, legs, and feet, P3, with temperature T_3,

$$\rho_3 \, c_{v3} \, V_3 \, \frac{dT_3}{dt} = Q_{g3} + Q_3 - U_3 \, S_3 \, (T_3 - T_4) \quad (19)$$

For the external portion of the peripheral part, legs, and feet, P4, with temperature T_4,

$$\rho_4 \, c_{v4} \, V_4 \, \frac{dT_4}{dt} = Q_{g4} + Q_4 - E_{t4} + U_3 \, S_3 (T_3 - T_4) - h_e \, S_4 \, (T_4 - T_e) \quad (20)$$

For the internal portion of the peripheral part, arms, and hands, P5, with temperature T_5,

$$\rho_5 \, c_{v5} \, V_5 \, \frac{dT_5}{dt} = Q_{g5} + Q_5 - U_5 \, S_5 \, (T_5 - T_6) \quad (21)$$

For the external portion of the peripheral part, arms, and hands, P6, with temperature T_6,

$$\rho_6 \, c_{v6} \, V_6 \, \frac{dT_6}{dt} = Q_{g6} + Q_6 - E_{t6} + U_5 \, S_5 \, (T_5 - T_6) - h_e \, S_6 \, (T_6 - T_e) \quad (22)$$

The metabolic heat is indicated with Q_g, while the subscripts are relative to each portion. The heat transfers exchanged through blood perfusion and capillary circulation are Q_2, Q_3, Q_4, Q_5, Q_6 and refer to the heat exchangers from the internal portion of the central part, where the heart distributes the blood, or P1, to the external portion of the central part, P2, Q_2, and the remaining four portions of the peripheral parts. Q_3 is the heat distributed to the internal portion of the peripheral part, legs, and feet, P3; Q_4 is the heat distributed to the external portion of the peripheral part, legs, and feet, P4. Q_5 is the heat distributed to the internal portion of the peripheral part, arms, and hands, P5; Q_6 is the heat distributed to the external portion of the peripheral part, arms, and hands, P6. The overall heat transfer coefficient between the internal and external portions of each part is indicated with U, while the subscripts are relative to each portion. Finally, R_s is the heat lost in respiration, E_t is the heat lost in evaporation from the external surfaces, and h_e is the convective heat transfer coefficient of the external surfaces with the outside air.

The heat transfer between internal and external portions of each part occurs mainly due to conduction; therefore, the global heat transfer coefficient U is reduced to the single term $\frac{k_{\text{eff}}}{l}$, with k_{eff} being the effective thermal conductivity of the tissue under examination and l being the thickness in which the conductive phenomenon occurs. The effective thermal conductivity of the body, k_{eff}, is calculated based on the theoretical model proposed in [4, 5] and then applied in [6] to frozen meat. Based on the model [5], it is assumed that the layer in which heat conduction occurs is a two-phase medium, with fat considered as the dispersed phase and a composition of fat and proteins as the continuous phase. From the tables of [6], the conductivity of fat is $\lambda_f = 0.18\,\frac{W}{m\,K}$, the conductivity of water is $\lambda_w = 0.56\,\frac{W}{m\,K}$, the conductivity of proteins is $\lambda_p = 0.20\,\frac{W}{m\,K}$, the volumetric fraction of fat is $\Theta_f = 0.851$, the volumetric fraction of water is $\Theta_w = 0.126$, and the volumetric fraction of proteins is $\Theta_p = 0.023$. To consider the biphasic medium, water and proteins are considered as a single phase, with conductivity $\lambda_{w,p} = \frac{\Theta_w \lambda_w + \Theta_p \lambda_p}{\Theta_w + \Theta_p} = 0.5\,\frac{W}{m\,K}$ and volumetric fraction $\Theta_{w,p} = \Theta_w + \Theta_p = 0.149$. The porosity of the medium is $\varepsilon = 0.149$ and $\beta = \sqrt[3]{\frac{1}{1-\varepsilon}} = 1.05$.

In the hypothesis of parallel isotherms [5], we obtain

$$k_{\text{eff},1} = \left(\frac{\beta - 1}{\beta\,\lambda_{w,p}} + \frac{\beta}{\lambda_{w,p}\,(\beta^2 - 1) + \lambda_f} \right)^{-1} = 0.214\,\frac{W}{m\,K}, \tag{23}$$

whereas, in the hypothesis of parallel heat flows [5], we have

$$k_{\text{eff},2} = \frac{\lambda_{w,p}\,(\beta^2 - 1)}{\beta^2} + \frac{1}{\frac{\beta^2 - \beta}{\lambda_{w,p}} + \frac{\beta}{\lambda_f}} = 0.213\,\frac{W}{m\,K}. \tag{24}$$

Since the two values are very similar, we assume $k_{\text{eff}} = 0.213$. Assuming that the thicknesses of fat, in which heat transfer occurs by conduction, are in the range 1.93–2.13 cm, we obtain, for the overall heat transfer coefficients used in the simulations, a value of about $U_1 = U_3 = U_5 = 10$ W/m^2/K. The heat transfer in each heat exchanger based on blood circulation is calculated as that present in a thermostat, and only the convection coefficient, h, in each heat exchanger and not the conduction of heat is taken into account.

The heat transfer due to the heat exchanger between the internal portion of the central part, P1, and the external one, P2, is

$$Q_2 = c_b \, G_2(T_1 - T_2)\left(1 - e^{\frac{-h \, A_2}{G_2 \, c_b}}\right) \tag{25}$$

The heat transfer between the internal portion of the central part, P1, and the internal portion of the peripheral part, legs, and feet, P3, is

$$Q_3 = c_b \, G_3 \, (T_1 - T_3)\left(1 - e^{\frac{-h \, A_3}{G_3 \, c_b}}\right) \tag{26}$$

The heat transfer between the internal portion of the central part, P1, and the external portion of the peripheral part, legs, and feet, P4, is

$$Q_4 = c_b \, G_4 \, (T_1 - T_4)\left(1 - e^{\frac{-h \, A_4}{G_4 \, c_b}}\right) \tag{27}$$

The heat transfer between the internal portion of the central part, P1, and the internal portion of the peripheral part, arms, and hands, P5, is

$$Q_5 = c_b \, G_5 \, (T_1 - T_5)\left(1 - e^{\frac{-h \, A_5}{G_5 \, c_b}}\right) \tag{28}$$

The heat transfer between the internal portion of the central part, P1, and the external portion of the peripheral part, arms, and hands, P6, is

$$Q_6 = c_b \, G_6(T_1 - T_6)\left(1 - e^{\frac{-h \, A_6}{G_6 \, c_b}}\right) \tag{29}$$

The system of the six differential equations, Eqs. 17–22, cannot be easily solved analytically because the equations are not separable due to the heat transfers, which contain the unknown temperatures of the other portions. However, if we consider the heat transfer constant with time based on the assumption that the temperatures are evaluated at time, t_o, i.e. at the beginning of the anaesthetic treatment, the system of six equations can be reduced to three pairs of equations, which can be solved as in the case of the two concentric spheres. In this way, the following temperatures are obtained for the three internal parts:

$$T_1 = A_1 \, e^{\lambda_{1,1} \, t} + B_1 \, e^{\lambda_{2,1} \, t} + C_1 \tag{30}$$

$$T_3 = A_2 \, e^{\lambda_{1,2} \, t} + B_2 \, e^{\lambda_{2,2} t} + C_2 \tag{31}$$

$$T_5 = A_3 \, e^{\lambda_{1,3} \, t} + B_3 \, e^{\lambda_{2,3} \, t} + C_3 \tag{32}$$

and for the three external parts

$$T_2 = T_1 - \frac{m_1}{k_1} + \frac{1}{k_1}\left(A_1\,\lambda_{1,1}\,e^{\lambda_{1,1}\,t} + B_1\,\lambda_{2,1}\,e^{\lambda_{2,1}t}\right) \tag{33}$$

$$T_4 = T_3 - \frac{m_3}{k_4} + \frac{1}{k_4}\left(A_2\,\lambda_{1,2}\,e^{\lambda_{1,2}\,t} + B_2\lambda_{2,2}\,e^{\lambda_{2,2}\,t}\right) \tag{34}$$

$$T_6 = T_5 - \frac{m_5}{k_7} + \frac{1}{k_7}\left(A_3\lambda_{1,3}\,e^{\lambda_{1,3}\,t} + B_3\,\lambda_{2,3}\,e^{\lambda_{2,3}t}\right) \tag{35}$$

where the heat transfers are given by

$$Q_2 = c_b\,G_2(T_{10} - T_{20})\left(1 - e^{\frac{-h\,A_2}{G_2\,c_b}}\right) \tag{36}$$

$$Q_3 = c_b\,G_3\,(T_{10} - T_{30})\left(1 - e^{\frac{-h\,A_3}{G_3\,c_b}}\right) \tag{37}$$

$$Q_4 = c_b\,G_4\,(T_{10} - T_{40})\left(1 - e^{\frac{-h\,A_4}{G_4\,c_b}}\right) \tag{38}$$

$$Q_5 = c_b\,G_5(T_{10} - T_{50})\left(1 - e^{\frac{-h\,A_5}{G_5\,c_b}}\right) \tag{39}$$

$$Q_6 = c_b\,G_6\,(T_{10} - T_{60})\left(1 - e^{\frac{-h\,A_6}{G_6\,c_b}}\right) \tag{40}$$

The constants k are

$$\begin{aligned}
k_1 &= \frac{h_1 S_1}{M_1 c_{v1}},\ k_2 = \frac{h_1 S_1}{M_2 c_{v2}},\ k_3 = \frac{h_e S_2}{M_2 c_{v2}},\ k_4 = \frac{h_3 S_3}{M_3 c_{v3}},\\
k_5 &= \frac{h_3 S_3}{M_4 c_{v4}},\ k_6 = \frac{h_e S_4}{M_4 c_{v4}},\ k_7 = \frac{h_5 S_5}{M_5 c_{v5}},\ k_8 = \frac{h_5 S_5}{M_6 c_{v6}},\ k_9 = \frac{h_e S_6}{M_6 c_{v6}},
\end{aligned} \tag{41}$$

the constants m are

$$\begin{aligned}
m_1 &= \frac{Q_{g1} - R_s - Q_2 - Q_3 - Q_4 - Q_5 - Q_6}{M_1 c_{v1}},\ m_2 = \frac{Q_{g2} - E_{t2} + Q_2}{M_2 c_{v2}},\ m_3 = \frac{Q_{g3} + Q_3}{M_3 c_{v3}},\\
m_4 &= \frac{Q_{g4} - E_{t4} + Q_4}{M_4 c_{v4}},\ m_5 = \frac{Q_{g5} + Q_5}{M_5 c_{v5}},\ m_6 = \frac{Q_{g6} - E_{t6} + Q_6}{M_6 c_{v6}},
\end{aligned} \tag{42}$$

the constants λ are

$$\lambda_{1,1} = \frac{-(k_1 + k_2 + k_3) - \sqrt{(k_1 + k_2 + k_3)^2 - 4k_1k_3}}{2},$$

$$\lambda_{2,1} = \frac{-(k_1 + k_2 + k_3) + \sqrt{(k_1 + k_2 + k_3)^2 - 4k_1k_3}}{2},$$

$$\lambda_{1,2} = \frac{-(k_4 + k_5 + k_6) - \sqrt{(k_4 + k_5 + k_6)^2 - 4k_4k_6}}{2},$$

$$\lambda_{2,2} = \frac{-(k_4 + k_5 + k_6) + \sqrt{(k_4 + k_5 + k_6)^2 - 4k_4k_6}}{2}, \tag{43}$$

$$\lambda_{1,3} = \frac{-(k_7 + k_8 + k_9) - \sqrt{(k_7 + k_8 + k_9)^2 - 4k_7k_9}}{2},$$

$$\lambda_{2,3} = \frac{-(k_7 + k_8 + k_9) + \sqrt{(k_7 + k_8 + k_9)^2 - 4k_7k_9}}{2},$$

the constants C are

$$C_1 = T_e + \frac{m_2}{k_3} + \frac{(k_2 + k_3)}{k_1k_3}m_1, \quad C_2 = T_e + \frac{m_4}{k_6} + \frac{(k_5 + k_6)}{k_4k_6}m_3, \quad C_3 = T_e + \frac{m_6}{k_9} + \frac{(k_8 + k_9)}{k_7k_9}m_5, \tag{44}$$

the constants B are

$$B_1 = \frac{m_1 - k_1(T_{10} - T_{20}) + C_1\lambda_{1,1} - \lambda_{1,1}T_{10}}{\lambda_{2,1} - \lambda_{1,1}},$$

$$B_2 = \frac{m_3 - k_4(T_{30} - T_{40}) + C_3\lambda_{1,2} - \lambda_{1,2}T_{30}}{\lambda_{2,2} - \lambda_{1,2}}, \tag{45}$$

$$B_3 \frac{m_5 - k_7(T_{50} - T_{60}) + C_3\lambda_{1,3} - \lambda_{1,3}T_{50}}{\lambda_{2,3} - \lambda_{1,3}},$$

and the constants A are

$$A_1 = T_{10} - B_1 - C_1, \quad A_2 = T_{30} - B_2 - C_2, \quad A_3 = T_{50} - B_3 - C_3 \tag{46}$$

2.3.2 Results and Comparison with Experimental Data

The results of the numerical simulations, obtained with the approximate analytical solution and with the *ODE45 program* of *MATLAB*, are presented and compared with the experimental data [1]. In addition, the simulations with reduced capillary circulation or even without a heat exchanger are also presented and discussed.

List of the MATLAB Code

The *MATLAB* calculation code of the approximate analytical solution of the body model divided into three parts is reported below [2].

```matlab
%%%NOTE: 1 indicates the internal part of the trunk+head and 2 its
external part,
%%% 3 indicates the inner part of the arms and 4 its external part,
%%% 5 indicates the inner part of the legs and 6 its external part.
functiondT=myfun4(~,T)
%Define global variables
global G2 G3 G4 G5 G6 Cb hb A2 A3 A4 A5 A6 Rs u1 u2 u3 u4 u5 u6 Et1 ...
Et2 Et3 Et4 Et5 Et6 S1 S2 S3 S4 S5 S6 he Qg1 Qg2 Qg3 Qg4 Qg5 Qg6 Te
%Blood perfusion in the five parts of the body:
G2=0.00098;
G3=0.00071;
G4=0.00069;
G5=0.0019;
G6=0.0016;
%Specific heat of blood:
Cb=3744;
%Specific heats of the various parts:
Cv1=3204;
Cv2=3105;
Cv3=3204;
Cv4=3105;
Cv5=3204;
Cv6=3105;
%Density:
rho=1000;
%Defining the weight and height of the average man:
H=1.70;
W=70;
%Volumes of the various parts:
V1=0.4831*0.001*W;
V2=0.1217*0.001*W;
V3=0.08*0.001*W;
V4=0.024*0.001*W;
V5=0.2369*0.001*W;
V6=0.0543*0.001*W;
%Masses of the various parts:
M1=rho*V1;
M2=rho*V2;
M3=rho*V3;
M4=rho*V4;
M5=rho*V5;
M6=rho*V6;
%Total external surface:
S=0.203*(H^0.725)*(W^0.425);
%External surfaces
```

```
S2=0.43*S;
S4=0.184*S;
S6=0.386*S;
%External radii:
R2=2*(V1+V2)/S2;
R4=2*(V3+V4)/S4;
R6=2*(V5+V6)/S6;
%Lengths:
L2=S2/(2*pi*R2);
L4=S4/(2*pi*R4);
L6=S6/(2*pi*R6);
%Internal radii:
R1=sqrt(V1/(pi*L2));
R3=sqrt(V3/(pi*L4));
R5=sqrt(V5/(pi*L6));
%Internal surfaces:
S1=2*pi*R1*L2;
S3=2*pi*R3*L4;
S5=2*pi*R5*L6;
%Heat generated:
Qg=1.07*W;
Qg1=0.88*Qg;
Qg2=0.007*Qg;
Qg3=0.03*Qg;
Qg4=0.0024*Qg;
Qg5=0.077*Qg;
Qg6=0.0052*Qg;
%Heat loss through respiration:
Rs=10.45*W/74.4;
%Evaporation losses:
Et=10.55/1.89;
Et2=Et*S2;
Et4=Et*S4;
Et6=Et*S6;
%Overall heat transfer coefficient:
u1=10;
u3=10;
u5=10;
%Convective heat transfer coefficient in the blood:
hb=1.9*10^5;
%Convective heat transfer coefficient with the outside (assuming natural
convection):
he=5;
%To define the number of capillaries per unit volume, the average length
of each capillary and the average diameter of a capillary:
```

```
Lc=1.5e-3;
D=13e-6;
Nc=1.5*10^5;
%Capillary surface area in various parts [m^2]
A2=Nc*pi*D*Lc*V2;
A3=Nc*pi*D*Lc*V3;
A4=Nc*pi*D*Lc*V4;
A5=Nc*pi*D*Lc*V5;
A6=Nc*pi*D*Lc*V6;
%External temperature
Te=25;
%Define initial temperatures:
Ti_1=36.71;
Ti_2=33;
Ti_3=35.9;
Ti_4=32.2;
Ti_5=36;
Ti_6=32.8;
%Define Q functions
Q2=G2*Cb*(T(1)-T(2))*(1-exp(-hb*A2/(G2*Cb)));
Q3=G3*Cb*(T(1)-T(3))*(1-exp(-hb*A3/(G3*Cb)));
Q4=G4*Cb*(T(1)-T(4))*(1-exp(-hb*A4/(G4*Cb)));
Q5=G5*Cb*(T(1)-T(5))*(1-exp(-hb*A5/(G5*Cb)));
Q6=G6*Cb*(T(1)-T(6))*(1-exp(-hb*A6/(G6*Cb)));
dT=[(Qg1-Rs-Q2-Q3-Q4-Q5-Q6-u1*S1*(T(1)-T(2)))/(M1*Cv1);(Qg2+Q2-
Et2+u1*S1*(T(1)-T(2))-he*S2*(T(2)-Te))/(M2*Cv2);(end
Here is the main:
clear all
close all
clc
%t=[0:600:18000];
t=[1800:600:12000]
%Solution with Matlab function ode45:
[t T]=ode45('myfun4',t,[36.71 33 35.9 32.2 36 32.8]);
global G2 G3 G4 G5 G6 Cbhb A2 A3 A4 A5 A6 Rs u1 u2 u3 u4 u5 u6 Et1 ...
Et2 Et3 Et4 Et5 Et6 S1 S2 S3 S4 S5 S6 he Qg1 Qg2 Qg3 Qg4 Qg5 Qg6 Te
%Temperature trend as a function of time:
figure(1)
semilogx(t,T(:,1),'r',t,T(:,2),'b',t,T(:,3),'c',t,T(:,4),'g',t,T
(:,5),'y',t,T(:,6),'m','linewidth',2)
grid on
hold on
legend('T1','T2','T3','T4','T5','T6')
%title('Solution obtained with ode')
%axis([10^2.5 10^5 31.5 37])
```

```
%xlim([10^2.5 10^10])
%ylim([25 50])
set(gca,'fontsize',14)
set(gcf,'color','w')
hold on
Q2=G2*Cb*(T(:,1)-T(:,2))*(1-exp(-hb*A2/(G2*Cb)));
Q3=G3*Cb*(T(:,1)-T(:,3))*(1-exp(-hb*A3/(G3*Cb)));
Q4=G4*Cb*(T(:,1)-T(:,4))*(1-exp(-hb*A4/(G4*Cb)));
Q5=G5*Cb*(T(:,1)-T(:,5))*(1-exp(-hb*A5/(G5*Cb)));
Q6=G6*Cb*(T(:,1)-T(:,6))*(1-exp(-hb*A6/(G6*Cb)));
%Trend of heat exchanged by the capillary heat exchangers in
%function of time, considering the temperatures obtained with ODE:
figure(2)
subplot(1,2,1)
semilogx(t,Q2,'b',t,Q3,'c',t,Q4,'g',t,Q5,'y',t,
Q6,'m','linewidth',2)
legend('Q2','Q3','Q4','Q5','Q6')
set(gca,'fontsize',14)
set(gcf,'color','w')
grid on
hold on
subplot(1,2,2)
%Trend of the thermal power exchanged by the six parts, taking as
%temperatures those obtained with ODE:
Q_part1=Qg1-Rs-Q2-Q3-Q4-Q5-Q6-u1*S1*(T(:,1)-T(:,2));
Q_part2=Qg2+Q2-Et2+u1*S1*(T(:,1)-T(:,2))-he*S2*(T(:,2)-Te);
Q_part3=Qg3+Q3-u3*S3*(T(:,3)-T(:,4));
Q_part4=Qg4+Q4-Et4+u3*S3*(T(:,3)-T(:,4))-he*S4*(T(:,4)-Te);
Q_part5=Qg5+Q5-u5*S5*(T(:,5)-T(:,6));
Q_part6=Qg6+Q6-Et6+u5*S5*(T(:,5)-T(:,6))-he*S6*(T(:,6)-Te);
semilogx(t,Q_part1,'r',t,Q_part2,'b',t,Q_part3,'c',t,Q_part4,'g',
t,Q_part5,'y',t,Q_part6,'m','linewidth',legend('Q_{part1}','Q_
{part2}','Q_{part3}','Q_{part4}','Q_{part5}','Q_{part6}')
set(gca,'fontsize',14)
set(gcf,'color','w')
grid on
hold on
```

The numerical values of the parameters used in the numerical simulations are taken in accordance with [1]. The densities are

$$\rho_1 = \rho_2 = \rho_3 = \rho_4 = \rho_5 = \rho_6 = 1000 \ \frac{\text{kg}}{\text{m}^3}, \tag{47}$$

the specific heats of the internal portions of the body parts are

$$c_{v1} = c_{v3} = c_{v5} = 3204 \ \frac{\text{J}}{\text{kg K}}, \tag{48}$$

and the specific heats of the external portions of the body parts are

$$c_{v2} = c_{v4} = c_{v6} = 3150 \ \frac{\text{J}}{\text{kg K}} \tag{49}$$

The volumes of the six portions are evaluated on the basis of the body weight W, according to the proportion used in [3]:

$$V_1 = 0.4831 \ W \ 0.001; \ V_2 = 0.1217 \ W \ 0.001; \ V_3 = 0.08 \ W \ 0.001;$$
$$V_4 = 0.024 \ W \ 0.001; \ V_5 = 0.2369 \ W \ 0.001; \ V_6 = 0.0543 \ W \ 0.001. \tag{50}$$

The total metabolic heat generated by the body is assumed as $Q_g = 1.07 \ W$, while the metabolic heats of the various portions are calculated according to the proportion used in [3]:

$$Q_{g1} = 0.88 \ Q_g; \ Q_{g2} = 0.007 \ Q_g; \ Q_{g3} = 0.03 \ Q_g;$$
$$Q_{g4} = 0.0024 \ Q_g; \ Q_{g5} = 0.077 \ Q_g; \ Q_{g6} = 0.0052 \ Q_g. \tag{51}$$

The heat lost through respiration from the internal portion of the central part is proportional to the body weight, according to

$$R_s = 10.45 \frac{W}{74.4}. \tag{52}$$

while the heat lost through evaporation from each portion is proportional to the total body surface,

$$S = 0.203 H^{0.725} \ W^{0.425} \tag{53}$$

according to the following proportions

$$S_2 = 0.43 \ S; \ S_4 = 0.184 \ S; \ S_2 = 0.386 \ S. \tag{54}$$

The heat lost through evaporation from the body per unit area is given by

$$E_t = \frac{10.55}{1.89}, \tag{55}$$

while the heat lost through evaporation from each external portion is

$$E_{t2} = E_t\, S_2; E_{t4} = E_t\, S_4; E_{t6} = E_t\, S_6 \tag{56}$$

The overall heat exchange coefficient between the internal and external portions is equal to

$$U_1 = U_3 = U_5 = 10\,\frac{W}{m^2\,K}. \tag{57}$$

The surfaces of each body portion are evaluated, assuming each part as a cylinder, whose outer radius is given by

$$R_2 = 2\,\frac{V_1 + V_2}{S_2}; R_4 = 2\,\frac{V_3 + V_4}{S_4}; R_6 = 2\,\frac{V_5 + V_6}{S_6} \tag{58}$$

The relative length is given by

$$L_2 = \frac{S_2}{2\pi R_2}; L_4 = \frac{S_4}{2\pi R_4}; L_6 = \frac{S_6}{2\pi R_6}, \tag{59}$$

The inner radius is

$$R_1 = \sqrt{\frac{V_1}{\pi L_2}}; R_2 = \sqrt{\frac{V_3}{\pi L_4}}; R_5 = \sqrt{\frac{V_5}{\pi L_6}}, \tag{60}$$

The surfaces of the internal portions are given by

$$S_1 = 2\,\pi\,R_1\,L_2; S_3 = 2\,\pi\,R_3\,L_4; S_5 = 2\,\pi\,R_5\,L_6, \tag{61}$$

The convective heat transfer coefficient from the external surfaces of the peripheral portions is in the range

$$h_e = 3 - 5\,\frac{W}{m^2\,K} \tag{62}$$

The specific heat of the blood is

$$c_b = 3744\,\frac{J}{kg\,K} \tag{63}$$

The blood perfusion in the five different portions of the body is given by

$$G_2 = 0.00098\left(\frac{Kg}{s}\right); G_3 = 0.00071\left(\frac{Kg}{s}\right); G_4 = 0.0069\left(\frac{Kg}{s}\right);$$
$$G_5 = 0.0019\left(\frac{Kg}{s}\right); G_6 = 0.0016\left(\frac{Kg}{s}\right). \tag{64}$$

The convective heat transfer coefficient of the blood in the capillaries (which act as heat exchangers) is calculated, assuming laminar flow with a Nusselt number equal to 4; therefore,

$$h = \frac{Nu\,k}{D} = \frac{4^*0.628}{13^*10^{-6}} = 1.9^*10^5\,\frac{W}{m^2\,K} \tag{65}$$

where D is the diameter of the capillary.

Finally, the heat exchange surface of each capillary is given by

$$A = V\,N_c\,\pi\,D\,L_c \tag{66}$$

where the number of capillaries per unit volume is

$$N_c = 10^5\,(m^{-3}) \tag{67}$$

while the average length of each capillary is

$$L_c = 1.5^*10^{-3}\,(m). \tag{68}$$

Comparison with the Experimental Measurements on a Woman of Height 1.60 m and Weight 66 Kg (Case D_1)

The initial temperatures of the various portions of the body parts are assumed as T_{10} (blue line) = 36.7 °C, T_{20} (red line) = 33.2 °C, T_{30} (green line) = 35.9 °C, T_{40} (yellow line) = 32.6 °C, T_{50} (purple line) = 36.0 °C, and T_{60} (sky blue line) = 32.8 ° C, in accordance with the experimental values of case D_1 [1].

The temperatures of the various portions of the body, given by Eqs. 17–22, are obtained numerically with the *ODE45* program of the *MATLAB* software for $N_c = 10^5\,(m^{-3})$ and reported in Fig. 10 up to the steady state. They are compared with the experimental measurements of case D_1 [1], obtained until the end of anaesthesia, which occur after about 10^4 s, or about 3 h. The temperatures of the internal portions, T_1 (blue line), T_3 (green line), *and* T_5 (purple line), decrease over time to reach values of about $T_1 = 36.4°C$, $T_3 = 35.0°C$, and $T_5 = 35.4°C$ after 3 h. The temperatures T_1 (blue line) are in good agreement with the experimental measurements of T_1 (blue points), measured until the end of anaesthesia. The temperatures of the external portions T_2 (red line), T_4 (yellow line), *and* T_6 (sky blue line), on the other hand, increase during the operation, with a maximum that depends on the portion but can be placed between 2000 s for $T_4 = 33.0°C$ and $T_6 = 33.2°C$ and 5000 s for $T_4 = 33.5°C$. The experimental measurements, T_2 (red points), T_4 (yellow points), *and* T_6 (sky blue points), are quite dispersed, but a general trend to increase and decrease can be detected. The numerical solutions then predict a decrease to the steady state, where the temperatures reached are $T_1 = 35.1°C$, $T_2 = 32.4°C$, $T_3 = 33.7°C$, $T_4 = 31.7°C$, $T_5 = 34.0°C$, and $T_6 = 32.0°C$, respectively.

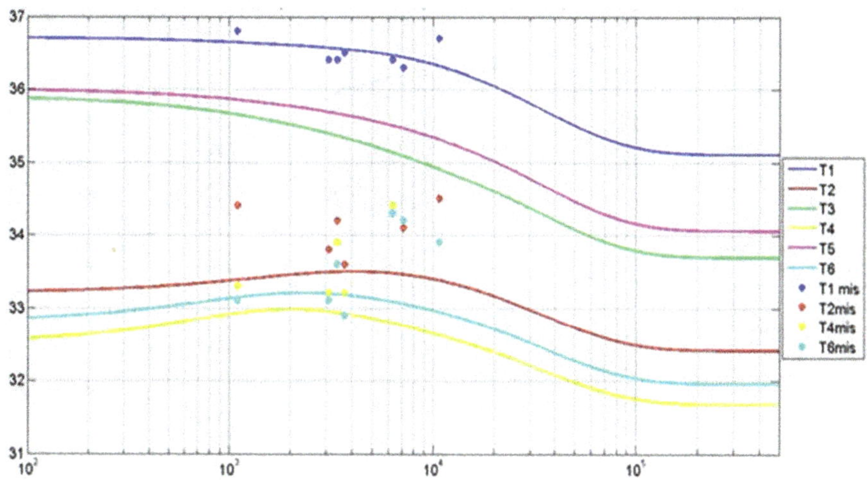

Fig. 10 Numerical temperatures (°C, lines) versus time (s) of the six portions of the body, obtained with the *ODE45* program of the *MATLAB* software for $N_c = 10^5$ (m^{-3}), compared with the experimental values (dots) of the case D_1 [2]. T_1: blue line and dots. T_2: red line and dots. T_3: green line. T_4: yellow line and dots. T_5: purple line. T_6: sky blue line and dots

In conclusion, the numerical solutions predict a decrease of T_1 and an increase, followed by a decrease, for T_2, T_4, and T_6, in agreement with the experimental measurements for T_1, while the experimental measurements of T_2, T_4, and T_6 are much more dispersed. The numerical solutions predict a decrease of T_3 and T_5.

The temperatures of the various body portions, obtained with the approximate analytical solutions of Eqs. 30–35 for $N_c = 10^5$ (m^{-3}) are reported in Fig. 11 up to the steady state. They are compared with the experimental measurements of case D_1 [1], obtained until the end of anaesthesia, after 10^4 s, i.e. about 3 h. The temperatures of the internal portions, T_1 (blue line), T_3 (green line), *and* T_5 (purple line), decrease with time and reach values of $T_1 = 36.4$°C, $T_3 = 34.7$°C, and $T_5 = 35.3$°C after 3 h, in general agreement with the experimental measurements of T_1 (blue points), measured until the end of anaesthesia. The temperatures of the external portions, T_2 (red line), T_4 (yellow line), *and* T_6 (sky blue line), on the other hand, increase during the operation, with a maximum that depends on the portion but can be placed between 3000 *s* for $T_4 = 33.2$°C and $T_6 = 33.4$°C and 5000 s for $T_4 = 33.6$°C. The experimental measurements, T_2 (red points), T_4 (yellow points), *and* T_6 (sky blue points), are quite dispersed, but a general trend to increase and decrease can be detected. The approximate analytical solutions predict a decrease of T_3 and T_5. The approximate analytical solution then predicts a decrease until the steady state, where the temperatures reached are $T_1 = 34.9$°C, $T_2 = 32.6$°C, $T_3 = 32.8$°C, $T_4 = 31.3$°C, $T_5 = 33.7$°C, *and* $T_6 = 32.0$°C.

The comparison among the temperatures obtained with the approximate analytical solution, that is, with the heat transfer constant during anaesthesia and equal to the initial ones, and the numerical solution, obtained with *ODE45* of the *MATLAB*

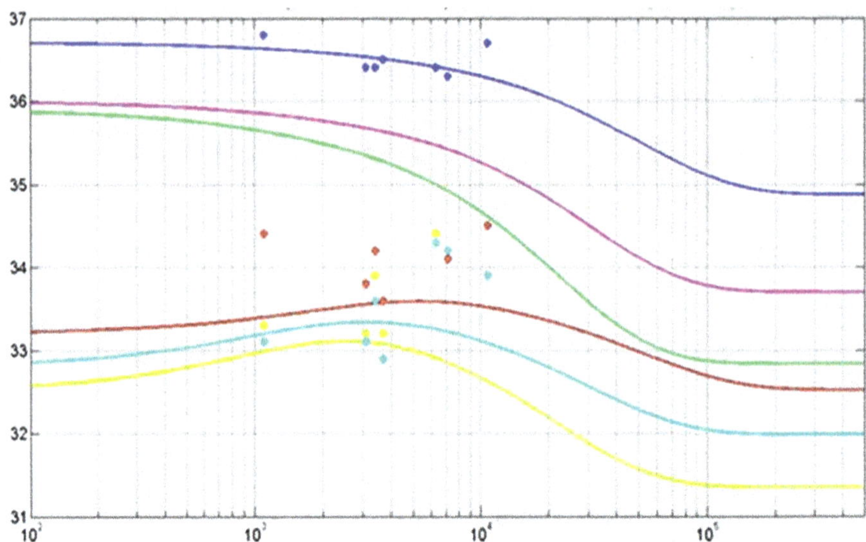

Fig. 11 Analytical temperatures (°C, lines) versus time (s) of the six portions of the body, obtained with the approximate solution for $N_c = 10^5$ (m^{-3}), compared with the experimental values (dots) of the case D_1 [2]. T_1: blue line and dots. T_2: red line and dots. T_3: green line. T_4: yellow line and dots. T_5: purple line. T_6: sky blue line and dots

software, is positive, especially during the anaesthesia transient. Even the steady-state temperatures reached with the approximate analytical solution are not too different from the numerical solution. In conclusion, the approximate analytical solution is able to predict a decrease of T_1 and an increase, followed by a decrease, for T_2, T_4, and T_6.

The heat transfer per unit of time *(W)*, exchanged among the internal central part and the other internal and external portions, versus time *(s)*, given by Eqs. 25–29, and calculated with *ODE45* of the *MATLAB* software for $N_c = 10^5$ (m^{-3}) are shown in Fig. 12. The values of the heat transfer, which, at the initial time, are Q_2 (blue line) $= 12$ W, Q_3 (red line) $= 2$ W, Q_4 (green line) $= 5.5$ W, Q_5 (yellow line) $= 5$ W, *and* Q_6 (purple line) $= 12$ W, are not constant during anaesthesia but tend to decrease or increase until they reach the steady-state values. The heat, Q_2 (blue line) and Q_6 (purple line), decrease during anaesthesia and reach a value of around 11 W at its end and then increase, up to the steady state, when they reach about 14.5 W. Also, Q_4 (green line) decreases during anaesthesia, with a minimum of about 5 W, and reaches almost the same initial value at its end, after which it increases until the steady state, when it reaches about 7 W. The heat exchanged by the inner portions increases until the steady state: Q_3 (red line) from 2 W to over 4 W at the end of anaesthesia and 7 W at the steady state and Q_5 (yellow line) from 5 W to 9 W at the end of anaesthesia and about 18 W at the steady state. These variations, however, do not seem to have too much influence on the temperatures, as shown previously.

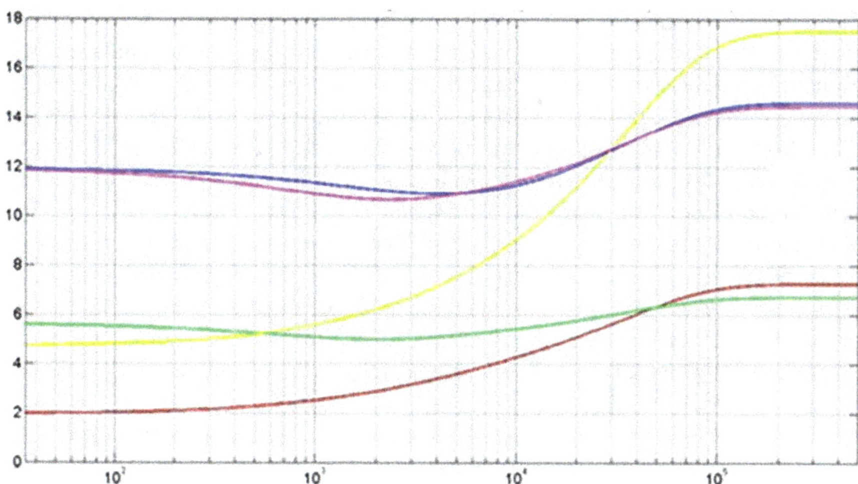

Fig. 12 Heat transfer (W) versus time (s) among the internal central portion, P1, and the other five portions, calculated with *ODE45* of the *MATLAB* software for $N_c = 10^5$ (m^{-3}) [2]. Q_2: blue line, Q_3: red line, Q_4: green line, Q_5: yellow line, and Q_6: purple line

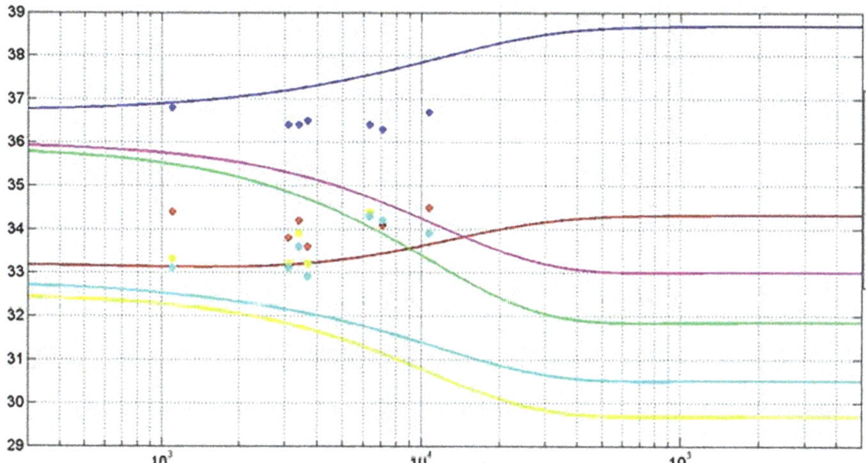

Fig. 13 Numerical temperatures (°C, lines) versus time (s) of the six portions of the body, obtained with the *ODE45* program of the *MATLAB* software for $N_c = 10^4$ (m^{-3}), compared with the experimental values (dots) of the case D_1 [2]. T_1: blue line and dots. T_2: red line and dots. T_3: green line. T_4: yellow line and dots. T_5: purple line. T_6: sky blue line and dots

The temperatures of the various portions of the body, given by Eqs. 17–22 and solved numerically with the *ODE45* program of the *MATLAB software* for a reduced blood perfusion, corresponding to a number of capillaries of $N_c = 10^4$ (m^{-3}), instead of $N_c = 10^5$(m^{-3}), as in the previous case, are reported in Fig. 13 up to the steady state. They are compared with the experimental measurements of case D_1 [1], obtained until the end of anaesthesia, after about 10^4 s, or about 3 h. The temperature

T_1 (blue line) does not decrease, as in the previous case, because the heat exchanger has a smaller heat transfer surface, having a reduced number of capillaries, and exchanges less heat from the internal central portion, reaching $T_1 = 37.8°C$ at the end of anaesthesia, after 3 h, and continuing to grow until the steady-state value of $T_1 = 38.7°C$. The temperatures of the internal portions, T_3 (green line) and T_5 (purple line) decrease over time, reaching the values of about $T_3 = 33.4°C$ and $T_5 = 34.3°C$ at the end of anaesthesia and $T_3 = 31.9°C$ and $T_5 = 33.0°C$ at the steady state. The temperature of the outer portion T_2 (red line) increases in agreement with the experimental measurements and reaches $33.5°C$ at the end of anaesthesia and $T_2 = 34.3°C$ at the steady state. The temperatures of the outer portions T_4 (yellow line) and T_6 (sky blue line), on the other hand, decrease during anaesthesia when they reach $T_4 = 30.8°C$ and $T_6 = 31.4°C$ and then arrive at the steady state at $T_4 = 29.7°C$ and $T_6 = 30.5°C$.

In conclusion, the numerical solution for a reduced blood perfusion, corresponding to a number of capillaries of $N_c = 10^4$ (m^{-3}), instead of $N_c = 10^5 (m^{-3})$, does not predict a decrease of T_1 but, rather, only an increase, while it predicts only an increase of T_2 and only a decrease of T_4 and T_6.

The temperatures of the various portions of the body, obtained with the approximate analytical solution, Eqs. 30–35 for a reduced blood perfusion and corresponding to a number of capillaries of $N_c = 10^4$ (m^{-3}), instead of $N_c = 10^5 (m^{-3})$, are reported in Fig. 14 up to the steady state. They are compared with the experimental measurements of case D_1 [1], obtained until the end of anaesthesia, after about 10^4 s, or about 3 h. The temperature T_1 (blue line) does not decrease because the heat exchanger has a smaller heat transfer surface, having a reduced

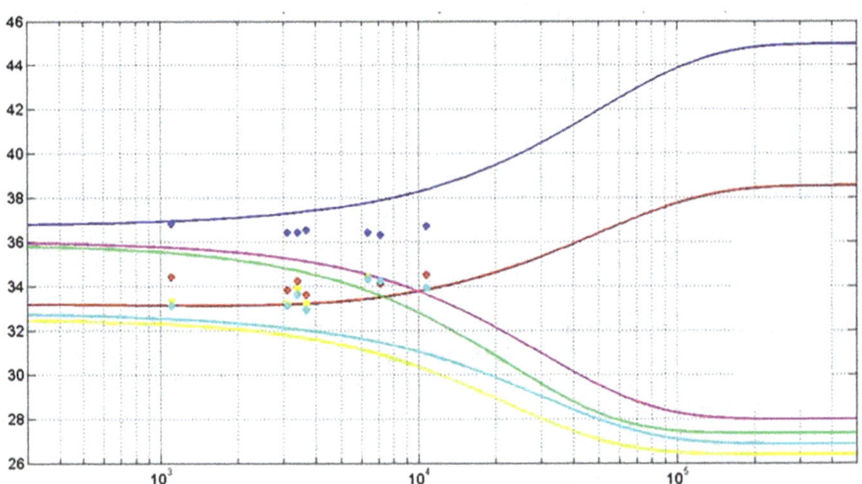

Fig. 14 Analytical temperatures (°C, lines) versus time (s) of the six portions of the body, obtained with the approximate solution for $N_c = 10^4$ (m^{-3}), compared with the experimental values (dots) of the case D_1 [2]. T_1: blue line and dots. T_2: red line and dots. T_3: green line. T_4: yellow line and dots. T_5: purple line. T_6: sky blue line and dots

number of capillaries, and the heat transfer from the internal central portion is smaller. Then T_1 increases, reaching 38.1 °C at the end of anaesthesia, after 3 h, and continuing to increase until the steady-state value of $T_1 = 45.0$°C. The temperatures of the internal portions, T_3 (green line) and T_5 (purple line) decrease over time, reaching values of about $T_3 = 32.8$°C and $T_5 = 33.8$°C at the end of anaesthesia, after about 3 h, and $T_3 = 27.3$°C and $T_5 = 28.0$ °C at the steady state. The temperature of the outer portion T_2 (red line) increases in agreement with the experimental measurements and reaches 34.0°C at the end of anaesthesia and $T_2 = 38.4$°C at the steady state. The temperatures of the outer portions T_4 (yellow line) and T_6 (sky blue line), conversely, decrease during anaesthesia and reach $T_4 = 30.3$°C and $T_6 = 31.0$°C at the end of anaesthesia, after about 3 h, to then reach the steady state at $T_4 = 26.5$°C and $T_6 = 27.0$°C.

In conclusion, the approximate analytical solution for a reduced blood perfusion, corresponding to a number of capillaries of $N_c = 10^4$ (m^{-3}), instead of $N_c = 10^5 (\mathrm{m}^{-3})$, does not predict a decrease of T_1 but, rather, only an increase, while it predicts only an increase of T_2 and only a decrease of T_4 and T_6, in agreement with the numerical solutions obtained with the *ODE45* program of the *MATLAB software*.

The variations of heat transfers per unit of time (W) between the internal central portion and the other portions versus time (s), given by Eqs. 25–29 and calculated with *ODE45* of the *MATLAB* software, with a reduced number of capillaries, equal to $N_c = 10^4$ (m^{-3}), are reported in Fig. 15. The values of heat transfers at the initial time are much smaller than in the case with $N_c = 10^5$ (m^{-3}), i.e. $Q_2 = Q_6 = 12$ W, $Q_3 = 2$ W, $Q_4 = 5.5$ W, and $Q_5 = 5$ W, showing that the heat transfers greatly reduce due to the reduction in the number of capillaries. The initial heat Q_2 (blue line) is

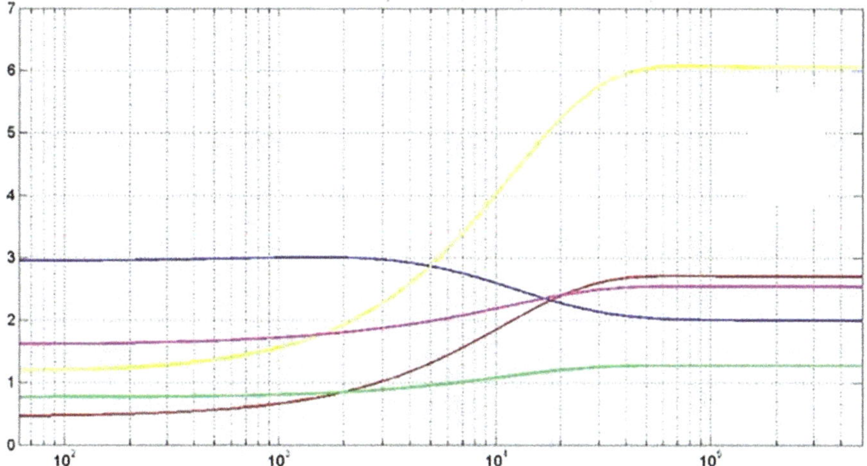

Fig. 15 Heat transfer (W) versus time (s) among the internal central portion, P1, and the other five portions, calculated with *ODE45* of the *MATLAB* software for $N_c = 10^4$ (m^{-3}) [2]. Q_2: blue line, Q_3: red line, Q_4: green line, Q_5: yellow line, and Q_6: purple line

now reduced to 3 W, while Q_6 (purple line) is reduced to 1.6 W, Q_4 (green line) 0.8 W, Q_5 (yellow line) to 1.2 W, and Q_3 (red line) *to* 0.5 W.

Figure 15 indicates that the heat transfers are not constant during anaesthesia. Q_3 (red line) increases to about 2.7 W, Q_4 (green line) increases to 1.2 W, Q_5 (yellow line) increases to 6 W, and Q_6 (purple line) increases to about 2.5 W. Only Q_2 (blue line) reduces to 2 W. These variations do not seem to affect the temperatures too much.

The temperatures of the various portions of the body, given by Eqs. 17–22 and solved numerically with the *ODE45* program of the *MATLAB software*, without blood perfusion, i.e. corresponding to a null number of capillaries, $N_c = 0$ (m^{-3}), that is, in the absence of the heat exchanger that redistributes the heat between the internal central portion and the other portions, are reported in Fig. 16, up to the steady state. They are compared with the experimental measurements of case D_1 [1], obtained until the end of anaesthesia, after about 10^4 s, or 3 h. The temperature T_1 (blue line) increases, as in the previous cases, since the heat exchanger is not present because it has a null heat transfer surface. It reaches 38.0°C at the end of anaesthesia, after 3 h, continuing to increase until the steady-state value of $T_1 = 47.2$°C. The temperatures of the internal portions, T_3 (green line) and T_5 (purple line), decrease over time, reaching the values of about $T_3 = 32.5$°C and $T_5 = 33.0$°C at the end of anaesthesia and $T_3 = 26.2$°C and $T_5 = 26.7$°C at the steady state. The temperature of the outer portion T_2 (red line) increases and reaches 33.5°C at the end of anaesthesia and $T_2 = 39.7$°C at the steady state. The temperatures of the outer portions T_4 (yellow line) and T_6 (light blue line), on the other hand, decrease and, at the end of

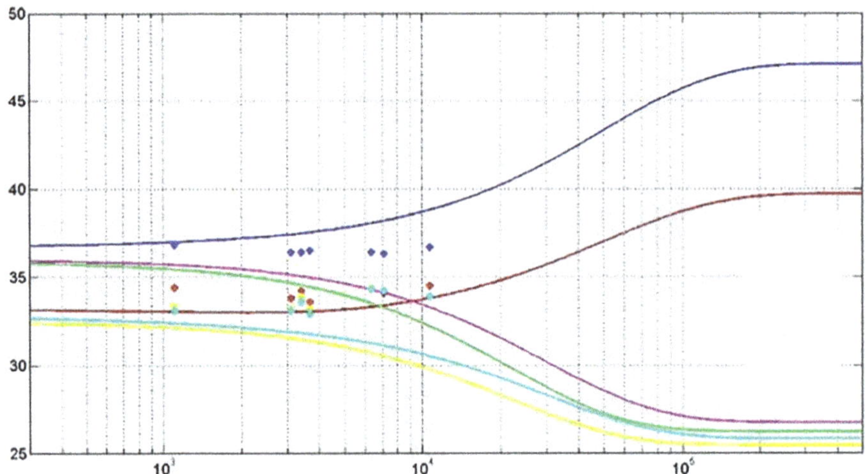

Fig. 16 Numerical temperatures (°C, lines) versus time (s) of the six portions of the body, obtained with the *ODE45* program of the *MATLAB* software for $N_c = 0$ (m^{-3}), compared with the experimental values (dots) of the case D_1 [2]. T_1: blue line and dots. T_2: red line and dots. T_3: green line. T_4: yellow line and dots. T_5: purple line. T_6: sky blue line and dots

anaesthesia, reach $T_4 = 30.0°C$ and $T_6 = 31.0°C$, then arriving at the steady state at $T_4 = 25.4°C$ and $T_6 = 25.8°C$.

In conclusion, the numerical solution in the absence of blood perfusion, i.e. corresponding to a number of capillaries of $N_c = 0$ (m^{-3}), instead of $N_c = 10^5$(m^{-3}), does not predict a decrease of T_1 but, rather, only an increase, while it predicts only an increase of T_2 and only a decrease of T_3, T_4, T_5, and T_6. The temperatures of the various portions of the body, obtained with the approximate analytical solution of Eqs. 30–35 and with a number of null capillaries, $N_c = 0$ (m^{-3}), coincide with the numerical temperatures due to the absence of heat exchangers.

Comparison with the Experimental Measurements on a Woman of Height 1.50 m and Weight 50 Kg (Case D_2)

The temperatures of the various portions of the body, given by Eqs. 17–22 and solved numerically with the *ODE45* program of the *MATLAB* software for $N_c = 10^5$ (m^{-3}), are reported in Fig. 17 up to the steady state. The initial temperatures of the various portions of the body parts in case D_2 are assumed to be $T_{10} = 37.2\ °C$, $T_{20} = 33.6\ °C$, $T_{30} = 36.4\ °C$, $T_{40} = 33.2\ °C$, $T_{50} = 36.5\ °C$, *and* $T_{60} = 33.4\ °C$, according to the experimental values measured in case D_2. They are compared with the experimental measurements of case D_2 [1], obtained until the end of anaesthesia, which occurred after $1.2·10^4$ s, or about 3.3 h. The temperatures of the internal portions, T_1 (blue line), T_3 (green line), *and* T_5 (purple line), decrease over time and reach values of about $T_1 = 36.5\ °C$, $T_3 = 35.2\ °C$, and $T_5 = 35.5\ °C$ after about 3.3 h. The temperature decrease of T_1 (blue line) is in good agreement

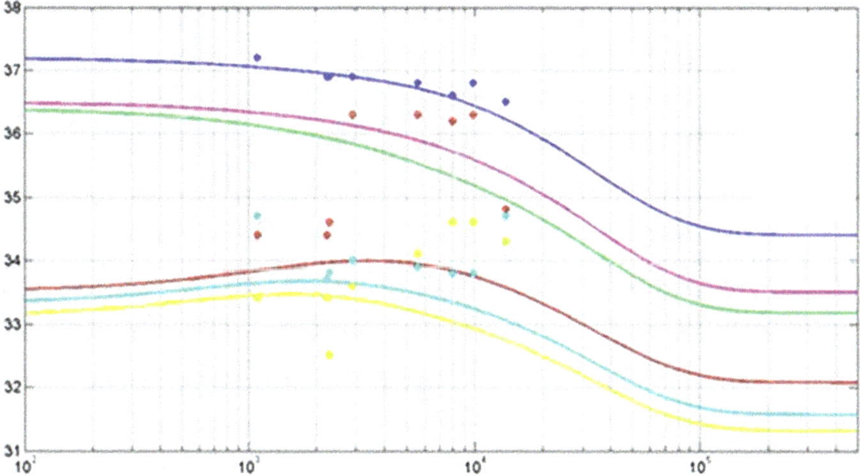

Fig. 17 Numerical temperatures (°C, lines) versus time (s) of the six portions of the body, obtained with the *ODE45* program of the *MATLAB* software for $N_c = 10^5$ (m^{-3}), compared with the experimental values (dots) of the case D_2 [2]. T_1: blue line and dots. T_2: red line and dots. T_3: green line. T_4: yellow line and dots. T_5: purple line. T_6: sky blue line and dots

with the experimental decrease of the measurements of T_1 (blue points). The temperatures of the external portions T_2 (red line), T_4 (yellow line), and T_6 (sky blue line), on the other hand, increase during the operation with a maximum, which depends on the portion but can be placed between 4000 s for $T_2 = 34.0$ °C and 2000 s for $T_6 = 33.6$ °C and $T_4 = 33.4$ °C. The experimental measurements, T_2 (red points), T_4 (yellow points), and T_6 (sky blue points), are quite dispersed, but a general trend to increase and decrease can be detected. The numerical model then predicts a decrease until the steady state, where the temperatures reach $T_1 = 34.4$ °C, $T_2 = 32.1$ °C, $T_3 = 33.2$ °C, $T_4 = 31.4$ °C, $T_5 = 33.5$ °C, and $T_6 = 31.5$ °C, respectively.

In conclusion, the numerical solution is able to predict a decrease of T_1 and an increase, followed by a decrease, for T_2, T_4, and T_6, in agreement with the trend of the experimental measurements. The numerical solutions predict a decrease of T_3 and T_5.

The temperatures of the various body portions, obtained with the approximate analytical solution of Eqs. 30–35, for $N_c = 10^5$ (m^{-3}), are reported in Fig. 18 up to the steady state. They are compared with the experimental results of case D_2 [1], obtained until the end of anaesthesia, which occurred after $1.2 \cdot 10^4$ s, or about 3.3 h. The temperatures of the internal portions, T_1 (blue line), T_3 (green line), and T_5 (purple line), decrease over time and reach values of approximately $T_1 = 36.2$ °C, $T_3 = 35.0$ °C, and $T_5 = 35.6$ °C after about 3.3 h, in general agreement with the experimental measurements of T_1 (blue points), measured until the end of anaesthesia. The temperatures of the outer portions T_2 (red line), T_4 (yellow line), and T_6 (sky blue line), on the other hand, increase during anaesthesia with a maximum, which

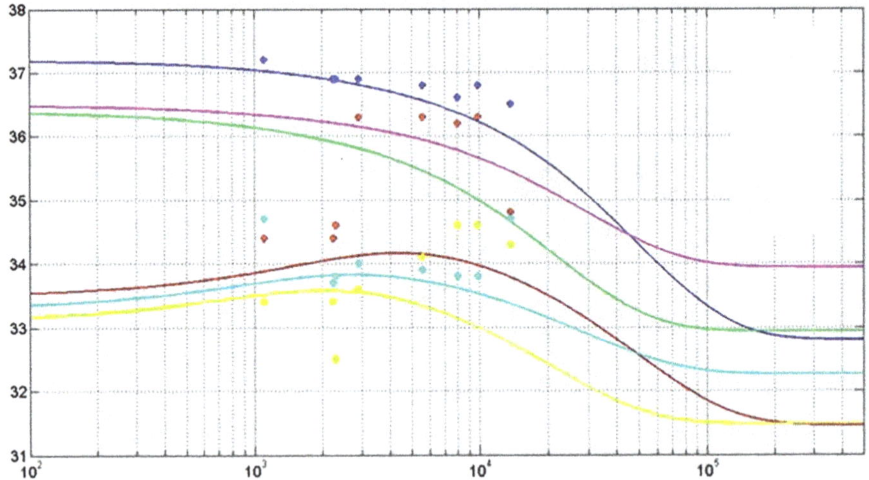

Fig. 18 Analytical temperatures (°C, lines) versus time (s) of the six portions of the body, obtained with the approximate solution for $N_c = 10^5$ (m^{-3}), compared with the experimental values (dots) of the case D_2 [2]. T_1: blue line and dots. T_2: red line and dots. T_3: green line. T_4: yellow line and dots. T_5: purple line. T_6: sky blue line and dots

depends on the portion but can be located between 4000 s for $T_2 = 34.2\ °C$ and 3000 s for $T_6 = 33.8\ °C$ and 2000 s for $T_4 = 33.6\ °C$, in reasonable agreement with the experimental measurements, T_2 (red points), T_4 (yellow points), *and* T_6 (sky blue points), which are quite dispersed. The approximate analytical model then predicts a decrease until the steady state, where the temperatures reached are $T_1 = 32.8\ °C$, $T_2 = 31.5\ °C$, $T_3 = 32.9\ °C$, $T_4 = 31.5\ °C$, $T_5 = 33.9\ °C$, and $T_6 = 32.3\ °C$, respectively.

In conclusion, the approximate analytical solution is able to predict a decrease in T_1 and an increase, followed by a decrease, for T_2, T_4, and T_6, in agreement with the trend of the experimental measurements. The approximate analytical solution predicts a decrease in T_3 and T_5. The comparison among the temperatures obtained with the approximate analytical solution, that is, with the heat transfer constant during the anaesthesia and equal to the initial ones, and the solution obtained with *ODE45* of the *MATLAB* software is quite positive, especially during the transient of the anaesthesia, even if the steady-state temperatures reached with the approximate analytical solution are different from the previous solution on the order of 1.5 °C at most.

The heat transfer per unit of time (W) between the internal central portion and the other portions versus time (s), given by Eqs. 25–29 and calculated with *ODE45* of the *MATLAB software* for $N_c = 10^5$ (m^{-3}), are shown in Fig. 19. The values of the heat transfer, which, at the initial time, are Q_2 (blue line) $= 12$ W, Q_3 (red line), $= 2$ W, Q_4 (green line) $= 4.5$ W, Q_5 (yellow line) $= 4.4$ W, and Q_6 (purple line) $= 9.5$ W, are not constant during anaesthesia but tend to decrease or increase until they reach the steady-state values. The heat Q_2 (blue line) decreases during

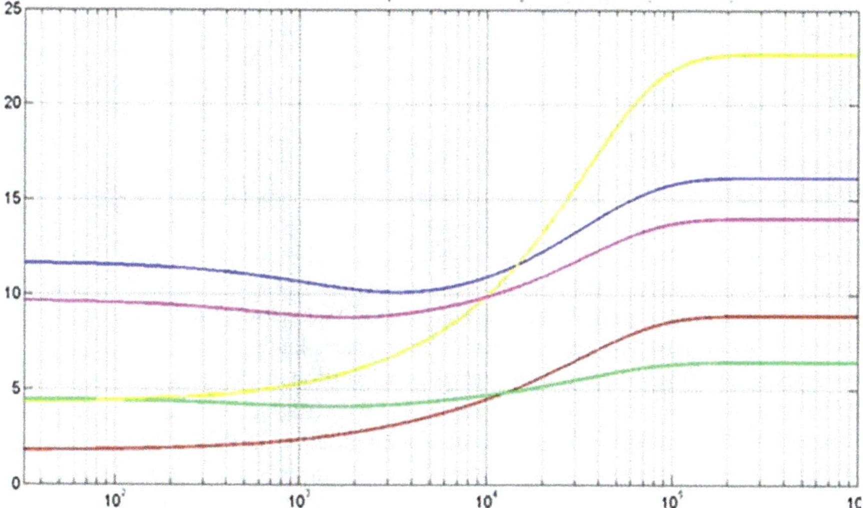

Fig. 19 Heat transfer (W) versus time (s) among the internal central portion, P1, and the other five portions, calculated with *ODE45* of the *MATLAB* software for $N_c = 10^5$ (m^{-3}) [2]. Q_2: blue line, Q_3: red line, Q_4: green line, Q_5: yellow line, and Q_6: purple line

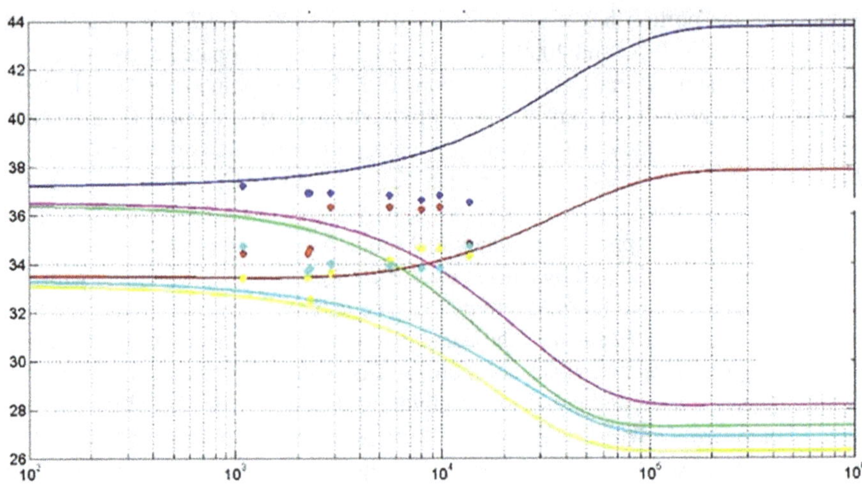

Fig. 20 Numerical temperatures (°C, lines) versus time (s) of the six portions of the body, obtained with the *ODE45* program of the *MATLAB* software for $N_c = 10^3$ (m^{-3}), compared with the experimental values (dots) of the case D_2 [2]. T_1: blue line and dots. T_2: red line and dots. T_3: green line. T_4: yellow line and dots. T_5: purple line. T_6: sky blue line and dots

anaesthesia, with a minimum of around 10 W, and then returns to its initial value at the end of anaesthesia to reach 17 W at steady state. The heat Q_6 (purple line) has a similar trend because, after a minimum of about 8 W, it returns to the initial value at the end of anaesthesia and reaches about 13 W at the steady state. Also, the heat Q_4 (green line) has a similar trend because, after a slight minimum, it returns to the initial value at the end of anaesthesia and reaches about 7 W at steady state. The heats, Q_3 (red line) and Q_5 (yellow line), do not have a minimum but increase up to about 8 *W* and about 23.0 *W* at the steady state, respectively. These variations, however, do not seem to affect the temperatures too much, as shown previously.

The temperatures of the various portions of the body, given by Eqs. 17–22 and solved numerically with the *ODE45* program of the *MATLAB software* for a reduced blood perfusion corresponding to the number of capillaries of $N_c = 10^3$ (m^{-3}), are reported in Fig. 20 up to the steady state. They are compared with the experimental measurements of case D_2 [1], obtained until the end of anaesthesia, after $1.2 \cdot 10^4$ s, about 3.3 h. The temperature T_1 (blue line) does not decrease, as in the previous case, because the heat exchanger has a smaller surface, having a reduced number of capillaries, and the heat transfer from the central portion of the central part is smaller. The temperature T_1 (blue line) reaches about 39.0 °C at the end of anaesthesia and continues to increase until the steady-state value of $T_1 = 43.9$ °C. The temperatures of the internal portions, T_3 (green line) and T_5 (purple line), decrease over time, reaching the values of about $T_3 = 32.0$ °C and $T_5 = 33.0$ °C at the end of anaesthesia and $T_3 = 27.3$ °C and $T_5 = 28.1$ °C at the steady state. The temperature of the outer portion T_2 (red line) increases in accordance with the experimental measurements and reaches 34.5 °C at the end of anaesthesia and $T_2 = 38.0$ °C at the steady state.

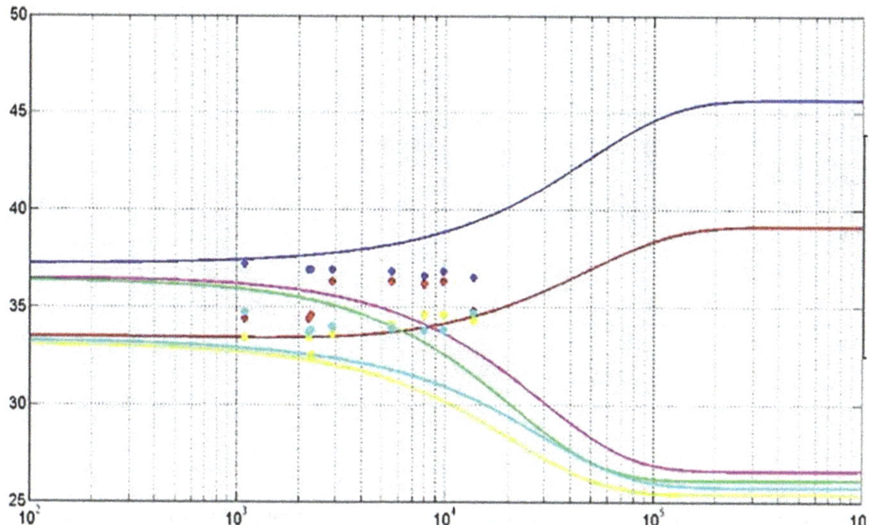

Fig. 21 Analytical temperatures (°C, lines) versus time (s) of the six portions of the body, obtained with the approximate solution for $N_c = 10^3$ (m^{-3}), compared with the experimental values (dots) of the case D_2 [2]. T_1: blue line and dots. T_2: red line and dots. T_3: green line. T_4: yellow line and dots. T_5: purple line. T_6: sky blue line and dots

The temperatures of the outer portions T_4 (yellow line) and T_6 (sky blue line), on the other hand, decrease during anaesthesia when they reach $T_4 = 29.0$ °C and $T_6 = 30.0$ °C, and then arrive at the steady state at $T_4 = 26.3$ °C and $T_6 = 27.0$ °C.

In conclusion, the numerical solution for a reduced blood perfusion, corresponding to a number of capillaries of $N_c = 10^3$ (m^{-3}), instead of $N_c = 10^5$(m^{-3}), does not predict a decrease of T_1 but, rather, only an increase, while it predicts only an increase of T_2 and only a decrease of the other temperatures.

The temperatures of the various portions of the body, obtained with the approximate analytical solution, Eqs. 30–35, for a reduced blood perfusion, corresponding to a number of capillaries of $N_c = 10^3$ (m^{-3}), are reported in Fig. 21 up to the steady state. They are compared with the experimental measurements of case D_2 [1], obtained until the end of anaesthesia, after $1.2 \cdot 10^4$ s, about 3.3 h. The temperature T_1 (blue line) does not decrease because the heat exchanger has a smaller heat transfer surface, having a reduced number of capillaries, and the heat transfer from the internal central portion is smaller; so T_1 increases, reaching 39.5 °C at the end of anaesthesia and continuing to increase to the steady-state value of $T_1 = 45.6$ °C. The temperatures of the internal portions, T_3 (green line) and T_5 (purple line), decrease over time, reaching values of about $T_3 = 31.0$ °C and $T_5 = 32.0$ °C at the end of anaesthesia and $T_3 = 26.0$ °C and $T_5 = 26.6$ °C at the steady state. The temperature of the outer portion T_2 (red line) increases and reaches 34.0 °C at the end of anaesthesia and $T_2 = 39.0$ °C at the steady state. The temperatures of the outer portions T_4 (yellow line) and T_6 (light blue line), on the other hand, decrease during

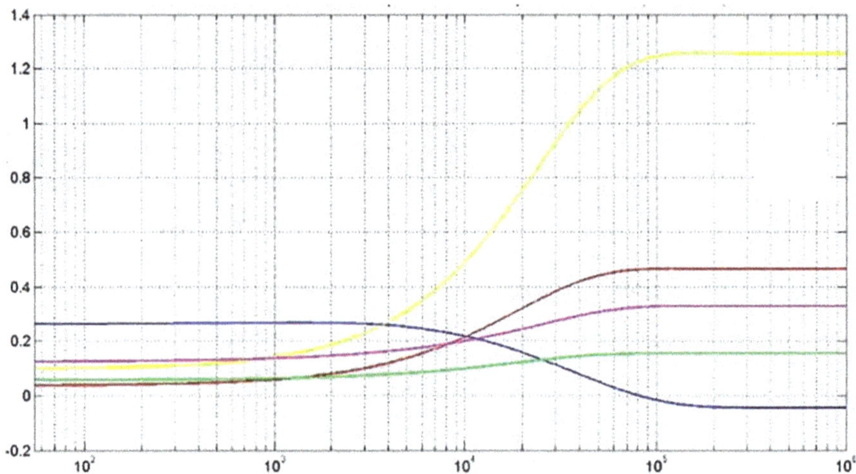

Fig. 22 Heat transfer (W) versus time (s) among the internal central portion, P1, and the other five portions, calculated with *ODE45* of the *MATLAB* software for $N_c = 10^3$ (m^{-3}) [2]. Q_2: blue line, Q_3: red line, Q_4: green line, Q_5: yellow line, and Q_6: purple line

the operation and reach $T_4 = 28.0\,°C$ and $T_6 = 30.0\,°C$ at the end of anaesthesia, then arriving at the steady state at $T_4 = 25.3\,°C$ and $T_6 = 25.7\,°C$.

In conclusion, the approximate analytical solution for a reduced blood perfusion, corresponding to a number of capillaries of $N_c = 10^3$ (m^{-3}), instead of $N_c = 10^5$ (m^{-3}), does not predict a decrease of T_1 but, rather, only an increase, while it predicts only an increase of T_2 and only a decrease of the other temperatures.

The variations of the heat transfer per unit of time (W) between the internal central portion and the other portions versus time (s), given by Eqs. 25–29 and calculated with *ODE45* of the *MATLAB* software, with a reduced number of capillaries, equal to $N_c = 10^3$ (m^{-3}), are shown in Fig. 22 . The values of the heat transfer at the beginning are much lower compared to the case with $N_c = 10^5$ (m^{-3}): Q_2 (blue line) = 0.28 W, Q_3 (red line), = 0.04 W, Q_4 (green line) = 0.05 W, Q_5 (yellow line) = 0.1 W, and Q_6 (purple line) = 0.12 W, showing that the heat transfer greatly reduces due to the reduction in the number of capillaries. Figure 22 indicates that the heat transfers are not constant over time. The heat Q_2 (blue line) decreases and even becomes negative at steady state. The other heat transfers tend to increase until steady state: like Q_3 (red line), which increases up to about 0.48 W; Q_4 (green line), which increases up to 0.15 W; Q_5 (yellow line), which increases up to about 1.25 W; and Q_6 (purple line), which increases up to about 0.32 W. These variations, however, do not seem to affect the temperatures too much.

The temperatures of the various portions of the body, given by Eqs. 17–22 and numerically solved with the *ODE45* program of the *MATLAB* software, for an absent blood perfusion, corresponding to a null number of capillaries, $N_c = 0$ (m^{-3}), that is, in the absence of the heat exchanger that redistributes the heat between the internal central portion and the other portions, are reported in Fig. 23, up to the steady state.

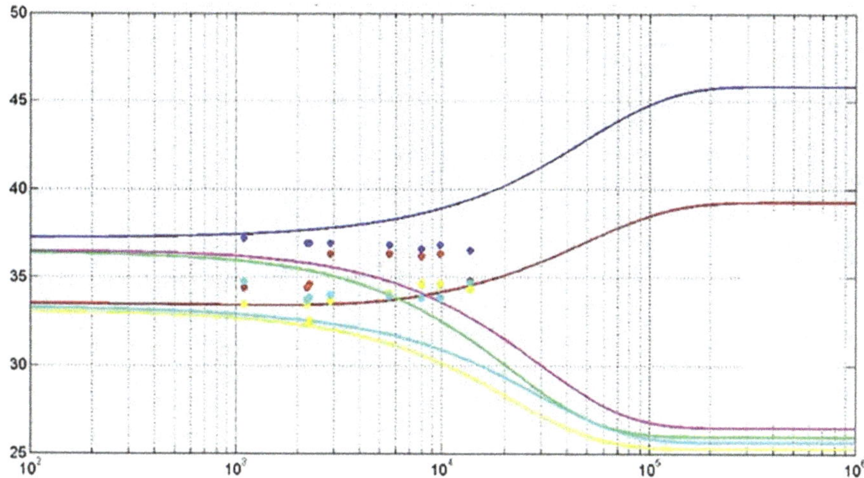

Fig. 23 Numerical temperatures (°C, lines) versus time (s) of the six portions of the body, obtained with the *ODE45* program of the *MATLAB* software for $N_c = 0$ (m^{-3}), compared with the experimental values (dots) of the case D_2 [2]. T_1: blue line and dots. T_2: red line and dots. T_3: green line. T_4: yellow line and dots. T_5: purple line. T_6: sky blue line and dots

They are compared with the experimental measurements of case D_2 [1], obtained until the end of anaesthesia, after $1.2 \cdot 10^4$ s, about 3.3 h. The temperature T_1 (blue line) does not decrease because the heat exchanger is not present since it has a null heat exchange surface. It reaches 39.0 °C at the end of anaesthesia, continuing to grow until the steady-state value of $T_1 = 45.9$ °C. The temperatures of the internal portions, T_3 (green line) and T_5 (purple line), decrease over time, reaching the values of about $T_3 = 32.5$ °C and $T_5 = 33.0$ °C at the end of anaesthesia and $T_3 = 25.9$ °C and $T_5 = 26.4$ °C at the steady state. The temperature of the outer portion T_2 (red line) increases and reaches 34.0 °C at the end of anaesthesia and $T_2 = 39.7$ °C at the steady state. The temperatures of the outer portions, T_4 (yellow line) and T_6 (light blue line), decrease and, at the end of anaesthesia, reach $T_4 = 29.0$ °C and $T_6 = 30.0$ °C and then arrive at the steady state at $T_4 = 25.4$ °C and $T_6 = 25.8$ °C.

In conclusion, the theoretical numerical solution for the absent heat exchanger, corresponding to a number of capillaries of $N_c = 0$ (m^{-3}), instead of $N_c = 10^5$(m^{-3}), does not predict a decrease of T_1 but, rather, only an increase, while it predicts only an increase of T_2 and only a decrease of all the other temperatures. The temperatures of the various portions of the body, obtained with the approximate analytical solution of Eqs. 30–35, with a null number of capillaries, $N_c = 0$ (m^{-3}), coincide with the previous temperatures due to the absence of the heat exchanger.

Comparison with the Experimental Measurements on a Woman of Height 1.68 m and Weight 66.5 Kg (Case D_3)

The temperatures of the various portions of the body, given by Eqs. 17–22 and solved numerically with the *ODE45* program of the *MATLAB* software for

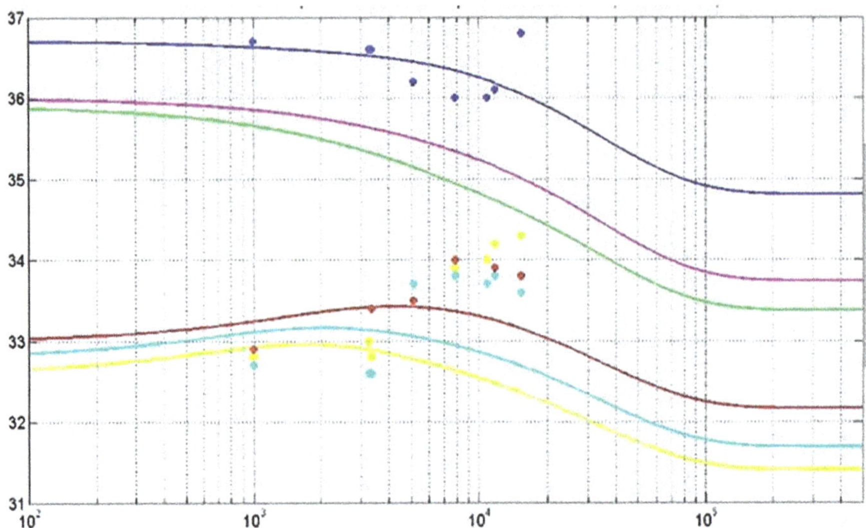

Fig. 24 Numerical temperatures (°C, lines) versus time (s) of the six portions of the body, obtained with the *ODE45* program of the *MATLAB* software for $N_c = 10^5$ (m^{-3}), compared with the experimental values (dots) of the case D_3 [2]. T_1: blue line and dots. T_2: red line and dots. T_3: green line. T_4: yellow line and dots. T_5: purple line. T_6: sky blue line and dots

$N_c = 10^5$ (m^{-3}), are reported in Fig. 24 up to the steady state. They are compared with the experimental measurements of the case D_3 [1], obtained until the end of anaesthesia, which occur after $1.5 \ 10^4$ s, about 4.2 h. The initial temperatures of the various portions of the body parts are assumed to be T_{10} (blue line) $= 36.8$ °C, T_{20} (red line) $= 33.1$ °C, T_{30} (green line) $= 35.9$ °C, T_{40} (yellow line) $= 32.7$ °C, T_{50} (purple line) $= 36.0$ °C, *and* T_{60} (sky blue line) $= 32.8$ °C, in accordance with the experimental values measured in case D_3. The temperatures of the internal portions, T_1 (blue line), T_3 (green line), *and* T_5 (purple line), decrease over time and reach values of about $T_1 = 36.0$ °C, $T_3 = 34.6$ °C, and $T_5 = 35.0$ °C after about 4.2 h. The temperature decrease of T_1 (blue line) is in good agreement with the experimental decrease of the measurements of T_1 (blue points). The temperatures of the external portions T_2 (red line), T_4 (yellow line), *and* T_6 (sky blue line), on the other hand, increase during anaesthesia, with a maximum depending on the portion but can be placed between 5000 s for $T_2 = 33.4$ °C and 2000 for $T_4 = 33.0$ °C and $T_6 = 33.2$ °C, in reasonable agreement with the experimental measurements, T_2 (red points), T_4 (yellow points), *and* T_6 (sky blue points), which are quite dispersed. The numerical model then predicts a decrease until the steady state, where the steady-state temperatures reached are $T_1 = 34.8$ °C, $T_2 = 32.2$ °C, $T_3 = 33.4$ °C, $T_4 = 31.4$ °C, $T_5 = 33.7$ °C, and $T_6 = 31.7$ °C, respectively.

In conclusion, the numerical solution is able to predict a decrease of T_1 and an increase, followed by a decrease, for T_2, T_4, and T_6, in agreement with the trend of the experimental measurements.

The temperatures of the various body portions, obtained with the approximate analytical solution of Eqs. 30–35, for $N_c = 10^5$ (m^{-3}), are reported in Fig. 25 up to

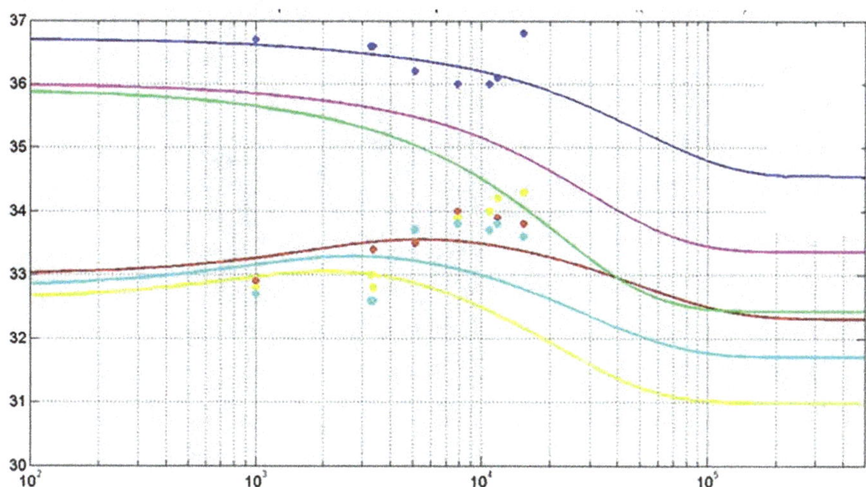

Fig. 25 Analytical temperatures (°C, lines) versus time (s) of the six portions of the body, obtained with the approximate solution for $N_c = 10^5$ (m^{-3}), compared with the experimental values (dots) of the case D_3 [2]. T_1: blue line and dots. T_2: red line and dots. T_3: green line. T_4: yellow line and dots. T_5: purple line. T_6: sky blue line and dots

the steady state and are compared with the experimental results of case D_3 [1]. They are shown until the end of anaesthesia, after $1.5\ 10^4$ s, about 4.2 h. The temperatures of the internal portions, T_1 (blue line), T_3 (green line), *and* T_5 (purple line), decrease over time and reach values of approximately $T_1 = 36.0$ °C, $T_3 = 34.0$ °C, and $T_5 = 34.8$ °C after about 4.2 h, at the end of anaesthesia, in general agreement with the experimental measurements of T_1 (blue points), measured until the end of anaesthesia. The temperatures of the external portions T_2 (red line), T_4 (yellow line), *and* T_6 (sky blue line), on the other hand, increase during anaesthesia, with the maximum that depends on the portion but can be placed between 6000 s for $T_2 = 33.5$ °C and 3000 s for $T_6 = 33.3$ °C and 2000 *s* for $T_4 = 33.0$ °C, in reasonable agreement with the experimental measurements, T_2 (red points), T_4 (yellow dots), *and* T_6 (sky blue dots), which are quite dispersed. The approximate analytical model then predicts a decrease to the steady state, where the steady temperatures reached are $T_1 = 34.6$ °C, $T_2 = 32.3$ °C, $T_3 = 32.4$ °C, $T_4 = 31.0$ °C, $T_5 = 33.4$ °C, and $T_6 = 31.7$ °C, respectively.

In conclusion, the approximate analytical solution is able to predict a decrease of T_1 and an increase, followed by a decrease, for T_2, T_4, and T_6, in agreement with the trend of the experimental measurements. The comparison between the temperatures obtained with the approximate analytical solution, that is, with the heat transfer constant during anaesthesia, and equal to the initial ones, and the solution obtained with *ODE45* of the *MATLAB* software is quite positive, especially during the transient of the anaesthesia, even if the steady-state temperatures reached with this approximate analytical solution are slightly different from the previous solution.

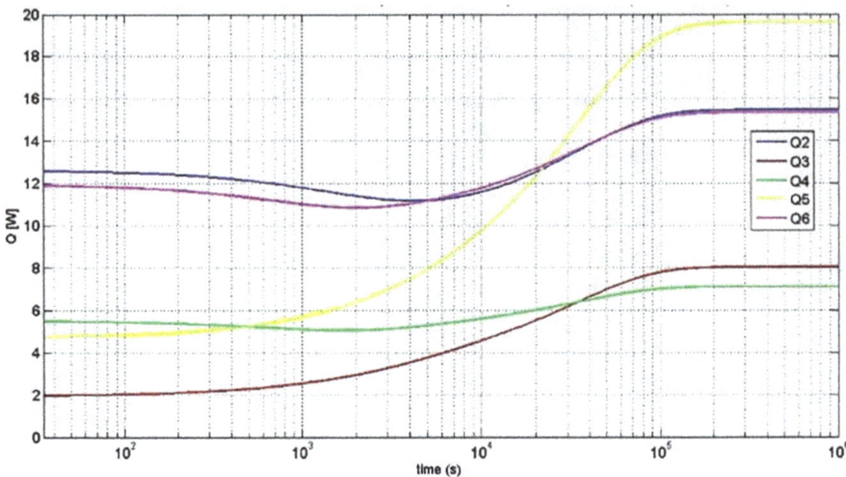

Fig. 26 Heat transfer (W) versus time (s) among the internal central portion, P1, and the other five portions, calculated with *ODE45* of the *MATLAB* software for $N_c = 10^5$ (m^{-3}) [2]. Q_2: blue line, Q_3: red line, Q_4: green line, Q_5: yellow line, and Q_6: purple line

The heat transfer per unit of time (W) between the internal central part and the other portions versus time (s), given by Eqs. 25–29 and calculated with *ODE45* of *MATLAB* for $N_c = 10^5$ (m^{-3}), are shown in Fig. 26. The values of the heat transfer, which, at the initial time, are Q_2 (blue line) = 12.5 W, Q_3 (red line), = 2.0 W, Q_4 (green line) = 5.5 W, Q_5 (yellow line) = 5 W, and Q_6 (purple line) = 12 W, are not constant during anaesthesia but tend to decrease or increase until they reach the steady-state values. The heat Q_2 (blue line) decreases during anaesthesia, with a minimum of around 11.0 W and then returns to its initial value at the end of anaesthesia to reach 15.5 W at the steady state. The heat Q_6 (purple line) has a similar trend because, after a minimum of about 11 W, it returns to the initial value at the end of the anaesthesia and reaches about 15.5 W at the steady state. Also, the heat Q_4 (green line) has a similar trend because, after a slight minimum, it returns to the initial value at the end of anaesthesia and reaches about 7 W at the steady state. The heats, Q_3 (green line) and Q_5 (purple line), do not have a minimum but increase up to about 7 W and about 19.8 W at the steady state, respectively. These variations, however, do not seem to affect the temperatures too much.

The temperatures of the various portions of the body, given by Eqs. 17–22 and solved numerically with the *ODE45* program of the *MATLAB software*, for a reduced blood perfusion, corresponding to a number of capillaries of $N_c = 10^4$ (m^{-3}), instead of $N_c = 10^5$(m^{-3}), as in the previous case, are reported in Fig. 27 up to the steady state. They are compared with the experimental measurements of case D_3 [1], obtained until the end of anaesthesia, after 1.5 10^4 s, about 4.2 h. The temperature T_1 (blue line) does not decrease, as in the previous case, because the heat exchanger has a smaller heat exchange surface, having a reduced number of capillaries, and the heat transfer from the internal central portion is smaller. The

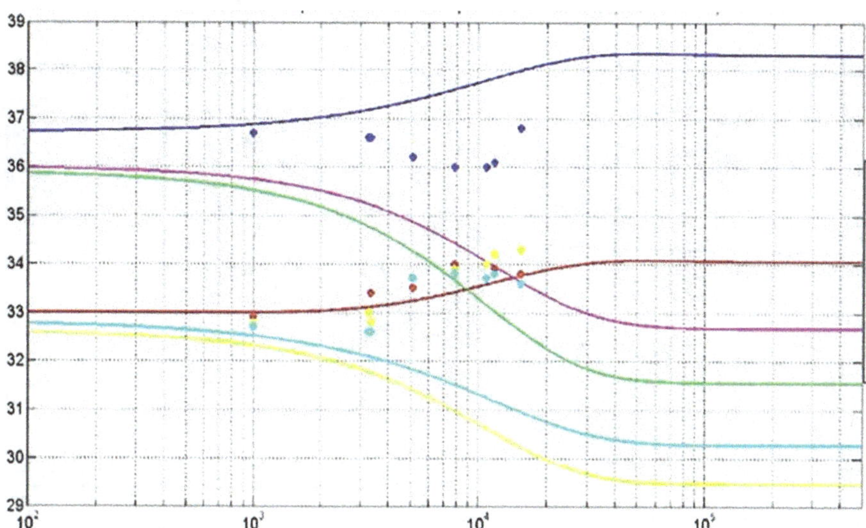

Fig. 27 Numerical temperatures (°C, lines) versus time (s) of the six portions of the body, obtained with the *ODE45* program of the *MATLAB* software for $N_c = 10^4$ (m^{-3}), compared with the experimental values (dots) of the case D_3 [2]. T_1: blue line and dots. T_2: red line and dots. T_3: green line. T_4: yellow line and dots. T_5: purple line. T_6: sky blue line and dots

temperature T_1 (blue line) reaches 38.0 °C at the end of anaesthesia, after about 4.2 h, and continues to grow until the steady-state value of $T_1 = 38.3$ °C. The temperatures of the internal portions, T_3 (green line) and T_5 (purple line), decrease over time, reaching the values of about $T_3 = 32.8$ °C and $T_5 = 33.7$ °C at the end of anaesthesia and $T_3 = 31.6$ °C and $T_5 = 32.7$ °C at the steady state. The temperature T_2 (red line) increases, in accordance with the experimental measurements, and reaches 33.8 °C at the end of anaesthesia and $T_2 = 34.0$ °C at the steady state. The temperature T_4 (yellow line) and T_6 (sky blue line), on the other hand, decrease during anaesthesia when they reach $T_4 = 30.5$ °C and $T_6 = 31.0$ °C and then arrive at the steady state at $T_4 = 29.5$ °C and $T_6 = 30.2$ °C.

In conclusion, the theoretical numerical solution for a reduced blood perfusion, corresponding to a number of capillaries of $N_c = 10^4$ (m^{-3}), instead of $N_c = 10^5$ (m^{-3}), does not predict a decrease of T_1 but, rather, only an increase, while it predicts only an increase of T_2 and only a decrease of all the other temperatures.

The temperatures of the various portions of the body, obtained with the approximate analytical solution, Eqs. 30–35, for a reduced blood perfusion, corresponding to a number of capillaries of $N_c = 10^4$ (m^{-3}), instead of $N_c = 10^5$ (m^{-3}), as in the previous case, are reported in Fig. 28 up to the steady state. They are compared with the experimental measurements of case D_3 [1], obtained until the end of anaesthesia, after about 4.2 h. The temperature T_1 (blue line) does not decrease, as in the previous case, because the heat exchanger has a smaller heat transfer surface, having a reduced number of capillaries, and the heat transfer from the internal central portion is smaller. Then, T_1 increases, reaching 38.8 °C at the end of anaesthesia, after 4.2 h,

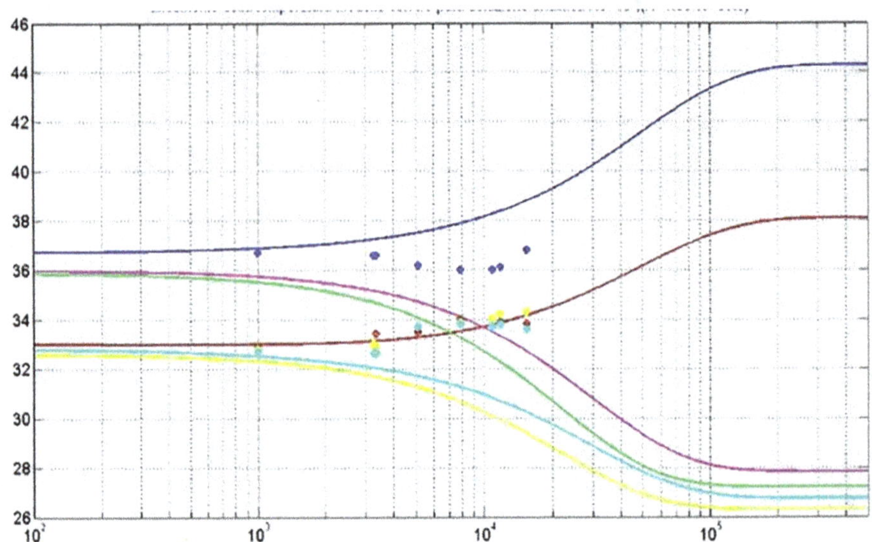

Fig. 28 Analytical temperatures (°C, lines) versus time (s) of the six portions of the body, obtained with the approximate solution for $N_c = 10^4$ (m^{-3}), compared with the experimental values (dots) of the case D_3 [2]. T_1: blue line and dots. T_2: red line and dots. T_3: green line. T_4: yellow line and dots. T_5: purple line. T_6: sky blue line and dots

continuing to grow up to the steady-state value of $T_1 = 44.3$ °C. The temperatures of the internal portions, T_3 (green line) and T_5 (purple line), decrease over time, reaching values of about $T_3 = 31.6$ °C and $T_5 = 32.8$ °C at the end of anaesthesia and $T_3 = 27.2$ °C and $T_5 = 27.8$ °C at the steady state. The temperature of the outer portion T_2 (red line) increases in accordance with the experimental measurements and reaches 34.1 °C at the end of anaesthesia and $T_2 = 38.1$ °C at the steady state. The temperatures of the outer portions T_4 (yellow line) and T_6 (sky blue line), on the other hand, decrease during anaesthesia and reach $T_4 = 29.5$ °C and $T_6 = 30.5$ °C at the end of anaesthesia, then arriving at the steady state at $T_4 = 26.3$ °C and $T_6 = 26.7$ °C.

In conclusion, the approximate analytical solution for a reduced blood perfusion, corresponding to a number of capillaries of $N_c = 10^4$ (m^{-3}), instead of $N_c = 10^5$ (m^{-3}), does not predict a decrease of T_1 but, rather, only an increase, while it predicts only an increase of T_2 and only a decrease of all the other temperatures.

The variation of the heat transfer per unit of time (W) between the internal central portion and the other portions, central and peripheral, versus time (s), given by Eqs. 25–29, calculated with *ODE45* of *MATLAB* and with a reduced number of capillaries, equal to $N_c = 10^4$ (m^{-3}), are shown in Fig. 29. The values of the heat transfer at the beginning are much lower compared to the case with $N_c = 10^5$ (m^{-3}), showing that the heat transfer greatly reduces due to the reduction in the number of capillaries. The heat Q_2 (blue line) reduced to 3.2 W, while Q_3 (red line) reduced to 0.5 W, Q_4 (green line) *reduced* to 0.8 W, Q_5 (yellow line) *reduced* to 1.2 W, *and* Q_6 (purple line) *reduced* to 1.5 W. Figure 29 indicates that the heat transfer is not

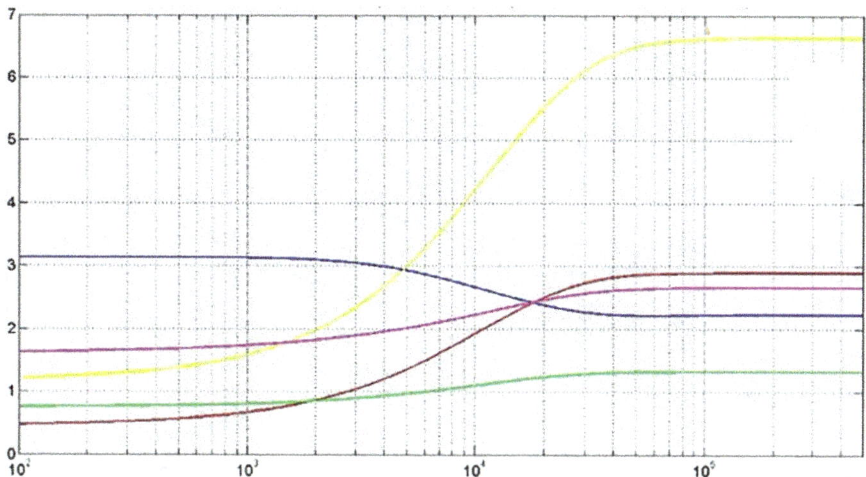

Fig. 29 Heat transfer (W) versus time (s) among the internal central portion, P1, and the other five portions, calculated with *ODE45* of the *MATLAB* software for $N_c = 10^4$ (m^{-3}) [2]. Q_2: blue line, Q_3: red line, Q_4: green line, Q_5: yellow line:, and Q_6: purple line

constant over time during anaesthesia but tends to increase all except Q_2 (blue line), which reduces to 2.3 W. The heat Q_3 (red line) increases up to about 2.8 W, Q_4 (green line) increases up to 1.3 W, Q_5 (yellow line) increases up to about 6.7 W, and Q_6 (purple line) increases to about 2.6 W. These variations, however, do not seem to affect the temperatures too much.

The temperatures of the various portions of the body, given by Eqs. 17–22 and numerically solved with the *ODE45* program of the *MATLAB* software, for absent blood perfusion, corresponding to a null number of capillaries, $N_c = 0$ (m^{-3}), that is, in the absence of the heat exchanger that redistributes the heat between the internal central portion and the other portions, are reported in Fig. 30, up to the steady state. They are compared with the experimental measurements of case D_3 [1], obtained until the end of anaesthesia, after $1.5 \ 10^4$ s, about 4.2 h. The temperature T_1 (blue line) does not decrease because the heat exchanger is not present since it has a null heat exchange surface. It reaches 39.0 °C at the end of anaesthesia after 4.2 h, continuing to grow until the steady-state value of $T_1 = 46.4$ °C. The temperatures of the internal portions, T_3 (green line) and T_5 (purple line), decrease over time, reaching the values of about $T_3 = 32.0$ °C and $T_5 = 33.0$ °C at the end of anaesthesia and $T_3 = 26.1$ °C and $T_5 = 26.6$ °C at the steady state. The temperature of the outer portion T_2 (red line) increases and reaches 34.0 °C at the end of anaesthesia and $T_2 = 39.2$ °C at the steady state. The temperatures of the outer portions T_4 (yellow line) and T_6 (light blue line), on the other hand, decrease and, at the end of anaesthesia, reach $T_4 = 28.5$ °C and $T_6 = 30.0$ °C and then arrive at the steady state at $T_4 = 25.4$ °C and $T_6 = 25.7$ °C .

In conclusion, the theoretical numerical solution for absent blood perfusion, corresponding to a number of capillaries of $N_c = 0$ (m^{-3}), instead of $N_c = 10^5$ (m$^-$

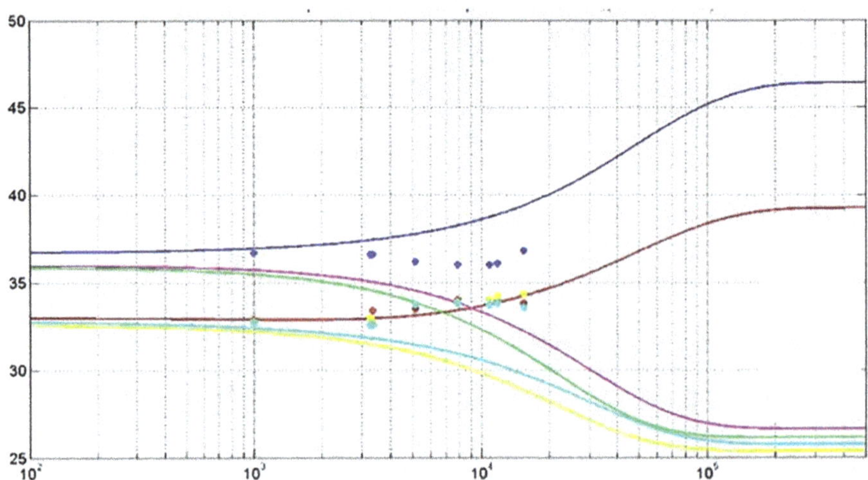

Fig. 30 Numerical temperatures (°C, lines) versus time (s) of the six portions of the body, obtained with the *ODE45* program of the *MATLAB* software for $N_c = 0$ (m^{-3}), compared with the four experimental values (dots) of the case D_3 [2]. T_1: blue line and dots. T_2: red line and dots. T_3: green line. T_4: yellow line and dots. T_5: purple line. T_6: sky blue line and dots

^{3}), does not predict a decrease of T_1 but, rather, only an increase, while it predicts only an increase of T_2 and only a decrease of all the other temperatures.

3   Conclusions

The results of all the numerical and approximate analytical solutions are that, in the case of correct capillary circulation, the typical trend of the central temperature is to decrease over time during the anaesthetic treatment. On the other hand, the external temperatures (legs, arms, and nipples) have a trend that is influenced by the introduction of the anaesthetic treatment; initially, the temperature rises due to the vasodilating effect of the anaesthetic drug, and then, it decreases in a similar way to the central temperature.

The experimental data were measured at the Santa Maria Annunziata Hospital in Florence by the group of E.Pagni [7]. Two different types of anaesthetic drugs were used for the three women; for the woman of height 1.6 m and weight 60 kg (case D_1), halothane was used, while for the other two women, isoflurane was used. Among the advantages of isoflurane are cardiovascular stability and muscle relaxation, while halothane is more dilating. In fact, the peripheral parts of the woman treated with halothane (case D_1) reached slightly higher average temperatures, as the blood perfusion was greater. The central temperature was measured in the left ear and in the oesophagus. Naturally, the experimental data were much more dispersed and depended on many other conditions that could not be evaluated with the model. However, the trend of the central temperature seemed to be in good agreement with the experimental data, while the decrease of the peripheral temperatures was less clear compared to the experimental data.

In the figures concerning the trend with absent or reduced capillary circulation, the fundamental role of the latter is immediately noticeable. In fact, with the loss of a way for the central part to dissipate heat towards the other parts and then consequently towards the outside, its temperature increases irreparably, while for legs and arms, the temperature will obviously have a strongly decreasing trend. The inevitable consequence of these situations is malignant hyperthermia.

References

1. Gori F (1991) Temperature variation of the human body during anaesthetic treatments. Macroscopic and microscopic heat and mass transfer in biomedical engineering. In: Diller K, Shitzer A (eds) 1992, Int. Symposium on Heat and Mass Transfer in Biomedical Engineering. ICHMT, Athens, September p. 119–33
2. Di Venuta I (2014-2015) Cooling of the human body during anaesthetic treatment, homework/dissertation of the course "Thermo-fluid-dynamics of biological systems", Master's degree in medical engineering. University of Rome Tor Vergata, A.Y
3. Stolwijk JAJ, Hardy JD (1966) Temperature regulation in man- a theoretical study. Pflugers Archiv 291:129–162
4. Gori F (1983) Prediction of effective thermal conductivity of two-phase media (including saturated soils). In: XVI International Congress of Refrigeration, Paris, vol B1-26, pp 525–531
5. Gori F (1983) A theoretical model for predicting the effective thermal conductivity of unsaturated frozen soils. In: The 4th international conference on permafrost, Fairbanks, AK, U.S.A., pp 363–368
6. Gori F et al (2005) Extension of soil thermal conductivity models to frozen meats with low and high fat content. Int J Refrig 28(6):840–850
7. Pagni E. Anestesia e termoregolazione (Anaesthesia and Thermoregulation). Editrice La Garangola, Padova

Dr. Eng. Ivan Di Venuta (Salerno, 5 March 1990) graduated in Medical Engineering from the University of Rome "Tor Vergata" in 2015, discussing his Master's Thesis on the Thermo-Fluid-Dynamics of stenotic coronary vessels titled "Influence of the degree of stenosis on the non-Newtonian blood flow in a stented coronary artery." He decided to continue his academic journey at "Tor Vergata" by enrolling in the Research Doctorate (Italian *Ph.D.*) course in Industrial Engineering. He was welcomed by the Technical Physics research group of Fabio Gori Ammannati, with whom he started a project on the Thermo-Fluid-Dynamics of submerged turbulent jets. Thanks to his valuable teachings and support during his Research Doctorate, Ivan Di Venuta published several scientific articles in international journals and carried out part of his research at the University of Cranfield, UK. He obtained his Research Doctorate in Industrial Engineering, with *European Label,* in April 2019 with the thesis titled "The new fluid dynamics, heat and mass transfer of submerged free jets." Currently, Ivan Di Venuta works as a business analyst at one of the leading business consulting firms, working on projects for the redesign of business processes in public administration.

Fabio Gori Ammannati was born in Montale (Pistoia) on 5 August 1947 to Emilio Gori and Cesarina Ammannati. On 16 July 2021, he added his mother's surname, Ammannati. He graduated in Chemical Engineering on 11 November 1971 with honours. In December 1971, he won a scholarship for young graduates at the Faculty of Engineering in Bologna, which he undertook from January 1972. In 1974, he became assistant professor and, in 1982, associate professor of Technical Physics at the Faculty of Engineering of the University of Florence, where he taught, also as a contract professor, until 1990. In 1978, he won a scholarship from the *British Council,* which he carried out at the *Imperial College* in London, where he collaborated with *Prof. D. Brian Spalding* on the topic of numerical analysis in the turbulent flow of liquid metals. During his time at the University of Florence, he established international scientific collaborations with *Prof. R. Echigo, Tokyo Institute of Technology, Tokyo, Japan; Prof. D.R. Chaudhary, Department of Physics, University of Rajasthan, Jaipur, India;* International Centre for Theoretical Physics *(ICTP)* in Trieste; *Centralny Osrodek Techniki Medycznej, Warsaw, Poland and Prof. T. Aihara, Institute of Fluid Science, Sendai, Japan.* In 1986, he won a CNR scholarship, which he carried out at *Cornell University, Ithaca, New York,* where he collaborated with *Prof. R. Miller* on the topic of ground freezing. In the same year, he became professor in the Faculty of Engineering at the University of Reggio Calabria. In 1988, he was appointed as *professor* at the *University of New York at Stony Brook,* teaching the course, *Introduction to Fluid Dynamics,* in the autumn semester of the same year. During his stay, he collaborated with *Prof. T.F. Irvine Jr.* on the method of measuring thermal conductivity with a thermal probe and on the measurement of the isobaric thermal expansion coefficient of non-Newtonian fluids. In 1990, he moved to the Milano Polytechnic, where he taught Technical Physics and Systems until 1993. From 1991 until 1998, he was an adjunct professor at the Faculty of Engineering of the University of Siena. From 1992 to 2017, he was a professor of Technical Physics in the Faculty of Engineering at the University of Rome Tor Vergata, and from 2017, he is a contract professor. In 1994, he proposed the Research Doctorate (Italian *Ph.D.*) in Energy-Environment Engineering, of which he was the coordinator until 2011. In 1995, he proposed and directed the second-level Master's in Thermo-fluid-dynamics until 2016. From 1998 to 2001, he was the director of the Department of Mechanical Engineering. In 2001, he proposed and coordinated the programme, *Master of Science, Energy Engineering and Thermal and Fluid Dynamics,* in collaboration with the *Department of Mechanical Engineering, College of Engineering, University of Illinois at Chicago, USA,* which takes place at the University of Rome Tor Vergata but awards the title of *Master of Science* from the same American University. Since the late 90s, he has been a coordinator of the joint *Ph.D.* research programme with *Prof. R.J. Goldstein, Department of Mechanical Engineering, University of Minnesota, Minneapolis, USA,* and with *Prof. J.P. Hartnett, Prof. L. Kennedy, Prof. W. Minkowycz and Prof. W.M. Worek, Department of Mechanical Engineering, College of Engineering, University of Illinois at Chicago, Chicago, USA.* Since the mid-90s, he has proposed and directed the *Socrates–Erasmus* programme with *Prof. Mayinger, Technical University of Munich, Germany; Prof. van Steenhoven, Eindhoven Technical University, the Netherlands and Prof. C. Caro, Imperial College of Science, Technology and Medicine.* During his stay at Tor Vergata, he established scientific collaborations with the *Department of Mechanical Engineering, University of Minnesota, Minneapolis, USA,* collaborating with *Prof. R.J. Goldstein* on Thermo-fluid-dynamics and mass transport in gas turbine blades; the *Energy Resources Center, University of Illinois at Chicago,* collaborating with *Prof. J.P. Hartnett, Prof. W. Minkowycz* and *Prof. W.M. Worek*; the *Department of Mechanical Engineering, College of Engineering, University of Illinois at Chicago,* collaborating with *Prof. L. Kennedy and Prof. W.M. Worek*; *Duke University,* collaborating with *Prof. A. Bejan* and *S. Mary's University, Halifax, Nova Scotia, Canada,* collaborating with *Prof. W. R. Tarnawski.* The bibliographic review of documents and citations, titled *PlosBiology Career,* places him within the top 2% of researchers in the field of *Mechanical Engineering and Transports.* Since 1992, he has been a tutor for over 30 PhD theses, supervisor for over 40 second-level university Master's theses, supervisor for over 70 degree theses (five-year, Specialist and Master's), supervisor for over ten *Master's degree in Mechanical Engineering* from the *University of Illinois at Chicago* and supervisor for over 20 theses *Erasmus-Socrates.*

Since the early 2000s, he has been a reviewer of international research projects on behalf of *Portuguese Science and Technology Foundation (FCT); Czech Science Foundation (GACR); European Science Foundation, Strasbourg, France,* on behalf of *FCT* and the *Shota Rustaveli National Science Foundation, Georgia.* Since 2017, the year of his retirement due to age limits, he has been a contract professor at the University of Rome Tor Vergata, where in the academic year 2024–2025, he has taught Technical Physics for the Master's Degree Course in Medical Engineering. M.D. Salvatore Mangiafico (Florence), scientific head of the NeuroVascular Base Camp, 2021, invited him to give a scientific presentation titled "*The engineering approach to the study of the Circle of Willis*" on 23 September 2021 at the Congress Centre, University of Rome La Sapienza, thus opening up a possible future collaboration.

Thermo-Haemo-Dynamics (THD) in Coronary Stents

Numerical Newtonian Thermo-Haemo-Dynamics (THD) in Coronary Stents

Mattia Amitrano, Andrea Boghi, and Fabio Gori Ammannati

1 Introduction

This chapter, based on [1], proposes numerical simulation as a method to predict changes in blood flow due to the presence of the implanted device. However, experimental checks are necessary in order to be sure of the results obtained through numerical simulation. Computer simulations, used to quantify some haemodynamic quantities in models that describe the anatomy of the specific patient, can provide the surgeon with additional information to assist him in planning the operation [2].

In this chapter, attention is focused on the numerical analysis of the fluid dynamics of the coronary stent *CYPHER*, produced by the company *Cordis (Johnson & Johnson)*. This is a drug-eluting stent (DES) (sirolimus) of the new generation and is one of the most used in the world, as well as one of the most studied, but the studies (generally clinical) are all aimed at drug release. Recent reviews, [3–5], suggest that drug-eluting stents may simply delay restenosis and are not able to facilitate the healing of the intima damaged by the implantation of the stent. Therefore, the restenosis rate of drug-eluting stents could be similar to that of simple metal stents. There are no specific fluid-dynamics studies on this stent in the literature, hence the importance of studying the fluid dynamics of the stent *CYPHER* regardless of drug release.

There are two variants of this stent: *"CYPHER Sirolimus-eluting coronary Stent"* and *"CYPHER SELECT Plus Sirolimus-eluting coronary Stent." CYPHER SELECT*

M. Amitrano
Strada delle Spiasce 2, Amelia, Terni, Italy

A. Boghi
Computational Science Ltd., Southampton, UK
e-mail: a.boghi@computationalscience.co.uk

F. G. Ammannati (✉)
University of Rome Tor Vergata, Rome, Italy
e-mail: gori@uniroma2.it

is the most recent (on the market since the second half of 2006) and is defined as a third-generation stent. In particular, it is designed for the treatment of difficult blood vessels. Both stent variants release the same drug, but they differ in geometry; in particular, they have a different connection among the meshes. The different geometric shapes are compared in order to understand the most effective one from a fluid dynamics point of view. In the literature, there are many studies aimed at characterizing coronary stents through fluid dynamic simulations. Some authors put their attention on the search for the best stent design. Attention is focused on the geometry of the stents, in particular on the size and number of meshes, as well as on the shape of the structural elements that make up the mesh. It can be said that all studies only consider simple meshes, formed by straight elements, which intersect at a constant angle, a strongly simplifying assumption, as the real geometry of a coronary stent (e.g., *CYPHER*) is much more complex.

Berry et al. [6] studied, through a two-dimensional (2D) fluid dynamics model, how the dimensions of the meshes influence blood flow. The mentioned authors carried out a series of fluid dynamic simulations by varying the ratio between the diameter of the structural elements of the stent meshes and the characteristic size of the mesh. A particular section is chosen as the cross-section of the structural elements. Before the simulations, they carry out in vivo experiments in which they highlight that the blood flow is only locally disturbed by the presence of the stent; in particular, they qualitatively see that the areas of stagnant flow are only immediately before and after the structural elements of the mesh. With these experiments, they justify the use of a 2D model for fluid dynamic simulations. From these simulations, by varying the ratio of structural elements' diameter to the characteristic size of the mesh, they highlight that if the size of the mesh is less than 6 diameters, the stagnation zone extends from one element to another without interruption throughout the cardiac cycle; but if, instead, the size of the mesh is greater than 6 diameters, two distinct stagnation zones form within a mesh for part or for the entire cardiac cycle. Since the residence time of the particles is much higher in the stagnation zones than in those of normal flow, these results suggest that the stent design plays an important role in determining how long the particles remain near the wall of the vessel (a low *wall shear rate* is associated with the stagnation zones, which could be the cause of thrombus formation and thickening of the intima). The authors show that the most important parameter influencing the stagnation zones is the ratio of diameter to elements-mesh size, rather than the diameter alone. In addition, they note that the stagnation zone, upstream of an element, is always smaller than the stagnation zone downstream, and they do not notice any substantial difference either in the first element or in the last. The authors therefore conclude that the dimensions of the mesh should be at least 6 times the diameter of the structural elements.

LaDisa et al. [7–10] carried out a number of studies aimed at determining the best stent design using fluid dynamic models, both 2D and 3D. LaDisa et al. carried out in [7, 8] a series of simulations to investigate the influence of the number, thickness and width of the structural elements of the mesh and the number of intersections of the elements per circumference on the blood fluid dynamics. In these simulations, they vary these parameters, and from the results of the simulations, they highlight that the highest values of the *wall shear stress* (*WSS*) are located, in all the simulations, on the surface of the structural elements and in the stent-vessel distal transition. If the thickness of the structural elements is reduced, the percentage of area exposed to low *WSS* decreases; if the width is reduced, the area affected by low *WSS* slightly increases and if the number of intersections of the elements per circumference is doubled, the area with low *WSS* more than doubles. In contrast to these results, the stagnation zones that surround the structural elements seem independent of the thickness, width and number of the elements. From these 3D computational fluid dynamics (*CFD*) simulations, the authors conclude that the stent influences the distribution of *WSS* within the arteries and suggest an optimal geometry that minimizes the number of intersections of the elements and the dimensions of the same elements. LaDisa et al. hypothesized in [9] that the implanted stent deforms the vessel wall, modelling this deformed wall with polygons. Then, through fluid dynamic simulations, they evaluate the effect of these possible deformations on the distribution of *WSS*. From these simulations, the authors highlight that the deformed circumference introduces a high *WSS* and high gradient of *WSS* among the structural elements, which is absent in the non-deformed circumference. Moreover, they note that the area exposed to low *WSS* depends on the degree of deformation of the circumference. With these simulations, the authors therefore demonstrate that not only does the geometry of the stent influence the *WSS* and, therefore, the possible formation of restenosis but also the deformation that the stent induces on the vessel wall plays an important role in the distribution of the *WSS*. LaDisa et al. evaluated in [10] the hypothesis that the angle formed among the structural elements and the blood flow can influence the distribution of the *WSS* on the vessel wall. With these simulations, the authors highlight that the intersection between the structural elements always creates a low *WSS* and high spatial gradient of *WSS* (*WSSG*), regardless of the angle of intersection (and therefore the angle with the direction of flow). Increasing this angle, the elements progressively misalign with respect to the direction of the flow, causing an alteration of the behaviour of the blood velocity vector near the wall and causing a decrease in the *WSS*. If this angle is kept narrow, the problem of misalignment decreases, but this causes an increase in the number of intersections of the elements per circumference, which, as previously demonstrated, should be minimized. From these results, the authors conclude that an optimal stent design must find a compromise that minimizes the areas of low *WSS*. LaDisa et al. used in [7–10], as a physiological reference, the canine anterior descending coronary artery, of which they measured the physical dimensions and the blood flow velocity [11]. The data obtained are used to reconstruct the artery by modelling it as a cylinder, and they chose, as stent mesh, a regular mesh formed by straight elements that always intersect at the same angle. In particular, they investigate various aspects of the stent design, such as the number, thickness and width of the mesh's structural elements; the number of intersections between the elements; the angle created by the elements with respect to the direction of blood flow; and the possible deformations of the vessel's circumference caused by the implanted stent.

The authors then carry out a series of simulations, both in a steady and unsteady state, and evaluate the *WSS* and the spatial gradient of the *WSS* (*WSSG*). As a threshold for low *WSS*, they define 5 dyn/cm, and as a threshold for high *WSS*, they define 8 dyn/cm, in agreement with [12–14]. With these thresholds, they calculate the regions of low and high *WSS* as a percentage of the total surface of the vessel.

Lanoye et al. [15] conducted a fluid dynamic (*CFD*) 2D study aimed at characterizing the shape of the cross-section of the structural elements of a stent mesh. In particular, they analyse three different geometries for the cross-section: rectangular, circular and rectangular with the side opposite the vessel wall in the shape of an arc of circumference. With a series of simulations, the authors evaluate the *WSS*, the oscillatory shear index (*OSI*) and the velocity field. From these simulations, the authors highlight that the velocity field is strongly modified by the presence of the stent, but recirculation zones are formed only in the presence of rectangular sections and not circular sections. The *WSS* is smaller near the rectangular sections than in the circular section, and, finally, the *OSI* is significantly higher in the two rectangular sections compared to the circular section. From these results, the authors conclude that the straight circular section should ensure a lower percentage of restenosis compared to the other geometries.

The analysis of the literature on more specific issues, such as the curvatures and bifurcations of the coronary arteries, shows few studies of computational fluid dynamics (*CFD*). Also, in this case, the meshes analysed are all very simple. Seo et al. [16] conducted a 2D study on the design of stents applied near coronary bends and compared this configuration with that related to stents applied on a straight section of the artery. In particular, they study the dimensions of the structural elements of the mesh. With the simulations done, they analyse the *WSS* and the stagnation zones of the blood flow. Their results are essentially in agreement with the conclusions set out in the previous studies, namely, that the size of the structural elements must be minimized.

LaDisa et al. [17] carried out a *CFD* 3D study aimed at investigating the effect that a stent can have on a curvature. In particular, they analyse two possibilities: a rigid stent that alters the natural curvature and a flexible stent that is able to reproduce this curvature. They, then, compare the effects that these two stents have on the blood flow. With these simulations, the authors show how, in fact, the rigid stent alters the blood fluid dynamics more significantly compared to the flexible stent. The velocity profile in the sections of the rigid stent is different from the profile that is found in the curvature, and the distribution of the *WSS* on the wall of the vessel with the rigid stent undergoes significant variations, with areas of smaller and greater *WSS* compared to the flexible stent. From these simulations, the authors conclude that a flexible stent, applied on a curvature, can ensure better blood flow and a lower percentage of restenosis compared to a rigid stent.

Deplano et al. [18] conducted a study in which they modelled a coronary T-bifurcation (90°) with the presence of two stents, one on the main branch and the other on the secondary one. They assume that the stent on the secondary branch is implanted first and the stent on the main branch is implanted straddling the bifurcation in order to repair both the distal and proximal sections of the artery. In this configuration, a final expansion can be performed to fully reopen the secondary branch, as it is partially obstructed by the second stent. This final expansion is not always performed, as it can cause damage to the implanted stents and remains a subject of discussion among cardiologists. The mentioned authors simulate both the stents with the final dilation and without, and, in the latter case, they simulate different mesh arrangements relative to the ostium of the vessel, as the actual mesh arrangement is random. Finally, they compare the fluid dynamic results of these configurations with the healthy bifurcation without a stent. From these simulations, the authors conclude that the percentage of restenosis is much smaller in the case where the final dilation is performed because the blood flow is very similar to the flow of the healthy artery only if this final dilation is performed. From the analysis of the literature, it emerges that the problem of the fluid dynamics of coronary stents is a current problem. The studies analysed are essentially in agreement in considering which parameters are useful to evaluate the percentage of restenosis, the *WSS*, the *WSSG*, the *OSI*, the stagnation zones of the flow and the velocity field. He et al. [19] studied a realistic geometry considering three geometric parameters of the structures, showing that the design of the stent is very important for the blood flow in the arteries.

The geometry of the chosen stent is designed with *CFD* software [1], modelling the coronary arteries with a simple cylinder. The reason for this choice does not lie so much in the computational difficulties resulting from the geometric reconstruction, as there are software programs capable of constructing 3D geometries from medical images, but rather in the desire to carry out a general study, not referred to individual cases, and to highlight the effects of the stent design. This choice is in agreement with the studies present in the literature on the influence of stent design on the fluid dynamics of the coronary arteries. The simulations are performed with numerical software. Steady-state simulations are carried out to test the independence from the grid. Finally, unsteady-state simulations are carried out, using the physiological flow of blood within the coronary arteries, to be able to obtain this physiological trend; a program is developed, in *MATLAB*, capable of sampling an image of a physiological signal and reconstructing the continuous signal throughout the discrete Fourier transform (*FFT*). In the simulations, some simplifying assumptions are made: the wall of the vessel is considered rigid, and the blood is a Newtonian fluid in laminar flow. As far as the physical properties are concerned, the blood is modelled as liquid water. The computational difficulties impose the use of simplified models. The substantial difference that this study presents lies in the shape of the mesh. All the studies analysed use a very simple mesh, while, in this study, the mesh of the geometry is reconstructed in a more realistic way in order to be able to analyse, with greater precision, the actual influence of the stent on the fluid dynamics of the coronary arteries.

2   Atherosclerosis and Coronary Stents

2.1   Introduction to the Pathophysiology of Atherosclerosis

Atherosclerosis is the most common cause of death and serious morbidity in the Western world. The World Health Organisation predicts that in the near future, it will become the leading cause of death worldwide. Atherosclerosis is a disease of the elastic arteries (such as the aorta, carotids and iliac) and of the large and medium muscular arteries (such as the coronary and popliteal), while it only rarely affects the smaller arteries. It is part of a group of arteriopathies characterized by the thickening and loss of elasticity of the vascular wall. The term commonly used for these diseases is that of arteriosclerosis, which literally means hardening of the arteries. Other diseases included in this group are arteriolosclerosis, characterized by the proliferation and thickening of the walls of small arteries and arterioles, and the sclerosis of also *Monckenberg*, characterized by the calcification of the tunica media of the muscular arteries. Atherosclerosis is, by far, the most common and important form of arteriosclerosis, so much so that at times these terms are used as synonyms.

The pathogenic process of atherosclerosis is mainly located at the level of the intima of the arterial wall, which is infiltrated by lipids and inflammatory cells, in addition to developing progressive degrees of fibrosis [20]. This observation leads to the belief that atherosclerosis is, at least, partly secondary to the activation at the vascular level of cellular repair processes. The arterial injury induces a scar-like reaction that involves the phenotypic modulation of the smooth muscle cells of the tunica media, with their transformation into repair-like fibroblastic cells, which migrate into the intima, where they begin to proliferate and produce extracellular matrix. It is believed that the accumulation of lipids derived from lipoproteins, including some components of low-density lipoproteins (LDLs), modified by oxidative and/or enzymatic means, is able to damage the arterial wall, inducing local inflammation and activating scarring processes [21, 22]. This process leads to the formation of intimal lesions that can, over time, become atheromas. Although the atherosclerotic process seems so strongly localized at the level of the intima, the other layers are not immune to the disease. For example, the tunica media, underlying an atherosclerotic plaque, often shows atrophy with loss of smooth muscle cells. The result of such atrophy is the dilation of the artery. However, even before this final stage, a remodelling of the intima is observed, which causes a progressive dilation of the vessel, which, in this way, manages to make room for the atherosclerotic plaque while maintaining the size of the vascular lumen. As a result, the artery may appear completely normal on angiographic evaluation, despite being the site of a severe atherosclerotic process. This represents a rather serious problem in the angiographic evaluation of atherosclerosis. It can generally be assumed that, when a plaque becomes visible by angiography, it is not a new plaque but rather the emergence of the tip of the iceberg.

Atherosclerosis does not uniformly alter the arteries, as it is a focal disease. This phenomenon is highlighted by the term plaque, used by early pathologists. The focal nature of this disease is in apparent contrast with the fact that most of the risk factors for atherosclerosis, such as hyperlipidaemia, arterial hypertension, smoking and diabetes, are, by their nature, systemic and, therefore, should have a similar effect on all the different sections of the arterial tree. This clearly demonstrates how the different systemic risk factors must act in concert with local factors. One of these local factors is the *WSS*, which is the tangential stress exerted by the blood flow at the level of the arterial wall. Atherosclerotic plaques do not develop randomly at different zones of the arterial system but follow the course of the *WSS*. For example, they are located, preferably, at the level of arterial bifurcations, i.e. in areas characterized by low values of *WSS* and, therefore, by a longer interaction time between particles carried by the blood flow (like the *low-density lipoproteins, LDLs*) and the luminal surface of the artery. This results in a greater diffusion of lipoproteins through the endothelium and, in the case of hyperlipidaemia, an accumulation of lipids in the sub-endothelial matrix. Another risk factor that can affect vascular permeability is represented by homocysteinaemia, as high concentrations of homocysteine (a sulphurated amino acid derived from the essential amino acid methionine by loss of a methyl group) can damage endothelial cells.

In recent years, researchers' attention has also focused on the possible role of microorganisms in the pathophysiology of atherosclerosis, as numerous epidemiological studies demonstrate the existence of an association between infection by *Chlamydia pneumoniae* and cardiovascular diseases [23]. The analysis of vascular tissue samples, surgically removed, shows the presence of *Chlamydia pneumoniae* in 50–70% of all atherosclerotic plaques [24]. It is likely that the local action of these microorganisms may contribute to the development of atherosclerotic plaques; this could explain the focal nature of this disease. Secondary symptoms from atherosclerosis depend not so much on the development of the plaque but rather on the degeneration and rupture of already present plaques. In the case of acute myocardial infarction and unstable angina, these can, almost invariably, be attributed to the formation of a thrombus above a cracked atherosclerotic lesion, as the formation of an atherosclerotic plaque is a slow phenomenon that, as such, allows for remodelling of the blood circulation to ensure the necessary influx of blood, such as the development of small collateral vessels. Moreover, the rapid growth of a plaque, sometimes observable on sequential angiographies, is probably caused by the rupture of the plaque, followed by scar-like repair processes that incorporate the thrombus and other blood cells. Consequently, it is likely that the treatment of atherosclerosis in coronary patients should focus primarily on plaque stabilization.

2.2 Anatomy and Cellular Biology of the Arterial Wall

2.2.1 Histological Organization

The anatomy of a normal artery is very simple. It has a wall made up of three layers: the tunica intima, which forms a barrier between the arterial wall and the circulating blood; the tunica media, which is a muscular state; and the tunica adventitia, which merges with the connective tissue of the surrounding organs. The tunica media represents the thickest layer of the arterial wall. It is composed of a single cell type, the vascular smooth muscle fibre, which constitutes the majority of the cellular component of the artery and produces the elements of the extracellular matrix of the tunica media. Smooth muscle cells are elongated, spindle-shaped and adhere to each other through junctional complexes, forming a syncytium. The cells are organized in circular layers that surround the arterial lumen in concentric circles. They produce large amounts of elastic fibres, which form the lamellae between the different layers of muscle cells. Therefore, the tunica media presents a "multilayer" organization in which two elastic lamellae surround a layer of smooth muscle cells, forming a lamellar unit. In elastic arteries like the aorta, the tunica media is made up of 20–50 lamellar units. In smaller muscular arteries, the organization is less developed, although layers of smooth muscle and elastic fibres can still be distinguished. The adventitia is a connective tissue structure that has a certain continuity with the surrounding stroma (interstitial tissue). Its inner part is fibrous and dominated by the presence of collagen and elastin, which gradually transform into loose connective tissue as one moves away from the media. In addition to fibres, the adventitia contains fibroblasts, mast cells, adipocytes and the terminations of sympathetic nerves. There are also blood and lymphatic vessels that partially penetrate into the media. In a normal artery, the inner part of the media and the intima are devoid of vessels, while, in pathological conditions, such as atherosclerosis, vascularization extends to these areas due to angioma factors.

The intima consists of a continuous single-cell layer, the endothelium; its basement membrane and a layer of connective tissue with some rare primitive mesenchymal cells. In the newborn, this layer is often only a few *micrometres*; over the course of life, it gradually thickens, reaching depths of several hundred *micrometres* in an adult's aorta due to a continuous and progressive accumulation of connective tissue fibres, proteoglycans and mesenchymal cells. At the bifurcation points of the arterial tree, there can be actual "cushions" that increase the thickness of the intima with a less organized appearance than the normal structure of the arterial wall. In this location, the endothelium shows increased permeability and a higher rate of proliferation. The intima is thickened, the lamellar organization of the media is altered and smooth muscle cells proliferate at a higher rate. These are "hot spots" of cell division and tissue renewal. However, it is also possible that they represent a tissue response to increased haemodynamic stress, as they are often found at points where the flow is particularly disturbed. It is interesting to note that the anatomical arrangement of atherosclerotic lesions overlaps with that of the intimal cushions. It is unclear whether the cushions are the substrate for the formation of lesions or whether the factors that induce the formation of cushions (e.g., haemodynamic stress) are also capable of promoting the atherosclerotic process [25].

2.2.2 The Cells of the Arterial Wall

Endothelial Cells

The endothelial cell is a thin and elongated epithelial cell, specialized in creating a barrier and controlling the permeability between blood and artery. Its cytoplasm is filled with pinocytotic vesicles, while, at the points of contact with adjacent endothelial cells, junctional complexes develop with specialized structures capable of increasing mechanical resistance and controlling the permeability to macromolecules. The endothelial cells of the arteries and most of the capillaries form a single layer of polygonal cells that fit perfectly together through such junctions. The endothelium has three important functions: determining blood-tissue permeability, controlling vascular tone and regulating the properties of the vascular surface related to haemostasis and inflammation. There is fine control of trans-endothelial permeability, which depends on the size and physico-chemical characteristics of different molecules. Small gaseous molecules, without an electric charge, such as oxygen, can diffuse without any restriction through the endothelium, following a concentration gradient. Many other small molecules, like glucose, can freely cross the endothelium. On the contrary, the endothelial penetration of macromolecules is much more limited, strictly depending on their size. In fact, while smaller proteins manage to pass through intra-endothelial gaps, larger proteins and particles can reach the sub-endothelial space only through transport involving endocytotic vesicles. The *lipoproteins*, which play a fundamental role in the development of atherosclerosis, penetrate the endothelium precisely through this mechanism.

Vascular tone is determined by the degree of concentration of smooth muscle cells present in the arterial wall, which is controlled by the endothelium through the release of paracrine vasoactive mediators. Among these mediators, the best characterized is nitric oxide (NO), an inorganic gas that is produced from *L-arginine*, thanks to an endothelial enzyme, NO synthase. Under normal conditions, this production occurs constitutively but can be significantly increased in response to an increase in the intracellular concentration of calcium in endothelial cells, as a result of the action of acetylcholine and other circulating mediators. Nitric oxide diffuses through the endothelial plasma membrane, reaches the extracellular space and activates an enzymatic cascade in the smooth muscle fibre. This results in a relaxation of the smooth muscle with a reduction in muscle tone. There are other endothelial vasoactive factors capable of counteracting this effect, such as the vasoconstrictor peptide endothelin-1 and angiotensin-II. The first one derives from the pre-pro-endothelin polypeptide, which is found within endothelial cells, and the second one derives from angiotensinogen, which is present in the bloodstream. These two vasoconstrictor factors are activated by specific enzymes produced by endothelial cells. The endothelium also performs important functions for haemostasis and inflammation. The endothelial surface contains a series of factors that regulate platelet adhesion, coagulation and fibrinolysis. Many of these factors are contained in specialized intracellular organelles, the *Weibel–Palade* bodies, which empty their content onto the endothelial surface when the cell is activated by thrombin or some other mediator. Similarly, endothelial cells control the inflammatory properties of the vascular surface, expressing chemokines, which promote the recruitment of leukocytes, and cytokines that activate immune cells.

Smooth Muscle Cells

The smooth muscle fibre is by far the most represented cell type in the arterial wall, constituting over 95% of all cells. These cells contain myosin and actin filaments, although their contractile apparatus is less developed than that of the striated muscle fibres of the cardiac and skeletal muscles. The vascular smooth muscle fibre is particularly primitive, combining the ability to modify its own tone with a fibroblast-like role in producing an extracellular matrix. In the arteries of adults, almost all the smooth muscle cells are located in the tunica media. They are joined together by junction complexes that include tighter adhesion zones and the so-called gap junctions. This not only increases mechanical resistance but also allows the rapid transfer of cell-signalling molecules through all the smooth muscle cells. The contractile apparatus of smooth muscle cells is dominated by actin filaments. Although the contractile filaments are associated with each other, forming specialized structures called "*dense bodies*," they are not as well developed as in striated muscle fibres; therefore, it is not possible to identify the sarcomeres. This reduces the contractile ability of the smooth muscle cell compared to the striated one. In most of the arterial tree, the contractile repertoire of the smooth muscle fibre is limited to modifications of vascular tone, which are fundamental not only in normal fluctuations of arterial pressure but also in the regulation of perfusion in different organs and tissues. Smooth muscle tone is regulated by numerous mechanisms. As mentioned earlier, the local regulation exercised by endothelial cells is of fundamental importance. Furthermore, muscle tone is influenced by metabolites produced by the surrounding tissue, by the autonomic nervous control exercised by sympathetic nerve endings and by various circulating mediators. Together, all these stimuli contribute to the fine regulation of vascular tone, blood pressure and blood perfusion.

The matrix produced by the smooth muscle is composed of two main types of fibres: elastic and collagen, cemented by a basic substance that contains a loose network of proteoglycans. These collagen fibres are particularly important for the mechanical properties of the tunica media. In general, smooth muscle cells, like endothelial cells, are quiescent cells that do not divide. However, a vascular injury triggers a proliferative response whereby smooth muscle cells in the tunica media begin to divide, then migrate into the intima, where they divide repeatedly to form an intimal thickening. This process, which resembles the restenosis observable after angioplasty and vascular surgery, is under the control of growth factors [26].

Leucocytes and Other Non-vascular Cells

Macrophages represent a small but significant portion of the normal artery's cellular population. They derive from blood monocytes, penetrate the arterial wall after interaction with the endothelium through leukocyte adhesion molecules and are located in the intima and adventitia. Lymphocytes and mast cells can also be found in the artery, although in smaller numbers than macrophages. Finally, adipocytes are common in the adventitia and represent an important cellular component of the loose connective tissue surrounding the artery. They derive from fibroblasts, and their size and quantity obviously depend on the nutritional state.

2.2.3   Histopathology of Atherosclerosis

Three types of atherosclerotic plaque have been identified:

- Lipid streaks.
- Fibrous plaques.
- Complicated lesions.

Lipid streaks consist of accumulations of macrophages filled with numerous lipid droplets (foam cells), actually composed of cholesterol esters derived from oxidized or aggregated *LDLs*, that are absorbed by a specific family of scavenger receptors. Macroscopically, they are visible as yellow stripes that follow the direction of blood flow. Lipid streaks do not hinder blood flow in the slightest.

In fibrous plaques, lipids are present both in macrophage foam cells and in the extracellular matrix. The intima is thickened due to the accumulation of smooth muscle cells and extracellular matrix proteins. T lymphocytes and, occasionally, B lymphocytes and mast cells may also be present. Smooth muscle cells and the extracellular matrix are more abundant in the sub-endothelial region, often forming a fibrous cap that covers the lipid and inflammatory cells in the deeper wall of the plaque. In the coronary arteries, fibrous plaques are often eccentric and, therefore, only cover part of the vessel. Although such fibrous plaques can grow to reduce the vascular lumen significantly, it is believed that they are not the main cause of clinical symptoms, at least until they remain intact. By their nature, fibrous plaques are nevertheless heterogeneous. Plaques with a fibrous cap and a large core of lipids and inflammatory cells are at high risk of rupture. This risk does not seem to depend on the size of the plaque.

Complicated lesions are plaques that, in addition to lipids, inflammatory cells and fibrous tissue, also contain a haematoma or haemorrhage with thrombotic deposits. Complicated lesions primarily develop as a result of the rupture of a fibrous plaque. Another possible cause could be the bleeding of capillaries that enter the plaque from the adventitial vessels. Other common features are the presence of fissures, erosions and ulcerations of the fibrous cap and luminal surface. The morbidity and mortality of coronary atherosclerosis mainly derive from these lesions. Often, in the elderly, these lesions contain calcium deposits. These deposits probably make the plaque more fragile, increasing the likelihood of rupture in response to mechanical stress.

Based on their morphological characteristics and in light of the experience accumulated in various experimental animal models, it is presumed that the three types of plaque represent different stages of atherosclerosis, developing in chrono-logical order: from lipid streaks to fibrous plaques and finally to complicated lesions. However, with the deepening of knowledge about the atherosclerotic process, it is clear that this hypothesis does not fully describe the complexity of this disease. In fact, only a certain type of lipid streak seems to be at risk of transforming into more advanced lesions. The rapid changes in plaque characteristics in response to rupture have important clinical implications that must be duly considered.

The *Committee on Vascular Lesions of the American Heart Association* has proposed a new classification, dividing the progression process of the lesions into eight different stages [27]. Although, at first, the terminology appears complicated, it has the advantage of integrating morphological changes and clinical consequences useful for both researchers and clinicians. Moreover, it takes into account the role of what are considered physiological modifications of the vascular wall, such as adaptive intimal thickening. Usually, human arteries have both thin and thick sections. These differences are present at birth and reflect physiological variations to tangential and longitudinal mechanical stresses. The thicker intimal sections are found near bifurcations and are defined as adaptive intimal thickenings. Their growth is self-limiting, and they almost never obstruct blood flow. Adaptive intimal thickening consists of a sub-endothelial layer of extracellular matrix, rich in proteoglycans, that contains smooth muscle cells (SMCs).

Type I lesion is the earliest one, characterized by scarce lipid deposits and rare foam cells. In the coronary arteries, type I lesions are generally co-localized with adaptive intimal thickenings, indicating that the same mechanical factors underlying such modifications are also responsible for the formation of atherosclerotic plaques.

In type II lesions, macrophage foam cells are more numerous and organized than those traditionally called lipid streaks. Type II lesions can occasionally contain T cells, mast cells and lipid-filled smooth muscle cells. It is generally accepted that more advanced lesions develop from a subpopulation of such lesions, defined as "type II-a lesions." Usually, they are present in some particular sites, and arterial segments with adaptive intimal thickening are those at greatest risk. Type II-b lesions are found in arterial tracts with relatively thin intima and rarely become more advanced plaques, even in the presence of risk factors such as hyperlipidaemia.

Type III lesion is the first stage classically recognized as an atherosclerotic plaque. The most important difference, compared to type II lesions, is the presence of small extracellular lipid deposits. The presence of type III lesions seems to be predictive of the development of clinical disease.

In type IV lesions, the amount of extracellular lipids is greater, forming a continuous accumulation, free from cells, of cholesterol deposits. These lipids can derive either from the degradation of foam cells or from direct deposition. According to the old classification, the disease has reached the stage of advanced lesion at this point. The lipid core is surrounded by inflammatory cells and is covered by a thin layer of smooth muscle cells and connective tissue. Generally, type IV lesions are crescent-shaped and increase the thickness of the vascular wall. At this point, the artery remodels itself in order to maintain the original lumen. The external profile of the vessel becomes oval, and consequently, these lesions are difficult to observe with angiography. These lesions, although clinically silent, are capable of rapidly undergoing rupture, generating symptoms and clinical manifestations of the disease.

Type V lesions are characterized by an increase in the fibrous tissue that covers the central lipid core of a type IV lesion. Often collagen becomes the main component of the plaque, constituting most of its volume. Generally, type V lesions are too large for the artery to compensate for them through remodelling, and this leads to a narrowing of the lumen. These lesions are therefore detectable by angiographic analysis. Like type IV lesions, type V lesions can also rupture.

Type VI lesions are plaques that contain thrombotic or haemorrhagic deposits. The main cause of the development of type VI lesions is the rupture of a plaque; often fissures, erosions and ulcerations of the sub-endothelial fibrous tissue are observed. Type VI lesions are, with few exceptions, those at which clinical events occur, such as acute myocardial infarction and unstable angina.

Type VII and VIII lesions are advanced plaques that do not contain lipids, except in minimal quantities, while they present significant deposits of calcium (type VII lesions) or are mainly composed of collagen (type VIII lesions). It is thought that these lesions represent the terminal stage of the disease. Calcification is an age-dependent phenomenon and is widely present in the coronary arteries of individuals over 70 years of age. Calcification generally does not contribute to plaque growth, and its clinical relevance is not entirely clear, although it is likely that it makes the lesions less elastic and more sensitive to mechanical stresses.

Type VIII lesions are more stable than type V and VI lesions. From a clinical point of view, the transformation of type V and VI lesions into type VIII lesions represents a significant advantage.

2.2.4 Pathogenesis of Atherosclerosis

Historical Notes

The mechanisms of formation and progression of atherosclerotic lesions have puzzled scientists for over 150 years. In 1856, the brilliant German pathologist Rudolf Virchow suggested the hypothesis that atherosclerosis is caused by an inflammatory response triggered by plasma components (including fluids) at the level of the arterial wall. Another pathologist, von Rokitansky, hypothesizes that the formation of the atherosclerotic lesion is secondary to the organization of thrombotic material on the surface of the arteries. In the early twentieth century, a large piece was added to the puzzle when Anitschkow, in St. Petersburg, highlighted the presence of significant lipid deposits in atherosclerotic plaques. A few years later, two Russian researchers, Starokadomskij and Sobolev, demonstrated that a mechanical lesion of the aorta is capable of causing intimal lesions very similar to atherosclerosis. These data are in line with Virchow's hypothesis, as a mechanical lesion leads to an increase in the infiltration of plasma components within the artery. In the 1950s, Florey and his study group linked some of these observations, demonstrating that an injury capable of causing de-endothelialization increases the accumulation of lipids and macrophages in the arterial wall [28].

With the advent of molecular medicine, it has been possible to formulate more specific hypotheses for the pathogenesis of atherosclerosis. Ross et al. [29] proposed the thesis that artery injury can be able to cause the local release of *platelet-derived growth factor (PDGF)* from adherent platelets and/or other cells. This could trigger a proliferative response in the population of smooth muscle cells, capable of leading to atherosclerosis. An alternative hypothesis, suggested by Benditt et al. [30], argues that atherosclerosis is caused by an uncontrolled proliferation of smooth muscle cells similar to that which can be noticed in a benign tumour. The discovery by Brown and Goldstein [31] of *LDL* receptors and cholesterol metabolism mechanisms allows the verification of the *"cholesterol hypothesis"* through efficient pharmacological and genetic tools. It is thus demonstrated with certainty, both in humans and in experimental animal models, the existence of a direct correlation between cholesterol levels (especially *LDL* cholesterol) and the extent of atherosclerosis. Based on these results, it becomes clear that any hypothesis on the pathogenesis of atherosclerosis must seek to explain the role of cholesterol. The new genetic models of the disease, based on alterations of lipid metabolism in mice that present specific *knockout* genes, allow for a detailed analysis of the different pathogenic stages and have significantly improved our understanding of atherosclerosis over the past 10 years [32].

2.2.5 Atherosclerosis: A Current Hypothesis

In humans, studies on atherosclerotic lesions lead to the identification of molecules and cells that participate in the physio-pathogenic process. The recent developments of murine experimental models of genetic engineering make it possible to examine the role of some specific factors in the formation and development of atherosclerotic lesions. Based on these studies, the unifying hypothesis of the induction of atherosclerosis can be proposed. *Low-density lipoproteins* (*LDLs*) enter the intima through an intact endothelium. In hypercholesterolemia, the uptake of *LDLs* exceeds the elimination capacity, forming an extracellular reservoir of *LDLs*. This is amplified by the association of *LDL* with the proteoglycans of the extracellular matrix. The *LDLs* in the intima are oxidized by the action of oxygen-free radicals. As a result, pro-inflammatory lipids are generated that induce endothelial expression of cell adhesion molecules, activate complement (C′) and stimulate the secretion of chemokines (CC). All these factors cause the adhesion and entry of mononuclear leukocytes, particularly monocytes (MC) and T lymphocytes (T).

Monocytes differentiate into macrophages. The macrophages stimulate the expression of scavenger receptors (*ScR*) that internalize the oxidized *LDLs*, which transform into foam cells. The uptake of oxidized *LDLs* by macrophages also leads to the presentation of fragments of these to antigen-specific T cells. This induces an autoimmune reaction that leads to the production of pro-inflammatory cytokines capable of acting on endothelial cells, stimulating the expression of adhesion molecules and pro-coagulant activity, on macrophages, activating proteases, endocytosis and the production of nitric oxide (*NO*) and cytokines and on smooth muscle cells (*SMCs*), inducing the production of *NO* and inhibiting growth and the expression of collagen and actin.

Accumulation and Modification of Low-Density Lipoproteins

The first detectable modifications in a laboratory animal subjected to hypercholesterolemia are represented by the appearance, in the sub-endothelial intima, of lipids derived from the blood and the expression of leukocyte adhesion molecules on the endothelial surface. Following the increase in plasma levels of *LDL* cholesterol, there is an increase in its passage through the endothelium to the inner portions of the tunica intima. The ability to remove *LDLs* from the intima is limited due to the lack of vascularization of this region. Therefore, the possibility of removing excess *LDL* is quickly exceeded, resulting in the accumulation of *LDLs* in the extracellular matrix [33]. The proteoglycans of the matrix have a high affinity for *LDLs*, which results in the formation of an *LDL* deposit [34, 35]. In the intima, *LDLs* undergo a series of modifications that include aggregation, oxidation and degradation of their components. These can largely be explained as oxidative attacks on the *LDL* particle, possibly carried out by oxygen-free radicals generated by tissue macrophages [36].

Recruitment of Inflammatory Cells

The oxidation of *LDLs* leads to the release of modified lipids. Many of these lipid species can act as signalling molecules, activating smooth muscle cells and endothelial cells [37, 38]. This leads to the expression of the leukocyte adhesion molecule and the vascular cell adhesion molecule type 1 (VCAM-1), which is a receptor for monocytes and T lymphocytes [39]. These cells express a "counter-receptor" VLA-4 (*very late activation antigen-4*), which can bind to VCAM-1 and some matrix molecules on the vascular surface. Together with other interactions with adhesion molecules, binding to VCAM-1 leads monocytes and T cells to adhere to the endothelial surface at sites of lipid accumulation and modification. Chemokines (i.e. chemotactic cytokines) are produced by macrophages, endothelial cells and smooth muscle cells. Their induction seems to depend on lipid accumulation and oxidation, although the exact mechanisms have not yet been clarified. In addition, aggregates of oxidized cholesterol induce the activation of complement, which, in turn, generates chemotactic signals [40]. Both these stimuli can promote the migration of mononuclear cells from the endothelial surface, through the intercellular spaces of the endothelial layer, into the sub-endothelial intima.

It is possible that, in addition to lipids, other stimuli are capable of activating the endothelium and triggering the recruitment of leukocytes in the intima. Once present in the intima, monocytes differentiate into macrophages. This process is triggered by the macrophage colony-stimulating factor (*M-CSF*) produced by activated vascular cells. Since the monocyte is a relatively inert precursor of the active macrophage, the differentiation process is of critical importance for the development of the disease. This fact is highlighted by the observation that genetically *M-CSF*-deficient mice do not develop atherosclerosis when exposed to hypercholesterolemia or crossed with mice that develop atherosclerosis [41].

Formation of Foam Cells

The macrophage plays a fundamental role in the formation of the atherosclerotic lesion. Thanks to its ability to internalize oxidized lipoproteins, it accumulates cholesterol, transforming into a lipid-laden foam cell, which is nothing more than the prototype cell of atherosclerosis. Brown and Goldstein [31], awarded the Nobel Prize for their discoveries on receptors and cholesterol metabolism, observed that macrophages are unable to absorb significant amounts of unmodified native *LDLs*, while they can internalize huge amounts of oxidized ones through a series of scavenger receptors [35, 42]. These cell surface receptors recognize "macromolecular patterns" that contain clusters of negative charges. Such patterns are present both on the oxidized *LDLs* and on bacterial endotoxins and numerous other macromolecules [43]. These ligands are recognized by the scavenger receptors and are then internalized and degraded by lysosomes. The cholesterol esters present in the oxidized *LDL* are hydrolysed and the free cholesterol is poured into the cytoplasm to be then re-esterified by cytosolic enzymes, initiating an accumulation of cholesterol esters with the formation of lipid droplets in the cytoplasm, transforming the macrophage into a lipid-laden foam cell. A lipid streak is essentially made up of a collection of foam cells with some T cells and an accumulation of extracellular cholesterol (generally associated with lipoproteins) beneath an area of intact endothelium. The family of scavenger receptors may have evolved during evolution for the purpose of providing protection from endotoxemia, rather than for the metabolism of oxidized *LDLs* [43]. The number of scavenger receptors on the macrophage surface is not controlled by the content of intracellular cholesterol, and, therefore, the macrophage continues to internalize oxidized *LDLs* until the cytoplasm is overloaded with cholesterol esters. Scavenger receptors are regulated by immune system cytokines and metabolic factors other than cholesterol [44, 45].

2.2.6 Progression from Lipid Streak to Atherosclerotic Plaque

The lipid streak has no clinical relevance, as it mostly disappears spontaneously. Some, however, progress into true atherosclerotic fibrofatty plaques, especially at the level of areas where there is a mechanical haemodynamic stimulus. Smooth muscle cells migrate into the sub-endothelial space, divide and synthesize extracellular matrix. The result is the formation of a fibrous cap that separates the central core of the lesion, full of lipids, from the endothelial surface. It is composed of smooth muscle cells, similar to fibrocytes, covered by thick layers of their own matrix. The stimuli that induce the formation of the fibrous cap probably act by inducing the activation of smooth muscle. The result is cell migration and proliferation with the synthesis of collagen and proteoglycans. In light of the focal localization of plaques, local factors of the arterial wall are likely able to activate smooth muscle cells. Almost all attention has focused on growth factors, and the results of studies on experimental vascular injury models show how the fibroblast growth factor, *basic fibroblast growth factor (bFGF)*, and the *platelet-derived growth factor (PDGF)* are able to induce the proliferation of quiescent smooth muscle cells in vivo. Furthermore, the *PDGF* acts as a chemotactic factor for smooth muscle cells. While the *bFGF* is deposited in the arterial extracellular matrix and can be released in case of injury, the expression of *PDGF* by endothelial cells can be up-regulated in response to flow changes through the stimulus exerted by variations in mechanical stress on the *PDGF* promoter [46].

Haematopoietic cells, and, in particular, platelets and macrophages, can also be important sources of the growth factor *PDGF*. The secretion of *PDGF* by macrophages is induced by their activation and is controlled by cytokines produced by T cells and some vascular cells. A plausible scenario for the transition from lipid streak to fibrous plaque is therefore constituted by the release of *PDGF* by platelets and/or macrophages, stimulated by haemodynamic stress and/or inflammatory activation. The *PDGF* stimulates the migration and division of smooth muscle cells, as well as the formation of the fibrous cap. The lipid core is thus physically separated from the endothelial surface, and the plaque stabilizes. Obviously, the price of this change is a narrowing of the artery lumen.

2.2.7 Plaque Activation and Triggering of Clinical Syndromes

The progressive reduction of the arterial lumen can lead to clinical syndromes, such as *exertional angina* and *intermittent claudication*, even though it is surprising to note that plaques of considerable size can be completely asymptomatic. The development of acute clinical complications generally requires the occurrence of a further pathogenic event, such as the formation of a thrombus on the plaque [47]. When this occurs, acute coronary syndromes, such as myocardial infarction, can be observed, while in other vascular districts, thrombosis at the level of a plaque can cause transient ischaemic attacks, cerebral infarctions and acute ischaemic gangrene.

Numerous lines of research have established the role of thrombosis in precipitating acute ischaemia. The injection of radioactive fibrinogen in patients with acute coronary syndromes shows how this is incorporated into fresh coronary thrombi, while histopathological analysis of the coronary arteries makes it possible to identify mural thrombi at the level of the *"culprit lesions."* It is interesting to note that, in these lesions, it is almost always possible to demonstrate the presence of superficial damage, which, in 80% of cases, are small fissures that cross the endothelium to reach the plaque and, in the remaining 20%, are coarse areas of endothelial desquamation [47–49]. These surface defects lead to the export of sub-endothelial prothrombotic material that rapidly induces the development of thrombosis.

The superficial fissures found in most of the culpable lesions are considered evidence of plaque rupture. This does not necessarily indicate the presence of an opening through the entire endothelial wall, but, rather, that the structure of the intimal plaque, or part of it, has been damaged. Generally, plaque fissures extend from the endothelial surface through the fibrous cap to the lipid core of the plaque. Why do plaques undergo rupture? An important clue is derived from the observation of the abundant presence of macrophages, T cells and activated mast cells near the rupture site [50, 51].

Immunohistochemical, biochemical and cellular biology studies make it possible to describe a scenario that illustrates the activation and rupture of the plaque [52]. According to this hypothesis, the inflammatory cells of the plaque are activated by pro-inflammatory lipids, cytokines, antigens or microbes. The resulting immune/inflammatory activation leads to the activation of macrophages. As part of this, macrophages release metalloproteases, enzymes capable of digesting the collagen of the fibrous cap. The pro-inflammatory cytokines, present in the inflammatory area, prevent the reconstruction of collagen fibres, inhibiting the expression of the collagen gene. These cytokines also inhibit the proliferation of smooth muscle and the expression of the actin gene, thus eliminating the mechanisms capable of inducing the repair of the fibrous cap. Cytotoxic oxygen species and *NO* radicals released by macrophages contribute to the process, causing cell death by apoptosis. All these events reduce the resistance to rupture of the fibrous cap to the point of making it succumb to the mechanical stress exerted by the pulsatile blood flow.

When blood components come into contact with the sub-endothelium, immediate activation of platelets occurs, which adhere to the surface and begin to aggregate. A series of surface receptors mediate adhesion between platelets and tissue; these include glycoprotein IIb/IIIa, which binds fibrinogen, and glycoprotein Ib, which binds von Willebrand factor. *Adenosine diphosphate* (*ADP*) and other factors, released by platelets after their adhesion, induce the activation of other platelets, while the exposure of platelet membrane structures promotes the activation of the humoral coagulation cascade. The latter is triggered by the exposure of the protein tissue factor thromboplastin, which can be induced in macrophages, endothelial cells and smooth muscle cells by pro-inflammatory cytokines. Therefore, immune/inflammatory activation promotes thrombosis in two ways: with plaque rupture and with the stimulation of tissue factor expression. As for humoral coagulation, in addition to these factors, some co-factors, such as membrane phospholipids and the complex of factors VIII-von Willebrand, are important. The end result is the formation of a fibrin clot that surrounds and stabilizes the platelet thrombus. With the formation of the solid thrombus, an acute coronary syndrome is usually triggered. However, a series of intrinsic blood and vessel wall defence mechanisms are able to counteract thrombosis and prevent or eliminate complete vessel occlusion. The most important of these defence mechanisms is the fibrinolytic system, a proteolytic cascade that is triggered by plasminogen activators that can be expressed by endothelial cells.

An inhibitor of fibrinolysis, plasminogen activator inhibitor type 1 (PAI-1), is also expressed by endothelial cells; therefore, the balance between fibrinolysis and antifibrinolysis is crucial for the control of thrombus formation. By modulating the expression of the plasminogen activator and *PAI-1* genes, pro-inflammatory cytokines become important regulatory factors in this delicate balance.

2.3   Ischaemic Coronary Diseases: Revascularization Treatments

Introduction

Atherosclerosis has, as secondary clinical syndromes, ischaemic coronary diseases, such as acute myocardial infarction and unstable angina, mainly due to the formation of a thrombus on the plaque. The treatment of coronary diseases aims to reduce the incidence of further ischaemic episodes. Current therapy is not only based on pharmacological measures but, increasingly, on coronary revascularization and, in particular, on the use of stents.

Coronary insufficiencies can therefore be treated in the following ways:

- Pharmacological therapy.
- Percutaneous coronary intervention (*PCI*).
- Balloon angioplasty.
- Intracoronary stent (*Stenting*).
- Surgery through aorto-coronary bypass (*coronary artery bypass grafting, CABG*).

Pharmacological Therapy

Pharmacological therapy is often used to prevent an acute ischaemic attack in at-risk subjects or as support to other therapies. Pharmacological therapy is based on drugs such as aspirin (reduces the incidence of cardiac events); beta-blockers (reduce myocardial oxygen consumption by decreasing heart rate, blood pressure and myocardial contractility; in particular, they reduce mortality in patients with a history of infarction); nitrates (increase venous capacitance, which reduces ventricular volume and pressure and improves sub-endocardial perfusion; this effect is increased by coronary vasodilation, improved collateral flow and reduced afterload); and calcium antagonists (have the effect of direct coronary vasodilation and reduction of peripheral vascular resistances). Pharmacological therapy is generally associated with the control of the patient's lifestyle to avoid, or at least limit, risk factors, so the patient is often subjected to a low-fat diet and controlled physical exercises.

Percutaneous Coronary Intervention

Percutaneous coronary intervention is a minimally invasive procedure that is performed under local anaesthesia and involves the introduction of a catheter via an artery (usually the femoral artery) that is able to reach practically any stenotic vessel. The balloon angioplasty procedure involves a small balloon being inserted into the occluded artery via a catheter and inflated in such a way as to push the atherosclerotic plaque against the arterial wall. This procedure may be followed by the implantation of a stent, via a catheter, which can be a normal metal stent or a drug-eluting stent (*DES*).

Surgery Through Aorto-Coronary Bypass

Surgery through aorto-coronary bypass is an open-heart surgery. It is therefore a highly invasive procedure that involves the use of a heart-lung machine and, consequently, extracorporeal circulation. The purpose of this procedure is to bypass the blocked section of the artery in such a way as to restore blood flow. For this purpose, a section of the saphenous vein taken from the patient is used (recently a section of the mammary artery has begun to be used, as the use of an arterial vessel seems to guarantee better results). One end of this vessel is sutured downstream of the coronary obstruction, while the other end is sutured onto the ascending aorta. Being an invasive procedure, this technique, widely used in the past, is now used only in cases of failure of the percutaneous procedure or in those cases where it is not possible to perform angioplasty.

2.3.1 Angioplasty Procedure

Coronary angioplasty with a balloon, introduced in 1977 by Gruentzig [53], has a dramatic impact on cardiological practice, changing the traditional role of invasive cardiology from a predominantly diagnostic to a therapeutic activity. Initially, angioplasty is limited to obstructions of a single coronary vessel in patients with stable angina. Progressively, the role of angioplasty has evolved to the entire spectrum of coronary syndromes, including also the treatment of acute ischaemic syndromes, such as acute myocardial infarction and unstable angina. The indications for angioplasty are contained in the guidelines of the most important international cardiology societies. Generally, the presence of at least one angiographic stenosis, presumably responsible for signs or symptoms of acute or chronic ischaemia, is required. However, angioplasty is also performed in asymptomatic patients suffering from stenosis that has induced signs of ischaemia in provocative tests. Angioplasty involves the dilation of a balloon that produces what is called a controlled injury to the coronary wall. The intimal plaque is destroyed, and a lesion is produced that tears the intima, propagates to the media and can reach the adventitia. Freed from the atherosclerotic plaque, the vessel wall expands and the lumen increases. Once the balloon is deflated, the vascular lumen tends to restore the initial diameter (recoil) with a loss of 15–30% of the diameter obtained during inflation. This phenomenon is recorded from 5 to 15 min after deflation and persists, albeit less intensely, for the following 24 h. The damage caused by the balloon on the arterial wall facilitates two processes: acute coronary occlusion and restenosis. Stents are used in order to provide a scaffold to the arterial wall, thus limiting the recoil phenomenon and making the intimal dissections adhere to the wall.

Phases of Angioplasty

The phases of angioplasty can be viewed in the relative Figure of [54].

1. In this first phase, a guide wire is pre-emptively passed through the occlusion.
2. The balloon passes over the guide wire and is then advanced through the stenosis and positioned on it.
3. Once in position, the balloon is inflated.
4. When it is believed that the plaque has been sufficiently compressed and the artery sufficiently opened, the catheter with the deflated balloon is removed.
5. The stent is introduced into the blood vessel, mounted on a catheter with a second balloon.
6. The second balloon is inflated, and this causes the expansion of the stent, which presses against the wall of the vessel.
7. The balloon is then deflated and removed, and the stent remains in position, keeping the vessel open and restoring the blood flow.

2.3.2 Complications of Angioplasty: Restenosis

Angioplasty involves the dilatation of the balloon, which produces so-called controlled damage. Following this mechanical trauma, the artery may undergo, generally within the 4–6 months following, a new stenosis due to neo-intimal proliferation and, therefore, the formation of a new atherosclerotic plaque, which again leads to a progressive reduction of the diameter of the vessel. This event can occur in up to 50% of balloon angioplasty cases and is favoured by haemodynamic factors, such as *WSS*, from the small diameter of the vessel, from the length and complexity of the stenosis and from some characteristics of the patient, for example, the presence of diabetes or other factors. Restenosis is therefore a pathology caused by angioplasty, and the resolution of this problem has been one of the most important objectives of research in interventional cardiology in recent decades.

Up to now the tool that has proven most effective is the stent [55, 56]. The introduction of the stent in the practice of angioplasty has reduced to 10–30% the cases of restenosis. In the presence of a stent, we speak of intrastent restenosis. Stents have therefore allowed the use of angioplasty in a larger number of patients and lesions. Despite stents having significantly reduced the percentage of restenosis, the absolute number of intrastent restenosis remains high. The objective of research in interventional cardiology has therefore shifted to the design of stents with the aim of lowering the percentages of restenosis.

Intrastent Restenosis

The first reaction that follows the implantation of a stent is thrombosis, which consists of the formation of a thin layer of clot due to the presence of foreign material and endothelial damage. Acute thromboses were one of the major causes of clinical failure of intravascular implants before the introduction of anti-thrombogenic drugs. Thrombosis occurs within a few days and begins with the adhesion of platelets. The detachment of these cells from the intravascular wall strongly depends on local fluid dynamic parameters, which depend on the geometry of the implanted stent. Inflammation of the arteries can occur a few days after implantation and involves the deposition on the surface and the infiltration of monocytes into the tunica intima. Also in this case, the detachment of these cells is governed by the parameters of local blood flow. After a few weeks from implantation, some smooth muscle cells migrate from the tunica media to the intima. This response of the arterial wall is mainly due to the chronic damage caused by the high tension associated with the stent and, as shown in [57], by the increase in the production of *endothelin-1* by endothelial cells subjected to low *WSS* (*endothelin-1*, in addition to being a vasoconstrictor, is also a strong mitogenic agent). As seen previously, smooth muscle cells produce the proteins of the extracellular matrix that constitute an important portion of the volume of the atherosclerotic plaque. Blocking the proliferation of smooth muscle cells with the use of drugs encapsulated on stents (drug-eluting stents) seems to significantly reduce the rate of in-stent restenosis [58, 59]. If these reactions, which occur to varying degrees in all stent implants, do not cause excessive neo-intimal hyperplasia, it is likely that the patient will not have significant clinical symptoms associated with restenosis. Despite the ongoing chronic damage caused by altered and therefore non-physiological tension, the artery can adapt to this new condition thanks to remodelling. The reactions just outlined are mediated by fluid dynamic parameters. Based on blood flow, blood cells can be retained, or not, on the vessel wall. Furthermore, it is shown that many functions of the cells of the arterial wall are directly or indirectly influenced by the need to maintain the *WSS* within a certain range [60].

3   Coronary Stents

3.1   Introduction

Stents are intravascular endoprostheses made up of small, generally metallic, hollow tubes. Stents reduce the phenomenon of elastic recoil and lower the percentage of restenosis compared to balloon angioplasty alone. The first pioneering study on stents was conducted by Dotter, [61], who conducted experiments with the implantation of stents on the canine popliteal artery with a device made from a particular shape memory alloy, *nitinol* (*nickel–titanium alloy*), used in naval applications, thus devising the precursor of the stents that today are called self-expanding stents, thereby opening up new clinical possibilities. Since then, these devices have undergone significant development.

In the early 1980s, Palmaz [62] conducted experiments and patented the first "modern" stent, Fig. 1B of [62], with thin walls, expandable with a balloon (*balloon-expandable stent*). The first human implant of the Palmaz device was placed in the aorta in 1987. This revolutionary event launched a real technological challenge in the design of stents, which continues today. From the first sale of the Palmaz stent, in 1994, the market for these devices has grown enormously, exceeding 4 billion dollars a year in the United States. To date, the *U.S. Food and Drug Administration* has approved for clinical use more than 100 different designs of balloon-expandable and self-expanding stents.

In the early 2000s, the first experiments with *drug-eluting stents* (*DESs*), also known as medicated stents, were successfully completed. These stents are capable of gradually releasing a drug, usually *sirolimus* (an immunosuppressive cytostatic, also used for the prophylaxis of kidney transplant rejection), which seems to be able to lower the percentage of restenosis. The first *DES* put on sale was the *CYPHER* stent in 2002, followed by the *TAXUS* stent. The latest frontier of intravascular endoprostheses is *biodegradable stents*. These devices, made of special bio-absorbable polymers in 8–9 months, are currently under clinical study. The initial clinical data are encouraging, but it does not seem that these stents significantly lower the percentage of restenosis.

3.2   Stent Design

The increasing use of these devices in recent years, supported by the significant reduction in the percentage of restenosis compared to simple angioplasty, has led to the development of different designs. Despite the existence of numerous stent designs, all aim to reduce the thickness of the stent wall, and therefore its profile, and to increase flexibility, both to facilitate implantation and to facilitate removal in case of failure. In clinical practice, the operator must decide the most appropriate stent for a particular patient and for the lesion that needs to be treated.

The general characteristics of the ideal stent are as follows:

- Flexibility.
- Traceability.
- Radiopacity.
- Thrombo-resistance.
- Biocompatibility.
- Ease and reliability of expansion.
- High radial resistance.
- Low profile.
- Minimal surface area.
- Hydrodynamic compatibility.

Stents can be classified based on various design variables [63–66]:

- Expansion mechanism (balloon-expandable or self-expandable).
- Material (stainless steel, cobalt alloys, tantalum, nitinol, etc.)
- Manufacturing technique (laser, water jet, photoetching, etc.)
- Biological response (with inert layer, with active layer or biodegradable).
- Geometric design (mesh structure, spiral, hollow tube, single ring, sequence of rings or custom design).
- Addition on the stent: graft, markers radio-opaque, surface layer (*coating*).

3.2.1 Expansion Mechanism

Generally, stents can be divided, based on the expansion mechanism, into two main categories: balloon-expandable and self-expandable. Balloon-expandable stents are made with a material capable of plastically deforming as a result of the pressure imposed by the inflating balloon. When the balloon deflates, these stents remain in their expanded position (unless there is recoil elasticity). Self-expandable stents, on the other hand, are manufactured in their expanded configuration and then compressed into a catheter, which, when removed, allows the stent to regain its expanded shape, thanks to elastic return. This second class also includes stents made from shape memory materials that exploit properties other than elasticity to recover the expanded shape.

3.2.2 Materials

The constituent materials of metallic stents must, first and foremost, have excellent corrosion resistance and biocompatibility. These materials should be radio-opaque and capable of not creating artefacts in magnetic resonance. For balloon-expandable stents, the ideal material should have a low yield stress (to be deformed under the pressure of the balloon) and a high elastic modulus (to minimize the recoil elasticity). Conversely, self-expandable stents, which rely on elastic properties, require a material with a low elastic modulus and a high yield limit to have large elastic deformations. Alternatively, the effects of shape memory materials, such as nitinol, can be used. In this case, large deformations can be achieved thanks to the super-elasticity or thermal memory of the material. The most commonly used material for stents is stainless steel, typically 316 L, with very low carbon content and with the addition of molybdenum and niobium. This steel has excellent corrosion resistance. After complete annealing, stainless steel 316 L is easily deformable and is the standard material for balloon-expandable stents. Without the annealing treatment, however, stainless steel has sufficient elasticity to be used in self-expandable stents. Other materials used for balloon-expandable stents are tantalum, platinum alloys, niobium alloys and cobalt alloys. These materials are used for their excellent properties, such as radio-opacity, corrosion resistance and compatibility with magnetic resonance, and have high resistance. Improved radio-opacity and high resistance allow for the design of stents with a smaller profile.

The materials for self-expandable stents must be able to undergo large elastic deformations. The most commonly used material in this case is nitinol, which is capable of elastically deforming beyond 10%. This unusual range of elasticity, called super-elasticity, is the result of a thermo-elastic martensitic transformation [67]. The limited elastic range of conventional materials, including stainless steel or cobalt and nickel–titanium alloys, limits the possibilities of the design of self-expandable stents.

3.2.3 Manufacturing Techniques

The choice of the manufacturing method for a stent essentially depends on the shape of the raw material used (tubular, filamentous or laminar). With a filamentous material, conventional processing techniques can be used to obtain spirals and shapes interwoven or knitted and welded. The simplest form for a filamentous stent is the spiral one. All spiral stents, to date, are made of nitinol and are self-expanding.

The most common self-expanding stent is the *WallStent*, by *Boston Scientific*, shown in Fig. 1, [63], a braided stent fabricated from cobalt alloy wire.

By welding the nitinol wire to form a closed-cell structure, it is possible to obtain a mesh stent, such as the self-expanding *Symphony stent* by *Boston Scientific*, shown in Fig. 2, [63].

Most coronary stents, and probably also most peripheral vessel stents, are produced by laser cutting from a tubular material. The lasers currently used can achieve cuts with a precision of less than 2 μm. Using tubes of 0.5 mm, very complex shapes can be obtained. Balloon-expandable stents are cut in the compressed form and only require post-cutting control and surface treatment (typically an electronic wash).

Fig. 1 WallStent (Boston Scientific), Fig. 8 of [63]

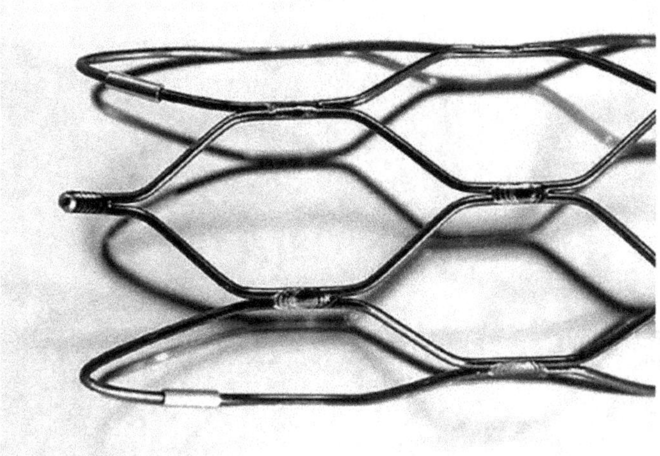

Fig. 2 Symphony Stent (Boston Scientific), Fig. 7 of [63]

Self-expanding stents can be manufactured in both compressed and expanded forms. If manufactured in the compressed form, in addition to control and surface washing, post-cutting expansion is also required. Laser cutting produces areas near the cutting lines with microstructural imperfections due to the high heat. These areas must be removed to achieve the required performance. An alternative cutting method that does not produce these areas is water jet cutting. A focused water jet with the addition of abrasive additives can be used to carve stent geometries. Another interesting manufacturing method is photochemical etching. Although this method is beginning to be used to produce stents from tubular material because it allows a large number of stents to be produced in a single manufacturing process, its real benefits are still under study.

3.2.4 Geometries

The first stent designs are, generally, either simple hollow tubes, like, for example, the first *Palmaz* stent, [62], or simple spirals. Simple hollow tubes have excellent radial resistance but very poor flexibility, while spiral stents have the opposite behaviour. The subsequent evolution of stents produced a rich variety of geometries, which can be classified into five major categories: spiral, helical spiral, mesh, single ring and sequence of rings.

Spiral

The spiral is the most used geometry in non-vascular applications, as it allows for easy explantation. An example of a stent for oesophageal applications is *Esophacoil*, a coil stent fabricated from nitinol ribbon, shown in Fig. 3, [63]. This geometry is very flexible, but its strength is limited. It is also subject to undesired elongations or compressions during the implantation phases, consequently generating cells of irregular sizes.

Fig. 3 Esophacoil Stent, Medtronic, Fig. 13 of [63]

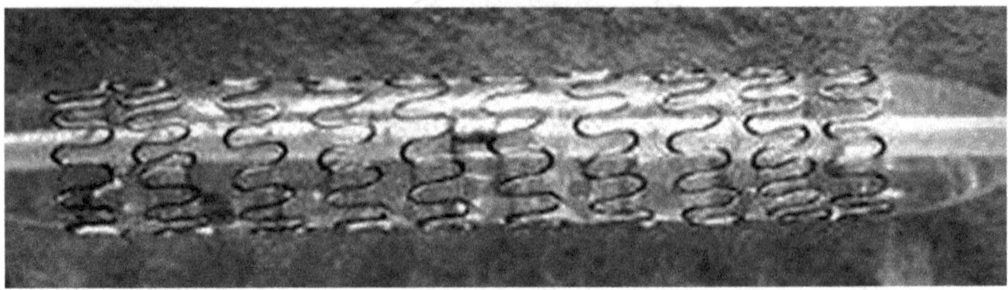

Fig. 4 Crossflex: a minimally connected helical spiral stent fabricated from stainless-steel wire, Fig. 14 of [63]

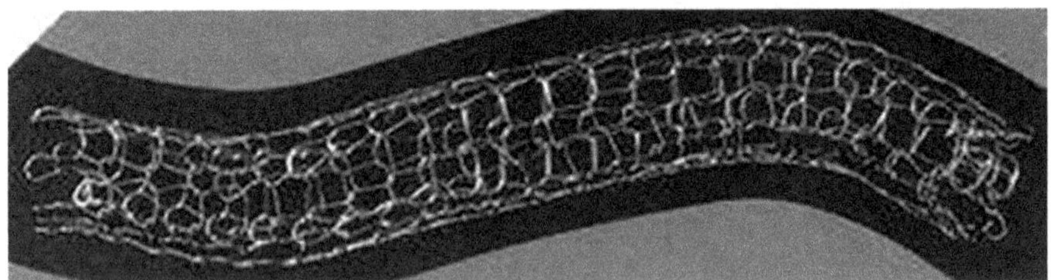

Fig. 5 Strecker stent made of knitted tantalum wire, Fig. 3 of [63]

3.2.5 Helical Spiral

This geometry is generally chosen for its flexibility. Without or with few internal connections, it is very flexible but lacks longitudinal support. This geometry can also undergo undesired elongations or compressions. With some internal connections, a bit of flexibility is sacrificed to achieve high longitudinal stability and better control of cell sizes. An example of a *Crossflex* stent, a minimally connected helical spiral stent fabricated from stainless-steel wire, is shown in Fig. 4, [63].

Mesh This category includes a wide variety of designs, made up of one or more metal wires. Examples of mesh stents are shown in Fig. 5, [63], (*Strecker*) and Fig. 6, (*Cook ZA*) [63], as well as in [68]. The braided geometry is often used for self-

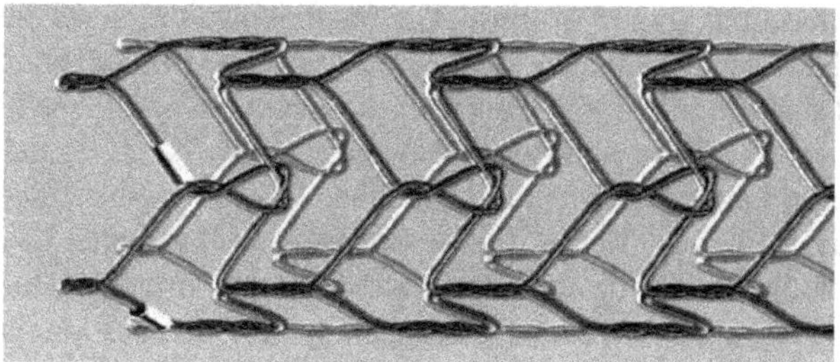

Fig. 6 Cook ZA: knitted nitinol wire design, featuring sleeve-type gold marker, Fig. 15 of [63]

expanding structures. This geometry offers excellent support but generally undergoes significant shortening during expansion.

Single Ring

Single "Z"-shaped rings are generally used as support grafts or similar prostheses. They can be individually sutured or, otherwise, joined to the support material during manufacturing. These structures are not generally used alone as vascular stents.

Sequence of Rings

This category includes stents made up of a sequence of expandable elementary structures, shaped like "Z" (struts, structural elements), joined together by connection elements. This type of geometry is the most used in stents currently on the market. This category can be further subdivided based on how the structural elements are connected and the resulting type of cells that are obtained:

- Regular connections, made with bridge elements at all flex points along the circumference of the structural elements (as in *CYPHER by Cordis*, Fig. 3 of [69], which shows a stent made up of a sequence of Z-shaped rings to form a regular peak-to-peak connection).
- Periodic connections, made with bridge elements only at some flex points along the circumference of the structural elements. The flex points with connections are alternated with flex points without connections (as in *Express TAXUS by Boston Scientific*, Fig. 5 of [69]).
- Peak-to-peak connection (Fig. 3 of [69]) or peak-to-valley connection (Fig. 5 of [69]) defines the connection points of the bridge elements with the structural elements.

Based on the type of connection, two different types of cells can be obtained: closed cell and open cell.

Closed Cell

When in the sequence of rings there are regular connections on the structural elements, we have closed cells. This configuration is only possible with peak-to-peak type connections (Fig. 3 of [69]). The first hollow tube stents are resistant but not flexible. New designs, like the stent *NIR TAXUS by Medinol* (Fig. 11 of [69]) or the *CYPHER* stent (Fig. 3 of [69]), improve these properties with the introduction of flexible connectors. These connectors have shapes like *U*, *V*, *S* and *N*, and they plastically deform during the stent implantation. This allows the structural elements to move away from, or closer to, the blood vessel wall to better adapt. The main advantages of closed-cell designs are the excellent scaffolding and the uniform surface. However, these structures are generally less flexible longitudinally than similar structures but with open cells.

Open Cell

This category includes ring sequence stents that have periodic connections. It allows peak-to-peak connections, peak-to-valley connections, intermediate-level connections of the structural elements and all possible hybrid combinations. In the open-cell structure, the unconnected elements contribute to increasing longitudinal flexibility. Periodic peak-to-peak connections are common in both self-expanding stents, such as the *SMART Cordis* stent (Fig. 7, [63], and balloon-expandable stents, such as the *CYPHER* stent (Fig. 3 of [69]) and the *AVE S7* stent (Fig. 8, [63]). In peak-to-valley connections, such as in the *TAXUS* stent (Fig. 5 of [69]) or the *ACS Multi-Link Vision, Abbott* stent (Fig. 9, [63]), the adjacent structural elements align in a peak-to-valley manner during stent expansion, optimizing scaffolding and support characteristics.

Fig. 7 SMART stent: self-expanding open-cell sequential ring design with periodic peak-to-peak non-flex connections, Fig. 18 of [63]

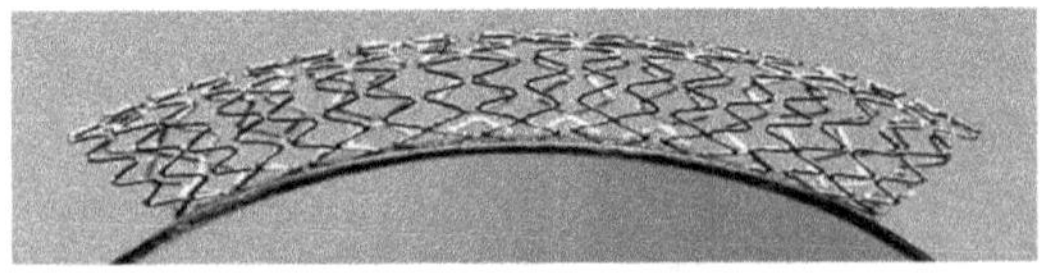

Fig. 8 AVE S7 stent: balloon-expandable open-cell sequential ring design with periodic peak-to-peak non-flex connections, Fig. 19 of [63]

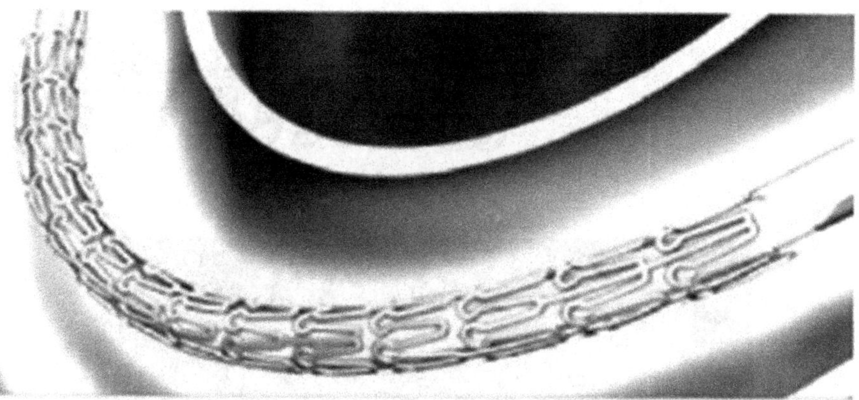

Fig. 9 ACS Multilink. Balloon-expandable open-cell sequential ring design with peak-to-valley connections, Fig. 20 of [63]

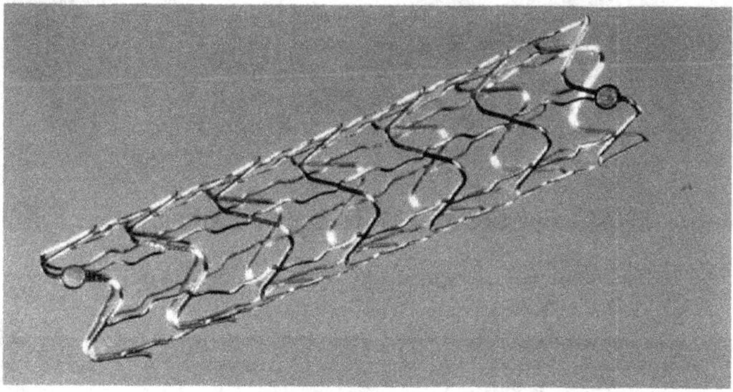

Fig. 10 Balloon-expandable BeStent open cell, with integral gold markers and mid-strut-to-mid-strut connections sequential ring design, Fig. 21 of [63]

On the other hand, peak-to-valley connections require more material that could otherwise be used for the structural elements. Consequently, these peak-to-valley structures are generally not as resistant as peak-to-peak structures. Peak-to-peak connections and peak-to-valley connections are the most common, but there are examples of recent stents, such as the *CYPHER SELECT* stent and the *BeStent, Abbott* stent (Fig. 10, [63]), that have intermediate-level connections of the structural elements.

Finally, there are stents with unique ring designs that are difficult to categorize into one of the aforementioned categories. An example is the *Navius* ZR1 ratcheting stent design, fabricated from a stainless-steel sheet and reported in Fig. 11, [63].

Radio-Opacity

Metallic stents, once implanted, are generally difficult to see with normal imaging techniques, especially if small or with thin structural elements. To improve visibility

Fig. 11 Navius, ZR1, ratcheting stent design, fabricated from stainless-steel sheet, Fig. 22 of [63]

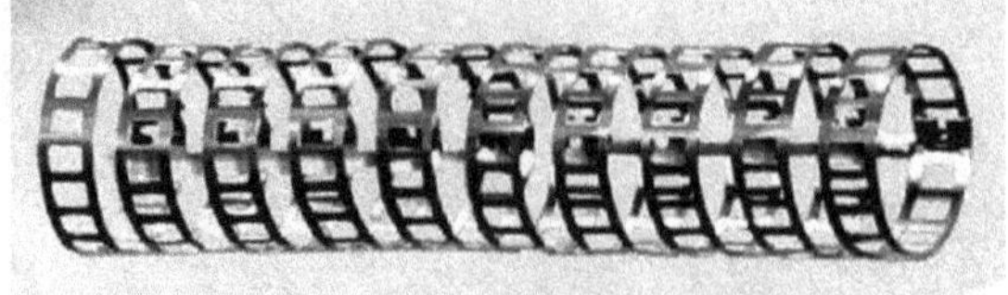

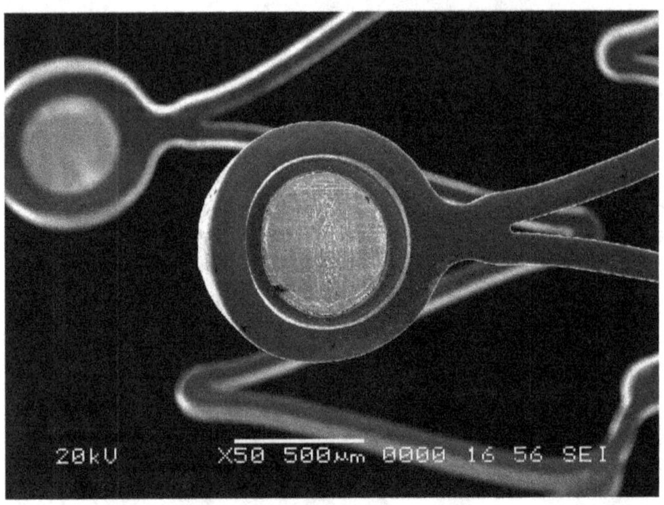

Fig. 12 SMARTeR tantalum radio-opaque marker in nitinol, Fig. 24 of [63]

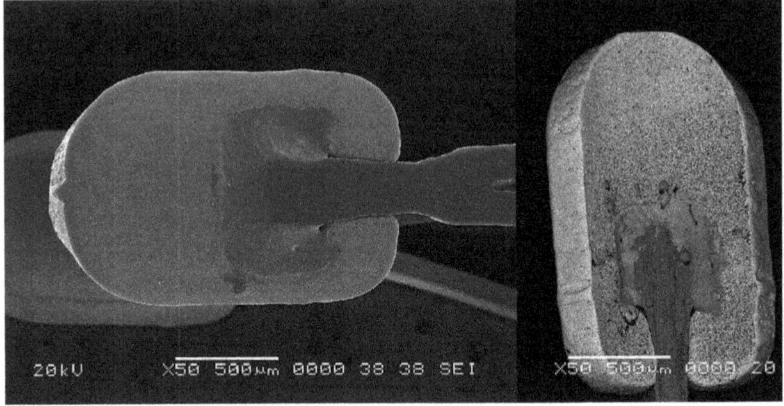

Fig. 13 Luminexx radio-opaque tantalum marker welded onto nitinol, Fig. 25 of [63]

to X-rays, stents are often added with markers. These additions are generally of gold, platinum or tantalum and can completely wrap around the stent, or just the ends, or they can be integrated by welding to the ends of the stent, as shown in Figs. 12 and 13, [63].

Coatings

Various active compounds are used to coat the stent with a surface layer and to improve biocompatibility and interactions with the different cells and molecules present in the blood and on the vascular wall. The aim is to achieve a significant reduction in the activity of these cells and molecules, thus increasing the safety and efficacy of the stents. Among the different compounds tested, heparin is one of the first, since it acts by reducing the coagulative cascade (and thus the thrombogenic risk) after implantation. Other compounds, such as phosphorylcholine and silicon carbide, are used to limit the activation and interaction of platelets, thus controlling their adhesion to the stent elements during the acute phase of stent re-endothelialization. Even passive layers have proven their effectiveness, e.g. stents coated with a surface layer of polytetrafluoroethylene (*PTFE*), since this layer of *PTFE* seems capable of trapping the material that could protrude from the stent elements.

Drugs

Drug-eluting stents require a fine combination between the design of the metallic stent and the drug delivery system. This fine combination is the standard approach of most drug-eluting stent (*DES*) designs. Drug delivery systems are essentially based on three different mechanisms:

- In bio-resorbable polymeric stents, the drug can be directly incorporated into the material and thus be released progressively over time.
- In metallic stents, the drug can be bound to the surface layer, or inserted into macroscopic fenestrations, or in microscopic nanopores, allowing a faster release.
- Metallic stents can be coated with a polymeric state (bio-resorbable or non-bio-resorbable) that incorporates the drug. This method allows a more regular release that should ensure a better drug–tissue interaction. In addition, there is greater control of the system.

Recent results [70] suggest that the configuration of the structural elements of the stents directly determines the pattern and degree of drug release obtained from the stent. Despite the drugs used being highly soluble and diffusive, the concentration of the drug in the vessel and in the various districts of the vascular wall is not uniform. The stent design, which maintains regular spaces within the meshes despite expansion in different anatomical circumstances, should ensure the predictable and regular release of the drug. For drugs with broad-spectrum immunosuppressive action, such as sirolimus and its derivatives, the geometric regularity of the meshes seems not to affect the efficacy of the drug. An adequate dose of the drug, applied to the stent, despite the wide local variability, ensures a uniform release and an adequate action on the entire vascular wall. On the contrary, drugs with limited-spectrum immuno-suppressive action, like paclitaxel, seem to be affected by the geometry of the stent. In areas where the structural elements of the stent are distant, it can have an insufficient dose of the drug, while in areas where the mesh is tighter, there is a risk of having an overdose or even toxic doses.

Among the different drug-eluting stents currently approved and on sale in the USA and Europe, the *CYPHER* stent (releasing sirolimus) from *Cordis* (*Johnson & Johnson*) and the *TAXUS* stent (releasing paclitaxel) from *Boston Scientific* have a relatively similar base: both are made up of a sequence of rings with an inert surface polymeric layer. These stents differ substantially in the drug released: the sirolimus was initially developed to reduce transplant rejections, while the paclitaxel is widely used in chemotherapy for the treatment of tumours. Many other new-generation drug-eluting stents are currently under study, both in terms of the constituent material and the drug release system. For example, the *Endeavor* stent (*Medtronic*), which uses the polymer phosphorylcholine for the release of sirolimus, is the first medicated stent, not by steel but by a cobalt–chrome alloy.

3.3   CYPHER Stent

The *CYPHER Sirolimus-eluting coronary Stent*, shown in Fig. 14, [71], is the first drug-eluting stent, launched in 2002, which caused a real revolution in interventional cardiology. This stent is produced by the *Cordis Corporation* (*Johnson & Johnson*), the world leader in the development and manufacturing of technologies for vascular interventions. The *CYPHER* stent consists of two components: a mechanical device and a drug with a release system. The mechanical device consists of the *Bx Velocity* stent, pre-mounted on a balloon catheter (*Raptor PTCA Dilatation Catheter*). The diameter of the stent varies from 2.5 to 3.5 mm. It is made of stainless steel 316 L and has a ring sequence geometry with closed cells and peak-to-peak connections. There are six cells in a circumference.

The drug sirolimus, also known as rapamycin, is a natural macrolide produced by Streptomyces hygroscopic, with potent immunosuppressive action, used for the prevention of rejection after kidney transplant. The inactive components are parylene C and two non-biodegradable polymers: polyethylene vinyl acetate (*PEVA*) and

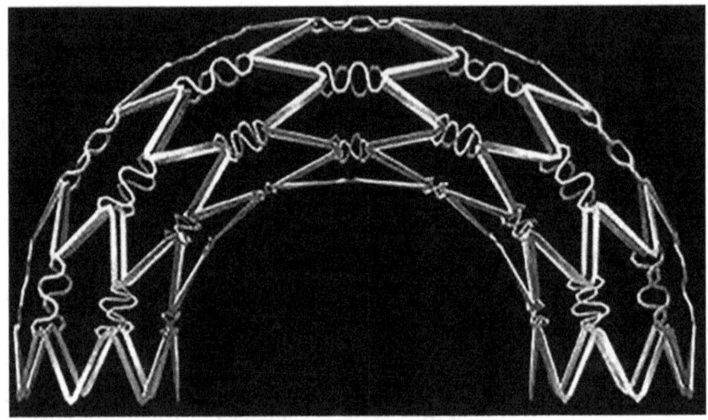

Fig. 14 CYPHER Sirolimus-eluting coronary Stent, Fig. 4a of [71]

polybutylmethacrylate (*PBMA*). A combination of these two polymers with the sirolimus (67%/33%) covers the stent treated with parylene C. On the stent, there is an additional layer of *PBMA* without a drug. The drug-polymeric layer is applied to the entire surface of the stent. The drug release occurs by diffusion and is completed in about 30 days.

The *CYPHER* stent is to date the most used coronary stent, with over two million implants worldwide. It is also one of the most studied stents, with more than 60 clinical studies, including independent studies, examining the effectiveness of this stent in a wide range of patients. All the trials conducted on the *CYPHER* stent are aimed at studying the effectiveness of the drug sirolimus. The first pilot study was the First-In-Man, [72], which included 45 patients suffering from ischaemic heart disease and non-complex atherosclerotic disease. In no case is intrastent restenosis observed. The evaluation of the results over time, carried out with intracoronary ultrasound, demonstrates the almost total absence of neo-intimal hyperplasia. After this initial study, many other trials are conducted comparing the *CYPHER* stent with traditional stents without drug release. The results of four trials, [73–75], show that the sirolimus-eluting stent significantly lowers the percentage of restenosis.

In September 2006, *Cordis* launched a new stent: the *CYPHER SELECT Plus Sirolimus-eluting Coronary Stent*, shown in Fig. 15, [1]. This stent is the evolution of the previous stent and is considered the first third-generation stent because it has an innovative drug delivery system. The delivery system is patented under the name of *CYPH $_2$ ONIC*. This system uses hydrophilic coating technology. The surface layer obtained with this technology seems to guarantee a more efficient drug release

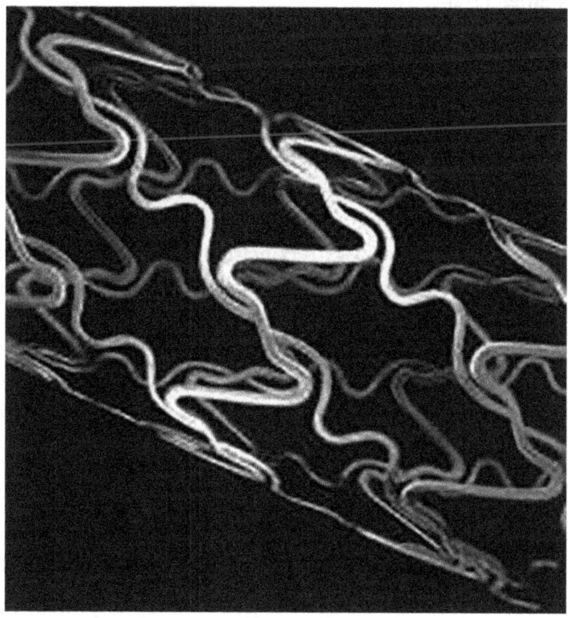

Fig. 15 CYPHER SELECT Plus Sirolimus-eluting Coronary Stent [1]

and better lubrication of the stent wall compared to the delivery systems used previously. From a mechanical point of view, the *CYPHER SELECT* stent is extremely flexible and has a very thin profile. The combination of all these factors is designed with the aim of making this stent suitable for the treatment of the most complicated lesions. The *CYPHER SELECT* stent is therefore created for the treatment of lesions located in tortuous and difficult-to-reach vessels. As this stent has only been on the market for a year, results from clinical studies on the effectiveness of this stent have not yet been published. Despite this, the *CYPHER SELECT* stent is rapidly spreading with great success and is completely replacing its predecessors.

4 Numerical Thermo-Haemo-Dynamics (THD) in CYPHER Coronary Stents

4.1 Introduction

The coronary stent *CYPHER*, in its two versions, *CYPHER* and *CYPHER SELECT*, is reconstructed as faithfully as possible with the *GAMBIT* software because it is the most used and most studied coronary stent in the world, but the studies (clinical) are all focused on the effects of drug release [1]. Concluding in about 30 days, the release of the drug is not the only aspect to consider in the design of a stent. Therefore, the goal of this work is to characterize its fluid dynamic behaviour, with particular attention to the geometry, disregarding other parameters. Computational fluid dynamics studies on coronary stents, present in literature, use 2D models [6, 15, 16] or use very simple geometries, reminiscent of the first stents, [62], like the studies [7–11, 18].

Gori et al. [76] presented the first computational analysis that uses a 3D model with a complex geometry. Because the geometry is periodic, it is possible, using appropriate boundary conditions, to use in the simulations only a part of the geometry. In particular, the *CYPHER* stents have six meshes per circumference, so it is sufficient to use a portion of the circumference of 60° to study the phenomenon. The simulations are carried out with numerical software, both in a steady and unsteady state. In the unsteady state, the physiological waveform of the flow within the coronary arteries is reconstructed with *MATLAB* software. A large number of simulations are carried out in order to evaluate the most suitable inlet and boundary conditions, the length of the stent and the dimensions of the mesh. The data analysis is carried out with the *Tecplot software*. Particular attention is paid to the stresses acting on the wall, as it is believed that these stresses are responsible for the phenomenon of restenosis. Therefore, the *WSS*, the *WSSG* and the *OSI* are analysed.

4.2 Creation of the Geometry and Grid

The geometry of the model is drawn with the *GAMBIT 2.2.30* software. This software is suitable for the generation of geometries and the grid and for the pre-processing of simulations carried out with other numerical software. It is structured at subsequent geometric levels: points, lines, areas and volumes. In the creation of geometry, it is necessary to start from the lowest level and then move on to the next upper level. *GAMBIT* has all the characteristics of drawing software and implements all the main functions but at a lower programming level compared to classic drawing software. *GAMBIT* is also a mesh generator, and, in this case, it is based on the principle of hierarchical levels. It is therefore necessary to first generate the meshes on the edges, then the surface meshes and finally the volume ones. The software gives a lot of freedom in the creation of meshes: it is possible to choose the type and size of the elements; moreover, it is possible to carry out non-uniform meshes, deciding the spacing for all the meshes (1D, 2D, 3D).

A real 3-D geometry of a coronary stent, made of sequential rings and connected with longitudinal flexible connectors, is studied. Two different connections are investigated: regular peak-to-peak (*CYPHER* or *S2*) and mid-strut-to-mid-strut (*CYPHER SELECT* or *S1*). The first type of stent has parallel connectors, and the second one has transverse connectors. As a first step, the basic geometric elements of the two stents, shown in Fig. 16, are drawn. To create the complex shape of the stents, the *non-uniform rational B-spline* (*NURBS*) curves are employed. These curves are obtained through algorithms that use parametric functions of a certain degree. In *GAMBIT*, it is possible to define the points of passage of the curve but not the tangents at these points.

The greatest difficulties in creating the geometry do not lie in the shape of the stent but in its definition on a cylindrical surface, as it is not possible to define the points of tangency of the *NURBS*. By defining only the points of passage on a cylindrical surface, it is not said that the relative curve belongs completely to the surface. To overcome this difficulty and achieve the desired result, a series of

Fig. 16 Basic elements of the two reconstructed stents, CYPHER SELECT on the left, S1; and CYPHER on the right, S2 [1]

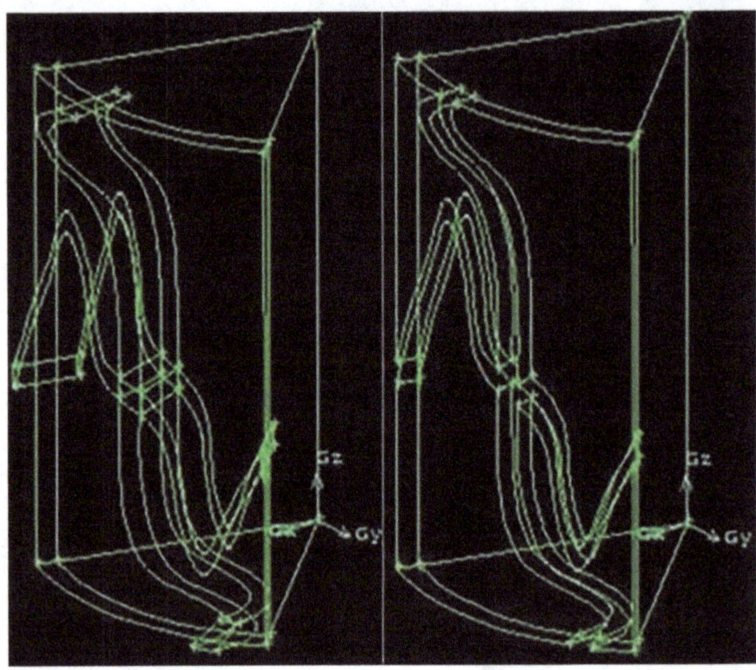

Fig. 17 Creation of the basic element of CYPHER SELECT [1]. (**a**) Before the Boolean cut. (**b**) After the Boolean cut

Boolean operations between volumes are carried out. In particular, the curves of *NURBS* that form the stent are drawn on a cylinder larger than the blood vessel, and then, through intersections between volumes, the stent on the cylinder that represents the blood vessel is obtained (Fig. 17).

To establish the dimensions of the geometry, measurements are made on a *CYPHER SELECT* stent in the expanded configuration. The cylinder representing the vessel is evaluated to have a diameter of 2.6 mm (in agreement with [11] on the size of coronary arteries), the thickness of the stent is evaluated to be 0.1 mm and the height of the basic element is 2.3 mm. Once the basic elements are created, it is sufficient to translate and rotate them to obtain the complete geometry of the stents (top of Fig. 18) [77]. The *CYPHER SELECT* stent is represented on the left of the top of Fig. 18, while the *CYPHER* stent is on the right of the top of Fig. 18. To simulate the stent inside the blood vessel, a simple cylinder area is added at the beginning and at the end of the stent. Given the symmetry of the geometry, it is sufficient to carry out the simulations with a portion of circumference of 60° (bottom of Fig. 18). The *CYPHER SELECT* stent is represented on the top of the bottom of Fig. 18, while the *CYPHER* stent is on the bottom of the bottom of Fig. 18. Given the geometric peculiarities, it is decided to use a non-uniform mesh. The stent is very thin compared to the diameter of the vessel; it is therefore necessary to use a very dense mesh near the wall.

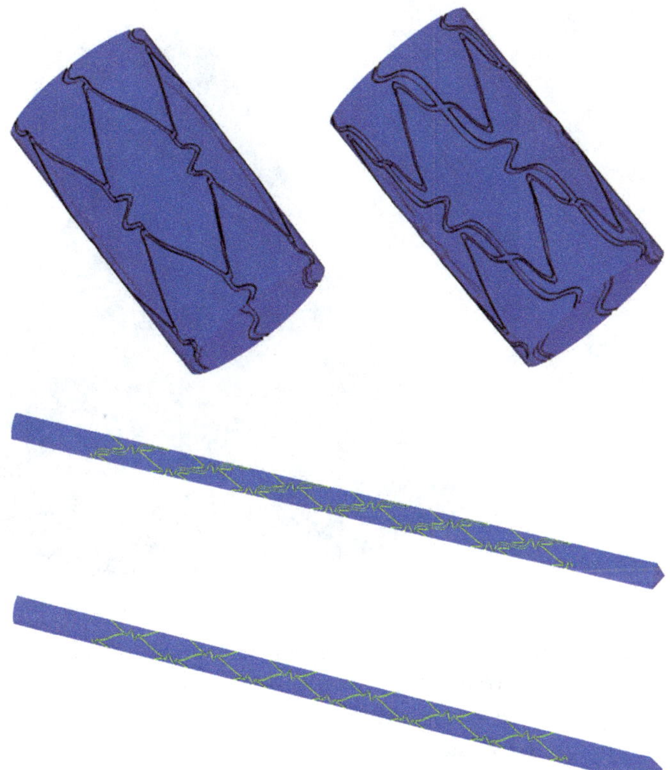

Fig. 18 Example of complete reconstruction and geometries used in the simulations, Fig. 1 of [77]

In the centre of the vessel, such a dense grid is not necessary, as we are sufficiently far from the wall. The size of the grid gradually increases from the vessel wall until it reaches a size that is maintained in the centre of the vessel (Fig. 19). Several simulations are carried out to evaluate the right size of the grid. The mesh chosen for the simulations is composed of 768,008 elements. The wall is assumed to be as rigid as the stent. This assumption does not significantly influence the results because the studies carried out on animals ensure that the implantation of the stent reduces the compliance of the stent to zero within the region of the stent itself, and the deformability of the wall does not alter the velocity field too much under normal conditions.

4.3 Numerical Thermo-Haemo-Dynamics (THD)

4.3.1 Simplifying Assumptions

The first simplifying assumption naturally concerns the geometry. As mentioned, the vessel wall is modelled with a simple cylinder. This assumption is made not so much

Fig. 19 Grid, Fig. 2 of [77]

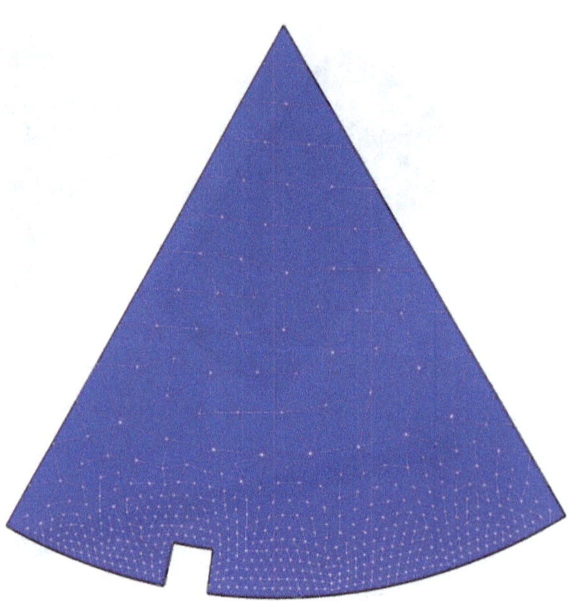

for the geometric difficulties of reconstructing the vessel, as there are *software* programs capable of reconstructing geometries from biomedical images, but in order to highlight the general effects of the stent geometry on fluid dynamics, regardless of individual clinical cases. In addition, the wall and the stent are considered rigid. The other simplifying assumptions concern the blood. The blood is modelled as an incompressible Newtonian fluid with the physical properties of water (density equal to 998.2 Kg/m^3 and dynamic viscosity equal to 1.003 10^{-3} Kg/ms). The physical properties of water are chosen because the blood models existing in the literature are very diversified, so the use of water does not lose generality. Moreover, the properties of these models, used in the computational fluid dynamics studies of the literature, are very similar to those of water. Laminar flow is assumed, in agreement with the literature, justified by the low velocities present in the coronary arteries, which consequently present a low Reynolds number. The Reynolds number used in the stationary simulations is about 272, while in the unsteady-state simulations, it varies between about 122 and 440.

4.3.2 Equations

The fundamental equations of fluid dynamics that are used are the principle of conservation of mass (scalar equation) and the second principle of dynamics, or law of conservation of momentum (vector equation). The first principle of thermodynamics (scalar equation) is not taken into consideration because there are no temperature variations in the simulations carried out.

The balance of mass conservation, with constant density, is

$$\nabla \cdot \vec{v} = 0 \tag{1}$$

while that of momentum is

$$\rho \frac{D\vec{v}}{Dt} = -\nabla p + \nabla \cdot (2\mu(\dot{\gamma})[D]) \tag{2}$$

The stress tensor is

$$[D] = \frac{\left(\left[\nabla \vec{v}\right] + {}^{t}\left[\nabla \vec{v}\right] \right)}{2} \tag{3}$$

with the shear stress defined by

$$\dot{\gamma} = \sqrt{2[D] : [D]} \tag{4}$$

4.3.3 Boundary and Initial Conditions

As an entry condition, a constant mass flow is assigned for the steady-state simulations and a physiological trend for the unsteady-state one. A null pressure is imposed at the exit. Given the periodicity of the geometry, the simulations are carried out only on a portion of the cylinder. To do this, appropriate boundary conditions are necessary. In the commercial software employed, it is possible to define a condition of rotational periodicity, provided that the geometry is appropriately pre-processed. The lateral surfaces of the cylinder portion must be connected to each other in *GAMBIT* before the generation of the mesh. With this operation, the mesh generated on the two faces is identical, and this allows the use of the boundary condition of periodicity. In the steady-state simulations, the initial solution is set equal to an estimate of the solution made by the numerical software. In the unsteady-state simulations, the solution is initialized to the steady-state solution.

4.4 Inlet Flow: Physiological Waveform

In order to model the pulsatile blood flow, the theory of Womersley–Evans, [78], is used to obtain the velocity profile

$$u(t, r) = 2u_0 \left(1 - \left(\frac{r}{R}\right)^2 \right) + 2 \sum_{m=1}^{N} \text{Re}\left(U_m \Psi(\tau_m, r) e^{j\omega_m t} \right) \tag{5}$$

where

$$\Psi(\tau_m, r) = \frac{J_0(\tau_m) - J_0(\tau_m r/R)}{J_0(\tau_m) - J_1(\tau_m)/\tau_m} \tag{6}$$

and

$$\tau_m = j^{\frac{3}{2}} R \sqrt{\frac{\rho}{\mu}} \omega_m = j^{\frac{3}{2}} \alpha_m \tag{7}$$

with J_0 and J_1, which are, respectively, the zero- and first-order Bessel functions of the first kind. α_m are the Womersley numbers of order m, Re() is the real part of a complex number, $j = \sqrt{-1}$ and U_m are the Fourier coefficients of the pulsatile mean velocity profile. By using the Fast Fourier Transform (*FFT*) algorithm, the mean value and the first six harmonics are extracted and used to reconstruct the velocity profile. The same procedure is applied to reconstruct the pressure waveform on the outlet.

The most significant fluid dynamics factor that affects vessel remodelling is wall shear stress (*WSS*). Other parameters, which, according to the literature, are related to intimal hyperplasia (*IH*) and are useful in locating critical haemodynamic points, are studied as well. The time-averaged wall shear stress (*TAWSS*) can be used to measure the cumulative effects of *WSS* within cardiac cycles and is defined as:

$$\text{TAWSS} = \frac{1}{T_o} \int_{T_o} \tau_w dt \tag{8}$$

The oscillatory shear index (OSI), defined as

$$\text{OSI} = \frac{1}{2} \left(1 - \left| \int_{T_o} \tau_w dt \right| \Big/ \int_{T_o} |\tau_w| dt \right) \tag{9}$$

is employed to measure the deviation of *WSS* from its average during the cycle.

The relative residence time (*RRT*), measured in Pa^{-1}, identifies the regions of the wall subjected to a stagnant flow and is defined as:

$$\text{RTT} = \left(\frac{1}{T_o} \int_{T_o} |\tau_w| dt \right)^{-1} \tag{10}$$

The wall is more prone to infiltrate cells and macromolecules where the relative residence time, *RRT*, is higher, but it gives a relative measure of the time spent by the fluid in the vicinity of the wall. Hence, the absolute values of *RRT* are not important by themselves but give an indication of the regions prone to infiltration. *OSI* and *RRT* are mathematically related by the following relation:

$$RRT = ((1 - 2OSI)TAWSS)^{-1} \tag{11}$$

which means that the *RRT* has greater values in regions where *OSI* is closer to 0.5 and where the time-averaged *WSS* is low. Thus the *RRT* is useful in locating low and oscillating regions for the *WSS*, while the *OSI* gives information about the *WSS* oscillation. The same information can be extracted from *RRT* maps: high values appear in the same location of the *OSI*, between the strut tips and in the distal vessel. The *RRT* also carries information about time-averaged *WSS* and can be useful in locating both oscillating and low shear stresses that seem to represent regions prone to atherosclerosis.

The physiological values used are experimentally measured in vivo in the left anterior descending coronary artery of a dog [11]. In particular, the aforementioned authors evaluate the average blood speed throughout the cardiac cycle. The dimensions of the arteries in which the measurement is carried out vary between 2.51 mm and 2.74 mm (in this study, the diameter is set to 2.6 mm). The derived waveform has a maximum value of 0.17 m/s and a minimum value of 0.05 m/s. The average value is 0.105 m/s. With the assumptions of fluid incompressibility, it is possible to transform this speed waveform into a mass flow waveform. With the diameter and density used, the average mass flow is 104.85 kg/m^2 s, and the maximum and minimum values are 47.73 kg/m^2 s and 162.28 kg/m^2 s, as shown in Fig. 20 as mass flow versus time. The period of the wave is set to 0.8 s, corresponding to a heart rate of 75 beats per minute.

4.4.1 List of the MATLAB Code for the Reconstruction of the Physiological Waveform

In order to be used in the simulations, the waveform of Fig. 20 must be defined by an analytical function. To achieve this, specific *software* is developed in *MATLAB*, which is capable of sampling the waveform image, calculating the Fast Fourier Transform (*FFT*) and, from this, extracting the sine and cosine coefficients to be

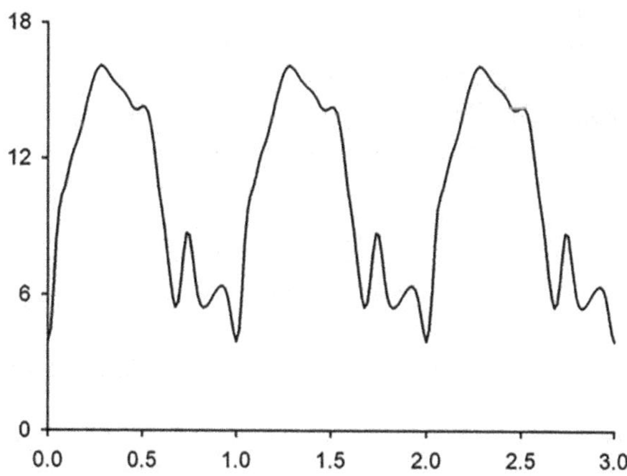

Fig. 20 Waveform of the mass flow (kg/m^2 s) versus time (s), Fig. 3 of [77]

used for the analytical reconstruction of the function. The function is then reconstructed through a summation of sines and cosines.

```matlab
MATLAB code
clear all
%input data
name='velcor2.jpg';
N=64;
minimum=47.7281;
maximum=162.2757;
T=0.8;
Fs=1/T;
t=0:(T/(N-1)):T;
% sampling of the sphygmic wave
x=sampling(name,N,minimum,maximum);
x=x';
%sampled image
figure(1)
hold on
plot(t, x, 'k--o', 'LineWidth', 1, 'MarkerEdgeColor','b',...
'MarkerFaceColor','b');
xlim([0,3]);
xlabel('time [s]');
ylabel('Mass flux [Kg/m^2s]');
hold off
%FFT of the vector containing the samples
X=fft(x);
%extraction of the sine and cosine coefficients
a=real(X);
b=-imag(X);
%initialisation of the new time vector
t=0:0.001:T;
%initialisation of the analytic function
xf=zeros(1,length(t));
%definition of the analytic function through
%a summation of sines and cosines.
for k=1:(N-1)/2
xf=xf+(1/N)*(a(k)*
cos(2*pi*(k-1)*(t*(N-1)/T)/N)+...
b(k)*sin(2*pi*(k-1)*(t*(N-1)/T)/N));
end
%image of the waveform obtained with the analytic function a
figure(2)
hold on
plot(t,xf);
```

```matlab
xlim([0,3]);
xlabel('time [s]');
ylabel('Mass flux [Kg/m^2s]');
hold off
%writing to file the sine and cosine coefficients
fid = fopen('a.txt', 'wt');
fprintf(fid, '%12.8f\n', a);
fclose(fid);
fid = fopen('b.txt', 'wt');
fprintf(fid, '%12.8f\n', b);
fclose(fid);
Sampling Function
function y=sampling(name,n,minimum,maximum);
%reading the wave image
C=imread(name);
%imread uses the three-colour code, the image is in black
%and white, so the three codes are not needed, one is enough.
%black=0,0,0. white=255,255,255
c=C(:,:,1);
%search for black pixels (where the wave is).
%the x vector indicates the sample number,
%the y vector indicates the sample value
[y,x]=find(c==0);
%flipping of the y axis
y=-y;
%the graph is not thin, to an x corresponds
%more y. it is necessary to eliminate these excesses.
%cycle on the x vector
for j=1:max(size(x))
%search for repeated samples
i=find(x==x(j));
if max(size(i))>=1
%calculation of the average of the samples
m=mean(y(i));
%nullification of repeated values
x(i(2:max(size(i))))=0;
y(i(2:max(size(i))))=0;
%insertion of the average in the unique sample
y(i(1))=m;
end
end
%removal of added zeros
[x,IX]=sort(x);
y=y(IX);
i=find(x);
```

```
x=x(i);
y=y(i);
%the number of samples obtained is high
%but different from the desired one. The excess samples are removed
%resulting in a vector of length n.
y=y(1:length(y)/n:length(y));
%transformation of sample values to achieve
%the maximum and minimum desired.
%%translation to have the minimum value equal to zero
%% y=y-(min(y));
% %scaling and translation
%% a= (maximum-minimum)/max(y);
%% y=y*a+minimum;
y=(y-(min(y)))*((maximum-minimum)/max(y-(min(y))))+minimum;
```

4.4.2   Results

Figure 21 reports the graph of the sampled waveform. In the numerical software employed, there is the possibility to use *user-defined functions* (*UDFs*) as boundary conditions or to define other properties, but these functions must be written in *C* language. This function is used for unsteady-state simulations, in which it is, therefore, possible to use a physiological waveform.

Fig. 21 Sampled mass flow waveform (kg/m^2 s) [1]

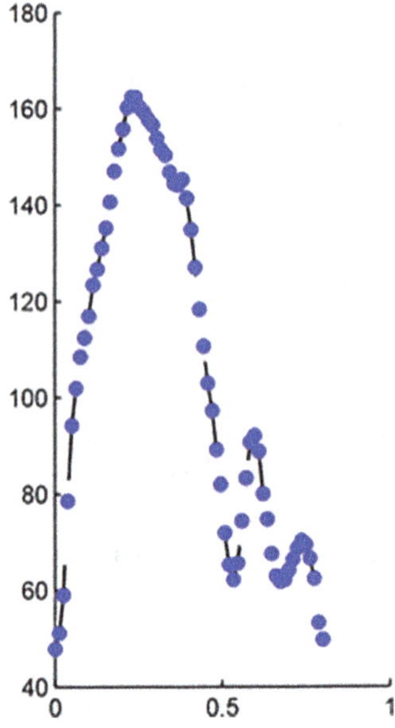

4.5 Grid Independence

The initial simulations carried out are aimed at testing the independence of the solution from the created mesh. In particular, steady-state simulations are carried out with the *CYPHER SELECT* stent, *S1*, varying the number of elements. All simulations are carried out with tetrahedral elements of increasing size. The grid with 768,008 elements is chosen for subsequent simulations, as it guarantees excellent results with the fewest number of elements.

Table 1 and Fig. 22 report the results of the main simulations carried out on the *CYPHER SELECT*, *S2*, which demonstrate the independence of the solution from the grid. The left of Fig. 22 shows the mesh with 523,849 elements, the centre of Fig. 22 with 768,008 elements and the right of Fig. 22 with 1,348,634 elements. Analysing these results, it is clear that the average values coincide with each other up to the third decimal place for speed and up to the first decimal place for pressure only in the

Table 1 Grid independence. Results of the simulations after 1000 iterations. The highest residuals are in all simulations of the order of 10^{-6}. The exit pressure is not reported, as it is set to zero by the boundary conditions

Number of elements	Average V (m/s) Entry	Average V (m/s) Exit	Average V (m/s) $Z = 0.0184$ mm	Average P (Pa) Entry	Average P (Pa) $Z = 0.0184$ mm
523,849	0.10176913	0.10308045	0.10625915	32.528706	11.598187
768,008	0.10237551	0.10338871	0.10628417	32.642593	11.671381
1,348,634	0.10284831	0.10363439	0.10627534	32.651765	11.640269

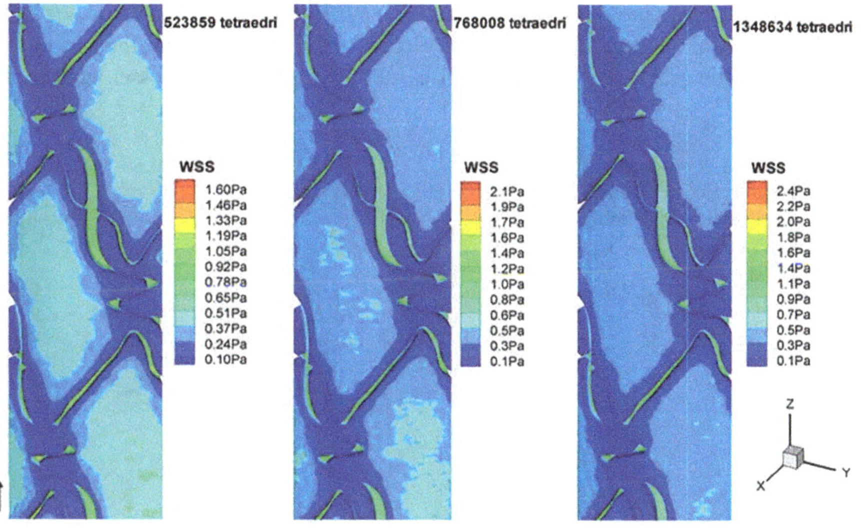

Fig. 22 Wall shear stress fields in the CYPHER SELECT stent, S2, in three simulations with a different number of elements. Zoom on a cell [1]

last two simulations. The solution of the first simulation deviates from the others by an order of magnitude more. Also, analysing the trend of the *WSS*, it is clear that the solution of the simulation with the fewest number of elements deviates from the others. The mesh with 523,849 elements (on the left of Fig. 22) is therefore discarded. The mesh with 768,008 elements (in the centre of Fig. 22) is instead chosen for subsequent simulations, as it guarantees excellent results with the fewest number of elements.

4.6   Steady-State Analysis

For both reconstructed stents, *CYPHER SELECT* and *CYPHER*, steady and unsteady-state simulations are performed. In the steady state, the mass flow at the inlet is set equal to the average value of the physiological waveform. In the steady state, about 1000 iterations are performed, and the highest value of the residuals is in all cases of the order of 10^{-6}.

In Fig. 23, the trend of the residuals in a generic simulation in the *CYPHER SELECT* stent is shown. The first curve at the top is related to the conservation of total mass, the second to the speed in the direction z, the third to the speed in the direction x and the last one, only noted after 1000 iterations, to the speed in the direction y. As can be seen, the residuals decrease regularly until they reach a level after which they decrease very slowly. The sudden increase that occurs during the simulation is due to the transition from first- to second-order accuracy. The results obtained from these simulations are exported to *Tecplot* to be visualized and analysed. In *Tecplot*, in addition to viewing the results, it is possible to perform post-processing that allows the definition of new variables.

Figure 24 shows the *WSS* for both stents, *CYPHER* on top and *CYPHER SELECT* on bottom on the entire simulated wall section, with the flow from the left to the right. From the graph, some considerations can be drawn: on the vessel wall, in all

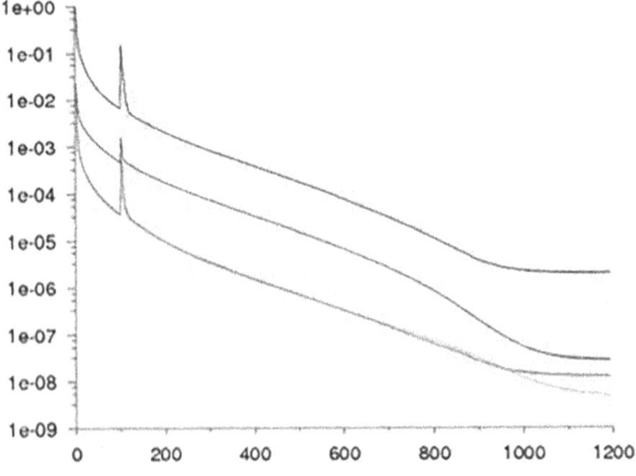

Fig. 23 Trend of residuals with the number of iterations in the steady-state simulation with the CYPHER SELECT stent, S2 [1]

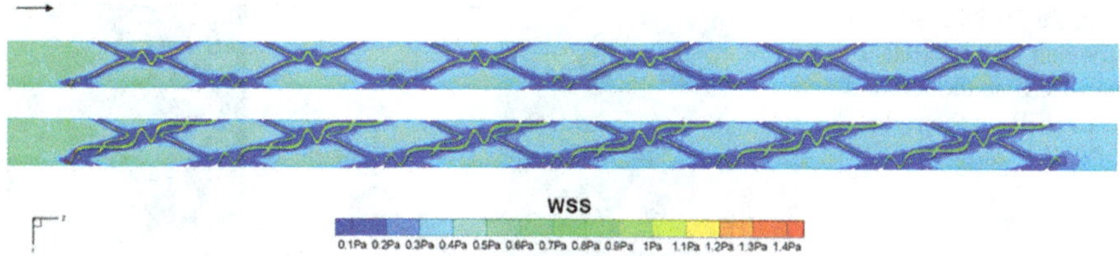

Fig. 24 WSS field over the entire simulated wall section for both stents [1]

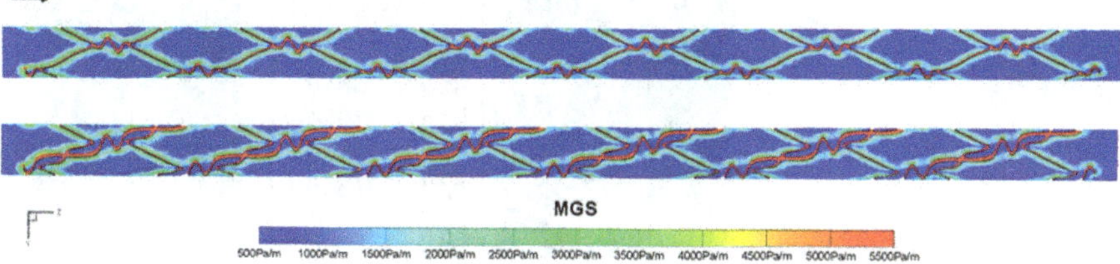

Fig. 25 MGS fields over the entire simulated wall section for both stents [1]

areas adjacent to the stents, there are very low values of the *WSS* (below 0.2 Pa). This area is uniform throughout the entire section. As for the comparison between the two stents, there are no substantial differences in the trend of the *WSS*, but the areas with low *WSS* are greater in the *CYPHER SELECT* stent (which is made of more material).

Another parameter derived from the *WSS* and important for the development of neo-intimal hyperplasia is the *MGS* [79]. Being the spatial gradient of a vector, its gradient is a tensor, so the modulus of the gradient of the wall shear stress, *MGS*, is defined, which is an index of neo-intimal hyperplasia when it assumes high values. Figure 25 shows the trend of the *MGS* for both stents, *CYPHER* on top and *CYPHER SELECT* on bottom, over the entire simulated wall section.

Figure 26, [77], shows a zoom of the *WSS* on a cell of the two stents with the flow from below. From this graph, it can be noticed the trend of the *WSS* on the stent and how evident is the area with very low *WSS* on the vessel wall near the stent.

Figure 27, [77], shows a zoom on a cell of the *MGS* in the two stents with the flow from below. In both graphs, the maximum value of the scale is set equal to the maximum value of the *MGS* on the vessel wall because the *MGS* on the stent is an order of magnitude higher than that on the vessel. From these graphs, it can be seen that the areas with high *MGS* are, similarly to the *WSS*, located near the stent but not uniformly; in particular, they are located near the stent elements that oppose the flow. Therefore, these areas are at high risk of neo-intimal hyperplasia as they present a low *WSS*, associated with a high *MGS*. Furthermore, it can be seen that in these areas the *CYPHER SELECT* stent has a higher *MGS* compared to the *CYPHER* stent.

Fig. 26 WSS field in a
zoomed cell, CYPHER
SELECT on the left and
CYPHER on the right, Fig. 6
of [77]

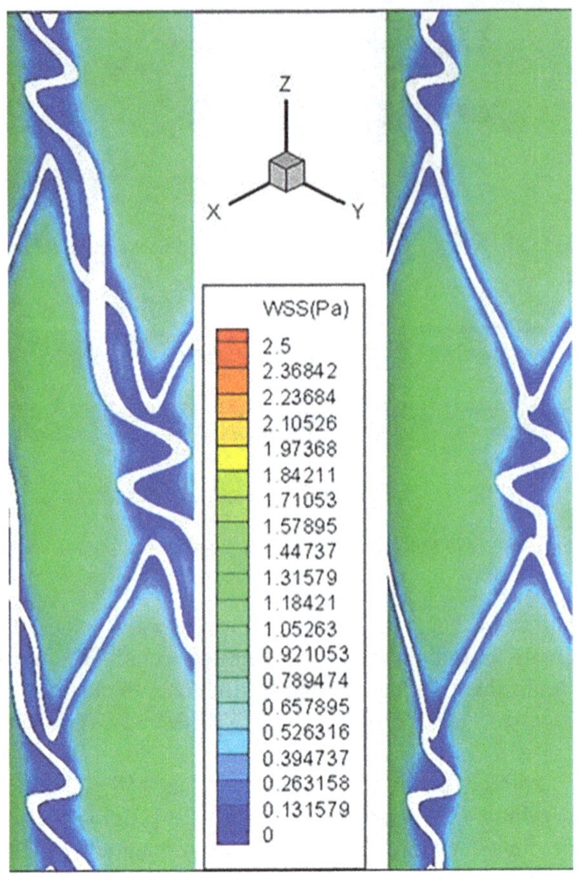

4.7   Optical Density: Parameter for Assessing Endothelial Permeability

Many studies demonstrate the influence of *WSS* on neo-intimal hyperplasia and
therefore on the formation of the atherosclerotic plaque [13, 14, 79–82]. In particular, the *WSS* and its gradient *MGS* seem to have an effect on endothelial permeability. An increase in endothelial permeability causes a greater accumulation of lipids
and, therefore, a higher probability of developing atherosclerosis.

4.7.1   Evans Blue Staining

Endothelial permeability can be assessed through the *Evans blue staining/dye*
(*EBD*), also known as *T-1824*. The *EBS* is an azo-dye that forms a strong bond
with albumin [83]. The extraordinary affinity of the *EBD* with albumin allows the
use of this dye as a marker of albumin in a wide variety of physiological, diagnostic
and pharmacological applications. The parenteral administration of *EBD* causes blue
spots on the intimal surface of the arterial tree. The accumulation of *EBD* is not

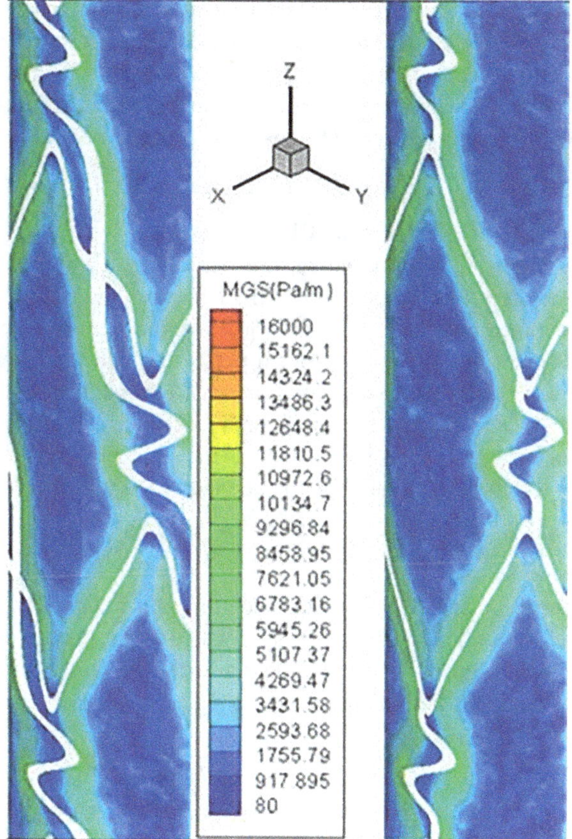

Fig. 27 MGS field in a zoomed cell, CYPHER SELECT on the left and CYPHER on the right, Fig. 7 of [77]

uniform but tends to be more intense in some areas than others. The accumulation is so characteristic that actual topographic accumulation maps are made for many animal species [84]. Due to the close bond between *EBD* and albumin, the areas of the arterial tree with a strong blue colour represent those regions where the transvascular transport of albumin is increased, and, consequently, also the endothelial permeability of other macromolecules, such as lipoproteins, globulins and fibrinogen, is increased, as some studies show [85]. Therefore, the EBD, as a marker of albumin, is a potential index of assessment of endothelial permeability. This method has many practical advantages over other methods, such as, for example, its cost-effectiveness and speed of execution and evaluation. Moreover, it does not require the use of radioactive materials, and, thanks to the visible nature of the marker, it is possible to use optical reflectance techniques, which allow a quick scan of the arteries.

The disadvantages of this method are essentially two: like all other methods of assessing endothelial permeability, it is highly invasive, and direct measurement can only be performed on guinea pigs. Moreover, the distribution of the *EBD* is not uniform along the arterial tree and strongly depends on the region; therefore, it would be necessary to have a detailed topographic map. Despite these difficulties, it is possible to derive relationships between the accumulation of *EBD* and optical factors, such as the optical density (*OD*), [86], which can be used as an index of high endothelial permeability. Finally, there are studies that demonstrate the existence of relationships between *OD* and fluid dynamic parameters, such as the *WSS* and *MGS* [87, 88].

4.7.2   Definition of OD

Endothelial walls reflect most of the radiation with wavelengths in the visible range; therefore, the accumulation of *EBD* can be measured using the optical technique outlined in relative Fig. 2 of [86]. A source generates a beam of visible light with a certain intensity I_1. The beam passes through the layer of dye (of thickness δ and with a volumetric concentration of *EBD* equal to c) and comes out with an intensity I_2. Once it reaches the uncoloured inner layers, the beam of light is reflected with an intensity I_3, passes through the dye layer again and is captured by the sensor with an intensity I_4. The ratio I_3/I_1 is defined as the reflection coefficient.

$$\sigma = I_3/I_1 \tag{12}$$

for the tissue. This reflection coefficient is assumed to be constant and independent of the amount of dye deposited, which implies that the thickness of the dye layer is small compared to the thickness of the vessel wall and that the thickness of the vessel is constant. In a first approximation, Lambert–Beer's law can be used,

$$I_2 = I_1 e^{-\alpha c \delta}, \tag{13}$$

and

$$I_4 = I_3 e^{-\alpha c \delta}, \tag{14}$$

to obtain

$$I_4 = \alpha I_2 e^{-\alpha c \delta} = \alpha I_1 e^{-2\alpha c \delta}, \tag{15}$$

where α is the optical absorbance coefficient of the *EBD* (measured in cm^2/nmol). When there is no deposition of *EBD* (c and $\delta = 0$), the reflected intensity I_4 is defined as equal to a value I_0. Substituting, we get

$$I_0 = \alpha I_1 \tag{16}$$

while the ratio between the reflected intensity in the presence ($I_4 = I$) and in the absence of *EBD:*

$$\frac{I}{I_0} = e^{-2\alpha c \delta}. \tag{17}$$

Applying logarithms to this equation finally defines the optical density OD (or optical absorbance) as

$$OD = -\ln\left(\frac{I}{I_0}\right) = 2\alpha c\delta. \tag{(18)}$$

The accumulation of dye M (amount of dye per unit of area, in nmol/cm^2 is the product $c\,\delta$; thus, the relationship between optical density and accumulation of EBD is

$$M = OD/2\alpha \tag{19}$$

4.7.3   Relationship between OD and Fluid Dynamic Parameters

Himburg et al. [87], and LaMack et al. [88], conducted studies aimed at finding relationships between the optical density OD and fluid dynamic parameters, such as the WSS and its spatial gradient MGS. The aforementioned authors have firstly carried out measurements of the OD on laboratory guinea pigs; in particular, they used the porcine aorta. Once the measurements are made, they reconstruct the arteries on a computer from images obtained from laser scanners and carry out simulations reproducing physiological conditions. With the results obtained, they carry out a point-by-point comparison between the fluid dynamic parameters (WSS and MGS) and the normalized optical density OD (divided by its average value). From this analysis, they derived some mathematical relationships that link these parameters.

In [87], they derive a relationship between OD and only WSS:

$$OD = 1.39\ WSS^{-0.118}. \tag{20}$$

In the subsequent study, [88], they integrate the spatial gradient module of the WSS as:

$$OD = 1.2\ WSS^{-0.11}MGS^{0.044}. \tag{21}$$

The physiological values of the OD (normalized), measured in [87, 88], are between 0.9 and 1.3. As stated, the accumulation of EBD is not uniform but varies significantly in different regions of the arterial tree. The OD, therefore, cannot be used in an absolute sense and does not have a threshold beyond which there is a strong correlation with neo-intimal hyperplasia, but it can still be used as an index of neo-intimal hyperplasia in a relative sense. The areas with the highest OD are most likely the areas at greatest risk of atherosclerosis. With this in mind, the results obtained from the steady-state simulations allow us to calculate the OD based on the last relationship.

Figure 28 reports the fields of the OD, zoomed in on a cell of both stents, in a steady state. The values of OD are slightly higher than the values measured in

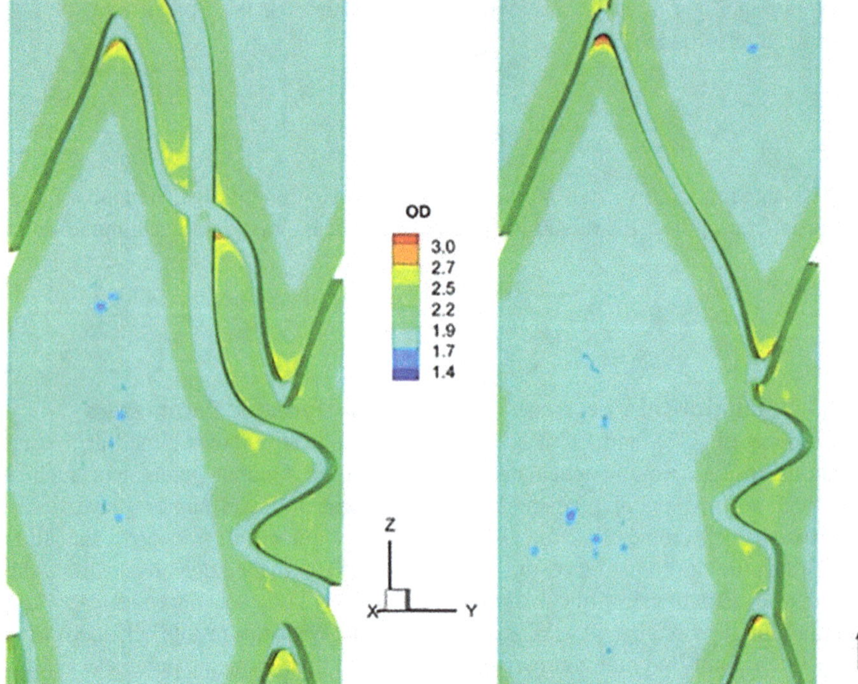

Fig. 28 Fields of OD in a zoomed cell of the two stents: CYPHER SELECT, S2, on the left and CYPHER, S1, on the right, in steady-state simulations [1]

[87, 88], but, as mentioned, the *OD* is only evaluated as a relative index. As clearly seen from Fig. 28, all the wall areas adjacent to the stents are subject to a higher value of *OD* compared to areas far from the stent. Furthermore, in both stents, there are circumscribed areas where the *OD* reaches very high values. These areas are more widespread in the vessel with the *CYPHER SELECT* stent.

4.8 Unsteady-State Analysis

As far as the unsteady-state simulations are concerned, the already-defined waveform is used. The time integration step is chosen to be 0.001 s, and, for each of these steps, 20 iterations are performed. A period of the wave (0.8 s) is therefore divided into 800 time steps. In order to obtain results with residuals of the order of $10^{-5} - 10^{-6}$, the simulations are carried out for two time periods, and the first period is discarded. The solutions are saved and displayed every 10 time integration steps. The graphs of the results are reported, displaying only 9 time instants per period, sufficient to describe the phenomenon.

Unsteady-state simulations, performed during two periods of the cardiac cycle, show large variations of *WSS* and *MGS* with time. Figure 29, [77], reports the time

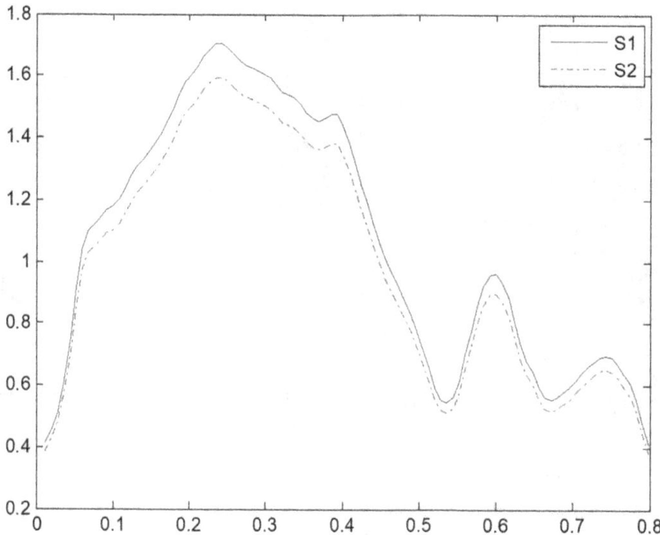

Fig. 29 WSS (Pa) variation with time (s) on the endothelium for CYPHER stent, S1, and CYPHER SELECT stent, S2, Fig. 8 of [76]

variation of the average *WSS* over the endothelium, showing that the average *WSS* on the two stents has a similar pattern, but higher values are present in the *CYPHER* stent, S1. Moreover, the differences are more significant on the systolic peak due to the stent geometry. In the *CYPHER SELECT* stent, S2, struts of adjacent cells become parallel in the proximity of the intersecting points of the struts, where the flow is trapped and *WSS* is quasi-time-independent. The absence of these zones in the *CYPHER* stent, S1, makes the *WSS* increase with the flow rate more evident, i.e. leading to the differences visible in Fig. 29.

Figure 30, [77], reports the patterns of the *WSS*, and Fig. 31, [77], that of the *MGS* at six time instants for the two stents. *WSS* and *MGS* increase with the systolic wave because velocity increases with the pressure, but it must be null on the stent due to the no-slip condition. Stagnation zones are present for all cardiac cycles, although their extension decreases with the increase of velocity. Areas adjacent to the stent struts present low *WSS* values for all cardiac cycles because of the stagnation regions. *WSS* has large variations close to the vessel wall where the velocity is free to increase. Low *MGS* values are present on the vessel wall during the entire cardiac cycle. *WSS* increases on the vessel wall with values spanning between 0 and 2.5 Pa, while *MGS* spans in the range between 0 and 16,000 Pa/m. *MGS* increases with the flow rate. Regions of high *MGS* are adjacent to the stent struts, except around the curved ones, which surround the stagnation regions where uniform and low *WSS* can be noticed. Uniform values of *WSS* mean low *MGS*. A different pattern is observed on the other struts because of the high *WSS* gradient, which increases during the flow rate, reaching a maximum on the edge of the curved struts. A high increase of *WSSG*

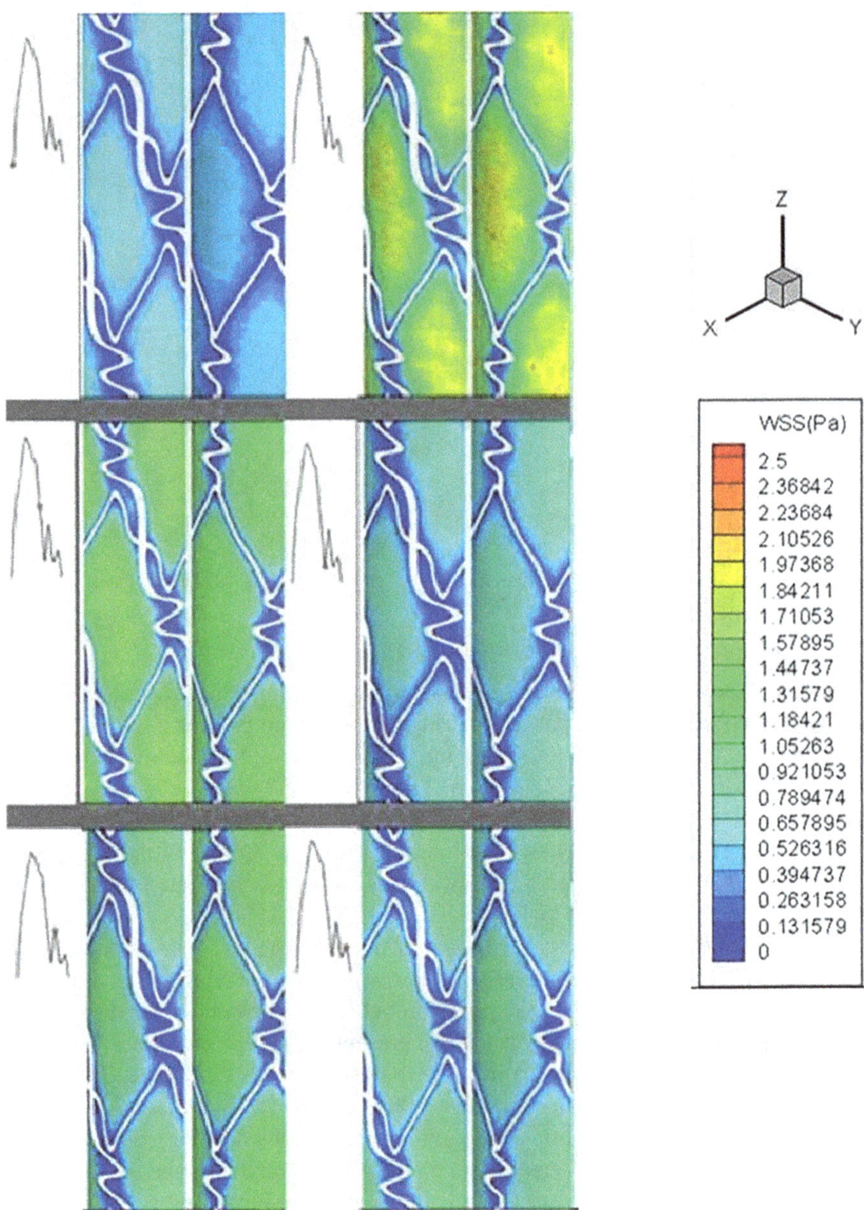

Fig. 30 Patterns of WSS during the cardiac cycle for the two stents, Fig. 9 of [77]

along the edge of the curved struts is produced due to the strong redirection of the velocity profile, which is caused by angular struts changing *WSS* direction.

As can be seen, both *WSS* and *MGS* vary significantly during the cardiac cycle. As for the *WSS*, the vascular wall areas far from the stent undergo apparently

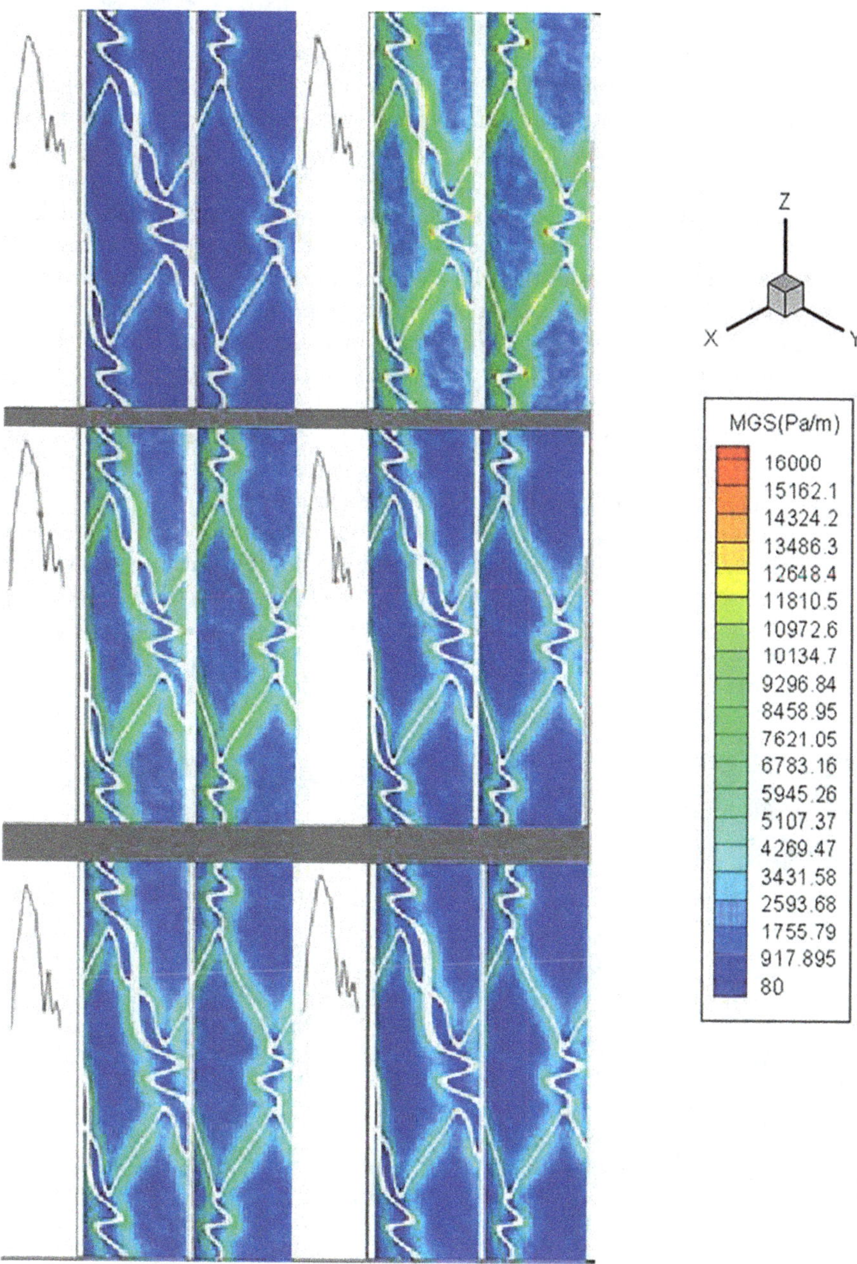

Fig. 31 Patterns of MGS (Pa/m) during the cardiac cycle for the two stents, Fig. 10 of [77]

significant variations. These variations are physiological, as higher speeds correspond to higher *WSS*; in fact, the values of the *WSS* in these areas correspond to the physiological values measured in the coronary arteries. The areas adjacent to the low *WSS* stent, instead, undergo only small variations, so these areas are permanently exposed to adverse conditions. The areas with high *MGS* undergo significant variations during the cardiac cycle. At the time instants when the flow speed is high, the areas with high *MGS* are very widespread, while when the speed is low, these areas are significantly reduced. This implies that, as far as *MGS* is concerned, the areas subjected to adverse conditions for the entire period are limited. There are areas exposed to unfavourable conditions only for some moments in time.

Even with the results obtained from these simulations, the *OD* is calculated to evaluate the combined effect of the *WSS* with its gradient *MGS*. Figure 32 shows the trend of the *OD* at various instants of time. As can be seen from these graphs, the trend of the *OD* undergoes a few variations during the entire cardiac cycle. The areas with high *OD* remain very concentrated, and it is interesting to note that these areas are more intense at the instants of time when the speed is lower.

From the unsteady-state simulations, it is possible to obtain another important parameter: the *OSI*, which is a measure of the reversal of the direction of the *WSS* over time. It is possible to define the *OSI* for all spatial directions, but, for the case under consideration, only the variation in the direction of the flow is significant.

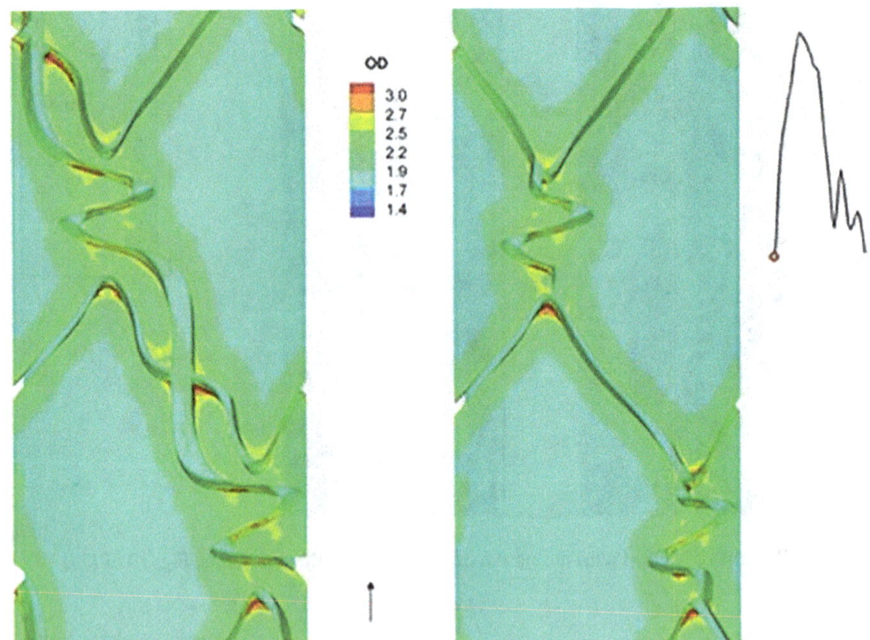

Fig. 32 Patterns of OD during the cardiac cycle for the two stents [1]

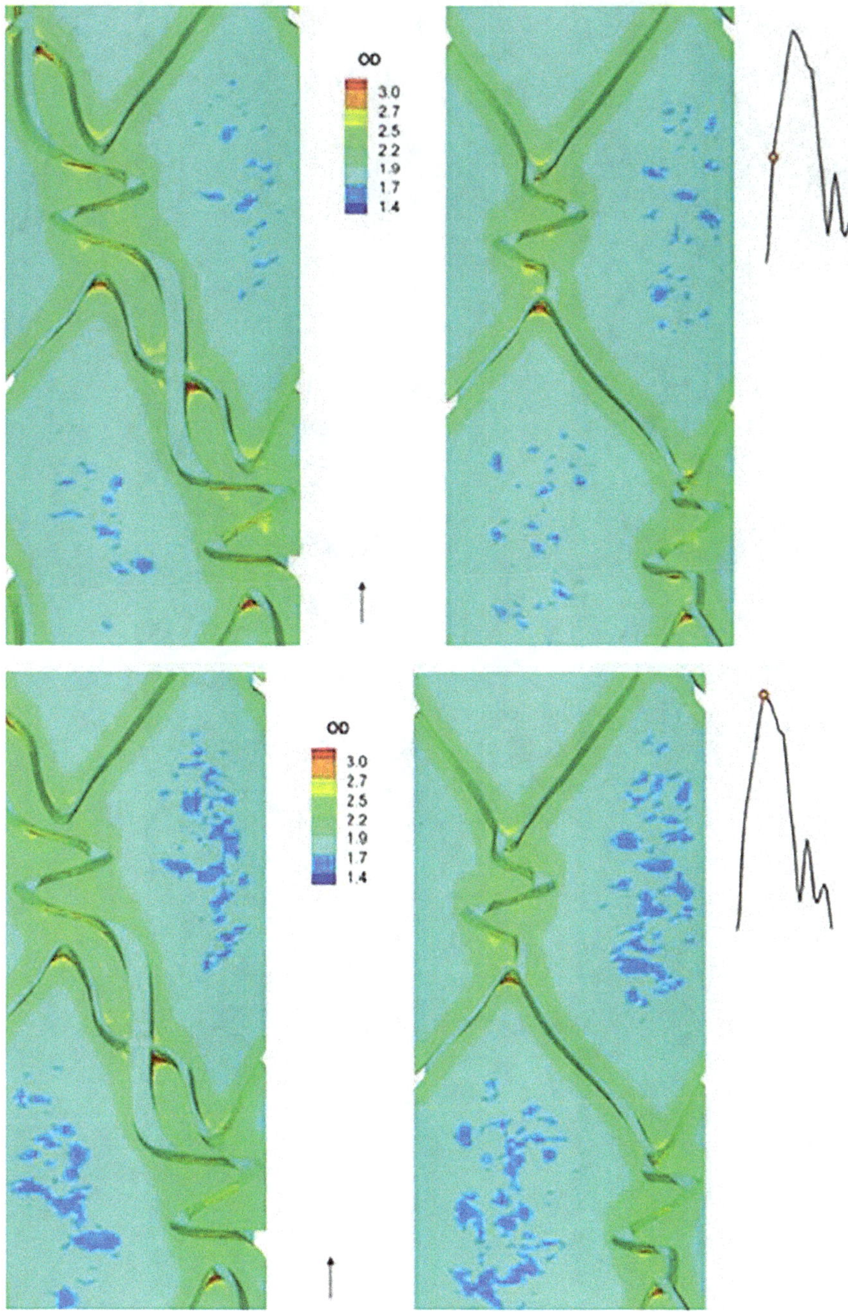

Fig. 32 (continued)

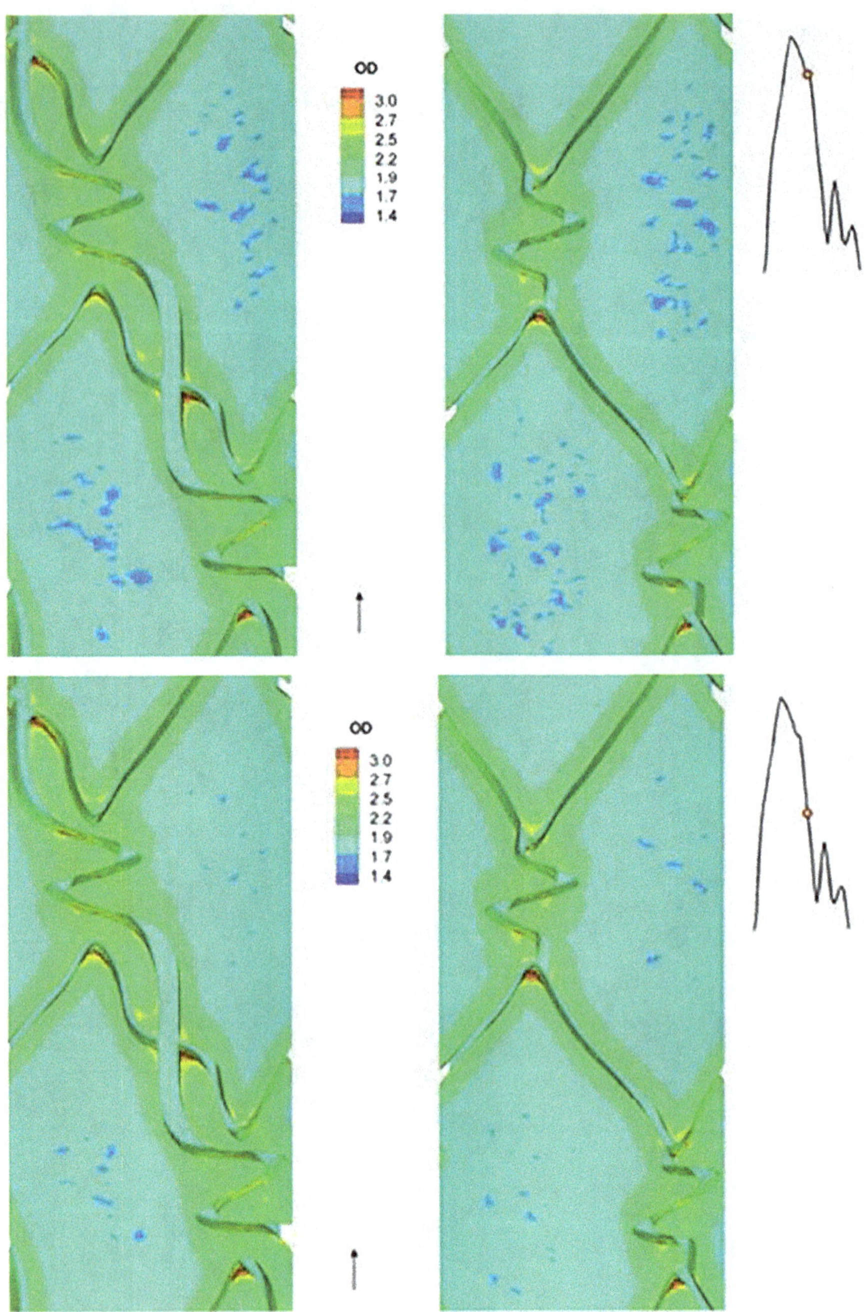

Fig. 32 (continued)

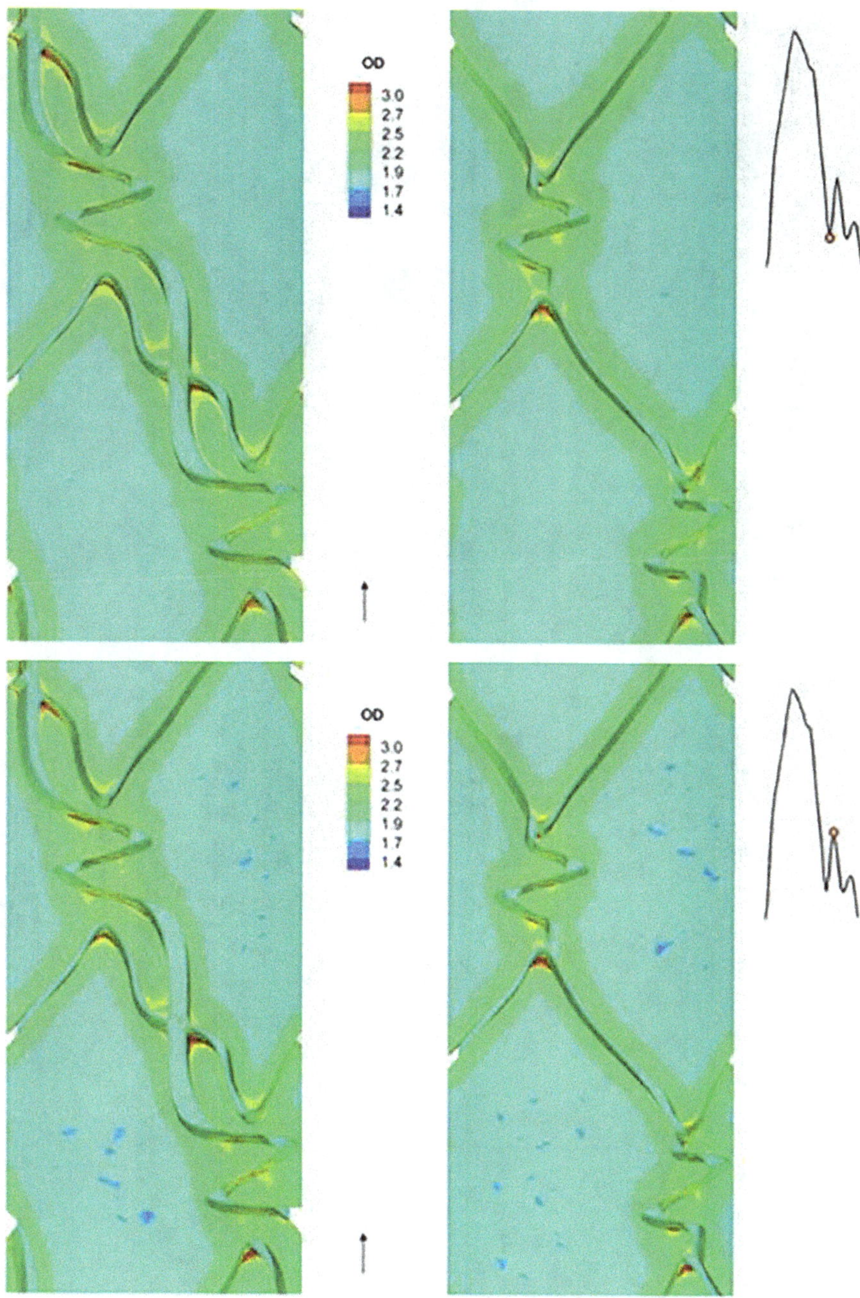

Fig. 32 (continued)

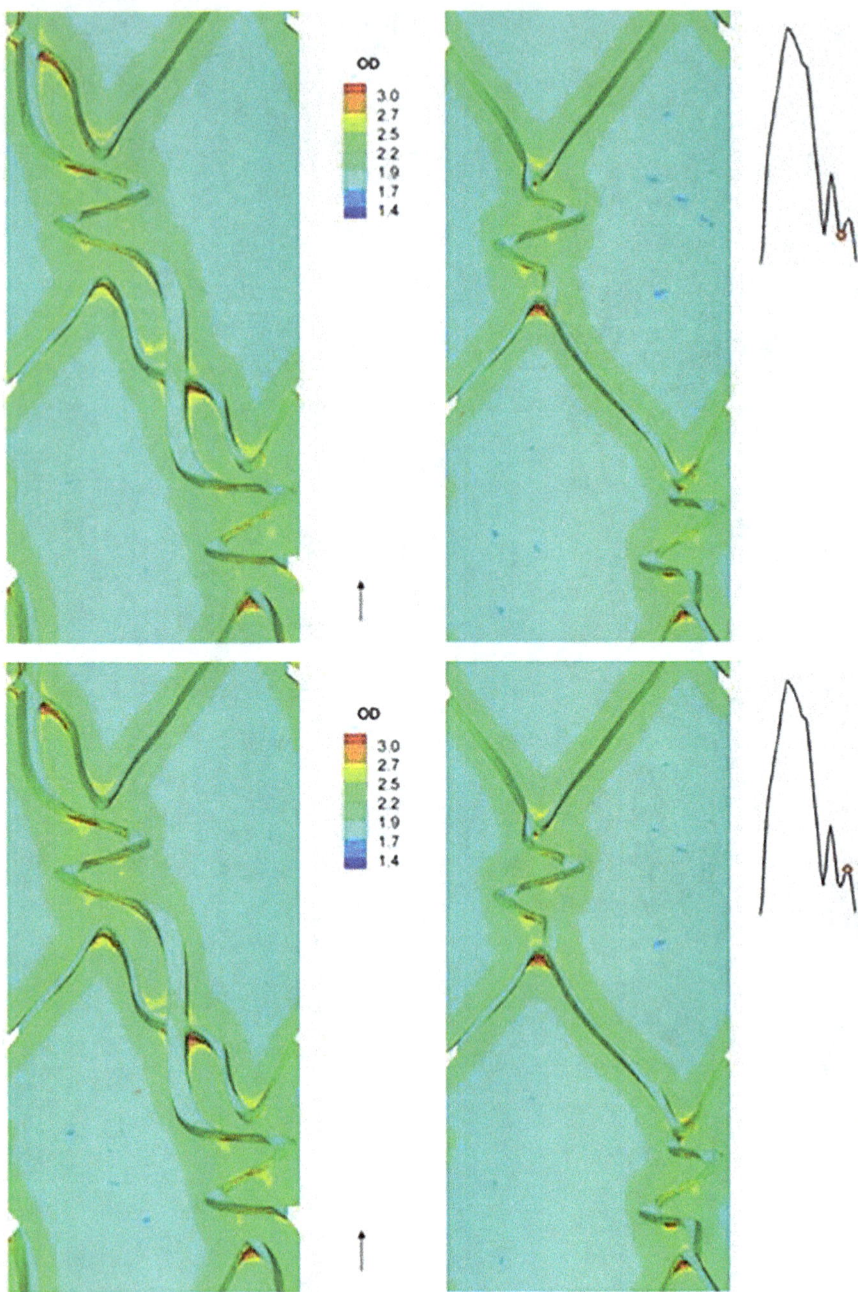

Fig. 32 (continued)

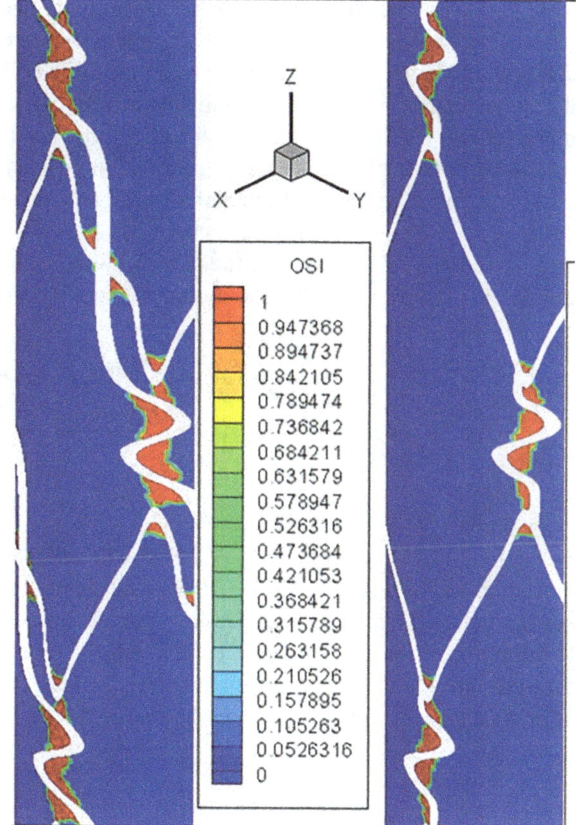

Fig. 33 Patterns of OSI for the two stents: S2 on the left and S1 on the right, Fig. 11 of [77]

$$\text{OSI} = \frac{1}{2}\left(1 - \frac{\displaystyle\int_0^T \text{WSS}_Z\, dt}{\displaystyle\int_0^T \|\text{WSS}_z\|\, dt}\right) \tag{((22)}$$

Figure 33, [77], shows the trend of the *OSI* calculated for both stents. From these graphs, it is clearly evident that the areas exposed to an *OSI* different from zero are located between the loops of the connecting elements and near the angular points. Comparing the trend of the two configurations, it is noted that the vessel with the *CYPHER SELECT* stent, S2, which has more angular points, is subjected to a greater number of areas with *OSI* different from zero. OSI results, presented in Fig. 33, show that the *CYPHER SELECT* stent, S2, presents large regions with OSI greater than 0.5. Regions with nonzero OSI are present between the handles of the struts and in the regions surrounding the points with tangent discontinuity. The *CYPHER*

SELECT stent, S2, has the highest curvature on the angular struts connecting repeating stent units, and its axis has a transverse direction that disturbs the flow and presents larger regions with nonzero OSI. In the *CYPHER* stent, S1, the axis of the connecting struts is parallel to the flow direction, and the angular struts are longer in the *CYPHER SELECT* stent, S2, disturbing the flow by a similar amount. The *OSI* parameter, which gives additional information besides *WSS* and *MGS*, shows that large differences in the fluid dynamics of the two stents are present in an unsteady state.

It is also possible to estimate the percent of critical intra-strut area, P, exposed to low WSS, i.e. lower than 0.5 Pa, during the cardiac cycle, defining the parameter P

$$P = \text{percArea}(\%) = 100 \frac{\displaystyle\int_{\text{vesel\_wall}} \frac{1}{2}\left(1 - \left(\frac{\text{WSS} - 0.5}{\|\text{WSS} - 0.5\|}\right)\right) dA}{\displaystyle\int_{\text{vesel\_wall}} dA} \tag{23}$$

Figure 34, [77], shows that the *CYPHER SELECT* stent, S2, has lower values of percent critical intra-strut area, P. The P value on the stented region, exposed to WSS smaller than 0.5 Pa, reduces as the flow rate increases, reaching the minimum value at the point of the cardiac cycle corresponding to peak flow velocity. Since *WSS* increases with the velocity, the endothelium surface is exposed averagely to higher *WSS* during the systole, decreasing the risk of developing neo-intimal hyperplasia.

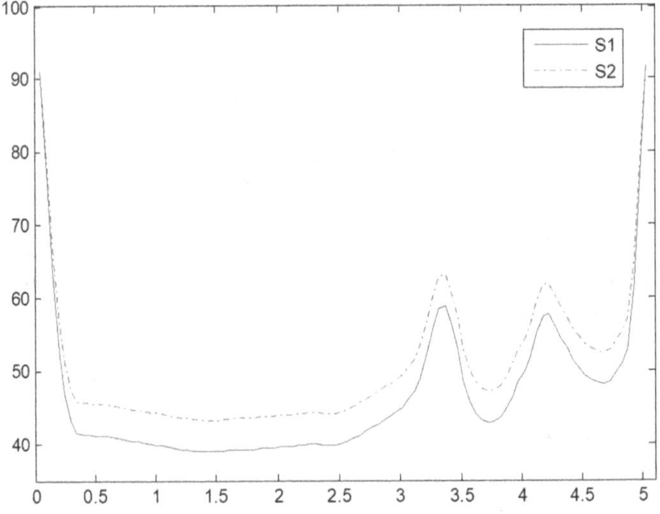

Fig. 34 Critical percent intra-strut area P (%) versus the arterial axis (mm) for CYPHER stent, S1, and CYPHER SELECT stent, S2, Fig. 12 of [77]

5 Numerical Thermo-Haemo-Dynamics (THD) of Two Carotid Self-Expanding Bare Metal Stents in Non-stenosed and Stenosed Conditions

5.1 Introduction

The carotid artery is one of the most common locations of cranio-cervical atheromatous diseases, accounting for 20–30% of strokes. Often, these conditions are the result of the detachment of thrombo-embolic material from the atheromatous plaque [89]. This problem is investigated numerically in a realistic geometry in [90]. Although carotid angioplasty and stenting recently emerged as valid therapeutic alternatives to carotid endarterectomy, risks associated with embolic complications are still high, ranging from 4% to 33%, and slowing down the introduction of the technique in the carotid circulation, as compared with the coronary and peripheral arteries [91, 92].

The aim of the present work is to investigate the blood flow in two carotid self-expanding bare metal stents (*SE-BMSs*), under physiological conditions, and in a non-completely restored vessel lumen by *3D CFD* simulations. The latter situation is likely to occur in *SE-BMS* grafts.

5.2 Geometry and Mesh

The simulations are carried out on two different stents, shown in Fig. 35, [70], in order to assess the influence of the different configurations on the fluid dynamics. The stent named S1, on the left of Fig. 35, has the strut tips arranged in the same direction, according to a configuration called peak-to-valley, while the stent named S2, on the right of Fig. 35, has the struts arranged in the opposite way and is called peak-to-peak. Both stents are applied to the carotid artery (8 mm in diameter), modelled as a cylindrical tube; in one case with a constant section and in the other with a slight reduction of the radius in the middle section.

Geometrical parameters of the stents are chosen from [93] to match an artery with a diameter of 8 mm; stent radial thickness is 0.122 mm, while the two lengths are 24.8 and 22.9 mm, respectively. The simulations are not carried out in the entire cylindrical domain because of the radial symmetry but only in one-sixth of the geometry, allowing significant savings in computational time. The choking in the middle section has a radius of 6.8 mm, corresponding to an area reduction of 30%.

Fig. 35 Stent configurations: (**a**) Peak-to-valley (S1); (**b**) Peak-to-peak (S2), Fig. 1 of [70]

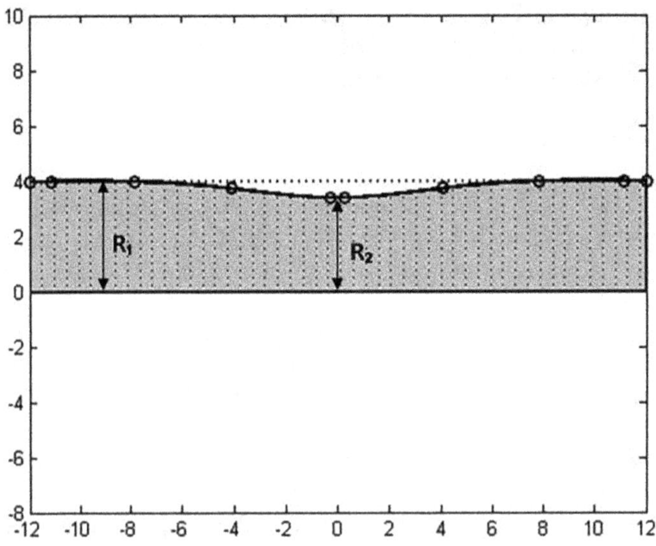

Fig. 36 Geometry of a stenosed vessel reconstructed by polynomial approximation, Fig. 2 of [70]

$$P(\%) = 100 \cdot \left(1 - \left(\frac{R_2}{R_1} \right)^2 \right) \tag{24}$$

The stenosis profile, represented in Fig. 36, [70], is reproduced by a piecewise cubic spline interpolation of ten control points. The height on the ordinate is in mm. The axial length on the abscissa is in mm.

The mesh used, generated with numerical software and refined with another software, is reported in Fig. 37, [70], and is made of tetrahedral elements in the core of the vessel, while near the wall a prismatic layer is employed. The grid is not uniform but coarser near the centreline and finer towards the wall and near the struts.

Blood is assumed as a homogeneous Newtonian fluid, having constant density and dynamic viscosity respectively equal to 1060 kg/m^3 and 0.0033 kg/m·s. The equations are solved under the boundary conditions of no-slip on the vessel wall and stent surface and prescribed velocity at the inlet and pressure at the outlet. The simulations are performed using numerical software, which solves the fluid dynamics equations through the finite volume method. The *SIMPLE* algorithm is employed to solve the pressure–velocity coupling, and the simulations are carried on until convergence is reached. A second-order implicit time-stepping method is used for the unsteady-state simulations, which are done for two cardiac cycles (2 s), with a fixed time step of 5 ms, which guarantees a Courant number smaller than 0.5 for all the time steps. Only the results of the second cycle are recorded, while those of the first cycle are only used to initialize the solution.

Fig. 37 Detail of mesh,
Fig. 3 of [70]

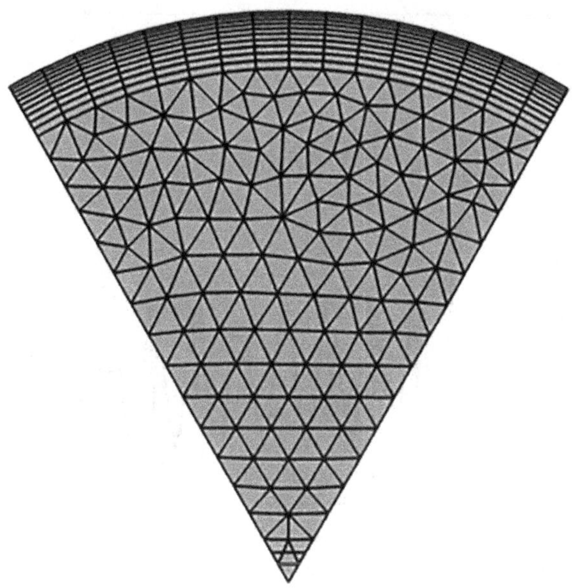

5.3   Steady-State Analysis

In steady-state simulations, the mean velocity of 0.25 m/s is assumed at the inlet. Since the Reynolds number associated with this configuration is about 640, the flow regime is laminar and a Poiseuille velocity profile is prescribed at the inlet. More than 200 slices are extracted by intersecting the vessel wall with parallel axial planes, and the resulting data are interpolated by splines. In order to show the mesh independence of WSS, the steady-state simulations are performed for each stent with three different meshes (named mesh-1, mesh-2 and mesh-3) of $5\ 10^5$, 10^6 and $2\ 10^6$ cells, respectively. Since the last two meshes offer comparable results, the intermediate mesh is used for the unsteady-state simulations.

5.3.1   Without Residual Stenosis

The results of the *WSS* (*Pa*), averaged over the vessel section, are shown in Figs. 38 and 39, [70], for the two stents versus the axial length (m) for a vessel without residual stenosis. The solution is validated against three different grids.

Figures 38 and 39, [70], show a repeating pattern for the *WSS*, with peaks and valleys corresponding, respectively, to the planes intersecting mostly the intra-strut area (where *WSS* is greater) and to the planes intersecting the regions of low *WSS*, located on the struts. Indeed, for a Newtonian fluid, the *WSS* only depends on the symmetric part of the velocity gradient at the wall, which strongly diminishes in proximity to the struts. This is due to the deceleration of the fluid impinging on an obstacle leaning on a wall. When the blood flow meets the edge of the strut, the streamlines, originally adhering to the wall, detach from it at a certain distance. A recirculation region is formed between the strut wall and the detachment point,

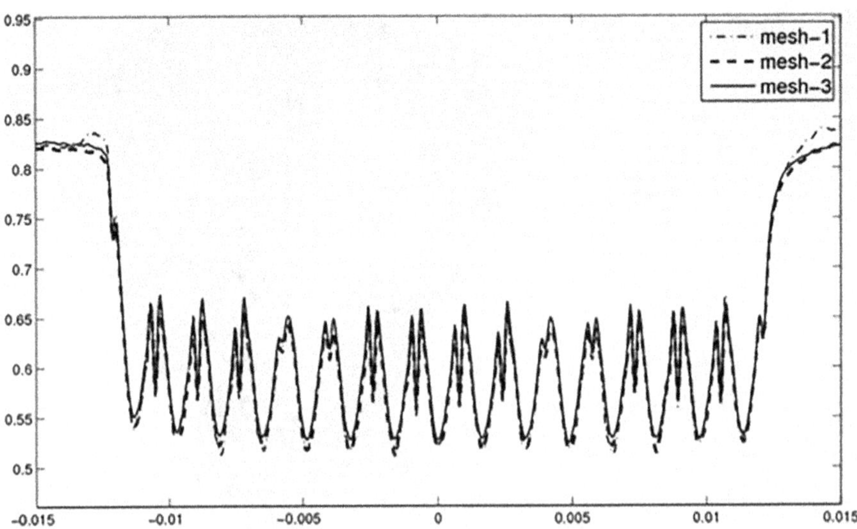

Fig. 38 WSS spatial distribution in S1 stent without residual stenosis for the three meshes, Fig. 5 of [70]

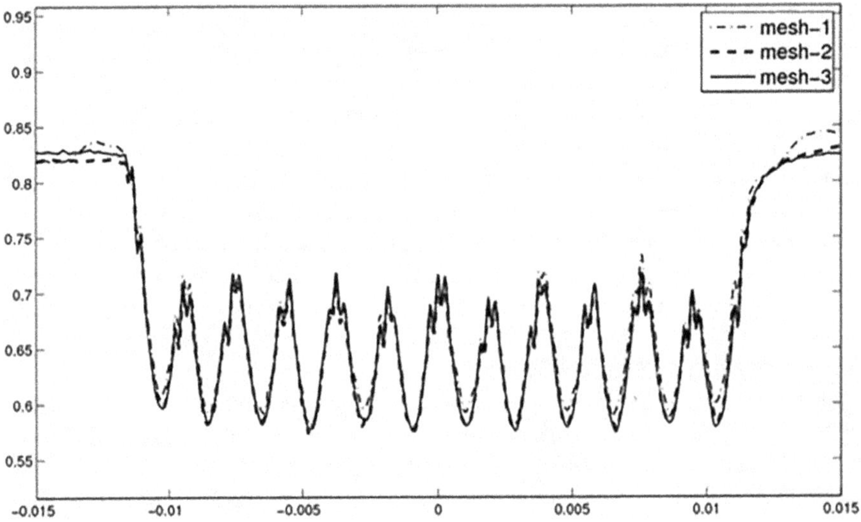

Fig. 39 WSS spatial distribution in S2 stent without residual stenosis for the three meshes, Fig. 6 of [70]

whose amplitude depends on the Reynolds number: the smaller it is, the wider the recirculation region. The stent causes a geometrical discontinuity in the endothelium wall and a consequent loss of charge. The presence of these recirculation regions increases the residence time of macrophages and macromolecules on the endothelium and increases the probability of uptake by intimal smooth muscle cells.

5.3.2 With Residual Stenosis

The results concerning the averaged *WSS* in the presence of a residual stenosis are shown in Figs. 40 and 41, [70], as well. The reduction of the diameter provokes a pressure drop and a consequent acceleration of the fluid. Since the *WSS* is, in first

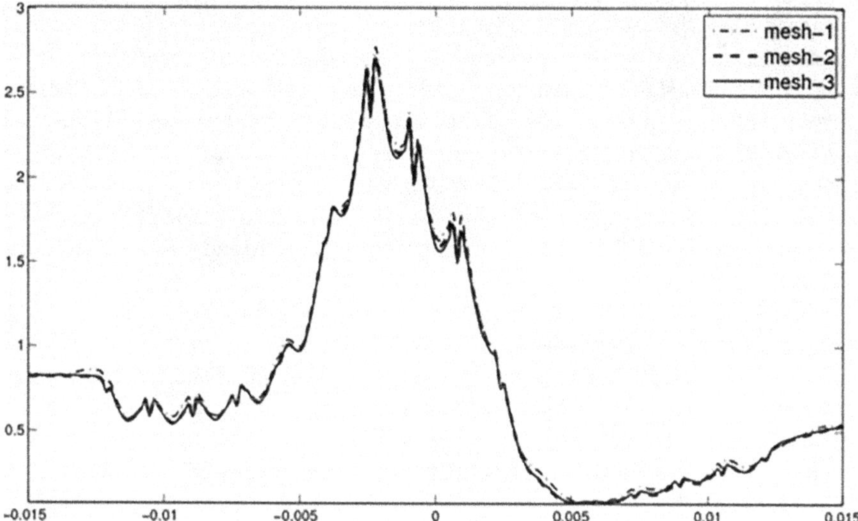

Fig. 40 WSS spatial distribution in S1 stent with residual stenosis for the three meshes, Fig. 7 of [70]

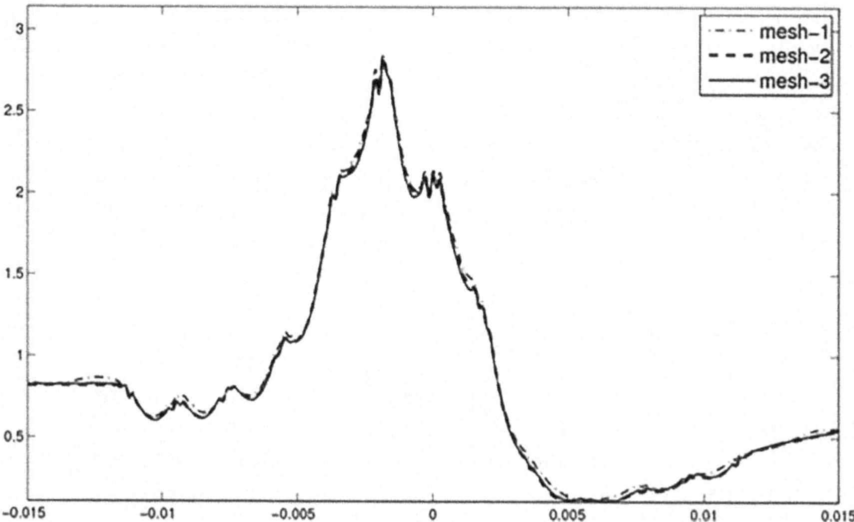

Fig. 41 WSS spatial distribution in S2 stent with residual stenosis for the three meshes, Fig. 8 of [70]

approximation, inversely proportional to the section diameter and directly proportional to the fluid speed, the *WSS* increases in the choked section. Downstream of this region, the diameter increases and so the *WSS* diminishes. If the slope of the wall downstream of the choking section is high enough, a streamline detachment is present and a consequent recirculation region, which contributes to lowering the *WSS* besides the geometrical changes.

Figure 40, [70], shows that at a distance of approximately 8 mm from the centre of the stent S1, the *WSS* starts to increase, as compared to the case without stenosis. It should be noted that the maximum *WSS*, which has a peak value of 2.66 Pa, does not correspond to the top of the stenosis but is 2.3 mm in front, because both the axial and transversal velocities change more rapidly before the maximum height of the stenosis. After this point, the *WSS* decreases with some local peaks and valleys due to the struts, which disappear 1.5 mm after the vessel centre. The *WSS* reaches a minimum value of 0.081 Pa at 5.1 mm from the vessel centre and gradually increases after that.

A similar behaviour can be observed in Fig. 41, [70], for the stent S2. The *WSS* starts to increase at 9 mm before the centre of the vessel, then it reaches the maximum value of 2.8 Pa at 1.9 mm in front of the vessel centre and then it decreases to the minimum monotonically, except for a small plateau around the centre of the vessel. The minimum of 0.11 Pa is reached at 6.5 mm as well. It must be remarked that the *WSS* in the stent S2 has a smoother profile compared to S1. The similarity in the behaviour of the two stents suggests that in the presence of a stenosis, the degree of stenosis plays a more important role in the fluid dynamics compared to the stent geometry. The stenosis has a significant effect in terms of *WSS* reduction, causing an abrupt drop below the critical value of 0.5 Pa. This takes place downstream and is due to the presence of a stagnation region after the stenosis. The narrowing has the effect of amplifying the *WSS* in the central region, without exceeding 7 Pa, which is the value above which thrombogenic effects may take place [13, 14]. In both fully expanded stents, the mean *WSS* value is kept in the physiological range of 0.5–7 Pa, which is usually found in medium and large arteries [13, 14, 94]. Nevertheless, from the computational data, the stent S2, *peak-to-peak*, seems to have better behaviour compared to the stent S1, *peak-to-valley*, since the *WSS* pattern is smoother, its average value is higher and the variance is lower, indicating that this design alters the fluid flow less.

5.4   Unsteady-State Analysis

Unsteady-state simulations, performed on an entire cardiac cycle, show large variations in terms of *WSS* minimum and maximum values. However, except for the systolic peak, the *WSS* pattern does not exhibit great differences, and thus time-averaged physiological parameters (*OSI, RRT*) are presented to take into account the cumulative effects of the fluid dynamics over the endothelium. *WSS* maps of the endothelium reveal the presence of stagnation regions around the struts, where the *WSS* is small.

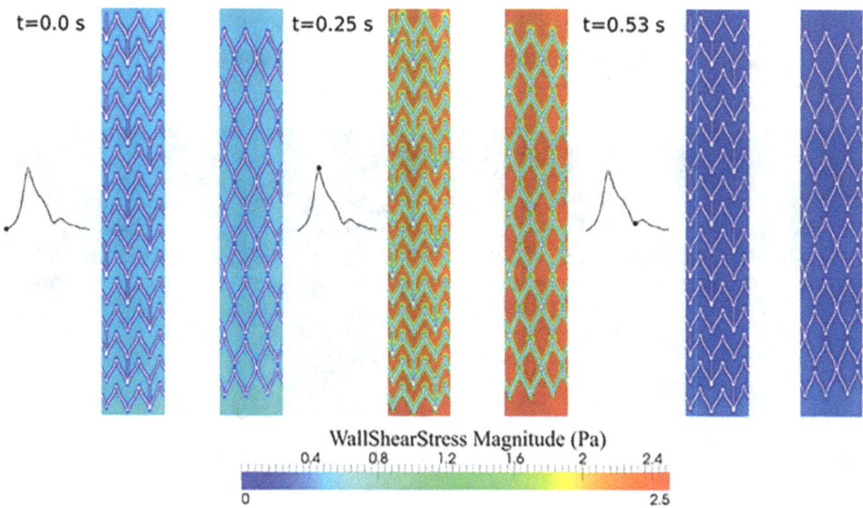

Fig. 42 WSS spatial distribution for S1 and S2 in a non-stenosed vessel at three different time instants, Fig. 9 of [70]

5.4.1 Without Residual Stenosis

In Fig. 42, [70], the spatial distribution of *WSS* magnitude in a vessel without residual stenosis is reported at three time steps: 0.0 s, 0.25 s and 0.53 s. At the systolic peak, the pattern changes as well, showing that the region downstream of the stenosis is always subjected to a relatively low wall shear stress, while the *WSS* increases upstream. This is because the increase in flow rate broadens the extent of the recirculation region as well, in order that the *WSS* changes in direction but its modulus remains quite low.

OSI is used to measure the oscillation of *WSS* from its mean direction. The *OSI* assumes a value equal to 0 where the *WSS* always has the same direction during the cycle and 0.5 when the *WSS* mean value is 0 Pa. The simulations in the two straight vessels are shown in Fig. 43, [70], and reveal a low value of *OSI* only near the struts, while in the intra-strut area, this value is close to 0. It is shown in [13, 95] that there is a correlation between atherosclerosis pathogenesis and *OSI* only for values between 0.1 and 0.5; thus, in this case, *OSI* is far from the critical range. Higher values of *OSI* can be observed in the small curvature formed by strut tips and connectors, where the fluid is trapped, as in Fig. 43, [70]. In the *peak-to-peak* configuration, the peaks are arranged opposite one another, and this makes the stent mesh made from a hexagonal pattern. Between the vertices of the mesh, there is a distance such that a region in which the motion is relatively undisturbed occurs. The configuration *peak-to-valley* is more rugged than the *peak-to-peak* one. The stent mesh has an "arrowhead" design, which provides an approximately constant distance between the opposite edges. Nevertheless, if this distance is not too long and the velocity is great enough, the mesh can act as roughness and favours the presence of recirculating regions and, so, inverted flow. This is in disagreement with Berry et al. [6], who suggested that a

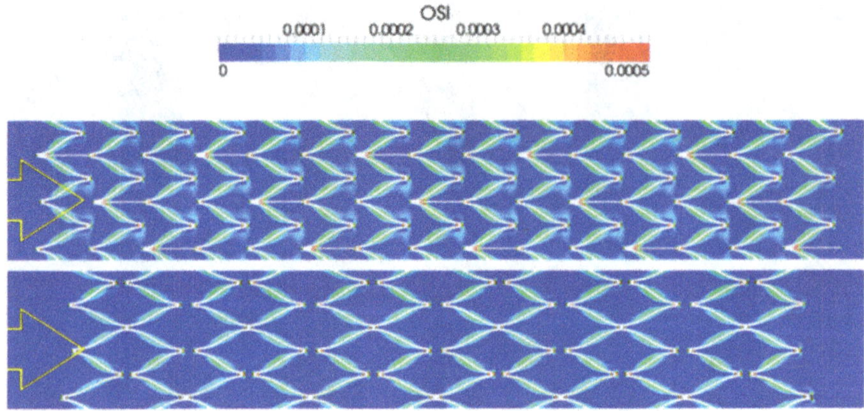

Fig. 43 OSI distribution without residual stenosis for S1 (top) and S2 (bottom), Fig. 11 of [70]

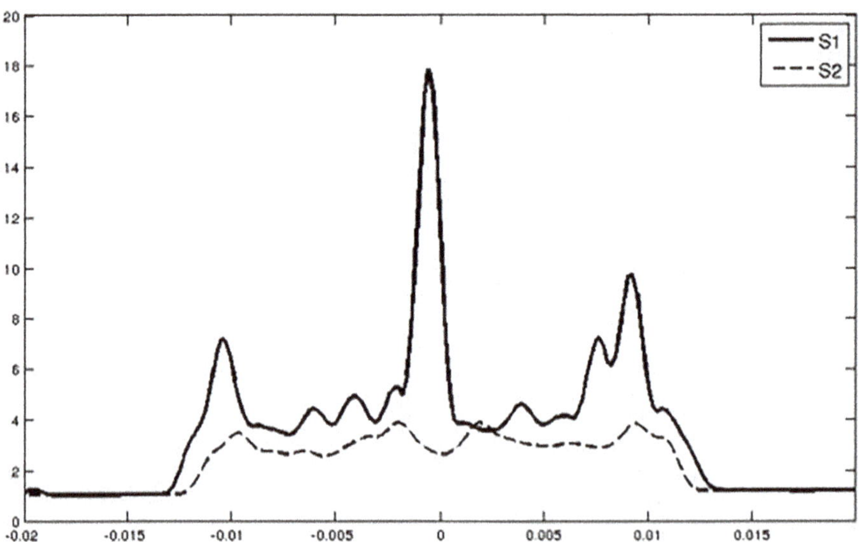

Fig. 44 RRT distribution without residual stenosis, Fig. 13 of [70]

finer mesh could disturb the flow more than a coarser one. From our results, it appears that, in physiological conditions, i.e. Re = 640, the stent S1 produces more recirculating regions compared to S2.

Figure 44, [70], shows the RRT(1/Pa) versus the axial length (m) of the two stents. Figure 44, [70], presents the *RRT* of the two stents in the vessel without stenosis, showing that the *RTT* increases where the stent is present and decreases on the vessel wall because the *RTT* is proportional to the reciprocal of the time-averaged wall shear stress, and this variable is considerably lower in the region where the stent

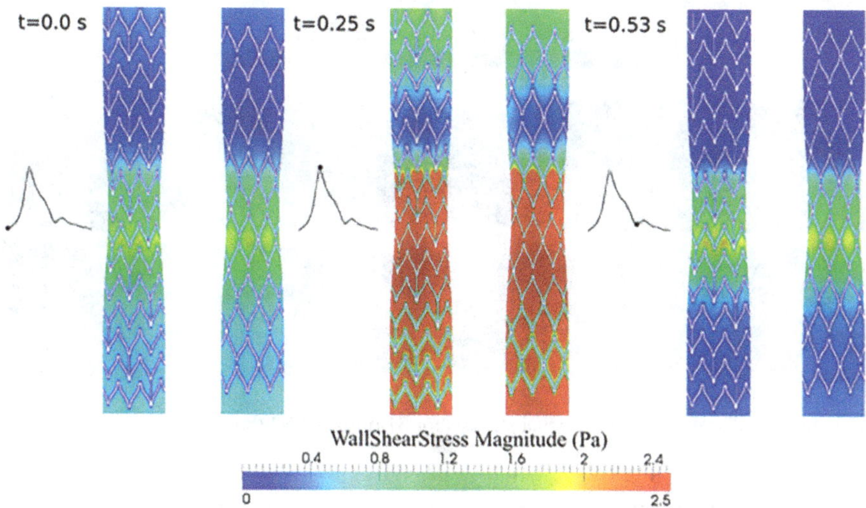

Fig. 45 WSS spatial distribution for S1 and S2 in a stenosed vessel, at three different time instants, Fig. 10 of [70]

is inserted. Moreover, since the *WSS* pattern is wavy in the stented region, *RRT* shows some peaks. The *RRT* presents higher values for the *peak-to-valley* configuration than the *peak-to-peak* one. This is due to the low *TAWSS* present in the region covered by the stent S1 and by higher values of *OSI* that are uniform in the stent S2.

5.4.2 With Residual Stenosis

In the presence of the residual stenosis, Fig. 45, [70], reports the *WSS* spatial distribution for the stents S1 and S2 at three different time instants. The *WSS* in the choking region has the same pattern as in the stents without residual stenosis and exhibits a high peak, corresponding to the minimum section, and a much lower value downstream of the narrowing. The *WSS* starts to grow again only when the flow attaches to the wall. The most significant role in the haemodynamic disturbance seems to be still played by the narrowing, while the different stent designs influence the *WSS* patterns only to a minor extent.

The *OSI* becomes more relevant when a stenosis exists, as shown by Fig. 46, [70], with high values between 0.2 and 0.5. The results show that a 30% reduction in the cross-sectional area is sufficient to provoke a boundary layer detachment after the stenosis. A wide recirculation region is present downstream of the stenosis. There are no major differences in the *OSI* pattern of the two stents when a 30% stenosis is over-imposed, although even in this case the stent S2 appears to have slightly smaller *OSI* values compared to the stent S1.

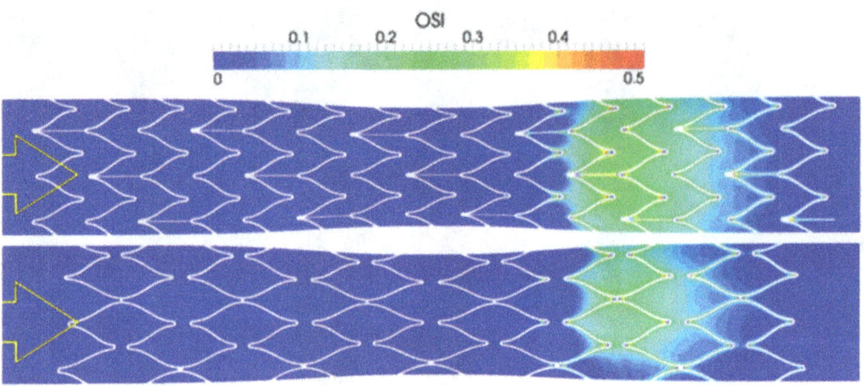

Fig. 46 OSI distribution with residual stenosis for S1 (top) and S2 (bottom), Fig. 12 of [70]

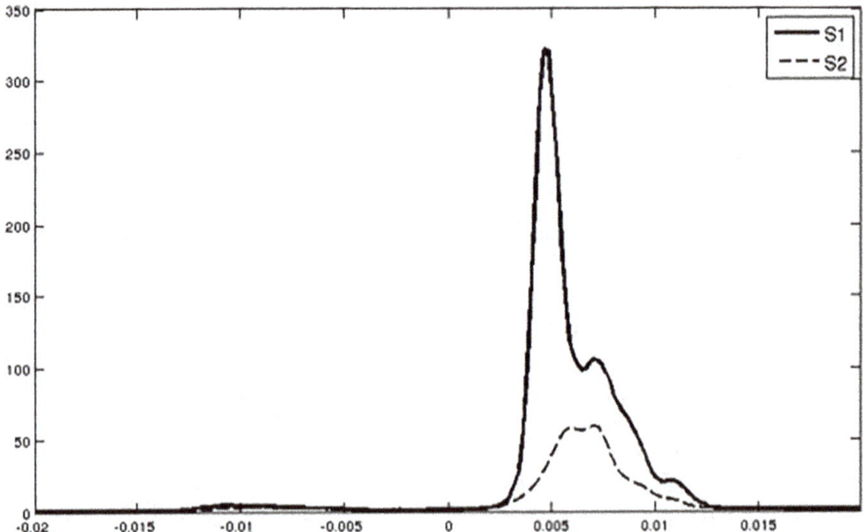

Fig. 47 OSI distribution with residual stenosis, Fig. 14 of [70]

Figure 47, [70], shows the *RRT* in the stenosed vessel. The variable *RRT* shows evident differences in the performances of the two stent configurations. The *RRT* for stent S1 is much higher than that for S2. Both the lowering of *TAWSS* and the increase of *OSI*, due to the stagnation region, contribute to its increase. The previous results about the *WSS* and the *OSI* already suggested that the *peak-to-peak* configuration had better fluid dynamic behaviour compared to the *peak-to-valley*, but the results were not as evident as in the *RRT* case.

6   Conclusions

6.1   Thermo-Haemo-Dynamics (THD) Analysis of CYPHER Coronary Stents

The aim of this study is the characterization of the fluid dynamics of coronary stents through numerical simulations. Understanding the fluid dynamics of coronary stents is very important for clinical success, as the phenomenon of restenosis, which causes implant failure, is strongly influenced by haemodynamic factors. After a careful study of the problems of atherosclerosis and existing coronary stents, a first numerical analysis of the fluid dynamics of a stent is carried out in this work. This analysis is compared with the studies present in the literature. The present one has important innovations; first of all, the realistic reconstruction of a stent currently on the market. All computational fluid dynamics studies on coronary stents present in literature use very simple geometries that recall the first stents used in the 80's.

The chosen stent is the *CYPHER* in its two versions marketed. The stent studied is a drug-eluting stent, and thanks to this, it has very low restenosis rates. There are many clinical studies on this stent aimed at studying the drug release, but there are no fluid dynamics studies. The drug release is exhausted in about 30 days, and recent long-term studies show that the drug is probably only able to delay restenosis. Therefore, the restenosis rate of a drug-eluting stent could be similar to that of a classic stent if evaluated over long periods. Hence, the importance of studying the fluid dynamics of the *CYPHER* stent regardless of the drug release.

The numerical analysis carried out includes the steady state and the unsteady one. For the latter, a physiological waveform is used, which is sampled and reconstructed with specially developed software. The fluid dynamic parameters that are taken into consideration for the evaluation of the probability of restenosis of the stents studied, in accordance with the literature, are the wall shear stress (*WSS*), the mean spatial gradient of the WSS (*MGS*) and the oscillatory shear index (*OSI*). These parameters are strongly correlated with endothelial permeability, which can cause neo-intimal hyperplasia and, therefore, restenosis. The different factors provide different and complementary information among them; moreover, the association of adverse conditions for several factors is an important index.

Remembering the simplifying assumptions used, from the analysis of the results obtained, some interesting considerations can be drawn. The parameter most certainly studied and used in the literature is the *WSS*. There are physiological values of this parameter measured experimentally in the coronary arteries. As a threshold value, the value of 0.5 Pa is commonly accepted (the physiological values vary between 0.5 Pa and 1.0 Pa). The present study shows that in some areas of the vessel wall, values lower than 0.2 Pa are obtained. These very low values are a strong indication of a high probability of restenosis. The low *WSS* values are widespread in all the vessel wall areas adjacent to the stents.

The *MGS* values detected in some areas are significantly high. Even though there is no threshold value for this parameter, it has been shown that high *MGS* values are correlated with neo-intimal hyperplasia. The high *MGS* areas are located near the stent, but they are not uniformly distributed around them like the *WSS* but are mainly located on the sides of the stent that oppose the flow. These areas, which are also low in *WSS*, are at high risk of neo-intimal hyperplasia and therefore restenosis.

Given the importance of the association of several adverse factors and the strong correlation with endothelial permeability, a model is studied that links these parameters (*WSS* and *MGS* with endothelial permeability). This model uses the optical density (*OD*) parameter, obtained from the optical measurements of the deposition of the *EBD* dye on the arterial walls. Using this model, the *OD* is calculated from the results obtained from the numerical simulations. The values of *OD* obtained are higher than the physiological values measured in some experiments, but there are no measurements in the coronary arteries, and the values strongly depend on the location within the arterial tree. Therefore, the *OD* is only considered as a relative index. The areas with the highest *OD* are located on the angular points. The unsteady-state analysis provides additional information compared to the steady-state one. In particular, it is seen that both the *WSS* and the *MGS* vary significantly during the cardiac cycle. There are areas that are subjected to adverse conditions throughout the cycle and others only at some instants of time. The *OD* instead remains practically constant during the entire cycle.

From the results of the unsteady-state analysis, it is possible to calculate another important factor for the evaluation of the probability of restenosis: the *OSI*. This parameter shows that the areas that undergo the greatest variations during the cardiac cycle are those areas adjacent to the loops of the connecting elements. The stents can undergo significant deformations during the implantation, based on the geometric conformation of the artery; therefore, the critical areas detected in this study can be reduced or amplified. The existence of these areas should be taken into consideration.

Comparing the results of the two configurations of the studied stent, *CYPHER SELECT* and *CYPHER*, it is noted that in the vessel section containing the stent *CYPHER SELECT*, the critical areas are more widespread. This suggests that this stent could be less suitable, as it may have a higher risk of restenosis. The stent *CYPHER SELECT* is a very recent stent (on the market since the second half of 2006) and is the successor of the stent *CYPHER* (on the market since 2002), which has had huge success, becoming the most used coronary stent in the world. The stent *CYPHER SELECT* is created for the treatment of lesions in difficult arteries, such as tortuous vessels that are difficult to reach and in which to untangle, and is defined as third generation, as it presents important innovations in drug release. The stent *CYPHER SELECT*, following the success of its predecessor, is having a rapid spread and is completely replacing the stent *CYPHER*, despite the few independent clinical studies that demonstrate its effectiveness.

From this study it emerges that, at least from a fluid dynamic point of view, the stent *CYPHER SELECT* is not superior to the stent *CYPHER;* in fact, it seems to present a higher risk of restenosis. Being specifically created for the treatment of difficult lesions, it would probably be useful not to use it indiscriminately, at least until the necessary independent clinical studies are carried out, and continue to use the stent *CYPHER* for the lesions that do not present particular difficulties, as its effectiveness is demonstrated with a large number of clinical studies.

6.2 Thermo-Haemo-Dynamics (THD) Analysis of Two Carotid Self-Expanding Bare Metal Stents

From the present numerical simulations, it can be concluded that the application of self-expandable stents has a twofold effect: on one hand, it can reduce the graft failure rate, limiting artery injury caused by balloon inflation (at high pressures of about 12–16 bar), and on the other hand, the residual narrowing could represent a critical region for the haemodynamics. This stenosis, even if not harmful for the patient, could disrupt the blood flow downstream of the narrowing, exposing the endothelium to hazardous *WSS*, which could result in the long term in neo-intimal proliferation, promoting the restenosis of the artery.

Long-term medical follow-up for this fairly new implant and in vivo experiments could help to point out whether the negative effects on fluid dynamics due to residual stenosis are more significant than the beneficial effects produced by the less traumatic implant technique. Moreover, the numerical simulation is a powerful tool to evaluate which is the best haemodynamic configuration. In this work, it appears that the better haemodynamic configuration is the less compliant one. This contradiction may be overcome in a more realistic configuration where the struts are deformed in a non-symmetric way. A study on a realistic geometry may give completely different results in terms of mechanical and haemodynamic properties. The effort in the stent design should be by conjugating good mechanical and haemodynamic properties. Further studies are required to investigate other configurations and a more realistic arterial geometry.

Acknowledgements for Figures The author thanks the *publisher Taylor and Francis* for the kind permission to reproduce images 3, 7, 8, 13, 14, 15, 18, 19, 20, 21, 22, 24 and 25 of Ref. [63]; the images 1, 2, 3, 5, 6, 7, 8, 9, 10, 11, 12, 13 and 14 of Ref. [70] and images 1, 2, 3, 6, 7, 8, 9, 10, 11 and 12 of Ref. [77].

The author thanks the *publisher MDPI* for the free reproduction of image 4a of Ref. [71].

References

1. Amitrano M (2006–2007) Analisi numerica della fluidodinamica di uno stent coronarico. Tesi di Laurea Magistrale in Ingegneria Medica, Università di Roma Tor Vergata, A.A

2. Ku JP, Elkins CJ, Taylor CA (2005) Comparison of CFD and MRI flow and velocities in an in-vitro large artery bypass graft model. Ann Biomed Eng 33(3):257–269

3. Palmaz JC (2004) Intravascular stents in the last and the next 10 years. J EndTher Suppl 2:200–206

4. Virmani R, Farb A, Guagliumi G, Kolodgie FD (2004) Drug-eluting stents: caution and concerns for long-term outcome. Coron Artery Dis 15:313–318

5. Shulz C, Herrmann RA, Beilharz C, Pasquantonio J, Alt E (2000) Coronary stent symmetry and vascular injury determine experimental restenosis. Heart 83:462–467

6. Berry JL, Santamarina A, Moore JE Jr, Chowdhury SR, Routh WD (2000) Experimental and computational flow evaluation of coronary stents. Ann Biomed Eng 28(4):386–398

7. LaDisa JF Jr, Olson LE, Guler I, Hettrik DA, Audi SH (2004) Stent design properties and deployment ratio influence indexes of wall shear stress: a three-dimensional computational fluid dynamics investigation within a normal artery. J Appl Physiol 97(1):424–430
8. LaDisa JF Jr, Guler I, Olson LE, Hettrik DA, Kersten JR, Warltier DC, Pagel PS (2003) Three-dimensional computational fluid dynamics modeling of alterations in coronary artery wall shear stress produced by stent implantation. Ann Biomed Eng 31:972–980
9. LaDisa JF Jr, Olson LE, Guler I, Hettrik DA, Kersten JR, Warltier DC, Pagel PS (2005) Circumferential vascular deformation after stent implantation alters wall shear stress evaluated with time-dependent 3D computational fluid dynamics models. J Appl Physiol 98:947–957
10. LaDisa JF Jr, Olson LE, Hettrik DA, Warltier DC, Kersten JR, Pagel PS (2005) Axial stent strut angle influences wall shear stress after stent implantation: analysis using 3D computational fluid dynamics models of stent foreshortening. Biomed Eng Online 4:59
11. LaDisa JF Jr, Hettrik DA, Olson LE, Guler I, Gross ER, Kress TT, Kersten JR, Warltier DC, Pagel PS (2002) Coronary stent implantation alters coronary artery hemodynamics and wall shear stress during maximal vasodilation. J Appl Physiol 93:1939–1946
12. Ku DN (1997) Blood flow in arteries. Ann Rev Fluid Mech 29:399–434
13. Ku DN, Giddens DP, Zarins CK, Glagov S (1985) Pulsatile flow and atherosclerosis in the human carotid bifurcation, positive correlation between plaque location and low oscillating shear stress. Arteriosclerosis 5:293–302
14. Malek AM, Alper SL, Izumo S (1999) Hemodynamic shear stress and its role in atherosclerosis. JAMA 282:2035–2042
15. Lanoye L, De Beule M, Dewijngaert C, Segers P, Van Impe P, Verdonck P (2006) The influence of the strut section shape on the flow field in a newly stented right coronary artery, the 56th National Congress of theoretical and applied mechanics. NCTAM
16. Seo T, Shahter LG, Barakat AI (2005) Computational study of fluid mechanical disturbance induced by endovascular stents. Ann Biomed Eng 33(4):444–456
17. LaDisa JF Jr, Olson LE, Hettrik DA, Warltier DC, Kersten J, Pagel P (2006) Alterations in regional vascular geometry produced by theoretical stent implantation influence distributions of wall shear stress: analysis of a curved coronary artery using 3D computational fluid dynamics modeling. Biomed Eng 5:40
18. Deplano V, Bertolotti C, Barragan P (2004) Three-dimensional numerical simulations of physiological flows in a stented coronary bifurcation. Med Biol Eng Comput 42:650–659
19. He Y, Duraiswamy N, Frank AO, Moore JE Jr (2005) Blood flow in stented arteries: a parametric comparison of strut design patterns in three dimensions. J Biomech Eng 127:637–647
20. Ross R (1999) Atherosclerosis-an inflammatory disease. New Engl J Med 340:1436–1442
21. Glass CK, Witztum JL (2001) Atherosclerosis: the road ahead. Cell 104:503–516
22. Hansson GK (2001) Immune mechanisms in atherosclerosis. Arterioscl ThrombVas Biol 21: 1876–1890
23. Seikku P, Leinonen M, Mattila K (1988) Serological evidence of an association of a novel Chlamydia, TWAR, with chronic coronary heart disease and acut emyocardial infarction. Lancet 2:983–986
24. Seikku P (1997) Chlamydia pneumoniae and atherocslerosis-an update. Scan J Infect Dis Suppl 104:53–56
25. Shwartz SM, Heimark RL, Majesky MW (1990) Developmental mehanisms underlying pathology of arteries. Physio Rev 70:1177–1209
26. Reidy MA, Fingerle J, Lindner V (1992) Fators ontrolling the development of arterial lesions after injury. Circulation 86:III43–III46
27. Stary HC, Chandler A, Dinsmore R (1995) A definition of advanced types of atherosclerotic lesions and histological classification of atherosclerosis. A report from the Committee on Vascular Lesions of the Council on Atherosclerosis, American Heart Association. Circulation 92:1355–1413

28. Poole J, Florey HW (1958) Changes in the endothelium of the aorta and the behaviour of macrophages in experimental atheroma of rabbits. J Pathol Bacteriol 75:245–252

29. Ross R, Glomset JA, Kariya B, Harker LA (1974) A platelet dependent serum factor that stimulates the proliferation of arterial smooth musle cells in vitro. Proc Natl Acad Sci U S A 71: 1207–1211

30. Benditt EP, Benditt JM (1973) Evidence for a monoclonal origin of human atheroslerotic plaques. Proc Natl Acad Sci U S A 70:1753–1756

31. Brown MS, Goldstein JL (1986) A receptor-mediated pathway for cholesterol homeostasis. Science 232:34–47

32. Breslow JL (1996) Mouse models of atherosclerosis. Science 272:685–688

33. Nordestgaard BG (1998) The vascular endothelial barrier – selective retention of lipoproteins. J Clin Inves 102:145–152

34. Boren J, Gustafsson M, Skalen K, Flood C, Innerarity TL (2000) Role of extracellular retention of low density lipoproteins in atherosclerosis. Cur Opi Lipido 11:451–456

35. Steinberg D (1997) Low density lipoprotein oxidation and its pathologial significance. J Biol Chem 272:20963–20966

36. Navab M, Berliner JA, Watson AD (1996) The Yin and Yang of oxidation in the development of the fatty streak. A review based on the 1994 George Lyman Duff Memorial Leture. Arterios Throm Vasc Bio 16:831–842

37. Niemann-Jonsson A, Dimayuga P, Jovinge S, Calara F, Ares MP, Fredrikson GN, Nilsson J (2000) Accumulation of LDL in rat arteries is associated with activation of tumor necrosis factor-alpha expression. Atheroscle Throm Vasc Bio 20:2205–2211

38. Cybulsky MI, Gimbrone MA (1991) Endothelial expression of a mononulear leukocyte adhesion moleule during atherosclerosis. Science 251:788–791

39. Quinn MT, Parthasarathy S, Steinberg D (1985) Endothelial cell-derived hemostatic activity for mouse peritoneal macrophages and the effects of modified forms of low density lipoprotein. Proc Natl Acad Sci U S A 82:5949–5953

40. Seifert PS, Hugo F, Hansson GK, Bhakdi S (1989) Prelesional complement activation in experimental atherosclerosis. Terminal C5b-9 complement deposition coincides with cholesterol accumulation in the aortic intima of hypercholesterolemic rabbits. Lab Inves 60:747–754

41. Smith JD, Trogan E, Ginsberg M, Grigaux C, Tian J, Miyata M (1995) Decresed atherosclerosis in mice deficient in both macrophages colony-stimulating factor (op) and apolipoprotein. Proc Natl Acad Sci U S A 92:8264–8268

42. Brown MS, Goldstein JL (1983) Lipoprotein metabolism in the macrophage: implications for cholesterol deposition in atherosclerosis. Annu Rev Biochem 52:223–261

43. Krieger M, Aton S, Ashkenas J, Pearson A, Penman M, Resnik D (1983) Molecular flypaper, host defense, and atheroslerosis. J Bio Chem 268:4569–4572

44. Geng YJ, Hansson GK (1992) Interferon-γ inhibits scavenger receptor expression and foam cell formation in human monocyte-derived macrophages. J Clin Inves 89:1322–1330

45. Tontonoz P, Nagy L, Alvarez JGA, Thomazy VA, Evans RM (1998) PPARg promotes monocyte/macrophage differentiation and uptake of oxidized LDL. Cell 93:241–252

46. Gimbrone MA Jr, Nagel T, Topper JN (1997) Perspectives series: cell adhesion in vascular biology. J Clin Inves 99:1809–1813

47. Davies MJ, Thomas A (1984) Thrombosis and acute coronary-artery lesions in sudden cardiac ischemic death. N Engl J Med 310:1137–1140

48. Davies MJ (1996) Stability and instability: two faces of coronary atherosclerosis. Circulation 94:2013–2020

49. Falk E, Shah P, Fuster V (1995) Coronary plaque disruption. Circulation 92:657–671

50. Van der Wal AC, Beker AE, Van der Loss CM, Das PK (1994) Site of intimal rupture or erosion of thrombosed coronary atherosclerotic plaques is charaterized by an inflammatory process irrespective of dominant plaque morphology. Circulation 89:36–44

51. Kovanen PT, Kaartinen M, Paavoncn T (1995) Infiltrates of activated mast cells at the site of coronary atheromatous erosion or rupture in mycoardial infartion. Circulation 92:1084–1088

52. Libby P (1995) Molecular bases of the acute coronary syndromes. Circulation 91:2844–2850
53. Gruentzig A, Schneider HJ (1977) The percutaneous dilatation of chronic coronary stenosis – experiments and morphology. Schweiz Med Wochenschr 107(44):1588
54. Berne RM, Levy MN (2001) Cardiovascular physiology. Mosby Inc, St. Louis
55. Serruys PW, de Jaegere P, Kiemeneij F, Maaya C, Rutsh W, Heyndrikx G, Emanuelsson H, Maro J, Legrand V, Materne P (1994) A comparison of balloon-expandable-stent implantation with balloon angioplasty in patients with coronary artery disease, Benestent study group. N Engl J Med 331:489–495
56. Fishman DL, Leon MB, Baim DS, Shatz RA, Savage MP, Penn I, Detre K, Veltri L, Rii D, Nobuyoshi M (1994) A randomized comparison of coronary stent placement and balloon angioplasty in the treatment of coronary artery disease, Stent restenosis study investigators. N Engl J Med 331:496–501
57. Yoshizumi M, Kurihara H, Sugiyama T, Takaku F, Yanagisawa M, Masaki T, Yazaki Y (1989) Hemo-dynamic shear stress stimulates endothelion production by cultured endothelial cells. Biochem Biophy Res Commun 161(2):859–864
58. Morice MC, Serruys PW, Sousa JE, Fajadet J, Ban Hayashi E (2002) A randomized comparison of a sirolimus-eluting stent with a standard stent for coronary revascularization. N Engl J Med 346(23):1773–1780
59. Moses JW, Leon MB, Popma JJ, Fitzgerald PJ, Holmes DR (2003) Sirolimus eluting stents versus standard stents in patients with stenosis in a native coronary artery. N Engl J Med 349(14):1315–1323
60. Zarins CK, Zatina MA, Giddens DP, Ku DN, Glagov S (1987) Shear stress regulation of artery lumen diameter in experimental atherogenesis. J Vasc Res 5(3):413–420
61. Dotter CT (1969) Transluminally-placed coilspring endoarterial tube grafts. Long term patency in canine popliteal artery. Invest Radiol 4(5):329–332
62. Palmaz J (1988) Expandable intraluminar graft, and method and apparatus for implanting an expandable intraluminar graft, United States Patent, Patent number: 4739762. Palmaz J (1985) Expandable intraluminal graft: a preliminary study. Radiology 156:73–77
63. Stoekel D, Bonsignore C, Duda S (2002) A survey of stent designs. Min Invas Ther Allied Technol 11:137–147
64. Serruys PW, Rensing BJ (2002) Handbook of coronary stents
65. Machraoui A, Grewe P, Fisher A (2001) Koronarstenting
66. Sangiorgi G, Melzi G, Agostoni P, Cola C, Clementi F, Romitelli P, Virmani R, Colombo A (2007) Engineering aspects of stents design and their translation into clinical practice. Ann Ist Super Sanità 43:89–100
67. Gori F et al (2006) A new hysteretic behavior in the electrial resistivity of flexinol shape memory alloys versus temperature. Int J Thermophys 27:866–879
68. Hernandez-Antoln RA, Suarez A, Corros C (2008) Case report. Diagnosis and treatment of coronary stent entanglement complicated by extreme stent distortion. J Invasive Cardiol 20(3): E67–E70
69. Butany J, Carmichael K, Leong SW, Collins MJ (2005) Coronary artery stents: identification and evaluation. J Clin Pathol 58:795–804. https://doi.org/10.1136/jcp.2004.024174
70. Boghi A et al (2015) Numerical simulation of blood flow through different stents in stenosed and non-stenosed vessels. Numer Heat Transf A Appl 68:225–242
71. Pan C, Han Y, Lu J (2021) Structural design of vascular stents: a review. Micromachines 12:770
72. Sousa JE, Costa MA, Abizaid A (2001) Lack of neointimal proliferation after implantation of sirolimus-coated stents in human coronary arteries: a quantitative coronaric angiography and three-dimensional intravascular ultrasound study. Circulation 103:192–195
73. Schofer M, Gershlik AH (2003) Sirolimus-eluting stents for treatment of patients with long atheroscleroti lesions in small coronary arteries: double-bind randomized controlled trial. Lancet 362:1093–1099

74. Shampaert E, Cohen EA, Shluter M (2004) The Canadian study of the sirolimus eluting stent in the treatment of patients with long de novo lesions in small native coronary arteries (C-SIRIUS). J Am Coll Cardiol 43:1110–1115
75. Kastrati A, Mehilli J, Dirshinger J, Pahe J, Ulm K, Shuhlen H, Seyfarth M, Shmitt C, Blasini R, Neumann FJ, Shomig A (2001) Restenosis after coronary placement of various stent types. Am J Cardiol 87:34–39
76. Gori F et al (2009) Three-dimensional numerical simulation of the fluid dynamics in a coronary stent. In: ASME international mechanical engineering congress and exposition proceedings, vol 2. Lake Buena Vista, FL, pp 407–411
77. Gori F et al (2011) Three-dimensional numerical simulation of blood flow in two coronary stents. Numer Heat Transf A Appl 59:231–246
78. Womersley JR (1955) Method for the calculation of velocity, rate of flow and viscous drag in arteries when the pressure gradient is known. J Physiol 127(3):553
79. Lei M, Kleinstreuer C, Truskey GA (1995) Numerical investigation and prediction of atherogenic sites in branching arteries. J Biomeh Eng 117:350–357
80. Buhanan JR Jr, Kleinstreuer C, Truskey GA, Le M (1999) Relation between non-uniform hemodynamics and sites of altered permeability and lesion growth at the rabbit aorto-celiac junction. Atherosclerosis 143:27–40
81. Butler PJ, Norwih G, Weinbaum S, Chien S (2001) Shear stress induces a time and position-dependent increase in endothelial cell membrane fluidity. Am J Physiol Cell Physiol 280:962–969
82. DePaola N, Gimbrone MA Jr, Davies PF, Dewey CF Jr (1992) Vascular endothelium responds to fluid shear stress gradients. Arterioscler Thromb 12:1254–1257
83. Freedman FB, Johnson JA (1969) Equilibrium and kinetic properties of the Evansblue-albumin system. Am J Physio 216:675–681
84. Bell FP, Adamson IL, Shwartz CJ (1974) Aortic endothelial permeability to albumin: foal and regional patterns of uptake and transmural distribution of 1311-albumin in the young pig. Exptl Mol Pathol 20:57–68
85. Pakham MA, Rowsell HC, Jorgensen L, Mustard JF (1967) Localized protein accumulation in the wall of the aorta. Exptl Mol Pathol 7:214–232
86. Fry DL (1977) Aortic Evans blue dye accumulation: its measurement and interpretation. Am J Phys 232:204–222
87. Himburg HA, Grzybowski DM, Hazel AL, LaMak JA, Li XM, Friedman MH (2004) Spatial comparison between wall shear stress measures and porcine arterial endothelial permeability. Am J Physiol Heart Circ Physiol 286:1916–1922
88. LaMack JA, Himburg HA, Li XM, Friedman MH (2005) Interaction of wall ShearStress magnitude and gradient in the prediction of arterial macromolecular permeability. Ann Biomed Eng 33:457–464
89. Riles T, Lieberman A, Kopelman I, Imparato AM (1981) Symptoms, stenosis, and bruit: interrelationships in carotid artery disease. Arch Surg 116(2):218–220
90. Gori F et al (2009) Image-based computational fluid dynamics in a carotid artery. In: ASME International Mechanical Engineering Congress and Exposition, Proceedings, vol 2, pp 123–128
91. Roubin GS (1999) The status of carotid stenting. AJNR Am J Neuroradiol 20:1378–1381
92. Bergeron P, De Chaumaray T, Gay J, Douillez V (2003) Endovascular treatment of thoracic aortic aneurysms. J Cardiovasc Surg 44(3):349–361
93. Muller-Hullsbeck S, Schafer PJ, Charalambous N, Schaffner SR, Heller M, Jahnke T (2009) Comparison of carotid stents: an in-vitro experiment focusing on stent Desig. J Endovasc Ther 16(2):168–177
94. Farb A, Weber DK, Kolodgie FD, Burke AP, Virmani R (2002) Morphological predictors of restenosis after coronary stenting in humans. Circulation 105:2974–2980
95. He X, Ku DN (1996) Pulsatile flow in the human left coronary artery bifurcation: average conditions. J Biomech Eng 118(1):74–82

Mattia Amitrano Born in 1982, Mattia Amitrano attended university with great passion, graduating with full marks. During his last years in university, he developed a passion for sailing. After finishing university, he started working for a company dealing with the maintenance of electromedical equipment. After about a year, he quit to embark on a round-the-world trip by boat. After returning to Italy following 2 years of travel, he resumed work with the same company, taking a position as a project supervisor, directly in the hospital. After just over a year, he changed companies, moving to a much larger entity and choosing to be the head technician to gain more practical experience. During this period, he also developed a passion for the mountains, particularly for snowboard mountaineering. After 5 years and following a personal growth journey based on rebirthing techniques, he decided to change his life. He, therefore, resigned from his permanent contract, choosing to collaborate occasionally with the same company, with which he still collaborates. In this collaboration, which has lasted several years, he has done various types of work, thus enhancing his curriculum. He moved near Livorno and lived for 3 years in a yurt, continuing his personal growth journey and where his first child was born. He currently lives in a farmhouse in the Umbrian countryside with his family and two children.

Andrea Boghi The Doctor of Research *(Ph. D.)*, Dr. Eng. Andrea Boghi, is the director of *Computational Science Ltd.*, a consultancy company in the field of software development and data analysis, which counts among its clients the British Air Traffic Agency *(NATS)* and the Bank of England, starting from 2018. He is the author of more than 30 international scientific publications indexed on the major scientific databases. For more than a decade, he has worked in academia in various European institutions, dealing with the modelling and simulation of transport phenomena, both for basic research and for industrial and biological applications. His academic journey began at the University of Rome "Tor Vergata," where, alongside research on the modelling and simulation of turbulent flows with variable laminar diffusivity and blood flow in arteries and *stents*, he also served as a contract lecturer for the courses of Technical Physics, Thermo-Fluid-Dynamics of Biological Systems and Numerical Calculation of Thermo-Fluid-Dynamic Systems (2006–2010). In the two-year period from 2010 to 2012, he was a contract researcher at the Department of Energy and Technology of the University of *Southampton* in the United Kingdom, where he dealt with modelling and simulation of bi-phase flows, characterised by spinodal decomposition. In 2012, he went to France to the Institute of Fluid Mechanics of Toulouse as a contract researcher for the OTE (Waves, Turbulence and Environment) research group, where he dealt with modelling and simulation of river and atmospheric fluid dynamics. He terminated his contract in 2013 to go to the University of *Cranfield* in the United Kingdom, where he was appointed *Senior Research Fellow*, becoming part of the academic staff. Here, for the following 5 years (2013–2018), he dealt with modelling and simulation of transport phenomena for various departments: Energy, Hydraulics, Environment and Agriculture, where he respectively dealt with transport of crude oil in pipelines, disinfection of tanks by chlorination and transport and diffusion of organic and inorganic material in soils, with associated absorption by plants. At the University of *Cranfield,* he also handled the course on modelling of atmospheric emissions in the two-year period from 2016 to 2017. Before leaving academia, he went again to the University of *Southampton* in 2018, where he dealt with the three-dimensional modelling of the transport of nitrates and carbonates at the air–water interface in soils. Throughout his period abroad, he maintained his collaboration with Fabio Gori Ammannati, supervising some students and collaborating on the production of scientific articles. In 2020, he obtained the national scientific qualification for the functions of associate university professor in the competitive sector of Technical Physics and Nuclear Engineering.

Fabio Gori Ammannati was born in Montale (Pistoia) on 5 August 1947 to Emilio Gori and Cesarina Ammannati. On 16 July 2021, he added his mother's surname, Ammannati. He graduated in Chemical Engineering on 11 November 1971 with honours. In December 1971, he won a scholarship for young graduates at the Faculty of Engineering in Bologna, which he undertook from January 1972. In 1974, he became assistant professor and, in 1982, associate professor of

Technical Physics at the Faculty of Engineering of the University of Florence, where he taught, also as a contract professor, until 1990. In 1978, he won a scholarship from the *British Council,* which he carried out at the *Imperial College* in London, where he collaborated with *Prof. D. Brian Spalding* on the topic of numerical analysis in the turbulent flow of liquid metals. During his time at the University of Florence, he established international scientific collaborations with *Prof. R. Echigo, Tokyo Institute of Technology, Tokyo, Japan; Prof. D.R. Chaudhary, Department of Physics, University of Rajasthan, Jaipur, India;* International Centre for Theoretical Physics *(ICTP)* in Trieste; *Centralny Osrodek Techniki Medycznej, Warsaw, Poland and Prof. T. Aihara, Institute of Fluid Science, Sendai, Japan.* In 1986, he won a CNR scholarship, which he carried out at *Cornell University, Ithaca, New York,* where he collaborated with *Prof. R. Miller* on the topic of ground freezing. In the same year, he became professor in the Faculty of Engineering at the University of Reggio Calabria. In 1988, he was appointed as *professor* at the *University of New York at Stony Brook,* teaching the course, *Introduction to Fluid Dynamics,* in the autumn semester of the same year. During his stay, he collaborated with *Prof. T.F. Irvine Jr.* on the method of measuring thermal conductivity with a thermal probe and on the measurement of the isobaric thermal expansion coefficient of non-Newtonian fluids. In 1990, he moved to the Milano Polytechnic, where he taught Technical Physics and Systems until 1993. From 1991 until 1998, he was an adjunct professor at the Faculty of Engineering of the University of Siena. From 1992 to 2017, he was a professor of Technical Physics in the Faculty of Engineering at the University of Rome Tor Vergata, and from 2017, he is a contract professor. In 1994, he proposed the Research Doctorate (Italian *Ph.D.*) in Energy-Environment Engineering, of which he was the coordinator until 2011. In 1995, he proposed and directed the second-level Master's in Thermo-fluid-dynamics until 2016. From 1998 to 2001, he was the director of the Department of Mechanical Engineering. In 2001, he proposed and coordinated the programme, *Master of Science, Energy Engineering and Thermal and Fluid Dynamics,* in collaboration with the *Department of Mechanical Engineering, College of Engineering, University of Illinois at Chicago, USA,* which takes place at the University of Rome Tor Vergata but awards the title of *Master of Science* from the same American University. Since the late 90s, he has been a coordinator of the joint *Ph.D.* research programme with *Prof. R.J. Goldstein, Department of Mechanical Engineering, University of Minnesota, Minneapolis, USA,* and with *Prof. J.P. Hartnett, Prof. L. Kennedy, Prof. W. Minkowycz and Prof. W.M. Worek, Department of Mechanical Engineering, College of Engineering, University of Illinois at Chicago, Chicago, USA.* Since the mid-90s, he has proposed and directed the *Socrates–Erasmus* programme with *Prof. Mayinger, Technical University of Munich, Germany; Prof. van Steenhoven, Eindhoven Technical University, the Netherlands and Prof. C. Caro, Imperial College of Science, Technology and Medicine.* During his stay at Tor Vergata, he established scientific collaborations with the *Department of Mechanical Engineering, University of Minnesota, Minneapolis, USA,* collaborating with *Prof. R.J. Goldstein* on Thermo-fluid-dynamics and mass transport in gas turbine blades; the *Energy Resources Center, University of Illinois at Chicago,* collaborating with *Prof. J.P. Hartnett, Prof. W. Minkowycz and Prof. W.M. Worek;* the *Department of Mechanical Engineering, College of Engineering, University of Illinois at Chicago, collaborating with Prof. L. Kennedy and Prof. W.M. Worek; Duke University, collaborating with Prof. A. Bejan and S. Mary's University, Halifax, Nova Scotia, Canada, collaborating with Prof. W. R. Tarnawski.* The bibliographic review of documents and citations, titled *PlosBiology Career,* places him within the top 2% of researchers in the field of *Mechanical Engineering and Transports.* Since 1992, he has been a tutor for over 30 PhD theses, supervisor for over 40 second-level university Master's theses, supervisor for over 70 degree theses (five-year, Specialist and Master's), supervisor for over ten *Master's degree in Mechanical Engineering* from the *University of Illinois at Chicago* and supervisor for over 20 theses *Erasmus-Socrates.*

Since the early 2000s, he has been a reviewer of international research projects on behalf of *Portuguese Science and Technology Foundation (FCT); Czech Science Foundation (GACR); European Science Foundation, Strasbourg, France,* on behalf of *FCT* and the *Shota Rustaveli National Science Foundation, Georgia.* Since 2017, the year of his retirement due to age limits, he has been a contract professor at the University of Rome Tor Vergata, where in the academic year 2024–2025, he has taught Technical Physics for the Master's Degree Course in Medical Engineering. M.D. Salvatore Mangiafico (Florence), scientific head of the NeuroVascular Base Camp, 2021, invited him to give a scientific presentation titled *"The engineering approach to the study of the Circle of Willis"* on 23 September 2021 at the Congress Centre, University of Rome La Sapienza, thus opening up a possible future collaboration.

Numerical Non-Newtonian Thermo-Haemo-Dynamics (THD) in Coronary Stents

Andrea Boghi, Ivan Di Venuta, and Fabio Gori Ammannati

1   Introduction

Cardiovascular diseases are the leading cause of death in the Western world, accounting for 50% of deaths occurring each year. Alongside the more traditionally known risk factors, there is strong evidence that haemo-dynamic factors are involved in the pathogenesis. Over the last decades, the use of computational fluid dynamics (*CFD*) to simulate blood flow has received increased attention because this technique can produce detailed three-dimensional (3D) information of the haemo-dynamics. The knowledge of local 3D flow fields can potentially support clinical diagnosis and treatment planning for several cardiovascular pathologies.

This chapter is based on [1–4]. The stenosis of the coronary arteries is a very widespread pathology, and the most common clinical practice is stent implantation. The failure of cardiovascular devices is due to haemo-dynamic factors, such as stagnation of blood flow and low wall shear stress (*WSS*) [5–10]. The stasis of blood flow and the detachment of the fluid promote hyperplasia of the tunica intima, atherosclerosis and thrombi formation. Therefore, a stent should prevent these adverse fluid dynamic conditions.

Both two- and three-dimensional *CFD* simulations, 2D and 3D, are performed in arteries restored by stents, mostly treating blood as a Newtonian fluid. Two-dimensional *CFD* studies are carried out to understand how these geometric features can change the haemo-dynamics, employing simple stent geometries to

A. Boghi
Computational Science Ltd., Southampton, UK
e-mail: a.boghi@computationalscience.co.uk

I. D. Venuta
Via Orazio Raimondo 43, Rome, Italy

F. G. Ammannati (✉)
University of Rome Tor Vergata, Rome, Italy
e-mail: gori@uniroma2.it

F. Gori Ammannati (ed.), *Thermo-Haemo-Dynamics in Medical Engineering*,
https://doi.org/10.1007/978-3-031-97214-0_6

investigate the influence on the blood flow of the mesh size [11] and the shape of the strut section [12]. LaDisa et al. [13, 14] performed *3D-CFD* simulations on different geometries and stent designs, analysing how the 3D geometric parameters can affect the distribution of plaque. He et al. [15] studied a realistic strut geometry, considering three geometric parameters and showed that the stent design is very important for blood flow. Duraiswamy et al. [16] provided a physiologic rating on the effects of stent geometry on platelet deposition. The localized platelet and the regions of deposition are dependent on flow convection, so arterial reaction to the stent can be modulated by altering the stent design. Gori et al. [17, 18] carried out *3D-CFD* simulations on commercial coronary stents, extrapolating different parameters strongly correlated with haemo-dynamics, suggesting design guidelines to avoid the failure of the implant. Blood flow in two carotid arteries, *SE-BMS*, under physiologic conditions and in an incompletely restored vessel lumen is investigated in [19].

Only a few studies investigate the effect of non-Newtonian behaviour on blood flow. Boghi [1–3] studied the non-Newtonian blood flow in a coronary and stenosed carotid artery. Seo et al. [20] investigated the influence of the stent design close to curvatures, assuming a Carreau non-Newtonian model. The study provides an understanding of the flow in the stent vicinity, suggesting strategies for the optimization of the stent in order to minimize the flow disturbance. Benard et al. [21] investigated numerically the blood flow in 3D steady, rigid-wall arteries, considering a non-Newtonian behaviour based on the Carreau–Yasuda relation. Non-Newtonian behaviour, based on the Carreau–Yasuda relation, in coronary stent flow seems to have important effects on the estimation of the wall shear stress (*WSS*) on the internal surface. Amblard et al. [22] analysed the phenomena of type I endo-leaks in a non-invasively stented abdominal aorta with Phan-Thien and Tanner models, derived from the rheology of polymer solutions. This study provides an evaluation of the stresses generated by the blood flow on the aorta wall. The artery is modelled as a cylinder with a rigid wall, and the blood is assumed to be an incompressible non-Newtonian fluid in laminar flow.

Non-Newtonian behaviour of the fluid is also investigated in turbulent flow in [23, 24]. Di Venuta [4] investigated the influence of the degree of stenosis on the non-Newtonian blood flow in a stented coronary artery. Non-Newtonian features are investigated in [25], and a 3D numerical simulation of a failed coronary stent implant is studied at different degrees of residual stenosis in [26, 27].

2 Medical Issues

2.1 Atherosclerosis

Atherosclerosis affects the elastic arteries, such as the aorta, and large-medium muscular arteries, such as coronaries. Despite its pathogenesis not yet being fully understood, various anatomical, physiological and behavioural risk factors, such as diabetes, advanced age and obesity [28], are identified. It is characterized by localized fibrosis thickening of the arterial walls, associated with infiltrated plaques of lipids that can also calcify and lead to loss of elasticity.

Lesions can occur in the sites of greater *WSS* as bending points and bifurcations. Endothelial damage causes the expression, by endothelial cells, of vascular cell adhesion molecules (*VCAMs*) that bind the monocytes, entering later into the sub-endothelial region. Even low-density lipoproteins (*LDL*) enter the vascular wall and are oxidized. Macrophages pick up the oxidized *LDL* and become the so-called foam cells, which, progressing at the site of the initial endothelial injury, raise the lipid streak, a characteristic feature of the atherosclerotic lesion, as shown in Fig. 1 of [29]. The oxidized *LDL* exerts a series of deleterious effects, including stimulation and release of cytokines and inhibition of nitric acid production.

The smooth muscle cells, close to the initial injury, are stimulated and migrate from the media to the intima [30], where they proliferate, deposit collagen and other matrix molecules and help increase the mass of the lesion. They accumulate even *T* cells, secreting one of the factors responsible for the transformation of monocytes into macrophages. Finally, the arterial wall, deformed and thickened, incorporates calcium and forms a crumbly plaque. The clot formation is favoured in areas of deformation of the vessels, caused by atherosclerotic plaques. Furthermore, calcified plaque may ulcerate and break, and further arterial damage promotes the formation of clots. So, the platelets can access sub-endothelial connective tissue, which is exposed, and form micro-thrombi. The platelets adhere to the connective tissue-free granules, whose constituents can penetrate the artery wall. The platelet factors interact with plasma components in the arterial wall and can stimulate muscle cell migration to the intima and regeneration of endothelial cells in an attempt to cover the exposed portion, which thickens rapidly due to the proliferation of smooth muscle cells and the formation of new connective tissue. Fragments of plaque, once broken, can tear away and form emboli. The presence of atherosclerotic plaque narrows the vessel and reduces blood flow.

2.2   Coronary Circulation

The coronary circulation supplies blood to the cardiac muscle, providing nourishment to the heart, and includes an extensive network of capillaries. The coronary arteries originate from the aortic root, above the right and left cusps of the aortic valve, respectively, representing the first branch of this vessel. Blood pressure is higher here than in the entire systemic circulation, ensuring a continuous blood flow, which is necessary for the metabolic demands of the cardiac muscle. The right coronary supplies blood to the right atrium and the electrical conductivity system of the heart, including the sino-atrial and atrio-ventricular nodes, and part of both ventricles [31].

The left coronary artery supplies blood to the left ventricle, left atrium and interventricular septum. The main factor responsible for myocardial perfusion is the aortic pressure. Changes in the aortic pressure generally result in parallel changes in coronary flow, partly due to changes in the perfusion pressure of the coronary arteries. However, to regulate the coronary blood flow, it is possible to change the arteriolar resistance as a result of increased metabolic activity of the heart that causes a reduction of the coronary resistance, while a reduction of cardiac metabolism causes an increase in these resistances. In addition to providing enough pressure to push the blood through the coronary vessels, the heart affects blood flow even in an action of squeezing exerted by the contraction of the myocardium on the same vessels. This squeeze is high on the left coronary during the first phase of systole when the blood flow is transiently stopped or reversed. The maximum flow in the left coronary artery occurs in the first phase of diastole when ventricles are relaxed and the extravascular compression on the coronary vessels is completely absent. After an initial period of reversal, in the first phase of systole, the flow in the left coronary artery follows the pressure of the aorta up to the first phase of diastole, when it suddenly increases and then slowly declines along with the aortic pressure, which is shrinking in the remaining part of the diastole. The trend of the flow in the right coronary artery is similar to the left coronary artery. However, because of the lower pressure developed during systole by the thin wall of the right ventricle, there is no reversal of blood flow in the first phase of systole [31].

2.3 Surgical Treatments

It is necessary to promptly restore the blood flow, reduced as a result of the occlusion of the coronary artery. In addition to a combination of pharmacological interventions and lifestyle changes, coronary stenosis is treated surgically. Nowadays, bypass surgery (which consists of taking a small stretch of another small artery or a peripheral vein and using it to create a bridge around the obstructed coronary artery) is very rare, and it is preferred to perform a minimally invasive treatment, for which a single plaque is removed with the aid of a catheter. In balloon angioplasty, an inflatable balloon is positioned at the end of the catheter. The expansion of the balloon produces "controlled injury" to the coronary artery wall. Freed from athero-sclerotic plaque, the vessel wall and the lumen increase. Once deflated the balloon, the lumen tends to restore the initial diameter (recoil) with a loss of 15–30% of the diameter obtained during inflation. Stents, with caps made of metal or polymer that help the lumen of the vessel remain lying, are used to provide scaffolding to the arterial wall, limiting the phenomenon of recoil, and adhere to the wall of the intimate dissections.

2.4 Stents

The adoption of intravascular stents to restore the lumen of stenotic vessels has brought great improvements in terms of quality of life in atherosclerotic patients. However, there are still some issues to be resolved. A risk, although infrequent, is thrombosis, which can be avoided by treating patients with anticoagulant therapy and employing biocompatible materials in stent construction. Nonetheless, even when thrombosis does not occur, the stent, recognized as a foreign body, starts to be covered with endothelial cells, whose uncontrolled proliferation can lead to restenosis of the vessel, returning to the starting point.

This problem can be tackled with drug-eluting stents (*DESs*), which show good clinical response with a low restenosis rate [32], but the same clinical outcome can be seen in the 12-month follow-up as with bare metal stents (*BMS*) and self-expanding bare metal stents (*SE-BMS*), as far as restenosis is concerned [33]. These devices are unable to repair the intima, damaged by the stent implantation procedure, which can be limited by slightly oversized *SE-BMS* [34]. Nonetheless, the residual stenosis after the implant, even if not significant in terms of reduction in flow rate, may represent a critical location for the haemo-dynamics. Moreover, there is good evidence that arterial injuries, caused by both balloons and stents, lead to an inflammatory response and activate a proliferative repair process, which can provoke luminal narrowing and in-stent restenosis [35, 36].

3 Thermo-Haemo-Dynamics (THD) of Non-Newtonian Blood

3.1 Model of Fluid

The blood is assumed to be an incompressible non-Newtonian fluid with $\rho = 1060\,\text{kg/m}$, flowing with the Reynolds number in the range from Re $= 122$ to 440. In a non-Newtonian fluid, the viscosity, a function of the shear rate, depends on the plasma viscosity with a complex relation with haematocrit. The non-Newtonian viscosity is modelled with the Casson relation [37]:

$$\mu(\dot{\gamma}) = \left(\mu_\infty + 2\sqrt{\frac{\tau_0 \mu_\infty}{\dot{\gamma}}} + \frac{\tau_0}{\dot{\gamma}} \right) \tag{1}$$

where τ_0 is the yield stress and μ_∞ is the asymptotic Newtonian viscosity, both functions of the blood haematocrit. The difficulty of using the Casson model in a numerical scheme lies in its divergence at the zero shear rate limit. To avoid this divergence, it used the regularization technique proposed in [38], obtaining:

$$\mu(\dot{\gamma}) = \left(\mu_\infty + 2\sqrt{\frac{\tau_0 \mu_\infty}{\dot{\gamma}}} \left(1 - e^{\sqrt{\frac{\dot{\gamma}}{\gamma_0}}} \right) + \frac{\tau_0}{\dot{\gamma}} \left(1 - e^{\sqrt{\frac{\dot{\gamma}}{\gamma_0}}} \right)^2 \right) \tag{2}$$

At zero shear rate, it is assumed that:

$$\mu(\dot{0}) = \left(\mu_\infty + 2\sqrt{\frac{\tau_0 \mu_\infty}{\dot{\gamma}_0}} + \frac{\tau_0}{\dot{\gamma}_0} \right) \tag{3}$$

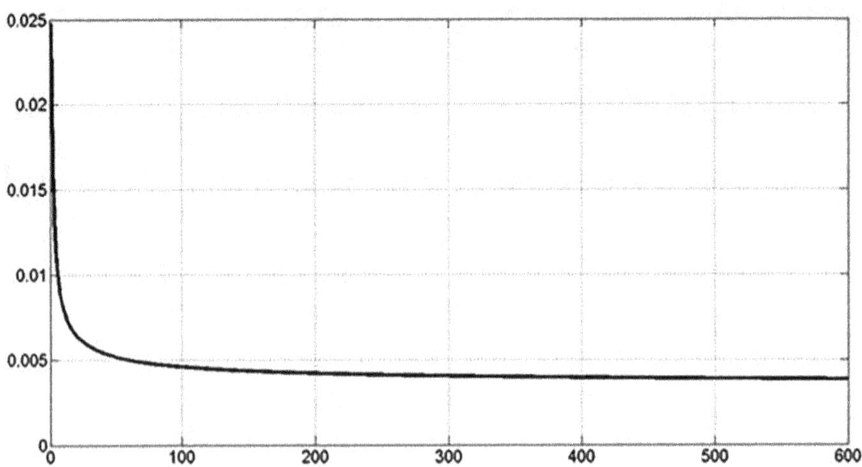

Fig. 1 Casson viscosity (Pa*s) versus shear rate (1/s), Fig. 1 of [27]

where the parameters are $\dot{\gamma}_0 = \frac{\tau_0}{\mu_0}\,s^{-1}$, $\tau_0 = 0.01$ Pa, $\mu_\infty = 0.00333$ Pa*s, $\mu_0 = 100\,\mu_\infty$, according to [24, 39].

The Casson viscosity model, introduced in the software with a *user-defined function (UDF)*, is evaluated as in the following and is represented in Fig. 1, [27].

3.1.1 List of the Code UDF for Casson Viscosity

```
#include "udf.h"
DEFINE_PROPERTY(cell_viscosity,c,t)
{
realmu_C;
realmu_inf;
real mu_0;
real S_0;
real tau_0;
real S=C_STRAIN_RATE_MAG(c,t);
tau_0=0.01;
mu_inf=0.00333;
mu_0=100*mu_inf;
S_0= (tau_0/mu_inf)/pow((sqrt(mu_0/mu_inf)-1.0),2.0);
if (S<=0.0)
mu_C=mu_inf*pow((1.0+sqrt(tau_0/(mu_inf*S_0))),2.0);
else
mu_C=mu_inf*pow((1.0+sqrt(tau_0*(1.0 -exp(-S/S_0))/
(mu_inf*S))),2.0);
returnmu_C;
}
```

3.2 Steady State

The THD equations are solved under the boundary conditions of no slip on the artery and stent walls, with a prescribed mass flow rate (constant or with a physiologic waveform) on the inlet and a constant pressure on the outlet. The two lateral surfaces of the cylinder portion need rotational periodic boundary conditions. The simulations are performed by using numerical software, which solves the fluid dynamics equations through the finite volume method. The SIMPLE algorithm is employed to solve the pressure–velocity coupling, and the simulations are carried out until the convergence is reached. A second-order implicit time-stepping method is used for the unsteady simulations, which run for two cardiac cycles (1.6 s), with a fixed time step of 8×10^{-4} s. The first cycle is used to initialize the solution, and the results of the second cycle are recorded. A steady-state velocity of 0.16 m/s is prescribed on the inlet, according to [31]. The associated Reynolds number is about 132, i.e. laminar flow, and a Casson velocity profile is prescribed on the inlet.

3.2.1 List of the Code UDF for Casson Velocity Profile and Results

A MATLAB program, implemented to solve, with the Newton–Raphson algorithm, the non-linear relationship between ξ_0 and the Casson number, is reported in the following.

```
#include "math.h"
DEFINE_PROFILE(inlet_x_velocity, thread, index)
{
real x[ND_ND]; /* this will hold the position vector */
real y;
real z;
realu_m;
real csi_0;
realcsi;
face_t f;
begin_f_loop(f, thread) /*loops over all faces in the thread passed in
the DEFINE macro argument*/
{
F_CENTROID(x,f,thread);
y=x[1];
z=x[2];
u_m=0.161;
csi_0=0.005112557661847;
csi = pow((pow(y,2)+pow(z,2)),0.5)/0.0013;
if (csi<=csi_0)
F_PROFILE(f,thread,index) = 2*u_m*(21-7*csi_0*csi_0+42*csi_0-
56*pow(csi_0,0.5))/(21-pow(csi_0,4)+28*csi_0-48*pow(csi_0,0.5));
else
F_PROFILE(f,thread,index) =2*u_m*(21*(1-pow(csi,2))-56*pow
```

```
(csi_0,0.5)*(1-pow(csi,3/2))+42*csi_0*(1-csi))/(21-pow(csi_0,4)
+28*csi_0-48*pow(csi_0,0.5));
}
end_f_loop(f,thread)
}
```

3.2.2 Velocity Profiles

The flow profile is given by

$$u(\xi) = 2u_m \frac{\left(21\left(1 - \xi^2\right) - 56\xi_0\left(1 - \xi^{\frac{3}{2}}\right) + 42\xi_0(1 - \xi)\right)}{\left(21 - 48\sqrt{\xi_0} + 28\xi_0 - \xi_0^4\right)} \tag{4}$$

where $\xi_0 = \frac{R_0}{R}$, $\xi = \frac{r}{R}$, with R radius of the cylinder, r radial coordinate, R_0 radius where the profile assumes a constant value. The profile implemented with a *UDF* is

$$u(\xi_0) = 2u_m \frac{\left(21 - 7\xi_0^2 + 42\xi_0 - 56\sqrt{\xi_0}\right)}{\left(21 - 48\sqrt{\xi_0} + 28\xi_0 - \xi_0^4\right)} \tag{5}$$

The Casson number, Ca, is defined as

$$Ca = \frac{D\tau_0}{\mu_\infty u_m} \tag{6}$$

with D diameter of the pipe, u_m mean velocity and the following relation with ξ_0

$$Ca = \frac{168\xi_0}{\left(21 - 48\sqrt{\xi_0} + 28\xi_0 - \xi_0^4\right)} \tag{7}$$

Figure 2 shows the non-linear relationship between Ca and ξ_0, while Fig. 3 presents the relation with the inlet velocity profile; for the highest Ca numbers, the non-Newtonian character is more pronounced.

3.3 Unsteady State

The pulsatile (unsteady state) blood flow is modelled with the Womersley–Evans theory [40]

$$u(t,r) = 2u_0 \left(1 - \left(\frac{r}{R}\right)^2\right) + 2\,\mathrm{Re}\left(\sum_n U_n \Phi(\tau_n, r)e^{j\omega_n t}\right) \tag{8}$$

with the relation

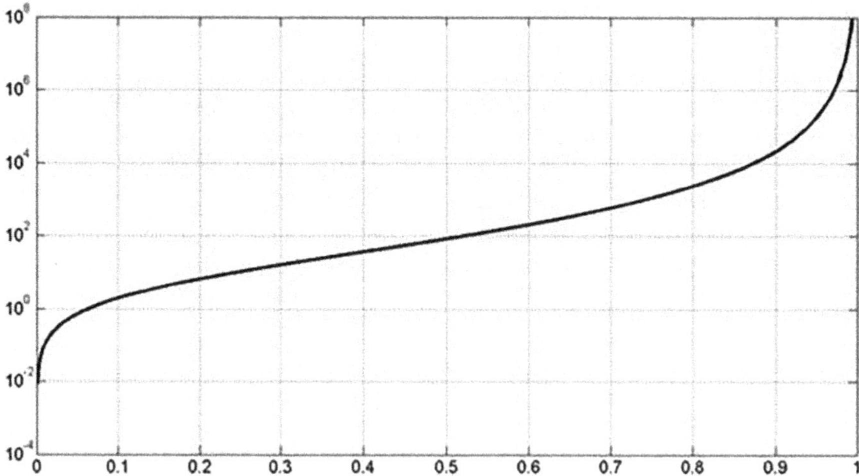

Fig. 2 Casson number, Ca, versus ξ_0 [4]

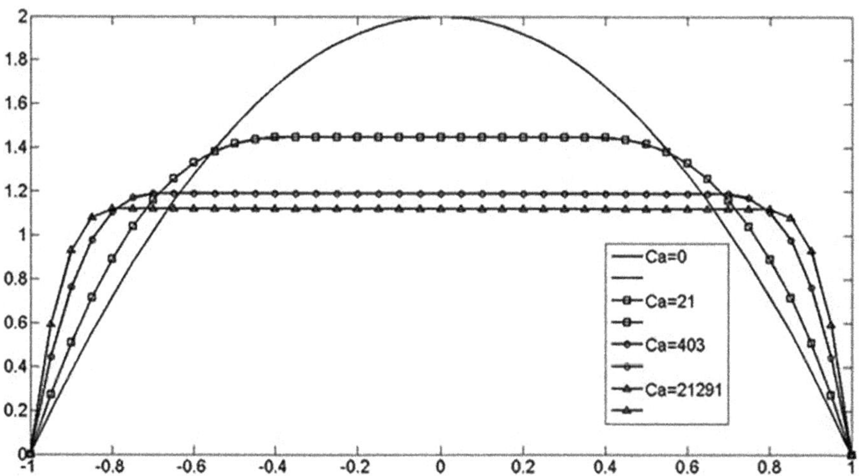

Fig. 3 Inlet velocity profile, u/u_m, versus ξ at various Ca numbers [4]

$$\Phi(\tau_n, r) = \frac{J_0(\tau_n) - J_0(\tau_n r/R)}{J_0(\tau_n) - 2J_1(\tau_n)/\tau_n} \tag{9}$$

and

$$\tau_n = j^{\frac{3}{2}} R \sqrt{\frac{\rho}{\mu}} \omega_n = j^{\frac{3}{2}} \alpha_n \tag{10}$$

where J_0 and J_1 are the 0th and first-order Bessel functions of the first kind, α_n are the Womersley numbers of order n, Re is the real part of a complex number, $j = \sqrt{-1}$

and U_n are the Fourier coefficients of the pulsatile mean velocity profile. By using the *Fast Fourier Transform* (*FFT*) algorithm, the mean value and the first 15 harmonics are extracted and used to reconstruct the velocity profile.

The physiological waveform is reported in [31]. The wave period is set equal to 0.8 s, corresponding to a heart rate of 75 bpm. As far as velocity is concerned, the maximum value is 0.31 m/s, the minimum is −0.015 m/s and the mean is 0.16 m/s (the same value used for the steady-state simulations). The waveform can be defined by an analytical expression, to be used in the simulations, by using a program developed in *MATLAB* capable of sampling the image of the waveform, Figs. 4 and 5, calculating the *FFT*, and extracting the Fourier coefficients.

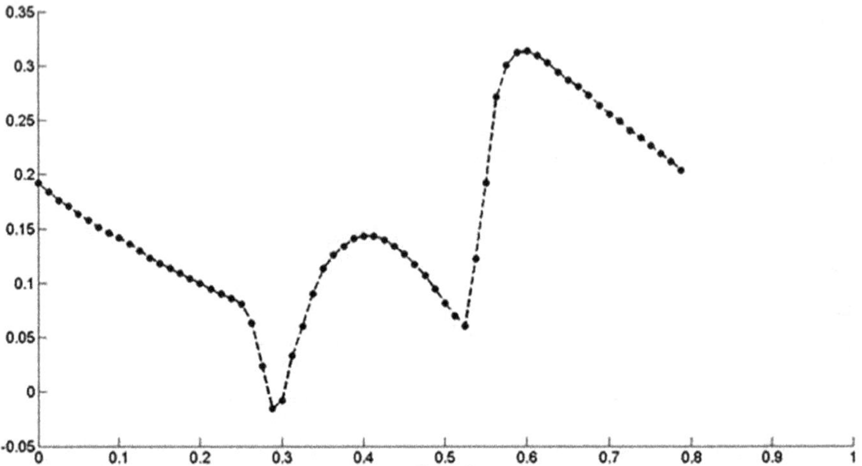

Fig. 4 Sampled velocity (m/s) versus time (s) [4]

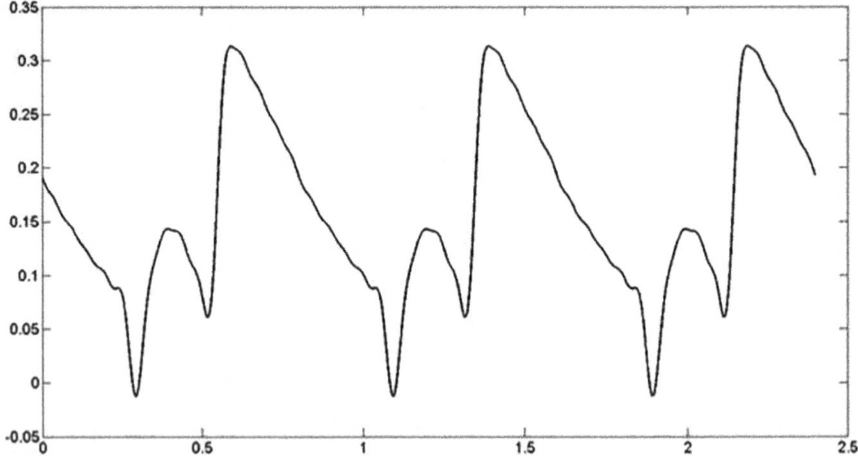

Fig. 5 Reconstructed velocity (m/s) versus time (s), in three periods [4]

3.3.1 List of the Code UDF for Womersley Velocity Profile

```c
/*womer-sine.c */
#include "udf.h"
#include <math.h>
/*** typedefs and defines you will need ***/
typedefstruct DCOMPLEX {double r,i;} dcomplex;
#ifndef PI
#define PI 3.1415926535
#endif
doublewomervel(intnfour, double *ccoef, double *scoef, double r,
double t);
double Cabs();
dcomplexCadd();
dcomplexCsub();
dcomplexCmul();
dcomplex Complex();
dcomplexCdiv();
dcomplexRCmul();
dcomplexzbes(int n, dcomplex y);
DEFINE_PROFILE(womersley, thread, position)
{
double x[ND_ND];
double a, r, y, z, t;
face_t f;
double ccoef[20] = {0.316109689495572,0.058844996267599,
-0.013581316692307,-0.031411468789966,0.035556212583219,
-0.012422133319352,-0.009706892736523,0.014263581935598,
-0.005711495430855,-0.003226304377924,0.005154898893547,
-0.002802474055863,-0.0009095752483694909,0.001968692504238,
-0.001088095307906,-0.0002240779637738993};
double scoef[20] = {0,-0.080458242624169,
-0.014632777907363,0.016040824829638,0.0004496917123870107,
-0.008257259391017,-0.00008615172106660450,
0.006796664615692,-0.006229915026003,
-0.001356935990269,0.005752810824752,-0.004185914111868,
-0.0007209062792056274,0.003043882126720,
-0.002587650749845,-0.0004307136230139006};
begin_f_loop(f, thread)
{
real q = RP_Get_Real("flow-time");
t = q/0.8;
F_CENTROID(x, f, thread);
y = x[1];
z = x[2];
a = sqrt(z*z + y*y);
```

```c
r = a/0.0013;
F_PROFILE(f, thread, position) =womervel(15, ccoef, scoef, r, t);
}
end_f_loop(f, thread)
}
/************************************************************
* returns velocity given:
* nfour = number of fourier coefficients
* ccoef = vector of n=0..nfour cosine fourier coefficients
* scoef = vector of n=0..nfour sine fourier coefficients
* r = radial coordinate/radius
* t = time
************************************************************/
doublewomervel(intnfour, double *ccoef, double *scoef, double r,
double t)
{
doublevel;
doublekt;
int k;
vel = ccoef[0]*(1-r*r);
for (k=1;k<=nfour;k+ +) {
dcomplex lambda, zexp, zvel, phi;
lambda=RCmul(0.0013*sqrt(1060*PI*k/(0.00333*0.8)),Complex(-1,1));
kt = 2.0*PI*k*t;
phi=Cdiv(Csub(zbes(0,lambda),zbes(0,RCmul(r,lambda))),Csub(zbes
(0,lambda),Cdiv(RCmul(2,zbes(1,lambda)),lambda)));
zexp = Complex(cos(kt),sin(kt));
zvel=Cmul(Complex(ccoef[k],-scoef[k]),Cmul(phi,zexp));
vel +=zvel.r;
}
returnvel;
}
/* the following routines are from Numerical Recipes in C */
dcomplexCadd(a,b)
dcomplexa,b;
{ dcomplex c;
c.r=a.r+b.r;
c.i=a.i+b.i;
return c;
}
dcomplexCsub(a,b)
dcomplexa,b;
{ dcomplex c;
c.r=a.r-b.r;
c.i=a.i-b.i;
```

```
return c;
}
dcomplexCmul(a,b)
dcomplexa,b;
{ dcomplex c;
c.r=a.r*b.r-a.i*b.i;
c.i=a.i*b.r+a.r*b.i;
return c;
}
dcomplex Complex(re,im)
doublere,im;
{ dcomplex c;
c.r=re;
c.i=im;
return c;
}
dcomplexCdiv(a,b)
dcomplexa,b;
{ dcomplex c;
doubler,den;
if (fabs(b.r) >= fabs(b.i)) {
r=b.i/b.r;
den=b.r+r*b.i;
c.r=(a.r+r*a.i)/den;
c.i=(a.i-r*a.r)/den;
} else {
r=b.r/b.i;
den=b.i+r*b.r;
c.r=(a.r*r+a.i)/den;
c.i=(a.i*r-a.r)/den;
}
return c;
}
double Cabs(z)
dcomplex z;
{ doublex,y,ans,temp;
x=fabs(z.r);
y=fabs(z.i);
if (x == 0.0)
ans=y;
else if (y == 0.0)
ans=x;
```

```
else if (x > y) {
temp=y/x;
ans=x*sqrt(1.0+temp*temp);
} else {
temp=x/y;
ans=y*sqrt(1.0+temp*temp);
}
returnans;
}
dcomplexRCmul(x,a)
double x;
dcomplex a;
{ dcomplex c;
c.r=x*a.r;
c.i=x*a.i;
return c;
}
/* this is something we wrote to compute the nth order Bessel
 * function given a complex argument */
dcomplexzbes(int n, dcomplex y)
{
dcomplexz,zarg,zbes;
inti;
zarg = RCmul(-0.25,Cmul(y,y));
z = Complex(1.0,0.0);
zbes = Complex(1.0,0.0);
i = 1;
while (Cabs(z)>1e-20 && i<=10000) {
z = Cmul(z,RCmul(1.0/i/(i+n),zarg));
if (Cabs(z)<=1.e-20) break;
zbes = Cadd(zbes,z);
i++;
}
zarg = RCmul(0.5,y);
for (i=1;i<=n;i++) zbes = Cmul(zbes,zarg);
returnzbes;
}
```

3.3.2　Womersley Velocity Profile

The pulse profile is implemented with a UDF to be used in the unsteady-state simulations. Womersley velocity profiles, prescribed at the inlet at different times, are reported in Fig. 6.

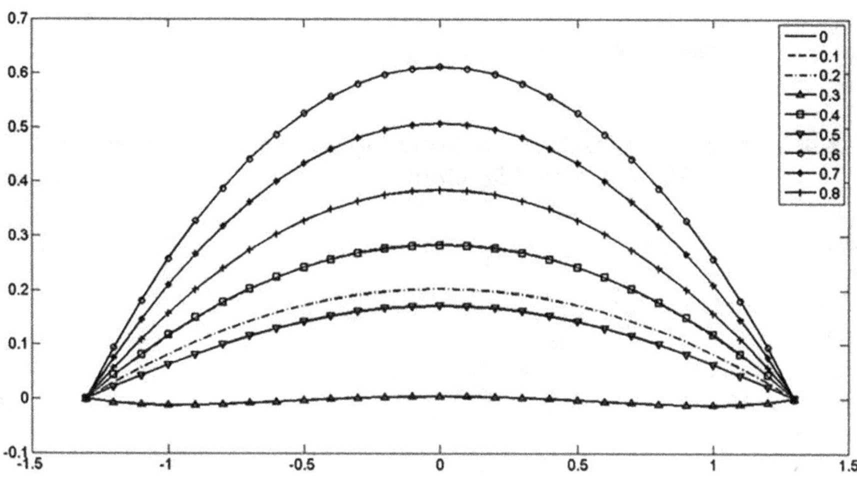

Fig. 6 Womersley velocity (m/s) profile versus radius (mm) at the inlet [4]

3.3.3 List of MATLAB Code for the Newton–Raphson Method

```
clear all
close all
clc
% Newton-Raphson method for solving non-linear equations.
%
% Calculating the zero of the function in the file fun.m
% with the iterative Newton-Raphson method performed by the function
newton.m
%INPUT
x0=0.05;
%OUTPUT
[zero]=newton(@fun, x0)
function [x] = newton(fun, x0)
cont_iter = 1;
x(cont_iter)=x0;
max_n_iter = 100;
whilecont_iter<max_n_iter
[nfun,dfun] = fun(x(cont_iter));
x(cont_iter+1)=x(cont_iter)-nfun/dfun;
diff=(x(cont_iter+1)-x(cont_iter));
cont_iter= cont_iter +1;
end
end
function [nfun,dfun] = fun(x)
Ca=21291;
nfun = 168*x/(Ca*(21-48*sqrt(x)+28*x-x^4))-1;
```

```
dfun= (168*Ca*(21-48*sqrt(x)+28*x-x^4)-168*x*Ca*(-24*x^(-1/2)+28-
4*x^3))/(((Ca*(21-48*sqrt(x)+28*x-x^4))^2));
return
```

3.3.4   List of MATLAB Code for Sampling and Reconstruction of the Physiological Waveform

```
close all
clear all
clc
name= 'waveform.jpg';
N_i=65;
minimus=-5/pi/(0.0013)^2/60/10^6;
maximus=100/pi/(0.0013)^2/60/10^6;
T=0.8;
N=N_i-1;
t_i=0:(T/N):T;
t=t_i(1,1:N);
x_i=sampling(name,N_i,minimus,maximus);
x_i=x_i';
x=x_i(1,1:N);
figure(1)
hold on
plot(t,x,'k--
o','LineWidth',2,'MarkerEdgeColor','b','MarkerFaceColor','b');
xlim([0 1])
xlabel('time [s]','FontSize',16,'FontWeight','bold','Color','k')
ylabel('velocity
[m/s]','FontSize',16,'FontWeight','bold','Color','k')
hold off
set (gca,'fontsize',14)
set (gcf,'color','w')
X=fft(T/N*x);
xf=zeros(1,N);
for n=1:N
for k=1:N
xf(n)=xf(n)+1/T*(X(k)*exp(1i*2*pi*(k-1)*(n-1)/N));
end
end
X=X(1,1:33);
a=real(2*X)/T;
b=-imag(2*X)/T;
fid=fopen('a.txt', 'wt');
fprintf(fid, '%12.8f',a);
fclose(fid);
fid=fopen('b.txt', 'wt');
```

```matlab
fprintf(fid, '%12.8f',b);
fclose(fid);
figure(2)
hold on
plot(t,xf,'o')
xlim([0 1])
xlabel('time [s]','FontSize',16,'FontWeight','bold','Color','k')
ylabel('velocity
[m/s]','FontSize',16,'FontWeight','bold','Color','k')
set(gca,'fontsize',14)
set(gcf,'color','w')
t=0:0.00001:3*T;
velocity=a(1)/2;
for k=2:16
velocity=velocity+(a(k)*cos(2*pi*(k-1)/T*t)+b(k)*sin(2*pi*(k-1)/
T*t));
end
figure(3)
plot(t,velocity,'linewidth',2)
xlabel('time [s]','FontSize',16,'FontWeight','bold','Color','k')
ylabel('velocity
[m/s]','FontSize',16,'FontWeight','bold','Color','k')
set(gca,'fontsize',14)
set(gcf,'color','w')
function y=sampling(name,n,minimum,maximum)
C=imread(name);
c=C(:,:,1);
[y,x]=find(c==0);
y=-y;
for j=1:max(size(x))
i=find(x==x(j));
if max(size(i))>=1
m=mean(y(i));
x(i(2:max(size(i))))=0;
y(i(2:max(size(i))))=0;
y(i(1))=m;
end
end
[x,IX]=sort(x);
y=y(IX);
i=find(x);
x=x(i);
y=y(i);
y=y(1:length(y)/n:length(y));
y=(y-(min(y)))*((maximum-minimum)/max(y-(min(y))))+minimum;
```

4   Thermo-Haemo-Dynamics (THD) of a Drug-Eluting Stent

The role of haemo-dynamic factors in the development of post-stenting restenosis in coronary arteries is investigated. A model of the artery, completely restored after stent implantation, is considered and, furthermore, three simplified models of the non-completely restored coronary artery, with different *degrees of residual stenosis* (*DOR* = 30%, 60% and 90% percentage of area reduction). The geometry is reconstructed as closely as possible to reality with the software *SOLIDWORKS*.

The geometry considered is relative to a patent of a drug-eluting stent [41], without considering the mass transport in the drug release. The simulations are performed with numerical software in steady and unsteady states. In the unsteady state, the physiological waveform in the coronary arteries is reconstructed with the software *MATLAB*. The post-processing is performed with the software *Tecplot*. Special attention is given to the stress acting on the wall, as it is believed to be responsible for restenosis.

4.1   Geometry and Mesh

The vessel wall is modelled as a cylinder; the wall and stent are assumed rigid because the goal is to focus on the general aspects of the stent geometry on the fluid dynamics. Besides, the presence of the stent reduces the deformability of the wall. As far as the fluid flow is concerned, laminar flow is assumed, in agreement with the literature, which is justified by the low velocity in the coronary arteries. The size of the stent is based on the data in the literature. The cylinder, representing the vessel, has a diameter equal to 2.6×10^{-3}m, according to the dimension of a coronary vessel [13], the thickness of the stent is 10^{-4}m, according to the thickness of coronary stents [42]. The length of the stent is about 15×10^{-3}m.

The more complicated part of the stent modelling is to wrap the sketch around a cylinder and to give thickness to the surface. The geometries of the four stents, without restenosis and with *DOR* 30%, 60% and 90%, are shown in Fig. 7, [4].

The second part is to subtract the stent from a solid cylinder of the same radius to obtain the fluid part. Given the symmetry of the geometry, it is sufficient to carry out simulations with a portion of a circumference of 90 degrees. To simulate the residual stenosis, three geometries, shown in Fig. 8, [4], are made with a choking in the middle section, with a radius of 1.09×10^{-3}m, 0.82×10^{-3}m and 0.45×10^{-3}m, respectively. They correspond to *DOR* 30%, 60% and 90%, given by

$$P(\%) = 100 \left(1 - \left(\frac{R_1}{R_2} \right)^2 \right) \tag{11}$$

In the latter geometry, the simple cylinder is longer at the outlet compared to the previous cases because there may be reverse flow due to the strong necking. Given the special geometry, a non-uniform mesh is used because the stent is much thinner than the diameter of the vessel, and a very dense mesh near the wall is employed. At the centre of the vessel, a thin mesh is not necessary as it is sufficiently far away from the stent. The size of the mesh increases gradually from the vessel wall. The periodicity of the vessel is set around the axis of the two periodic faces.

Three simulations are carried out to evaluate the most convenient size of the mesh, which are shown in Fig. 9, [26], with 8×10^5 elements (a), with 1.2×10^6 elements (b) and with 2.3×10^6 elements. The final choice of the mesh is that with 1.2×10^6 elements.

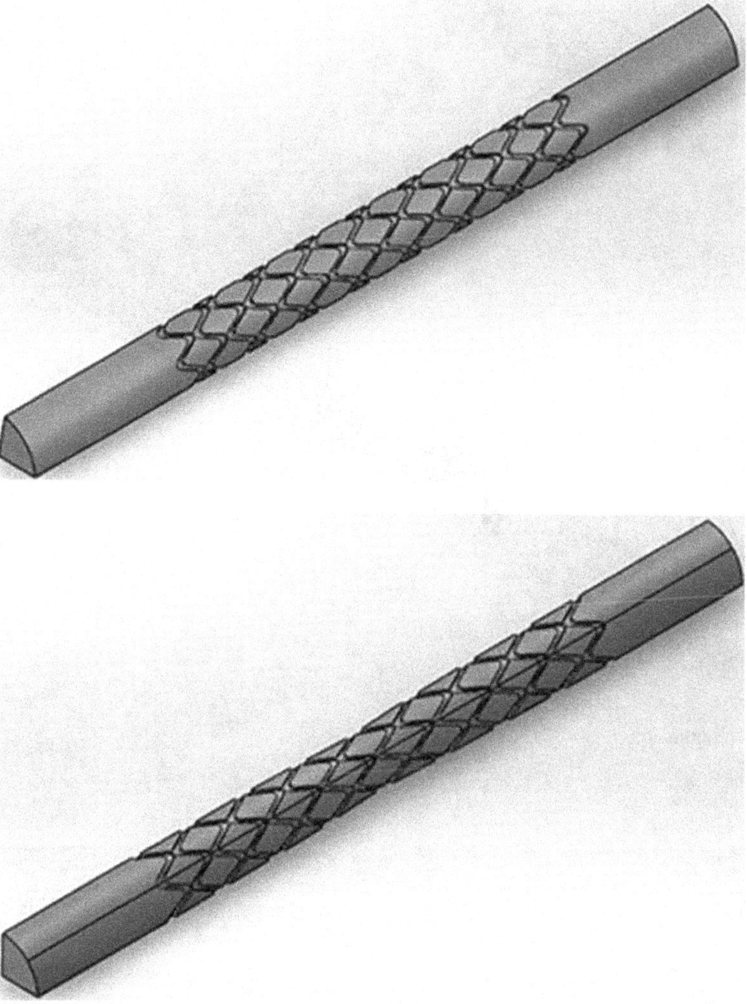

Fig. 7 Geometry of the stent. Top: without residual stenosis, DOR $=$ 0%; second from top: DOR $=$ 30%; second from bottom: DOR $=$ 60%; bottom: DOR $=$ 90% [4]

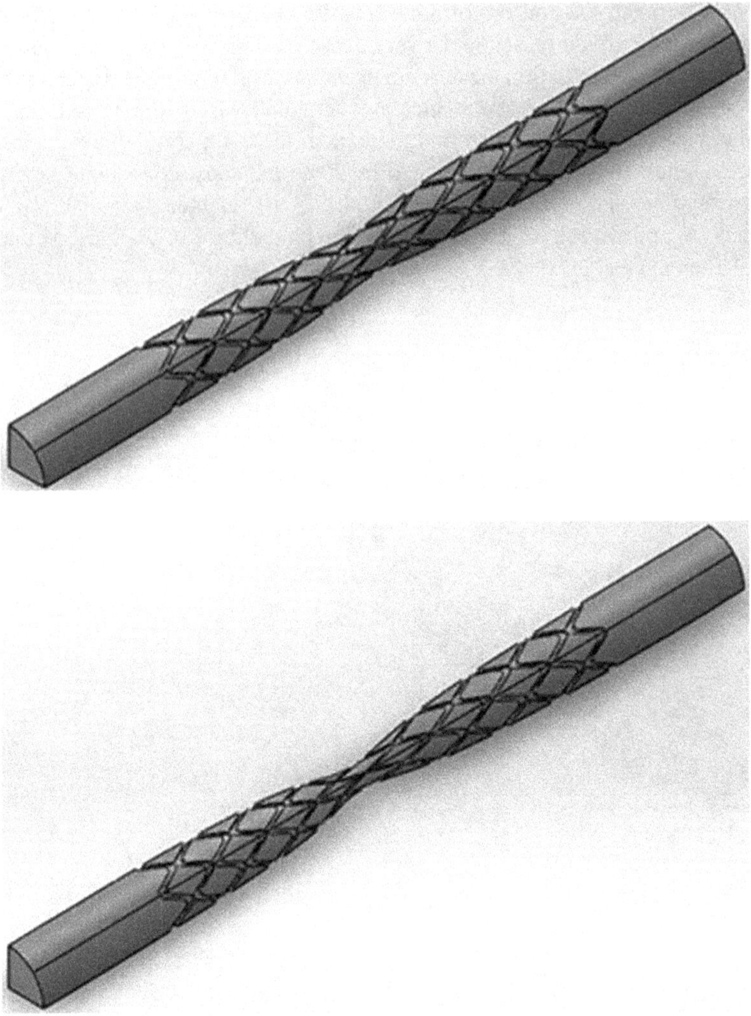

Fig. 7 (continued)

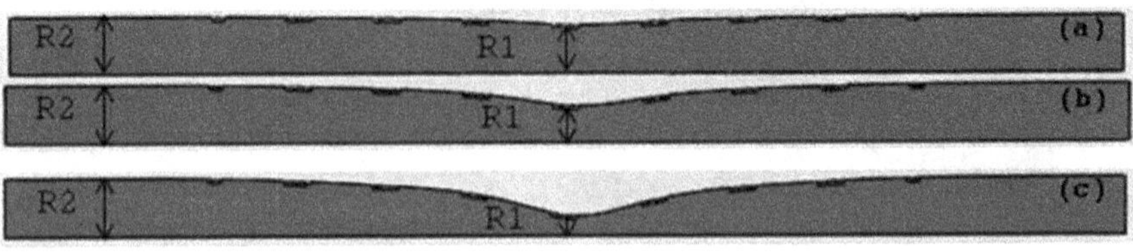

Fig. 8 The three geometries with residual stenosis: (**a**) DOR = 30%, (**b**) DOR = 60%, (**c**) DOR = 90%, [4]

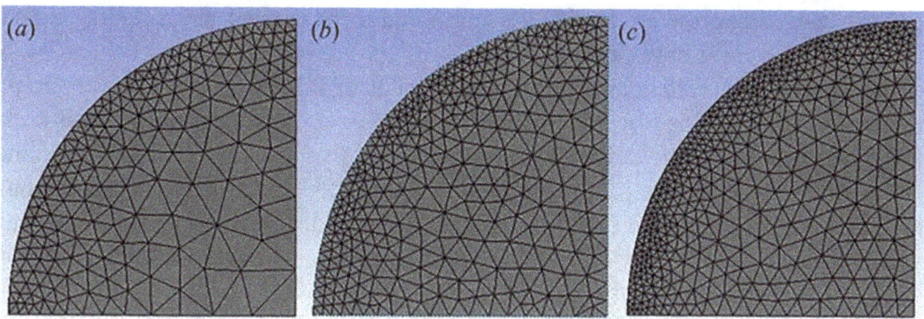

Fig. 9 Computational grid with: (**a**) 8×10^5 elements; (**b**) 1.2×10^6 elements; (**c**) 2.3×10^6 elements, Fig. 2 of [26]

4.2   Results

4.2.1   Grid Independence

The results of WSS (dyn/cm^2) versus the axial length (m), averaged over the vessel perimeter, are shown in Fig. 10, [26], for the four geometries. The results on the top are for $DOR = 0\%$, those on the second from the top for $DOR = 30\%$, those on the second from the bottom for $DOR = 60\%$, and the ones on the bottom for $DOR = 90\%$. The solutions are validated against three different grids. More than 300 slices are extracted by intersecting the vessel wall with parallel axial planes, and the resulting data are interpolated by splines. To show the mesh independence from WSS, the steady-state simulations are performed for each stent with three different meshes, made of approximately 800 k, 1200 k and 2300 k elements, respectively. Since the last two meshes offer similar results, the intermediate mesh, with 1200 k, is used for the unsteady-state simulations as a compromise between accuracy and computation time.

The top chart of Fig. 10, [26], for $DOR = 0\%$, shows a repeating pattern for WSS. Peaks and valleys correspond to slices intersecting mostly the intra-strut area (where WSS is higher) and those intersecting regions of low WSS, located on the struts. WSS depends on the symmetric part of the velocity gradient, and the velocity strongly diminishes in proximity to the struts. This is due to the deceleration of the fluid impinging on an obstacle. When the blood flow meets the strut, the streamlines, originally adhering to the wall, detach at a certain distance. A recirculation region forms between the strut wall and the detachment point, whose amplitude depends on the Reynolds number. The presence of these recirculation regions increases the residence time of macrophages and macromolecules on the endothelium.

The other charts of Fig. 10, [26] present the results for the average WSS in the presence of residual stenosis. The reduction of the diameter provokes a pressure drop and a consequent acceleration of the fluid. Since WSS is, as a first approximation, inversely proportional to the section diameter and directly proportional to the fluid speed, it increases in the restricted section. Downstream of this region, the diameter increases and WSS decreases. If the slope of the wall, downstream of the restriction, is sufficiently high, the streamline detaches, and a consequent recirculation region appears, which contributes to lowering the *WSS*, besides the geometrical changes. In all geometries with restenosis, the maximum value of *WSS* is registered upstream of the maximum shrinkage of the section (in the middle of the vessel) because both axial and transversal velocities change more rapidly upstream of the maximum height of the stenosis. The stenosis has a significant effect in terms of *WSS* reduction, and, in the geometries with $DOR = 60\%$ and 90%, it causes a drop below the critical value of 5 dyne/cm^2 (0.5 Pa). This takes place downstream and is due to the presence

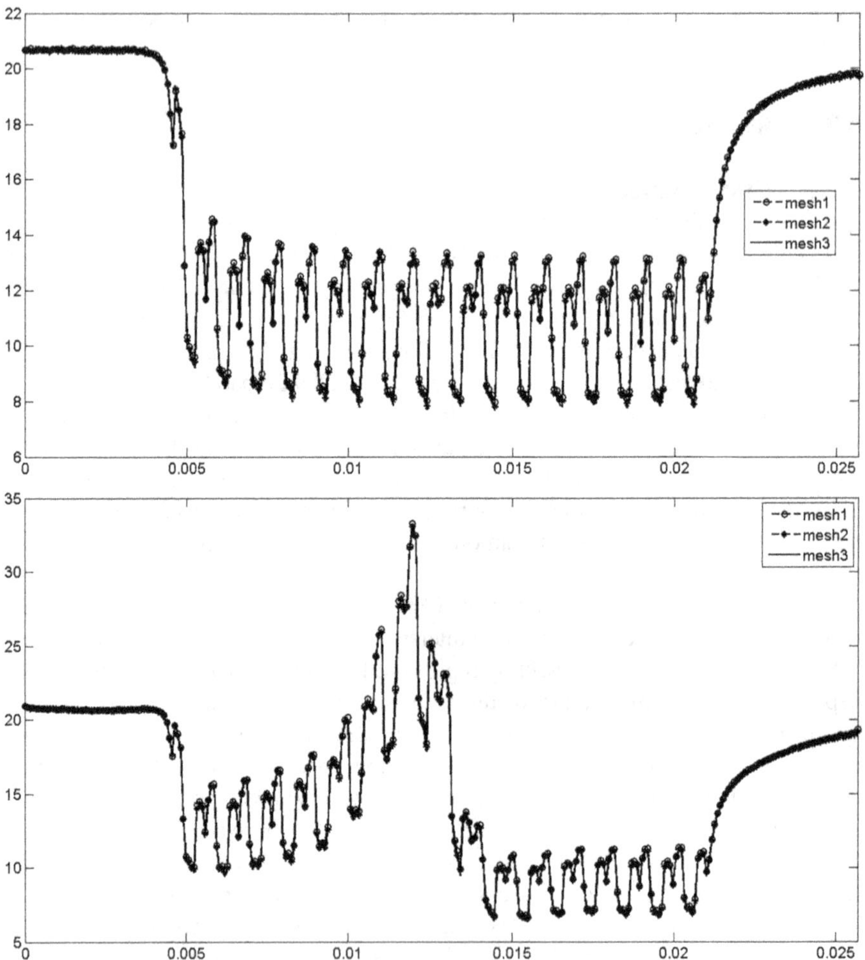

Fig. 10 Average WSS (dyn/cm^2) versus the axial length (m) for the three meshes: top, DOR $= 0\%$; second from top, DOR $= 30\%$; second from bottom, DOR $= 60\%$; bottom, DOR $= 90\%$, Fig. 4 of [26]

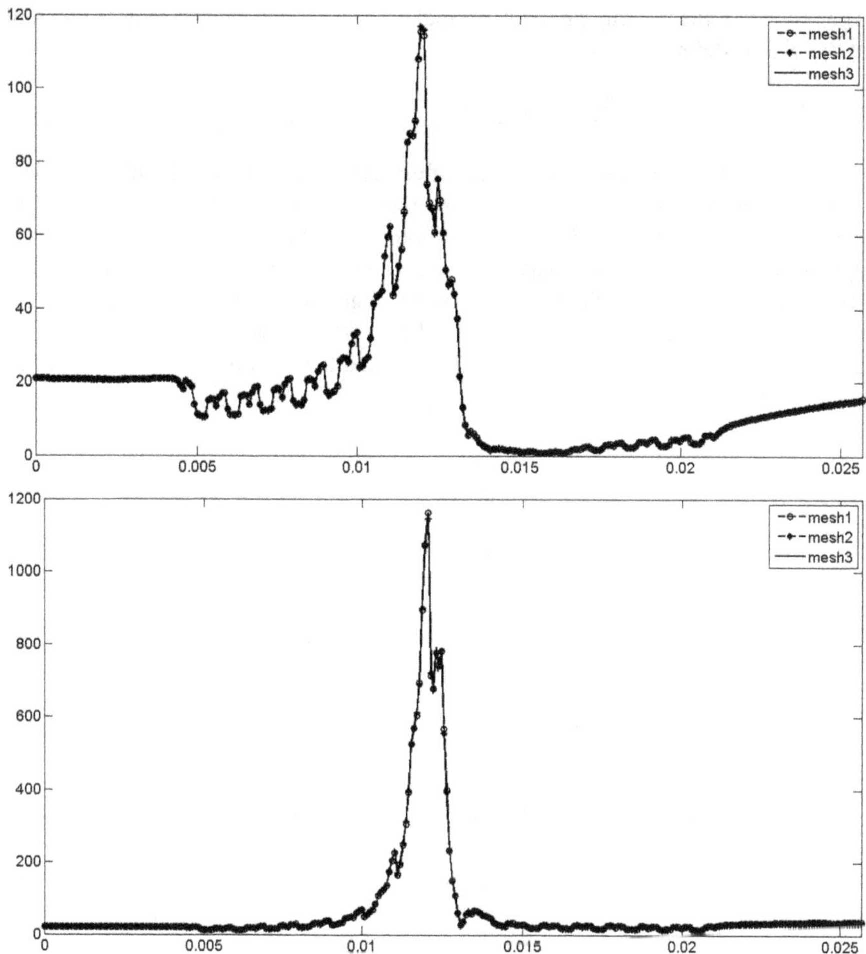

Fig. 10 (continued)

of a stagnation region beyond the stenosis. The narrowing of the section has the effect of amplifying the *WSS* in the central region, and still, with $DOR = 60\%$ and 90%, the averaged *WSS* exceeds 70 dyne/cm^2 (7 Pa), a critical value above which the thrombogenic effects may take place.

4.2.2 Steady State

In the steady state, the iterations are carried out until the highest value of the residuals is 10^{-6}. The results obtained are exported to *Tecplot.case format* to be analysed and displayed. It is also possible to do post-processing with the definition of new variables. This work is particularly interested in the *WSS*, which is the tangential drag force produced by blood moving across the endothelial surface.

The product of the velocity gradient, normal to the wall, and the viscosity of the fluid at the wall is defined as

$$\mathrm{WSS}_i = 2\mu\left(S_{ij} - \left(\widehat{n}_p S_{pq} \widehat{n}_q\right)\delta_{ij}\right)\widehat{n}_j \tag{12}$$

The unit of measurement is Pa, but, often, dyne/cm^2 is used. The physiological range for the mean *WSS* value is 5–70 dyne/cm^2 (0.5–7 Pa), usually found in medium and large arteries [36, 43, 44]. Another important parameter for the development of neo-intimal hyperplasia is the spatial gradient of the wall shear stress (*WSGG*) [45] and its absolute value, module of gradient stress (*MGS*). The unit of measurement is Pa/m, often used as dyne/cm^3. *MGS* values on the stents are less interesting because high values of *MGS* are related to restenosis, but their range has not been established yet.

$$[\mathrm{WSSG}] = \begin{pmatrix} \dfrac{\partial \mathrm{WSS}_x}{\partial x} & \dfrac{\partial \mathrm{WSS}_x}{\partial y} & \dfrac{\partial \mathrm{WSS}_x}{\partial z} \\[2mm] \dfrac{\partial \mathrm{WSS}_y}{\partial x} & \dfrac{\partial \mathrm{WSS}_y}{\partial y} & \dfrac{\partial \mathrm{WSS}_z}{\partial z} \\[2mm] \dfrac{\partial \mathrm{WSS}_z}{\partial x} & \dfrac{\partial \mathrm{WSS}_z}{\partial y} & \dfrac{\partial \mathrm{WSS}_z}{\partial z} \end{pmatrix} \tag{13}$$

$$\mathrm{MGS} = \sqrt{\sum_{i,j=1}^{3} \mathrm{WSSG}_{ij} \mathrm{WSSG}_{ij}} \tag{14}$$

Figure 11 presents the results for the average *MGS*. The pattern is similar to *WSS* but with some important differences. *MGS* is very low in the absence of the stent; it is mainly due to the developing velocity profile and grows in the presence of the stent. In the geometries with area reduction, higher values of *WSS* and *MGS* are present in correspondence with necking. Downstream of this point, lower values of the *MGS* are registered, counterbalanced by the negative impact of other parameters (specifically, low *WSS*).

Figures 12 and 13 show very low values of *WSS* on the vessel wall in all the zones adjacent to the stent, while areas with high *MGS* are localized close to the stent but not uniformly. They are located close to the struts, which are the parts of the stent that oppose most of the flow.

Another parameter to be considered is the optical density (*OD*), which evaluates the endothelial permeability. For macromolecules, such as *LDL*, a high endothelial permeability may be associated with the development of atherosclerotic lesions. The spatial variation of macromolecular permeability can be assessed by *Evans blue dye* (*EBD*), which has a great affinity for albumin, being a good marker [46]. The parenteral administration of *EBD* causes the formation of spots of blue colour on the surface intima of the arterial tree, where the trans-vascular transport and the permeability of other macromolecules, such as globulin, fibrinogen and lipoproteins, increase [47]. The relationship between the accumulation of *EBD* and the optical density, indicative of high endothelial permeability [48], is

$$OD = 2M\alpha \tag{15}$$

where *M* is the dye deposit per area unit (nmol/cm) and α is the absorption coefficient of *EBD* (cm^2/nmol).

Several studies have been performed to obtain a direct relation between fluid dynamic parameters (*WSS* and *GWSS*) and *OD* [46–51]. A composite parameter, including *MGS* and *WSS*, is related to endothelial permeability,

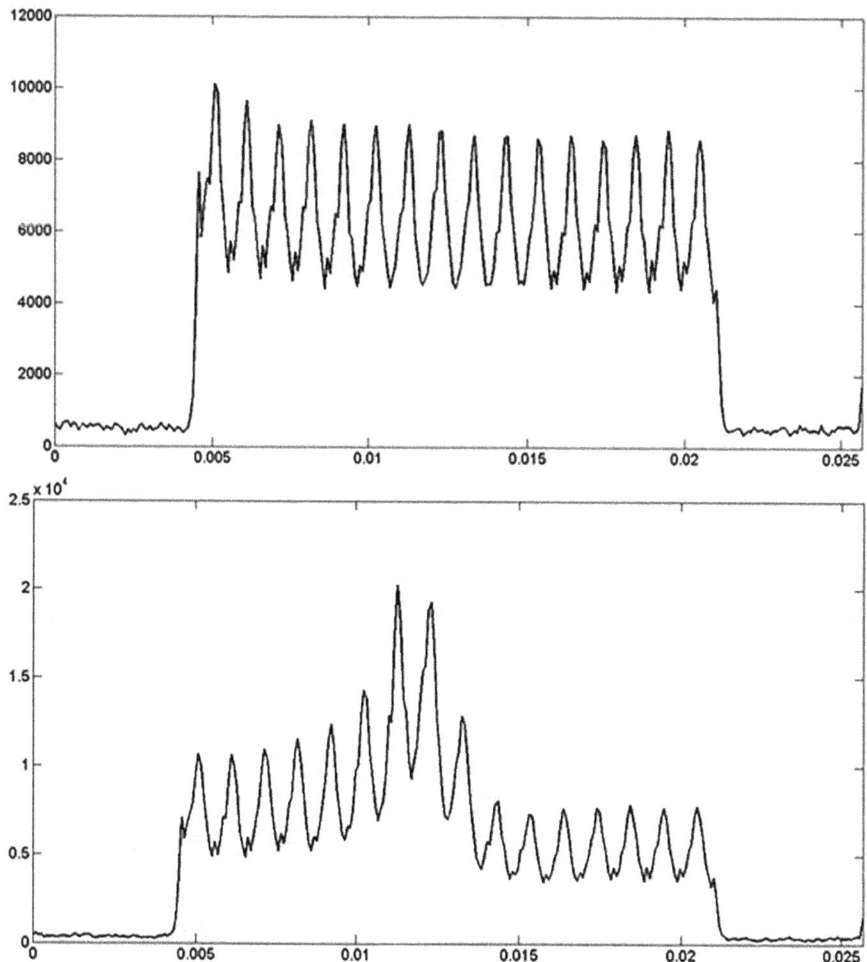

Fig. 11 Average MGS (dyne/cm^3) versus axial length (m). Top: DOR = 0%; second from top: DOR = 30%; second from bottom: DOR = 60%; bottom: DOR = 90% [4]

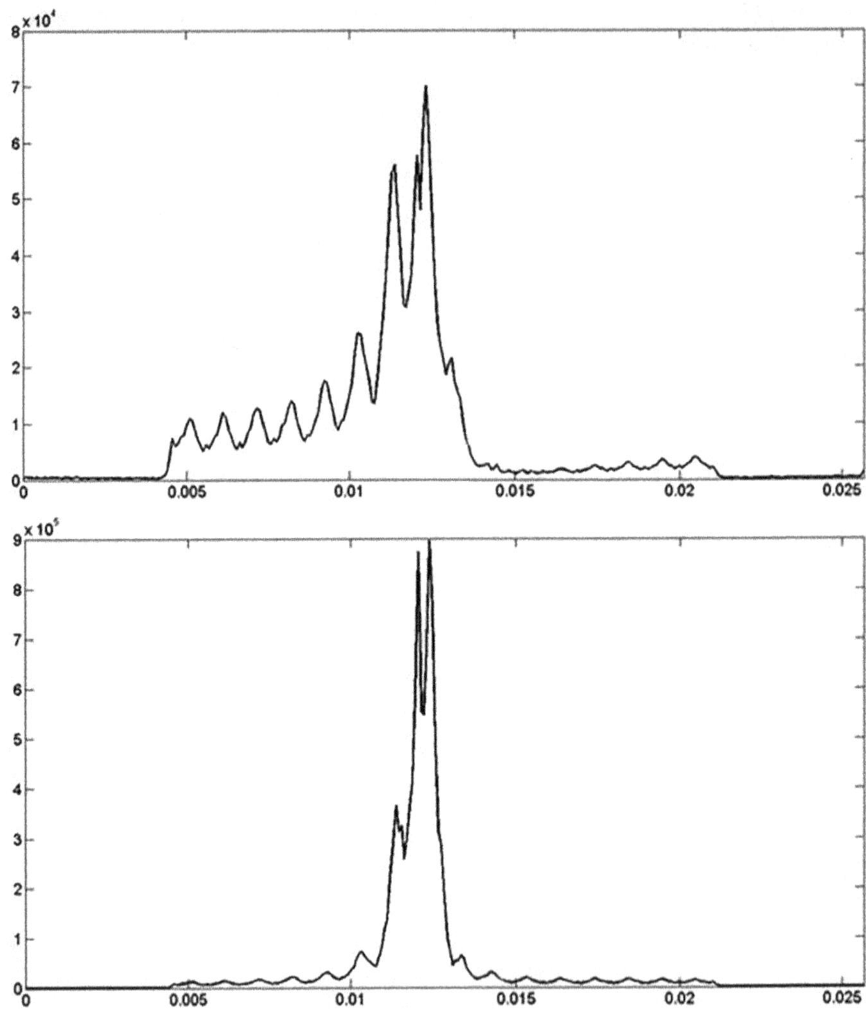

Fig. 11 (continued)

$$OD = 1.2WSS^{-0.11}MGS^{0.044} \tag{16}$$

The physiological values of the *OD* are between 0.9 and 1.3. The accumulation of *EBD* is highly variable in different regions; then, *OD* can only be used as a comparative parameter.

Figure 14 presents the average *OD* on the vessel wall before the stent. The *OD* assumes physiological values with the increase of the stent endothelial permeability, and the average *OD* has peaks and valleys corresponding to areas with greater or smaller curvature. In the presence of residual stenosis, permeability decreases in the direction of necking due to very high *MGS*; then, downstream of the stenosis, the permeability increases again, assuming higher values. An exception is the geometry with *DOR* = 90%, where the wide area of recirculation makes the *OD* values similar to those prior to the stenosis.

Figure 15 shows that all the areas of the vessel wall adjacent to the stent are associated with an *OD* value greater than those far from the stent. There are small areas where the *OD* reaches a very high value in the vicinity of higher curvature.

The streamlines obtained from the instantaneous velocity fields at four meaningful times and corresponding to four points of the mean velocity profile are shown in Figs. 16, 17, and 18, [26], for the geometries with *DOR* = 30%, 60% and 90%.

Figure 16, [26], presents the streamlines for *DOR* = 30%. The streamlines appear mostly parallel to the direction of the mean flow, even in proximity to the struts and

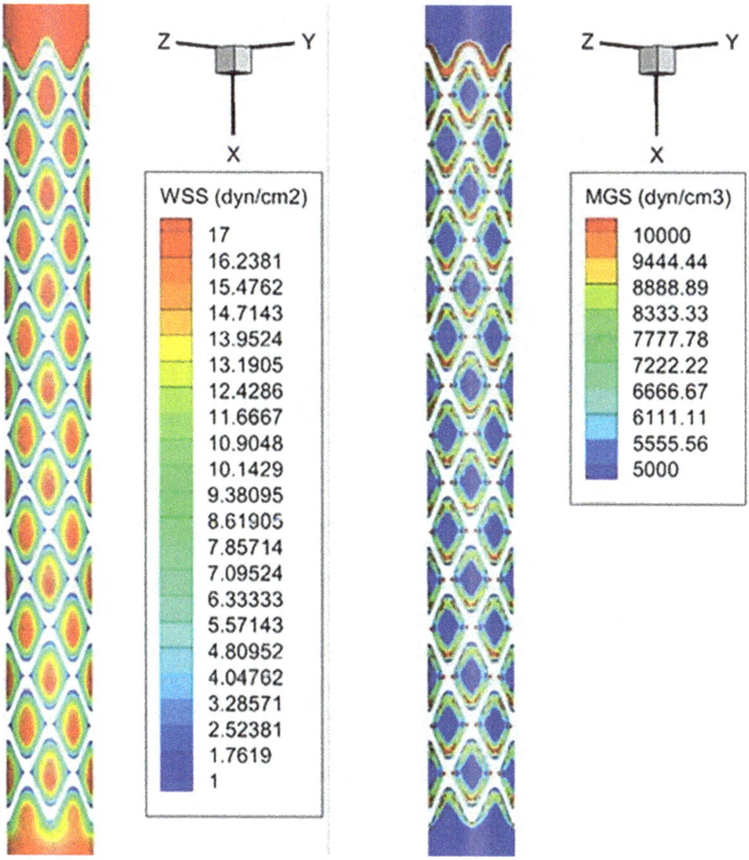

Fig. 12 WSS (dyn/cm^2) and MGS (dyn/cm^3) fields. Top: DOR = 0%; second from top: DOR = 30%; second from bottom: DOR = 60%; bottom: DOR = 90% [4]

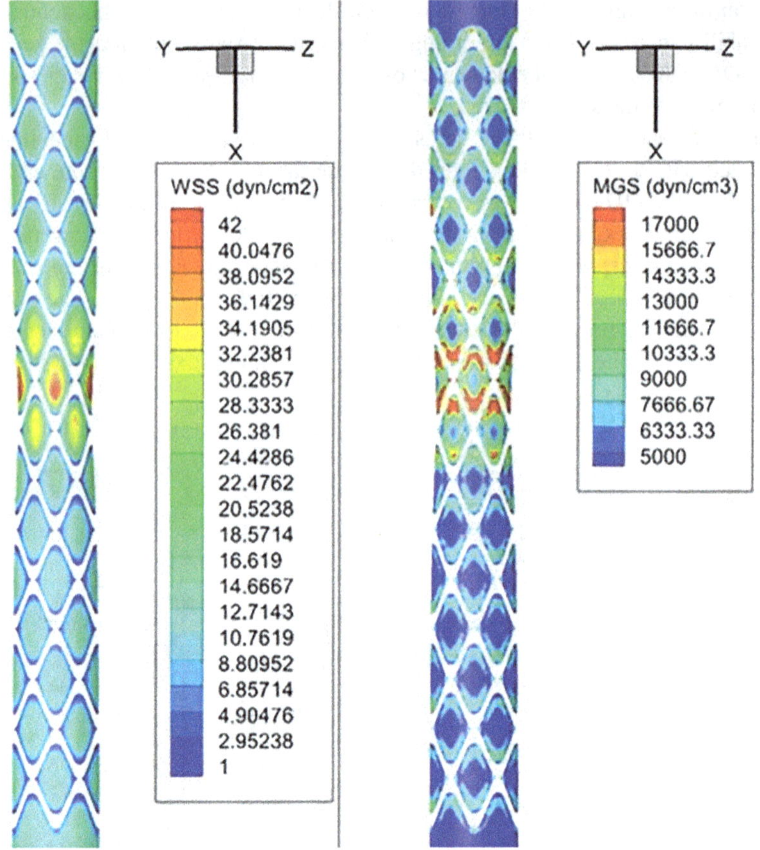

Fig. 12 (continued)

during the systole (Fig. 16d), meaning that the recirculation, if it exists, is very small. An exception to this behaviour is at the diastole (Fig. 16a), where the fluid proceeds backwards everywhere except downstream of the stenosis at the centre of the vessel, where a train of vortices can be seen. However, no vortex is seen on the vessel wall.

Figure 17, [26], presents the streamlines for $DOR = 60\%$. The streamline pattern is similar to the previous case, except downstream of the stenosis during the diastole (Fig. 17a), since the region subject to the train of vortices is wider. Figure 17b and c show streamlines mostly parallel to the direction of the mean flow, and no near-wall vortex is visible. The near-wall vortices are present during the systole (Fig. 17d), downstream of the stenosis, because the fluid at the narrowing is faster and therefore more prone to generating near-wall vortices when the section expands. However, this effect disappears in 1.5 diameters, and the streamlines return parallel to the mean flow.

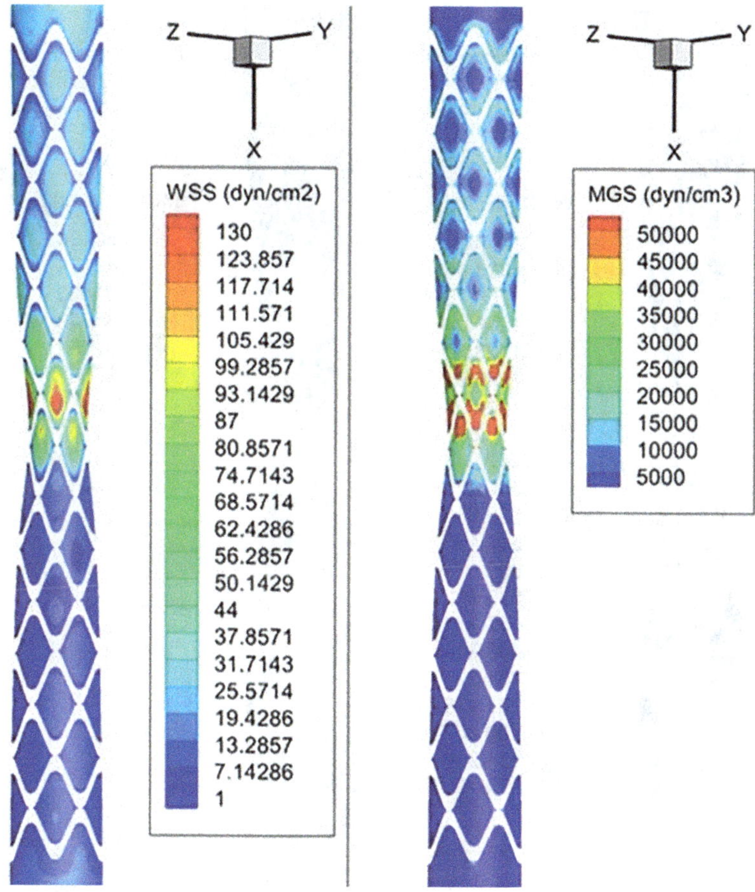

Fig. 12 (continued)

Figure 18, [26] presents the streamlines for $DOR = 90\%$. The results in this case are different from the previous one because a large recirculation region is visible in the diastole but with irregular contours. Figure 18b–d shows that at the other three time steps, $t = 0.4$ s, $t = 0.5$ s and $t = 0.6$ s, a wide recirculation region is present near the vessel wall and beyond the stenosis, and the flow on the vessel wall is directed upstream. This region of recirculation terminates several diameters downstream of the end of the stent and is not visible in the figures.

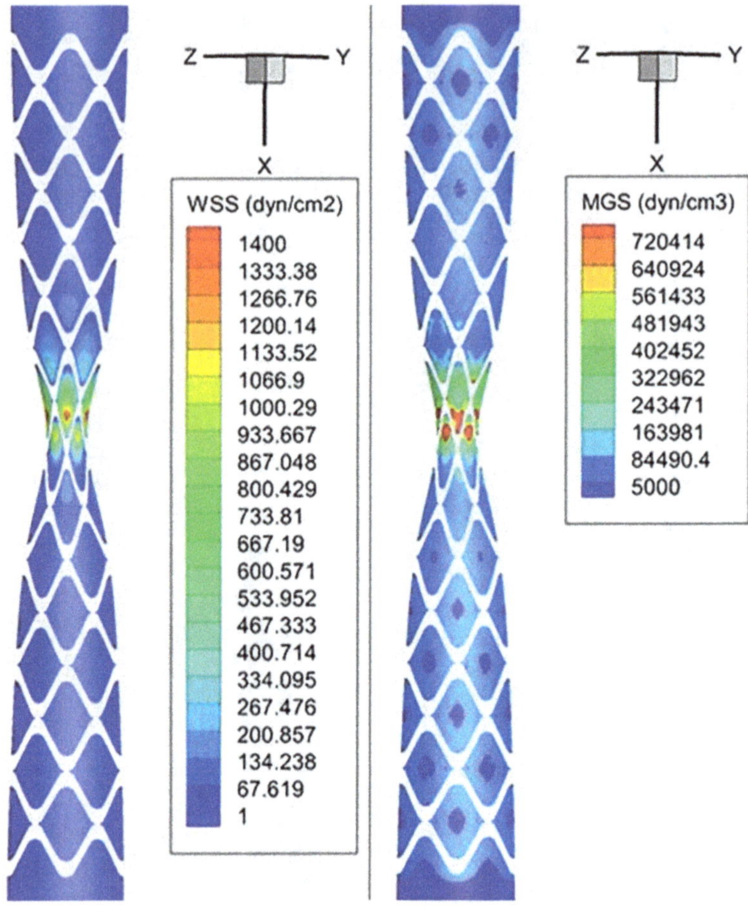

Fig. 12 (continued)

4.2.3 Unsteady State

Other haemodynamic parameters can be investigated, over those examined for the stationary case, in the unsteady-state analysis. The oscillatory shear index (*OSI*) [43] is defined as

$$OSI = 0.5 \left(1 - \frac{\left| \int_0^T WSS_i dt \right|}{\int_0^T WSS dt} \right) \tag{17}$$

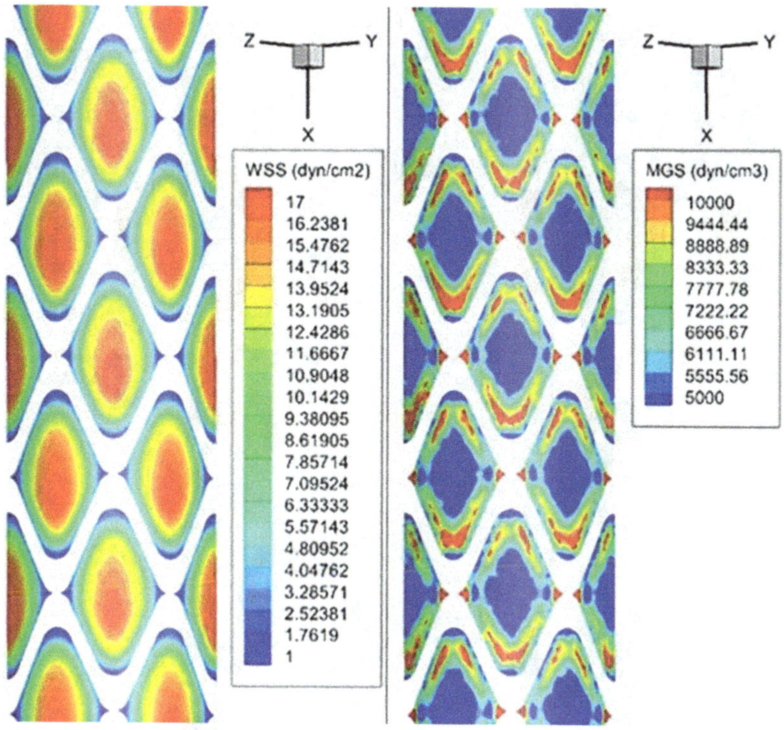

Fig. 13 More detailed WSS (dyn/cm^2) and MGS (dyn/cm^3) fields. Top: DOR = 0%; second from top: DOR = 30%; second from bottom: DOR = 60%; bottom: DOR = 90% [4]

The *OSI* clarifies, during the cardiac cycle, the *WSS* vector deflection from the blood flow predominant direction. Its value can vary from zero, for no cyclic variation, to 0.5 for 180° deflection of the *WSS* direction.

The time-averaged wall shear stress (*TAWSS*) magnitude, defined as

$$\text{TAWSS} = \frac{1}{T} \int_0^T \text{WSS} dt \tag{18}$$

can compare the cumulative effects of *WSS* over an entire cardiac cycle.

The relative residence time (*RTT*) [52], measured in Pa^{-1},

$$\text{RTT} = [(1 - 2 \cdot \text{OSI}) \cdot \text{TAWSS}]^{-1} \tag{19}$$

identifies the regions subjected to stagnant flow. The values higher than those on the endothelium are more prone to cell and macromolecule infiltration. However, this parameter gives only a relative evaluation of the residence time of the fluid close to the wall.

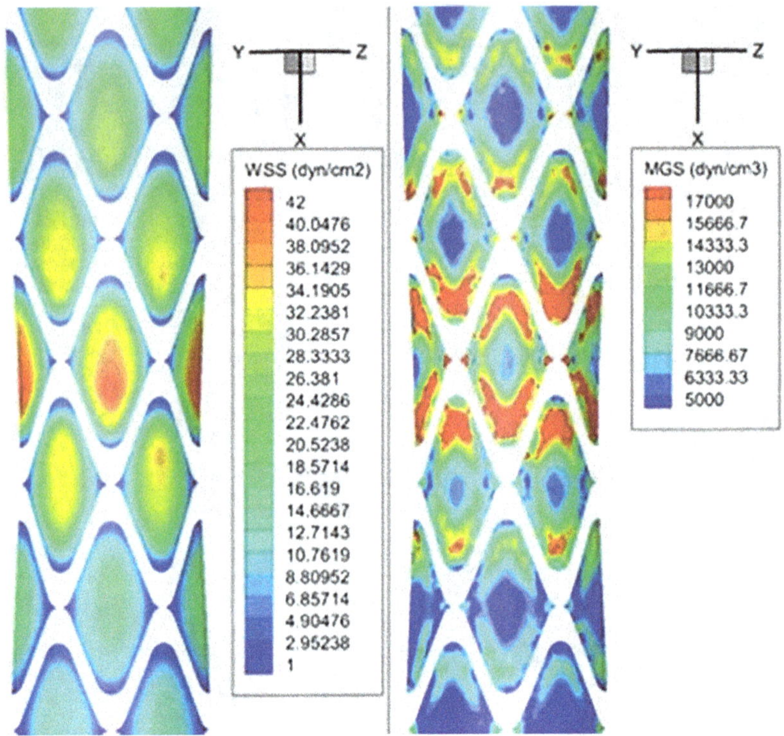

Fig. 13 (continued)

Geometry with DOR = 0%

Figure 19, [26], shows the trend of local *WSS* at different instants of time, which vary significantly during the cardiac cycle. As far as *WSS* is concerned, the zones of the vascular wall undergo relevant variations far from the stent, which is physiological, since higher speeds correspond to higher *WSS*. On the other hand, the zones adjacent to the stent undergo small variations during the cardiac cycle and remain permanently exposed to adverse conditions.

Figure 20, [4], shows the trend of local *MGS* at different instants of time, which vary significantly during the cardiac cycle. The local *MGS* field shows greater variations than *WSS*. When the flow velocity is high, *MGS* spreads in large areas, whereas when it is small, the areas are greatly reduced. In conclusion, as far as *MGS* is concerned, areas are exposed to adverse conditions only for some instants of time.

Figures 21 and 22 show the average *WSS* and *MGS* on the vessel wall at different instants of time during the cardiac cycle.

The *OD* during the cardiac cycle is calculated and analysed to evaluate the combined effect of *WSS* and its gradient. Figure 23 presents the average *OD* on the vessel wall, while Fig. 24 shows the local *OD* field on the vessel wall at different times. The trend of the time-averaged *OD* does not show significant variations

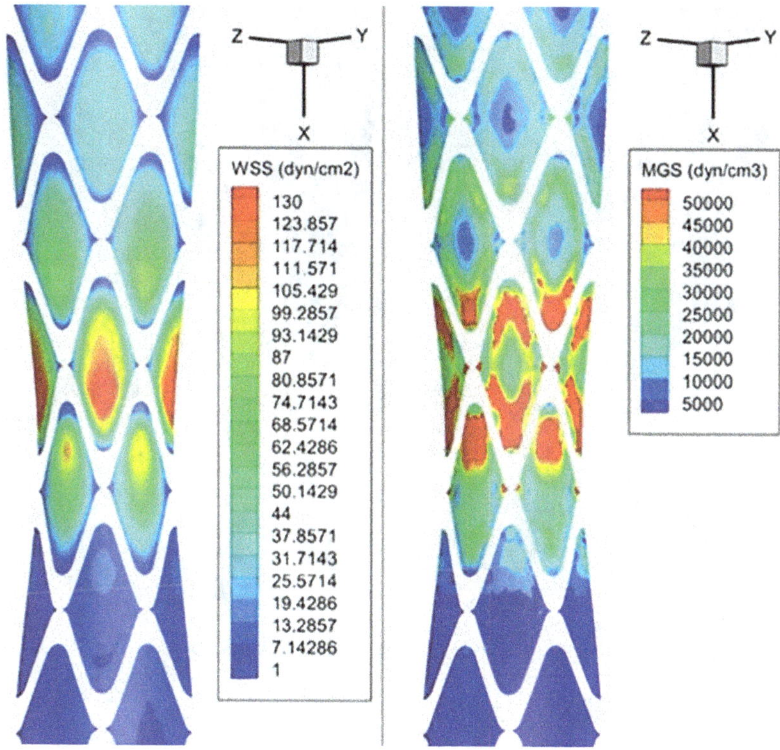

Fig. 13 (continued)

during the cardiac cycle; the higher OD values remain in the same positions, i.e. close to the struts, as shown in Fig. 24 as well, and they increase with the decreasing flow rate, indicating a higher wall permeability. The OSI is used to measure the deviation of WSS from its mean direction. Although it is possible to define the OSI for all spatial directions, for the geometry with $DOR = 0\%$, only the change in the mean flow direction is significant.

The OSI is used to measure the deviation of WSS from its mean direction. Although it is possible to define the OSI for all spatial directions, for the geometry with $DOR = 0\%$, only the change in the mean flow direction is significant. As can be seen from Fig. 25, the areas exposed to non-zero OSI are located near the struts and close to the points with greater curvature, while in the intra-strut area, the values are close to zero. Since there is a correlation between atherosclerosis pathogenesis and OSI only for values between 0.1 and 0.5, areas subject to these values are limited for this geometry.

RTT measures the relative time that the fluid spends in a certain region. RTT has higher values where OSI is closer to 0.5 and $TAWSS$ is small; therefore, RTT allows us to identify regions of low and oscillatory WSS, which are deemed to indicate areas

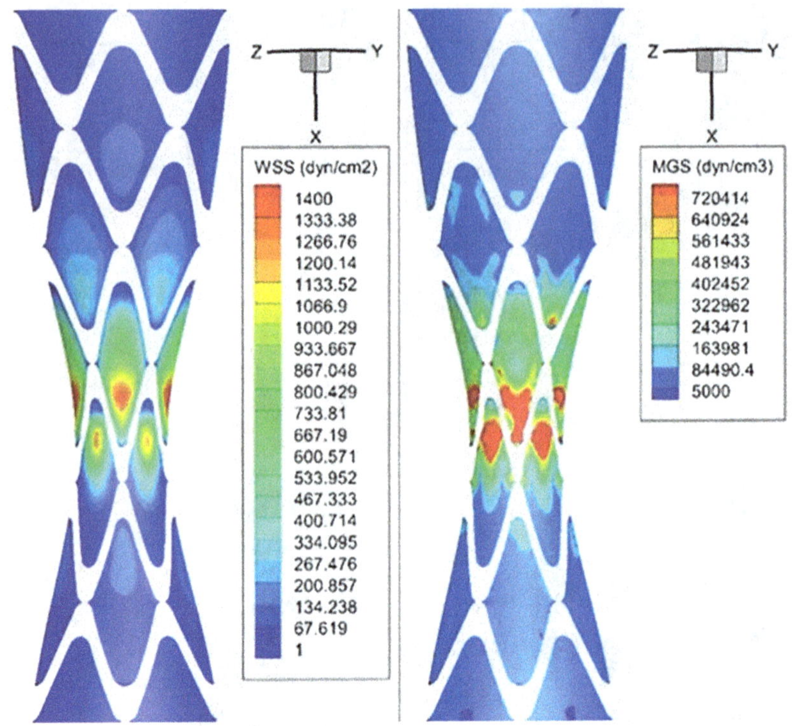

Fig. 13 (continued)

prone to atherosclerosis. Figures 26 and 27 show the trends of *RTT* on the vessel wall for the geometry with *DOR* = 0% and the spatial distribution of the average *RTT* on the vessel wall. High values appear in the same regions as *OSI*. *RTT* increases where the stent is present and decreases on the vessel wall because *RTT* is inversely proportional to *TAWSS* (smaller in proximity to the stent).

Geometry with DOR = 30%
In the geometry with DOR = 30%, the results change considerably. As shown in Fig. 28, [26] and Fig. 29, [4], respectively, for local and average *WSS*, the region downstream of the stenosis is subjected to relatively small *WSS*, while upstream *WSS* increases. Downstream of the choking, *WSS* increases when the flow rate returns to physiological conditions, far enough from the stenosis, as a result of an increase in flow rate, in correspondence to the section narrowing. This effect occurs when the flow rate is higher, while, at low velocity, its pattern is similar to the absence of residual stenosis.

The same considerations can be made for local and average *MGS* (Figs. 30 and 31).

The *OD*, shown by Figs. 32 and 33 (average and local), has a trend similar to those obtained in the absence of residual stenosis during diastole, while, during systole, *OD* decreases upstream of the stenosis and increases slightly downstream. In conclusion, downstream of the stenosis, endothelial permeability increases and *OD* has higher values for a longer time interval during the cardiac cycle as compared to the case with $DOR = 0\%$.

Figure 34, [4] shows that local *OSI* increases slightly, in correspondence to and downstream of the stenosis, but does not reach critical values, and the pattern is quite similar to the previous case. The area reduction is 30%, $DOR = 30\%$, and the flow

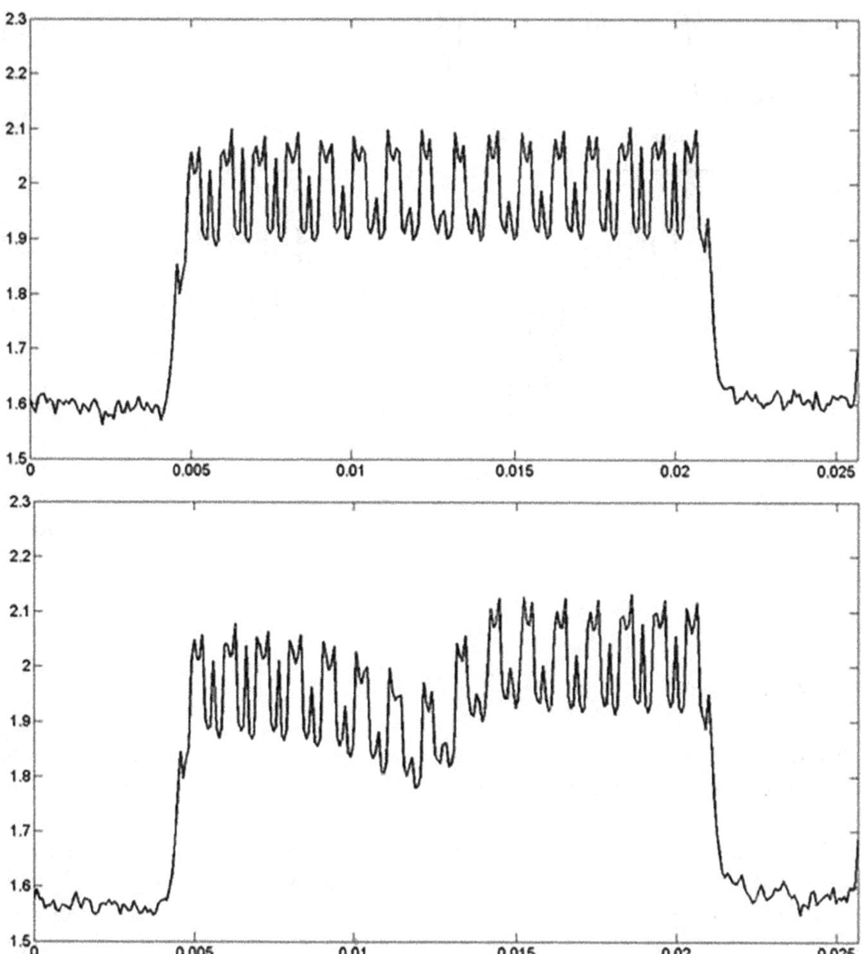

Fig. 14 Average OD versus axial length (m). Top: DOR = 0%; second from top: DOR = 30%; second from bottom: DOR = 60%; bottom: DOR = 90% [4]

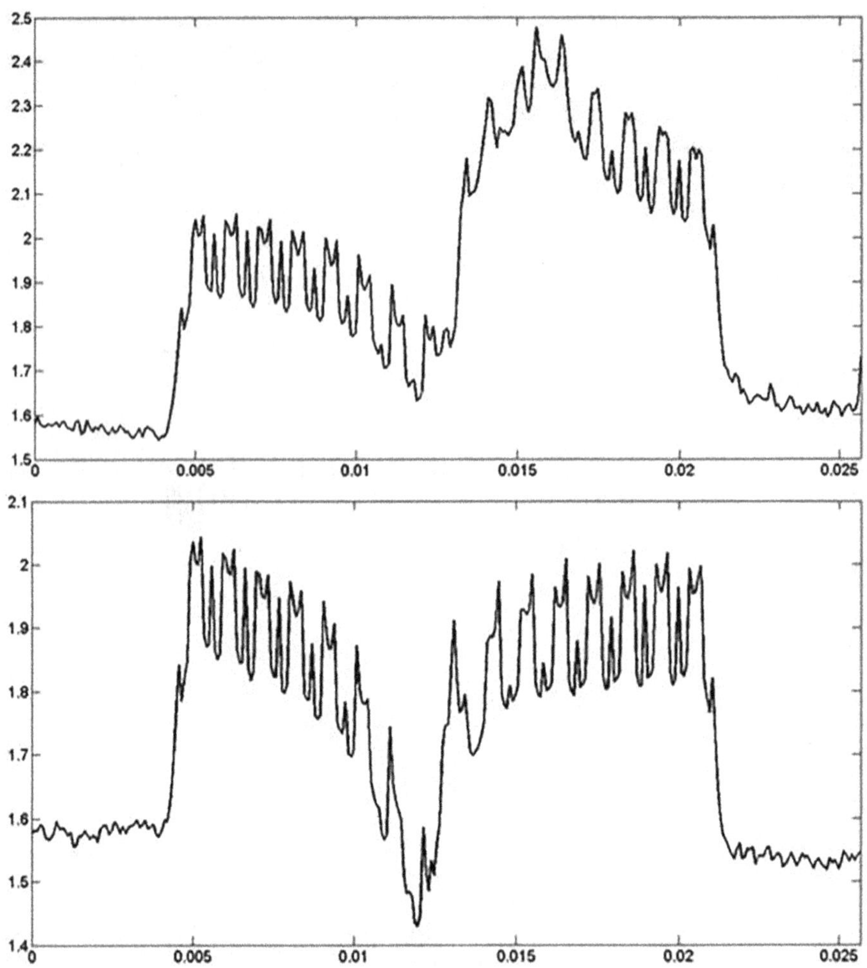

Fig. 14 (continued)

conditions are not able to cause the detachment of the fluid vein after the contraction, a situation that would certainly lead to a proliferation of regions affected by critical values of *OSI*.

Figures 35 and 36 show that the residence time, *RTT*, has no substantial and worrisome changes from the case without residual stenosis, $DOR = 0\%$. As might be expected, *RTT* decreases in the proximity of the stenosis, where blood velocity increases. Then, the *RTT* of the macromolecules decreases but increases slightly afterwards, downstream of the stenosis, due to the lower *TAWSS*.

Geometry with DOR = 60%

As far as the geometry with $DOR = 60\%$ is concerned, the patterns of *WSS* and *MGS* are similar to those of the previous case with $DOR = 30\%$, but the effect is more pronounced, as shown by Fig. 37, [4], and Fig. 38, [26], for average and local *WSS* and by Figs. 39 and 40 for average and local *MGS*. The local *WSS* and *MGS* assume very high values, specifically during the systole, in the section of minimum area, but this is not enough to cause thrombogenic effects. Downstream of this point, both *WSS* and *MGS* assume very small values. Indeed, the geometry is such that, in the area of the vascular endothelium, it forms a recirculation zone, delimited by the periodic pattern of the struts. The areas where the fluid is more stagnant are critical and especially prone to massive post-stent restenosis.

Figures 41 and 42, which are relative to average and local *OD*, confirm that downstream of the stenosis, the endothelial permeability increases because of the recirculation zones formed in areas of the vascular endothelium delimited by the geometry of the stent.

Fig. 15 Average OD fields. Top: DOR = 0%; second from top: DOR = 30%; second from bottom: DOR = 60%; bottom: DOR = 90% [4]

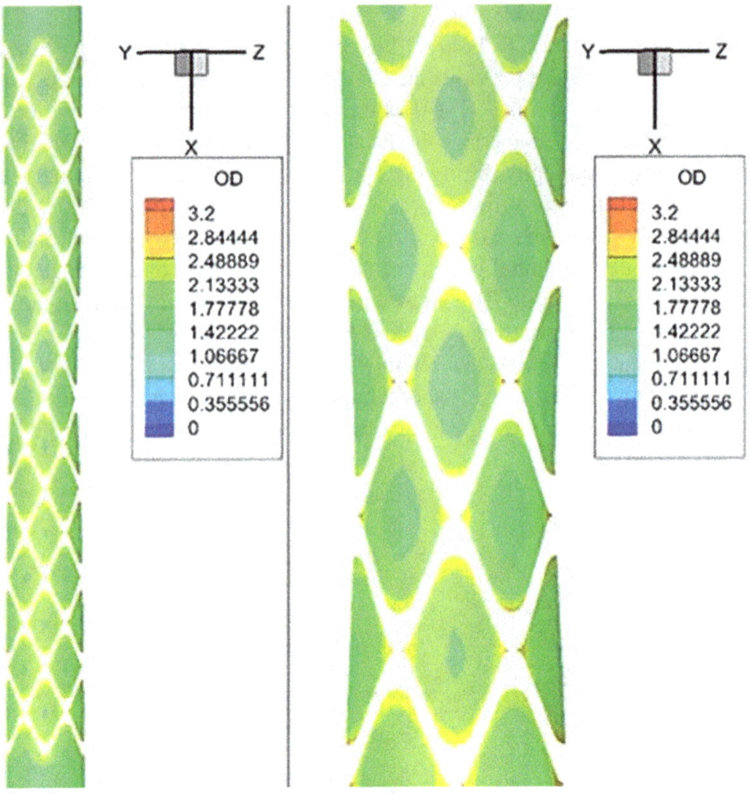

Fig. 15 (continued)

Figure 43 shows that the *OSI* becomes more relevant with *DOR* = 60% because it is sufficient to provoke a slight boundary layer detachment beyond the stenosis at the systolic peak. *OSI*, as noted in Fig. 43, assumes greater values, 0.2–0.5, in areas where the fluid is detached from the wall and where, consequently, a recirculation region is present. This happens also in regions previously analysed, where there is a local recirculation in correspondence with the endothelium delimited by the struts.

Figures 44 and 45 show the average and local *RTT*. High *OSI* and low *TAWSS* values increase the residence time beyond the stenosis.

Geometry with DOR = 90%

Finally, we show the results for the geometry with *DOR* = 90%. Figures 46, [4], and Fig. 47, [26], for *WSS*, and Figs. 48 and 49, [4], for *MGS*, show that *WSS* and *MGS* assume hazardous values for most of the cardiac cycle, which may lead to the occurrence of thrombogenic effects. Beyond the stenosis, a wide region of recirculation is formed; the wider it is, the higher the blood flow rate in correspondence with the necking. This is due to the large detachment of the fluid vein from the wall, which reattaches to the wall several diameters after the contraction point. This is due to the

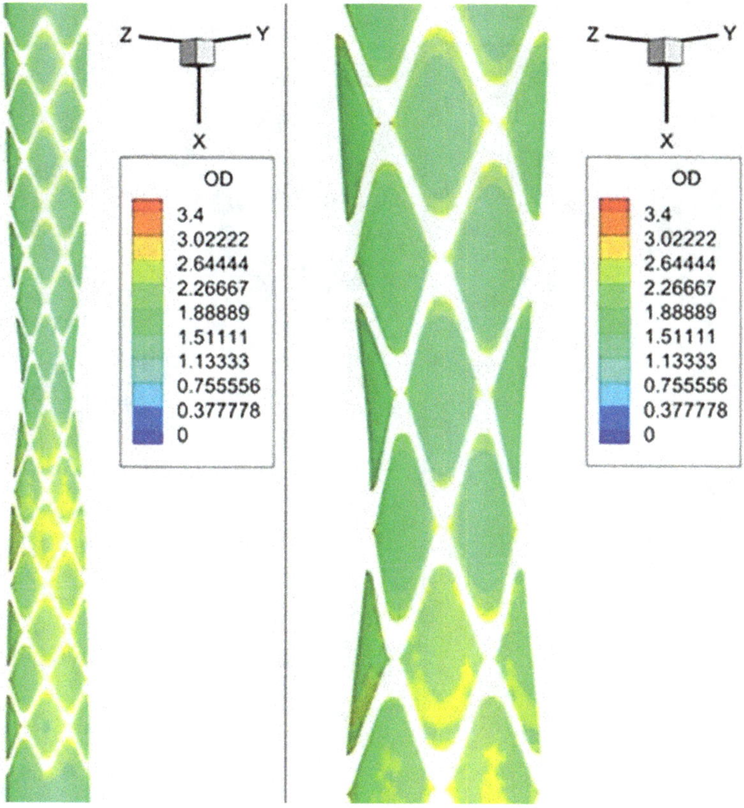

Fig. 15 (continued)

fact that we are considering a straight tube for a long stretch, but in reality, the vessel has a certain curvature.

Figures 50 and 51 show that the OD seems to improve in the region downstream of the stenosis, despite the presence of extensive recirculation, which, from the physiological point of view, is a hazardous factor.

Figure 52 shows that OSI is high right downstream of the section of the minimum area, where the WSS vector is oscillating, because of the recirculation region. Beyond the stenosis, the OSI has a trend similar to that for the geometry without residual stenosis, but, in those regions, the WSS vector is oriented, for most part of the cardiac cycle, in the opposite direction with respect to the main flow.

Figures 53 and 54 show that even the trend of RTT seems to improve compared to the case with $DOR = 60\%$, but the same considerations done for the OD can be applied to this case as well.

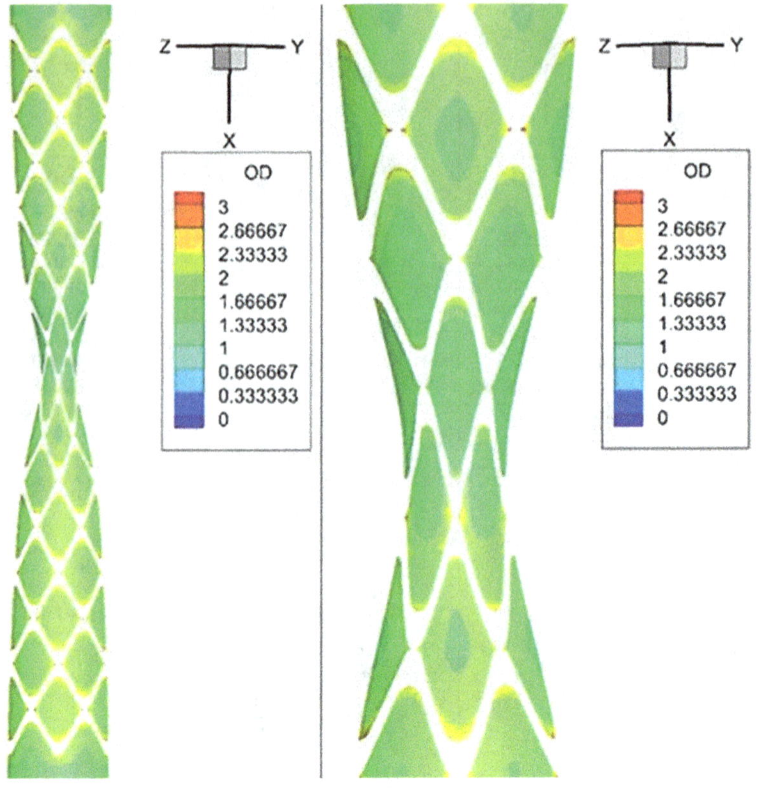

Fig. 15 (continued)

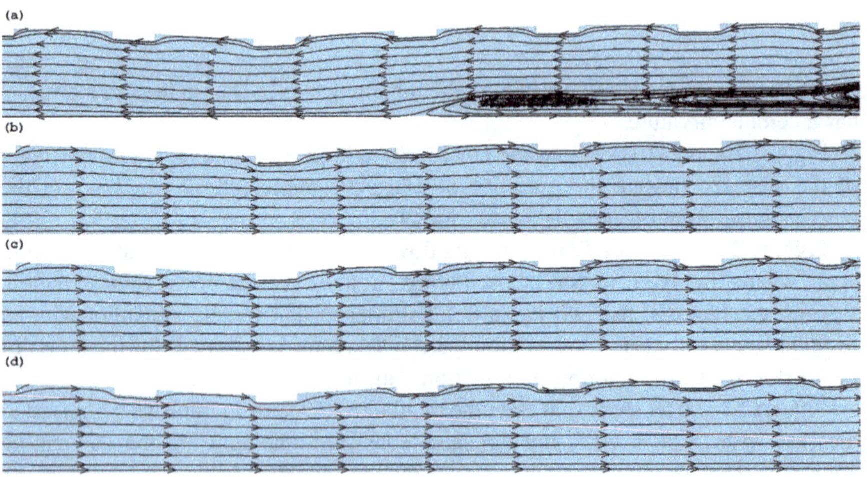

Fig. 16 Streamlines for the geometry with DOR $= 30\%$ during the cardiac cycle after: (**a**) $t = 0.3$ s (diastole), (**b**) $t = 0.4$ s, (**c**) $t = 0.5$ s, (**d**) $t = 0.6$ s (systole), Fig. 5 of [26]

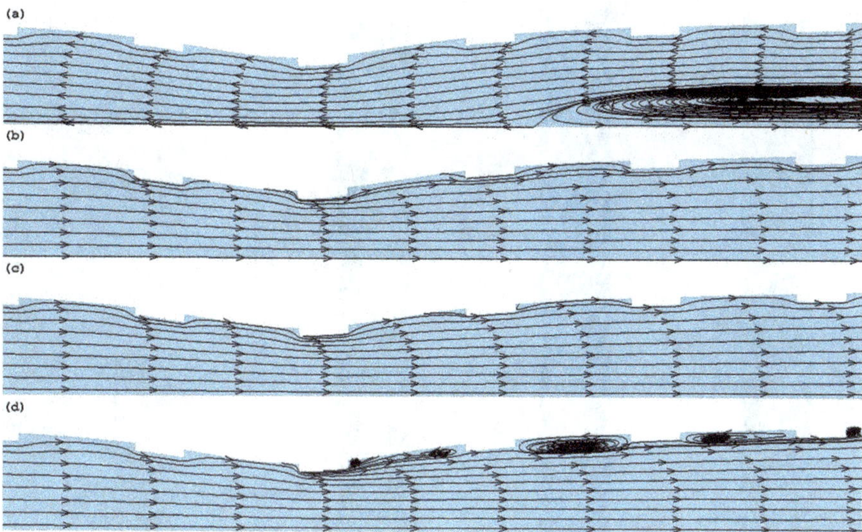

Fig. 17 Streamlines for the geometry with DOR = 60% during the cardiac cycle after: (a) $t = 0.3$ s (diastole), (**b**) $t = 0.4$ s), (**c**) $t = 0.5$ s, (**d**) $t = 0.6$ s (systole), Fig. 6 of [26]

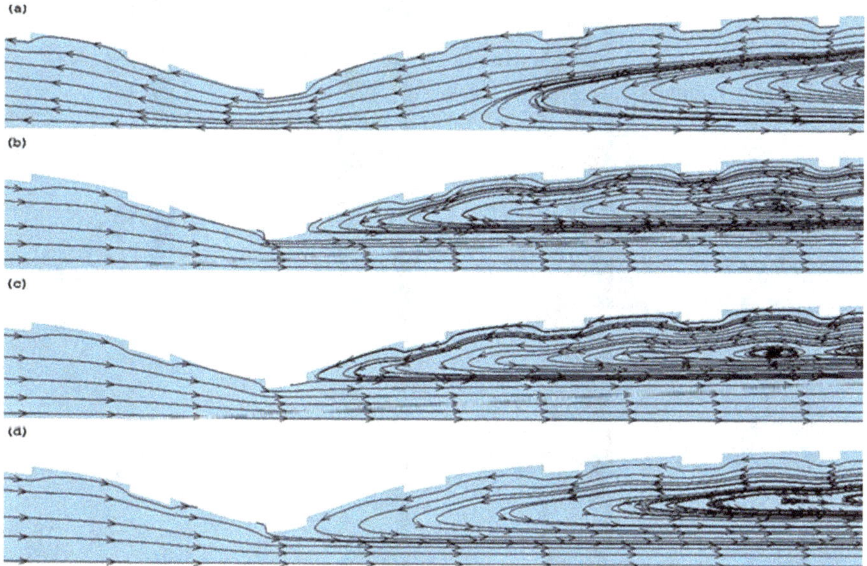

Fig. 18 Streamlines for the geometry with DOR = 90% during the cardiac cycle after: (**a**) 0.3 s (diastole); (**b**) 0.4 s; (**c**) 0.5 s; (**d**) 0.6 s (systole), Fig. 7 of [26]

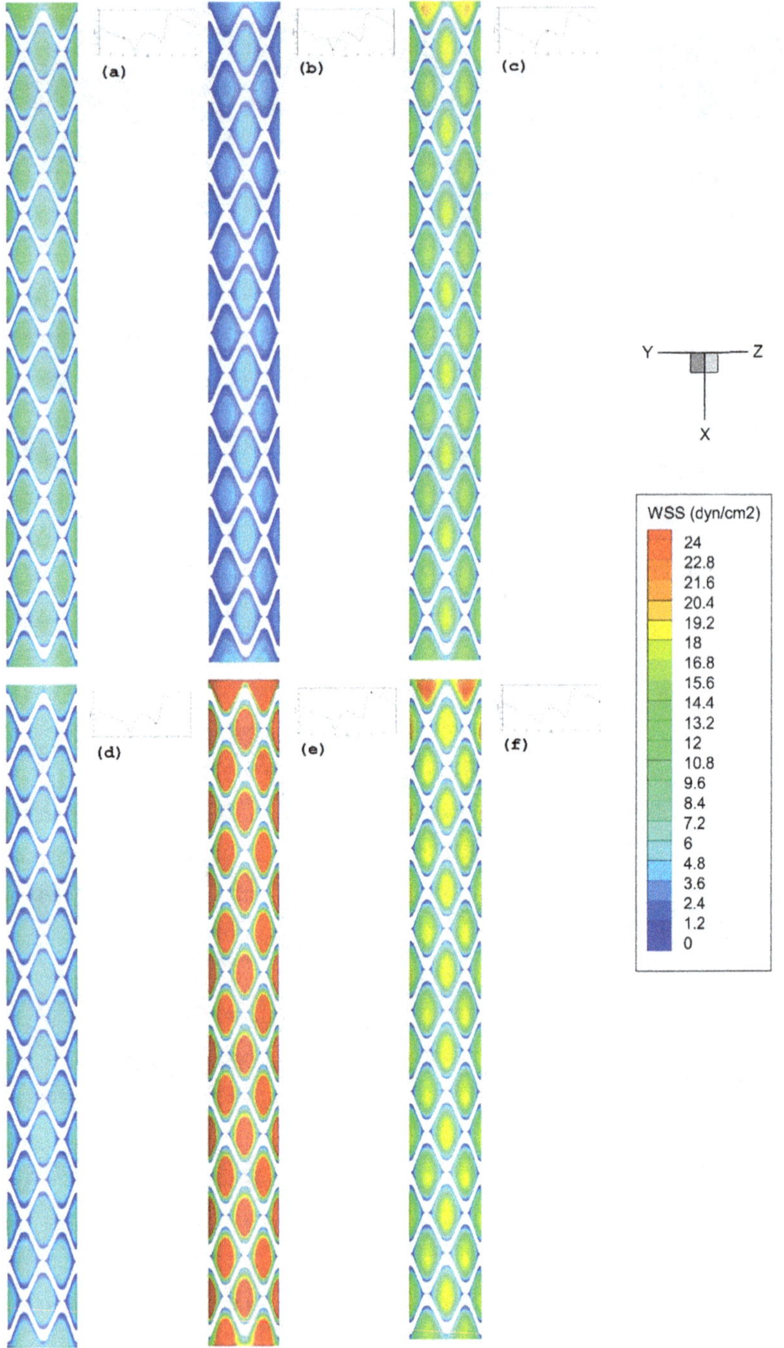

Fig. 19 Local WSS during the cardiac cycle with DOR = 0% after: (**a**) $t = 0.2$ s, (**b**) $t = 0.3$ s (diastole), (**c**) $t = 0.4$ s, (**d**) $t = 0.5$ s, (**e**) $t = 0.6$ s (systole), (**f**) $t = 0.8$ s, Fig. 8 of [26]

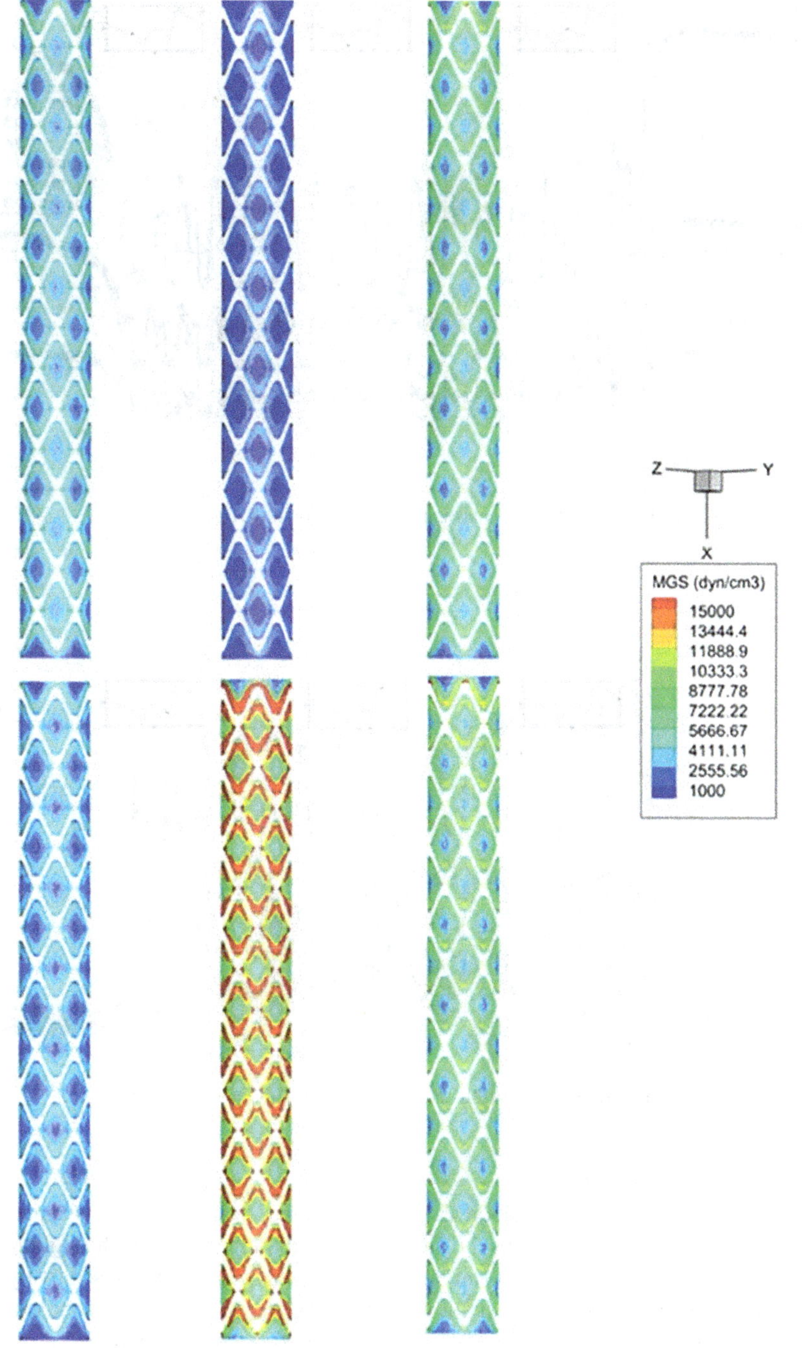

Fig. 20 Local MGS with DOR = 0% [4]

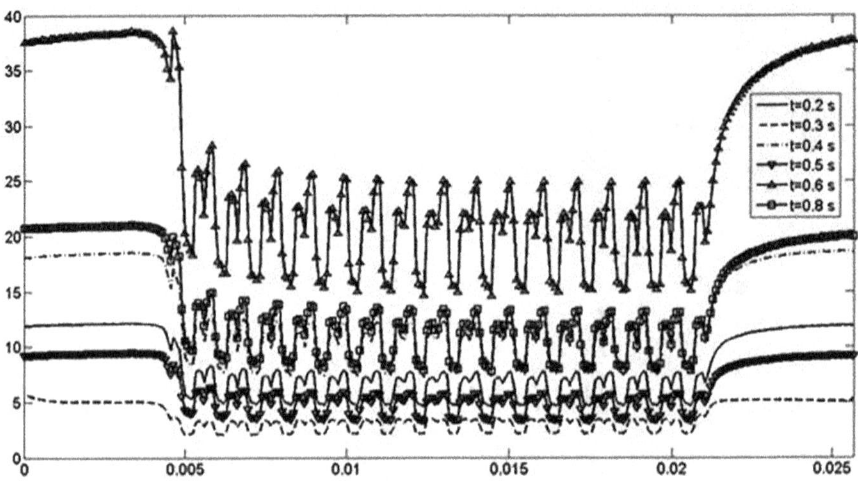

Fig. 21 Average WSS (dyne/cm^2) versus axial length (m) with DOR $= 0\%$ [4]

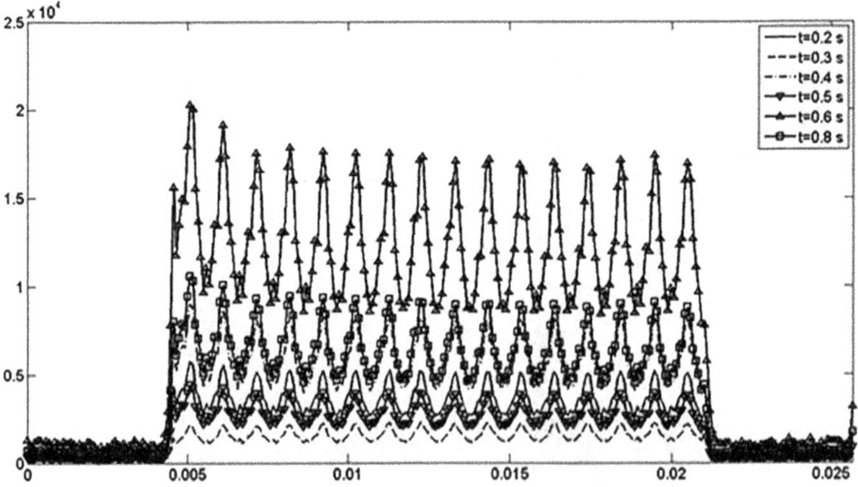

Fig. 22 Average MGS (dyne/cm^3) versus axial length (m) with DOR $= 0\%$ [4]

4.2.4 Blood Viscosity

A strong correlation between blood viscosity and cardiovascular events is evidenced in [53]. Blood viscosity explains the anatomic distribution of lesions throughout the body, provides a role for platelet activation by turbulent blood flow caused by hyperviscosity, and includes an explanation of the protective role of *HDL* cholesterol, which appears at lower viscosity [54]. Therefore, viscosity is a fundamental parameter to be taken into account. In a Casson fluid, viscosity is inversely proportional to the shear rate; then, non-Newtonian behaviour can be significant only where the fluid

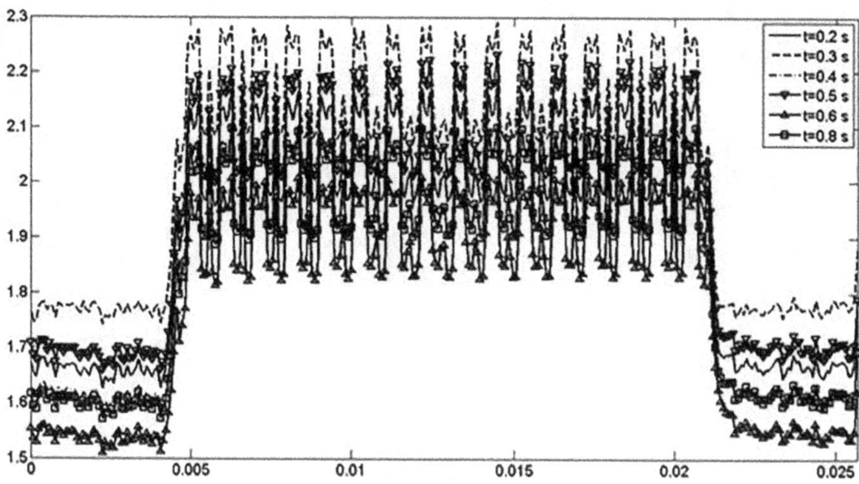

Fig. 23 Average OD versus axial length (m) with DOR $= 0\%$ [4]

is quite slow. In a rigid pipe, the shear rate increases with the flow rate, and consequently, the viscosity decreases, while a decrease in the flow and in the shear rate leads to an increase in viscosity. Moreover, the shear rate is also a function of the geometry and is higher where the duct diameter is smaller and vice versa.

The apparent viscosity is averaged across the section and is shown along the axial length in the following Fig. 55 for the cases with $DOR = 0\%$, 30%, 60% and 90% at four instants: (a) $t = 0.3$ s, (b) $t = 0.4$ s, (c) $t = 0.5$ s and (d) $t = 0.6$ s. The apparent viscosity is higher where the flow rate is smaller, and the presence of the stent influences the viscosity. Local maxima are present in correspondence with the sections with higher strut density.

Figure 56 presents the trend of viscosity with the increase of the DOR. Viscosity decreases where shrinkage occurs, and it increases beyond the stenosis. Downstream of the stenosis, the viscosity assumes the highest value, with a more pronounced effect in the case with $DOR = 60\%$. In the presence of $DOR = 90\%$, viscosity stabilizes at a lower value due to the recirculation region, which is an unexpected result and, to the best of our knowledge, has never been reported before in the literature. Figure 56 presents the viscosity with different DOR at four significant time instants.

Figure 57 shows a higher apparent viscosity at the centre of the stent tube, which corresponds to the positions of higher erythrocyte concentration, according to the Casson model. Figure 57d shows that with $DOR = 90\%$, the higher viscosity is concentrated close to the vessel wall instead of in the centre. This means that, for a high degree of stenosis, the erythrocytes tend to be closer to the vessel wall, showing an alteration of the physiological conditions, as the erythrocytes tend to be concentrated at the centre of the pipe.

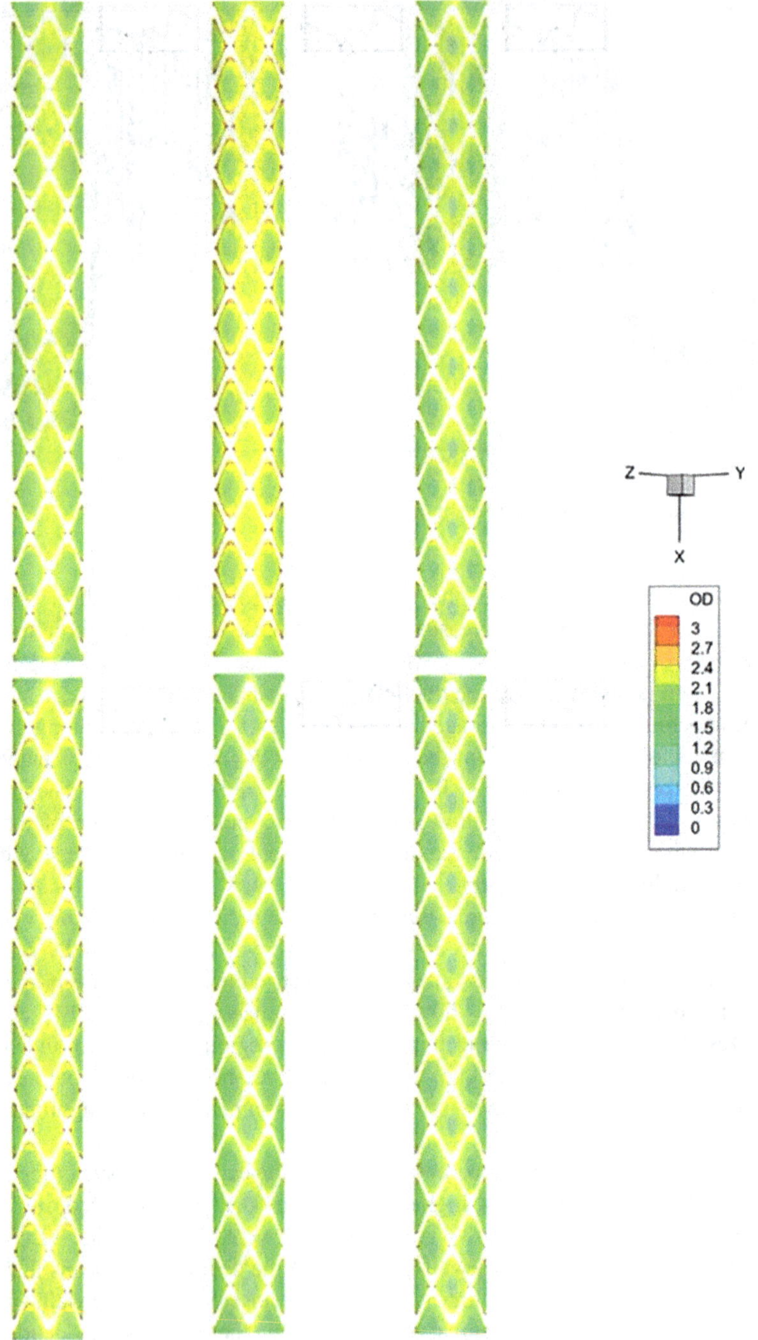

Fig. 24 Local OD with DOR = 0% [4]

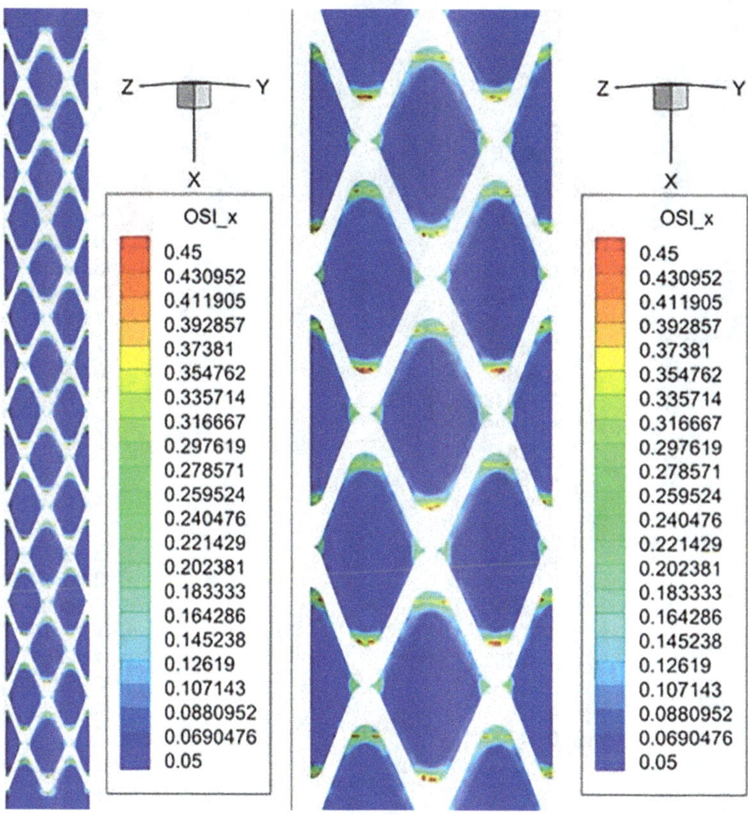

Fig. 25 Local OSI field for the geometry with DOR $= 0\%$ [4]

Figures 58, 59, and 60, [27] show the results of the apparent viscosity at three instants of time, $t = 0.3$ s (diastole), $t = 0.4$ s and $t = 0.6$ s (systole), on the periodic face of the stents with different *DOR*, 0%, 30%, 60% and 90%. Because of the steep variation in the shear rate, the numerical results are reported as the logarithm of the viscosity. In a Casson fluid, the viscosity is inversely proportional to the shear rate; therefore, the non-Newtonian behaviour can be significant only in slow-flow conditions. In a rigid tube, the shear rate is higher where the pipe diameter is smaller and vice versa. Therefore, the apparent viscosity is smaller in correspondence to the stenosis and greater where the flow rate is lower. From the results observed in [6, 15, 19, 20], it is possible to infer that the presence of the stent reduces the shear rate, and the WSS, proportional to this variable, is reduced as well.

Figure 58 reports the viscosity profile in the diastolic phase, after $t = 0.3$ s. Figure 58a shows the result with $DOR = 0\%$, which is higher at the centre of the pipe, in correspondence with the low shear rate caused by the pulsatile nature of the flow. The viscosity assumes another peak value at a distance from the wall corresponding to one-third of the radius due to the rheology of the blood. As far as

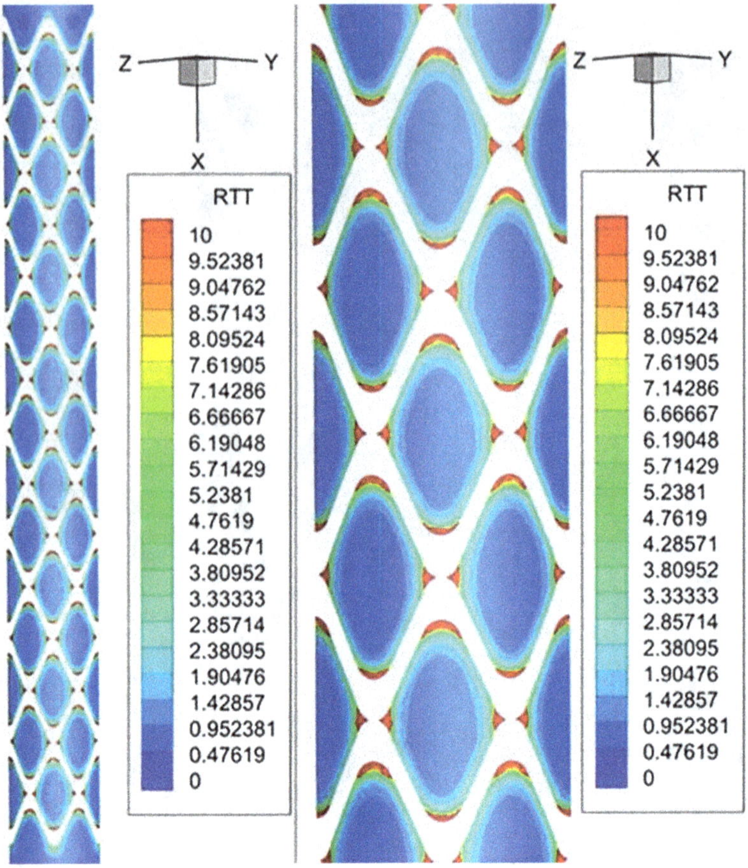

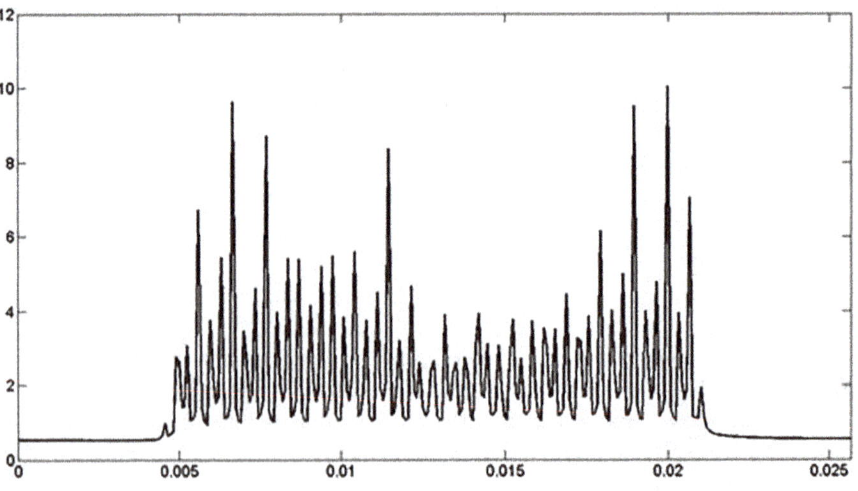

Fig. 26 Local RTT with DOR = 0% [4]

Fig. 27 Average RTT (1/Pa) versus axial length (m) with DOR = 0% [4]

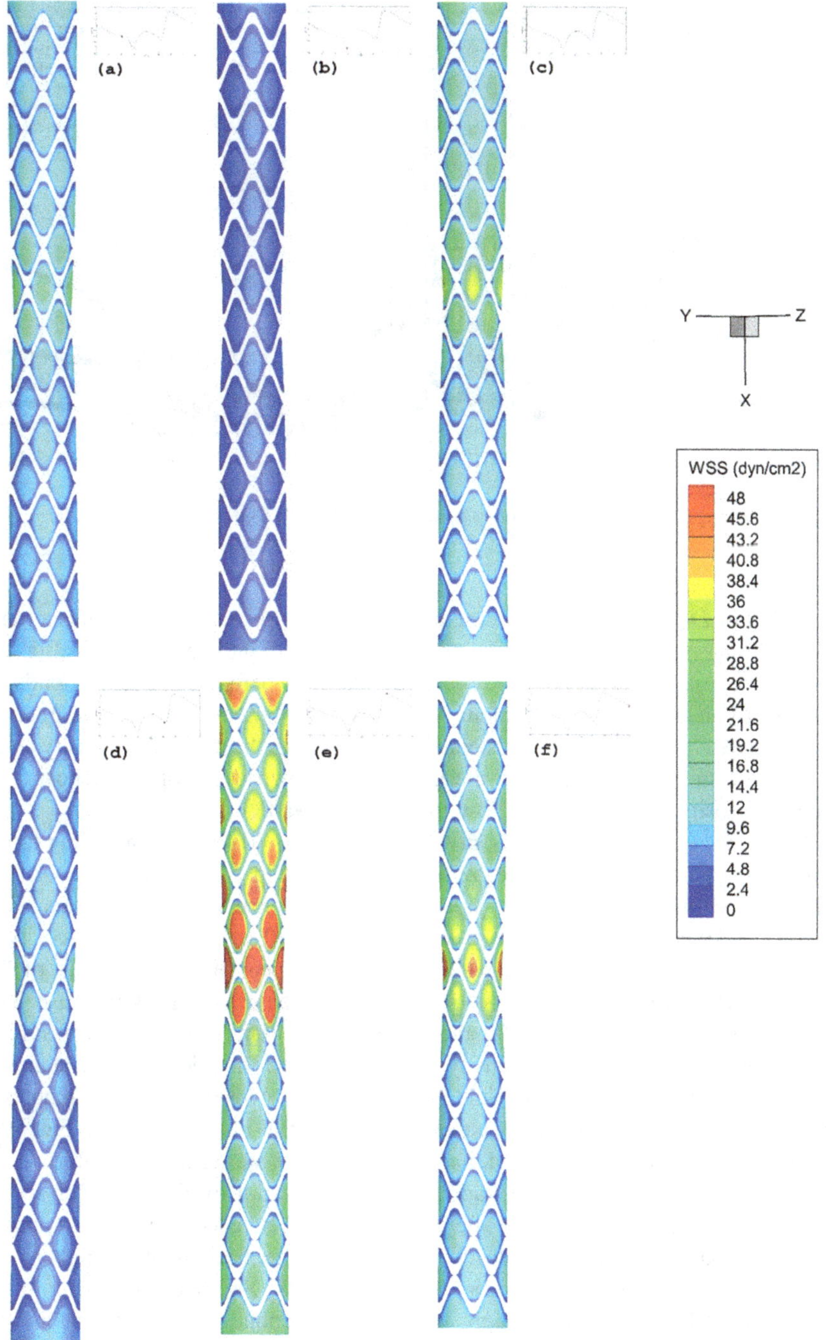

Fig. 28 Local WSS field (dyne/cm^2) during the cardiac cycle with DOR $= 30\%$ after: (**a**) $t = 0.2$ s, (**b**) $t = 0.3$ s (diastole), (**c**) $t = 0.4$ s, (**d**) $t = 0.5$ s, (**e**) $t = 0.6$ s (systole), (**f**) $t = 0.8$ s, Fig. 9 of [26]

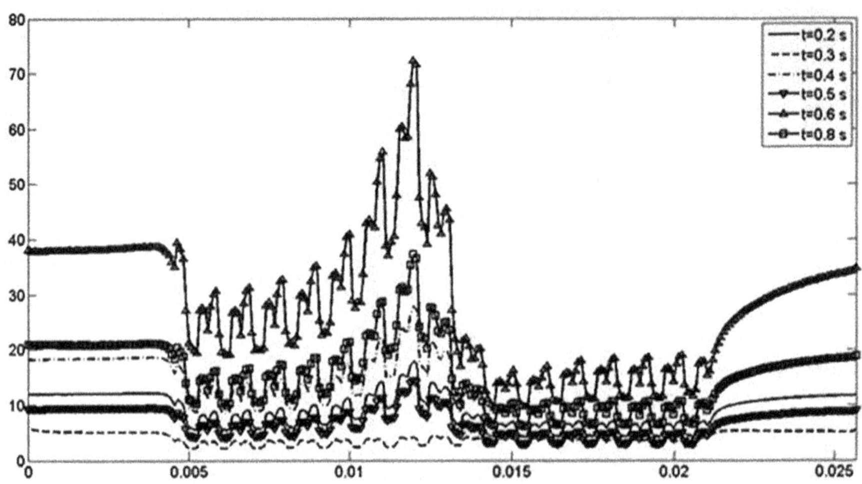

Fig. 29 Average WSS (dyne/cm^2) versus axial length (m) with DOR = 30% [4]

the effect of $DOR = 30\%$ is concerned, Fig. 58b, the apparent viscosity has a trend fairly similar to the case without stenosis, except slightly before and after the stenosis, particularly in the centre of the vessel, where the viscosity tends to decrease. This effect is more evident for the cases with $DOR = 60\%$, Fig. 58c. As far as the case with $DOR = 90\%$ is concerned, Fig. 58d, the apparent viscosity tends to decrease in correspondence to the neck, while downstream of the stenosis, the higher viscosity is concentrated close to the vessel wall.

Figure 59, [27] shows the viscosity patterns after $t = 0.4$ s, corresponding to the central time peak of the flow rate. With $DOR = 0\%$, Fig. 59a, the high viscosity is concentrated in the centre of the pipe, in correspondence with the low shear rate regions, and there is no secondary peak. With $DOR = 30\%$, Fig. 59b, the viscosity in the centre of the pipe decreases while it assumes a peak just downstream of the choking, where the velocity starts to gradually decrease. As far as the case with $DOR = 60\%$ is concerned, Fig. 59c, the field upstream of the stenosis is fairly similar to the previous case, but the viscosity decreases in correspondence to the stenosis, where a peak is present just after the necking, while downstream of the stenosis it assumes lower values, even in the centre of the vessel. With $DOR = 90\%$, Fig. 59d, these effects are more pronounced, and a peak near the vessel wall is visible.

Similar considerations can be made from the observation of Fig. 60, which shows the viscosity pattern in the systole after $t = 0.6$ s. The differences are small compared to the previous cases, but the viscosity values are smaller because of the higher flow rate, and the secondary peak appears for $DOR = 60\%$, Fig. 60c. According to the Casson model, the areas of high viscosity correspond to higher erythrocyte concentrations.

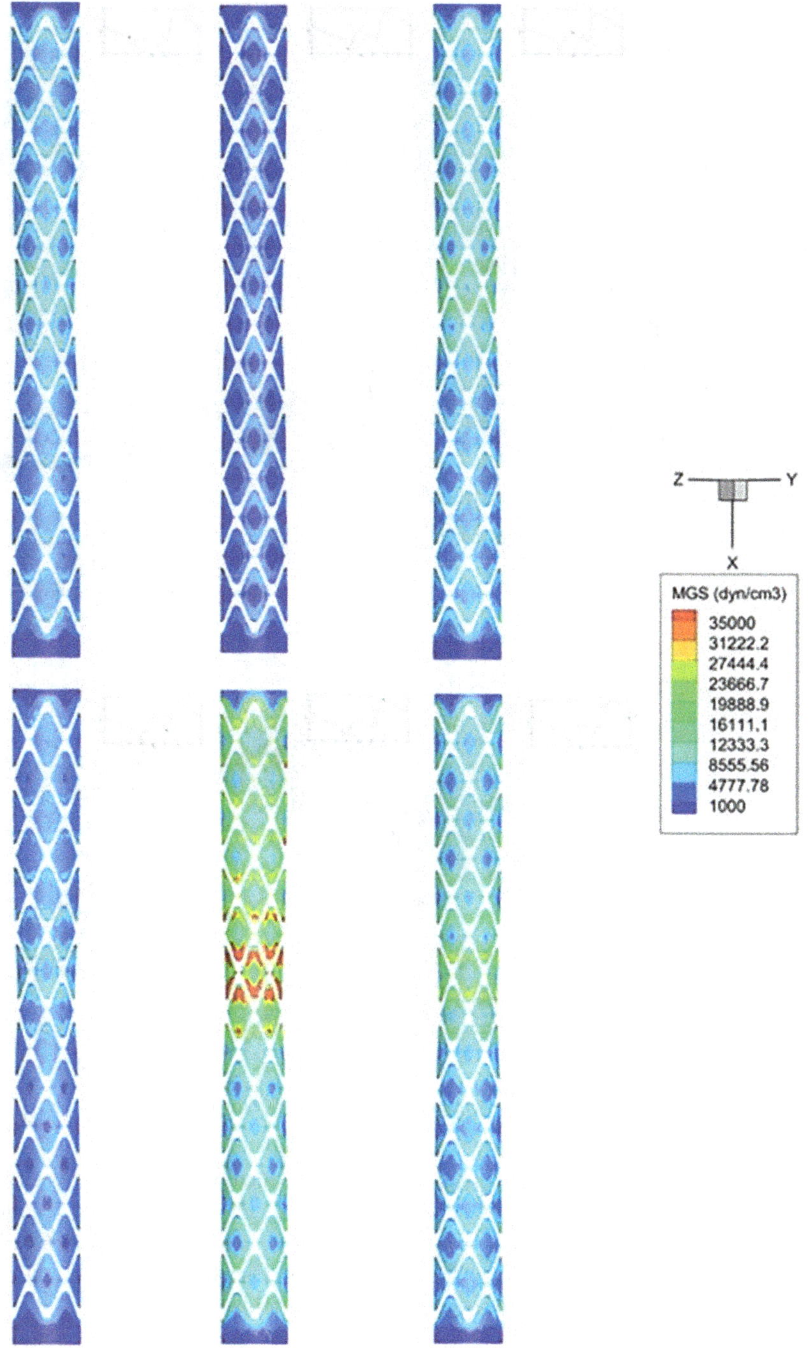

Fig. 30 Local MGS with DOR = 30% [4]

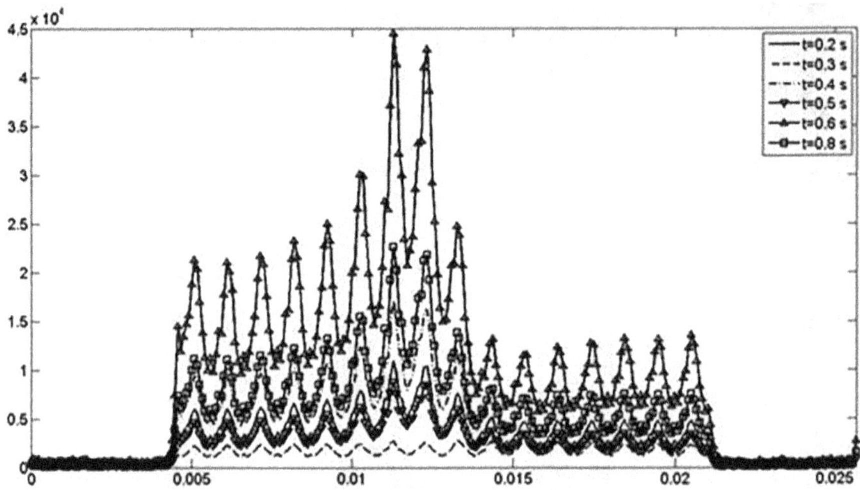

Fig. 31 Average MGS (dyne/cm^3) versus axial length (m) with DOR = 30% [4]

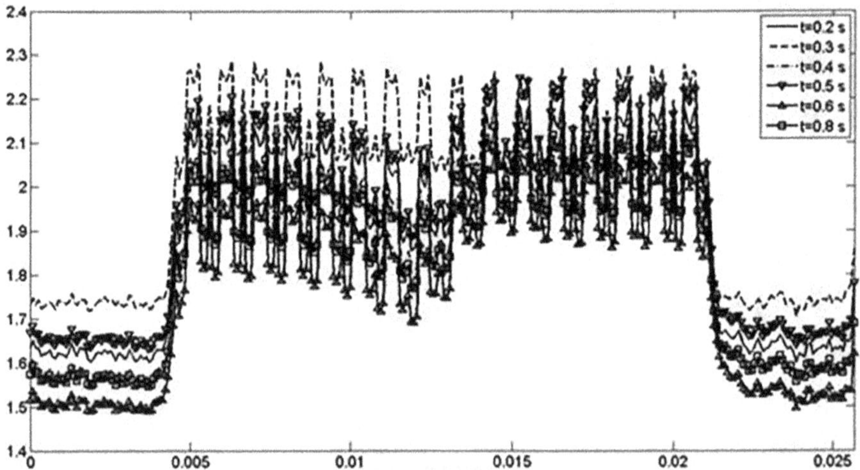

Fig. 32 Average OD versus axial length (m) with DOR = 30% [4]

In conclusion, the results with different *DORs* show that the stenosis increases the erythrocytes' concentration towards the wall, which is not a pathological situation, as the erythrocyte can move close to the wall in the diastole, as shown in Fig. 57. However, the presence of the stenosis increases the time spent by the red blood cells close to the wall, which can lead to pathological conditions.

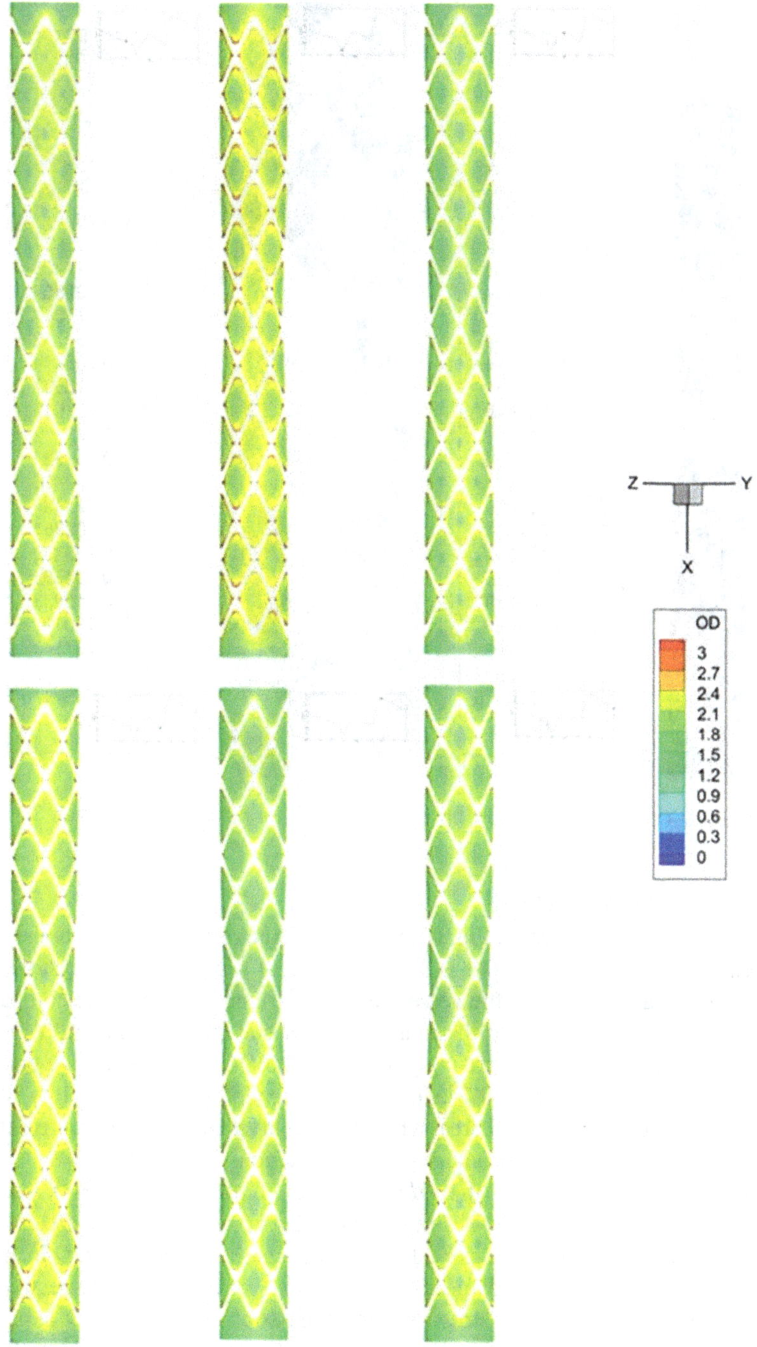

Fig. 33 Local OD with DOR = 30% [4]

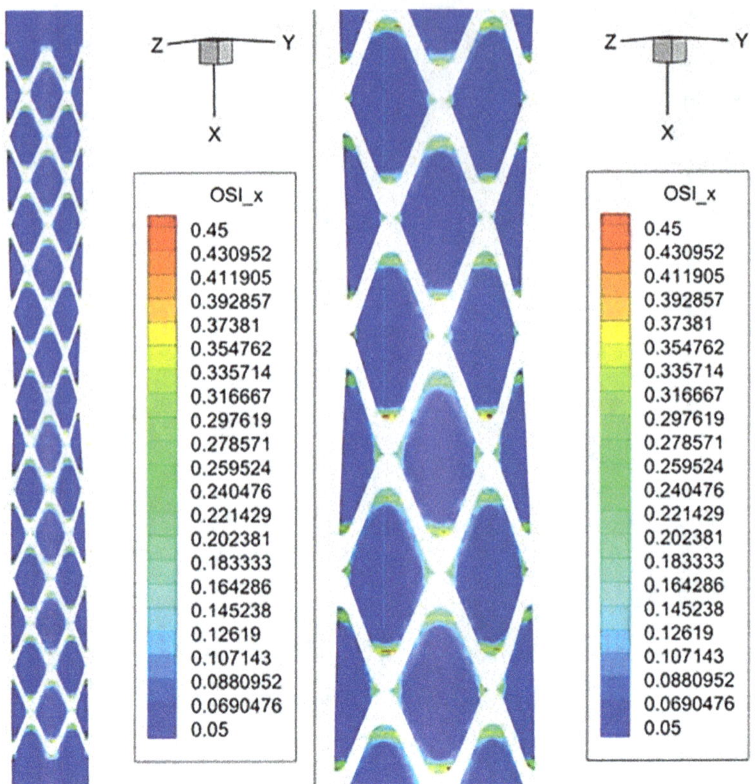

Fig. 34 Local OSI with DOR $= 30\%$ [4]

4.2.5 OD, OSI and RTT

As far as the *OD* is concerned, Fig. 61, [27], shows the results for the average *OD* at different instants of time: $t = 0.3$ s (diastole), $t = 0.4$ s and $t = 0.6$ s (systole). The *OD* results are averaged over the vessel perimeter. Figure 61 shows the comparison between the results of *OD* with the four *DOR* values at three instants: (a) $t = 0.3$ s, (b) $t = 0.4$ s, (c) $t = 0.6$ s. Surprisingly, the *OD* with *DOR* $= 90\%$ has lower values than the cases with *DOR* $= 0\%$ and 30%. The endothelial permeability should increase in a wide recirculation area, while the *OD* takes into account more extreme conditions, up to a certain percentage of area reduction.

The top of Fig. 61 shows the *OD* pattern in diastole, after $t = 0.3$ s, with *DOR* $= 0\%$, 30%, 60% and 90%. In the area of the vascular wall, the *OD* assumes physiological values before the stent. In the presence of the stent, the *OD* increases, assuming typical patterns with peaks and valleys, corresponding respectively to slices intersecting mostly the struts and the intra-strut areas. The *OD* assumes a peak where the *WSS* assumes very low values. The *OD* values are higher in diastole because of the lower velocity, and the presence of stenosis seems to have no effect, except with *DOR* $= 90\%$.

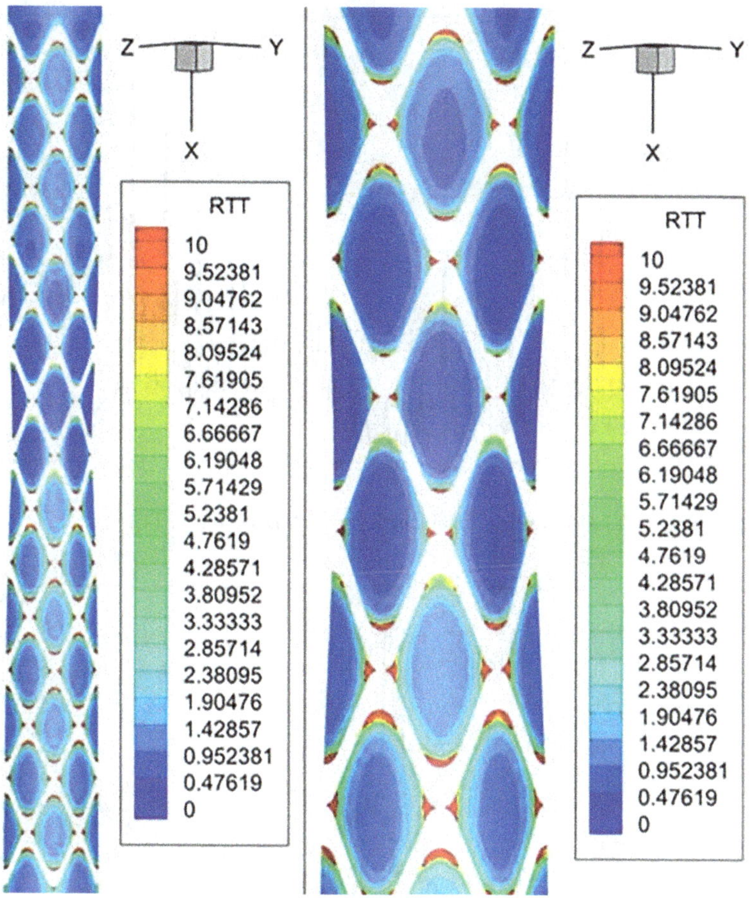

Fig. 35 Local RTT with DOR $= 30\%$ [4]

The second from the top of Fig. 61 shows the OD pattern after $t = 0.4$ s, with $DOR = 0\%$, 30%, 60%, 67.5%, 75% and 90%. The OD starts decreasing in correspondence with the necking, due to the very high MGS, while downstream, it assumes a different pattern, depending on the value of the DOR. In particular, it seems that with DOR greater than 60%, the OD pattern assumes a peak downstream of the stenosis, which moves along the axis with the increasing DOR, in correspondence to the end of the wide recirculation region. With $DOR = 90\%$, the peak is not visible, which is due to the longer recirculation region. In this case, the local flow is always directed upstream, and the WSS direction does not change. Moreover, the WSS is spatially more uniform, which contributes to a reduction of MGS and permeability. With $DOR = 30\%$, the OD returns to physiological values downstream of the stenosis.

The second from the bottom of Fig. 61 shows the OD pattern after $t = 0.6$ s, with $DOR = 0\%$, 30%, 60%, 67.5%, 75% and 90%.

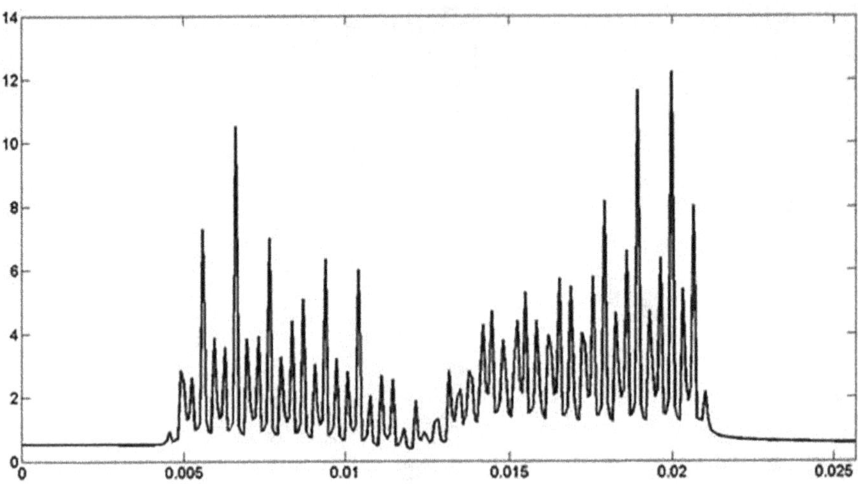

Fig. 36 Average RTT (1/Pa) versus axial length (m) with DOR = 30% [4]

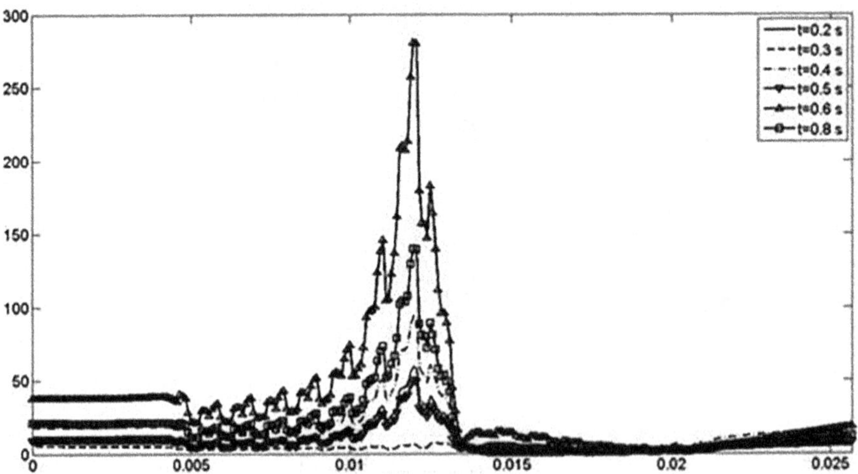

Fig. 37 Average WSS (dyne/cm^2) versus axial length (m) with DOR = 60% [4]

Figure 61c shows that in the systole, after $t = 0.6$ s, the most relevant *DORs* are 0%, 30%, 52.2%, 60%, 67.5% and 90%. At this instant of time, the *OD* assumes the lowest average values upstream of the stenosis because of the highest velocity reached by the blood flow during the cardiac cycle. Downstream of the stenosis, even with *DOR* = 52.2%, the local recirculation regions are present in correspondence with the meshes, where the *OD* assumes higher values. With *DOR* = 60%, a peak in correspondence to the end of the stent is present, and a wide recirculation region closes. The last peak is for *DOR* = 67.5%, while for *DOR* = 90%, the recirculation region is longer than the domain length.

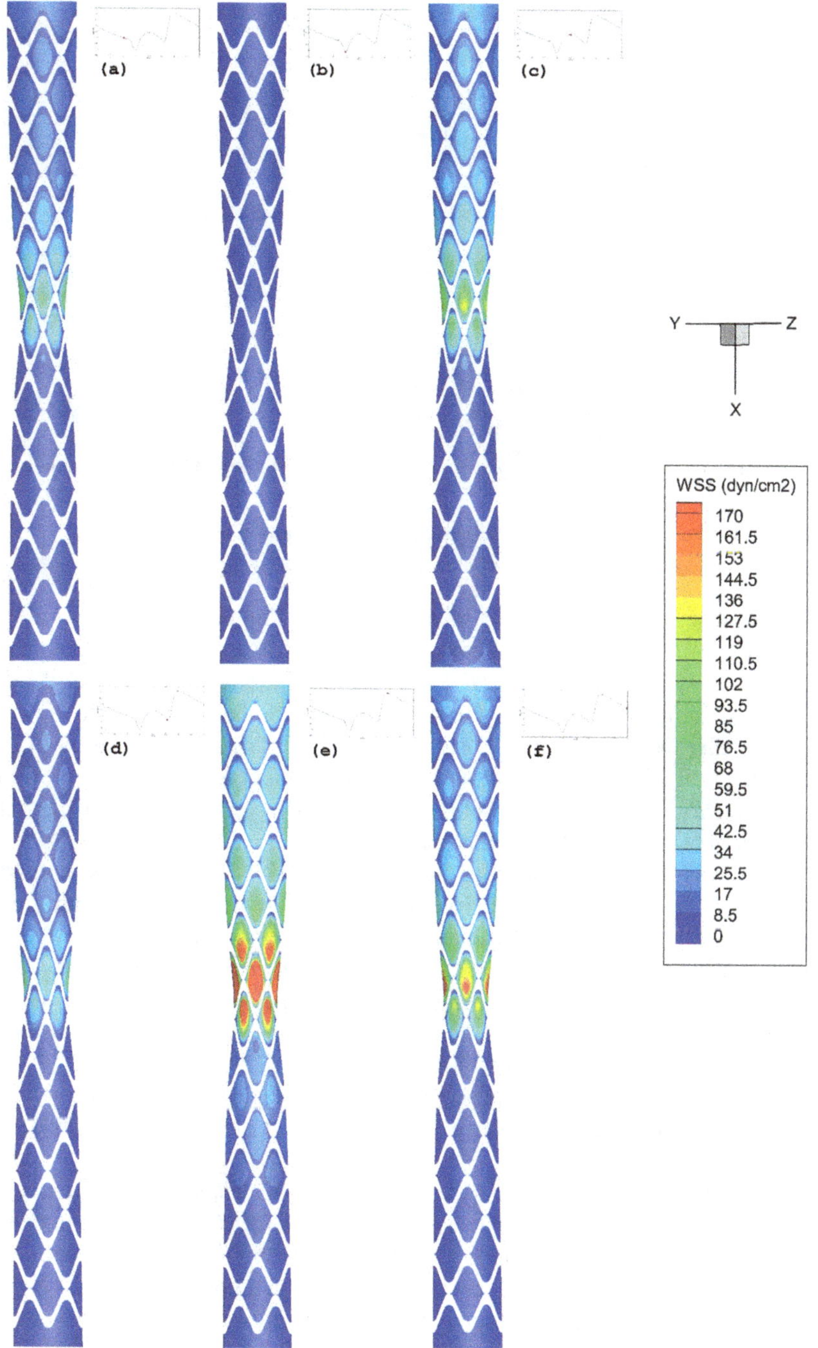

Fig. 38 Local WSS field during the cardiac cycle with $DOR = 60\%$ after: (**a**) $t = 0.2$ s, (**b**) $t = 0.3$ s (diastole), (**c**) $t = 0.4$ s, (**d**) $t = 0.5$ s, (**e**) $t = 0.6$ s (systole), (**f**) $t = 0.8$ s, Fig. 10 of [26]

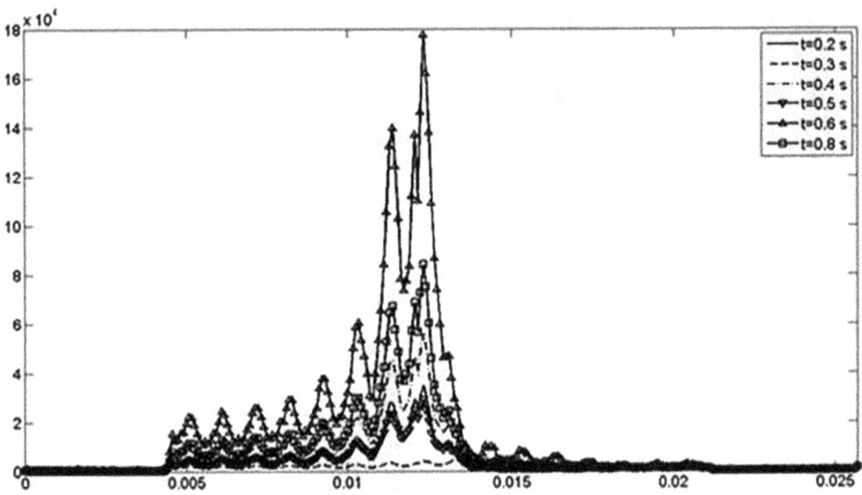

Fig. 39 Average MGS (dyne/cm^3) versus axial length (m) with DOR = 60% [4]

Similar considerations can also be made for *OSI*, Fig. 62, [27], because, during the cardiac cycle, it measures the deviation of the *WSS* vector from the main blood flow direction. In the case with *DOR* = 90%, the *WSS* vector, on the wall, is directed opposite to the main flow. Then, the situation seems to be better than in the cases with a smaller *DOR*, while it is an extreme situation. The *OSI*, defined by Eq. 17, measures the *WSS* deviation from its mean direction and is reported in Fig. 62 for different *DORs*. Although it is possible to define the OSI for all spatial directions, the change in the mean flow direction is a significant one.

Figure 62a presents the results for the case with *DOR* = 0%. The areas exposed to the non-zero *OSI* are located near the struts and close to the points with greater curvature, while in the intra-strut areas, the values approach zero. Since the *OSI* values that trigger atherosclerosis span between 0.1 and 0.5 [44], the case with *DOR* = 0% represents a safe scenario, as there are few areas in which the *OSI* assumes these values.

Figure 62b–d report the *OSI* values with *DOR* = 52.5%, 60% and 67.5%. The *OSI* assumes greater values, in the range of 0.2–0.5, where the fluid is detached from the wall and where, consequently, a recirculation region is present. This is evident in the centre of each mesh.

Figure 62e shows that, for the greatest *DOR* investigated, 90%, the *OSI* assumes hazardous values mostly at the beginning and the end of the recirculation zone. In these two regions, the *OSI* trend is similar to that with *DOR* = 0%, but the *WSS* vectors are oriented in the direction opposite to the main flow.

With *DOR* = 90%, *OSI* assumes greater values only immediately beyond the stenosis, where the recirculation region originates, but the values are very similar to the case with *DOR* = 0%, as evidenced by Fig. 63.

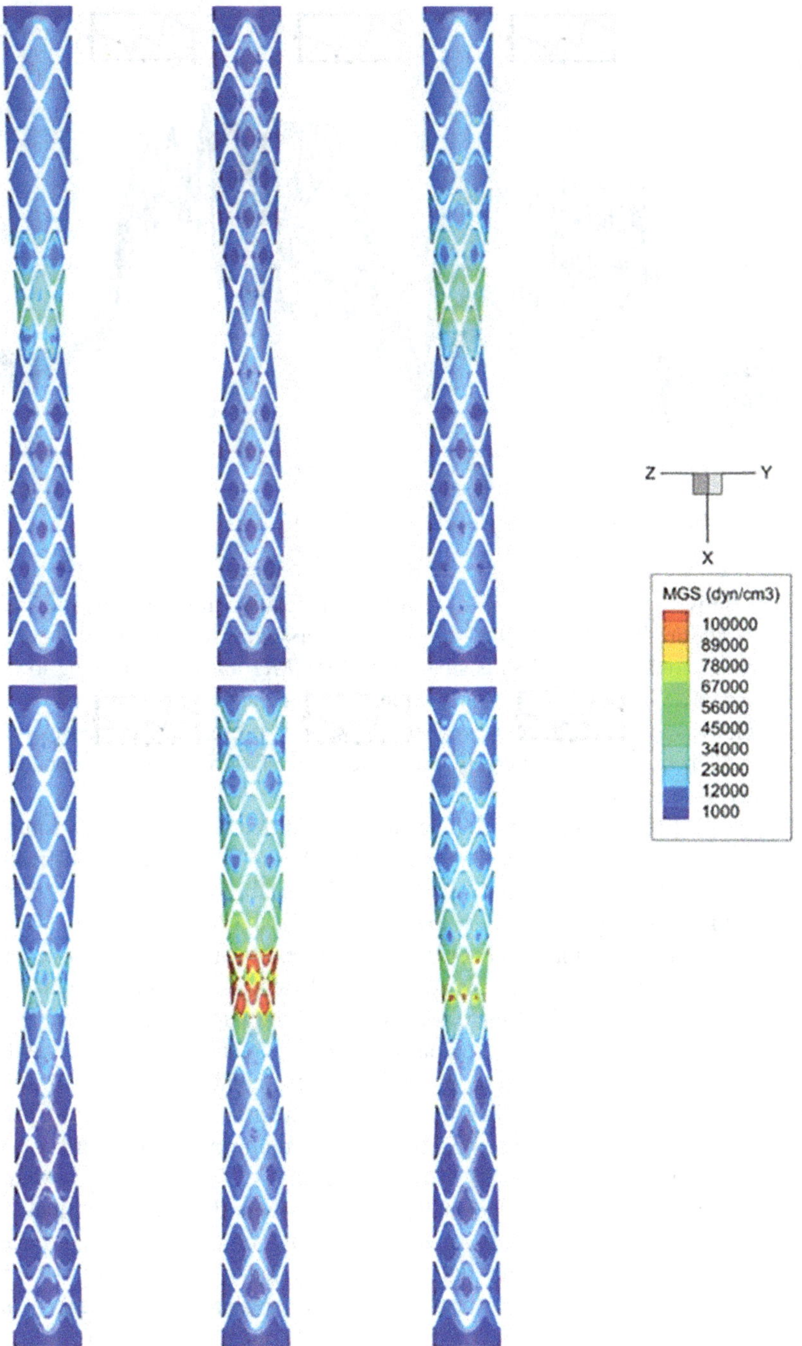

Fig. 40 Local MGS with DOR = 60% [4]

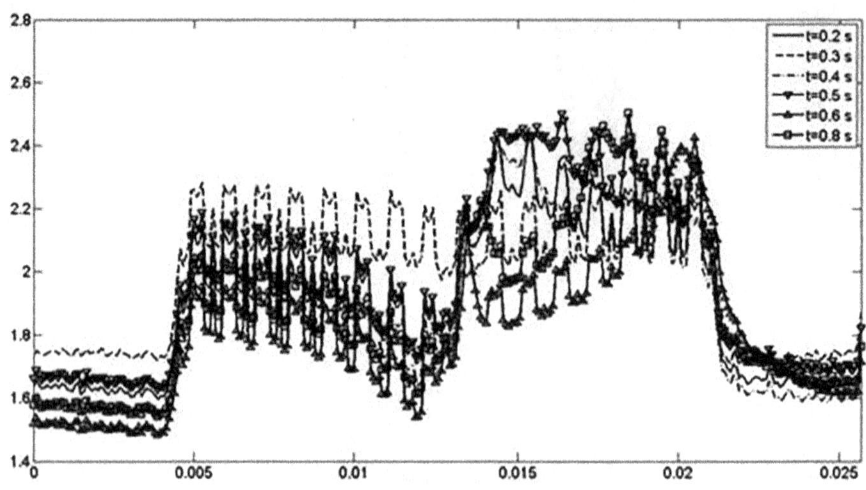

Fig. 41 Average OD versus axial length (m) with DOR = 60% [4]

The relative residence time (*RRT*) measures the relative time spent by the fluid in a certain region during the flow. This parameter assumes higher values where the *OSI* is closer to 0.5 and *TAWSS* is small; therefore, *RRT* allows identifying low and oscillating *WSS* regions, which are deemed to indicate areas prone to develop atherosclerosis. The *RRT* increases where the stent is present and decreases on the vessel wall because it is inversely proportional to *TAWSS*, which is smaller in proximity to the stent.

Figure 64, [27], reports the RRT versus the axial length with several values of *DOR*: 0%, 52.2%, 60%, 67.5%, 75% and 90%. The *RRT* increases downstream of the stenosis, with *DOR* = 52.2%, because of the local recirculation in correspondence to the centre of the mesh. With *DOR* = 60%, the *RRT* beyond the stenosis assumes higher values than with *DOR* = 52.2%, except in a region immediately downstream of the first peak, due to the detachment of the fluid stream. With *DOR* = 67.5% and 75%, *RRT* increases with a peak moving along the axial length, which is due to the stable recirculation region downstream of the stenosis, whose width is proportional to the value of *DOR*. In these conditions, the local flow is always directed upstream, and therefore, the *WSS* direction does not change, except close to the struts. Moreover, the *WSS* field is more uniform, which contributes to a reduction of *MGS* and permeability. Nevertheless, the *WSS* assumes very low values at the end of the recirculation region. The regions where the vortex terminates are susceptible to atherosclerotic lesions, as indicated by non-uniform haemo-dynamic indicators, such as the near-zero *WSS* and the elevated *WSS* gradients, as well as blood particle accumulation and deposition. This explains why the peak of *RRT* assumes the trend shown in Fig. 64. With the increase of *DOR*, the recirculation zone ends further downstream of the stenosis, and with *DOR* = 90%, this happens several diameters downstream of the end of the stent and is not visible in Fig. 64. *RTT* should

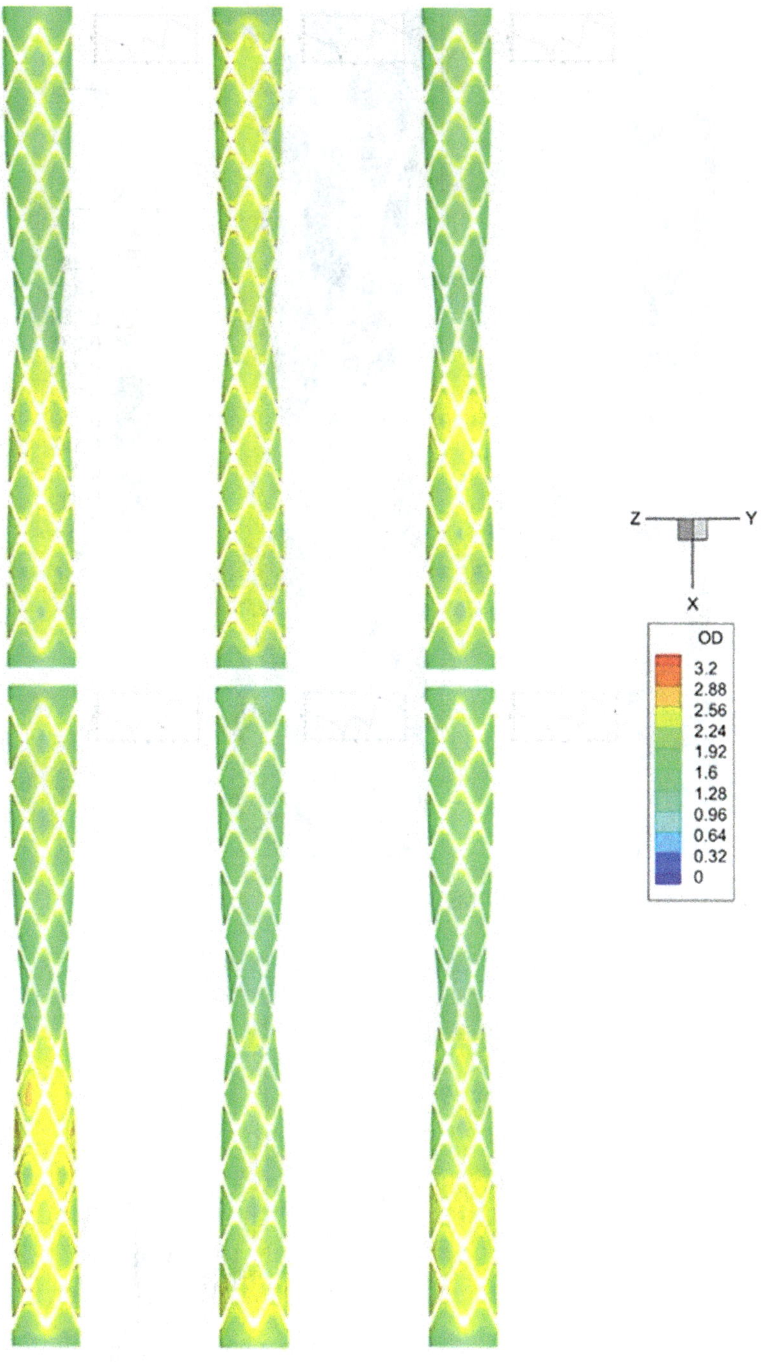

Fig. 42 Local OD with DOR = 60% [4]

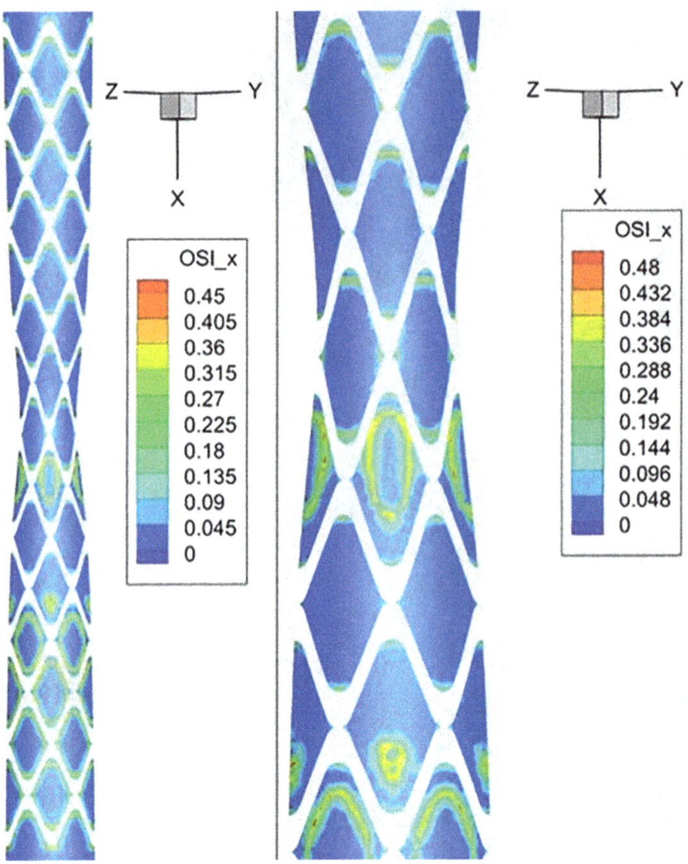

Fig. 43 Local OSI with DOR = 60% [4]

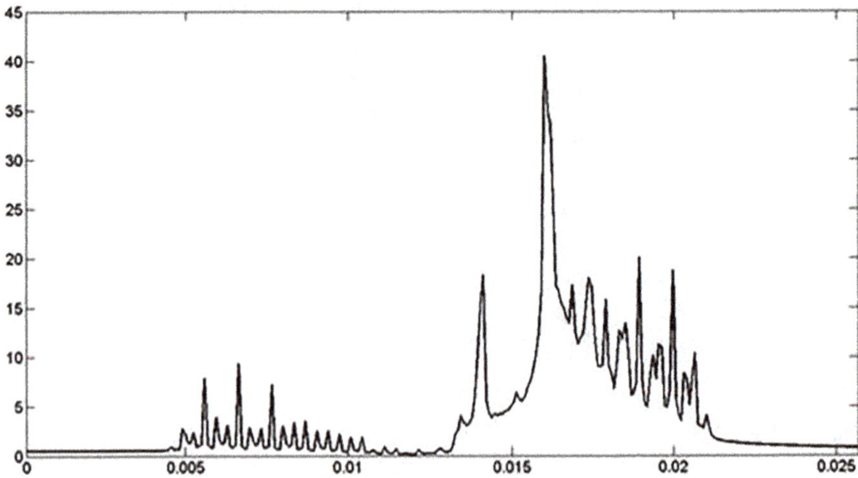

Fig. 44 Average RTT (1/Pa) versus axial length (m) with DOR = 60% [4]

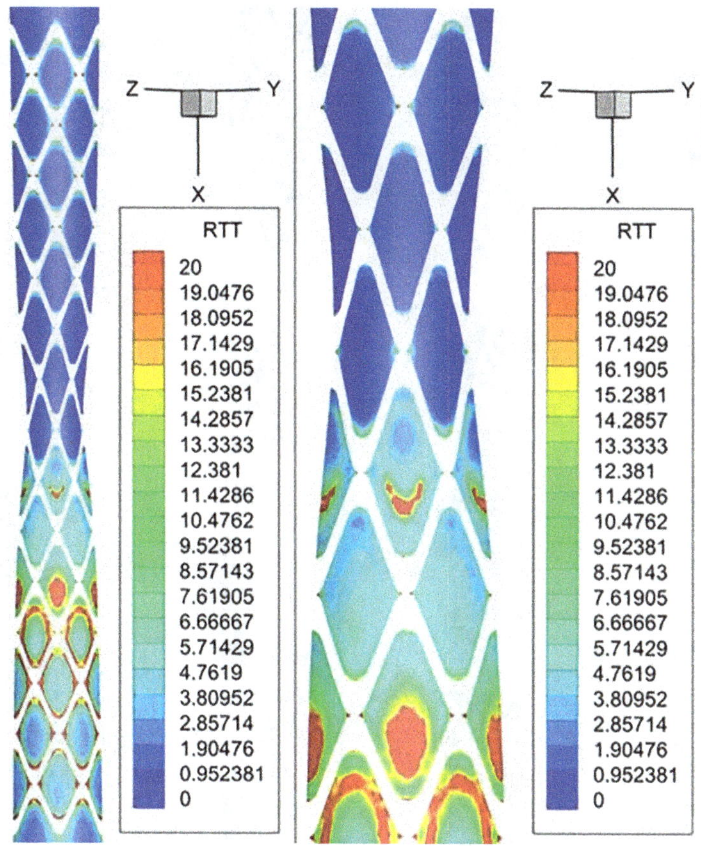

Fig. 45 Local RTT with DOR = 60% [4]

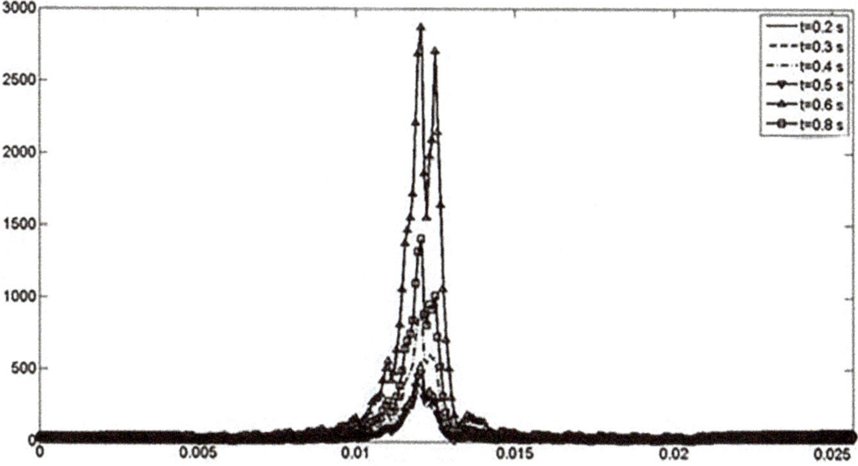

Fig. 46 Average WSS (dyne/cm^2) versus axial length (m) with DOR = 90% [4]

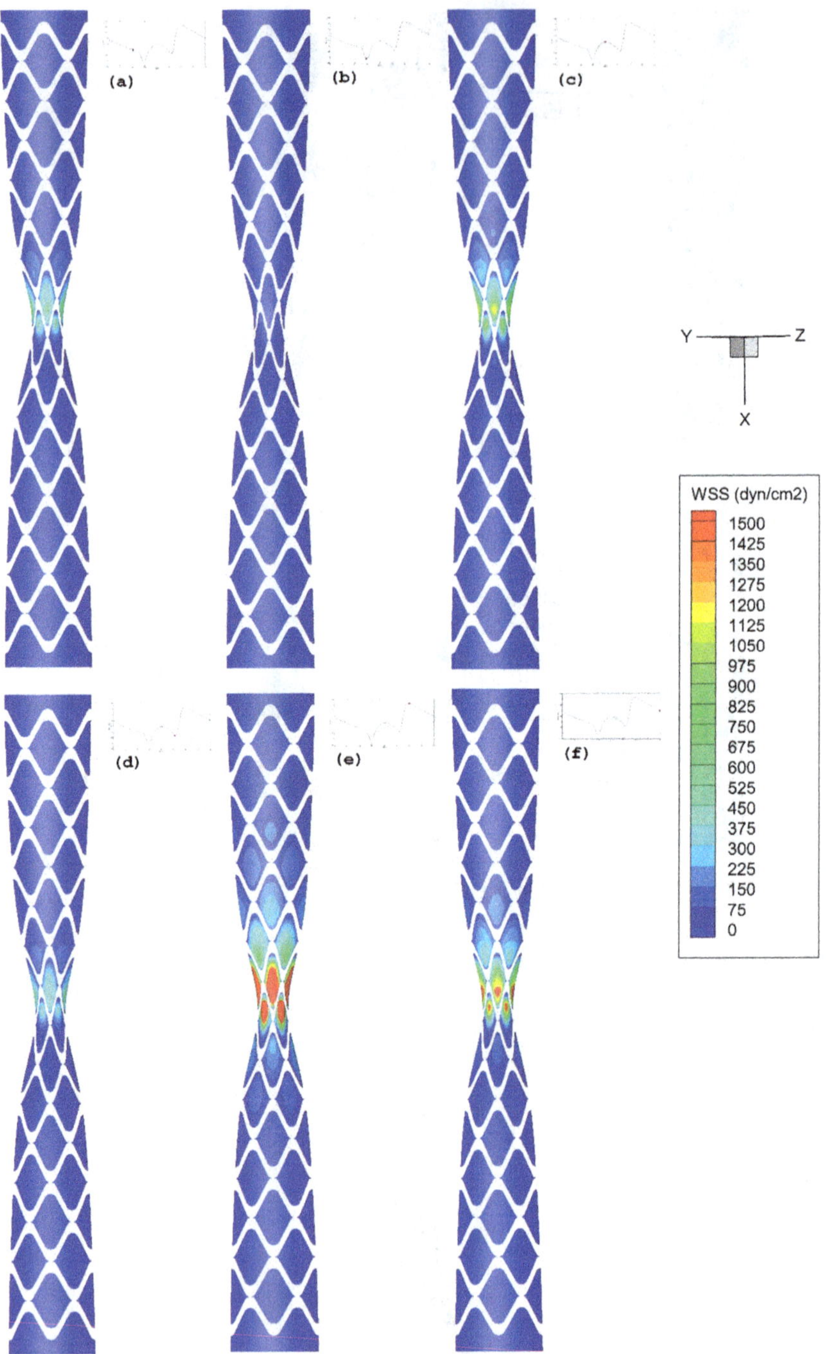

Fig. 47 Local WSS field during the cardiac cycle with DOR $= 90\%$ after: (**a**) $t = 0.2$ s, (**b**) $t = 0.3$ s (diastole), (**c**) $t = 0.4$ s, (**d**) $t = 0.5$ s, (**e**) $t = 0.6$ s (systole), (**f**) $t = 0.8$ s, Fig. 11 of [26]

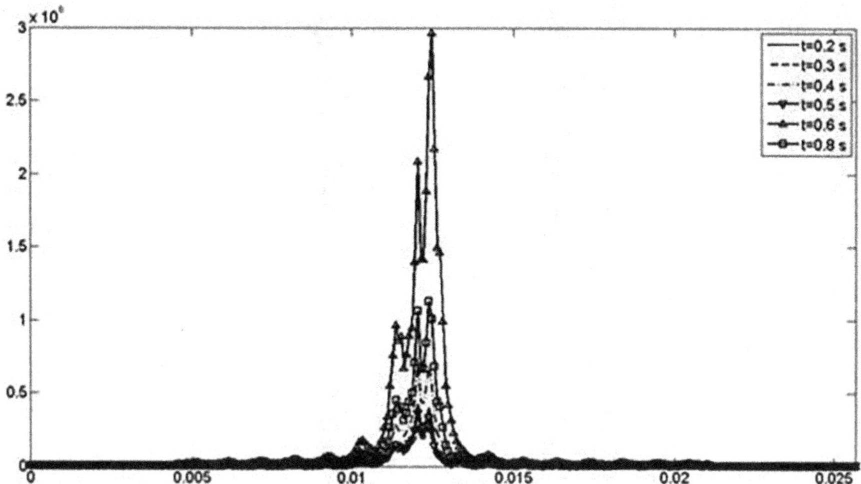

Fig. 48 Average MGS (dyne/cm^3) versus axial length (m) with DOR = 90%, [4]

be higher where the endothelium is more prone to infiltrate cells and macromolecules, which is expected where a recirculation region is present. *RTT* takes into account increasingly adverse conditions with *DOR* = 30% and 60%, but with *DOR* = 90%, it provides better values, as shown by Fig. 64.

The results obtained for *OD*, *OSI* and *RTT*, which describe the endothelial permeability, present a better situation with *DOR* = 90% compared to *DOR* = 60%. The importance of these parameters lies in their ability to predict post-stent restenosis. With a greater *DOR*, a stable recirculation region forms downstream of the stenosis, the local flow always directs upstream, and, therefore, the *WSS* direction does not change, except close to the struts. Moreover, *WSS* is flatter, contributing to a reduction of both *MGS* and permeability. Certainly, the degree of danger associated with stenosis is not due only to haemo-dynamics but also to structural factors. Unlike the abovementioned parameters, *WSS* increases monotonically with the degree of stenosis, and, therefore, the plaque is subject to higher stresses, which increases the likelihood of breaking it. In conclusion, there seems to be a *DOR*, between 30% and 90%, for which there is a maximum of endothelial permeability, and *OSI*, *OD* and *RTT* assume a maximum value.

5   Thermo-Haemo-Dynamics of Cypher Coronary Stents

The aim of this part of the chapter, based on [4], is to perform a numerical *CFD* simulation, in a realistic 3D geometry, of the two *Cypher* coronary stents already investigated in the previous chapter with Newtonian blood, IV.I, acting under physiological conditions, and to investigate the influence of the non-Newtonian behaviour of the blood, modelled as a Casson fluid. The artery is modelled as in

A. Boghi et al.

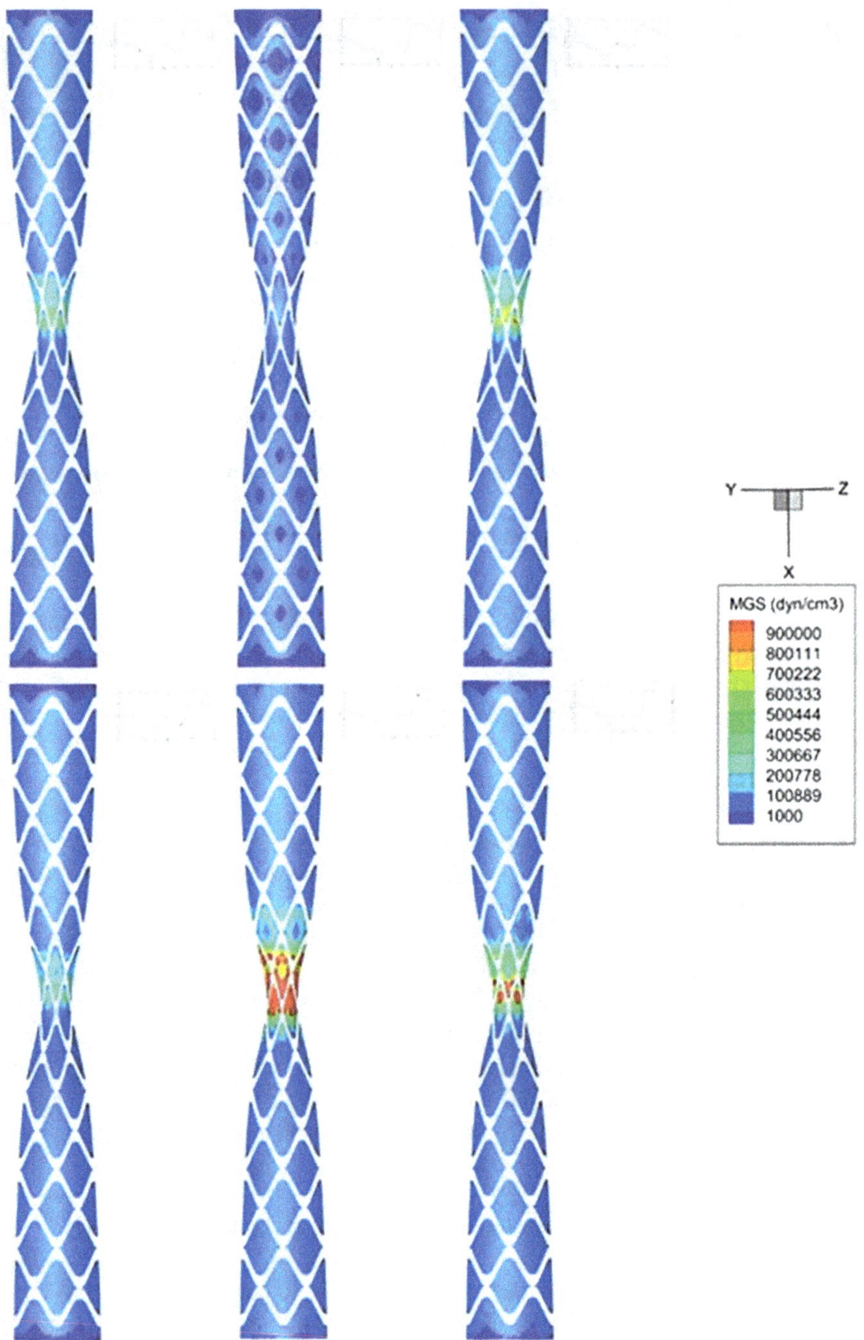

Fig. 49 Local MGS with DOR = 90% [4]

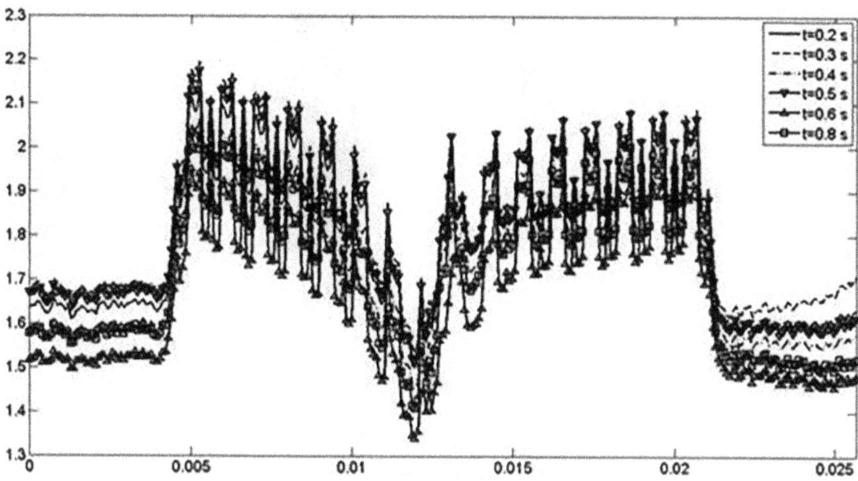

Fig. 50 Average OD versus axial length (m) with DOR = 90% [4]

the previous chapter, and the same numerical code is employed on the same calculation grid. The simulations are carried out in steady-state and unsteady-state conditions.

5.1   Results

The residuals of the numerical simulations, after 2100 iterations, for the conservation of mass and the three components of velocity are less than 10^{-6} in all cases. In the unsteady state, the total investigated time (0.8 s) is divided into 800 time intervals, and 20 iterations are made for each period.

5.1.1   Steady State

The results of the *WSS* are presented in Fig. 65, [4], for each of the two stents, and compared with the similar results obtained with Newtonian blood in the previous chapter. Stagnation zones are present around and between the stent structures with lower *WSS* than in the Newtonian case in both stents, whereas, away from the structures, there are areas with higher *WSS*. The diameter of the vein is reduced on the surface of the stent, producing convergence of the current lines and higher *WSS*, whereas, after the stent structure, larger diameters are present, and diverging current lines reduce the *WSS*. The divergence effect is smaller and disappears in a frame diameter, with higher *WSS* in the centre of each stent. The greater values of the *WSS* are present on the surface of the stent, and the smaller values are around the stent structure, i.e. in the transition region between the vein and stent. The conclusion is a greater difference between *WSS* in a small region.

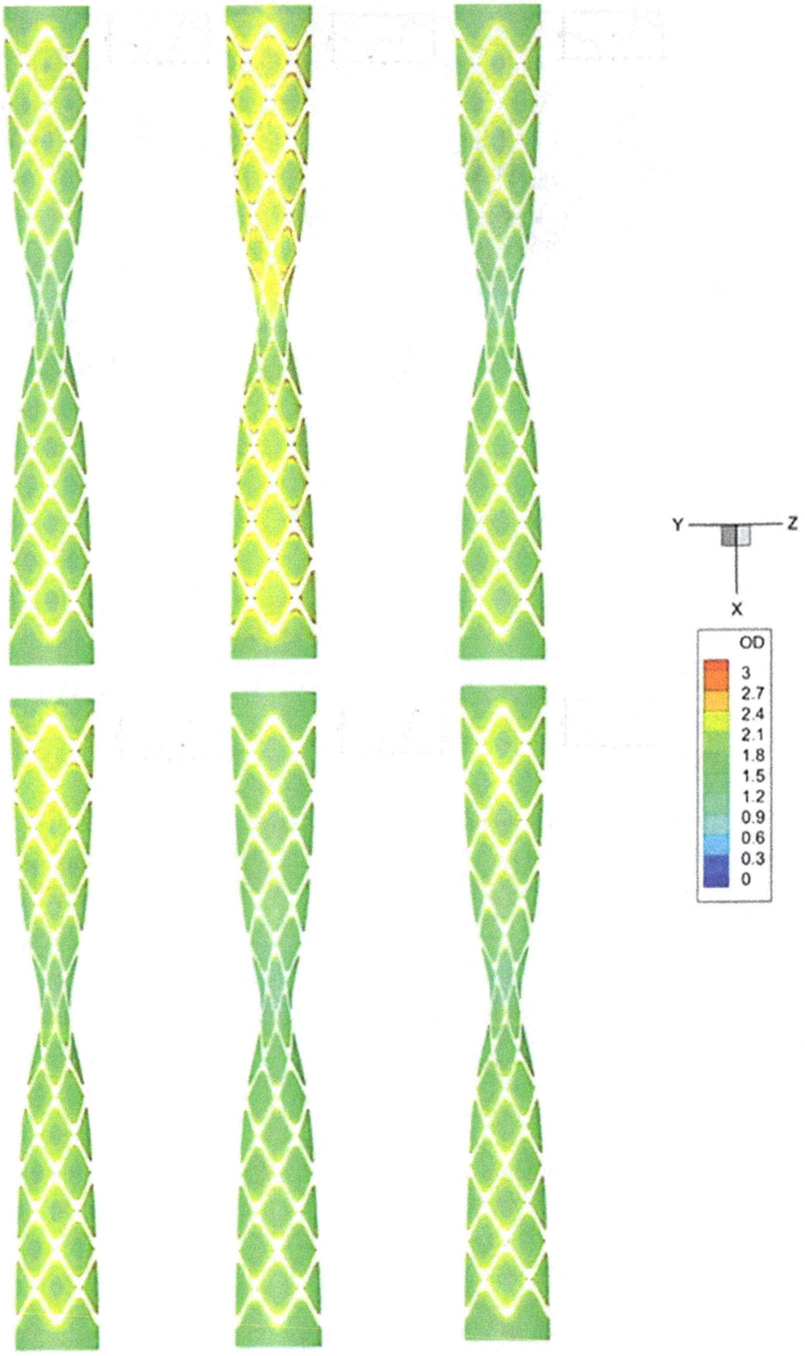

Fig. 51 Local OD with DOR = 90% [4]

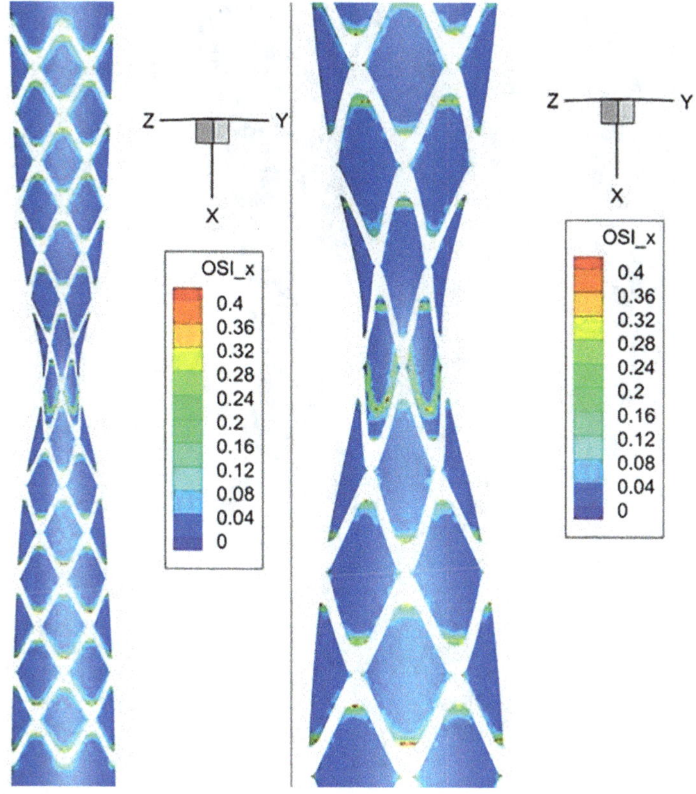

Fig. 52 Local OSI with DOR $= 90\%$ [4]

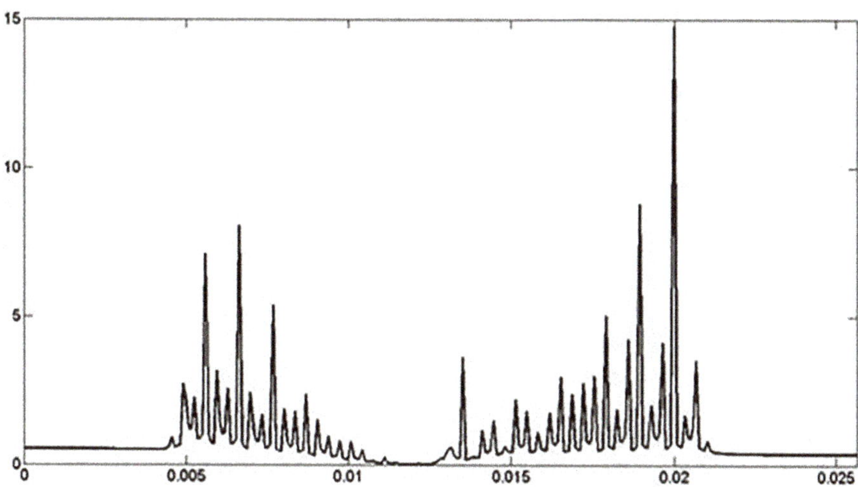

Fig. 53 Average RTT (1/Pa) versus axial length (m) with DOR $= 90\%$ [4]

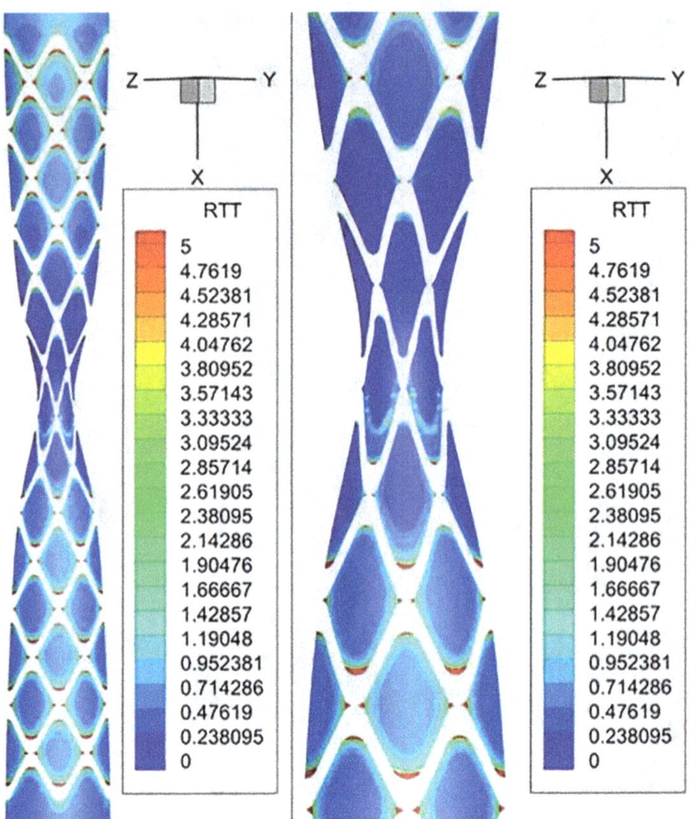

Fig. 54 Local RTT with DOR $= 90\%$ [4]

5.1.2 Unsteady State

Unsteady-state numerical simulations are performed during two periods of the cardiac cycle, and Fig. 66, [4], shows the distribution of *WSS* at the instants of time 0.84 s, 0.88 s, 1.04 s, 1.28 s, 1.44 s and 1.6 s. The velocity increases with pressure and is zero on the stent and the wall. In the duct, the velocity increases with the systolic wave, causing an increase in the velocity gradient and *WSS*. The mean value of the *WSS* on the wall is described by a wave with the same pattern as the systolic wave. The stent on the artery wall is a fluid dynamic obstacle, which is small in comparison to the diameter of the artery but large enough to create a stagnation zone around it. These stagnation zones are present in all cardiac cycles, while their extent decreases with increasing speed. The areas around the stent structures exhibit low *WSS* values for all cardiac cycles due to the stagnation zones, while the *WSS* has large variations during the cycle near the artery wall, where the *WSS* ranges between 0 and 2.5 Pa.

The flow of the non-Newtonian fluid behaves differently from the Newtonian fluid [17]. The maximum values of the *WSS* are present in the centre of the stent structure, and the minimums are on the surface close to the stent wires. One can then estimate the intra-structural areas exposed to low *WSS* (i.e. less than 0.5 Pa) during the cardiac cycle. Percentage area, *P*, is defined as the area of the artery wall exposed to *WSS* values that are below the critical value, according to

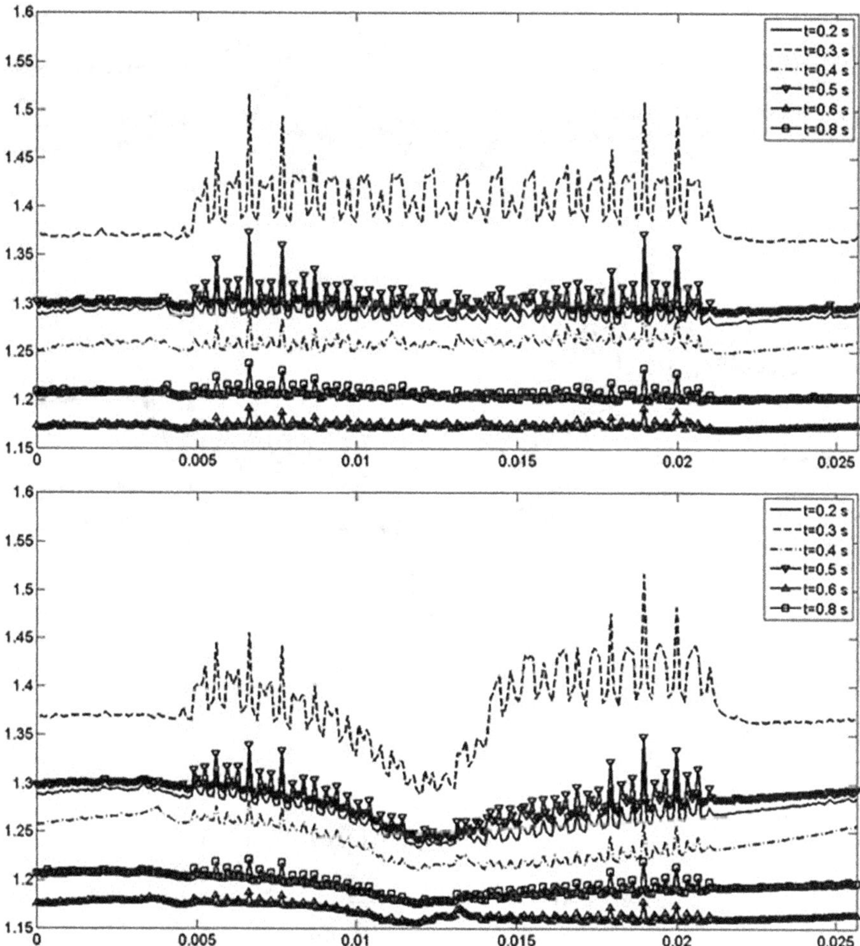

Fig. 55 Apparent viscosity averaged across the section versus axial length (m) at four instants: (**a**) $t = 0.3$ s, (**b**) $t = 0.4$ s, (**c**) $t = 0.5$ s, (**d**) $t = 0.6$ s. Top: DOR = 0%; second from top: DOR = 30%; second from bottom: DOR = 60%; bottom: DOR = 90% [4]

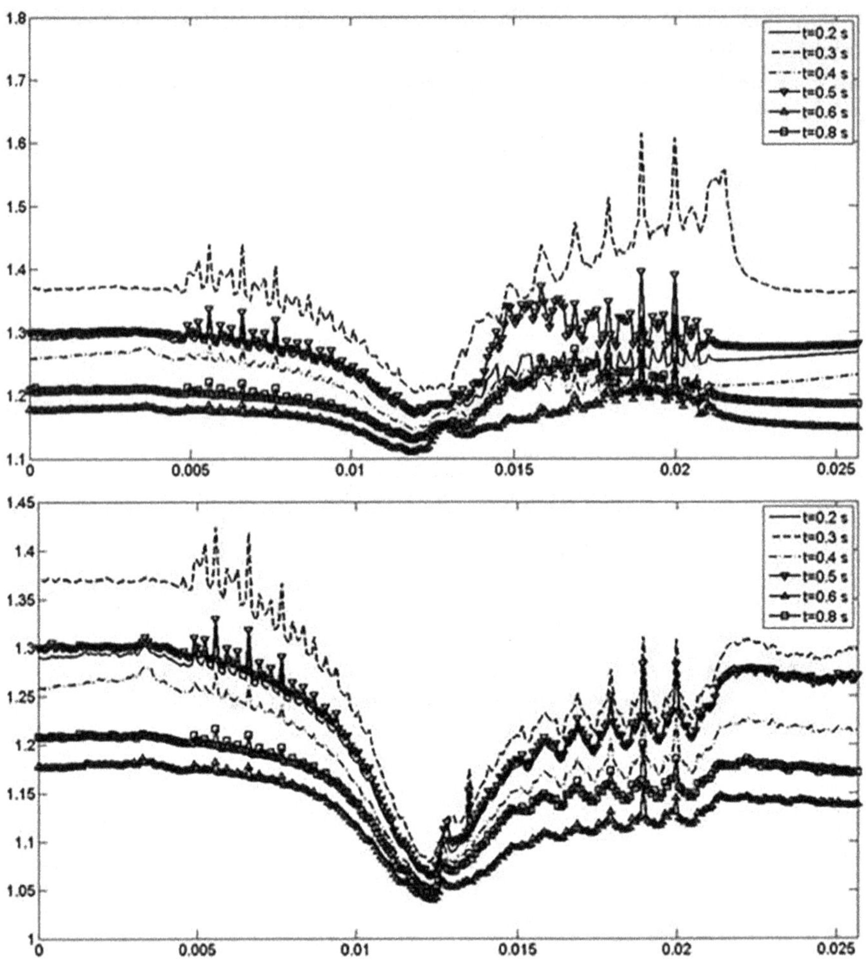

Fig. 55 (continued)

$$P(\%) = 100 \, \frac{\displaystyle\int_{\text{vesel wall}} \tfrac{1}{2}\left(1 - \left(\frac{\text{WSS} - 0.5}{\|\text{WSS} - 0.5\|}\right)\right) dA}{\displaystyle\int_{\text{vesel wall}} dA} \tag{20}$$

The percentage area is assessed for both stents (S1 and S2) at each time interval of the cardiac cycle. The results are shown in Fig. 67 and indicate that the closed-cell stent S1 has a lower P with low *WSS*.

The values of P in the stent region, exposed to WSS smaller than 0.5 Pa, are reduced as the flow rate increases and reach the minimum at the point of the cardiac cycle corresponding to the peak velocity. As WSS increases with velocity, the endothelium surface is exposed to higher WSS on average during systole, decreasing the risk of developing neo-intimal hyperplasia. Comparing these results with those of [17] and of the previous chapter, it can be concluded that simulations with Newtonian fluid have the effect of overestimating the critical area. The results shown in Fig. 67 indicate a significant percentage area with WSS below the critical value for the two stents investigated, in agreement with [55].

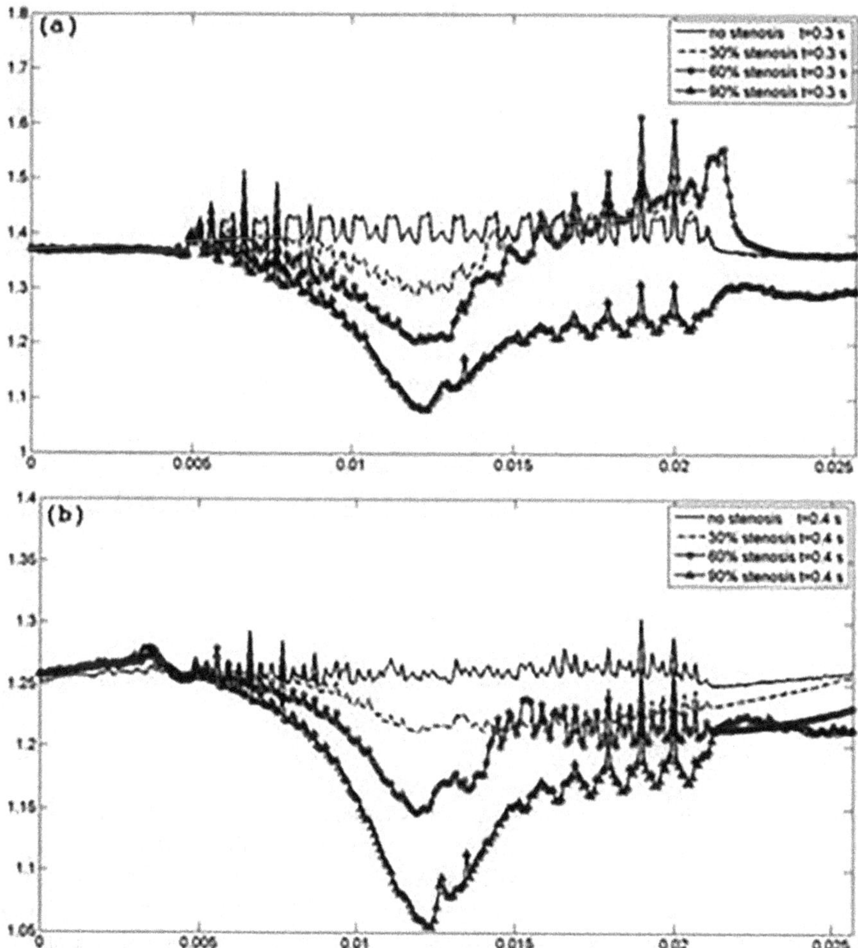

Fig. 56 Apparent viscosity μ/μ_∞ versus axial length (m) at four instants: (**a**) $t = 0.3$ s, (**b**) $t = 0.4$ s, (**c**) $t = 0.5$ s, (**d**) $t = 0.6$ s. Top: DOR $= 0\%$; second from top: DOR $= 30\%$; second from bottom: DOR $= 60\%$; bottom: DOR $= 90\%$ [4]

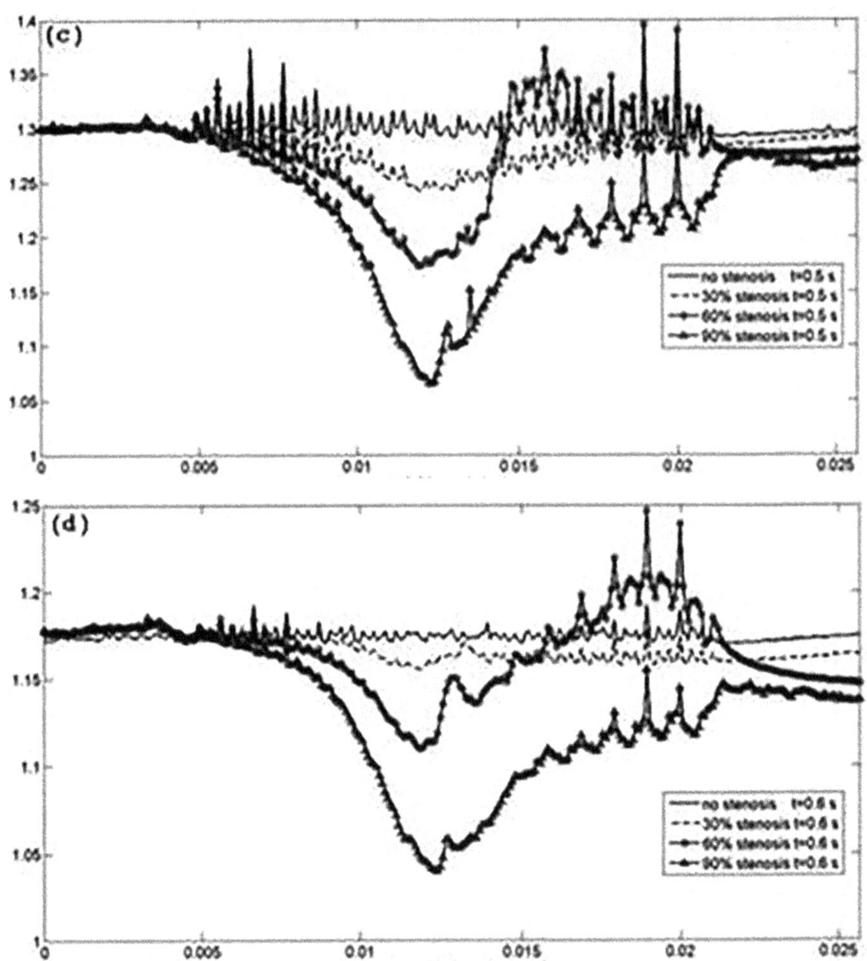

Fig. 56 (continued)

5.2 Discussion

Physiological values of *WSS* range from 0.5 Pa to 0.8–1 Pa [43, 44, 56]. Lower *WSS* values are strongly correlated with endothelial permeability and may promote neo-intimal hyperplasia, while higher values are unrelated to neo-intimal hyperplasia. The steady-state results, presented in Fig. 65, show that in the artery wall regions, there is a zone where the *WSS* is uniformly less than 0.3 Pa. There are no substantial differences between the two stents, but these areas have a greater extent in the transverse direction. The differences in the critical values of both stents are around 5%, but most of the artery wall is subject to *WSS* below the critical value.

The unsteady state results shown in Fig. 66 show a large variation in *WSS* during the cardiac cycle. Regions far from the stent have greater variation, while regions close to the stent have smaller variation, allowing it to be permanently exposed to low critical *WSS* values.

The comparison between these results and those obtained in [17] and in the previous chapter for Newtonian fluid allows us to conclude that the Newtonian behaviour underestimates the *WSS*, with the clinical consequence of overestimating the area at risk of restenosis, in agreement with the results of [20].

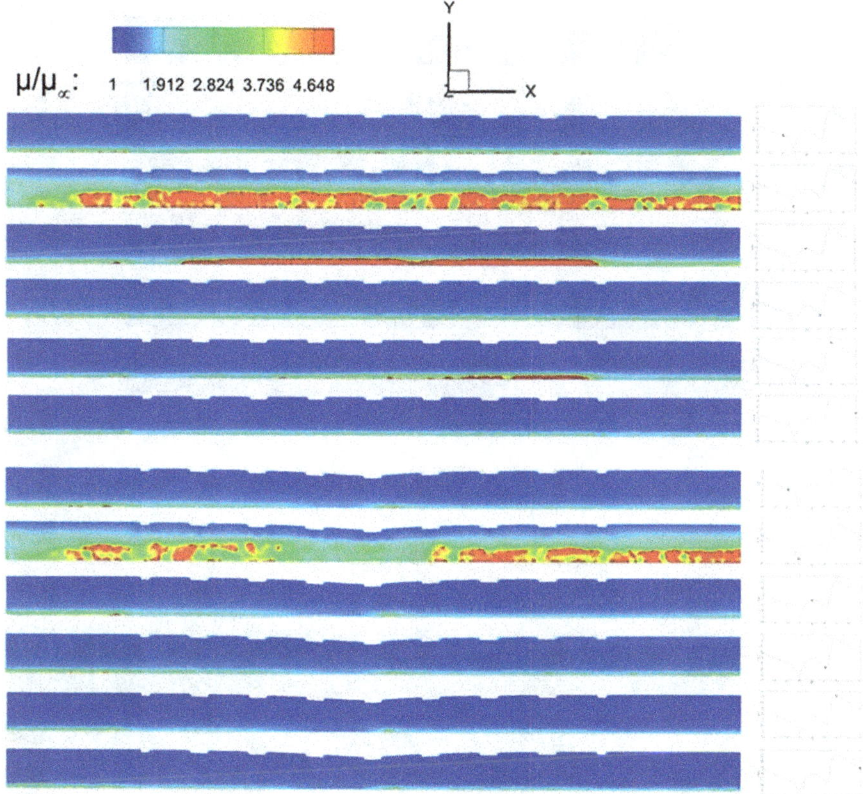

Fig. 57 Apparent viscosity μ/μ_∞ during the cardiac cycle. Top: DOR $= 0\%$; second from top: DOR $= 30\%$; second from bottom: DOR $= 60\%$; bottom: DOR $= 90\%$ [4]

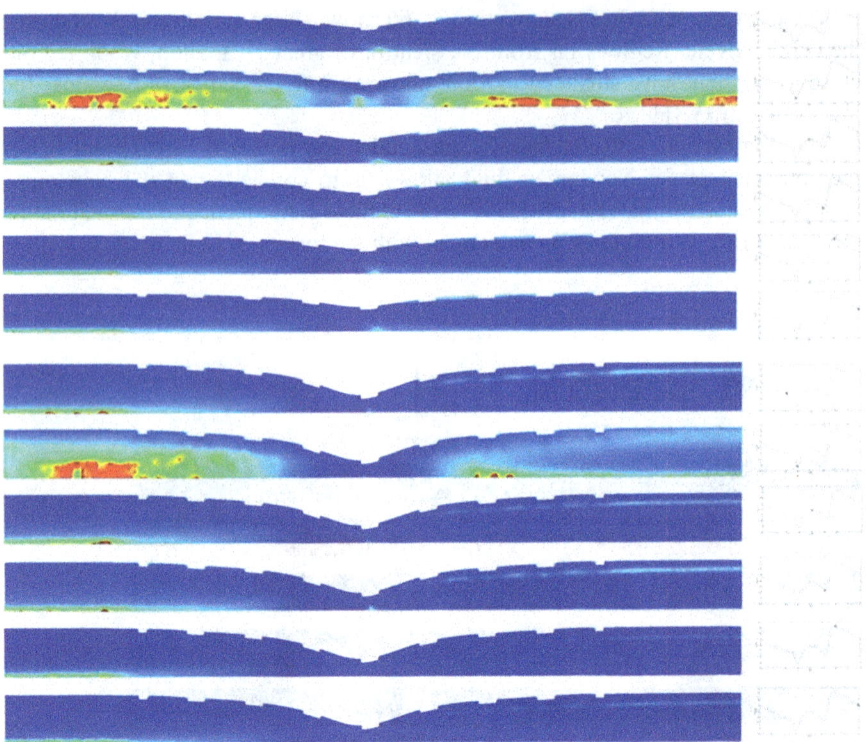

Fig. 57 (continued)

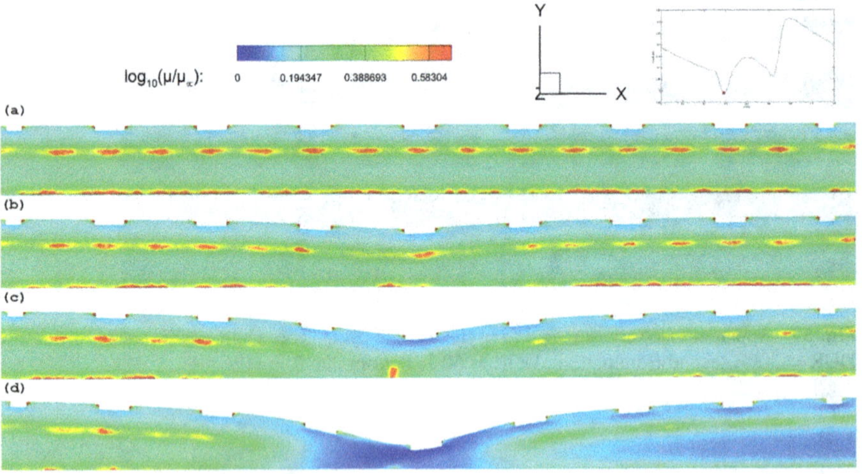

Fig. 58 Comparison of apparent viscosity log_10(μ/μ_∞) after $t = 0.3$ s (diastole). DOR values: (**a**) 0%, (**b**) 30%, (**c**) 60%, (**d**) 90%, Fig. 2 of [27]

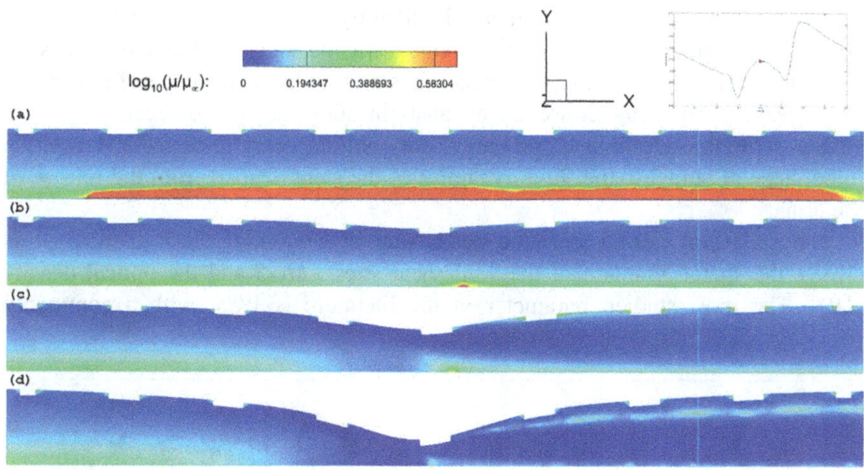

Fig. 59 Comparison of apparent viscosity log_10(μ/μ_∞) at $t = 0.4$ s. DOR values: (**a**) 0%, (**b**) 30%, (**c**) 60%, (**d**) 90%, Fig. 3 of [27]

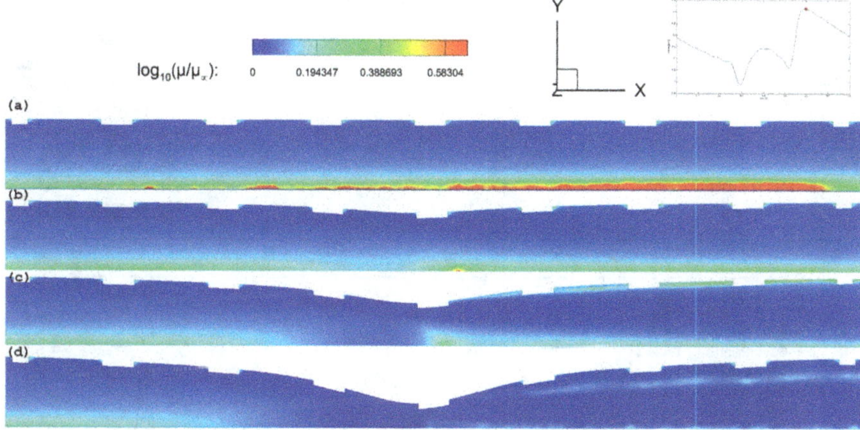

Fig. 60 Comparison of apparent viscosity log_10(μ/μ_∞) after $t = 0.6$ s (systole). DOR values: (**a**) 0%, (**b**) 30%, (**c**) 60%, (**d**) 90%, Fig. 4 of [27]

6   Conclusions

6.1   Conclusions on a Drug-Eluting Stent

This chapter investigates numerically the fluid dynamics in four configurations of a drug-eluting coronary stent [41], i.e. without residual stenosis and with 30%, 60% and 90% degrees of residual stenosis (DOR). Steady and unsteady numerical simulations are carried out to evaluate the fluid dynamics conditions, which could, potentially, promote atherosclerotic events with the stent application. The stent investigated is a drug-eluting one, which has a very low percentage of restenosis. The release of the drug, however, exhausts in 30/40 days, and recent long-term studies show that the drug is, probably, only able to delay restenosis. Hence, the importance of studying the fluid dynamics of the stent is apart from the release of the drug.

The analysis is carried on in the steady state by examining the wall shear stress (*WSS*), the spatial gradient of the wall shear stress, *MGS* and the optical density (*OD*). The most studied parameter in the literature is *WSS*, with a commonly

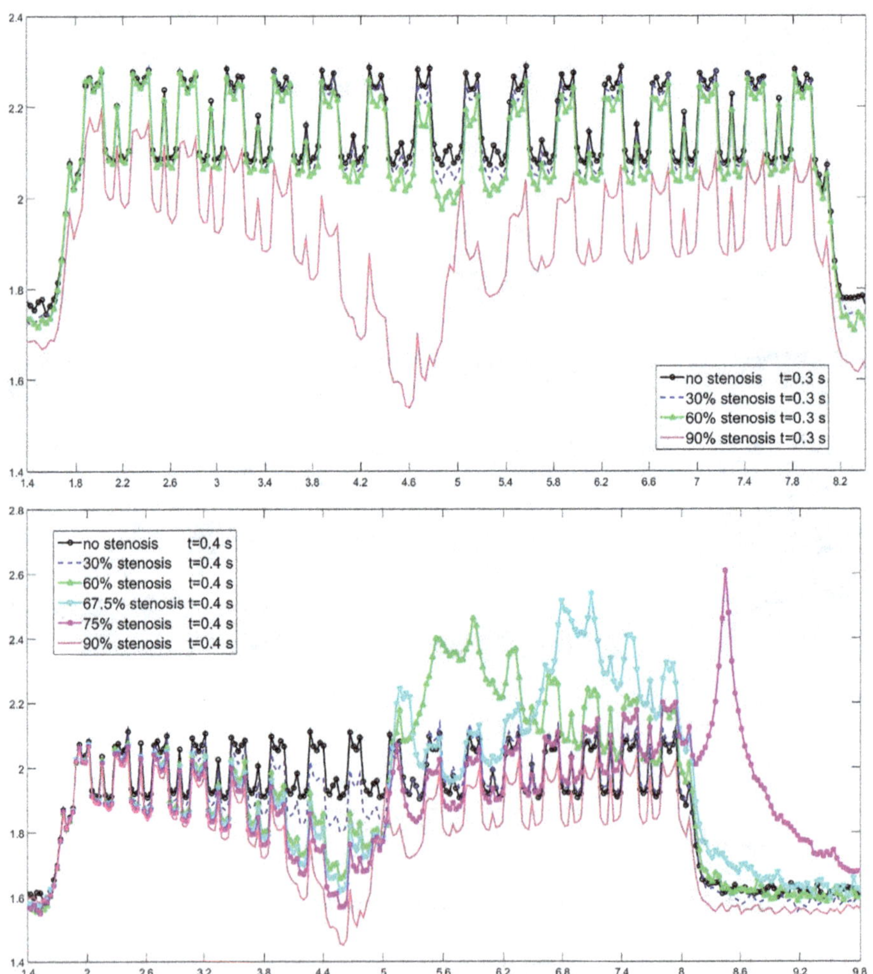

Fig. 61 Average OD versus axial length (m) at four instants: (**a**) $t = 0.3$ s, Fig. 5, [27]; (**b**) $t = 0.4$ s, Fig. 6, [27]; (**c**) t = 0.6 s, Fig. 7, [27]

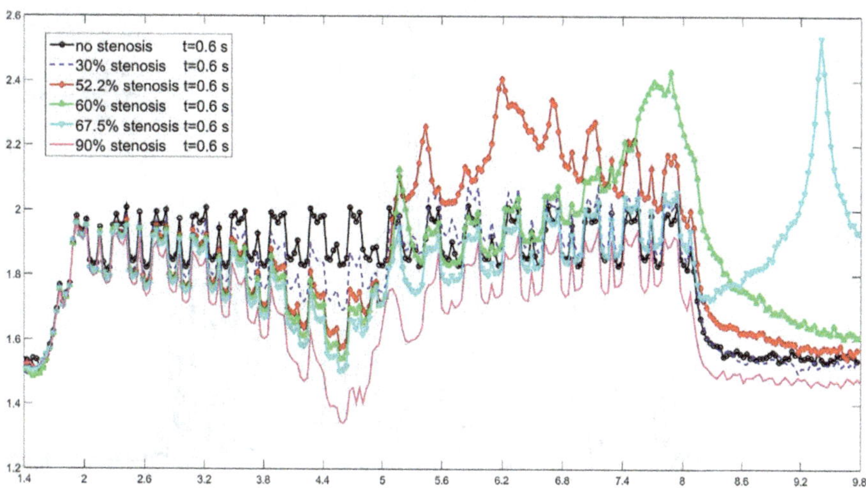

Fig. 61 (continued)

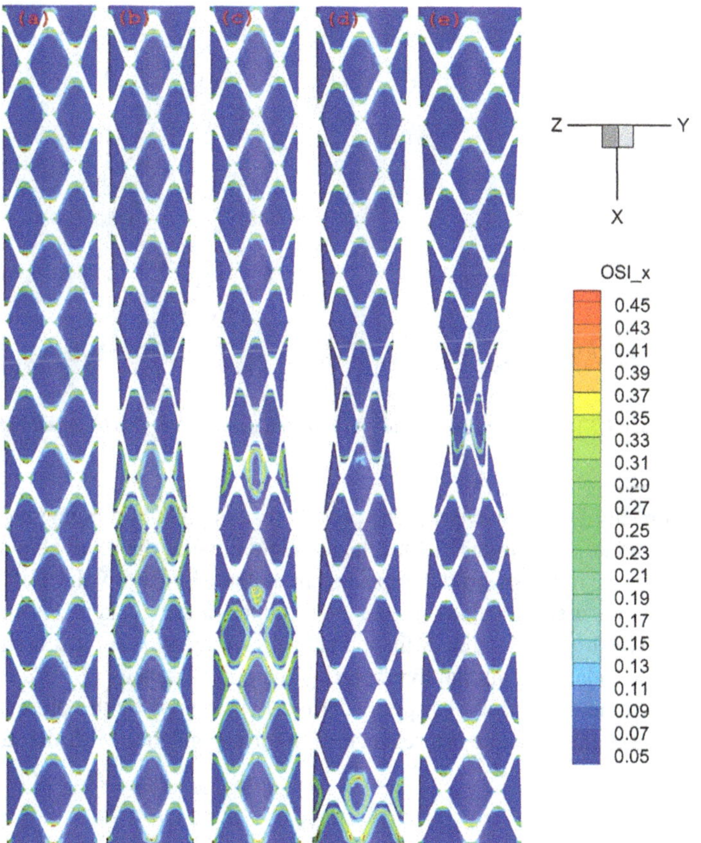

Fig. 62 OSI field with several DOR: (**a**) 0%, (**b**) 52.5%, (**c**) 60%, (**d**) 67.5%, (**e**) 90%, Fig. 8 of [27]

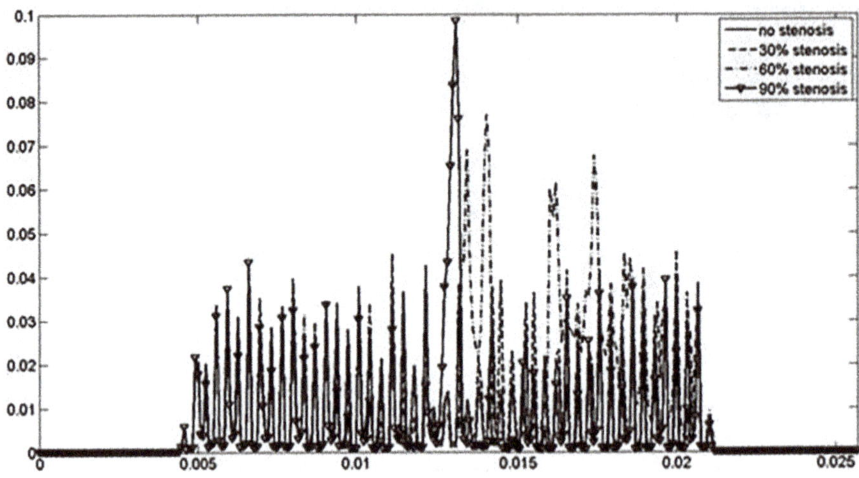

Fig. 63 Average OSI versus axial length (m) with the four values of DOR [4].

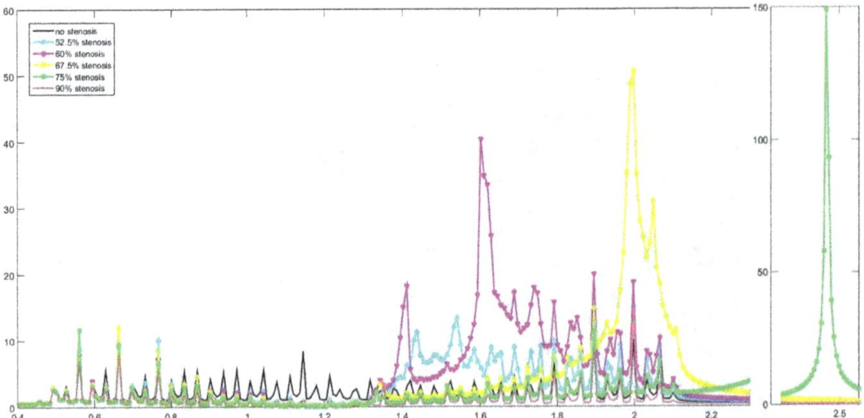

Fig. 64 Average RTT (1/Pa) versus axial length (m) with the different DOR values, Fig. 9 of [27]

accepted threshold of 5 dyne/cm^2. The present results predict, in some areas of the vessel wall and adjacent to the stent, critical values smaller than 5 dyne/cm^2, which can give a strong index of probable post-stent restenosis. High values of *MGS* are predicted in some areas, localized where the presence of the stent creates an obstacle to the flow. The threshold value of this parameter is unknown, but it was shown that high values are related to intimal hyperplasia. The optical density (*OD*), a factor that takes into account the endothelial permeability, is increased by the presence of the stent, and the highest values are registered in correspondence with the areas with greater curvature.

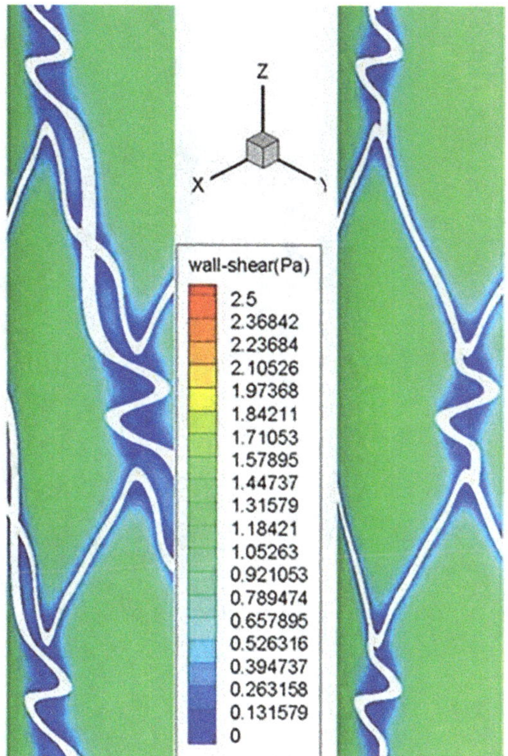

Fig. 65 Steady-state WSS field, S2 stent on the left and S1 on the right [4]

In the unsteady state, besides *WSS*, *MGS* and *OD*, the oscillatory shear index (*OSI*) and the relative residence time (*RRT*) are investigated. The parameters can vary during the cardiac cycle, with areas under critical conditions during the whole cycle and others only in some instants of time. The non-zero values of *OSI* are located near the struts and close to the points with greater curvature, which are the areas opposed to the flow, where the *WSS* vector can change direction. The presence of the stent also increases the *RRT* of the macromolecules. The results, obtained in the stenosed configurations, show that residual stenosis could expose the endothelium to hazardous conditions for longer during the cardiac cycle. Even a small $DOR = 30\%$, which is asymptomatic in terms of blood delivery to the left ventricle, atrium and inter-ventricular septum, is such to cause smaller values of *WSS* and to increase endothelial permeability beyond the stenosis. With $DOR = 60\%$, there is a worrying deterioration in *OSI* and *RTT* beyond the stenosis due to the formation of recirculation zones located in a pattern of spatial repetition of the stent geometry. The analysis of the geometry with $DOR = 90\%$ shows that *WSS* assumes very high values at the stenosis level, above which thrombogenic effects may take place. Also, beyond the section contraction, the detachment of the fluid vein causes a wide region of recirculation, where the fluid is stagnant. In these conditions, *OD*, *OSI* and *RTT* describe a better situation with $DOR = 90\%$, compared to the case with

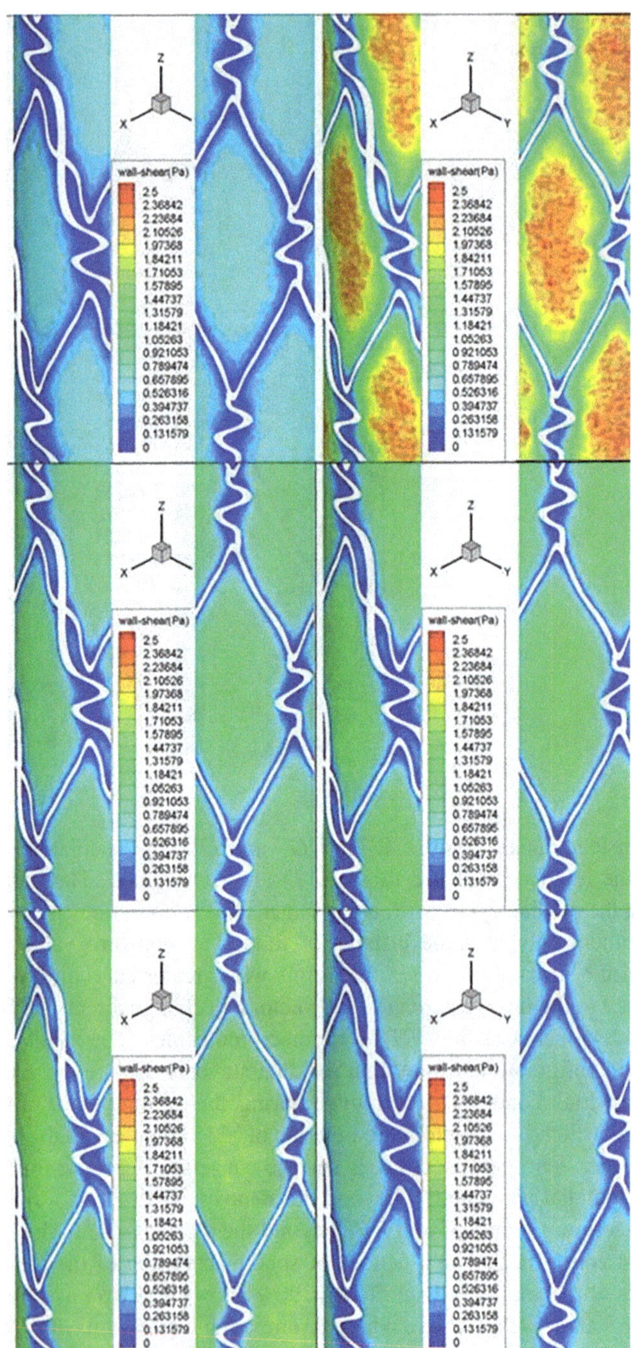

Fig. 66 Unsteady-state WSS field, S2 stent on the left and S1 on the right of each fig [4].

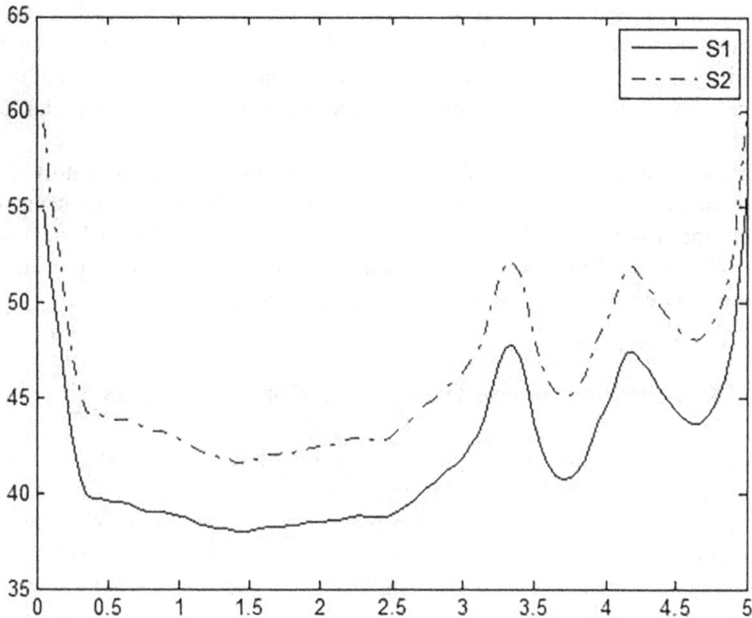

Fig. 67 Percentage area, $P(\%)$, with WSS < 0.5 Pa, versus the axial coordinate (mm) [4]

$DOR = 60\%$. There seems to be a DOR, between 30% and 90%, where a maximum of endothelial permeability is present and where OSI, OD and RTT assume a maximum value.

Blood viscosity is high in patients who have experienced cardiovascular diseases or are possibly subject to them. Residual stenosis increases blood viscosity beyond the stenosis, while a high degree of stenosis stabilizes the viscosity at lower values due to the recirculation region. This is an unexpected result and, to the best of our knowledge, has never been reported before in the literature. With $DOR = 90\%$, the higher viscosity is concentrated close to the vessel wall instead of being in the centre. This means that for a high value of DOR, the erythrocytes tend to be closer to the wall of the vessel. Certainly, this is an alteration of the physiological conditions, as the erythrocytes, usually, tend to be concentrated in the centre of the vessel.

The numerical simulations are performed by assuming the artery as a straight and rigid vessel, while diseased carotid arteries have a non-zero compliance and a more complex geometry. The rigid vessel assumption should not affect the results because stent implants limit the wall compliance to zero. As shown in the literature, the complexity of the geometry affects the fluid dynamics, but the study of a simplified geometry allows the understanding of the influence of each parameter. If a given DOR adversely affects the fluid dynamics conditions in a simplified and symmetrical geometry, it is possible to assert, with good confidence, that this effect should be even worse in a complex and asymmetrical geometry.

From the present numerical simulations, it can be concluded that the stent implantation increases the endothelial permeability compared to a vessel without a stent, in agreement with the literature, and this can favour the development of post-stent restenosis. Residual stenosis could expose the endothelium to hazardous conditions.

The most important result of this work is the discovery of a non-monotonic pattern for endothelial permeability as a function of the degree of stenosis. In addition, the results with $DOR = 90\%$ seem to indicate that the erythrocytes flow close to the endothelium, downstream of the stenosis, whereas under physiological conditions, they flow mostly in the centre of the vessel.

6.2 Conclusions on the Two Cypher Coronary Stents

The two stents with different designs, used as drug-eluting stents ($DESs$), in clinical practice, are investigated at steady state and unsteady state with the same 3D numerical simulation to study their fluid dynamic behaviour. Numerical simulations show that there are regions near the stent where the WSS is below the critical value. Unsteady-state simulations show large differences between mean and instantaneous values. By comparing the WSS and the percentage area of the two stents, it is possible to establish that the stent with the structure parallel to the mean flow has better fluid dynamics compared to the stent with the transverse structure.

Moreover, the comparison of the results with the blood as a non-Newtonian fluid to those with the blood as a Newtonian fluid [17] allows us to conclude that the use of a Newtonian fluid for the estimation of the WSS leads to an overestimation of the percentage of restenosis.

Acknowledgements for Figures The author thanks the *publisher Taylor and Francis* for the kind permission to reproduce images *2, 4, 5, 6, 7, 8, 9, 10 and 11 of Ref.* [26] and images *1, 2, 3, 4, 5, 6, 7, 8 and 9 of Ref.* [27].

References

1. Boghi A (2003–2004) Impiego della CFD nella diagnosi e cura delle stenosi vascolari. Tesi di Laurea in Ingegneria Medica, Università di Roma Tor Vergata, A.A
2. Boghi A (2005–2006) Termofluidodinamica in stenosi carotidee. Studio del carattere non-newtoniano del fluido e del regime di moto. Tesi di Laurea Specialistica in Ingegneria Medica, Università di Roma Tor Vergata, A.A
3. Boghi A (2009–2010) New transport equations for turbulent flow with variable transport properties. Biomedical applications of non-Newtonian blood flow in coronary stent and stenosed carotid artery. Ph. D., Faculty of Engineering, University of Rome Tor Vergata, A.A
4. Di Venuta I (2014–2015) Influence of the degree of stenosis on the non-Newtonian blood flow in a stented coronary artery. Tesi di Laurea Magistrale in Ingegneria Medica, Università di Roma Tor Vergata, A.A

5. Caro CG, FitzGerald JM, Schroter RC (1969) Arterial wall shear and distribution of early atheroma in man. Nature 223:1159–1160
6. Caro CG, FitzGerald JM, Schroter RC (1971) Atheroma: a new hypothesis. Br Med J 2:651
7. Caro CG, FitzGerald JM, Schroter RC (1971) Atheroma, and arterial wall shear. Observation, correlation and proposal of a shear dependent mass transfer mechanism for atherogenesis. Proc R Soc Lond B Biol Sci 177:109–159
8. Davies PF (1995) Flow-mediated endothelial mechano-transduction. Physiol Rev 75(3): 519–560
9. Gimbrone MAJ, Garca-Cardena G (2013) Vascular endothelium, hemodynamics, and the pathobiology of atherosclerosis. Cardiovasc Pathol 22(1):9–15
10. Himburg HA, Grzybowski DM, Hazel AL, LaMak JA, Li XM, Friedman MH (2004) Spatial comparison between wall shear stress measures and porcine arterial endothelial permeability. Am J Physiol Heart Circ Physiol 286:1916–1922
11. Berry JL, Santamarina A, Moore JE Jr, Chowdhury SR, Routh WD (2000) Experimental and computational flow evaluation of coronary stents. Ann Biomed Eng 28(4):386–398
12. Lanoye L, De Beule M, Dewijngaert C, Segers P, Van Impe P, Verdonck P (2006) The influence of the strut section shape on the flow field in a newly stented right coronary artery. 56th National Congress of Theoretical and Applied Mechanics, NCTAM
13. LaDisa JF Jr, Hettrik DA, Olson LE, Guler I, Gross ER, Kress TT, Kersten JR, Warltier DC, Pagel PS (2002) Coronary stent implantation alters coronary artery hemodynamics and wall shear stress during maximal vasodilation. J Appl Physiol 93:1939–1946
14. LaDisa JF Jr, Olson LE, Guler I, Hettrik DA, Audi SH (2004) Stent design properties and deployment ratio influence indexes of wall shear stress: a three-dimensional computational fluid dynamics investigation within a normal artery. J Appl Physiol 97(1):424–430
15. He Y, Duraiswamy N, Frank AO, Moore JE Jr (2005) Blood flow in stented arteries: a parametric comparison of strut design patterns in three dimensions. J Biomech Eng 127:637–647
16. Duraiswamy N, Cesar JM, Schoephoerster RT, Moore JE Jr (2008) Effects of stent geometry on local flow dynamics and resulting platelet deposition in an in vitro model. Biorheology 45:547–561
17. Gori F, Boghi A, Amitrano M (2009) Three-dimensional numerical simulation of the fluid dynamics in a coronary stent. ASME International Mechanical Engineering Congress and Exposition Proceedings, 2, Lake Buena Vista, FL, pp 407–411
18. Gori, F., & Boghi, A. (2011). Three-Dimensional Numerical Simulation of Blood Flow in Two Coronary Stents. Numerical Heat Transfer, Part A: Applications, 59(4), 231–246. https:// urldefense.com/v3/__https://doi.org/10.1080   /10407782.2011.541147__;!!O5Bi4QcV! EzQbvIn1n_9cyu2-9cTCYWaZTPFursHJ8FZYLftwrkzM2PC_EGNfiZKgdUQI5tWnPc_81NU3DfghL6egunoQ-AKSjtSg5aQ$
19. Boghi, A., & Gori, F. (2015). Numerical Simulation of Blood Flow through Different Stents in Stenosed and Non-Stenosed Vessels. Numerical Heat Transfer, Part A: Applications, 68(3), 225–242. https://urldefense.com/v3/__https://doi.org/10.1080 /10407782.2014.977151__;!! O5Bi4QcV!EzQbvIn1n_9cyu2-9cTCYWaZTPFursHJ8FZYLftwrkzM2PC_EGNfiZKgdUQI5tWnPc_81NU3DfghL6egunoQ-AKTJB9gvvg$
20. Seo T, Shahter LG, Barakat AI (2005) Computational study of fluid mechanical disturbance induced by endovascular stents. Ann Biomed Eng 33(4):444–456
21. Benard N, Perrault R, Coisne D (2006) Computational approach to estimating the effects of blood properties on changes in intra-stent flow. Ann Bio Eng 34(8):1259–1271
22. Amblard A, Le Berre HW, Bou-Saïd B, Brunet M (2009) Analysis of type Iendoleaks in a stented abdominal aortic aneurysm. Med Eng Phys 31:27–33
23. Gori F, Boghi A (2011) Two new differential equations of turbulent dissipation rate and apparent viscosity for non-Newtonian fluids. Int Comm Heat Mass Transfer 38:696–703

24. Gori F, Boghi A (2012) A three-dimensional exact equation for the turbulent dissipation rate of generalized Newtonian fluids. Int Comm Heat Mass Transfer 39:477–485
25. Gori F, Boghi A (2010) Three-dimensional numerical simulation of non-Newtonian blood in two coronary stents. 14th International Heat Transfer Conference, IHTC14, 1, Washington, D.C., pp 109–114
26. Di Venuta, I., Boghi, A., & Gori, F. (2017). Three-dimensional numerical simulation of a failed coronary stent implant at different degrees of residual stenosis. Part I: Fluid dynamics and shear stress on the vascular wall. Numerical Heat Transfer, Part A: Applications, 71(6), 638–652. https://urldefense.com/v3/__https://doi.org/10.1080/10407782.2017.1293985__;!!O5Bi4QcV! DxrsVSQR3YaYOk6aJHXty8SQpWFjaFLio5W-Ytz9MSa8eUO8pVg7INhlAVHLlFcnT8cdMsBiS0d7yqggqflvFO_DyGlo0Q$
27. Boghi, A., Di Venuta, I., & Gori, F. (2017). Three-dimensional numerical simulation of a failed coronary stent implant at different degrees of residual stenosis. Part II: Apparent viscosity and wall permeability. Numerical Heat Transfer, Part A: Applications, 71(6), 653–665. https://urldefense.com/v3/__https://doi.org/10.1080/10407782.2017.1293976__;!!O5Bi4QcV! DxrsVSQR3YaYOk6aJHXty8SQpWFjaFLio5W-Ytz9MSa8eUO8pVg7INhlAVHLlFcnT8cdMsBiS0d7yqggqflvFO85FPkJJw$
28. Ross R (1999) Atherosclerosis-an inflammatory disease. New Engl J Med 340:1436–1442
29. Visentin S, Bertin M, Rampon M, Trevisanuto D, Zanardo V, Cosmi E (2012) Infants born with intrauterine growth restriction: renal and cardiovascular follow-up. In: Ozdemir O (ed) Contemporary pedriatics. IntechOpen
30. Himburg HA, Grzybowski DM, Hazel AL, Lamak JA, Li XM, Friedman MH (2004) Spatial comparison between wall shear stress measures and porcine arterial endothelial permeability. Am J Physiol HeartCir Physiol 286:1916–1922
31. Berne RM, Levy MN (2000) Fisiologia, Casa Editrice Ambrosiana. Berne RM, Levy MN (2001) Cardiovascular physiology. Mosby Inc. St. Louis
32. Park S, Lee D, Chung W, Lee DH, Suh DC (2013) Long-term outcomes of drug-eluting stents in symptomatic intracranial stenosis. Neurointervention 8(1):9–14
33. Mwipatayi BP, Thomas S, Angel D, Wong J, Vijayan V (2013) Stent outcomes for infrapopliteal arterial occlusive disease. Vascular 21(3):121–128
34. Lownie SP, Pelz DM, Lee DH, Men S, Gulka I (2005) Kalapos, efficacy of treatment deliberate use of angioplasty balloons. AJNR Am J Neuroradiol 26:1241–1248
35. Schillinger M, Exner M, Mlekusch W, Haumer M, Ahmadi R, Rumpold H, Wagner O, Minar E (2002) Inflammatory response to stent implantation: differences in femoropopliteal, iliac, and carotid arteries. Vasc Interv Radiol 224(2):529–535
36. Farb A, Weber DK, Kolodgie FD, Burke AP, Virmani R (2002) Morphological predictors of restenosis after coronary stenting in humans. Circulation 105:2974–2980
37. Casson N (1959) A flow equation for pigment-oil suspensions of the printing ink type. In: Rheology of disperse systems, p 84
38. Mukundakrishnan K, Ayyaswamy PS, Eckmann DM (2008) Finite-sized gas bubble motion in a blood vessel: non-Newtonian effects. Phys Rev E 78:036303
39. Gori F, Boghi A (2010) Image-based computational fluid dynamics in a carotid artery. ASME 2009 International Mechanical Engineering Congress and Exposition. American Society of Mechanical Engineers, 2, Lake Buena Vista, FL, pp 123–128. 80879
40. Womersley JR (1955) Method for the calculation of velocity, rate of flow and viscous drag in arteries when the pressure gradient is known. J Physiol 127:553
41. Robaina S, Hitzman C, Robertson K, Davis L, Lenz JT (2011) Drug-eluting stent. US Patent 7,951,193
42. Kastrati A, Mehilli J, Dirshinger J, Pahe J, Ulm K, Shuhlen H, Seyfarth M, Shmitt C, Blasini R, Neumann FJ, Shomig A (2001) Restenosis after coronary placement of various stent types. Am J Cardiol 87:34–39

43. Ku DN, Giddens DP, Zarins CK, Glagov S (1985) Pulsatile flow, and atherosclerosis in the human carotid bifurcation, positive correlation between plaque location and low oscillating Shear stress. Arteriosclerosis 5:293–302

44. Malek AM, Alper SL, Izumo S (1999) Hemodynamic shear stress and its role in atherosclerosis. JAMA 282:2035–2042

45. DePaola N, Gimbrone MA Jr, Davies PF, Dewey CF Jr (1992) Vascular endothelium responds to fluid shear stress gradients. Arterioscler Thromb 12:1254–1257

46. Freedman FB, Johnson JA (1969) Equilibrium and kinetic properties of the Evansblue-albumin system. Am J Physio 216:675–681

47. Pakham MA, Rowsell HC, Jorgensen L, Mustard JF (1967) Localized protein accumulation in the wall of the aorta. Exptl Mol Pathol 7:214–232

48. Fry DL, Mahley RW, Weinsgraber KH, Sy O (1977) Simultaneus accumulation of Evans blue dye and albumin in the canine aortic wall. Am J Physiol Heart Circ Physiol 233:H66–H79

49. LaMack JA, Friedman MH (2005) Interactive effects of spatial shear stress gradient on endothelial cell behaviour in vivo and in vitro. Summer Bioengineering Conference

50. Bell FP, Adamson IL, Shwartz CJ (1974) Aortic endothelial permeability to albumin: foal and regional patterns of uptake and transmural distribution of 1311-albumin in the young pig. Exptl Mol Pathol 20:57–68

51. Fry DL (1977) Aortic Evans blue dye accumulation: its measurement and interpretation. Am J Phys 232:204–222

52. Murray CD (1926) The physiological principle of minimum work. I. The vascular system and the cost of blood volume. Proc Natl Acad Sci 12:207–214

53. Lowe GD, Lee AJ, Rumley A et al (1997) Blood viscosity and risk of cardiovascular events: the Edinburgh Artery Study. Br J Haematol 96:168–173

54. Sloop GD (1996) A unifying theory of atherogenesis. Med Hypotheses 47:321–325

55. Mejia J, Ruzzeh B, Mongrain R, Leask R, Bertrand OF (2009) Evaluation of the effect of stent strut profile on shear stress distribution using statistical moments. Biomed Eng Online 8:8

56. Ku DN (1997) Blood flow in arteries. Ann Rev Fluid Mech 29:399–434

Andrea Boghi The Doctor of Research *(Ph. D.)*, Dr. Eng. Andrea Boghi, is the director of *Computational Science Ltd.*, a consultancy company in the field of software development and data analysis, which counts among its clients the British Air Traffic Agency *(NATS)* and the Bank of England, starting from 2018. He is the author of more than 30 international scientific publications indexed on the major scientific databases. For more than a decade, he has worked in academia in various European institutions, dealing with the modelling and simulation of transport phenomena, both for basic research and for industrial and biological applications. His academic journey began at the University of Rome "Tor Vergata," where, alongside research on the modelling and simulation of turbulent flows with variable laminar diffusivity and blood flow in arteries and *stents*, he also served as a contract lecturer for the courses of Technical Physics, Thermo-Fluid-Dynamics of Biological Systems and Numerical Calculation of Thermo-Fluid-Dynamic Systems (2006–2010). In the two-year period from 2010 to 2012, he was a contract researcher at the Department of Energy and Technology of the University of *Southampton* in the United Kingdom, where he dealt with modelling and simulation of bi-phase flows, characterised by spinodal decomposition. In 2012, he went to France to the Institute of Fluid Mechanics of Toulouse as a contract researcher for the OTE (Waves, Turbulence and Environment) research group, where he dealt with modelling and simulation of river and atmospheric fluid dynamics. He terminated his contract in 2013 to go to the University of *Cranfield* in the United Kingdom, where he was appointed *Senior Research Fellow*, becoming part of the academic staff. Here, for the following 5 years (2013–2018), he dealt with modelling and simulation of transport phenomena for various departments: Energy, Hydraulics, Environment and Agriculture, where he respectively dealt with transport of crude oil in pipelines, disinfection of tanks by chlorination and transport and diffusion of organic and inorganic material in soils, with associated absorption by plants. At the University of *Cranfield,* he also handled the course on modelling of atmospheric emissions in the two-year period from 2016 to 2017. Before leaving academia, he went again to the University of *Southampton* in 2018, where he dealt with the three-dimensional modelling of the transport of nitrates and carbonates at the air–water interface in soils. Throughout his period abroad, he maintained his collaboration with Fabio Gori Ammannati, supervising some students and collaborating on the production of scientific articles. In 2020, he obtained the national scientific qualification for the functions of associate university professor in the competitive sector of Technical Physics and Nuclear Engineering.

Dr. Eng. Ivan Di Venuta (Salerno, 5 march 1990) graduated in Medical Engineering from the University of Rome "Tor Vergata" in 2015, discussing his Master's Thesis on the Thermo-Fluid-Dynamics of stenotic coronary vessels titled "Influence of the degree of stenosis on the non-Newtonian blood flow in a stented coronary artery." He decided to continue his academic journey at "Tor Vergata" by enrolling in the Research Doctorate (Italian *Ph.D.*) course in Industrial Engineering. He was welcomed by the Technical Physics research group of Fabio Gori Ammannati, with whom he started a project on the Thermo-Fluid-Dynamics of submerged turbulent jets. Thanks to his valuable teachings and support during his Research Doctorate, Ivan Di Venuta published several scientific articles in international journals and carried out part of his research at the English University of Cranfield. He obtained his Research Doctorate in Industrial Engineering, with *European Label,* in April 2019 with the thesis titled "The new fluid dynamics, heat and mass transfer of submerged free jets." Currently, Ivan Di Venuta works as a business analyst at one of the leading business consulting firms, working on projects for the redesign of business processes in public administration.

Fabio Gori Ammannati was born in Montale (Pistoia) on 5 August 1947 to Emilio Gori and Cesarina Ammannati. On 16 July 2021, he added his mother's surname, Ammannati. He graduated in Chemical Engineering on 11 November 1971 with honours. In December 1971, he won a scholarship for young graduates at the Faculty of Engineering in Bologna, which he undertook from January 1972. In 1974, he became assistant professor and, in 1982, associate professor of Technical Physics at the Faculty of Engineering of the University of Florence, where he taught, also as a contract professor, until 1990. In 1978, he won a scholarship from the *British Council,* which he carried out at the *Imperial College* in London, where he collaborated with *Prof. D. Brian Spalding* on the topic of numerical analysis in the turbulent flow of liquid metals. During his time at the University of Florence, he established international scientific collaborations with *Prof. R. Echigo, Tokyo Institute of Technology, Tokyo, Japan; Prof. D.R. Chaudhary, Department of Physics, University of Rajasthan, Jaipur, India;* International Centre for Theoretical Physics *(ICTP)* in Trieste; *Centralny Osrodek Techniki Medycznej, Warsaw, Poland and Prof. T. Aihara, Institute of Fluid Science, Sendai, Japan.* In 1986, he won a CNR scholarship, which he carried out at *Cornell University, Ithaca, New York,* where he collaborated with *Prof. R. Miller* on the topic of ground freezing. In the same year, he became professor in the Faculty of Engineering at the University of Reggio Calabria. In 1988, he was appointed as *professor* at the *University of New York at Stony Brook,* teaching the course, *Introduction to Fluid Dynamics,* in the autumn semester of the same year. During his stay, he collaborated with *Prof. T.F. Irvine Jr.* on the method of measuring thermal conductivity with a thermal probe and on the measurement of the isobaric thermal expansion coefficient of non-Newtonian fluids. In 1990, he moved to the Milano Polytechnic, where he taught Technical Physics and Systems until 1993. From 1991 until 1998, he was an adjunct professor at the Faculty of Engineering of the University of Siena. From 1992 to 2017, he was a professor of Technical Physics in the Faculty of Engineering at the University of Rome Tor Vergata, and from 2017, he is a contract professor. In 1994, he proposed the Research Doctorate (Italian *Ph.D.*) in Energy-Environment Engineering, of which he was the coordinator until 2011. In 1995, he proposed and directed the second-level Master's in Thermo-fluid-dynamics until 2016. From 1998 to 2001, he was the director of the Department of Mechanical Engineering. In 2001, he proposed and coordinated the programme, *Master of Science, Energy Engineering and Thermal and Fluid Dynamics,* in collaboration with the *Department of Mechanical Engineering, College of*

Engineering, University of Illinois at Chicago, USA, which takes place at the University of Rome Tor Vergata but awards the title of *Master of Science* from the same American University. Since the late 90s, he has been a coordinator of the joint *Ph.D.* research programme with *Prof. R.J. Goldstein, Department of Mechanical Engineering, University of Minnesota, Minneapolis, USA,* and with *Prof. J.P. Hartnett, Prof. L. Kennedy, Prof. W. Minkowycz and Prof. W.M. Worek, Department of Mechanical Engineering, College of Engineering, University of Illinois at Chicago, Chicago, USA.* Since the mid-90s, he has proposed and directed the *Socrates–Erasmus* programme with *Prof. Mayinger, Technical University of Munich, Germany; Prof. van Steenhoven, Eindhoven Technical University, the Netherlands and Prof. C. Caro, Imperial College of Science, Technology and Medicine.* During his stay at Tor Vergata, he established scientific collaborations with the *Department of Mechanical Engineering, University of Minnesota, Minneapolis, USA,* collaborating with *Prof. R.J. Goldstein* on Thermo-fluid-dynamics and mass transport in gas turbine blades; the *Energy Resources Center, University of Illinois at Chicago,* collaborating with *Prof. J.P. Hartnett, Prof. W. Minkowycz* and *Prof. W.M. Worek;* the *Department of Mechanical Engineering, College of Engineering, University of Illinois at Chicago,* collaborating with *Prof. L. Kennedy and Prof. W.M. Worek; Duke University, collaborating with Prof. A. Bejan* and *S. Mary's University, Halifax, Nova Scotia, Canada, collaborating with Prof. W. R. Tarnawski.* The bibliographic review of documents and citations, titled *PlosBiology Career,* places him within the top 2% of researchers in the field of *Mechanical Engineering and Transports.* Since 1992, he has been a tutor for over 30 PhD theses, supervisor for over 40 second-level university Master's theses, supervisor for over 70 degree theses (five-year, Specialist and Master's), supervisor for over ten *Master's degree in Mechanical Engineering* from the *University of Illinois at Chicago* and supervisor for over 20 theses *Erasmus-Socrates.*

Since the early 2000s, he has been a reviewer of international research projects on behalf of *Portuguese Science and Technology Foundation (FCT); Czech Science Foundation (GACR); European Science Foundation, Strasbourg, France,* on behalf of *FCT* and the *Shota Rustaveli National Science Foundation, Georgia.* Since 2017, the year of his retirement due to age limits, he has been a contract professor at the University of Rome Tor Vergata, where in the academic year 2024–2025, he has taught Technical Physics for the Master's Degree Course in Medical Engineering. M.D. Salvatore Mangiafico (Florence), scientific head of the NeuroVascular Base Camp, 2021, invited him to give a scientific presentation titled *"The engineering approach to the study of the Circle of Willis"* on 23 September 2021 at the Congress Centre, University of Rome La Sapienza, thus opening up a possible future collaboration.

Hydraulic Circuit Design for the Study of Vascular Problems

Hydraulic Circuit Design for the Study of Vascular Problems

Ivano Petracci, Andrea Boghi, and Fabio Gori Ammannati

1 Introduction

In this chapter, based on [1], a hydraulic circuit capable of simulating blood flow in the arteries of the human body is planned and designed. Through a numerical simulation implemented in *MATLAB*, its one-dimensional behaviour is simulated, which changes according to the initial conditions of the sample of interest. The medical aspect is presented, with particular emphasis on diseases affecting the arteries and, in particular, bifurcations. The chapter also focuses on the possibility of simulating in vitro blood flow in particular areas of the human body in order to understand its most important aspects. The technique named *Particle Image Velocimetry (PIV)*, which allows us to study the flow behaviour with pulsation, is also presented. The state of the art of research is summarised. Of particular relevance are the examples of hydraulic circuits, used and developed in order to create a workbench useful for solving the problem. The design of the circuit is then described. The numerical results, obtained with the simulation in *MATLAB* of the circuit, are then shown, reporting the variations of pressures and velocities over time, correlated with the input velocity.

I. Petracci · F. G. Ammannati (✉)
University of Rome Tor Vergata, Rome, Italy
e-mail: ivano.petracci@uniroma2.it; gori@uniroma2.it

A. Boghi
Computational Science Ltd., Southampton, UK
e-mail: a.boghi@computationalscience.co.uk

F. Gori Ammannati (ed.), *Thermo-Haemo-Dynamics in Medical Engineering*,
https://doi.org/10.1007/978-3-031-97214-0_7

2   Medical Aspects

2.1   Structure of the Arteries

The walls of the arteries are divided into three layers called tunics. The tunica intima, the innermost layer, is in contact with the basal membrane of the endothelium in which the blood flows; it is composed of connective tissue, with fibrocytes and rare muscle fibrous cells, and it is separated from the internal elastic membrane by the tunica media, in which smooth muscle fibro-cells are found, arranged circularly. In turn, the tunica media is divided by the external elastic membrane (which is made of elastic fibres) from the tunica adventitia; it is composed of collagen and elastic fibres, oriented along the axis of the vessel. The muscular component is prevalent in small-calibre arteries, which, with autonomous contractile movements, must advance the blood, for which the heart's push is no longer sufficient. In large-calibre arteries, the elastic component prevails, which has the ability to dilate with each new emission of blood from the heart. Arteries are subject to a natural ageing process, characterised by an increase in connective tissue. This results in a decrease in elasticity, evidenced by an increase in pressure values, as is normally found in elderly people. Endothelial lesions are predisposed to the formation of thrombi due to agglutination on them of platelets.

Endothelial tissue is a particular type of simple squamous epithelial lining tissue, derived from embryonic mesoderm, which lines the inner surface of blood vessels and the heart (endocardium). The endothelial cells that make it up are flat, polygonal and elongated in the direction of flow; the apical face is oriented towards the lumen of the vessels, and the nucleus protrudes towards it. Endothelial cells contain relatively few intracellular organelles (Golgi, mitochondria, endoplasmic reticulum and free ribosomes), numerous pinocytotic vesicles for the transport of substances through the endothelium and the characteristic electron-dense granules of *Weibel–Palade*, also present in platelets. These are secretory granules, in which the membrane contains P-selectin, inside which is contained the factor of *von Willebrand* (factor VIII of coagulation).

Considered in the past as a simple lining of vessels, currently, the endothelium is seen as a true organ, capable of processing a vast amount of active substances, able to modulate the activity, both of the various structures of the vascular wall covered by it and of the blood cells and proteins of the coagulation system, which come into contact with its luminal surface. Parts of these substances are secreted by endothelial cells in the immediate vicinity (paracrine secretion) to exert their effects on the vascular wall or are released into circulation (endocrine secretion) to perform their action at a distance, as in the case of substances that contribute to the control of blood pressure (for example, nitric oxide and endothelin). Other molecules, produced by the endothelium, exert their action by remaining bound to the surface of endothelial cells, as happens for the adhesion molecules, for leukocytes, or for those that influence coagulation. In summary, at the wall level, the endothelium modulates the vascular tone and the same vascular structure, playing a leading role in remodelling, which is observed in hypertension, in stenosis after angioplasty and in atherosclerosis. At the luminal level, the endothelium modulates coagulation and interactions with blood cells, leukocytes and platelets. Some of these substances are produced constitutively (i.e. also under basal conditions); others, instead, are elaborated only when endothelial cells are activated by appropriate stimuli, as happens, for example, during inflammation. Endothelial dysfunction is spoken of when the ability of endothelial cells to process those substances that are produced under physiological conditions is compromised. Endothelial activation means the stimulation to the synthesis of molecules that are not produced under physiological conditions.

There are two main mechanical forces exerted by blood flow on the vascular wall: shear stress and tensile stress. The shear stress is produced by the friction of laminar flow on the endothelium and exclusively affects endothelial cells, while tensile stress is produced by hydrostatic pressure within the vessel and affects the entire vessel wall (endothelium, fibroblasts and smooth muscle cells). The shear stress activates endothelial cells and promotes the release of vaso-dilator mediators, while tensile stress directly stimulates smooth muscle cells, inducing their contraction, and causes the stretching of endothelial cells. The net effect on vascular tone is the result of the interaction between myogenic contraction, induced by pressure, and endothelium-dependent dilatation induced by flow. The role of the endothelium in modulating the response to flow variations is highlighted by observing how flow-induced dilatation is dependent, in vitro and in vivo, on the integrity of the endothelium.

2.2 Arterial Pressure

Arterial pressure (*AP*) expresses the intensity of the force with which the blood (content) pushes on the arterial walls (container), divided by the area of the wall. This pressure is the result of the following factors: (*a*) heart contraction force; (*b*) stroke volume, or the amount of blood expelled for each ventricular contraction (systole); (*c*) peripheral resistances, or resistance opposed to the progression of blood from the state of constriction of small arteries; and (*d*) elasticity of the aorta and large arteries. During contraction, or ventricular systole, it determines the systolic pressure (maximum pressure), while, during relaxation or ventricular diastole, it determines the diastolic pressure (minimum pressure). It is currently believed that under normal conditions, the values of arterial pressure, measured in resting conditions, at the level of the humerus with the sphygmomanometer, should not exceed the value of 130 mm Hg as systolic pressure and that of 85 mm Hg as diastolic pressure.

2.3 Electrical Stimulus of Heart Contraction

The stimulus that generates contraction is electrical in nature and originates involuntarily from control centres located in the central nervous system in the brain and spinal cord. It is transported from the central nervous system to the heart through the parasympathetic and sympathetic efferent pathways. This stimulator, called the sino-atrial node, produces the nerve impulse, which, like an electric shock, generates the contraction of the heart. It is this property, called automaticity, that spontaneously produces the stimulus that regulates the heart rate (sinus rhythm), which is a real pacemaker of the heart. It is located in the right atrium, near the opening of the superior vena cava. The sino-atrial node rhythmically emits an impulse that depolarises the adjacent cardiac muscle. The resulting waves propagate through the atria until they reach the second specific conduction structure, called the atrioventricular node, which is located in the floor of the right atrium, to the left of the orifice of the coronary sinus. Its end is continuous with the atrial myocardium and with the fibres of the internodal tracts.

The stimulus, arriving in the atrioventricular node, slows down so that the depolarisation of the two atria can be completed; it then regains speed, spreading through the tissue specialised in conduction, which is the continuation of the atrioventricular node and is located in the membranous and proximal portions of the interventricular septum. It divides into the right and left branches, which run under the endocardium (the floor of the ventricular cavities) along the two surfaces of the heart septum:

- The left branch quickly divides, forming a wide streak of bundles that are arranged on the septal surface of the left ventricle.
- The right branch extends for a greater stretch, usually up to crossing the distal portion of the right ventricle with a moderator band, while the other parts extend on the endocardial surface of the right ventricle.

Peripherally, both branches of the common bundle divide and form the subendocardial network of *Purkinje* fibres, which extend into the ventricular walls, in direct relation with the fibres of the ventricular musculature. Therefore, as soon as the electrical impulse, which started from the sino-atrial node, arrives at the ventricles, the heart beats, and the blood flows until it reaches all parts of the body.

2.4   Coronary Arteries

The heart pumps blood into all the tissues of the body and performs this work incessantly, contracting and relaxing rhythmically. To live and perform its pump function, the heart muscle (myocardium) needs the continuous, uninterrupted supply of blood that it receives through the coronary arteries. There are two arteries, the *right coronary artery* and the *left coronary artery*, which originate from the aorta, just above the aortic valve. The coronary arteries run on the external surface of the heart (in the epicardium, indicated as epicardial arteries) or in the septum that separates the two ventricles, emitting secondary branches, third- and fourth-order branches and so on, composing a dense network of vessels.

The *left coronary artery*, almost from the start, bifurcates into the *anterior interventricular (AIV) branch*, which provides irrigation to the left ventricle and, therefore, has strategic importance, Table 1. Because of this early bifurcation, the

Table 1 Radius, Reynolds, and Womersley numbers [1]

Artery	R(cm)	Re	Wo
Aorta	2.5	1500	22
Main left coronary	0.43	270	6.1
Right coronary	0.1	230	1.8

main coronary vessels are three, and usually, when talking about the coronaries, reference is made to three vessels: the *right coronary*, the *AIV* branch and the *circumflex artery*, which is the name that the left coronary assumes after the detachment of the *AIV*. The *right coronary artery* originates from the right sinus of *Valsalva* of the aorta and runs along the right atrioventricular groove, surrounded by fat. It is differently branched: one branch, usually well developed and large, that runs along the acute edge of the heart; another, the posterior descending interventricular branch, and a series of other small branches that irrigate the part of the right ventricle that rests on the diaphragm. Each coronary artery has its own defined vascularisation district, although with variations from person to person. Some parts of the heart, like the posterior papillary muscle of the left ventricle, can have a double right–left vascularisation. The right atrial branches of the right coronary artery assume a particular importance. It originates from the right coronary artery and rises along the anteromedial wall of the right atrium. It enters the upper part of the interatrial septum and reappears as a branch of the superior vena cava.

During systole, the subendocardial coronary vessels (those that enter deep into the myocardium) are compressed due to high intraventricular pressures. Instead, the epicardial coronary vessels (which run on the surface of the heart) are not subject to these pressures. Because of this, the subendocardial flow stops. Most of the perfusion of the heart muscle occurs during diastole, when the heart is relaxed, and then, in case of decreased diastolic duration, the subendocardial tissue may undergo ischaemia.

The coronary circulation is completed by the coronary veins, which discharge their blood into the coronary sinus, which, in turn, discharges it into the right atrium. The heart must contract regularly to perform its pumping activity, and this is possible due to a delicate system of self-regulation of the cardiac rhythm. The heart, in addition to connections with the nervous system and central and autonomous systems, has its own intrinsic conduction structure. Unlike the other muscles of the body, whose activity depends on the brain and the spinal cord (which is also called voluntary musculature because it is controllable by our will), the heart is self-sufficient, as it has its own stimulator that generates the electrical impulse that determines the cardiac contraction (beat).

2.5 Carotid Artery

The carotids originate from the aortic arch asymmetrically on both sides, as drawn by Leonardo da Vinci in Fig. 1 [2], mirrored to be read from left to right.

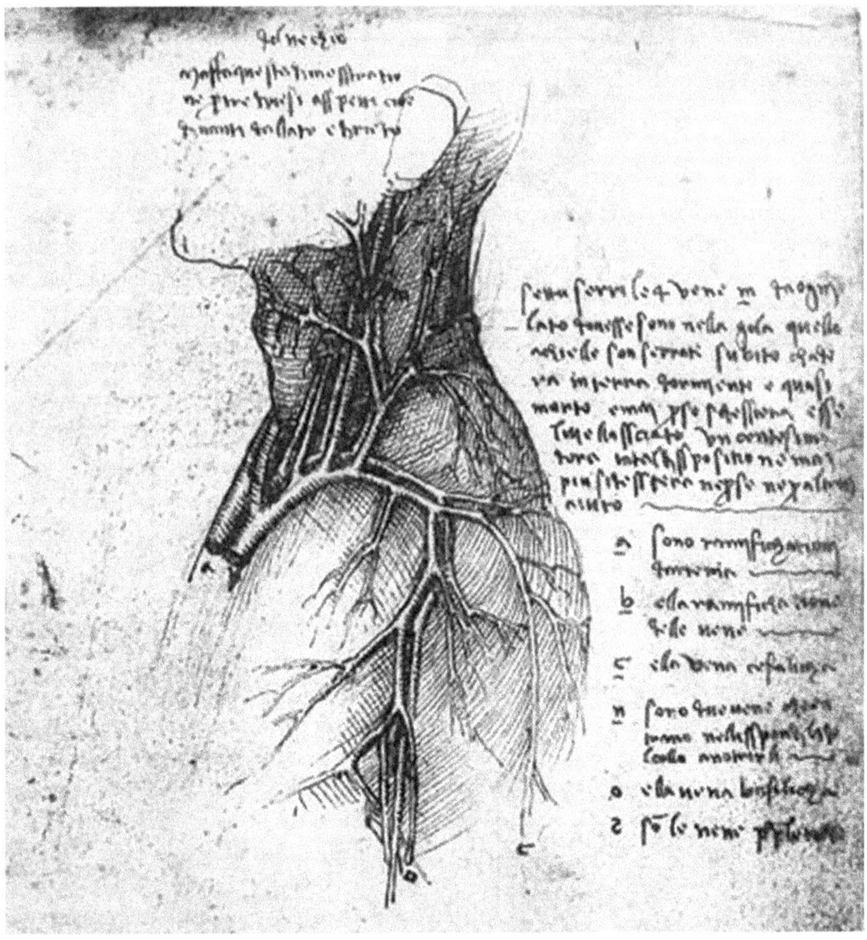

Fig. 1 Leonardo da Vinci: study of the arteries and veins of the head, (mirrored image of Sheet B 32 v) [2]

The left carotid artery originates directly from the aortic arch, while the right one originates from the division of the brachiocephalic trunk, and from these points, it is named the common carotid artery. The common carotid artery points upwards and is covered by the sternocleidomastoid muscle, and in the neck region, it becomes part of the vasculo-nervous bundle of the neck, together with the internal jugular vein and the vagus nerve. It reaches up to the upper edge of the thyroid cartilage of the larynx, where it divides into its two terminal branches: the internal carotid artery (ICA) and the external carotid artery (ECA). The internal carotid is located laterally, and the external one is located medially, but then, the internal carotid curves medially and backwards to point towards the cranial base and enter inside the skull, while the external one remains outside the skull and provides the irrigation of the face. Before the origin of the two terminal branches, the lumen of the common carotid presents a fusiform dilation, the carotid sinus, whose wall is very rich in nerve endings with a pressure-sensing function, which, stimulated by the pressure of the blood circulating in the artery, regulates blood pressure, heart rate and respiratory rate. At the point where the carotid bifurcates, there is the *Carotid Glomus*, which has chemoreceptor functions.

The internal carotid artery is a large and important vessel that carries its branches to the brain, the arachnoid, the eye and the organs of the orbital cavity. The internal carotid artery originates from the common carotid artery, just above the upper edge of the thyroid cartilage of the larynx, where the common carotid divides into a sling in the external and internal carotid. At its origin, it has a diameter of 8 mm, a larger diameter than that of the external carotid. Generally, the left internal carotid is larger than the right one. The cervical portion is in relation to the internal jugular vein and the vagus nerve, which accompany it, and with the stylohyoid, stylopharyngeal and digastric (posterior belly) muscles, which cross it laterally. The glossopharyngeal nerve also crosses it externally. It rises upwards along the lateral wall of the pharynx and then penetrates the carotid canal, carved in the petrous rock of the temporal bone. Having travelled the entire carotid canal (intrapetrous tract), it is found inside the cranial cavity, where it bends forwards and penetrates the cavernous sinus (intracavernous tract). Inside this characteristic venous sinus, it is not directly bathed by the blood contained therein but is externally covered by the endothelium of the sinus itself.

Inside the cavernous sinus, it forms a curve directed backwards and upwards (carotid siphon). In this tract, it is accompanied by various nerve structures, which are the following:

- Oculomotor nerve (III).
- Trochlear nerve (IV).
- Ophthalmic branch of the trigeminal nerve (V).
- Abducens nerve (VI).

In the neck region, no lateral branch detaches from the artery; in the carotid canal, it gives rise to a small arterial branch, the carotid-tympanic branch, which irrigates the mucosa of the tympanic box, and the pterygoid artery. In the intracavernous tract, it detaches branches for the cavernous sinus, branches for the semilunar ganglion of Gasser, hypophyseal branches (for the pituitary and the ventral parts of the hypothalamus) and meningeal branches (for the dura mater of the anterior cranial fossa). Just after exiting the cavernous sinus (intracranial tract), it detaches its second lateral branch, which is represented by the ophthalmic artery, destined for the eyeball, which, among its collaterals, presents the central artery of the retina that penetrates the optic nerve and then opens at the level of the optic papilla.

Finally, the internal carotid bends medially in front of the clinoid processes and divides into its four terminal branches, which are the following:

- Anterior cerebral artery.
- Middle cerebral artery.

- Anterior choroidal artery.
- Posterior communicating artery.

The external carotid artery is an artery that originates from the common carotid, which bifurcates, giving rise to the internal and the external carotids. The external carotid irrigates some structures of the head and neck. Shortly after its origin, it bends forwards, moving away from the internal carotid to move upwards and return backwards, passing between the mastoid process and the angle of the jaw. Beyond this point, it divides into the maxillary artery and the superficial temporal artery. It is in relation to the skin, the subcutaneous tissue, and the edge of the sternocleidomastoid muscle and is crossed by various satellite veins; outside the carotid triangle, it is in relation with the digastric muscle and the stylohyoid muscle that it follows until it enters the parotid gland.

Medially, below, the artery is in relation to the larynx; it is divided from the internal carotid, from the glossopharyngeal nerve and from the stylopharyngeus muscle. Its branches are the following:

- Superior thyroid artery.
- Ascending pharyngeal artery.
- Lingual artery.
- Facial artery.
- Occipital artery.
- Posterior auricular artery.
- Superficial temporal artery.
- Maxillary artery.

2.6   Pathological Conditions of the Arteries

Numerous pathological conditions (such as atherosclerosis, hypercholesterolemia and diabetes) and smoking can damage the endothelium and alter its function in controlling the vascular tone. Endothelial dysfunction involves the reduction of flow-mediated vaso-dilation and reduces vaso-dilation (or even paradoxical vaso-constriction) to agonists, which normally determine endothelium-dependent vaso-dilation, such as bradykinin, histamine, substance P and, especially, acetylcholine. These mediators induce vaso-dilation through the endothelial production of NO. Due to the anatomical characteristics it possesses, the carotid is a preferential site for the formation of atherosclerotic plaques; in fact, at the bifurcation of internal and external carotids, a turbulence of the blood flow is formed, which ceases to be a laminar flow, generating vortices. These vortices, associated with arterial hypertension and hypercholesterolemia, are the major risk factors for the genesis of carotid atherosclerosis. The formation of an atheromatous plaque produces an obstruction to the passage of blood, which is no longer free to pass and reach the peripheral irrigation districts.

In general, unilateral carotid obstructions, with contralateral carotid patent, are asymptomatic because the existing anastomoses between the internal carotid, external carotid and vertebral artery manage to ensure an adequate blood supply to the central nervous system. In general, surgical intervention to remove the plaque is resorted to in case of obstructions greater than 70% of the vascular lumen. The consequences of carotid obstruction can vary: generally, the obstruction establishes over a long period of time, which allows other arteries to modulate cerebral flow, but sometimes, a thrombotic event can acutely worsen the symptoms, and from the atherosclerotic site, emboli can be released that cause stroke events.

2.7   Atherosclerosis and Stenosis

The epicardial arteries, along with the interventricular septum arteries, are the most well-known and easily explored because they are well-outlined by coronary angiography (the radiological examination that allows their visualisation). From these major (macroscopic) vessels, almost at a right angle, branches depart that penetrate deeply into the cardiac muscle and give rise to the fine network of cardiac capillaries (microscopic) that constitute the microcirculation. The correct nourishment of the heart and the normality of its function depend on the correct functioning of both the macrocirculation and the coronary microcirculation. When the blood supply to the myocardium becomes defective, in a more or less extensive part, the heart suffers. The most common diseases due to some obstacle to the flow in the coronary arteries are myocardial ischaemia (transitory but critical reduction of blood flow in some coronary segments) and myocardial infarction, in which case the small afflux of blood persists beyond 20–30 min. At the base of these defects, segmental or diffuse, of coronary blood flow, ranging from simple narrowing (stenosis) to obstruction of the vascular lumen, there is generally a pathology of the arterial walls that also affects the coronary arteries: atherosclerosis. This is the most widespread disease that represents the main cause of death in our industrialised society.

Atherosclerosis is a chronic inflammatory disease of large- and medium-calibre arteries, which is established due to cardiovascular risk factors: smoking, hypercholesterolemia, diabetes mellitus, hypertension, obesity and hyperhomocysteinemia. It is suspected that there may also be other causes, particularly of an infectious and immunological nature. Anatomically, the characteristic lesion of atherosclerosis is the atheroma, or atherosclerotic plaque, that is, a thickening of the intima of the arteries due mainly to the accumulation of lipid (fat) material and proliferation of connective tissue. Clinically, atherosclerosis can be asymptomatic or manifest, usually from 40 to 50 years onwards, with acute or chronic ischaemic phenomena, which mainly affect the heart, brain, lower limbs and intestine. The term atherosclerosis was proposed to underline the presence of the atheroma (from the Greek *athere*, which means *gruel*, to indicate the fatty, pasty material contained in the plaques). The lesions, which have, as a specific characteristic, a more or less abundant lipid component, evolve over time: they start in childhood as lipid streaks (reversible in nature) and tend to become real atherosclerotic plaques, which, in advanced stages, can narrow (stenosis) the arterial lumen or ulcerate and complicate with a superimposed thrombosis, which can lead to an occlusion of the artery.

Due to the anatomical characteristics it possesses, the carotid is a preferential site for the formation of atherosclerotic plaques; in fact, at the bifurcation between internal and external carotids, a turbulence of the blood flow is formed, generating vortices. These vortices, associated with arterial hypertension and hypercholesterolemia, are the major risk factors for the genesis of carotid atherosclerosis. The formation of an atheromatous plaque produces an obstruction to the passage of blood, which, therefore, is no longer free to pass and reach the peripheral irrigation districts. In general, unilateral carotid obstructions, with contralateral carotid pervious, are asymptomatic because the existing anastomoses between the internal carotid, external carotid and vertebral artery, manage to ensure an adequate blood supply to the central nervous system.

2.8   Pathogenesis

The initial events in the formation of atherosclerosis (atherogenesis) are identified in the damage of the endothelium (functional damage or endothelial dysfunction) and in the accumulation and subsequent modification (aggregation, oxidation and/or glycosylation) of low-density lipoproteins (*LDL*) in the intima of the arteries, two events that influence each other. The accumulation of *LDL* is due not only to the increase in the permeability of the endothelium, functionally or anatomically damaged, but also to their binding to the constituents of the extracellular matrix of the intima: a bond that increases the residence time of lipoproteins in situ. An important factor causing an increase in the connective intimal matrix (intimal thickening) is represented by the friction of the blood flow on the vascular surface (haemodynamic stress), which is particularly accentuated at the branches and curvatures of the vessels, sites that are particularly predisposed to the development of atherosclerotic lesions. It starts from a small laceration of the artery; then fatty material is deposited at the level of the artery wall, obstructing the lumen of the artery; finally, there is the complete obstruction of the lumen caused by a thrombus.

The dysfunction/activation of the endothelium by cardiovascular risk factors is followed by the adhesion and migration of monocytes and *T* lymphocytes into the intima in response to the expression on the endothelial surface of adhesive molecules (Selectins, VCAM-1 and ICAM-1) and chemotactic signals (MCP-1) emitted by the damaged endothelium. Figure 2 shows the histological image of a coronary artery, occluded to almost 50%, where a dead thrombus has formed in the remaining part of the lumen. The dashed line indicates the outer limit of the artery, while the continuous line indicates the thrombus [3].

Fig. 2 Histological image of a coronary artery with nearly 50% area occluded, Fig. 1 of [3]

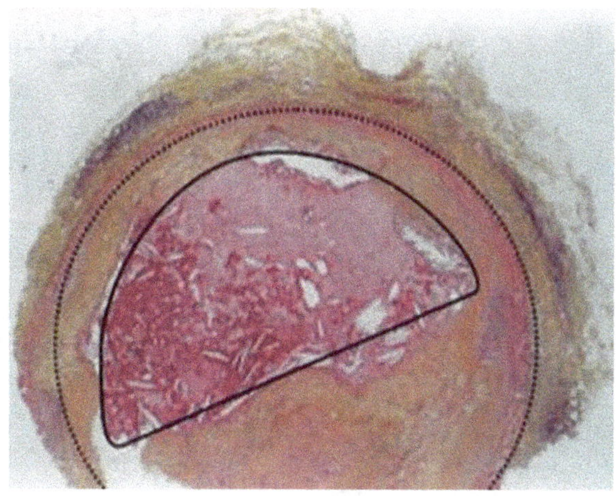

3 State of the Art

Various blood flow study techniques have been introduced, such as, for example, *Laser Doppler Anemometry (LDA), Ultrasound Doppler (US) or Magnetic Resonance Imaging (MRI)*. An alternative and complementary approach to studying arterial haemodynamics uses numerical simulations. The development of efficient numerical methods and new generations of computers allows the calculation of local behaviours in physiological and anatomically realistic conditions.

3.1 Particle Image Velocimetry (PIV)

3.1.1 Introduction

The *PIV* technique is one of the most recent methods of measuring the velocity of a fluid, which arises from the need to measure velocities simultaneously throughout the field of interest. Figure 3 [4] shows the *PIV* technique, which is a non-intrusive optical laser method. The principle of operation of the *PIV* technique is as follows: the flow under investigation is seeded with small particles (solid particles or liquid droplets, depending on the nature of the fluid and the temperature range within the flow). The speed of the fluid is determined by measuring the displacement of the particles over a certain time interval. Two laser beams, separated by an appropriate time interval, illuminate, with consecutive pulses, the tracer particles. The light, scattered by the particles present on the laser blade (*laser sheet*), is collected by a *charge-coupled device (CCD)* camera, where the various frames are recorded in pairs. The displacement of the illuminated particles is then easily determined by calculating a displacement function between two consecutive images. The velocity vector projected onto the *laser sheet* is then determined, as in [5–9].

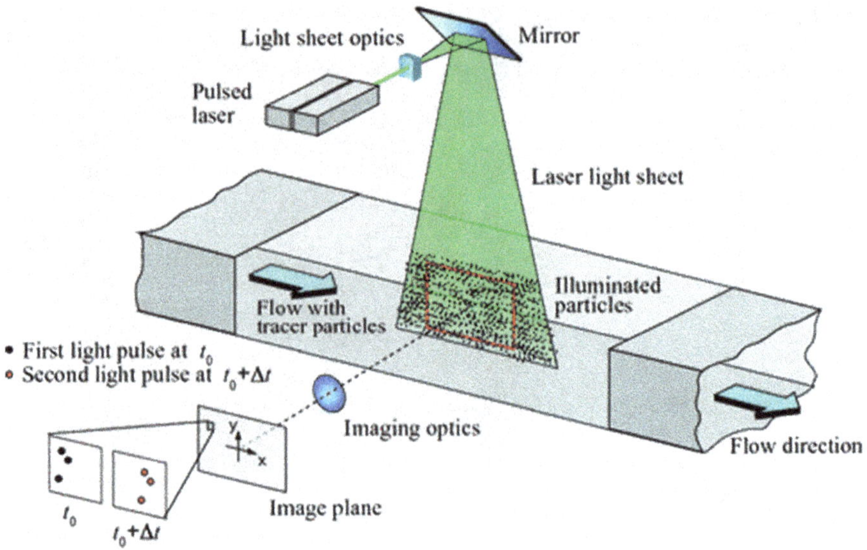

Fig. 3 PIV technique, Fig. 1.9 of [4]

The instrumentation present in the Technical Physics and Energy Engineering Laboratory of the University of Rome Tor Vergata for the *PIV* experiments is a system produced by *Intelligent Laser Applications, GmbH (ILA)*. It consists of a computer, a monitor, a synchroniser, a laser power supply, a laser head, an articulated arm with a light sheet adapter and a camera. Except for the camera, all the mentioned elements are mounted on a piece of furniture that allows the system to be easily positioned in the laboratory.

3.1.2 Laser

The *Light Amplification by Stimulated Emission of Radiation (LASER)* is an optical source that emits photons in a coherent beam. Unlike common light sources, such as the light bulb, which emits incoherent photons in all directions, laser light is typically almost monochromatic; that is, it consists of a single wavelength emitted in a limited beam. Electrons absorb the energy that is supplied to the atom, thus reaching an excited state, which allows them to occupy the outermost orbitals. An electron that absorbs energy and jumps to a certain orbital finds itself in an unnatural and unstable situation; so after a period of time that can range between *milliseconds* and *nanoseconds*, it returns to the ground state in the form of quanta of light (photons), with a defined frequency and dependent on the element to which it belongs and the energy it had absorbed.

In the case of conventional light sources, the transfer of energy occurs spontaneously due to the randomness of time and direction of emission of the photon. The *LASER*, on the other hand, uses a stimulated emission through energy pumping. In this case, a photon, acting as an energy source, hits an already excited particle. The result is the emission of a photon with the same frequency and the same phase as the stimulating one and, therefore, an amplification of the light emission. To have a stimulated amplification, it is necessary that the number of excited atoms is greater than that of the atoms that remain in the ground state (that is, population inversion must occur); otherwise, there should be an absorption of the pumped energy and a stimulated emission. Only materials defined as active allow this. Since stimulated emission produces identical photons, in phase and direction, to the stimulating ones, we are able to obtain a coherent amplification of the incident electromagnetic wave. The fact that everything happens between two parallel reflective surfaces, that is, in a resonant optical cavity, means that the wave, whose wavelength is a multiple of the distance between the two mirrors, is automatically selected, dispersing the others. This selected wave will reflect multiple times within the cavity, stimulating other coherent emissions and gradually amplifying itself. One of the mirrors being partially reflective, the laser beam thus formed and characterised by coherent, iso-directional and monochromatic photons can pass through it and exit. Simultaneously, each event of stimulated emission brings a particle back from the excited state to its fundamental state, reducing the capacity of the active medium for further amplification. The balance between the power supplied by the pumping lamp and saturation and the losses of the medium produces a power value that determines the operating point of the *LASER*.

The *LASER* can be classified by the active medium used: gas, chemical, solid or semiconductors. The *LASER* most commonly used in *PIV* applications is the solid state one, in which the crystalline material is doped with ions that provide the required energy states. The *LASER* used in this laboratory is the *Nd:YAG LASER*, which uses a neodymium-doped yttrium aluminium crystal as the active medium. This *LASER* can operate both in continuous wave and in pulses; in the latter mode, it is generally used in Q switching, that is, with an optical switch inserted in the resonant cavity, which remains closed until the crystal has reached the maximum population inversion, at which point, by opening, it allows a single pulse of very high power to be discharged. The normal wavelength λ of a *Nd:YAG LASER* is 1064 nm or in the infrared range. In order for the beam to hit the camera's spectrum, it must pass through a second harmonic generator to double its frequency ($\lambda = 532$ nm).

The power supply and the head of the two *LASERs* used are produced by *New Wave Research*, a company specialising in *LASER* for multiple applications. The model is *Solo PIV 120XT*, with an output energy of 120 mJ for an emission with a wavelength of 532 nm. The head of the *LASER* is the component that contains the beam-generating parts, while the power supply contains all the electronic parts and the water used for cooling. These *LASERs* can operate at a repetition rate of 15 Hz, while the distance between the pulses used can vary from 40 ns up to a few seconds.

3.1.3 Articulated Arm and Light Sheet

After leaving the head of the *LASER*, the beam must be conducted to the measurement plane. For this purpose, an articulated arm is fixed to the *PIV* unit, which directs the light towards the lens that produces the laser beam. This arm consists of a support

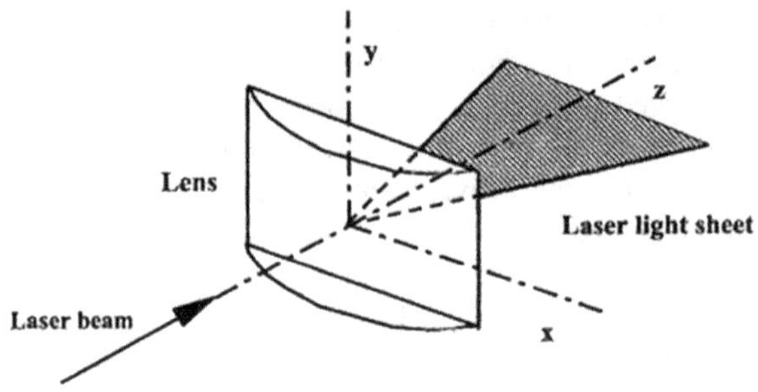

Fig. 4 Laser beam [1]

block, several metal tubes connected by counterweight joints. The light leaves the head of the *LASER* and, for a short stretch, crosses free air, before entering the support block, where a mirror directs the light at an angle of 90° inside the articulated arm. The beam then passes inside the first tube until the mirror of the first joint deflects it by 45°. In this way, the beam is directed through seven rods (ranging from 20 to 1000 mm) and seven joints to the lens adapter of the *light sheet*. The seven joints correspond to seven degrees of freedom, as each joint allows a rotation, but a configuration cannot be locked. To stabilise the *LASER* beam in one position, the lens is fixed to an additional arm on a tripod. In order to illuminate the experimental region, the laser beam is transformed into a light sheet by a cylindrical lens, Fig. 4. The adjustment of the focal distance allows the thickness of the blade to be set, while the divergence angle can be chosen from among 8°, 16°, 30° and 50°.

3.1.4 Tracers

The *LASER* light is not visible in vacuum; therefore, it is necessary for the beam to hit particles in order to be scattered and create a light beam. This type of scattering, called *Mie Scattering*, involves the deviation of the electromagnetic radiation from spherical particles with a diameter larger than the wavelength of the radiation. By increasing the number of particles present in the air, it is possible to make the beam more visible. When analysing a submerged flow, it is absolutely necessary to seed the fluid with fine particles so that the *LASER* light is sufficiently reflected. There are various methods of seeding the flow, depending on the type and speed of the fluid or the wavelength of the laser beam. In PIV applications, polyethylene glycol smoke or the atomisation of fluids, such as water and oil, are commonly used. Other methods include fluidisation of particles, sublimation, chemical reactions or combustion.

The choice of particles must meet the following needs: firstly, a particle of the right size must be available to scatter the light sufficiently; it must also be easily generated and should not require an excessive number of additional tools; furthermore, it should not leave impurities inside the apparatus that would result in a waste of time in cleaning the same, and finally, the seeding of the flow must appear as uniform as possible. It was chosen as tracer for this study the *ether speedometer*. For the production of the particles, an aerosol generator, *PivPart 30*, produced by *PIVTEC, GmbH*, is used. This generator consists of a closed cylindrical container with an inlet for compressed air and an outlet for the aerosol. Four air ducts, immersed in the liquid, are present inside the container. Each of them is connected to the air distributor via a valve. These tubes are closed at their lower ends but have lateral holes (*Laskin nozzle*) that are 1 μm in diameter. Compressed air is applied, with a pressure difference of 0.5 bar compared to the aerosol outlet pressure, through these air jets, and the bubbles inside the liquid are broken. Due to the shear forces generated by the jets, small liquid droplets are created and carried by the air bubbles to the surface. A circular plate (*impactor plate*) is located above the liquid level, thus forming a lateral passage of 2 mm and allowing the passage of the smaller particles to reach the aerosol outlet duct. The four valves determine the number of particles, while their size depends on the type of atomised liquid.

3.1.5 Camera

The cameras used for *PIV* visualisations are, in most cases, cameras with *charge-coupled device* (CCD) sensors. The operating principle of a *CCD* sensor is as follows: an image is projected by a lens onto a grid of capacitive elements capable of accumulating an electric charge proportional to the intensity of the electromagnetic radiation that hits them. These elements are coupled in order to transmit each of their own charge to the adjacent element. The last capacitor discharges in favour of an amplifier that converts the charge into voltage. By repeating this process, a timed sequence of pulses is created, which results in an electrical signal, through which it is possible to reconstruct and digitise the matrix of the pixels composing the image. The chosen camera is a *PCO Sensicam QE* with a single *CCD* sensor, a high-resolution camera (1376 × 1040 pixels) capable of capturing a maximum of *ten frames per second* at full resolution, that is, five pairs of frames.

3.1.6 Synchronisation

Among the different components of the system, such as the camera, the *LASER* and the Q switches, a perfect synchronisation is required in *PIV* measurements. Therefore, the presence of a synchroniser that executes programmed sequences of commands for the above-mentioned devices and for other external devices is required. The first to be activated is the camera; then, the pumping lamp of the first *LASER* is activated. After reaching the required energy level, the Q switch is activated so as to coincide the *LASER* pulse with the first frame of the camera. Then, the lamp of the second resonant cavity is activated, depending on the interval between the pulses. The closure of the camera's shutter determines the time of acquisition of the second frame, and after the pumping, the second Q switch is activated. The interval between the first and second switching is defined by the flow speed, which imposes the distance between the pulses. The minimum repetition time of the sequence is about 250 s, corresponding to 4000 Hz.

3.1.7 Computer and Software

Through an optical fibre connection, the camera is connected to the *PC* on which the *VidPIV* and *PIVSync* programs are installed, both produced by *ILA*. The *VidPIV* is the main program used for the calibration and execution of the experiment. The *PIVSync* controls the synchroniser and sets the energy levels of the *LASER*. Once the images from the camera are acquired, the software is able to calculate the velocity vectors of the various pairs of photos, calculating the displacement of the particles with the help of a cross-correlation. This operation consists of dividing the image into *interrogation spots* (*IS*), that is, into square matrices in greyscale with a size of 2 pixels (e.g. 32×32 or 64×64), and then, defining, for each square, the correlation function and, therefore, the displacement and velocities of the particles, thus providing a field of vectors containing a vector for each *IS*.

3.2   Circulatory Apparatus

As far as the simulation of the circulatory system is concerned, the objective is to analyse the behaviour of a fluid, in this case blood, in particular areas of the body. The focus is not only on the bifurcations of the coronary arteries but also on the studies which are using the *PIV* technique, through which one can trace the haemodynamic behaviour of the fluid under examination.

The objective of [10] is to describe the haemodynamic losses in a bypass anastomosis using instantaneous velocities obtained through *PIV*. The circuit is driven by the *Berlin Heart*, which has ensured a pulsatile flow of about 200 ml/ min. The working fluid is a mixture of glycerine and water with a viscosity equal to 4 mPa s. Hollow glass spheres, with a diameter of about 913 μm, are used as tracer particles. The velocity field is obtained through *PIV*, and the *shear stress* is calculated; *ten* measurements are made at intervals of 100 ms per cardiac cycle. The pressures are measured at the entrance and exit of the circuit so as to be able to calculate the pressure losses.

The study [11] describes the numerical simulation of a pulsatile flow in a carotid artery bifurcation. The pressure and flow rate of the flow under examination in the carotid artery are obtained using a non-invasive technique, while the geometry of the computational model was reconstructed through angiography *Magnetic Resonance Imaging* (MRI). The shear stress, the velocity field and the stress tension are also analysed. The inconsistency with previous results, obtained through an ideal geometric model, is due to the fact that the flow in the carotid is dominated by a helical flow, accompanied by a secondary vortex. This type of flow is induced by the asymmetry and the curvature of the real geometry. The flow simulation is carried out assuming firstly rigid walls and then deformable ones. The comparison of the results demonstrates the quantitative influence of the movement of the vessel walls. Generally, a reduction of the shear stress is proven, also depending on the position in the artery and the phase of the cardiac output. The regions where the flow is slower in the deformable model, the global flow and stress characteristics remain unchanged. The analyses of stress mechanics on the surface of the artery walls show a very complicated tension field, not homogeneous, with areas more subject to stress than others. The comparison of the movement of the walls between the real method and the estimated one shows good agreement. Subsequently, the study deals with simulating the blood flow and the stress mechanics of the walls in a real human carotid artery, combining the simulation with an algorithm. To trace the pressure and flow rate in the two bifurcations of the carotid, they used, in vivo, the *Doppler ultrasound* method, while the mechanical stress was measured by *B-mode* ultrasound. In this study, the blood is assumed to be incompressible, homogeneous and with Newtonian behaviour.

In [12], *Computational Fluid Dynamics (CFD)* and *Computational Structural (CS)* are used to evaluate flows and shear stress, which are difficult to calculate in vivo. Therefore, *CFD* and *CS* offer many more advantages compared to in vivo and in vitro models, as long as these methods are based on observations and comparisons with experimental and clinical data. New technologies for clinical imaging, such as Magnetic Resonance and Computed Tomography, have opened a new path for the specific description, for each patient, of the haemodynamic and structural behaviour of living tissues. Therefore, by coupling *CFD/CS* with clinical images, the diagnosis of problems of all kinds can be evaluated. The study focuses on the analysis of fluid and structural dynamics of anastomoses.

The authors of [13] also focus on the haemodynamic problems related to the damage of the carotid arteries. They address the problem with a computational analysis, simulating the blood flow precisely in the areas most affected by tissue damage. They consider two cases: in the first case, they uniformly reduce the peripheral resistance to maintain constant arterial pressure, resulting in an increase in blood flow, and in the second case, they leave the resistance unchanged. The results show that the flow does not change significantly, but the values of the shear stress on the walls increase sharply in proportion to the flow, except for the minimum flow, when the shear stress in the carotid bulb increases threefold. The method used is to simulate two beats per minute: 72 bpm and 108 bpm.

The study [14] emphasises the fact that most of the studies are carried out in rigid or simplified conduits. For a better resemblance to reality, they use a scale (1:1) and exact geometric measurements. Figure 5 [14] shows the circuit diagram for measurements with a *LASER Doppler anemometer*. The liquid is pushed from a pressure tank by compressed air into a reservoir (3), from which it flows into a container (4). To maintain a constant pressure, the excess liquid returns to the initial tank (1) and passes through the model (6), and a tap makes it flow back into the container (1). The flow rate can be increased or decreased, and different Reynolds numbers can be simulated, without reflections caused by the valves. The average time per pulse is measured. The liquid is transported from the container by reflux into the pressure tank, from which it is pressed with a compressor into the reservoir (3). At this point, a computer (14) comes into play that operates a piston pump that superimposes an oscillatory impulse in order to obtain an oscillating flow. The stroke of the piston can be changed, creating different forms of impulse and flow waveforms. To avoid shock caused by piston pressure, a reserve tank (7) is installed in front of the model as a filter. The temporal flow is measured with inductive flow meters, while the pressure drop is measured with inductive-type pressure transducers.

The study [15] uses the *Echo Particle Image Velocimetry (Echo PIV)*, which shows the results for the measurements of the characteristics of blood flow in vitro. A carotid bifurcation model is created using the biplane angiography data of an adult.

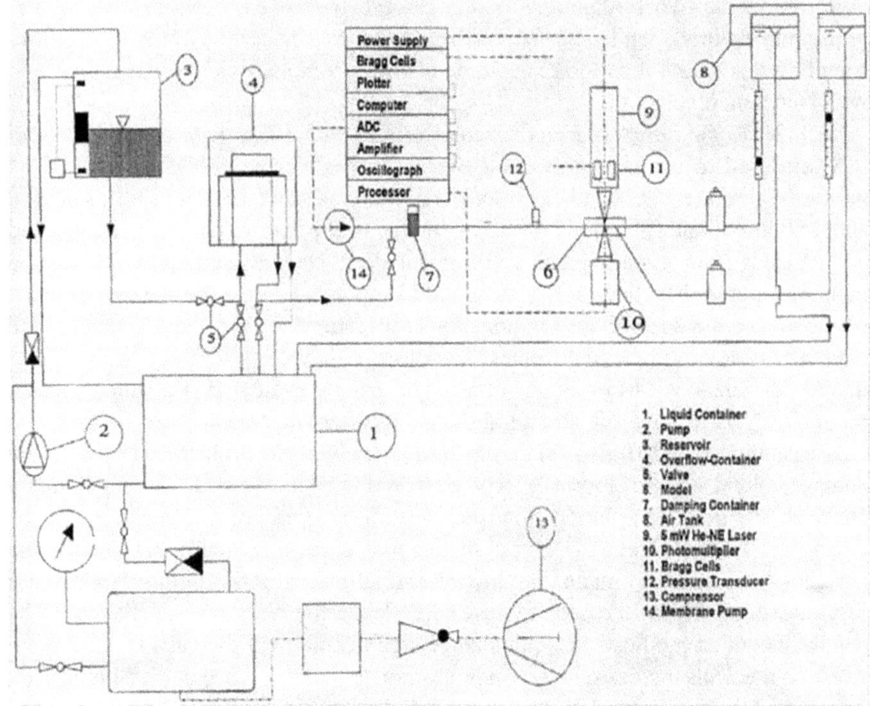

Fig. 5 Hydraulic circuit for the simulation of blood flow, Fig. 1 of [14]

The model reproduces precise vascular geometry, allowing the visualisation of the same flow, both with *PIV* optical and with *Echo PIV*. Figure 1a of [15] shows that a pulsatile pump and a compliance chamber are used to produce conditions of realistic physiological flow, with peak speeds of about 70 cm/s, a heart rate of 75 beats/min, an average Reynolds number of about 1500 and a Womersley number of 16.1. Deionised water at 20 °C is used as a working fluid. The water is drawn from a tank with a pulsatile pump and sent into a compliance chamber in the carotid model. Subsequently, the bifurcation converges, after 40 cm, flowing into a tank that maintains a constant pressure downstream. From the head tank, it flows back to the water tank. The water provides a good coupling between the transducer and the carotid models. The carotid bifurcation model, built in silicone, is shown in Fig. 1b of [15]. The length of the model is about 120 mm, and the distance from the entrance of the common artery to the bifurcation sinus is about 50 mm. The internal diameters of the common carotid artery (*CCA*), the internal carotid artery (*ICA*) and the external carotid artery (*ECA*) are 11.1 mm, 8.9 mm and 8.5 mm, respectively. The use of silicone in this model allows the analysis of the same flow field, both with *Echo PIV* and *Optical PIV*. The flow condition, adopted for the physiological model, has a peak speed of 70 cm/s in the common carotid of normal people, even though the average flow rate of 1.4 l/min and the peak flow rate of 3.0 l/min are much higher than human physiological values. The average Reynolds number calculated is 1484, with an average diameter of 11.5 mm and an average speed of 29 cm/s per cycle.

In [16], the velocity profiles and the shear stresses within human arteries are studied, as there is no simple method to obtain such information in vivo. The analytical solution of a constant flow and optical measurements *PIV* (for pulsatile flow) are used for comparison. Compared to the analytical solution, both *Echo PIV* and *Optical PIV* resolve the velocity profile quite well. The error in the velocity gradient measured with *Echo PIV*, 8%, is comparable to that measured with *optical PIV*, 6.5%. In pulsatile flow, the velocity profiles, measured with *Echo PIV*, agree with those of *optical PIV*. In vivo, instead, through the method of *MRI*, it is possible to measure the azimuthal variance of the wall shear stress (WSS) of the arteries, and recently, it has become possible to resolve the asymmetric distribution of velocity along the circumferential boundary of a non-circular vessel wall. However, it has not reached a temporal precision to acquire these data. In this study, an *Echo PIV* is analysed in detail as an investigation method in comparison to the *optical PIV* method. The work scheme is shown in Fig. 2b of [16].

The study, [17], comparing a low speed with the plaque distribution in the human carotid, shows that on the outer wall of the carotid sinus, where the wall shear stress and the speed are low, the plaque thickness is maximum, and instead, where shear stress and speeds are high, the plaque thickness is minimum. The goal is to verify that numerical simulations are comparable to real simulations. Pulsatile flow velocity measurements are performed in an anatomically realistic replica of a human carotid bifurcation (*true-to-model* on an elastic scale) made of silicone rubber, cast in a box full of 1.5% agarose gel, Fig. 6 [17]. With this procedure, a nearly identical model is reconstructed. To simulate the rheological properties of blood, a water–glycerin mixture is used, with kinematic viscosity, $\nu = 0.037$ cm^2/s. The experiment of

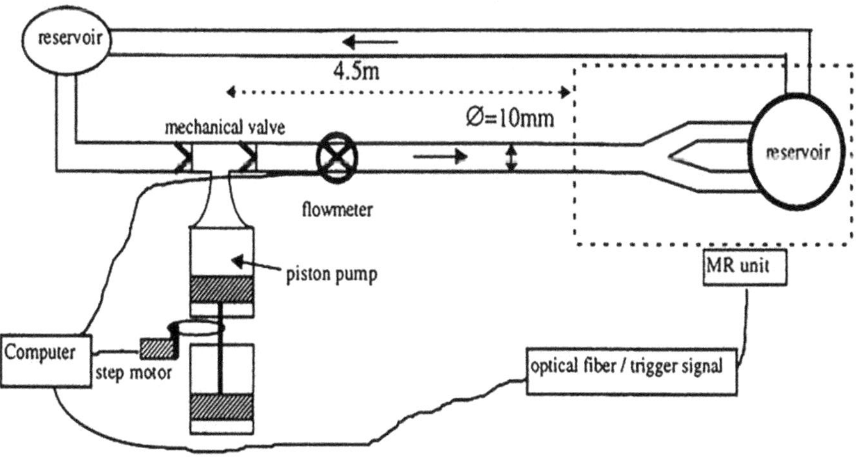

Fig. 6 Working circuit, Fig. 2 of [17]

[17] is conducted under unsteady flow conditions, with an average systolic flow rate of 28 ml/s (corresponding to the maximum Reynolds number, Re = 1454) in the common segment of the carotid. A programmable motor allows driving a pump in such a way as to reproduce physiological flow conditions. The direction of the liquid flow is controlled by two mechanical valves. During the dead times of the cycle, the liquid is sucked from a water tank into the piston and pumped towards the bifurcation during the working cycle. To ensure the flow profiles, the pump is connected to the bifurcation with a rigid plexiglass tube, with a diameter of 10 mm and a length of 4.5 m. The long-term stability of the pulsation wave is monitored during the experiment with an electromagnetic flow meter installed near the pump. The sampling frequency is 25 Hz. The average hourly flow rate is 9.4 ml/s, corresponding to a Reynolds number of 488.

The diameters of the sections examined (common section, *midsinus* and *endsinus*, internal and external of the carotid) are (A) 6.6 mm, (B) 6.2 mm, (C) 5.3 mm and (D) 4.8 mm (Fig. 7, [17]), with a flow ratio of 0.5 for both daughter tubes. The length of the sinus is 7 mm, and the bifurcation angle is about 45°.

The researchers of [18] studied the dependence of the blood flow on the angulation present between the two branches of models of proximal anastomosis with angles of 30°, 45°, 75° and 90°, made of Pyrex glass, and on the transplant conditions, facing forwards or backwards. The aorta and the graft have an internal diameter of 20 and 6 mm, respectively, and are shown in Fig. 8 [18]. The working fluid is a mixture of 30% glycerine and 70% ammonium thiocyanate (NH_4SCN). The aqueous solution of ammonium thiocyanate is composed of equal parts of ammonium thiocyanate salt and distilled water, by weight. The solution has a refractive

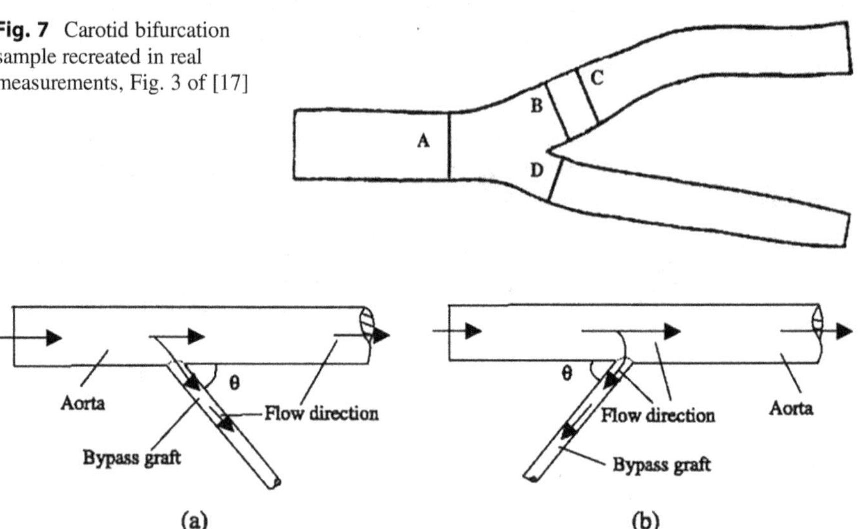

Fig. 7 Carotid bifurcation sample recreated in real measurements, Fig. 3 of [17]

Fig. 8 Recreated bifurcation samples in Pyrex glass, Fig. 1 of [18]

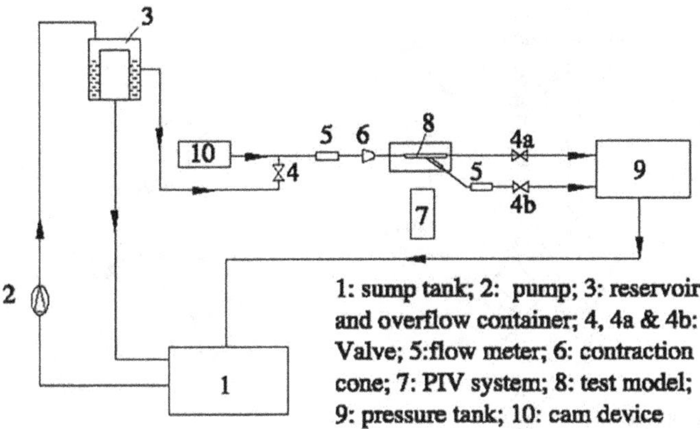

Fig. 9 Working circuit, Fig. 2 of [18]

index of 1.47 and dynamic viscosity of $4.08 \, 10^{-3}$ Pa s at a temperature of 22 °C, measured with a commercial refractometer and a controlled-rate rheometer. *Polyamide seeding* particles are inserted into the liquid to highlight the flow field.

The test circuit is shown in Fig. 9 [18]. The fluid, contained in a cup tank (*1*), is set in motion by a centrifugal pump (*2*) and passes into the overflow tank of the container (*3*). To maintain constant static pressure, the excess liquid flows back into the tank (*1*). The fluid passes through the contraction cone (*6*) and then the test model (*8*). The three valves (4, 4a and 4b) control the inflow and the outflow. The PIV system consists of a *Q-switched* double pulsed cavity; *Nd: YAG* is used as a light source, with a repetition rate of 10 Hz, providing two thin (0.3–1 mm) green laser sheets ($E = 532$ nm). A *DoubleImage 80C42 700* camera with *Nikon AF Micro-Nikkor 60/2.8* is used to capture two consecutive images. Since clinical studies show that the average flow in aorta–saphenous vein–coronary trigger is around ±25 ml; the Reynolds number, based on a diameter of 6 mm, is about 169. Consequently, the Reynolds numbers of 100, 169 and 250 are selected as control parameters for the study of the flow characteristics under different peripheral resistance conditions after the aorta. Subsequently, to discover the effect of the resistance conditions at the distal end (4a–4b), the flow characteristics due to the different ratio between the two points of exit are studied with a flow rate of 5 l/min.

To study the fluid dynamics within a bifurcation, the researchers of [19] use a transparent model of the carotid, as shown in Fig. 1 of [19], in which the geometry is constructed in order to use the *PIV* in a complex three-dimensional geometry. The velocity data are acquired through studies on the cross sections of the sample and combined in such a way as to provide a three-dimensional velocity field. Flows that occur within complex arterial geometries are, however, difficult to observe. An alternative is to examine the flow phenomena through in vitro studies. The failure of the graft due to an intimal thickening is the main concern, and the coronary revascularisation remains unresolved even after surgery. Similar pathological processes are often found in nature or are reproducible at bifurcations and arterial

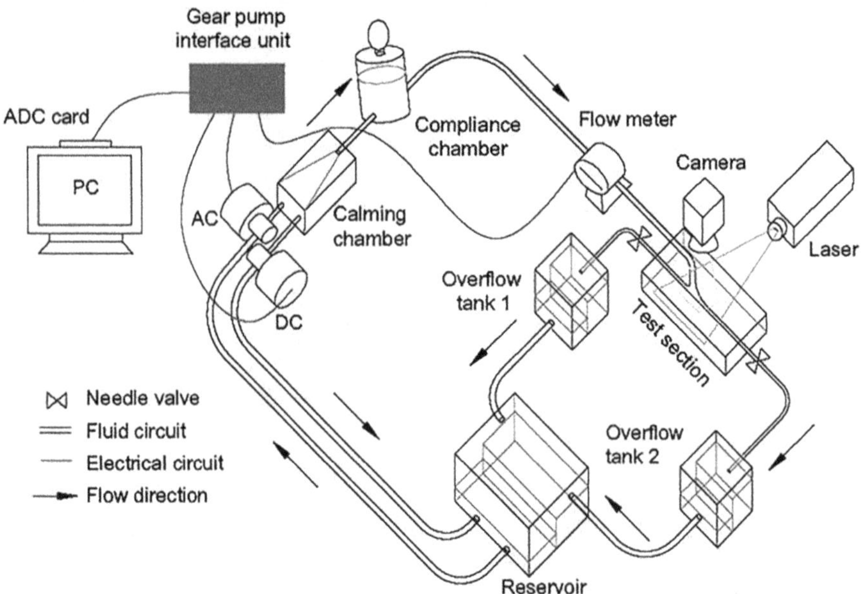

Fig. 10 Hydraulic circuit for flow study with PIV, Fig. 3 of [20]

curvatures. Some studies show that these regions are associated with disturbed flows and abnormal haemodynamic forces, suggesting a possible mechanical role of the geometry in regulating the haemodynamics in the genesis and, therefore, the progression of the diseases. In particular, low wall shear stress (*WSS*) and high spatial gradient shear stress (*SGWSS*) are critical negative factors for the disease.

The researchers of [20] proposed the working circuit shown in Fig. 10 [20]. A conventional coronary distal anastomosis is created using a computer-aided design software. The graft and the host artery are circular conduits with the internal diameters of $d(g) = 5$ mm and $d(a) = 3.3$ mm, respectively, typical of distal coronary anastomoses. The geometric data are exported to a numerical control milling machine in order to produce an aluminium model, which is connected with stainless steel connectors and suspended in a rectangular acrylic container. A degassed mixture of silicone and hardener is then inserted, allowing it to be treated at 60 °C for 12 h. A transparent silicone flow passage replica is obtained after the removal of the alloy cast. The working fluid is a mixture of glycerol (42% by weight) of dynamic viscosity 4.0 mPa s and density 1.1 g/cm^3, comparable with the characteristics of blood at 25 °C. A gear pump system is used to generate the physiological flow input waveform. The average Reynolds number is 230 based on the diameter of the host artery, representative of the flow in the left anterior descending artery. The heart rate is 75 beats/min, simulating a resting condition. The flow immediately upstream of the test section is measured and controlled by an electromagnetic flow meter. The flow division ratio in the proximal and distal segments of the artery is set at 20:80 based on the average flow by using the time collection method; that is, the average flow rate in each outflow segment is determined by measuring the volume of liquid collected in a known time interval. The pulsatility index ($PI = 2.4$) and the acquired waveform are controlled and adjusted according to the required waveform by modifying the resistance, which is regulated by needle valves placed downstream of each flow segment, and the compliance of the circuit, controlled by the compliance chamber, as well as the gain of the pump system.

The study [21] proposes a circuit from which important information is drawn regarding the one proposed in the present chapter. The Pyrex glass geometric models used in the *PIV* experiments are similar to the numerical simulation models, which are designed to represent the proximal anastomosis, based on clinical data for patients at the *National Heart Centre in Singapore*. The internal diameters of the aorta and the graft are 20 and 6 mm, respectively, and the study is focused on the influence of the graft angle. In constant flow conditions, the inflow to the aorta is assumed to be 5 l/min, while the flow within the graft is varied to simulate the resistance to flow, which can represent the level of stenosis, or blockage, within the graft or coronary arteries. In this study, the assumptions of incompressible Newtonian fluid and rigid arterial walls and graft are made. To be consistent with the working fluid used in the *PIV* measurements, a dynamic viscosity of 4.08 10^{-3} Pa s and a density of 1055 kg/m^3 are assumed for the blood. Subsequently, a commercial software is used to solve the conservation equations of mass and momentum.

Figure 11 [21] shows the scheme of the experimental test. The first phase of the experiments is carried out by applying different resistances at the graft, with a

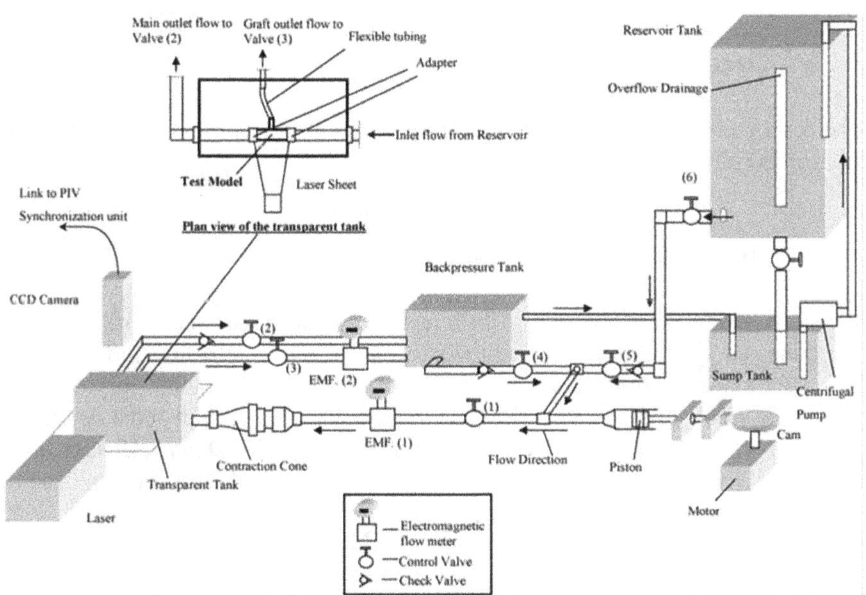

Fig. 11 Hydraulic circuit, Fig. 3 of [21]

constant inflow of 5 l/min in the aorta, and adjusting the opening of the control valve (3) (55%, 75%, 85% and 95%), as illustrated. In the second phase, pulsatile flow conditions are applied to mimic physiological flow. Initially, the fluid of the tank fills the entire system in order to eliminate any trapped air bubbles. After the system is completely filled, the control valve (5) is closed, and the engine starts. The special *cam* activates the piston to create the required waveform, and an external control circuit synchronises the *PIV* system for capturing full-field images. At the viewing section, the *Nd: YAG* is adopted to illuminate the section of the test model plane, which is immersed in a transparent tank filled with the same liquid as the working fluid. The *CCD* camera, mounted in a vertical position above the test model, immediately captures two digital images at a short time interval, which are processed to obtain the velocity vectors in the measurement plane. In order to include the refractive index of the Pyrex glass model and mimic the properties of blood, the working fluid is a mixture of *ammonium thiocyanate, glycerine and water*, which has a refractive index and dynamic viscosity of 1.47 and 4.08 10^{-3} Pa s at 22 °C, respectively. Seeding particles of *polyamide*, with an average diameter of 20 μm, are selected based on a compromise between a particle tracing response and a high *signal-to-noise ratio (SNR)* of the light signals. Experimental parameters, such as the interval time between pulses, the duration of the pulses and the separation, are carefully determined and verified during the measurements. The experimental calculation of the *WSS* is based on the detection of the flow velocity near the walls within the viscous boundary layer. A curve is then constructed to obtain a velocity profile, and the slope of the curve at the wall is determined as a measure of the velocity gradient.

4 Circuit Design

The design criteria for the circuit that must simulate the blood flow within coronary and carotid bifurcations are now illustrated.

4.1 Fluid to Be Used

Blood, in accordance with [18, 20], is replaced by a mixture composed of 30% glycerine and 70% aqueous ammonium thiocyanate, with a refractive index of 1.47 10^{-3} Pa s and dynamic viscosity of 4.08 10^{-3} N s/m^2. Within this mixture, *polyamide* seeding particles are inserted, with a diameter of 10 μm and a density of 1.1 g/cm^3 *(Polyamide Seeding Particle 38A2–121 PSP-50, Dantec Measurement Technology)*, so that the laser light is scattered, but the particles do not significantly influence the motion of the fluid under examination.

4.2 Model to Be Investigated

The model to be studied is the entire coronary and carotid bifurcation, and the material to be used to simulate a human artery is silicone, according to which a rigid wall model underestimates the temporal and spatial characteristics of the stresses in areas that actually present a lower blood flow in laminar motion. The circuit is sized on the basis of the spaces available in the laboratory and an estimate of the concentrated head losses in each section. In addition, a numerical method is implemented in *MATLAB*, reported in the following, that simulates the desired trend inside the specimen based on the already-obtained experimental data in order to obtain the main frequencies of the pump to be inserted. The pump needs to be reprogrammable, depending on the speeds present inside the specimen, which need to be analysed. The circuit, at this point, is suitable for the study of any artery or bifurcation, provided that its physical characteristics and the speed trend in it are known. It is, therefore, enough to change the specimen and reprogram the pump's output pressure, which can be calculated simply by changing the input data of the specimen.

4.3 Circuit Elements

The basic elements of the circuit are the main tank, the peristaltic pump, the compensation chamber (*compliance*) and the study chamber.

Figure 12 [1] shows the view of the circuit from the top. The circuit is built on two levels to ensure the return of the liquid under examination to the main tank, which is open, and is used as a collection basin and as a means to cancel the frequencies of the circuit that come from the return section. In addition, arranged in this way, it guarantees a constant pressure at the pump inlet. The peristaltic pump is the mechanical means that guarantees the required speeds inside the specimen. It must have a constant inlet pressure, so its positioning is at the foot of the main tank. The compliance chamber, used in these circuits, has the main function of preventing

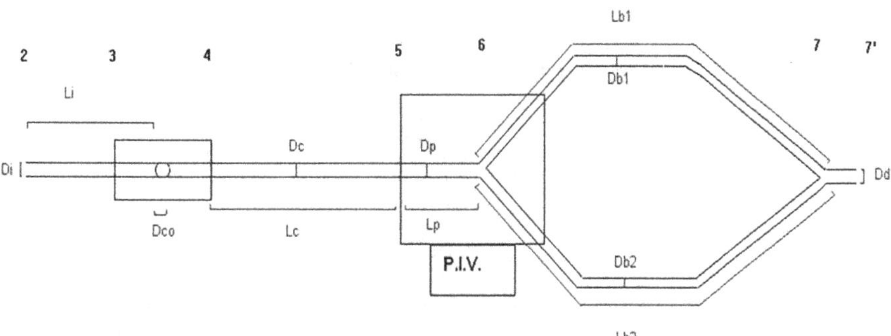

Fig. 12 Top view of the circuit [1]

Table 2 Dimensions of the hydraulic circuit [1]

Section	Description	Diameter [cm]	Length [cm]	Coefficient λ'
1–2	Initial part	2	70	0
2–3	First part	2	25	0.5 (90 curve)
3–4	Compliance	2–10	0	0
4–5	Calm part	1	25	1 (calm entrance)
5–6	Sample	1	25	0
6–7	Bifurcation	To be defined	25	0.5–1 1.2 valve 1 0.5 bifurcation 2 1.2 valve 2
7–7'	Return 1–1	2	25	1 reconnection
7'-8	Return 1–2	2	70	0.5 (90 curve)
8–9	Return 2	2	125	0.5 (90-degree curve)

breakages in the specimen and ensuring that the elastic oscillations remain unchanged. The equations of varied motion of currents can be used in pressure conduits for the study of two substantially different forms of oscillations that occur in the transients of hydraulic systems: elastic and mass. The former are pressure waves that propagate with high speed due to the compressibility of the fluid and the elasticity of the conduit itself. The cause of these elastic oscillations lies in speed variations, caused, for example, by opening or closing manoeuvres of the end valves of a branch or pumping conduit, or, in the latter case, also directly from the connection or disconnection of the pump. Mass oscillations, on the other hand, are block displacements that the liquid column, contained in the conduit connecting two free-surface tanks, undergoes, almost rigidly, when the level of at least one of the tanks varies over time. Mass oscillations have periods and durations of an order of magnitude much greater than the former, to be considered as subsequent, and independent oscillations from the elastic ones.

The details of the circuit, with the indication of the number assigned to each section, the name, the diameter, the length and the friction coefficient of the concentrated losses, are reported in Table 2.

5   Simulation of the Circuit Operation

Starting from the sample, *part 5–6*, the circuit is followed, upstream and downstream, calculating for each section the pressure loss and the speed at every instant of time.

5.1   Pressure Losses in the Various Parts of the Circuit

The pressure drop within each section of the circuit is given by

$$\Delta p_{m-n} = -\rho \frac{dU_i}{dt} L_i - \left(\rho \frac{U_i |U_i|}{2} \right) \left(\lambda_i \frac{L_i}{D_i} + \lambda' \right) \tag{1}$$

where the subscript i changes for each part of the circuit and assumes the relative symbol, while m and n assume the numbers relative to the various sections. The Reynolds number is defined by

$$\mathrm{Re}_i = \rho U_i \frac{D_i}{\mu} \tag{2}$$

and in laminar flow, we have

$$\lambda_i = \frac{64}{\mathrm{Re}_i} \tag{3}$$

5.1.1 Part 5–6 (Sample)

The pressure drop within the sample (Sects. 5 and 6) is given by

$$\Delta p_{5-6} = -\rho \frac{dU_p}{dt} L_p - \left(\rho \frac{U_p |U_p|}{2} \right) \left(\lambda_p \frac{L_p}{D_p} + \lambda'_p \right) \tag{4}$$

5.1.2 Part 4–5 (Calm)

Upstream of the sample, we encounter the calm part, and from the mass balance, we obtain the speed

$$U_c = U_p \left(\frac{D_p}{D_c} \right)^2 \tag{5}$$

The pressure drop within the calm part (4–5) is given by

$$\Delta p_{4-5} = -\rho \frac{dU_c}{dt} L_c - \left(\rho \frac{U_c |U_c|}{2} \right) \left(\lambda_c \frac{L_c}{D_c} + \lambda'_c \right) \tag{6}$$

5.1.3 Part 3–4 (Compliance Chamber)

As for the mass balance, we have

$$U_i \left(\frac{D_i}{2} \right)^2 = \frac{dZ}{dt} \left(\frac{D_{co}}{2} \right)^2 + U_c \left(\frac{D_c}{2} \right)^2 \tag{7}$$

but the change in level, balancing the pressures in the compliance chamber, is

$$Z = \frac{P_{\text{sc}} + (P_{\text{rif}} - P_{\text{atm}})}{\rho g} \tag{8}$$

through which we can calculate the speed in the part preceding the compliance chamber.

5.1.4　Part 2–3

Having already calculated the speed in part 2–3, from the balance in the compliance chamber, we only need to calculate the pressure drop, which is

$$\Delta p_{2-3} = -\rho \frac{dU_i}{dt} L_i - \left(\rho \frac{U_i |U_i|}{2}\right)\left(\lambda_i \frac{L_i}{D_i} + \lambda'_i\right) \tag{9}$$

5.1.5　Part 1–2 (Initial Part or Exit from the Pump: Uphill)

From the mass balance, we have

$$U_s = U_i \left(\frac{D_i}{D_s}\right)^2 \tag{10}$$

and the relative pressure drop is given by

$$\Delta p_{1-2} = -\rho g L_s \sin \varphi_2 - \rho \frac{dU_s}{dt} L_{is} - \left(\rho \frac{U_s |U_s|}{2}\right)\left(\lambda_s \frac{L_s}{D_s} + \lambda'_s\right) \tag{11}$$

5.1.6　Part 6–7 (Bifurcation)

Downstream of the sample, we find the bifurcation, as shown in Fig. 12, and it is reasonable to assume that

$$U_p \left(\frac{D_p}{2}\right)^2 = U_{b1}\left(\frac{D_{b1}}{2}\right)^2 + U_{b2}\left(\frac{D_{b2}}{2}\right)^2 \tag{12}$$

while the pressure drop in the first branch is equal to that in the second

$$\Delta p_{b1} = \Delta p_{b2} \tag{13}$$

where

$$\Delta p_{b1} = -\rho \frac{dU_{b1}}{dt} L_{b1} - \left(\rho \frac{U_{b1} |U_{b1}|}{2}\right)\left(\lambda_{b1} \frac{L_{b1}}{D_{b1}} + \lambda'_{b1}\right) \tag{14}$$

$$\Delta p_{b2} = -\rho \frac{dU_{b2}}{dt} L_{b2} - \left(\rho \frac{U_{b2} |U_{b2}|}{2}\right)\left(\lambda_{b2} \frac{L_{b2}}{D_{b2}} + \lambda'_{b2}\right) \tag{15}$$

Equating these expressions, we obtain the speed U_{b1}, using the implicit method of *Crank–Nicholson* to arrive at the solution. Once U_{b1} is known, we are able to calculate the pressure losses in the various branches.

5.1.7 Part 7–7′ (Return 1–1)

The reconnection of the bifurcation leads to the speed

$$U_d = U_{b1} \left(\frac{D_{b1}}{D_d}\right)^2 + U_{b2} \left(\frac{D_{b2}}{D_d}\right)^2 \tag{16}$$

and to the pressure drop

$$\Delta p_d = -\rho \frac{dU_d}{dt} L_d - \left(\rho \frac{U_d|U_d|}{2}\right)\left(\lambda_d \frac{L_d}{D_d} + \lambda'_d\right) \tag{17}$$

5.1.8 Part 7′–8 (Return 1–2)

For the part of the return downhill, only the pressure drop changes, as there is also the component related to gravitational acceleration

$$\Delta p_d = -\rho g L_d \sin \varphi_3 - \rho \frac{dU_d}{dt} L_{isd} - \left(\rho \frac{U_d|U_d|}{2}\right)\left(\lambda_d \frac{L_d}{D_d} + \lambda'_d\right) \tag{18}$$

5.1.9 Part 8–9 (Return 2)

From the mass balance, we have

$$U_r = U_d \left(\frac{D_d}{D_r}\right)^2 \tag{19}$$

The pressure drop in the last part, before the main tank, where the main frequencies of the pump are nullified, is

$$\Delta p_r - -\rho \frac{dU_r}{dt} L_r - \left(\rho \frac{U_r|U_r|}{2}\right)\left(\lambda_r \frac{L_r}{D_r} + \lambda'_r\right) \tag{20}$$

6 Results

In this paragraph, the results of the *MATLAB* simulation in the various parts of the circuit are shown. Mainly, the speed and the pressure in the various parts are illustrated and, finally, the total power of the circuit and the level difference in the tower, as a function of time, in *seconds*.

6.1   List of MATLAB Code for the Simulation

The program in *MATLAB* is the following.

```
function [DP_b1,DP_b2,Ub1,Ub2] = bifurcation(Up,Nt,dt,rho,mu,Db1,
Db2,Lb2,Lb1,Dp)
DP_b1= zeros(1,Nt);
Ub1= zeros(1,Nt);
DUb1DT= zeros(1,Nt);
lambda_b1= zeros(1,Nt);
Re_b1= zeros(1,Nt);
DP_b2= zeros(1,Nt);
Ub2= zeros(1,Nt);
DUb2DT= zeros(1,Nt);
lambda_b2= zeros(1,Nt);
Re_b2= zeros(1,Nt);
g=9.81; %m/s2
ucm=sum(Up(1,:))/Nt;
%%%% WE INITIALISE THE VARIABLES
%Z(1,1)= 0.1;
Ub1(1,:)= Up(1,:)*(Db1/(Db1+Db2))^2;
Ub2(1,:)= Up(1,:)*(Db2/(Db1+Db2))^2;
Re_b1(1,:)= rho*Ub1(1,:)*Db1/mu;
lambda_b1(1,:)= 64./Re_b1(1,:);
lambda_apex_b1=0.5;
lambda_apex_b2=0.5;
[DUpDT] = second_order_derivative(Up,Nt,dt);
%%%% FIRST TEMPORAL CYCLE
alphaU=0.8; %alphaU= [0:1];
Nite=80;
for It=2:Nt;
% Ub1_new(1,It)=Ub1(1,It-1);
Ub1(1,It)= Ub1(1,It-1);
for K=1:Nite;
% Ub1(1,It)= Ub1_new(1,It);
Re_b1(1,It)= rho*Ub1(1,It)*Db1/mu;
Ub2(1,It)= ((Dp/Db2)^2)*Up(1,It) - ((Db1/Db2)^2)*Ub1(1,It);
Re_b2(1,It)= rho*Ub2(1,It)*Db2/mu;
lambda_b1(1,It)= 64/Re_b1(1,It);
lambda_b2(1,It)= 64/Re_b2(1,It);
% Ub1(1,It)= Ub1(1,It-1)+(1/2)*dt*((1/((rho*Lb1)+rho*((Db1/Db2)^2)
*Lb2))*((rho*DUpDT(1,It-1)*((Dp/Db1)^2)*Lb2) - (Ub1(1,It-1)*(abs
(Ub1(1,It-1)))*lambda_b1(1,It-1)*(Lb1/Db1)))+....
```

```
% (1/((rho*Lb1)+rho*((Db1/Db2)^2)*Lb2))*(1/2)*((Up(1,It-1)*
(Dp/Db2)^2-Ub1(1,It-1)*(Db1/Db2)^2)*abs(Up(1,It-1)*(Dp/Db2)^2-Ub1
(1,It-1)*(Db1/Db2)^2)*(lambda_b2(1,It-1)*(Lb2/Db2))))+....
% +(1/2)*dt*((1/((rho*Lb1)+rho*((Db1/Db2)^2)*Lb2))*((rho*DUpDT(1,
It)*((Dp/Db1)^2)*Lb2) - (Ub1(1,It)*(abs(Ub1(1,It)))*lambda_b1(1,It)
*(Lb1/Db1))))+....
% (1/((rho*Lb1)+rho*((Db1/Db2)^2)*Lb2))*(1/2)*((Up(1,It)*(Dp/Db2)
^2-Ub1(1,It)*(Db1/Db2)^2)*abs(Up(1,It)*(Dp/Db2)^2-Ub1(1,It)*
(Db1/Db2)^2)*(lambda_b2(1,It)*(Lb2/Db2))));
% Ub1(1,It)= Ub1(1,It-1)+dt*(1/2)*((1/(Lb2*((Db1/Db2)^2) +Lb1))*
(Lb2*((Dp/Db2)^2)*DUpDT(1,It) - ((Ub1(1,It)*abs(Ub1(1,It)))/2)*
(lambda_b1(1,It)*Lb1/Db1)+(((((Dp/Db2)^2)*Up(1,It) - ((Db1/Db2)^2)
*Ub1(1,It))*abs(((Dp/Db2)^2)*Up(1,It) - ((Db1/Db2)^2)*Ub1(1,It)))/
2)*(lambda_b2(1,It)*Lb2/Db2))+...
% (1/(Lb2*((Db1/Db2)^2) +Lb1))*(Lb2*((Dp/Db2)^2)*DUpDT(1,It-1) -
((Ub1(1,It-1)*abs(Ub1(1,It-1)))/2)*(lambda_b1(1,It-1)*Lb1/Db1)
+...
% (((((Dp/Db2)^2)*Up(1,It-1) - ((Db1/Db2)^2)*Ub1(1,It-1))*abs
(((Dp/Db2)^2)*Up(1,It-1) - ((Db1/Db2)^2)*Ub1(1,It-1)))/2)*
(lambda_b2(1,It-1)*Lb2/Db2)));
FCNU(1,It)= ((1/(Lb2*((Db1/Db2)^2) +Lb1))*(Lb2*((Dp/Db2)^2)*DUpDT
(1,It) - ((Ub1(1,It)*abs(Ub1(1,It)))/2)*(lambda_b1(1,It)*Lb1/
Db1+lambda_apex_b1)...
+(((((Dp/Db2)^2)*Up(1,It) - ((Db1/Db2)^2)*Ub1(1,It))*abs(((Dp/Db2)
^2)*Up(1,It) - ((Db1/Db2)^2)*Ub1(1,It)))/2)*(lambda_b2(1,It)*Lb2/
Db2+lambda_apex_b1)));
Ub1(1,It)= Ub1(1,It-1)+dt*(1/2)*(FCNU(1,It-1)+FCNU(1,It));
% Ub1(1,It)= Ub1_new(1,It);
% Ub1_old(1,It) = Ub1_new(1,It);
% Ub1_new(1,It) = alphaU*Ub1(1,It)+(1-alphaU)*Ub1_old(1,It);
end;
end;
%%%% END OF THE ITERATIVE CYCLE
[DUb1DT] = second_order_derivative(Ub1,Nt,dt);
for I=1:Nt;
Re_b1(1,I)= rho*Ub1(1,I)*Db1/mu;
end;
for I=1:Nt;
lambda_b1(1,I)= 64./Re_b1(1,I);
end;
for I=1:Nt;
DP_b1(1,I)= -rho*DUb1DT(1,I)*Lb1 - (rho*(Ub1(1,I)*abs(Ub1(1,I)))/2)
*(lambda_b1(1,I)*Lb1/Db1+lambda_apex_b1);
end;
for I=1:Nt;
```

```matlab
Ub2(1,I)=((Dp/Db2)^2)*Up(1,I) -(Ub1(1,I)*((Db1/Db2)^2));
end;
[DUb2DT] = second_order_derivative(Ub2,Nt,dt);
for I=1:Nt;
Re_b2(1,I)= rho*Ub2(1,I)*Db2/mu;
end;
for I=1:Nt;
lambda_b2(1,I)= 64./Re_b2(1,I);
end;
for I=1:Nt;
DP_b2(1,I)= -rho*DUb2DT(1,I)*Lb2 -(rho*(Ub2(1,I)*abs(Ub2(1,I)))/2)
*(lambda_b2(1,I)*Lb2/Db2);
end;
% DP_b1(1,I)= -rho*DUb1DT(1,I)*Lb1 -(rho*(Ub1(1,I)*abs(Ub1(1,I)))/
2)*(lambda_b1(1,I)*Lb1/Db1)=DP_b2(1,I)= -rho*DUb2DT(1,I)*Lb2 -
(rho*(Ub2(1,I)*abs(Ub2(1,I)))/2)*(lambda_b2(1,I)*Lb2/Db2);
% Ub2(1,It)=((Dp/Db2)^2)*Up(1,It) -((Db1/Db2)^2)*Ub1(1,It);
% DUb2DT(1,I)=((Dp/Db2)^2)*DUpDT(1,It) -((Db1/Db2)^2)*DUb1DT(1,It)
% -rho*DUb1DT(1,I)*Lb1 -(rho*(Ub1(1,I)*abs(Ub1(1,I)))/2)*(lambda_b1
(1,I)*Lb1/Db1)= -rho*DUb2DT(1,I)*Lb2 -(rho*(Ub2(1,I)*abs(Ub2(1,
I)))/2)*(lambda_b2(1,I)*Lb2/Db2);
% DUb1DT(1,It)=....
% (1/(Lb2*((Db1/Db2)^2) +Lb1))*(Lb2*((Dp/Db2)^2)*DUpDT(1,It) -(rho*
(Ub1(1,I)*abs(Ub1(1,I)))/2)*(lambda_b1(1,I)*Lb1/Db1) ....
% (((((Dp/Db2)^2)*Up(1,It)
% -((Db1/Db2)^2)*Ub1(1,It))*abs(((Dp/Db2)^2)*Up(1,It) -((Db1/Db2)
^2)*Ub1(1,It)))/2)*(lambda_b2(1,I)*Lb2/Db2))
% DUb1DT(1,It)=(1/(Lb2*((Db1/Db2)^2) +Lb1))*(Lb2*((Dp/Db2)^2)
*DUpDT(1,It) -(rho*(Ub1(1,It)*abs(Ub1(1,It)))/2)*(lambda_b1(1,It)
*Lb1/Db1)+...
% (((((Dp/Db2)^2)*Up(1,It) -((Db1/Db2)^2)*Ub1(1,It))*abs(((Dp/Db2)
^2)*Up(1,It) -((Db1/Db2)^2)*Ub1(1,It)))/2)*(lambda_b2(1,It)*Lb2/
Db2));
% Ub1(1,It)=Ub1(1,It-1)+dt*(1/2)*((1/(Lb2*((Db1/Db2)^2) +Lb1))*
(Lb2*((Dp/Db2)^2)*DUpDT(1,It) -(rho*(Ub1(1,It)*abs(Ub1(1,It)))/2)*
(lambda_b1(1,It)*Lb1/Db1)+...
% (((((Dp/Db2)^2)*Up(1,It) -((Db1/Db2)^2)*Ub1(1,It))*abs(((Dp/Db2)
^2)*Up(1,It) -((Db1/Db2)^2)*Ub1(1,It)))/2)*(lambda_b2(1,It)*Lb2/
Db2))+...
% (1/(Lb2*((Db1/Db2)^2) +Lb1))*(Lb2*((Dp/Db2)^2)*DUpDT(1,It-1) -
(rho*(Ub1(1,It-1)*abs(Ub1(1,It-1)))/2)*(lambda_b1(1,It-1)*Lb1/
Db1)+...
% (((((Dp/Db2)^2)*Up(1,It-1) -((Db1/Db2)^2)*Ub1(1,It-1))*abs
(((Dp/Db2)^2)*Up(1,It-1) -((Db1/Db2)^2)*Ub1(1,It-1)))/2)*
(lambda_b2(1,It-1)*Lb2/Db2)));
```

```
%
%
function [Ui,Z,DZDT,lambda_apex_i]= compliancechamber(Uc,Nt,dt,
rho,mu,Di,Li,Dco,Dc,P_sc)
Ui=zeros(1,Nt);
lambda_i=zeros(1,Nt);
Re_i=zeros(1,Nt);
Z=zeros(1,Nt);
DZDT=zeros(1,Nt);
%%%% COMPLIANCE CHAMBER
%%%% MASS BALANCE
%%%% Ui*Si*dt=Sco*dz+Uc*Sc*dt
%%%% H1-H2=-z
%%%% Ui=(Sco/Si)*dz/dt+(Sc/Si)*Uc
%%%% DUi/DT=(Sco/Si)*d2z/dt2+(Sc/Si)*DUc/DT
%%%% -z = (1/g)*Li*((Sco/Si)*d2z/dt2+(Sc/Si)*DUc/DT)+(1/g)
*lambda_i*(Li/Di)*(1./2)*((Sco/Si)*dz/dt+(Sc/Si)*Uc)^2
%%%% d2z/dt2 +lambda_i*(1./Di)*(1./2)*(Sco/Si)*(dz/dt)^2 +lambda_i*
(1./Di)*(Sc/Si)*Uc*dz/dt +(Si/Sco)*(g/Li)*z +(Sc/Sco)*DUc/DT
+(Si/Sco)*((Sc/Si)*Uc)^2 = 0
%%%% -z =
%%%% (1/g)*Li*DUi/DT+(1/g)*lambda_i*(Li/Di)*(1./2)*((Sco/Si)
*Ui*abs(Ui)
lambda_apex_i=3;
g=9.81; %m/s2
ucm=sum(Uc(1,:))/Nt;
%%%% INITIALISE THE VARIABLES
Z(1,1)= 0.0;
Ui(1,1)= 0.00001;
Re_i(1,1)= rho*Ui(1,1)*Di/mu;
lambda_i(1,1)= 64/Re_i(1,1);
%%%% TIME LOOP FIRST
for It=2:Nt;
%%%% WE USE THE CRANK-NICHOLSON METHOD
Z(1,It)=Z(1,It-1);
Ui(1,It)=Ui(1,It-1);
for IK=1:50;
Re_i(1,It)= rho*Ui(1,It)*Di/mu;
lambda_i(1,It)= 64/Re_i(1,It);
Z(1,It)=Z(1,It-1)+dt*(1/2)*((((Di/Dco)^2)*Ui(1,It-1)-((Dc/Dco)^2)
*Uc(1,It-1))+(((Di/Dco)^2)*Ui(1,It)-((Dc/Dco)^2)*Uc(1,It)));
Ui(1,It)=Ui(1,It-1)+dt*(g/Li)*(1/2)*((-(Z(1,It-1)-P_sc(1,It-1)/
(rho*g))-(lambda_i(1,It-1)*(Li/Di)+lambda_apex_i)*(1/g)*(1./2)*Ui
(1,It-1)*abs(Ui(1,It-1)))+...
(-(Z(1,It)-P_sc(1,It)/(rho*g))-(lambda_i(1,It)*(Li/Di)
```

```
+lambda_apex_i)*(1/g)*(1./2)*Ui(1,It)*abs(Ui(1,It))));
end;
%%%% END OF ITERATIVE LOOP
%%% TEST
[DP_p,Up] = piv(Nt,dt,wo,rho,mu,Dp,Lp,To);
%%% CALM_SECTION
[DP_c,Uc] = third_section(Up,Nt,dt,rho,mu,Dc,Lc,Dp);
%BIFURCATION
[DP_b1,DP_b2,Ub1,Ub2] = bifurcation(Up,Nt,dt,rho,mu,Db1,Db2,Lb2,
Lb1,Dp);
% OUTPUT TANKS following the bifurcation
%First Tank
[Us1,Zs1,DZs1DT]= tankob1(Ub1,Nt,dt,rho,mu,Dco,Lb1,Ds,Db1);
% Second Tank
[Us2,Zs2,DZs2DT]= tankob2(Ub2,Nt,dt,rho,mu,Dco,Lb2,Ds,Db2);
% rejoining
[Ud] = rejoining(Nt,dt,Dd,Db1,Db2,Ub2,Ub1);
% First RETURN Section downhill
[DP_d]=downhill_section(Nt,dt,rho,mu,Dd,Ld,g,phi3,Ud);
% Last return section into the main tank B1
[DP_r,Ur] = return_section(Ud,Nt,dt,rho,mu,Dr,Lr,Dd);
%%%% COMPLIANCE CHAMBER
%%%% MASS BALANCE
%%%% Ui*Si*dt=Sco*dz+Uc*Sc*dt
%%%% H1-H2=-z
%%%% Ui= (Sco/Si)*dz/dt+(Sc/Si)*Uc
%%%% DUi/DT= (Sco/Si)*d2z/dt2+(Sc/Si)*DUc/DT
%%%% -z = (1/g)*Li*((Sco/Si)*d2z/dt2+(Sc/Si)*DUc/DT)+(1/g)
*lambda_i*(Li/Di)*(1./2)*((Sco/Si)*dz/dt+(Sc/Si)*Uc)^2
%%%% d2z/dt2 +lambda_i*(1./Di)*(1./2)*(Sco/Si)*(dz/dt)^2 +lambda_i*
(1./Di)*(Sc/Si)*Uc*dz/dt +(Si/Sco)*(g/Li)*z +(Sc/Sco)*DUc/DT
+(Si/Sco)*((Sc/Si)*Uc)^2 = 0
for I=1:Nt;
P_dc(1,I) = (DP_r(1,I)); %Pressure in the return section
end;
for I=1:Nt;
P_bc(1,I) = (DP_d(1,I)+DP_r(1,I)); %Pressure at the start of the
descent
end;
for I=1:Nt;
P_pc(1,I) = (DP_b1(1,I)+DP_d(1,I)+DP_r(1,I)); %Pressure at the start
of the bifurcation
end;
for I=1:Nt;
P_cc(1,I) = (DP_p(1,I)+DP_b1(1,I)+DP_d(1,I)+DP_r(1,I)); %Pressure at
```

```
the inlet to the sample
end;
for I=1:Nt;
P_ic(1,I) = (DP_c(1,I)+DP_p(1,I)+DP_b1(1,I)+DP_d(1,I)+DP_r(1,I)); %
Pressure at the inlet to the calm (in the tower)
end;
for I=1:Nt;
P_sc(1,I) = (DP_i(1,I)+DP_c(1,I)+DP_p(1,I)+DP_b1(1,I)+DP_d(1,I)
+DP_r(1,I)); %Pressure at the inlet to the section containing the tower
end;
%%% COMPLIANCE_SECTION
[Ui,Z,DZDT,lambda_apex_i]= cameradicompliance2(Uc,Nt,dt,rho,Di,
Dco,Dc,P_sc);
%%% INITIAL_SECTION
[DP_i]=first_section(Ui,Nt,dt,rho,mu,Di,Li,phi1,g,lambda_apex_i);
[DP_s,Us]=initial_section(Ui,Nt,dt,rho,mu,Ds,Ls,Di,g,phi2);
for I=1:Nt;
P_itc(1,I) = (DP_s(1,I)+DP_i(1,I)+DP_c(1,I)+DP_p(1,I)+DP_b1(1,I)
+DP_d(1,I)+DP_r(1,I)); % Total Pressure
end;
L_dot= P_itc.*Us*(pi*(Ds/2)^2);
a=Nt/2;
FoP_ic=fft(P_ic(a:Nt));
aP_ic=abs(FoP_ic);
phiP_ic=angle(FoP_ic);
FoL_dot=fft(L_dot(a:Nt));
aL_dot=abs(FoL_dot);
phiL_dot=angle(FoL_dot);
wmax=2*pi/dt;
wmin=2*pi/To;
dw=(wmax-wmin)/Nt;
w=[wmin:dw:wmax-dw];
w=w(1:Nt-a+1);
figure(1)
plot(t,Up,'linewidth',2)
xlabel('time (s)')
ylabel('velocity (m/s)')
title('velocity inside the sample piv')
xlim([10, 12])
figure(2)
plot(t,(6.5-(10^-3)*P_cc)*3.7/8,'linewidth',2)
xlabel('time (s)')
ylabel('Pressure (kPa)')
title('pressure at the sample inlet')
xlim([10, 12])
```

```
figure(3)
plot(t,Uc,'linewidth',2)
xlabel('time (s)')
ylabel('velocity (m/s)')
title('velocity in the third section (calm section)')
xlim([10, 12]) end;
%%%% END OF TIME CYCLE
[DZDT] = second_order_derivative(Z,Nt,dt);
for I=1:Nt;
Ui(1,I)= ((Dco/Di)^2)*DZDT(1,I)+((Dc/Di)^2)*Uc(1,I);
end;
function [Ui,Z,DZDT,lambda_apex_i]= cameradicompliance2(Uc,Nt,dt,
rho,Di,Dco,Dc,P_sc)
Ui=zeros(1,Nt);
Z=zeros(1,Nt);
DZDT=zeros(1,Nt);
%%%% COMPLIANCE CHAMBER
%%%% MASS BALANCE
%%%% Ui*Si*dt=Sco*dz+Uc*Sc*dt
%%%% H1-H2=-z
%%%% Ui=(Sco/Si)*dz/dt+(Sc/Si)*Uc
%%%% DUi/DT=(Sco/Si)*d2z/dt2+(Sc/Si)*DUc/DT
%%%% -z = (1/g)*Li*((Sco/Si)*d2z/dt2+(Sc/Si)*DUc/DT)+(1/g)
*lambda_i*(Li/Di)*(1./2)*((Sco/Si)*dz/dt+(Sc/Si)*Uc)^2
%%%% d2z/dt2 +lambda_i*(1./Di)*(1./2)*(Sco/Si)*(dz/dt)^2 +lambda_i*
(1./Di)*(Sc/Si)*Uc*dz/dt +(Si/Sco)*(g/Li)*z +(Sc/Sco)*DUc/DT
+(Si/Sco)*((Sc/Si)*Uc)^2 = 0
%%%% -z =
%%%% (1/g)*Li*DUi/DT+(1/g)*lambda_i*(Li/Di)*(1./2)*((Sco/Si)
*Ui*abs(Ui)
lambda_apex_i=3;
g=9.81; %m/s2
%%%% INITIALISE VARIABLES
for I=1:Nt;
Z(1,I)= P_sc(1,I)/(rho*g);
end;
[DZDT] = second_order_derivative(Z,Nt,dt);
for I=1:Nt;
Ui(1,I)= ((Dco/Di)^2)*DZDT(1,I)+((Dc/Di)^2)*Uc(1,I);
end;
%main_circuit
close all
clear all
clc
Nt=20000;
```

```
Di=0.02; % diameter of the first section (before the compliance)
Li=0.25; % length of the first section (before the compliance)
Dc=0.01; % diameter of the third section (calm section)
Lc=0.25; % length of the third section (calm section)
Dp=0.01; % sample diameter
Lp=0.25; % sample length
Dco=0.1; % diameter of the turret
Ds=0.02; % diameter of the initial section
Ls =0.7; % length of the initial section
Db1=0.01; % diameter of the first branch of the bifurcation
Db2=0.005; % diameter of the second branch of the bifurcation
Lb1=0.25; % length of the first branch of the bifurcation
Lb2=0.25; % length of the second branch of the bifurcation
Dd=0.02; % diameter of the section following the rejoining of the
bifurcation
Ld=0.95; % length of the section following the rejoining of the
bifurcation
Dr=0.02; % diameter of the final return section
Lr=1.25; % length of the final return section
DP_b1=zeros(1,Nt); %pressure drop in the first branch of the bifurcation
DP_b2=zeros(1,Nt); %pressure drop in the second branch of the
bifurcation
DP_i=zeros(1,Nt); %pressure drop in the first section
DP_c=zeros(1,Nt); %pressure drop in the calm section
DP_p=zeros(1,Nt); %pressure drop in the sample
DP_d=zeros(1,Nt); %pressure drop in the section following the
reconnection of the bifurcation
DP_r=zeros(1,Nt); %pressure drop in the last return section
Us=zeros(1,Nt); % speed in the initial section
Ui=zeros(1,Nt); % speed in the first section
Uc=zeros(1,Nt); % speed in the calm section
Up=zeros(1,Nt); % speed in the sample
Ub1=zeros(1,Nt);% speed in the first branch of the bifurcation
Ub2=zeros(1,Nt);% speed in the second branch of the bifurcation
Ud=zeros(1,Nt); % speed in the section following the reconnection
Ur=zeros(1,Nt); % speed in the last return section
P_dc=zeros(1,Nt); % pressure in the last return section
P_bc=zeros(1,Nt); % pressure at the entrance of the reconnection
section of the bifurcation
P_pc=zeros(1,Nt); % pressure at the entrance of the bifurcation
P_cc=zeros(1,Nt); % pressure at the entrance of the sample
P_ic=zeros(1,Nt); % pressure at the entrance to the third section (calm
section)
P_sc=zeros(1,Nt); % pressure at the entrance to the compliance chamber
P_itc=zeros(1,Nt);% Total Pressure
```

```
rho=1000.0; %kg/m3
mu=0.001; %Pa*s
To=1; % period
Toss=12*To; % observation time
dt=Toss/Nt;
t=[dt:dt:Toss];
wo=(2*pi/To); % frequency
%H=1.0;
%Hd=0.8;
g=9.81; % gravitational acceleration
phi1=0*(pi/4);
phi2=0*(-pi/4);
phi3=0*(-pi/2);
%%%%%%%%%%%%%%%%%%%%%%%%%%%%%%%%%%%%%%%%%%%%%%%%%%%%%%%%%%%%%%%%%%%%%%%%%%%%%%
%%%%%%%%%%%%
%%%%WARNING: NEVER INITIALISE THE INPUT VARIABLES TO THE FUNCTION
%%%%%%%%%%%%%%%%%%%%%%%%%%%%%%%%%%%%%%%%%%%%%%%%%%%%%%%%%%%%%%%%%%%%%%%%%%%%%%%
%%%%%%%%%%%%
figure(4)
plot(t,3-(10^-3)*P_ic*5.5/9.5,'linewidth',2)
xlabel('time (s)')
ylabel('Pressure (kPa)')
title('pressure at the entrance of the calm stretch (in the tower)')
xlim([10, 12])
figure(5)
plot(t,Z,'linewidth',2)
xlabel('time (s)')
ylabel('Level difference in the tower (m)')
title('compliance chamber')
xlim([10, 12])
figure(6)
plot(t,-(0.025/500)*Ui,'linewidth',2)
xlabel('time (s)')
ylabel('velocity (m/s)')
title('first stretch (before the compliance chamber)')
xlim([10, 12])
figure(7)
plot(t,2.94-(10^-3)*P_sc*0.13/9.5,'linewidth',2) %3-(10^-3)
*P_sc*5.5/9.5
xlabel('time (s)')
ylabel('Pressure (kPa)')
title('pressure at the entrance of the section containing the
compliance chamber')
xlim([10, 12])
figure(8)
```

```
plot(t,-(0.025/500)*Us,'linewidth',2)
xlabel('time (s)')
ylabel('velocity (m/s)')
title('initial section (uphill section, heading towards the first
section)')
xlim([10, 12])
figure(9)
plot(t,10.75+(10^-8)*P_itc*5.1/3,'linewidth',2)
xlabel('time (s)')
ylabel('Pressure (kPa)')
title('Pressure at the pump outlet')
xlim([10, 12])
figure(10)
plot(t,(10^-3)*DP_b1,'b',t,(10^-3)*DP_b2,'r','linewidth',2)
xlabel('time (s)')
ylabel('Pressure (kPa)')
legend('large branch','small branch')
title('bifurcation 1')
xlim([10, 12])
figure(11)
plot(t,Ub1,'linewidth',2)
xlabel('time (s)')
ylabel('velocity (m/s)')
title('branch 1')
%xlim([10, 12])
figure(12)
plot(t,Ub2,'linewidth',2)
xlabel('time (s)')
ylabel('velocity (m/s)')
title('branch 2')
%xlim([10, 12])
figure(13)
plot(t,Ud,'linewidth',2)
xlabel('time (s)')
ylabel('velocity (m/s)')
title('velocity in the downhill section of the return')
xlim([10, 12])
figure(14)
plot(t,9.81-(10^-3)*P_bc*0.08/4.3,'linewidth',2)
xlabel('time (s)')
ylabel('Pressure (kPa)')
title('pressure at the entrance of the downhill section of the return')
xlim([10, 12])
figure(15)
plot(t,Ur,'linewidth',2)
```

```
xlabel('time (s)')
ylabel('velocity (m/s)')
title('velocity in the last section of the return')
xlim([10, 12])
figure(16)
plot(t,9.81-(10^-3)*P_dc*0.08/2.4,'linewidth',2)
xlabel('time (s)')
ylabel('Pressure (kPa)')
title('pressure at the entrance of the last section')
xlim([10, 12])
N=3
figure(17)
loglog(w(1:Nt/2),smooth(aP_ic(1:Nt/2)/aP_ic(1),N),'linewidth',2)
xlabel('frequency (Hz)')
ylabel('Pressure (Pa)')
title('fft P')
figure(18)
plot(t,L_dot,'linewidth',2)
xlabel('time (t)')
ylabel('Power (W)')
N=3
figure(19)
loglog(w(1:Nt/2),smooth(aL_dot(1:Nt/2)/aL_dot(1),
N),'linewidth',2)
xlabel('frequency (Hz)')
ylabel('Pressure (Pa)')
title('fft P')
figure(20)
plot(t,(10^-3)*P_dc,'b',t,(10^-3)*P_bc,'r',t,(10^-3)*P_pc,'c',t,
(10^-3)*P_cc,'g',t,(10^-3)*P_ic,'y',t,(10^-3)*P_sc,'k',t,(10^-3)
*P_sc,'m',t,(10^-3)*P_itc,'b','linewidth',2)
xlabel('time (s)')
ylabel('Pressure (kPa)')
legend('Pressure in the return section','Pressure at the start of the
descent','Pressure at the start of the bifurcation','Pressure at the
entrance to the sample',....
'Pressure at the entrance to the calm (in the tower)','Pressure at the
entrance to the section containing the tower','Total Pressure')
xlim([10, 12])
figure(21)
plot(t,0.3-3.5*Zs1,'linewidth',2)
xlabel('time (s)')
ylabel('Level difference in tower 1 (m)')
title('fork 1 reservoir')
xlim([10, 12])
```

```
figure(22)
plot(t,Zs2,'linewidth',2)
xlabel('time (s)')
ylabel('Level difference in tower 2 (m)')
title('fork 2 reservoir')
xlim([10, 12])
figure(23)
plot(t,-(6.8/400)*Ui,'linewidth',2)
xlabel('time (s)')
ylabel('Power (W)')
%title('first section (before the compliance chamber)')
xlim([10, 12])
['the necessary pressure difference is ', num2str(max(abs(P_ic(1,
I)))),'Pa']
function [DP_p,Up] = piv(Nt,dt,wo,rho,mu,Dp,Lp,To)
DP_p = zeros(1,Nt);
Up= zeros(1,Nt);
DUpDT= zeros(1,Nt);
lambda_p= zeros(1,Nt);
Re_p= zeros(1,Nt);
lambda_apex_p=1;
for I=1:Nt;
% Up(1,I)= (1/rho)*(3327.37908178-862.89309093*cos
(wo*I*dt)-328.04844576*cos(2*wo*I*dt)-240.65304266*cos
(3*wo*I*dt)....
% +68.14896524*cos(4*wo*I*dt)-224.35074725*cos
(5*wo*I*dt)-176.39609014*cos(6*wo*I*dt)-22.92131212*cos
(7*wo*I*dt).....
% -8.71197406*cos(8*wo*I*dt)-68.45635802*cos
(9*wo*I*dt)-0.70622372*cos(10*wo*I*dt)-5.26754494*cos
(11*wo*I*dt).....
% -2.81406010*cos(12*wo*I*dt)+7.72282589*cos(13*wo*I*dt)
+11.20952170*cos(14*wo*I*dt)+13.15340842*cos(15*wo*I*dt).....
% -0.30416579*cos(16*wo*I*dt)+9.43509700*cos(17*wo*I*dt)
+5.74480893*cos(18*wo*I*dt)+11.97121956*cos(19*wo*I*dt).....
% +5.29051745*cos(20*wo*I*dt)+3.42362751*cos(21*wo*I*dt)
+4.97112842*cos(22*wo*I*dt)+4.50096886*cos(23*wo*I*dt)......
% -0.74952339*cos(24*wo*I*dt)-5.09646128*cos
(25*wo*I*dt)-6.56022745*cos(26*wo*I*dt)-4.21634370*cos
(27*wo*I*dt).....
% +1.46834049*cos(28*wo*I*dt)-4.48597970*cos
(29*wo*I*dt)-6.19885283*cos(30*wo*I*dt)-3.52181539*cos
(31*wo*I*dt) -5.05432935*cos(32*wo*I*dt).....
% +1339.70206107*sin(wo*I*dt)-214.70689347*sin(2*wo*I*dt)
+106.21907018*sin(3*wo*I*dt)-17.76139433*sin(4*wo*I*dt).....
```

```
% -109.64895883*sin(5*wo*I*dt)+25.34068175*sin(6*wo*I*dt)
+44.31085321*sin(7*wo*I*dt)-58.76816196*sin(8*wo*I*dt)......
% +1.35383329*sin(9*wo*I*dt)-13.98336895*sin
(10*wo*I*dt)-14.85734149*sin(11*wo*I*dt)-17.26854689*sin
(12*wo*I*dt)......
% -5.40827340*sin(13*wo*I*dt)-8.78640674*sin(14*wo*I*dt)
+4.89415093*sin(15*wo*I*dt)-4.40069649*sin(16*wo*I*dt).....
% -0.67340986*sin(17*wo*I*dt)+5.18011647*sin(18*wo*I*dt)
+0.31732217*sin(19*wo*I*dt)+1.20673568*sin(20*wo*I*dt)......
% -0.48786692*sin(21*wo*I*dt)+6.48434504*sin(22*wo*I*dt)
+4.73294659*sin(23*wo*I*dt)+0.92599160*sin(24*wo*I*dt)......
% -0.09268785*sin(25*wo*I*dt)-1.74417804*sin
(26*wo*I*dt)-2.27987463*sin(27*wo*I*dt)+2.42239393*sin
(28*wo*I*dt).....
% -8.19430263*sin(29*wo*I*dt)+2.54675588*sin(30*wo*I*dt)
+2.54969842*sin(31*wo*I*dt)-0.00000000*sin(32*wo*I*dt))/32;
Up(1,I)=0.1440+0.0916*cos(wo*I*dt-1.3236)+0.0488*cos
(2*wo*I*dt-1.9969)+0.0257*cos(3*wo*I*dt-1.9415)+....
0.0403*cos(4*wo*I*dt-2.4972)+0.0316*cos(5*wo*I*dt+2.5717)
+0.0174*cos(6*wo*I*dt+1.4304)+...
0.0121*cos(7*wo*I*dt+0.4314)+0.0041*cos(8*wo*I*dt-1.5113)
+0.0047*cos(9*wo*I*dt-1.3014)+0.0032*cos(10*wo*I*dt-1.3213);
end
for I=1:Nt;
Re_p(1,I)= rho*Up(1,I)*Dp/mu;
end;
for I=1:Nt;
lambda_p(1,I)= 64/Re_p(1,I);
end;
[DUpDT] = second_order_derivative(Up,Nt,dt);
for I=1:Nt;
DP_p(1,I)= -rho*DUpDT(1,I)*Lp -(rho*(Up(1,I)*abs(Up(1,I)))/2)*
(lambda_p(1,I)*Lp/Dp+lambda_apex_p);
end;
function [DP_i]=first_section(Ui,Nt,dt,rho,mu,Di,Li,phi1,g,
lambda_apex_i)
DP_i =zeros(1,Nt);
DUiDT=zeros(1,Nt);
lambda_i=zeros(1,Nt);
Re_i=zeros(1,Nt);
for I=1:Nt;
Re_i(1,I)= rho*Ui(1,I)*Di/mu;
end;
for I=1:Nt;
lambda_i(1,I)= 64/Re_i(1,I);
```

```
end;
[DUiDT] = second_order_derivative(Ui,Nt,dt);
lambda_apex_i=0.5
for I=1:Nt;
DP_i(1,I)= -rho*g*Li*sin(phi1) -rho*DUiDT(1,I)*Li - (rho*(Ui(1,I)
*abs(Ui(1,I)))/2)*(lambda_i(1,I)*Li/Di+lambda_apex_i);
end;
function [Ud] = reconnection(Nt,dt,Dd,Db1,Db2,Ub2,Ub1)
Ud = zeros(1,Nt);
for It=1:Nt;
Ud(1,It)=Ub1(1,It)*((Db1/Dd)^2)+Ub2(1,It)*((Db2/Dd)^2);
end;
function [DADT]=second_order_derivative(A,Nt,dt)
%%% CALCULATION OF DA/Dt
DADT(1,1)= (-3*A(1,1)+4*A(1,2)-A(1,3))/(2*dt);
for I=2:Nt-1;
DADT(1,I)= (A(1,I+1)-A(1,I-1))/(2*dt);
end;
DADT(1,Nt)= (3*A(1,Nt)-4*A(1,Nt-1)+A(1,Nt-2))/(2*dt);
function [Us1,Zs1,DZs1DT]= tankob1(Ub1,Nt,dt,rho,mu,Dus1,Lb1,Ds1,
Db1)% Dusb1 is the diameter of the tank's vertical outlet, Ds1 instead is
the diameter of the tank
Us1=zeros(1,Nt);
Zs1=zeros(1,Nt);
DZs1DT=zeros(1,Nt);
DUb1DT=zeros(1,Nt);
Re_b1=zeros(1,Nt);
lambda_b1=zeros(1,Nt);
%%%% COMPLIANCE CHAMBER
%%%% MASS BALANCE
%%%% Ui*Si*dt=Sco*dz+Uc*Sc*dt
%%%% H1-H2=-z
%%%% Ui=(Sco/Si)*dz/dt+(Sc/Si)*Uc
%%%% DUi/DT=(Sco/Si)*d2z/dt2+(Sc/Si)*DUc/DT
%%%% -z = (1/g)*Li*((Sco/Si)*d2z/dt2+(Sc/Si)*DUc/DT)+(1/g)
*lambda_i*(Li/Di)*(1./2)*((Sco/Si)*dz/dt+(Sc/Si)*Uc)^2
%%%% d2z/dt2 +lambda_i*(1./Di)*(1./2)*(Sco/Si)*(dz/dt)^2 +lambda_i*
(1./Di)*(Sc/Si)*Uc*dz/dt +(Si/Sco)*(g/Li)*z +(Sc/Sco)*DUc/DT
+(Si/Sco)*((Sc/Si)*Uc)^2 = 0
g=9.81; %m/s2
ucm=sum(Ub1(1,:))/Nt;
%%%% WE INITIALISE THE VARIABLES
[DUb1DT] = second_order_derivative(Ub1,Nt,dt);
for I=1:Nt;
Re_b1(1,I)= rho*Ub1(1,I)*Db1/mu;
```

```matlab
end;
for I=1:Nt;
lambda_b1(1,I)= 64./Re_b1(1,I);
end;
for It=2:Nt;
Zs1(1,It)=-(1/g)*DUb1DT(1,It)*Lb1-(1/g)*((Ub1(1,It)*abs(Ub1(1,
It))))/2)*(lambda_b1(1,It)*Lb1/Db1);
end;
[DZs1DT] = second_order_derivative(Zs1,Nt,dt);
for It=2:Nt;
Us1(1,It)=Ub1(1,It)*(Db1/Ds1)^2-DZs1DT(1,It)*(Dus1/Ds1)^2;
end;
function [Us2,Zs2,DZs2DT]= serbatoiob2(Ub2,Nt,dt,rho,mu,Dus2,Lb2,
Ds2,Db2)% Dusb2 is the diameter of the vertical outlet of the tank, Ds2
instead is the diameter of the tank
Us2=zeros(1,Nt);
Zs2=zeros(1,Nt);
DZs2DT=zeros(1,Nt);
DUb2DT=zeros(1,Nt);
Re_b2=zeros(1,Nt);
lambda_b2=zeros(1,Nt);
%%%% COMPLIANCE CHAMBER
%%%% MASS BALANCE
%%%% Ui*Si*dt=Sco*dz+Uc*Sc*dt
%%%% H1-H2=-z
%%%% Ui= (Sco/Si)*dz/dt+(Sc/Si)*Uc
%%%% DUi/DT= (Sco/Si)*d2z/dt2+(Sc/Si)*DUc/DT
%%%% -z = (1/g)*Li*((Sco/Si)*d2z/dt2+(Sc/Si)*DUc/DT)+(1/g)
*lambda_i*(Li/Di)*(1./2)*((Sco/Si)*dz/dt+(Sc/Si)*Uc)^2
%%%% d2z/dt2 +lambda_i*(1./Di)*(1./2)*(Sco/Si)*(dz/dt)^2 +lambda_i*
(1./Di)*(Sc/Si)*Uc*dz/dt +(Si/Sco)*(g/Li)*z +(Sc/Sco)*DUc/DT
+(Si/Sco)*((Sc/Si)*Uc)^2 = 0
g=9.81; %m/s2
ucm=sum(Ub2(1,:))/Nt;
%%%% INITIALISE VARIABLES
[DUb2DT] = second_order_derivative(Ub2,Nt,dt);
for I=1:Nt;
Re_b2(1,I)= rho*Ub2(1,I)*Db2/mu;
end;
for I=1:Nt;
lambda_b2(1,I)= 64./Re_b2(1,I);
end;
for It=2:Nt;
Zs2(1,It)=-(1/g)*DUb2DT(1,It)*Lb2-(1/g)*((Ub2(1,It)*abs(Ub2(1,
It))))/2)*(lambda_b2(1,It)*Lb2/Db2);
```

```
end;
[DZs2DT] = second_order_derivative(Zs2,Nt,dt);
for It=2:Nt;
Us2(1,It)=Ub2(1,It)*(Db2/Ds2)^2-DZs2DT(1,It)*(Dus2/Ds2)^2;
end;
function [DP_c,Uc] = third_section(Up,Nt,dt,rho,mu,Dc,Lc,Dp)
DP_c= zeros(1,Nt);
Uc= zeros(1,Nt);
DUcDT= zeros(1,Nt);
lambda_c= zeros(1,Nt);
Re_c= zeros(1,Nt);
lambda_apex_c =1;
for I=1:Nt;
Uc(1,I)=Up(1,I)*(Dp/Dc)^2; %mass balance
end;
for I=1:Nt;
Re_c(1,I)= rho*Uc(1,I)*Dc/mu;
end;
for I=1:Nt;
lambda_c(1,I)= 64/Re_c(1,I);
end;
[DUcDT] = second_order_derivative(Uc,Nt,dt);
for I=1:Nt;
DP_c(1,I)= -rho*DUcDT(1,I)*Lc - (rho*(Uc(1,I)*abs(Uc(1,I))))/2)*
(lambda_c(1,I)*Lc/Dc+lambda_apex_c);
end;
function [DP_d]=descent_section(Nt,dt,rho,mu,Dd,Ld,g,phi3,Ud)
DP_d =zeros(1,Nt);
DUdDT=zeros(1,Nt);
lambda_d=zeros(1,Nt);
Re_d=zeros(1,Nt);
lambda_apex_d=1;
for I=1:Nt;
Re_d(1,I)= rho*Ud(1,I)*Dd/mu;
end;
for I=1:Nt;
lambda_d(1,I)= 64/Re_d(1,I);
end;
[DUdDT] = second_order_derivative(Ud,Nt,dt);
for I=1:Nt;
DP_d(1,I)=-rho*g*Ld*sin(phi3) -rho*DUdDT(1,I)*Ld - (rho*(Ud(1,I)
*abs(Ud(1,I))))/2)*(lambda_d(1,I)*Ld/Dd+lambda_apex_d);
end;
function [DP_s,Us]=initial_section(Ui,Nt,dt,rho,mu,Ds,Ls,Di,g,
phi2)
```

```
DP_s =zeros(1,Nt);
Us= zeros (1,Nt);
DUsDT=zeros(1,Nt);
lambda_s=zeros(1,Nt);
Re_s=zeros(1,Nt);
lambda_apex_s=1.2;
for I=1:Nt;
Us(1,I)=Ui(1,I)*(Di/Ds)^2;
end;
for I=1:Nt;
Re_s(1,I)= rho*Us(1,I)*Ds/mu;
end;
for I=1:Nt;
lambda_s(1,I)= 64/Re_s(1,I);
end;
[DUsDT] = second_order_derivative(Us,Nt,dt);
for I=1:Nt;
DP_s(1,I)=-rho*g*Ls*sin(phi2) -rho*DUsDT(1,I)*Ls - (rho*(Us(1,I)
*abs(Us(1,I)))/2)*(lambda_s(1,I)*Ls/Ds+lambda_apex_s);
end;
function [DP_r,Ur] = return_section(Ud,Nt,dt,rho,mu,Dr,Lr,Dd)
DP_r= zeros(1,Nt);
Ur= zeros(1,Nt);
DUrDT= zeros(1,Nt);
lambda_r= zeros(1,Nt);
Re_r= zeros(1,Nt);
lambda_apex_r=0.5;
for I=1:Nt;
Ur(1,I)=Ud(1,I)*(Dd/Dr)^2; %mass balance
end;
for I=1:Nt;
Re_r(1,I)= rho*Ur(1,I)*Dr/mu;
end;
for I=1:Nt;
lambda_r(1,I)= 64/Re_r(1,I);
end;
[DUrDT] = second_order_derivative(Ur,Nt,dt);
for I=1:Nt;
DP_r(1,I)= -rho*DUrDT(1,I)*Lr - (rho*(Ur(1,I)*abs(Ur(1,I)))/2)*
(lambda_r(1,I)*Lr/Dr+lambda_apex_r);
end;
```

6.2   Numerical Results

Figure 13 [1] shows the speed (m/s) inside the sample *(top)* and in the calm part *(bottom)*, as a function of time *(seconds)*. The speed in the sample corresponds to the desired physiological shape, and the circuit is sized to achieve this shape. The speeds are identical because the two sections are contiguous.

Figure 14 [1] shows the speeds in the first part (before the compliance chamber) *(top)* and in the initial uphill part (directed towards the first part) *(bottom)*.

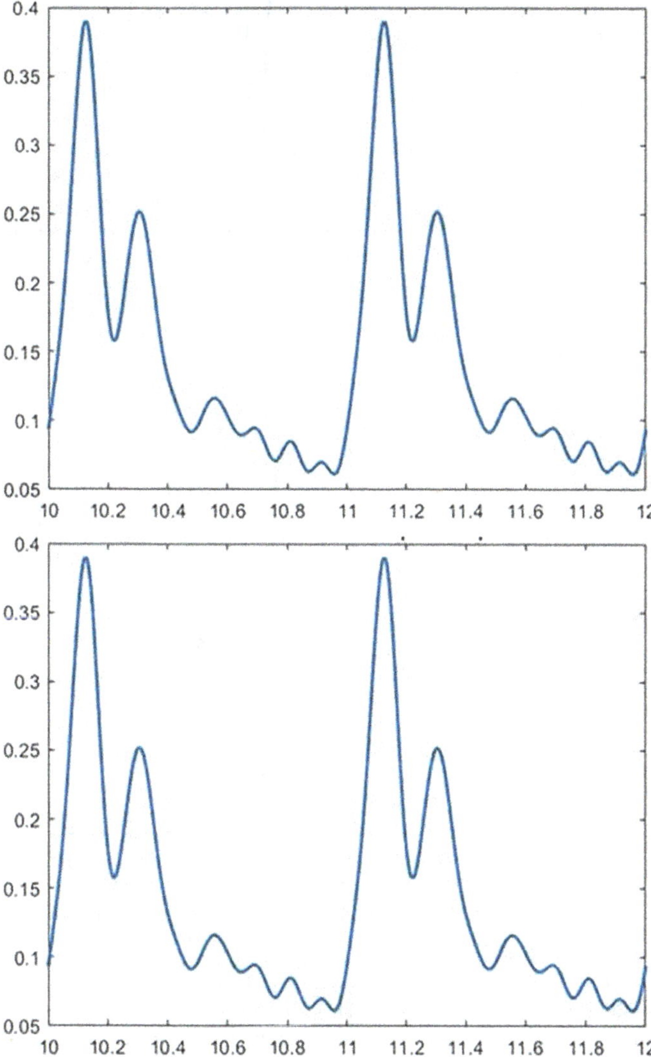

Fig. 13 Speed (m/s) versus time (s): inside the PIV sample (top) and in the calm part (bottom) [1]

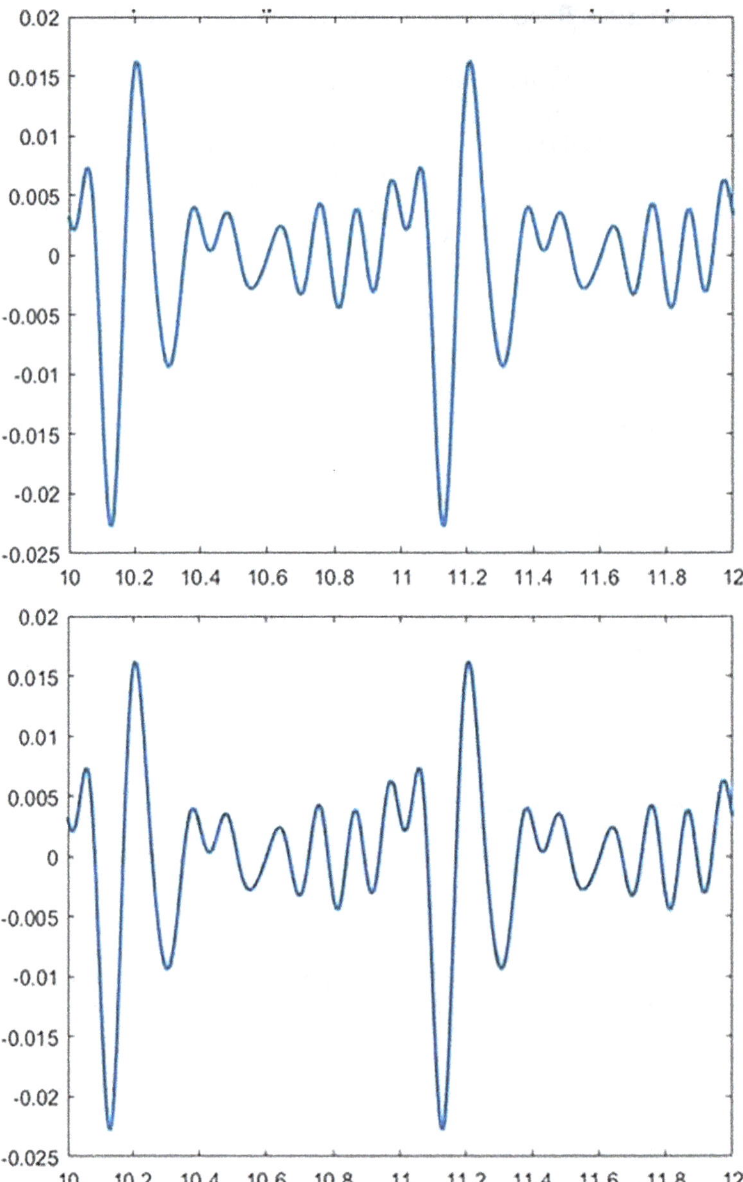

Fig. 14 Speed (m/s) versus time (s): in the first section (top) and in the initial uphill section (bottom) [1]

Comparing the waveforms of the figures, Figs. 13 and 14, it can be noted that they are different. The waveform in the initial part necessarily has a greater frequency because the tower acts as a low-pass filter. The speeds in the two parts are equal because they are contiguous.

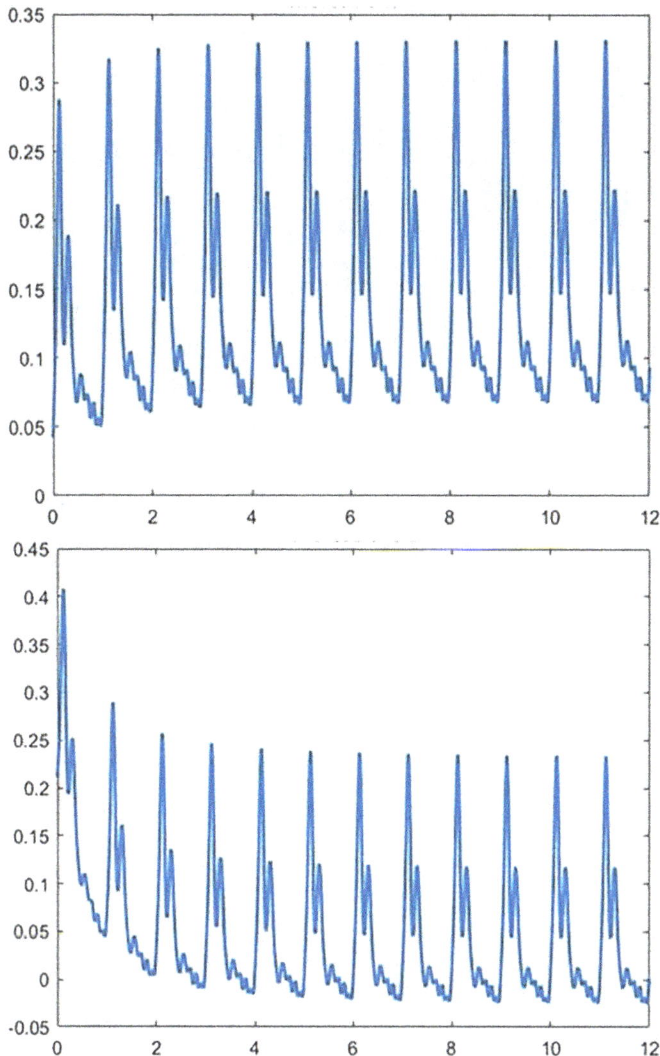

Fig. 15 Speed (m/s) versus time (s): in bifurcation 1 (top) and in bifurcation 2 (bottom) [1]

Figure 15 [1] shows the speed profiles in the two branches of the bifurcation. It is noted that in both branches, the effect of the initial transient is present, for which the flow is underestimated in branch 1 and overestimated in branch 2. The time scale is extended to show the influence of the transient. The bifurcation does not alter the frequency content of the waveform. However, the waveforms are reduced in amplitude in the two branches due to the flow reduction in each branch.

Figure 16 [1] shows the speeds in the downhill return part and in the last part. The speeds are identical to each other and equal to the speed in the sample and in the calm

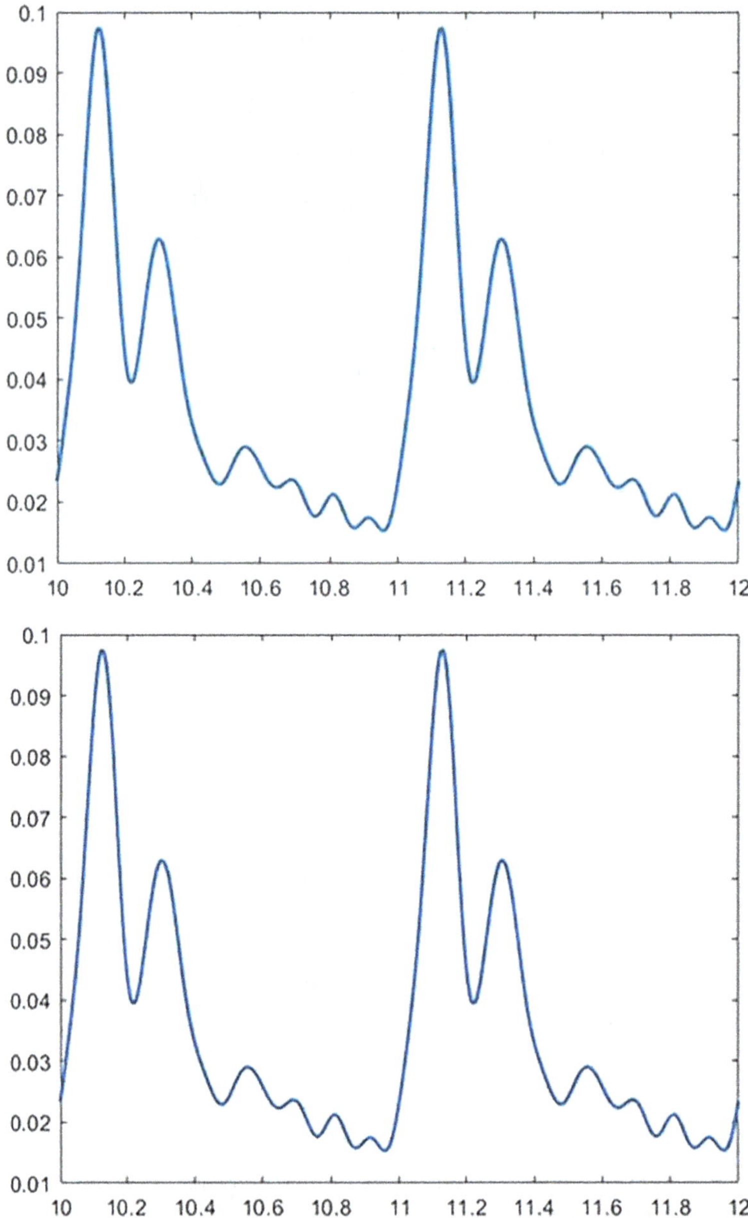

Fig. 16 Speed (m/s) versus time (s): in the downhill part of the return (top) and in the last part of return (bottom) [1]

part for continuity and because there are no elements acting as a filter, like the piezometric tower, downstream of the calm chamber.

Figure 17 [1] shows the pressure drop in the sample *(top)* and in the compliance chamber *(bottom)*. The two waveforms are almost identical, with negligible variations in amplitude, due to the different resistances in the two parts. Note that the waveform in the compliance chamber is not altered. The piezometric tower has a capacitive effect, so it alters the waveform of the flow rate but not of the potential (pressure).

Figure 18 [1] shows the pressure in the calm part *(top)* and in the two branches of the bifurcation *(bottom)*. The waveform in the calm part is the same as in Fig. 17 because there are no reactive elements in this part. The pressure drop in the two branches of the bifurcation is the same because the two parts are in parallel. For this reason, only one of the two curves (the red one) is visible; the other is superimposed.

Figure 19 [1] shows the pressure drop in the downhill part of the return *(top)* and in the last part *(b)*. The two parts are in series, and the two pressure drops are identical because the resistive effects are comparable.

Figure 20 [1] shows the pressure drop in the compliance chamber *(a)* and on the pump outlet *(b)*. The waveforms are slightly different. In particular, on the pump outlet, the pressure is higher, and there is a slight high-pass effect in the compliance chamber due to the presence of the tower.

Figure 21 [1] shows the total power in the circuit *(top)* and the level difference of the tower *(bottom)*. Power is the product of the total pressure drop with the volumetric flow rate in the circuit. The two quantities are out of phase due to the inductive effects caused by the fluid inertia, and the power frequency spectrum is wider than pressure and flow rate. The results show that the circuit requires a low operating power, with the effective value estimated between 3 and 4 W. As for the level difference in the piezometric tower, we can note that *(i)* the waveform is the same as the pressure because there is a linear relationship between the variables, and *(ii)* the maximum level difference in the tower is about 60 cm, compared to 3 m of average level difference, indicating that the tower height must be about 4 m.

7   Conclusions

The design of a hydraulic circuit is analysed from the point of view of blood flow in bifurcations, particularly in the carotid. The experimental technique to be used is the *Particle Image Velocimetry (PIV)*, which provides a three-dimensional representation of blood flow, allowing us to study vortices and tangential stresses on the vascular wall. In order to represent the flow in the bifurcation as accurately as possible, the hydraulic circuit is sized to represent the physiological flow rate in the carotid. To ensure this speed, a pump is needed that provides the desired pulsation associated with the speed. To obtain the desired flow rate, the pressure to be delivered by the pump must be calculated from the information obtained with the mathematical modelling of the circuit.

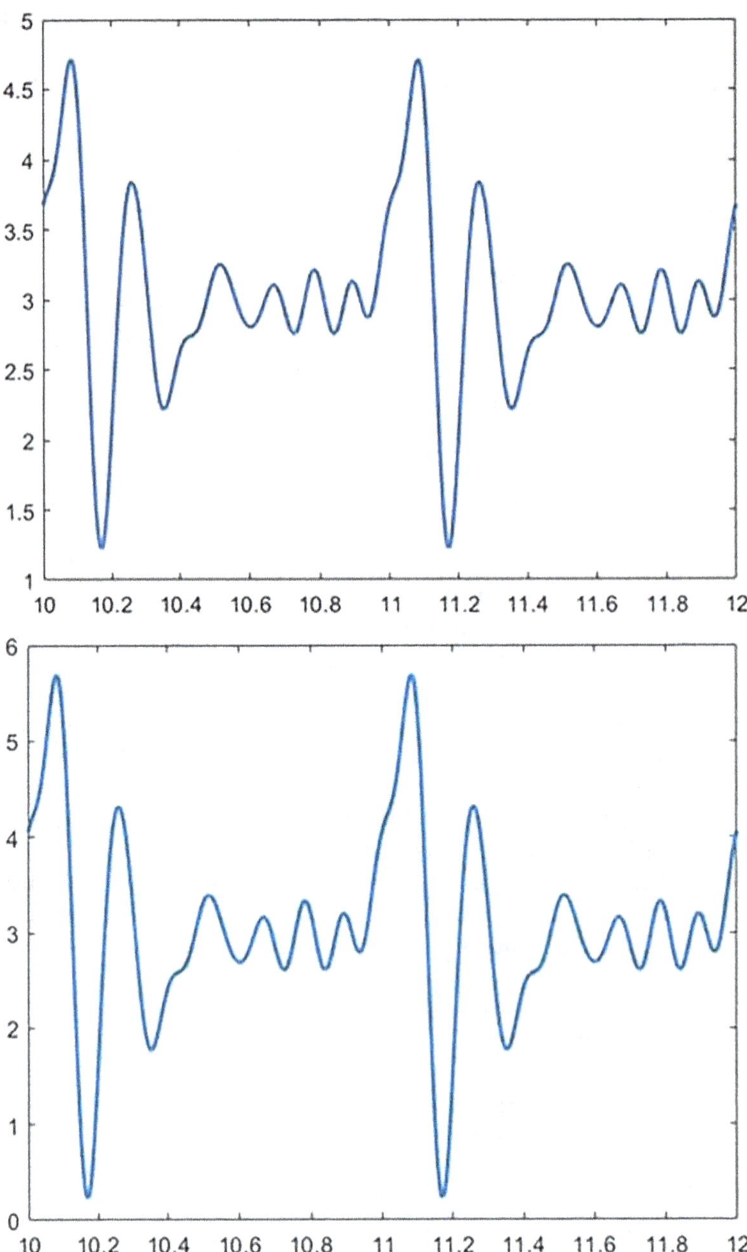

Fig. 17 Pressure (k Pa) versus time (s): in the sample (top) and in the part containing the compliance chamber (bottom) [1]

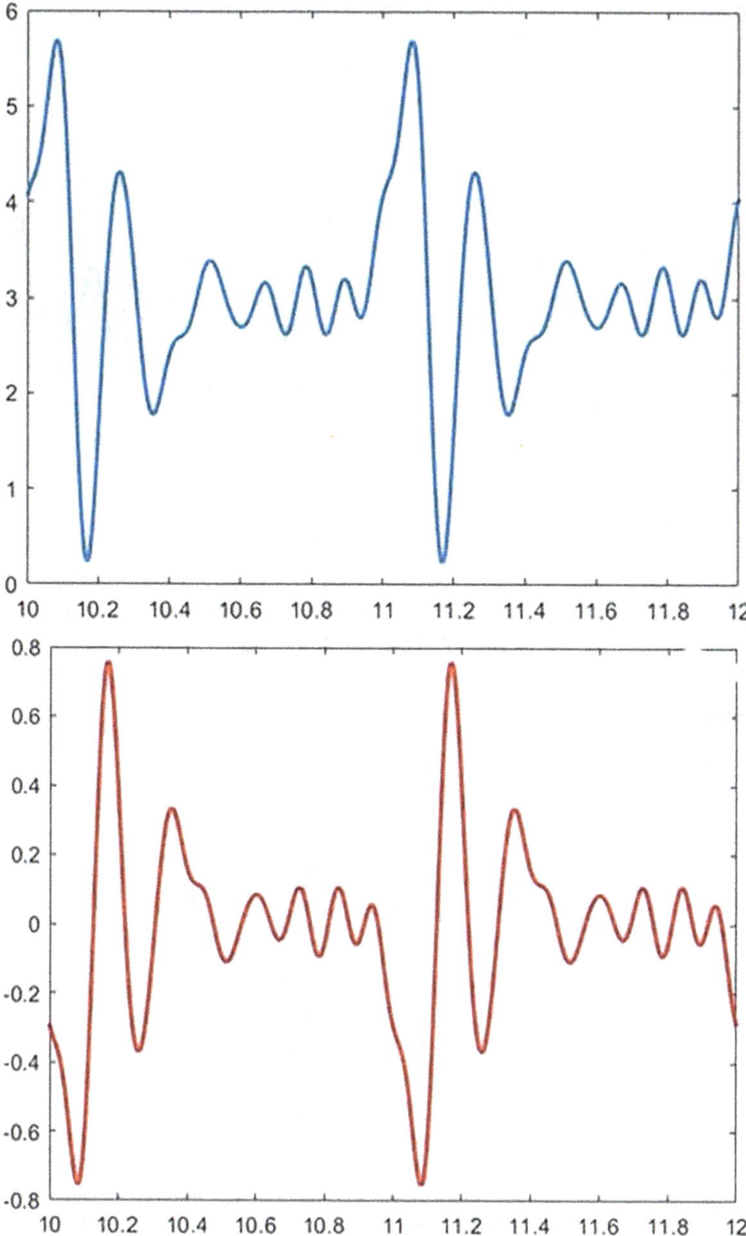

Fig. 18 Pressure (k Pa) versus time (s): in the calm part (top) and in the bifurcation (bottom) [1]

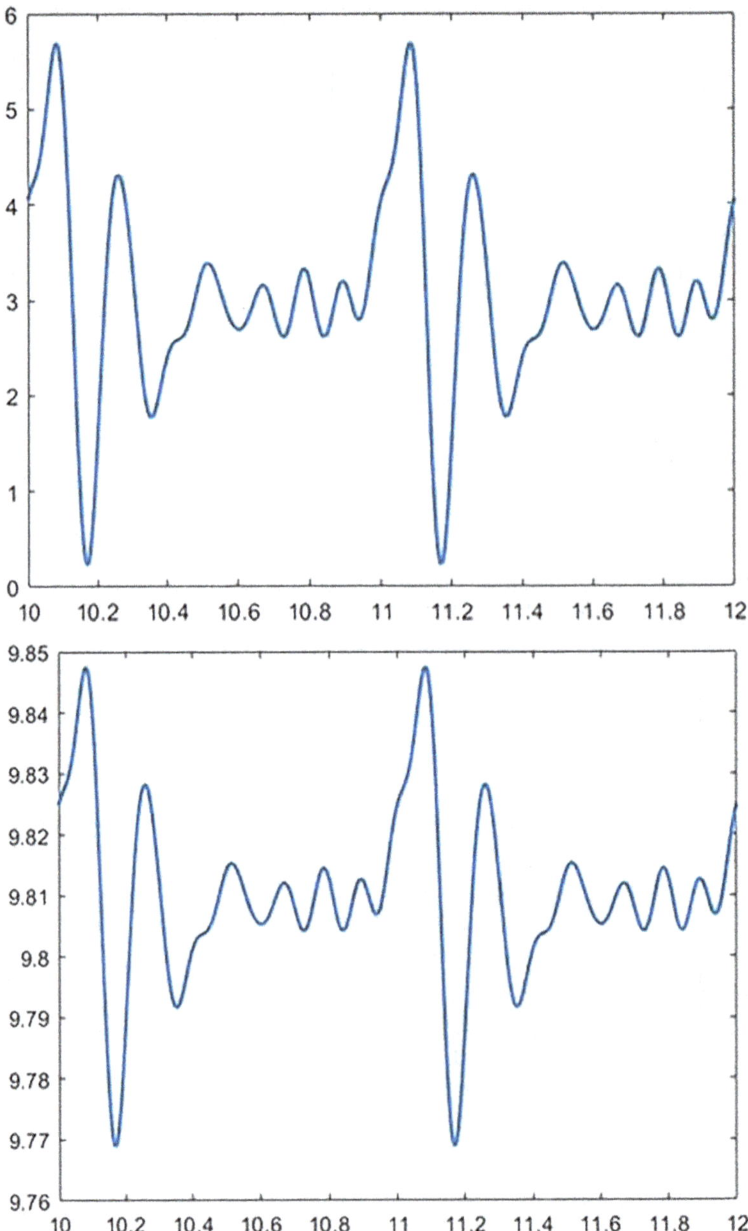

Fig. 19 Pressure (k Pa) versus time (s): in the downhill part of the return (top) and in the last part (bottom) [1]

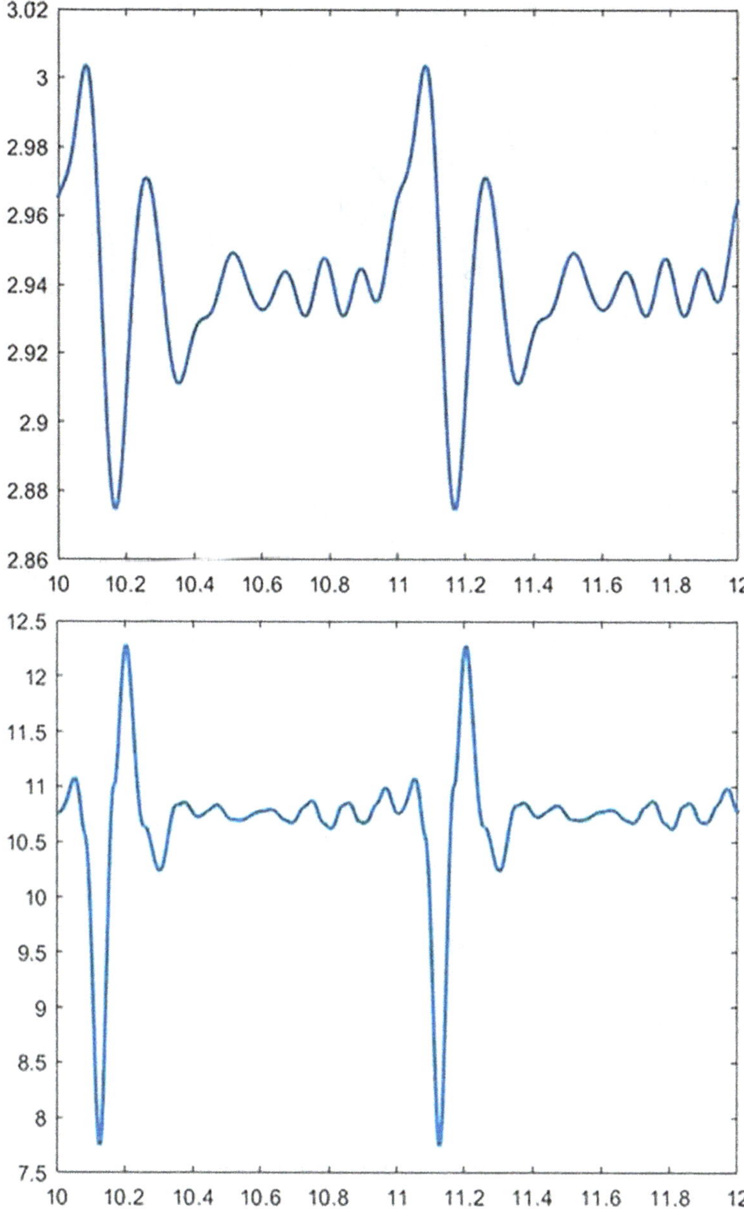

Fig. 20 Pressure (k Pa) versus time (s): in the compliance chamber (top) and on the pump outlet (bottom) [1]

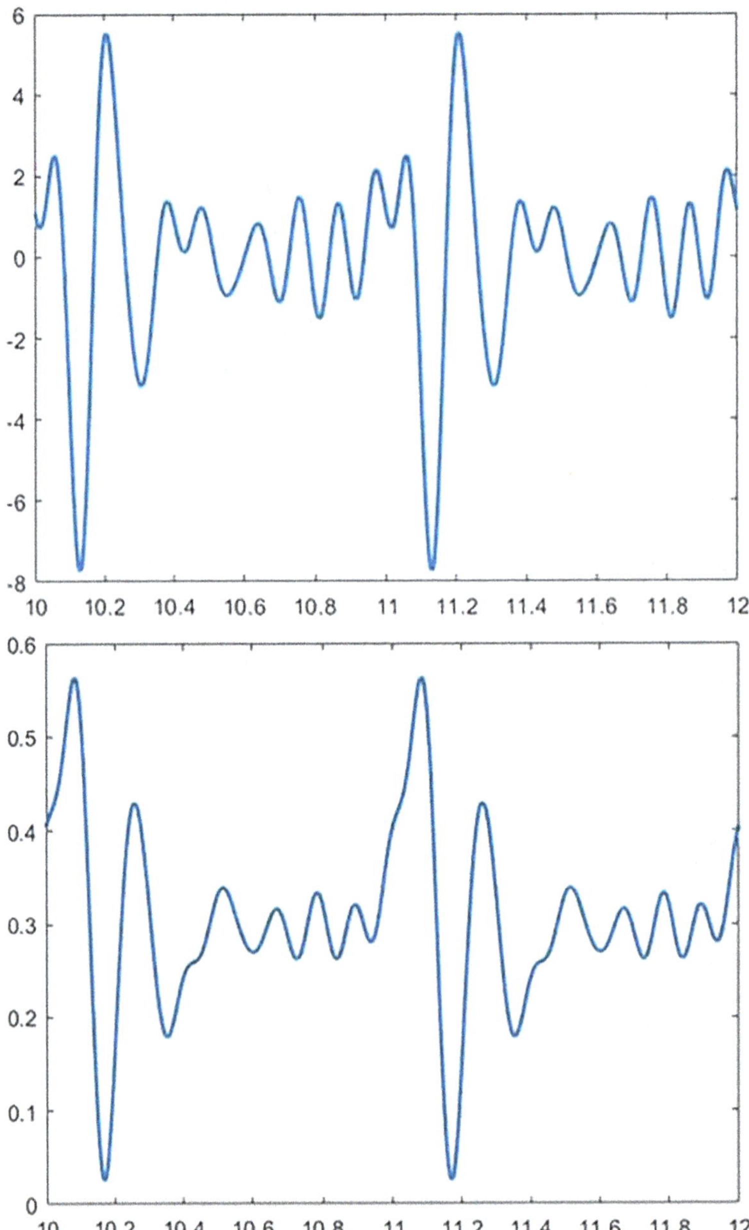

Fig. 21 Power (W) in the tower (top) and level difference (m) of the tower (bottom) in the circuit as a function of time [1]

The model is based on the solution of the fluid dynamic equations, averaged over the volume of the considered section. The viscous stresses are modelled with dimensional analysis, while the convective term is neglected because the section of each part is almost constant. The model is implemented in a *MATLAB* code that simulates the flow throughout the circuit and calculates speed and pressure in each part as a function of time. From the numerical simulation of the equations associated with the circuit, it is possible to determine the pressure that must be provided at the pump outlet. The circuit is composed primarily of conduits (resistive elements), but a piezometric tower (capacitive element) is also inserted to prevent breakages in the sample and ensure that the elastic oscillations remain unchanged.

The sizing of the circuit reveals a low operating power, 3–4 W, but a large size of the piezometric tower, about 4 m. This can pose a problem in laboratories without the necessary height. A problem that the circuit can present is the realisation of the pressure waveform delivered by the pump, the spectrum of the latter being relatively wide, and the not-easy or, in any case, expensive procurement of a suitable pump in the market.

Acknowledgements for Figures The author thanks the *editor, Anthelios* (anthelios@anthelios.it), for the kind concession to the reproduction of *image B32v,* from the volumes of the Tables of the Royal Library of Windsor, edited by Teodoro Sabachnikoff. The images are mirrored to be read more easily.

The author thanks the *publisher, Elsevier,* for the kind permission to reproduce images *1 of the reference [3], 1 of the reference* [14], *2 and 3 of the reference* [17], *1 and 2 of the reference* [18], *3 of the reference* [20] *and 3 of the reference* [21].

The author thanks the *publisher, Springer,* for the kind permission to reproduce image *1.9 of the reference* [4].

References

1. Galgano V (2008–2009) Circuito idraulico per lo studio di problemi vascolari. Progettazione preliminare di un sistema fluidodinamico in regime non stazionario, (Hydraulic circuit for the study of vascular problems. Preliminary design of a non-steady state fluid dynamic system). Tesi di Laurea in Ingegneria Medica (Three-years Degree Thesis in Medical Engineering), University of Rome Tor Vergata, A. A
2. Sabachnikoff T (1901) I Manoscritti di Leonardo da Vinci della Reale Biblioteca di Windsor, Dell'anatomia - Fogli B, Torino, Roux e Viarengo Editori. (The Manuscripts of Leonardo da Vinci of the Royal Library of Windsor, Anatomy - Sheets B, Turin, 1901, Roux and Viarengo Publishers)
3. Brunette J, Mongrain R, Laurier J, Galaz R, Tardif JC (2008) 3D flow study in a mildly stenotic coronary artery phantom using a whole volume PIV method. Med Eng Phys 30:1193–1200
4. Raffel M, Willert CE, Scarano F, Kähler CJ, Wereley ST, Kompenhans J (2018) Particle image velocimetry, a practical guide, 3rd edn. Springer
5. Gori F et al (2013) Flow evolution of a turbulent submerged two-dimensional rectangular free jet of air. Average Particle Image Velocimetry (PIV) visualizations and measurements. Int J Heat Fluid Flow 44:764–775
6. Gori F et al (2014) Influence of the Reynolds number on the instant flow evolution of a turbulent rectangular free jet of air. Int J Heat Fluid Flow 50:386–401

7. Di Venuta I et al (2018) Numerical simulation of mass transfer and fluid flow evolution of a rectangular free jet of air. Int J Heat Mass Transf 117:235–251
8. Petracci I et al (2019) Experiments and numerical simulations of mass transfer and flow evolution in transient rectangular free jet of air. Int Commun Heat Mass Transf 108:104290
9. Petracci I et al (2022) Forced convective heat transfer in a metallic foam cylinder cooled by a slot jet flow and comparison with a smooth cylinder and a full flow. Int J Heat Mass Transf 183: 122118
10. Heise M, Schmidt S, Kruger U, Ruckert R, Rosler S, Neuhaus P, Settmacher U (2004) Flow pattern and shear stress distribution of distal end-to-side anastomoses. A comparison of the instantaneous velocity fields obtained by particle image velocimetry. J Biomech 37:1043–1051
11. Zhao SZ, Xu XY, Hughes AD, Thom SA, Stanton AV, Ariff B, Long Q (2000) Blood flow and vessel mechanics in a physiologically realistic model of a human carotid arterial bifurcation. J Biomech 33:975–984
12. Migliavacca F, Dubini G (2005) Computational modeling of vascular anastomoses. Biomech Model Mechanobiol 3:235–250
13. Younis HF, Kaazempur-Mofrad MR, Chung C, Chan RC, Kamm RD (2003) Computational analysis of the effects of exercise on hemodynamics in the carotid bifurcation. Ann Biomed Eng 31:995–1006
14. Liepsch D (2002) An introduction to biofluid mechanics-basic models and applications. J Biomech 35:415–435
15. Zhang F, Lanning C, Mazzaro L, Rech B, Chen J, Chen SJ, Shandas R (2008) Systematic validation of the echo particle image velocimetry technique using a patient specific carotid bifurcation model. IEEE International Ultrasonics Symposium Proceedings, Digital Object Identifier. https://doi.org/10.1109/ULTSYM.2008.0003
16. Kim H-B, Hertzberg J, Lanning C, Shandas R (2004) Noninvasive measurement of steady and pulsating velocity profiles and shear rates in arteries using echo PIV: *In Vitro* validation studies. Ann Biomed Eng 32:1067–1076
17. Botnar R, Rappitsch G, Scheidegger MB, Liepsch D, Perktold K, Boesiger P (2000) Hemodynamics in the carotid artery bifurcation: a comparison between numerical simulations and in vitro MRI measurements. J Biomech 33:137–144
18. Chua LP, Ji W, Zhou T (2005) In vitro study on the steady flow characteristics of proximal anastomotic models. Int Commun Heat Mass Transf 32:464–472
19. Vetel J, Garon A, Pelletier D (2009) Lagrangian coherent structures in the human carotid artery bifurcation. Exp Fluids 46:1067–1079
20. Xiong FL, Chong CK (2007) PIV-validated numerical modelling of pulsatile flows in distal coronary end-to-side anastomoses. J Biomech 40:2872–2881
21. Zhang J-M, Chua LP, Ghista DN, Zhou T-M, Tan YS (2008) Validation of numerical simulation with PIV measurements for two anastomosis models. Med Eng Phys 30:226–247

Ivano Petracci , born in Marino (RM) on 3 March 1974, is currently an associate professor in the scientific discipline of ING-IND/10 at the Department of Industrial Engineering, University of Rome "Tor Vergata." On 6 July 2001, he obtained a degree (5-year) in Mechanical Engineering (100/100 with honours) from the University of Rome "Tor Vergata" with a thesis titled "Experimental Thermo-Fluid-Dynamic research and numerical simulations on the cooling of electrically heated cylinders, using submerged jets." In 2002, he attended and obtained the title of Magister in Thermo-Fluid Dynamics with the thesis "Fluid dynamics of submerged air jets: measurements in hot anemometry, shadowgraph visualisations and numerical simulations." On 30 June 2005, he obtained the Research Doctorate degree in Energy-Environment Engineering from the University of Rome "Tor Vergata" with a thesis titled "Thermo-Fluid-Dynamics of submerged air jets. Measurements in hot anemometry and shadowgraph visualisations of free, stationary and pulsed jet. Optimisation of heat transfer with smooth and finned cylinders," supervised by Prof. Fabio Gori Ammannati. He was a visiting researcher at the University of Minnesota, Prof. R.J. Goldstein, in

June 2009. In April 2017, he obtained the national scientific qualification (ASN) as an associate professor. His research activity, mainly of an experimental type, is focused on methods of increasing the convective heat transfer with both active and passive techniques, with extended and treated surfaces, reticular ducts, acoustic fields applied to fluids, jet flows and composite solutions. The research is supported by experimental thermodynamic and fluid dynamic investigations (hot wire anemometry and flow visualisation techniques, such as Particle Image Velocimetry) and by numerical Computational Fluid Dynamics, CFD, studies in thermo-fluid dynamic applications. He is a referee for international scientific journals, such as *Applied Thermal Engineering* and *International Journal of Thermal Sciences*. He was a member of the Local Scientific Committee of the 12th International Conference on Computational Heat, Mass and Momentum Transfer (ICCHMT 2019), Rome, 3–6 September 2019. He has participated in research projects supported by public funds: University Scientific Research Project "Mission-Sustainability" titled "Jet Flow and Heat Transfer upon/from Cylinders/Wind Blades," 2018–2019; PRIN project "Thermophysical and fluid dynamic properties of materials for electrodes and electrolyte of high-temperature fuel cells," 2005–2007; PRIN project "Fundamental and applicative aspects in innovative heat exchange techniques," 2002–2004 and the winner of research grant in "Thermo-fluid dynamics of air jets," 2002 He has participated in research projects supported by private funds: co-responsible for scientific research activity with Elettronica S.p.a. in the project "Experiments on the Lattice Channel Section," 2017–2019, and co-responsible for scientific research activity with Elettronica S.p.a. in the project " Thermal conductivity test campaign on epoxy adhesives," 2021. He currently teaches "Technical Physics" for Bachelor's and Master's degrees in Mechanical and Energy Engineering; "Thermo-technics" for the Master's degrees in Energy Engineering and "Technical Physics," together with Fabio Gori Ammannati, for the Master's degree in Medical Engineering. He has been a supervisor and co-supervisor for 55 Bachelor's, Master's and Doctoral theses. In the A.Y. 2015–2016, he was the director of the second-level Master's in "Thermo-Fluid-Dynamics."

Andrea Boghi The Doctor of Research *(Ph. D.),* Dr. Eng. Andrea Boghi, is the director of *Computational Science Ltd.,* a consultancy company in the field of software development and data analysis, which counts among its clients the British Air Traffic Agency *(NATS)* and the Bank of England, starting from 2018. He is the author of more than 30 international scientific publications indexed on the major scientific databases. For more than a decade, he has worked in academia in various European institutions, dealing with the modelling and simulation of transport phenomena, both for basic research and for industrial and biological applications. His academic journey began at the University of Rome "Tor Vergata," where, alongside research on the modelling and simulation of turbulent flows with variable laminar diffusivity and blood flow in arteries and *stents*, he also served as a contract lecturer for the courses of Technical Physics, Thermo-Fluid-Dynamics of Biological Systems and Numerical Calculation of Thermo-Fluid-Dynamic Systems (2006–2010). In the two-year period from 2010 to 2012, he was a contract researcher at the Department of Energy and Technology of the University of *Southampton* in the United Kingdom, where he dealt with modelling and simulation of bi-phase flows, characterised by spinodal decomposition. In 2012, he went to France to the Institute of Fluid Mechanics of Toulouse as a contract researcher for the OTE (Waves, Turbulence and Environment) research group, where he dealt with modelling and simulation of river and atmospheric fluid dynamics. He terminated his contract in 2013 to go to the University of *Cranfield* in the United Kingdom, where he was appointed *Senior Research Fellow*, becoming part of the academic staff. Here, for the following 5 years (2013–2018), he dealt with modelling and simulation of transport phenomena for various departments: Energy, Hydraulics, Environment and Agriculture, where he respectively dealt with transport of crude oil in pipelines, disinfection of tanks by chlorination and transport and diffusion of organic and inorganic material in soils, with associated absorption by plants. At the University of *Cranfield,* he also handled the course on modelling of atmospheric emissions in the two-year period from 2016 to 2017. Before leaving academia, he went again to the University of *Southampton* in 2018, where he dealt with the three-dimensional modelling of the transport of nitrates and carbonates at the air–water interface in soils. Throughout his period abroad, he maintained his collaboration with Fabio Gori Ammannati, supervising some students and collaborating on the production of scientific articles. In 2020, he obtained the national scientific qualification for the functions of associate university professor in the competitive sector of Technical Physics and Nuclear Engineering.

Fabio Gori Ammannati was born in Montale (Pistoia) on 5 August 1947 to Emilio Gori and Cesarina Ammannati. On 16 July 2021, he added his mother's surname, Ammannati. He graduated in Chemical Engineering on 11 November 1971 with honours. In December 1971, he won a scholarship for young graduates at the Faculty of Engineering in Bologna, which he undertook from January 1972. In 1974, he became assistant professor and, in 1982, associate professor of Technical Physics at the Faculty of Engineering of the University of Florence, where he taught, also as a contract professor, until 1990. In 1978, he won a scholarship from the *British Council,* which he carried out at the *Imperial College* in London, where he collaborated with *Prof. D. Brian Spalding* on the topic of numerical analysis in the turbulent flow of liquid metals. During his time at the University of Florence, he established international scientific collaborations with *Prof. R. Echigo, Tokyo Institute of Technology, Tokyo, Japan; Prof. D.R. Chaudhary, Department of Physics, University of Rajasthan, Jaipur, India;* International Centre for Theoretical Physics *(ICTP)* in Trieste; *Centralny Osrodek Techniki Medycznej, Warsaw, Poland and Prof. T. Aihara, Institute of Fluid Science, Sendai, Japan.* In 1986, he won a CNR scholarship, which he carried out at *Cornell University, Ithaca, New York,* where he collaborated with *Prof. R. Miller* on the topic of ground freezing. In the same year, he became professor in the Faculty of Engineering at the University of Reggio Calabria. In 1988, he was appointed as *professor* at the *University of New York at Stony Brook,* teaching the course, *Introduction to Fluid Dynamics,* in the autumn semester of the same year. During his stay, he collaborated with *Prof. T.F. Irvine Jr.* on the method of measuring thermal conductivity with a thermal probe and on the measurement of the isobaric thermal expansion coefficient of non-Newtonian fluids. In 1990, he moved to the Milano Polytechnic, where he taught Technical Physics and Systems until 1993. From 1991 until 1998, he was an adjunct professor at the Faculty of Engineering of the University of Siena. From 1992 to 2017, he was a professor of Technical Physics in the Faculty of Engineering at the University of Rome Tor Vergata, and from 2017, he is a contract professor. In 1994, he proposed the Research Doctorate (Italian *Ph.D.*) in Energy-Environment Engineering, of which he was the coordinator until 2011. In 1995, he proposed and directed the second-level Master's in Thermo-fluid-dynamics until 2016. From 1998 to 2001, he was the director of the Department of Mechanical Engineering. In 2001, he proposed and coordinated the programme, *Master of Science, Energy Engineering and Thermal and Fluid Dynamics,* in collaboration with the *Department of Mechanical Engineering, College of Engineering, University of Illinois at Chicago, USA,* which takes place at the University of Rome Tor Vergata but awards the title of *Master of Science* from the same American University. Since the late 90s, he has been a coordinator of the joint *Ph.D.* research programme with *Prof. R.J. Goldstein, Department of Mechanical Engineering, University of Minnesota, Minneapolis, USA,* and with *Prof. J.P. Hartnett, Prof. L. Kennedy, Prof. W. Minkowycz and Prof. W.M. Worek, Department of Mechanical Engineering, College of Engineering, University of Illinois at Chicago, Chicago, USA.* Since the mid-90s, he has proposed and directed the *Socrates–Erasmus* programme with *Prof. Mayinger, Technical University of Munich, Germany; Prof. van Steenhoven, Eindhoven Technical University, the Netherlands and Prof. C. Caro, Imperial College of Science, Technology and Medicine.* During his stay at Tor Vergata, he established scientific collaborations with the *Department of Mechanical Engineering, University of Minnesota, Minneapolis, USA,* collaborating with *Prof. R.J. Goldstein* on Thermo-fluid-dynamics and mass transport in gas turbine blades; the *Energy Resources Center, University of Illinois at Chicago,* collaborating with *Prof. J.P. Hartnett, Prof. W. Minkowycz and Prof. W.M. Worek;* the *Department of Mechanical Engineering, College of Engineering, University of Illinois at Chicago, collaborating with Prof. L. Kennedy and Prof. W.M. Worek; Duke University, collaborating with Prof. A. Bejan and S. Mary's University, Halifax,*

Nova Scotia, Canada, collaborating with Prof. W. R. Tarnawski. The bibliographic review of documents and citations, titled *PlosBiology Career,* places him within the top 2% of researchers in the field of *Mechanical Engineering and Transports.* Since 1992, he has been a tutor for over 30 PhD theses, supervisor for over 40 second-level university Master's theses, supervisor for over 70 degree theses (five-year, Specialist and Master's), supervisor for over ten *Master's degree in Mechanical Engineering* from the *University of Illinois at Chicago* and supervisor for over 20 theses *Erasmus-Socrates.*

Since the early 2000s, he has been a reviewer of international research projects on behalf of *Portuguese Science and Technology Foundation (FCT); Czech Science Foundation (GACR); European Science Foundation, Strasbourg, France,* on behalf of *FCT* and the *Shota Rustaveli National Science Foundation, Georgia.* Since 2017, the year of his retirement due to age limits, he has been a contract professor at the University of Rome Tor Vergata, where in the academic year 2024–2025, he has taught Technical Physics for the Master's Degree Course in Medical Engineering. M.D. Salvatore Mangiafico (Florence), scientific head of the NeuroVascular Base Camp, 2021, invited him to give a scientific presentation titled *"The engineering approach to the study of the Circle of Willis"* on 23 September 2021 at the Congress Centre, University of Rome La Sapienza, thus opening up a possible future collaboration.

Thermo-Haemo-Dynamics (THD) and Magneto-Hydro-Dynamics (MHD)

Bases of Thermo-Haemo-Dynamics (THD) and Numerical Magneto-Hydro-Dynamics (MHD)

Flavia Russo, Andrea Boghi, and Fabio Gori Ammannati

1 Magneto-Hydro-Dynamics (MHD)

1.1 Origin of Magneto-Hydro-Dynamics

Magneto-Hydro-Dynamics *(MHD)* concerns the flow of electrically conducting fluids in the presence of a magnetic field, to be applied externally to the fluid or inside by inductive action, [2]. Its origin comes from the pioneering discoveries made by Hartmann, Alfven and Northrup. The first study, done by Alfven, [3], concerns electromagnetic phenomena in astrophysics, which led him to discover the so-called Alfven waves. From his study, two important features come out: the concept of frozen-in magnetic field lines in a highly conducting medium and the *MHD* waves, where these field lines can be seen as an elastic string in a dynamic process. Hartmann and Lazarus, [4], studied the *MHD* flow, focusing on pressure-driven channel flow. The results show that the magnetic field and the resulting induced electric currents have a strong influence on the velocity profile, with boundary layers of strong velocity gradients and enhanced drag force.

The first analytical characterisation of the problem is that of Shercliff, [5], who, later on, studies high Hartmann number flows, [6]. At the beginning of the 60 s Gold, [7], obtained the solution for the steady one-dimensional flow of an incompressible, viscous, electrically conducting fluid through a circular pipe in the presence of an applied uniform magnetic field. Other studies obtain the exact solutions, as Hunt, [8], Hunt and Stewartson, [9], who analyse rectangular ducts. Different geometries, such as cylindrical and toroidal square ducts, are investigated analytically, [10], as the toroidal geometry for nuclear applications. Liquid metals are employed as a coolant as well as a breeder fluid, [11].

F. Russo · A. Boghi · F. G. Ammannati (✉)
University of Rome Tor Vergata, Rome, Italy
e-mail: a.boghi@computationalscience.co.uk; gori@uniroma2.it

F. Gori Ammannati (ed.), *Thermo-Haemo-Dynamics in Medical Engineering*,
https://doi.org/10.1007/978-3-031-97214-0_8

1.2 Thermo-Haemo-Dynamics (THD) Equations

Fluid dynamics studies the flow under the hypothesis of continuum mechanics, which states that a body is completely defined by continuous functions of spatial and time variables, $\vec{x}$ and t, as mass density $\rho\left(t, \vec{x}\right)$ and velocity $\vec{v}\left(t, \vec{x}\right)$. The continuous functions are the average of the corresponding property over a volume that is small compared to the spatial variations of the fluid but large compared to the distance between the individual molecules.

1.2.1 Conservation of Mass

The equation of mass conservation is obtained by considering an arbitrarily small amount of fluid, which occupies a volume Ω in space. By following the fluid flow, its mass remains constant in time if no mass sources are present in the volume Ω.

The previous statement leads to

$$\frac{dM}{dt} = \frac{d}{dt}\int_{\Omega}\rho\left(t, \vec{x}\right)dV = 0 \tag{1}$$

The size of the integration domain can vary in time and must be taken into account by bringing the time derivative inside the integral by the transport theorem

$$\int_{\Omega}\left(\frac{D\rho}{Dt} + \rho\,\mathrm{div}\left(\vec{v}\right)\right)dV = 0 \tag{2}$$

Moreover, the positions $\vec{x}$ of the fluid element depend on time, so the total time derivative can be decomposed as

$$\frac{D\rho\left(t, \vec{x}\right)}{Dt} = \frac{\partial\rho}{\partial t} + \frac{d\vec{x}}{dt}\cdot\nabla\rho = \frac{\partial\rho}{\partial t} + \vec{v}\cdot\nabla\rho \tag{3}$$

giving

$$\int_{\Omega}\left(\frac{\partial\rho}{\partial t} + \mathrm{div}\left(\rho\,\vec{v}\right)\right)dV = 0 \tag{4}$$

It is possible to leave aside the volume integration since the expression is applicable for any volume Ω, giving the mass conservation equation

$$\frac{\partial\rho}{\partial t} + \mathrm{div}\left(\rho\,\vec{v}\right) = 0 \tag{5}$$

If the fluid has a velocity smaller than the speed of sound in the medium, it is possible to assume that the volume of a small amount of fluid molecules does not change with time. The fluid is defined as incompressible, obeying to

$$\frac{d}{dt} \int_{\Omega} dV = 0 \tag{6}$$

which is equivalent to a solenoidal limit on the velocity

$$\operatorname{div}\left(\vec{v}\right) = 0 \tag{7}$$

1.2.2   Conservation of Momentum

The conservation of momentum states that the rate of change of the momentum of a body equals the net forces exerted on that body

$$\frac{d\vec{P}}{dt} = \frac{d}{dt} \int_{\Omega} \rho\left(t, \vec{x}\right) \vec{v}\left(t, \vec{x}\right) dV = \sum_{i=1}^{N} \vec{F}_{i,ext} + \sum_{j=1}^{M} \vec{F}_{j,int} \tag{8}$$

The left-hand side of Eq. 8 can be developed as

$$\frac{d}{dt} \int_{\Omega} \rho\left(t, \vec{x}\right) \vec{v}\left(t, \vec{x}\right) dV = \int_{\Omega} \left(\frac{D\rho\vec{v}}{Dt} + \rho\vec{v}\operatorname{div}\left(\vec{v}\right)\right) dV$$

$$= \int_{\Omega} \left(\rho\frac{D\vec{v}}{Dt} + \vec{v}\left(\frac{D\rho}{Dt} + \rho\operatorname{div}\left(\vec{v}\right)\right)\right) dV \tag{9}$$

The second term of the right-hand side of Eq. 9 is the mass conservation equation. Therefore, it is possible to express the momentum conservation equation as

$$\frac{d}{dt} \int_{\Omega} \rho\left(t, \vec{x}\right) \vec{v}\left(t, \vec{x}\right) dV = \int_{\Omega} \rho\frac{D\vec{v}}{Dt} dV \tag{10}$$

The net forces, $\sum_{i=1}^{N} \vec{F}_{i,ext} + \sum_{j=1}^{M} \vec{F}_{j,int}$, can be written as the sum of the mass, or body, forces, $\rho\vec{f}$, and surface forces, $\vec{t}_n = [T]\hat{n}$. They are evaluated in a fluid volume, and all their internal contributions cancel each other because every elementary force is accompanied by an opposite reaction force, i.e., $\sum_{j=1}^{M} \vec{F}_{j,int}$. In this way, it is possible to write the resulting total force as a volumetric force by using the Gauss divergence theorem

$$\sum_{i=1}^{N} \vec{F}_{i,ext} = \int_{\Omega} \rho\vec{f} \, dV + \int_{\partial\Omega} [T]\hat{n} dA = \int_{\Omega} \left(\rho\vec{f} + \operatorname{div}([T])\right) dV \tag{11}$$

Finally, the momentum conservation equation becomes

$$\rho\frac{D\vec{v}}{Dt} = \operatorname{div}([T]) + \rho\vec{f} \tag{12}$$

The conservation of the angular momentum, not reported here, requires a symmetric [T]. In a fluid at rest, the tensor is diagonal and isotropic, [T] $= -p[I]$, where p is the static pressure and [I] the identity tensor. It is still possible to decompose the stress tensor into a multiple of the unit tensor, [T] $= -p[I] + [\tau]$, where p is the isotropic stress, called mechanical pressure, while [τ] is a deviatoric tensor, linked to the internal friction mechanism. We assume that the stress tensor is an isotropic and linear function of the velocity gradient tensor, i.e. it is a Newtonian fluid, with the form

$$[\tau] = \eta\left(\left[\nabla \vec{v}\right] + \left[\nabla \vec{v}\right]^{T} - \frac{2}{3}\operatorname{div}\left(\vec{v}\right)[I]\right) \tag{13}$$

where η is the dynamic viscosity. If the flow is incompressible, the equation becomes

$$[\tau] = \eta\left(\left[\nabla \vec{v}\right] + \left[\nabla \vec{v}\right]^{T}\right) \tag{14}$$

with the kinematic viscosity

$$\nu = \frac{\eta}{\rho} \tag{15}$$

Finally, the momentum conservation equations are

$$\frac{D\vec{v}}{Dt} = -\frac{1}{\rho}\nabla p + \nu\nabla^{2}\vec{v} + \vec{g} \tag{16}$$

where the generic volume force field is replaced by the gravitational acceleration.

1.2.3 Conservation of Energy

If the fluid is incompressible, the thermal energy balance does not influence the dynamics of the system. The kinetic energy per unit volume is defined as $\rho\frac{\vec{v}\cdot\vec{v}}{2}$; then, to define an equation for this variable, it is possible to apply the scalar product between the velocity and the equation of conservation of momentum. Under the assumption of incompressible flow, it is possible to write:

$$\frac{\partial}{\partial t}\left(\frac{\vec{v}\cdot\vec{v}}{2}\right) + \operatorname{div}\left(\left(\frac{\vec{v}\cdot\vec{v}}{2} + \frac{p}{\rho} + gz\right)\vec{v}\right) = \nu\nabla^{2}\left(\frac{\vec{v}\cdot\vec{v}}{2}\right) - \nu\left[\nabla\vec{v}\right]$$
$$: \left[\nabla\vec{v}\right] \tag{17}$$

The last two terms on the right-hand side concern viscous effects; the first one is about energy diffusion, while the last one, always negative, represents a loss of kinetic energy. The effect of the viscous interaction is to dissipate kinetic energy.

1.3   Magneto-Hydro-Dynamics (MHD) Equations

The coupling between fluid dynamics and magneto-dynamics is effective because electromagnetic body forces influence the momentum balance, while, on the other hand, the particle motion of a conducting medium determines the complex behaviour of the electromagnetic field. The constitutive laws of electromagnetism are combined to get the conservation equation of the magnetic field, called the induction equation, which can be described in terms of dimensionless parameters.

1.3.1   Classical Electromagnetism

The fundamental equations for electromagnetism are a combination of the Maxwell equation with the Lorentz force, which acts on a charged particle in the presence of a magnetic field. The latter law states that a mass, carrying a charge, q, and moving with a velocity, $\vec{v}_p$, in an electric field, $\vec{E}$, and/or in a magnetic field, $\vec{B}$, undergoes the following force

$$\vec{F}_L = q\left(\vec{E} + \vec{v}_p \times \vec{B}\right) \tag{18}$$

We introduce a charge density, ρ_e, and an electric current density, $\vec{J}$, for a continuum medium, which, for an arbitrary volume, obeys the following relationships

$$q = \int_\Omega \rho_e dV \tag{19}$$

$$\vec{J} = \rho_e \vec{v} \tag{20}$$

where $\vec{v}$ is the velocity. It is then possible to derive the Lorentz force density as

$$\vec{f}_L = \rho_e \vec{E} + \vec{J} \times \vec{B} \tag{21}$$

The Maxwell equations allow to compute electric and magnetic fields for a specified charge and current distribution with the following equations

$$\mathrm{div}\left(\vec{D}\right) = \rho_e \tag{22}$$

$$\mathrm{curl}\left(\vec{E}\right) = -\frac{\partial \vec{B}}{\partial t} \tag{23}$$

$$\mathrm{div}\left(\vec{B}\right) = 0 \tag{24}$$

$$\mathrm{curl}\left(\vec{H}\right) = \vec{J} + \frac{\partial \vec{D}}{\partial t} \tag{25}$$

Applying the divergence operator to Eq.25, we obtain the electrical charge conservation equation:

$$\frac{\partial \rho_e}{\partial t} + \mathrm{div}\left(\vec{J}\right) = 0 \tag{26}$$

The Maxwell equations are expressed in function of the electric induction field, $\vec{D}$, electric field, $\vec{E}$, magnetic induction field, $\vec{B}$, and magnetic field, $\vec{H}$. The relations linking fields and the respective inductions are, $\vec{D} = \varepsilon_0 \vec{E}$, and $\vec{B} = \mu_0 \vec{H}$, where ϵ_0 and μ_0 are the vacuum electric permittivity and magnetic permeability.

The previous set of equations is not closed, and it is necessary to supply a constitutive relation linking the electric field and current density. In steady isotropic conducting media, and for low-frequency magnetic fields, the empirical evidence is the proportionality between current density and electric field, with a constant, σ, called electric conductivity. This brings us to Ohm's law

$$\vec{J} = \sigma \vec{E} \tag{27}$$

A conductor in motion needs to take into account the Lorentz force. If the medium velocity is much smaller than the speed of light, it is possible to use an approximation of the Lorentz transformation law

$$\vec{J} = \sigma\left(\vec{E} + \vec{v} \times \vec{B}\right) \tag{28}$$

If Eq. 28 is replaced in Eq. 26, then

$$\frac{\partial \rho_e}{\partial t} + \frac{\sigma}{\varepsilon_0}\rho_e + \sigma\mathrm{div}\left(\vec{v} \times \vec{B}\right) = 0 \tag{29}$$

1.3.2 Induction Equation

To eliminate the electric field, the Ohm law is substituted into the Faraday law, as

$$\frac{\partial \vec{B}}{\partial t} = -\mathrm{curl}\left(\vec{E}\right) = -\mathrm{curl}\left(\frac{1}{\sigma}\vec{J}\right) + \mathrm{curl}\left(\vec{v} \times \vec{B}\right) \tag{30}$$

The hypothesis that the current induced by the electric induction field (capacitor effect) is negligible gives

$$\mathrm{curl}\left(\vec{B}\right) \cong \mu_0 \vec{J} \tag{31}$$

The final equation is

$$\frac{\partial \vec{B}}{\partial t} = \mathrm{curl}\left(\vec{v} \times \vec{B}\right) - \mathrm{curl}\left(\frac{1}{\sigma \mu_0}\mathrm{curl}\left(\vec{B}\right)\right) \tag{32}$$

Considering the solenoidal character of the magnetic induction field, and after some algebra, we obtain

$$\frac{\partial \vec{B}}{\partial t} + \left(\vec{v} \cdot \nabla\right)\vec{B} = \left(\vec{B} \cdot \nabla\right)\vec{v} + \frac{1}{\sigma \mu_0}\nabla^2 \vec{B} \tag{33}$$

1.4 Thermo-Haemo-Dynamics (THD) and Magneto-Hydro-Dynamics (MHD) Equations

The previous relations allow us to derive the governing equations of Magneto-Hydro-Dynamics (*MHD*) for incompressible flow of a viscous, conducting liquid, which combines Navier-Stokes and Maxwell equations, including the induction equation.

The dimensional equations are

$$\mathrm{div}\left(\vec{v}\right) = 0 \tag{34}$$

$$\rho\frac{\partial \vec{v}}{\partial t} + \rho\left(\vec{v} \cdot \nabla\right)\vec{v} = -\nabla(p + \rho g z) + \eta\nabla^2 \vec{v} + \frac{1}{\mu_0}\mathrm{curl}\left(\vec{B}\right) \times \vec{B} \tag{35}$$

$$\frac{\partial \vec{B}}{\partial t} + \left(\vec{v} \cdot \nabla\right)\vec{B} = \left(\vec{B} \cdot \nabla\right)\vec{v} + \frac{1}{\sigma \mu_0}\nabla^2 \vec{B} \tag{36}$$

$$\mathrm{div}\left(\vec{B}\right) = 0 \tag{37}$$

while the dimensionless ones are

$$\mathrm{div}\left(\tilde{\vec{v}}\right) = 0 \tag{38}$$

$$\frac{\partial \tilde{\vec{v}}}{\partial t} + \left(\tilde{\vec{v}} \cdot \nabla\right)\tilde{\vec{v}} = -\nabla\left(\tilde{p} + \frac{1}{\mathrm{Fr}^2}\tilde{z}\right) + \frac{1}{Re}\nabla^2 \tilde{\vec{v}} + \frac{St}{Re_m}curl\left(\tilde{\vec{B}}\right) \times \tilde{\vec{B}} \tag{39}$$

$$\frac{\partial \vec{\tilde{B}}}{\partial t} + \left(\vec{\tilde{v}} \cdot \nabla \right) \vec{\tilde{B}} = \left(\vec{\tilde{B}} \cdot \nabla \right) \vec{\tilde{v}} + \frac{1}{Re_m} \nabla^2 \vec{\tilde{B}} \tag{40}$$

$$\mathrm{div}\left(\vec{\tilde{B}} \right) = 0 \tag{41}$$

The dimensionless equations allow the *MHD* flow to be described with the dimensionless groups: Reynolds number, *Re;* interactive parameter, or Stuart number, *St;* magnetic Reynolds number, *Re_m*; Froude number, *Fr;* Hartmann number, *Ha*, defined as

$$Re = \frac{UL}{\nu} \tag{42}$$

$$Re_m = \mu_0 \sigma UL \tag{43}$$

$$St = \frac{\sigma B_0^2 L}{\rho U} \tag{44}$$

$$Fr = \frac{U}{\sqrt{gL}} \tag{45}$$

$$Ha = B_0 L \sqrt{\frac{\sigma}{\eta}} \tag{46}$$

The Hartmann number measures the ratio between electromagnetic force and viscosity and is useful in laminar flow because the convective term is negligible. The *MHD* flow depends strongly on the magnetic Reynolds number because it governs the induction equation, and, when the magnetic field is small, its fluctuations relax quickly. It is then possible to notice the dissipative nature of the phenomenon since the kinetic energy of the fluid is dissipated due to the Joule effect. The Stuart number measures the ratio between electromagnetic and inertial forces, or, equivalently, the ratio between two timescales. On the one hand, the Joule damping time, defined as

$$t_J = \frac{\rho}{\sigma B_0^2} \tag{47}$$

is the time necessary for the Lorentz force to damp a vortex, while the eddy-turnover time, defined as

$$t_e = \frac{L}{U} \tag{48}$$

is the time needed for a vortex to move over a typical length scale.

2   Thermo-Haemo-Dynamics (THD) and Magneto-Hydro-Dynamics (MHD) Solvers in *OpenFOAM*

2.1   Introduction

OpenFOAM is an open-source *CFD* software package, written in *C++* libraries, used to create executables, known as applications. Two categories of applications are present: solvers and utilities. The solvers are designed to solve specific problems in continuum mechanics, while the utilities are designed to perform tasks involving data manipulation. Both applications can be created by the users. The software is supplied with pre- and post-processing environments, respectively, *BlockMesh* and *ParaView*, [12], as illustrated in Fig. 1.

2.1.1   Structures of OpenFOAM

Every single case is organised in a precise way inside the software. There are three folders: the first one, for the time directories, is called *0;* the second one, for the geometry of the problem, is called *constant*; and the last one, for the schemes and the libraries used in the numerical solution of the problems, is called *system*. The name of each time directory is based on the simulation time, as shown in Fig. 2. The folder *time directories* include individual files of data for specific fields. The data include both initial values and boundary conditions which can be specified by the user to define the problem. They can also be written upon a file by the same *OpenFOAM* during its run.

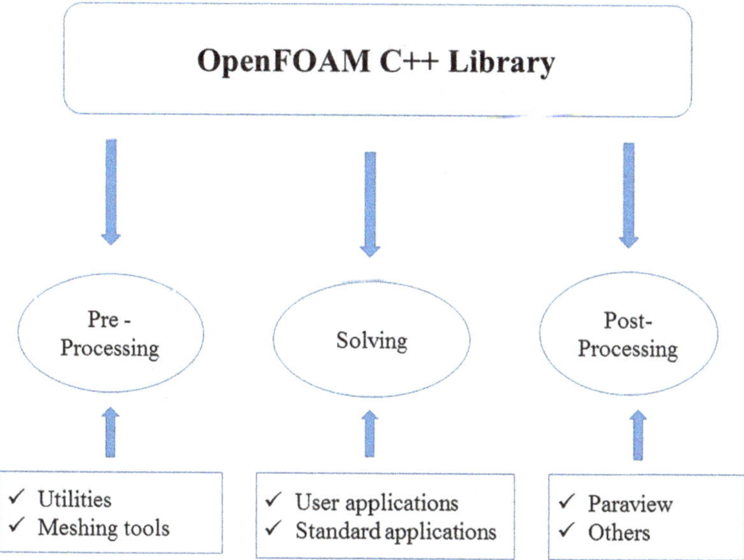

Fig. 1 Overview of OpenFOAM structure, [1]

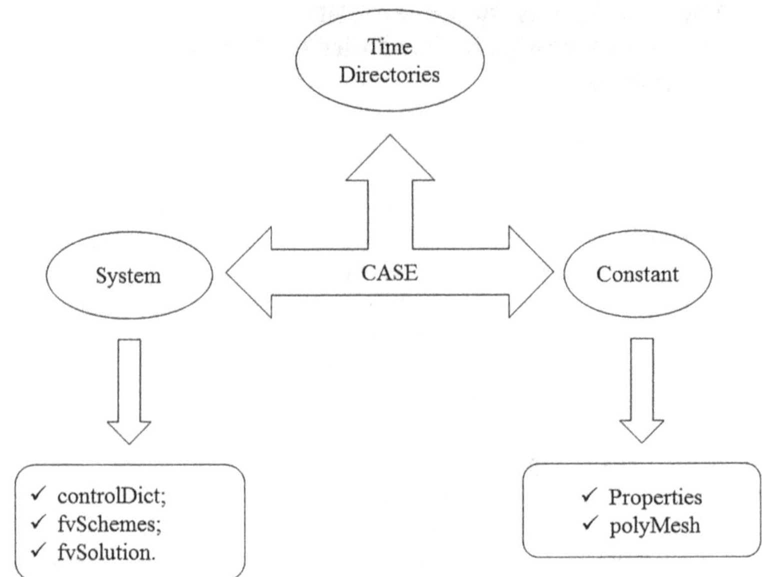

Fig. 2 Overview of OpenFOAM case structure, [1]

The folder *constant* contains a full description of the case mesh in a specific subdirectory and files, specifying the physical properties of the applications involved. The folder *system,* necessary for setting parameters associated with the method of solution, contains at least three files:

- *controlDict:* the user can set up the parameters necessary to run the simulation, such as start and end time, time-step and parameters for data output.
- *fvSchemes:* the user can specify the discretisation schemes.
- *fvSolution:* the solvers for the algebraic system, tolerances and other algorithm controls are set up for the simulation.

2.1.2   Mesh Generation and Properties

In the folder *constant,* there are all the necessary specifications for the geometry domain and the properties for the specific problem; its peculiarity is to decompose the domain geometry into blocks. The blocks can be realised by using hexahedral elements, or wedges, prisms, pyramids and tetrahedra. The edges of the blocks can be lines, arcs or splines. Each block of the geometry is defined by eight vertices, one at each corner of the hexahedra. Each block has a local coordinate system, generally right-handed.

The local coordinate system is defined by the order of the vertices presented in the block definition:

- The axis origin is the first entry in the block definition.
- The x-direction is described by moving from vertex 0 to vertex 1.
- The y-direction is described by moving from vertex 1 to vertex 2.
- The vertices from 0 to 3 define the plane z = 0.
- The z-direction is described by moving from vertex 0 to vertex 4.
- The vertices from 5 to 7 are found by moving in the z-direction from vertices 1 to 3.

Two vertices are joined by an edge, which can be:

- Arc
- Line
- Simple Spline
- Poly Spline
- Poly Line

After the characterisation of vertices and edges, it is possible to define the blocks. Inside the block definition, the user can set up: the number of cells in each of the x, y and z directions for that block and the cell expansion ratios. The last specification gives the cell expansion ratios for each direction in the block, enabling the mesh to be graded, or thickened, in specific directions. The cell expansion ratio is calculated as the ratio of the width of the end cell, along one edge of a block, to the width of the start cell along that edge. The ratio can be split into two types:

- *simpleGrading,* which specifies uniform expansions in the local x, y, z directions;
- *edgeGrading,* which gives a ratio for each edge of the block.

The most trivial part in defining the geometry is related to the boundaries definition. Boundaries are necessary for applying boundary conditions; generally, a boundary is broken up into a set of patches, and one patch can include one or more enclosed areas of the boundary, which do not necessarily need to be physically connected. Three types of attributes can be associated with a patch:

- *Base Type:* the patch is described in terms of geometry;
- *Primitive Type:* the numerical patch condition is assigned to a field variable on that patch;
- *Derived Type:* the complex patch condition, derived from the primitive, is assigned to a field variable on the patch.

Since *OpenFOAM* has a large range of solvers, each designed for a specific class of problems, the equations and algorithms differ from one solver to another. The selection of a solver involves the user making some initial choices, such as parameters and physical properties. These last two conditions can be specified in a properties file inside the constant folder.

2.1.3 Fields

The input and output data in *OpenFOAM* can be Tensor, Vector or Scalar fields, and they must be read and written in the time directories. In the *0* folder, every field of the problems is written in a detailed way using specific keywords:

- *Dimensions:* to specify the dimensions of a field;
- *InternalField:* to specify the value of the internal field;
- *BoundaryField:* to assign the boundary conditions to a well-known patch.

The *Dimensions* are associated with the physical properties of the variables used in the problem.

The *InternalField* can be divided into two types:

- *Uniformfield:* a single value is assigned to all the elements within the field;
- *Nonuniformfield:* each field element is assigned to a unique value from a list.

The *BoundaryField* is a dictionary containing the names of the boundary patches, listed in the boundary file in the *polyMesh* directory. The compulsory keyword *Type* describes the patch field conditions specified for that field. The other entries correspond to the type of patch field condition selected and can include field data specifying initial conditions on patch faces.

2.1.4 Time and Data Control

Every *OpenFOAM* solver begins all runs by setting up a database, which controls input and output at specific time intervals during the run. Generally, time is an essential part of the database. The *controlDict* file is the dictionary where the essential parameters for the creation of the database are set up. The parameters for the time control are related to the beginning/end of the runtime of the simulation. The parameters for the data writing are related to the writing format, *ASCII* or *BINARY,* and the format for the file compression, uncompressed or compressed. Moreover, the user can also set up the graph format by specifying how the solver can save the data in the time directories.

As depicted in Fig. 2, there are two other files in the system folder: *fvSchemes* and *fvSolution*. The *fvSchemes* file is the dictionary where the user sets the numerical schemes for terms, such as derivatives in equations. Generally, the terms that the user has to specify range from derivatives to interpolations of values from one set of points to another. In the *ddtSchemes,* the first time derivative terms are specified; it is possible to choose the discretisation scheme for each term. When we specify the time scheme, it must be noted that an application designed for transient problems will not run as a steady state and vice versa. Any second derivative with respect to time is specified in the *d2dt2Schemes*. For steady state, the time derivative terms have been turned off.

The *gradSchemes* sub-dictionary contains the gradient schemes, and the default setting is *Gauss linear*. The *Gauss* keyword specifies that most spatial derivative terms are first integrated over a cell volume, V, then converted to integrals over the cell surface bounding the volume using the *Gauss theorem*. *Linear* is the choice of the interpolation scheme. The *divSchemes* sub-dictionary contains the divergence terms. The *Gauss* scheme is the only choice of discretisation and requires a selection of the interpolation scheme for the dependent field. The choice of the interpolation scheme is critical for the determination of the numerical behaviour, i.e. linear implies a second-order unbounded for the numerical behaviour. The *laplacianSchemes* sub-dictionary contains Laplacian terms. The only possible choice is to use a *Gaussian* scheme, which requires a selection of both an interpolation scheme for the diffusion coefficient and a surface normal gradient scheme. In detail, for our purpose, we have imposed a *Gauss Linear Corrected*, whose numerical behaviour is unbounded, second-order and conservative.

The *interpolationScheme* sub-dictionary contains terms that are interpolations of values typically from cell to face centres. The selection of interpolation schemes can be divided into four categories: Centred schemes, Upwinded convection schemes, TVD and NVD schemes. The convection-specific schemes calculate the interpolation based on the flux of the flow velocity. The specification of these schemes requires the name of the flux field on which the interpolation is based; in most *OpenFOAM* applications, this is *phi*, the name adopted for the *surfaceScalarField* velocity flux, ϕ. In our simulations, we have used a centred scheme with a *linear* interpolation.

The *snGradSchemes* sub-dictionary contains surface normal gradient terms. A surface normal gradient is evaluated at a cell face; it is the component, normal to the face, of the gradient of the values at the centres of the 2 cells that the face connects. A surface normal gradient is also required to evaluate a Laplacian term using Gaussian integration. The user can choose the right type of scheme. In our simulations, we have used a *corrected* scheme. The flux requires sub-dictionary lists for which the flux is generated in the application. In our simulations, we are interested in the fluxes generated after solving the pressure equation and the equation of the magnetic field, which evaluates the magnetic pressure: p, pB.

The *fvSolution* contains a set of sub-dictionaries that are specific to the solver and include relaxation factors, tolerances and pressure-velocity coupling algorithms, such as *Pressure-Implicit with Splitting of Operators (PISO)*, *Semi-Implicit Method for Pressure-Linked Equations (SIMPLE)* and *BPISO* for the magnetic field. For the pressure equation, a *GAMG (Generalised Geometric-Algebraic Multi-Grid)* solver is used, along with a preconditioner, a smoother and the definition of its tolerance and relative one. The *GAMG* uses the principle of generating a quick solution on a mesh with a small number of cells, mapping this solution onto a finer mesh, and using it as an initial guess to obtain an accurate solution on the fine mesh. It aims to be faster than standard methods when the increase in speed by solving first on coarser meshes outweighs the additional costs of mesh refinement and mapping of field data. The switching from a coarse mesh to the fine mesh is handled by *agglomeration* of cells, either by a *geometric agglomeration* where cells are joined together or by an *algebraic agglomeration* where matrix coefficients are joined, [12]. This algorithm allows a faster convergence for the preconditioned. For the *GAMG*, the suggested preconditioner is *FDIC (Faster Diagonal Incomplete Cholesky)*. As the name suggests, this is a faster version of the *DIC* (*Diagonal Incomplete Cholesky*) for symmetric matrices, in which the reciprocal of the preconditioned diagonal and the upper coefficients, divided by the diagonal, are calculated and stored, [13]. Although the preconditioner can reduce the number of iterations, it does not normally reduce the mesh dependency on the number of iterations. For this purpose, it is necessary to specify the type of smoother; for the *GAMG*, the default smoother is *Gauss-Siedel*. In general, this is a technique used to solve linear systems of equations and is the most reliable option. The method is an improved version of the Jacobi method and is defined on matrices with non-zero diagonals, but the convergence is guaranteed only if the matrix is either diagonally dominant or symmetric and positive definite.

In general, matrix solvers are iterative, and they are based on reducing the equation residual over a succession of iterations. The residual is a measure of the error in the solution, so the smaller it is, the more accurate the solution is. The residual is evaluated by substituting the current solution into the equation and taking the magnitude of the difference between the left and the right-hand sides; it is normalised to make it independent of the scale of the problem analysed. Before solving an equation, the initial residual is evaluated based on the current value of the field. After each iteration, the residual is re-evaluated. The solver stops if either two of the following conditions are reached:

- Residual falls below the solver tolerance and is defined as the tolerance parameter;
- Ratio of current to initial residual falls below the solver's relative tolerance and is defined as the *relTol* parameter;
- Number of iterations exceeds the maximum number of iterations, which is defined as the *maxIter* parameter and is optional.

The solver's tolerance should represent the level at which the residual is so small that the solution can be estimated sufficiently accurately. The solver's relative tolerance limits the relative improvement from the initial to the final solutions. In transient simulation, to force the convergence in each time-step, it is normal to set this parameter to zero.

For the velocity and the intensity of the magnetic field, we use the *PBiCG* (*Preconditioned Bi-Conjugate Gradient*) solver. It is a solver for symmetric *LDU* matrices using a runtime selectable preconditioner. In our case, the suggested *DILU* preconditioner is used. This is a simplified diagonal-based incomplete *LU* preconditioner for asymmetric matrices. The reciprocal of the preconditioned diagonal is calculated and then subsequently stored. After setting up the solvers for the variables involved in the problems, in the *fvSolution* file, there are two other important sub-dictionaries to analyse:

- Relaxation factors;
- Correctors for algorithms.

The relaxation factors control the under-relaxation coefficients, which are used to improve the stability of a computation, particularly in steady-state problems. The under-relaxation coefficients work by limiting the amount by which a variable changes from one iteration to the next, either by modifying the solution matrix and source before solving for a field or by directly modifying the field. The under-relaxation factor is a real number α ranging between 0 and 1. A null under-relaxation factor represents a solution that does not change at all with the successive iterations, while $\alpha = 1$ corresponds to a solution independent from the previous state. The under-relaxation factor should be the highest value in the [0,1] interval, which ensures a stable computation.

About the solver, in general, the *PISO* (*Pressure-Implicit with Splitting of Operators*) is mostly used in transient, compressible problems; it is a *predictor-corrector* algorithm, which requires more than one correction but less than four. The *BPISO* is similar to the *PISO* (*Pressure-Implicit with Splitting of Operators*) algorithm, in which there are also correctors for the magnetic field. In *BPISO*, it is compulsory to specify the number of mesh correctors. The *SIMPLE* algorithm is of the *predictor-corrector* type; it evaluates some initial solutions and then corrects them. In the *SIMPLE* case, there is only one correction. Inside these, the number of correctors, both orthogonal and non-orthogonal, is specified. In general, a mesh is orthogonal if, for each face, the normal face is parallel to the vector between the centres of the cells to which the face connects. The *nNonOrthogonalCorrectors* takes into account the non-orthogonality of the mesh. The number of non-orthogonal correctors should correspond to the mesh for the case being solved: for an orthogonal mesh, it is zero, while it increases with the degree of the mesh's non-orthogonality.

2.1.5 MapFields Utility

This utility can map one or more fields relative to a given geometry onto the corresponding fields for another geometry. It is completely generalised, so it does not need any similarity between the geometries to which the fields are related. For cases with *consistent* geometries, it can be executed with a special option that simplifies the mapping process. To use this utility, it is necessary to define some terms. In general, the data is mapped from the *source* to the *target*. The fields are considered *consistent* if the geometry and boundary types, or conditions, of both *source* and *target* fields are identical. The field data that *mapFields* maps are those fields within the time directory specified by *startFrom/startTime* inside the *controlDict* of the *target* case. The data is read from the equivalent time directory of the *source* case and mapped onto the equivalent time directory of the *target* case.

2.1.6 Groovy BC

The *groovyBC* offers the possibility to specify expressions involved in the fields. It is possible to set up the parameters in every patch by some expressions:

- *valueExpression:* a string used if a Dirichlet condition is needed;
- *value:* used for the first time-step/iteration, when *valueExpression* is specified;
- *gradientExpression:* a string used if a Neumann condition is needed;

- *fractionExpression:* to distinguish between Neumann or Dirichlet conditions;
- *variables:* a list of all temporary variables;
- *timelines:* a list with sub-dictionaries that specifies interpolation tables over time.

Moreover, *groovyBC* allows accessibility to other fields from remote patches. Since *groovyBC* is used, the necessary libraries for the setting of the new boundary condition are specified in *controlDict*. The same formulation, used for the previous test case, is used in *fvSchemes* and *fvSolution*.

2.1.7 SIMPLE Algorithm

The *SIMPLE (Semi-Implicit Method for Pressure-Linked Equations)* is a guess-and-correct procedure for the calculation of pressure on the grid arrangement. To initiate the *SIMPLE* calculation process, a pressure field p^* is guessed, and the discretised momentum equations are solved using the guessed pressure field to yield the velocity components u^*, v^*:

$$a_{i,J}u^*_{i,J} = \sum a_{nb}u^*_{nb} + \left(p^*_{I-1,J} - p^*_{I,J}\right)A_{i,J} + b_{i,J} \tag{49}$$

$$a_{i,J}v^*_{i,J} = \sum a_{nb}v^*_{nb} + \left(p^*_{I-1,J} - p^*_{I,J}\right)A_{i,J} + b_{i,J} \tag{50}$$

Afterwards, we define the correction p' as the difference between the correct pressure field p and the guessed field p^*, so that

$$p = p^* + p' \tag{51}$$

In a similar way we define velocity corrections u', v' and relate the correct velocities u, v to the guessed u^*, v^*:

$$u = u^* + u' \tag{52}$$

$$v = v^* + v' \tag{53}$$

Substitution of the correct pressure field, p, into the momentum equations yields the correct velocity field, u, v. Subtraction of Eqs. (49) and (50) from the discretised momentum equations, respectively, gives:

$$a_{i,J}u'_{i,J} = \sum a_{nb}u'_{nb} + \left(p'_{I-1,J} - p'_{I,J}\right)A_{i,J} \tag{54}$$

$$a_{i,J}v'_{i,J} = \sum a_{nb}v'_{nb} + \left(p'_{I-1,J} - p'_{I,J}\right)A_{i,J} \tag{55}$$

The terms $\sum a_{nb}u'_{nb}$ and $\sum a_{nb}v'_{nb}$ are dropped to simplify the Eqs. (54) and (55) for the velocity corrections. The omission of these terms is the main approximation of the *SIMPLE* algorithm, to obtain:

$$u'_{i,J} = d_{i,J}\left(p'_{I-1,J} - p'_{I,J}\right) \tag{56}$$

$$v'_{i,J} = d_{i,J}\left(p'_{I-1,J} - p'_{I,J}\right) \tag{57}$$

being $d_{i,j} = \frac{A_{i,J}}{a_{i,J}}, d^{A_{I,j}}_{a_{I,j}\,I,j}$.

Equations. (56) and (57) describe the corrections to be applied to velocities through Eqs. (52) and (53), which gives:

$$u_{i,J} = u^*_{i,J} + d_{i,J}\left(p'_{I-1,J} - p'_{I,J}\right) \tag{58}$$

$$v_{I,j} = v^*_{I,j} + d_{I,j}\left(p'_{I,J-1} - p'_{I,J}\right) \tag{59}$$

Similar expressions exist for $u_{i+1,\,J},\,v_{I,\,j+1}$:

$$u_{i+1,J} = u^*_{i+1,J} + d_{i+1,J}\left(p'_{I,J} - p'_{I+1,J}\right) \tag{60}$$

$$v_{I,j+1} = v^*_{I,j+1} + d_{I,j+1}\left(p'_{I,J} - p'_{I,J+1}\right) \tag{61}$$

Also, the continuity equation is satisfied in the discretised form of the scalar control volume:

$$\left[(\rho uA)_{i+1,J} - (\rho uA)_{i,J}\right] + \left[(\rho vA)_{I,j+1} - (\rho vA)_{I,j}\right] = 0 \tag{62}$$

Substituting the corrected velocities from Eqs. (58–59) into the discretised continuity Eq. (62) and identifying the coefficients of p', the final equation can be written as:

$$a_{I,J}p'_{I,J} = a_{I+1,J}p'_{I+1,J} + a_{I-1,J}p'_{I-1,J} + a_{I,J+1}p'_{I,J+1} + a_{I,J-1}p'_{I,J-1} + b'_{I,J} \tag{63}$$

This represents the equation for pressure correction p'. The source term b' in the equation is the continuity imbalance arising from the incorrect velocity field u^*, v^*. By solving Eq. (63), the pressure correction p' can be obtained at all points. Once the pressure correction field is known, the correct pressure field may be obtained using Eq. (51) and velocity components through the correction Eqs. (58–61).

The pressure correction equation is susceptible to divergence unless some under-relaxation factors are used during the iterative process and new, improved pressures p^{new} are obtained with:

$$p^{new} = p^* + \alpha_p p' \tag{64}$$

where α_p is the pressure under-relaxation factor. If we select α_p equal to 1, the guessed pressure field p^* is corrected by p'; in specific, when the guessed field p^* is far away from the final solution, it is often too large for stable computations. A value of α_p equal to 0 would apply no correction at all. Taking α_p between 0 and 1 allows us to add to the guessed field p^* a fraction of the correction field p' that is large enough to ensure stable computation.

The velocities are also under-relaxed. The iteratively improved velocity components u^{new} and v^{new} are obtained from:

$$u^{new} = \alpha_u u + (1 - \alpha_u) u^{(n-1)} \tag{65}$$

$$v^{new} = \alpha_v v + (1 - \alpha_v) v^{(n-1)} \tag{66}$$

where α_u, α_v are the u and v velocity under-relaxation factors with values between 0 and 1; u, v are the corrected velocity components without relaxation factors, and $u^{(n-1)}$, $v^{(n-1)}$ represent their values attained in the previous iteration. A correct choice of under-relaxation factors α is essential for cost-effective simulations. Too large values of α may lead to oscillatory or even divergent iterative solutions, and a value that is too small will cause slow convergence. Unfortunately, the optimum values of under-relaxation factors are flow-dependent and must be sought on a case-by-case basis, [14].

2.1.8 PISO Algorithm

The *PISO algorithm (Pressure-Implicit with Splitting Operators)* is a pressure-velocity calculation procedure, developed originally for the non-iterative computation of unsteady compressible flows. *PISO* involves one predictor and two corrector steps, so it can be considered an extension of *SIMPLE*, with a further corrector step to enhance it. In the *predictor step,* the discretised momentum equations are solved with an initial guess for the pressure field p^* to give the velocity components u^*, v^*. The last two fields will not satisfy continuity unless the pressure field p^* is correct. The *first corrector step* is used to give a velocity field u^{**}, v^{**} that satisfies the discretised continuity equation.

The resulting equations are:

$$p^{**} = p^* + p' \tag{67}$$

$$u^{**} = u^* + u' \tag{68}$$

$$v^{**} = v^* + v' \tag{69}$$

From Eqs. (68–69) it is possible to define corrected velocities u^{**}, v^{**}:

$$u_{i,J}^{**} = u_{i,J}^* + d_{i,J}\left(p'_{I-1,J} - p'_{I,J}\right) \tag{70}$$

$$v_{i,J}^{**} = v_{i,J}^* + d_{i,J}\left(p'_{I-1,J} - p'_{I,J}\right) \tag{71}$$

Equations (70–71) are substituted into the discretised continuity equation to yield the pressure correction equation with its coefficients and source term:

$$a_{I,J}p'_{I,J} = a_{I+1,J}p'_{I+1,J} + a_{I-1,J}p'_{I-1,J} + a_{I,J+1}p'_{I,J+1} + a_{I,J-1}p'_{I,J-1} + b'_{I,J} \qquad (72)$$

Equation (72) is called the *first pressure correction equation*. It is solved to yield the first pressure correction field p'. Once the pressure corrections are known, the velocity components u^{**}, v^{**} can be obtained through Eqs. (70–71).

To enhance the *SIMPLE* procedure, the *PISO* performs a *second corrector step*. The discretised momentum equations for u^{**}, v^{**} are

$$a_{i,J}u^{**}_{i,J} = \sum a_{nb}u^{*}_{nb} + \left(p^{**}_{I-1,J} - p^{**}_{I,J}\right)A_{i,J} + b_{i,J} \qquad (73)$$

$$a_{i,J}v^{**}_{i,J} = \sum a_{nb}v^{*}_{nb} + \left(p^{**}_{I-1,J} - p^{**}_{I,J}\right)A_{i,J} + b_{i,J} \qquad (74)$$

A twice-corrected velocity field u^{***}, v^{***} may be obtained by solving the momentum equations once more:

$$a_{i,J}u^{***}_{i,J} = \sum a_{nb}u^{**}_{nb} + \left(p^{***}_{I-1,J} - p^{***}_{I,J}\right)A_{i,J} + b_{i,J} \qquad (75)$$

$$a_{i,J}v^{***}_{i,J} = \sum a_{nb}v^{**}_{nb} + \left(p^{***}_{I-1,J} - p^{***}_{I,J}\right)A_{i,J} + b_{i,J} \qquad (76)$$

The summation terms are evaluated using the velocities u^{**}, v^{**}, calculated in the *first corrector step*. Subtraction of Eq. (73) from (75) and Eq. (74) from (76) gives:

$$u^{***}_{i,J} = u^{**}_{i,J} + \frac{\sum a_{nb}\left(u^{**}_{nb} - u^{*}_{nb}\right)}{a_{i,J}} + d_{i,J}\left(p''_{I-1,J} - p''_{I,J}\right) \qquad (77)$$

$$v^{***}_{i,J} = v^{**}_{i,J} + \frac{\sum a_{nb}\left(v^{**}_{nb} - v^{*}_{nb}\right)}{a_{i,J}} + d_{i,J}\left(p''_{I-1,J} - p''_{I,J}\right) \qquad (78)$$

where p'' is the second pressure correction so that p^{***} may be obtained by:

$$p^{***} = p^{**} + p'' \qquad (79)$$

Substitution of u^{***}, v^{***} into the discretised continuity equation yields a *second pressure correction equation*:

$$a_{I,J}p''_{I,J} = a_{I+1,J}p''_{I+1,J} + a_{I-1,J}p''_{I-1,J} + a_{I,J+1}p''_{I,J+1} + a_{I,J-1}p''_{I,J-1} + b'_{I,J} \qquad (80)$$

Deriving the previous equation, the source term is zero since the velocity components u^{**}, v^{**} satisfy the continuity. Eq. (80) is solved to obtain the second pressure correction field p'' and a twice-corrected pressure field is obtained from:

$$p^{***} = p^{**} + p'' = p^{*} + p' + p'' \qquad (81)$$

Finally, the twice-corrected velocity field is obtained from Eqs. (77–78). For the non-iterative calculation of unsteady flows, the pressure field p^{***} and the velocity field u^{***}, v^{***} are considered to be the correct u, v, p. Furthermore, all time-dependent terms are retained in the momentum and continuity equations; this gives additional contributions to the momentum and pressure correction equations.

The *PISO* procedure is carried out at each time level to calculate the velocity and pressure fields; the temporal accuracy achieved by the predictor-corrector process for pressure and momentum is of order 3 and 4, respectively. Therefore, the pressure and velocity fields obtained at the end of the *PISO* process with suitably small time steps are considered to be accurate to proceed to the next time-step immediately. Since the algorithm relies on the higher-order temporal accuracy gained by the splitting technique, small time steps are recommended to ensure accurate results, [14].

2.1.9 MHDFOAM

The default solver for Magneto-Hydro-Dynamics is *MHDFOAM*, used for the incompressible, laminar flow of a conducting fluid under the influence of a magnetic field. An applied magnetic field, $\vec{B}$, acts as a driving force. At present, boundary conditions cannot be set via the electric field, $\vec{E}$, or current density, $\vec{J}$. The fluid magnetic permeability, μ_0, the electric conductivity, σ, and the kinematic viscosity, ν, are read as uniform constants.

The equations considered by the solver are the following

$$\mathrm{div}\left(\vec{v}\right) = 0 \tag{82}$$

$$\frac{\partial \vec{v}}{\partial t} + \left(\vec{v} \cdot \nabla\right)\vec{v} = -\frac{1}{\rho}\nabla(p + \rho gz) + \nu\nabla^2\vec{v} + \frac{1}{\rho\mu_0}\mathrm{curl}\left(\vec{B}\right) \times \vec{B} \tag{83}$$

$$\frac{\partial \vec{B}}{\partial t} + \left(\vec{v} \cdot \nabla\right)\vec{B} = \left(\vec{B} \cdot \nabla\right)\vec{v} + \frac{1}{\sigma\mu_0}\nabla^2\vec{B} \tag{84}$$

$$\mathrm{div}\left(\vec{B}\right) = 0 \tag{85}$$

The magnetic stress tensor is split into a hydrostatic and a deviatoric part for the convenience of solution. The hydrostatic part is identified by the so-called magnetic pressure, p_B, which enforces the solenoidal nature of the magnetic field. A fictitious magnetic flux pressure, pH, is introduced to compensate for discretisation errors and to create a magnetic face flux field, which is divergence-free, as required by the Maxwell equations.

2.1.10 ParaView

After the simulations have run, it is possible to post-process the results to obtain plots and maps of the interested variables. We analyse the results with the *ParaView* open-source software, which allows the post-processing of the data. *ParaView* has extensive scripting and batch-processing capabilities. The standard scripting interface uses the widely used *Python* programming language for scripted control.

2.2   THD and MHD Equations in Cartesian Coordinates

To study *2D-3D* channels, we need to simplify the *MHD* equations in Cartesian coordinates, as

$$\frac{\partial u}{\partial x} + \frac{\partial v}{\partial y} + \frac{\partial w}{\partial z} = 0 \tag{86}$$

The momentum equations are written, for each component, as

$$\frac{\partial u}{\partial t} + u\frac{\partial u}{\partial x} + v\frac{\partial u}{\partial y} + w\frac{\partial u}{\partial z} = -\frac{\partial p}{\partial x} + \frac{1}{Re}\left(\frac{\partial^2 u}{\partial x^2} + \frac{\partial^2 u}{\partial y^2} + \frac{\partial^2 u}{\partial z^2}\right) + \frac{St}{Re_m}$$
$$\times \left(B_z\left(\frac{\partial B_x}{\partial z} - \frac{\partial B_z}{\partial x}\right) - B_y\left(\frac{\partial B_y}{\partial x} - \frac{\partial B_x}{\partial y}\right)\right) \tag{87}$$

$$\frac{\partial v}{\partial t} + u\frac{\partial v}{\partial x} + v\frac{\partial v}{\partial y} + w\frac{\partial v}{\partial z} = -\frac{\partial p}{\partial y} + \frac{1}{Re}\left(\frac{\partial^2 v}{\partial x^2} + \frac{\partial^2 v}{\partial y^2} + \frac{\partial^2 v}{\partial z^2}\right) + \frac{St}{Re_m}$$
$$\times \left(B_x\left(\frac{\partial B_y}{\partial x} - \frac{\partial B_x}{\partial y}\right) - B_z\left(\frac{\partial B_z}{\partial y} - \frac{\partial B_y}{\partial z}\right)\right) \tag{88}$$

$$\frac{\partial w}{\partial t} + u\frac{\partial w}{\partial x} + v\frac{\partial w}{\partial y} + w\frac{\partial w}{\partial z} = -\frac{\partial p}{\partial z} + \frac{1}{Re}\left(\frac{\partial^2 w}{\partial x^2} + \frac{\partial^2 w}{\partial y^2} + \frac{\partial^2 w}{\partial z^2}\right) + \frac{St}{Re_m}$$
$$\times \left(B_y\left(\frac{\partial B_z}{\partial y} - \frac{\partial B_y}{\partial z}\right) - B_x\left(\frac{\partial B_x}{\partial z} - \frac{\partial B_z}{\partial x}\right)\right) \tag{89}$$

The transport equations for the magnetic induction field are

$$\frac{\partial B_x}{\partial t} + u\frac{\partial B_x}{\partial x} + v\frac{\partial B_x}{\partial y} + w\frac{\partial B_x}{\partial z} = B_x\frac{\partial u}{\partial x} + B_y\frac{\partial u}{\partial y} + B_z\frac{\partial u}{\partial z} + \frac{1}{Re_m}$$
$$\times \left(\frac{\partial^2 B_x}{\partial x^2} + \frac{\partial^2 B_x}{\partial y^2} + \frac{\partial^2 B_x}{\partial z^2}\right) \tag{90}$$

$$\frac{\partial B_y}{\partial t} + u\frac{\partial B_y}{\partial x} + v\frac{\partial B_y}{\partial y} + w\frac{\partial B_y}{\partial z} = B_x\frac{\partial v}{\partial x} + B_y\frac{\partial v}{\partial y} + B_z\frac{\partial v}{\partial z} + \frac{1}{Re_m}$$
$$\times \left(\frac{\partial^2 B_y}{\partial x^2} + \frac{\partial^2 B_y}{\partial y^2} + \frac{\partial^2 B_y}{\partial z^2}\right) \tag{91}$$

$$\frac{\partial B_z}{\partial t} + u\frac{\partial B_z}{\partial x} + v\frac{\partial B_z}{\partial y} + w\frac{\partial B_z}{\partial z} = B_x\frac{\partial w}{\partial x} + B_y\frac{\partial w}{\partial y} + B_z\frac{\partial w}{\partial z} + \frac{1}{Re_m}$$
$$\times \left(\frac{\partial^2 B_z}{\partial x^2} + \frac{\partial^2 B_z}{\partial y^2} + \frac{\partial^2 B_z}{\partial z^2}\right) \tag{92}$$

2.3 Numerical Simulation of 2D Laminar Steady-State THD–MHD Duct Flow

The solution of *2D* laminar steady *MHD* duct flow is obtained with the assumption of fully developed flow and with each variable independent from the axial coordinate, except the pressure, which decays linearly, [1]. The equations involved are

$$0 = -Re\,\frac{\partial p}{\partial x} + \frac{d^2 u}{dy^2} + \frac{Ha^2}{Re_m}\frac{dB_x}{dy} \tag{93}$$

$$0 = \frac{\partial p}{\partial y} + \frac{St}{Re_m}B_x\frac{dB_x}{dy} \tag{94}$$

$$0 = \frac{du}{dy} + \frac{1}{Re_m}\frac{d^2 B_x}{dy^2} \tag{95}$$

Integrating Eq. (95)

$$[c_1 - u(y)] = \frac{1}{Re_m}\frac{dB_x}{dy} \tag{96}$$

and substituting into Eq. (96)

$$\frac{d^2 u}{dy^2} - Ha^2 u(y) + \left(Ha^2 c_1 + Re\,\frac{\Delta p}{L}\right) = 0 \tag{97}$$

we obtain, from Eq. (94),

$$0 = \frac{d}{dy}\left(p + \frac{St}{Re_m}\frac{B_x^2}{2}\right) \tag{98}$$

Equation (98) shows that pressure is not uniform along the vertical direction, being

$$p(x, y) = P(x) - \frac{St}{Re_m}\frac{B_x^2(y)}{2} \tag{99}$$

The solution of Eq. 97 gives the velocity

$$u(y) = \frac{\left(1 - \frac{cosh(Ha\cdot y)}{cosh\left(\frac{Ha}{4}\right)}\right)}{\left(1 - \frac{4}{Ha}\,tanh\left(\frac{Ha}{4}\right)\right)} \tag{100}$$

under the assumption that the non-dimensional average velocity is 1

$$1 = \left(c_1 + \frac{Re}{Ha^2 \frac{\Delta p}{L}} \left(1 - \frac{4}{Ha} \tanh\left(\frac{Ha}{4}\right) \right) \right) \tag{101}$$

Integration of Eq. 99 gives

$$B_x(y) = \left(c_1 - \frac{1}{\left(1 - \frac{4}{Ha}\tanh\left(\frac{Ha}{4}\right)\right)} \right) Re_m y + \frac{Re_m}{Ha\left(1 - \frac{4}{Ha}\tanh\left(\frac{Ha}{4}\right)\right)} \frac{\sinh(Ha \cdot y)}{\cosh\left(\frac{Ha}{4}\right)} \tag{102}$$

with

$$c_1 = \frac{1}{Re_m} \left. \frac{\partial B_x}{\partial y} \right|_w \tag{103}$$

2.3.1 Two-Dimensional Duct Flow with Uniform Magnetic Induction

Consider the incompressible and unidirectional flow, $\vec{u} = u(x, y)$, of an electrically conducting fluid through a rectangular duct, Fig. 3. An applied magnetic field, $\vec{B} = B_0(y)$ is imposed, and the flow is driven by a uniform pressure gradient along the x-axis. In the $2D$ geometry, two blocks are realised, and a patch type is set up for the whole edge. Moreover, the left edge *Inlet*, as well as the right edge *Outlet*, the lower and upper edges, respectively *Lower* wall and *Upper* wall, are defined. The cells are *600* along the x-axis and *40* along the y-axis. As far as cell ratios are concerned, a *simpleGrading* is used.

In the *transportProperties* the default parameters are: $\rho = 1000$ kg·m^{-3}, $\sigma = 1$ A·s·m^{-2}, $\eta = 10^3$ kg/m/s, $\mu = 1.26 \cdot 10^{-10}$ $T \cdot m \cdot A^{-1}$. The velocity profile is obtained for several Hartmann numbers, from 0.1 to 100. The correct time-step is set up in the

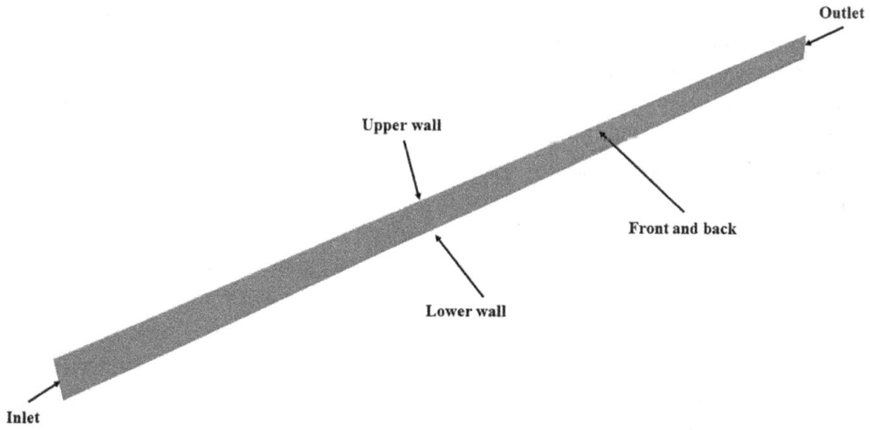

Fig. 3 Grid detail, [1]

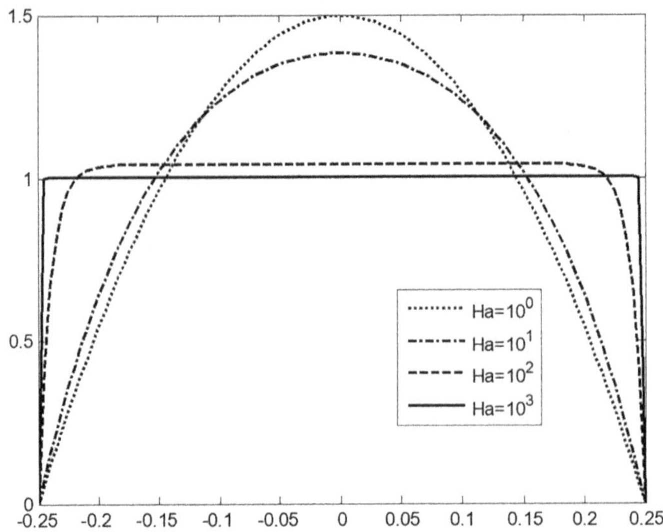

Fig. 4 Velocity profile U/U$_0$ versus y (cm) at several Ha numbers, [1]

controlDict to obtain a Courant number smaller than 1. The parameters to run the simulation are set up in *fvSchemes,* for the whole variables involved in the problems and in *fvSolution*. The initial conditions for velocity, pressure and magnetic field are fixed in the first time directory, 0, as well as the boundary conditions on every surface, previously defined in the *blockMesh* file. The intensity of the magnetic field, perpendicular to the velocity component, is set up in the *B* file to compare numerical and Hartmann solutions.

The first test is the validation of the Hartmann problem for insulating walls. The results are reported in Figs. 4 and 5, respectively, for velocity and magnetic field profiles.

Figure 4 shows that the velocity profile is flat in the middle of the duct at high Hartmann numbers because high-velocity gradients appear near the walls. Two *MHD* boundary layers, called Hartmann layers, are present on the walls, perpendicular to the direction of the applied magnetic field. The thickness is variable with Ha^{-1}. Two secondary boundary layers, called *Side* layers, are present on the walls, parallel to the direction of the applied magnetic field. Their thickness varies with $Ha^{-0.5}$.

Figure 5 presents the magnetic field profiles in *Tesla (T)* for insulating walls versus y (cm) at several Hartmann numbers. The shape of the profile shows that the interaction between the magnetic field and fluid flow induces, in the middle of the duct, an electric field, $\vec{u} \times \vec{B}$, and an electric density current, $\vec{J}$, resulting in a voltage difference between the side layers. The hypothesis of an electrically insulating material determines that any current can leave the geometric domain, and the current paths need to be closed, from the equation of charge conservation,

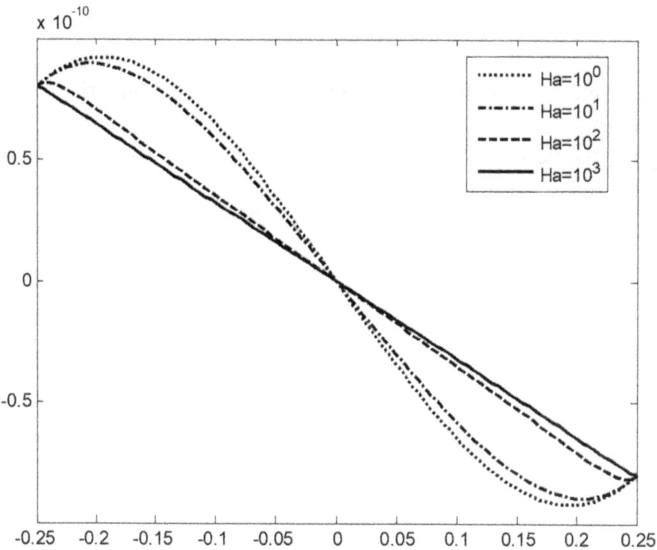

Fig. 5 Magnetic field B_x (T) versus y (cm) at several Ha numbers, [1]

$\nabla \cdot \vec{J}$. The electric current is driven along the wall by a potential gradient in the viscous layer, where the velocity is small. The Lorentz force is specified by the component of the current perpendicular to the magnetic field. In the middle of the duct, this force opposes the flow, reducing the velocity intensity. The momentum balance is established between the Lorentz force and the pressure gradient in the region where the profile is flattened. Velocity drops sharply, near the Hartmann layers, to satisfy the no-slip condition.

2.3.2 Two-Dimensional Duct Flow with Non-uniform Magnetic Induction

The two-dimensional rectangular duct flow, under the influence of a non-uniform magnetic field, is tested in [1]. A magnetic dipole, with a sine function modulator, is generated when a coil electrically conducting, is embedded in a magnetic field. The field generates a torque, allowing the rotation along a specific axis, generally corresponding to the centre of mass of the whole coil.

The torque is

$$\vec{M} = \vec{m} \times \vec{B} \tag{104}$$

where $\vec{B}$ is the vector of the magnetic field, and $\vec{m}$ the magnetic dipole vector

$$\vec{m} = i\,\vec{S} \tag{105}$$

The current intensity, flowing in the coil, is I, and $\vec{S}$ is the orientated surface of the coil. The expression for the intensity of the magnetic field derives from the magnetic potential vector

$$\varphi = \frac{\mu_m}{4\pi}\frac{\vec{m}\cdot\vec{r}}{|\vec{r}|^3} \tag{106}$$

$$\vec{A} = \frac{\mu_m}{4\pi}\frac{\vec{m}\times\vec{r}}{|\vec{r}|^3} = -\frac{\mu_m}{4\pi}iS\frac{\sin\theta}{|\vec{r}|^3}\widehat{\varphi}$$

$$= \frac{\mu_m}{4\pi}iS\frac{\widehat{e}_x(z-z_0)-\widehat{e}_z(x-x_0)}{\left((x-x_0)^2+(y-y_0)^2+(z-z_0)^2\right)^{\frac{3}{2}}} \tag{107}$$

$$\vec{B} = \mathrm{curl}\left(\vec{A}\right) = \frac{\widehat{e}_x 3(x-x_0)(y-y_0)-\widehat{e}_y\left((x-x_0)^2-2(y-y_0)^2+(z-z_0)^2\right)-\widehat{e}_z 3(z-z_0)(y-y_0)}{\left((x-x_0)^2+(y-y_0)^2+(z-z_0)^2\right)^{\frac{5}{2}}} \tag{108}$$

where μ_0 is the magnetic permeability of the vacuum and $\vec{m}$ the magnetic dipole vector.

The grid for the numerical simulation is the same as the *2D* duct with the uniform magnetic field, while the driven force is the pressure gradient along the *x*-axis. A magnetic field, generated by a magnetic dipole, is imposed along the *y*-axis with an intensity modulated with a sine function. A peculiar boundary condition, *groovyBC*, is imposed to be implemented in *OpenFOAM*, which, in its last releases, takes place in *swak4Foam*. The non-uniform magnetic field condition is fixed by *groovyBC* to the lower and upper walls in the time directory *0*, inside the B file. Positive and negative currents, *I*, are analysed. Constant velocity along the *y*-axis and constant pressure drop are used.

Figures 6, 7, 8, 9 and 10 present the results obtained with a positive current, I. Figures 6, 7 and 8 present velocity maps *(m/s)*, showing that the maximum velocity is reached in the middle of the duct, near the lower and upper walls, where the magnetic dipole also acts.

Figure 7 reports the *x* component of the velocity profile, U/U_0, versus y/y_0, which is the fully developed Hagen-Poiseuille type.

Figure 8 shows the *y* component of the velocity profile, V/V_0, versus y/y_0, which follows the sine function present in the magnetic field. In conclusion, the velocity profile is in agreement with the analytical results of the literature.

The results of the magnetic field are shown in Figs. 9 and 10.

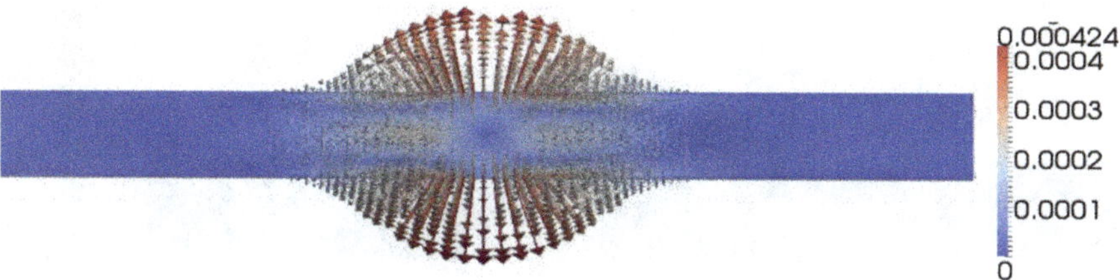

Fig. 6 Velocity magnitude U (m/s) for positive current, [1]

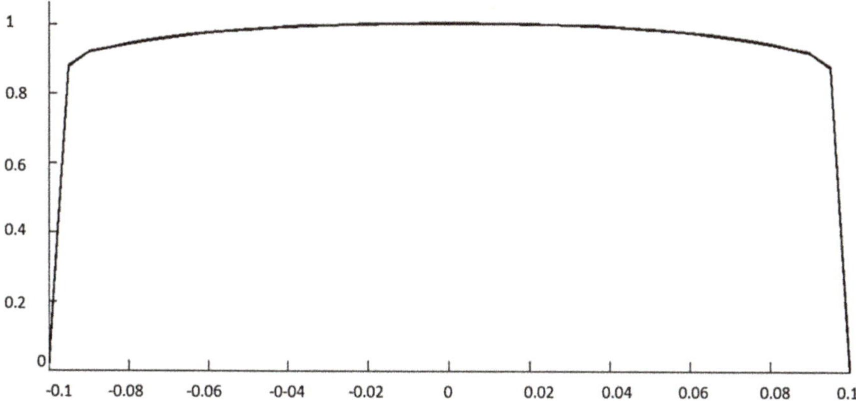

Fig. 7 Velocity profile U/U_0 versus y/y_0 for positive current, [1]

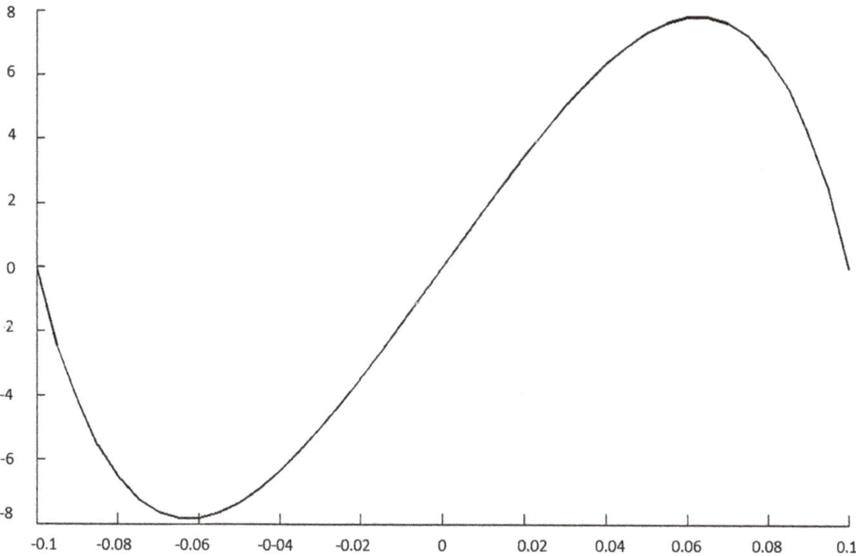

Fig. 8 Velocity profile V/V_0 versus y/y_0 for positive current, [1]

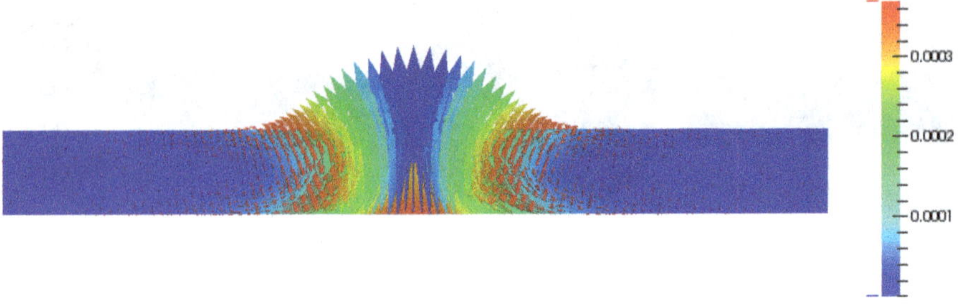

Fig. 9 Magnetic field B (T) for positive current, [1]

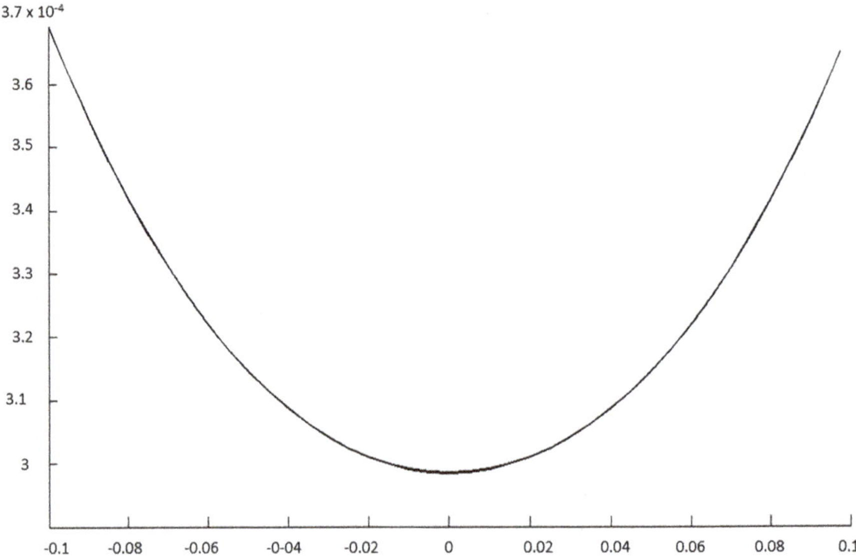

Fig. 10 Magnetic field B_y, (T), versus y/y_0 for positive current, [1]

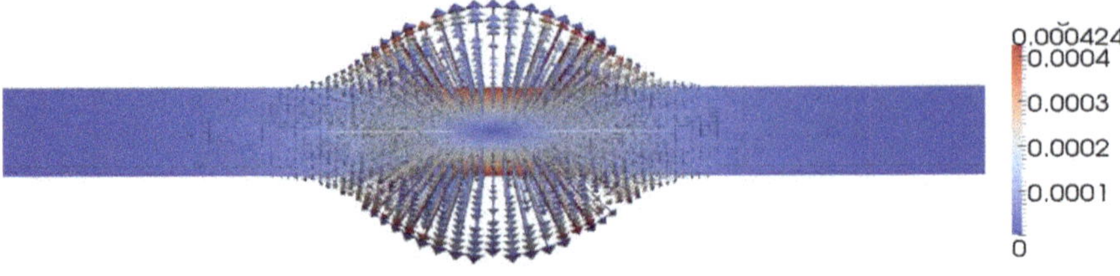

Fig. 11 Velocity magnitude U (m/s) for negative current, [1]

Figure 10 presents a positive parabolic profile for the y component of the magnetic field, which comes from the positive current value used in the numerical simulation.

The results with a negative current intensity I are shown in Figs. 11, 12, 13, 14 and 15.

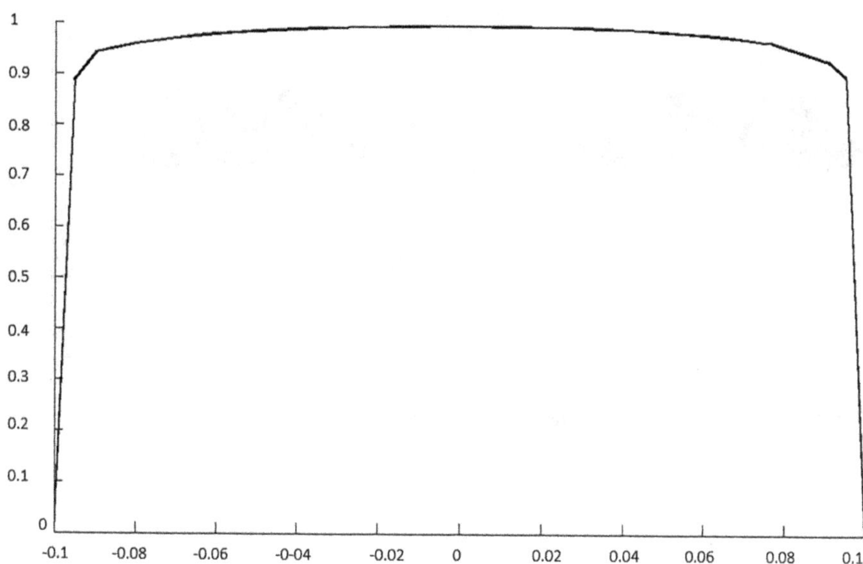

Fig. 12 Velocity U/U_0 versus y/y_0 for negative current, [1]

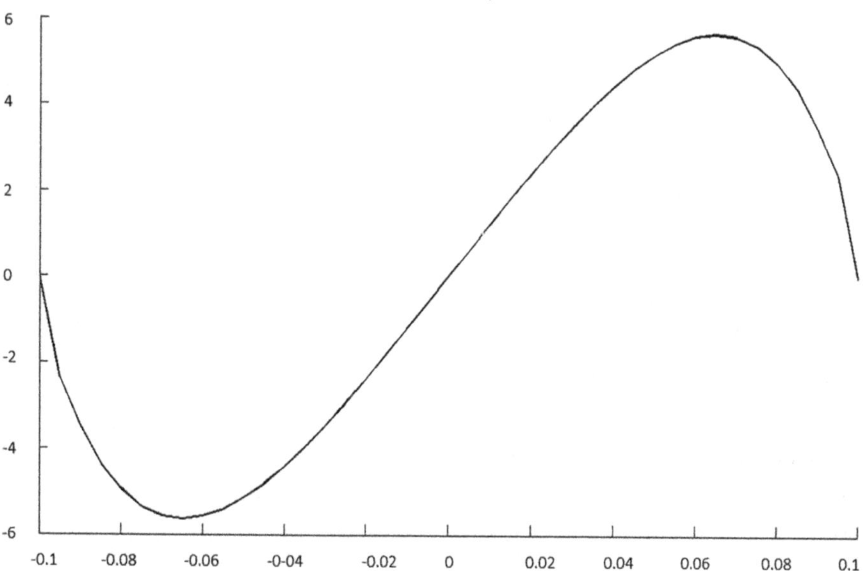

Fig. 13 Velocity V/V_0 versus y/y_0 for negative current, [1]

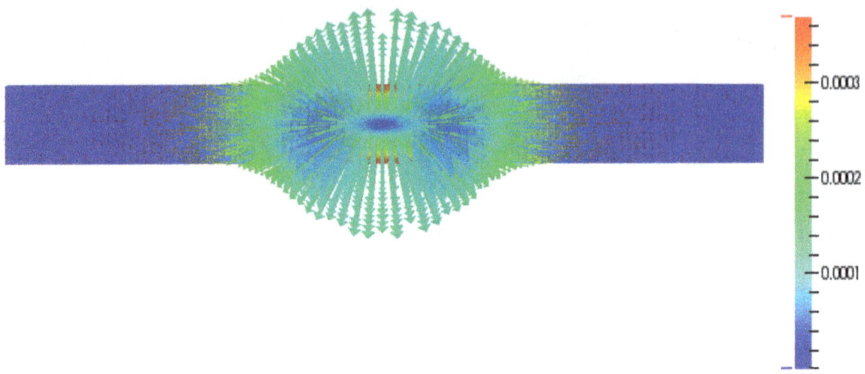

Fig. 14 Magnetic field B (T) for negative current value, [1]

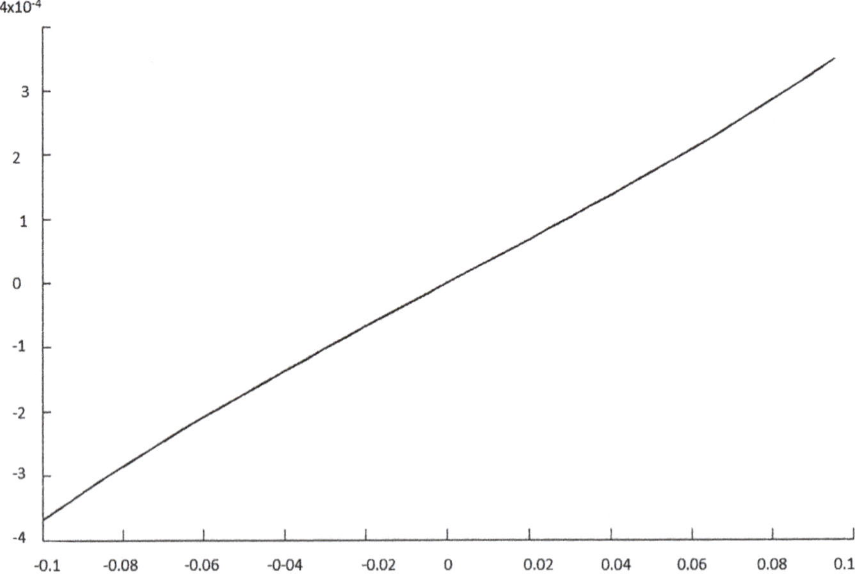

Fig. 15 Magnetic field profile B_y, (T), versus y/y_0 for negative current value, [1]

The velocity maps are reported in Figs. 12 and 13. The maximum velocity is not in the centre of the duct but near the borders of the upper and lower walls because the electric current is negative and the net force is towards the outside. Velocity profiles along x and y axes have the same trend as the case with a positive current.

The results for the magnetic field are presented in Figs. 14 and 15. The profile of the magnetic field shows a symmetric profile, increasing towards the centre of the duct.

2.4   Numerical Simulation of 3D Laminar Steady-State THD–MHD Duct Flow

Several tests are also performed in *3D* geometries, [1], with different types of geometries, such as rectangular ducts and cylinders, different types of walls, such as conducting and insulating, and different types of boundary conditions, such as uniform and non-uniform magnetic fields.

2.4.1   3D Rectangular Duct Flow

The coordinate system in *3D* is made of a horizontal *x*-axis, a vertical *y*-axis, and a span-wise *z*-axis. In the planar case, the derivatives along the *x*-direction are zero, $\frac{\partial}{\partial x} = 0$, as $\frac{\partial u}{\partial x} = 0$. Due to the boundary conditions

$$\left(B_y, B_z\right) = (1, 0)\forall(x, y, z) \in \Omega \tag{109}$$

the system of equations is

$$v = \frac{u}{\left(Re\,\frac{\Delta p}{L}\right)} \tag{110}$$

$$b = \frac{\frac{Ha}{Re_m}B_x}{\left(Re\,\frac{\Delta p}{L}\right)} \tag{111}$$

$$\frac{\partial^2 v}{\partial y^2} + \frac{\partial^2 v}{\partial z^2} + Ha\frac{\partial b}{\partial y} + 1 = 0 \tag{112}$$

$$\frac{\partial^2 b}{\partial y^2} + \frac{\partial^2 b}{\partial z^2} + Ha\frac{\partial v}{\partial y} = 0 \tag{113}$$

The final solution depends on the boundary conditions of the insulating wall, $b = 0$, or the conducting wall, $\frac{\partial b}{\partial n} = 0$.

2.4.2   3D Rectangular Duct Flow with Insulating Walls

Under the assumption of $\vec{B} = 0$ the equations are

$$b(y, z) = \sum_{k=0}^{\infty} b_k(y)\cos(\lambda_k z) \tag{114}$$

$$v(y, z) = \sum_{k=0}^{\infty} v_k(y)\cos(\lambda_k z) \tag{115}$$

$$1 = \sum_{k=0}^{\infty} a_k \cos(\lambda_k z) \tag{116}$$

$$a_k = 2\frac{(-1)^k}{\lambda_k} \tag{117}$$

The y component equations are

$$\begin{cases} \dfrac{\partial^2 v_k(y)}{\partial y^2} - \lambda_k^2 v_k(y) + Ha\dfrac{\partial b_k(y)}{\partial y} + a_k = 0 \\[2ex] \dfrac{\partial^2 b_k(y)}{\partial y^2} - \lambda_k^2 b_k(y) + Ha\dfrac{\partial v_k(y)}{\partial y} = 0 \end{cases} \quad \forall k \tag{118}$$

The decoupling of the equations, with

$$\begin{cases} \eta_k = v_k + b_k \\ \xi_k = v_k - b_k \end{cases} \Rightarrow \begin{cases} v_k = \dfrac{\eta_k + \xi_k}{2} \\[2ex] b_k = \dfrac{\eta_k - \xi_k}{2} \end{cases} \tag{119}$$

leads to

$$\begin{cases} \dfrac{\partial^2 \xi_k(y)}{\partial y^2} - Ha\dfrac{\partial \xi_k(y)}{\partial y} - \lambda_k^2 \xi_k(y) + a_k = 0 \\[2ex] \dfrac{\partial^2 \eta_k(y)}{\partial y^2} + Ha\dfrac{\partial \eta_k(y)}{\partial y} - \lambda_k^2 \eta_k(y) + a_k = 0 \end{cases} \quad \forall k \tag{120}$$

Their integration gives

$$\begin{cases} \xi_k(y) = \dfrac{a_k}{\lambda_k^2} + a_{\xi,k}\, exp(\gamma_{1,k}y) + b_{\xi,k}\, exp(\gamma_{2,k}y) \\[2ex] \eta_k(y) = \dfrac{a_k}{\lambda_k^2} + a_{\eta,k}\, exp(-\gamma_{2,k}y) + b_{\eta,k}\, exp(-\gamma_{1,k}y) \end{cases} \quad \forall k \tag{121}$$

The substitution into Eq. 119 gives

$$\begin{cases} v_k(y) = \dfrac{a_k}{\lambda_k^2} + a_{\xi,k}\, cosh(\gamma_{1,k}y) + b_{\xi,k}\, cosh(\gamma_{2,k}y) \\[2ex] b_k(y) = -a_{\xi,k}\, sinh(\gamma_{1,k}y) + b_{\xi,k}\, sinh(\gamma_{2,k}y) \end{cases} \quad \forall k \tag{122}$$

where the coefficients $a_{\zeta,\kappa}$ and $b_{\zeta,\kappa}$ are

$$\begin{cases} a_{\xi,k} = -\dfrac{a_k}{\lambda_k^2}\dfrac{sinh\left(\gamma_{2,k}\dfrac{L_y}{2L}\right)}{sinh\left(\gamma_{2,k}\dfrac{L_y}{2L}\right)cosh\left(\gamma_{1,k}\dfrac{L_y}{2L}\right) + cosh\left(\gamma_{2,k}\dfrac{L_y}{2L}\right)sinh\left(\gamma_{1,k}\dfrac{L_y}{2L}\right)} \\[4ex] b_{\xi,k} = -\dfrac{a_k}{\lambda_k^2}\dfrac{sinh\left(\gamma_{1,k}\dfrac{L_y}{2L}\right)}{sinh\left(\gamma_{2,k}\dfrac{L_y}{2L}\right)cosh\left(\gamma_{1,k}\dfrac{L_y}{2L}\right) + cosh\left(\gamma_{2,k}\dfrac{L_y}{2L}\right)sinh\left(\gamma_{1,k}\dfrac{L_y}{2L}\right)} \end{cases} \tag{123}$$

$$\begin{cases} v_k(y) = \dfrac{a_k}{\lambda_k^2}\left(1 - \dfrac{\sinh\left(\gamma_{2,k}\frac{L_y}{2L}\right)\cosh(\gamma_{1,k}y) + \sinh\left(\gamma_{1,k}\frac{L_y}{2L}\right)\cosh(\gamma_{2,k}y)}{\sinh\left(\gamma_{2,k}\frac{L_y}{2L}\right)\cosh\left(\gamma_{1,k}\frac{L_y}{2L}\right) + \cosh\left(\gamma_{2,k}\frac{L_y}{2L}\right)\sinh\left(\gamma_{1,k}\frac{L_y}{2L}\right)}\right) \\[3em] b_k(y) = \dfrac{a_k}{\lambda_k^2}\dfrac{\sinh\left(\gamma_{2,k}\frac{L_y}{2L}\right)\sinh(\gamma_{1,k}y) - \sinh\left(\gamma_{1,k}\frac{L_y}{2L}\right)\sinh(\gamma_{2,k}y)}{\sinh\left(\gamma_{2,k}\frac{L_y}{2L}\right)\cosh\left(\gamma_{1,k}\frac{L_y}{2L}\right) + \cosh\left(\gamma_{2,k}\frac{L_y}{2L}\right)\sinh\left(\gamma_{1,k}\frac{L_y}{2L}\right)} \end{cases} \forall k \tag{124}$$

In conclusion, the equations for velocity, magnetic field and current density for an insulating wall are

$$u(y,z) = \dfrac{\sum_{k=0}^{\infty}\frac{a_k}{\lambda_k^2}\cos(\lambda_k z)\left(1 - \frac{\sinh\left(\gamma_{2,k}\frac{L_y}{2L}\right)\cosh(\gamma_{1,k}y) + \sinh\left(\gamma_{1,k}\frac{L_y}{2L}\right)\cosh(\gamma_{2,k}y)}{\sinh\left(\gamma_{2,k}\frac{L_y}{2L}\right)\cosh\left(\gamma_{1,k}\frac{L_y}{2L}\right) + \cosh\left(\gamma_{2,k}\frac{L_y}{2L}\right)\sinh\left(\gamma_{1,k}\frac{L_y}{2L}\right)}\right)}{\left(\frac{2D_h}{L_z}\right)\sum_{k=0}^{\infty}\frac{2}{\lambda_k^4}\left(1 - \frac{\left(\frac{2D_h}{L_y}\right)\frac{(\gamma_{1,k}+\gamma_{2,k})}{\gamma_{2,k}\gamma_{1,k}}\tanh\left(\gamma_{1,k}\frac{L_y}{2L}\right)\tanh\left(\gamma_{2,k}\frac{L_y}{2L}\right)}{\tanh\left(\gamma_{1,k}\frac{L_y}{2L}\right) + \tanh\left(\gamma_{2,k}\frac{L_y}{2L}\right)}\right)} \tag{125}$$

$$b(y,z) = \dfrac{Re_m}{Ha}\dfrac{\sum_{k=0}^{\infty}\frac{a_k}{\lambda_k^2}\cos(\lambda_k z)\left(\frac{\sinh\left(\gamma_{2,k}\frac{L_y}{2L}\right)\sinh(\gamma_{1,k}y) - \sinh\left(\gamma_{1,k}\frac{L_y}{2L}\right)\sinh(\gamma_{2,k}y)}{\sinh\left(\gamma_{1,k}\frac{L_y}{2L}\right)\cosh\left(\gamma_{2,k}\frac{L_y}{2L}\right) + \sinh\left(\gamma_{2,k}\frac{L_y}{2L}\right)\cosh\left(\gamma_{1,k}\frac{L_y}{2L}\right)}\right)}{\left(\frac{2D_h}{L_z}\right)\sum_{k=0}^{\infty}\frac{2}{\lambda_k^4}\left(1 - \frac{\left(\frac{2D_h}{L_y}\right)\frac{(\gamma_{1,k}+\gamma_{2,k})}{\gamma_{2,k}\gamma_{1,k}}\tanh\left(\gamma_{1,k}\frac{L_y}{2L}\right)\tanh\left(\gamma_{2,k}\frac{L_y}{2L}\right)}{\tanh\left(\gamma_{1,k}\frac{L_y}{2L}\right) + \tanh\left(\gamma_{2,k}\frac{L_y}{2L}\right)}\right)} \tag{126}$$

The current density components

$$j_y = \frac{Ha^2}{Re\left(\frac{1}{Re_m}\frac{\partial b}{\partial z}\right)}; j_z = -\frac{Ha^2}{Re\left(\frac{1}{Re_m}\frac{\partial b}{\partial y}\right)} \tag{127}$$

$$j_y(y,z) = -\dfrac{Ha}{Re\dfrac{\sum_{k=0}^{\infty}\frac{a_k}{\lambda_k}\sin(\lambda_k z)\left(\frac{\sinh\left(\gamma_{2,k}\frac{L_y}{2L}\right)\sinh(\gamma_{1,k}y) - \sinh\left(\gamma_{1,k}\frac{L_y}{2L}\right)\sinh(\gamma_{2,k}y)}{\sinh\left(\gamma_{1,k}\frac{L_y}{2L}\right)\cosh\left(\gamma_{2,k}\frac{L_y}{2L}\right) + \sinh\left(\gamma_{2,k}\frac{L_y}{2L}\right)\cosh\left(\gamma_{1,k}\frac{L_y}{2L}\right)}\right)}{\left(\frac{2D_h}{L_z}\right)\sum_{k=0}^{\infty}\frac{2}{\lambda_k^4}\left(1 - \frac{\left(\frac{2D_h}{L_y}\right)\frac{(\gamma_{1,k}+\gamma_{2,k})}{\gamma_{2,k}\gamma_{1,k}}\tanh\left(\gamma_{1,k}\frac{L_y}{2L}\right)\tanh\left(\gamma_{2,k}\frac{L_y}{2L}\right)}{\tanh\left(\gamma_{1,k}\frac{L_y}{2L}\right) + \tanh\left(\gamma_{2,k}\frac{L_y}{2L}\right)}\right)}} \tag{128}$$

$$j_z(y,z) = -\dfrac{Ha}{Re\dfrac{\sum_{k=0}^{\infty}\frac{a_k}{\lambda_k^2}\cos(\lambda_k z)\left(\frac{\gamma_{1,k}\sinh\left(\gamma_{2,k}\frac{L_y}{2L}\right)\cosh(\gamma_{1,k}y) - \gamma_{2,k}\sinh\left(\gamma_{1,k}\frac{L_y}{2L}\right)\cosh(\gamma_{2,k}y)}{\sinh\left(\gamma_{1,k}\frac{L_y}{2L}\right)\cosh\left(\gamma_{2,k}\frac{L_y}{2L}\right) + \sinh\left(\gamma_{2,k}\frac{L_y}{2L}\right)\cosh\left(\gamma_{1,k}\frac{L_y}{2L}\right)}\right)}{\left(\frac{2D_h}{L_z}\right)\sum_{k=0}^{\infty}\frac{2}{\lambda_k^4}\left(1 - \frac{\left(\frac{2D_h}{L_y}\right)\frac{(\gamma_{1,k}+\gamma_{2,k})}{\gamma_{2,k}\gamma_{1,k}}\tanh\left(\gamma_{1,k}\frac{L_y}{2L}\right)\tanh\left(\gamma_{2,k}\frac{L_y}{2L}\right)}{\tanh\left(\gamma_{1,k}\frac{L_y}{2L}\right) + \tanh\left(\gamma_{2,k}\frac{L_y}{2L}\right)}\right)}} \tag{129}$$

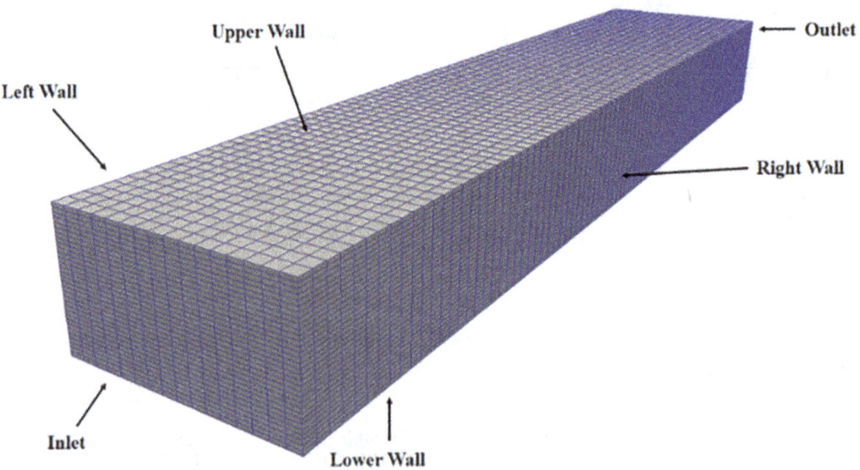

Fig. 16 3D geometry, [1]

All the patches are fixed in the *blockMesh* file, naming and setting the grading for each axis and the number of elements for each edge.

Figure 16 depicts the resulting mesh. The parameters are set in the *transportProperties* file, leaving the default values for all of them, as well as for the system directory, *fvSchemes* and *fvSolutions*. A runtime equal to *4 s* is set in the *controlDict* file to have a fully developed velocity profile. The y component, in the upper and lower wall patches, is fixed in the *0* time directory of the B file, depending on its value on the Hartmann number, from *1* to *100*. Moreover, the values of B_y are considered as initial conditions since the fluid is immersed in the magnetic field, achieving a perfect insulating case. A constant pressure gradient is set up in the p file, while, in the U file, a constant value along the x-axis is fixed on the inlet patch. The U value is chosen to have a laminar flow with a Reynolds number equal to *100*.

The results are shown in Figs. 17, 18, 19, 20, 21, 22, 23, 24 and 25. Figures 17 and 18 show the velocity profiles at several Hartmann numbers for insulating walls along the y (cm) (Fig. 17) and z *(cm)* (Fig. 18) axes. As far as velocity is concerned, the profile for low Hartmann numbers follows the Poiseuille flow, as obtained in the *2D* case, whereas, for high Hartmann numbers, it becomes flat in the middle of the duct.

The profiles of the magnetic field (T) are shown in Figs. 19 and 20 at several Hartmann numbers for insulating walls along the y (cm) (Fig. 19) and z (cm) (Fig. 20) axes. They present a decrease in amplitude for high Hartmann numbers, whereas the amplitude is higher for low Hartmann numbers.

The y component profile of the density current, j_y (A/m^2), is shown in Figs. 21 and 22, along the y (cm) and z (cm) directions, respectively. The trend is symmetric with respect to the origin and alongside the vertical and span-wise directions. Moreover, it is possible to notice that the profile has a sinusoidal tendency with the increase of the Hartmann number.

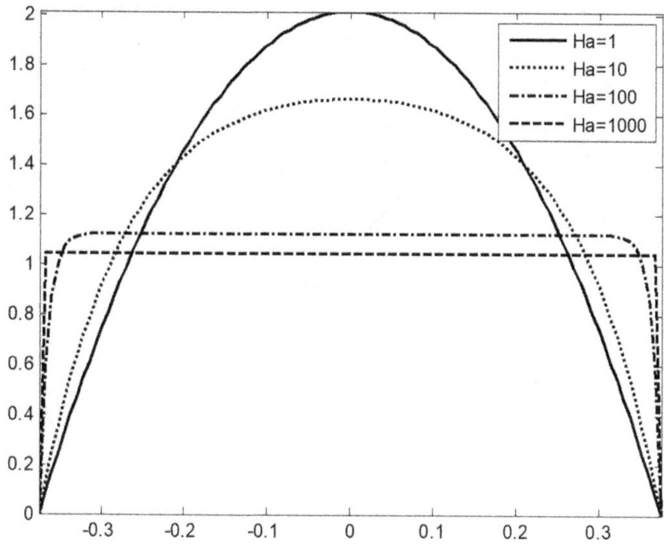

Fig. 17 Velocity profile U/U_0 versus y (cm) at several Ha numbers, [1]

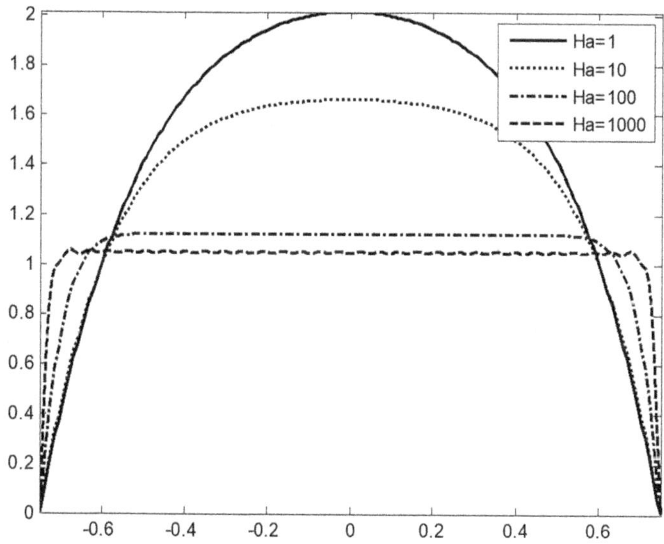

Fig. 18 Velocity profile U/U_0 versus z (cm) at several Ha numbers, [1]

Figures 23 and 24 present the z-component of the current density, j_z (A/m^2), along the y (cm) and z (cm) directions, respectively. Figure 23 shows a constant behaviour alongside the vertical direction, y, for every Hartmann number. Only the magnitude of the vector changes as the Hartmann number increases, while the trend is parabolic for increasing Hartmann number, Fig. 24.

F. Russo et al.

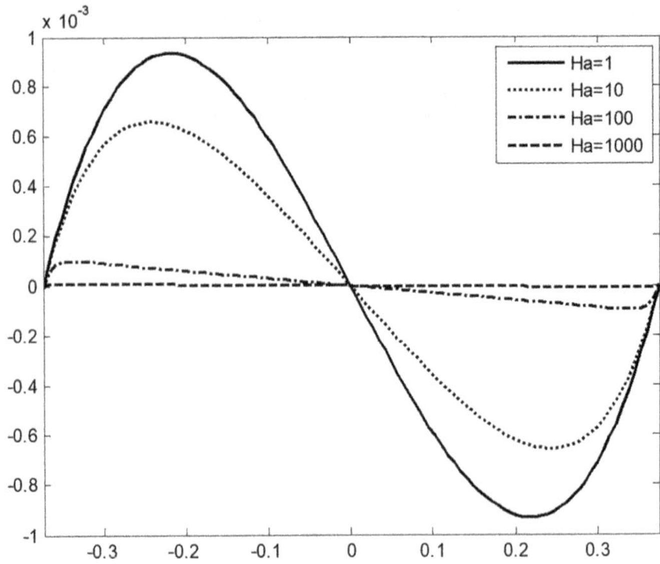

Fig. 19 Magnetic field profile B (T) versus y (cm) for insulating walls, [1]

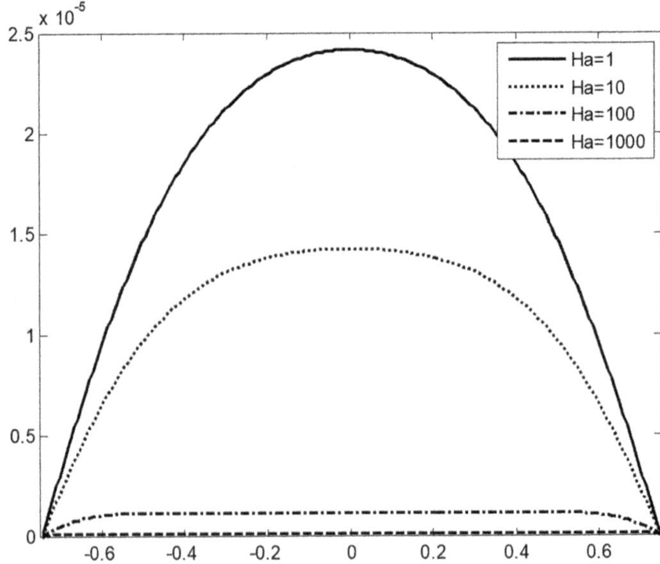

Fig. 20 Magnetic field profile B, (T), versus z, (cm), for insulating walls, [1]

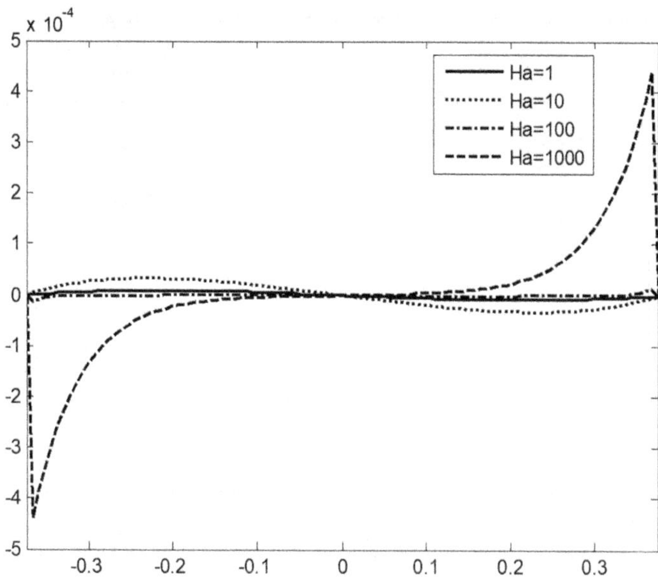

Fig. 21 Current density, j_y (A/m^2), versus y (cm), [1]

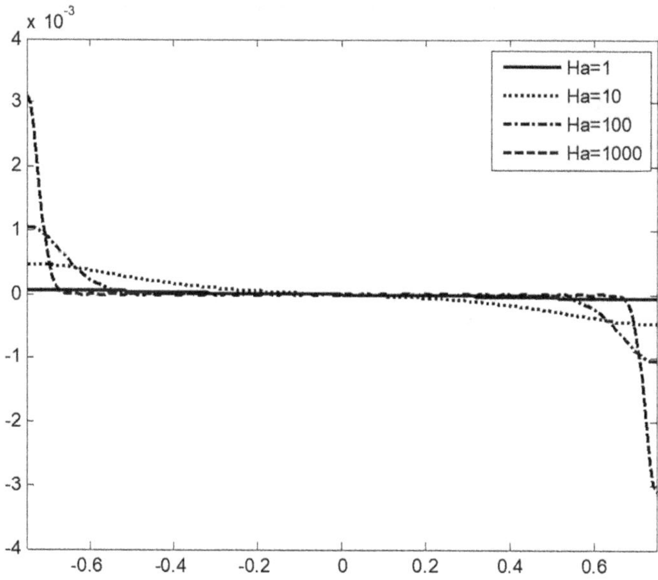

Fig. 22 Current density, j_y (A/m^2), versus z (cm), [1]

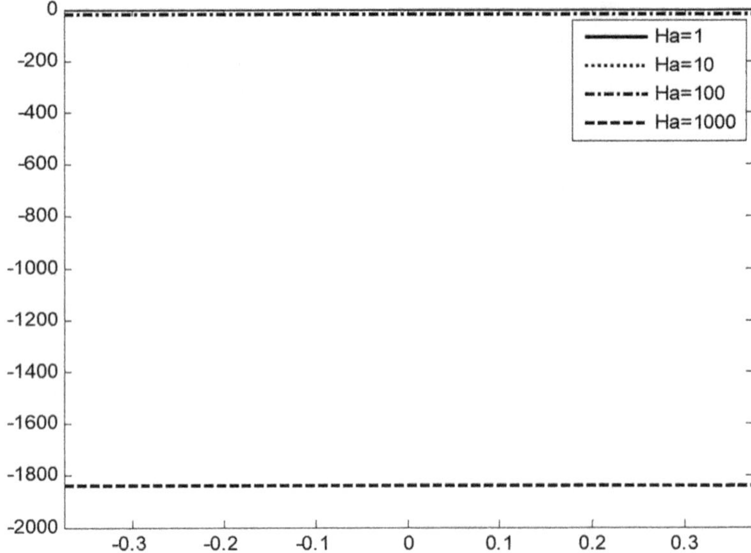

Fig. 23 Current density, j_z (A/m^2), versus z (cm), [1]

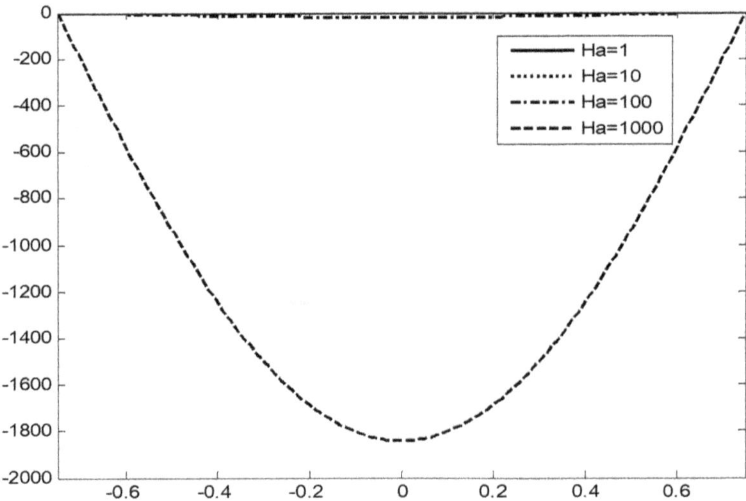

Fig. 24 Current density, j_z (A/m^2), versus z (cm), [1]

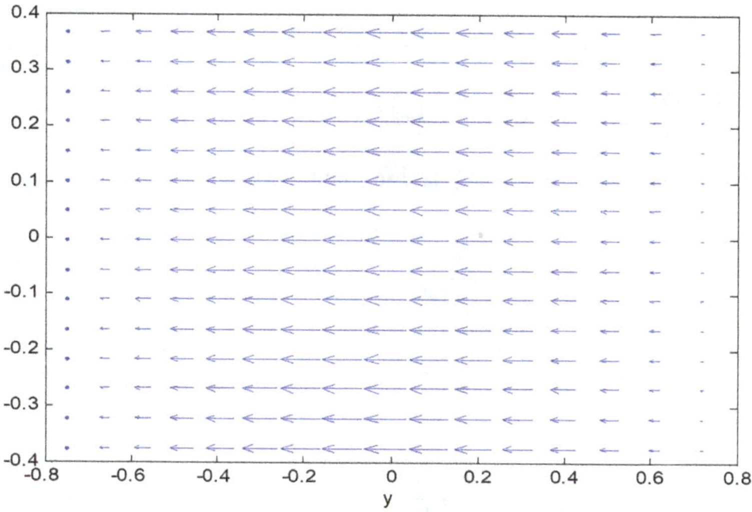

Fig. 25 Current density, $\vec{J}$ (A/m²), in the plane y-z (cm-cm), [1]

Figure 25 shows an enlargement of the vector $\vec{J}$ of the density current in the plane y-z.

2.4.3   3D Rectangular Duct Flow with Conducting Walls

The three-dimensional rectangular duct, with conducting walls, is studied by changing, in the setup, the boundary condition for the B field, assuming $\frac{\partial b}{\partial n} = 0$, [1], are:

$$b(y,z) = \sum_{i=0}^{\infty} b_i(z) \, sin(\alpha_i y) \tag{130}$$

$$v(y,z) = \sum_{k=0}^{\infty} v_i(z) \, cos(\alpha_i y) \tag{131}$$

$$1 = \sum_{i=0}^{\infty} a_i \, cos(\alpha_i y) \tag{132}$$

$$a_i = (2i+1)\frac{\pi}{2}\frac{2L}{L_y} \tag{133}$$

The equations along the z-axis are

$$\begin{cases} \dfrac{d^2 v_i(z)}{dz^2} - \alpha_i^2 v_i(z) + Ha\alpha_i b_i(z) + a_i = 0 \\[2mm] \dfrac{d^2 b_i(z)}{dz^2} - \alpha_i^2 b_i(z) - Ha\alpha_i v_i(z) = 0 \end{cases} \forall i \tag{134}$$

which are solved by applying a second derivative

$$\begin{cases} \dfrac{d^4 v_i(z)}{dz^4} - 2\alpha_i^2 \dfrac{d^2 v_i(z)}{dz^2} + \alpha_i^2\left(\alpha_i^2 + Ha^2\right)v_i(z) - \alpha_i^2 a_i = 0 \\[2mm] \dfrac{d^4 b_i(z)}{dz^4} - 2\alpha_i^2 \dfrac{d^2 b_i(z)}{dz^2} + \alpha_i^2\left(\alpha_i^2 + Ha^2\right)b_i(z) + a_i\alpha_i Ha = 0 \end{cases} \qquad \forall i \qquad (135)$$

The final equations, for velocity and magnetic fields, are

$$v_i(z) = \frac{a_i}{\left(\alpha_i^2 + Ha^2\right)} + - \frac{a_i}{\left(\alpha_i^2 + Ha^2\right)} \frac{\left(p_i \sinh\left(p_i\frac{L_z}{2L}\right)\cos\left(q_i\frac{L_z}{2L}\right) - q_i \cosh\left(p_i\frac{L_z}{2L}\right)\sin\left(q_i\frac{L_z}{2L}\right)\right)}{\left(p_i \cosh\left(p_i\frac{L_z}{2L}\right)\sinh\left(p_i\frac{L_z}{2L}\right) - q_i \cos\left(q_i\frac{L_z}{2L}\right)\sin\left(q_i\frac{L_z}{2L}\right)\right)} \cosh(p_i z)\cos(q_i z)$$

$$+ - \frac{a_i}{\left(\alpha_i^2 + Ha^2\right)} \frac{\left(p_i \cosh\left(p_i\frac{L_z}{2L}\right)\sin\left(q_i\frac{L_z}{2L}\right) + q_i \sinh\left(p_i\frac{L_z}{2L}\right)\cos\left(q_i\frac{L_z}{2L}\right)\right)}{\left(p_i \cosh\left(p_i\frac{L_z}{2L}\right)\sinh\left(p_i\frac{L_z}{2L}\right) - q_i \cos\left(q_i\frac{L_z}{2L}\right)\sin\left(q_i\frac{L_z}{2L}\right)\right)} \sinh(p_i z)\sin(q_i z)$$

$$(136)$$

$$b_i(z) = - \frac{a_i Ha}{\alpha_i\left(\alpha_i^2 + Ha^2\right)} + \frac{a_i}{\left(\alpha_i^2 + Ha^2\right)} \frac{\left(p_i \cosh\left(p_i\frac{L_z}{2L}\right)\sin\left(q_i\frac{L_z}{2L}\right) + q_i \sinh\left(p_i\frac{L_z}{2L}\right)\cos\left(q_i\frac{L_z}{2L}\right)\right)}{\left(p_i \cosh\left(p_i\frac{L_z}{2L}\right)\sinh\left(p_i\frac{L_z}{2L}\right) - q_i \cos\left(q_i\frac{L_z}{2L}\right)\sin\left(q_i\frac{L_z}{2L}\right)\right)} \cosh(p_i z)\cos(q_i z)$$

$$+ - \frac{a_i}{\left(\alpha_i^2 + Ha^2\right)} \frac{\left(p_i \sinh\left(p_i\frac{L_z}{2L}\right)\cos\left(q_i\frac{L_z}{2L}\right) - q_i \cosh\left(p_i\frac{L_z}{2L}\right)\sin\left(q_i\frac{L_z}{2L}\right)\right)}{\left(p_i \cosh\left(p_i\frac{L_z}{2L}\right)\sinh\left(p_i\frac{L_z}{2L}\right) - q_i \cos\left(q_i\frac{L_z}{2L}\right)\sin\left(q_i\frac{L_z}{2L}\right)\right)} \sinh(p_i z)\sin(q_i z)$$

$$(137)$$

The y component, along the upper and the lower walls, is set as *zeroGradient*, and the internal field is initialised with the scalar value, which depends on the Hartmann number, 1–100. The *zeroGradient* condition states that the variable is fully developed in the patches where it is set up.

The results of the velocity profiles for conducting walls are shown along the z-axis (cm) in Fig. 26 and along the y-axis (cm) in Fig. 27 at several Hartmann numbers. A Hagen-Poiseuille flow profile is shown for low Hartmann numbers, as for the insulating walls case, while the profile is called an *M-profile* for high Hartmann numbers. This characteristic profile is present because high-velocity jets appear near the walls, i.e. parallel to the magnetic field. The jet formation occurs because of the high flow-opposing vertical Lorentz force in the bulk, while no force appears near the parallel walls. Moreover, the *M-shaped* profile has inflection points, and therefore, under certain conditions, the flow becomes unstable.

The magnetic field profile, reported in Fig. 28, presents a less significant decrease, compared to the insulating case, when the Hartmann number increases. The continuous line is for $Ha = 1$, the dashed-dot one for $Ha = 10$ and the dashed one for $Ha = 100$.

2.4.4   3D Cylinder Flow with Uniform Magnetic Field

The three-dimensional cylinder flow, with conducting and insulating walls, is investigated under the condition of uniform and non-uniform magnetic fields. The equations for velocity and magnetic fields are

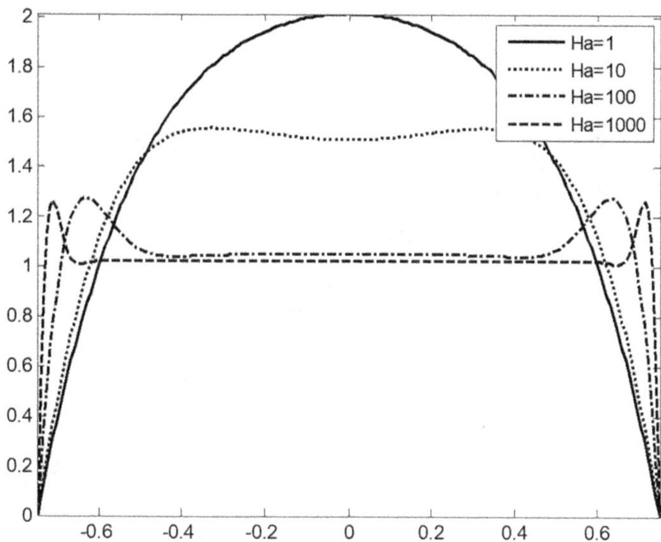

Fig. 26 Velocity profile, U/U_0, versus z (cm), [1]

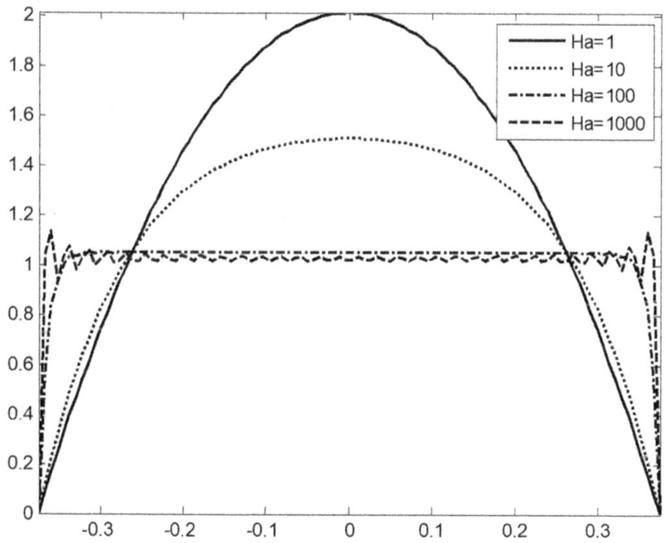

Fig. 27 Velocity profile, U/U_0, versus y (cm), [1]

$$\frac{1}{r}\frac{\partial(rv_r)}{\partial r} + \frac{1}{r}\frac{\partial v_\theta}{\partial \theta} + \frac{\partial v_z}{\partial z} = 0 \tag{138}$$

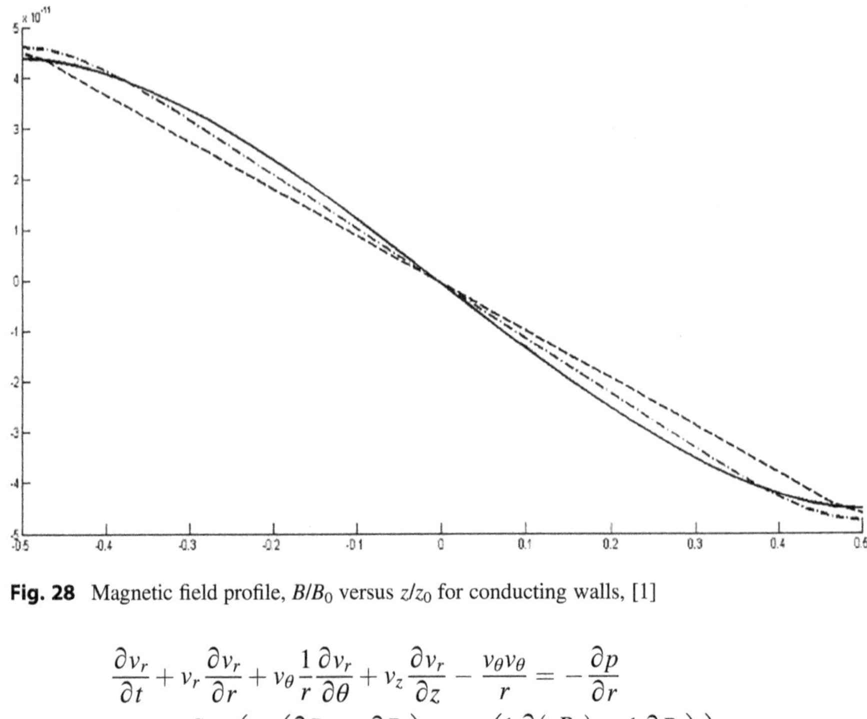

Fig. 28 Magnetic field profile, B/B_0 versus z/z_0 for conducting walls, [1]

$$\frac{\partial v_r}{\partial t} + v_r \frac{\partial v_r}{\partial r} + v_\theta \frac{1}{r}\frac{\partial v_r}{\partial \theta} + v_z \frac{\partial v_r}{\partial z} - \frac{v_\theta v_\theta}{r} = -\frac{\partial p}{\partial r}$$
$$+ \frac{St}{Re_m}\left(B_z\left(\frac{\partial B_r}{\partial z} - \frac{\partial B_z}{\partial r}\right) - B_\theta\left(\frac{1}{r}\frac{\partial (rB_\theta)}{\partial r} - \frac{1}{r}\frac{\partial B_r}{\partial \theta}\right)\right) \tag{139}$$
$$+ \frac{1}{Re}\left(\frac{\partial^2 v_r}{\partial r^2} + \frac{1}{r}\frac{\partial v_r}{\partial r} - \frac{v_r}{r^2} + \frac{1}{r^2}\frac{\partial^2 v_r}{\partial \theta^2} + \frac{\partial^2 v_r}{\partial z^2} - \frac{2}{r^2}\frac{\partial v_\theta}{\partial \theta}\right)$$

$$\frac{\partial v_\theta}{\partial t} + v_r \frac{\partial v_\theta}{\partial r} + v_\theta \frac{1}{r}\frac{\partial v_\theta}{\partial \theta} + v_z \frac{\partial v_\theta}{\partial z} + \frac{v_\theta v_r}{r} = -\frac{1}{r}\frac{\partial p}{\partial \theta}$$
$$+ \frac{St}{Re_m}\left(B_r\left(\frac{1}{r}\frac{\partial (rB_\theta)}{\partial r} - \frac{1}{r}\frac{\partial B_r}{\partial \theta}\right) - B_z\left(\frac{1}{r}\frac{\partial B_z}{\partial \theta} - \frac{\partial B_\theta}{\partial z}\right)\right) \tag{140}$$
$$+ \frac{1}{Re}\left(\frac{\partial^2 v_\theta}{\partial r^2} + \frac{1}{r}\frac{\partial v_\theta}{\partial r} - \frac{v_\theta}{r^2} + \frac{1}{r^2}\frac{\partial^2 v_\theta}{\partial \theta^2} + \frac{\partial^2 v_\theta}{\partial z^2} + \frac{2}{r^2}\frac{\partial v_r}{\partial \theta}\right)$$

$$\frac{\partial v_z}{\partial t} + v_r \frac{\partial v_z}{\partial r} + v_\theta \frac{1}{r}\frac{\partial v_z}{\partial \theta} + v_z \frac{\partial v_z}{\partial z} = -\frac{\partial p}{\partial z} + \frac{St}{Re_m}\left(B_\theta\left(\frac{1}{r}\frac{\partial B_z}{\partial \theta} - \frac{\partial B_\theta}{\partial z}\right) - B_r\left(\frac{\partial B_r}{\partial z} - \frac{\partial B_z}{\partial r}\right)\right)$$
$$+ \frac{1}{Re}\left(\frac{\partial^2 v_z}{\partial r^2} + \frac{1}{r}\frac{\partial v_z}{\partial r} + \frac{1}{r^2}\frac{\partial^2 v_\theta}{\partial \theta^2} + \frac{\partial^2 v_z}{\partial z^2}\right)$$

$$\tag{141}$$

$$\frac{\partial B_r}{\partial t} + v_r \frac{\partial B_r}{\partial r} + v_\theta \frac{1}{r}\frac{\partial B_r}{\partial \theta} + v_z \frac{\partial B_r}{\partial z} = B_r \frac{\partial v_r}{\partial r} + B_\theta \frac{1}{r}\frac{\partial v_r}{\partial \theta} + B_z \frac{\partial v_r}{\partial z} +$$
$$\frac{1}{Re_m}\left(\frac{\partial^2 B_r}{\partial r^2} + \frac{1}{r}\frac{\partial B_r}{\partial r} - \frac{B_r}{r^2} + \frac{1}{r^2}\frac{\partial^2 B_r}{\partial \theta^2} + \frac{\partial^2 B_r}{\partial z^2} - \frac{2}{r^2}\frac{\partial B_\theta}{\partial \theta}\right)$$
$$(142)$$

$$\frac{\partial B_\theta}{\partial t} + v_r \frac{\partial B_\theta}{\partial r} + v_\theta \frac{1}{r}\frac{\partial B_\theta}{\partial \theta} + v_z \frac{\partial B_\theta}{\partial z} - \frac{v_r B_\theta}{r} = B_r \frac{\partial v_\theta}{\partial r} + B_\theta \frac{1}{r}\frac{\partial v_\theta}{\partial \theta} - \frac{v_\theta B_\theta}{r}$$
$$+ B_z \frac{\partial v_\theta}{\partial z} + \frac{1}{Re_m}\left(\frac{\partial^2 B_\theta}{\partial r^2} + \frac{1}{r}\frac{\partial B_\theta}{\partial r} - \frac{B_\theta}{r^2} + \frac{1}{r^2}\frac{\partial^2 B_\theta}{\partial \theta^2} + \frac{\partial^2 B_\theta}{\partial z^2} + \frac{2}{r^2}\frac{\partial B_r}{\partial \theta}\right)$$
$$(143)$$

$$\frac{\partial B_z}{\partial t} + v_r \frac{\partial B_z}{\partial r} + v_\theta \frac{1}{r}\frac{\partial B_z}{\partial \theta} + v_z \frac{\partial B_z}{\partial z} = B_r \frac{\partial v_z}{\partial r} + B_\theta \frac{1}{r}\frac{\partial v_z}{\partial \theta} + B_z \frac{\partial v_z}{\partial z} +$$
$$\frac{1}{Re_m}\left(\frac{\partial^2 B_z}{\partial r^2} + \frac{1}{r}\frac{\partial B_z}{\partial r} + \frac{1}{r^2}\frac{\partial^2 B_z}{\partial \theta^2} + \frac{\partial^2 B_z}{\partial z^2}\right)$$
$$(144)$$

The geometry of the cylinder is created with the *O-grid,* which, in the *blockMesh* file, has a central square and outer circumference elements, as shown in Fig. 29.

The detail of the *O-grid* is reported in Fig. 30. Three patches are named *Inlet, Outlet* and *Wall.* The mesh quality is checked by the skewness of the grid, an indicator of its suitability. Generally, a large skewness compromises the accuracy of the interpolated regions. In a mesh with hexahedron and tetrahedron cells, the skewness should not exceed 0.85, and, in the present simulations, the maximum skewness is 0.75.

The applied magnetic field is set, as usual, along the *y*-axis in the case of a cylinder with an insulating wall and a uniform magnetic field. The runtime is set to have a fully developed velocity profile. The magnetic field intensity, shown in Fig. 31 for insulating walls, decreases with the increase of the Hartmann number. The continuous line is for $Ha = 1$, the dashed-dot one for $Ha = 10$ and the dashed one for $Ha = 100$.

The velocity profiles for insulating walls are presented in Fig. 32 versus x/x_0. The velocity profiles are in agreement with the literature, i.e. a Hagen-Poiseuille flow for low Hartmann numbers and a flat profile for high Hartmann numbers. The continuous line is for $Ha = 1$, the dashed-dot one for $Ha = 10$ and the dashed one for $Ha = 100$.

Figure 33 shows the velocity profile for the conducting wall. It has two peaks near the wall, and the profile has the typical *M-shape.* The continuous line is for $Ha = 1$, the dashed-dot one for $Ha = 20$ and the dashed one for $Ha = 100$.

The magnetic field intensity for the conducting wall is shown in Fig. 34. The continuous line is for $Ha = 1$, the dashed-dot one for $Ha = 20$ and the dashed one for $Ha = 100$. Its profile does not change significantly for Ha greater than 20.

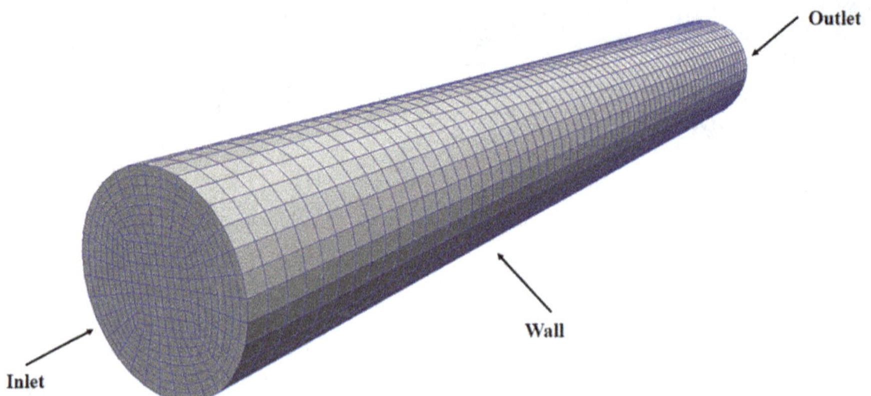

Fig. 29 3D cylinder geometry grid, [1]

Fig. 30 O-grid detail, [1]

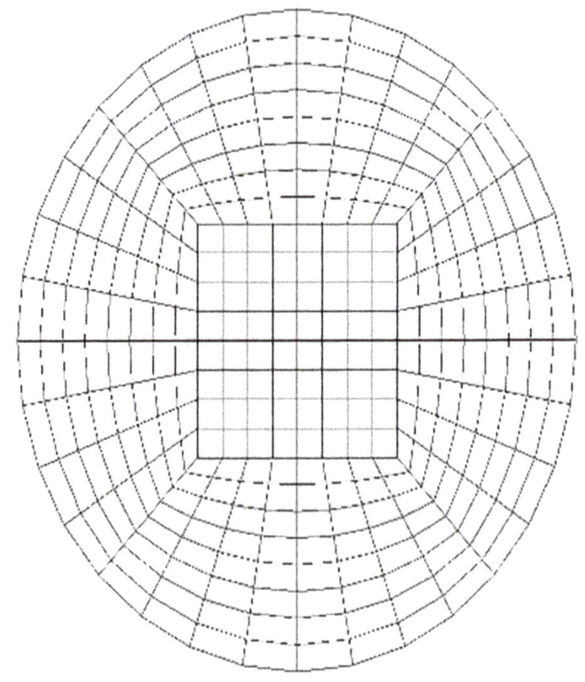

2.4.5   3D Cylinder Flow with Non-uniform Magnetic Field

The numerical simulations, in the case of a cylinder with a non-uniform magnetic field, use a magnetic dipole with a negative intensity by setting up a *groovyBC* condition for the magnetic dipole, [1].

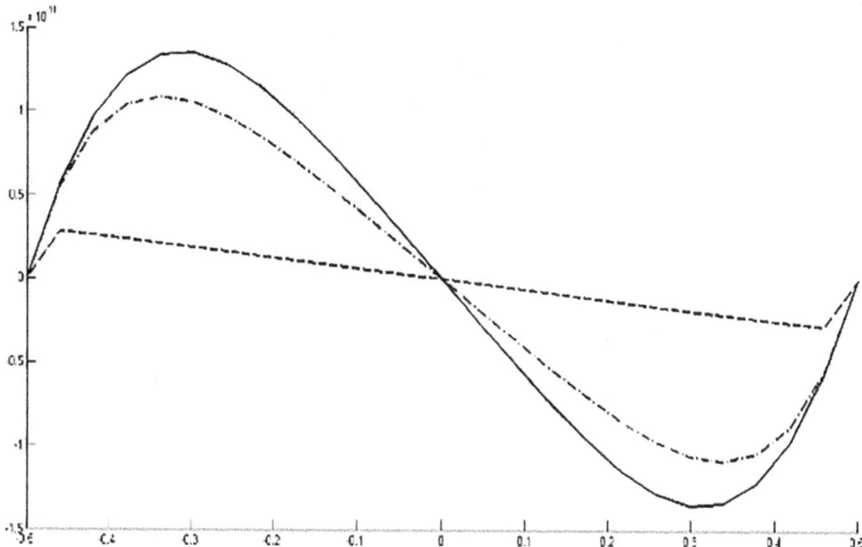

Fig. 31 Magnetic field profile, B/B_0, versus y/y_0 for insulating walls, [1]

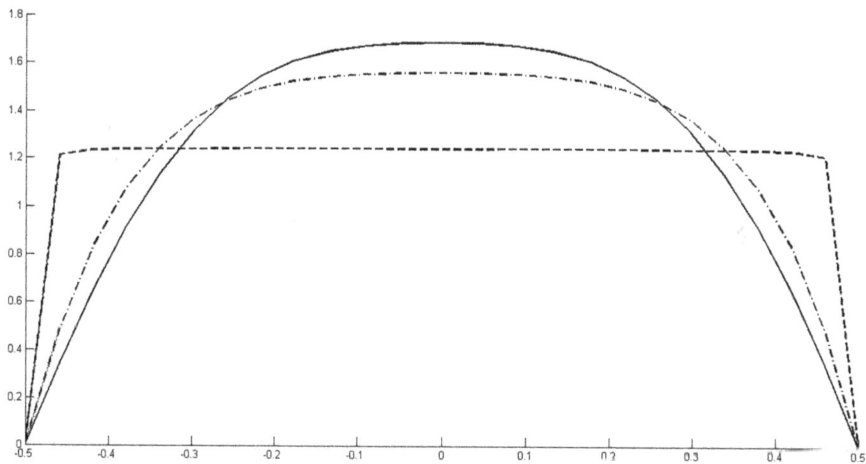

Fig. 32 Velocity profile, U/U_0, versus x/x_0 for insulating walls, [1]

Figures 35, 36, 37 and 38 present the velocity profile (*m/s*) and magnetic field (*T*) for the positive current value *I* along the span-wise *z* (cm) and vertical directions *y* (cm). Velocity, zero at the wall, varies as a sine function. The magnetic field has the maximum value in the middle of the duct, where the probe is present, while near the wall, its intensity is zero.

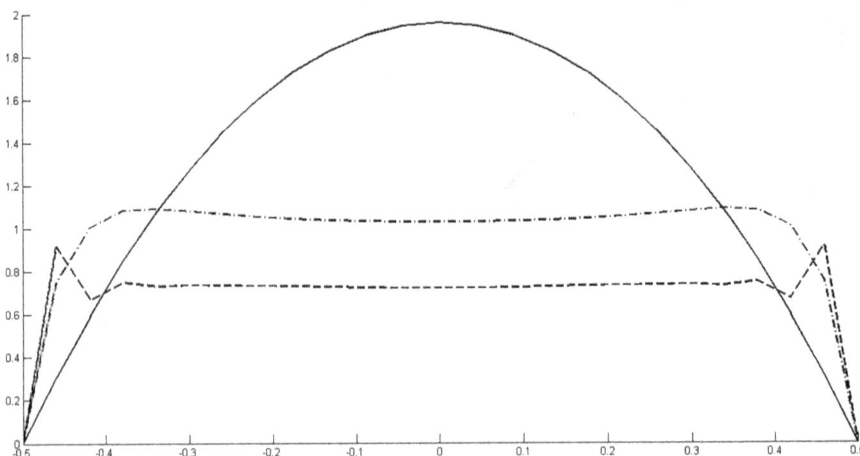

Fig. 33 Velocity profile, U/U_0, versus y/y_0 for conducting wall, [1]

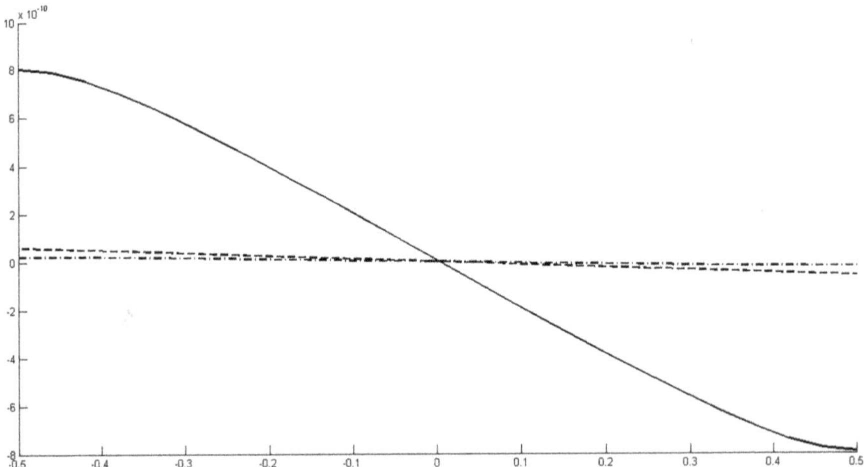

Fig. 34 Magnetic field profile B/B_0 versus y/y_0 for conducting wall, [1]

2.5 Pulsatile Flow

The flow of blood and air is pulsed through the circulatory system because the heart has two phases. The diastole, occurring when the two atriums fill with venous blood, and the systole, occurring when the blood is ejected from the two ventricles. Air has an inspiratory and an expiratory phase. The blood flow in the arterial system is investigated in [15].

Velocity, flow rate and viscous drag, in the presence of a time-dependent periodic pressure gradient, are given by

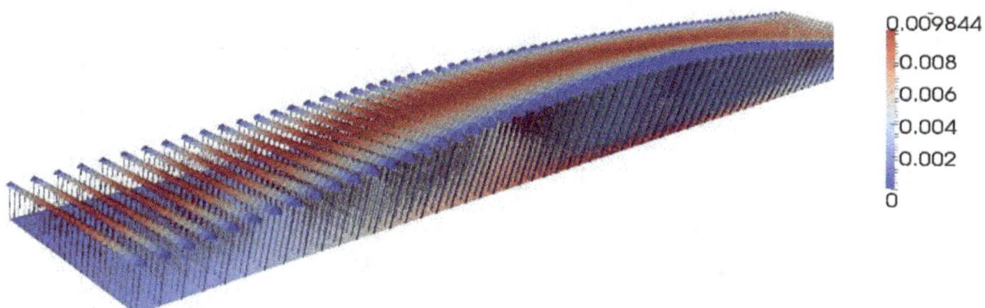

Fig. 35 Velocity field profile U (m/s), along z-axis (cm), [1]

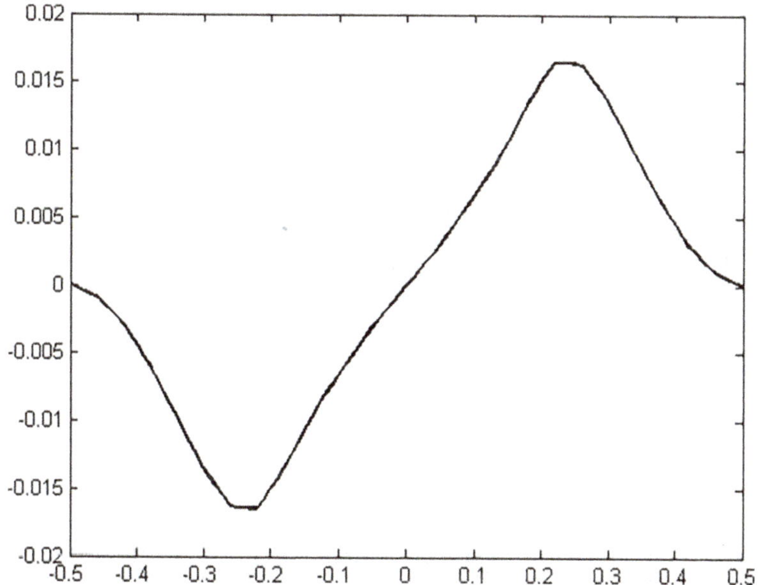

Fig. 36 Velocity profile V/V_0 versus y/y_0, [1]

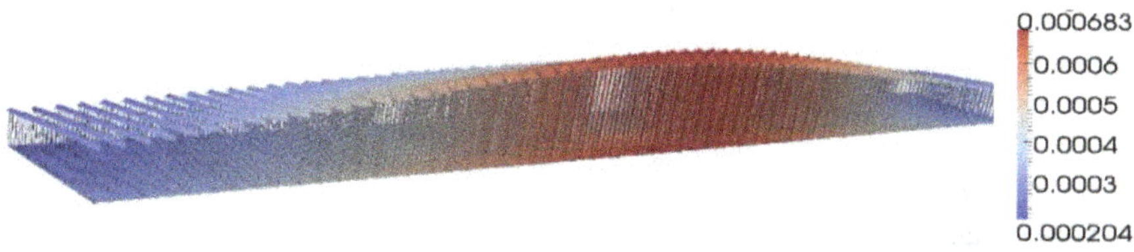

Fig. 37 Magnetic field profile B (T), versus z (cm), [1]

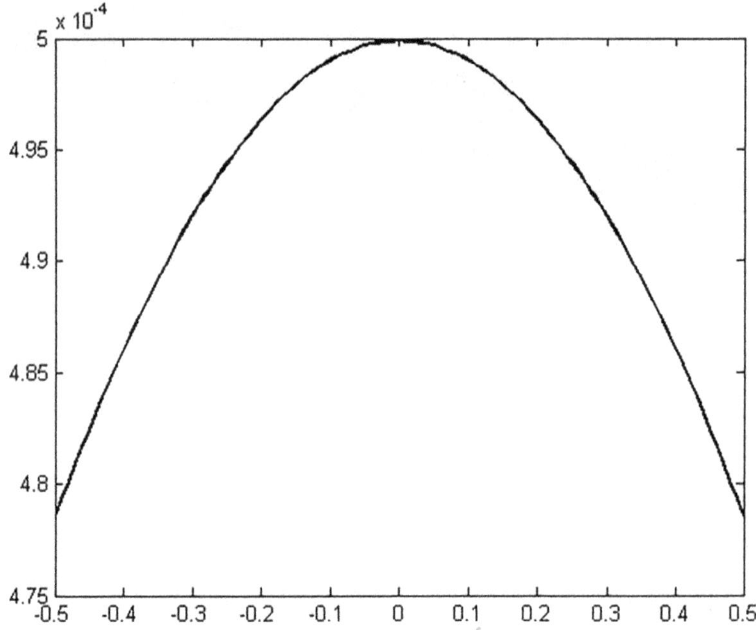

Fig. 38 Magnetic field profile B_y (T), versus y/y_0, [1]

$$\frac{p_1 - p_2}{l} = A e^{j2\pi f \cdot t} \tag{145}$$

where f is the frequency,

$$f = \frac{1}{T_o} \tag{146}$$

j the imaginary unit and T_o the pulse period. In this way, any time a periodic function can be expressed as the sum of a series of terms of this form.

The velocity equation for axial, fully developed flow can be written as

$$\frac{\partial^2 w}{\partial r^2} + \frac{1}{r}\frac{\partial w}{\partial r} - \frac{\rho}{\mu}\frac{\partial w}{\partial t} = -\frac{A}{\mu} e^{j2\pi f \cdot t} \tag{147}$$

The condition on the velocity w is

$$w = u(r) e^{j2\pi f \cdot t} \tag{148}$$

The equation for u, depending on r only, is

$$\frac{\partial^2 u}{\partial r^2} + \frac{1}{r}\frac{\partial u}{\partial r} - j2\pi f \frac{\rho}{\mu} u = -\frac{A}{\mu} \tag{149}$$

with the solution

$$u(r) = \frac{A}{\rho}\frac{1}{j2\pi f}\left(1 - \frac{J_0\left(rj^{\frac{3}{2}}\sqrt{2\pi f \frac{\rho}{\mu}}\right)}{J_0\left(Rj^{\frac{3}{2}}\sqrt{2\pi f \frac{\rho}{\mu}}\right)}\right) \tag{150}$$

where J_0 is the Bessel function of the first kind of order zero and $\alpha = R\sqrt{2\pi f \frac{\rho}{\mu}}$ a dimensionless parameter. Defining $y = \frac{r}{R}$ velocity is

$$w(r,t) = \frac{A}{\rho}\frac{1}{j2\pi f}\left(1 - \frac{J_0\left(yj^{\frac{3}{2}}\alpha\right)}{J_0\left(j^{\frac{3}{2}}\alpha\right)}\right)e^{j2\pi f \cdot t} \tag{151}$$

This is the complex form of the velocity equation. The real part of the equation is taken as $Ae^{j2\pi f \cdot t}$. The Bessel function is

$$J_0\left(xj^{\frac{3}{2}}\right) = M_o(x)e^{J\theta_0(x)} \tag{152}$$

If the real part of $Ae^{j2\pi f \cdot t}$ is $M\cos(2\pi f \cdot t + \varphi)$, the corresponding velocity, after the comparison with the steady-state one, is

$$w(t,r) = \frac{M}{\rho}\frac{1}{2\pi f}\left[\sin(2\pi f \cdot t) - \frac{M_0(y)}{M_0}\sin(2\pi f \cdot t + \varphi - \delta_0)\right] \tag{153}$$

The equation for the flow rate can be written as

$$Q(t) = \frac{M\pi R^4}{\mu\alpha^2}\left[\sin(2\pi f \cdot t + \varphi) - \frac{2M_1}{\alpha M_0}\sin(2\pi f \cdot t + \varphi - \delta_{10})\right] \tag{154}$$

The previous equations are valid for the pulsatile airflow, with the pressure gradient referred to as the difference between atmospheric and alveolar pressure. When the atmospheric pressure exceeds the alveolar one, the inspiratory phase is present; on the contrary, when the alveolar pressure is higher than the atmospheric one, we have the expiratory phase.

3   Particle Tracking

3.1   Introduction

Medical studies demonstrated that the use of magnetic drugs, guided by a magnetic field, improves the absorption of the same drugs in most carcinomas, increasing the efficacy of the treatment [16–18]. The behaviour of magnetic nanoparticles under the influence of an external magnetic field in the human body is investigated in [19, 20]. The particles are laden in both air and blood, circulating inside the airways and the hepatic artery. To achieve this goal, the *mhdFOAM* solver of the *OpenFOAM* package is modified. The specific solver for particles and their interaction, called *solidParticleFOAM,* is not suitable for the present approach and is not integrated into the MHD equations.

3.2   Equations in the Frame of Reference

The equations for the incompressible fluid flow in the Eulerian frame are briefly reviewed. The momentum equation of an incompressible, viscous fluid, subject to gravity, is

$$\rho_f \frac{\partial \vec{u}}{\partial t} + \rho_f \vec{u} \cdot \nabla \vec{u} = -\nabla p + \mu_f \nabla^2 \vec{u} + \rho \vec{g} - \vec{S}_p \tag{155}$$

with $\vec{u}$ is velocity; p, pressure; μ_f, viscosity. The additional source term, $\vec{S}_p$, is due to the influence of the particles. Considering a particle P of mass, m_p, and velocity, $\vec{u}_p$, the force exerted by a particle on a unit volume of fluid is proportional to the difference in the particle momentum between the time of the inlet, t_{in}, and the outlet, t_{out}, of the control volume

$$\vec{F}_{pf} \propto m_p \frac{\left(\vec{u}_p\big|_{t_{out}} - \vec{u}_p\big|_{t_{in}}\right)}{t_{out} - t_{in}} \tag{156}$$

A momentum source contribution is generated by the particle P in each cell, visited along its path, during one Eulerian time-step, dt, which is used to evaluate the time variation of the continuum phase. The contribution of all particles in cell k, of volume V_k, during the Eulerian time-step dt, is

$$\vec{S}_{p,k} = \frac{1}{V_k dt} \sum_P m_p \left(\vec{u}_p\big|_{t_{out,k}} - \vec{u}_p\big|_{t_{in,k}}\right) \tag{157}$$

In the Lagrangian frame of reference, it is possible to evaluate the previous equation in a slightly different way. A spherical particle P, with diameter, D_p, velocity, $\vec{u}_p$, density, ρ_p, mass, $m_p = \frac{1}{6}\rho_p \pi D_p^3$, is defined by the position of its centre, x_p. The position vector $\vec{x}_p$ of each particle, in a Lagrangian frame of reference, is

$$\frac{d\vec{x}_p}{dt} = \vec{u}_p \tag{158}$$

The motion of the particle is governed by Newton's equation

$$m_p \frac{d\vec{u}_p}{dt} = \sum_{i=1}^{N} \vec{F}_i \tag{159}$$

The dominant force acting on the particle in a dilute flow is the drag, $\vec{F}_d$, from the fluid phase. Other forces, such as the Magnus force (assuming that particle rotation is small compared to particle translation), the added mass and buoyancy force, are relatively small, and the Basset history term is neglected. The right-hand side term of Eq. 152 is

$$\sum_{i=1}^{N} \vec{F}_i = \vec{F}_d + m_p \vec{g} \tag{160}$$

The drag force is expressed as

$$\vec{F}_d = -m_p \frac{\left(\vec{u}_p - \vec{u}\right)}{\tau_p} \tag{161}$$

The relaxation time, τ_P, of the particle is the time that the particle takes to respond to changes in the local flow velocity

$$\tau_p = \frac{4}{3} \frac{\rho_p D_p}{\rho_f C_d \left|\vec{u}_p - \vec{u}\right|} \tag{162}$$

where the standard definition of the drag coefficient is

$$C_d = \begin{cases} \dfrac{24}{\mathrm{Re}_p} & \mathrm{Re}_p < 0.1 \quad \text{Stokes regime} \\[2ex] \dfrac{24}{\mathrm{Re}_P}\left(1 + \dfrac{1}{6}\mathrm{Re}_P^{2/3}\right) & 0.1 \leq \mathrm{Re}_P \leq 1000 \ \text{Transition regime} \\[2ex] 0.44 & \mathrm{Re}_P > 1000 \ \text{Newtonian regime} \end{cases} \tag{163}$$

The particle Reynolds number is defined as

$$Re_p = \frac{\rho_f D_p \left|\vec{u}_p - \vec{u}\right|}{\mu_f} \tag{164}$$

The drag coefficient for a solid sphere, at low Reynolds numbers, is evaluated with the Stokes law, which is extended to high Re_p with an empirical non-linear correction, which takes into account both viscous and inertial effects, [21]. At high values of Re_p inertial effects dominate the viscous ones, the drag coefficient becomes independent of the Reynolds number, and a constant value is normally used. Since the fluid velocity, $\vec{u}$, calculated in the Eulerian frame, is necessary for the calculation of the drag force in the Lagrangian frame, it needs to be interpolated in the position of the particle from the neighbourhood grid points.

Finally, the velocity of the particle is calculated using Eqs. 159–160 and Eq. 162,

$$m_P \frac{\vec{u}_p^{t+\Delta t} - \vec{u}_p^{t}}{\Delta t} = -m_P \frac{\vec{u}_p^{t+\Delta t} - Int\left[\vec{u}\right]_P^{t}}{\tau_P} + m_P \vec{g} \tag{165}$$

The particle velocity is updated after the first Lagrangian time-step, Δt, used to evaluate the time variation of the discrete phase, in the following way

$$\vec{u}_p^{t+\Delta t} = \frac{\tau_p \vec{u}_p^{t} + Int\left[\vec{u}\right]_P^{t}\Delta t + \tau_p \vec{g} \Delta t}{\tau_p + \Delta t} \tag{166}$$

The position of the particle is evaluated according to Eq. (158)

$$\vec{x}_P^{t+\Delta t} = \vec{x}_P^{t} + \vec{u}_P^{t}\Delta t \tag{167}$$

If the Eulerian time-step is accomplished in several Lagrangian time steps, velocity and position are evaluated at the n-th Lagrangian time-step, [22],

$$\vec{u}_p^{t+\sum_{i=1}^{n}\Delta t_i} = \frac{\vec{x}_p^{t+\sum_{i=1}^{n-1}\Delta t_i} + Int\left[\vec{u}\right]^{t}\frac{\Delta t}{\tau_p} + \vec{g}\,\Delta t}{1 + \frac{\Delta t}{\tau_p}} \tag{168}$$

$$\vec{x}_p^{t+\sum_{i=1}^{n}\Delta t_i} = \vec{x}_p^{t+\sum_{i=1}^{n-1}\Delta t_i} + \vec{u}_p^{t+\sum_{i=1}^{n-1}\Delta t_i}\Delta t_n \tag{169}$$

3.3   Solid Particle Development

The *solidParticle* class in *OpenFOAM* is a one-way coupling Lagrangian Particle Tracking, which tracks a *defaultCloud*, a collection of Lagrangian particles. It is possible to define a one-way coupling when the continuum phase influences the discrete phase via drag and turbulence, and the discrete phase does not influence the continuum phase. Moreover, the collision between two particles and the wall can be specified. Only the collision between particles and the wall is implemented in the *solidParticle*. For this reason, it has been specified as the restitution coefficient, e, which is a measure of the *bounciness* of a collision between two objects. It indicates how much of the kinetic energy remains for the objects to rebound from one another versus how much is lost as heat, or work done, deforming the objects.

The coefficient e is defined as the ratio of the relative speeds, after and before an impact, taken along the impact line

$$\vec{e} \cdot \hat{n} = \frac{\vec{u}_2 \cdot \hat{n}}{\vec{u}_1 \cdot \hat{n}} \tag{170}$$

$$\vec{e} \cdot \hat{t} = \frac{\vec{u}_2 \cdot \hat{t}}{\vec{u}_1 \cdot \hat{t}} \tag{171}$$

The coefficient e is usually a positive real number between 0 and 1.

- If $e = 0$ the collision is perfectly inelastic. The objects do not move apart after the collision, but they coalesce. Kinetic energy is converted into heat or work done in deforming the objects.
- If $0 < e < 1$ the collision is a real-world inelastic collision, where some kinetic energy is dissipated.
- If $e = 1$ the collision is perfectly elastic, where no kinetic energy is dissipated, and the objects rebound from one another with the same relative speed they approached.
- If $e < 0$ the coefficient of restitution is smaller than zero, it represents a collision where the separation velocity of the objects has the same direction as the closing velocity, implying that the objects pass through one another without full engagement. This may be thought of as an incomplete transfer of momentum. An example of this might be a small, dense object passing through a large, less dense one.
- If $e > 1$ the collision is energy released. Moreover, some recent articles describe super-elastic collisions, where it is argued that the coefficient of restitution can take a value greater than 1 in a special case of oblique collisions, [23–25]. These phenomena are due to the change of rebound trajectory caused by friction, and it does not mean that the collisions generate kinetic energy.

Figure 39 summarises the previous considerations.

Another important parameter in the *solidParticle* formulation is the viscosity, μ, which comes from the frictional force, i.e. the force exerted by a surface when an object moves across it, or it makes an effort to move across it. The frictional force is

$$\vec{F}_f = \mu \vec{N} \tag{172}$$

where $\vec{F}_f$ is the frictional force, μ the viscosity and $\vec{N}$ the normal force.

Fig. 39 Restitution coefficient

3.3.1 Particle Injection

Some important modifications are necessary inside the solver because the aim is to add the injection of a cloud of particles. Firstly, the addition of a function *inject*, which allows the injection of the particles. The injector sends the *nbInjByDt* particles every time-step between *tStart* and *tEnd* in a rectangle, of centre and side r_0, defined by the user in the dictionary (*constant/particleProperties*). The initial velocity of the particle is

$$\vec{u}_p = \vec{u}_0 + rand(n) \cdot \vec{u}' \tag{173}$$

The parameters $\vec{u}_0$, and $\vec{u}'$, the diameter, d, the density, ρ_P, the coefficient of restitution, e, and viscosity, μ, are defined by the user, [22]. Furthermore, the simulations run in a domain where there is also an applied magnetic field, and the final aim is to evaluate the influence of the magnetic field on the particles involved. Then, we need to modify the source file to take into account this last effect as well. We define three different variables for the acceleration, $\vec{a}_{em}$, $\vec{a}_F$ and $\vec{a}_V$, which take into account, respectively, the electromagnetic, the inertial and the viscous effects:

$$\vec{a}_F = \frac{\partial \vec{u}}{\partial t} + \vec{u} \cdot \nabla \vec{u} \tag{174}$$

$$\vec{a}_V = -\nu \nabla^2 \vec{u} \tag{175}$$

$$\vec{a}_{em} = -\nabla \cdot \left(\vec{B} \otimes \frac{\vec{B}}{\mu\rho} \right) + \nabla \left(\frac{\left\| \vec{B} \right\|^2}{\mu\rho} \right) \tag{176}$$

These variables are expressed in Eq. 168, and the particle velocity has the following form

$$\vec{u}_p^{n+1} = \frac{(2\rho_p + \rho_f)\,\vec{u}_p^n + \Delta t D_c \left(2\rho_p D_c \vec{u}_f^n + 2(\rho_p - \rho_f)\,\vec{g} + 3\rho_f \vec{a}_F^n - 2\rho_f \left(\vec{a}_{em}^n + \vec{a}_V^n + \vec{a}_0^n \right) \right)}{2\rho_p + \rho_f + 2\rho_p \Delta t D_c} \tag{177}$$

The lift force, Ω, produced perpendicularly to the flow, is specified in the main code *mhdFOAM*. Bernoulli's law for the flow past an obstacle predicts that the streamlines are compressed at one boundary and expand on the opposite one, resulting in a pressure difference, which causes the lift force

$$\Omega = \nabla \times \vec{u} \tag{178}$$

In the end, the solver has to move the injected particles, which is incorporated in the function *move*. We have to add some folders and files to the case directory. All the files necessary for the continuous phase must be in the folder *0*, and in a folder called *Lagrangian*, where another folder is present, called *defaultCloud*, with three more files: U, d, and *positions*. These last three files contain the information about the velocity, diameter and position of the particle cloud. In the folder *constant*, there is the *particleProperties* dictionary, where we have specified all the parameters necessary for the injection: density, restitution coefficient, friction coefficient, number of particles injected per time-step, particle diameter, injector dimension, *startTime* and *endTime* of injection.

3.4 New Boundary Condition Development

We are interested in investigating the behaviour of a nanoparticle cloud under the effect of an externally applied magnetic field. Then, we need a suitable new boundary condition to apply to file B, inside folder 0 in *OpenFOAM*. We need to modify the solver, which has to take into account the effect of the Lagrangian cloud of particles and the effect of the magnetic field.

3.4.1 Magnetic Field Produced by a Rectangular Coil

Once modified, the *mhdFoam* solver, including the physics related to the solid particle class, needs a new boundary condition for the magnetic field, generated by a coil with electric current. The rectangular geometry is chosen because the three spatial components of the flux density are in closed form, [26].

The first formulation was developed in [27], considering the vector potential for a rectangular loop of wire in the x-y plane, A_x and A_y, and calculating the vector components of the magnetic flux density, according to

$$\vec{B} = \mathrm{curl}\left(\vec{A}\right) \tag{179}$$

The components of the vector potential, under the assumption of a single rectangular loop of wire of negligible wire cross-section and side dimensions $2a_1$ and $2b_1$, are

$$A_x = \frac{\mu_0 I_1}{4\pi} \ln\left(\frac{r_1 + a_1 + x}{r_2 - a_1 + x} \frac{r_3 - a_1 + x}{r_4 + a_1 + x}\right) \tag{180}$$

$$A_y = \frac{\mu_0 I_1}{4\pi} \ln\left(\frac{r_2 + b_1 + y}{r_3 - b_1 + y} \frac{r_4 - b_1 + y}{r_1 + b_1 + y}\right) \tag{181}$$

where μ_0 is the magnetic permeability in the vacuum, and I_1 the intensity of the current in the loop. It must be noted that $A_z = 0$. The parameters r_1, r_2, r_3, r_4 are the distances from the corners of the loop to a point $\vec{P} = (x_P, y_P, z_P)$, where the magnetic flux density is evaluated. The z-component of the magnetic flux density at $\vec{P}$ has the expression

$$B_z = \frac{\mu_0 I_1}{4\pi} \sum_{a=1}^{4} \left[\frac{d_a(-1)^a}{r_a\left[r_a + C_a(-1)^{a+1}\right]} - \frac{C_a}{r_a[r_a + d_a]} \right] \tag{182}$$

where

$$C_1 = -C_4 = a_1 + x; C_2 = -C_3 = a_1 - x; d_1 = d_2 = y + b_1; d_3 = d_4 = y - b_1;$$

$$r_1 = \sqrt{\left((a_1 + x)^2 + (b_1 + y)^2 + z^2\right)}; r_2 = \sqrt{\left((a_1 - x)^2 + (b_1 + y)^2 + z^2\right)};$$

$$r_3 = \sqrt{\left((a_1 - x)^2 + (y - b_1)^2 + z^2\right)}; r_4 = \sqrt{\left((a_1 + x)^2 + (y - b_1)^2 + z^2\right)}$$

$$\tag{183}$$

The expressions for the x and y components of the magnetic flux density can be derived from Eqs. 179 and 181, as

$$B_x = \frac{\mu_0 I_1}{4\pi} \sum_{a=1}^{4} \left[\frac{z(-1)^{a+1}}{r_a[r_a + d_a]} \right] \tag{184}$$

$$B_y = \frac{\mu_0 I_1}{4\pi} \sum_{a=1}^{4} \left[\frac{z(-1)^{a+1}}{r_a\left[r_a + C_a(-1)^{a+1}\right]} \right] \tag{185}$$

4   Conclusions

The Magneto-Hydro-Dynamic, *MHD*, equations are investigated numerically, for *2D* and *3D* geometries, by using the open-source software *OpenFOAM*, which possesses, among others, an appropriate solver for the *MHD* equations. Different boundary conditions are tested for velocity, walls and applied magnetic field. The simulations can be performed in steady-state conditions. After the most convenient mesh is found, the unsteady fluid dynamic simulations can be carried on by imposing an inlet velocity profile on the flow of air or blood.

A specific Lagrangian solver is added to the *MHD* module, available in *OpenFOAM*, to simulate the fluid dynamics of nanoparticles in the flow. Then, the boundary conditions for the magnetic field are set up. Since the most common way to create a magnetic field outside the body is to use a coil, a rectangular coil is employed to solve the magnetic field generated by a current *I*. The last part of the implementation of the new boundary condition is to find the coordinates of the target and the source. The simulations are carried on with null and different from zero magnetic fields to assess the influence of the magnetic field on the adsorption of the particles. The behaviour of magnetic nanoparticles under the influence of an external magnetic field is investigated by modifying the *mhdFOAM* solver of the *OpenFOAM* package.

References

1. Russo F (2014/2015) Numerical simulation of magneto-hydro-dynamics inside cardiovascular and respiratory systems. PhD Thesis, University of Rome Tor Vergata, A.A
2. Molokov S, Moreau R, Moffatt H (2007) Magnetohydrodynamics. Springer, New York
3. Alfven H (1942) Existence of electromagnetic-hydrodynamic waves. Nature 150:405–406
4. Hartmann J, Lazarus F (1937) Hg-dynamics II. Experimental investigations. Det Kongelige Danske Videnskabernes Selskab-Mat Phys 7:1–45
5. Shercliff JA (1956) The flow of conducting fluids in circular pipes under transverse magnetic fields. J Fluid Mech 6:644
6. Shercliff JA (1962) Magnetohydrodynamic pipe flow. Part 2. High Hartmann number. J Fluid Mech 13:513–518
7. Gold R (1962) Magnetohydrodynamic pipe flow. Part 1. J Fluid Mech Agosto:505–512
8. Hunt JCR (1965) Magnetohydrodynamic flow in rectangular ducts. J Fluid Mech 5:577
9. Hunt JCR, Stewartson K (1965) Magnetohydrodynamic flow in rectangular ducts, II. J Fluid Mech 5:563
10. Vantieghem S, Albets-Chico X, Knaepen B (2009) The velocity profile of laminar MHD flows in circular conducting pipes. Theor Comput Fluid Dyn 525-533
11. Mistrangelo C, Bühler L (2009) Disturbances produced by an electric potential probe on MHD flows in rectangular ducts, vol 48. s.l.: FZKA, p 157
12. Open FOAM the open source CFD toolbox. User Guide Open FOAM, 2012
13. Behrens T (2009) Openfoam's basic solvers for linear systems of equations. s.l.:s.n
14. Versteeg H, Malalaskera W (1995) An introduction to computtational fluid dynamic. The finite volume method. Longman Scientific & Technical, Longman House, Burnt Mill, Harlow
15. Womersley J (1955) Method for the calculation of velocity, rate of flow and viscous drag in arteries when the pressure gradient is known. J Physiol 127:553–563
16. Voltairas P, Fotiadis D, Michalis L (2002) Hydrodynamics of magnetic drug targeting. J Biomechan:813–821
17. Mody V, Coz A, Shah S, Singh A, Bevins W, Parihar H (2014) Magnetic nanoparticle drug delivery systems for targeting tumor. Appl Nanosci 4:385
18. Wilson M, Kerlan RK Jr, Fidelman N, Venook AP, LaBerge JM, Koda J, Gordon RL (2004) Hepatocellular carcinoma: regional therapy with a magnetic targeted carrier bound to doxorubicin in a dual MR imaging/conventional angiography suite—initial experience with four patients. Radiology 4:287–293
19. Boghi A, Russo F, Gori F (2017) Numerical simulation of magnetic Nano drug targeting in a patient-specific coeliac trunk. J Magn Magn Mater 437:86–97
20. Russo F, Boghi A, Gori F (2018) Numerical simulation of magnetic Nano drug targeting in a patient-specific lower respiratory tract. J Magn Magn Mater 451:554–564
21. Naumann Z, Schiller L (1935) A drag coefficient correlation. Z Ver Deutsch Ing 77:318–332
22. Vallier A (2010) Tutorial iso lagrangian foam/solid particle foam
23. Louge M, Adams ME (2002) Anomalous behavior of normal kinematic restitution in the oblique impacts of a hard sphere on an elastoplastic plate. Physical Review E 65(2)
24. Kuninaka H, Hayakawa H (2004) Anomalous behavior of the coefficient of normal restitution in oblique impact. Phys Rev Lett 93(15)
25. Calsamiglia J, Kennedy SW, Chatterjee A, Ruina A, Jenkins JT (1999) Anomalous frictional behavior in collisions of thin disks. J Appl Mech 66(1):146–152
26. Misakian M (2000) Equations for the magnetic field produced by one or more rectangular loops of wire in the same plane. J Res Nat Inst Stand Technol 105(4):557–564
27. Weber E (1950) Electromagnetic theory-theory and applications. John Wiley & Sons, Inc, New York; Chapman & Hall, Limited, London

Flavia Russo, a Medical Engineer, obtained her Research Doctorate degree in Industrial Engineering from the University of Rome "Tor Vergata." She completed a six-month fellowship at the Center for Biomolecular Nano Technologies of the IIT in Lecce for the study of optical fibre cochlear prostheses. She has worked at a medical software house as a product specialist and in a research and development start-up as a product owner of wearable medical devices. She currently holds the position of project manager at a company that deals with telemedicine.

Dr. Eng. Andrea Boghi, The Doctor of Research *(Ph. D.)*, is the director of *Computational Science Ltd.*, a consultancy company in the field of software development and data analysis, which counts among its clients the British Air Traffic Agency *(NATS)* and the Bank of England, starting from 2018. He is the author of more than 30 international scientific publications indexed on the major scientific databases. For more than a decade, he has worked in academia in various European institutions, dealing with the modelling and simulation of transport phenomena, both for basic research and for industrial and biological applications. His academic journey began at the University of Rome "Tor Vergata," where, alongside research on the modelling and simulation of turbulent flows with variable laminar diffusivity and blood flow in arteries and *stents*, he also served as a contract lecturer for the courses of Technical Physics, Thermo-Fluid-Dynamics of Biological Systems and Numerical Calculation of Thermo-Fluid-Dynamic Systems (2006–2010). In the two-year period from 2010 to 2012, he was a contract researcher at the Department of Energy and Technology of the University of *Southampton* in the United Kingdom, where he dealt with modelling and simulation of bi-phase flows, characterised by spinodal decomposition. In 2012, he went to France to the Institute of Fluid Mechanics of Toulouse as a contract researcher for the OTE (Waves, Turbulence and Environment) research group, where he dealt with modelling and simulation of river and atmospheric fluid dynamics. He terminated his contract in 2013 to go to the University of *Cranfield* in the United Kingdom, where he was appointed *Senior Research Fellow*, becoming part of the academic staff. Here, for the following 5 years (2013–2018), he dealt with modelling and simulation of transport phenomena for various departments: Energy, Hydraulics, Environment and Agriculture, where he respectively dealt with transport of crude oil in pipelines, disinfection of tanks by chlorination and transport and diffusion of organic and inorganic material in soils, with associated absorption by plants. At the University of *Cranfield,* he also handled the course on modelling of atmospheric emissions in the two-year period from 2016 to 2017. Before leaving academia, he went again to the University of *Southampton* in 2018, where he dealt with the three-dimensional modelling of the transport of nitrates and carbonates at the air–water interface in soils. Throughout his period abroad, he maintained his collaboration with Fabio Gori Ammannati, supervising some students and collaborating on the production of scientific articles. In 2020, he obtained the national scientific qualification for the functions of associate university professor in the competitive sector of Technical Physics and Nuclear Engineering.

Fabio Gori Ammannati was born in Montale (Pistoia) on 5 August 1947 to Emilio Gori and Cesarina Ammannati. On 16 July 2021, he added his mother's surname, Ammannati. He graduated in Chemical Engineering on 11 November 1971 with honours. In December 1971, he won a scholarship for young graduates at the Faculty of Engineering in Bologna, which he undertook from January 1972. In 1974, he became assistant professor and, in 1982, associate professor of Technical Physics at the Faculty of Engineering of the University of Florence, where he taught, also as a contract professor, until 1990. In 1978, he won a scholarship from the *British Council*, which he carried out at the *Imperial College* in London, where he collaborated with *Prof. D. Brian Spalding* on the topic of numerical analysis in the turbulent flow of liquid metals. During his time at the University of Florence, he established international scientific collaborations with *Prof. R. Echigo, Tokyo Institute of Technology, Tokyo, Japan; Prof. D.R. Chaudhary, Department of Physics, University of Rajasthan, Jaipur, India;* International Centre for Theoretical Physics *(ICTP)* in Trieste; *Centralny Osrodek Techniki Medycznej, Warsaw, Poland and Prof. T. Aihara, Institute of Fluid Science, Sendai, Japan.* In 1986, he won a CNR scholarship, which he carried out at *Cornell University, Ithaca, New York,* where he collaborated with *Prof. R. Miller* on the topic of ground freezing. In the same year, he became professor in the Faculty of Engineering at the University of Reggio Calabria. In 1988, he was appointed as *professor* at the *University of New York at Stony Brook*, teaching the course, *Introduction to Fluid Dynamics,* in the autumn semester of the same year. During his stay, he collaborated with *Prof. T.F. Irvine Jr.* on the method of measuring thermal conductivity with a thermal probe and on the measurement of the isobaric thermal expansion coefficient of non-Newtonian fluids. In 1990, he moved to the Milano Polytechnic, where he taught Technical Physics and Systems until 1993. From 1991 until 1998, he was an adjunct professor at the Faculty of Engineering of the University of Siena. From 1992 to 2017, he was a professor of Technical Physics in the Faculty of Engineering at the University of Rome Tor Vergata, and from 2017, he is a contract professor. In 1994, he proposed the Research Doctorate (Italian *Ph.D.*) in Energy-Environment Engineering, of which he was the coordinator until 2011. In 1995, he proposed and directed the second-level Master's in Thermo-fluid-dynamics until 2016. From 1998 to 2001, he was the director of the Department of Mechanical Engineering. In 2001, he proposed and coordinated the programme, *Master of Science, Energy Engineering and Thermal and Fluid Dynamics,* in collaboration with the *Department of Mechanical Engineering, College of Engineering, University of Illinois at Chicago, USA,* which takes place at the University of Rome Tor Vergata but awards the title of *Master of Science* from the same American University. Since the late 90s, he has been a coordinator of the joint *Ph.D.* research programme with *Prof. R.J. Goldstein, Department of Mechanical Engineering, University of Minnesota, Minneapolis, USA,* and with *Prof. J.P. Hartnett, Prof. L. Kennedy, Prof. W. Minkowycz and Prof. W.M. Worek, Department of Mechanical Engineering, College of Engineering, University of Illinois at Chicago, Chicago, USA.* Since the mid-90s, he has proposed and directed the *Socrates–Erasmus* programme with *Prof. Mayinger, Technical University of Munich, Germany; Prof. van Steenhoven, Eindhoven Technical University, the Netherlands and Prof. C. Caro, Imperial College of Science, Technology and Medicine.* During his stay at Tor Vergata, he established scientific collaborations with the *Department of Mechanical Engineering, University of Minnesota, Minneapolis, USA,* collaborating with *Prof. R.J. Goldstein* on Thermo-fluid-dynamics and mass transport in gas turbine blades; the *Energy Resources Center, University of Illinois at Chicago,* collaborating with *Prof. J.P. Hartnett, Prof. W. Minkowycz* and *Prof. W.M. Worek*; the *Department of Mechanical Engineering, College of Engineering, University of Illinois at Chicago, collaborating with Prof. L. Kennedy and Prof. W.M. Worek*; *Duke University, collaborating with Prof. A. Bejan* and *S. Mary's University, Halifax, Nova Scotia, Canada, collaborating with Prof. W. R. Tarnawski.* The bibliographic review of documents and citations, titled *PlosBiology Career,* places him within the top 2% of researchers in the field of *Mechanical Engineering and Transports.* Since 1992, he has been a tutor for over 30 PhD theses, supervisor for over 40 second-level university Master's theses, supervisor for over 70 degree theses (five-year, Specialist and Master's), supervisor for over ten *Master's degree in Mechanical Engineering* from the *University of Illinois at Chicago* and supervisor for over 20 theses *Erasmus-Socrates.*

Since the early 2000s, he has been a reviewer of international research projects on behalf of *Portuguese Science and Technology Foundation (FCT); Czech Science Foundation (GACR); European Science Foundation, Strasbourg, France,* on behalf of *FCT* and the *Shota Rustaveli National Science Foundation, Georgia.* Since 2017, the year of his retirement due to age limits, he has been a contract professor at the University of Rome Tor Vergata, where in the academic year 2024–2025, he has taught Technical Physics for the Master's Degree Course in Medical Engineering. M.D. Salvatore Mangiafico (Florence), scientific head of the NeuroVascular Base Camp, 2021, invited him to give a scientific presentation titled "*The engineering approach to the study of the Circle of Willis*" on 23 September 2021 at the Congress Centre, University of Rome La Sapienza, thus opening up a possible future collaboration.

Numerical Thermo-Haemo-Dynamics (THD) and Magneto-Hydro-Dynamics (MHD) in Respiratory Airways

Flavia Russo, Andrea Boghi, and Fabio Gori Ammannati

1 Introduction and Medical Information

Lung cancer, a malignant tumour, is the leading cause of death among common cancers, with the World Health Organisation reporting a number of deaths of 1.5 million during 2012 [2]. The disease strikes mainly men, 16.7% of the total, with the highest estimated age-standardised incidence rate in Central, Eastern Europe and Eastern Asia. The incidence is smaller among women, with the highest estimated rates in Northern America and Northern Europe [3]. It appears impossible to prevent this disease, although the incidence can be reduced by avoiding the main risk factors, such as smoking and air pollution. Treatment and long-term outcomes rely on the type of cancer, the stage when it is treated and the health of the overall person. Chemotherapy, surgery and radiotherapy are the most commonly used treatments, despite their numerous hazardous side effects.

1.1 Cardiovascular System

The cardiovascular system is a closed loop, through which the blood is transported throughout the whole body [4]. The system is made of two parts. The pulmonary circulation starts in the right ventricle and ends in the left atrium. The blood volume inside this loop is about 9% of the total volume. This system is necessary for the

F. Russo
Rome, Italy

A. Boghi
Computational Science Ltd., Southampton, UK

F. G. Ammannati (✉)
University of Rome Tor Vergata, Rome, Italy
e-mail: gori@uniroma2.it

F. Gori Ammannati (ed.), *Thermo-Haemo-Dynamics in Medical Engineering*,
https://doi.org/10.1007/978-3-031-97214-0_9

correct oxygenation of the blood, which eliminates carbon dioxide in the lungs and gets charged with oxygen, which is distributed to the whole body after it returns to the heart. The systemic circulation, which accounts for 84% of the total blood issued from the heart, starts from the left ventricle and ends in the right atrium. It transports blood to the whole body. Its main arteries are: ascending aorta, aortic arch, descending aorta and common iliac arteries. The ascending aorta is the first part of the vessel. The aortic arch is divided into three other arteries: the brachiocephalic trunk, the common carotid and the left subclavian. The descending aorta is divided into two parts: the thoracic and the abdominal aorta. The common iliac arteries bring blood to the legs and feet. As mentioned, the system closes in the right atrium since the blood arrives through the venous system. The most important vessels for this system are two: the superior and the inferior vena cava.

The walls of arteries and veins are made of three layers:

- Tunica intima is the inner layer, which includes a thick layer of elastic fibres, called the inner elastic lamina;
- Tunica media has smooth muscle cells and elastic fibres, and an outer elastic lamina;
- Tunica adventitia has a thick layer outside the vessel, made of connective tissue, collagen and elastic fibres. It is possible to stretch the arteries in order to prevent an overextension, due to excessive pressure exerted by blood on the walls.

The total blood volume is unequally divided between arteries, capillaries and veins. Generally, the heart, arteries and capillaries contain 30–35% of the blood volume, while the venous system contains about 65–70% of the remaining blood volume. Moreover, the venous system acts as a venous reserve, with an amount of blood of 21%, compared to the total blood volume.

1.2 Respiratory System

The human gas-exchanging organ, the lung, is located in the thorax, where its delicate tissues are protected by the bony and muscular thoracic cage [5]. The lung provides the organism with a continuous flow of oxygen and clears the blood of the gaseous waste product, carbon dioxide. Athmospheric air is pumped in and out regularly through a system of pipes, called conducting airways, which join the gas-exchange region with the outside. The airways can be divided into the upper and lower airway systems. The transition between the two systems is located where the pathways of the respiratory and digestive systems cross, just at the top of the larynx.

The upper airway system includes the nose and paranasal cavities, called sinuses; the pharynx, or throat; and partly the oral cavity, since it may be used for breathing. The lower airway system consists of the larynx, trachea, stem bronchi and all the airways branching intensively within the lungs, such as the intrapulmonary bronchi, the bronchioles and the alveolar ducts. For respiration, the collaboration of other organs is clearly essential. The diaphragm, as the main respiratory muscle, and the intercostal muscles of the chest play an essential role by generating, under the control of the central nervous system, the pumping action on the lungs. The muscles expand and contract the internal space of the thorax, whose bony framework is formed by the ribs and the thoracic vertebrae [6].

1.2.1 Trachea and Stem Bronchi

Below the larynx lies the trachea, a tube about 10–12 cm long and 2 cm wide. Its wall is stiffened by 16–20 characteristic horseshoe-shaped incomplete cartilage rings that open towards the back and are embedded in dense connective tissue. The dorsal wall contains a strong layer of transverse smooth muscle fibres that span the gap of the cartilage. The interior of the trachea is lined by the typical respiratory epithelium. The mucosal layer contains mucous glands. At its lower end, the trachea divides in an inverted Y into the two stem (or main) bronchi, one each for the left and right lungs. The right main bronchus has a larger diameter, is orientated more vertically and is shorter than the left main bronchus. Due to this arrangement, foreign bodies passing beyond the larynx usually slip into the right lung. The structure of the stem bronchi closely matches that of the trachea.

1.2.2 Lung Anatomy

The lungs occupy most of the intrathoracic space. The space between them is filled by the mediastinum, a region of connective tissue containing the heart, the major blood vessels, the trachea with the stem bronchi, the oesophagus and the thymus gland. The right lung accounts for 56% of the total lung volume and is composed of three lobes, a superior, a middle and an inferior one, separated from each other by a deep horizontal and an oblique fissure. The left lung, smaller in volume because of the asymmetric position of the heart, has only two lobes separated by an oblique fissure. In the thorax, the two lungs rest with their bases on the diaphragm, while their apexes extend above the first rib. They are connected to the mediastinum at the hilum, a limited area where airways, blood and lymphatic vessels, and nerves enter or leave the lungs. The inside of the thoracic cavities and the lung surface are covered with serous membranes, respectively the parietal pleura and the visceral pleura, which are in direct continuity with the hilum. Because of the presence of pleural recesses, which form a kind of reserve space, the pleural cavity is larger than the lung volume.

During inspiration, the recesses are in part opened by the expanding lung, thus allowing the lung to increase in volume. Although the hilum is the only place where the lungs are secured to surrounding structures, they are maintained in close apposition to the thoracic wall by a negative pressure between the visceral and parietal pleurae. A thin film of extracellular fluid between the pleurae enables the lungs to move smoothly along the walls of the cavity during breathing. If the serous membranes become inflamed (pleurisy), respiratory movements can be painful. If air enters a pleural cavity (pneumothorax), the lung immediately collapses owing to its inherent elastic properties, and breathing is abolished on this side.

The lung lobes are subdivided into smaller units, the pulmonary segments. There are 10 segments in the right lung and, depending on the classification, 8 to 10 segments in the left lung. Unlike the lobes, the pulmonary segments are not delimited from each other by fissures but by thin membranes of connective tissue containing veins and lymphatics; the arterial supply follows the segmental bronchi. These anatomical features are important because pathological processes may be limited to discrete units, and the surgeon can remove single diseased segments instead of whole lobes.

In the intrapulmonary bronchi, the cartilage rings of the stem bronchi are replaced by irregular cartilage plates; furthermore, a layer of smooth muscle is added between the mucosa and the fibrocartilaginous tunic. The bronchi are unsheathed by a layer of loose connective tissue that is continuous with the other connective tissue elements of the lung and hence is part of the fibrous skeleton spanning the lung from the hilum to the pleural sac. This outer fibrous layer contains, besides lymphatics and nerves, small bronchial vessels to supply the bronchial wall with blood from the systemic circulation. Bronchioles are small conducting airways ranging in diameter from three to less than one millimetre. The walls of the bronchioles lack cartilage and seromucous glands. Their lumen is lined by a simple cuboidal epithelium with ciliated cells and *Clara cells*, which produce a chemically ill-defined secretion. The bronchiolar wall also contains a well-developed layer of smooth muscle cells, capable of narrowing the airway. Abnormal spasms of this musculature cause the clinical symptoms of bronchial asthma.

1.3   Magnetic Therapy and Drug Delivery

In magnetic therapy, the drugs are guided directly into the interested parts of the body. The advantage of this technique is the reduction of the drug absorption in other organs, not interested by the tumour. Lungs represent an ideal target for drug delivery due to the direct access and the large area exposed to drugs [7]. Different types of nanoparticles to enhance drug delivery are studied in [8]. Mouse lungs are investigated in [9], where a large number of leukocytes are found in the lungs' parenchyma and in the bronchiole lumen.

The development of drug delivery starts at the beginning of the twentieth century, when Paul Ehlrich states that if an agent could selectively target a disease-causing organism, a toxin for that specific organism could be delivered with the selectivity agent [10]. In these last two decades, *micro* (10^{-6} m) and *nano* (10^{-9} m) magnetic particles are used especially as contrast agents in Magnetic Resonance Imaging (*MRI*) and for magnetic cell sorting in pathology laboratories [11]. Other experimental works aimed to develop magnetic particle/fluid hyperthermia treatment for tumours [12]. Their potential is really exciting, specifically in the treatment of cancer through tumour targeting. In fact, the most important disadvantage of chemotherapeutic approaches to cancer is that they are not specific.

Several Computational Fluid Dynamics (CFD) studies are carried out to investigate blood flow [13–19], and the effect of nanoparticles embedded in the air flow

[20]. Nano and microparticles deposition, after the construction of idealised airway geometries, are studied in [21] in order to find the optimal diameter of the particle for drug targeting. The transport and deposition of nanoparticles for cyclic and steady flow at low Reynolds numbers are studied in [22] by evaluating the mass transfer due to the nanoparticles' dispersion. The inspiratory flow in a three-generation symmetric bifurcation and under the assumption of low Reynolds numbers is investigated in [23], while the turbulent flow is investigated with the k-ω model in [24]. Two breathing conditions, the resting/normal and the maximal one, are studied in [25] by employing a patient-specific geometry. The secondary flow fields and the inertial effects in patient-specific lung geometries, obtained from a *Computed Tomography (CT)* data set, are studied in [26]. Other studies focus on subject-specific boundary conditions [27], and on the application of *Computational Fluid Dynamics (CFD)* to the surgery field in order to evaluate the flow rates in patients with bidirectional anastomosis [28]. The particle deposition in the lungs is investigated in [29] by using two different geometry models in order to analyse the best regions where the deposition mechanism is higher.

The reason for magnetic micro- and nanoparticle-based targeting lies in the potential to reduce the side effects of chemotherapy drugs by decreasing their systemic distribution. At the beginning, magnetic nanoparticles are proposed as contrast agents for localised radiation therapy and to induce vascular closure in tumours. Their development starts in the 1970s [30, 31], when the first researchers proposed to use this type of delivery system [32]. Specifically, they use magnetic erythrocytes for delivering cytotoxic drugs. At first, the researchers focused on applying this delivery in the presence of tumours as an anticancer drug, and they bound reversibly to magnetic fluids. They concentrate on specific advanced tumours by magnetic fields arranged on the surface outside the tumour; in this case, the toxicity might be minimised and the efficacy on the local tumour might be increased [33]. A drug delivery system is based on binding established anticancer drugs with fluids that concentrate the drug in the interested area by means of magnetic fields [34]. With this technique, the side effects linked to anticancer drugs decrease, as observed in some clinical trials [35]. In general, this kind of system is used to modify the pharmacokinetics and biodistribution of the associated drug. The dimensions of the particles used to make this system are 200 nm or less. They can be used to treat fungal infections, various types of carcinomas, aneurysmatic vessels, and in stenotic vessels where stents are inserted [36–38]. Moreover, magnetic particles can be used in special medical techniques like *Magnetic Resonance Imaging (MRI)* to check the effectiveness of the drug released in a specific injured area [39].

In the past few years, nanoparticles have been developed to be used as carriers for drug delivery. The advantages of using this new system are:

- Ability to target specific locations in the body;
- Decrease the drug quantity needed by the area of the targeted body;
- Decrease the concentration of the drug in non-target areas.

In some applications, the nanoparticles are made with the help of magnetic materials that enhance their capability to release drugs in the affected areas [40]. This new system is made of a radionuclide bound to a magnetic compound, which is inserted into the body. It reaches the target area by using a magnetic field, which can be externally applied. The drug can be released by simple diffusion, by enzymatic activity, or, also by changes in physiological conditions like: pH, osmolality or temperature.

The advantages of using magnetic nanoparticles are:

- Visualisation of the particles;
- Guidance in a specific place by means of a magnetic field;
- Heating in a magnetic field.

The parameters to be taken into account for the behaviour of the magnetic nanoparticles are: the surface chemistry, the size and the magnetic properties. The application fields for this system are: cancer treatment, anaemic, chronic kidney disease and disorders associated with the musculoskeletal system. For cancer treatment, this delivery system is developed in order to reduce the damaging side effects coming from chemotherapy [41]. The first studies about drug targeting with magnetic particles were carried out on animals and in particular on swine. The magnetic targeted carriers are focused on the liver, lungs and other affected areas to analyse their retentions and releases and the positive effects they have in those areas [42].

Several important biomedical applications of different types of magnetically drug carrier systems are investigated; in specific, the possibility of using iron-fluids for drug localisation in blood vessels and in deep organs [43]. The magnetic target principles are based on the attraction of magnetic nanoparticles to an external magnetic field source. In the presence of a magnetic field gradient, a translational force is generated and is exerted on the drug compound. The result is the trap in the field at the target area.

The magnetic force, which is established in the target area, is

$$\vec{F}_{mag} = (\chi_2 - \chi_1)\, V\, \frac{1}{\mu_0}\, \vec{B} \cdot \nabla \vec{B} \tag{1}$$

where: χ_1 is the magnetic susceptibility of the medium; χ_2 the magnetic susceptibility of the magnetic particle; $\vec{B}$ the magnetic induction field and ∇ the special gradient operator. From the above equation, it is possible to underline that the important parameters for the capture of the nanoparticles are the magnetic properties and the volume of the particles, the magnetic field strength and the magnetic field gradient. From these, it is clear that, as the field strength decreases, the ability to capture the nanoparticles also decreases, and this is the main reason to explain the difficulty of magnetic targeting from small animals to humans.

The most important factors for a correct delivery are: the volume of the particles, the magnetic properties and the magnetic field strength and its gradient. About the magnetic properties of the materials, researchers prove how the aggregation of blood cells is dependent on these properties [44]. In particular, it can be noticed that as the field strength decreases, the capability to take the nanoparticles decreases as well. The magnetic strength at the target area should be about 200–700 mT, with a gradient along the z-axis of about 8–100 T/m. Other researchers demonstrate that magnetic targeting is the most effective system for delivering drugs, especially on surfaces really close to the body [45].

The applied magnetic field can be located internally or externally of the target area; moreover, inside the interested area is located a ferromagnetic wire implant in order to observe the correct release of the drugs. Recent studies evaluate three other types of magnetic drug targeting systems: a permanent magnet, combined with an implanted wire; a single permanent magnet and a homogenous magnetic field, combined with an implanted wire. Specifically, the implanted magnetic field can have the shape of wires, seeds or stents in the blood vessels [46, 47].

The first numerical studies on this problem were developed for cylindrical geometries in order to test the efficiency of the delivery in the presence of a magnetic field [48]. Furthermore, some models are developed in order to simulate a real bloodstream with the magnetic drugs inside. At first, the blood flow is modelled in a 2D way, injecting iron fluid to study the problem from a numerical point of view [49]. Subsequently, the behaviour of the blood is studied as a non-Newtonian fluid, analysing the coupling between Navier-Stokes and Maxwell equations [50]. Other studies are developed in order to analyse different artery configurations, in particular: a straight blood vessel and an asymmetric bifurcation [51]. The results show that it is possible to slow down drug particles, thus having, in this way, a better distribution of the drugs in the intended area. Other researchers simulate the blood flow with magnetic particles in motion. Specifically, one paper studied numerically the left coronary and carotid arteries, using the properties of magnetic carriers and strong superconducting magnets [52]. Another study simulates the blood flow and the diffusion of the drug particles when the magnetic field is induced by a permanent magnet embedded inside the muscular volume [53]. Another trial is done analysing the efficiency of the delivery in a hepatic carcinoma, under the supervision of an *MRI*, in order to observe the correct delivery of the drugs employed [54]. In the presence of deep tumours, it is also important to note the distance between the affected area and the magnet. This aspect is investigated with different techniques to focus small magnetic particles on the microvasculature of tumours [55]. The distribution of the nanoparticles is also discussed in other studies, due to the fact that this is an important point to analyse before starting the design and the use of the magnetic system targeting [56].

Biological effects of electromagnetic fields are investigated in [57], while the application of the magnetic technique for the transport of drugs and tracers in specific targets is developed in [58]. Few *CFD* studies are conducted to simulate the magnetic drug targeting in blood vessels, both in idealised [59–62] and patient-specific [20–63] geometries. A mathematical model is proposed in [64] to investigate the deposition of magnetic particle aerosol in the lung alveolus by considering only one alveolus with a simplified spherical geometry. A particle's diameter of 5 µm and a quadrupolar Halbach permanent magnet array are used as a magnetic probe.

2 Three-Dimensional Modelling of Anatomic Structures

2.1 Introduction

All the detailed anatomic data about vascular structures are provided by angiographic image acquisition techniques. Graphic workstations are linked to *CT* or *MR* scanners, and they produce a 3D patient-specific image on the basis of the acquired data. The current techniques used for these purposes are not suitable for geometric analysis and for *CFD* computation. A first step in the modelling of a vascular structure is the extraction of the vascular wall position from medical images. Nowadays, it is possible to find several techniques that are able to do this first step, in particular, the parametric and the implicit deformable models, which allow building 3D models in a very accurate way.

In order to reconstruct the 3D vessel geometry, an *Open-Source Software, VMTK (The Vascular Modelling Toolkit),* is used. This software, developed in [65–67], relies on two other, already existing, software: *VTK* and *ITK*. The first one provides data structures for images, polygonal surfaces and volumetric meshes, whereas the second one is a *C++* toolkit for image processing, segmentation and registration. *ITK* becomes a standard resource in medical imaging analysis.

In the following paragraphs, the different imaging acquisition techniques are analysed, and it is discussed how to reconstruct a medical image.

2.2 Image Acquisition

The acquisition of 3D images of the anatomy of vascular segments is done by using several techniques, starting with Computed Tomography *(CT)*.

2.2.1 Computed Tomography

Computed Tomography *(CT)* is a technology which uses computer-processed *X-rays* to obtain tomographic images of specific areas of the scanned body. To generate a 3D image of the inside of the human body, a digital geometry procedure is used, which starts from a large series of two-dimensional radiographic images. It was launched in the 1970s, and it has become an important tool in medical imaging. It has several advantages in comparison with traditional 2D medical radiography. X-ray slice data is generated by an X-ray source which rotates around the object. X-ray sensors are positioned on the opposite side of the circle from the X-ray source. At the beginning, the machines rotate the X-ray source and the detectors around the body. In the newest machines, the rotation is continuous around the body, and they are called *Helical* or *Spiral CT*. A new improvement of these machines is the *Multi-slice CT*, with multiple rows of detectors which are used to capture multiple cross sections at the same time.

This kind of diagnostic tool has several advantages. First of all, *CT* eliminates the superimposition of images of structures outside the area of interest. Secondly, differences between tissues, which differ by less than 1%, can be distinguished due to the inherent high-contrast resolution of *CT*. Lastly, data from a single *CT* scan procedure can be viewed as images in the axial, coronal and sagittal sections. A visual representation of the raw data is called a sinogram. When the scan data is acquired, it must be processed by using a form of tomographic reconstruction, which produces a series of cross sectional images. The final images are obtained by using the inverse Radon transform. In the reconstructed images, pixels are displayed in terms of radio density. Moreover, the pixels are displayed according to the mean attenuation of the tissues, from the most attenuated, +3071, to the least attenuated, − 1024, according to the Hounsfield scale. In this specific scale, water has an attenuation of 0 *HU*, air of −1000 HU, and bone material can reach 2000 HU, or more, which can cause artefacts. Moreover, the attenuation of metallic implants depends on the atomic number of the elements used. For example, titanium has a value of 1000 HU, while iron-steel can completely extinguish the X-ray and is responsible, in some cases, for the presence of artefacts.

2.2.2 Image Representation and Handling

The medical images are stored on workstations and then transferred to *PCs* for processing. Generally, the images are represented by numbers of grey levels greater than 256. In order to transfer all the information linked to the images, such as patient and investigator data, image number, position, resolution, acquisition time and modality, and scan parameters, a standard for communication and storage of medical image data, *DICOM* (*Digital Communication in Medicine*), was developed. The *DICOM* image has a header, made of all the information necessary to reconstruct the 3D image. The scalar fields of the grey levels, sampled into 3D images, are defined as

$$I\left(\vec{x}\right);\ \vec{x}\in\mathfrak{R}^3 \tag{2}$$

2.3   Contouring

Different tissues correspond to different grey levels on the radiological images. For this reason, the first and easiest approach to reconstruct a *3D* geometry is to create a surface in correspondence with a given grey level. This method of proceeding is known as *Contouring,* and the most popular algorithm to obtain the desired geometry is the *Marching Cubes.* This algorithm has some limitations during the reconstruction of blood vessels from angiographic images. However, it is used both for the initialisation and the final step of the *Level Sets* algorithm, the principal instrument for vessel reconstruction.

2.3.1 Marching Cubes Algorithm

The *Marching Cubes* algorithm is proposed in [68] to create triangle models of constant density surfaces from 3*D* medical data. The algorithm processes the 3*D* medical data in a scan-line order and calculates triangle vertices using linear interpolation. The algorithm is made up of two main steps. The first one is to create the surface, corresponding to the user-specified value. The second one is to calculate the normal vectors of the surface at each vertex of each triangle. The algorithm locates the surface in a logical cube, created from eight pixels, four from each adjacent slice. In order to find out at which point the surface intersects the logical cube, the algorithm compares the values assumed by each vertex of the cube with the two slices. The algorithm defines the vertex, above and below, by comparing its grey level with a reference one. Since there are eight vertices and two states, inside and outside, there are $2^8 = 256$ ways a surface can intersect the cube; nevertheless, from the two different symmetries, it is possible to reduce the cases from 256 to 15.

In order to prevent ambiguity, a set of rules is set-up to select the proper case according to the configuration of the surrounding cubes. The final step of the algorithm evaluates a normal unit for each triangle vertex. A surface with constant density has a zero gradient component along the tangential direction to the surface, and, in this way, the direction of the gradient vector is normal to the surface. The vector, $\vec{g}$, is the gradient of the density function

$$\vec{g}(x, y, z) = \nabla D(x, y, z) \tag{3}$$

In order to estimate the gradient on the surface of interest, the *Marching Cubes* algorithm estimates the gradient at the cube's vertices and linearly interpolates it at that point of intersection. The gradient on the cube vertex, *(i, j, k)*, is estimated using central differences along the three axes

$$g_x(i,j,k) = \frac{D(i+1,j,k) - D(i-1,j,k)}{2\Delta x} + o\left(\Delta x^2\right) \tag{4}$$

$$g_y(i,j,k) = \frac{D(i,j+1,k) - D(i,j-1,k)}{2\Delta y} + o\left(\Delta y^2\right) \tag{5}$$

$$g_z(i,j,k) = \frac{D(i,j,k+1) - D(i,j,k-1)}{2\Delta z} + o\left(\Delta z^2\right) \tag{6}$$

where $D(i, j, k)$ is the density at pixel *(i, j)* in slice k, and Δx, Δy, Δz are the lengths of the cube edges. The division of the gradient by its length produces the normal unit at the vertex, required for rendering.

2.3.2   Limitations of Iso-Surface Extraction

The method to create the $3D$ image reconstruction has some restrictions, especially for *CFD* simulations, because it is based on medical image grey levels. The resulting surface depends on the choice of the contouring level, normally made by the operator. Moreover, since image grey level varies more near interfaces, a slight change in the contouring level can produce large changes in geometric and even topological features of the resulting surface.

In the angiographic images, especially in the presence of a contrast medium, the section can have different opacity levels due to the time variation in the concentration of the contrast medium. Another problem involved in this algorithm is that the *Marching Cubes* algorithm does not allow the deletion of branching vessels, which is not useful in the reconstruction. In order to solve these limitations, the *Parametric Deformable* models are developed.

2.4   Parametric Deformable Models

The *Parametric Deformable* models are developed to obtain a reconstructed image based on geometrical features. These models evolve in the image space, whose deformations are described from a Lagrangian point of view. Deformation refers to the non-deformed configuration so that the material points are followed during the evolution.

2.4.1   Snakes

The first model is the active contours, a common solution in the field of $2D$ image segmentation. The active contours, called snakes, are a parameterised curve, embedded in the image plane, $(x, y) \in \mathfrak{R}^2$ [69]. The contour is represented as

$$\vec{v}(s) = \{x(s), y(s)\}^T \tag{7}$$

where x and y are the coordinates and $s \in [0, 1]$ the parametrical domain. The shape of the contour, subject to an image, is dependent on the following functional

$$\varepsilon\left(\vec{v}\right) = S\left(\vec{v}\right) + Q\left(\vec{v}\right) \tag{8}$$

which can be seen as a representation of the energy of the contour. The final shape of the contour corresponds to the minimum of its energy, and the first term of Eq. 8 can be seen as

$$S\left(\vec{v}\right) = \int_0^1 \left(w_1(s)\left|\frac{d\vec{v}}{ds}\right|^2 + w_2(s)\left|\frac{d^2\vec{v}}{ds^2}\right|^2\right) ds \tag{9}$$

This is the internal deformation energy, which characterises the deformation of a stretchy, flexible contour. Two physical parameter functions prescribe the simulated physical characteristics of the contour: $w_1(s)$ controls the tension of the contour, with the effect of reducing the snake length, while $w_2(s)$ controls the rigidity, producing a smoothing effect. The second term of Eq. 8 couples the snake to the image in the following way

$$Q\left(\vec{v}\right) = \int_0^1 w_3(s)P\left(\vec{v}\right) ds \tag{10}$$

where $P\left(\vec{v}\right)$ represents a scalar potential function, which accounts for the image features. External potentials are designed to apply snakes to images in order to connect the intensity of an image with its interesting features. In accordance with the calculus of variations, the contour, $\vec{v}(s)$, which minimises the energy, $\varepsilon\left(\vec{v}\right)$, must satisfy the Euler-Lagrange equation

$$-w_1(s)\frac{d^2\vec{v}}{ds^2} + w_2(s)\frac{d^4\vec{v}}{ds^4} + w_3(s)\nabla P\left(\vec{v}\right) = \vec{0} \tag{11}$$

The first two terms represent the internal stretching and bending forces, while the third one represents the external forces coupling the snake to the image data.

Since in a $2D$ geometry any closed curve shrinks under the effect of its curvature and evolves into a circle before collapsing into one point, snakes are usually initialised as closed lines surrounding the regions of interest. In some cases, especially for modelling blood vessels, snakes must be inflated inside the areas of interest until image features are encountered. The advantages of this model are the ease of gaining results and the high control over them, as demonstrated by the results of some research groups [70–72]. In general, the construction of a parametric surface from a set of $2D$ contours is not always a trivial task, especially near a bifurcation.

2.4.2 Balloons

In a $3D$ geometry, snakes take the name of balloons, which are parametric deformable surfaces, and can be represented by

$$\vec{S} : U_r \times U_s \times \mathfrak{R}_t^+ \to \mathfrak{R}^3 \tag{12}$$

where, $U \subset \mathfrak{R}$ and $r, s \in U$ are parameterised surface.

Generally, it is not true that a surface, shrinking under the effect of its curvature, evolves into a sphere before collapsing into one point in $3D$. The lack of this result on the regularity of the surface evolution under curvature makes it more convenient to inflate the $3D$ surface by means of an internal pressure term rather than to shrink it under curvature. It is possible to write this idea into an equation where a term takes into account image-driven balloon inflation

$$E_{\text{baloon}}\left(\vec{S}\right) = E_{\text{inf late}}\left(\vec{S}\right) + E_{\text{smoot}}\left(\vec{S}\right) + E_{\text{image}}\left(\vec{S}\right) \tag{13}$$

The corresponding evolution equation is

$$\frac{\partial \vec{S}}{\partial t} = w_1 G\left(\vec{S}\right)\hat{n} + w_2\left(\frac{\partial^2 \vec{S}}{\partial s^2} + \frac{\partial^2 \vec{S}}{\partial r^2}\right) - w_3 \nabla P\left(\vec{S}\right) \tag{14}$$

Where: $G\left(\vec{S}\right)$ is the scalar inflation speed, which can be constant or depending on image features; $\hat{n}$ is the outward surface normal; and $\left(\frac{\partial^2 \vec{S}}{\partial s^2} + \frac{\partial^2 \vec{S}}{\partial r^2}\right)$ is a smoothing term, which can be substituted by more complex regularising expressions to take into account the expected shape of the result.

Equation 14 can be solved by using numerical methods that describe the deformation of the initial model through its parameterisation [73]. The use of balloons in vessel modelling makes it possible to deal with the 3D geometry without processing single 2D frames, gaining speed and operator independence, even if the user interaction can be much more difficult.

2.5   Level Sets Technique

The *Level sets* technique, introduced for the first time in [74], allows for moving interfaces. This technique, also used in free boundary problems, is quite useful to process the image segmentation; it allows overcoming the limitations on topology and great deformations affecting parametric deformable models [75]. It is then possible to describe the surface motion, observed through the evolution of scalar field values at fixed points in space. Since then, the problem of reconstructing a 3D surface involves the resolution of the so-called *Evolution Equation*.

2.5.1   Evolution Equation

A moving surface, $\vec{S}(t)$, defined as

$$\vec{S} : \mathfrak{R}^2 \times \mathfrak{R}^+ \to \mathfrak{R}^3 \tag{15}$$

can be described using an iso-surface of level k of a time-dependent scalar function, as in [76],

$$\vec{S}(t) = \left\{ \vec{x} \in \mathfrak{R}^3 : F\left(t, \vec{x}\right) = k \right\} \tag{16}$$

Since $\vec{S}$ remains the k-level set of the function F over time, it is possible to evaluate the previous equation as

$$\frac{\partial F\left(t, \vec{S}\right)}{\partial t} = -\nabla F\left(t, \vec{S}\right) \cdot \frac{d\vec{S}}{dt} = -\left\|\nabla F\left(t, \vec{S}\right)\right\| \hat{n} \cdot \frac{d\vec{S}}{dt} \tag{17}$$

where $\hat{n} = \dfrac{\nabla F\left(t, \vec{S}\right)}{\left\|\nabla F\left(t, \vec{S}\right)\right\|}$ is the outward normal to level sets.

The advantage of using this approach is that the description of $\vec{S}(t)$ does not require a global parameterisation, but it relies on the geometric properties of $F\left(t, \vec{x}\right)$. The embedding of the inflation term in Eq. 14 can be rewritten as

$$\|\nabla F\| w_1 G\left(\vec{x}\right) \hat{n} \cdot \hat{n} = w_1 G\left(\vec{x}\right) \|\nabla F\| \tag{18}$$

For the smoothing term it is necessary to express $\frac{\partial^2 \vec{S}}{\partial r^2}$ and $\frac{\partial^2 \vec{S}}{\partial s^2}$ in terms of a differential expression on $F\left(t, \vec{x}\right)$. To obtain this condition, the derivative of $F\left(t, \vec{x}\right)$, with respect to r, is calculated under the assumption $\left|\frac{\partial \vec{S}}{\partial r}\right| = 1$, as

$$\frac{\partial F}{\partial r} = \nabla F \cdot \frac{d\vec{S}}{dr} = 0 \tag{19}$$

Equation 19 is derived once more

$$\frac{\partial^2 F}{\partial r^2} = \frac{d\vec{S}}{dr}^T \cdot \left[H_F\left(\vec{x}\right)\right] \cdot \frac{d\vec{S}}{dr} + \nabla F \cdot \frac{d^2\vec{S}}{dr^2} = \frac{\partial^2 F}{\partial u^2} + \nabla F \cdot \frac{d^2\vec{S}}{dr^2} = 0 \tag{20}$$

where $\left[H_F\left(\vec{x}\right)\right]$ is the Hessian matrix of F, and $\frac{\partial^2 F}{\partial u^2}$ the second directional derivative of F in the direction $\frac{\partial \vec{S}}{\partial r}$. The final expression for the smoothing term is

$$w_2 \nabla F \cdot \left(\frac{d^2\vec{S}}{dr^2} + \frac{d^2\vec{S}}{ds^2}\right) \hat{n} = -w_2 \left(\frac{d^2 F}{du^2} + \frac{d^2 F}{dv^2}\right) = -2w_2 H\left(\vec{x}\right) \|\nabla F\| \tag{21}$$

where $H\left(\vec{x}\right)$ is the mean curvature of level sets,

$$H\left(\vec{x}\right) = \mathrm{div}\left(\frac{\nabla F}{\|\nabla F\|}\right) = \frac{\nabla^2 F \|\nabla F\|^2 - \|\nabla F\|^T \cdot \|\nabla F\|}{\|\nabla F\|^3} = \frac{1}{\|\nabla F\|}\left(\frac{\partial^2 F}{\partial u^2} + \frac{\partial^2 F}{\partial v^2}\right) \tag{22}$$

A generic orthonormal reference frame is x, y, z, and u, v, w is the reference frame, relative to the tangent plane of the level set, where w is set along the normal direction. It follows that $H\left(\vec{x}\right)$ can be computed from the second-order differential operators of F. The embedding of the image driving term, $w_3 \nabla P\left(\vec{S}\right)$, is

$$\|\nabla F\| w_3 P\left(\vec{x}\right) \cdot \hat{n} = w_3 \nabla P\left(\vec{x}\right) \cdot \nabla F \tag{23}$$

The substitution of the terms of Eqs. 21–23 into Eq. 17 allows obtaining the evolution equation for the level sets

$$\frac{\partial F\left(t, \vec{x}\right)}{\partial t} = -w_1 G\left(\vec{x}\right) \|\nabla F\| + 2w_2 H\left(\vec{x}\right) \|\nabla F\| + w_3 \nabla P\left(\vec{x}\right) \cdot \nabla F \tag{24}$$

Equation 24 describes a deformable surface, embedded as a level set of a scalar field, evolving in time. Moreover, Eq. 24 can be solved directly, on the image regular grid, with the approach of the finite differences when dealing with $3D$ images. The image-based evolution terms is

$$G\left(\vec{x}\right) = \frac{\text{const}}{1 + \left\|\nabla I\left(\vec{x}\right)\right\|} \tag{25}$$

Inflation speed is smaller when the image gradient is greater, giving the valley potential

$$P\left(\vec{x}\right) = -\left\|\nabla I\left(\vec{x}\right)\right\| \tag{26}$$

All definitions are studied in [77] in the context of medical images, and they are well suited for all $3D$ modelling of blood vessels.

An alternative formulation for the evolution equation is

$$\frac{\partial F\left(t, \vec{x}\right)}{\partial t} = -w_1 G\left(\vec{x}\right)\|\nabla F\| + 2w_2 G\left(\vec{x}\right)H\left(\vec{x}\right)\|\nabla F\| + w_3 \nabla P\left(\vec{x}\right) \cdot \nabla F \tag{27}$$

The curvature term is weighted by the function $G\left(\vec{x}\right)$, in order to have a stronger smoothing effect in regions with smaller image gradients, where fewer image features are present, whereas more surface details are allowed where $\|\nabla I\|$ is higher.

The last step is the initialisation of the embedding function, $F\left(t_0, \vec{x}\right)$, which can be modelled as the signed distance function

$$F\left(t, \vec{x}\right) = \begin{cases} -D_{\vec{S}(t_0)}\left(\vec{x}\right)\_if\_\vec{x} \in \vec{S}(t_0) \\ D_{\vec{S}(t_0)}\left(\vec{x}\right)\_if\_\vec{x} \notin \vec{S}(t_0) \end{cases} \tag{28}$$

where $D_{\vec{S}(t_0)}\left(\vec{x}\right) = \min\left\{\left\|\vec{x} - \vec{S}(t_0)\right\|\right\}$ is the distance function from $\vec{S}(t_0)$, the embedded surface defined in Eq. 15 with $k = 0$.

2.5.2 Finite Difference Equations

In order to solve the aforementioned equation, several methods are available. One possibility is to solve the equation with the finite difference method, which consists of replacing the derivatives of the differential equations with finite difference approximations. This method leads to a large algebraic system of equations to be solved instead of the differential equations.

The evolution equation described before belongs to the Hamilton-Jacobi class, which has the general form

$$\frac{\partial F\left(t, \vec{x}\right)}{\partial t} + H\left(\vec{x}, \nabla F\right) = 0 \tag{29}$$

It could be generalised for the level sets in the following way:

$$\frac{\partial F\left(t, \vec{x}\right)}{\partial t} + H\left(t, \vec{x}\right)\left\|\nabla F\left(t, \vec{x}\right)\right\| = 0 \tag{30}$$

The forward scheme for the numerical approximation of this equation is given by

$$F\left(t + \Delta t, \vec{x}\right) = F\left(t, \vec{x}\right) + \frac{\partial F\left(t, \vec{x}\right)}{\partial t}\Delta t = F\left(t, \vec{x}\right) - H\left(t, \vec{x}\right)\left\|\nabla F\left(t, \vec{x}\right)\right\|\Delta t \tag{31}$$

Because of the behaviour of the hyperbolic equations, the use of central finite differences leads to instabilities. Approximating $\nabla F\left(t, \vec{x}\right)$ with central finite differences in this region produces incorrect results in spite of the arbitrary grid refinement due to the fact that information from each side of the first-order discontinuity is averaged in calculating gradient value. To overcome this problem, the use of upwind finite differences for the approximation of odd-order derivatives is suggested.

The central difference expression for the first derivative of $F\left(t, \vec{x}\right)$ in the grid point (i, j, k), on a grid of space h, can be derived from the Taylor expansion of $F\left(t, \vec{x}\right)$, around $\vec{x}_{i,j,k}$, up to the first order:

$$F_x\big|_{\vec{x}_{i,j,k}} \approx D^0_{i,j,k} = \frac{F(i+1,j,k) - F(i-1,j,k)}{2} \tag{32}$$

The above expression proves to be second-order accurate and gets contributions from both forwards and backwards directions.

In order to take into account only upstream contributions, the following upwind finite difference schemes are considered:

$$F_x\big|_{\vec{x}_{i,j,k}} \approx D^{+x}_{i,j,k} = \frac{F(i+1,j,k) - F(i,j,k)}{2} \tag{33}$$

$$F_x\big|_{\vec{x}_{i,j,k}} \approx D^{-x}_{i,j,k} = \frac{F(i,j,k) - F(i-1,j,k)}{2} \tag{34}$$

Though these expressions are first-order accurate, they are able to correct the approximate first derivatives in regions where the solution presents cusps. The solution converges to the exact solution as the grid spacing $h \to 0$, even if the convergence is slow.

Using the upwind finite difference for the level set approximation leads to:

$$F\left(t + \Delta t, \vec{x}\right) = F\left(t, \vec{x}\right) - \Delta t\left[\max\left\{H\left(\vec{x}\right), 0\right\}\nabla^+ + \min\left\{H\left(\vec{x}\right), 0\right\}\nabla^-\right]$$

(35)

where

$$\nabla^+ = \left[\left(\max\left\{D_{i,j,k}^{-x}, 0\right\}\right)^2 + \left(\min\left\{D_{i,j,k}^{+x}, 0\right\}\right)^2 + \left(\max\left\{D_{i,j,k}^{-y}, 0\right\}\right)^2\right.$$

$$\left. + \left(\min\left\{D_{i,j,k}^{+y}, 0\right\}\right)^2 + \left(\max\left\{D_{i,j,k}^{-z}, 0\right\}\right)^2 + \left(\min\left\{D_{i,j,k}^{+z}, 0\right\}\right)^2\right]^{1/2}$$

(36)

$$\nabla^- = \left[\left(\max\left\{D_{i,j,k}^{+x}, 0\right\}\right)^2 + \left(\min\left\{D_{i,j,k}^{-x}, 0\right\}\right)^2 + \left(\max\left\{D_{i,j,k}^{+y}, 0\right\}\right)^2\right.$$

$$\left. + \left(\min\left\{D_{i,j,k}^{-y}, 0\right\}\right)^2 + \left(\max\left\{D_{i,j,k}^{+z}, 0\right\}\right)^2 + \left(\min\left\{D_{i,j,k}^{-z}, 0\right\}\right)^2\right]^{1/2}$$

(37)

Moreover, the evolution equation needs the computation of the second-order derivatives for the approximation of level sets curvature, $K\left(\vec{x}\right)$, for the smoothing term.

Concerning the temporal discretisation, and in order to have stability, Δt is required to be equal to or lower than a limit time-step, since forward time is not unconditionally stable. The limit time-step for explicit time integration is given by the *Courant-Friedrichs-Lewy (CFL)* condition, which ensures that the rate of change in the solution does not exceed the maximum velocity that can be modelled with a given grid of spacing h:

$$\max\left\{G\left(\vec{x}\right)\right\} \leq \frac{h}{\Delta t}$$

(38)

3 Fluid Dynamic Analysis of the Airflow in the Respiratory Airways and Three-Dimensional Modelling of the Trachea

3.1 Introduction

An interesting field of study, from a medical point of view, is the behaviour of air inside the airways of the human respiratory system. In this paragraph we analyse the portion of the system made by the *trachea* and the first bifurcations of the *bronchioles.* Once medical images are used to obtain the grid for the geometry involved in the numerical simulations, we can study the grid independence and evaluate the *wall shear stress* (WSS) for four different types of grids. In this kind of analysis, we set-up a steady-state simulation; afterwards, in order to observe the fluid dynamic behaviour of a real airflow profile, an unsteady simulation is carried on. Because our field of interest is the *Magneto-Hydro-Dynamic, MHD,* we impose a fixed value of the magnetic field intensity. Finally, the effect of the addition of drug *nanoparticles* under the effect of an external magnetic field is investigated.

3.2 Airflow inside the Respiratory Airways

Air moves in and out of the lungs in response to the pressure gradient. When the air pressure within the alveolar spaces falls below the atmospheric pressure, air enters the lungs during the so-called *inspiration*, provided the larynx is open; when the air pressure within the alveoli exceeds the atmospheric pressure, air is blown out of the lungs during the so-called *expiration*. The flow of air is fast or slow, depending on the magnitude of the pressure gradient. Because of the relatively constant atmospheric pressure, the flow direction is determined by the increasing or decreasing pressure relative to the atmosphere. Alveolar pressure fluctuations are caused by the expansion and contraction of the lungs resulting from the tensing and relaxing of the muscles of the chest and abdomen. Each small increment of expansion transiently increases the space enclosed in the lung. Moreover, there is less air per unit volume in the lungs, and the pressure falls. A difference in air pressure between the atmosphere and the lungs is created, and air, flowing in equilibrium with the atmospheric pressure, is restored at a higher lung volume. When the muscles of inspiration relax, the volume of the chest and lungs decreases, lung air becomes transiently compressed, its pressure rises above the atmospheric pressure, and the flow into the atmosphere stays the same until the pressure equilibrium is reached at the original lung volume. This is the sequence of events during each normal respiratory cycle: the lung volume changes leading to a pressure difference, resulting in the flow of air inside or outside the lung and the establishment of a new lung volume [78].

From the above observations, it is possible to deduce that the profile of the pressure is periodic. The main effort in this study is to find suitable boundary conditions for fluid flow simulation of the *trachea* and its first bifurcations. This means that a flow profile, reflecting the complete flux inside the whole lungs, must be modelled. For the sake of simplicity, given the fact that we are coping with an *MHD* problem, the airflow in the bronchioles is considered *laminar*; in this way, the Poiseuille law is used to model the outlet boundary condition for the pressure. Moreover, the pressure in each alveolus is considered the same, and therefore, this value is used as a reference. The flow rate in a laminar pipe flow follows the Poiseuille law:

$$Q = \frac{\pi D^4}{128\mu} \frac{\Delta p}{\Delta L} \tag{39}$$

being $\frac{\Delta p}{\Delta L}$ the pressure gradient along the airway, D the mean airway diameter, μ the air dynamic viscosity. From the above equation, it is possible to deduce the resistance to flow rate, which can be defined as the driving pressure divided by the volumetric flow rate. Poiseuille's law can be rearranged to calculate the resistance, as shown in the following equation:

$$R = \frac{128\mu\Delta L}{\pi D^4} \tag{40}$$

Although the resistance to flow associated with the pipe is inversely related to the fourth power of the radius, the airway contribution is not predominant in the smallest diameter *bronchioles*, because the branches become narrower, but also more numerous. The major site of airway resistance is the medium-sized *bronchi*. Small bronchioles contribute relatively little resistance because of their increased numbers. Most resistance in airways occurs up to the seventh generation of branching vessels. Less than 20% of the resistance is attributable to airways that are less than 2 mm in diameter because of the large number of vessels. Between 25% and 40% of total resistance is found in the upper airways, including the mouth, nose, pharynx, larynx and trachea. In the lungs, fully developed laminar flow probably occurs only in small airways at low Reynolds numbers. Flow in the trachea may be fully turbulent. Much of the flow in intermediate-sized airways is transitional flow, in which it is difficult to predict whether the flow is laminar or turbulent [79]. Nevertheless, to simplify this study, we assume that the flow is laminar, which is verified for a *Reynolds number (Re)* smaller than 2000, where *Re* is defined as

$$\text{Re} = \frac{\rho U D}{\mu} \tag{41}$$

being D the pipe diameter, ρ the air density, μ the air viscosity and U the mean fluid velocity.

A real flow rate profile of air during a respiratory cycle can be found in the literature [80]; secondly, the profile is digitised and saved in a *csv* file format, and finally, it is imported into the *Matlab* package. We define a *linspace* for the temporal variable and an array with the values of the plot for the function to study. We execute the *Fast Fourier Transform (FFT)* routine and compute the coefficients of the Fourier series, c_n, φ_n, i.e. the modulus and the phase of the complex Fourier coefficients. Once we assess the parameters of the series, we find the coefficients necessary to reproduce the flow rate. In detail, every coefficient has the following form:

$$f_{\text{rec}} = \sum_{n=0}^{N-1} c_n \cos(n\omega t + \varphi_n) \tag{42}$$

being f_{rec} the reconstructed profile, c_n the modulus of the Fourier coefficient of the n-th harmonic, φ_n the phase of the Fourier coefficient of the n-th harmonic, ω the fundamental pulsation of the periodic function.

Figure 1 [1], presents the reconstructed profile of the flow rate inside the airways.

The inlet velocity profile, reconstructed from the Fourier analysis, has the following form and is shown in Fig. 2 [1],

$$U_{\text{in}}(t, r) = 2U_0\left(1 - \left(\frac{r}{R}\right)^2\right) + 2\sum_{n=1}^{N-1} \Re\left(U_n \Psi(\tau_n, r) e^{jn\omega t}\right) \tag{43}$$

where

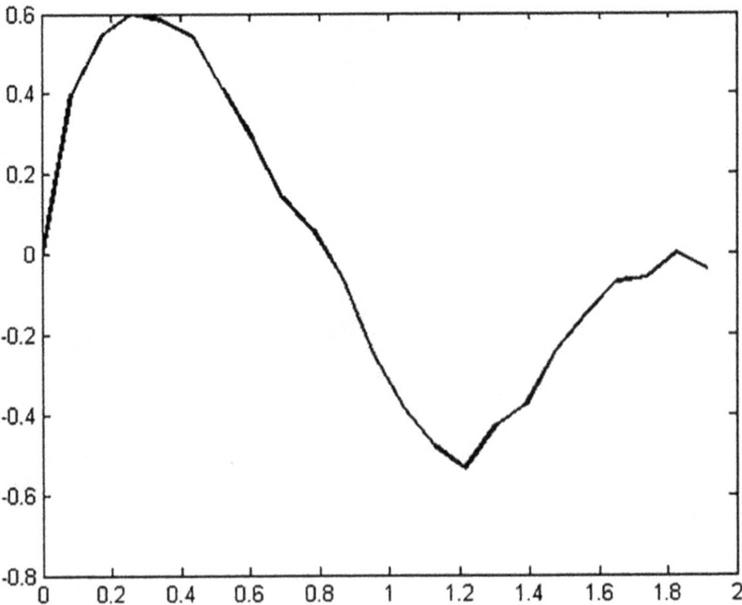

Fig. 1 Pulmonary airflow profile (l/s) versus time (s) [1]

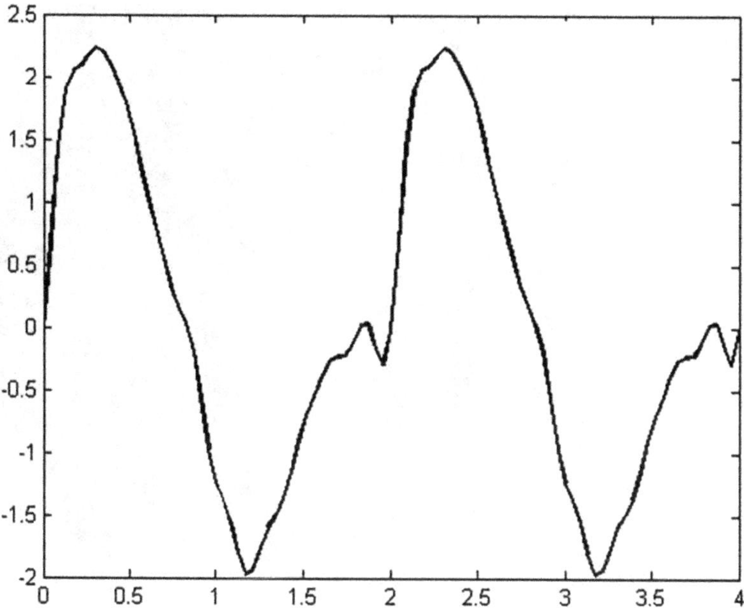

Fig. 2 Velocity (m/s) of the airflow versus time (s) [1]

$$\Psi(\tau_n, r) = \frac{J_0(\tau_n) - J_0\left(\tau_n \frac{r}{R}\right)}{J_0(\tau_n) - 2\frac{J_1(\tau_n)}{\tau_n}} \tag{44}$$

and

$$\tau_n = j^{\frac{3}{2}}R\sqrt{\frac{\rho}{\mu}\omega_n} = j^{\frac{3}{2}}\alpha_n \tag{45}$$

being U_0 the average flow velocity, R the inlet radius, $\Re()$ the real part of a given value, U_n the complex Fourier coefficients of the velocity. It must be noted that $U_n = \frac{c_n}{A\,in}$, being A_{in} the inlet area.

3.3   Trachea and Respiratory Airways Geometry

The first step, in order to assess the *CFD* simulations, is to have the right geometry. In order to perform this task, we use *DICOM images* of the whole-lung low-dose *CT images* of the chest, coming from the public database of [80], and we use *VMTK* to generate the necessary *3D* geometry. In the *VMTK package*, described previously, we modify some settings specifically for the airways. Since in medical images the air is 'black' because of its low density, in order to visualise the airways, the first thing

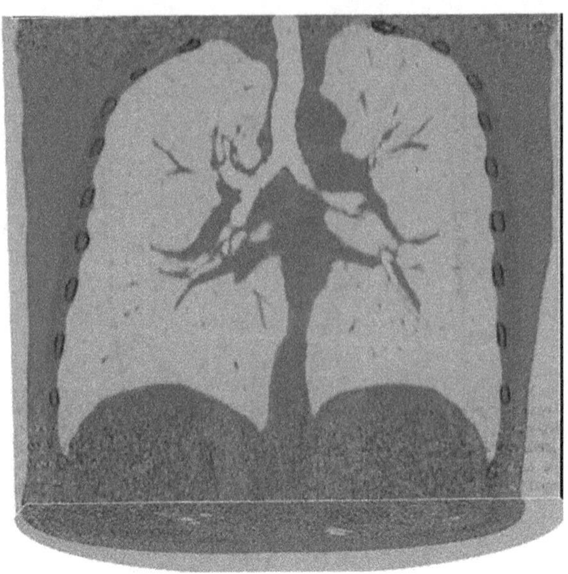

Fig. 3 CT lung grey scale shifted [1]

to do is to shift the grey scale of the same images in order to gain brightness, as Fig. 3 [1], shows.

Afterwards, a volume of interest, the so-called *VOI,* is chosen. This is a small portion of the previous *CT* image, in which there are structures useful for the reconstruction of the geometry. The *Level Sets Method* is applied to generate an image-based surface model from a *vti* image. In order to build the desired geometry, we handle *Colliding fronts*; this type of initialisation consists of placing two seeds on the image. Two fronts propagate from the seeds, with their speed proportional to the image intensity. The region where the two fronts cross is the initial deformable model. This type of initialisation is very effective when it is necessary to initialise the tract of a vessel, and in this case, the tract of the airways. By using *level sets,* it is possible to reconstruct a vessel in separate chunks and then merge the result into a single final model. It is possible to proceed in steps, mixing initialisation types, which can be quite useful. Additionally, it is also possible to save level sets into a file and resume the segmentation later on, adding the result to the first model [81]. For the present geometry, we add a lower threshold, equal to 800, because the airways are soft, and in order to visualise the structures, we need to use a low threshold value. For the upper threshold, a zero value is set-up.

There are also other important parameters to set-up for the correct representation of the geometry:

- *Number of iterations* is the number of deformation steps the model performs. For numerical reasons, the distance scanned by the model depends on some parameters, among which are *voxel* size and image gradient modulus intensity.
- *Propagation scaling* is the weight assigned to model inflation.

- *Curvature scaling* is the weight assigned to model surface regularisation (this can eventually make the model collapse and vanish if it is too strong).
- *Advection scaling* is the most important weight. It regulates the attraction of the surface of the image gradient modulus ridges, which is ultimately what is desired.

From experience, it is recommended that propagation and curvature should be set to 0, and advection to 1. This is possible if the initialisation is sufficiently close to where you want the surface to converge. The number of iterations should be large enough for the level set not to move anymore. It must be noted that setting advection to 1 and propagation and curvature to 0 will robustly lead to reproducible results [81]. In Fig. 4 [1], it is possible to depict how the Colliding fronts work.

When the *Level Sets Method* has converged, the zero-level image is generated, and this can be used to generate the surface by the *Marching Cubes* algorithm. Figure 5 [1], shows the final result of the procedure.

Once the surface is obtained, the next step is to prepare the surface for meshing. This step is made of three parts:

- *Smoothing*
- *Opening the surfaces*
- *Extension flows*

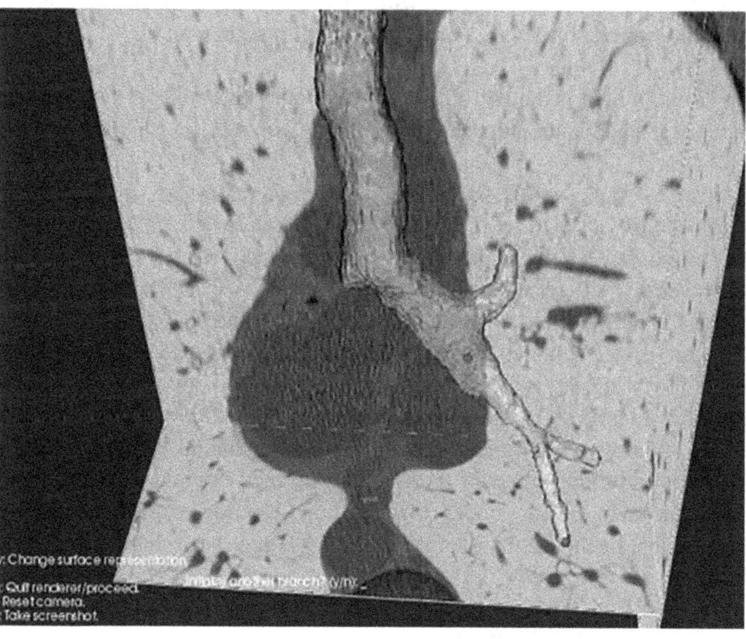

Fig. 4 Colliding fronts [1]

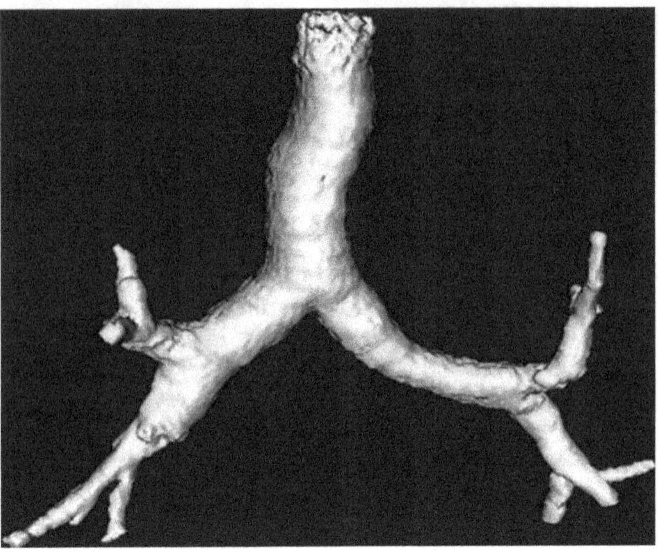

Fig. 5 Marching Cubes algorithm [1]

The smoothing is a necessary operation because in the reconstructed images, there can be some artificial elements, which can affect the *WSS* distributions. During the smoothing operation, there are two important parameters:

- *Pass-Band:* which is the cut-off spatial frequency of the low-pass filter
- *Iterations:* which is the number of smoothing passes

For this geometry, we use two different types of smoothing: in the first smoothing, a pass-band value equal to 0.005 and 40 iterations is imposed, while, in the second smoothing, we fix a pass-band equal to 0.001 and 30 iterations. This is necessary because after the first smoothing, we notice that a branch has too many bumps. The result of these two smoothings is shown in Fig. 6 [1].

If the geometry is generated using a deformable model, the surfaces at inlets and outlets are likely to be closed; therefore, it is necessary to open them by clipping the endcaps [20]. Finally, it is necessary to add the inlet and outlet flow extensions to ensure that the flow entering and leaving the computational domain is fully developed. This is important in order that fully developed boundary conditions do not force the solution in the tract under consideration. The flow extensions are cylinders, and it is possible to choose their length. For this geometry, we impose 6 diameters in length, as can be seen in Fig. 7 [1].

3.3.1 Grids for the Trachea

In order to obtain the final grid, a radio-adaptive element mesh, in which the dimensions of the elements of the grid are related to the dimensions of the radii, is

Fig. 6 Smoothing [1]

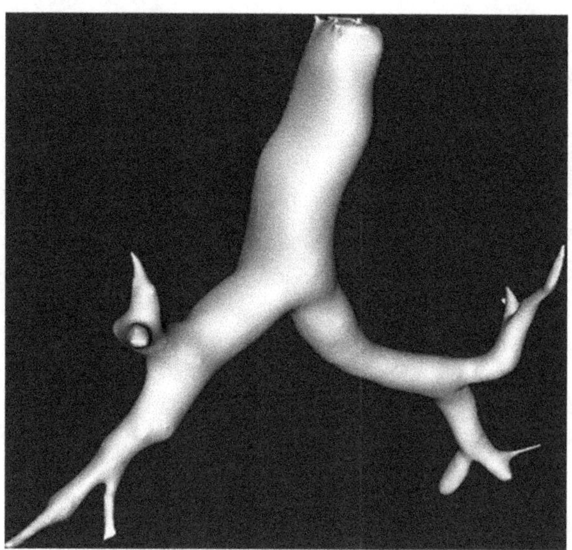

Fig. 7 Flow extensions [1]

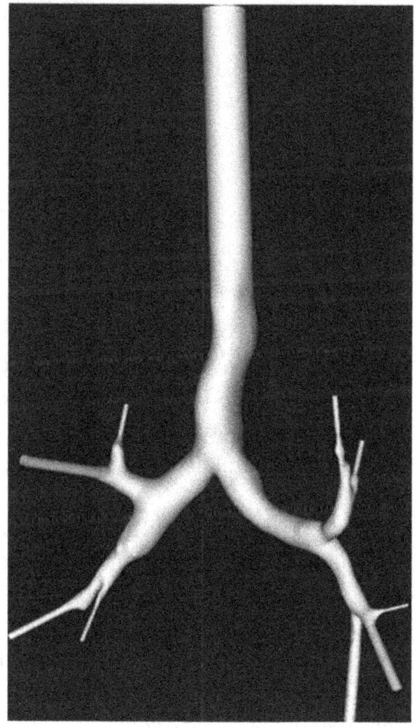

Fig. 8 Radio-adaptive element mesh [1]

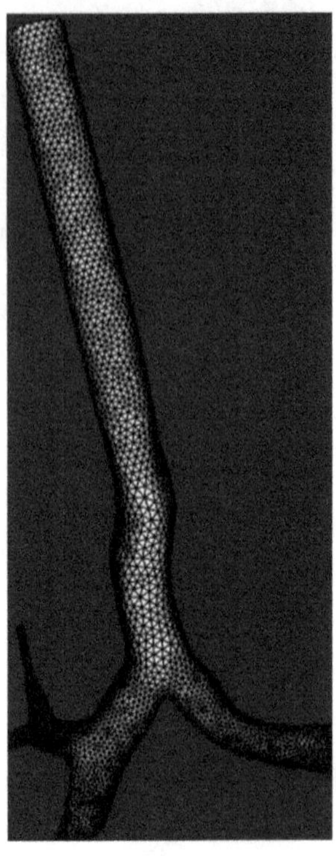

Table 1 Grid characteristics

	Grid 1	Grid 2	Grid 3	Grid 4
Elements	144,712	514,723	853,982	1,245,423
Nodes	30,620	100,241	162,332	233,728
Skew	0.737467	0.733254	0.735652	0.826272

generated. We realise four different grids with different dimensions of the tetrahedron's edge length; in particular, we fix the length equal to 0.25, 0.28, 0.33 and 0.5. Finally, we adjust the value of the skewness by fixing the value of the maximum dihedral angle to 75°. Figure 8 [1], illustrates the result of the mentioned procedure.

We implement four different grids, and Table 1 shows the features of each one.

Figures 9 and 10 [1], show the geometries with their four grids. As already discussed in the previous chapters, *VMTK* creates a grid in a format that is not supported by *OpenFOAM*, so in order to run a simulation, we convert the geometry into an appropriate file, which can be read by *blockMeshDict* in *OpenFOAM*. Afterwards, we define in the boundary file the name of the patches so that we can later give the boundary conditions for those geometry zones.

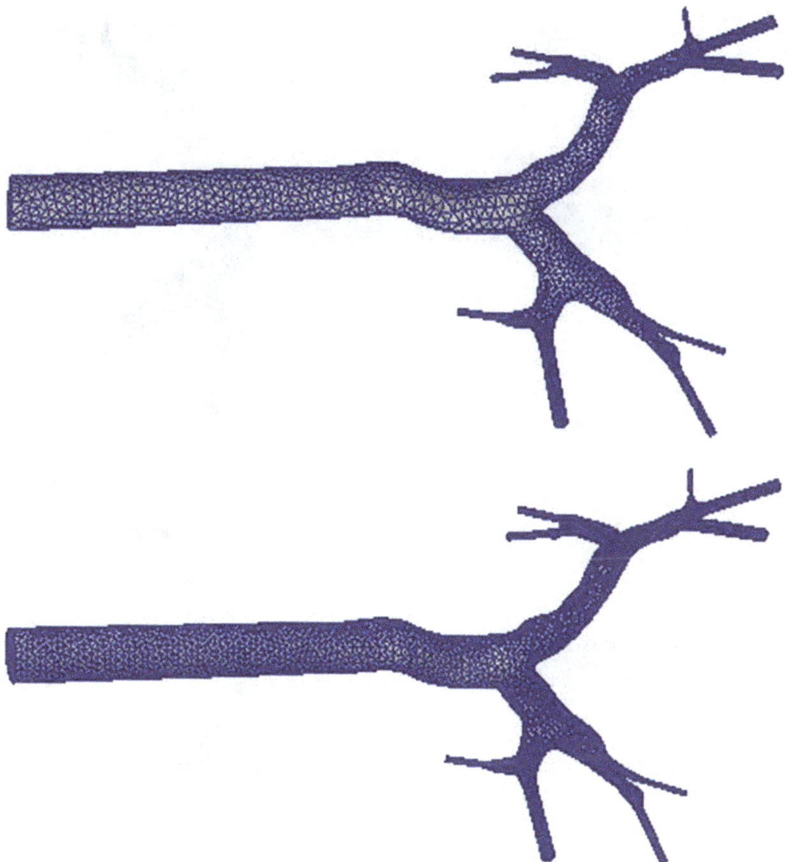

Fig. 9 Grid 1 (top) and Grid 2 (bottom) [1]

4 Numerical Thermo-Haemo-Dynamics (THD) and Magneto-Hydro-Dynamics (MHD) Analysis of the Respiratory Airways

4.1 Trachea Steady-State Analysis

In order to decide which of the four grids is convenient to use, we run steady-state simulations. First of all, we specify the boundary and the initial conditions and the schemes necessary for those specific cases. Secondly, the steady-state simulations for all four grids are performed and the results compared, based on the *WSS*. Finally, once the most convenient grid is found, we set-up the conditions for the unsteady simulation.

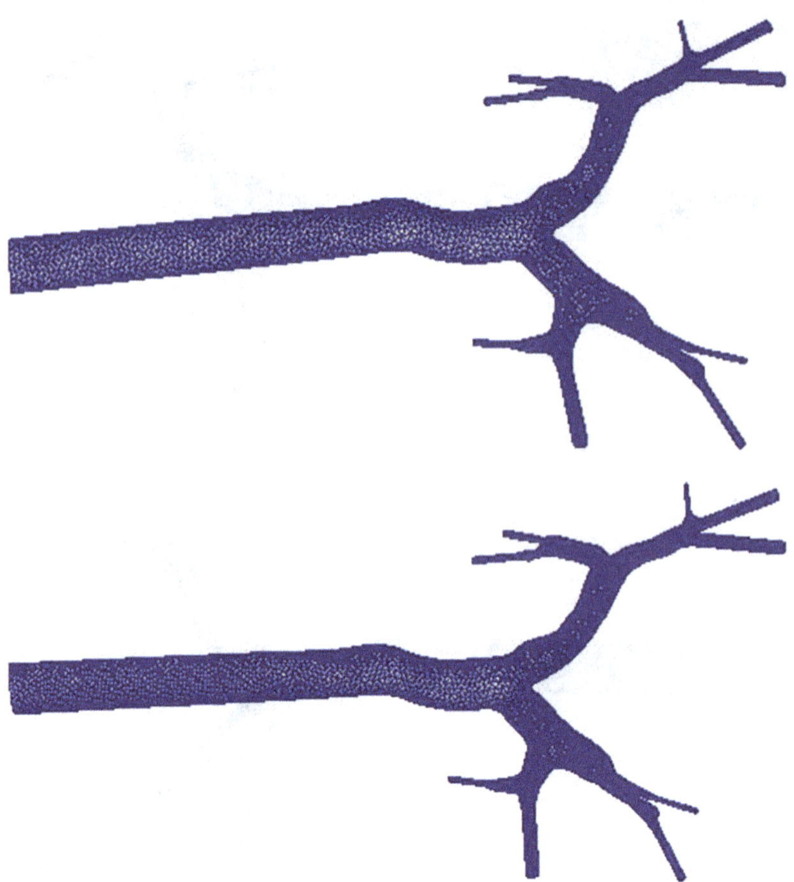

Fig. 10 Grid 3 (top) and Grid 4 (bottom) [1]

4.1.1 Parameter Settings

In the constant folder, there is the boundary file, in which the geometry patches are specified. The patches are renamed as shown in Fig. 11 [1].

In the first-time directory, the boundary and initial conditions are set-up. We want to investigate the fluid dynamics in the trachea and in the first branches of the bronchioles. For the pressure, a *uniform field* is initialised, while for every patch, a *zeroGradient* boundary condition is imposed. The velocity field is initialised to zero everywhere. As a boundary condition for the walls, we use *no-slip conditions*. At the inlet, we set-up a *groovyBC* condition because we want to impose a parabolic profile. For the remaining patches, we impose the flow rate relative to every branch. In *OpenFOAM*, there is a way in which this type of time-varying boundary condition can be assigned, and it is called *flow Rate Inlet Velocity*. In general, for the previous boundary condition, it is important to specify the value of the flow rate, and it is possible to do that with the *volumetric Flow Rate* condition. To evaluate the

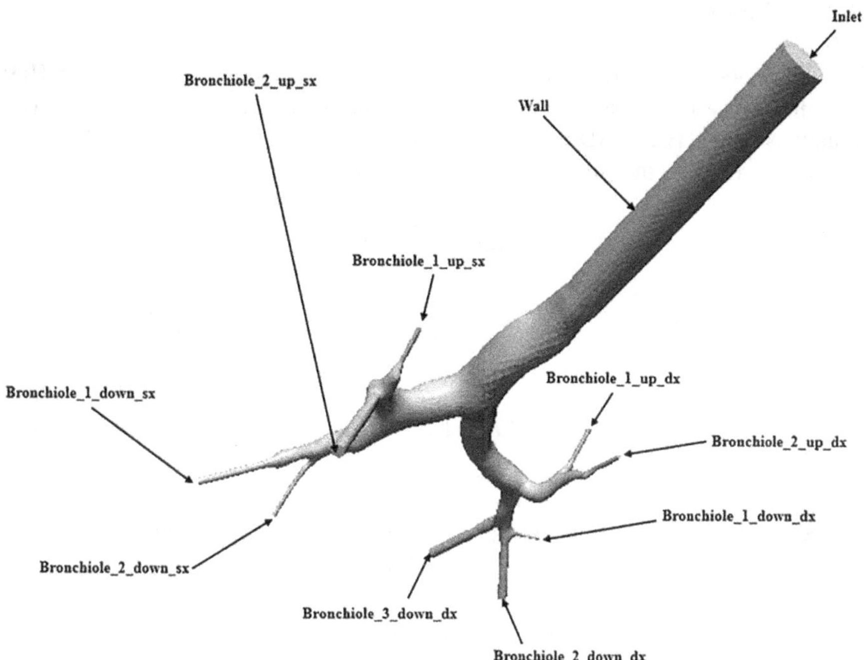

Fig. 11 Definitions of the patches for the boundary conditions [1]

volumetric flow rate at each outlet, we use the code described previously, from which we obtain, for the first temporal instant and for each outlet, the value of the velocity. From the continuity equation, we assess the value of the volumetric flow rate, knowing the value of all cross sections and of the velocity. In the end, we set-up these values inside the velocity file of *OpenFOAM*, and subsequently, the simulation can be done.

Finally, in the *system* folder, the time parameters are set. In the *fvSchemes* sub-dictionary, we have to set-up the temporal schemes; in this first simulation, we set-up a steady-state analysis. In the *fvSolution* sub-dictionary, we set-up all the parameters necessary for the solvers for the steady-state study, such as tolerances and correctors for the involved algorithms. The steps in which the algorithm solves the pressure equation, applies the under-relaxation factors and corrects the mass fluxes can be repeated for a prescribed number of times to correct the non-orthogonality [82]. Since it is not a transient simulation, we fix the *startTime* and the *endTime* of the run and decide the number of time steps to save. In this simulation, just the last two time steps are saved, and the simulation runs for a total of 4000 *time-step*.

4.2 Eulerian Model

The Thermo-Haemo-Dynamics (THD) and Magneto-Hydro-Dynamic (MHD) systems are based on the coupling between the Navier-Stokes and the Maxwell equations. The THD-MHD equations can be written for incompressible flow of a conducting fluid, in this specific case, air, as follows:

$$\mathrm{div}\left(\vec{v}_a\right) = 0 \tag{46}$$

$$\frac{\partial \vec{v}_a}{\partial t} + \left(\vec{v}_a \cdot \nabla\right)\vec{v}_a = -\frac{1}{\rho_a}\nabla p + \nu_a \nabla^2 \vec{v}_a + \vec{g} + \frac{1}{\rho_a \mu_0}\mathrm{curl}\left(\vec{B}\right) \times \vec{B} \tag{47}$$

$$\frac{\partial \vec{B}}{\partial t} + \left(\vec{v}_a \cdot \nabla\right)\vec{B} = \left(\vec{B} \cdot \nabla\right)\vec{v}_a + \frac{1}{\sigma_a \mu_0}\nabla^2 \vec{B} \tag{48}$$

where the subscript a denotes 'air', ν_a is the kinematic viscosity, ρ_a the density, μ_0 the magnetic permeability in the vacuum, σ_a the electric conductivity, $\vec{g}$ the gravity acceleration, p the static pressure, $\vec{v}_a$ the velocity field, and $\vec{B}$ the magnetic induction field.

The air momentum equation, Eq. (2), does not take into account the particle-air momentum transfer. This is due to the fact that the volume fraction of the nanoparticles is negligible, and so is their contribution to the momentum of the air.

Air is assumed to be incompressible because, in physiological conditions, its velocity is much smaller than the speed of sound, and, therefore, the Mach number is much less than unity, which is the condition required to treat a gas as incompressible [83].

The Eulerian conservation equations, i.e. Eqs. (1)–(3), do not take into account the influence of particles because for particle volume fractions smaller than 10^{-6} the disperse phase does not influence the continuum phase, in agreement with [84, 85]. The effect of the magnetic field on the free ions and the erythrocytes is taken into account by the Lorentz force in Eq. (2). The particle diameter used in this work is 5 nm, which is typical of gold/iron oxide nanoparticles [10, 86], and the particle geometry is spherical. Other shapes can be used, and their effect on the fluid flow is documented in the literature [87–89], but they are not considered here, as the different terms in Eq. (5) should be multiplied by unknown correction coefficients in order to account for shape effects.

4.3 Boundary Conditions (BC)

The solution of the Eulerian system requires appropriate boundary conditions (BC) for each field. A non-slip BC has been imposed for the velocity on the wall. As far as the outlets are concerned, a mixed BC has been used. If the air leaves the domain, the velocity normal derivative is set to zero, whereas if air enters through the boundary, the tangential velocity is set to zero.

The velocity profile is imposed on the inlet. In the steady state, a parabolic profile is used, whereas in the unsteady state, a Womersley-Evans profile is employed, in analogy with [17, 82].

$$v_a(t, \xi) = \frac{8Q}{\pi D^2} \left(1 - \xi^2\right) + 2\Re\left(\sum_{n=1}^{N} V_n \Phi(\tau_n, \xi) e^{j\omega_n t}\right) \tag{49}$$

where

$$\Phi(\tau_n, \xi) = \frac{J_0(\tau_n) - J_0(\tau_n, \xi)}{J_0(\tau_n) - 2J_1(\tau_n)/\tau_n} \tag{50}$$

and

$$\tau_n = j^{\frac{3}{2}} \frac{D}{2} \sqrt{\frac{\rho}{\mu_\infty} \omega_n} = j^{\frac{3}{2}} \alpha_n \tag{51}$$

being $\xi = \frac{2r}{D}$ J_0 and J_1 the zero th and first-order Bessel functions of the first kind, α_n the Womersley numbers of order n, $\Re()$ the real part of a complex number, $j = \sqrt{-1}$ the imaginary unit, V_n the Fourier coefficients of the pulsatile mean velocity profile and N the number of harmonics used to reproduce the flow rate. By using the Fast Fourier Transform (FFT) algorithm, the first ten Fourier coefficients of the flow rate in the trachea, derived from experimental data [25–27], are employed to reconstruct the velocity profile.

The flow rate is such that the Reynolds number varies from 0 to 2536. This means that the flow is laminar for most of the time and becomes transitional only at the peaks of the inhalation and exhalation phases. For this reason, the flow has been treated as laminar in agreement with [22, 23].

As far as the pressure is concerned, in the unsteady state, on all the outlets, the 'resistive' BC is imposed, derived in [16, 28],

$$p = p_b + RQ \tag{52}$$

being p_b the reference pressure inside the branches, Q the volumetric flow rate, and R the hydraulic resistance. The value of the resistance and the reference pressure are extrapolated from the steady-state simulations by imposing the volumetric flow rates and a zero normal derivative condition for the pressure on the outlets.

4.3.1   Wall Shear Stress Analysis

In order to choose one of the four meshes, the results from four different grids are compared by using the *mapFields* utility. We map the *wall shear stress, WSS*, fields for the four geometries in order to compare the grids from the coarsest to the finest one and decide which one is the best compromise between accuracy and speed of the solution. For the *WSS* field, we estimate the differences among the absolute values of each grid. This variable measures the shear stress with this specified boundary. In general, for a Newtonian fluid, it is defined as

$$\tau_{\text{wall}} = \widehat{i}_{\text{axis}} \cdot \mu \left([I] - \widehat{n}_{\text{wall}} \bigotimes \widehat{n}_{\text{wall}} \right) \left(\left[\nabla \vec{v} \right] + \left[\nabla \vec{v} \right]^{T} - \frac{2}{3} \text{div} \left(\vec{v} \right) [I] \right)_{\text{wall}} \widehat{n}_{\text{wall}}$$

$$(53)$$

As shown by the maps, the *WSS* reaches the maximum value in the small branches of the airway tree. Indeed, the *shear stress* is influenced by the geometry and the flow rate. The trachea cross section is larger than the other branches, so, under the assumption that the flow rate is constant, for the continuity equation, the velocity is smaller; consequently, the *wall shear stress* is smaller and is dependent only on the air viscosity. On the contrary, in the small branches of the tree, the wall shear stress is greater due to a reduction of their radii. Moreover, in the proximity of a bifurcation, or in the presence of a bend, the *WSS* increases due to the fact that the laminar flow becomes a reverse flow.

The grid with 514,723 elements, i.e. Grid 2, is deemed a good compromise between accuracy and speed of execution. Moreover, the last steady-state time directory is applied as the initial condition for the unsteady simulation.

4.4 Trachea Unsteady-State Analysis

Once the Grid 2 is chosen, the unsteady-state simulations can be performed. We set-up the initial conditions of the unsteady state from the last steady-state time directory. In the 0 folder, we impose the boundary conditions for the unsteady-state simulation. In order to have a fully developed profile on the *inlet* and at the beginning of the unsteady-state simulation, we use the value of U and p coming from the last time directory of the steady one. The only thing to change is the type of boundary conditions for every patch of the aforementioned fields. For the pressure, we use for the wall and the inlet a *zero Gradient* condition, while for the outlets, we set a *groovy BC*. This last type of condition requires some input value, such as: *value, variables, fraction Expression* and *value Expression*. We write a *Matlab code* to calculate the necessary variables to add to *groovyBC*.

Firstly, from the steady state for each outlet and for the inlet, we get the.*csv* file with the value of velocity, pressure and the cross section area. The previous.*csv* file was created with *Paraview;* once we import the geometry, we use the specific filter *Integrate Variables* to obtain the necessary variables. With *Paraview* we evaluate the components of the normal vector n_x, n_y, n_z, by the filter *normalglyph* for each outlet and for the inlet. In *Matlab*, it is possible to read the.*csv* file by a *csvreader* command. For each patch of the geometry, we evaluate the pressure, the velocity, the flow rate and the resistance. About pressure, we normalise its value, dividing it by the value of the cross section. For the velocity, we do the same thing as for the pressure; each velocity component is divided by the value of the cross section. The flow rate is the integral of the velocity over the surface, so its value is equal to the product of each velocity component for the equivalent component of the normal vector. Finally, the resistance comes from Eq. 41. Once the code has run, we obtain the values of the pressure and of the resistance that we write inside *groovy BC*. In *value,* for each outlet, we write the value of the pressure coming from the code. In *variables*, we declare the parameters used in the definition of the equation to evaluate the pressure, which are the resistance, the flow rate and the pressure for every outlet. In particular, the resistance is evaluated by Eq. 41, and the flow rate is evaluated at each outlet and updated every time-step.

FractionExpression distinguishes between the *Dirichlet* and *Neumann* boundary conditions. In the *valueExpression,* we define the *Dirichlet* condition with the equation for the evaluation of the pressure at each outlet. The general form of the final equation is of the following type:

$$p = p_b + RQ \tag{54}$$

As far as the velocity is concerned, for all the outlets an *inlet Outlet* boundary condition is used. In particular, this condition behaves as a *zero Gradient* when the flow is directed outwards from the boundary and a *fixed Value* when the flow is inwards. For this type of condition, we specify some entries: *inlet Value* in which the direction of the flow is defined. *Value* is the actual value of the field at the boundary. At time 0, it is the initial value of the boundary condition. At the first iteration, *OpenFOAM* does not know if the boundary is an inlet or an outlet, so it needs some initial value to start the simulation. The values of the velocity components, to write in this condition, are evaluated with the code written in *Matlab* and used to calculate the pressure values. For the inlet, a *groovy BC* boundary condition is imposed because the velocity profile is periodic. The implemented profile has been discussed previously.

Finally, we set-up the time parameters in the *system* folder. In the *fvSchemes* sub-dictionary, we set-up the temporal schemes using an *Euler* scheme. In the *fvSolution* sub-dictionary, we set-up all the parameters necessary for the solvers for the unsteady state, such as tolerances and correctors for the involved algorithms. In detail, we do not have to specify the corrector for the *SIMPLE* algorithm because, in the transient case, we need only the *PISO* and *BPISO* correctors.

4.4.1 Zero Magnetic Field

The first simulation is concerned with the case in which the magnetic probe is turned off. We do this to understand the fluid dynamics of the problem without the influence of an external source. The time of observation is set equal to 4 s because it is equal to twice the duration of a respiratory cycle. In order to evaluate the convergence of a time-dependent differential equation, the *Courant-Friedrichs-Lewy* (CFL) criterion is used. It is derived in numerical analysis of explicit time integration schemes when they are used in some numerical solutions, so the time-step must be less than a certain value in many explicit time-marching simulations; otherwise, the simulation

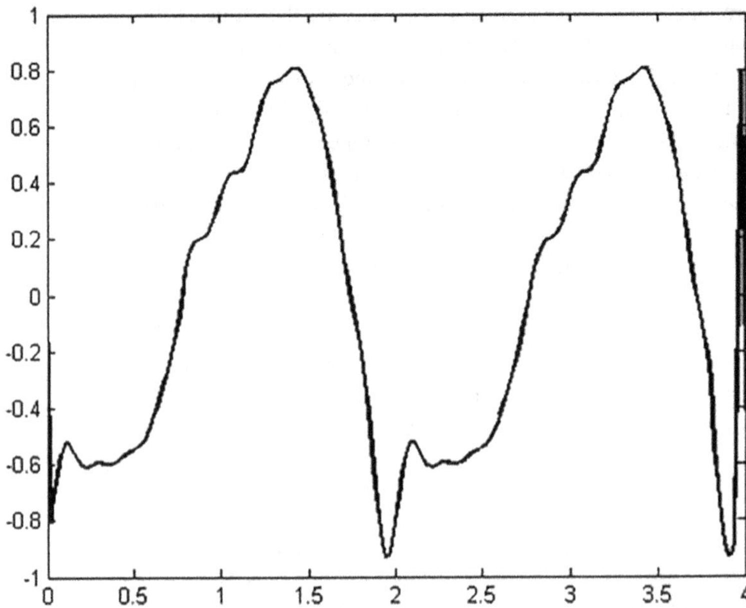

Fig. 12 Flow rate (l/s 10^{-3}) at the inlet versus time (s) [1]

can be unstable. The *CFL condition* relates the length of the time-step to the grid stencil and to the maximum speed at which the information can travel in the grid. The *CFL condition* is prescribed for those terms of the *finite difference approximation* of general *partial differential equations,* which model the *advection* phenomenon. This condition in *OpenFOAM* is necessary to achieve temporal accuracy and numerical stability. The Courant number is defined for one cell as

$$Co = \frac{U\Delta t}{\Delta x} \tag{55}$$

being Δt the time-step, U the value of the velocity orthogonal to the grid boundary, Δx, the grid stencil. The flow velocity varies across the domain, and we must ensure $Co < 1$ everywhere. We therefore choose Δt based on the worst case: the maximum Co corresponding to the combined effect of a large flow velocity and small cell size.

In Fig. 12 [1], we show the time evolution of the *flow rate* on the inlet. In Fig. 12, it is possible to depict that the *inspiratory* and *expiratory phases* have a similar duration in time. In the plot, it is possible to verify the existence of a small amount of negative flow because the trachea, as well as bronchi and bronchioles, do not participate in respiratory exchange, and the gas that fills these structures occupies an anatomical dead space. Moreover, the radius of the trachea is bigger than the branches' radii, so the hydrodynamic resistance decreases, and for the fluid, it is easier to enter the airways due to the existence of a pressure gradient between the atmospheric environment and the respiratory airways.

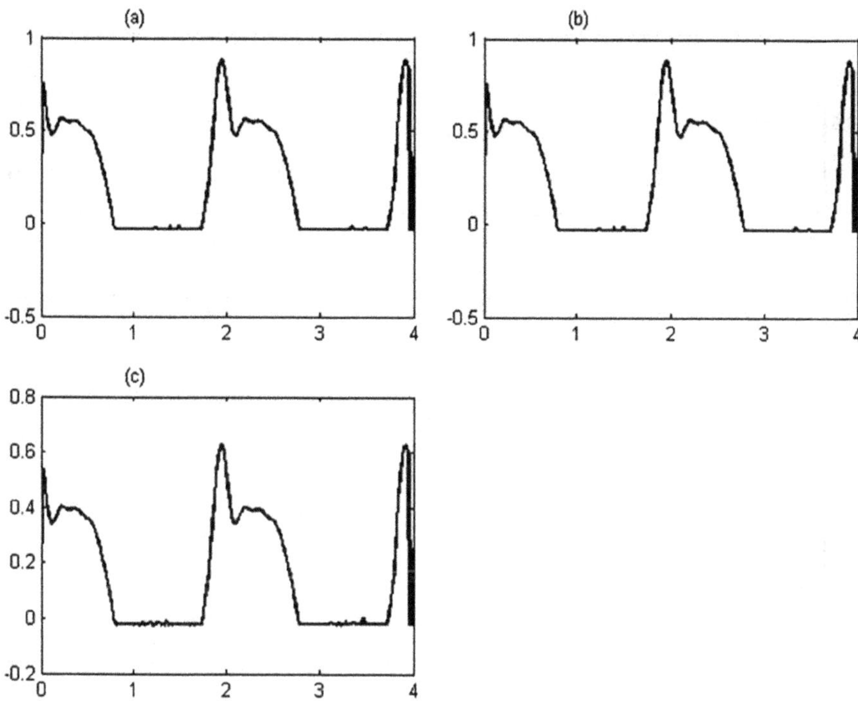

Fig. 13 Pressure profile (cm H_2O) versus time (s) at the down right side. Bronchiole: (**a**) 1; (**b**) 2; (**c**) 3 [1]

In the following, it is possible to see what happens at the outlets of this geometry. We analyse all the outlets, and the plots are obtained, dividing the geometry into four zones, as shown in Fig. 11. Figure 13 [1], presents the pressure profile (cm H_2O) on the outlet of the three bronchioles versus time (s) at the down right side. As can be seen, we have two *respiratory cycles*; the duration of the *inspiratory* phase is approximately 1 s, and the other second is for the *expiratory phase*. The profile of the *pressure* for all four groups of outlets is the same. Between time 0 and 1 s, we are in the *inspiratory phase*, so the airflow can enter, thanks to a *pressure gradient* between the atmosphere and the inside of the thorax. In fact, the atmospheric pressure is higher than the thoracic pressure, and the air can flow from a high-pressure zone to a low-pressure zone. Also, for the *expiratory phase*, a pressure gradient is necessary, but, in this case, it is between the inside of the thorax and the atmospheric environment. At the end of the *expiratory phase*, the pressure inside the thorax is higher than outside, and an outward flow is generated. Moreover, in bifurcations, where the radius decreases, the *hydro-dynamic resistance* increases and consequently also increases the *pressure gradient*. Meanwhile, on the big pipe, the *resistance* is lower as well as the *pressure gradient*. Furthermore, the *pressure* is also related to the velocity, and, inside the small branches, the velocity is higher than

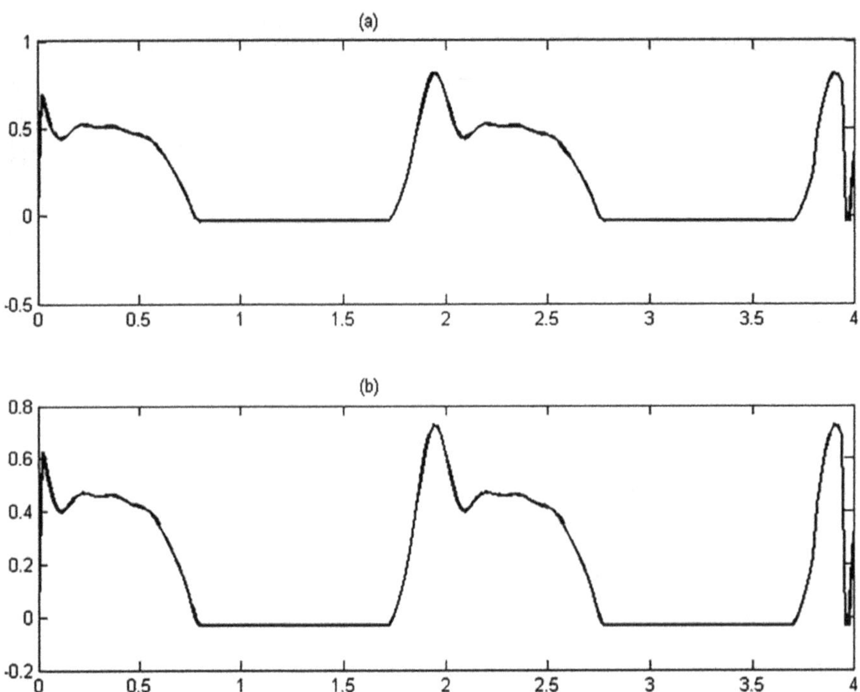

Fig. 14 Pressure profile (cm H_2O) versus time (s) at the up right side. Bronchiole: (**a**) 1; (**b**) 2 [1]

the pressure, compared to the inlet pressure. In the trachea, the cross section is great, the velocity is low, and the pressure is too.

Figures 14, 15, and 16 [1], present the pressure profile (cm H_2O) on the outlet of the two bronchioles, 1 and 2, versus time (s). Similar considerations can be carried on as for Fig. 13.

Figure 17 [1], presents the profile of the *flow rate* (l/s 10^{-3}) on the outlet of the three bronchioles versus time (s) at the down right side. As for the *pressure*, we can recognise a periodic trend. The first cycle starts at 0 and finishes at 2 s; the *inspiratory phase* lasts 1 s, and the *expiratory phase* the remaining 1 s. In the first part of the cycle, on the outlets, due to a reduction of the cross section and of the radii, the velocity increases, and for the continuity equation, the *flow rate* has to increase. In the second part of the previous cycle, we are in the *expiratory phase*, and the flow rate is equal to zero inside the lung because there is a reverse flow towards the atmospheric environment.

Figures 18, 19, and 20 [1], present the pressure profile (cm H_2O) on the outlet of the two bronchioles, 1 and 2, versus time (s). Similar considerations can be carried on as for Fig. 17.

Figure 21 [1], presents the profile of the *velocity (m/s)* at the outlet of the three bronchioles versus time (s) on the down right side. As previously said, the profile shows a periodic trend for the second phase of the respiratory cycle. In the small

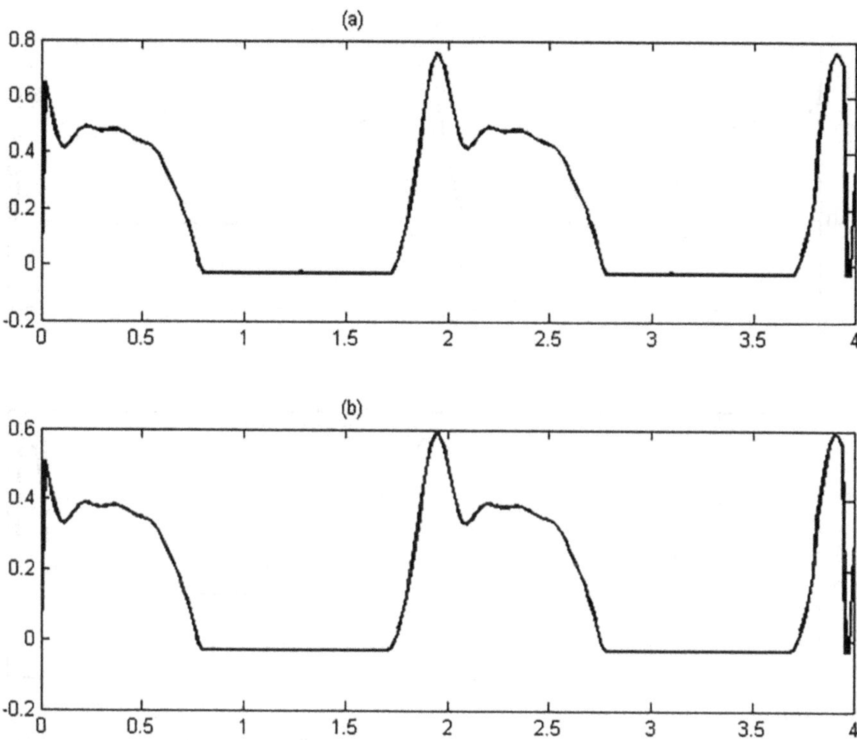

Fig. 15 Pressure profile (cm H_2O) versus time (s) at the down left side. Bronchiole: (**a**) 1; (**b**) 2 [1]

branches of the arterial tree, we observe a cross sectional reduction, and consequently, the velocity increases. During the *inspiratory phase*, we observe a peak for the velocity, whereas during the expiratory phase, we observe a plateau because the flow is towards the outside.

Figures 22, 23, and 24 [1], present the pressure profile (cm H_2O) on the outlet of the two bronchioles, 1 and 2, versus time (s). Similar considerations can be carried on as for Fig. 21.

In the following figures, the maps of *pressure, velocity* and *wall shear stress* are reported. Figures 25, 26, 27, 28, 29, and 30 [1], report the *pressure* maps, (cm H_2O). At time 0, Fig. 25, we are in a rest condition, so everything is steady, and the pressure is the same everywhere in the respiratory tree. At 0.25 s, Fig. 26, we are at the beginning of the *inspiratory phase*, so the pressure is higher towards the inlet and lower inside the bifurcations. The pulmonary volume increases, the pressure decreases, and, due to this pressure gradient, the air flows from outside to inside the thorax. At 0.82 s, Fig. 27, we are at the end of the *inspiratory phase,* and it is possible to see a spot of high pressure due to the high velocity of the inlet flow, while, in the remaining part of the system, the pressure creates the necessary gradient for the respiratory process. At time 1 s, Fig. 28, and 2 s, Fig. 29, we are at the end of the *inspiratory* and *expiratory phase*; the pressure is higher in the large branches

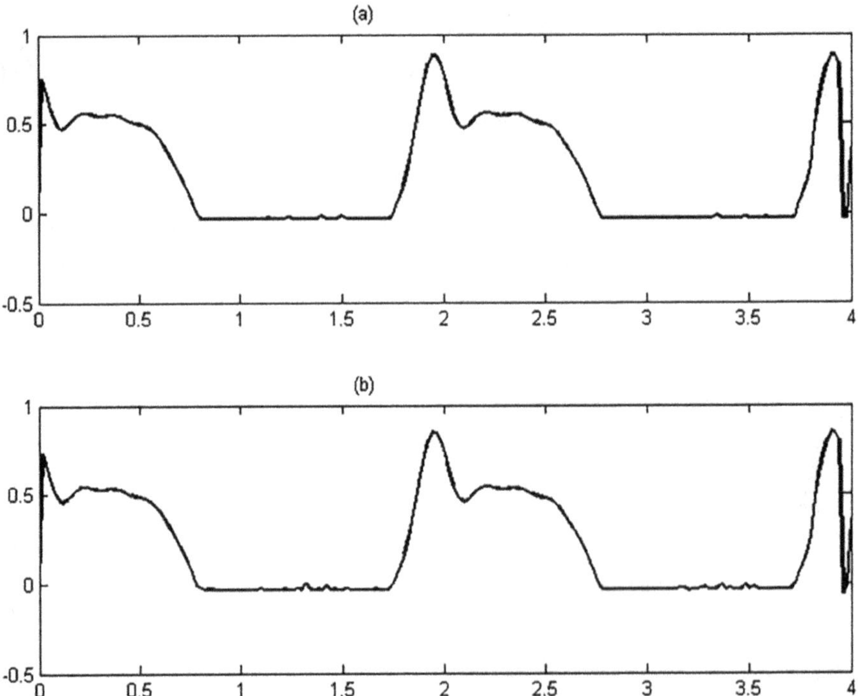

Fig. 16 Pressure profile (cm H_2O) versus time (s) at the up left side. Bronchiole: (**a**) 1; (**b**) 2 [1]

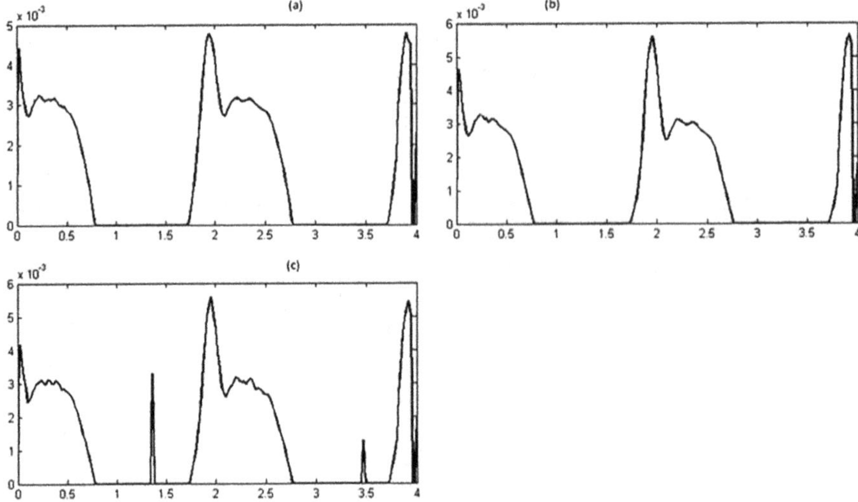

Fig. 17 Flow rate profile (l/s 10^{-3}) versus time (s) at the down right side. Bronchiole: (**a**) 1; (**b**) 2; (**c**) 3 [1]

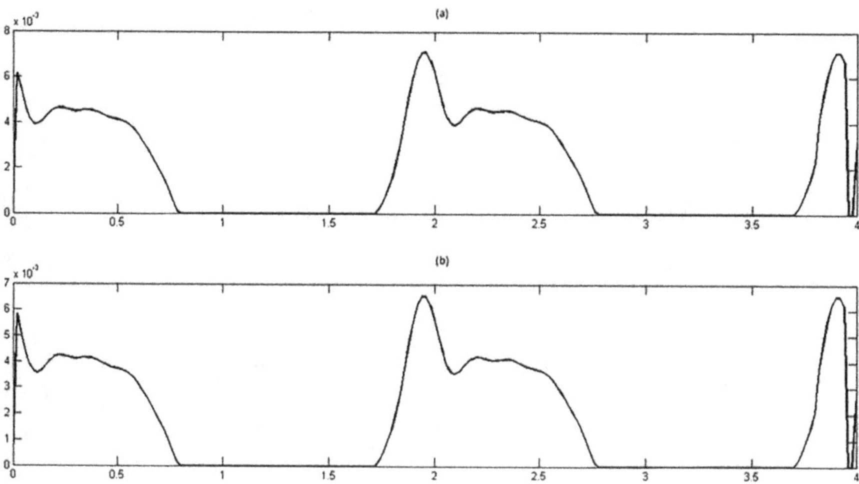

Fig. 18 Flow rate profile (l/s 10^{-3}) versus time (s) at the up right side. (**a**) 1; (**b**) 2 [1]

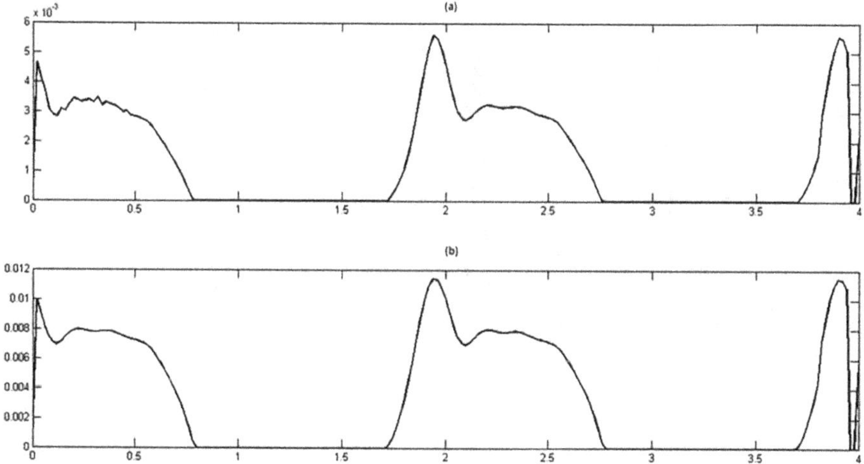

Fig. 19 Flow rate profile (l/s 10^{-3}) versus time (s) at the down left side. (**a**) 1; (**b**) 2 [1]

than in the small bifurcations because the air is in a steady state and there is no transition of the gas between inside and outside. At 1.35 s, Fig. 30, we are in the *expiratory phase*; the pressure at the beginning is equal in all the structures; subsequently, due also to the movement of the expiratory muscles, the thorax is contracted, its volume decreases, the pressure increases, and the air flows outside.

The map fields of the *wall shear stress*, (dyne/cm^2), are discussed without showing the relative figures. At time 0 s, we are at the beginning of the *inspiratory phase*, so in a rest condition, there is no flow through the trachea; also, the velocity is zero, and the wall shear stress is at its minimum value. As the inspiratory phase flows

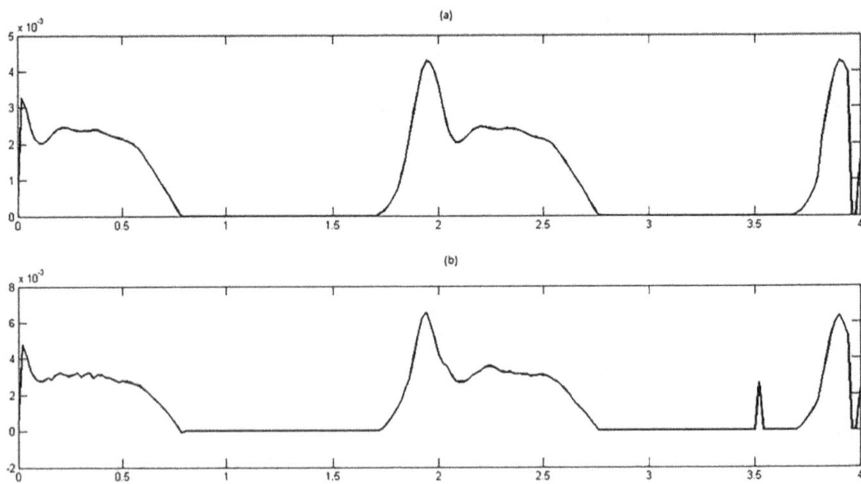

Fig. 20 Flow rate profile (l/s 10^{-3}) versus time (s) at the up left side. (**a**) 1; (**b**) 2 [1]

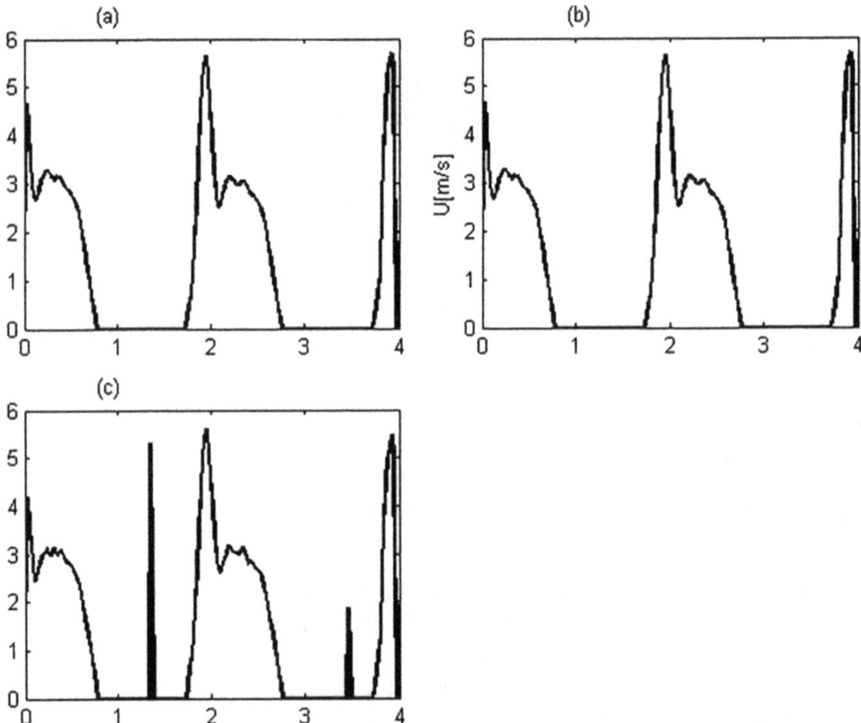

Fig. 21 Velocity profile (m/s) versus time (s) at the down right side. Bronchiole: (**a**) 1; (**b**) 2; (**c**) 3 [1]

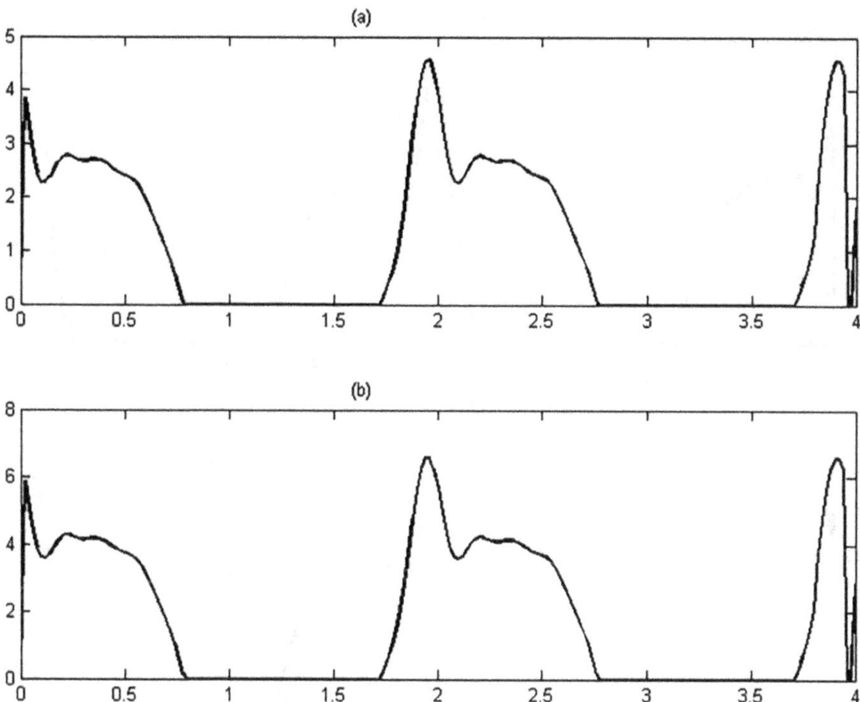

Fig. 22 Velocity profile (m/s) versus time (s) at the up right side. Bronchiole: (a) 1; (b) 2 [1]

by, at time 0.5 s, we are at the peak of the inspiratory phase, so the flow rate is maximum inside the trachea, and this explains the high value at the wall of the inlet. In the small bronchioles, the wall shear stress starts to increase due to the fact that the flow is already turbulent and the radii decrease. At time 1 s, we are at the end of the inspiratory phase; the flow rate inside the trachea is now zero, and the wall shear stress is minimum in this region and maximum in the small branches for the previous reasons. Now the expiratory phase starts; at time 1.25 s, we are in the opposite situation of the inspiratory phase. The flow in the bronchioles is at its highest value, and the wall shear stress, due to the geometry and to the air velocity, is also at its highest value. In the trachea, the flow is zero, and the velocity is smaller, so the wall shear stress is lower too. As the expiratory phase flows by for a time equal to 1.5 s, the wall shear stress starts to decrease, reaching its minimum value inside the trachea. At the end of the expiratory phase, 2 s, the flow in the trachea is equal to zero; the wall shear stress is already smaller, whereas it is higher near the bifurcations of the branches due to the high velocity of the flow, which is moving towards the outside.

Figures 31, 32, 33, 34, 35, and 36 [1], depict the map for the velocity streamlines, (*m/s*). At the beginning of the inspiration, time 0 s, Fig. 31, we are in a rest condition, and the velocity is equal in all respiratory trees. As the cycle goes on, at time 0.4 s, Fig. 32, the flow is turbulent even if the velocity is low, due to the geometry of the

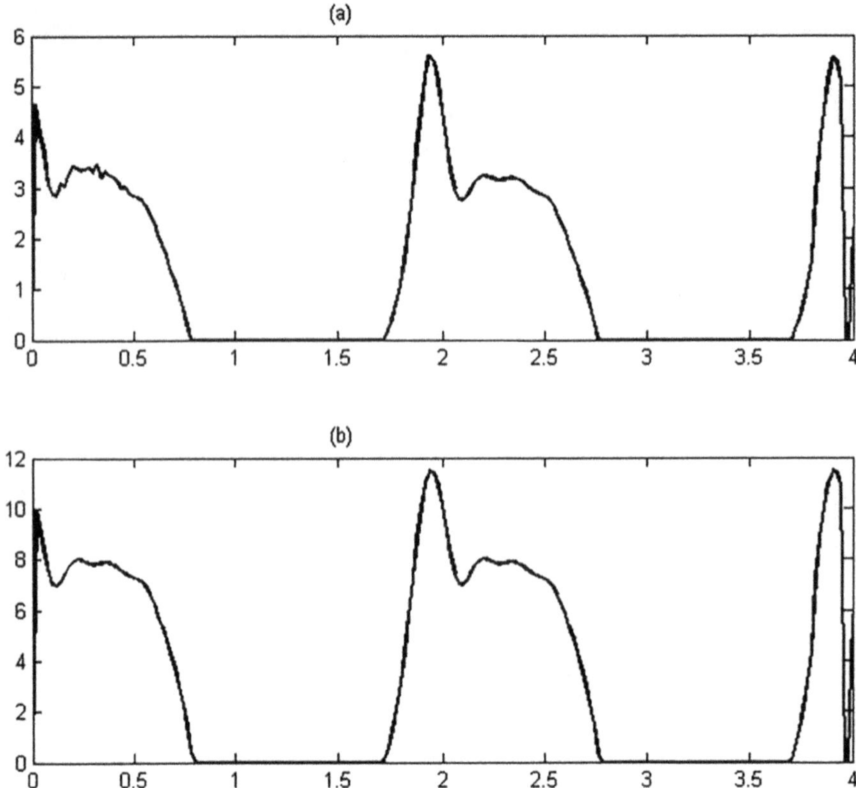

Fig. 23 Velocity profile (m/s) versus time (s) at the down right side. Bronchiole: (**a**) 1; (**b**) 2 [1]

respiratory system. At the end of the inspiratory phase, at 0.8 s, Fig. 33, the flow is turbulent, with the appearance of some vortices, especially near the bifurcations, where the velocity gradient becomes higher due to a reduction in the cross section. The highest value of the velocity is reached at the inlet because the air coming from outside has high kinetic energy. During the expiratory phase, at 1.4 s, Fig. 34, there are vortices near the bifurcations due to the changes in the geometry. At time 1.8 s, Fig. 35, we are at the end of the expiratory cycle; we can observe a reverse flow, and, again, the highest velocity is reached at the inlet. At the end of the respiratory cycle, at 2 s, Fig. 36, we have a maximum value of the velocity inside the small branches and a null velocity at the inlet.

In conclusion, we can say that for the velocity, it is possible to deduce that the profile is turbulent at the beginning of the system and becomes laminar inside the bigger channels. In the proximity of a bifurcation, the flow becomes turbulent even if the Re is smaller than 2000, due to the geometric properties of the respiratory system. It goes back to laminar in the distal airways.

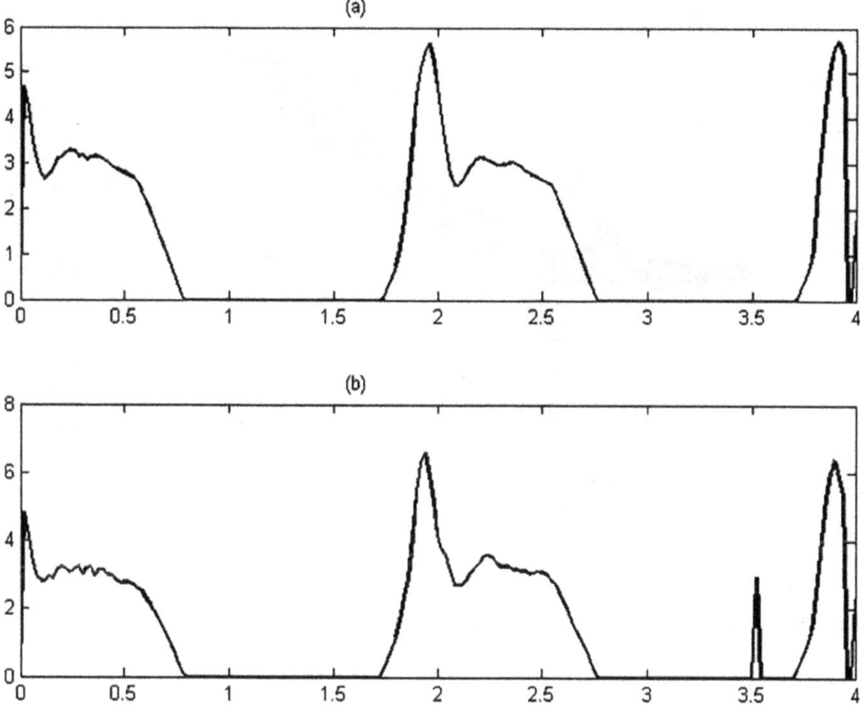

Fig. 24 Velocity profile (m/s) versus time (s) at the up right side. Bronchiole: (**a**) 1; (**b**) 2 [1]

5   Particle Tracking

5.1   Lagrangian Model

Let us consider a particle of diameter d_p, velocity $\vec{v}_p$, and density ρ_p, whose centre position is $\vec{x}_p$. In a Lagrangian frame of reference, the position of each particle is obtained by the integration of its velocity,

$$\frac{d\vec{x}_p}{dt} = \vec{v}_p \tag{56}$$

which is evaluated from the momentum conservation equation, written as follows

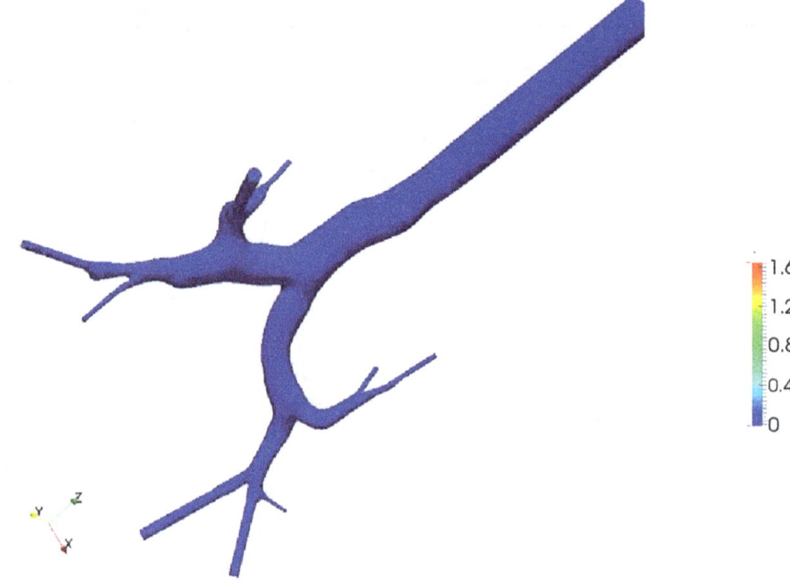

Fig. 25 Pressure map (cm H_2O) at $t = 0$ s [1]

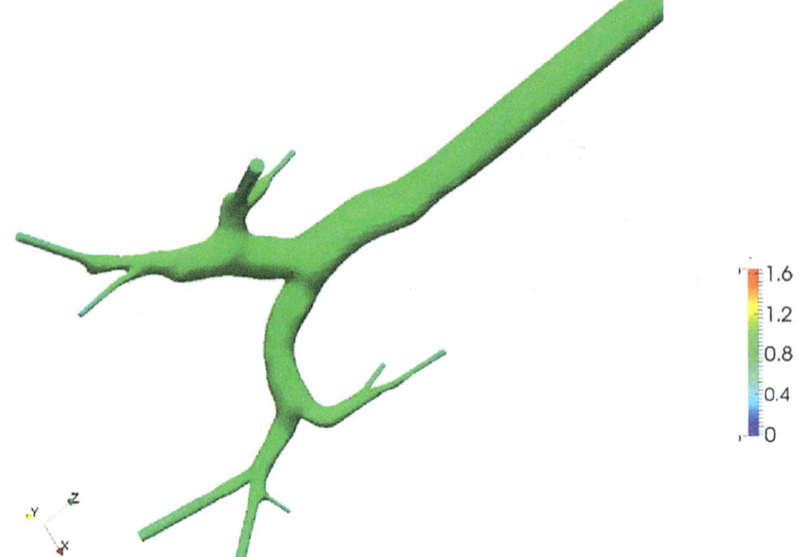

Fig. 26 Pressure map (cm H_2O) at $t = 0.25$ s [1]

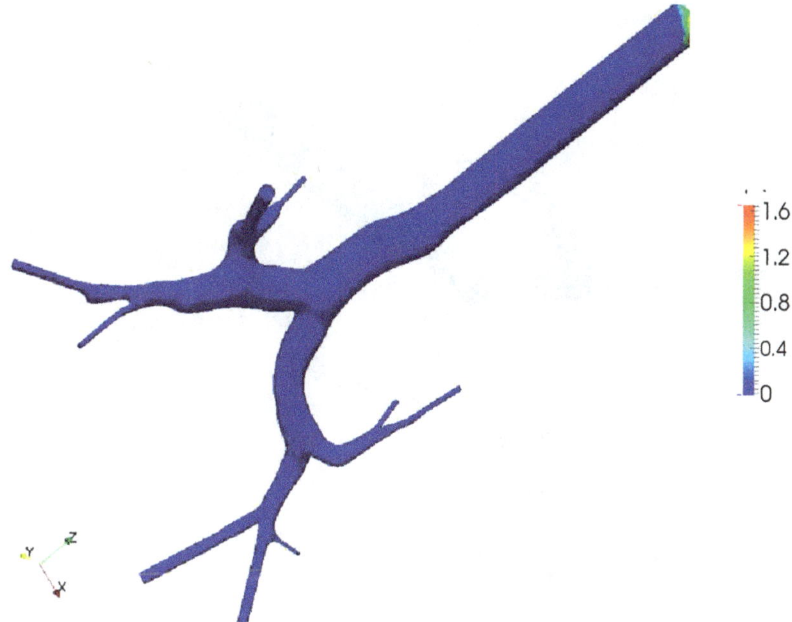

Fig. 27 Pressure map (cm H_2O) at $t = 0.82$ s [1]

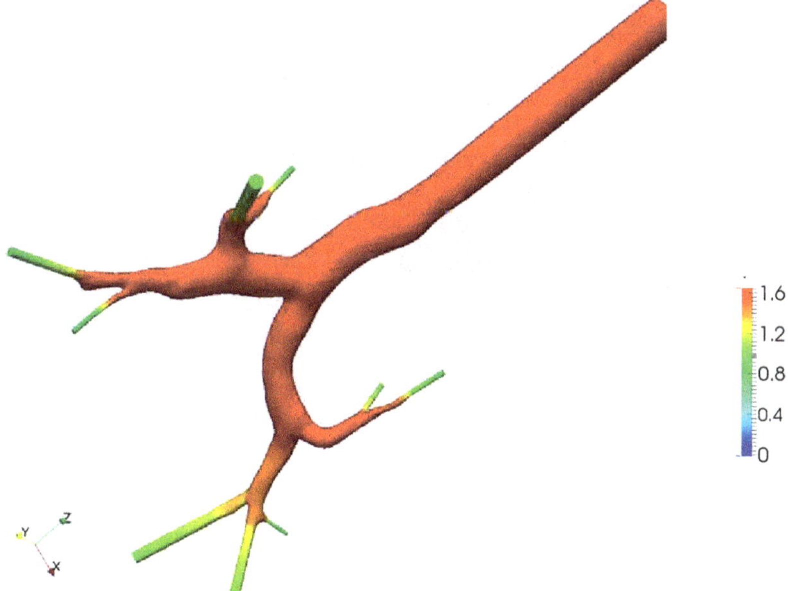

Fig. 28 Pressure map (cm H_2O) at $t = 1$ s [1]

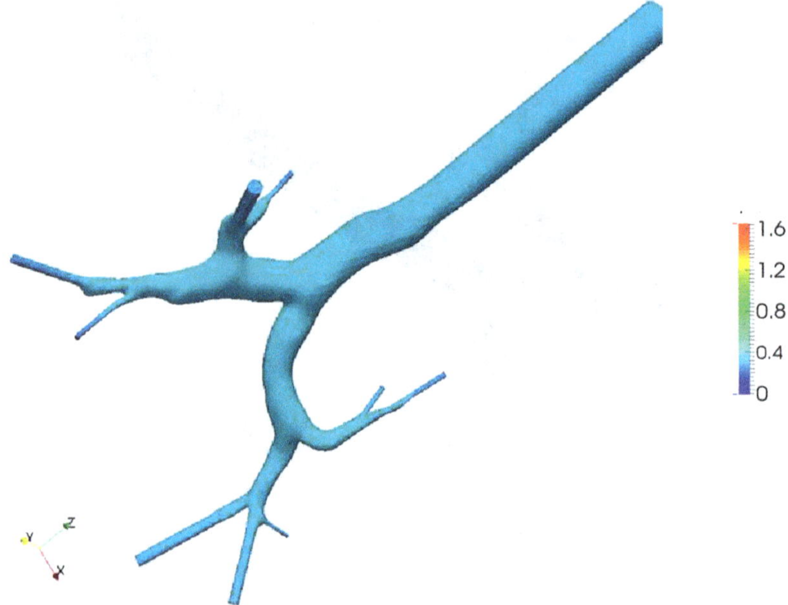

Fig. 29 Pressure map (cm H_2O) at $t = 1.35$ s [1]

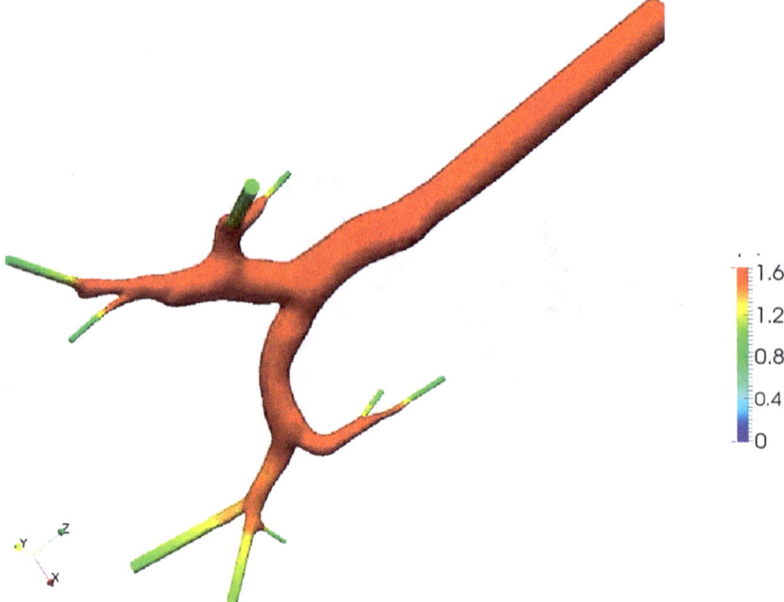

Fig. 30 Pressure map (cm H_2O) at $t = 2$ s [1]

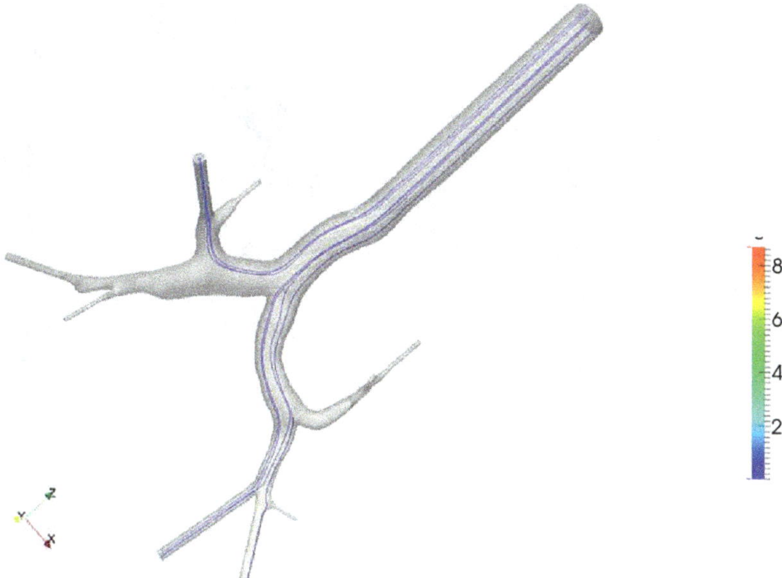

Fig. 31 Velocity streamlines, (m/s), map at $t = 0$ s [1]

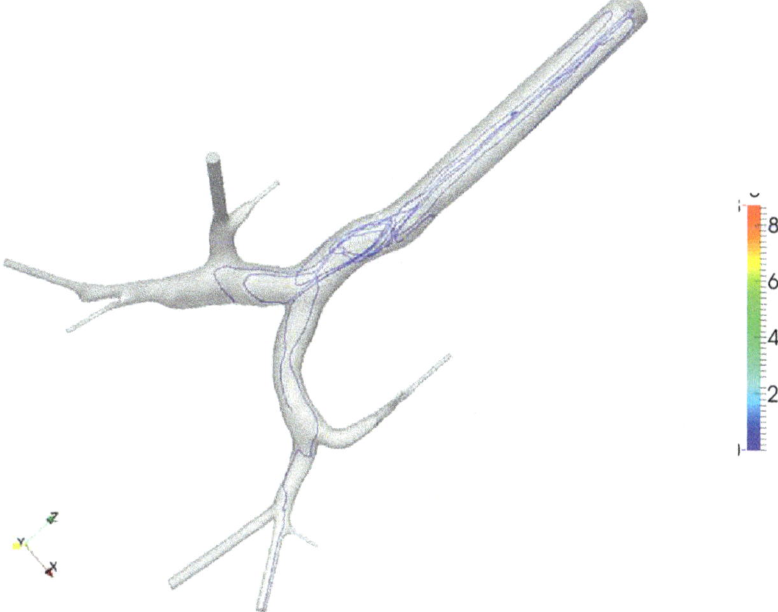

Fig. 32 Velocity streamlines, (m/s), map at $t = 0.4$ s [1]

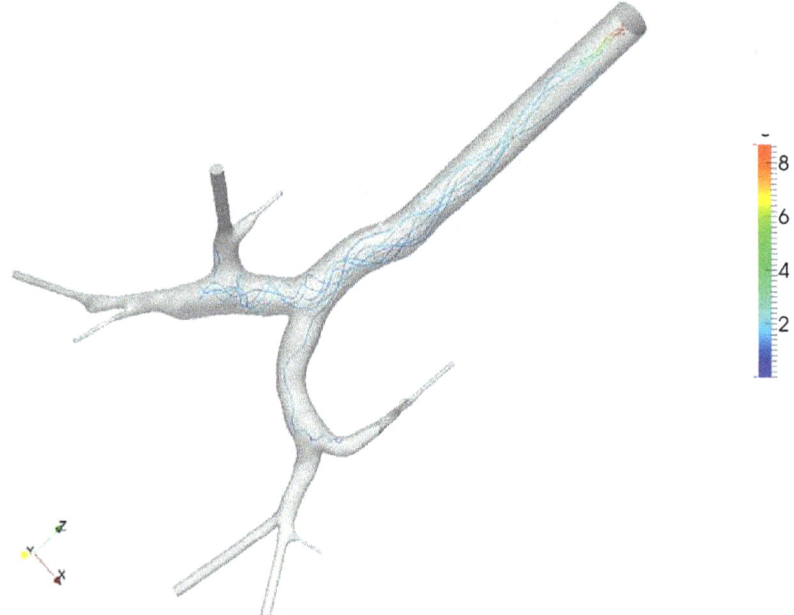

Fig. 33 Velocity streamlines, (m/s), map at $t = 0.8$ s [1]

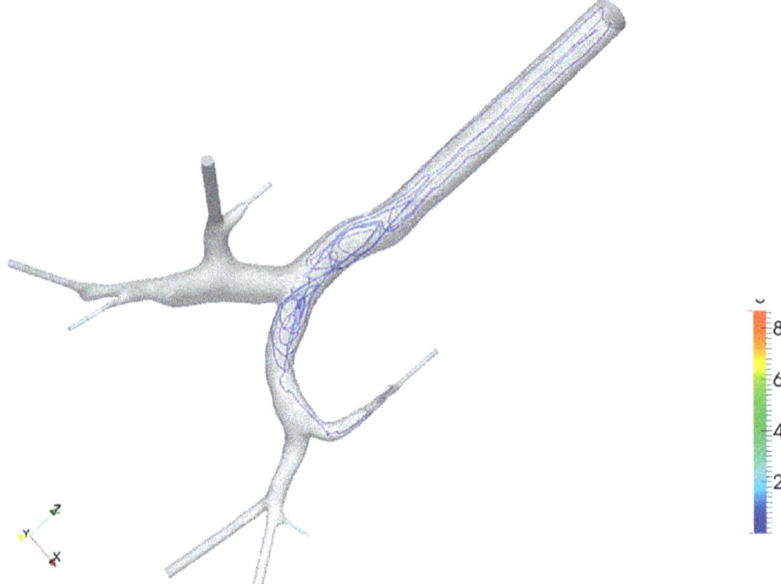

Fig. 34 Velocity streamlines, (m/s), map at $t = 1.4$ s [1]

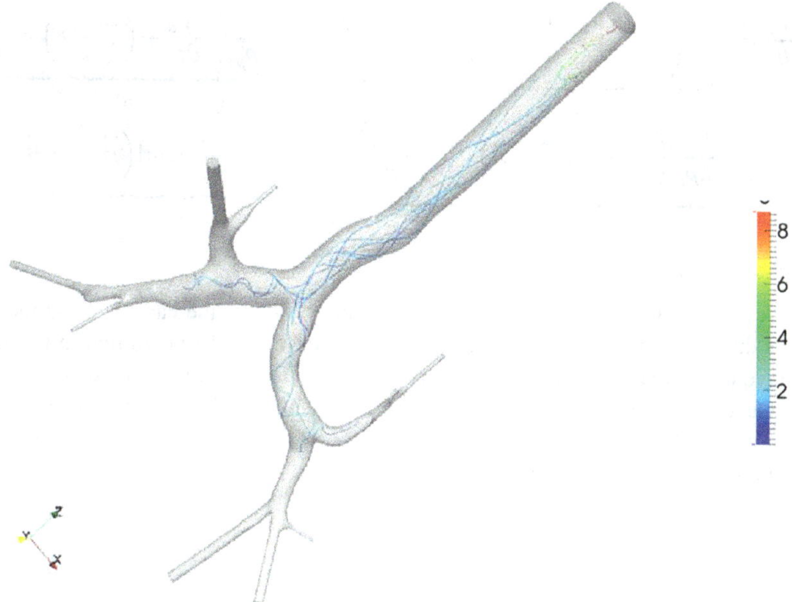

Fig. 35 Velocity streamlines, (m/s), map at $t = 1.8$ s [1]

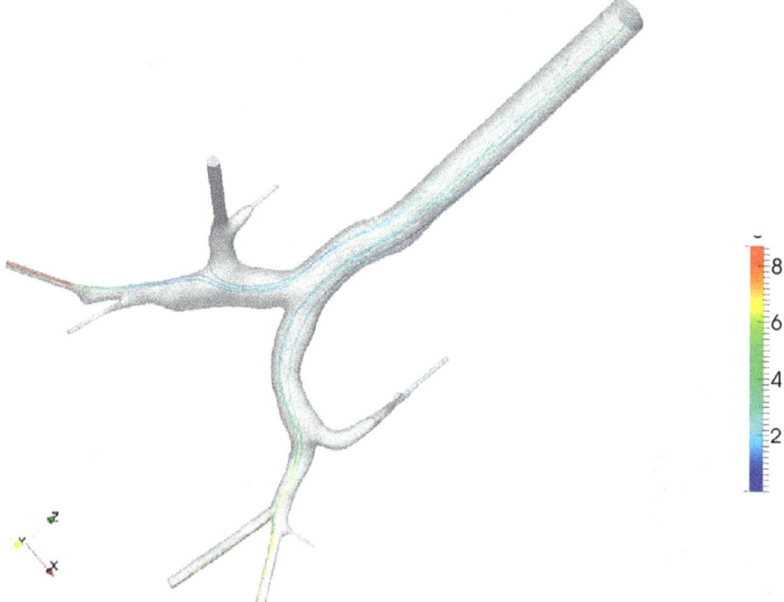

Fig. 36 Velocity streamlines, (m/s), map at $t = 2$ s [1]

$$\frac{d\vec{v}_p}{dt} = -\underbrace{\frac{1}{\tau_p}\left(\vec{v}_p - \vec{v}_a + \frac{d_p^2}{12}\nabla^2\vec{v}_a\right)}_{(I)} + \underbrace{\left(1 - \frac{\rho_a}{\rho_p}\right)\vec{g}}_{(II)} + \underbrace{\frac{\rho_a}{\rho_p}\left(\frac{\partial\vec{v}_a}{\partial t} + \left(\vec{v}_a\cdot\nabla\right)\vec{v}_a\right)}_{(III)}$$

$$+\underbrace{\frac{1}{2}\frac{\rho_a}{\rho_p}\left(\frac{\partial\vec{v}_a}{\partial t} + \left(\vec{v}_a\cdot\nabla\right)\vec{v}_a - \frac{d\vec{v}_p}{dt}\right)}_{(IV)} + \underbrace{\left(\frac{q_p}{m_p}\vec{v}_p - \frac{1}{\rho_p\mu_0}\mathrm{curl}\left(\vec{B}\right)\right)\times\vec{B}}_{(V)}$$

$$(57)$$

Where (I) is the drag term, (II) is the buoyancy, (III) is the carrier phase inertia, (IV) is the added mass and (V) is the Lorentz force. It must be noted that in Eq. (56) we have neglected the particle-particle interactions. This assumption has been made because of the small volume of the particles, which reduces the probability of collision. Furthermore, q_p is the electric charge of the particle, m_p its mass and τ_p the relaxation time, defined as

$$\tau_p = \frac{4}{3}\frac{\rho_p d_p}{\rho_a C_d \left|\vec{v}_a - \vec{v}_p\right|} \tag{58}$$

The standard definition of the drag coefficient is the following

$$C_d = \begin{cases} \dfrac{24}{\mathrm{Re}_p} & \mathrm{Re}_p < 0.1 \\[2mm] \dfrac{24}{\mathrm{Re}_P}\left(1 + \dfrac{1}{6}\mathrm{Re}_p^{2/3}\right) & 0.1 \leq \mathrm{Re}_p \leq 1000 \\[2mm] 0.44 & \mathrm{Re}_p > 1000 \end{cases} \tag{59}$$

The particle Reynolds number is defined as

$$\mathrm{Re}_p = \frac{d_p\left|\vec{v}_a - \vec{v}_p\right|}{\nu_a} \tag{60}$$

5.2   Magnetic Induction

As far as the magnetic induction field is concerned, zero normal derivatives are set to zero everywhere, except on the wall, where the magnetic field of the probe is imposed. The external magnetic field is generated by a single rectangular coil with a negligible cross section of the wire where an electric current flows. This geometry is quite common in clinical practice [90–93].

The analytical expression for the magnetic induction field has been derived in [94]. A point in the coil reference frame, whose origin is at its centre, is identified by the coordinates (x', y', z'). The coil dimensions are $2a_1$ along the x' axis and $2b_1$ in the

y' direction, and its section is 2.5 cm^2. The axis z' is normal to the coil surface. The components of the magnetic induction field are

$$B_{x'} = \frac{\mu_0 I_1}{4\pi} \sum_{a=1}^{4} \left[\frac{z'(-1)^{a+1}}{r_a [r_a + d_a]} \right]$$ (61)

$$B_{y'} = \frac{\mu_0 I_1}{4\pi} \sum_{a=1}^{4} \left[\frac{z'(-1)^{a+1}}{r_a \left[r_a + C_a(-1)^{a+1} \right]} \right]$$ (62)

$$B_{z'} = \frac{\mu_0 I_1}{4\pi} \sum_{a=1}^{4} \left[\frac{d_a(-1)^a}{r_a \left[r_a + C_a(-1)^{a+1} \right]} - \frac{C_a}{r_a [r_a + d_a]} \right]$$ (63)

with

$$\begin{cases} C_1 = -C_4 = a_1 + x' \\ C_2 = -C_3 = a_1 - x' \\ d_1 = d_2 = y' + b_1 \\ d_3 = d_4 = y' - b_1 \\ r_a = \sqrt{C_a^2 + d_a^2 + z'^2} \end{cases}$$ (64)

The magnetic probe is located 1 cm above the patient's skin, in order for its modulus to be smaller than 1.5 T, which is the limit allowed in clinical treatments [7, 8]. In fact, higher values can cause damage to the patient. The distance between the origin of the probe and the target is set parallel to the z' axis. The target of the magnetotherapy is located at the upper right lobe of the lungs; the maximum magnetic field must be concentrated on it.

The external chest surface of the patient is reconstructed using VMTK, with the procedure previously illustrated for the trachea, and the two geometries are located in the same reference frame in order to calculate the coordinates of the probe and the target. The dimensions of the rectangular coil, the current intensity and the positions of the coil and target can be found in Table 2, while the main simulation parameters are listed in Table 3.

Table 2 Source and target coordinates; values for the normal to the chest surface of the patient; intensity of the magnetic field along the z-axis

	Normal	Coil Coords(m)	Target Coords(m)	Bz|1 cm(mT)
1st *run*	$n_x = -0.082166$	$x_c = -0.054377$	$x_t = 0.0314$	46,21
	$n_y = 0.93057$	$y_c = -0.0891$	$y_t = -0.3554$	
	$n_z = 0.35677$	$z_c = -0.21683$	$z_t = 0.9342$	

Table 3 Simulation parameters: H_D (domain height), W_D (domain width), D_D (domain depth), $D_{M\text{-}T}$ (magnet–tumour distance), T_0 (observation time)

H_D(cm)	W_D(cm)	D_D(cm)	$D_{M\text{-}T}$(cm)	T_0(s)
31.6	10	2	12.32	2

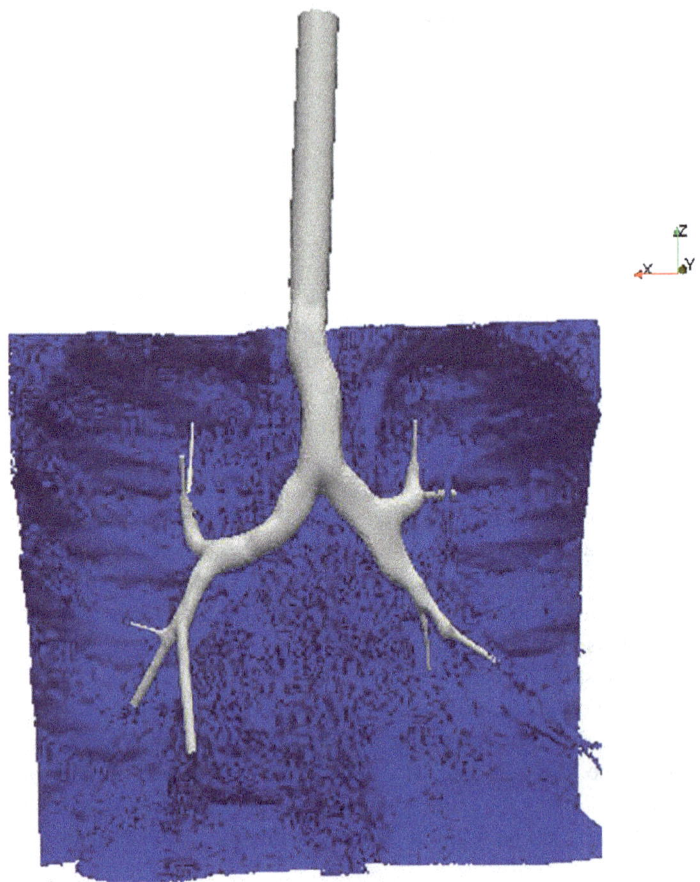

Fig. 37 Reconstruction of the thorax and the trachea by VMTK [1]

5.3 Set-Up for the Trachea

We implement the equations inside the solver for setting up the boundary conditions for the trachea. We need to find the vector normal to the thorax plane, the coordinates of a point on the target surface and the coordinates of a point on the source surface. We do the reconstruction of the thorax, and we work on it in order to extrapolate only the very external surface, as can be seen in Fig. 37 [1]. Once the necessary coordinates of the target, the source and the normal are found, the file B, in the folder 0, can be modified, and the numerical simulation begins. In the aforementioned file, we use the new boundary condition for the wall patch, specifying the value of the current I used during the simulation, the dimensions of the rectangular coil and the coordinates of the target, source and normal vector. For the remaining patches, we fix a *zeroGradient* condition, because the magnetic field, in this case, acts like a probe, and we are interested in analysing its behaviour for the whole geometry. For the U file and the wall patch, we prescribe a *fixedValue* equal to zero, while, for the outlets, the condition is *InletOutlet*. For the inlet patch, we use a *groovyBC* condition for a periodic profile. For the p file, at the wall and the inlet, we fix a *zeroGradient* condition that is equivalent to a *Neumann condition*, where the gradient, normal to those boundaries, is specified. For the outlets, we impose a *grovyBC*, in which the value is evaluated, as described previously. As usual, before starting the simulation, we evaluate the right *Courant number* in order to ensure stability.

5.4   Post-processing of the Trachea

The simulation runs for 2 s of model time. The post-processing is done with *ParaView*. We analyse the vector quantities, $\overline{U}, \overline{B}, \overline{U}_p$ and the scalar quantities, p, Position$_p$. Moreover, we evaluate how the magnetic field influences the particles cloud inside the first airways of the respiratory tree and the mechanism of absorption of the particles, caused by the presence of the external magnetic field, generated by a rectangular coil.

5.4.1   Unsteady-State Simulation with Null Magnetic Field

The first simulation with the addition of nanoparticles is done with a null magnetic field. The set-up of the boundary conditions is already explained. Figures 38, 39, 40, and 41 [1], show the maps of the pressure (cmH$_2$O) at different instants of time. First

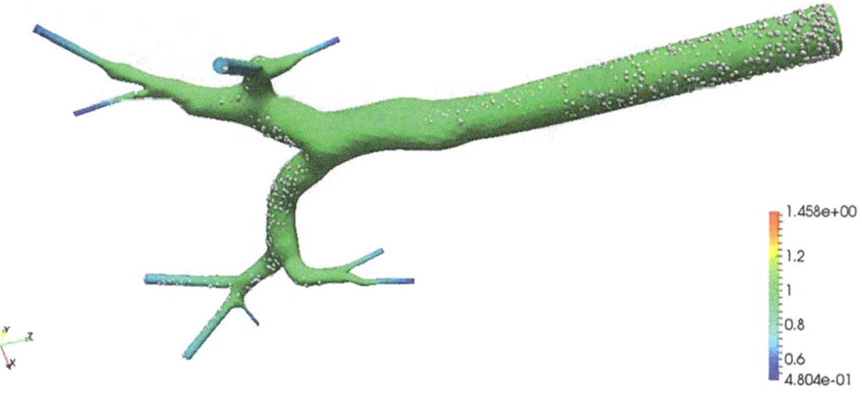

Fig. 38 Pressure map (cmH$_2$O) with nanoparticles at $t = 0.15$ s [1]

Fig. 39 Pressure map (cmH$_2$O) with nanoparticles at $t = 0.95$ s [1]

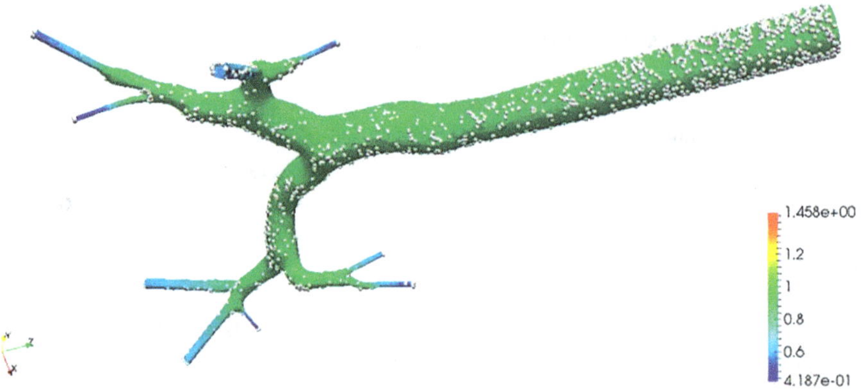

Fig. 40 Pressure map (cmH$_2$O) with nanoparticles at $t = 1$ s [1]

Fig. 41 Pressure map (cmH$_2$O) with nanoparticles at $t = 2$ s [1]

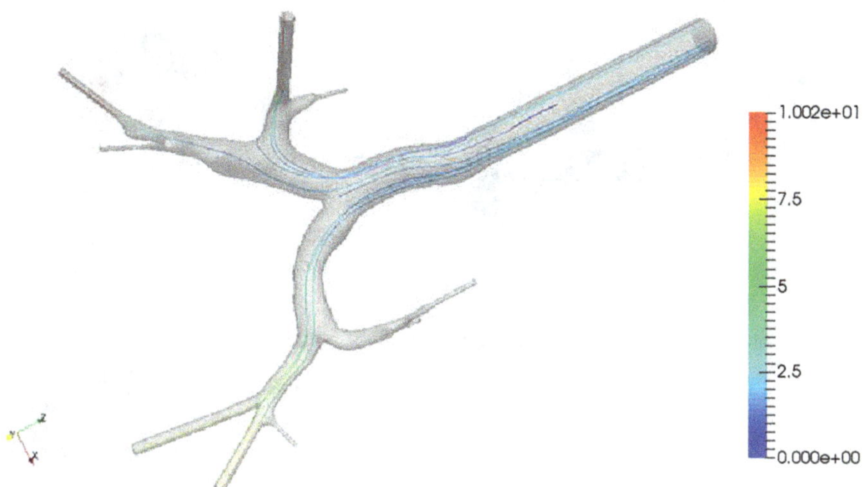

Fig. 42 Velocity streamlines, (m/s), with nanoparticles at $t = 0$ s [1]

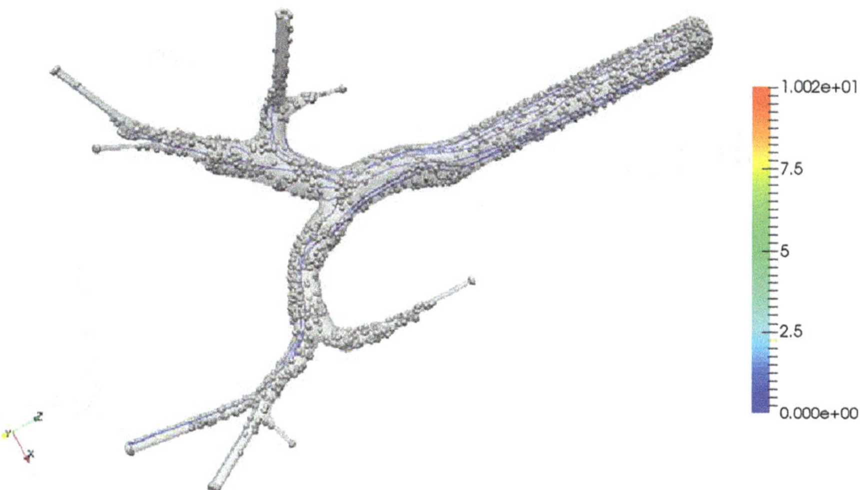

Fig. 43 Velocity streamlines, (m/s), with nanoparticles at $t = 0.40$ s [1]

of all, it is possible to observe that the pressure is particle-independent for the trachea and the first bifurcations of the respiratory tree. Secondly, we have the same trend at the beginning of each phase of the respiratory cycle. All the considerations are still valid from an Eulerian point of view.

Figures 42, 43, 44, 45, 46, and 47 [1], depict the velocity streamlines, (*m/s*), at different instants of time, with the addition of nanoparticles, obtained with the coupling between the velocity equations of the fluid and the equations of the velocity particles. At time 0 s, Fig. 42, we are in a rest condition, and the particles are not

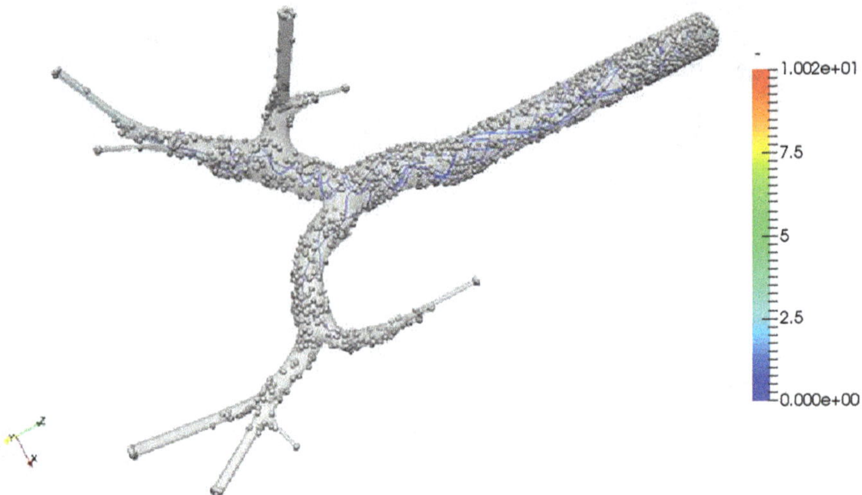

Fig. 44 Velocity streamlines, (m/s), with nanoparticles at $t = 0.8$ s [1]

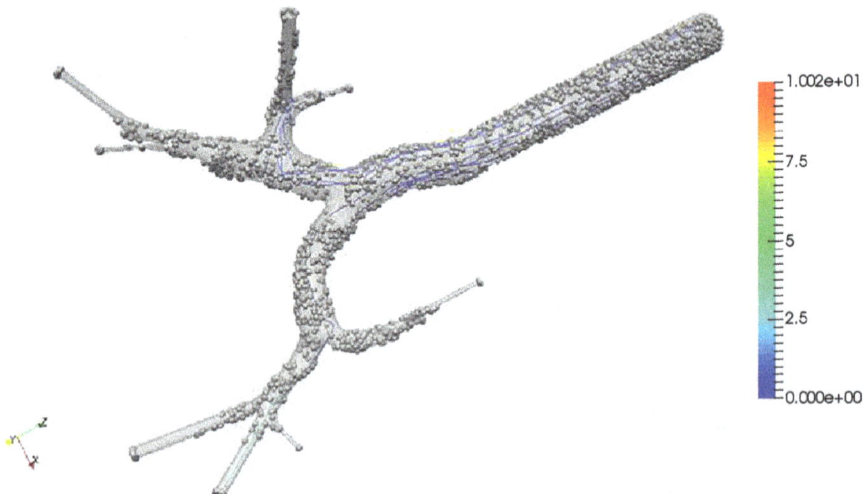

Fig. 45 Velocity streamlines, (m/s), with nanoparticles at $t = 1.40$ s [1]

injected. As the cycle proceeds, the flow becomes transitional even if the velocity is low, due to the geometry of the respiratory system. The particles try to occupy all the possible space, especially in the proximity of the main vortex. At the end of the inspiratory phase, 0.4 s, Fig. 43, the path lines are highly tangled, with the presence of some vortices, especially near the bifurcations, where the velocity gradient becomes higher due to a reduction of the cross section. The highest velocity is reached in the common trachea because the air coming from outside has a high

Fig. 46 Velocity streamlines, (m/s), with nanoparticles at $t = 1.80$ s [1]

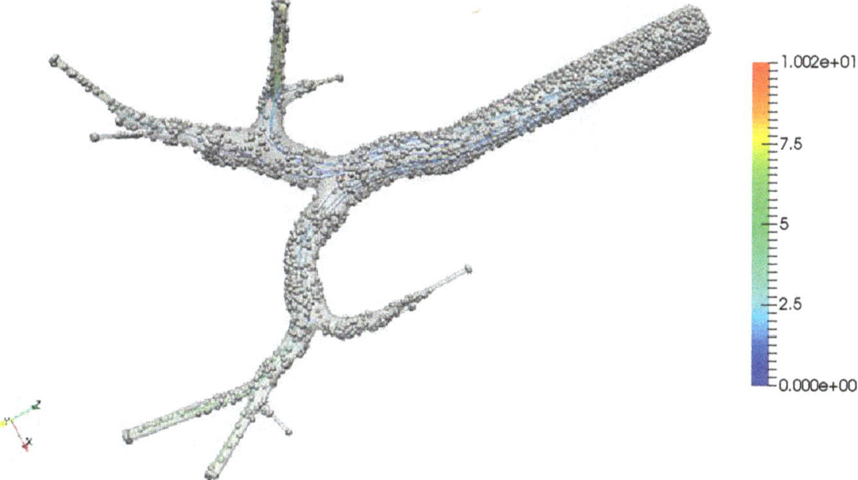

Fig. 47 Velocity streamlines, (m/s), with nanoparticles at $t = 2$ s [1]

kinetic energy. During the expiratory phase, 1.4 s, Fig. 45, there are vortices near the bifurcations due to the changes in the geometry. At the time 1.8 s, Fig. 46, we can observe a reverse flow, and the highest velocity is reached at the inlet. At the end of the respiratory cycle, 2 s, Fig. 47, we have a maximum of the velocity inside the small branches and a null velocity at the inlet. As far as the particles are concerned, it is possible to see how their flow follows the air motion, and they move towards the trachea from the branches.

Figures 48, 49, 50, 51, and 52 [1], show the viscous acceleration term, m/s^2, with nanoparticles. In general, this term is higher in the small branches because it depends

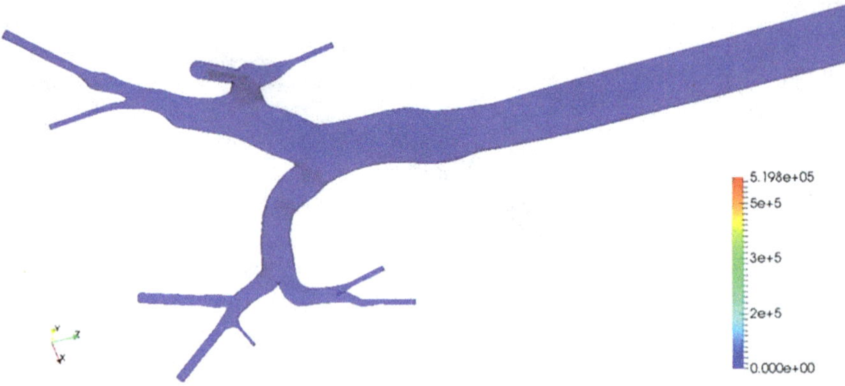

Fig. 48 Viscous acceleration term, (m/s^2), with nanoparticles at $t = 0$ s [1]

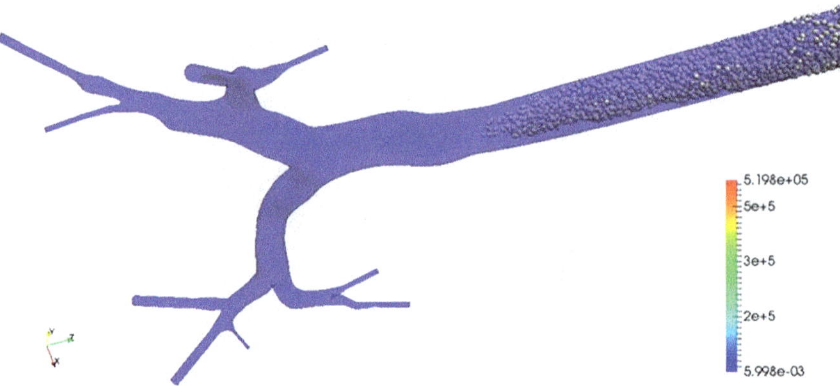

Fig. 49 Viscous acceleration term, (m/s^2),with nanoparticles at $t = 0.04$ s [1]

Fig. 50 Viscous acceleration term, (m/s^2),with nanoparticles at $t = 0.75$ s [1]

Fig. 51 Viscous acceleration term, (m/s^2), with nanoparticles at $t = 1$ s [1]

Fig. 52 Viscous acceleration term with nanoparticles at $t = 2$ s [1]

directly on the velocity gradient. Since it reaches its peak in the inspiratory phase, this term in the small branches increases and has the same value at the end of the inspiratory and expiratory phases because airflow changes its motion, and the velocity is higher than in the other parts of the respiratory tree.

Figures 53, 54, 55, 56, and 57 [1], depict the inertial term, or Newtonian acceleration term, (m/s^2), which is high in the last part of the small branches, at $t = 0.7$ s, Fig. 54, and $t = 1.7$ s, Fig. 56, of the inspiratory and expiratory phase. Moreover, it is equal to zero at the beginning and end of each phase of the respiratory cycle because it depends on the velocity. In general, inside the first generations of the respiratory tree, the flow is turbulent.

Figures 58, 59, 60, 61, and 62 [1], show the lift acceleration term, Ω, (m/s^2), which is the force component perpendicular to the oncoming flow direction. The lift is orthogonal to the drag force, which, in general, is the component parallel to the

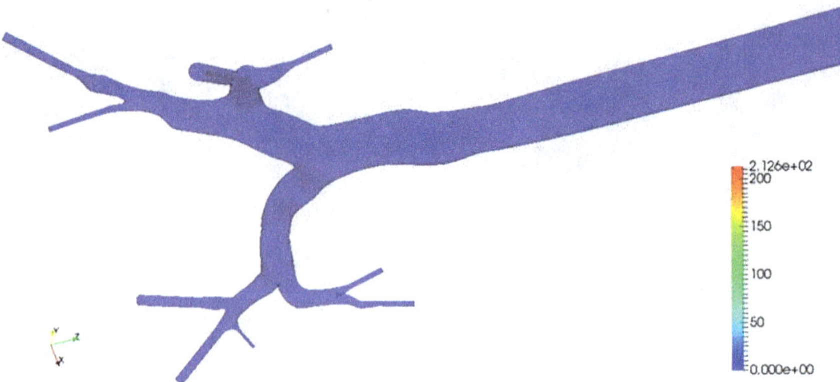

Fig. 53 Newtonian acceleration term, (m/s^2), with nanoparticles at $t = 0$ s [1]

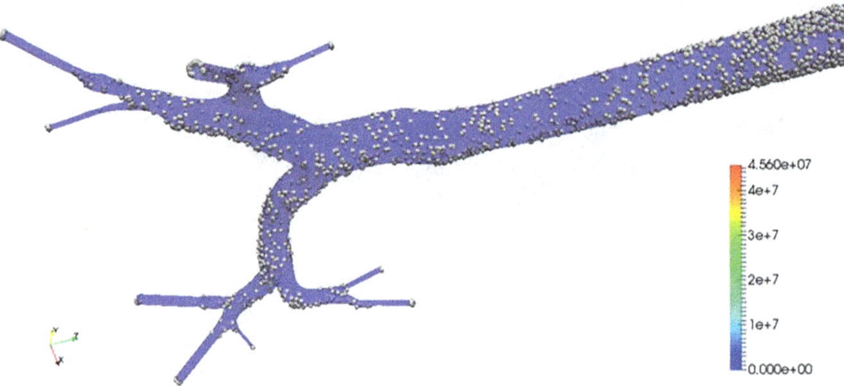

Fig. 54 Newtonian acceleration term, (m/s^2), with nanoparticles at $t = 0.7$ s [1]

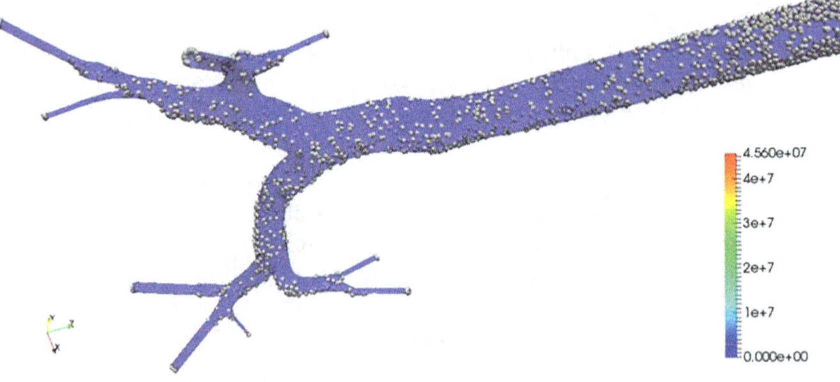

Fig. 55 Newtonian acceleration term, (m/s^2), with nanoparticles at $t = 1$ s [1]

Fig. 56 Newtonian acceleration term, (m/s^2), with nanoparticles at $t = 1.7$ s [1]

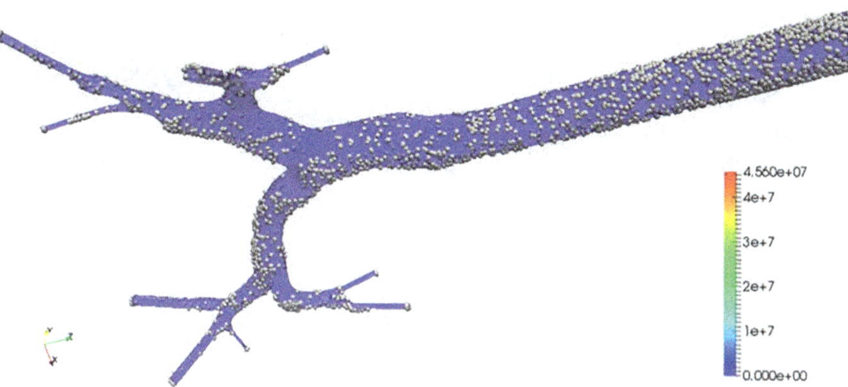

Fig. 57 Newtonian acceleration term, (m/s^2), with nanoparticles at $t = 2$ s [1]

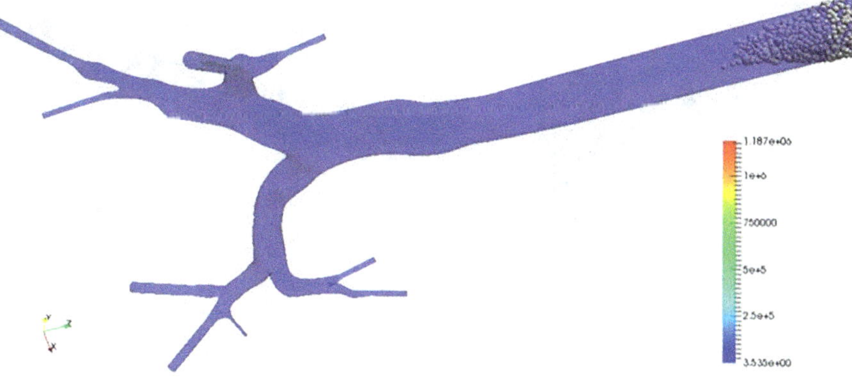

Fig. 58 Lift acceleration term, (m/s^2), with nanoparticles at $t = 0.01$ s [1]

Fig. 59 Lift acceleration term, (m/s^2), with nanoparticles at $t = 0.5$ s [1]

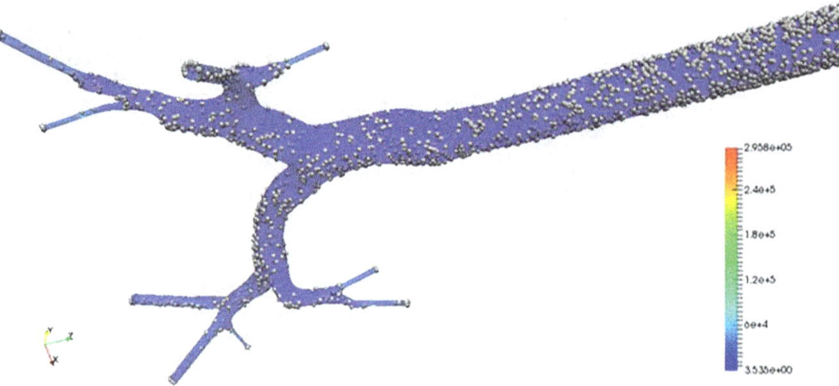

Fig. 60 Lift acceleration term, (m/s^2), with nanoparticles at $t = 1$ s [1]

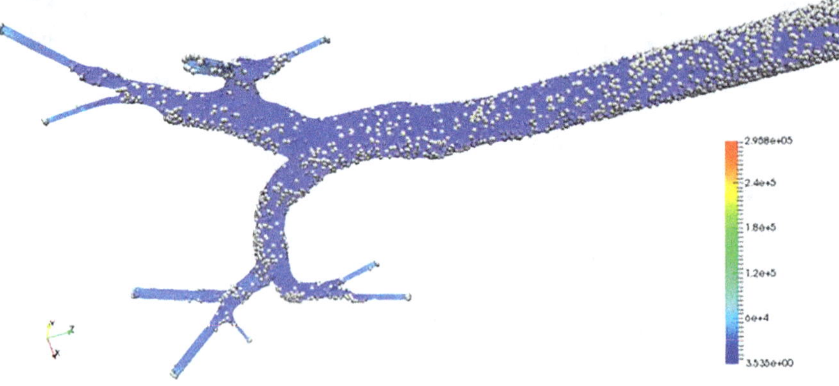

Fig. 61 Lift acceleration term, (m/s^2), with nanoparticles at $t = 1.7$ s [1]

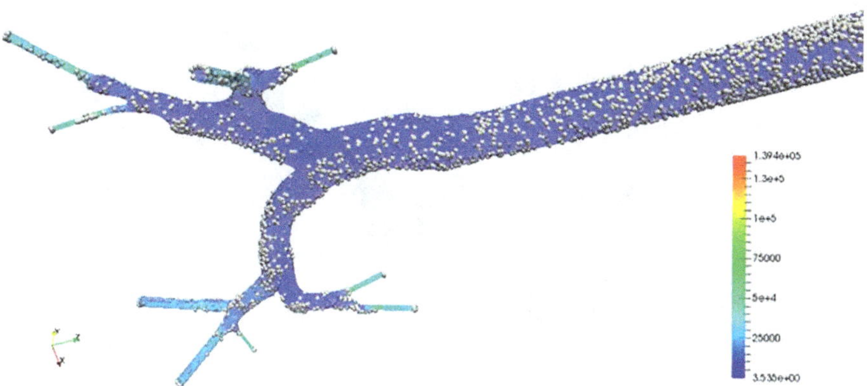

Fig. 62 Lift acceleration term, (m/s^2), with nanoparticles at $t = 2$ s [1]

Table 4 Values used for the simulation

Normal	Coordinate source	Coordinate target	$Bz|_{1cm}$[T]
−0.082166	−0.054377	0.0314	0.04621[T]
0.93057	−0.0891	−0.3554	
0.35677	−0.21683	0.9342	

flow direction and is proportional to the air density and to the curl of velocity as expressed in the following equation

$$\Omega = \nabla \times \vec{u} \tag{65}$$

For the last reason, the maximum value of the lift is reached in the small branches and, especially, in proximity to the peaks of the inspiratory and expiratory phases and at the beginning of the cycle.

5.4.2 Unsteady-State Simulation with Magnetic Field

We run one simulation where we evaluate the intensity of the magnetic field along the axis at 1 cm from the probe. In Table 4 below, the characteristics used for this simulation are listed.

Figures 63, 64, 65, and 66 [1], show the maps of the pressure field (cm H$_2$O) at different times, which is particle independent. At the beginning of each phase of the respiratory cycle, we have the same trend. All the previous considerations are still valid from an Eulerian point of view.

Figures 67, 68, 69, 70, 71 and 72 [1], depict the velocity streamlines, (m/s), with the movement of the nanoparticles. As far as the fluid dynamic analysis is concerned, the conclusions are similar to the previous paragraph. The difference is that, in this case, the particles are also under the effect of the magnetic field generated by an

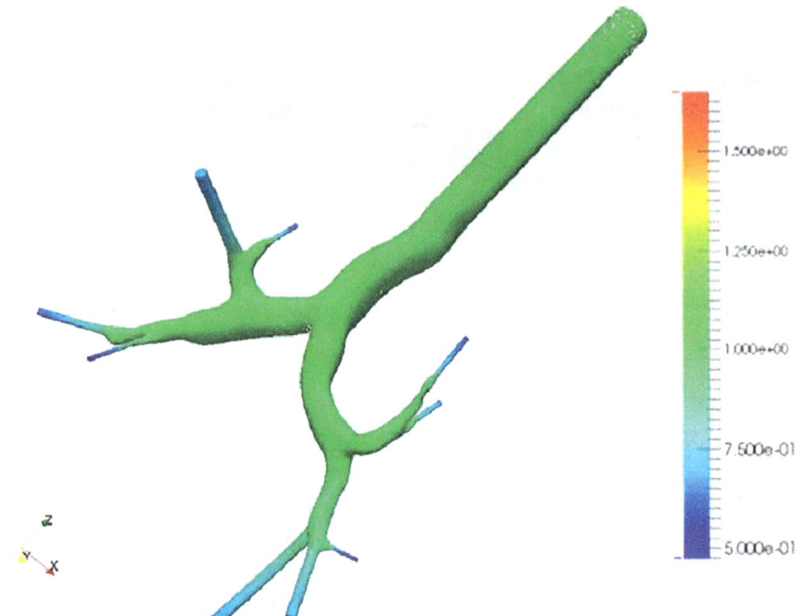

Fig. 63 Pressure field (cm H_2O) with nanoparticles at $t = 0.15$ s [1]

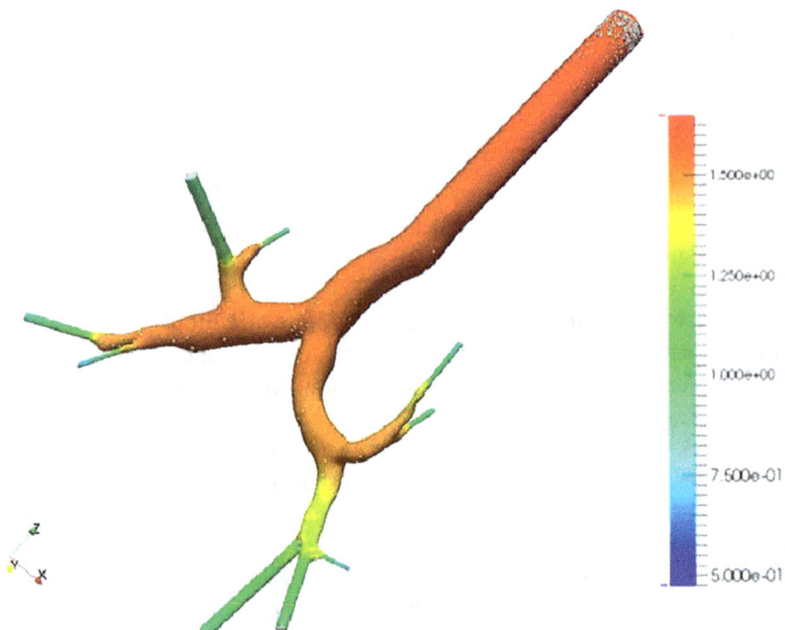

Fig. 64 Pressure field (cm H_2O) with nanoparticles at $t = 0.95$ s [1]

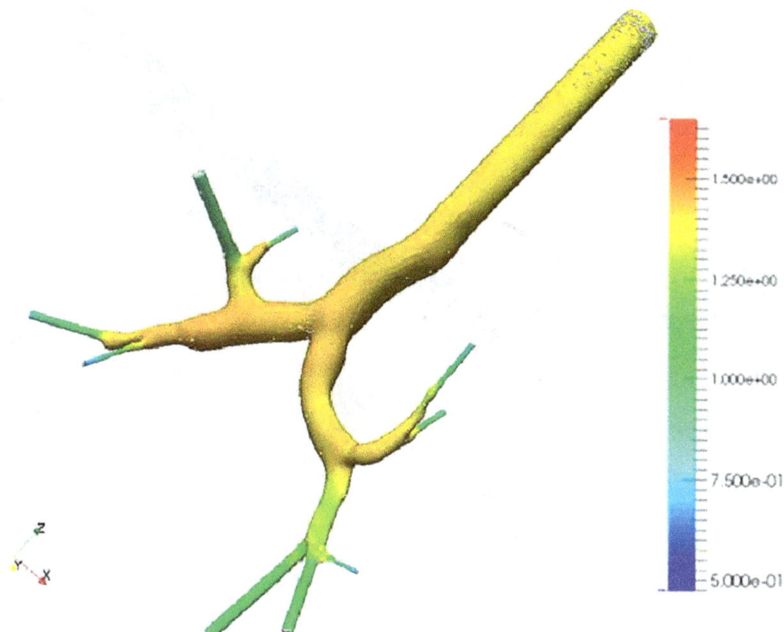

Fig. 65 Pressure field (cm H_2O) with nanoparticles at $t = 1$ s [1]

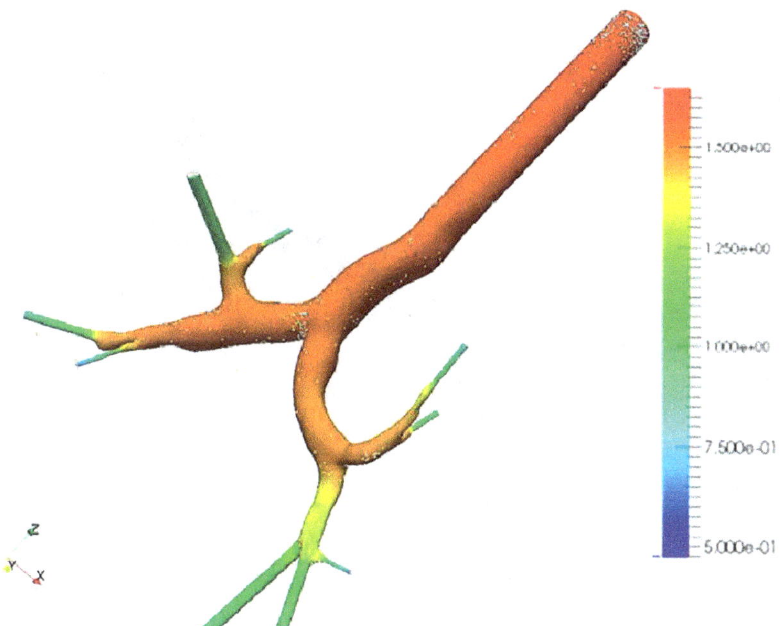

Fig. 66 Pressure field (cm H_2O) with nanoparticles at $t = 2$ s [1]

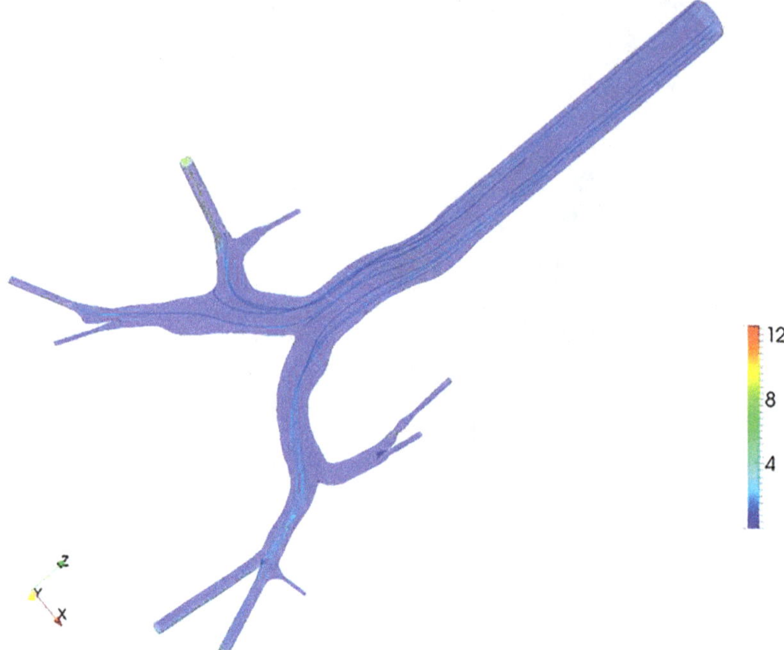

Fig. 67 Velocity streamlines, (m/s), with nanoparticles at $t = 0$ s [1]

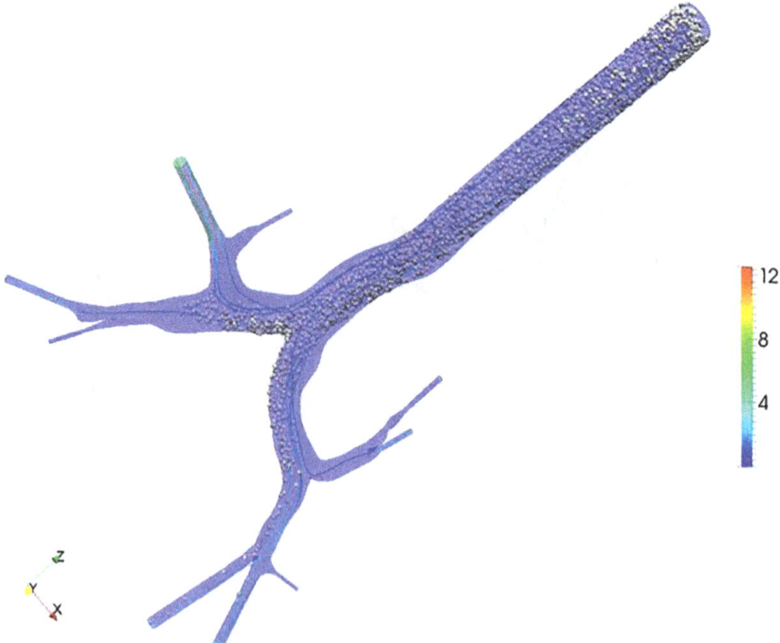

Fig. 68 Velocity streamlines, (m/s), with nanoparticles at $t = 0.15$ s [1]

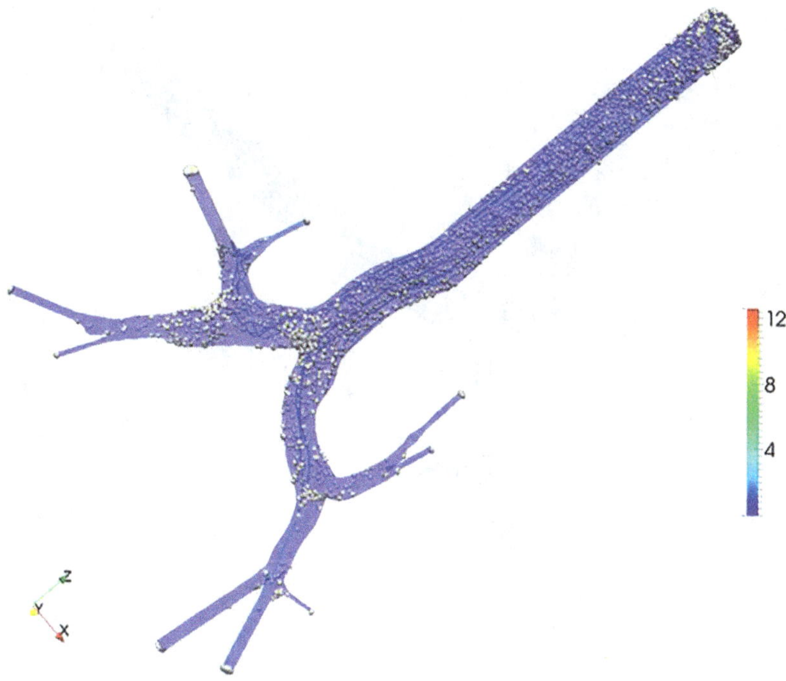

Fig. 69 Velocity streamlines, (m/s), with nanoparticles at $t = 0.5$ s [1]

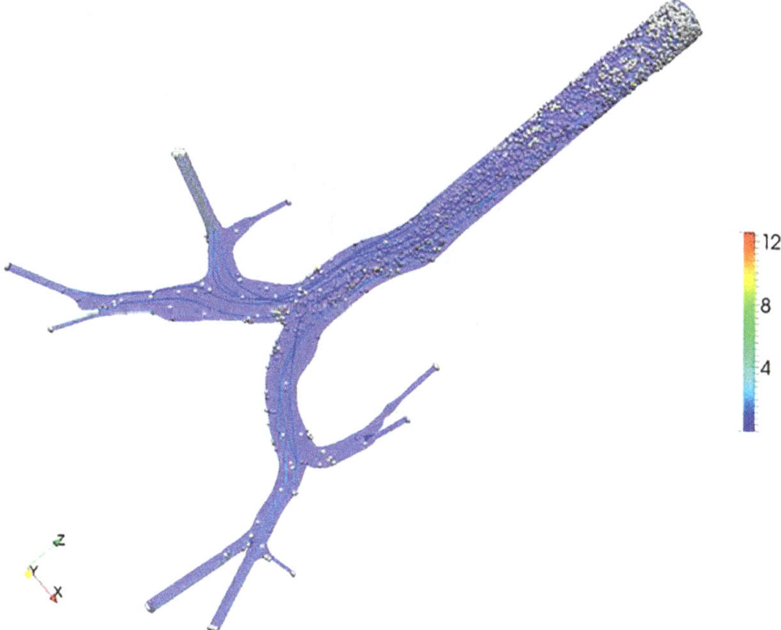

Fig. 70 Velocity streamlines, (m/s), with nanoparticles at $t = 1$ s [1]

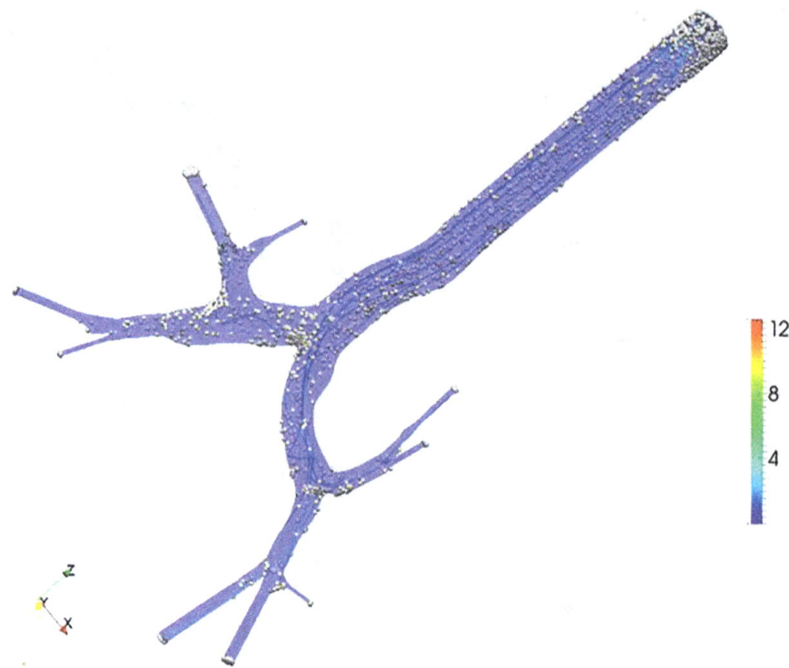

Fig. 71 Velocity streamlines, (m/s), with nanoparticles at $t = 1.5$ s [1]

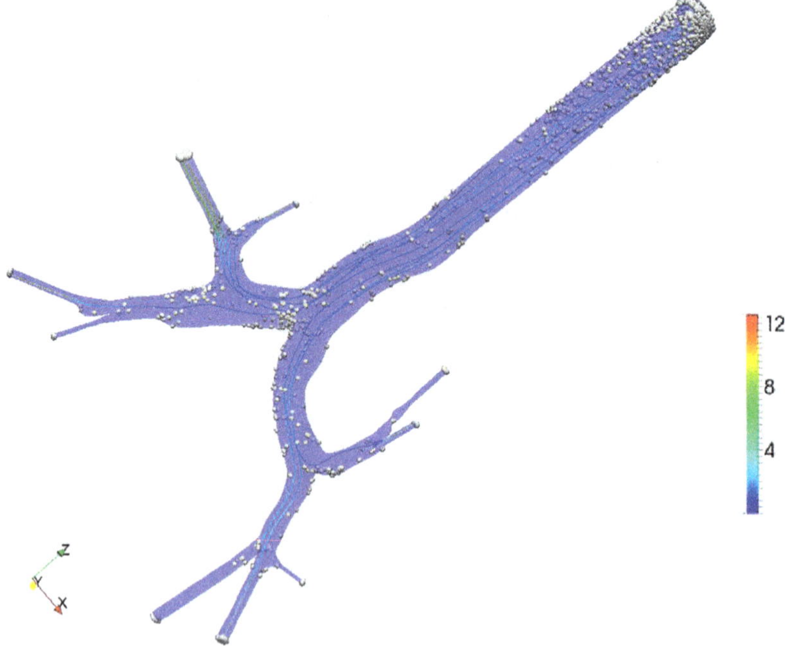

Fig. 72 Velocity streamlines, (m/s), with nanoparticles at $t = 2$ s [1]

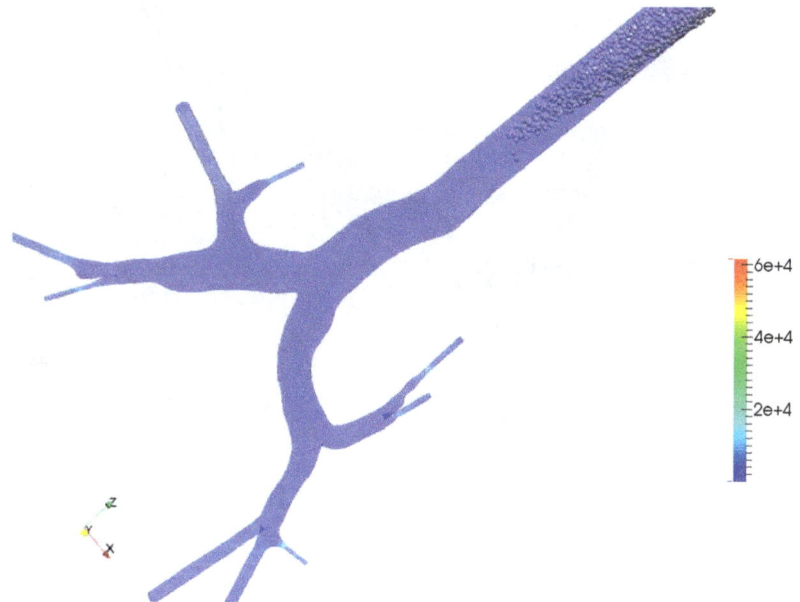

Fig. 73 Viscous acceleration term, (m/s^2), with nanoparticle at $t = 0.04$ s [1]

external coil. The effect of the magnetic field is visible in an acceleration in the proximity of the probe.

Figures 73, 74, and 75 [1], illustrate the viscous acceleration term, (m/s^2), with the nanoparticle, which is higher in the small branches because it depends directly on the velocity gradient. As the inspiratory phase reaches its peak, this term increases in the small branches, and it has the same value at the end of the inspiratory and expiratory phases because the airflow changes its motion, and the velocity is higher than in the other parts of the respiratory tree.

Figures 76 and 77 [1], show the inertial term, or Newtonian acceleration term, (m/s^2). As previously seen, also this term is high in the last part of the small branches and in specific instants of the inspiratory and expiratory phase, i.e. $t = 1.7$ s; moreover, at the beginning and at the ending of each phase of the respiratory cycle, it is equal to zero. As we already explained, it strictly depends on the velocity. In general, inside the first generations of the respiratory tree, the flow is transitional unless the Re is less than 2100.

Figures 78, 79, and 80 [1], depict the electromagnetic acceleration, (m/s^2), which derives from the application of the magnetic field. This acceleration describes how the particles change their velocity under the effect of an external magnetic field, and, as is possible to see, it is higher closer to the magnetic probe. The particles are sped up by the magnetic field, and we can notice, looking at three different time steps, the increase in the number of particles adhering to the wall.

Figures 81, 82, and 83 [1], depict the effect of the lift acceleration, Ω, (m/s^2), which is the component of the force perpendicular to the oncoming flow direction.

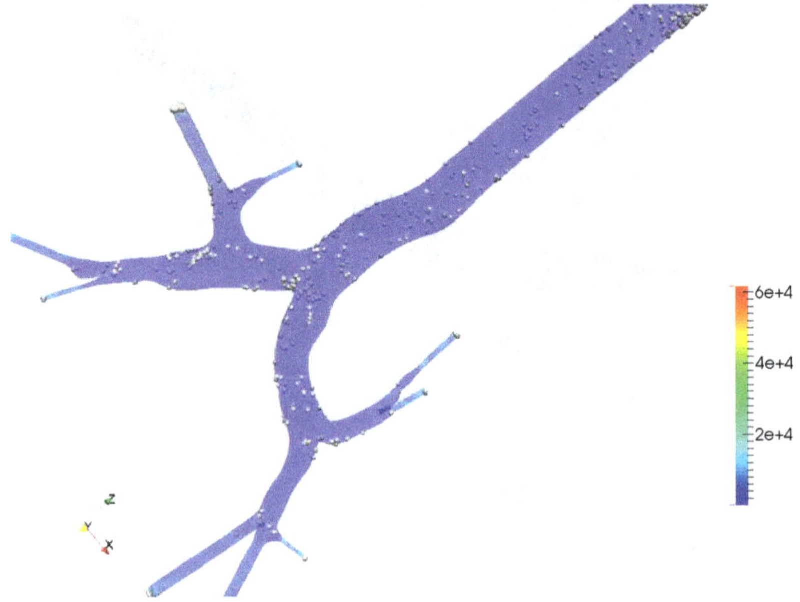

Fig. 74 Viscous acceleration term, (m/s^2), with nanoparticle at $t = 0.75$ s [1]

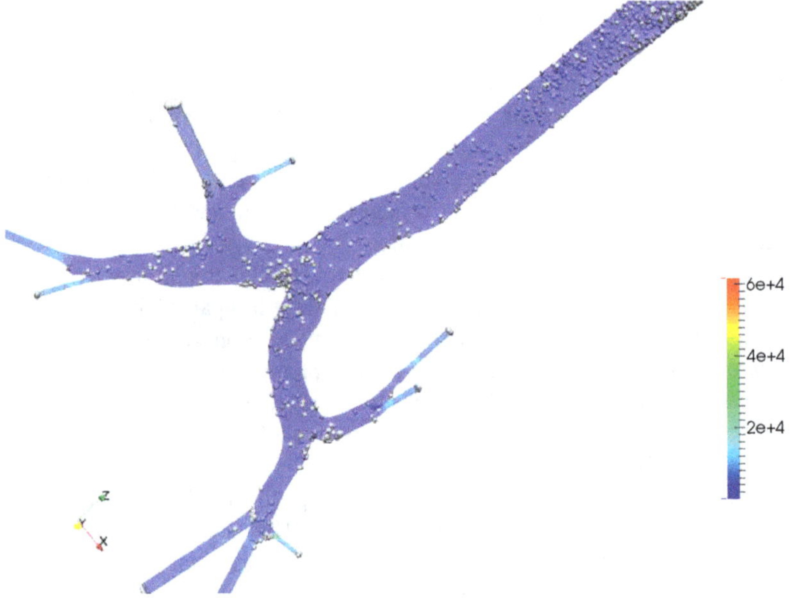

Fig. 75 Viscous acceleration term, (m/s^2), with nanoparticle at $t = 2$ s [1]

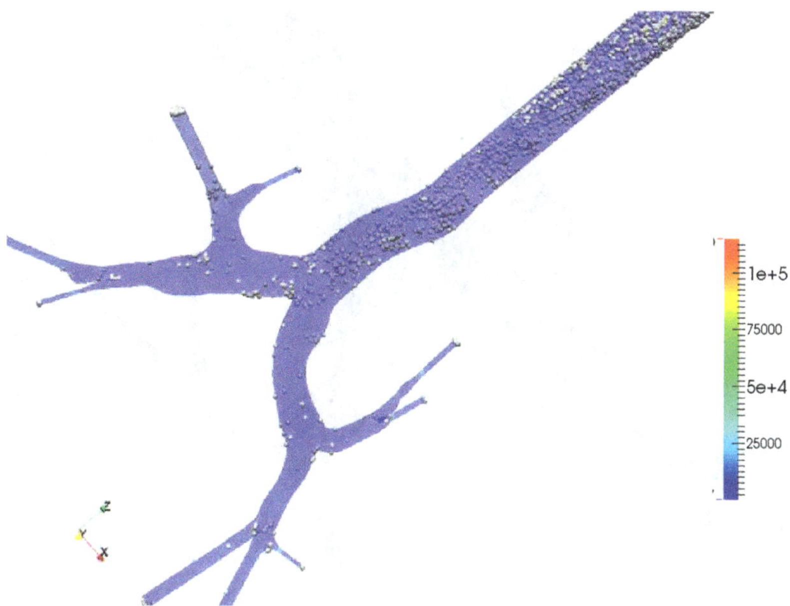

Fig. 76 Newtonian acceleration term, (m/s^2), with nanoparticles at $t = 1$ s [1]

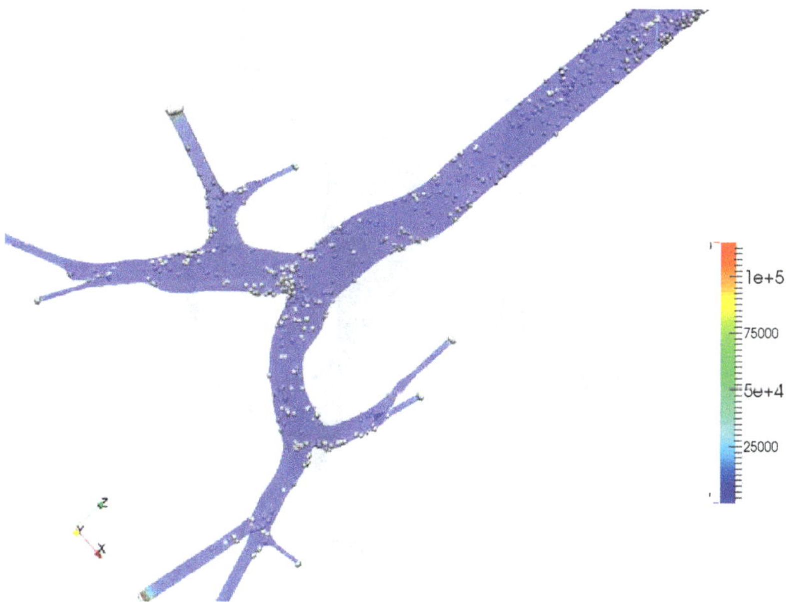

Fig. 77 Newtonian acceleration term, (m/s^2), with nanoparticles at $t = 1.7$ s [1]

Fig. 78 Electromagnetic acceleration term (m/s^2) with nanoparticles at $t = 1.15$ s [1]

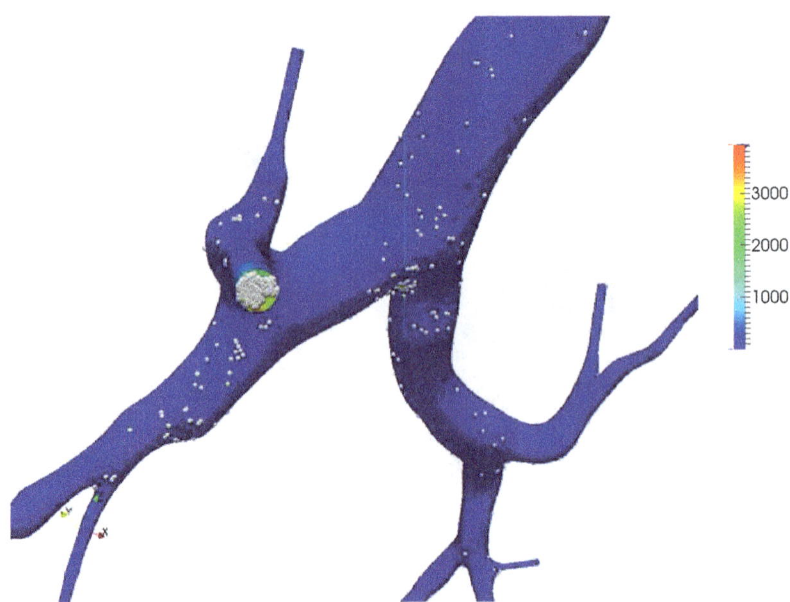

Fig. 79 Electromagnetic acceleration term (m/s^2) with nanoparticles at $t = 1.75$ s [1]

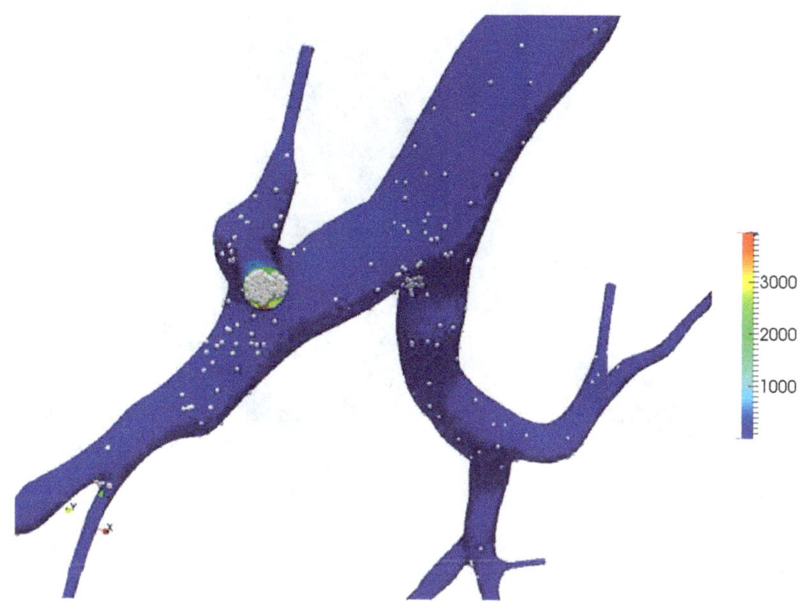

Fig. 80 Electromagnetic acceleration term (m/s^2) with nanoparticles at $t = 2$ s [1]

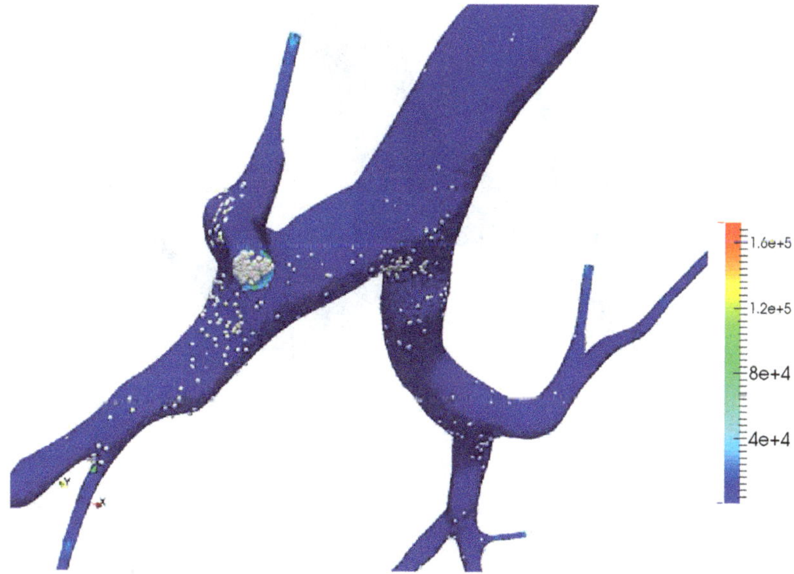

Fig. 81 Lift acceleration term (m/s^2) with nanoparticles at $t = 0.5$ s [1]

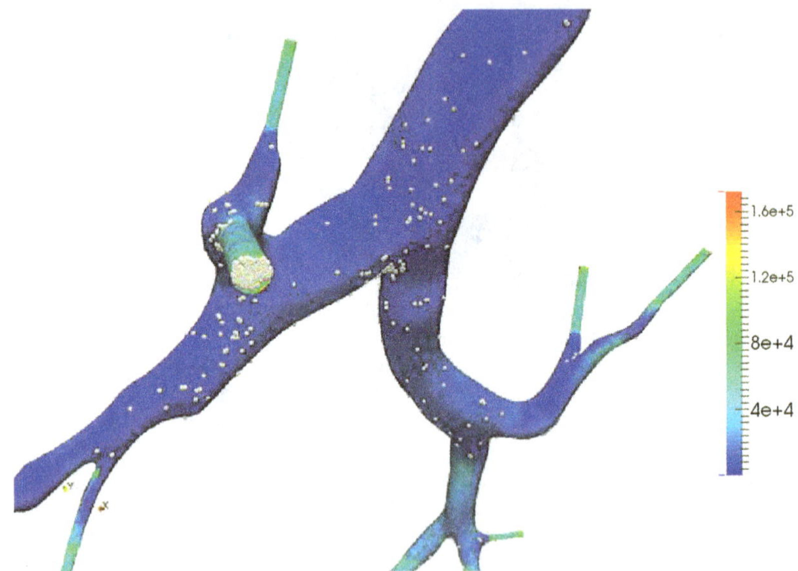

Fig. 82 Lift acceleration term (m/s^2) with nanoparticles at $t = 1.7$ s [1]

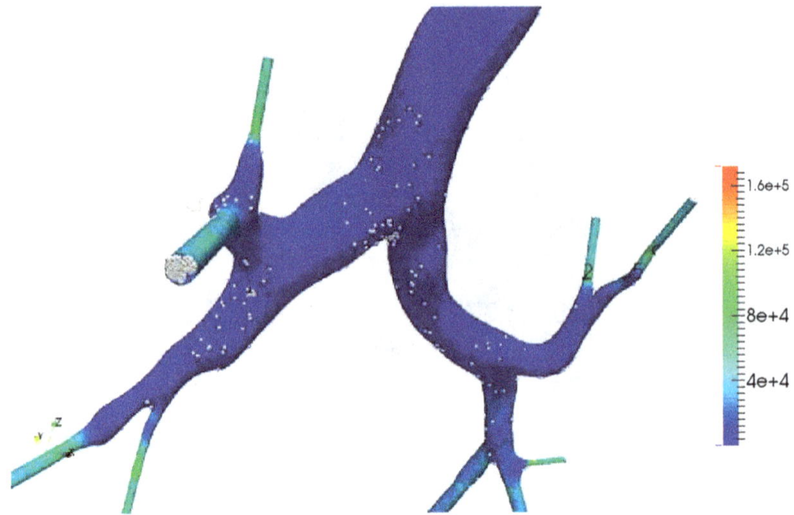

Fig. 83 Lift acceleration term (m/s^2) with nanoparticles at $t = 2$ s [1]

The lift is proportional to the air density and to the curl of the velocity. For the last reason, the maximum value of this field is present in the small branches and, especially, in proximity to the peaks of the inspiratory and expiratory phases and at the beginning of the cycle.

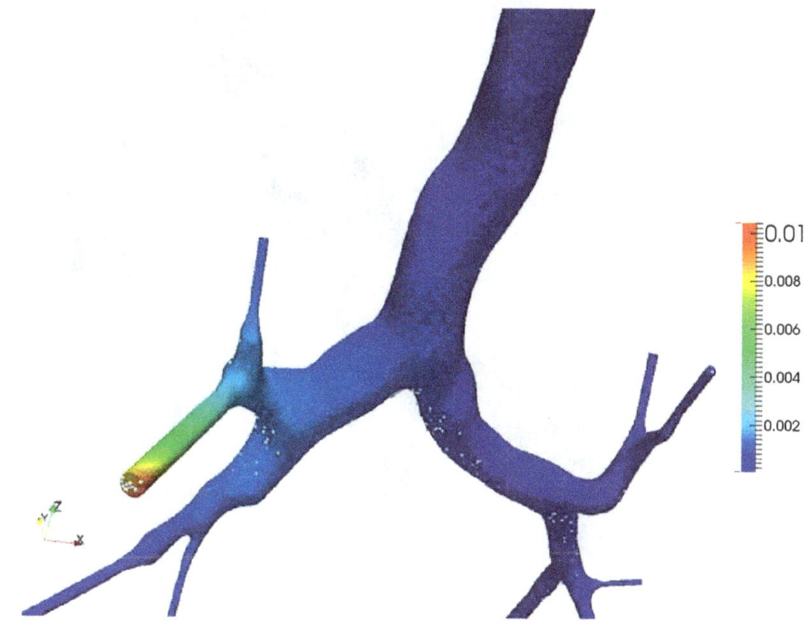

Fig. 84 Magnetic field, (T), with nanoparticles at $t = 0.2$ s [1]

Figures 84, 85, and 86 [1], depict the magnetic field maps, (T), at different instants of time with the nanoparticles. Since the magnetic field is pointed towards a precise branch and its coordinates are set-up inside the boundary conditions file, it is higher on that specific branch and constant during the simulation.

The effect of the magnetic field is to drag the motion of the nanoparticles towards this area, due to their magnetic properties. Consequently, the magnetic field promotes the absorption of the nanoparticles in the same zone. As can be seen from the maps, the number of particles, which adhere to the outlet of the branches and along their walls, increases during the run-time. The main goal of these simulations is the analysis of the nanoparticles' absorption under the effect of an applied magnetic field. The absorption is independent of the sign of the intensity current, whereas the choice of the target and source coordinates is more important. Indeed, this last condition establishes the extinction of the magnetic field and, consequently, a good absorption of the nanoparticles.

5.5   Influence of the Magnetic Field on the Particle Absorption

The influence of the magnetic field on the particles' absorption is shown in Figs. 87 and 88 [20]. The effect of the magnetic field is underlined by reporting the results with the magnetic probe turned off as well.

Fig. 85 Magnetic field, (T), with nanoparticles at $t = 1.15$ s [1]

Fig. 86 Magnetic field, (T), with nanoparticles at $t = 2$ s [1]

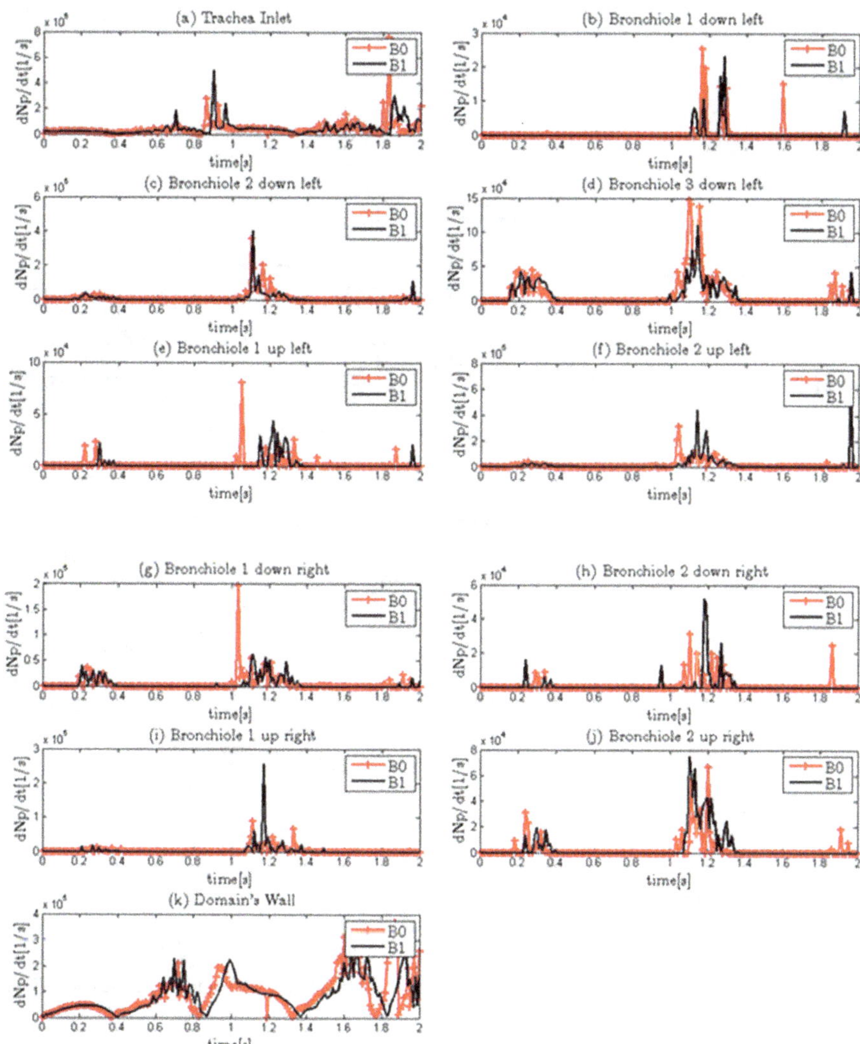

Fig. 87 Particle volumetric flow rate per unit volume (l/s) versus time (s) in the absence of a magnetic field (B0) and in presence (B1), Fig. 8 of [20]

Figure 87 [20], reports the particle flow rate per unit volume (l/s) versus time (s) on the different boundaries during the respiratory cycle. In the trachea and on the wall of the domain, the particle flow rates are of the same order of magnitude. The peak of the exhalation process occurs between 1 s and 1.4 s, when the highest flow rate from the lungs to the environment occurs.

Figure 88 [20], reports the number of particles versus time (s) on the different boundaries during the respiratory cycle. Figure 88b, e, f, h, j shows the

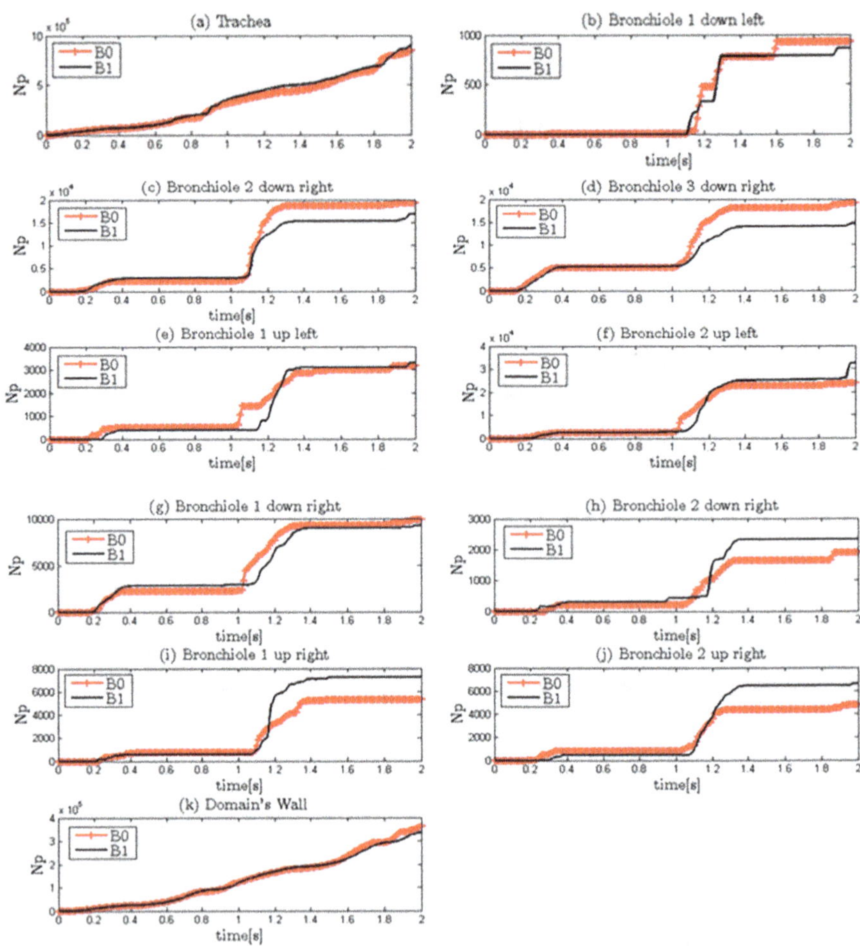

Fig. 88 Particle number versus time (s) in absence of magnetic field (B0) and in the presence (B1), Fig. 9 of [20]

correspondence with an increase in the particle uptake in bronchioles, probably due to the formation of several low-speed recirculation regions during exhalation. The diameter of the bronchi decreases from the trachea to the alveoli. Therefore, since during exhalation the air moves from the alveoli to the trachea, a flow detachment occurs, causing regions of recirculation and promoting the uptake of nanoparticles. Since the particle flow rate may not show the effectiveness in the use of a rectangular coil in the magnetic targeting, its time integration, corresponding to the total number of particles that cross a given section from the beginning of the process, is performed. The uptake of nanoparticles increases in the branches of the lower respiratory tract bent in the frontal direction, Fig. 88b, e, f, h, j, while it decreases in the branches bent towards the spine, Fig. 88c, d, g, i. A slight reduction in the particle's adhesion to the domain wall is also evidenced. Despite the increase in particle concentration in the targeted regions, shown in Fig. 88b, the effect of the magnetic field is not significant. More than 50% of the injected particles adhere to the domain wall when the magnetic probe is turned off, while it is only 47% when the magnetic probe is turned on. Similarly, the increase of the nanoparticles in the different regions is less than 1%. This behaviour can be due to the high surface-volume ratio of the lower respiratory tract, which inherently increases the probability of adhesion, regardless of the intensity of the magnetic field, and to the geometry of the probe. The magnetic field on the axis of a single square loop decreases quickly from its origin, about 100 *times*, at a distance equal to 2.5 times its width. Considering that the distance between the source and the target is about 12.32 cm, it is clear that this makes the modulus of the magnetic induction considerably low.

6   Conclusions

The present chapter investigates the nanoparticles air flow during magnetic therapy of a lung tumour, which is an emerging alternative to chemotherapy because of the reduced side effects. To the best of the author's knowledge, this work represents the first numerical investigation of this technique in the patient-specific lower respiratory tract. The lack of numerical studies is probably due to the complexity of the problem, which couples the air flow with the nanoparticle dynamics under the influence of an external magnetic field. The purpose of this work is to verify numerically the effectiveness of this technique in the treatment of lung cancer.

The effect of Magneto-Hydro-Dynamic, *MHD*, on the fluid dynamics of airflow is investigated numerically for *2D* and *3D* geometries. The numerical simulations are performed by using the open-source software *OpenFOAM*, which possesses, among others, an appropriate solver for the *MHD* equations. Different boundary conditions are tested for velocity, walls and applied magnetic field. The application of a magnetic probe on the external surface of the human body can improve the absorption of drugs, especially in some tumour treatments, if they are electrically charged. To test numerically this technique on a patient-specific geometry, the open-source software *VMTK*, which allows the reconstruction of a *3D* geometry of arteries or other organs starting from medical images saved in *DICOM* format, is used. Once the final geometry is obtained, the grid independence of the results from the mesh used is tested. The simulations are performed in steady-state conditions, taking into account only the fluid dynamics. After the most convenient mesh is found, the unsteady-state fluid dynamic simulations are carried on by imposing an inlet velocity profile coherently with the experimental pulsatile flow rate. A specific Lagrangian solver is added to the *MHD* module, available in *OpenFOAM*, in order to simulate the fluid dynamics of nanoparticles in the flow.

Then, the boundary conditions for the magnetic field are set-up. Since the most common way to create a magnetic field outside the body is to use a coil, a rectangular coil is employed to solve the magnetic field generated by a current *I*. The last part of the implementation of the new boundary condition is to find the coordinates of the target and source. The target, inside the human body, corresponds to a specific branch of the respiratory tree. The magnetic field, generated by the probe, decays quite quickly with distance, and, at a certain time, the axial component becomes comparable to the radial and angular ones. All the body is interested in is a weak magnetic field, while the magnetic field is higher in correspondence to the target.

The numerical simulations, for steady and unsteady flow, are carried out in laminar flow, with a Newtonian viscosity for the air and absence of inter- and intraparticle forces, due to the small size of the drug carriers. Two simulations are set-up, with a null and a different from zero magnetic field, in order to assess the influence of the magnetic field on the adsorption of the particles and to establish the maximum value to which a person can be exposed. In general, experiments proved that short-term exposure to electromagnetic fields does not cause any apparent effects. It is considered a starting point for the magnetic field intensity, the value coming from the datasheet of a portable magnetic resonance (0.15 T). We have adjusted this value in order to find a suitable value that fits our geometries. One simulation with the magnetic probe turned off and one with the probe turned on are carried out. The results are compared to assess the increase of drug uptake in the lung due to the magnetotherapy and consequently the reduced dispersion in other locations. The rectangular coil, pointing perpendicularly to the tumour, is positioned at a specific point, 1 cm above the upper right lobe of the lungs.

The results of the numerical simulations show that, despite the induced magnetic field increasing the particle uptake, a small fraction of the total number of particles injected reaches the target. This can be due to the high surface-volume ratio of the lower respiratory tract and the design of the magnetic probe. This conclusion is in contrast with the mathematical model developed for lung alveolus in [64] and is probably due to the different probe employed, the larger particle diameter ($5 \ \mu m$) and the higher magnetic field (0.2–2.2 T). Magnetic drug delivery is extremely sensitive to these parameters. Moreover, the drug distribution in different branches is not investigated in [64]. The low performance of the magnetic drug delivery in the present study can be due to three factors: (i) the nanoparticle size; (ii) the magnet–tumour distance; (iii) the probe design. As far as the nanoparticle size is concerned, as shown in [82], a low particle diameter increases the acceleration due to the Lorentz force, but it also increases the friction. However, because of the small magnetic field applied [95, 96], the effect on the friction is predominant. Therefore, larger particles should increase the effectiveness of the technique, as in [64]. The probe-tumour distance is another factor which limits the effectiveness of the technique. Unfortunately, this parameter cannot be tuned at will, and in a superficial tumour, the magnetic drug targeting is likely to be more effective.

The design of the magnetic probe is another factor which influences the outcome of the procedure. The modulus of the magnetic induction field of a single rectangular coil decreases considerably with the distance from the target, as in the right lung. Being the right lung about 12.32 cm from the source, the modulus of the magnetic induction is considerably small. A solution to this problem could be to increase the value of the electric current intensity flowing in the probe, but this is not possible because higher values of the magnetic field could affect the health of the patient, which is forbidden by law [95, 96]. A solution could be to modify the design of the probe in order to reduce the rate of decay of the magnetic field. A possible configuration could include multiple probes, positioned in appropriate locations.

CFD allows detailed visualisation of biological fluid flows, which increases our understanding of natural phenomena, but has some limitations due to the computational resources required to simulate a process in a complex domain. The question then arises as to whether or not the observation time (2 s) is sufficient to judge the effectiveness of the technique. It would certainly be better to extend the simulations to a few minutes, but the particle volumetric flow rate, shown by Fig. 88 in different cases, shows negligible differences. It seems unlikely that the percentage of injected particles that reach the right lung will significantly increase over time.

In conclusion, the present simulations validate the theory and the clinical practice, which suggests that the guide, with an appropriate magnetic field from outside, of the drugs, inhaled or injected, allows better absorption with a reduction of side effects for the patient.

Acknowledgements for the Figures The author thanks the *publisher Elsevier* for the kind permission to reproduce *the full paper* with the images *8 and 9* of the *Ref.* [20].

References

1. Russo F (2014/2015) Numerical simulation of MagnetoHydroDynamics inside cardiovascular and respiratory systems, PhD Thesis. University of Rome Tor Vergata, A.A
2. Parkin DM, Bray F, Ferlay J, Pisani P (2005) Global cancer statistics, 2002. CA Cancer J Clin 55:74–108
3. IARC, Globocan 2012: estimated cancer incidence, mortality and prevalence worldwide in 2012
4. Rogers M. Encyclopaedia Britannica, [Online] available at: http://media-3.web.britannica.com/eb-media/49/115249-004-14D1F54A.jpg
5. Anon (n.d.) Humanity First Serving mankind. [Online] available at: http://medicinembbs.blogspot.it/2011/04/respiratory-system.html
6. Burri P. Encyclopaedia Britannica, [Online] available at: http://www.britannica.com/science/human-respiratory-system
7. Labiris NR, Dolovich MB (2003) Pulmonary drug delivery. Part I: physiological factors affecting therapeutic effectiveness of aerosolized medications. Br J Clin Pharmacol 56(6):588–599
8. Chandolu V, Dass CR (2013) Treatment of lung cancer using nanoparticle drug delivery systems. Curr Drug Discov Technol 10(2):170–176
9. Garcia MP et al (2005) Morphological analysis of mouse lungs after treatment with magnetite-based magnetic fluid stabilized with DMSA. J Magn Magn Mater 293(1):277–282
10. Arruebo M, Fernández-Pacheco R, Ibarra MR, Santamaría J (2007) Magnetic nanoparticles for drug delivery. NanoToday 2(3):22–32
11. Dobson J (2006) Magnetic nanoparticles for drug delivery. Drug Dev Res 67(1):55–60
12. Pankhurst QA, Connolly J, Jones SK, Dobson J (2003) Applications of magnetic nanoparticles in biomedicine. J Phys D: Appl Phys Luglio 36(13):R167–R181

13. Gori F, Boghi A, Amitrano M (2009) Three-dimensional numerical simulation of the fluid dynamics in a coronary stent. In: ASME 2009 international mechanical engineering congress and exposition. American Society of Mechanical Engineers, pp 407–411
14. Gori F, Boghi A (2009) Image-based computational fluid dynamics in a carotid artery. In: ASME 2009 international mechanical engineering congress and exposition. American Society of Mechanical Engineers, pp 123–128
15. Gori F, Boghi A (2010) Three-dimensional numerical simulation of non-Newtonian blood in two coronary stents. In: 2010 14th international heat transfer conference. American Society of Mechanical Engineers, pp 109–114
16. Gori F, Boghi A (2011) Three-dimensional numerical simulation of blood flow in two coronary stents. Numer Heat Transf A Appl 59:231–246
17. Boghi A, Gori F (2015) Numerical simulation of blood flow through different stents in stenosed and non-stenosed vessels. Numer Heat Transf A Appl 68:225–242
18. Di Venuta I, Boghi A, Gori F (2017) Three-dimensional numerical simulation of a failed coronary stent implant at different degrees of residual stenosis. Part I: Fluid dynamics and shear stress on the vascular wall. Numer Heat Transf A Appl 71(6):638–665
19. Boghi A, Di Venuta I, Gori F (2017) Three-dimensional numerical simulation of a failed coronary stent implant at different degrees of residual stenosis. Part II: Apparent viscosity and wall permeability. Numer Heat Transf A Appl 71(6):653–665
20. Russo F, Boghi A, Gori F (2018) Numerical simulation of magnetic Nano drug targeting in a patient-specific lower respiratory tract. J Magn Magn Mater 451:554–564
21. Kleinstreuer C, Zhang Z, Li Z (2008) Modeling airflow and particle transport/deposition in pulmonary airways. Respir Physiol Neurobiol 163(1–3):128–138
22. Zhang Z, Kleinstreuer C (2004) Airflow structures and nano-particle deposition in a human upper airway model. J Comput Phys 198(1):178–210
23. Liu Y, So RMC, Zhang CH (2003) Modeling the bifurcating flow in an asymmetric human lung airway. J Biomech 36(7):951–959
24. Luo HY, Liu Y (2008) Modeling the bifurcating flow in a CT-scanned human lung airway. J Biomech 41(12):2681–2688
25. Calay RK, Kurujareon J, Holdø AE (2002) Numerical simulation of respiratory flow patterns within human lung. Respir Physiol Neurobiol 130(2):201–221
26. De Backer JW, Vos WG, Gorlé CD, Germonpré P, Partoens B, Wuyts FL, Parizel PM, De Backer W (2008) Flow analyses in the lower airways: patient-specific model and boundary conditions. Med Eng Phys 30(7):872–879
27. Yin Y, Choi J, Hoffman EA, Tawhai MH, Lin C-L (2010) Simulation of pulmonary air flow with a subject-specific boundary condition. J Biomech 43(11):2159–2163
28. Pennati G, Corsini C, Cosentino D, Hsia T-Y, Luisi VS, Dubini G, Migliavacca F (2011) Boundary conditions of patient-specific fluid dynamics modelling of cavopulmonary connections: possible adaptation of pulmonary resistances results in a critical issue for a virtual surgical planning. Interface Focus 1(3):297–307
29. Nowak N, Kakade PP, Annapragada AV (2003) Computational fluid dynamics simulation of airflow and aerosol deposition in human lungs. Ann Biomed Eng 31(4):374–390
30. Senyei A, Widder K, Czerlinski G (1978) Magnetic guidance of drug-carrying microspheres. J Appl Phys 49(6):3578–3583
31. Mosbach K, Schroder U (1979) Preparation and application of magnetic polymers for targeting of drugs. FEBS Lett 102(1):112–116
32. Zimmerman U, Pilwat G (1975) Organ specific application of drugs by means of cellular capsule systems. Z Naturforsch C Biosci:732–736
33. Lübbe A, Bergemann C, Riess H (1996) Clinical experiences with magnetic drug targeting: a phase I study with 4'-Epidoxorubicin in 14 patients with advanced solid tumors. Cancer Res 56:4686–4693
34. Lubbe A, Alexiou C, Bergemann C (2001) Clinical applications of magnetic drug targeting. J Surg Res 95(2):200–206

35. Alexiou C, Arnold W, Klein RJ, Parak FG, Hulin P, Bergemann C, Erhardt W, Wagenpfeil S, Lubbe AS (2000) Locoregional cancer treatment with magnetic drug targeting. Cancer Res 60(23):6641–6648
36. Allen T, Cullis P (2004) Drug delivery system, entering the mainstream. Science 303(5665): 1818–1822
37. Chorny M, Fishbein I, Adamo RF, Forbes SP, Folchman-Wagner Z, Alferiev IS (2012) Magnetically targeted delivery of therapeutic agents to injured blood vessels for prevention of in-stent restenosis. Methodist De Bakey Cardiovasc J 8(1):23–27
38. Forbes Z, Halverson DS, Fridman G, Yellen BB, Chorny M, Friedman G, Barbee KA (2004) Locally targeted drug delivery to magnetic stents for therapeutic applications, pp 1–6
39. Shinkai M (2002) Functional magnetic particles for medicine. J Biosci Bioeng 94(6):606–613
40. Voltairas P, Fotiadis D, Michalis L (2002) Hydrodynamics of magnetic drug targeting. J Biomechanics:813–821
41. Sayed Z, Telang S, Ramchand C (2003) Application of magnetic techniques in the field of drug discovery and biomedicine. Biomagn Res Technol
42. Goodwin S, Peterson C, Hoh C, Bittner C (1999) Targeting and retention of magnetic targeted carriers (MTC'S) enhancing intra-arterial chemoterapy. J Magn Magn Mater Aprile:132–139
43. Ruuge E, Rusetski A (1993) Magnetic fluids as drug carriers: targeted transport of drugs. J Magn Magn Mater 122(1–3):335–339
44. Li X, Yao K, Liu Z (2008) CFD study on the magnetic fluid delivering in the vessel in high-gradient magnetic field. J Magn Magn Mater 320:1753–1758
45. Grief A, Richardson G (2005) Mathematical modeling of magnetically targeted drug delivery. J Magn Magn Mater:455–463
46. Cregg P, Murphy K, Mardinoglu A (2008) Calculation of nanoparticle capture efficiency in magnetic drug targeting. J Magn Magn Mater 320(23):3272–3275
47. Cregg P, Murphy K, Mardinoglu A (2009) Inclusion of magnetic dipole-dipole and hydrodynamic interactions in implant-assisted magnetic drug targeting. J Magn Magn Mater 321(23): 3893–3898
48. Haverkort J, Kenjereš S (2009) Optimizing drug delivery using non-uniform magnetic fields: a numerical study. In: 4th European conference of the International Federation for Medical and Biological Engineering. Springer, Berlin Heidelberg, pp 26323–22627
49. Ganguly R, Gaind A, Sen S, Puri I (2005) Analyzing ferrofluid transport for magnetic drug targeting. J Magn Magn Mater Marzo:331–334
50. Reza Habibi M, Ghasemi M (2011) Numerical study of magnetic nanoparticles concentration in biofluid (blood). J Magn Magn Mater 323(1):32–38
51. Cherry E, Eaton J (2014) A comprehensive model of magnetic particle motion during magnetic drug targeting. Int J Multhiphase Flow Febbraio:173–185
52. Haverkort J, Kenjeres S, Kleijn C (2009) Computational simulations of magnetic particle capture in arterial flows. Ann Biomed Eng 37(12):2436–2448
53. Morega A, Dobre A, Morega M (2011) Magnetic field-flow interactions in drug delivery through an arterial system. Rev Roumaine Sci Techn Electrotech et Energ 56(2):199–120
54. Pouponneau P, Leroux JC, Martel S (2009) Magnetic nanoparticles encapsulated into biodegradable microparticles steered with an upgraded magnetic resonance imaging system for tumor chemoembolization. Biomaterials 30(31):6327–6332
55. Rotariu O, Strachan N (2005) Modelling magnetic carrier particle targeting in the tumor. J Magn Magn Mater 293(1):639–646
56. Nacev A, Beni C, Bruno B, Shapiro B (2010) Magnetic nanoparticle transport within flowing blood and into surrounding tissue. Nanomedicine 1(November):1459–1466
57. Adey WR (1993) Biological effects of electromagnetic fields. J Cell Biochem 51:410–410
58. Saiyed ZM, Telang SD, Ramchand CN (2003) Application of magnetic techniques in the field of drug discovery and biomedicine. Biomagn Res Technol 1:2
59. Larimi M, Ramiar A, Ranjbar A (2014) Numerical simulation of magnetic nanoparticles targeting in a bifurcation vessel. J Magn Magn Mater 362:58–71

60. Tehrani MD, Yoon J-H, Kim MO, Yoon J (2015) A novel scheme for nanoparticle steering in blood vessels using a functionalized magnetic field. IEEE Trans Biomed Eng 62:303–313
61. Larimi M, Ramiar A, Ranjbar A (2016) Numerical simulation of magnetic drug targeting with eulerian-lagrangian model and effect of viscosity modification due to diabetics. Appl Math Mech 37:1631–1646
62. Momeni Larimi M, Ramiar A, Ranjbar AA (2016) Magnetic nanoparticles and blood ow behavior in non-newtonian pulsating ow within the carotid artery in drug delivery application. Proc Inst Mech Eng H J Eng Med 230:876–891
63. Kenjeres S, Righolt B (2012) Simulations of magnetic capturing of drug carriers in the brain vascular system. Int J Heat Fluid Flow 35:68–75
64. Krafcik A, Babinec P, Frollo I (2015) Computational analysis of magnetic field induced deposition of magnetic particles in lung alveolus in comparison to deposition produced with viscous drag and gravitational force. J Magn Magn Mater 380:46–53
65. Antiga L, Ene-Iordache B, Caverni L, Cornalba GP, Remuzzi A (2002) Geometric reconstruction for computational mesh generation of arterial bifurcations from CT angiography. Comput Med Imaging Graph 26(4):227–235
66. Steinman D (2002) Image-based computational fluid dynamics modeling in realistic arterial geometries. Ann Biomed Eng 30(4):483–497
67. Antiga L, Piccinelli M, Botti L, Ene-Iordache B, Remuzzi A, Steinman DA (2008) An image-based modeling framework for patient-specific computational hemodynamics. Med Biol Eng Comput 46(11):1097–1112
68. Lorensen W, Cline H (1987) Marching cubes: a high resolution 3D surface construction algorithm. Computer Graphics 21(4):163–169
69. McInerney T, Terzopoulos D (1996) Deformable models in medical image analysis: a survey. Med Image Anal 1(2):91–108
70. Long Q, Xu XY, Ariff B, Thomas SA, Hughes AD, Stanton AV (2000) Reconstruction of blood flow patterns in a human carotid bifurcation: a combined CFD and MRI study. J Magn Reson Imaging 11(3):299–311
71. Redaelli A, Rizzo G, Arrigoni S, Di Martino E, Origgi D, Fazio F, Montevecchi F (2002) An assisted automated procedure for vessel geometry reconstruction and hemodynamic simulations from clinical imaging. Comput Med Imaging Graph 26(3):143–152
72. Bulpitt A, Efford N (1996) An efficient 3D deformable model with a self-optimising mesh. Image Vis Comput 14(8):573–580
73. Sethian J (1999) Advancing interfaces: level set and fast marching methods. Dept. of Mathematics, University of California, Berkeley
74. Sifakis E, Garcia C, Tziritas G (2002) Bayesian level sets for image segmentation. J Vis Commun Image Represent 13(1–2):44–64
75. Whitaker R (1995) Algorithms for implicit deformable models. In: Proceedings of the Fifth International Conference on Computer Vision, pp 822–827
76. Malladi R, Sethian J (1998) A real-time algorithm for medical shape recovery. In: Proceedings International Conference on Computer Vision, pp 304–310
77. Siebens A (n.d.) Encyclopaedia Britannica. [online] available at: http://www.britannica.com/science/human-respiratory-system/Control-of-breathing#toc66147
78. Waite L, Fine J (2007) Applied biofluid mechanics. McGraw-Hill, s.l.
79. Hall & Guyton (2012) Fisiologia Medica. XII ed. Elsevier-Masson, s.l.
80. VIA-GROUP (n.d.) ELCAP Public Lung Image Database. [Online] available at: http://www.via.cornell.edu/databases/lungdb.html
81. Antiga L, Manini S (n.d.) vmtk. [Online] available at: http://www.vmtk.org/
82. Boghi A, Russo F, Gori F (2017) Numerical simulation of magnetic Nano drug targeting in a patient-specific coeliac trunk. J Magn Magn Mater 437:86–97
83. Young DF, Munson BR, Okiishi TH, Huebsch WW (2010) A brief introduction to fluid mechanics. John Wiley & Sons
84. Elghobashi S (1994) On predicting particle-laden turbulent flows. Appl Sci Res 52:309–329

85. Tzirtzilakis E, Sakalis V, Kafoussias N, Hatzikonstantinou P (2004) Biomagnetic fluid flow in a 3d rectangular duct. Int J Numer Meth Fluids 44:1279–1298
86. Seino S, Matsuoka Y, Kinoshita T, Nakagawa T, Yamamoto TA (2009) Dispersibility improvement of gold/iron-oxide composite nanoparticles by polyethylenimine modification. J Magn Magn Mater 321:1404–1407
87. Hassan M, Zeeshan A, Majeed A, Ellahi R (2017) Particle shape effects on ferrofuids flow and heat transfer under influence of low oscillating magnetic field. J Magn Magn Mater 443:36–44
88. Zeeshan A, Hassan M, Ellahi R, Nawaz M (2017) Shape effect of nanosize particles in unsteady mixed convection flow of nanofluid over disk with entropy generation, vol 231. SAGE Publications Sage, UK, London, pp 871–879
89. Majeed A, Zeeshan A, Hayat T (2017) Analysis of magnetic properties of nanoparticles due to applied magnetic dipole in aqueous medium with momentum slip condition. Neural Comput & Applic 1–9:189
90. Metzger GJ, van de Moortele P-F, Akgun C, Snyder CJ, Moeller S, Strupp J, Andersen P, Shrivastava D, Vaughan T, Ugurbil K, Adriany G (2010) Performance of external and internal coil configurations for prostate investigations at 7 T. Magn Reson Med 64(6):1625–1639
91. Nemkov V, Ruffini R, Goldstein R, Jackowski J, DeWeese TL, Ivkov R (2011) Magnetic field generating inductor for cancer hyperthermia research. COMPEL – Int J Comput Math Electric Electron Eng 30(5):1626–1636
92. Lübbe AS, Alexiou C, Bergemann C (2001) Clinical applications of magnetic drug targeting. J Surg Res 95(2):200–206
93. Lübbe AS, Bergemann C, Riess H, Schriever F, Reichardt P, Possinger K, Matthias M, Dörken B, Herrmann F, Gürtler R, Hohenberger P, Haas N, Sohr R, Sander B, Lemke A-J, Ohlendorf D, Huhnt W, Huhn D (1996) Clinical experiences with magnetic drug targeting: a phase I study with 4′-Epidoxorubicin in 14 patients with advanced solid tumors. Cancer Res 56(20):4686–4693
94. Misakian M (2000) Equations for the magnetic field produced by one or more rectangular loops of wire in the same plane. J Res-Natl Inst Standards Technol 105(4):557–564
95. E. Council, Recommendation, 519/e of 12 July 1999 on the limitation of exposure of the general public to electromagnetic fields (0 Hz to 300 GHz). Official J. Eur. Communities L 59 (1999) (July 1999), pp 59–70
96. F.C. Commission (1997) Evaluating compliance with FCC guidelines for human exposure to radiofrequency electromagnetic fields, vol 65. OET Bullettin

Flavia Russo , a Medical Engineer, obtained her Research Doctorate degree in Industrial Engineering from the University of Rome "Tor Vergata." She completed a six-month fellowship at the Center for Biomolecular Nano Technologies of the IIT in Lecce for the study of optical fibre cochlear prostheses. She has worked at a medical software house as a product specialist and in a research and development start-up as a product owner of wearable medical devices. She currently holds the position of project manager at a company that deals with telemedicine.

Andrea Boghi The Doctor of Research *(Ph. D.)*, Dr. Eng. Andrea Boghi, is the director of *Computational Science Ltd.*, a consultancy company in the field of software development and data analysis, which counts among its clients the British Air Traffic Agency (*NATS*) and the Bank of England, starting from 2018. He is the author of more than 30 international scientific publications indexed on the major scientific databases. For more than a decade, he has worked in academia in various European institutions, dealing with the modelling and simulation of transport phenomena, both for basic research and for industrial and biological applications. His academic journey began at the University of Rome "Tor Vergata," where, alongside research on the modelling and simulation of turbulent flows with variable laminar diffusivity and blood flow in arteries and *stents*, he also served as a contract lecturer for the courses of Technical Physics, Thermo-Fluid-Dynamics of Biological Systems and Numerical Calculation of Thermo-Fluid-Dynamic Systems (2006–2010).

In the two-year period from 2010 to 2012, he was a contract researcher at the Department of Energy and Technology of the University of *Southampton* in the United Kingdom, where he dealt with modelling and simulation of bi-phase flows, characterised by spinodal decomposition. In 2012, he went to France to the Institute of Fluid Mechanics of Toulouse as a contract researcher for the OTE (Waves, Turbulence and Environment) research group, where he dealt with modelling and simulation of river and atmospheric fluid dynamics. He terminated his contract in 2013 to go to the University of *Cranfield* in the United Kingdom, where he was appointed *Senior Research Fellow*, becoming part of the academic staff. Here, for the following 5 years (2013–2018), he dealt with modelling and simulation of transport phenomena for various departments: Energy, Hydraulics, Environment and Agriculture, where he respectively dealt with transport of crude oil in pipelines, disinfection of tanks by chlorination and transport and diffusion of organic and inorganic material in soils, with associated absorption by plants. At the University of *Cranfield,* he also handled the course on modelling of atmospheric emissions in the two-year period from 2016 to 2017. Before leaving academia, he went again to the University of *Southampton* in 2018, where he dealt with the three-dimensional modelling of the transport of nitrates and carbonates at the air–water interface in soils. Throughout his period abroad, he maintained his collaboration with Fabio Gori Ammannati, supervising some students and collaborating on the production of scientific articles. In 2020, he obtained the national scientific qualification for the functions of associate university professor in the competitive sector of Technical Physics and Nuclear Engineering.

Fabio Gori Ammannati was born in Montale (Pistoia) on 5 August 1947 to Emilio Gori and Cesarina Ammannati. On 16 July 2021, he added his mother's surname, Ammannati. He graduated in Chemical Engineering on 11 November 1971 with honours. In December 1971, he won a scholarship for young graduates at the Faculty of Engineering in Bologna, which he undertook from January 1972. In 1974, he became assistant professor and, in 1982, associate professor of Technical Physics at the Faculty of Engineering of the University of Florence, where he taught, also as a contract professor, until 1990. In 1978, he won a scholarship from the *British Council,* which he carried out at the *Imperial College* in London, where he collaborated with *Prof. D. Brian Spalding* on the topic of numerical analysis in the turbulent flow of liquid metals. During his time at the University of Florence, he established international scientific collaborations with *Prof. R. Echigo, Tokyo Institute of Technology, Tokyo, Japan; Prof. D.R. Chaudhary, Department of Physics, University of Rajasthan, Jaipur, India;* International Centre for Theoretical Physics *(ICTP)* in Trieste; *Centralny Osrodek Techniki Medycznej, Warsaw, Poland and Prof. T. Aihara, Institute of Fluid Science, Sendai, Japan.* In 1986, he won a CNR scholarship, which he carried out at *Cornell University, Ithaca, New York,* where he collaborated with *Prof. R. Miller* on the topic of ground freezing. In the same year, he became professor in the Faculty of Engineering at the University of Reggio Calabria. In 1988, he was appointed as *professor* at the *University of New York at Stony Brook,* teaching the course, *Introduction to Fluid Dynamics,* in the autumn semester of the same year. During his stay, he collaborated with *Prof. T.F. Irvine Jr.* on the method of measuring thermal conductivity with a thermal probe and on the measurement of the isobaric thermal expansion coefficient of non-Newtonian fluids. In 1990, he moved to the Milano Polytechnic, where he taught Technical Physics and Systems until 1993. From 1991 until 1998, he was an adjunct professor at the Faculty of Engineering of the University of Siena. From 1992 to 2017, he was a professor of Technical Physics in the Faculty of Engineering at the University of Rome Tor Vergata, and from 2017, he is a contract professor. In 1994, he proposed the Research Doctorate (Italian *Ph.D.*) in Energy-Environment Engineering, of which he was the coordinator until 2011. In 1995, he proposed and directed the second-level Master's in Thermo-fluid-dynamics until 2016. From 1998 to 2001, he was the director of the Department of Mechanical Engineering. In 2001, he proposed and coordinated the programme, *Master of Science, Energy Engineering and Thermal and Fluid Dynamics,* in collaboration with the *Department of Mechanical Engineering, College of Engineering, University of Illinois at Chicago, USA,* which takes place at the University of Rome Tor Vergata but awards the title of *Master of Science* from the same American University. Since the late 90s, he has been a coordinator of the joint *Ph.D.* research programme with *Prof. R.J. Goldstein,*

Department of Mechanical Engineering, University of Minnesota, Minneapolis, USA, and with *Prof. J.P. Hartnett, Prof. L. Kennedy, Prof. W. Minkowycz and Prof. W.M. Worek, Department of Mechanical Engineering, College of Engineering, University of Illinois at Chicago, Chicago, USA.* Since the mid-90s, he has proposed and directed the *Socrates–Erasmus* programme with *Prof. Mayinger, Technical University of Munich, Germany; Prof. van Steenhoven, Eindhoven Technical University, the Netherlands and Prof. C. Caro, Imperial College of Science, Technology and Medicine.* During his stay at Tor Vergata, he established scientific collaborations with the *Department of Mechanical Engineering, University of Minnesota, Minneapolis, USA,* collaborating with *Prof. R.J. Goldstein* on Thermo-fluid-dynamics and mass transport in gas turbine blades; the *Energy Resources Center, University of Illinois at Chicago,* collaborating with *Prof. J.P. Hartnett, Prof. W. Minkowycz* and *Prof. W.M. Worek*; the *Department of Mechanical Engineering, College of Engineering, University of Illinois at Chicago, collaborating with Prof. L. Kennedy and Prof. W.M. Worek; Duke University, collaborating with Prof. A. Bejan* and *S. Mary's University, Halifax, Nova Scotia, Canada, collaborating with Prof. W. R. Tarnawski.* The bibliographic review of documents and citations, titled *PlosBiology Career,* places him within the top 2% of researchers in the field of *Mechanical Engineering and Transports.* Since 1992, he has been a tutor for over 30 PhD theses, supervisor for over 40 second-level university Master's theses, supervisor for over 70 degree theses (five-year, Specialist and Master's), supervisor for over ten *Master's degree in Mechanical Engineering* from the *University of Illinois at Chicago* and supervisor for over 20 theses *Erasmus-Socrates.*

Since the early 2000s, he has been a reviewer of international research projects on behalf of *Portuguese Science and Technology Foundation (FCT); Czech Science Foundation (GACR); European Science Foundation, Strasbourg, France,* on behalf of *FCT* and the *Shota Rustaveli National Science Foundation, Georgia.* Since 2017, the year of his retirement due to age limits, he has been a contract professor at the University of Rome Tor Vergata, where in the academic year 2024–2025, he has taught Technical Physics for the Master's Degree Course in Medical Engineering. M.D. Salvatore Mangiafico (Florence), scientific head of the NeuroVascular Base Camp, 2021, invited him to give a scientific presentation titled *"The engineering approach to the study of the Circle of Willis"* on 23 September 2021 at the Congress Centre, University of Rome La Sapienza, thus opening up a possible future collaboration.

Numerical Thermo-Haemo-Dynamics (THD) and Magneto-Hydro-Dynamics (MHD) in the Hepatic Artery

Flavia Russo, Andrea Boghi, and Fabio Gori Ammannati

1 Introduction and Medical Information

The liver tumour results from a quick proliferation of the cells. The primary hepatic tumours, called hepatocellular tumours, are born inside the hepatic cells and spread towards the bones and lungs. It is a rare condition, but its mortality is quite high, as the World Health Organization has reported a number of deaths equal to 745,000 during 2012, which is the second cause of death among common cancers. Moreover, it is mostly widespread in Asia, accounting for 782,000 new cancer cases in 2012. A smaller number of cases per year is reported in the USA and Europe, e.g. in Southern Europe the incidence is 9.5% and in Northern Europe only 4.6%, which can be compared to 2% of overall tumours [4]. Nowadays, it is impossible to prevent this disease except by avoiding the main risk factors, such as alcohol excess and hepatitis viruses. Chemotherapy can ensure a good quality of life, avoiding the spread of the metastasis. However, these drug treatments can have numerous hazardous side effects. In the last few years, new technologies have been developed in order to decrease these undesired side effects.

In magnetic drug delivery, the drugs are guided directly to the interested regions of the body, and the absorption in other organs or systems not interested by the tumour decreases, which results in a higher quality of life for the patient. Magnetic nano-drug targeting, through the use of an external magnetic field, is a new technique for the treatment of several diseases, which can potentially avoid the dispersion of drugs in undesired locations of the body. Nevertheless, due to the limitations on the intensity of the magnetic field applied, the hydrodynamic forces can reduce the effectiveness of the procedure. This technique is studied with the *Thermo-Haemo-Dynamics (THD)*, focusing on the influence of the magnetic probe position and the direction of the circulating electric current.

F. Russo · A. Boghi · F. G. Ammannati (✉)
University of Rome Tor Vergata, Rome, Italy
e-mail: a.boghi@computationalscience.co.uk; gori@uniroma2.it

F. Gori Ammannati (ed.), *Thermo-Haemo-Dynamics in Medical Engineering*,
https://doi.org/10.1007/978-3-031-97214-0_10

1.1 Magnetic Therapy and Drug Delivery

The study of *Magneto-Hydro-Dynamics (MHD)* started in the late 1930s by Hartmann and Lazarus [5], who investigated the flow of an incompressible Newtonian fluid inside a duct under the influence of an external magnetic field. The first analytical characterisation of the problem was done by Shercliff at the end of the 1950s [6], who, later on, investigated the flow at high Hartmann numbers [7]. At the beginning of the 1960s, Gold [8] obtained the solution for the steady one-dimensional flow of an incompressible, viscous, electrically conducting fluid through a circular pipe in the presence of an applied uniform magnetic field. The early studies were performed for liquid metals, and only in the 1990s was the research extended to biological fluids [9, 10].

Several applications in bioengineering and medicine have been developed since then. Among them, the targeted transport of drugs by external magnetic fields and magnetic tracers is of particular interest [6, 11].

1.2 Blood as Fluid

The blood is the most important biological fluid, composed of plasma and erythrocytes, which contain haemoglobin, whose main molecule is iron. The presence of this element allows the blood to be magnetised in the presence of an electromagnetic field. In the deoxygenated state, the blood behaves as a paramagnetic material with a magnetic susceptibility of $\chi = 3.5 \ 10^{-6}$, while, in the oxygenated state, the blood is diamagnetic and its susceptibility is $\chi = -6.6 \ 10^{-7}$, [12].

Several studies have been conducted to investigate the blood velocity profile inside the human circulatory system, both for healthy and non-healthy patients. In the last decades, some experimental data have been compared with numerical results obtained with *Computational Fluid Dynamics (CFD)*, a technique well established in biomedical applications [13–19]. The influence of geometrical factors on pathological conditions is investigated, suggesting that the most important factor to be taken into account is the velocity field, whose measurements can be provided by Doppler ultrasound and MRI [20, 21]. Long et al. [22] reconstruct the blood flow in the carotid bifurcation from MRI images. Moreover, Moulinier et al. [21] use Doppler echocardiography to get information on the blood flow in the aortic arch.

Some studies investigate the effect of the magnetic field on metal nanoparticles embedded in the blood flow and their use as drug carriers [23–25]. Other investigations on drug delivery are performed on non-biological fluid [20] and on clinical trials of cancer patients [26, 27]. A few CFD studies have been conducted to simulate the magnetic drug targeting. In [28], the behaviour of magnetic particles in blood vessels and surrounding tissue is analysed in a cylindrical geometry by solving the particle concentration and including a "magnetic drift" term. In [29], numerical simulations of blood flow and magnetic drug carrier distributions in a patient-specific brain vascular system are reported. The blood is represented as a non-Newtonian power-law fluid, neglecting the influence of the Lorentz force. A Lagrangian approach is used to simulate the particle tracking, neglecting the buoyancy and the carrier phase inertia. The Lorentz force is introduced in [30], with a Lagrangian particle tracking algorithm to simulate magnetic drug delivery in a two-dimensional bifurcation. In [31], the nanoparticle sticking-steering relationship is investigated using Lagrangian particle tracking. In [32, 33], magnetic drug delivery is studied in a three-dimensional bifurcation using Lagrangian particle tracking.

The present study investigates the *THD* of blood in the coeliac trunk, where nanoparticles, used for drug targeting, are dragged into the liver by an external magnetic field. A patient-specific geometry is reconstructed from a data set of CT scan images of a middle-aged healthy man. The blood is treated as a continuum medium, and therefore, an Eulerian formulation is employed to describe it, while the nanoparticles are treated with a Lagrangian approach. A rectangular coil is employed as a source of the external magnetic field. The influence of the probe position and the current intensity is investigated.

2   Three-Dimensional Modelling of the Hepatic Artery

2.1   Hepatic Artery Geometry

The first step to do, in order to reconstruct a *3D* surface using the *Vascular Modelling Toolkit (VMTK)* [34], is to extract the so-called *Volume Of Interest (VOI)* from a dataset (*VMTK user-guide*), as in Fig. 1, [1]. This operation is really important because medical images contain structures which are not of interest.

Once the structures of no interest are deleted, it is possible to apply the level set algorithm, [35], to reconstruct the surface, [36]. In general, level sets are deformable models in which the deformable surface is not represented by a set of points or triangles but rather described by a *3D* function, whose contour, at level zero, is the surface in question. The output of a level set segmentation is an image. The advantage of using a deformable model is that the location of the surface does not depend on the level chosen, but it will locally conform to the peaks of the gradient modulus of the image levels.

In conclusion, the final surface is located on the regions corresponding to the steepest change of intensity across the vessel wall, which is a solid objective criterion. *VMTK* gives the possibility to the operator to choose the way the deformable model is initialised. The aim of the initialisation is to have a model as close to the vessel wall as possible. *VMTK* enables four different ways of initialisation:

- Colliding fronts
- Fast marching
- Iso-surface
- Threshold

Fig. 1 VOI selector tool [1]

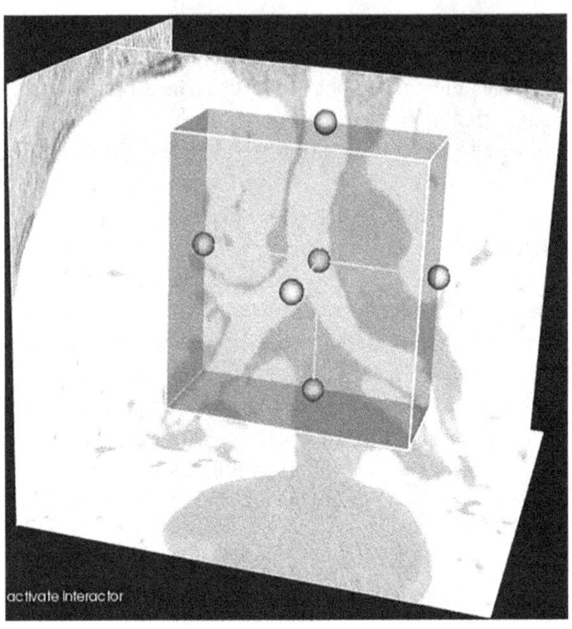

The best choice for reconstructing blood vessels is the *Colliding Fronts*, where two fronts propagate from the seeds (one front from each), with their speed proportional to the image intensity. The region where the two fronts collide is the initial deformable model. For reconstructing a tract of the vessel, in general, a seed is placed on each of the two extremities, and the region between the seeds is then selected. An advantage of this method is that the side branches are ignored. Moreover, it is possible to reconstruct vessels in separate chunks with level sets and then merge the results of the different chunks into a single, cohesive, final model. Once the whole set of the interested segment is initialised, the evolution equation is solved for each segment. In this case, it is possible to choose four main parameters:

- Number of iterations: the number of deformation steps that the model will perform
- Propagation scaling: the weight to assign to the model inflation
- Curvature scaling: the weight to assign to the model surface in order to regularise it
- Advection scaling: the most important weight, which regulates the attraction of the surface to the image gradient

It is observed that propagation and curvature should be set to 0.0, while advection is set to 1.0. This is possible if the initialisation is really close to where the surface has to converge. The number of iterations should be set large enough: in general, if the region is not too big, 300 iterations are correct. The output of the level set algorithm is an image, and, in order to obtain a surface, it is necessary to use

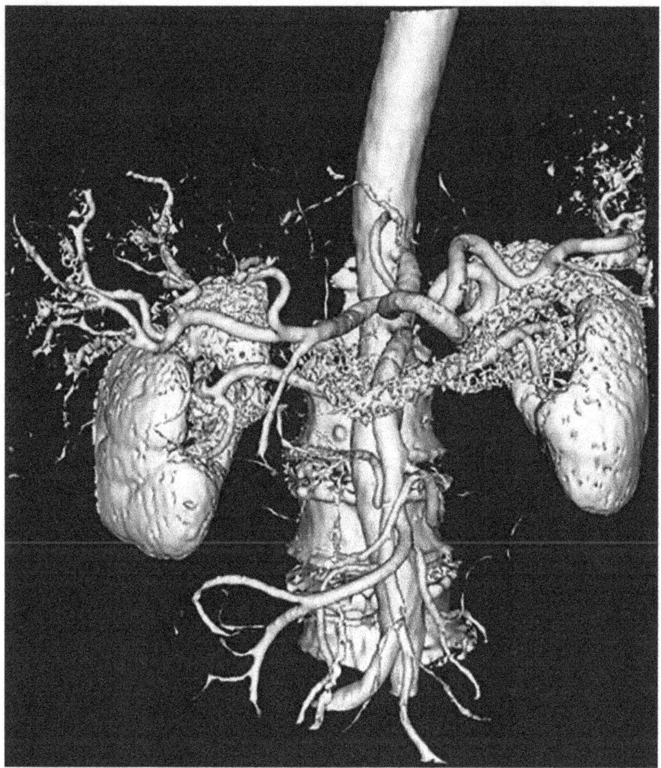

Fig. 2 Marching cubes algorithm [1]

the marching cubes algorithm. In this case, to reconstruct the surface, the *Marching Cubes algorithm* is used due to the depth of the interested vessel, as reported in Fig. 2, [1]. In order to get a mesh from the reconstructed surface, to be used in a *CFD* simulation, four more steps are necessary.

The first one is to compute the centerlines, [37]; they are powerful descriptors of the shape of the vessel. Several studies have been proposed in the literature for their computations, all from angiographic images and *3D* models. In *VMTK*, the algorithm to compute the centerlines deals with their computation, starting from surface models, and the advantage is the stability from the perturbations in the surface. Centerlines can be determined as weighted shortest paths, traced between two extreme points. In order to ensure that the final lines are central, the paths cannot lie anywhere in space, but they are bound to run on the *Voronoi* diagram of the vessel model, as in Fig. 3, [1]. In the *Voronoi* diagrams, the paths can be described as the places where the centres of maximal inscribed spheres are defined.

About the centerlines algorithm, it takes as input a surface and gives back centerlines, the *Voronoi* diagrams and the *Delaunay* tessellation, as in Fig. 4, [1]. It is important to give the algorithm the right source and target points for the inlet and outlet of the surface in order to obtain its centrelines.

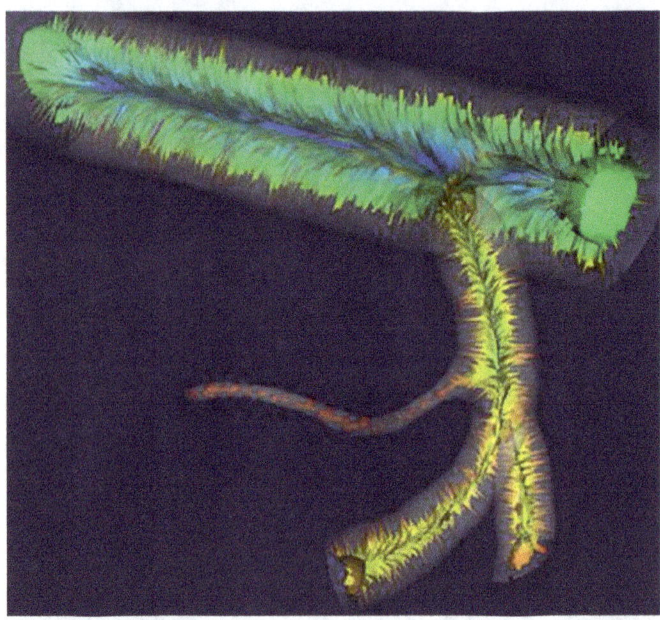

Fig. 3 Voronoi diagram [1]

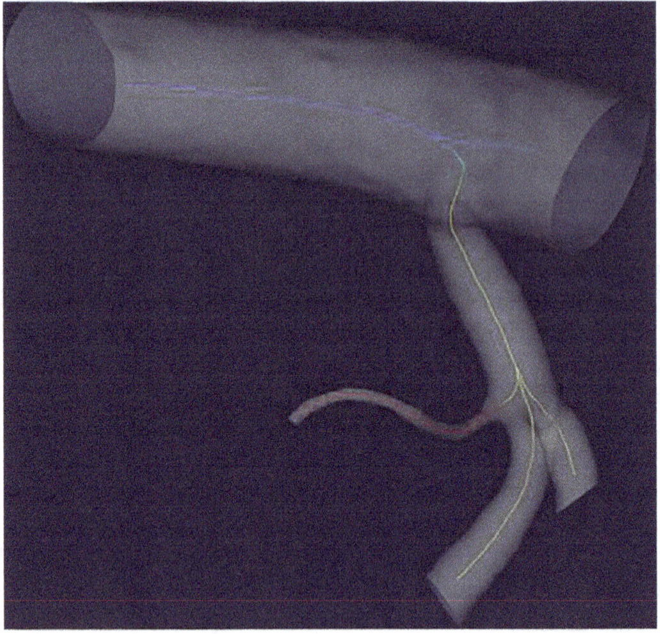

Fig. 4 Centerlines [1]

The second step is a *Smoothing* of the obtained surface. This step is necessary because sometimes the image segmentation can result in bumpy surfaces, especially if the image quality is not high. Since artefacts can result in some flow features, and they can affect wall shear stress distributions, it is necessary to smooth the surface before building a mesh. In general, there are two important parameters for smoothing a surface:

- The type of filter
- The number of iterations

In general, smoothing is implemented with a low-pass filter that is based on the *Taubin* algorithm, and the spatial cut-off frequency is quite low, about 0.1. As far as the number of iterations, they represent the number of smoothing passes, and, for blood vessels, 30 are sufficient. A large number of iterations could lead to a shift of the reconstructed surface in space.

The third step is the opening of the surface inlet and outlet; in general, it happens when a surface is created by using a deformable model. To open a surface, it is possible to choose between two options:

- Clip the surface with a plane
- Clip end-caps automatically, but this method requires the computed centerlines

The fourth and last step is to add flow extensions; they have a cylindrical shape, and, generally, they are added to the inlets and outlets of a model, as in Fig. 5, [1]. Their use is important because they ensure that the flow, entering and leaving the computational domain, is fully developed in order to have boundary conditions that

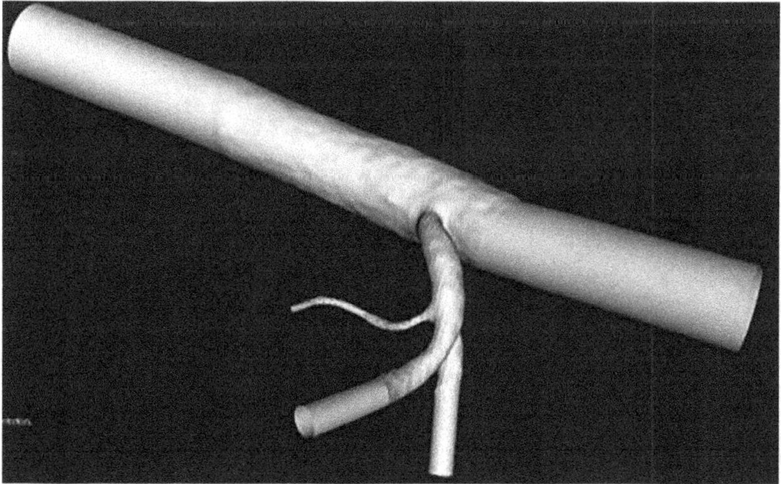

Fig. 5 Flow extensions [1]

do not force the solution in the actual vessel. Adding flow extensions relies on centerlines, which is a typical problem in *CFD* modelling. The automatic procedure ensures the reproducibility and the speeding up of the modelling phase. *VMTK* makes it possible to choose the length of the added extensions, and, in general, the algorithm computes the length of each flow extension, proportional to the mean profile radius. After adding flow extensions to the surface, it is possible to generate a mesh for the final model in order to use it for numerical *THD* simulations, [38].

2.2 Grids for the Hepatic Artery

In order to study the fluid dynamics of a human hepatic artery in a patient-specific geometry, we need the *DICOM* images taken from a *CT* scan. The subject under investigation is a healthy middle-aged man. The slices of the abdomen are supplied by the Department of Imaging of the Hospital "Tor Vergata", and the hepatic artery is reconstructed by using *VMTK*. When dealing with *THD* problems, the first thing to do after the grid creation is to analyse the grid's independence. The final step is to generate an adaptive mesh in order to have a much more refined grid only in the part where the interest is focused, as in Fig. 6, [1]. The step needs to generate an adaptive mesh, which has a much more refined grid only where the interest is focused. Before converting the mesh to a format supported by the *THD* solvers, it is possible to inspect the elements of the final volume. After this step, it is possible to export the mesh and run a *THD* simulation for the reconstructed vessel [39].

In our case, we build three grids with different numbers of elements. In the following Table 1, the characteristics for each grid are listed.

In the following Figs. 7, 8 and 9, [1], it is possible to see the three different geometries: the first one is the coarsest grid, *Grid 1*, Fig. 7; the second one is a more

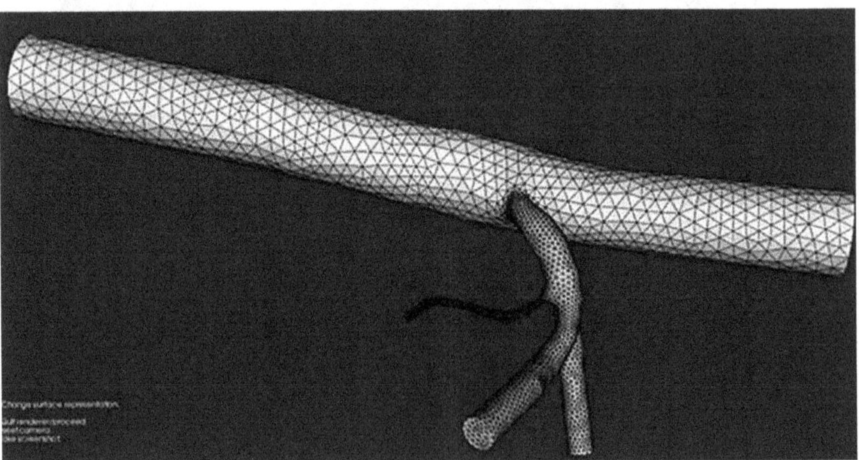

Fig. 6 Adaptive mesh [1]

Table 1 Grid characteristics

	Grid 1	Grid 2	Grid 3
Elements	174619	375972	724779
Nodes	34283	71193	134133
Skew	0.666132	0.684929	0.757877

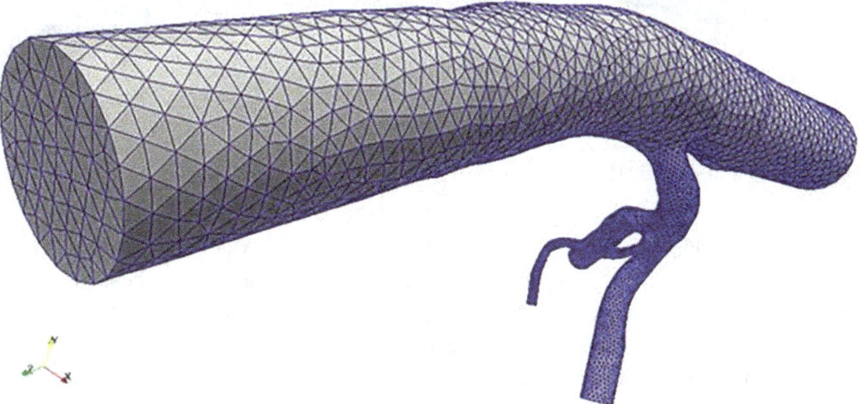

Fig. 7 Coarse grid, Grid 1 [1]

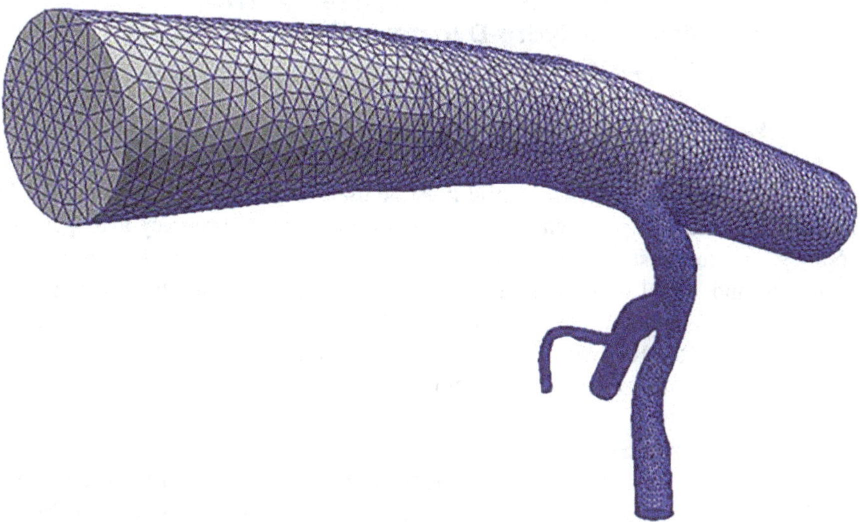

Fig. 8 Intermediate grid, Grid 2 [1]

refined version, the intermediate grid, *Grid 2*, Fig. 8; and the last one is the finest, the fine grid, *Grid 3,* Fig. 9. *VMTK* can create a radio-adaptive grid, which means that, as the cross-sectional area decreases, the software increases the number of elements in the desired region. In Figs. 7, 8 and 9, it is possible to observe this specific feature in

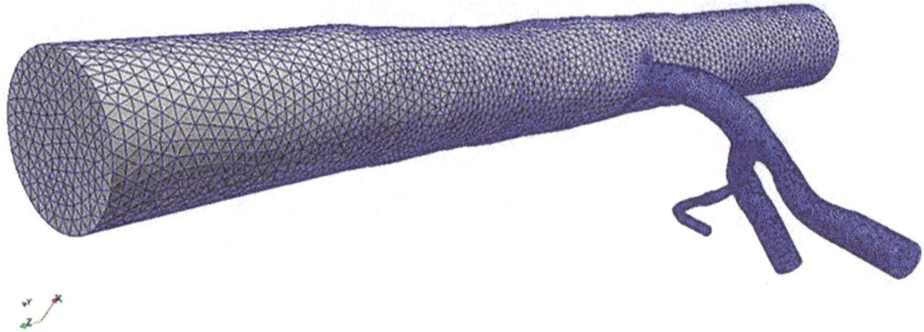

Fig. 9 Fine grid, Grid 3 [1]

the coeliac trunk and in the hepatic artery bifurcation. *VMTK* generates a mesh file in a format which is not supported by *OpenFOAM*, so in order to use these grids, we have to save the geometry in an appropriate format, which can be read by the *blockMesh utility* in *OpenFOAM*. Afterwards, we define in the boundary file the name of each patch in order to assign the appropriate boundary conditions for these regions.

3 Numerical Thermo-Haemo-Dynamics (THD) and Magneto-Hydro-Dynamics (MHD) Analysis of the Hepatic Artery

3.1 Hepatic Artery Steady-State Analysis

At first, we need to decide which grid must be used for the steady-state analysis. In order to find the best grid, which is a compromise between accuracy and speed of solution, we run a steady-state simulation for each grid. We fix the appropriate boundary and initial conditions and the schemes necessary for the calculations. Secondly, the simulations are run and the results compared. In particular, we analyse the pressure and the *Wall Shear Stress (WSS)*. After the right grid is found, we set up the conditions for the unsteady simulation.

3.1.1 Parameter Setting

In the constant folder, there is the boundary file, where the geometry patches are specified. The boundary regions are renamed, as displayed in Fig. 10, [1]. In the start-time directory, the boundary and initial conditions for the case are set up. The fluid dynamics of the magnetic nanoparticles, laden in the blood flow and moving under the influence of a magnetic field, is the object of this work. The non-uniform magnetic field is generated by a probe, which behaves as a magnetic dipole. The probe is located externally, on the abdomen of the patient. The current intensity is set equal to zero in the first steady-state simulation since we want to understand how the magnetic field changes the fluid dynamics.

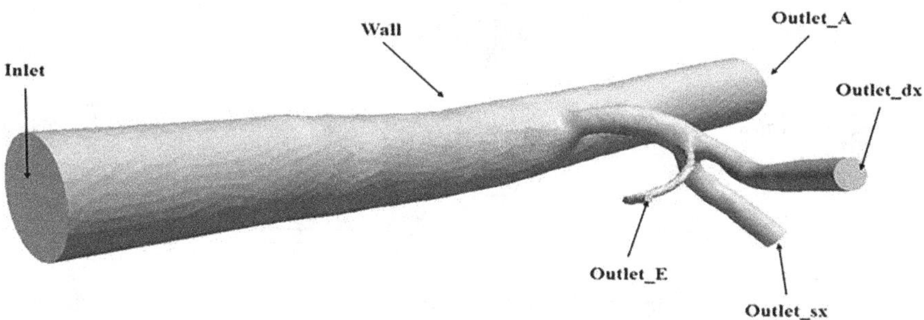

Fig. 10 Definition of the patches for the boundary conditions [1]

As far as the inlet velocity is concerned, a non-uniform profile is set up by using a *groovyBC* utility. In the steady-state simulation, the profile is parabolic. For the outlets, we have to decide whether to specify the velocity or the flow rate. In clinical practice, it is difficult to know the exact value of the blood speed in the sections of interest, while it is quite easy to know the value of the flow rate. Normally, the total volume of blood in resting condition is between 5.5 and 6 litres/min; the variation depends on many factors, as, for instance, gender. There are plenty of studies in the literature in which the velocity and the flow rate of blood are measured, [40–42], in rest and under exercise conditions, [43, 44], using some techniques, like the Doppler effect, [45, 46]. Moreover, some researchers investigated the blood velocity and the flow rate in the hepatic artery, [47], as well as by using ultrasounds [48].

We assume the flow rate in the hepatic artery and in the two gastric arteries, left and right, from the aforementioned papers. For three of the outlets, *outlet_E*, *outlet_dx* and *outlet_sx*, respectively, relative to the hepatic, right and left gastric arteries, we impose the volumetric flow rate. In *OpenFOAM*, this is done with the *flowRateInletVelocity* boundary condition. In particular, this boundary condition is mostly used with time-dependent flow rate. The flow rate value is specified by the *volumetricFlowRate key-word*. For the *outlet_A*, relative to the descending aorta, we impose a *zeroGradient* boundary condition. This is done in order to avoid mass-balance problems due to the specification of the flow rate at the other outlets. For the pressure field in the *outlet_A* the value of the pressure is used as a reference because it is the relatively constant aortic pressure to impose a *fixedValue* boundary condition. For the other boundaries, we consider a *zeroGradient* condition due to the fact that we assume those constant values for the flow rate. In the end, we fix the parameters for the simulation in terms of time schemes.

3.1.2 Eulerian Model

The mathematical model is based on the coupling between the Navier-Stokes equations and the Maxwell equations. It is known that an external magnetic field can change the direction of a moving charged particle, and the force responsible for it is known as the Lorentz force [49]. The *THD-MHD* equations can be written for incompressible flow of a viscous, conducting liquid

$$\mathrm{div}\left(\vec{v}_b\right) = 0 \tag{1}$$

$$\frac{\partial \vec{v}_b}{\partial t} + \left(\vec{v}_b \cdot \nabla\right)\vec{v}_b = -\frac{1}{\rho_b}\nabla p + \nu_b\nabla^2\vec{v}_b + \vec{g} + \frac{1}{\rho_b\mu_0}\mathrm{curl}\left(\vec{B}\right) \times \vec{B} \tag{2}$$

$$\frac{\partial \vec{B}}{\partial t} + \left(\vec{v}_b \cdot \nabla\right)\vec{B} = \left(\vec{B} \cdot \nabla\right)\vec{v}_b + \frac{1}{\sigma_b\mu_0}\nabla^2\vec{B} \tag{3}$$

where the subscript b denotes blood, ν_b is the kinematic viscosity, ρ_b the density, μ_0 the magnetic permeability in the vacuum, which is approximately the same as that of the blood, σ_b the electric conductivity, g the gravity acceleration, p the static pressure, v_b the velocity field and B the magnetic induction field.

The momentum conservation equation does not take into account the momentum which the particles transfer to the blood, and the last term represents the Lorentz force in the context of continuum mechanics. According to [50, 51], for particle volume fractions less than 10^{-6} particle motion is influenced by continuous phase properties, while, practically, there is no feedback from the dispersed phase. The particle diameter used in this work is 5 nm, which is typical of gold/iron oxide nanoparticles [52, 53]. Therefore, we assume that the inertia of the magnetic particles is much smaller than that of the blood, which is justified by their small volume of the order of 65 nm^3.

3.1.3   Boundary Conditions (BC)

The solution of the Eulerian system requires appropriate boundary conditions (BC) for each field. A non-slip BC is imposed for the velocity on the wall. As far as the outlets are concerned, a mixed BC is used. If the blood leaves the domain, the velocity normal derivative is set to zero, whereas if blood enters through the boundary, the tangential velocity is set to zero.

The velocity profile is imposed on the inlet. In the steady state, a parabolic profile is used, whereas in the unsteady state, a Womersley-Evans profile is employed, in analogy with [17].

$$v_b(t,\xi) = \frac{8Q}{\pi D^2}\left(1 - \xi^2\right) + 2\Re\left(\sum_{n=1}^{N} V_n\Phi(\tau_n,\xi)e^{j\omega_n t}\right) \tag{4}$$

where

$$\Phi(\tau_n,\xi) = \frac{J_0(\tau_n) - J_0(\tau_n,\xi)}{J_0(\tau_n) - 2J_1(\tau_n)/\tau_n} \tag{5}$$

and

$$\tau_n = j^{\frac{3}{2}} \frac{D}{2} \sqrt{\frac{\rho}{\mu_\infty}} \, \omega_n = j^{\frac{3}{2}} \alpha_n \tag{6}$$

being r the radial coordinate, D the diameter of the ascending aorta, Q the volumetric flow rate, $\xi = \frac{2r}{D}$, J_0 and J_1 the zero[th] and first-order Bessel functions of the first kind, α_n the Womersley numbers of order n, $\Re()$ the real part of a complex number, $j = \sqrt{-1}$ the imaginary unit, V_n the Fourier coefficients of the pulsatile mean velocity profile and N the number of harmonics used to reproduce the flow rate. By using the *Fast Fourier Transform (FFT) algorithm*, the first ten Fourier coefficients of the flow rate in the ascending aorta are used to reconstruct the velocity profile, according to the experimental data [54].

As far as the pressure is concerned, a constant total value, equal to the sum of the static and dynamic ones, is imposed on the outlet of the abdominal aorta (AA). On the other outlets, the resistive BC, derived in [55], is used

$$p = p_b + RQ \tag{7}$$

being p_b the reference pressure at the right atrium, Q the volumetric flow rate, and R the hydraulic resistance. The value of the resistance and the reference pressure are extrapolated from the steady-state simulations by imposing the volumetric flow rates and a zero normal derivative condition for the pressure on the outlets. The values of the flow rates, in steady state, for the different outlets are derived from the literature [46, 56, 57].

3.1.4   Results

The cardiac cycle is the sequence of the rhythmic activity of the heart, or, rather, a sequence of a *systolic* and a *diastolic* phase. In the *late diastolic phase*, the two atria and the two ventricles fill with blood coming from the veins. In the *atrial systolic phase*, the contraction of the atria pumps the blood inside the ventricles with the closure of the atrium-ventricles valves, but the pressure gradient inside the ventricles is not sufficient to open the semilunar valves. The ejection of blood from the ventricles is present only when the pressure inside them is higher than the pressure in the arteries. When the ventricles are not contracted, the ventricular pressure diminishes, the blood passes through the semilunar valves, and this movement closes the aforementioned valves. As can be noticed, the principal variable involved in this cycle is the pressure.

The pressure is the main variable that regulates the flow inside the blood system. In particular, the flow rate can be defined as:

$$Q = \frac{\Delta p}{R} \tag{8}$$

with Q flow rate, Δp pressure gradient and R flow resistance, which, in a cylindrical pipe, is expressed by the Poiseuille law

$$R = \frac{8L\eta}{\pi r^4} \tag{9}$$

where L is the length of the pipe, η the viscosity of the blood and r the radius of the pipe. As far as the radius decreases, the resistance increases. So, the mass flow rate is:

$$Q = \frac{\Delta p \pi r^4}{8L\eta} \tag{10}$$

After the simulation has run for a fixed number of time steps equal to *2000*, it is possible to post-process the result in order to find out the grid independence. We evaluate the results with *ParaView*.

Pressure and Wall Shear Stress Analysis

In order to choose the best mesh, we compare them with the *mapFields* utility. The data are read from the equivalent time directory of the *source* case and mapped onto the equivalent time directory of the *target* case, [58]. We map the *pressure* and the *WSS* for the three geometries in order to compare the fields from the coarsest to the thickest one.

The grid independence results are shown in Fig. 11, [1], highlighting the trend of the pressure in the three grids. The large tract of the artery is the descending aorta, and we can see that the value of the pressure in this part is the highest. The radius of the aorta is big, so the resistance is low, the flow rate is high, and the pressure gradient is high. Moreover, the aorta is the first artery in which the blood is pumped at the end of the *systolic phase*; then the maximum pressure at the entrance of this artery is the systolic pressure or the maximum pressure of the systemic circuit.

In the other three branches, the pressure diminishes, and the minimum value is reached inside the hepatic artery. This happens because the radii of the three arteries decrease, and this leads to an increase in the resistance R; the cross-section and the flow rate inside the arteries diminish too, due to conservation, and consequently, the pressure gradient decreases. The pressure values do not change significantly in the three grids, and therefore, the pressure field is grid independent. From these results, we can say that the coarsest grid is sufficient for the accuracy of the solution.

We perform the same analysis on the *WSS* to confirm this by comparing the differences between its modulus. In general, for a Newtonian fluid the *WSS* is defined as

$$\tau_{wall} = \widehat{\imath}_{axis}$$

$$\cdot \mu \left([I] - \widehat{n}_{wall} \otimes \widehat{n}_{wall} \right) \left(\left[\nabla \vec{v} \right] + \left[\nabla \vec{v} \right]^T - \frac{2}{3} \mathrm{div} \left(\vec{v} \right) [I] \right)_{wall} \widehat{n}_{wall} \tag{11}$$

The contours of WSS for the grid independence study are shown in Fig. 12, [59].

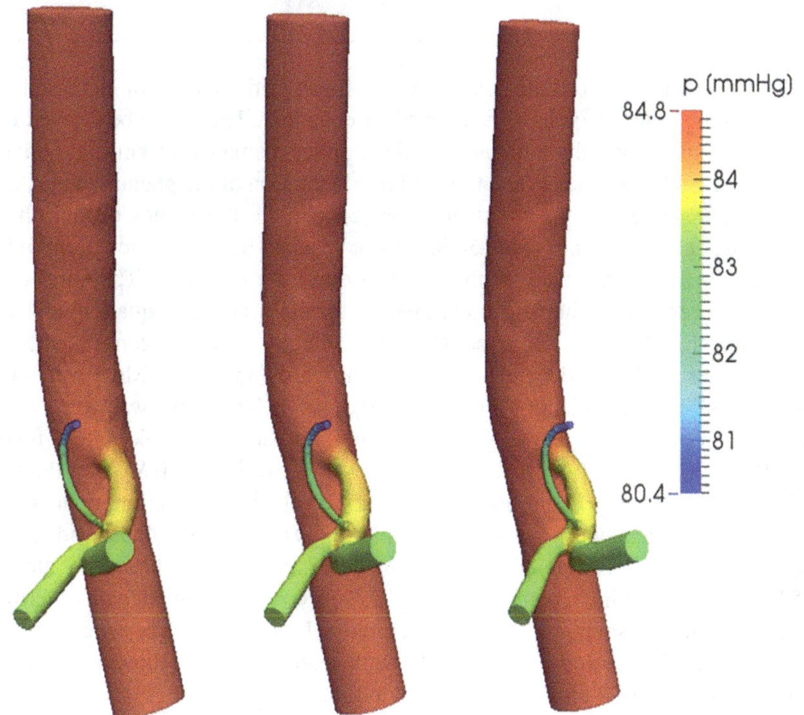

Fig. 11 Steady-state pressure (mmHg): left, Grid 1; centre, Grid 2; right, Grid 3 [1]

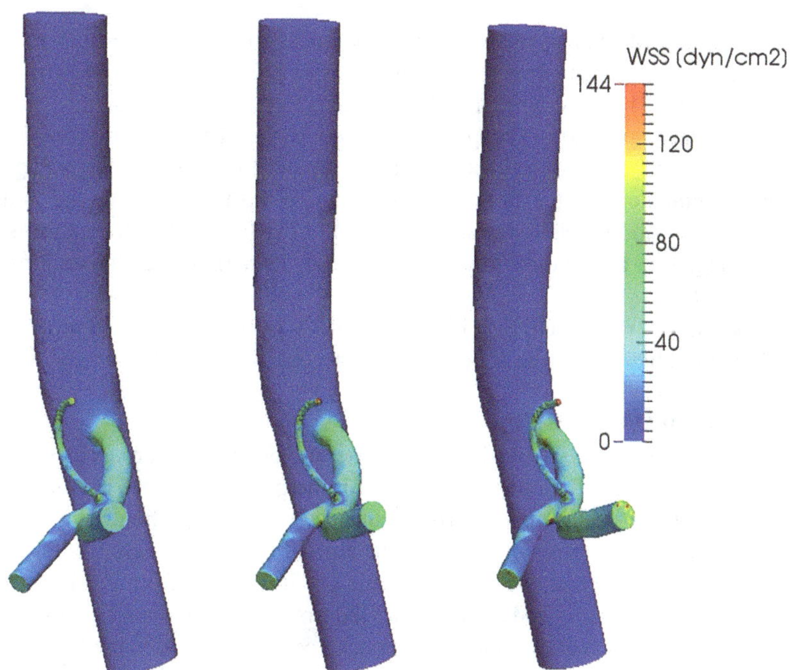

Fig. 12 Wall shear stress (dyn cm^{-2}): left, Grid 1; centre, Grid 2; right, Grid 3, Fig. 4 of [59]

It can be seen that from the coarsest to the finest grid, there is an increase in the *WSS*, meaning that this variable is not grid independent. This could be derived from the fact that increasing the number of nodes implies a smaller spacing in the mesh, and it could lead to a better evaluation and representation of the phenomenon. As far as the intermediate and the finest grids are concerned, the values do not change appreciably. The maximum value of the *WSS* is reached in the hepatic artery. In fact, the shear stress is influenced by the geometry and the flow rate. The aorta cross-section is larger than the hepatic artery one, so, for the continuity equation, under the assumption that the flow rate is constant, the velocity is lower; consequently, the *WSS* is lower, and it is dependent only on the blood viscosity. Moreover, in the proximity of a bifurcation or in the presence of a bend, the wall shear stress increases. In particular, near a bifurcation, the laminar flow breaks into a reverse flow. So, the inner wall of the bifurcation is subject to higher *WSS* than the outer wall. The hepatic artery has a small radius, and the velocity, for continuity, increases. The choice now is between the intermediate grid, *Grid 2*, and the finest grid, *Grid 3*, and, in order to save computational time, the best grid is *Grid 2*, because it has only *375,972* elements, which, compared to *724,779* of *Grid 3*, is a great gain in terms of computational time saving. The *Grid 2* is employed for the unsteady-state simulation with a pulsatile blood flow. Moreover, the last steady-state time directory is applied as the initial condition for the unsteady-state run.

3.2   Hepatic Artery Unsteady-State Analysis

For the unsteady-state analysis, we take the pulsatile profile from the studies of the literature, [60, 61], for the flow rate and the relative pressure versus time inside the ascending aorta. Then, we write a code in *Matlab* to evaluate the flow rate starting from the relative pressure. The procedure used is the same as described in the previous chapter for the reconstruction of the airflow profile. By using the *Womersley* equation for the flow rate, we evaluate the harmonics of the Fourier series. In detail, in our case, we take into account the first 10 harmonics to reconstruct the signal in a period of time equal to four of its periods, i.e. 4 s, each period being equal to 1 s.

The pressure is reported in Fig. 13, while Fig. 14, [59], presents the pattern of the inlet flow rate obtained by the code written in accordance with *Womersley* theory.

3.2.1   OpenFOAM Results

Once the pressure values for the different arteries are obtained in the steady state, we can set up the resistive boundary conditions for the pressure and the magnetic field generated by the probe. At first, we run an unsteady-state simulation in the absence of a magnetic field by fixing the value of the current intensity to zero. As far as the inlet velocity is concerned, we use a pulsatile profile derived from the *Womersley* equation; we implement a code to calculate the first ten Fourier coefficients for the velocity profile. After that, we import the Fourier coefficients inside the U boundary condition, and we set up a *groovyBC* in order to use the *Womersley* formulation. For the outlets, as in the steady state, we use the *pressureInletOutletVelocity* boundary

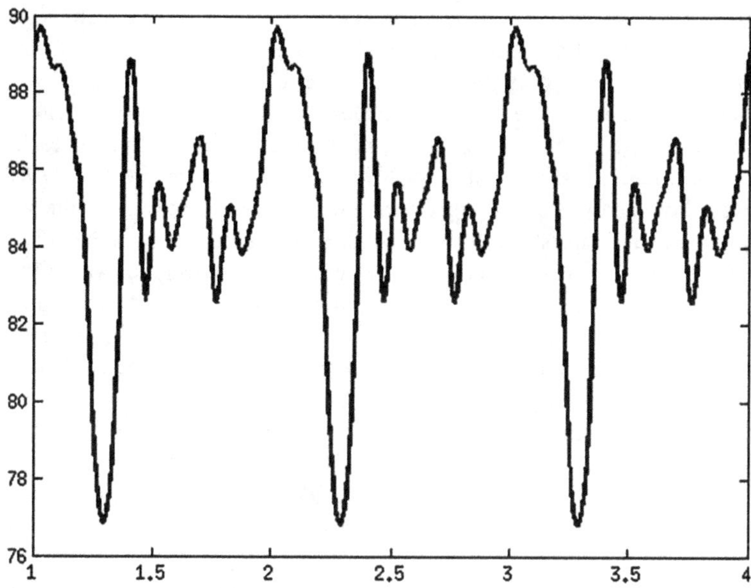

Fig. 13 Inlet pressure gradient (mmHg) versus time (s) in ascending aorta [1]

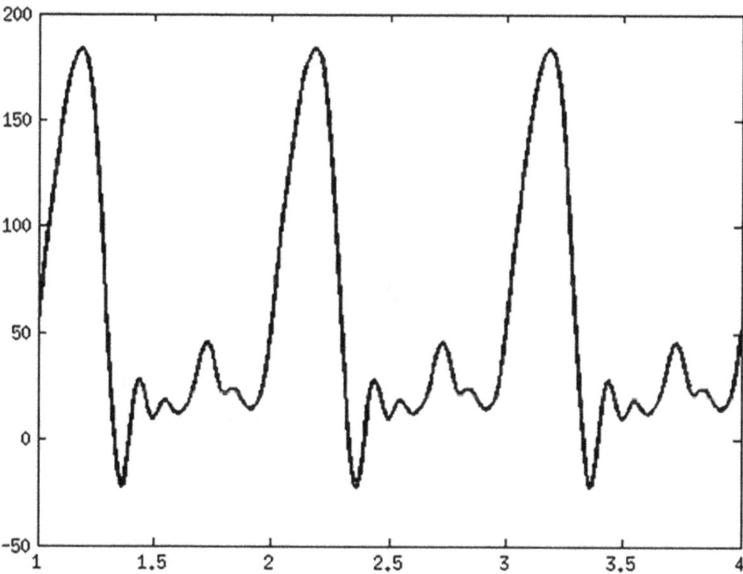

Fig. 14 Inlet flow rate (cm^3/s) versus time (s) in ascending aorta, Fig. 2 of [59]

condition, which switches between *zeroGradient* and *fixedValue,* depending on the direction of the flow. As far as the pressure is concerned, we impose a *groovyBC* condition at every patch. This last type of condition requires some input value such as: *value, variables, fractionExpression* and *valueExpression.*

From the steady state for each outlet and inlet, we get the *.csv* file with the value of velocity, pressure and cross-section area. The previous *.csv* file was created with *ParaView*; once we import the geometry, we use the specific filter *Integrate Variables* to obtain the necessary variables. With *ParaView*, we evaluate the components of the normal vector n_x, n_y, n_z, by the filter *glyph*for, for each outlet and for the inlet. In *Matlab,* it is possible to read the *.csv* file by a *csvreader* command. For each patch of the geometry, we evaluate the pressure, the velocity, the flow rate and the resistance. About pressure, we normalise its value, dividing it by the value of the cross-section. As far as velocity is concerned, we do the same thing as for the pressure, and each velocity component is divided by the value of the cross-section. The flow rate is the integral of the velocity over the surface, so its value is equal to the product of each velocity component for the equivalent component of the normal vector. Finally, the resistance comes from Eq. 2.

Once the code has completed the simulation, we obtain the values of pressure and resistance that we write inside *groovyBC*. In *value,* for each outlet, we write the value of the pressure coming from the code. In *variables,* we declare the parameters used in the definition of the equation to evaluate the pressure, which are the resistance, the flow rate and the pressure for every outlet. In particular, the resistance is evaluated by Eq. 2, as well as the flow rate at each outlet, and updated every time step. The *FractionExpression* distinguishes between *Dirichlet* and *Neumann* boundary conditions. In the *valueExpression,* we define the *Dirichlet* condition with the equation for the evaluation of the pressure at each outlet. The general form of the final equation is of the following type:

$$p = p_b + RQ \tag{12}$$

The simulation is carried out for two periods, each lasting 1 s. We opt for the application of an adjustable time step, in order to have a maximum Co equal to 0.5. We plot the pressure and the flow rate for the outlets over a time of 2 s. The pressure and the flow rate profile for the abdominal aorta, right coeliac branch, left coeliac branch and hepatic artery are shown respectively in Figs. 15 and 16. As can be seen, the profiles are periodic. We examine four periods for the case with a null magnetic field, and, as mentioned before, every period lasts 1 s.

In Fig. 17, [59], we show the time variations of the *pressure* field on the artery wall, which is representative of the pressure inside the whole artery. Because of the periodicity of the profile, we have chosen only some meaningful instants of time for the different images. From Fig. 17, it is possible to say that the highest *pressure* value is reached in the abdominal aorta, whereas the lowest *pressure* value is reached inside the hepatic artery. In the two coeliac branches, right and left, the pressure has the same intermediate values. These results come from the fact that as the radius of the artery decreases, the resistance increases due to a reduction in the cross-section of the artery, and the flow rate for the continuity equation decreases. So the pressure decreases in the small arteries and increases in the big ones. Moreover, the maps are taken at four different times within the same period of the cycle. At time 0 s, we can see that the pressure is maximum in the aorta because we are in the last part of the

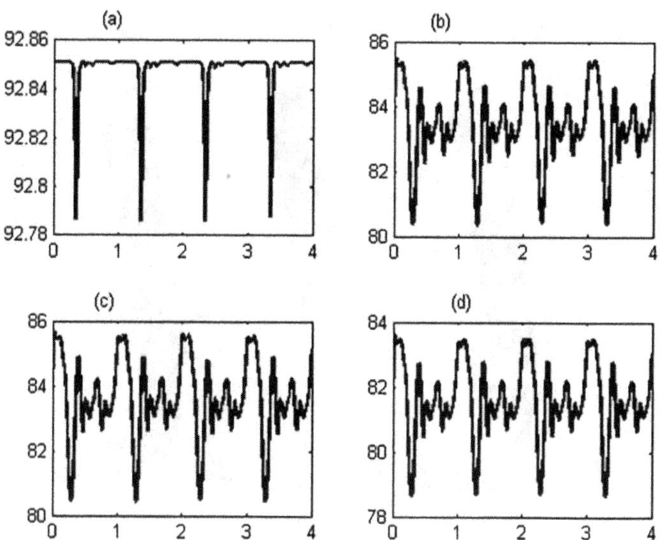

Fig. 15 Outlet pressure profile (mmHg) versus time (s) in the different arteries: (**a**) abdominal aorta; (**b**) right coeliac branch; (**c**) left coeliac branch; (**d**) hepatic artery [1]

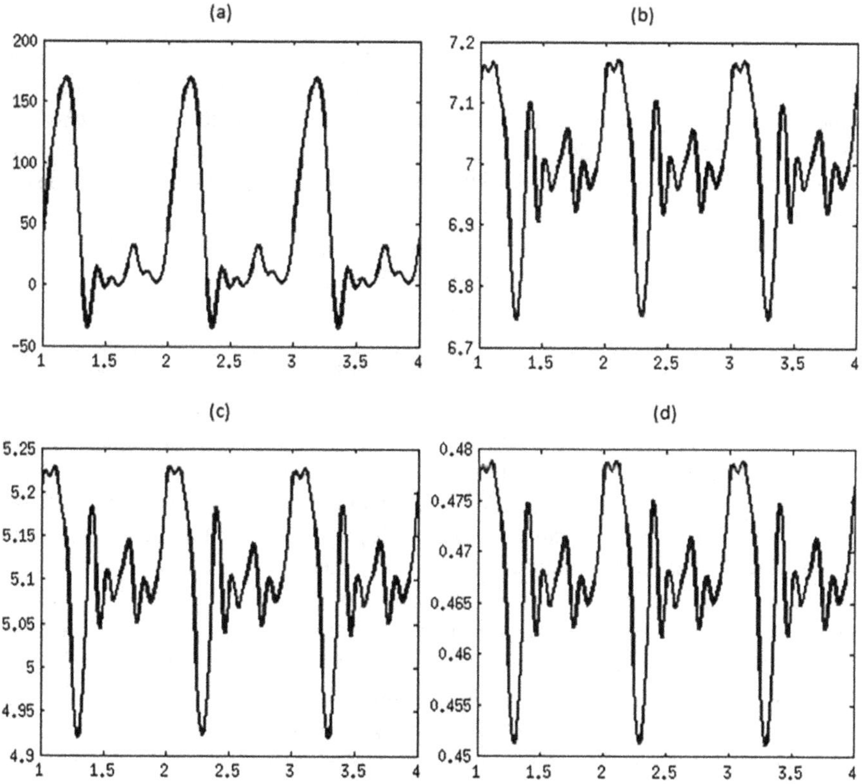

Fig. 16 Outlets Flow profile (cm^3/s) versus time (s) in the different arteries: (**a**) abdominal aorta; (**b**) right coeliac branch; (**c**) left coeliac branch; (**d**) hepatic artery [1]

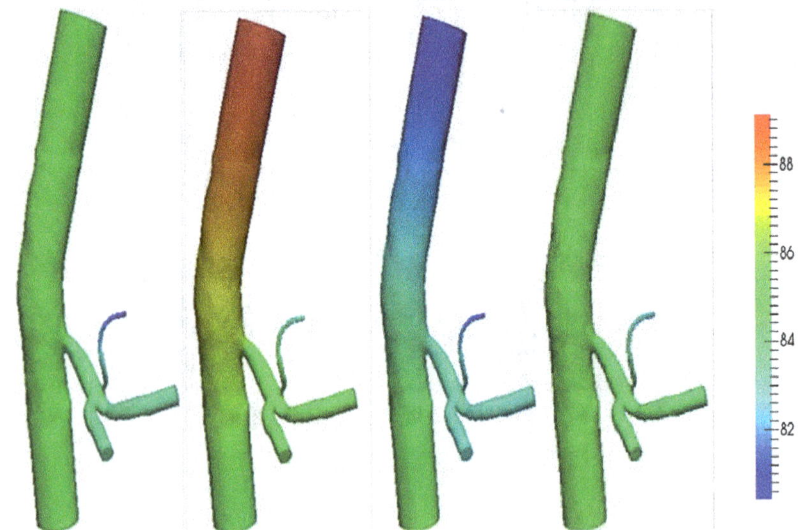

Fig. 17 Pressure map (mmHg) at 0 s, 1 s, 1.25 s, and 1.5 s, from the left, inside the whole artery, Fig. 5 of [59]

systolic phase; the blood flows from the atria to the ventricles. At time 1 s, the blood is ejected from the cardiac chambers, so the pressure in the aorta is the highest. At time 1.25 s, the pressure is lower in the aorta because we are in the *diastolic phase* of the cardiac cycle, and finally, at time 1.5 s, we are in the last part of the *diastolic phase*; the pressure increases again in the aorta and reaches the lower value inside the hepatic artery.

Two more variables are investigated: the *WSS* field, shown in Fig. 18, and the velocity streamlines in the hepatic artery, in Figs. 19 and 20, which is particularly interesting because the target of the magnetotherapy is the liver, and therefore, the magnetic particles are dragged into this organ.

The *WSS* for the whole geometry is visible in Fig. 18. The *WSS* presents a periodic pattern and is dependent on the diastolic and systolic phases of the cardiac cycle. Indeed, at time 0 s, we are in the last part of the systolic phase, and the ventricles are filling with blood. The velocity in the aorta is smaller, and the *WSS* is higher only near a bifurcation or inside bending branches. As the cardiac cycle flows by, the trend for the *WSS* remains the same. It is possible to verify that the maximum value is reached at the bifurcations and wherever there is a bend in the artery. In the presence of bending or bifurcations, the velocity of the blood changes from laminar to reverse flow, and consequently also the shear stress also changes. In a bifurcation, it is bigger near the inner wall than in the outer one, whereas, in bending arteries, the *WSS* is bigger in the outer wall than in the inner one.

The velocity streamlines (*m/s*) inside the hepatic artery are shown in Figs. 19 and 20 at different instants of time. From these images, it is possible to see that the flow,

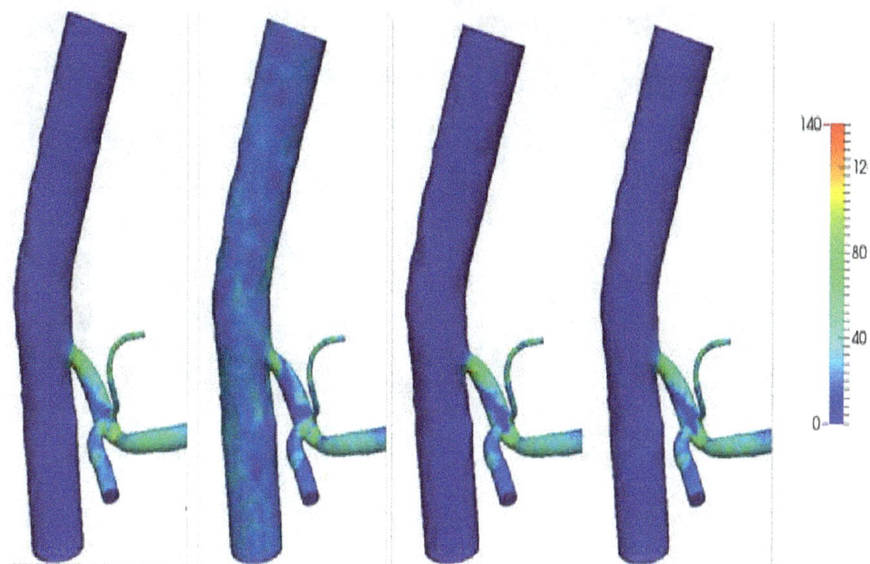

Fig. 18 Wall shear stress map (dyne/cm^2) at 0 s, 1.25 s, 1.5 s, and 2.0 s, from left, inside the whole artery, Fig. 6 of [59]

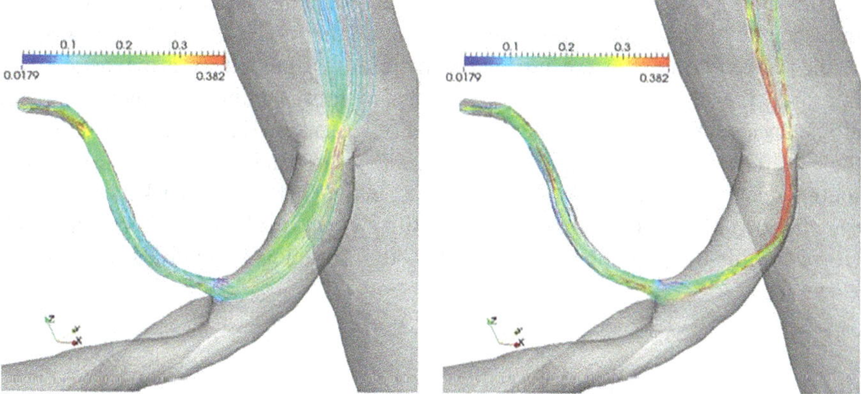

Fig. 19 Velocity streamlines (m/s) at 0 s and 1.25 s, from left, inside the whole artery [1]

transitional in the aorta, is subject to a laminarisation process. This is evident, especially at $t = 1.5$ s, Fig. 20.

Figure 19 evidences that the particles at the centre of the common coeliac branch tend to go into the hepatic artery, except in the period of maximum flow rate, i.e. $t = 1.25$ s, where the velocity is so high that the particles are first deflected towards the bottom side of the common coeliac artery and then transported in the hepatic artery. The streamlines also evidence a helical motion at the point where the hepatic artery originates. This contributes to increasing the local pressure drop.

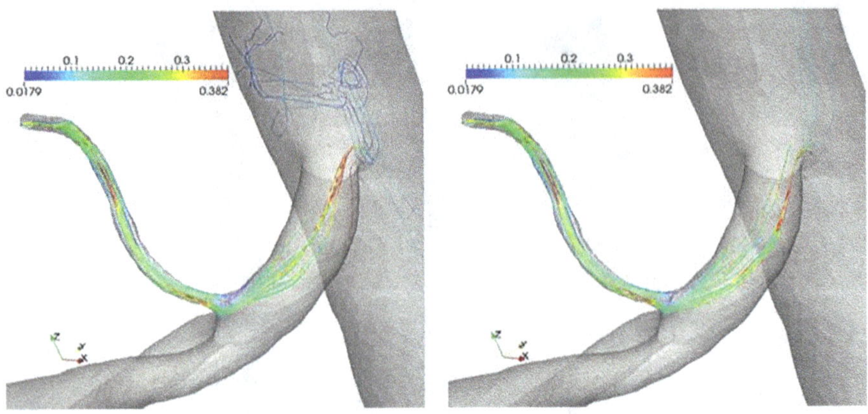

Fig. 20 Velocity streamlines (m/s) at 1.5 s and 2.0 s, from left, inside the whole artery [1]

4 Particle Tracking

4.1 Introduction

At this point, we evaluate the behaviour of magnetic nanoparticles under the influence of an external magnetic field in the human body. Indeed, some medical studies have shown that the use of magnetic drugs, guided by a magnetic field, improves the absorption of the same drugs in most carcinomas, increasing the efficiency of the treatment, [27, 62, 63]. In order to achieve this goal, we modify the *mhdFOAM* solver of the *OpenFOAM* package. In detail, we add to the code the routine related to the injection of the particles. In the *OpenFOAM* package, there is a specific solver for particles and their interaction, which is called *solidParticleFOAM*. However, this is not suitable for our study and is not integrated into the *THD-MHD* equations.

4.2 Lagrangian Model

A particle of diameter d_p, velocity v_p, and density ρ_p is defined by the position of its centre, x_p. In a Lagrangian frame of reference, the position of each particle is obtained by the integration of its velocity,

$$\frac{d\vec{x}_p}{dt} = \vec{v}_p \tag{13}$$

which is evaluated from the momentum conservation equation, written as follows

$$\frac{d\vec{v}_p}{dt} = \underbrace{-\frac{1}{\tau_p}\left(\vec{v}_p - \vec{v}_b + \frac{d_p^2}{12}\nabla^2\vec{v}_b\right)}_{(I)} + \underbrace{\left(1 - \frac{\rho_b}{\rho_p}\right)\vec{g}}_{(II)}$$

$$+ \underbrace{\frac{\rho_b}{\rho_p}\left(\frac{\partial\vec{v}_b}{\partial t} + \left(\vec{v}_b \cdot \nabla\right)\vec{v}_b\right)}_{(III)} + \underbrace{\frac{1}{2}\frac{\rho_b}{\rho_p}\left(\frac{\partial\vec{v}_b}{\partial t} + \left(\vec{v}_b \cdot \nabla\right)\vec{v}_b - \frac{d\vec{v}_p}{dt}\right)}_{(IV)}$$

$$+ \underbrace{\left(\frac{q_p}{m_p}\vec{v}_p - \frac{1}{\rho_p\mu_0}\mathrm{curl}\left(\vec{B}\right)\right) \times \vec{B}}_{(V)}$$

$$(14)$$

where (I) is the drag term, (II) is the buoyancy, (III) is the carrier phase inertia, (IV) is the added mass and (V) is the Lorentz force. It must be noted that we have neglected the particle-particle interactions. This assumption is made because of the small volume of the particles, which reduces the probability of collision. Moreover, the intra-particle interactions are computationally demanding and do not seem to be important in this study. Furthermore, q_p is the electric charge of the particle, m_p its mass and τ_p the relaxation time, defined as

$$\tau_p = \frac{4}{3}\frac{\rho_p d_p}{\rho_b C_d\left|\vec{v}_b - \vec{v}_p\right|} \tag{15}$$

The standard definition of the drag coefficient is the following

$$C_d = \begin{cases} \dfrac{24}{\mathrm{Re}_p} & \mathrm{Re}_p < 0.1 \\[2ex] \dfrac{24}{\mathrm{Re}_P}\left(1 + \dfrac{1}{6}\mathrm{Re}_p^{2/3}\right) & 0.1 \leq \mathrm{Re}_p \leq 1000 \\[2ex] 0.44 & \mathrm{Re}_p > 1000 \end{cases} \tag{16}$$

The particle Reynolds number is defined as

$$\mathrm{Re}_p = \frac{d_p\left|\vec{v}_b - \vec{v}_p\right|}{\nu_b} \tag{17}$$

4.3   Magnetic Induction

As far as the magnetic induction field is concerned, zero normal derivatives are set to zero everywhere, except on the wall, where the magnetic field of the probe is imposed. The external magnetic field is generated by a single rectangular coil with a negligible cross-section of the wire where an electric current flows. This geometry is quite common in clinical practice [64–67]. The analytical expression for the magnetic induction field is derived as in [68].

A point in the coil reference frame, whose origin is at its centre, is identified by the coordinates (x', y', z'). The coil dimensions are $2a_1$ along the x' axis and $2b_1$ in the y' direction. The axis z' is normal to the coil surface. The components of the magnetic induction field are

$$B_{x'} = \frac{\mu_0 I_1}{4\pi} \sum_{a=1}^{4} \left[\frac{z'(-1)^{a+1}}{r_a[r_a + d_a]} \right] \tag{18}$$

$$B_{y'} = \frac{\mu_0 I_1}{4\pi} \sum_{a=1}^{4} \left[\frac{z'(-1)^{a+1}}{r_a\left[r_a + C_a(-1)^{a+1}\right]} \right] \tag{19}$$

$$B_{z'} = \frac{\mu_0 I_1}{4\pi} \sum_{a=1}^{4} \left[\frac{d_a(-1)^a}{r_a\left[r_a + C_a(-1)^{a+1}\right]} - \frac{C_a}{r_a[r_a + d_a]} \right] \tag{20}$$

with

$$\begin{cases} C_1 = -C_4 = a_1 + x' \\ C_2 = -C_3 = a_1 - x' \\ d_1 = d_2 = y' + b_1 \\ d_3 = d_4 = y' - b_1 \\ r_a = \sqrt{C_a^2 + d_a^2 + z'^2} \end{cases} \tag{21}$$

The magnetic probe is located 1 cm above the patient's skin in order for its modulus to be smaller than 1.5 T, which is the limit allowed in clinical treatments [69], since higher values can cause damage to the patient. The liver is the target of the magnetotherapy, and the maximum magnetic field must be concentrated on it. Therefore, the distance between the origin of the probe and the target must be parallel to the z' axis.

The external abdominal surface of the patient is reconstructed using *VMTK* with the procedure previously illustrated for the arteries, and the two geometries are located in the same reference frame in order to calculate the coordinates of the probe and the target. The dimensions of the rectangular coil, the current intensity and the positions of the coil and target can be found in Table 1, while the main simulation parameters are reported in Tables 2 and 3.

4.4 Set-Up for the Hepatic Artery

We implement the previous equations inside the solver for a new type of boundary condition. In the folder 0 and in the B file, in order to use the aforementioned condition, we need to find the normal vector to the target, the coordinates of a

Table 2 Source and target coordinates; values for the normal to the abdomen surface of the patient; intensity of the magnetic field along the z-axis, Bz

	Normal	Coil Coords (m)	Target Coords (m)	Bz|1 cm(mT)
1st run	$n_x = -0.05899$ $n_y = 0.94519$ $n_z = 0.32114$	$x_c = 0.0368$ $y_c = -0.12346$ $z_c = -0.14145$	$x_t = 0.9983$ $y_t = -0.056$ $z_t = -0.019$	46,21
2nd run	$n_x = -0.05899$ $n_y = 0.94519$ $n_z = 0.32114$	$x_c = 0.0368$ $y_c = -0.12346$ $z_c = -0.14145$	$x_t = 0.9983$ $y_t = -0.056$ $z_t = -0.019$	
3rd run	$n_x = -0.05303$ $n_y = 0.97961$ $n_z = 0.19377$	$x_c = 0.001322$ $y_c = -0.12699$ $z_c = -0.12485$	$x_t = 0.9986$ $y_t = -0.052$ $z_t = -0.0103$	
4th run	$n_x = -0.05303$ $n_y = 0.97961$ $n_z = 0.19377$	$x_c = 0.001322$ $y_c = -0.12699$ $z_c = -0.12485$	$x_t = 0.9986$ $y_t = -0.052$ $z_t = -0.0103$	

Table 3 Simulation parameters: H_D (domain height), W_D (domain width), D_D (domain depth), $D_{M\text{-}T}$ (magnet—tumour distance), T_0 (observation time), Re_m (mean aortic number)

H_D (cm)	W_D (cm)	D_D (cm)	$D_{M\text{-}T}$ (cm)	T_0(s)	Re_m
17.1	4.0	8.37	12.0	2	1097

point on the target surface and the coordinates of a point on the source surface. For the hepatic artery, thanks to *VMTK*, we do a reconstruction of the abdomen in order to extrapolate only the very external abdominal surface, as depicted in Fig. 21. Subsequently, we add the geometry of the hepatic artery in order to have the two of them on the same plane. In this way, it is easy to find a normal start from the abdominal surface to the hepatic artery. Once we find the necessary coordinates of the target, the source and the normal, the file B, in the folder 0, can be modified to run the simulation, [59, 70].

In the mentioned file, the new boundary condition is employed on the wall patch; it is also necessary to specify the value of the current I used during the simulation and the dimensions of the rectangular coil. For the remaining patches, we use a *zeroGradient* condition, because the magnetic field, in this case, acts like a probe, and we are interested in analysing its behaviour for the whole geometry. For the U file and the wall patch, we prescribe a *fixedValue* equal to zero. For the four outlets, the condition was *pressureInletOutletVelocity*; this type of condition implies a *zeroGradient* on all components, except where there is an inflow, in which case a *fixedValue* is applied to the tangential component. An alternative way of defining the velocity profiles on the inlet is to map the data values at specified points along the height of the inlet patch. This can be done with the *timeVaryingMappedFixedValue* boundary condition, which is often used to map experimentally obtained values to the inlet. Firstly, we want to set different velocity values for each time step of the simulation. All these time steps are inside the directory *constant* and inside the directory *boundaryData*. Because we want to map data values in specified points along the inlet patch, in the same directory it will be placed the file *points*; this last

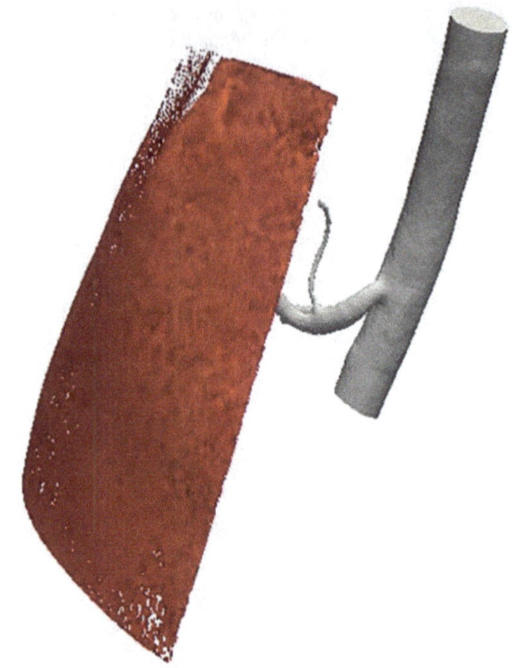

Fig. 21 Abdomen and hepatic artery reconstruction by VMTK [1]

file contains the points where the data are located; in this case, we consider the data on the inlet and the point referred to that patch.

For the p file, at the wall and on the inlet, we fix a *zeroGradient* condition that is equivalent to a *Neumann* condition, where the gradient normal to those boundaries is specified. For the so-called *outlet_A*, corresponding to the abdominal aorta, we set up a *totalPressure* condition, which is a *fixedValue* condition, calculated from the specified total pressure p_0 and local velocity, $\vec{u}$. The total pressure p_0 in *OpenFOAM* is evaluated as

$$p_0 = p + \frac{1}{2}\rho|\vec{u}|^2 \tag{22}$$

In this case, ρ is equal to zero, so the local velocity is also zero, and the total pressure is coincident with p. For the three remaining outlets, we have imposed a *grovyBC*, in which the value is taken from physiology. For the variables, we declare the parameters in the definition of the equation to evaluate the pressure as resistance, flow rate and pressure for every outlet. In detail, the resistance is given by the equation:

$$\Re = \frac{8\mu}{\pi R^4} \tag{23}$$

where R is the outlet radius. The flow rate is evaluated inside the geometry and updated every time step, and the pressure is considered equal in every outlet with a mean value. *Fraction Expression* can distinguish between *Dirichlet* and *Neumann* conditions. In the *value Expression,* we define the *Dirichlet* condition with the equation which evaluates the pressure at each outlet. The general form of the final equation is:

$$p = p_b + \Re Q \tag{24}$$

Before starting the simulation, we evaluate the right *Courant number, Co,* in order to ensure stability. However, if the *Co* number is too small and more computational time is requested, then it is possible to use *Co* to control the time step in a transient simulation by using some parameters. These parameters are usually imposed inside the *controlDict* file. When the option *adjustTimeStep* is enabled, the software changes the value of the time step during the simulation so that the value of the *CFL* condition is below the maximum specified in the *maxCo*. An output file is written every *writeInterval* seconds, and the simulation adjusts the time step to coincide with the *writeInterval* because, in this case, we use a time-step adjustment. Another modification is required inside the source code of the main program's solver. We include a *readTimeControls* because this command line is used to read the control parameters used by *setDeltaT,* which has to be defined always in the same main source code.

4.5   Post-processing of the Hepatic Artery

The post-processing is done with *ParaView.* We analyse the vector quantities, $\overline{U}, \overline{B}, \overline{U}_p$, the scalar quantities, p, *Position$_p$*. Moreover, we evaluate how the magnetic field influences the particle cloud inside the artery and the mechanism of absorption of the particles, caused by the presence of the external magnetic field, generated by a rectangular coil.

4.5.1   Unsteady State Simulation with Null Magnetic Field

The first simulation assumes a null magnetic field for 2 s, and the setting up of the boundary conditions is explained previously. In Figs. 22, 23, 24, 25 and 26, we report the pressure map in the presence of the particles. The pressure is particle independent, without coupling with the pressure equation in the Lagrangian frame. We can notice that the profile of the pressure, at different time steps, is the same as the fluid-dynamic analysis. Furthermore, it is possible to see that the particles increase with time, and, at the beginning, the profile is parabolic, occupying the full artery. The highest pressure value is reached in the abdominal aorta, whereas the lowest pressure value is reached inside the hepatic artery. In the two coeliac

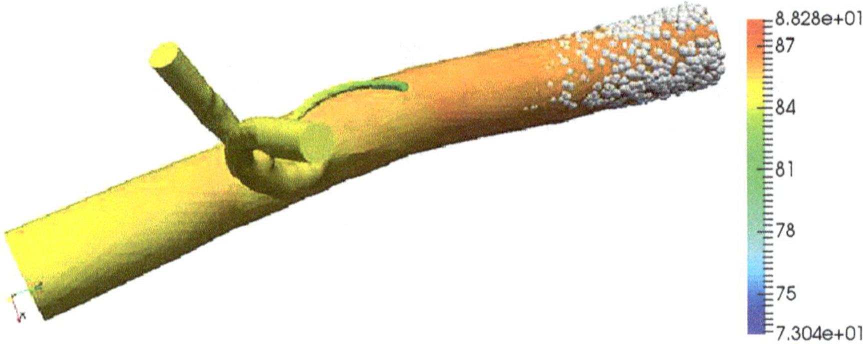

Fig. 22 Pressure map (mmHg) with nanoparticles at $t = 0.15$ s [1]

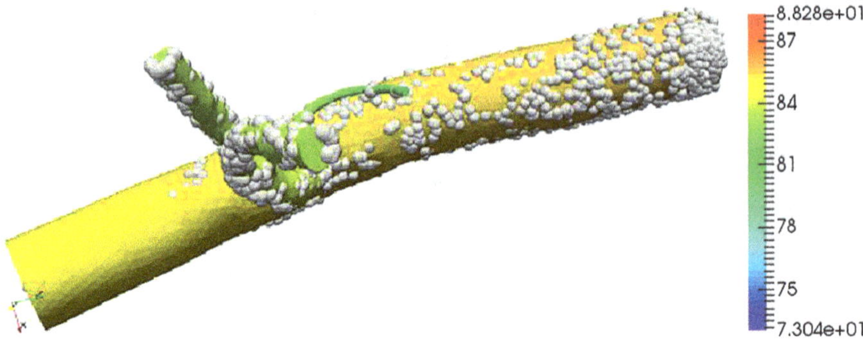

Fig. 23 Pressure map (mmHg) with nanoparticles at $t = 0.50$ s [1]

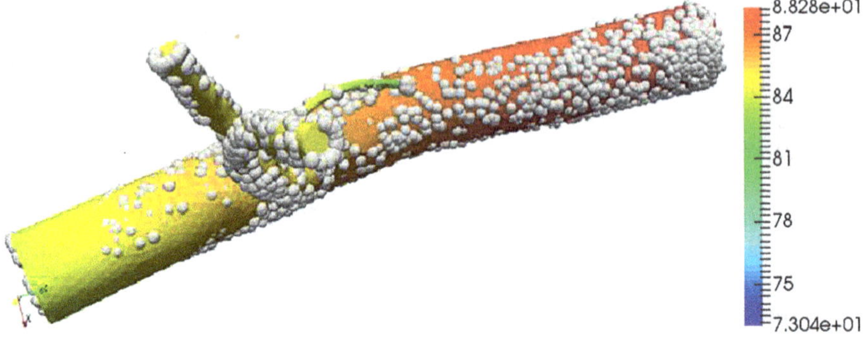

Fig. 24 Pressure map (mmHg) with nanoparticles at $t = 1$ s [1]

branches, right and left, the pressure has the same value. These results happen because when the radius of the artery decreases, the resistance increases due to a reduction of the cross-section of the artery, and the flow rate decreases because of the continuity equation. Then the pressure decreases in the small arteries and increases

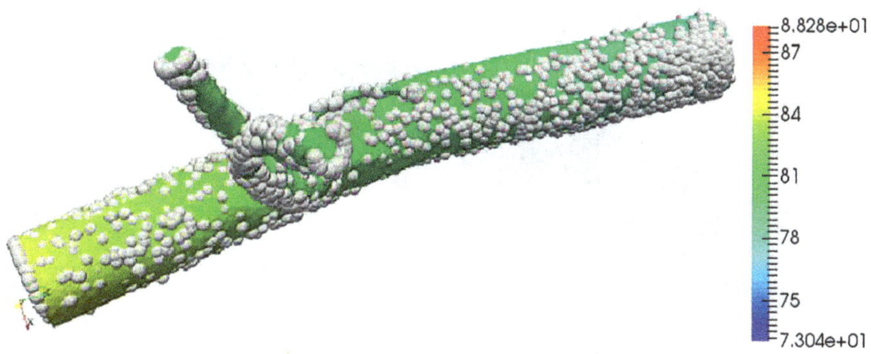

Fig. 25 Pressure map (mmHg) with nanoparticles at $t = 1.20$ s [1]

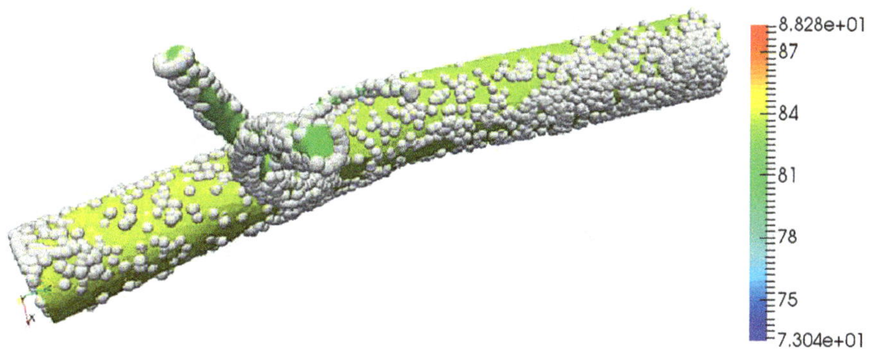

Fig. 26 Pressure map (mmHg) with nanoparticles at $t = 2$ s [1]

in the big ones. At the time equal to 1 s, the blood is ejected from the cardiac chambers, and the pressure in the aorta is the highest. At the time of 1.20 s, the pressure is small because we are in the diastolic phase of the cardiac cycle, and finally, at time 2 s, we are in the last part of the diastolic phase; the pressure increases again in the aorta and reaches the lower value inside the hepatic artery.

In Figs. 27, 28, 29, 30, 31 and 32, we present the maps of the velocity (*m/s*). In this case, there is a coupling between the particles and the fluid velocity, even if the particle's velocity is smaller than the fluid's velocity. The velocity is periodic, due to the two phases of the cardiac cycle. At the beginning, the blood is in a resting condition; the ventricles are filled with blood, and at the early stage of the systolic phase, the velocity is highest in the aorta. In this artery, the flow is transitional, as evident at $t = 0.15$ s and $t = 1.15$ s. Figures 27, 28, 29, 30, 31 and 32 show that the particle at the centre of the common coeliac branch tends to go into the hepatic artery, except in the period of maximum flow rate, i.e. $t = 0.15$ s and $t = 1.15$ s. Indeed, the velocity at these time steps is so high that the particles are first deflected

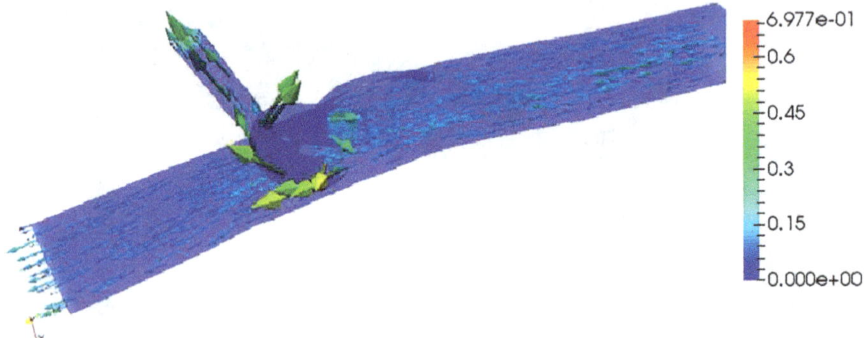

Fig. 27 Velocity vectors, (m/s), with nanoparticles at $t = 0$ s [1]

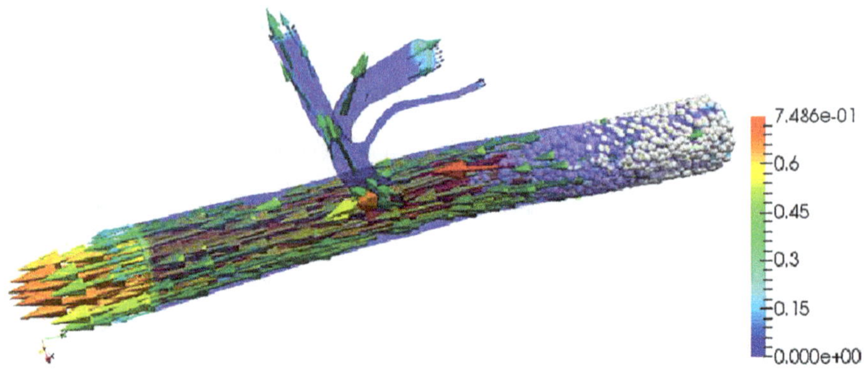

Fig. 28 Velocity vectors, (m/s), with nanoparticles at $t = 0.15$ s [1]

Fig. 29 Velocity vectors, (m/s), with nanoparticles at $t = 0.50$ s [1]

Fig. 30 Velocity vectors, (m/s), with nanoparticles at $t = 1$ s [1]

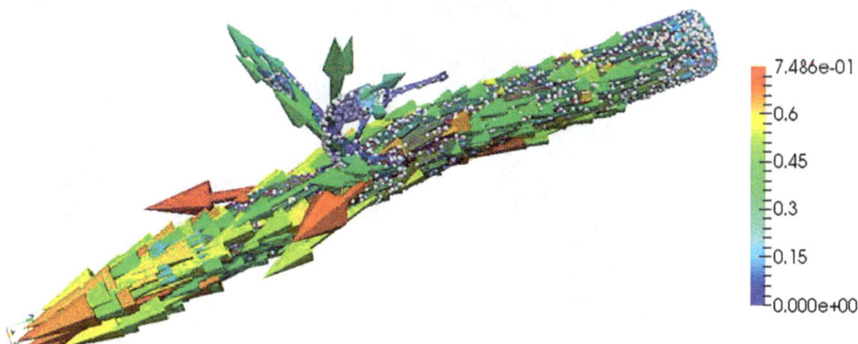

Fig. 31 Velocity vectors, (m/s), with nanoparticles at $t = 1.15$ s [1]

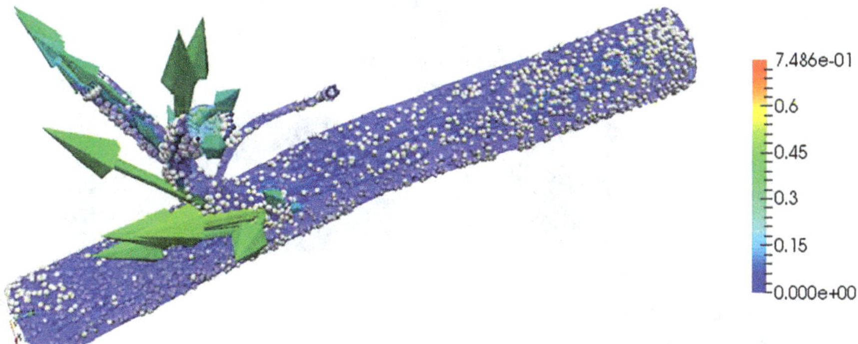

Fig. 32 Velocity vectors, (m/s), with nanoparticles at $t = 2$ s [1]

towards the bottom side of the common coeliac artery and then transported into the hepatic artery.

Figures 33, 34, 35, 36, 37, 38, 39, 40, 41 and 42 show the profiles for the so-called viscous and Newtonian acceleration term, (m/s^2), with nanoparticles, which takes into account the viscous stresses and the inertia of the fluid.

As far as the viscous term is concerned, Figs. 33, 34, 35 and 36 show that it is higher near a bifurcation and in the small branches at the beginning of the cardiac cycle and when the particles are far from these regions. This is because of the higher velocity gradient. In the neighbourhood of the bifurcation, velocity is high in a thin boundary layer, which causes high shear stress and, consequently, high viscous

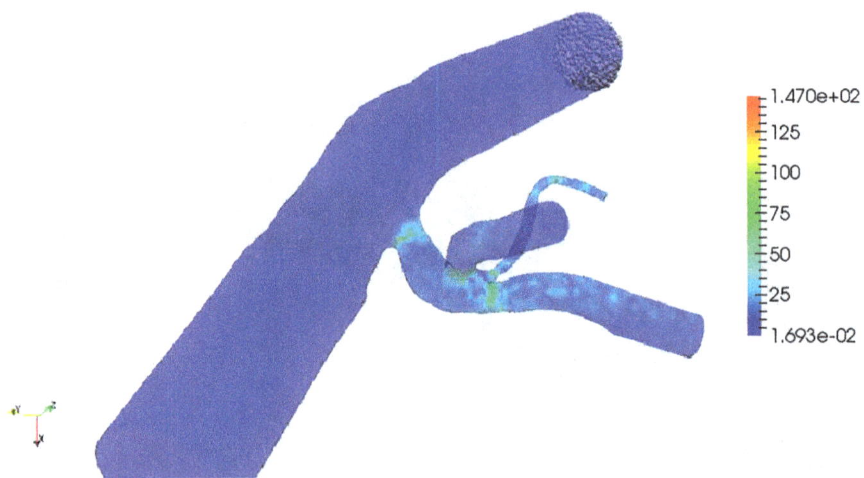

Fig. 33 Viscous acceleration term, (m/s^2), with nanoparticles at $t = 0.01$ s [1]

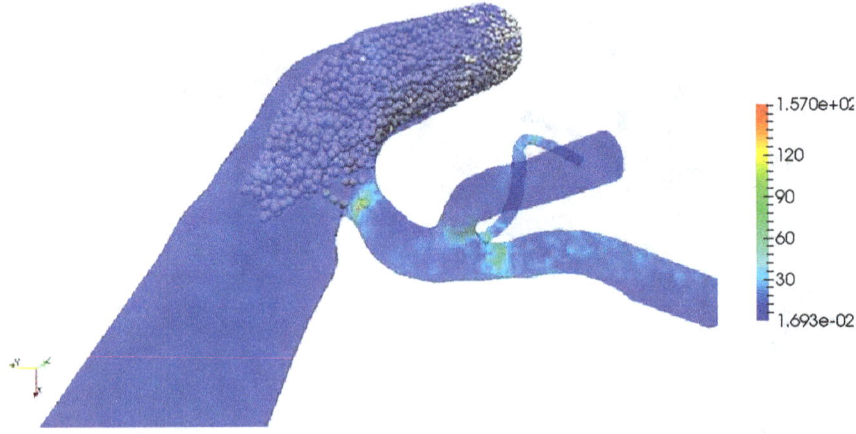

Fig. 34 Viscous acceleration term, (m/s^2), with nanoparticles at $t = 0.2$ s [1]

Fig. 35 Viscous acceleration term, (m/s^2), with nanoparticles at $t = 1$ s [1]

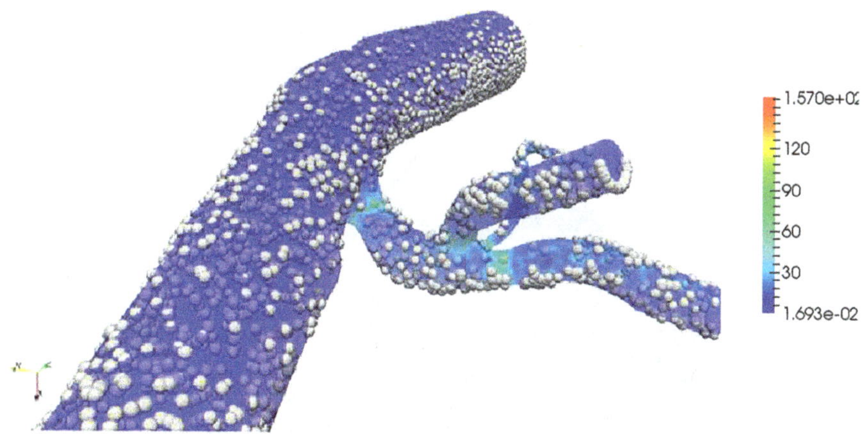

Fig. 36 Viscous acceleration term, (m/s^2), with nanoparticles at $t = 2$ s [1]

acceleration. Due to this, it is possible to observe high values in the mentioned regions.

Figures 37, 38, 39, 40, 41 and 42 show the contribution of the fluid inertia to the Newtonian acceleration (m/s^2). We can see that at $t = 0.01$ s, the term is high because it strictly depends on the velocity. Furthermore, at the bifurcations, the gradient velocity is high. Moreover, at $t = 0.2$ s, there is another region where the term is at its maximum value; because we are at the beginning of the blood ejection from the ventricles, there is a change in the curvature, and the velocity and its gradient are high.

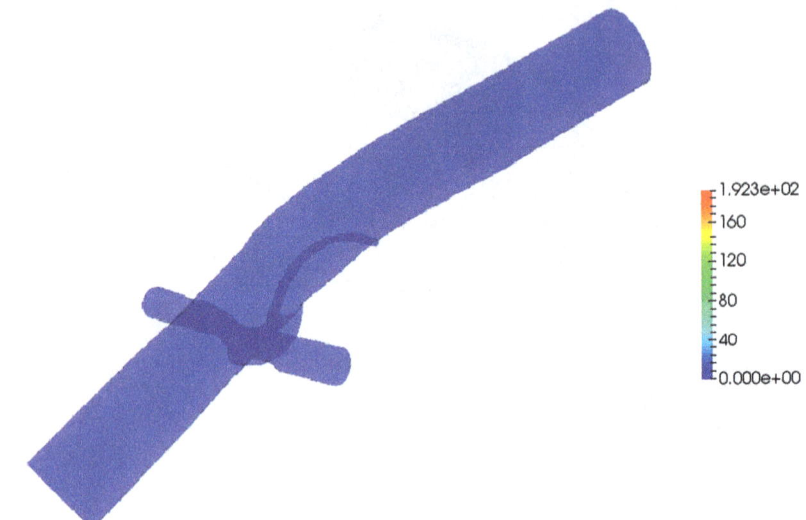

Fig. 37 Newtonian acceleration term, (m/s^2), with nanoparticles at $t = 0$ s [1]

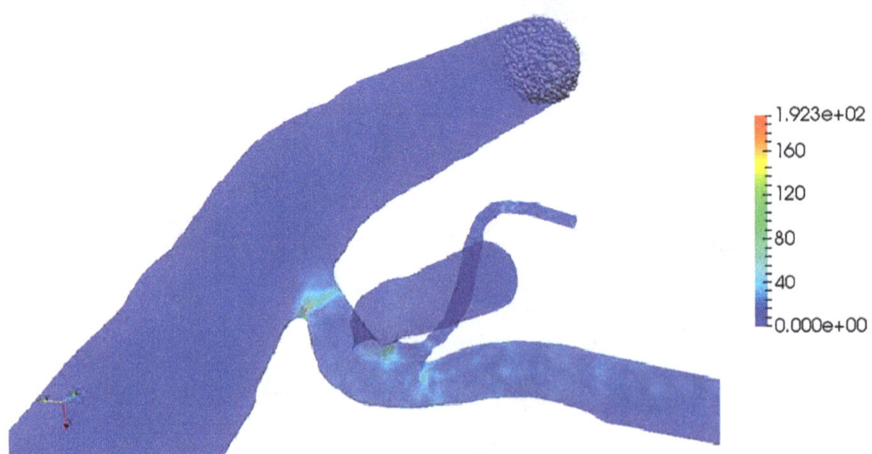

Fig. 38 Newtonian acceleration term, (m/s^2), with nanoparticles at $t = 0.01$ s [1]

4.5.2   Unsteady State Simulations with Magnetic Field

We carry out four different simulations where we evaluate the intensity of the magnetic field along the axis at 1 cm from the probe. We find two different places where the probe is positioned. In the following table, the characteristics used in each simulation are listed.

In Figs. 43, 44, 45, 46, 47, 48, 49, 50, 51 and 52, we compare the pressure fields (mmHg) for the first and third simulations of Table 4 at different instants of time with the nanoparticles. The pressure is particle independent but is not independent of the

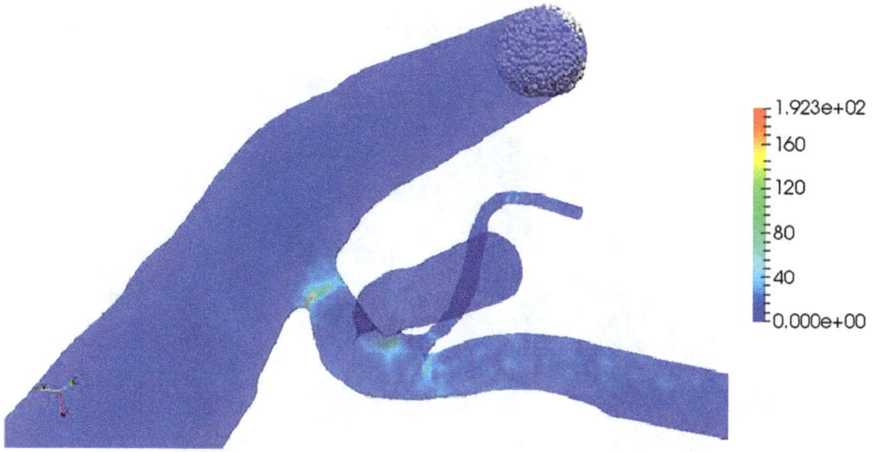

Fig. 39 Newtonian acceleration term, (m/s^2), with nanoparticles at $t = 0.04$ s [1]

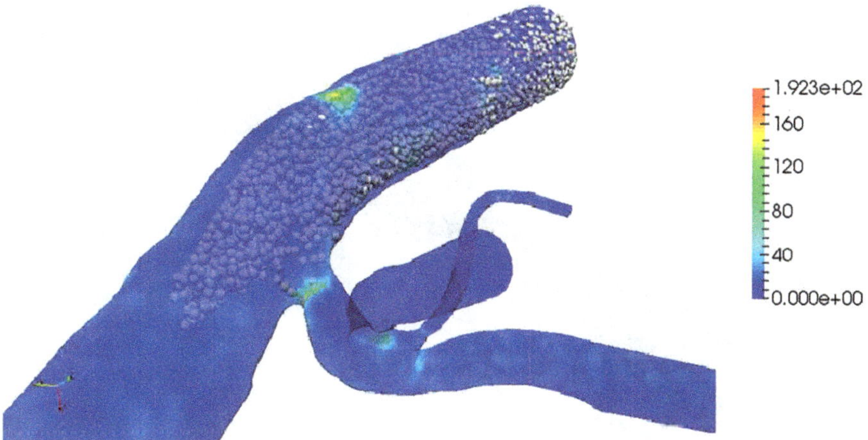

Fig. 40 Newtonian acceleration term, (m/s^2), with nanoparticles at $t = 0.2$ s [1]

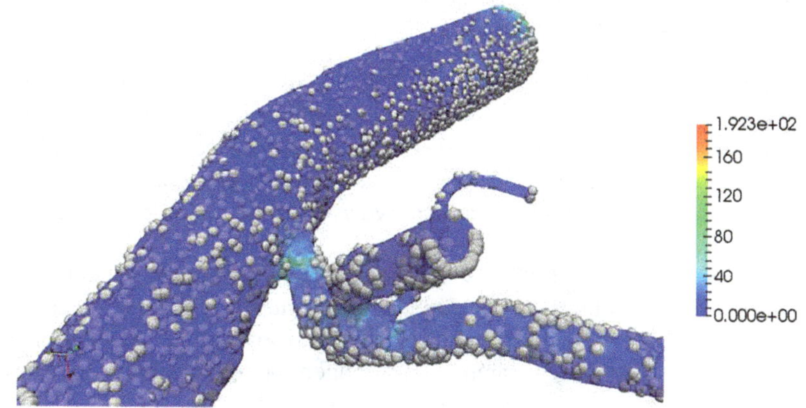

Fig. 41 Newtonian acceleration term, (m/s^2), with nanoparticles at $t = 1.34$ s [1]

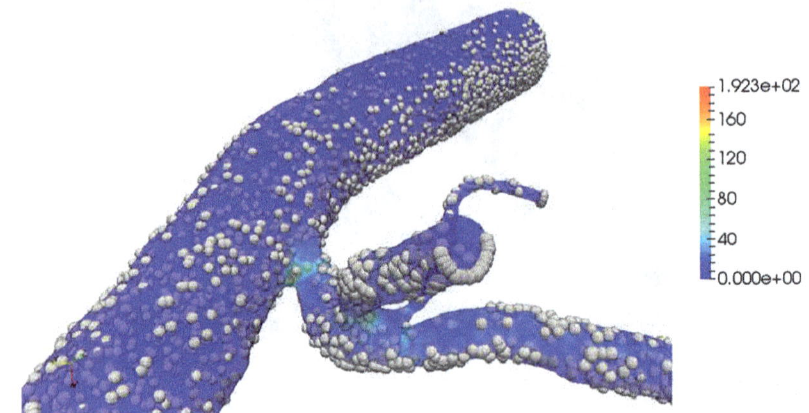

Fig. 42 Newtonian acceleration term, (m/s^2), with nanoparticles at $t = 2$ s [1]

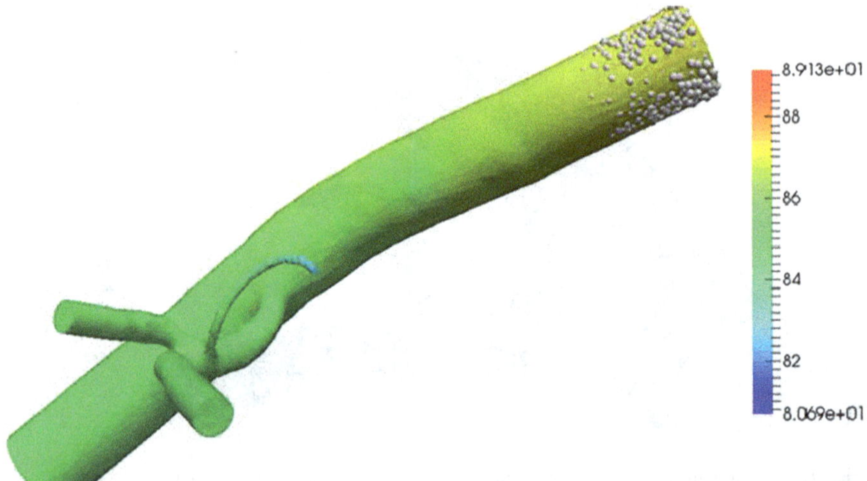

Fig. 43 Pressure field (mmHg) with nanoparticles for the first simulation at 0.15 s [1]

magnetic field. Indeed, the pressure drop is proportional to Ha^2 for conducting walls in a strong magnetic field, whereas it is proportional to Ha for non-conducting walls. The difference between the two cases is more evident in the coeliac trunk and in the right and left coeliac arteries. At the same instant, it is greater for the third simulation because the coordinates of the target are closer to the source, where the magnetic probe is located.

Figures 53, 54, 55 and 56 report the velocity fields *(m/s)* with the nanoparticles at two different instants of time for the first and third simulations of Table 4. As can be seen, the particles are influenced by the velocity profile of the fluid. At the beginning of the simulation, they have a parabolic distribution for both conditions; moreover,

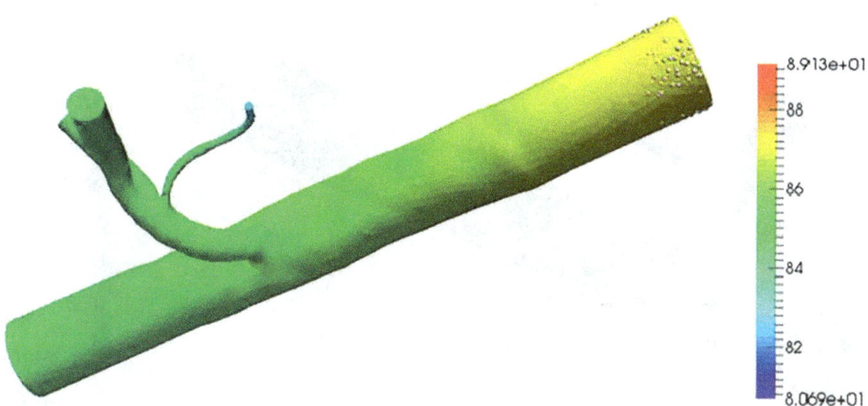

Fig. 44 Pressure field (mmHg) with nanoparticles for the third simulation at 0.15 s [1]

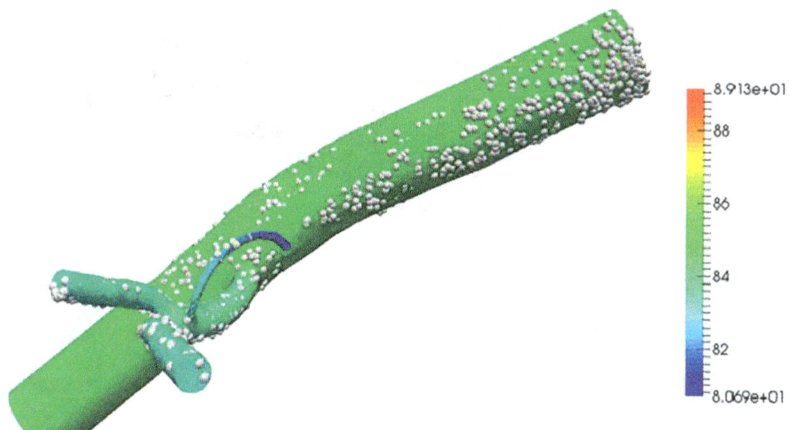

Fig. 45 Pressure field (mmHg) with nanoparticles for the first simulation at 0.5 s [1]

for the third simulation, the velocity is higher compared to the first one. At the end of the simulation, the velocity peak is reached inside the coeliac trunk for both cases because the cross-section decreases, and, due to the continuity equation, the velocity has to increase, and, consequently, also the particles increase their velocity because they are dragged.

Figures 57 and 58 report the magnetic field maps, (T), with the nanoparticles for the first and third simulations of Table 4 at $t = 2$ s. The difference between the two conditions is that for the first simulation, we have a higher intensity in the right and left coeliac arteries, while for the third simulation, the highest value is reached only in the right coeliac arteries, due to the choice of the target and source coordinates. The magnetic field is generated by a rectangular coil with a height equal to 1.5 cm.

Fig. 46 Pressure field (mmHg) with nanoparticles for the third simulation at 0.5 s [1]

Fig. 47 Pressure field (mmHg) with nanoparticles for the first simulation at 1.0 s [1]

Figures 59 and 60 present the map of the Newtonian acceleration term, (m/s^2), for the first and third simulations of Table 4 at $t = 0.15$ s. At the beginning, the acceleration is high because it depends on the velocity, and, at the bifurcations, the velocity gradient is high and remains high at all the time steps. Moreover, under the effect of the magnetic field, the velocity of the fluid with the particles is affected, and it is sped up in such a way that the inertial term is also affected, increasing its value.

Figures 61 and 62 present the viscous term, (m/s^2), with the nanoparticles for the first and third simulations of Table 4 at $t = 0.15$ s. The profile is the same for both simulations, and, furthermore, it has the same trend at all times. It is higher near the bifurcation and in the small branches at the beginning of the cardiac cycle, and when the particles are far from these regions. The particle's addition causes an increase in the total velocity, and near the bifurcation or near the reduction of the cross-section,

Fig. 48 Pressure field (mmHg) with nanoparticles for the third simulation at 1.0 s [1]

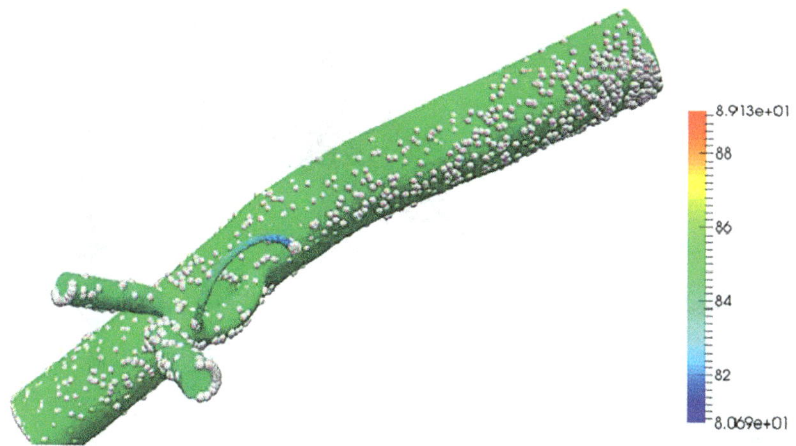

Fig. 49 Pressure field (mmHg) with nanoparticles for the first simulation at 1.2 s [1]

the value of the gradient increases and the stress between two adjacent fluid layers too, because it depends on the velocity gradient. Due to this, it is possible to observe high values in those regions.

Figures 63 and 64 report the electromagnetic acceleration term (m/s^2), which derives from the application of the magnetic field (T) with the nanoparticles for the first and third simulations of Table 4 at $t = 1.0$ s. It describes how the particles change their velocity under the effect of an external magnetic field, and, as it is possible to see, it is higher where the applied magnetic probe is present. The particles are sped up by the magnetic field, and under this driving force, they adhere to the wall of the arteries, where the magnetic field is greater.

Figures 65 and 66 report the lift acceleration term (m/s^2), which is the component of the force perpendicular to the flow direction with the nanoparticles, for the first

Fig. 50 Pressure field (mmHg) with nanoparticles for the third simulation at 1.2 s [1]

Fig. 51 Pressure field (mmHg) with nanoparticles for the first simulation at 2.0 s [1]

and third simulations of Table 4, at $t = 0.15$ s. It is opposite to the drag force, which, in general, is the component parallel to the flow direction. It depends on the fluid speed and density and is proportional to the density of the blood and to the square of the flow speed. For this reason, the maximum value of this field is reached in the small branches and near a bifurcation. The behaviour of this variable is the same for both the first and the third simulations. It is not magnetically independent because the velocity field, in the presence of the magnetic field, is sped up, and consequently, the lift increases.

We conduct another comparison for the second and the fourth simulations of Table 4, with the same variables as the previous case. Figures 67, 68, 69, 70, 71, 72, 73, 74, 75 and 76 report the pressure field (*mmHg*) with the nanoparticles, which is

Fig. 52 Pressure field (mmHg) with nanoparticles for the third simulation at 2.0 s [1]

Table 4 Values used for the four different simulations

	Normal	Coordinate source	Coordinate target	$Bz|_{1\ cm}$[T]
1st simulation	0.058986 0.94519 0.32114	0.003678 −0.12346 −0.14145	0.9983 −0.0559 −0.0190	0.04621
2nd simulation	0.058986 0.94519 0.32114	0.003678 −0.12346 −0.14145	0.9983 −0.0559 −0.0190	0.04621
3rd simulation	0.053028 0.97961 0.19377	0.0013223 −0.12699 −0.12485	0.9986 −0.0520 −0.0103	0.04621
4th simulation	0.0053028 0.97961 0.19377	0.0013223 −0.12699 −0.12485	0.9986 −0.0520 −0.0103	0.04621

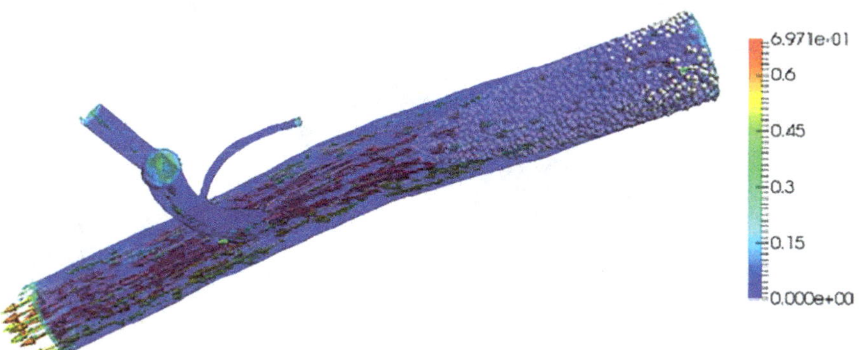

Fig. 53 Velocity field (m/s) with nanoparticles for the first simulation at $t = 0.15$ s [1]

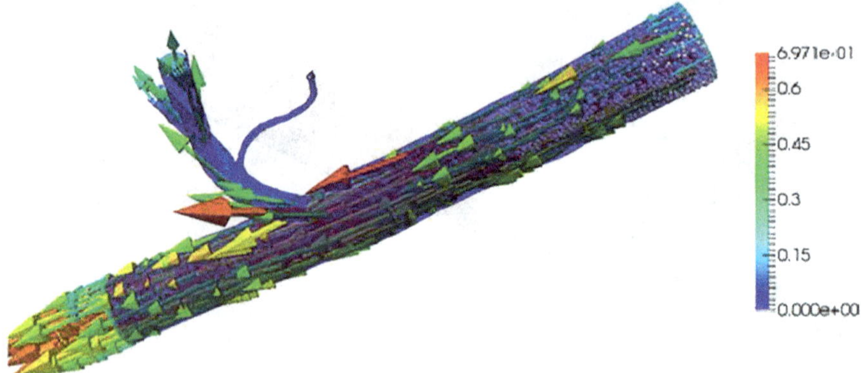

Fig. 54 Velocity field (m/s) with nanoparticles for the third simulation at $t = 0.15$ s [1]

Fig. 55 Velocity field (m/s) with nanoparticles for the first simulation at $t = 2.0$ s [1]

Fig. 56 Velocity field (m/s) with nanoparticles for the third simulation at $t = 2.0$ s [1]

Fig. 57 Magnetic field (T) with nanoparticles for the first simulation at $t = 2.0$ s [1]

Fig. 58 Magnetic field (T) with nanoparticles for the third simulation at $t = 2.0$ s [1]

not independent of the magnetic field. For these two simulations, at the different times and evolution of the cardiac cycle, the pressure has the same trend and values.

Figures 77, 78, 79 and 80 depict the velocity fields (*m/s*) with the nanoparticles for the second and the fourth simulations of Table 4. The particles are dragged, and their distribution profile at the beginning is parabolic. The maximum velocity is reached inside the coeliac trunk.

Figures 79 and 80 present the magnetic field intensity (T) for the second and the fourth simulations of Table 4, at $t = 2.0$ s. The value is equal at all times, and its maximum is reached in the right coeliac artery, due to the choice of the target and source coordinates. The magnetic field is generated by a rectangular coil of height equal to 1.5 cm (Figs. 81 and 82).

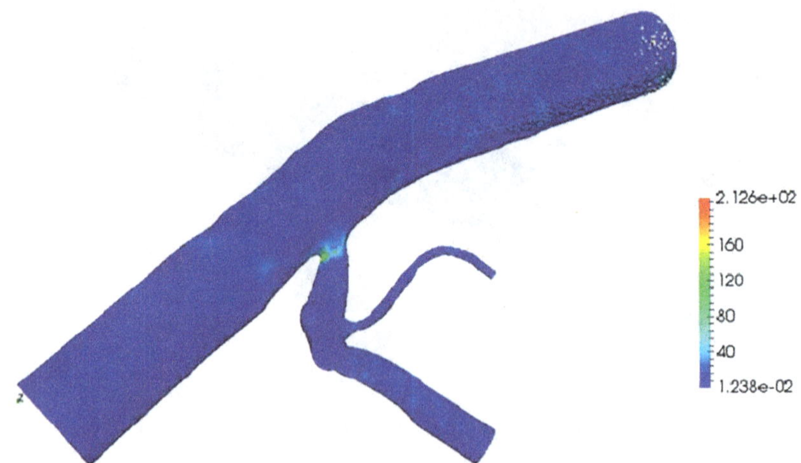

Fig. 59 Newtonian acceleration, (m/s^2), with nanoparticles for the first simulation at $t = 0.15$ s [1]

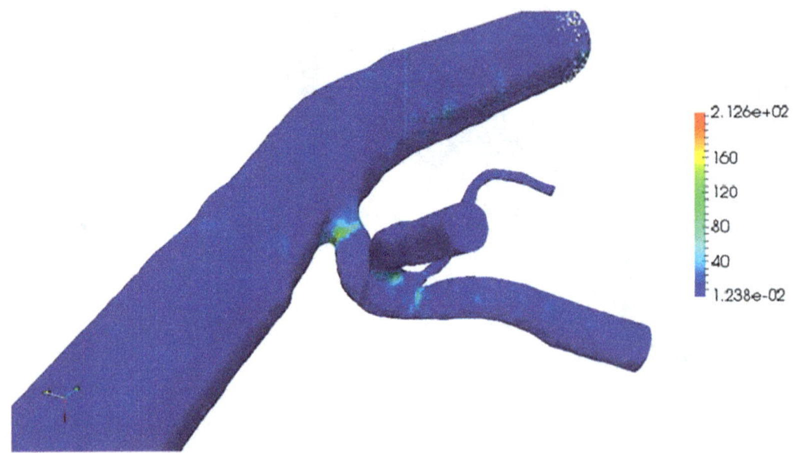

Fig. 60 Newtonian acceleration, (m/s^2), with nanoparticles for the third simulation at $t = 0.15$ s [1]

Figures 83 and 84 report the inertial term (m/s^2) for the second and the fourth simulations of Table 4 at $t = 0.15$ s. At the beginning, the inertial term is high because it depends on the velocity, and, in laminar flow, the relation is linear, whereas in a turbulent regime, the relation is quadratic. Furthermore, at the bifurcations, the gradient of velocity is high, remaining so at all times, because, in those regions, the gradient is always high. Moreover, under the effect of the magnetic field, the velocity of the fluid with the particles is affected by it, and it is sped up. Also, the inertial term is affected and increases its value. The different sign for the intensity current, or the different coordinates for target and source, are not

Fig. 61 Viscous acceleration field, (m/s^2), with nanoparticles for the first simulation at $t = 0.15$ s [1]

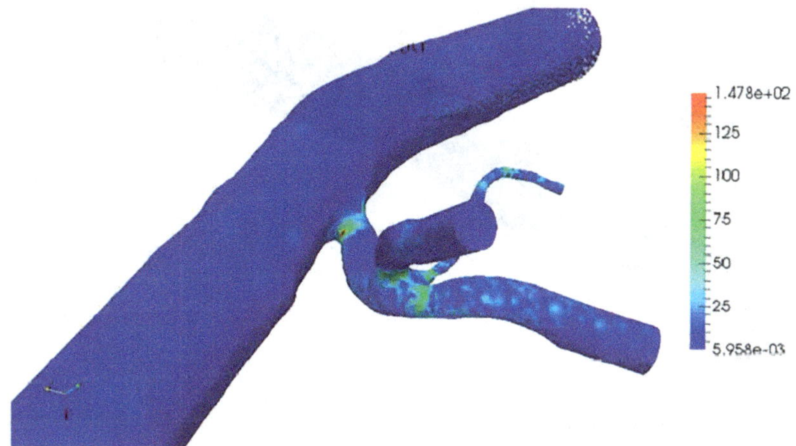

Fig. 62 Viscous acceleration field, (m/s^2), with nanoparticles for the third simulation at $t = 0.15$ s [1]

affected. In conclusion, we can say that the inertial acceleration is equal in all the simulations.

Figures 85 and 86 present the viscous term (m/s^2) for the second and the fourth simulations of Table 4 at $t = 0.15$ s. The viscous term is greater near a bifurcation and in the small branches at the beginning of the cardiac cycle, and when the particles are far from these regions. The particle's addition increases the total velocity near a bifurcation or near a cross-section reduction; its gradient increases, and the stress between two adjacent fluid layers too, because it depends on the

Fig. 63 Electromagnetic acceleration term field, (m/s^2), with nanoparticles for the first simulation at 1.0 s [1]

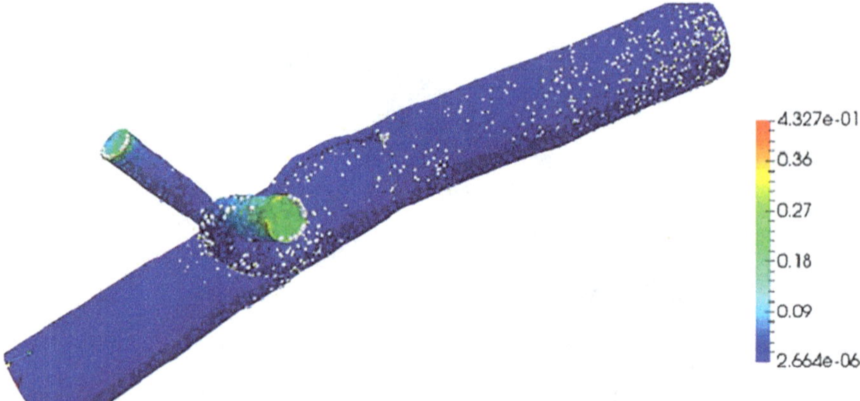

Fig. 64 Electromagnetic acceleration term field, (m/s^2), with nanoparticles for the third simulation, at 1.0 s [1]

second derivatives of the velocity. Also, for this field, the sign of the current intensity and the coordinates of the source and target do not affect the maps.

Figures 87 and 88 report the electromagnetic acceleration term (m/s^2), which derives from the application of the magnetic field, for the second and the fourth simulations of Table 4, at $t = 0.15$ s. This term describes how the particles change their velocity under the effect of the external magnetic field, and, as is possible to be seen, it is higher where the applied magnetic probe is present. The particles are sped up by the magnetic field, and, under this driving force, they adhere to the wall of the arteries, where the magnetic field is stronger. In conclusion, the behaviour of this variable is the same for the four different conditions.

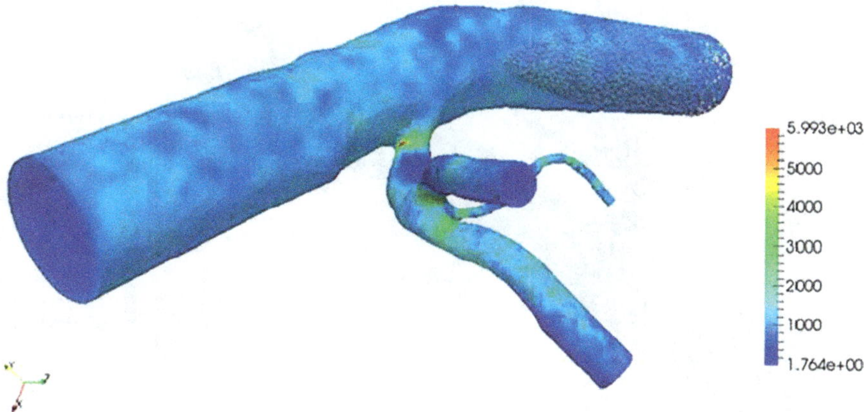

Fig. 65 Lift acceleration term field, (m/s^2), with nanoparticles for the first simulation at $t = 0.15$ s [1]

Fig. 66 Lift acceleration term field, (m/s^2), with nanoparticles for the third simulation at $t = 0.15$ s [1]

Figures 89 and 90 present the lift acceleration term (m/s^2), which is opposite to the drag force, which, in general, is parallel to the flow direction, for the second and the fourth simulations of Table 4, at $t = 1.15$ s. The lift acceleration term is proportional to the density of the blood and to the curl of the flow speed. Then, the maximum value of this field is reached in the small branches and near a bifurcation. The behaviour of this variable is the same for both the second and the fourth simulations and depends on the magnetic field because velocity, in the presence of the magnetic field, is sped up, and also the lift increases.

The main goal of these simulations is the analysis of the nanoparticles' absorption under the effect of an applied magnetic field. The absorption is independent of the

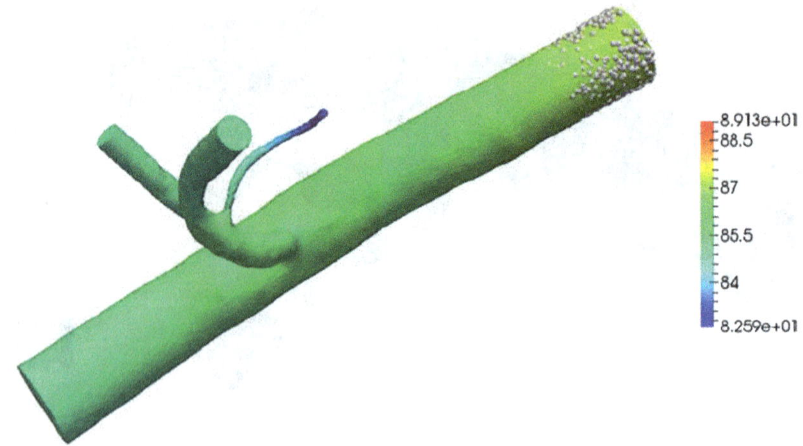

Fig. 67 Pressure field (mmHg) with nanoparticles for the second simulation at $t = 0.15$ s [1]

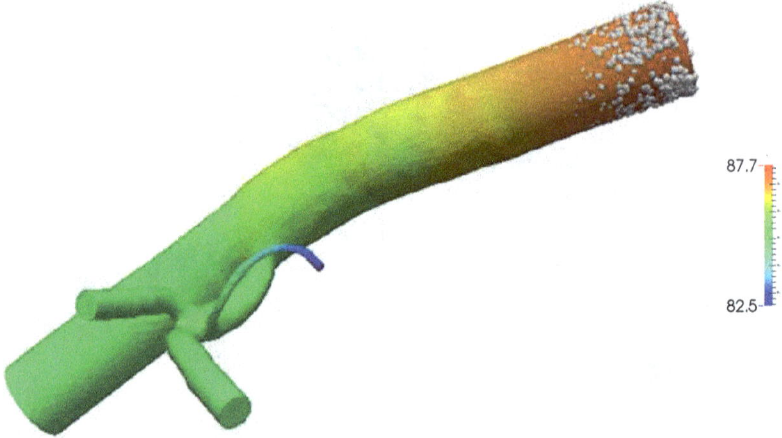

Fig. 68 Pressure field (mmHg) with nanoparticles for the fourth simulation at $t = 0.15$ s [1]

sign of the intensity current, whereas the choice of the target and the source coordinates is much more important. This last condition determines the extinction of the magnetic field and, consequently, a good absorption of the nanoparticles. From the previous results, according to a qualitative point of view, we can notice that the particles are conveyed by the continuous phase, which is influenced by the magnetic field. For this reason, in the arteries where the magnetic field is applied, we can notice better particle adhesion to the wall and, consequently, their absorption inside the wall of the same arteries. For the remaining arteries, the particle adhesion decreases because they are not affected by the magnetic field due to the fact that the field decreases its intensity with the square of the distance.

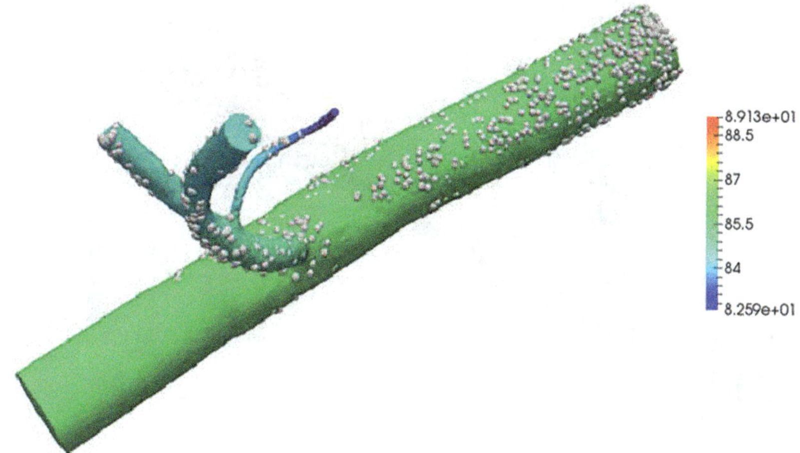

Fig. 69 Pressure field (mmHg) with nanoparticles for the second simulation at $t = 0.5$ s [1]

Fig. 70 Pressure field (mmHg) with nanoparticles for the fourth simulation at $t = 0.5$ s [1]

4.6   Influence of the Magnetic Field on the Particle Absorption

Figures 91 and 92, [59], show the influence of the magnetic field on the absorption of the particles. To put the phenomenon in evidence, the results with the magnetic probe turned off are also reported.

Figure 91 shows the particle flow rate per unit volume crossing a given surface per unit time during the cardiac cycle in the *Thoracic Aorta (TA)*, in the *abdominal aorta (AA)*, in the *Splenic Artery (SA)*, in the *Left Gastric Artery (LGA)*, in the *common hepatic artery (CHA)*, and in the *Domain's Wall*. In the *Thoracic Aorta (TA)* and in the *Domain's Wall*, the particle flow rates are of the same order of

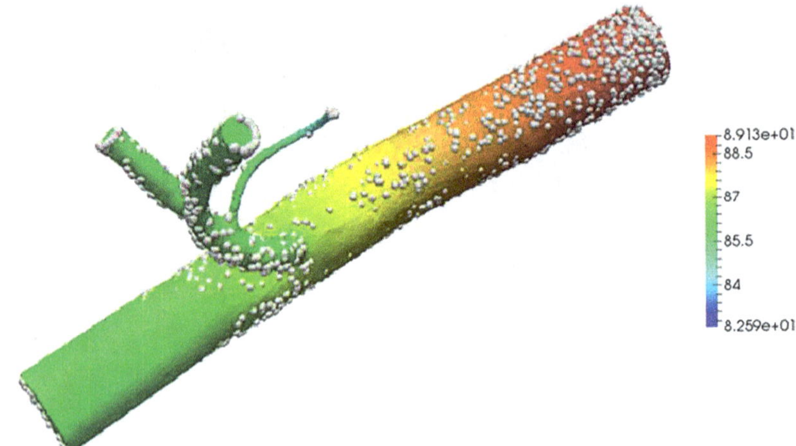

Fig. 71 Pressure field (mmHg) with nanoparticles for the second simulation at $t = 1.0$ s [1]

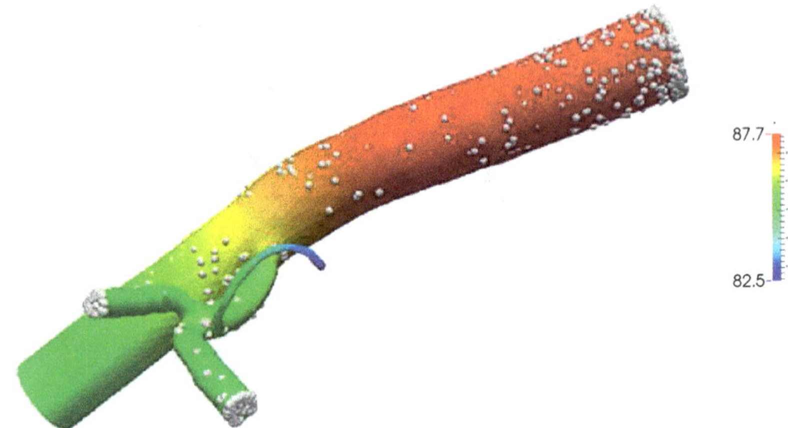

Fig. 72 Pressure field (mmHg) with nanoparticles for the fourth simulation at $t = 1.0$ s [1]

magnitude. A peak is present in the *TA* around 1 s because the blood is ejected from the ventricles into the *aorta,* and there is a higher number of particles inside the artery. Afterwards, the particle flow rate decreases due to the diastolic phase. The comparison of Fig. 91, a *(TA)* and Fig. 91, f *(Domain's Wall)* shows that the particle flow rates in those regions do not seem to be influenced by the parameters used in Table 1. On the contrary, on the outlets and especially on the *common hepatic artery (CHA),* it is possible to observe an increase in the particle flow, thanks to the applied magnetic field. The results concerning the particle flow rate may not be enough to judge the effectiveness of the use of a rectangular coil in magnetic targeting. Therefore, the time integration of the particle flow rate per unit volume is done,

Fig. 73 Pressure field (mmHg) with nanoparticles for the second simulation at $t = 1.2$ s [1]

Fig. 74 Pressure field (mmHg) with nanoparticles for the fourth simulation at $t = 1.2$ s [1]

which corresponds to the total number of particles that crossed a given section since the beginning of the process.

Figure 92 shows that the number of particles is approximately the same for the five different conditions inside the *TA* and on the *Domain's Wall*, confirming the results shown in Fig. 91. On the other hand, the results concerning the outlets show that the magnetic field influences the solution because the number of particles increases slightly in the *LGA* and *SA*, while it decreases in the *AA*. The *CHA* seems to be the branch which is mostly influenced by the presence of the probe. Depending on the position of the coil, the number of particles dragged into this vessel can increase twice or three times compared to the base case when the probe is turned off. In particular, the last two cases of Table 1 represent the optimal conditions for the therapy because they maximise the absorption of the nanodrugs

Fig. 75 Pressure field (mmHg) with nanoparticles for the second simulation at $t = 2.0$ s [1]

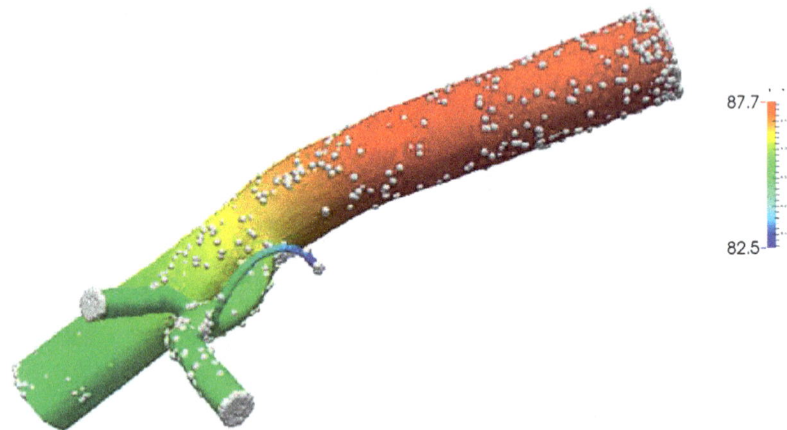

Fig. 76 Pressure field (mmHg) with nanoparticles for the fourth simulation at $t = 2.0$ s [1]

in the liver, and, consequently, they minimise the dispersion of the drugs in undesired locations, reducing the side effects of the therapy.

5   Conclusions

All the numerical simulations are performed using the open-source software: *OpenFOAM*. This program possesses, among others, an appropriate solver for the equations of *MHD*. We test different types of boundary conditions for velocity, walls and the applied magnetic field. The main point is the application of a magnetic probe on the external surface of the human body, which can improve the absorption of

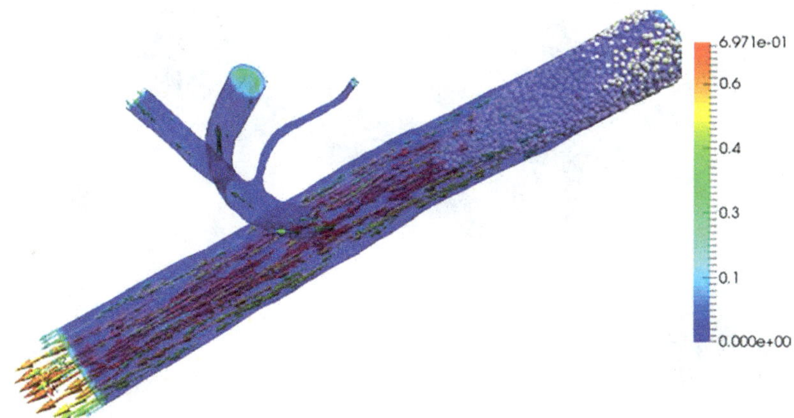

Fig. 77 Velocity field (*m/s*) with nanoparticles for the second simulation at $t = 0.15$ s [1]

Fig. 78 Velocity field (m/s) with nanoparticles for the fourth simulation at $t = 0.15$ s [1]

drugs, especially in some tumour treatments, if they are electrically charged. In order to test numerically this technique on a patient-specific geometry, we use another open-source software, *VMTK*, which allows the user to reconstruct a *3D* geometry of arteries or other organs, starting from medical images saved in *DICOM* format.

We investigate a magnetic treatment on the *hepatic artery*. Once the final geometry is obtained, the grid independence of the results from the mesh used is tested. To this aim, the first set of simulations is done in steady-state conditions, taking into account only the fluid dynamics. Once the most convenient mesh is found, which ensures the grid independence, the unsteady-state fluid-dynamic simulations are carried on. We impose an inlet velocity profile on the blood flow, based on the Womersley solution, which is coherent with the experimental pulsatile flow rate. In order to simulate the fluid dynamics of nanoparticles in the blood flow, we add to the *MHD* module, available in *OpenFOAM*, a specific Lagrangian solver.

Fig. 79 Velocity field (m/s) with nanoparticles for the second simulation at $t = 2.0$ s [1]

Fig. 80 Velocity field (m/s) with nanoparticles for the fourth simulation at $t = 2$ s [1]

Once we couple the two solvers, we set up the boundary conditions for the magnetic field. From the state of the art, the most common way to create a magnetic field outside the body is to use a coil. In our cases, we decide to implement a rectangular coil, based on Weber's equation, to solve the magnetic field generated by a current I. The last part of the implementation of the new boundary condition is to find the coordinates of the target and source. The target is inside the human body and, in our case, corresponds to the *coeliac trunk*. The magnetic field, generated by the probe, decays quite quickly with distance, and, at a certain time, the axial component becomes comparable, as far as the order of magnitude is concerned, to the radial and angular ones, so that all the body is interested in is a weak magnetic field. Nevertheless, the magnetic field is higher in correspondence with the target. We set up two simulations, one with a null magnetic field and the other with a magnetic field different from zero. This is done in order to assess the influence of the magnetic field on the adsorption of the particles and to establish the maximum value to which a

Fig. 81 Magnetic field (T) with nanoparticles for the second simulation at $t = 2.0$ s [1]

Fig. 82 Magnetic field (T) with nanoparticles for the fourth simulation at $t = 2.0$ s [1]

Fig. 83 Inertial acceleration term (m/s^2) with nanoparticles for the second simulation at $t = 0.15$ s [1]

Fig. 84 Inertial acceleration term (m/s^2) with nanoparticles for the fourth simulation at $t = 0.15$ s [1]

person can be exposed. In general, experiments proved that short-term exposure to electromagnetic fields does not cause any apparent effects. Whereas exposure to higher levels could be dangerous, they are regulated by national and international guidelines.

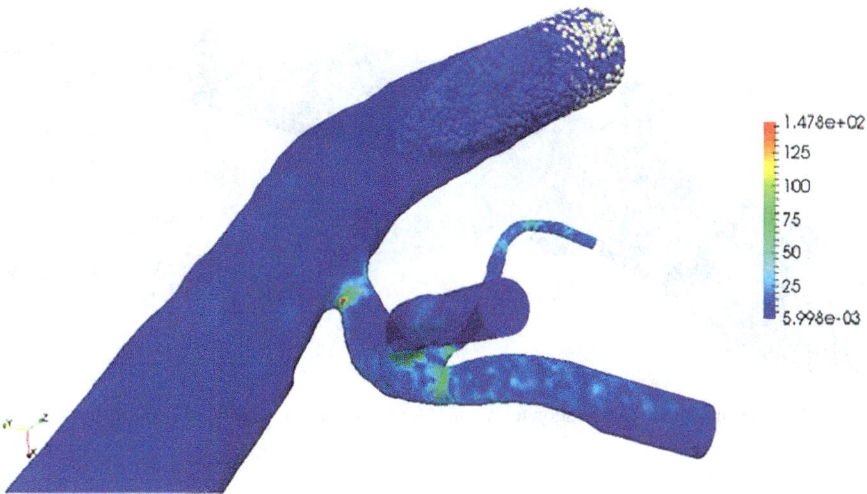

Fig. 85 Viscous acceleration term (m/s^2) with nanoparticles for the second simulation at $t = 0.15$ s [1]

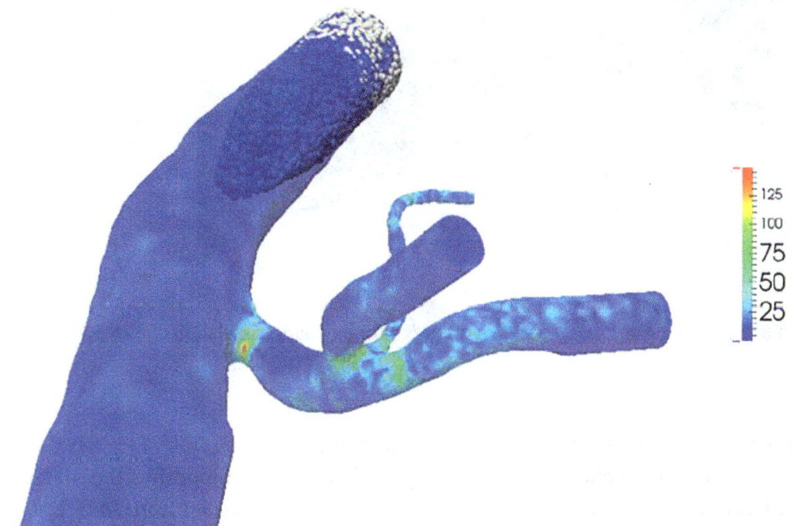

Fig. 86 Viscous acceleration term (m/s^2) with nanoparticles for the fourth simulation at $t = 0.15$ s [1]

The actual problem is whether long-term low exposure can stimulate biological effects influencing people's health. In our studies, we consider, as a starting point for the magnetic field intensity, the value which comes from the data sheet of a portable magnetic resonance (0.15 T). Afterwards, we adjust this value in order to find a suitable one that fits our geometries. From the qualitative analysis, it emerges that the

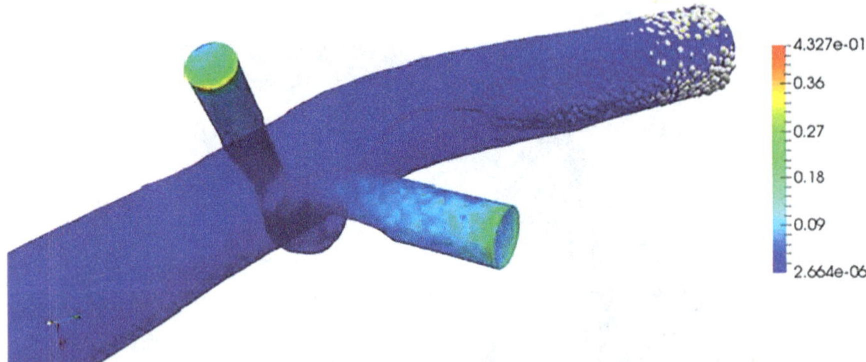

Fig. 87 Electromagnetic acceleration term (m/s^2) for the second simulation of Table 4, at $t = 0.15$ s [1]

Fig. 88 Electromagnetic acceleration term (m/s^2) for the fourth simulation of Table 4, at $t = 0.15$ s [1]

magnetic field allows better absorption directly in the areas of interest because the magnetic field interacts with the velocity. This interaction modifies the viscosity of the blood and the shear stress, especially at the wall. Furthermore, this last effect arises because in the biological fluid, there are ions that can be affected by a magnetic field. In conclusion, these simulations validate the theory and the clinical practice, which suggests that guiding with an appropriate magnetic field from outside the drugs inhaled or injected allows better absorption with a reduction of side effects for the patient. Future studies can involve different humans' organs and different pathologies, i.e. from aneurysms to other types of tumours. The geometry of the magnetic probe is an interesting research topic for a better focalisation of the magnetic field, which can increase the adsorption in the desired locations and, further on, reduce the side effects.

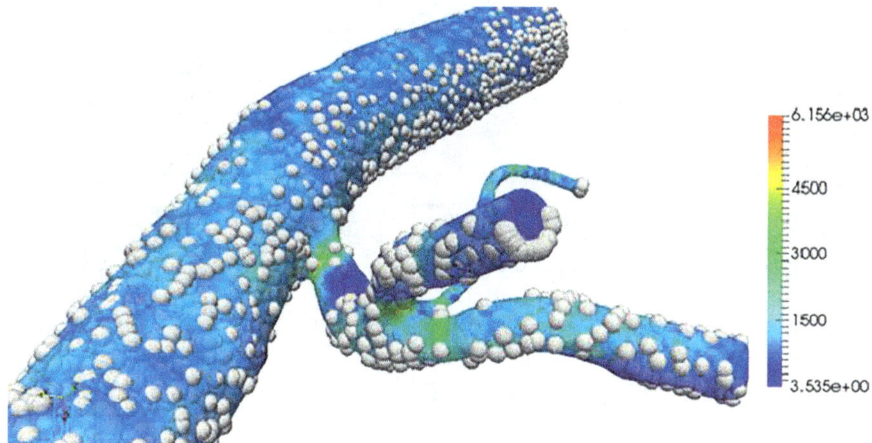

Fig. 89 Lift acceleration term (m/s^2) for the second simulation of Table 4, at $t = 1.15$ s [1]

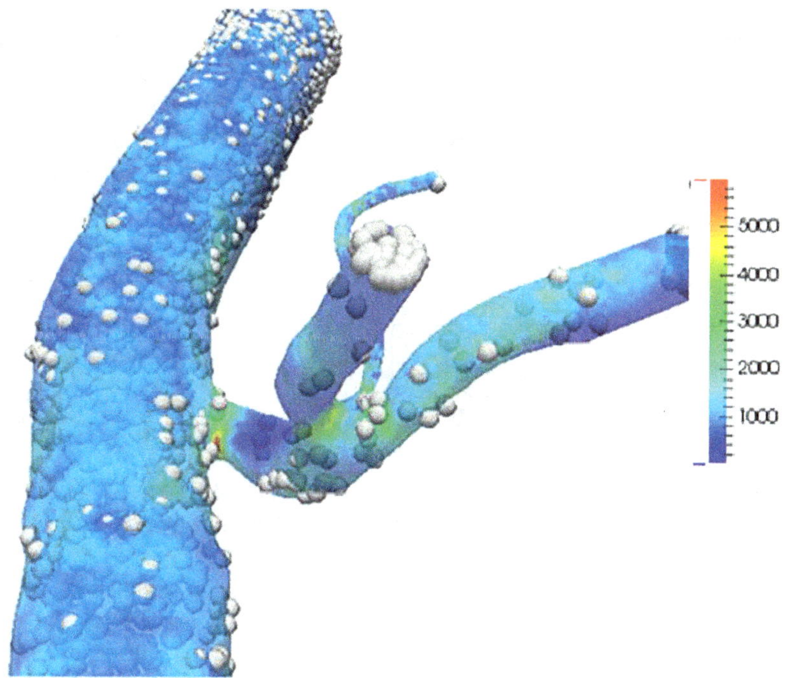

Fig. 90 Lift acceleration term (m/s^2) for the fourth simulation of Table 4, at $t = 1.15$ s [1]

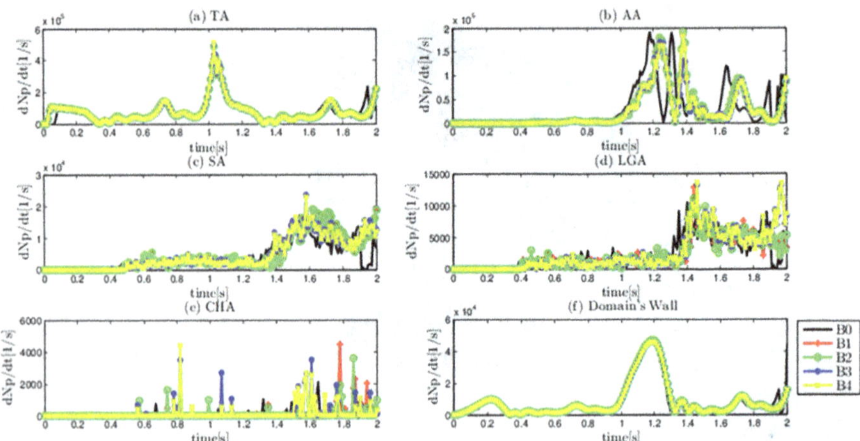

Fig. 91 Particle volumetric flow rate, per unit volume, versus time, for the different conditions of Table 1. B0 = no magnetic field; B1 = first run; B2 = second run; B3 = third run; B4 = fourth run, Fig. 9 of [59]

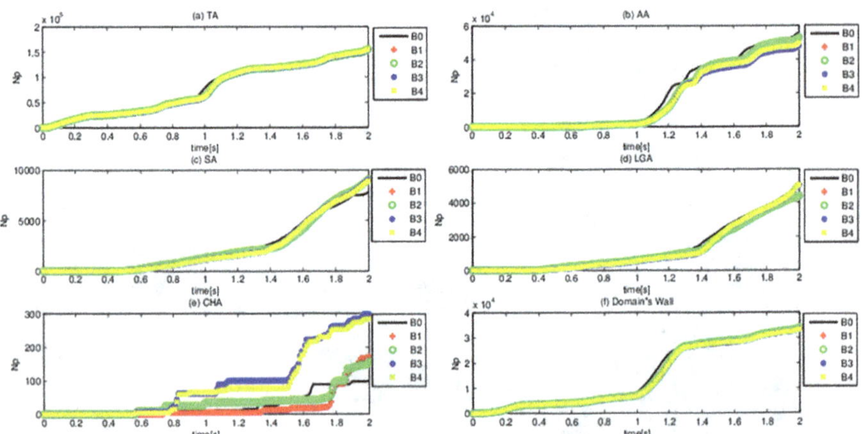

Fig. 92 Particle number vs time for the different conditions in Table 1. B0 = no magnetic field; B1 = first run; B2 = second run; B3 = third run; B4 = fourth run, Fig. 10 of [59]

One simulation with the magnetic probe turned off and four simulations with two different current intensities and two positions of the coil are carried out. The results are compared in order to find the best configuration which maximises the intake of the drug in the liver and consequently minimises the dispersion in other locations. The rectangular coil, pointing perpendicularly to the liver, is positioned at two points, 1 cm above the abdomen and 2 cm. The results of the present numerical simulations show that the position of the probe highly influences the intake in the *CHA*. Nevertheless, despite the absolute number of particles flowing in this vessel being able to double or triple, it is still a very small fraction of the total number of particles injected. In all the cases studied, only a minimal amount of the drug injected, about 0.2% in the *CHA*, 3.3% in the *LGA* and 6.6% in the *SA*, goes to the liver after 2 s from the beginning of the technique. About half of the injected drug is absorbed by the arterial wall, and the rest flows away through the *AA*. It was impossible to investigate all the possible positions of the target on the abdomen, but it seems unlikely to significantly increase the drug intake in the liver by changing the position of the probe. This conclusion is in agreement with [30, 31], but in contrast with the numerical findings of [29] for brain tumours. This is probably due to the closer distance between the magnet and the tumour (3 cm), the larger particle diameter (2 μm), and the higher magnetic field (1–4 T). Magnetic drug delivery is extremely sensitive to these parameters; it should not be surprising that the technique performs differently in different conditions. The low performance of the magnetic drug delivery in the present study can be due to three factors: (i) the nanoparticle size; (ii) the magnet-tumour distance; and (iii) the probe design. As far as the nanoparticle size is concerned, as shown in Eqs. (18 and 19) of [59], a small particle diameter increases the acceleration due to the Lorentz force, but it also increases the friction. However, because of the law limitation on the magnetic field applied [71, 72], the effect on the friction is predominant. Therefore, having larger particles should increase the effectiveness of the technique. The probe-tumour distance is another factor which limits the effectiveness of the technique. Unfortunately, this parameter cannot be tuned at will. If the tumour is superficial, the magnetic drug targeting is likely to be more effective.

Finally, the design of the magnetic probe is another factor which influences the outcome of the procedure. This conclusion was suggested in [31] as well, where it was found that the sticking of particles to vessels occurred because of low blood flow velocity near the vessel walls, and this was suggested to be the main reason for low targeting efficiency. The modulus of the magnetic induction field of a single rectangular coil decreases considerably with the distance from the target, i.e. the liver. Being the liver about 12 cm away from the source, it is clear that this makes the modulus of the magnetic induction considerably low. A solution to this problem could be to increase the value of the current intensity flowing in the probe, but this is not possible because higher values of the magnetic field could affect the health of the patient, which is why this is forbidden by law [71, 72]. A different solution could be the employment of a different type of probe, which must focalise most of the energy on its axis in order to reduce the rate of decay of the magnetic field. Moreover, a possible configuration with multiple probes, positioned in appropriate locations, could also focus a higher magnetic field inside the target and limit the dangerous values of *B* elsewhere. A different solution, based on the use of a time-varying magnetic field, is proposed in [31] and shows a significant decrease in particle adherence to walls.

THD-MHD allows detailed visualisation of biological fluid flows, which increases our understanding of natural phenomena. Yet, *THD–MHD* has some limitations due to the computational resources required to simulate a process in a complex domain. The question then arises as to whether or not the observation time (2 s) is sufficient to judge the effectiveness of the technique. It would certainly be better to extend the simulations to a few minutes, but the particle volumetric flow rate in Fig. 91 in the different cases shows negligible differences. It seems unlikely that the percentage of injected particles which reach the liver will significantly increase over time. Nevertheless, the liver uptake could increase if the aortic wall were saturated with nanoparticles. A few works have studied this problem [73–75], finding that the nanoparticle uptake is reduced for a concentration of a few µg/ml. In this work, we use a Lagrangian approach to simulate the non-particle flow, so the particle concentration is not directly available. However, we can derive an average concentration by multiplying the total number of particles injected by the mass of a single particle and dividing it by the domain volume. After 2 s, we obtain an average particle concentration of about 10^{-9} mg/ml, which is far below the saturation limit. For a particle with $d_p = 500$ nm, the saturation limit would be reached in 3 minutes, but with the particle diameter used in this work, the saturation would not occur during the time frame of the procedure. Moreover, the particle release from the wall is allowed for in a few numerical studies, e.g [30, 33]., but this did not result in a significant increase in particle penetration into the tumour. In conclusion, it is unlikely that extending the simulation time would significantly change the predicted efficiency of the technique.

Acknowledgements for the Figures The Author thanks the *publisher Elsevier* for the kind permission to reproduce the *full papers of Refs.* [65, 71].

References

1. Russo F (2014/2015) Numerical simulation of magneto-hydro-dynamics inside cardiovascular and respiratory systems. PhD Thesis, University of Rome Tor Vergata, A.A
2. Womersley J (1955) Method for the calculation of velocity, rate of flow and viscous drag in arteries when the pressure gradient is known. J Physiol 127:553–563
3. Tang B et al (2006) Abdominal aortic hemodynamics in young healthy adults at rest and during lower limb exercise: quantification using image-based computer modeling, vol 291. American Journal of Physiology—Heart and Circulatory Physiology, August, pp H668–H676
4. I. A. for Research on Cancer, W. H. Organization, et al (2014) Globocan: Estimated cancer incidence, mortality, and prevalence worldwide in 2012. IARC
5. Hartmann J, Lazarus F (1937) Hg-dynamics II. Experimental investigations, Det Kongelige Danske Videnskabernes Selskab-Mat Phys 7:1–45
6. Shercliff JA (1956) The flow of conducting fluids in circular pipes under transverse magnetic fields. J Fluid Mech 6:644
7. Shercliff JA (1962) Magnetohydrodynamic pipe flow. Part2. High Hartmann number. J Fluid Mech 13:513–518
8. Gold R (1962) Magnetohydrodynamic pipe flow. Part 1. J Fluid Mech Agosto:505–512
9. Adey WR (1993) Biological effects of electromagnetic fields. J Cell Biochem 51:410–416
10. Tzirtzilakis E (2005) A mathematical model for blood flow in magnetic field, Physics of Fluids (1994-present) 17: 077103
11. Saiyed Z, Telang S, Ramchand C (2003) Application of magnetic techniques in the field of drug discovery and biomedicine. Biomagn Res Technol 1:2
12. Ikbal MA, Chakravarty S, Wong KK, Mazumdar J, Mandal PK (2009) Unsteady response of non-newtonian blood flow through a stenosed artery in magnetic field. J Comput Appl Math 230:243–259

13. Gori F, Boghi A, Amitrano M (2010) Three-dimensional numerical simulation of the fluid dynamics in a coronary stent. ASME International Mechanical Engineering Congress and Exposition, Proceedings, Volume 2, Pages 407–411, 2009 ASME International Mechanical Engineering Congress and Exposition, IMECE2009; Lake Buena Vista, FL; United States; 13 November 2009 through 19 November 2009; Code 80879

14. Gori F, Boghi A (2010) Image-based computational fluid dynamics in a carotid artery. ASME International Mechanical Engineering Congress and Exposition, Proceedings, Volume 2, Pages 123–128, 2009 ASME International Mechanical Engineering Congress and Exposition, IMECE2009; Lake Buena Vista, FL; United States; 13 November 2009 through 19 November 2009; Code 80879

15. Gori F, Boghi A (2010) Three-dimensional numerical simulation of non newtonian blood in two coronary stents. 14th International Heat Transfer Conference, American Society of Mechanical Engineers, pp 109–114

16. Gori F, Boghi A (2011) Three-dimensional numerical simulation of blood flow in two coronary stents. Numer Heat Transf A Appl 59:231–246

17. Boghi A, Gori F (2015) Numerical simulation of blood flow through different stents in stenosed and non-stenosed vessels. Numer Heat Transf A Appl 68:225–242

18. Di Venuta I, Boghi A, Gori F (2017) Three-dimensional numerical simulation of a failed coronary stent implant at different degree of residual stenosis. Part I: fluid dynamics and shear stress on the vascular wall. Numer Heat Transf A Appl 71(6):638–652

19. Boghi A, Di Venuta I, Gori F (2017) Three-dimensional numerical simulation of a failed coronary stent implant at different degree of residual stenosis. Part II: apparent viscosity and wall permeability. Numer Heat Transf A Appl 71(6):653–665

20. Li X, Yao K, Liu Z (2008) Cfd study on the magnetic fluid delivering in the vessel in high-gradient magnetic field. J Magn Magn Mater 320:1753–1758

21. Moulinier L, Venet T, Schiller NB, Kurtz TW, Morris RC, Sebastian A (1991) Measurement of aortic blood flow by doppler echocardiography: day to day variability in normal subjects and applicability in clinical research. J Am Coll Cardiol 17:1326–1333

22. Long Q, Xu X, Collins M, Bourne M, Griffith T (1998) Magnetic resonance image processing and structured grid generation of a human abdominal bifurcation. Comput Methods Prog Biomed 56:249–259

23. Haverkort J, Kenjere˘s S, Kleijn C (2009) Computational simulations of magnetic particle capture in arterial flows. Ann Biomed Eng 37:2436–2448

24. Jain RK (2001) Delivery of molecular and cellular medicine to solid tumors, 508. Adv Drug Deliv Rev 46:149 168

25. Habibi MR, Ghasemi M (2011) Numerical study of magnetic nanoparticles concentration in biofluid (blood) under influence of high gradient magnetic field. J Magn Magn Mater 323:32–38

26. Lubbe AS, Alexiou C, Bergemann C (2001) Clinical applications of magnetic drug targeting. J Surg Res 95:200–206

27. Lubbe AS, Bergemann C, Riess H, Schriever F, Reichardt P, Possinger K, Matthias M, Dorken B, Herrmann F, Gurtler R et al (1996) Clinical experiences with magnetic drug targeting: a phase i study with 4- epidoxorubicin in 14 patients with advanced solid tumors. Cancer Res 56:4686–4693

28. Nacev A, Beni C, Bruno O, Shapiro B (2011) The behaviors of ferromagnetic nano-particles in and around blood vessels under applied magnetic fields. J Magn Magn Mater 323:651–668

29. Kenjere˘s S, Righolt B (2012) Simulations of magnetic capturing of drug carriers in the brain vascular system. Int J Heat Fluid Flow 35:68–75

30. Larimi M, Ramiar A, Ranjbar A (2014) Numerical simulation of magnetic nanoparticles targeting in a bifurcation vessel. J Magn Magn Mater 362:58–71

31. Tehrani MD, Yoon J-H, Kim MO, Yoon J (2015) A novel scheme for nanoparticle steering in blood vessels using a functionalized magnetic field. IEEE Trans Biomed Eng 62:303–313

32. Larimi M, Ramiar A, Ranjbar A (2016) Numerical simulation of magnetic drug targeting with eulerian-lagrangian model and effect of viscosity modification due to diabetics. Appl Math Mech 37:1631–1646

33. Momeni Larimi M, Ramiar A, Ranjbar AA (2016) Magnetic nanoparticles and blood flow behavior in non-newtonian pulsating flow within the carotid artery in drug delivery application. Proc Inst Mech Eng H J Eng Med 230:876–891

34. Steinman DA (2002) Image-based computational fluid dynamics modeling in realistic arterial geometries. Ann Biomed Eng 30:483–497

35. Baillard C, Barillot C (1935) Robust 3D segmentation of anatomical structures with level sets. MICCAI 2:236–245

36. Antiga L, Piccinelli M, Botti L, Ene-Iordache B, Remuzzi A, Steinman DA (2008) An image-based modeling framework for patient-specific computational hemodynamics. Med Biol Eng Comput 46(11):1097–1112

37. Attali D, Auchad D (2001) Delaunay conforming iso-surface, skeleton extraction and noise removal. Comput Geom 19(2–3):175–189

38. Antiga L, Ene-Iordache B, Caverni L, Cornalba GP, Remuzzi A (2002) Geometric reconstruction for computational mesh generation of arterial bifurcations from CT angiography. Comput Med Imaging Graph 26(4):227–235

39. Wang S, Liu H, Xu W (2008) Hydrodynamic modelling and CFD simulation of ferrofluids flow in magnetic targeting drug delivery. Int J Comput Fluid Dynamic 22(10):659–667

40. Zacek M, Krause E (1996) Numerical simulation of the blood flow in the human cardiovascular system. J Biomech 29(1):13–20

41. Perko M et al (1998) Mesenteric, coeliac and splanchnic blood flow in humans during exercise. J Physiol 513(3):907–913

42. Mano Y et al (2013) Hemodynamic assessment of Celiaco-mesenteric anastomosis in patients with pancreaticoduodenal artery aneurysm concomitant with celiac artery occlusion using flow-sensitive four-dimensional magnetic resonance imaging. Eur J Vasc Endovasc Surg 46(3): 321–328

43. Tang B et al (2001) Abdominal aortic hemodynamics in young healthy adults at rest and during lower limb exercise: quantification using image-based computer modeling. Am J Physiol Heart Circ Physiol 291(2):H668–H676

44. Steele B, Olufsen M, Taylor C (2007) Fractal network model for simulating abdominal and lower extremity blood flow during resting and exercise conditions. Comput Methods Biomech Biomed Engin 10(1):39–51

45. Gardin J, Burn C, Childs W, Henry W (1984) Evaluation of blood flow velocity in the ascending aorta and main pulmonary artery of normal subjects by Doppler echocardiography. Am Heart J 107(2):310–319

46. Gill R (1985) Measurement of blood flow by ultrasound: accuracy and sources of error. Ultrasound Med Biol 11(4):625–641

47. Greenway C, Stark R (1971) Hepatic vascular bed. Physiol Rev 51(1):23–65

48. Carlisle K, Halliwell M, Read A, Wells P (1992) Estimation of total hepatic blood flow by duplex ultrasound. Gut 33(1):92–97

49. Tenforde TS (2005) Magnetically induced electric fields and currents in the circulatory system. Prog Biophys Mol Biol 87(545):279–288

50. Elghobashi S (1994) On predicting particle-laden turbulent flows. Appl Sci Res 52:309–329

51. Tzirtzilakis E, Sakalis V, Kafoussias N, Hatzikonstantinou P (2004) Biomagnetic uid ow in a 3d rectangular duct. Int J Numer Meth Fluids 44(550):1279–1298

52. Arruebo M, Fernandez-Pacheco R, Ibarra MR, Santamaria J (2007) Magnetic nanoparticles for drug delivery. Nano Today 2:22–32

53. Seino S, Matsuoka Y, Kinoshita T, Nakagawa T, Yamamoto TA (2009) Dispersibility improvement of gold/iron-oxide composite nanoparticles by polyethylenimine modification. J Magn Magn Mater 321:1404–1407

54. Taylor CA, Hughes TJ, Zarins CK (1998) Finite element modeling of three dimensional pulsatile flow in the abdominal aorta: relevance to atherosclerosis. Ann Biomed Eng 26:975–987

55. Grinberg L, Karniadakis GE (2008) Outflow boundary conditions for arterial networks with multiple outlets. Ann Biomed Eng 36(564):1496–1514

56. Tang BT, Cheng CP, Draney MT, Wilson NM, Tsao PS, Herfkens RJ, Taylor CA (2006) Abdominal aortic hemodynamics in young healthy adults at rest and during lower limb exercise: quantification using image based computer modeling, American Journal of Physiology-Heart and Circulatory. Physiology 291:H668–H676

57. Zácek M, Krause E (1996) Numerical simulation of the blood flow in the human cardiovascular system. J Biomech 29:13–20

58. OPENFOAM (2012) OpenFOAM the open sourec CFD toolbox. s.l.:s.n. OPENFOAM, 2012. User Guide OpenFOAM v.2.2. s.l.:s.n

59. Boghi A, Russo F, Gori F (2017) Numerical simulation of magnetic nano drug targeting in a patient-specific coeliac trunk. J Magn Magn Mater 437:86–97

60. Azer K, Peskin C (2001) A one-dimensional model of blood flow in arteries with friction and convection based on the Womersley velocity profile. Cardiovasc Eng 7(2):51–73

61. Axner L et al (2009) Simulations of time harmonic blood flow in the mesenteric artery: comparing finite element and lattice Boltzmann methods. Biomed Eng Online 8(1):23

62. Mody V, Coz A, Shah S, Singh A, Bevins W, Parihar H (2014) Magnetic nanoparticle drug delivery systems for targeting tumor. Appl Nanosci 3:385

63. Wilson M, Kerlan RK Jr, Fidelman N, Venook AP, LaBerge JM, Koda J, Gordon RL (2004) Hepatocellular carcinoma: regional therapy with a magnetic targeted carrier bound to doxorubicin in a dual MR imaging/conventional angiography suite—initial experience with four patients. Radiology 2:287–293

64. Azpurua MA (2012) A semi-analytical method for the design of coil-systems for homogeneous magnetostatic field generation. Prog Electromagn Res B 37:171–189

65. Metzger GJ, van de Moortele P-F, Akgun C, Snyder CJ, Moeller S, Strupp J, Andersen P, Shrivastava D, Vaughan T, Ugurbil K et al (2010) Performance of external and internal coil configurations for prostate investigations at 7 t. Magn Reson Med 64:1625–1639

66. Dughiero F, Baake E, Forzan M, Nemkov V, Ruffini R, Goldstein R, Jackowski J, DeWeese T, Ivkov R (2011) Magnetic field generating inductor for cancer hyperthermia research. COMPEL 30:1626–1636

67. Takamura T, Ko PJ, Sharma J, Yukino R, Ishizawa S, Sandhu A (2015) Magnetic-particle-sensing based diagnostic protocols and applications. Sensors 15:12983–12998

68. Misakian M (2000) Equations for the magnetic field produced by one or more rectangular loops of wire. J Res Nat Inst Stand Technol 105:557

69. Wilson MW, Kerlan RK Jr, Fidelman NA, Venook AP, LaBerge JM, Koda J, Gordon RL (2004) Hepatocellular carcinoma: regional therapy with a magnetic targeted carrier bound to doxorubicin in a dual mr imaging/conventional angiography suiteinitial experience with four patients 1. Radiology 230:287–293

70. Russo F, Boghi A, Gori F (2018) Numerical simulation of magnetic nano drug targeting in a patient-specific lower respiratory tract. J Magn Magn Mater 451:554–564

71. Council Recommendation of 12 July (1999) On the limitation of exposure of the general public to electromagnetic fields (0 hz to 300 ghz). Off J Eur Commun L 59:59–70

72. Cleveland RF, Sylvar DM, Ulcek JL (1997) Evaluating compliance with FCC guidelines for human exposure to radiofrequency electromagnetic fields. Standards Development Branch, Allocations and Standards Division, Office of Engineering and Technology, Federal Communications Commission

73. Song C, Labhasetwar V, Cui X, Underwood T, Levy RJ (1998) Arterial uptake of biodegradable nanoparticles for intravascular local drug delivery: results with an acute dog model. J Control Release 54:201–211
74. Labhasetwar V, Song C, Humphrey W, Shebuski R, Levy RJ (1998) Arterial uptake of biodegradable nanoparticles: effect of surface modifications. J Pharm Sci 87:1229–1234
75. Davda J, Labhasetwar V (2002) Characterization of nanoparticle uptake by endothelial cells. Int J Pharm 233:51–59

Flavia Russo, a Medical Engineer, obtained her Research Doctorate degree in Industrial Engineering from the University of Rome "Tor Vergata." She completed a six-month fellowship at the Center for Biomolecular Nano Technologies of the IIT in Lecce for the study of optical fibre cochlear prostheses. She has worked at a medical software house as a product specialist and in a research and development start-up as a product owner of wearable medical devices. She currently holds the position of project manager at a company that deals with telemedicine.

Dr. Eng. Andrea Boghi, The Doctor of Research *(Ph. D.)*, is the director of *Computational Science Ltd.*, a consultancy company in the field of software development and data analysis, which counts among its clients the British Air Traffic Agency (*NATS*) and the Bank of England, starting from 2018. He is the author of more than 30 international scientific publications indexed on the major scientific databases. For more than a decade, he has worked in academia in various European institutions, dealing with the modelling and simulation of transport phenomena, both for basic research and for industrial and biological applications. His academic journey began at the University of Rome "Tor Vergata," where, alongside research on the modelling and simulation of turbulent flows with variable laminar diffusivity and blood flow in arteries and *stents*, he also served as a contract lecturer for the courses of Technical Physics, Thermo-Fluid-Dynamics of Biological Systems and Numerical Calculation of Thermo-Fluid-Dynamic Systems (2006–2010). In the two-year period from 2010 to 2012, he was a contract researcher at the Department of Energy and Technology of the University of *Southampton* in the United Kingdom, where he dealt with modelling and simulation of bi-phase flows, characterised by spinodal decomposition. In 2012, he went to France to the Institute of Fluid Mechanics of Toulouse as a contract researcher for the OTE (Waves, Turbulence and Environment) research group, where he dealt with modelling and simulation of river and atmospheric fluid dynamics. He terminated his contract in 2013 to go to the University of *Cranfield* in the United Kingdom, where he was appointed *Senior Research Fellow*, becoming part of the academic staff. Here, for the following 5 years (2013–2018), he dealt with modelling and simulation of transport phenomena for various departments: Energy, Hydraulics, Environment and Agriculture, where he respectively dealt with transport of crude oil in pipelines, disinfection of tanks by chlorination and transport and diffusion of organic and inorganic material in soils, with associated absorption by plants. At the University of *Cranfield*, he also handled the course on modelling of atmospheric emissions in the two-year period from 2016 to 2017. Before leaving academia, he went again to the University of *Southampton* in 2018, where he dealt with the three-dimensional modelling of the transport of nitrates and carbonates at the air–water interface in soils. Throughout his period abroad, he maintained his collaboration with Fabio Gori Ammannati, supervising some students and collaborating on the production of scientific articles. In 2020, he obtained the national scientific qualification for the functions of associate university professor in the competitive sector of Technical Physics and Nuclear Engineering.

Fabio Gori Ammannati was born in Montale (Pistoia) on 5 August 1947 to Emilio Gori and Cesarina Ammannati. On 16 July 2021, he added his mother's surname, Ammannati. He graduated in Chemical Engineering on 11 November 1971 with honours. In December 1971, he won a scholarship for young graduates at the Faculty of Engineering in Bologna, which he undertook from January 1972. In 1974, he became assistant professor and, in 1982, associate professor of Technical Physics at the Faculty of Engineering of the University of Florence, where he taught, also as a contract professor, until 1990. In 1978, he won a scholarship from the *British Council,* which he carried out at the *Imperial College* in London, where he collaborated with *Prof. D. Brian Spalding* on the topic of numerical analysis in the turbulent flow of liquid metals. During his time at the University of Florence, he established international scientific collaborations with *Prof. R. Echigo, Tokyo Institute of Technology, Tokyo, Japan; Prof. D.R. Chaudhary, Department of Physics, University of Rajasthan, Jaipur, India;* International Centre for Theoretical Physics *(ICTP)* in Trieste; *Centralny Osrodek Techniki Medycznej, Warsaw, Poland and Prof. T. Aihara, Institute of Fluid Science, Sendai, Japan.* In 1986, he won a CNR scholarship, which he carried out at *Cornell University, Ithaca, New York,* where he collaborated with *Prof. R. Miller* on the topic of ground freezing. In the same year, he became professor in the Faculty of Engineering at the University of Reggio Calabria. In 1988, he was appointed as *professor* at the *University of New York at Stony Brook,* teaching the course, *Introduction to Fluid Dynamics,* in the autumn semester of the same year. During his stay, he collaborated with *Prof. T.F. Irvine Jr.* on the method of measuring thermal conductivity with a thermal probe and on the measurement of the isobaric thermal expansion coefficient of non-Newtonian fluids. In 1990, he moved to the Milano Polytechnic, where he taught Technical Physics and Systems until 1993. From 1991 until 1998, he was an adjunct professor at the Faculty of Engineering of the University of Siena. From 1992 to 2017, he was a professor of Technical Physics in the Faculty of Engineering at the University of Rome Tor Vergata, and from 2017, he is a contract professor. In 1994, he proposed the Research Doctorate (Italian *Ph.D.*) in Energy-Environment Engineering, of which he was the coordinator until 2011. In 1995, he proposed and directed the second-level Master's in Thermo-fluid-dynamics until 2016. From 1998 to 2001, he was the director of the Department of Mechanical Engineering. In 2001, he proposed and coordinated the programme, *Master of Science, Energy Engineering and Thermal and Fluid Dynamics,* in collaboration with the *Department of Mechanical Engineering, College of Engineering, University of Illinois at Chicago, USA,* which takes place at the University of Rome Tor Vergata but awards the title of *Master of Science* from the same American University. Since the late 90s, he has been a coordinator of the joint *Ph.D.* research programme with *Prof. R.J. Goldstein, Department of Mechanical Engineering, University of Minnesota, Minneapolis, USA,* and with *Prof. J.P. Hartnett, Prof. L. Kennedy, Prof. W. Minkowycz and Prof. W.M. Worek, Department of Mechanical Engineering, College of Engineering, University of Illinois at Chicago, Chicago, USA.* Since the mid-90s, he has proposed and directed the *Socrates–Erasmus* programme with *Prof. Mayinger, Technical University of Munich, Germany; Prof. van Steenhoven, Eindhoven Technical University, the Netherlands and Prof. C. Caro, Imperial College of Science, Technology and Medicine.* During his stay at Tor Vergata, he established scientific collaborations with the *Department of Mechanical Engineering, University of Minnesota, Minneapolis, USA,* collaborating with *Prof. R.J. Goldstein* on Thermo-fluid-dynamics and mass transport in gas turbine blades; the *Energy Resources Center, University of Illinois at Chicago,* collaborating with *Prof. J.P. Hartnett, Prof. W. Minkowycz and Prof. W.M. Worek;* the *Department of Mechanical Engineering, College of Engineering, University of Illinois at Chicago, collaborating with Prof. L. Kennedy and Prof. W.M. Worek; Duke University, collaborating with Prof. A. Bejan and S. Mary's University, Halifax, Nova Scotia, Canada, collaborating with Prof. W. R. Tarnawski.* The bibliographic review of documents and citations, titled *PlosBiology Career,* places him within the top 2% of researchers in the field of *Mechanical Engineering and Transports.* Since 1992, he has been a tutor for over 30 PhD theses, supervisor for over 40 second-level university Master's theses, supervisor for over 70 degree theses (five-year, Specialist and Master's), supervisor for over ten *Master's degree in Mechanical Engineering* from the *University of Illinois at Chicago* and supervisor for over 20 theses *Erasmus-Socrates.*

Since the early 2000s, he has been a reviewer of international research projects on behalf of *Portuguese Science and Technology Foundation (FCT); Czech Science Foundation (GACR); European Science Foundation, Strasbourg, France,* on behalf of *FCT* and the *Shota Rustaveli National Science Foundation, Georgia.* Since 2017, the year of his retirement due to age limits, he has been a contract professor at the University of Rome Tor Vergata, where in the academic year

2024–2025, he has taught Technical Physics for the Master's Degree Course in Medical Engineering. M.D. Salvatore Mangiafico (Florence), scientific head of the NeuroVascular Base Camp, 2021, invited him to give a scientific presentation titled "*The engineering approach to the study of the Circle of Willis*" on 23 September 2021 at the Congress Centre, University of Rome La Sapienza, thus opening up a possible future collaboration.

Cardiac Acoustics

Basics of Acoustics for the Evaluation of Environmental and Fluid Dynamic Noise

Giulio La Bella and Fabio Gori Ammannati

1 Environmental Noise

Noise is defined as a set of unwanted sounds (the English word *noise* derives from the Latin word *noxia,* which means harmful). This definition does not allow to classify a sound as noise solely based on its own physical characteristics (frequency, intensity, etc.). Subjective factors come into play: it is pleasant to listen to the music from your own stereo. If the same music comes from the neighbour's stereo while you are resting, this is perceived as an annoying sound. In today's world, humans are constantly exposed to sources of noise, particularly for those living in large urban centres. Environmental noise is defined as the noise produced by all sources of noise in a given place and during a certain time. In an urban centre, there are numerous sources of noise that contribute to environmental noise: road, rail and air traffic; wind turbines; industrial machinery; noisy neighbours; fans; pipes; lifts and other mechanical equipment present in most buildings.

The type of noise that is predominant is the noise due to transport [1]. Specifically, the aerodynamic noise caused by air traffic is often the most annoying, being present in most cases even during the night. In areas such as wind farms, wind turbines cause significant disturbance due to the noise produced, both during the day and night.

G. La Bella
Rome, Italy

F. G. Ammannati (✉)
University of Rome Tor Vergata, Rome, Italy
e-mail: gori@uniroma2.it

F. Gori Ammannati (ed.), *Thermo-Haemo-Dynamics in Medical Engineering,*
https://doi.org/10.1007/978-3-031-97214-0_11

1.1 Aerodynamic Noise Due to Wind Turbine Blades

Wind turbines are a source that contributes to environmental noise by producing two types of noise, mechanical and aerodynamic. The mechanical noise is due to the generator and the reducer. Over the years, it has been possible to significantly reduce this type of noise, making it lower than that produced aerodynamically. Over time, the blades have become increasingly larger, which has little influence on the noise produced by the mechanical components but increases the aerodynamic noise. The aerodynamic noise is due to the passage of the rotor blades through the air. This increases with the speed of the blades, and being the latter directly related to the production of energy, an increase in efficiency implies an increase in speed. It is expected that with the improvement of the efficiency of the blades and with the advancement of the years, the noise produced will also increase.

The *WHO (World Health Organisation)* has drafted a document containing guidelines for the regulation of environmental noise, [2], listing the possible sources of noise and the possible effects on health. The document also includes limits in terms of LA_{eq} *T (A-weighted equivalent sound pressure level for period T)*, LA_{MAX} *(Maximum A-weighted sound pressure level in a stated interval)* and *SEL (Sound Exposure Level)*, recommended for various types of inhabited areas (residential areas, hospitals, schools, airports, etc.). While an agreement and international standardisation have been reached for air traffic with the *Chicago Convention*, through the *ICAO (International Civil Aviation Organisation)*, this is not the case for wind turbines. Each country indeed has its own rules or guidelines.

In Europe, for example, Sweden recommends a maximum of 40 dBA in inhabited areas, while Denmark has drafted a specific legislative decree *(Bekendtgrelse om stj fra vindmller BEK nr 304 af 14/05/1991)*, as have the Netherlands *(Besluit van 18 oktober 2001, houdende regels voor voorzieningen en installaties; Besluit voorzieningen en Installaties milieubeheer; Staatsblad van het Koninkrijk der Nederlanden 487)*, Germany *(Bundes-Immissionschultz-Gesetzes. BimSchG, Germany, 1974)*, France *(Loi n 92–1444 du 31 décembre 1992: Loi relative à la lutte contre le bruit)*, and Great Britain *(ETSU for DTI 1996)* [3]. In Italy, no specific regulations or specialised legislation for wind turbines have been developed yet, and the legal limits in force and any penalties are those of general validity, reported in the D.M.C.M. of 14 November 1997. However, regional regulations are issued, such as in Campania and in Molise, that impose predetermined distances of wind farms from individual inhabited centres. Given the lack of international regulations, not many studies are carried out to relate the noise produced by wind turbines to the effects on health, as done for aerodynamic noise due to air traffic. However, studies are carried out on the annoyance caused by the noise of wind turbines on groups of people living near wind farms.

1.2   Aerodynamic Noise Due to Air Traffic

Aerodynamic noise is an unwanted effect produced by aircraft. Initially, however, aerospace engineering was mainly focused on improving the performance of aircraft, without giving too much weight to the aerodynamic noise generated by them. At the beginning of the 60 s, having now reached good levels in the performance of aircraft, greater importance began to be given to emissions, safety and aerodynamic noise. In the 60 s, airport owners were forced to impose local limits on the noise produced. The problem of aerodynamic noise is also continuously developing, as the continuous increase in the performance of aircraft results in a continuous increase in the aerodynamic noise generated. The problem of reducing aerodynamic noise mainly concerns civil aviation, as in the field of military aviation, the focus is mainly on continuous performance improvement. In the 70 s, it was the manufacturers who had to guarantee sound emission standards to obtain the certifications suitable for marketing aircraft. The first standards concerning the noise produced were drafted in 1971 by the *ICAO* committee for aircraft noise and came into force in 1973.

1.2.1   ICAO

The *ICAO* is a specialised agency, established in 1944 by the United Nations, during the *Chicago Convention*. This agency is responsible for coordinating and regulating international air transport, issuing standardised guidelines (continuously updated) for global air transport. The *ICAO* works with the 191 member states adhering to the Chicago Convention (including Italy). The *ICAO* has also drafted the *Standard And Recommended Practices* (*SARPs*), containing more than 12,000 technical specifications to be adopted in the field of aviation, to achieve the highest degree of standardisation. These specifications are in accordance with Article 37 of the Chicago Convention, [4]: "*Each contracting State undertakes to collaborate to ensure the highest degree of uniformity in regulation, standards, procedures and organisation relating to aircraft, personnel, flight routes and auxiliary services and in all matters where such uniformity facilitates and improves air navigation.*"

The *ICAO* defines a standard as any specification for: physical characteristics, material, configuration, performance, personnel or procedure; the uniform application of which is recognised as necessary for safety or regulation of international air traffic and to which the member states will adhere according to the Chicago Convention. One of the main objectives of the *ICAO* is the reduction of noise produced by air traffic and the reduction of the number of people subjected to noise from air traffic. The policy of the *ICAO* regarding this topic is the *Balanced Approach to Aircraft Noise Management*, [5], which is based on four key points:

1. Reduction of noise at the source. Through the application of the *SARPs*, aircraft manufacturers ensure that the most advanced technologies have been used in the design (both regarding engines, wing profiles and fuselage).
2. Planning and management of land use, that is, ensuring that the activities near the airport are compatible with aviation. This goal is achieved by defining how the various areas of the airport are to be used and using the areas near the airport

appropriately to minimise the population subjected to high levels of noise. It would, in fact, make little sense to reduce the noise generated by an aircraft only to then build a large residential centre at a very short distance from an airport. In addition, airports must apply countermeasures regarding the noise present in the airport itself, for example, by installing soundproofing glass.

3. Noise abatement procedures. They mainly deal with reducing noise due to various operations carried out at airports (as for example, refuelling operations, wheel transport, etc.) and that of an aircraft during take-off and landing phases. This type of noise is controlled, mainly, by setting preferential routes for take-off and landing.

4. Operational restrictions. Operational restrictions impose certain behaviours on airports; for example, they do not allow transit to aircraft not certified by *ICAO*. One of the most common operational restrictions is to impose a ban on air traffic during night hours at airports located near residential areas.

1.3   Health Effects

The disturbances and damage caused to health by intense and/or continuous noise on humans are multiple, [6] such as effects on the cardiovascular system, sleep disturbances, damage to the auditory system, psychosocial effects, and effects on cognitive performance. The damage caused to an organism is strictly dependent on the context. Those who are in a disco are subjected to noises of the order of 100 dB, and the risk incurred is the damage to hearing (temporary or permanent tinnitus, damage to the outer or inner ear [7]); a family living near the disco will, for example, experience sleep disturbances but certainly will not report hearing damage.

1.3.1   WHO

The *WHO* is a specialised agency, established on 7 April 1948, whose main purpose is to safeguard the health of the world's population. The *WHO* has its own offices in 147 nations around the world. Among the active programmes within the association, one concerns the protection of the world's population from damage caused by excessive environmental noise. According to the *WHO,* a negative effect due to noise is defined as a temporary or long-term change in the physiology and morphology of an organism, resulting in damage to functional capacity, or in reducing the ability to compensate for additional stress, or in increasing the organism's susceptibility to other harmful effects present in the environment, [8].

The *WHO* has drawn up guidelines on the maximum levels of noise that should be present in certain environments, reported in Table 1. The maximum noise is reported in $LA_{eq}\,T$, with $T = 16$ h for noise during the day and $T = 8$ h for noise during the night. For some types of noise, for which a total energy sum is not sufficient but for which it is necessary to limit the noise associated with the single event (for example, the noise of a passing plane), the value LA_{MAX} is also reported. The various effects that noise causes on humans are analysed individually below.

Table 1 Guidelines for environmental noise in specific environments [9]

Type of environment	Health effects	$LA_{eq}T$	LA_{MAX}
Outside of dwelling	Severe annoyance	55 dB (16 h)	
Outside of dwelling	Moderate annoyance	50 dB (16 h)	
Apartments, inside of dwellings	Difficulty in understanding conversations, moderate annoyance	35 dB (16 h)	
Bedrooms	Sleep disturbances	30 dB (8 h)	45 dB
School classrooms and nurseries	Difficulty in understanding conversations, learning disturbances	35 dB stay time	
Rest areas in nurseries	Sleep disturbances	30 dB during rest	45 dB
School outdoor spaces	Annoyance	55 dB during recreation	
Hospitals,	Sleep disturbances	30 dB (8 h)	40 dB
Patient rooms	Annoyance	30 dB (16 h)	
Hospitals, operating rooms	Interference with work and recovery	As low as possible	
Industrial areas, commercial areas and generally busy areas	Hearing damage	70 dB (24 h)	110 dB
Venues, ceremonies and other events	Hearing damage	100 dB(4 h)	110 dB
Speaker systems, both indoor and outdoor	Hearing damage	85 dB (1 h)	110 dB
Listening to music through headphones	Hearing damage	85 dB (1 h)	110 dB
Impulsive sounds due to toys, firearms and fireworks	Hearing damage (adults) Hearing damage (children)	–	140 dB 120 dB

1.3.2 Effects on the Cardiovascular System

Numerous studies are carried out on groups of people living or working in noisy areas, and it is found that already in Europe alone, 50% of the population lives in environments that exceed the 55 dB LA_{eq} 16 h limit suggested by the *WHO*, [10].

In a study carried out by an international group of 13 researchers, [11], within the *HYENA* (*Hypertension and Exposure to Noise near Airports*) project, the blood pressure and heartbeats of 140 people living near four major airports—Milan-Malpensa (Italy), Athens (Greece), London-Heathrow (Great Britain) and Arlanda (Sweden)—are monitored. The subjects' blood pressure is measured using *ABPM* (*Ambulatory Blood Pressure Monitoring*), that is, a *Holter* pressure monitor, while the heartbeats are measured using an *ECG* dynamic, always according to *Holter*. On average, it is found that systolic pressure increases by 6.2 mmHg while diastolic pressure increases by 7.4 mmHg if blood pressure is measured within 15 min of a noisy event due to an aeroplane. The noise is measured using special microphones located in the subjects' rooms, and a noisy event is considered as such if it has an

$LA_{\mathrm{MAX}} > 35$ dB. An average increase of an additional 0.63 mmHg in diastolic pressure is recorded if LA_{MAX} is $40 \div 45$ dB (threshold at which awakenings occur during sleep). This indicates that the increase in blood pressure due to the noisy event is a consequence of the activation of the autonomic nervous system, and awakening slightly increases blood pressure. Other studies (particularly on anaesthetised animals) observe that consciousness is not necessary for an increase in blood pressure following a noisy event due to the connection between the auditory pathways and the autonomic nervous system (amygdala, hypothalamus, hippocampus). No noteworthy or statistically valid effects on heart rate variation due to the passage of planes during sleep are highlighted. The increase in blood pressure, as a consequence of noisy events due to air traffic during the night, is reported in numerous other studies, although the exact physiological mechanism underlying this phenomenon is not yet clear.

Another study, [12], also within the *HYENA* project, attempts to provide an answer. Three groups, each consisting of 75 people, are subjected to rotation during the night, to no noisy event (control group), and to 30 and to 60 noisy events due to the passage of a plane during sleep. The passage of the plane is simulated by reproducing the recorded noise, due to the real passage of a plane, with LA_{MAX} of 40 dB. On each of the three mornings following the noise exposure, a blood analysis is carried out to study any variation in plasma composition and the measurement of the *PTT (Pulse Transit Time),* which measures how long a pressure wave needs to transit between two arterial sites. A less *compliant* artery will therefore have a shorter *PTT*. From the results, reported in Figs. 2 and 3 of [12], it is seen that as the number of noisy events increases during the night, the *PTT* decreases and the vascular *compliance* decreases (flow-mediated dilatation) (Fig. 2 of [12]). This factor contributes to the development of endothelial dysfunction, leading to atherosclerosis and hypertension. By administering vitamin C (which has an antioxidant function) to subjects exposed to 30 or 60 noisy episodes at night, the *PTT* values are almost normal. Therefore, it is assumed that it is precisely the production of reactive oxygen species that promotes the onset of cardiovascular system diseases following excessive exposure to noisy episodes. On Fig. 3 of [12], it is also seen how, as the number of noisy episodes increases, the concentration of adrenaline in the blood plasma increases, a hormone activator of the sympathetic nervous system. During sleep, the action of the parasympathetic nervous system prevails, and a concentration of adrenaline in the blood plasma, higher than normal, indicates that the body is under stress. Being a hormone activator of the sympathetic nervous system, adrenaline, among its various effects, promotes vasoconstriction, an increase in heart rate and metabolic activities in general.

1.3.3 Sleep Disturbances

The acoustic stimuli that come from the ear, which, through the auditory pathways, reach the brain area dedicated to hearing, are then branched to both the cerebral cortex (the seat of cognitive mental functions) and to the autonomic nervous system. The body, therefore, responds to this type of stimulus even without the intervention of consciousness. Some types of noise are particularly harmful to sleep: a constant background noise is much less annoying than a sudden noise or a series of short-duration noisy events. The most common and annoying type of environmental noise detrimental to the quality of sleep is that derived from air traffic and wind turbines in exposed areas. Road and rail traffic, in some areas, is present even at night, but is often reduced in quantity at night compared to the day. As previously seen, the *ICAO* has imposed a ban on night air traffic for some airports.

In Table 1 the maximum recommended values for bedrooms are reported: LA_{eq} 8 h = 30 dB and LA_{MAX} = 45 dB. The effects of night noise are divisible into two categories: short-term and long-term, [1]. Short-term effects include:

1. The alteration of the number and duration of the various phases of sleep, compared to physiological values.
2. The presence of night awakenings and the fact that noise exceeds a certain *LA* $_{MAX}$ (the value in *dB* is subjective).
3. The alteration of physiological parameters, such as blood pressure, vasoconstriction and respiratory rate, following the modification of the *REM (rapid eye movement)* phase of sleep.
4. The decrease in total sleep, due to early awakenings in the morning or difficulty in falling asleep late in the evening.

Long-term effects include:

1. The fatigue the day after, due to deprivation of hours and quality of sleep. This can have a negative long-term impact on the social and working lives of individuals and, in the most extreme cases, be the cause of various types of accidents.
2. The need for compensation periods for lost rest.
3. The chronic increase in concentrations of stress hormones and reactive oxygen species in the blood plasma. Points 2 and 3 are caused by prolonged sleep deprivation due to noises that regularly occur at night. Noise from air traffic and from wind turbines certainly falls into these categories.

In the study [13], some subjects are divided into three groups, one as a control and two subjected to 64 recorded noisy events. One of the two groups, subjected to noisy events, receives them at a *Sound Pressure Level (SPL)* of 45 dBA and the other at an *SPL* of 65 dBA. It is seen how the electroencephalograms (*EEG*) of the subjects subjected to noise present a greater number of awakenings and *arousals* (cognitive states of alertness in which one is more responsive to external stimuli) and variations in the physiological duration of the various phases of sleep. In particular, the duration and number of *REM* phases are reduced, which seems to be the phase of sleep in which information acquired during the day is memorised. Also, the total duration of stage 4 of sleep is reduced, which is the stage used for the body's energy recovery.

In literature, two hypnograms are reported, one concerning a physiological, or undisturbed, sleep cycle, and one referring to a disturbed sleep cycle (by noise). In disturbed sleep, first of all, there is a delay in the appearance of stage 1 of sleep, which translates into greater difficulty in falling asleep. In physiological sleep, stage 4 prevails in the first part of the cycle, and the duration of the *REM* phases increases in the second part, while, in disturbed sleep, stage 4 is fragmented and the *REM* phases are shorter. In disturbed sleep, there are also some awakenings (*W*). In 1997, the *Health Council of the Netherlands*, following the numerous studies carried out in the sector, estimated the trend of the number of night awakenings in a year as a function of LA_{eq} 8 h.

1.3.4 Damage to the Auditory System

The auditory system is a particularly delicate system and, if overstimulated, can easily suffer damage, even permanent. For the auditory system to suffer damage, the sound exposure must be high, for example, with an exposure exceeding 70 dB LA_{eq} 24 h in busy areas, Table 1. Particularly risky are impulsive noises of the order of 140 dB LA_{MAX}, such as those due to explosions and gunshots. The main negative effects on the auditory system (which can be either transitory or permanent) are:

1. The decrease in the audible frequency spectrum.
2. The increase in the audibility threshold.
3. The *tinnitus* (the subject hears whistles, buzzing or hissing), which can also be caused by oxidative stress.

This can lead to serious problems at work and in everyday life, as reduced hearing ability is considered one of the most severe handicaps in social relationships.

1.3.5 Psychosocial Effects

The dominant psychosocial effect, caused by noise, is certainly the annoyance caused by continuous or intermittent traffic noise, wind generators and road and rail traffic, in addition to various other noises of various origins. By comparing the results obtained in various studies on the subject and through questionnaires filled out by subjects located in areas with different intensities and types of noise, a trend in the percentage of people strongly disturbed by noise is hypothesised. It is clear that the percentage of people who are strongly annoyed increases with the increase of noise, and that air traffic is the most annoying.

In the long term, the effects can be dramatic, particularly for sensitive individuals. An exposure to continuous noise can indeed cause annoyance to change into depressive episodes, mood swings, increased aggression, and headaches, [6]. It is also shown that there is a higher admission of people to facilities dedicated to the treatment of psychiatric diseases, [14], in areas exposed to high levels of noise. Large-scale studies are conducted to demonstrate the association between noise levels and consumption of tranquillisers and sleeping pills, [15].

1.3.6 Effects on Cognitive Performance

It is common sense that annoying noise can reduce concentration and compromise the quality of certain activities. The most affected category is certainly that of children at school, who are in the most delicate learning phase, in which they learn to read, write and study. A very important study, [16], was carried out in Munich when the old airport closed in 1992 to then inaugurate the new one, a unique opportunity to conduct studies on noise due to air traffic in the surrounding areas. In the study, three sessions (waves) with tests on a control group, not subjected to noise from air traffic, and on a group whose school was located near the old airport. The first session is carried out before the closure of the old airport; the second and third are carried out after the closure of the latter. In the literature, the results of the test for reading are reported, in which the children of the two groups are made to read texts of increasing difficulty, and the errors made are counted. It is evident that the number of errors is higher for the group of children subjected to air traffic compared to those made by the control group. This difference is only present in the first session; after the closure of the airport, the errors made in the two subsequent sessions are the same for the two groups. Both groups are also subjected to a memory test, in which they have to remember parts of texts and diagrams seen the day before. Again, the score made by the group of children subjected to air traffic is lower than that obtained by the control group in the first session, while in the second and third sessions, this difference is no longer present. Relationships are hypothesised between the intensity of the noise to which the mother is exposed during pregnancy and the weight of the newborn. The results obtained for the topic are still unclear.

2   Auditory System

The human auditory system is an extremely complex system. The tympanic membrane, for example, is able to perceive and translate vibrations with a wavelength of the order of 10^{-11} m. The reaction time of the auditory system is also very fast, about 1000 times faster than the responses of the visual photoreceptors. The magnitudes of a sound wave that interest the human ear are frequency and amplitude, which correspond to pitch and perceived intensity. The frequencies audible by the human ear are included in an interval that goes from 20 Hz to 20 kHz. This range varies greatly from individual to individual: a child can perceive frequencies even beyond 20 kHz, while the upper limit of frequencies audible for an adult is about 15–17 kHz. The auditory system is generally divided into the external, middle and inner ear.

2.1   Outer Ear

The external ear is formed by the pinna, the auricle and the auditory canal. The combination of these three structures aims to convey sound energy onto the tympanic membrane. The configuration of the ear is such that it selectively increases sound pressure from 30 to 100 times for frequencies around 3 kHz due to passive resonance effects. This makes the human ear particularly sensitive to frequencies from 800 Hz to 5 kHz. A sound wave of 40 dB of *SPL* is perceived differently by the human ear if oscillating at 200 Hz or 2 kHz. The inverse problem is interesting, that is, understanding what intensity *(SPL)* a sound wave must have, at different frequencies, to be perceived in the same way as a wave that has a reference frequency and intensity.

In 1933, *Fletcher and Munson*, [17], experimentally determined the audiograms, later called *Fletcher-Munson,* which show the *SPL (dB)* as a function of frequency *(Hz)*. Taking as a reference the frequency of 1 kHz, at various *SPL*, they determine a series of isophonic curves. Each isophonic curve presents a value in *phon*. For example, the isophonic curve at 40 *phon* indicates the value of *SPL* that each sound wave must have, at various frequencies, to be perceived in the same way as a sound wave of 40 dB at 1 kHz. The audiograms are corrected and refined over the years, and the current international standard is reported in the standard *ISO 226– 2003*.

The pinna of the external ear does not make a significant contribution to channelling the sound. Due to the ridges that characterise it, the pinna filters frequencies above 4 kHz differently depending on the point of origin of the sound. The sound is channelled through the auricle into the auditory canal, which resonates at a frequency equal to a wavelength of about four times the length of the auditory canal itself. The length of the auditory canal, on average, is about 2.4 cm, although it varies depending on some parameters, such as age and gender. The resonance frequency is, as already anticipated, around 3 kHz. Mehrgardt and Mellert [18] conduct a series of experiments to evaluate the transfer function of the outer ear and the impedance of the eardrum, with the source in front of the listener. The *head-related transfer function (HRTF)* is compared with the curves obtained previously.

2.2 Middle Ear

The middle ear consists of the eardrum membrane, the tympanic cavity, the ossicle chain, and the Eustachian tube (or auditory tube). The purpose of the middle ear is to adapt the impedance so that the sound transmitted through the outer ear can efficiently reach the inner ear. While in the outer ear, sound transmission occurs through air; the cochlea, in the inner ear, contains an aqueous solution, and the impedance of the cochlear fluid is about 4000 times higher than the impedance of air. Direct sound transmission between these two media results in a loss of 99.9% of the acoustic energy, transmitting only 0.01%, which corresponds to a loss of about 40 dB compared to the original signal. The middle ear solves this problem and ensures the transmission of sound energy across the air-liquid boundary, increasing the pressure exerted at the eardrum membrane by almost 200 times before it reaches the inner ear. The ossicle chain connects the eardrum membrane to the oval window of the cochlea. The eardrum membrane has a surface area of about $60 \div 80$ mm^2, while the oval window has a surface area of 3.2 mm^2. If the ossicle chain were a simple, rigid connection between the two membranes, the pressure would be amplified by a factor of 19–25. However, the ossicle chain is structured to minimise the inertia of the system and to act as a lever relative to its own axis. This system increases the amplification by a factor of 1.3.

The last characteristic that amplifies the pressure signal is the curved membrane mechanism. The resting eardrum membrane is extroverted outward, and upon the arrival of the sound wave, the pressure distribution on the surface generates a force F_1 that moves the eardrum membrane a distance D_1. The end of the hammer (the handle) is attached to the eardrum membrane, and when it undergoes a displacement D_1 the hammer moves by $D_2 < D_1$. For a lever system, it holds that $F_1 * D_1 = F_2 * D_2$, and therefore, the force F_2 that the hammer exerts on the ossicle chain is greater than F_1. The mechanism of the curved membrane increases the amplification by a factor of 2. The combination of various amplification factors leads to a total amplification of about 60, that is, about 36 dB, a value that varies from individual to individual and depends on the frequency of the signal. From the transfer function of the middle ear, it is noted that the resonance frequency is between 800 Hz and 1500 Hz. Up to 5–6 kHz, the transfer function still maintains its gain, indicating that energy transmission is maximum, more or less, up to these frequencies. A certain amount of sound energy can be transmitted through the bones of the skull directly to the cochlea.

In the middle ear, there are two very small muscles, the stapedius and the tensor tympani, responsible for the acoustic reflex, to which various functions are attributed. The stapedius has a protective function, as it prevents excessive movement of the stapes following very loud sounds, reducing the efficiency of transmission of the ossicle chain. The contraction of the stapedius and the tensor tympani muscles also increases the stiffness of the tympanic membrane, increasing its acoustic impedance and making the transmission of sound waves less effective. According to the theory of perception, these two muscles serve to modulate the frequency response of the middle ear in order to increase the attention of the organism, changing the intensity and frequency of characteristic environmental sounds. The acoustic reflex is able to attenuate very effectively low frequencies, improving the signal/noise ratio, as the physiological noises of the organism itself are mainly at low frequencies, such as the noises produced by chewing and by speaking. These latter activities seem, in fact, to selectively activate the tensor tympani muscle.

2.3   Inner Ear

The cochlea of the inner ear ensures that the energy of sound waves is transformed into nerve impulses, also mechanically analysing the frequencies involved and breaking down complex signals into simpler components. The base of the stapes transmits the acoustic pressure signals, recorded by the outer and middle ear, to the cochlea, leaning on the oval window, which, along with the round window, are the two regions of the cochlea not surrounded by bone. The spiral of the cochlea is divided into three scales: vestibular, tympanic, and middle. The vestibular and tympanic scales are connected by the end of the cochlea, the helicotrema, and both are thus bathed by the same fluid, the perilymph (rich in Na^+ and poor in K^+). The middle scale is instead bathed by the endolymph, poor in Na^+ and rich in K^+. The movement of the base of the stapes on the oval window causes a movement of the internal fluid, which causes a slight extroflexion of the round window and deforms the cochlear partition. The organ of Corti is responsible for translating this signal.

The organ of Corti is composed of a basilar membrane, where the ciliated cells, internal and external, are located and on which the tectorial membrane rests. The basilar membrane and the tectorial membrane have two different anchoring points, and the movement of fluid inside the cochlea, following a sound signal, causes a displacement of the basilar membrane so that the tectorial membrane slides (forward or backwards) on the stereocilia of the ciliated cells. The ciliated cells are able to detect movements of the dimensions of an atom and have response times of the order of microseconds. They also quickly adapt to constant stimuli, managing to filter out background noise to some extent. The stereocilia have a progressively higher height along the ciliated cell, and the tips of the stereocilia are connected to each other via filamentous structures, known as tip links. On the tips of the stereocilia are ion channels that are opened or closed, depending on the direction of the relative movement between the basilar and tectorial membranes. The ion channels allow the entry of ions K^+, so the ciliated cell depolarises or hyperpolarises based on the opening or closing of the ion channels. The oscillation, back and forth, of the stereocilia modulates the ionic flow, determining a graded receptor potential that follows the movements of the stereocilia. The endolymph is at a potential of 80 mV and the perilymph at 0 mV, while the resting potential of ciliated cells is between $-$ 45 mV $\div -60$ mV. When the channels of K^+ open, the depolarisation of the ciliated cell causes the opening of voltage-dependent calcium channels on the cell membrane, and the consequent entry of Ca^{2+} causes a release, at the basal end, of vesicles containing neurotransmitters that act on the terminations of the acoustic nerve. As some channels of K^+ are open in the resting state, their closure and the consequent hyperpolarisation make the potential of the ciliated cell biphasic; that is, to a sinusoidal stimulus, the ciliated cell responds with a sinusoidal potential. Up to about 3 kHz, the temporal course of the response faithfully follows that of the stimulus, while at higher frequencies, the response is a constant component.

The potentials of the endolymph and perilymph are such that they can restore the resting potential of the ciliated cells (the details of the process are reported in [19]). The basilar membrane is wider and more flexible at the apical end and is narrower and stiffer at the basal end. Regardless of where the energy is provided, the movement always starts from the rigid end and then propagates to the more flexible end. The acoustic stimulus generates in the cochlea a travelling wave of the same frequency, which propagates from the base towards the apex of the basilar membrane, increasing in amplitude and slowing down in speed until the point of maximum displacement is reached, determined by the frequency of the sound stimulus. The points that respond to high frequencies are at the base of the basilar membrane, where it is stiffer, while the points that respond to low frequencies are at the apex, giving rise to a topographic representation of frequencies, a phenomenon called tonotopy. Complex sounds produce vibrations equivalent to the superposition of the vibrations generated by individual sounds, achieving a spectral decomposition of sounds. This system is extremely useful in revealing the harmonic combinations that distinguish different natural sounds, including speech.

3   Basics of Acoustics

The following paragraph, with some of its elementary solutions and the main quantities used to describe noise, refers to [20].

3.1   Wave Equation

A sound wave is a small-amplitude oscillatory motion in a compressible fluid. At each point in the fluid, the passage of the sound wave front causes an alternation of compression and rarefaction. The variables under consideration (in bold, the vectors) can be broken down into a mean component and a fluctuating one, much smaller than the average,

$$p = p_0 + p', \quad \rho = \rho_0 + \rho', \quad \boldsymbol{u} = \boldsymbol{u_0} + \boldsymbol{u'} \tag{1}$$

Substituting Eq. 1 into the mass conservation equation, ignoring the second-order terms of the fluctuations and considering the fluid at rest ($\boldsymbol{u_0} = 0$) in which the sound waves are transmitted, we obtain

$$\frac{\partial \rho'}{\partial t} + \rho_0 \, \Delta \cdot \boldsymbol{u'} = 0 \tag{2}$$

For the conservation of momentum equation, similar considerations apply, and, since the fluid is at rest, it is possible to ignore the contribution due to viscous and the convective terms,

$$\rho_0 \frac{\partial \boldsymbol{u'}}{\partial t} + \Delta p' = 0 \tag{3}$$

The Reynolds number, associated with the acoustic wave with wavelength, $\lambda = \frac{c}{f}$, is

$$Re_\lambda = \frac{\lambda^2 f}{\upsilon} \tag{4}$$

A sound wave with a frequency of $f = 1000$ Hz in air has a Re_λ of the order of 10^7. This means that, for the viscous effects to be comparable with those inertial, the sound wave must propagate for at least 10^7 wavelengths, which corresponds to about 3000 km, distances enormously greater than those of interest in the simulations carried out. It is also noted that the lower frequencies are attenuated less than the higher frequencies.

Heat transfer due to the passage of a sound wave is negligible, and propagation can be assumed as an adiabatic transformation. This allows to relate the pressure fluctuation with the density fluctuation

$$p' = \left(\frac{\partial p}{\partial \rho_0}\right)_s \rho' \tag{5}$$

At this point, the potential function of the velocity fluctuation is introduced

$$\boldsymbol{u}' = \nabla \varnothing \tag{6}$$

which, substituted into Eq. 3, gives

$$p' = -\rho_0 \left(\frac{\partial \varnothing}{\partial t}\right) \tag{7}$$

Substituting Eq. 5 into Eq. 2, we obtain

$$\frac{\partial p'}{\partial t} + \rho_0 \left(\frac{\partial p}{\partial \rho_0}\right) \nabla \cdot \boldsymbol{u}' = 0 \tag{8}$$

Finally, the substitution of Eq. 7 into Eq. 8 gives

$$\frac{\partial \left(-\rho_0 \frac{\partial \varnothing}{\partial t}\right)}{\partial t} + \rho_0 \left(\frac{\partial p}{\partial \rho_0}\right) \nabla \cdot (\nabla \varnothing) = 0 \tag{9}$$

Remembering the link between the divergence of the gradient and the Laplacian, Δ, the homogeneous wave equation is obtained

$$\frac{\partial^2 \varnothing}{\partial t^2} - c^2 \Delta \varnothing = 0 \tag{10}$$

with c being the speed of sound. This equation has general validity, and, as the velocity potential is defined, it is valid for pressure, density and velocity fluctuations.

The speed of sound, for an adiabatic transformation, is given by

$$c = \sqrt{\left(\frac{\partial p}{\partial \rho_0}\right)_s} \tag{11}$$

and Eq. 5 becomes

$$p' = c^2 \rho' \tag{12}$$

The solution of the homogeneous wave equation, Eq. 10, depends only on the initial and boundary conditions. If there is a source term, q, Eq. 10 becomes

$$\frac{\partial^2 \varnothing}{\partial t^2} - c^2 \Delta \varnothing = q \tag{13}$$

3.1.1   Plane Wave

If the phenomenon depends only on the axial coordinate, x, as well as time, we are in the presence of a plane wave, and Eq. 10 becomes

$$\frac{\partial^2 \varnothing}{\partial t^2} - c^2 \frac{\partial^2 \varnothing}{\partial x^2} = 0 \tag{14}$$

which can be rewritten as

$$\left(\frac{\partial}{\partial t} - c\frac{\partial}{\partial x}\right)\left(\frac{\partial}{\partial t} + c\frac{\partial}{\partial x}\right)\varnothing = 0 \tag{15}$$

to indicate that there are two waves propagating in opposite directions. If two new variables are introduced, η and ξ,

$$\eta = t + \frac{x}{c} \tag{16}$$

$$\xi = t - \frac{x}{c} \tag{17}$$

from which

$$t = \frac{1}{2}(\eta + \xi) \tag{18}$$

$$x = \frac{c}{2}(\eta - \xi) \tag{19}$$

it follows that Eq. 14 becomes the following plane wave equation

$$\frac{\partial^2 \varnothing}{\partial \eta \partial \xi} = 0 \tag{20}$$

Integration of Eq. 20 with respect to ξ gives

$$\frac{\partial \varnothing}{\partial \eta} = F(\xi) \tag{21}$$

where $F(\xi)$ is an arbitrary function of ξ. Integrating again, we have

$$\phi = f_1(\eta) + f_2(\xi) \tag{22}$$

with $f_1(\eta)$ and $f_2(\xi)$ being arbitrary functions of η and ξ, respectively. Returning to the initial variables, x and t, Eq. 22 is

$$\phi = f_1\left(t + \frac{x}{c}\right) + f_2\left(t - \frac{x}{c}\right) \tag{23}$$

where it is evident that the solution is composed of two waves that propagate in opposite directions. In particular, f_1 propagates along the negative direction of x, while f_2 propagates along the positive direction of x. Only the component along x of the velocity fluctuation, $\mathbf{u}' = \nabla \phi$, is different from 0; therefore

$$u'_x = \frac{\partial \varnothing}{\partial x} \tag{24}$$

The direction of the velocity fluctuations is, therefore, in the same direction as the propagation one, which is why these waves are also called longitudinal. A relationship between pressure fluctuations and velocity can now be derived.

Consider only the wave that propagates in the positive direction of x

$$\phi = f\left(t - \frac{x}{c}\right) \tag{25}$$

From Eq. 24, we get

$$u'_x = -\frac{1}{c}f'\left(t - \frac{x}{c}\right) \tag{26}$$

while from Eq. 7

$$p' = -\rho_0 f'\left(t - \frac{x}{c}\right) \tag{27}$$

From Eqs. 26 and 27, we obtain the relationship between pressure fluctuations and velocity

$$u'_x = \frac{p'}{\rho_0 c} \tag{28}$$

In the case where the sound wave is monochromatic, all quantities are periodic functions of time. Using the Fourier series decomposition, a sound wave can be represented as a sum of monochromatic sound waves at different frequencies, expressed as the real part of a complex quantity.

For the potential of velocity

$$\phi = Re\left\{A \cdot e^{-i\omega\left(t - \frac{x}{c}\right)}\right\} \tag{29}$$

where A is the complex amplitude, defined as

$$A = a \cdot e^{j\alpha} \tag{30}$$

with a and α real constants. By making explicit the real part, we finally get

$$\phi = a \cdot \cos\left(\omega\frac{x}{c} - \omega t + \alpha\right) \tag{31}$$

in which a is the amplitude of the wave, and the argument of the cosine is the phase of the wave. We can introduce the wave vector

$$\boldsymbol{k} = k \cdot \boldsymbol{n} \tag{32}$$

where $\boldsymbol{n}$ is the unit vector in the direction of propagation and k is the wave number, defined as

$$k = \frac{\omega}{c} \tag{33}$$

The velocity potential can then be written as.

$$\phi = a \cdot \cos(\boldsymbol{k} \cdot \boldsymbol{x} - \omega t + \alpha) \tag{34}$$

If we consider the equation for pressure fluctuations, Eq. 34 can be written as.

$$p' = a \cdot \cos(\boldsymbol{k} \cdot \boldsymbol{x} - \omega t + \alpha) \tag{35}$$

3.1.2 Spherical Wave

If the phenomenon depends only on the radial coordinate, r, as well as time, we are dealing with a spherical wave. Eq. 10 becomes

$$\frac{\partial^2 \varnothing}{\partial t^2} - c^2 \frac{1}{r^2} \frac{\partial}{\partial r}\left(r^2 \frac{\partial \varnothing}{\partial r}\right) = 0 \tag{36}$$

In analogy to the previous case of a plane wave, we look for a solution of the type

$$\varnothing = \frac{f(r,t)}{r} \tag{37}$$

Equation 37 is a solution of Eq. 36, and, by substituting it into Eq. 36, we get

$$\frac{1}{r}\frac{\partial^2 f}{\partial t^2} - c^2 \frac{1}{r^2} \frac{\partial}{\partial r}\left[r^2 \frac{\partial\left(\frac{f}{r}\right)}{\partial r}\right] = 0 \tag{38}$$

from which

$$\frac{\partial^2 f}{\partial t^2} - c^2 \frac{\partial^2 f}{\partial r^2} = 0 \tag{39}$$

Equation 39 is analogous to Eq. 14, and the solution is of the type

$$f = f_1\left(t + \frac{r}{c}\right) + f_2\left(t - \frac{r}{c}\right) \tag{40}$$

Using Eq. 37, the solution for the potential of speed is derived

$$\varnothing = \frac{f_1\left(t + \frac{r}{c}\right)}{r} + \frac{f_2\left(t - \frac{r}{c}\right)}{r} \tag{41}$$

in which the first term represents a wavefront that propagates towards the origin of the coordinate r, while the second term represents a wavefront that propagates from the origin. Unlike the plane wave, in which the amplitude remains constant, in the spherical wave, the amplitude decreases with the distance from r. The expression of the outgoing spherical wave is

$$\varnothing = \frac{f\left(t - \frac{r}{c}\right)}{r} \tag{42}$$

In the case where the outgoing spherical wave is monochromatic, it can be written

$$\phi = \mathrm{Re}\left\{ \frac{A \cdot e^{-i\omega\left(t - \frac{r}{c}\right)}}{r} \right\} \tag{43}$$

By explicating the real part, with the same procedure as the monochromatic plane wave, the potential of speed can be rewritten as

$$\phi = \frac{a \cdot \cos(\boldsymbol{k} \cdot \boldsymbol{r} - \omega t + \alpha)}{r} \tag{44}$$

and the equation for pressure fluctuations as

$$p' = \frac{a \cdot \cos(\boldsymbol{k} \cdot \boldsymbol{r} - \omega t + \alpha)}{r} \tag{45}$$

In a symmetrical spherical wave there are no tangential motions, and the moment equation, linearised and in spherical coordinates, is

$$\rho_0 \left(\frac{\partial u'_r}{\partial t}\right) + \frac{\partial p'}{\partial r} = 0 \tag{46}$$

The relationship between radial velocity fluctuations and pressure can be derived

$$u'_r = \frac{p'}{\rho_0 c}\left(1 - \frac{1}{kr}\right) \tag{47}$$

This last relationship is analogous to Eq. 28 if

$$k\,r = \frac{2\pi r}{\lambda} \gg 1 \tag{48}$$

that is, when the distance of the wavefront from the origin is much greater than the wavelength considered. This condition is called the far field and will be verified in the simulations taken into account.

3.1.3 Green's Function for the Solution of Non-homogeneous Wave Equation

One of the most used methods for the solution of the non-homogeneous wave equation is the use of the *Green* function, $G(\mathbf{x}, t \mid \mathbf{y}, \tau)$, defined as the response of the flow to a point impulsive source in space and time

$$\frac{1}{c^2}\frac{\partial^2 G}{\partial t^2} - \Delta G = \delta(\mathbf{x} - \mathbf{y})\delta(t - \tau) \tag{49}$$

where $\mathbf{x}$ is the position of the receiver, $\mathbf{y}$ that of the source, $\delta(t - \tau)$ indicates that the signal arrives at the receiver with the delayed time τ, and δ is the Dirac impulse, defined by

$$\int_{-\infty}^{+\infty} F(t)\,\delta(t)\,dt = F(0) \tag{50}$$

with F any function not singular at 0. In particular,

$$\delta(\mathbf{x} - \mathbf{y}) = \delta(x_1 - y_1)\,\delta(x_2 - y_2)\,\delta(x_3 - y_3) \tag{51}$$

To solve Eq. 49, the Fourier transform and its related properties are used. For simplicity, the source is considered located at the origin of the reference system, and it is obtained

$$-k^2 G - \Delta G = \delta(\mathbf{x})e^{-i\omega\tau} \tag{52}$$

Using spherical coordinates, we have

$$-\left[\frac{1}{r^2}\frac{d}{dr}\left(r^2\frac{dG(r)}{dr}\right) + k^2\,G(r)\right] = \delta(r)e^{-i\omega\tau} \tag{53}$$

The use of the total derivative is possible because, in free space, a point source radiates like an acoustic monopole, and, as seen, the radiation does not depend on the direction θ. The new variable is used

$$\tilde{G} = r\, G \tag{54}$$

so that

$$\frac{d\tilde{G}}{dr} = r\,\frac{dG}{dr} + G \tag{55}$$

Multiplying everything by r

$$r^2\frac{dG}{dr} = r\frac{d\tilde{G}}{dr} - \tilde{G} \tag{56}$$

and substituting it into Eq. 53, for $r \neq 0$, we get

$$\frac{1}{r}\frac{d}{dr}\left(r\frac{d\tilde{G}}{dr} - \tilde{G}\right) + k^2\tilde{G} = 0 \tag{57}$$

Carrying out the derivative, we get

$$\frac{d^2\tilde{G}}{dr} + k^2\tilde{G} = 0 \tag{58}$$

and the solution is of the type

$$\tilde{G} = C_1 e^{-ikr} + C_2 e^{ikr} \tag{59}$$

therefore

$$G = C_1\frac{e^{-ikr}}{r} + C_2\frac{e^{ikr}}{r} \tag{60}$$

To have a physically sensible solution, at infinity, G must tend to 0, and therefore

$$C_2 = 0 \tag{61}$$

To determine C_1, we integrate Eq. 52 with respect to a sphere of volume V_0 and radius r_0, including the source:

$$-\int_{V_0} \Delta G\, dV - k^2\int_{V_0} G\, dV = e^{-i\omega\tau}\int_{V_0}\delta(r)dV = e^{-i\omega\tau} \tag{62}$$

Using a corollary of the divergence theorem, carrying out the integrals in spherical coordinates and making the radius r_0 of the sphere tend to 0, we get

$$C_1 = \frac{e^{-i\omega\tau}}{4\pi} \tag{63}$$

whose solution is

$$G = \frac{e^{-ikr}e^{-i\omega\tau}}{4\pi r} = \frac{e^{-i\omega(\tau+r/c)}}{4\pi r} \tag{64}$$

Anti-transforming and using the properties of the Fourier anti-transform, we finally get

$$G = \frac{1}{4\pi r}\,\delta\!\left(t - \tau - \frac{r}{c}\right) \tag{65}$$

In the case where

$$r = |\mathbf{x} - \mathbf{y}| \tag{66}$$

we get

$$G = \frac{1}{4\pi\,|\mathbf{x} - \mathbf{y}|}\,\delta\!\left(t - \tau - \frac{|\mathbf{x} - \mathbf{y}|}{c}\right) \tag{67}$$

Knowing the solution of Eq. 49 and introducing a generic function $F(\mathbf{x}, t)$, all the solutions of the equations of this type are known

$$\frac{1}{c^2}\frac{\partial^2 F}{\partial t^2} - \Delta F = q(\mathbf{x}, t) \tag{68}$$

Calling

$$L(D) = \left(\frac{1}{c^2}\frac{\partial^2}{\partial t^2} - \Delta\right) \tag{69}$$

Equation 49 can be rewritten as

$$L(D)\,G = \delta(\mathbf{x} - \mathbf{y})\delta(t - \tau) \tag{70}$$

while Eq. 68 as

$$L(D)\,F = q(\mathbf{x}, t) \tag{71}$$

So, if G is a solution of Eq. 70, then the solution of Eq. 71 is

$$F = \int_{-\infty}^{+\infty} \int_{-\infty}^{\tau} G\, q\,(\mathbf{y}, \tau)d\tau\, d\mathbf{y} \tag{72}$$

with

$$d\mathbf{y} = dy_1 dy_2 dy_3 \tag{73}$$

Recalling that

$$x(t) = \int_{-\infty}^{+\infty} \delta(t - \tau)x(\tau)\, d\tau \tag{74}$$

we obtain

$$L(D)\, F = \int_{-\infty}^{+\infty} \int_{-\infty}^{\tau} L(D)\, G\, q\,(\mathbf{y}, \tau)d\tau\, d\mathbf{y} \tag{75}$$

and therefore

$$L(D)\, F = \int_{-\infty}^{+\infty} \int_{-\infty}^{\tau} \delta(\mathbf{x} - \mathbf{y})\delta(t - \tau)\, q\,(\mathbf{y}, \tau)d\tau\, d\mathbf{y} = q(\mathbf{x}, t) \tag{76}$$

confirming what is asserted by Eq. 72. The use of the Green's function for free space is extremely useful for the solution of the non-homogeneous wave equation.

3.1.4 Generalised Functions

Some results obtained for generalised functions are introduced, that is, functions in which there are discontinuities. The theory of distributions is extremely vast, and attention is only paid to the derivation of generalised functions. Consider a function $f(x)$ that has a discontinuity in x_0, such that at x_0 there is a jump equal to

$$\Delta f = f(x_{0+}) - f(x_{0-}) \tag{77}$$

The generalised derivative of $f(x)$ is defined as (the bar over indicates a generalised variable)

$$\frac{\overline{df}}{dx} = \overline{f}'(x) = f'(x) + \Delta f\, \delta(x - x_0) \tag{78}$$

with $f'(x)$ ordinary derivative of $f(x)$ and $\delta(x)$ Dirac delta. Also, given

$$a < x_0 < x \tag{79}$$

the integration of Eq. 78 gives

$$\int_a^x \overline{f'}(x)\, dx = f(x) - f(a) \neq \int_a^x f'_x\, dx \tag{80}$$

This result is fundamental because it indicates that $\overline{f'}(x)$ takes into account the discontinuity in x_0, while f'_x does not. Consider the function of *Heaviside*, defined by

$$F(x) = \begin{cases} 0 & \text{for } x < 0 \\ 1 & \text{for } x > 0 \end{cases} \tag{81}$$

one has

$$F'(x) = 0 \tag{82}$$

$$\Delta F = 1 \tag{83}$$

in $x = 0$, and therefore

$$\overline{F'}(x) = \delta(x) \tag{84}$$

Similarly, one has

$$\int_a^x \overline{F'}(x)\, dx = \int_a^x \delta(x)\, dx = F(x) \tag{85}$$

very different from

$$\int_a^x F'(x)\, dx = 0 \tag{86}$$

Equation 78 expresses the process of differentiation for generalised functions, useful if a mathematical function describes a physical phenomenon in which discontinuities are present. It is important to extend the discussion to functions in three dimensions. Let $q(\mathbf{x})$ be a function of $\mathbf{x} = (x_1, x_2, x_3)$ that has a discontinuity along the surface $s(\mathbf{x}) = 0$. The jump Δq of $q(\mathbf{x})$ along $s(\mathbf{x}) = 0$ is

$$\Delta q = q(s = 0_+) - q(s = 0_-) \tag{87}$$

In analogy to Eq. 78, the generalised partial derivative is introduced as

$$\frac{\overline{\partial q}}{\partial x_i} = \frac{\partial q}{\partial x_i} + \Delta q \, \frac{\partial s}{\partial x_i} \, \delta(s) \tag{88}$$

with $\frac{\partial q}{\partial x_i}$ the ordinary partial derivative of $q(\mathbf{x})$. From Eq. 88, the generalised gradient of $q(\mathbf{x})$ is

$$\overline{\nabla} q = \nabla q + \Delta q \nabla s \, \delta(s) \tag{89}$$

There are three useful properties of generalised functions:

1. If $Q(\mathbf{x})$ is an arbitrary function,

$$\int_{\mathbf{x}} Q(\mathbf{x}) |\nabla s| \, \delta(s) \, d\mathbf{x} = \int_{s=0} Q(\mathbf{x}) \, dS \tag{90}$$

where the integral on the right of Eq. 90 is a surface integral, with $Q(\mathbf{x})$ calculated precisely on the surface $s = 0$;

2. For generalised functions, the operation of derivation can be inverted

$$\frac{\overline{\partial}^2 Q(\mathbf{x})}{\partial x_i \partial x_j} = \frac{\overline{\partial}^2 Q(\mathbf{x})}{\partial x_j \partial x_i} \tag{91}$$

3. In case the integral limits are not a function of x_i, it holds

$$\frac{\overline{\partial}}{\partial x_i} \int Q(\mathbf{x}, \mathbf{y}) \, d\mathbf{y} = \int \frac{\overline{\partial}}{\partial x_i} Q(\mathbf{x}, \mathbf{y}) \, d\mathbf{y} \tag{92}$$

3.2 Energy, Intensity and Momentum of Sound Wave

3.2.1 Energy

The energy, ε, per unit volume, is the sum of the kinetic and internal energy per unit of mass, ϵ,

$$\varepsilon = \frac{1}{2} \rho \, u^2 + \rho \, \epsilon \tag{93}$$

disregarding the potential energy, as it is not significant for the applications used in the present simulations. The internal energy, per unit of mass, ϵ, for oscillatory motions can be expressed as the sum of the average and fluctuating components

$$\epsilon = \epsilon_0 + \epsilon' \tag{94}$$

Developing $\rho \, \epsilon$ in a Taylor series up to the second order and neglecting the third order fluctuations of kinetic energy, we obtain, for the energy, ε

$$\varepsilon = \frac{1}{2}\rho_0 u'^2 + \rho_0\epsilon_0 + \rho'\frac{\partial(\rho\,\epsilon)}{\partial\rho_0} + \frac{1}{2}\rho'^2\frac{\partial^2(\rho\,\epsilon)}{\partial\rho_0^2} \tag{95}$$

If the propagation of the sound wave occurs as an adiabatic transformation, it can be written

$$\frac{\partial(\rho\,\epsilon)}{\partial\rho_0} = h_0 \tag{96}$$

with h_0 specific enthalpy at rest, and

$$\frac{\partial^2(\rho\,\epsilon)}{\partial\rho_0^2} = \frac{c^2}{\rho_0} \tag{97}$$

Equation 95 becomes

$$\varepsilon = \frac{1}{2}\rho_0 u'^2 + \rho_0\epsilon_0 + h_0\rho' + \frac{1}{2}\frac{c^2\rho'^2}{\rho_0} \tag{98}$$

The term $\rho_0\,\epsilon_0$ represents the internal energy of the fluid at rest, not due to the passage of the sound wave. The term $h_0\rho'$ represents the change in energy due to the change in local density, which disappears with the integration over the entire volume of fluid. In conclusion, a sound wave causes a series of compressions and rarefactions of the fluid, leaving the total mass unchanged.

The energy of a sound wave, in a volume of fluid, due to its passage is

$$E = \int_{V_0}\left(\frac{1}{2}\frac{c^2\rho'^2}{\rho_0} + \frac{1}{2}\rho_0 u'^2\right)dV \tag{99}$$

In the case of a plane wave, the sound energy per unit volume becomes

$$\varepsilon_s = \rho_0 u'^2 \tag{100}$$

If a non-monochromatic wave is considered to be composed of the sum of monochromatic waves, the average energy is equal to the sum of the average energies due to the various monochromatic components.

3.2.2 Intensity

The energy flux density vector is given by

$$\dot{\varsigma} = \rho\,u\left(\frac{1}{2}u^2 + h\right) \tag{101}$$

Integrating this quantity with respect to a surface that limits the volume of fluid considered, the power that flows through this surface is obtained. The power associated with the sound wave is due to the fluctuations of the quantities considered; therefore, also for the specific enthalpy, it is necessary to distinguish a mean component and a fluctuating one

$$h = h_0 + h'$$ (102)

Neglecting the second-order fluctuations, we obtain

$$\dot{\varsigma} = \rho_0 h_0 \boldsymbol{u} + \rho_0\, h'\, \boldsymbol{u}$$ (103)

For small fluctuations of h', it is found that

$$h' = \frac{p'}{\rho_0}$$ (104)

and Eq. 103 can be rewritten as

$$\dot{\varsigma} = \rho_0 h_0 \boldsymbol{u} + p'\, \boldsymbol{u}$$ (105)

If the fluid is at rest, $\boldsymbol{u_0} = 0$, and Eq. 105 becomes

$$\dot{\varsigma} = \rho_0 h_0 \boldsymbol{u}' + p'\, \boldsymbol{u}'$$ (106)

The integral of $\dot{\varsigma}$ on the surface of the fluid volume represents the power flowing through the surface

$$\dot{E} = \int_{S_0} (\rho_0 h_0 \boldsymbol{u}' + p'\, \boldsymbol{u}') \cdot \boldsymbol{n}\, dS$$ (107)

The average power is given by

$$\overline{\dot{E}} = \int_{S_0} \overline{(p'\,\boldsymbol{u}')} \cdot \boldsymbol{n}\, dS$$ (108)

The average intensity is

$$\overline{\boldsymbol{I}} = \overline{p'\,\boldsymbol{u}'}$$ (109)

while the instantaneous intensity is given by

$$\boldsymbol{I} = p'\boldsymbol{u}'$$ (110)

For a plane wave, the instantaneous intensity is rewritten as

$$\boldsymbol{I} = c\,\rho_0 u'^2 = c\,\varepsilon_s\,\boldsymbol{n}$$ (111)

where $\boldsymbol{n}$ is the unit vector normal to the direction of propagation.

3.2.3   Momentum

The momentum, per unit volume of fluid, is

$$\mathbf{q} = \rho\,\mathbf{u} \tag{112}$$

Considering the fluctuations and using Eq. 12, with $\mathbf{u_0} = 0$, we obtain

$$\mathbf{q} = \rho_0\mathbf{u}' + \frac{p'\mathbf{u}'}{c^2} \tag{113}$$

and, based on Eqs. 6 and 110, we have

$$\mathbf{q} = \rho_0\nabla\varnothing + \frac{\mathbf{I}}{c^2} \tag{114}$$

The total momentum is obtained with the integral of the momentum per unit volume in the fluid volume

$$\mathbf{Q} = \int_{V_0} \left(\rho_0\nabla\varnothing + \frac{\mathbf{I}}{c^2}\right) dV \tag{115}$$

Based on the corollary of the divergence theorem, the first term of the integral is

$$\int_{S_0} (\rho_0\varnothing)\cdot \mathbf{n}\, dS = 0 \tag{116}$$

as $\varnothing$ is null outside the volume of fluid traversed by the sound wave. The total momentum of the sound wave is therefore

$$\mathbf{Q} = \frac{1}{c^2}\int_{V_0} \mathbf{I}dV \tag{117}$$

3.3   Acoustic Monopoles, Dipoles and Quadrupoles

Acoustic sources can be modelled as combinations of ideal acoustic sources. The three main models of acoustic source are described.

3.3.1   Acoustic Monopole

The acoustic monopole is the simplest acoustic source, which emits sound waves equally in all directions. A point-like sphere, the radius of which oscillates infinitesimally, and a point, in which alternating additions and subtractions of infinitesimal mass occur, are examples of ideal acoustic monopoles. An acoustic source can be considered ideal if the characteristic dimension of the acoustic source is much smaller than the wavelength of the radiated sound. Therefore, the radiation of the acoustic monopole is independent of the direction, θ, and depends on the distance, r, from the source and the time, t. An ideal source can be described as an outgoing spherical wave. For the analytical description of the acoustic monopole, the complex field is used

$$\phi(r,t) = A \frac{e^{-i(\omega t - kr)}}{r} \tag{118}$$

The amplitude A is determined by the boundary conditions on the surface of the monopole.

Consider the general case of a sphere of radius, a, oscillating at a certain pulsation, ω. The oscillations have a velocity normal to the surface of the type

$$u'_a = U_a e^{-i\omega t} \tag{119}$$

with U_a as the average velocity of the oscillations of the surface of the sphere. From Eq. 6, the radial velocity fluctuations are

$$u'_r = \frac{\partial \phi}{\partial r} \tag{120}$$

Substituting Eq. 119 into Eq. 120, calculated at $r = a$, we obtain the complex amplitude

$$A = -U_a \left(\frac{a^2}{1 - i k a} \right) e^{-i k a} \tag{121}$$

and therefore

$$\phi(r,t) = -\frac{U_a}{r} \left(\frac{a^2}{1 - i k a} \right) e^{-i [\omega t - k(r - a)]} \tag{122}$$

In literature, the concept of source strength is often used, defined as

$$Q e^{i \omega t} = \int_S u'_a \cdot \mathbf{n} \, dS \tag{123}$$

which indicates the amount of fluid displaced by the oscillating sound source. In the case of the oscillating sphere, Eq. 123 becomes

$$Q = 4 \pi a^2 U_a \tag{124}$$

Equation 122 can be rewritten as

$$\phi(r,t) = -\frac{Q}{4\ \pi\ r}\left(\frac{1}{1-i\,k\,a}\right)e^{-i\,[\omega t - k(r-a)]} \tag{125}$$

from which we obtain the analogous relationship for pressure fluctuations

$$p' = -i\ \frac{1}{4\ \pi\ r}\left(\frac{1}{1-i\,k\,a}\right)e^{-i\,[\omega t - k(r-a)]} \tag{126}$$

In the far field, that is where

$$k\,r \gg 1 \tag{127}$$

Eq. 126 becomes

$$p' = -i\ \frac{1}{4\ \pi\ r}e^{-i\,[\omega t - k\,r]} \tag{128}$$

which confirms that the sound radiation of the acoustic monopole does not depend on the direction θ. The modulus of the pressure fluctuation of the acoustic monopole is worth

$$|p'| = \frac{1}{4\ \pi\ r} \tag{129}$$

The first two Figures of [21] show the diagram of a monopole radiation, normalised to the value obtained at the angle $\theta = 0°$, and the snapshot of a monopole's radiation.

3.3.2 Acoustic Dipole

Two acoustic monopoles with the same source strength but in phase opposition and separated by a small distance, d, such that

$$k\,d \ll 1 \tag{130}$$

constitute an acoustic dipole. Unlike the acoustic monopole, there are no additions or subtractions of fluid mass because the addition of fluid mass from one of the two monopoles is compensated by the subtraction of mass from the other monopole in phase opposition. The fluid around the dipole therefore moves back and forth between the two monopoles. It is therefore the net force applied to the fluid that causes the radiation of sound energy. Furthermore, as is analytically demonstrated, the radiation of the acoustic dipole, in addition to the distance from the source, r, and time, t, also depends on the direction θ.

Figure 1 shows the scheme of an acoustic dipole, where 1 and 2 are two monopoles, distant d; r_1 and r_2 are the distances of the monopoles from point P; and Θ is the angle between the segment d and the direction r.

The distances that separate the two monopoles from P are

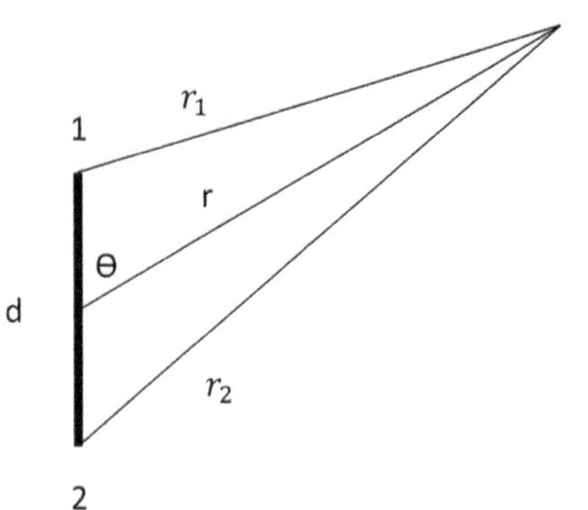

Fig. 1 Acoustic dipole

1. $r_1{}^2 = r^2 + \frac{d^2}{4} - r\,d\,\cos\theta$
2. $r_2{}^2 = r^2 + \frac{d^2}{4} + r\,d\,\cos\theta$

Considering P located in the far field and given that

$$k\,d \ll 1 \tag{131}$$

the two distances can be rewritten as

1. $r_1 \approx r - d\,\cos\theta$
2. $r_2 \approx r + d\,\cos\theta$

The pressure fluctuations are given by the sum of the fluctuations of the two individual acoustic monopoles of equal strength and in opposition in phase

$$p' = p'_1 + p'_2 \tag{132}$$

and therefore

$$p' = -i\,\frac{}{4\,\pi\,r_1}\,e^{-i\,[\omega t - k\,r_1 - \pi/2)]} - i\,\frac{}{4\,\pi\,r_2}\,e^{-i\,[\omega t - k\,r_2 - \pi/2)]} \tag{133}$$

Given that

$$r \gg d \tag{134}$$

it can be assumed

$$r_1 \approx r_2 \approx r \tag{135}$$

while, for the phase terms, this approximation is not acceptable. Equation 133 can therefore be rewritten as

$$p' = -i \frac{}{4\,\pi\,r} e^{-i\,(\omega t - k\,r)} \left[e^{i\,(k\,d/2\;\cos\theta - \pi/2)} + e^{-i\,(k\,d/2\;\cos\theta - \pi/2)} \right] \tag{136}$$

Using Euler's formulas, we obtain

$$p' = -i \frac{}{4\,\pi\,r} e^{-i\,(\omega t - k\,r)}\, 2\, \cos\left(\frac{k\,d}{2}\,\cos\theta - \frac{\pi}{2} \right) \tag{137}$$

and remembering that

$$\cos(\alpha - \pi/2) = \sin\alpha \tag{138}$$

it can be written

$$p' = -i \frac{}{4\,\pi\,r} e^{-i\,(\omega t - k\,r)}\, 2\, \sin\left(\frac{k\,d}{2}\,\cos\theta \right) \tag{139}$$

Finally, remembering that, for α small,

$$\sin\alpha \approx \alpha \tag{140}$$

and given that

$$k\,d \ll 1 \tag{141}$$

the expression of the pressure fluctuation for the acoustic dipole becomes

$$p'(r, t, \theta) = i \frac{}{4\,\pi\,r} e^{-i\,(\omega t - k\,r)} \cos\theta \tag{142}$$

It is evident how the sound radiation depends on the direction θ, while the magnitude of the pressure fluctuation of the acoustic dipole is

$$|p'| = \frac{}{4\,\pi\,r} e^{-i\,(\omega t - k\,r)} \cos\theta \tag{143}$$

The second two Figures of [21] show the radiation diagram of an acoustic dipole normalised to the value obtained for the angle $\theta = 0°$, and the snapshot of dipole radiation.

3.3.3 Acoustic Quadrupole

The acoustic quadrupole is, as an extension of the acoustic dipole, composed of two acoustic dipoles with the same source strength and out of phase, separated by a small distance, D, such that

$$k\,D \ll 1 \tag{144}$$

As in the case of the acoustic dipole, there are no additions or subtractions of fluid mass, but there is also no net force, and it is the fluctuation of stresses that causes the radiation of sound waves from the source. The radiation of the acoustic quadrupole depends on the distance from the source, r, the time, t, and the direction, θ. The acoustic quadrupole can be of two types: lateral and longitudinal. The lateral acoustic quadrupole is composed of two acoustic dipoles, whose axes are not located on the same line. The expression of the pressure fluctuation for the lateral acoustic quadrupole is

$$p'(r,t,\theta) = -i\,\frac{}{\pi\,r}\,e^{-i\,(\omega t - k\,r)}\cos\theta\sin\theta \tag{145}$$

with modulus

$$|p'(r,t,\theta)| = \frac{}{\pi\,r}\,e^{-i\,(\omega t - k\,r)}\cos\theta\sin\theta \tag{146}$$

In the third, two Figures of [21], the radiation diagram of a lateral acoustic quadrupole is shown, normalised with respect to the value obtained for the angle $\theta = 0°$, and the snapshot of lateral quadrupole radiation. For a longitudinal acoustic quadrupole, the axes of the two dipoles that make it up are located on the same line. The pressure fluctuation for the longitudinal acoustic quadrupole is

$$p'(r,t,\theta) = -i\,\frac{}{\pi\,r}\,e^{-i\,(\omega t - k\,r)}\cos^2\theta \tag{147}$$

with modulus

$$|p'(r,t,\theta)| = \frac{}{\pi\,r}\,e^{-i\,(\omega t - k\,r)}\cos^2\theta \tag{148}$$

The fourth, two Figures of [21] show the radiation diagram of a longitudinal acoustic quadrupole, normalised with respect to the value obtained for the angle, $\theta = 0°$, and the snapshot of longitudinal quadrupole radiation.

Comparing the expressions obtained for the pressure fluctuations of the three cases of acoustic sources and remembering that both d and D are very small distances, it is evident that the acoustic dipole radiates sound waves less effectively than an acoustic monopole, and the acoustic quadrupole is less effective than the acoustic dipole. This phenomenon is even more pronounced at low frequencies, as the term k does not appear in the expression of the acoustic monopole but as a first-order power in the acoustic dipole and as a second-order power in the acoustic quadrupole.

3.4   Units of Sound Measurement

The main units of measurement involved in the study of acoustic phenomena are described below.

3.4.1 Sound Pressure Level

The *Sound Pressure Level (SPL)* measures noise, in decibels, *dB*, relative to a reference value, p_{ref}

$$SPL = 10 \, \log_{10}\left(\frac{p'^2}{p_{ref}{}^2}\right) \tag{149}$$

equal to the human auditory threshold, at 1000 Hz in air,

$$p_{ref} = 20 \, \mu \, Pa \tag{150}$$

3.4.2 Type A Weighting

The human ear has different sensitivity for equal sound levels, depending on the frequency. To account for this, acoustic signals are filtered by the weighting curve, type *A*, which attenuates the value in *dB* of very low and excessively high frequencies, which are perceived less intensely by the human ear. The transfer function of the weighting curve *A* is

$$H(s) = \frac{4 \, \pi^2 \cdot 12200^2 s^3}{(s + 2 \, \pi \cdot 20,6)^2 \cdot (s + 2 \, \pi \cdot 12200)^2 \cdot (s + 2 \, \pi \cdot 107,7)(s + 2 \, \pi \cdot 737,9)} \tag{151}$$

with *s* being a complex variable.

Figure 2 shows the transfer function, $H(s)$, also known as type A weighting, which attenuates low frequencies more than high ones and provides a slight gain, in *dB,* at frequencies where the human ear has the highest sensitivity. Generally, the transfer function $H(s)$ is not used to evaluate signals in *dB* (*A*), but specific tables are used, with corrective factors reported relative to frequency values in octave or 1/3 octave bands.

Table 2 shows an example of applying the type *A* filter to a signal. There are other weighting curves (*B*, *C* and *D*) that will not be discussed here.

3.4.3 Equivalent Continuous Sound Level

The Equivalent Continuous Sound Level (LAeq) is the most suitable method for describing a sound whose intensity varies over time, providing a single value, in *dB,* that takes into account all the sound energy measured in the interval of interest. This corresponds to considering a hypothetical constant noise, which, if replaced by the real noise in the same time interval, results in the same total amount of sound energy.

LA_{eq} is defined as

$$LA_{eq} = 10 \, \log_{10} \frac{1}{T} \int_0^T \frac{p'^2(t)}{p_0{}^2} \, dt \tag{152}$$

where *T* is the time interval considered, and $p'(t)$ is the effective value of the acoustic pressure fluctuation. Generally, the periods *T* are 8 h, 16 h and 24 h. LA_{eq} 24 h is

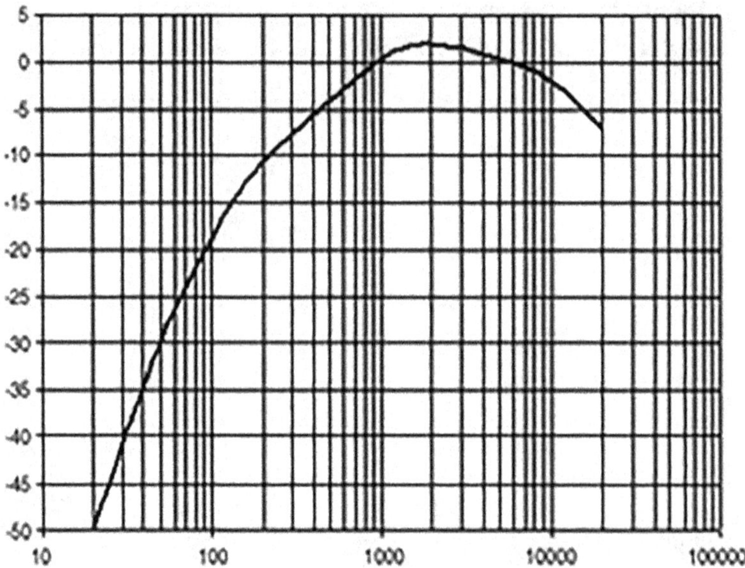

Fig. 2 H(s) (dBA) versus frequency (Hz) [9]

Table 2 Application of the type A filter

Hz	63	125	250	500	1000	2000	4000	8000	16,000
dB	85	88	77	75	70	65	63	60	58
Filter A	−26.2	−16.1	−8.6	−3.2	0	+1.2	+1	−1.1	−6.9
dB(A)	58.8	71.9	68.4	71.8	70	66.2	64	58.9	51.1

normally used to measure the noise to which one person is subjected during an entire day. In order to take into account the greater sensitivity of the human being to noise during rest hours, a corrective factor of +10 dB is added in the time interval between 23:00 and 7:00 and a corrective factor of +5 dB in the time interval between 19:00 and 23:00.

3.4.4 Single Event Level

The *Single Event Level (SEL)* is equivalent to LA_{eq} normalised with respect to 1 s and allows relating noisy phenomena of different durations. It is defined by

$$SEL = 10 \, \log_{10} \frac{1}{T_0} \int_{t_1}^{t_2} \frac{p'^2(t)}{p_{\text{ref}}^2} \, dt \tag{153}$$

with $T_0 = 1$ s, and $t_2 - t_1$ being the observation period of the phenomenon. *SEL* is the constant noise level that, if maintained for *1 s*, has the same acoustic energy as the noisy event observed.

3.4.5 Overall Sound Pressure Level

The *Overall Sound Pressure Level (OASPL)*, which represents the energy contained in the entire resolved frequency spectrum, is defined as

$$E = \int_{f_{res}} S(f)\, df \tag{154}$$

where $S(f)$ is the frequency spectrum of sound pressure fluctuations, normalised with respect to p_{ref}, while f_{res} is the resolved frequency spectrum. All these units of measurement can be calculated in *dB* or in *dB (A)*.

3.4.6 Octave Bands

Acoustic signals can be reported in octave bands or in bands of 1/3 of an octave. The entire energy contained in the whole band is attributed to the central frequency of each band. Each band has a lower cut-off frequency, f_i, and an upper cut-off frequency, f_s. The central band frequency is defined by

$$f_c = \sqrt{f_i f_s} \tag{155}$$

The bandwidth is

$$\Delta f = f_s - f_i \tag{156}$$

while the percentage bandwidth is

$$\%f = \frac{\Delta f}{f_c} \tag{157}$$

For octave bands, the percentage bandwidth is equal to $\sqrt{2}/2$. The cut-off frequencies of the bands are such that the lower cut-off frequency of an octave band is double that of the corresponding previous band. Similarly, the upper cut-off frequencies and the central band frequencies double. As a consequence, the bandwidths also double, moving from one band to the next. Using the octave bands (the discussion is similar for the bands of 1/3 of an octave), the total level, in *dB,* or in *dB (A)*, is expressed as

$$\mathrm{SPL}_{tot} = 10\,\log\left(\sum_{i=1}^{N} 10^{\mathrm{SPL}_i/10}\right) \tag{158}$$

with SPL_i, the value referred to each octave band is in *dB* or in *dB (A)*.

Table 3 reports the octave bands up to the frequencies of interest.

As for the bands of 1/3 octave, the lower cut-off frequency of a band is greater than the corresponding one of the previous band by the factor $\sqrt[3]{2}$. The other quantities follow discussions similar to the octave bands. In each octave band, there are 3 third-octave bands of increasing width.

Table 3 Octave bands

f_i (Hz)	f_c (Hz)	f_s (Hz)	f_i (Hz)	f_c (Hz)	f_s (Hz)
11	16	22	710	1000	1420
22	31.5	44	1420	2000	2840
44	63	88	2840	4000	5680
88	125	177	5680	8000	11,360
177	250	355	11,360	16,000	22,720
355	500	710			

Table 4 Bands of 1/3 octave

f_i (Hz)	f_c (Hz)	f_s (Hz)	f_i (Hz)	f_c (Hz)	f_s (Hz)	f_i (Hz)	f_c (Hz)	f_s (Hz)
14.1	16	17.8	178	200	224	2239	2500	2818
17.8	20	22.4	224	250	282	2818	3150	3548
22.4	25	28.2	282	315	355	3548	4000	4467
28.2	31.5	35.5	355	400	447	4467	5000	5623
35.5	40	44.7	447	500	562	5623	6300	7079
44.7	50	56.2	562	630	708	7079	8000	8913
56.2	63	70.8	708	800	891	8913	10,000	11,220
70.8	80	89.1	891	1000	1122	11,220	12,500	14,130
89.1	100	112	1122	1250	1413	14,130	16,000	17,780
112	125	141	1413	1600	1778	17,780	20,000	22,390
141	160	178	1778	2000	2239			

Table 4 reports the bands of 1/3 octave up to the frequencies of interest.

4   Thermo-Haemo-Dynamics (THD) and Acoustic Analogy

Aeroacoustics is a branch of acoustics and aerodynamics that deals with the aerodynamic noise produced by turbulent flow or by aerodynamic forces acting on a surface. An example of an aeroacoustic phenomenon is the wind tones produced by the wind passing over an obstacle. Just think about what happens when you leave the car window open while travelling: the air hits the window and assumes a turbulent regime, causing a noise perceived as very annoying by the passengers. This field of study is mostly neglected until *Lighthill* publishes the article [22], which does not contain references, testifying to the novelty of the subject. *Ffowcs Williams* and *Hawkings,* referred to in the following as *FW-H,* generalise the analogy of *Lighthill,* introducing, in their theory, the moving surfaces, [23]. Since then, *Computational Aeroacoustic* (*CAA*) has seen huge development.

4.1   Thermo-Haemo-Dynamics (THD) Equations

The equations of fluid dynamics are analysed, then *Lighthill's* analogy and the generalisation obtained by *FW-H* are derived.

4.1.1 Conservation of Mass

A fluid of density ρ and volume $V(t)$ has the mass

$$m = \int_{V(t)} \rho \, dV \qquad (159)$$

The Reynolds transport theorem allows, given an intensive quantity, ψ, and the corresponding extensive quantity, Ψ, to relate them

$$\Psi = \int_{V(t)} \rho \, \psi \, dV \qquad (160)$$

Applying the Reynolds transport theorem to ψ and Ψ, it is obtained

$$\frac{d\Psi}{dt} = \int_{V_0} \frac{\partial(\rho\,\psi)}{dt} \, dV + \int_{S_0} \rho\,\psi\,\mathbf{u}\cdot\mathbf{n}\, dS = \int_{V_0} \left[\frac{\partial(\rho\,\psi)}{\partial t} + \nabla\cdot(\rho\,\psi\,\mathbf{u}) \right] dV \qquad (161)$$

with S_0 surface of the control volume V_0. Applying the Reynolds transport theorem to Eq. 159, we obtain:

$$\frac{dm}{dt} = \int_{V_0} \left[\frac{\partial\rho}{\partial t} + \nabla\cdot(\rho\,\mathbf{u}) \right] dV = 0 \qquad (162)$$

The control volume V_0 is arbitrary, and Eq. 162 must be valid for any V_0. Thus, we obtain the mass conservation equation, in differential form,

$$\frac{\partial\rho}{\partial t} + \nabla\cdot(\rho\,\mathbf{u}) = 0 \qquad (163)$$

In the case where the fluid moves incompressible, Eq. 163 is reduced to

$$\nabla\cdot\mathbf{u} = 0 \qquad (164)$$

In general, Eq. 163 can be written as

$$\frac{\partial\rho}{\partial t} + \frac{\partial(\rho\,u_i)}{\partial x_i} = 0 \qquad (165)$$

4.1.2 Conservation of Momentum

The momentum of a fluid of density ρ and volume $V(t)$, moving at a speed $\mathbf{u}$, is

$$\mathbf{q} = \int_{V(t)} \rho\,\mathbf{u}\, dV \qquad (166)$$

If external forces are applied, with resultant $\mathbf{F}$, the temporal variation of the momentum is

$$\frac{d\mathbf{q}}{dt} = \mathbf{F} \tag{167}$$

The resultant of the forces, $\mathbf{F}$, can be decomposed into volume and surface forces. In this treatment, volume forces are neglected compared to surface forces, which can be expressed in terms of the surface stress tensor associated with them

$$\mathbf{F} = \int_{S(t)} \mathbf{P} \cdot \mathbf{n}\, dS \tag{168}$$

The surface stress tensor can be decomposed into pressure and viscous stresses

$$\mathbf{P} = -p \cdot \mathbf{I} + \tau \tag{169}$$

Applying the Reynolds transport theorem, we obtain

$$\frac{d\mathbf{q}}{dt} = \int_{V_0} \left[\frac{\partial(\rho\,\mathbf{u})}{\partial t} + \nabla \cdot (\rho\,\mathbf{uu}) \right] dV = \int_{V_0} (-\nabla p + \nabla \cdot \tau)dV \tag{170}$$

from which, the control volume V_0 being totally arbitrary, we obtain the momentum balance equation in differential form

$$\frac{\partial(\rho\,\mathbf{u})}{\partial t} + \nabla \cdot (\rho\,\mathbf{uu}) = -\nabla p + \nabla \cdot \tau \tag{171}$$

Equation 171 can be further developed, using the constitutive relation for Newtonian fluids (where the dynamic viscosity μ is not a function of position),

$$\tau = -\frac{2}{3}\mu\,(\nabla \cdot \mathbf{u}) \cdot \mathbf{I} + \mu\,(\nabla\mathbf{u} + \nabla\mathbf{u}^T) \tag{172}$$

which can also be written in tensorial form

$$\tau_{ij} = -\frac{2}{3}\mu\left(\frac{\partial u_k}{\partial x_k}\right)\delta_{ij} + \mu\left(\frac{\partial u_i}{\partial x_j} + \frac{\partial u_j}{\partial x_i}\right) \tag{173}$$

where δ_{ij} is the Kronecker delta, defined as $\delta_{ij} = \begin{cases} 1 \ if \ i=j \\ 0 \ if \ i\neq j. \end{cases}$

Substituting Eq. 172 into Eq. 173, we obtain

$$\frac{\partial(\rho\,\mathbf{u})}{\partial t} + \nabla \cdot (\rho\,\mathbf{uu}) = -\nabla p - \frac{2}{3}\mu\nabla \cdot (\nabla \cdot \mathbf{u}) \cdot \mathbf{I} + \mu\,(\nabla\mathbf{u} + \nabla\mathbf{u}^T) \tag{174}$$

which is written in the most common form

$$\frac{\partial(\rho\,\mathbf{u})}{\partial t} + \nabla\cdot(\rho\,\mathbf{uu}) = -\nabla p - \frac{\mu}{3}\nabla\cdot(\nabla\cdot\mathbf{u}) + \mu\left(\nabla^2\mathbf{u}\right) \qquad (175)$$

and, in the case of incompressible flow ($\nabla\mathbf{u} = 0$), becomes

$$\frac{\partial(\rho\,\mathbf{u})}{\partial t} + \nabla\cdot(\rho\,\mathbf{uu}) = -\nabla p + \mu\left(\nabla^2\mathbf{u}\right) \qquad (176)$$

To derive the acoustic analogies, the momentum balance equation of momentum is expressed, without specifying all the terms of the surface stress tensor, as

$$\frac{\partial(\rho\,u_i)}{\partial t} + \frac{\partial\left(\rho\,u_i u_j\right)}{\partial x_j} = \frac{\partial P_{ij}}{\partial x_j} \qquad (177)$$

or

$$\frac{\partial(\rho\,u_i)}{\partial t} + \frac{\partial\left(\rho\,u_i u_j - P_{ij}\right)}{\partial x_j} = 0 \qquad (178)$$

4.2 Lighthill Acoustic Analogy

The idea of *Lighthill* is to consider noise as the difference between the fluctuating flow and a reference at rest flow. The fluctuating flow will only occupy a small portion of the volume of the external fluid at rest. The difference between the equations governing the two flows is considered the effect of the application of an external fluctuating force field, known if the motion of the fluctuating flow is known. This force field is considered applied to the portion of fluid at rest, which will radiate the noise according to the laws of acoustics. This allows, in numerical simulations, to decouple the propagation of sound from the fluid dynamic motion that generated it.

The approach of *Lighthill's* analogy is exact, as it is the result of a rearrangement of the equations of *Navier-Stokes*. This is fundamental according to *Lighthill*; the acoustic energy is only a small part of the kinetic energy due to the turbulent flow that produced it. Making any kind of a priori approximation could lead to results very different from reality. A limitation in the acoustic analogy of *Lighthill* emerges, as it is valid if the sound is radiated into free space. If, instead, the flow is confined, phenomena of interaction between the emitted noise and the flow itself may arise, which, not being considered by the analogy, would lead to erroneous estimates. Also, the effects of reflection, refraction, absorption and scattering are not taken into consideration. These latter can, however, be considered a posteriori, unless these effects alter the flow itself.

Thanks to this approach, aeroacoustics becomes accessible to numerical calculation, as until then it was studied only experimentally and, when possible, at a theoretical level. The numerical techniques of fluid dynamics are applied to solve the associated aeroacoustic problems, giving rise to the birth of computational aeroacoustics (*CAA*). Some considerations about the orders of magnitude are necessary. Consider a fluid at rest(air, for example). The threshold of hearing, at the frequency of 1 kHz, is SPL $= 0$ dB, while the pain threshold is about SPL $= 130$ dB. This means that pressure fluctuations are worth(in effective value) about 0.00002 Pa and 100 Pa. Compared to the reference value of atmospheric pressure, $p_0 = 101{,}325$ Pa, the value of the fluctuation for the pain threshold is about 1000 times smaller.

4.2.1 Equations

The fluid dynamic quantities are decomposed into the average and the fluctuating components

$$p = p_0 + p', \quad \rho = \rho_0 + \rho', u_i = u_{i0} + u' \tag{179}$$

The following relationships between pressure, density and velocity, c, of sound (in air at *15 °C*, $c_0 = 340$ m/s) are

$$\rho' = \frac{p'}{c^2} \tag{180}$$

$$u'_i = \frac{p'}{\rho_0 c} \tag{181}$$

Therefore, for density and velocity, the fluctuating component is much smaller than the average.

The mass conservation equation, substituting density and velocity from Eq. 179, becomes

$$\frac{\partial (\rho_0 + \rho')}{\partial t} + \frac{\partial \left[(\rho_0 + \rho')(u_{i0} + u'_i) \right]}{\partial x_i} = 0 \tag{182}$$

Constant terms are eliminated in the differentiation, and $u_{i0} = 0$ as the medium is at rest. The term $\rho' u'_i$ is neglected, being of order $O\left(\rho_0 u'\right)_i$, therefore

$$\frac{\partial \rho'}{\partial t} + \rho_0 \frac{\partial u'_i}{\partial x_i} = 0 \tag{183}$$

Doing the same for the momentum balance, we get

$$\frac{\partial \left[(\rho_0 + \rho')(u_{i0} + u'_i) \right]}{\partial t} + \frac{\partial \left[(\rho_0 + \rho')(u_{i0} + u'_i)(u_{j0} + u'_j) - P_{ij} \right]}{\partial x_j} = 0 \tag{184}$$

Since $u_{i0} = u_{j0} = 0$, P_{ij} contains only the term due to hydrostatic pressure, $\rho_0 u_i$ 'u_j'; and ρu_i 'u_j' are negligible compared to hydrostatic pressure, we get

$$\rho_0 \frac{\partial u'_i}{\partial t} + \frac{\partial p'}{\partial x_j} = 0 \tag{185}$$

Deriving Eq. 183 with respect to time, we get

$$\frac{\partial^2 \rho'}{\partial t^2} + \rho_0 \frac{\partial^2 u'_i}{\partial t\, \partial x_i} = 0 \tag{186}$$

and subtracting the divergence of Eq. 185

$$\rho_0 \frac{\partial^2 u'_i}{\partial t\, \partial x_i} + \frac{\partial^2 p'}{\partial x_i\, \partial x_j} = 0 \tag{187}$$

we get

$$\frac{\partial^2 \rho'}{\partial t^2} - \Delta p' = 0 \tag{188}$$

Substituting Eq. 180 into Eq. 188, we get the homogeneous wave equation for pressure

$$\frac{1}{c} \frac{\partial^2 p'}{\partial t^2} - \Delta p' = 0 \tag{189}$$

Considering the turbulent flow portion, the mass conservation equations and momentum balance are now not linearisable. The mass conservation equation is therefore

$$\frac{\partial \rho'}{\partial t} + \frac{\partial (\rho\, u_i)}{\partial x_i} = 0 \tag{190}$$

while the momentum balance equation

$$\frac{\partial (\rho\, u_i)}{\partial t} + \frac{\partial \left(\rho\, u_i u_j - P_{ij}\right)}{\partial x_j} = 0 \tag{191}$$

Similarly to the previous case, deriving Eq. 190 with respect to time, we get

$$\frac{\partial^2 \rho'}{\partial t^2} + \frac{\partial^2 (\rho\, u_i)}{\partial t\, \partial x_i} = 0 \tag{192}$$

and subtracting the divergence of Eq. 191

$$\frac{\partial^2 (\rho\, u_i)}{\partial t\, \partial x_i} + \frac{\partial^2 \left(\rho\, u_i u_j - P_{ij}\right)}{\partial x_i\, \partial x_j} = 0 \tag{193}$$

we get

$$\frac{\partial^2 \rho'}{\partial t^2} + \frac{\partial^2 \left(\rho\, u_i u_j - P_{ij}\right)}{\partial x_i\, \partial x_j} = 0 \tag{194}$$

Subtracting from both sides of Eq. 194 the term

$$c^2 \frac{\partial^2 \left(\rho' \delta_{ij}\right)}{\partial x_i\, \partial x_j} \tag{195}$$

it is obtained

$$\frac{\partial^2 \rho'}{\partial t^2} - c^2 \Delta\rho' = \frac{\partial^2 \left(\rho\, u_i u_j - P_{ij} - c^2 \rho' \delta_{ij}\right)}{\partial x_i\, \partial x_j} = 0 \tag{196}$$

Introducing the *Lighthill* tensor

$$T_{ij} = \rho\, u_i u_j - P_{ij} - \rho' \delta_{ij} \tag{197}$$

and using Eq. 180, the equation for pressure is obtained

$$\frac{1}{c}\frac{\partial^2 p'}{\partial t^2} - \Delta p' = \frac{\partial^2 T_{ij}}{\partial x_i\, \partial x_j} \tag{198}$$

The non-homogeneous wave equation for pressure is obtained, which has as its source term the double divergence of the *Lighthill* tensor. Knowing the motion field also makes the source term of Eq. 198 known. In the case where the turbulent flow has a speed such that the Mach number, M, is small, the *Lighthill* tensor can be reasonably approximated by

$$T_{ij} \approx \rho_0 u_i u_j \tag{199}$$

and the error increases according to M^2. The solution of the equation for pressure is obtained by applying the Green function for free space, obtaining

$$p'(\mathbf{x}, t) = \int_{-\infty}^{+\infty} \int_{-\infty}^{\tau} \left[\frac{1}{4\pi |\mathbf{x} - \mathbf{y}|} \delta\left(t - \tau - \frac{|\mathbf{x} - \mathbf{y}|}{c} \right) \frac{\partial^2 T_{ij}(\mathbf{y}, \tau)}{\partial x_i\, \partial x_j} \right] \tag{200}$$

therefore

$$p'(\mathbf{x}, t) = \frac{1}{4\pi} \int_{-\infty}^{+\infty} \frac{\partial^2 T_{ij}\left(\mathbf{y}, t - \frac{|\mathbf{x}-\mathbf{y}|}{c}\right)}{\partial x_i \, \partial x_j} \frac{d\mathbf{y}}{|\mathbf{x}-\mathbf{y}|} \tag{201}$$

In Eq. 201, $\mathbf{y}$ indicates the position of the noise source, $\mathbf{x}$ the position of the receiver, $d\mathbf{y} = dy_1 \, dy_2 \, dy_3$ and $|\mathbf{x} - \mathbf{y}|$ the distance between receiver and source. It should be noted that the integral is calculated at the delayed time

$$\tau = t - \frac{|\mathbf{x}-\mathbf{y}|}{c} \tag{202}$$

The pressure fluctuation p is calculated at the receiver, and the signal must travel the distance separating the source from the receiver at the speed of sound in the medium considered.

4.3   Ffowcs Williams and Hawkings Acoustic Analogy

In 1969, *Ffowcs Williams* and *Hawkings (FW-H)* generalised the analogy of *Lighthill* by introducing moving surfaces, [23]. In the derivation of the equations, the original procedure is not used, but that obtained in [24], as it is the formulation used in the software employed. To derive the *FW-H* equation, the theory of generalised functions is used. The moving mathematical surface

$$f(\mathbf{x}, t) = 0 \tag{203}$$

encloses the region of fluid generating noise (including solid surfaces, such as a portion of a wind turbine blade).

The purpose of the *FW-H* formulation is to obtain a non-homogeneous wave equation, valid for the volume of fluid outside the mathematical surface. It is assumed that:

1. $f > 0$ outside the mathematical surface;
2. $f < 0$ within the mathematical surface.

The mathematical surface is also such that

$$\nabla f = \widehat{\mathbf{n}} \tag{204}$$

with $\widehat{\mathbf{n}}$ outgoing unit vector, normal to the mathematical surface. In order to obtain the most general treatment possible, the mathematical surface is considered permeable.

The mass conservation equation, considering the partial derivatives as generalised partial derivatives, as there is a discontinuity due to the surface, becomes

$$\frac{\overline{\partial}\rho}{\partial t} + \frac{\overline{\partial}\,\rho u_i}{\partial x_i} = \frac{\partial \rho}{\partial t} + (\rho - \rho_0)\frac{\partial f}{\partial t}\,\delta(f) + \frac{\partial(\rho u_i)}{\partial x_i} + (\rho u_i)\frac{\partial f}{\partial x_i}\,\delta(f) \tag{205}$$

where $(\rho - \rho_0)$ is the jump of ρ on $f = 0$ and ρu_i is the jump of ρu_i on $f = 0$. Calling:

1.

$$n_i = \frac{\partial f}{\partial x_i} \tag{206}$$

the components of the outgoing unit vector normal to the mathematical surface;

2.

$$u_n = u_i n_i \tag{207}$$

the local fluid velocity in the direction normal to the mathematical surface;

3.

$$v_n = -\frac{\partial f}{\partial t} \tag{208}$$

the local velocity normal to the mathematical surface, and using Eq. 165 and Eq. 205 with the generalised fluid dynamic quantities, becomes

$$\frac{\overline{\partial}\rho}{\partial t} + \frac{\overline{\partial}\,\rho u_i}{\partial x_i} = \left[\rho_0 v_n + \rho\,(u_n - v_n)\right]\delta(f) \tag{209}$$

It therefore appears that the discontinuity due to the mathematical surface has introduced a source term in the mass conservation equation, proportional to the local introduction of mass per unit of surface. Proceeding in a similar way for Eq. 178, we obtain

$$\frac{\overline{\partial}\rho u_i}{\partial t} + \frac{\overline{\partial}\rho u_i u_j - P_{ij}}{\partial x_j} = \frac{\partial(\rho u_i)}{\partial t} + (\rho u_i)\frac{\partial f}{\partial t}\,\delta(f) + \frac{\partial\left(\rho u_i u_j - P_{ij}\right)}{\partial x_j}$$

$$+ \left(\rho u_i u_j - \Delta P_{ij}\right)\frac{\partial f}{\partial x_j}\,\delta(f) \tag{210}$$

Reusing Eqs. 178 and 209, we obtain the momentum balance equation with the generalised fluid dynamic quantities:

$$\frac{\overline{\partial}\rho u_i}{\partial t} + \frac{\overline{\partial}\rho u_i u_j - P_{ij}}{\partial x_j} = \left[\rho u_i(u_n - v_n) - \Delta P_{ij}n_j\right]\delta(f) \tag{211}$$

The discontinuity, due to the mathematical surface, has introduced in the equation a source term, proportional to the local introduction of momentum per unit of surface, and to the sum of the intensity of the stresses located on the mathematical surface. At this point, the same steps employed for the derivation of the *Lighthill* equation are carried out, with the only difference that the quantities considered are generalised functions:

1. All quantities are replaced by the sum of their average and fluctuating component, and all necessary observations on the orders of magnitude involved are made.
2. Equation 209 is derived with respect to time, $\frac{\bar{\partial}}{\partial t}$.
3. To subtract the divergence $\bar{\partial}/\partial x_i$ from Eq. 211.
4. To subtract from both sides of the resulting equation

$$c^2 \Delta \rho'. \tag{212}$$

5. To replace the density fluctuation with the pressure fluctuation by using the linear relationship of Eq. 180.

At the end of these operations, the equation for the formulation of *FW-H* for pressure is obtained

$$\frac{1}{c} \frac{\bar{\partial}^2 p'}{\partial t^2} - \overline{\Delta} p' = \frac{\partial}{\partial t} \{ [\rho_0 v_n + \rho \, (u_n - v_n] \, \delta(f) \} + - \frac{\partial}{\partial x_i}$$

$$\times \left\{ \left[\rho u_i (u_n - v_n) + \Delta P_{ij} \widehat{n}_j \right] \delta(f) \right\} + \frac{\partial^2 T_{ij} H(f))}{\partial x_i \, \partial x_j} + - \frac{\partial}{\partial x_i}$$

$$\times \left\{ \left[\rho u_i (u_n - v_n) + \Delta P_{ij} \widehat{n}_j \right] \delta(f) \right\} + \frac{\partial^2 T_{ij} H(f))}{\partial x_i \, \partial x_j} \tag{213}$$

A non-homogeneous wave equation with three source terms is obtained. The generalised partial derivatives of the first two source terms do not have the bar over the derivation symbol, as, being present $\delta \, (f)$, they are generalised quantities. The first source term is called thickness or monopole and is due to the displacement of the fluid, caused by the relative motion with the surface. The second source term is called load, or dipole and is caused by the non-stationary distribution of the pressure field on the surface. The term $\delta \, (f)$, which appears in these two source terms, indicates that they are both due to the surface and disappear on the rest of the domain. The third source term is called the quadrupole term (T_{ij} is again the tensor of *Lighthill*), and $H \, (f)$ indicates that this term is calculated outside the surface.

If the mathematical surface coincides with a solid surface, the velocity u_n coincides with v_n, and the equation of *FW-H* for pressure becomes

$$\frac{1}{c} \frac{\bar{\partial}^2 p'}{\partial t^2} - \overline{\Delta} p' = \frac{\partial}{\partial t} \{ [\rho_0 v_n] \, \delta(f) \} - \frac{\partial}{\partial x_i} \{ [\Delta P_{ij} \widehat{n}_j] \, \delta(f) \} + \frac{\partial^2 T_{ij} H(f))}{\partial x_i \, \partial x_j} \tag{214}$$

In the formulation used, the quadrupole source term is ignored, as it is effectively negligible compared to the monopole and dipole source terms. *Brentner and*

Farassat, [24], report numerical results in which this hypothesis is verified, as well as in [25, 26]. Equation 214 is reduced to

$$\frac{1}{c}\frac{\overline{\partial}^2 p'}{\partial t^2} - \overline{\Delta}p' = \frac{\partial}{\partial t}\{[\rho_0 v_n]\,\delta(f)\} - \frac{\partial}{\partial x_i}\{[\Delta P_{ij}\widehat{n}_j]\,\delta(f)\} \tag{215}$$

Using the function of *Green* for free space, the solution of Eq. 215 is obtained, also called formulation 1 of *Farassat*

$$p'(\mathbf{x},t) = \frac{\partial}{\partial t}\int_{-\infty}^{+\infty}\int_{-\infty}^{\tau}\frac{\rho_0 v_n \delta(g)\delta(f)}{4\pi|\mathbf{x}-\mathbf{y}|}\,d\tau\,d\mathbf{y} - \frac{\partial}{\partial x_i}\int_{-\infty}^{+\infty}\int_{-\infty}^{\tau}\frac{l_i \delta(g)\delta(f)}{4\pi|\mathbf{x}-\mathbf{y}|}\,d\tau\,d\mathbf{y}\frac{\partial}{\partial x_i} \tag{216}$$

where $l_i = \Delta P_{ij}\,\hat{n}_j$ is the local stress acting on the impermeable surface, while

$$\delta(g) = \delta\left(t - \tau - \frac{r}{c}\right) \tag{217}$$

in order to make Eq. 215 more compact. To minimise computation times, maximise the accuracy of the numerical solution, and reduce spatial differentiation, the solution of [24], for the formulation of *FW-H*, is further reworked in [27]. The solution, also called formulation 1A of *Farassat*, is

$$p'(\mathbf{x},t) = p'_T(\mathbf{x},t) + p'_L(\mathbf{x},t) \tag{218}$$

where

$$4\pi p'_T(\mathbf{x},t) = \int_{f=0}\left[\frac{\rho_0(\dot{v}_n + v_{\dot{n}})}{r|1-M_r|^2}\right]_{ret}dS + \int_{f=0}\left[\frac{\rho_0 v_n(r\dot{M}_r + cM_r - cM^2)}{r^2|1-M_r|^3}\right]_{ret}dS \tag{219}$$

$$4\pi p'_L(\mathbf{x},t) = \frac{1}{c}\int_{f=0}\left[\frac{\dot{l}_r}{r|1-M_r|^2}\right]_{ret}dS + \int_{f=0}\left[\frac{l_r - l_M}{r|1-M_r|^2}\right]_{ret}dS$$

$$+\frac{1}{c}\int_{f=0}\left[\frac{l_r(r\dot{M}_r + cM_r - cM^2)}{r^2|1-M_r|^3}\right]_{ret}dS \tag{220}$$

while $p'_T(\mathbf{x},t)$, represents the contribution to the acoustic pressure due to the thickness source term; $p'_L(\mathbf{x},t)$ represents the contribution to the acoustic pressure due to the load source term. The quantities with subscripts indicate that they are the product between a vector and a unit vector ($\hat{r}$ indicates the unit vector in the direction of radiation, while $\hat{n}$ indicates the unit vector normal to the considered surface), for example

$$l_r = \hat{l} \cdot \hat{r} = l_i r_i \qquad (221)$$

M is the Mach number, and a dot above a variable denotes differentiation with respect to time. All integrals are calculated at the delayed time

$$\tau = t - \frac{|\mathbf{x} - \mathbf{y}|}{c} \qquad (222)$$

This formulation has great advantages from a numerical point of view. The disadvantage is in having introduced singularities in the integrals, which, however, can be a problem only for motions with main flow velocities close to that of sound.

5　Conclusions

The extensive literature search covers the effects of noise on human health, describing the damage that can result from excessive exposure to environmental noise. The damage due to environmental noise is multiple: effects on the cardiovascular system, psychosocial effects and effects on cognitive performance, in addition to damage to the auditory system and sleep disturbances. Therefore, the equations used in computational aeroacoustics, *Computational AeroAcoustic (CAA)*, are derived.

References

1. Muzet A (2007) Enviromental noise, sleep and health. Sleep Med Rev 11(2):135–142
2. Berglund B, Lindvall T, Schwela DH (1999) Guidelines for community noise. WHO
3. Pedersen E, Halmstad HI (2003) Noise annoyance from wind turbines: a review. Naturvårdsverket
4. United States of America (1944), Convention on international civil aviation, ICAO archives, Doc 7300
5. ICAO (2008) Guidance on the balanced approach to aircraft noise management, ICAO archives, Doc 9829
6. Vermeer WP, Passchier WF (2000) Noise exposure and public health. Environ Health Perspect 108(Suppl.1):123
7. Axelsson A, Prasher D et al (1999) Tinnitus: a warning signal to teenagers attending discotheques? Noise Health 1(2):1
8. King RP, Davis JR (2003) Community noise: health effects and management. Int J Hyg Environ Health 206(2):123–131
9. La Bella G (2015–16) Il rumore aerodinamico. Simulazioni numeriche con la formulazione di FW-H per lo studio del rumore di generatori eolici, Tesi di Laurea Magistrale in Ingegneria Medica, Università di Roma Tor Vergata, A.A
10. Eriksson C, Roselund M, Pershagen G, Hilding A, Östenson CG, Bluhm G (2007) Aircraft noise and incidence of hypertension. Epidemiology 18(6):716–721
11. Haralabidis AS, Dimakopoulou K, Vigna-Taglianti F, Giampaolo M, Borgini A, Dudley ML, Pershagen G, Bluhm G, Houthuijs D, Babisch W, Velonakis M, Katsouyanni K, Jarup L, the HYENA Consortium (2008) Acute effects of night-time noise exposure on blood pressure in populations living near airports. Eur Heart J 29(5):658–664

12. Schmidt FP, Basner M, Kröger G, Weck S, Schnorbus B, Muttray A, Sariyar M, Binder H, Gori T, Warnholtz A et al (2013) Effect of nighttime aircraft noise exposure on endothelial function and stress hormone release in healthy adults. Eur Heart J:269
13. Basner M, Glatz C, Griefahn B, Penzel T, Samel A (2008) Aircraft noise: effects on macro and microstructure of sleep. Sleep Med 9(4):382–387
14. Wickrama IA, a'Brook MF, Gattoni FEG, Herridge CF (1969) Mental-hospital admissions and aircraft noise. Lancet 294(7633):1275–1277
15. Stansfeld SA (1992) Noise, noise sensitivity and psychiatric disorder: epidemiological and psychophysiological studies. Psychol Med Monogr Suppl 22:1–44
16. Evans GW, Maxwell L (1997) Chronic noise exposure and reading deficits the mediating effects of language acquisition. Environ Behav 29(5):638–656
17. Fletcher H, Munson WA (1933) Loudness, its definition, measurement and calculation. J Acoust Soc Am 5:82–108
18. Mehgardt S, Mellert V (1977) Transformation characteristics of the external human ear. J Acoust Soc Am 61(6):1567–1576
19. Purves D, Augustine GJ, Fitzpatrick D, Hall WC, La Mantia AS, McNamara JO, White LE (2013) Neuroscienze
20. Landau LD, Lifshitz EM (1959) Fluid mechanics. Pergamon Press, New York
21. https://www.acs.psu.edu/drussell/demos/rad2/mdq.html
22. Lighthill MJ (1952) On sound generated aerodynamically I. General theory. Proc R Soc Lond A Math Phys Eng Sci 211:564–587
23. Ffowcs Williams JE, Hawkings DL (1969) Sound generation by turbulence and surfaces in arbitrary motion. Philos Trans R Soc Lond A Math Phys Eng Sci 264(1151):321–342
24. Brentner KS, Farassat F (2003) Modeling aerodynamically generated sound of helicopter rotors. Prog Aerosp Sci 39(2):83–120
25. Casalino D, Jacob M, Roger M (2003) Prediction of rod-airfoil interaction noise using the Ffowcs Williams-Hawkings analogy. AIAA J 41(2):182–191
26. Giret JC, Sengissen A, Moreau S, Sanjosé M, Jouhaud JC (2012) Prediction of the sound generated by a rod-airfoil configuration using a compressible unstructured les solver and FW-H analogy. In: 18th AIAA/CEAS aeroacustics conference, AIAA paper, volume 2058, pag. 4–6
27. Brentner KS (1986) Prediction of helicopter rotor discrete frequency noise: a computer program incorporating realistic blade motions and advanced acoustic analogy approach. NASA Tech Memo

Giulio La Bella was born in Rome on 28 July 1992. He attended the University of Rome "Tor Vergata," where he obtained both his Bachelor's and Master's degrees in Medical Engineering. During his Master's degree, he attended courses taught by Prof. Fabio Gori Ammannati, Thermo-Fluid-Dynamics of Biological Systems and Numerical Calculation for Thermo-Fluid-Dynamic Systems, then continuing the topics covered in the courses with his Master's thesis, titled "Aerodynamic noise. Numerical simulations with the FW-H formulation for the study of wind turbine noise." After graduation, he moved to Germany, Munich, where he worked for three years in the field of industrial automation, dealing with feasibility studies and implementation of artificial vision in industrial quality control. In 2020, he returned to Italy and now works in the field of life science and bioprocess.

Fabio Gori Ammannati was born in Montale (Pistoia) on 5 August 1947 to Emilio Gori and Cesarina Ammannati. On 16 July 2021, he added his mother's surname, Ammannati. He graduated in Chemical Engineering on 11 November 1971 with honours. In December 1971, he won a scholarship for young graduates at the Faculty of Engineering in Bologna, which he undertook from January 1972. In 1974, he became assistant professor and, in 1982, associate professor of Technical Physics at the Faculty of Engineering of the University of Florence, where he taught, also as a contract professor, until 1990. In 1978, he won a scholarship from the *British Council,* which he carried out at the *Imperial College* in London, where he collaborated with *Prof. D. Brian Spalding* on the topic of numerical analysis in the turbulent flow of liquid metals. During his time at the University of Florence, he established international scientific collaborations with *Prof. R. Echigo, Tokyo Institute of Technology, Tokyo, Japan; Prof. D.R. Chaudhary, Department of Physics, University of Rajasthan, Jaipur, India;* International Centre for Theoretical Physics *(ICTP)* in Trieste; *Centralny Osrodek Techniki Medycznej, Warsaw, Poland* and *Prof. T. Aihara, Institute of Fluid Science, Sendai, Japan.* In 1986, he won a CNR scholarship, which he carried out at *Cornell University, Ithaca, New York,* where he collaborated with *Prof. R. Miller* on the topic of ground freezing. In the same year, he became professor in the Faculty of Engineering at the University of Reggio Calabria. In 1988, he was appointed as *professor* at the *University of New York at Stony Brook,* teaching the course, *Introduction to Fluid Dynamics,* in the autumn semester of the same year. During his stay, he collaborated with *Prof. T.F. Irvine Jr.* on the method of measuring thermal conductivity with a thermal probe and on the measurement of the isobaric thermal expansion coefficient of non-Newtonian fluids. In 1990, he moved to the Milano Polytechnic, where he taught Technical Physics and Systems until 1993. From 1991 until 1998, he was an adjunct professor at the Faculty of Engineering of the University of Siena. From 1992 to 2017, he was a professor of Technical Physics in the Faculty of Engineering at the University of Rome Tor Vergata, and from 2017, he is a contract professor. In 1994, he proposed the Research Doctorate (Italian *Ph.D.*) in Energy-Environment Engineering, of which he was the coordinator until 2011. In 1995, he proposed and directed the second-level Master's in Thermo-fluid-dynamics until 2016. From 1998 to 2001, he was the director of the Department of Mechanical Engineering. In 2001, he proposed and coordinated the programme, *Master of Science, Energy Engineering and Thermal and Fluid Dynamics,* in collaboration with the *Department of Mechanical Engineering, College of Engineering, University of Illinois at Chicago, USA,* which takes place at the University of Rome Tor Vergata but awards the title of *Master of Science* from the same American University. Since the late 90s, he has been a coordinator of the joint *Ph.D.* research programme with *Prof. R.J. Goldstein, Department of Mechanical Engineering, University of Minnesota, Minneapolis, USA,* and with *Prof. J.P. Hartnett, Prof. L. Kennedy, Prof. W. Minkowycz* and *Prof. W.M. Worek, Department of Mechanical Engineering, College of Engineering, University of Illinois at Chicago, Chicago, USA.* Since the mid-90s, he has proposed and directed the *Socrates–Erasmus* programme with *Prof. Mayinger, Technical University of Munich, Germany; Prof. van Steenhoven, Eindhoven Technical University, the Netherlands* and *Prof. C. Caro, Imperial College of Science, Technology and Medicine.* During his stay at Tor Vergata, he established scientific collaborations with the *Department of Mechanical Engineering, University of Minnesota, Minneapolis, USA,* collaborating with *Prof. R.J. Goldstein* on Thermo-fluid-dynamics and mass transport in gas turbine blades; the *Energy Resources Center, University of Illinois at Chicago,* collaborating with *Prof. J.P. Hartnett, Prof. W. Minkowycz* and *Prof. W.M. Worek;* the *Department of Mechanical Engineering, College of Engineering, University of Illinois at Chicago,* collaborating with *Prof. L. Kennedy and Prof. W.M. Worek; Duke University, collaborating with Prof. A. Bejan* and *S. Mary's University, Halifax, Nova Scotia, Canada, collaborating with Prof. W. R. Tarnawski.* The bibliographic review of documents and citations, titled *PlosBiology Career,* places him within the top 2% of researchers in the field of *Mechanical Engineering and Transports.* Since 1992, he has been a tutor for over 30 PhD theses, supervisor for over 40 second-level university Master's theses, supervisor for over 70 degree theses (five-year, Specialist and Master's), supervisor for over ten *Master's degree in Mechanical Engineering* from the *University of Illinois at Chicago* and supervisor for over 20 theses *Erasmus-Socrates.*

Since the early 2000s, he has been a reviewer of international research projects on behalf of *Portuguese Science and Technology Foundation (FCT); Czech Science Foundation (GACR); European Science Foundation, Strasbourg, France,* on behalf of *FCT* and the *Shota Rustaveli National Science Foundation, Georgia.* Since 2017, the year of his retirement due to age limits, he has been a contract professor at the University of Rome Tor Vergata, where in the academic year

2024–2025, he has taught Technical Physics for the Master's Degree Course in Medical Engineering. M.D. Salvatore Mangiafico (Florence), scientific head of the NeuroVascular Base Camp, 2021, invited him to give a scientific presentation titled *"The engineering approach to the study of the Circle of Willis"* on 23 September 2021 at the Congress Centre, University of Rome La Sapienza, thus opening up a possible future collaboration.

Acoustics in Cardiac Thermo-Haemo-Dynamics

Alex Colucci and Fabio Gori Ammannati

1   Introduction

This chapter, based on [1], arises from the idea of examining acoustics in the cardiac *CTHD* field, with the intention of studying the physics of heart murmurs. For this scope, it combines what is known in the literature on *Computational Aero Acoustics (CAA)*, applied almost exclusively to the industrial field, to blood flow in the new field of *Computational Thermo-Haemo-Dynamics (CTHD)*. In order to arrive at a model to be investigated, we start from a clinical, physiological and pathophysiological analysis of the causes that lead to the production of noise by a turbulent blood flow. We evaluate which heart valve pathologies are associated, from a clinical point of view, with the generation of murmurs and how they are diagnosed. We evaluate the differences between the physiological and pathological cases, from which we choose to deepen the stenosis and insufficiency of the aortic valve.

The dysfunction of the aortic valve apparatus is mentioned in the literature as one of the greatest clinical interests, especially for the role of this valve in cardiac mechanics. Evaluating studies present in the literature on valvular stenosis and on the noise generated by turbulent flows, we make assumptions to create a model on which to carry out the numerical simulations with numerical software, using the approach of acoustic analogy for the study of noise. We make sure that these assumptions respect the pathophysiology in question and the numerical models to be implemented. In this way, we undertake a study in which a first approach to the application of *CAA* was made in the *CTHD* field to turbulent *Cardiac Thermo-Haemo-Dynamics*.

A. Colucci
Rome, Italy

F. G. Ammannati (✉)
University of Rome Tor Vergata, Rome, Italy
e-mail: gori@uniroma2.it

     755
F. Gori Ammannati (ed.), *Thermo-Haemo-Dynamics in Medical Engineering*,
https://doi.org/10.1007/978-3-031-97214-0_12

2 Physiology and Pathophysiology

2.1 Circulatory System and Heart Valves

The circulatory system can be thought of as the coupling of two hydraulic circuits in series with each other [2]. The first circuit, systemic, carries oxygenated blood from the left heart to all the tissues of the body through the systemic arteries and the blood to be oxygenated from the tissues to the right heart through the systemic veins. The second circuit, pulmonary, transports the blood to be oxygenated from the right heart to the lungs, via the pulmonary artery and its branches, and the oxygenated blood from the lungs to the left heart, via the pulmonary veins.

A fundamental element for the circulation of blood in the two circuits is the heart, a double pulsatile pump that constantly provides the energy necessary for the fluid to reach the tissues. The heart comprises four chambers: the two upper ones are the atria, and the two lower ones are the ventricles. The separation between atria and ventricles, and between ventricles and arteries, is guaranteed by four heart valves, structures designed to work unidirectionally, thus allowing the blood to flow from the atrium to the ventricle and from the ventricle to the artery and then close, to prevent the fluid from going back in the form of regurgitation flow. Two of the heart valves are called cuspid valves, or atrioventricular, as their function is to ensure the complete separation between the atrium and ventricle during ventricular contraction. In particular, the mitral valve, or bicuspid, separates the two left chambers of the heart, while the tricuspid valve separates the right cavities. The other two valves are called semilunar and connect the ventricles with the large efferent vessels. Of these, the aortic valve is present between the left ventricle and the aorta, and the pulmonary valve is between the right ventricle and the pulmonary artery.

2.2 Cardiac Cycle

To understand the physiology and pathophysiology of the sounds generated by the flow within the heart, it is important to refer to all those phenomena that occur from the beginning of a heartbeat to the beginning of the next. The set of events that follow in this time span constitutes the cardiac cycle. Each cycle begins with the spontaneous onset of an action potential in the sinoatrial node. *(Action potentials are rapid changes in the cell membrane potential that go from the normal negative value to a positive value and end with a change that restores the negative potential. As regards the excitation and conduction system of the action potential within the myocardium, we find two types of development of the electrical potential: one concerns the atrial and ventricular fibres, and another concerns the cells of the sinoatrial node, or cells of the pacemaker).* It is located in the upper lateral wall of the right atrium, on the outlet of the superior vena cava. The action potential from which it originated propagates through the atria and then into the ventricles via the atrioventricular bundle. Due to the particular arrangement of the atrioventricular conduction system, there is a delay of about 0.1 s between the passage of the electrical impulse from the atria to the ventricles. This allows the atria to contract in advance and to pump blood into the ventricles before the contraction occurs in them. In this way, the atria act as a primary pump for the ventricles, which generate the pressure necessary to push the blood into the vascular system.

The cardiac cycle consists of two main phases: an initial period of contraction, called systole, followed by a period of relaxation in which the heart fills, diastole. The latter, in turn, includes a phase of isovolumetric relaxation and one of ventricular filling, while the systolic phase is characterised by an isovolumetric contraction followed by ventricular ejection. Immediately after the beginning of ventricular contraction, the pressure rises rapidly, causing the mitral valve to close. Subsequently, the ventricle requires an additional time of about 0.03 s to generate enough pressure to open the aortic semilunar valve against the pressure present in the aorta. Therefore, during this period, although the contraction, there is no emptying of the ventricle. This phase is defined as isovolumetric contraction. This term indicates that in the myocardium, the increase in tension is not accompanied by a shortening of the fibres, which is present from the apex to the base of the heart, and a circumferential elongation. When the pressure in the left ventricle slightly exceeds the aortic pressure, the valve opens, and the blood begins to pour out of the ventricle. About 70% of the emptying occurs in the first third of the ejection period, while the remaining 30% occupies the other two-thirds. The first third is therefore defined as the period of rapid ejection, while the other two-thirds represent the period of slow ejection. During the systole phase, ventricular emptying results in a decrease in volume of about 70 ml, defined as systolic stroke. The remaining volume of blood present in each ventricle, approximately 40–50 ml, is defined as end-systolic. At the end of ventricular systole, relaxation begins, and ventricular pressure drops rapidly. The high-pressure values present in the aorta push the blood back towards the ventricles and cause the closure of the aortic valve. In the following 0.03–0.06 s, the ventricular myocardium continues to relax, although the volume does not change. Thus begins the period of isovolumetric relaxation, and the pressure in the ventricle drops to low diastolic values. Subsequently, the mitral valve opens, and the blood flows rapidly into the left ventricle.

All of this is evident from the analysis of the curve of ventricular volume changes, which shows a clear increase in this phase. This event is defined as the period of rapid ventricular filling and lasts about the first third of the diastolic phase. In the second third, instead, only a small amount of blood flows into the ventricle, represented by the flow coming from the pulmonary veins, which passes directly from the atrium to the ventricle. In the last third, finally, atrial contraction occurs, which contributes, for about 25%, to ventricular filling. The volume of blood present inside the left ventricle at the end of the diastolic phase is about 110–120 ml; a volume defined as end-diastolic. The end-diastolic volume that is ejected during the systole phase is defined as the ejection fraction and normally amounts to about 60% of the total filling volume.

2.3   Physiology of Tones and Pathophysiology of Heart Murmurs

From the phonocardiogram, it is noted that certain cardiac cycle events are associated with some heart sounds, produced by the turbulent flow of blood and by the vibration of the valve cusps in closure, [3]. Indeed, the flow of blood, predominantly laminar, does not produce any sound. If it becomes turbulent, then it produces a noticeable sound, Fig. 6.2 of [4]. The physiological heart tones are linked to the closure of the valves, atrioventricular and semilunar. The first tone corresponds to the closure of the mitral valve, M, and tricuspid, T, which close at the end of diastole and atrial systole and at the beginning of ventricular systole. This first tone, also known as S_1, is a low and prolonged sound and is due to several haemo-dynamic events: low-frequency vibrations originating from the flaps of the mitral and tricuspid valves during the phase of isovolumetric ventricular contraction and two larger amplitude vibrations, M_1 and T_1, due to the closure of the mitral and tricuspid valves. Under physiological conditions, M_1 precedes T_1, the frequency range is 25–45 Hz, the average duration is 0.15 s and S_1 is onomatopoeically represented by $LUBB$. After a brief pause, which in the cardiac cycle corresponds to the phase of blood ejection from the ventricles, follows the second tone, which corresponds to the mechanical closure of the semilunar valves, which close at the end of the ventricular systole and at the beginning of diastole. This second tone, also known as S_2, is short and sharp and is made up of high-frequency vibrations, A_2 and P_2, due to the closure of the semilunar aortic, A, and pulmonary, P, valves. Under physiological conditions, A_2 precedes P_2, the average frequency is 50 Hz, the average duration is 0.12 s, and it is onomatopoeically represented by DUP. The second tone is followed by a long pause that corresponds to diastole, which occupies most of the cardiac cycle.

The opening tones of the pulmonary valve, PO, and aortic, AO, give rise to the click of ejection, perceivable as splitting of the first tone; the opening snap, OS, of tricuspid, TO, and mitral, MO, followed by the third tone, S_3, due to the rapid filling of the ventricles, and finally the fourth tone, S_4, mechanically associated with atrial contraction. The tones S_3 and S_4, under physiological conditions, are not perceived, and a phonocardiogram is necessary. The noises, caused by a turbulent blood flow, are of longer duration compared to those of the tones and noises generated by vibrations of valves and vascular structures, normal or pathological. Based on the moment of the cardiac cycle in which they occur and in relation to cardiac dynamics, these are classified as systolic and diastolic, or systo-diastolic, and, in turn, a murmur that arises in systole or in diastole can be due to regurgitation or ejection, depending on which cardiac structure has generated the latter [5].

We do not consider, in this in-depth study, all the details of the classifications of murmurs, but we focus on some characteristics of two particular ones, to which we refer throughout the discussion. We compare the case called A, or normal, with B, which corresponds to a systolic ejection murmur due to an anatomical-pathological alteration of the aortic valve, defined as aortic stenosis, and with C, which schematises a proto-diastolic murmur, always due to an alteration of the semilunar aortic valve, which in this case is called aortic insufficiency [5].

2.3.1 Flow Through a Partial Obstruction of the Aortic Valve Apparatus

We now focus on the mechanisms of generation of the murmurs [6]. The aortic valve, in the phase of ventricular systole, should be fully open, but in the presence of a stenotic valve, it is not open enough, as it has a reduced valve orifice. In this case, the blood ejected from the ventricle faces a greater resistance to pass into the aorta, and a systolic ejection murmur is identified that grows from the beginning of systole to a mid-systolic peak, then decreases. This is due to the fact that at the beginning, less blood is ejected through the stenotic orifice, while, with the progress of ventricular contraction, the volume of blood passing through the stenosis is greater and the flow turbulent, with the accentuation of the murmur. As the ventricular contraction decreases, so does the blood flow and the murmur. In a rheumatic aortic valve, it is noted how the calcified leaflets create an obstacle to the outflow of blood from the ventricle, thus generating a turbulent flow. Aortic insufficiency is a pathology in which, during diastole, the valve should be fully closed, but not being perfect, there is a regurgitation because the blood flows back from the aorta to the ventricle, causing a murmur. The murmur at the beginning of diastole is particularly strong because the ventricle is empty and the pressure difference with the full aorta is maximum, enough to cause a large regurgitation. As diastole and ventricular filling continue, this pressure difference is minimised, and consequently, it is the regurgitation.

Figures 28-29 of [7] show two examples of regurgitation through two insufficient aortic valves, evaluated with echocardiography, [7]. On the top left, A, a mild aortic insufficiency (*AI*), is shown, while on the bottom right, B, a severe aortic insufficiency (*AI*) is shown. The colour flow Doppler of the aortic valve is done in the midesophageal long-axis view. The aortic insufficiency (*AI*) is graded using the relative ratio of the jet thickness to the diameter of the left ventricular outflow tract (*LVOT*). The measurements are performed at the same site, usually within 0.5–1 cm proximal to the aortic valve (*AV*) plane. The width of the *AI* jet is measured as it crosses the *AV* cusps (*vena contracta*). Figures 28-29 of [7] report that the *vena contracta* measures 0.2 cm in the mild *AI* but 0.5 cm in the severe *AI*. Further, the ratio *AI/LVOT* is smaller than 30% in the mild *AI* but is greater than 65% in the severe *AI*.

3 Fluid Dynamics of Valvular Stenosis

3.1 One-Dimensional Flow Model

The behaviour of blood flow through a stenotic valve can be represented as in Fig. 1, [8]. Compared to the real situation, the model captures the narrowing of the section, but the three bulges in the aortic root downstream of the valve, called *Valsalva* sinuses, one for each valve leaflet, do not appear. The narrowing of the flow section does not end at the orifice but continues until it reaches the minimum section, called *vena contracta*, v_c, or *Effective Orifice Area (EOA)*. From the section of *vena*

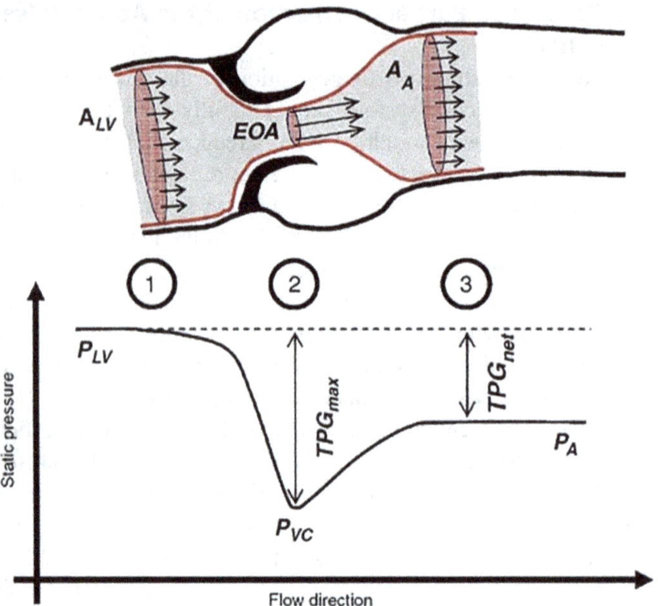

Fig. 1 Blood flow through a stenotic valve, Fig. 1 of [8]

contracta, the flow behaves like a submerged jet; that is, it gradually begins to widen until it occupies again, at the section 3, the entire available section. The length of the 2–3 section is called the reattachment length. There are vortices, adjacent to the jet, in the section where the current expands, which create localised pressure dissipations. The consequences of the presence of the section narrowing can be deduced from the conservation equations of mass and energy.

The volumetric flow rate, Q, is constant along the coordinate on the axis of the duct, s,

$$Q = \cos t \tag{1}$$

while the average speed V

$$V = Q/A \tag{2}$$

increases, in 1–2 section of Fig. 1, with the decrease of A, presenting the maximum speed right on the section of *vena contracta*. In sections 2–3 of Fig. 1, the jet decelerates until the section 3, where it resumes almost the same speed as the current upstream of the orifice, except for the distributed load losses.

The energy conservation equation in an unsteady-state regime, or Euler's equation, [9], between the section 1 and the generic section s is

$$\left(\frac{p_1}{\gamma}+h_1+\frac{V_1{}^2}{2g}\right)-\left(\frac{p_s}{\gamma}+h_s+\frac{V_s{}^2}{2g}\right)=\frac{1}{g}\int_1^s\frac{\partial V}{\partial t}\,ds+\Delta E_{1\rightarrow s} \qquad (3)$$

where the velocity profile is uniform, with V average velocity in the section and $\Delta E_{1\rightarrow s}$ sum of distributed and concentrated head losses. In the problem under consideration, it is assumed

$$h_1=h_S \qquad (4)$$

If we consider the steady-state flow, we can neglect the temporal acceleration. We also neglect the contribution of distributed head losses in the section 1-vc,

$$\Delta E_{1\rightarrow vc}=0,\quad s\in(1,vc) \qquad (5)$$

Conversely, in the section vc-3, the deceleration of the current induces the formation of vortices, which contribute to the losses of localised pressure

$$\Delta E_{vc\rightarrow 3}\neq 0,\quad s\in(vc,3) \qquad (6)$$

Therefore, from Eq. 3, we have

$$\frac{p_{vc}}{\gamma}=\frac{p_1}{\gamma}+\left(\frac{V_1{}^2-V_{vc}{}^2}{2g}\right),\quad s\in(1,vc) \qquad (7)$$

and

$$\frac{p_3}{\gamma}=\frac{p_{vc}}{\gamma}+\left(\frac{V_{vc}{}^2-V_3{}^2}{2g}\right)-\Delta E_{vc\rightarrow 3},\quad s\in(vc,3) \qquad (8)$$

from which it is deduced that the minimum energy is at $s=vc$. As the velocity increases in the length 1-vc, the pressure decreases, reaching its minimum value at the *vena contracta* section, where the speed is maximum. At this point, there is the maximum pressure difference between section 1 and vc, indicated with $\left(\frac{\Delta p}{\gamma}\right)_{max}$ or TPG_{max} (maximum transvalvular pressure gradient)

$$\left(\frac{\Delta p}{\gamma}\right)_{max}=\frac{p_1}{\gamma}-\frac{p_{vc}}{\gamma} \qquad (9)$$

In the length vc-3, the pressure increases again to the exit value, where the pressure difference, $\frac{p_1}{\gamma}-\frac{p_3}{\gamma}$, defined $\left(\frac{\Delta p}{\gamma}\right)_{net}$, or *TPGnet*, represents the net transvalvular pressure gradient between sections 1 and 3. Therefore, the pressure does not return to the initial value due to the presence of vortices that are generated downstream of the contractive vein section.

In the case of unsteady-state flow, the effect of temporal inertia is included in the fluid dynamic model of valvular stenosis. From Eq. 3, with $h_1 = h_3$ and $V_1 = V_3$, the difference in transvalvular pressure is

$$\left(\frac{\Delta p}{\gamma}\right)_{net} = \frac{(p_1 - p_3)}{\gamma} = \frac{1}{g}\int_1^3 \frac{\partial V}{\partial t}\, ds + \Delta E_{1\to3} \tag{10}$$

The energy dissipation between sections 1 and 3, is assumed to be equal to that between the *vena contracta* section, 2, and section 3

$$\Delta E_{1\to3} = \Delta E_{2\to3} = \frac{(p_2 - p_3)}{\gamma} + \frac{V_2^2 - V_3^2}{2\,g} - \frac{1}{g}\int_2^3 \frac{\partial V}{\partial t}\, ds \tag{11}$$

Applying the momentum theorem to the fluid contained in the control volume, between vc and 3, we have

$$(p_{vc} - p_3)A_2 + \rho\, Q\, (V_{vc} - V_3) - \rho\, A_2 \int_{vc}^3 \frac{\partial V}{\partial t}\, ds = 0 \tag{12}$$

where, the first term is related to the pressure forces, the second to the dynamic, or inertial, convective thrust that the incoming and outgoing current exerts on the control volume, and the third is the force related to the local inertia of the fluid. From Eq. 12, dividing by A_2, and considering $Q = V_2\, A_2$, we have

$$\frac{(p_{vc} - p_3)}{\gamma} - \frac{1}{g}\int_{vc}^3 \frac{\partial V}{\partial t}\, ds = \frac{V_2}{g}(V_3 - V_{vc}) \tag{13}$$

which, substituted into Eq. 11, becomes

$$\Delta E_{1\to3} = \Delta E_{vc\to3} = \frac{V_{vc}^2 - V_3^2}{2\,g} + \frac{V_2}{g}(V_3 - V_{vc}) = \frac{(V_{vc} - V_3)^2}{2\,g} \tag{14}$$

Substituting Eq. 14 into 10, we obtain

$$\left(\frac{\Delta p}{\gamma}\right)_{net} = \frac{(p_1 - p_3)}{\gamma} = \frac{(V_2 - V_3)^2}{2\,g} + \frac{1}{g}\int_1^3 \frac{\partial V}{\partial t}\, ds \tag{15}$$

and, remembering the link between flow rate, Q, and average speed in the section, V, Eq. 15 can be rewritten

$$\left(\frac{\Delta p}{\gamma}\right)_{net} = \frac{Q^2}{2\,g\,\mathrm{EOA}^2}\left(1 - \frac{\mathrm{EOA}}{A_3}\right)^2 + \frac{1}{g}\int_1^3 \frac{\partial\,Q/A}{\partial t}\, ds \tag{16}$$

where the area that appears under the integral sign is that of the generic section along the jet, from 1 to 3. Assuming that the geometry of the jet remains constant over time, that is, $A = A\,(s)$, and therefore $\frac{\partial A}{\partial t} = 0$, Eq. 17 can be rewritten as

$$\left(\frac{\Delta p}{\gamma}\right)_{net} = \frac{Q^2}{2\,g\,EOA^2}\left(1 - \frac{EOA}{A_3}\right)^2 + \frac{1}{g}\frac{\partial Q}{\partial t}\int_1^3 \frac{1}{A}\,ds \qquad (17)$$

where the derivative of Q over time is taken out of the integral sign, since the flow rate is independent of s. The execution of the integral itself, however, requires knowledge of the geometry of the jet, that is, to adopt a model for $A(s)$.

An in vitro experimental study is conducted in [10] on the evaluation of the *TPGnet* for the determination of the geometry of the jet, therefore of $A(s)$, to confirm Eq. 17, in which $A(s)$ is inserted as a parameter explained with dimensional analysis. The two constants, obtained from the dimensional analysis, are also calculated experimentally, obtaining

$$\left(\frac{\Delta p}{\gamma}\right)_{net} = \frac{Q^2}{2\,g\,EOA^2}\left(1 - \frac{EOA}{A_3}\right)^2 + \frac{1}{g}\frac{\partial Q}{\partial t}\frac{2\pi}{\sqrt{EOA}}\sqrt{1 - \frac{EOA}{A_2}} \qquad (18)$$

Figure 2 and Fig. 6 of [10] show the comparison between the measured (dashed lines) and calculated (solid lines) trend of instantaneous Δp in bioprostheses through rigid circular orifices for different values of *EOA* and systolic jet. The comparison is satisfactory. This result is also due to the fact that the rigidity of the orifices used is a condition respectful of the assumption, made in the theoretical model, that the geometry of the jet is almost independent of time.

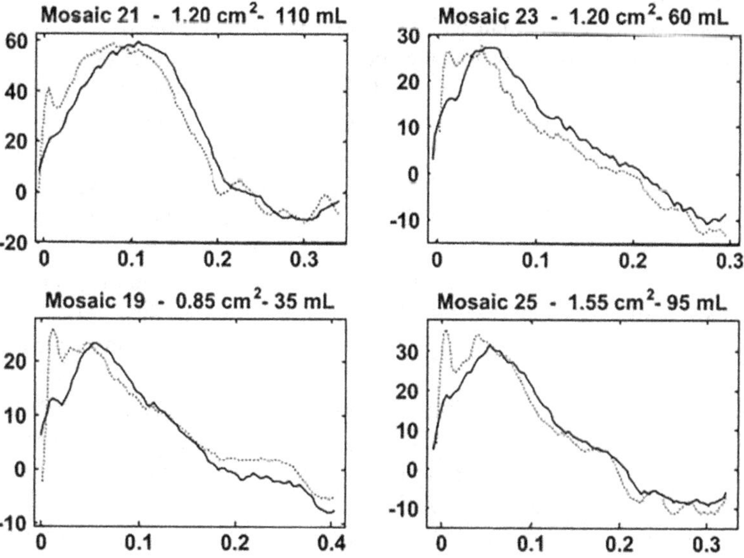

Fig. 2 Instantaneous Δp (mmHg) in bioprostheses as a function of time (s): estimated (solid line) and measured (dotted line), Fig. 6 of [10]

3.2   Behaviour of the Stenotic Valve

One of the most stringent hypotheses, and not at all found in reality, assumed in the fluid dynamic model already described is the staticity of the valve orifice. To evaluate this issue, reference is made to the stenosis of the aortic valve, a pathology most often treated in biomedical literature, but where the study of valve dynamics is quite recent. Significant differences are found between the dynamics of healthy and stenotic valves, mainly related to the duration of transvalvular flow and the phase of opening and closing of the valve. Indeed, a stenotic valve, due to degenerative processes, often has calcified valve leaflets, therefore stiffened. This results in a progressive reduction in the speed of movement of the valve leaflets so that the opening of the valve is always slower as the pathology worsens, reaching, in the ejective systole, a reduced value of the maximum opening of the orifice, which occurs in a longer time compared to that of a physiologically healthy valve.

A study on valve dynamics [11] evaluates the response of the valve to flow variations during systole, which can influence the calculation of the area of the aortic valve in the presence of stenosis and lead to inaccuracies in the evaluation of the latter. The area is evaluated directly by short-range planimetry with the possibility of performing serial planimetries during an ejection phase and analysing the dynamic behaviour of the stenotic aortic valve. The results are compared with measurements derived from the flow (*Doppler*). The average and maximum area, opening and closing rates, and ejection times are analysed on 40 patients with echocardiograms that presented different degrees of aortic stenosis through planimetry *frame-by-frame* of the valve, from the beginning of the opening until complete closure. It is found that the area of the valve varies during ejection. Stenotic valves open and close more slowly than normal, and the period of maximum opening is shorter. In patients with aortic stenosis, surprisingly, the average area is better correlated with the maximum anatomical area than with the average anatomical area.

Figure 3 and Fig. 2 of [11] show the opening and closure of the aortic valve (*AVA, aortic valve area*) of three patients as a function of time, expressed in numbers of video *frames* (each equal to 33 ms) for an ejection period. The valve area is smaller than 1 cm^2 on the left of Fig. 3; the area is between 1 and 2 cm^2 in the centre of Fig. 3; and the area is greater than 2 cm^2 on the right of Fig. 3. The variation of the valve area can be easily appreciated in these nine examples and seems more pronounced in patients with more severe aortic stenosis. The valves of patients with an area smaller than 1 cm^2 open and close more slowly and remain open for a very short period. In patients without significant stenosis, on the right of Fig. 3, the valve opens and closes very quickly, and there is a time lapse during which there is little variation in the area.

The study [12], which confirms the importance of the instantaneous behaviour of the aortic valve during ventricular ejection, examines 56 patients using three-dimensional echocardiography. The patients are divided into three categories: (1) Group *NL*: healthy subjects; (2) Group *CMP*: cardio-myopathic subjects with reduced left ventricular function; (3) Group *AS*: subjects with severe aortic stenosis but with good ventricular function. The comparison of the temporal trends of the *GOA* (*Geometrical Orifice Area*), representative of a patient for each group, is

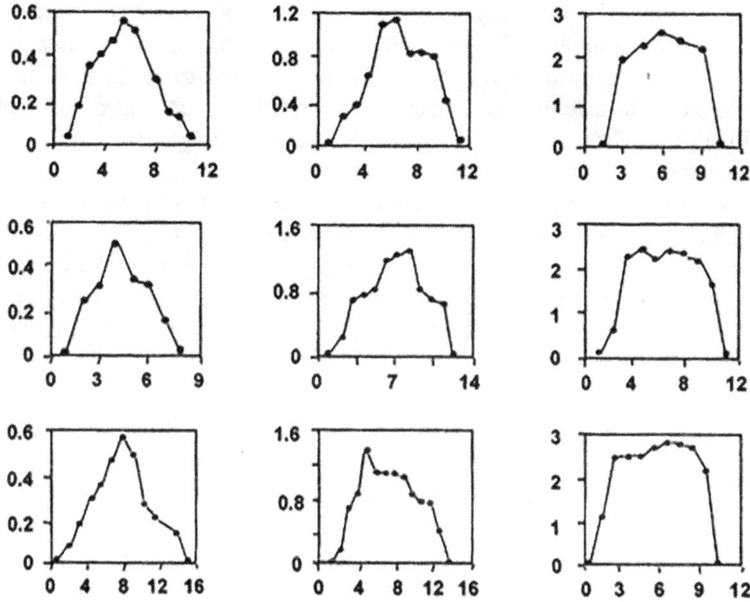

Fig. 3 Variations of the aortic valve area (cm^2) as a function of time (frames), Fig. 2 of [11]

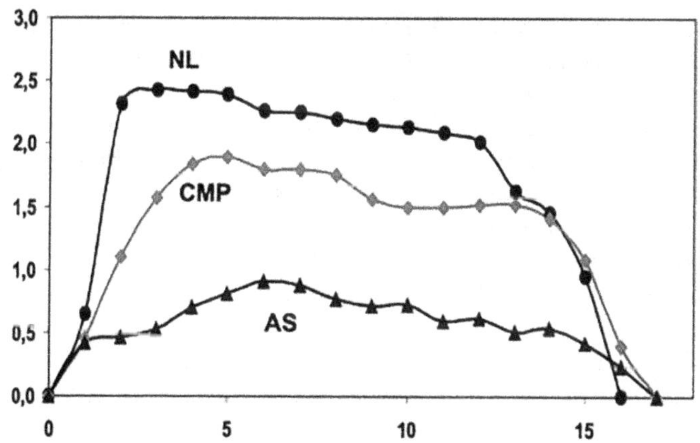

Fig. 4 Variations of the aortic valve area (cm^2) as a function of time (frames), Fig. 5 of [12]

shown in Fig. 4, Fig. 5 of [12], which presents the *AOA (Aortic Orifice Area)*, in cm^2, as a function of time, or images (*frames*) for the three categories of subjects studied. The presence of aortic stenosis and the comparison, shown in Fig. 4 with the non-pathological case (*NL*), allow the following considerations to be made: (i) the initial valve opening movement is slower; (ii) the maximum opening of the valve is reached later during systole; (iii) in the valve closing movement, it is often not possible to differentiate the slow and fast phases; (iv) the valve closing phase is characterised by *GOA* oscillations due to fluttering of the calcified valve flaps.

In the case of a healthy subject, the determination of the speed, area and time indices is carried out in [12], with which the trend of the valve area can be differentiated, based on the different phases of the cardiac cycle. In Fig. 2 of [12], the areas are indicated in cm^2; where $A1$ is relative to the maximum area of the orifice, $A2$ is the difference between $A1$ and $A3$, which, in turn, is the area of the orifice at the end of the slow closing movement. Figure 2 of [12] also reports the times, in *frames*, where $T1$ is the time required to reach $A1$, $T2$ is the time required for the slow closing movement, and $T3$ is the time required for the rapid closing movement. Finally, Fig. 2 of [12] indicates the speeds, where $V1$ is the speed of the rapid opening movement, $V2$ that of the slow closing movement, and $V3$ that of the rapid closing movement. In the healthy valve there is a phase of rapid initial opening (of duration $T1$) in which the valve reaches the maximum value of its area $A1$, or the maximum degree of orifice opening; the valve, subsequently, begins very soon, still in the phase of systole, the slow closing movement (of duration $T2$), where the shape of its orifice varies, assuming a triangular morphology. At the end of systole, there is a rapid closing movement (of duration $T3$) until the valve is completely closed.

A further important result, reached by the authors of [12], is that during the systole phase the aortic orifice not only changes in size but also in shape, which can appear star-shaped, circular, triangular, or with an intermediate shape between these variants (Fig. 5, Fig. 1 of [12]). The top of Fig. 5 shows the dynamic changes in the normal aortic valve. During systole, the aortic orifice takes an intermediate shape (between round and triangular) at the time of maximum opening, which is followed by a slow closing movement, and the shape clearly becomes triangular. The corresponding temporal scheme of the aortic orifice area, AOA, is the one reported at the bottom of Fig. 5 with a continuous line with dots.

Figure 6, Fig. 4 of [12], shows the dynamic changes in a stenosed aortic valve during the systole. The valve is functionally bicuspid; left and right coronary cusps are fused. The opening movement of the valve is lethargic compared with a normal valve, Fig. 5. The bottom of Fig. 6 shows the corresponding area-time diagram, AOA (cm^2), aortic orifice area versus time (frames).

The relationship between the geometric area of the orifice *(GOA)* and the area of the smallest cross-section of the jet, EOA, or the contraction coefficient $CC = EOA/GOA$, for heart valves is in the field $CC = 0.6 \div 1$. The value of CC depends on numerous factors, particularly: (i) geometric characteristics of the orifice, such as shape, inclination and curvature relative to the axis of the current, and eccentricity of the orifice; (ii) Reynolds number; (iii) whether the current is stationary or not. Despite their connection, the behaviour shown by the EOA still reflects that shown by the GOA, and it is therefore possible to integrate the information obtained from investigations focused on one or the other quantity with the aim of describing the dynamics of the aortic valve. The differences between the ejection profiles of EOA, found in subjects with healthy and severely stenotic aortic valves, are reported in Fig. 7, Fig. 1 of [13]. Figure 7 shows the instantaneous trends for a healthy subject, with an average value of $EOA = 3.2$ cm^2 during ejection, and for a subject with a severely stenotic aortic valve, with $EOA = 0.71$ cm^2. The trends are graphically represented as the ratio $EOA(t)/EOA_{max}$ (where $EOA(t)$ is the instantaneous value

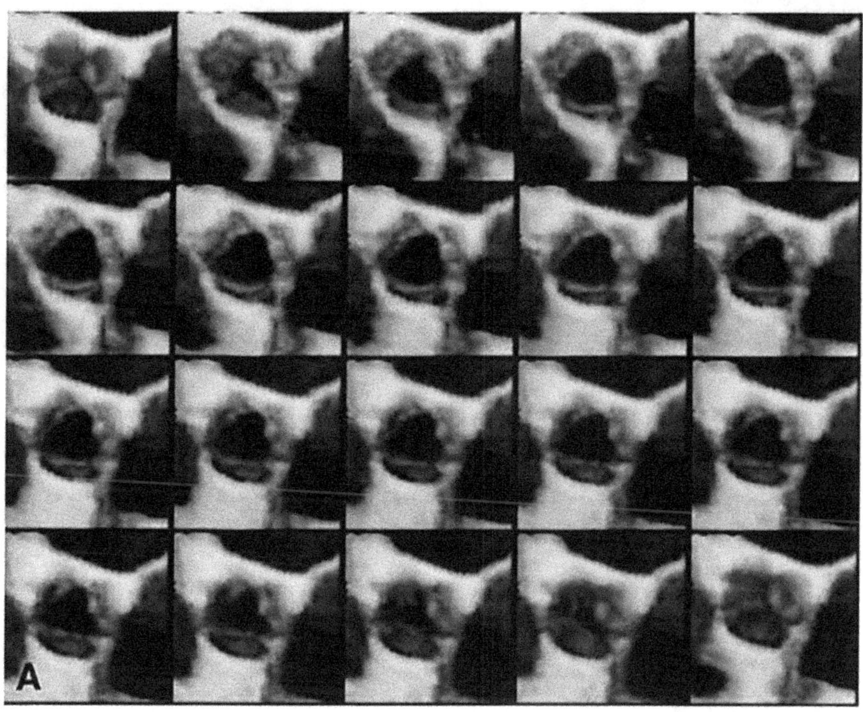

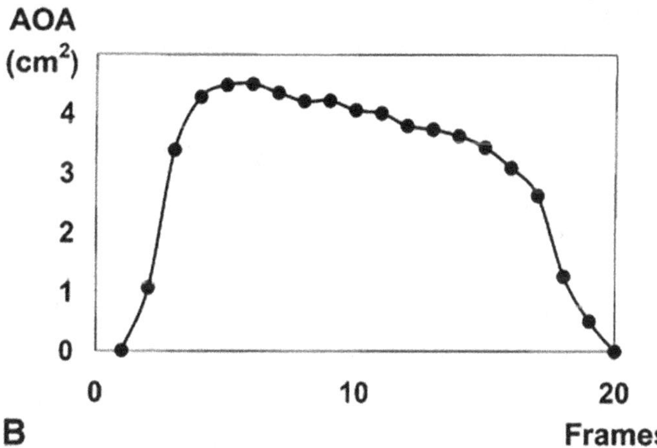

Fig. 5 Shape changes in normal aortic valve (top). Variations of a normal aortic valve area (cm²) as a function of time (frames) (bottom), Fig. 1 of [12]

and EOA_{max} is its maximum) as a function of t/T_{ejec} (where t is the generic instant in time and T_{ejec} is the ejection period). The characteristics already mentioned for the other studies are highlighted due to the different profiles of transvalvular flow speed (*VEOA*) and the flow in the outflow tract of the left ventricle (*LV*), which depend on

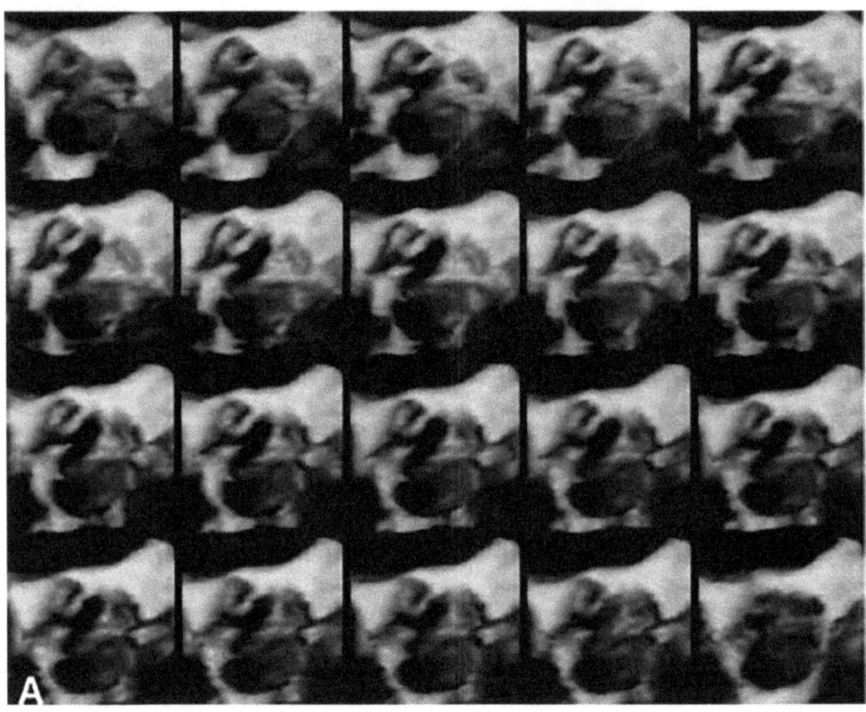

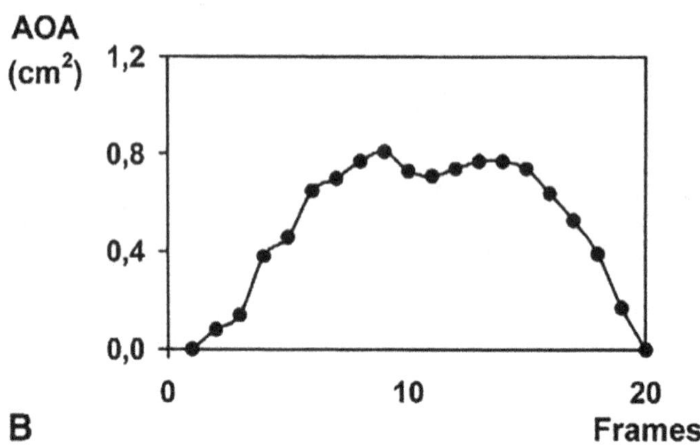

Fig. 6 Shape changes in a stenosed aortic valve (top). Variations of a stenosed aortic valve area (cm²) as a function of time (frames), (bottom), Fig. 4 of [12]

the absence or presence of pathology and the degree of severity of the same. The study shows that, while in a normal valve, *LV* and *VEOA* are practically simultaneous, in the presence of aortic stenosis, the maximum value, both of *LV* and of *VEOA*, is reached with a delay in ejection relatively greater than in healthy subjects and with a consequent delay in reaching the maximum value of *EOA*.

Fig. 7 EOA(t)/EOA$_{\mathrm{max}}$ (%) versus t/T$_{\mathrm{ejec}}$ (%). In a healthy subject (black dots) and in a severely stenotic aortic valve (white dots), Fig. 1 of [13]

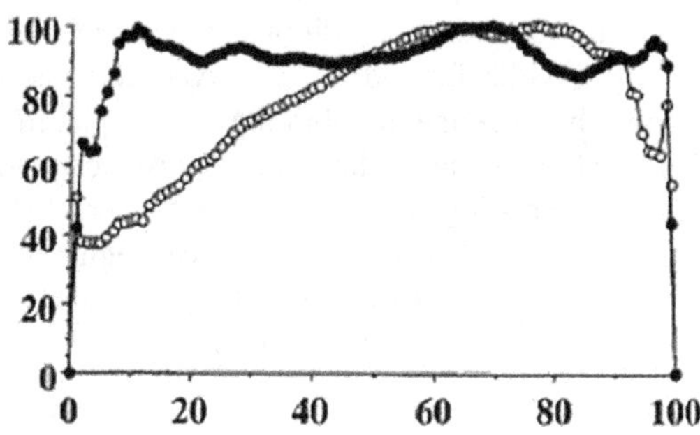

4   Hyperbolic Wave Equation

The analytical solution of the hyperbolic wave equation is illustrated to implement, in the programming language *Fortran* 90, the numerical solution and to discuss its stability.

4.1   Characteristics of the Wave Equation

We refer to the basic relationship of a second-order partial differential equation

$$A\,\frac{\partial^2 \phi}{\partial x^2} + B\,\frac{\partial^2 \phi}{\partial x\,\partial y} + C\,\frac{\partial^2 \phi}{\partial y^2} + D\,\frac{\partial \phi}{\partial x} + E\,\frac{\partial \phi}{\partial y} + F\,\phi + G(x,y) = 0 \qquad (19)$$

in which the constants are only functions of the variables, x and y; ϕ is the dependent variable; x and y are the independent variables; and $G(x, y)$ is a source term. Eq. 19, calculated at a point (x_0, y_0), is hyperbolic if $B^2 - 4\,A\,C > 0$, a condition that respects the wave equation since $B = 0$, $A = 1$ and $C = -c^2$.

4.2   Analytical Solution

4.2.1   Solution with the Method of the Separation of Variables

The wave equation in one dimension, on a finite interval, of a hyperbolic problem, with the initial and boundary values, is

$$\begin{cases} \dfrac{\partial^2 u}{\partial t^2} - c^2\,\dfrac{\partial^2 u}{\partial x^2} = 0,\, 0 < x < L, t > o \\[2ex] u(x,0) = f(x),\ \dfrac{\partial u(x,0)}{\partial t} = g(x),\ \le x \le L \\[2ex] u(0,t) = 0, u(L,t) = 0, t \ge 0 \end{cases} \qquad (20)$$

where the boundary conditions, $u(0, t) = 0$ and $u(L, t) = 0$, are called *null* or *Dirichlet*, while $f(x)$ and $g(x)$ are given functions and represent the initial conditions. Physically, imagine a vibrating string placed along the segment $(0, L)$. Once disturbed, the string arches, and the displacement in the plane (x, y), in the direction y, from the resting position is given by $u(x, t)$. This means that, if x is fixed, we are looking at what happens on a vertical segment, and the trend is dictated by $u(x, t)$. We can proceed to solve the differential equation using the separation of variables. We look for solutions of the form $u(x, t) = P(x) R(t)$, which we insert into Eq. (20) to get

$$P(x)R''(t) = c^2 P''(x)R(t) \tag{21}$$

being P and R functions of different variables. The only possibility is to have

$$R''(t) = c^2 \, \lambda \, R(t), P''(x) = \lambda \, P(x), \lambda \in R \tag{22}$$

The boundary conditions are $u(0, t) = P(0) R(t) = 0$ and $u(L, t) = P(L) R(t) = 0, \forall \, t$. Since $R(t) = 0$ is not a solution, as we would have $u(x, 0) = f(x) \neq 0$, we must impose $P(0) = P(L) = 0$. The function $P(x)$ satisfies the equation

$$P''(x) = \lambda \, P(x), P(0) = P(L) = 0 \tag{23}$$

1. Case 1: $\lambda = 0$; $P''(x) = 0$, from which $P(x) = \alpha + \beta x$; $P(0) = \alpha = 0$ and $P(L) = \beta L = 0$, that is, $\beta = 0$ and therefore, $P = 0$, which does not satisfy Eq. 20, as we have already seen.
2. Case 2: >0; $P(x) = \alpha \, e^{\sqrt{\lambda} x} + \beta \, e^{-\sqrt{\lambda} x}$, and the boundary conditions give $P(0) = \alpha + \beta = 0$ and $P(L) = \alpha \, e^{\sqrt{\lambda} L} + \beta \, e^{-\sqrt{\lambda} L}$, from which $2 \, \alpha \, \sin h\left(\sqrt{\lambda L}\right) = 0$ and therefore $\lambda = 0$, which we have excluded.
3. Case 3: $\lambda = -|\lambda| < 0$; $P(x) = \alpha \, \cos \sqrt{|\lambda|} \, x + \beta \, \sin \sqrt{|\lambda|} \, x$, and the boundary conditions give $P(0) = \alpha = 0$, where $P(0) = \alpha = 0$ and $P(L) = \beta \, \sin \sqrt{|\lambda|} \, L$, from which $L\sqrt{|\lambda|} = k \, \pi$ and therefore $|\lambda| = k^2 \frac{\pi^2}{L^2}$, that is $\lambda_k = -k^2 \frac{\pi^2}{L^2}, k \in N$. $k = 1, 2, 3$. As we can see, there are infinite solutions, and indeed Eq. 20 is called a boundary problem or a problem with given boundary conditions.

Once Eq. 23 is solved, we need to solve

$$R''(t) = c^2 \lambda_k R(t) = -c^2 |\lambda_k| R(t) \tag{24}$$

whose solution is

$$R_k(t) = A_k \cos\left(c\sqrt{|\lambda_k|} t\right) + B_k \sin\left(c\sqrt{|\lambda_k|} t\right) \tag{25}$$

In the end, we obtain

$$u_k(x,t) = P_k(t)R_k(t) = \left[A_k\cos\left(c\sqrt{|\lambda_k|}t\right) + B_k\sin\left(c\sqrt{|\lambda_k|}t\right)\right]\sin\left(\sqrt{|\lambda_k|}x\right) \quad (26)$$

Since the problem has null boundary conditions and is linear, the sum of $u_k(x,t)$ and of $u_{k1}(x,t)$ is still a solution, so the most general solution is given by

$$u(x,t) = \sum_{k=1}^{+\infty} u_k(x,t) \quad (27)$$

Now we need to satisfy the initial conditions of the solution. Since $u(0,t) = 0$ and $\frac{\partial u}{\partial t}(0,t) = 0$ for every $t \geq 0$, it follows that $f(0) = 0 = g(0)$. We extend the function $f(x)$ in an odd way to all R defining

$$F(x) = \begin{cases} f(x) \text{ for } 0 \leq x \leq L \\ -f(x) \text{ for } -L \leq x \leq 0 \end{cases} \quad (28)$$

Being odd, its Fourier development will be made only by sines

$$F(x) = \sum_{k=1}^{+\infty} f_k \sin\sqrt{|\lambda_k|}x = \sum_{k=1}^{+\infty} f_k \sin\frac{k\pi}{L}x, f_k = \frac{2}{L}\int_0^l f(x)\sin\left(\frac{\pi k}{L}x\right)dx \quad (29)$$

The equality $u(x,0) = f(x)$ gives us

$$u(x,0) = \sum_{k=1}^{+\infty} A_k \sin\sqrt{|\lambda_k|}x = \sum_{k=1}^{+\infty} f_k \sin\frac{k\pi}{L}x, \quad \text{da cui } A_k = f_k \quad (30)$$

We do the same for the function $g(x)$, defining

$$G(x) = \begin{cases} g(x) \text{ per } 0 \leq x \leq L \\ -g(x) \text{ per } -L \leq x \leq 0 \end{cases} \quad (31)$$

and therefore

$$\frac{\partial u}{\partial t}(x,0) = \sum_{k=1}^{+\infty} B_k \, c\sqrt{|\lambda_k|}\,\sin\sqrt{|\lambda_k|}x = \sum_{k=1}^{+\infty} g_k \sin\frac{k\pi}{L}x \quad (32)$$

from which $g_k = B_k\frac{ck\pi}{L}$

$$g_k = \frac{2}{L}\int_0^l g(x)\sin\left(\frac{\pi k}{L}x\right)dx \quad (33)$$

In the end, the solution is

$$u(x, t) = \sum_{k=1}^{+\infty} \left[f_k \cos\left(c\sqrt{|\lambda_k|}t\right) + g_k \frac{L}{ck\pi} \sin\left(c\sqrt{|\lambda_k|}t\right) \right] \sin\sqrt{|\lambda_k|}x \qquad (34)$$

If we assume that $g(x) = 0$ we get

$$\begin{aligned} u(x, t) &= \sum_{k=1}^{+\infty} f_k \cos\left(c\sqrt{|\lambda_k|}t\right) \sin\sqrt{|\lambda_k|}x \\ &= \sum_{k=1}^{+\infty} \frac{f_k}{2} \left\{ \sin\left[\frac{\pi k}{L}(x+ct) + \sin\frac{\pi k}{L}(x-ct)\right] \right\} \\ &= \frac{1}{2}[f(x+ct) + f(x-ct)] \end{aligned} \qquad (35)$$

and, as we can see, the effect is that of two travelling waves in opposite directions. If, instead,
$g(x) \neq 0$ we have

$$\frac{1}{2}[f(x+ct) + f(x-ct)] + \frac{1}{2c}\int_{x-ct}^{x+ct} g(y)dy \qquad (36)$$

which is D'Alembert's equation, or the classic solution of the wave equation.

4.3 Numerical Solution

4.3.1 Explicit Finite Difference Scheme for Initial and Boundary Value Problems

We construct a numerical scheme to approximate the solution of a problem of the type of Eq. 20. Before proceeding, we must focus on the solution that led us to D'Alembert's formula by referring to Fig. 8. From Eq. 19 it follows that, considering a generic point $P^*(x^*, t^*)$, the solution P^* depends only on the values of the data in $[x^* - ct^*, x^* + ct^*]$. The range $I = [x^* - ct^*, x^* + ct^*]$ is called the dependency interval of the point P^*; the triangle, with base I and opposite vertex P^*, is delimited by the characteristic lines for P^*, and is called the domain of continuous dependency

Fig. 8 Domain of integration

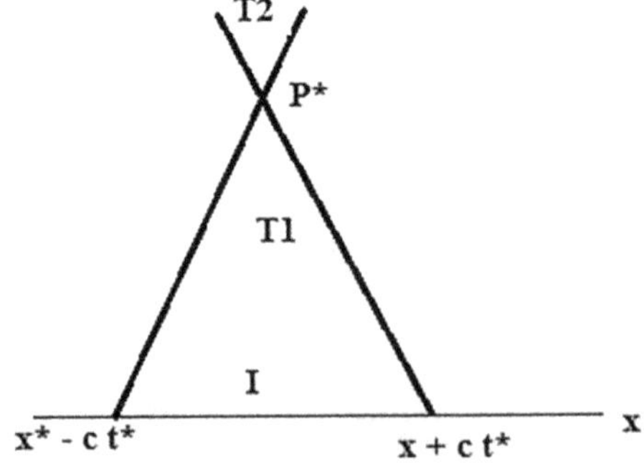

Fig. 9 Conditions on an
interval [0, a]

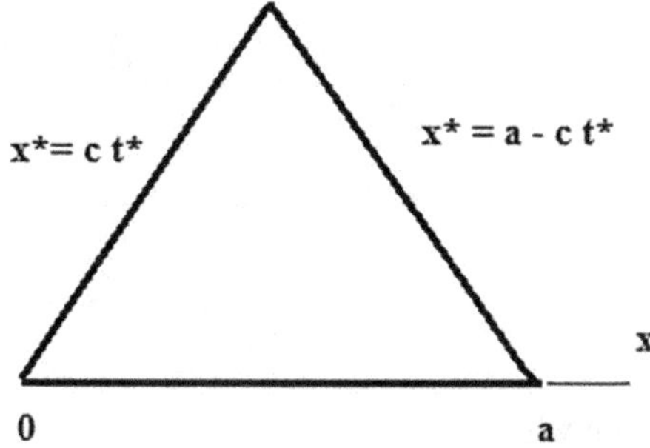

of the point P^*. The knowledge of the data on I uniquely determines the solution
only in T_1. On the other hand, the triangle T_2, with vertex P^*, also delimited by
characteristic lines, is called the domain of influence of P^*, because the value of the
solution in P^* influences the solution at all points in T_2. It can be said that an
observer located in P^* perceives the effects of what happens in T_1, but not of what
happens outside of T_1, and, at the same time, the effect of a disturbance in P^* can be
perceived only in the domain T_2.

However, if the conditions are assigned only on an interval, let's say (0, a) of the
axis x (Fig. 9), the solution can only be identified in the triangle with base (0, a),
delimited by the two characteristics for (0, 0) and (a, 0) respectively. If the
conditions are assigned on (0, $+\infty$), the solution is determined only below the
characteristic line passing through the origin; therefore, in both cases, to determine
the solution in the remaining areas, other conditions are needed. Finally, we note that
if φ, or ψ, has a discontinuity at a certain point x_0, it follows that the discontinuity
will appear, at time t, in the positions $(x^* - ct^*, x^* + ct^*)$; therefore, the
characteristics are the lines along which the discontinuities propagate. These
clarifications will be useful for the discretisation of the domain.

Now, returning to the numerical solution of the problem, we consider a change of
variables, $y = ct$, and the initial boundary conditions

$$\begin{cases} \dfrac{\partial^2 u}{\partial x^2} - \dfrac{\partial^2 u}{\partial xy} = 0, 0 < x < L, y > o \\[2mm] u(x,0) = \varphi(x), \dfrac{\partial u(x,0)}{\partial y} = \psi(x), 0 \leq x \leq 1 \\[2mm] u(0,y) = 0, u(L,y) = 0, y \geq 0 \end{cases} \qquad (37)$$

We consider, in the domain, a lattice of equispaced nodes. We approximate the
second derivatives at the nodes with numerical derivation to the central finite
differences, that is

$$\frac{\partial^2 u(x_i, y_j)}{\partial x^2} = \frac{u(x_{i+1}, y_j) - 2\,u(x_i, y_j) + u(x_{i-1}, y_j)}{h^2} + O(h^2) \qquad (38)$$

$$\frac{\partial^2 u(x_i, y_j)}{\partial y^2} = \frac{u(x_i, y_{j+1}) - 2\,u(x_i, y_j) + u(x_i, y_{j-1})}{k^2} + O(k^2) \tag{39}$$

instead, for the $\partial u/\partial y\,(x_i, 0)$, the forward derivative along y is applied

$$\frac{\partial u(x_i, y_j)}{\partial y} = \frac{u(x_i, y_{j+1}) - u(x_i, y_j)}{k} + O(k) \tag{40}$$

from which

$$\frac{\partial u(x_i, 0)}{\partial y} = \frac{u(x_i, y_1) - u(x_i, y_{-1})}{2k} + O(k^2) \tag{41}$$

Neglecting the error terms and indicating with u_{ij} the approximated values of the solution at the nodes and assuming $\varphi_i = \varphi(x_i)$, $\psi_i = \psi(x_i)$,, we obtain

$$\frac{u_{i+1,j} - 2\,u_{i,j} + u_{i-1,j}}{h^2} - \frac{u_{i,j+1} - 2\,u_{i,j} + u_{i,j-1}}{k^2} = 0 \tag{42}$$

where, by adding the assigned conditions and simplifying, we get

$$\begin{cases} u_{i,j+1} = 2\,u_{i,j} + u_{i,j-1} + \dfrac{k^2\left(u_{i+1,j} - 2\,u_{i,j} + u_{i-1,j}\right)}{h^2} \\[2mm] u_{i,0} = \varphi_i \\[1mm] \dfrac{u_{i,1} - u_{i,-1}}{2\,k} = \varphi_i, \quad i = 1, 2, \ldots, N-1 \\[2mm] u_{0j} = u_{Nj}, \quad j = 0, 1, 2 \ldots \end{cases} \tag{43}$$

Taking into account the order of magnitude of the neglected truncation errors, we thus have an explicit second-order scheme that generates the discrete solution at node $j + 1$, starting from the values at nodes j and $j - 1$. The third equation allows us to obtain the values $u_{i,-1}$, indeed

$$u_{i,-1} = u_{i1} - 2\,k\,\varphi_i \tag{44}$$

Given

$$\alpha = \frac{k}{h} \tag{45}$$

we obtain the following explicit scheme

$$\begin{cases} u_{i,j+1} = \alpha^2 u_{i-1,j} + 2\,(1 - \alpha^2)\,u_{i,j} + \alpha^2 u_{i+1,j} - u_{i,j-1}, \quad j = 1, 2, \ldots \\[1mm] u_{i1} = \dfrac{\alpha^2}{2}\,(\varphi_{i-1} + \varphi_{i+1}) + (1 - \alpha^2)\varphi_i + k\varphi_i, \quad i = 1, 2, \ldots, N-1 \\[1mm] u_{i0} = \varphi_i \end{cases} \tag{46}$$

Fig. 10 Discrete dependence domain

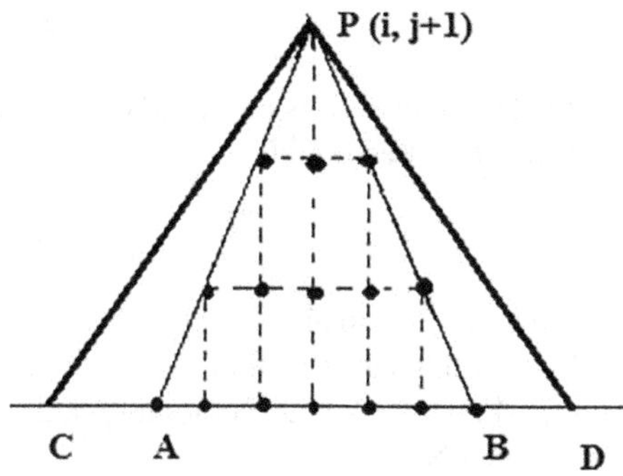

Referring to Fig. 10, from Eq. 46 it is evident that the discrete solution in $P_{i, j+1}$ depends on the nodes indicated in the figure and contained in the triangle delimited by the lines $P_{i, j+1} - A$ and $P_{i, j+1} - B$, having angular coefficients $\pm \frac{k}{h}$, called the discrete dependence domain. If the lines $P_{i, j+1} - C$ and $P_{i, j+1} - D$ are the characteristics for $P_{i, j+1}$, with angular coefficients ± 1, whereas $\alpha = k/h < 1$, it is clear that the variations of the data on AC and on BD modify the solution $u\,(P_{i, j+1})$, which depends on the data on all CD but does not influence the discrete solution, which depends on the data on AB, which remain unchanged. Therefore, there can be no convergence of the discrete solution to the continuous if the discrete dependence domain does not contain the continuous one.

The condition that the discrete dependence domain contains the continuous one is

$$\alpha = \frac{k}{h} \leq 1 \tag{47}$$

a condition only necessary for convergence, but, as demonstrated, it is also sufficient to have conditional stability.

We now analyse the stability by introducing the vectors

$$U_j\left[u_{1j}, u_{2j}, \ldots, u_{N-1,j}\right]^T; U_0\left[\varphi_1, \varphi_2, \ldots, \varphi_{N-1}\right]^T \tag{48}$$

and the matrix A, of dimensions $N - 1$, tridiagonal, symmetric

$$A = \begin{bmatrix} 2(1-\alpha^2) & \alpha^2 & 0 & 0 & 0 & 0 \\ \alpha^2 & & & & & \\ 0 & & & & & \\ 0 & & & & & \alpha^2 \\ 0 & & & & \alpha^2 & 2(1-\alpha^2) \end{bmatrix} \tag{49}$$

then the scheme of Eq. 48 can be rewritten in vectorial form

$$U_{j+1} = A\, U_j - U_{j-1} \tag{50}$$

If a disturbance E_0 is produced on the vector U_0, the error propagated to the stage $j + 1$ also satisfies Eq. 50

$$E_{j+1} = A\, E_j - E_{j-1} \tag{51}$$

or rather E_{j+1} satisfies a finite difference system, which is its vectorial form. In analogy to this last equation, the solution of Eqs. 50 and 51 must be sought in the form $\beta^j V$, where V is a vector and β a non-zero scalar. Imposing that $\beta^j V$ satisfies Eq. 51 we obtain

$$\beta^{j+1}\, V - A\, \beta^j\, V + \beta^{j-1}\, V = \tag{52}$$

from which

$$A\, V = \frac{\beta^2 + 1}{\beta}\, V \tag{53}$$

which means that V and $\frac{\beta^2+1}{\beta}$ are eigenvector and eigenvalue of the matrix A. Indicating with λ the generic eigenvalue of A, β must be such that

$$\frac{\beta^2 + 1}{\beta} = \lambda \tag{54}$$

Now, for the solution of Eq. 51 and, therefore, for the propagated error to remain limited, it must result $|\beta| \leq 1$. Since A, and consequently its eigenvalues, depends on, this condition will reflect in a condition on α and on the choice of discretisation steps. Eq. 54 implies

$$\beta = \frac{\lambda}{2} \pm \sqrt{\frac{\lambda^2}{4} - 1} \tag{55}$$

and, if results $|\lambda| > 2$, one of the two values β would have a modulus greater than 1. Therefore, it must be assumed $|\lambda| < 2$, since, as we will see, the case $|\lambda| = 2$ will be excluded. The eigenvalues of A are given by

$$\lambda_i = 2 - 4\,\alpha^2\, \sin^2 \frac{i\,\pi}{2\,N}, \quad i = 1, 2, \ldots N - 1 \tag{56}$$

from which it follows, not that $\lambda = 2$ only if $\alpha = 0$, (which is impossible), but also that $|\lambda_i| < 2$ implies

$$\alpha^2 < \frac{1}{\sin^2 \frac{i\,\pi}{2\,N}}; \quad i = 1, 2, \ldots N - 1 \tag{57}$$

Since $\sin^2 \frac{i\pi}{2N} < 1$, $\lim_{N \to \infty} \sin^2 \frac{i\pi}{2N} = 1$, and the conclusion is that the scheme is stable if

$$\alpha \le 1 \tag{58}$$

4.4 Case Studies

4.4.1 Hyperbolic Initial Value Problem

Analytical Solution The problem is proposed in the following form

$$\begin{cases} \dfrac{\partial^2 u}{\partial t^2} - c^2 \dfrac{\partial^2 u}{\partial x^2} = 0, x \in R, t > o \\[2mm] u(x,0) = \varphi(x) = \dfrac{x^2}{50}, x \in R \\[2mm] \dfrac{\partial u(x,0)}{\partial t} = \psi(x) = 0, x \in R \end{cases} \tag{59}$$

where, in the jargon of the vibrating string, the pair of data $u(x,0) = \varphi(x)$ and $\psi(x) = 0$ represents the problem of the plucked string, that is, the disturbance of the equilibrium configuration without disturbance of the velocity. According to D'Alembert's solution, this type of disturbance of the equilibrium configuration divides equally into a progressive wave and a regressive wave. In fact, substituting our data, we have

$$u(x,t) = \frac{1}{2}\left[\frac{1}{50}(x+ct)^2 + \frac{1}{50}(x-ct)^2\right] + \frac{1}{2c}\int_{x-ct}^{x+ct} 0 \, dy \tag{60}$$

which, when developed, leads to

$$u(x,t) = \frac{1}{50}\left(x^2 + [ct]^2\right) \tag{61}$$

which, in the case under examination, expresses the physical displacement of the string from its resting position. As is known, c represents the wave propagation speed, which, for a vibrating string, can assume very different values; for example, for some materials and geometries, like an elastic helix, it can be reduced to 1 m/s. For simplicity, therefore, we impose $c = 1$ m/s. Equation 61 becomes

$$u(x,t) = \frac{1}{50}\left(x^2 + [t]^2\right) \tag{62}$$

Numerical Solution

For the numerical solution of the case under consideration, the system of Eq. 43 is rewritten as

$$
\begin{cases}
u_{i,j+1} = 2\,u_{i,j} + u_{i,j-1} + \dfrac{k^2\left(u_{i+1,j} - 2\,u_{i,j} + u_{i-1,j}\right)}{h^2}\,, j=0,1,2,\ldots \\[2mm]
u_{i,0} = \varphi_i \\[2mm]
u_{i,1} - u_{i,-1}, i = 1,2,\ldots N
\end{cases}
\tag{63}
$$

which, in the form of Eq. 46, gives

$$
\begin{cases}
u_{i,j+1} = \alpha^2 u_{i-1,j} + 2\,(1-\alpha^2)u_{i,j} + \alpha^2 u_{i+1,j} - u_{i,j-1}, j=0,1,2,\ldots \\[2mm]
u_{i,1} = \dfrac{\alpha^2}{2}\left(\varphi_{i-1} + \varphi_{i+1}\right) + (1-\alpha^2)\varphi_i,\ i=1,2,\ldots N \\[2mm]
u_{i,0} = \varphi_i,
\end{cases}
\tag{64}
$$

where:

1. $\varphi_i = \varphi(x_i)$, $\varphi_{i+1} = \varphi(x_{i+1})$, $\varphi_{i-1} = \varphi(x_{i-1})$;
2. $x_i = i\,h$, with $h = L/N$ step of discretisation of the spatial domain;
3. $y_j = j\,k$, with k steps of discretisation of the variable y, remembering that $y = c\,t$;
4. $\alpha = h/k \le 1$, with α stability parameter of the numerical solution.

4.4.2 Hyperbolic Initial and Boundary Value Problem

Analytical Solution

The problem, in this second case, is always proposed in the form of Eq. 43, and we solve

$$
\begin{cases}
\dfrac{\partial^2 u}{\partial t^2} - c^2 \dfrac{\partial^2 u}{\partial x^2} = 0, x \in R, t > o \\[3mm]
u(x,0) = \varphi(x) = \dfrac{x^2}{50}, x \in R \\[3mm]
\dfrac{\partial u(x,0)}{\partial t} = \psi(x) = 0, x \in R
\end{cases}
\tag{65}
$$

that is, the perturbation at the initial instant is now both of the equilibrium configuration and of the speed. According to the D'Alembert solution, we have

$$
u(x,t) = \frac{1}{2}\left[\frac{1}{50}(x+ct)^2 + \frac{1}{50}(x-ct)^2\right] + \frac{1}{2c}\int_{x-ct}^{x+ct} \cos x\, dx
\tag{66}
$$

which, developed also using the addition and subtraction formulas for trigonometric functions, leads to

$$u(x,t) = \frac{1}{50}x^2 + \frac{1}{50}c^2t^2 + \frac{1}{c} + \cos x \sin(c\,t) \tag{67}$$

which always expresses the physical displacement of the string from its resting position.

Numerical Solution

For the numerical solution of the case under consideration, the system of Eq. 43 is rewritten as

$$\begin{cases} u_{i,j+1} = 2\,u_{i,j} + u_{i,j-1} + \dfrac{k^2\left(u_{i+1,j} - 2\,u_{i,j} + u_{i-1,j}\right)}{h^2}, j = 0, 1, 2, \dots, N \\[2mm] u_{i,0} = \varphi_i \\[1mm] u_{i,1} = u_{i,-1}, i = 0, 1, 2, \dots, N \end{cases} \tag{68}$$

which, in the form of Eq. 46, gives

$$\begin{cases} u_{i,j+1} = \alpha^2 u_{i-1,j} + 2\,(1-\alpha^2)u_{i,j} + \alpha^2 u_{i+1,j} - u_{i,j-1}, j = 0, 1, 2, \dots \\[2mm] u_{i,1} = \dfrac{\alpha^2}{2}\left(\varphi_{i-1} + \varphi_{i+1}\right) + \left(1-\alpha^2\right)\varphi_i, i = 1, 2, \dots N \\[1mm] u_{i,0} = \varphi_i, \end{cases} \tag{69}$$

where we add, compared to the first example, Eq. 63, the term $k\,\psi_i$, with $\psi_i = \psi(x_i)$.

4.5 Numerical Results

The numerical solutions of the two cases are implemented in *Fortran* 90 and then compared with the analytical ones just discussed.

4.5.1 Numerical Solution of the Homogeneous Hyperbolic Wave Eq. 90

Below we report the list of the code in *Fortran* 90, in which the numerical solution for the homogeneous hyperbolic wave equation is implemented.

List of the Code in Fortran 90.

```
1 program equation_wave
2 implicit none ! initialization of parameters
3double precision ,parameter:: L =30.0 D +00
4integer :: i ,j ,N ,t ,s,m
5double precision :: h,k ,c, delta_t
6double precision :: alpha, x_i,phi_i,phi_i_meno_1,phi_i_piu_1 ,psi_i
7double precision , dimension (20 ,1) :: T_j
8double precision , dimension (200 ,20) :: U_i_j
```

```fortran
9data U_i_j /4000*0.0 D +00/, T_j /20*0.0 D +00/
10! Insert number of nodes
11write(*,*) 'Numerical solution of hyperbolic equation of plane waves
12write(*,*)
13write(*,*) 'Insert number of nodes to use for spatial discretisation
14read(*,*) N
15if (N >200) then
16write(*,*) 'Too many nodes!'
17goto 100
18elseif (N ==1) then
19write(*,*) 'Only one node!'
20goto 100
21elseif (N <1) then
22write(*,*) 'Node is invalid !'
23goto 100
24endif
25h=L/N !Calculate spatial step!
26! Insert number of time steps!
27write(*,*) 'Insert number of time instants to evaluate transient
solution!
28read (* ,*)t
29if (t >20) then
30write(*,*) 'Too many instants! '
31goto 100
32elseif (t==0) then
33write(*,*) 'No time instant'
34write(*,*)
35goto 200
36elseif (t <1) then
37write(*,*) 'Number of time instants not valid'
38goto 100
39endif
40! Insert time step!
41write(*,*) 'Insert time step'
42write(*,*)
43read (* ,*) delta_t
44if (delta_t <0) then
45write (* , *) 'Time step not valid! '
46goto 100
47elseif (delta_t >10) then
48write (* , *) 'Time step not valid! '
49goto 100
50elseif ( delta_t ==0) then
51write(*,*) ' Solution is the same of the initial one '
52write(*,*)
```

```
53goto 200
54endif
55! Insert wave velocity !
56write(*,*) ' Insert wave velocity '
57write(*,*)
58write (* , *) 'Wave velocity is '
59read (* ,*) c
60 k= delta_t *c
61alpha=k/h ! Calculate stability parameters '
62 if (alpha >1) then
63 write(*,*) 'Numerical solution is unstable '
64 write(*,*)
65 goto 100
66 endif
67! Insert initial condition !
67write(*,*) 'Insert initial condition '
69write(*,*)
70write(*,*) 'Insert 1 for perturbation on equilibrium '
71write(*,*)
72write(*,*) 'Insert 2 for perturbation on equilibrium position and
velocity '
73read(*,*) m
74 if (m <1) then
75 write(*,*) 'Initial condition is invalid !'
76 goto 100
77 elseif (m >2) then
78 write(*,*) 'Initial condition is invalid !'
79 goto 100
80 endif
81! Fill the matrix U_i_j
82 do j =3,t
83 do i =1,N
84 if (j ==3) then
85 x_i=i*h
86 phi_i =( x_i **2) /50
87 U_i_j (i ,1) =phi_i ! t=o
88 phi_i_meno_1 =( ((i-1) *h) **2) /50
89 phi_i_piu_1 =( ((i +1) *h) **2) /50
90 if (m ==1) then
91 psi_i =0
92 elseif (m ==2) then
93 psi_i= cos (x_i)
94 endif
95 U_i_j (i ,2) =( (( alpha **2) * ( phi_i_meno_1 +
phi_i_piu_1 ) ) /2) +((1 - ( alpha **2) ) *phi_i) + (k* psi_i) ! t =1* k
```

```fortran
96 U_i_j (i,j) = ((alpha**2)*(U_i_j(i-1,j-1))) + (2*(1-(alpha**2))
*(U_i_j(i,j-1))) + ((alpha**2)*(U_i_j(i+1,j-1))) - (U_i_j(i,j-
2))
97 elseif (j >3) then
98 U_i_j (i,j) = ((alpha**2)*(U_i_j(i-1,j-1))) + (2*(1-(alpha
**2))*(U_i_j(i,j-1))) + ((alpha**2)*(U_i_j(i+1,j-1))) - (U_i_j
(i,j-2))
99 endif
100 enddo
101 enddo
102 ! check
103 !write(*,*) " Values of u(x) on the nodes are :"
104 !write (* ," ( F6.2) ") (U_i_j (s,1),s=1, N)
105 !write(*,*)
106 !write (* ," ( F6.2) ") (U_i_j (s,2),s=1, N)
107 !write(*,*)
108 ! write (* ," ( F6.2) ") (U_i_j (s,8),s=1, N)
109 !write(*,*)
110 !write (* ," ( F6.2) ") (U_i_j (0,2))
111 ! pause
112 do j =1, t -1
113 T_j (1,1) =0
114 T_j (j+1,1)= j* delta_t
115 enddo
116 open (unit =1, file=" results. dat")
117 write (1 ,*) " Vectors of solutions U_j "
118 do j =1 ,t
119 write (1 ,*) " Vector U_ " ,j -1
120 do i =1 ,N
121 write (1 , " (F6.2) ") U_i_j (i, j)
122 enddo
123 enddo
124 write (1 ,*) " Vector of nodal positions' "
125 do i =1 , N
126 x_i =i*h
127 write (1 , " (F6.2) ") x_i
128 enddo
129 write (1 ,*) " Vector of time instants ' "
130 do j =1 ,t
131 write (1 , " (F6.2) ") T_j (j ,1)
132 enddo
133 write (1 ,*) "N t m c"
134 write (1 ,*) N
135 write (1 ,*) t
136 write (1 ,*) m
```

```
137write (1 ,*) c
138return
139100 pause
140stop
141200 write (*,*) "' Solution is the declared one as initial condition ! "
142pause
143stop
144end program equation_wave
```

4.5.2 Hyperbolic Initial Value Problem

We now report the response obtained from the numerical solution and compare the numerical solutions with the analytical ones, Eq. 61. The parameters set for the solutions are: $L = 30$ for the considered spatial domain, with $0 \le x \le L$; $N = 10$ for the number of nodes on which to evaluate the numerical solution, from which $h = L/N = 3$ is the step of spatial discretisation; $\Delta t = 0.2$ the time discretisation step, from which $\alpha = \frac{k}{h} = 1/16$. It is noted that the stability parameter of the numerical solution, is <1, thus respecting the necessary and sufficient condition of conditional stability, discussed earlier.

In the various tests carried out during programming, it is, however, noticed that if the value of α approaches unity, the numerical solution, for subsequent time instants, is of the same form as the analytical solution but gradually moves away from the latter over time. Therefore, as expected, the solution is stable, but the error propagates too much. With the parameters listed above, which give $=1/16$, the numerical solution, at various time instants, coincides with the analytical solution. Figure 11 shows the numerical and analytical solutions of $u(x)$ as a function of x_i for $\alpha = 1/16$. On top is shown the solution after $t = 0.8$ s and on the bottom, after $t = 1.6$ s. We remind you that this solution is for a speed $c = 1$ m/s. We do not report the overlap of the various solutions at various time instants, as the oscillation over time is not well perceived; this will be more explicit in the results of the following case.

As already mentioned, the solution reported is similar to that of a sound wave propagating in a fluid since the equation is of the same form. The differences lie in the presence of source terms, as discussed in the chapter on the acoustic analogy, both in the wave speed, as the latter depends on the medium in which it propagates. For a fluid, the wave speed is defined as $c_{\text{fluid}} = \sqrt{\frac{K}{\delta}}$, where K is the compressibility modulus and δ the density of the fluid. However, as the wave speed varies, the solution remains similar, but only the speed at which it propagates in space varies. Referring to the discussion of the problem that we address throughout this chapter, we can say that for the determination of the speed of sound produced by a turbulent blood flow, one must take into account the characteristic parameters of this fluid. As a first approximation, given the composition of the blood itself, it is taken as a wave speed value that in water at atmospheric pressure and at a temperature of 37 °C, i.e. about $c = 1500$ m/s. This result allows us to make some considerations on the convergence and stability of the numerical methods that we use through an analogy

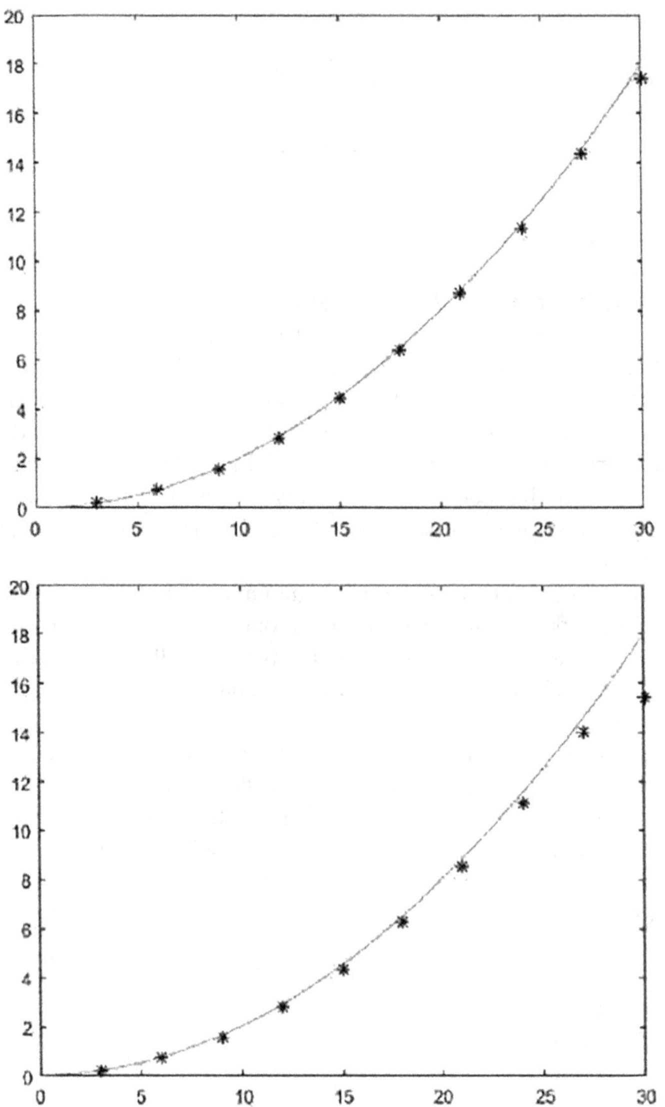

Fig. 11 Numerical (points) and analytical (line) solutions of $u(x)$ as a function of x_i for $\alpha = \frac{1}{16}$ and $c=1$ m/s. [1]

between the constraints on some parameters to be taken into account in the different numerical resolution approaches.

4.5.3 Hyperbolic Initial and Boundary Value Problem

We report the numerical solution and compare it to the analytical one, Eq. 67. The parameters L and N, set for the solutions, are identical to those used in the previous

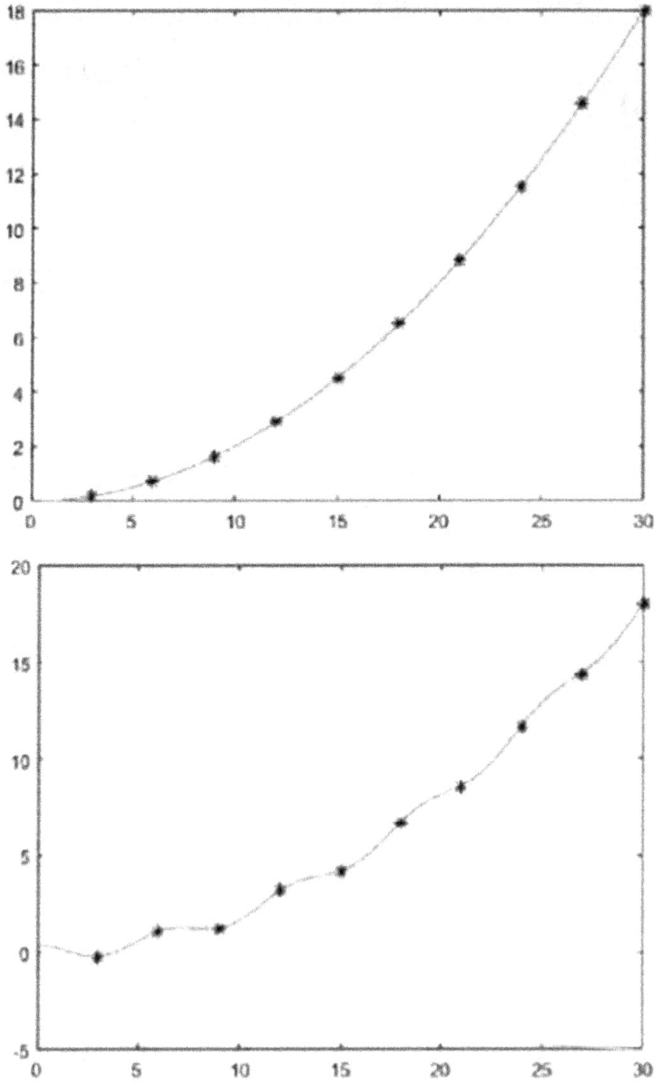

Fig. 12 Numerical solution (points) and analytical (line) of $u(x)$ as a function of x_i for $\alpha = \frac{1}{16}$ and $c = 1$ m/s. On the top at $t = 0$ and on the bottom at $t = 0.4$ s, [1]

problem. In this problem, however, we report the solutions for two speed values: $c = 1$ m/s, with $\delta_t = 0.2$ s, and $c = \frac{1}{2}$ m/s, with $\delta_t = 0.4$ s. In Figs. 12, 13 and 14, we report the trends of $u(x)$ at some temporal instants evaluated for the solution with $c = 1$ m/s, while in Figs. 15, 16 and 17, we report the solution with $c = 1/2$ m/s.

Figure 12 presents the solutions at $t = 0$ on the top and $t = 0.4$ s on the bottom. Figure 13 presents the solutions at $t = 0.8$ s on the top and $t = 1.2$ s on the bottom.

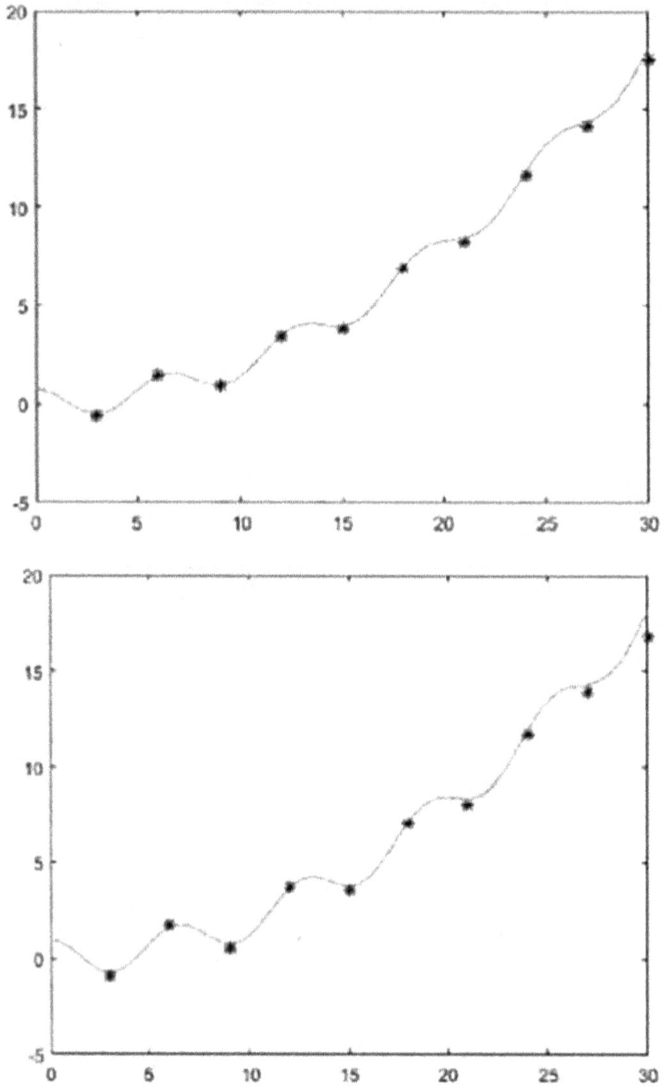

Fig. 13 Numerical solution (points) and analytical (line) of $u(x)$ as a function of x_i for $\alpha = \frac{1}{16}$ and $c = 1$ m/s. On the top at $t = 0.8$ and on the bottom at $t = 1.2$ s, [1]

Figure 14 presents the solutions at $t = 1.6$ s on the top and at all times on the bottom.

Figure 15 presents the solutions at $t = 0$ on the top and $t = 0.8$ s on the bottom.
Figure 16 presents the solutions at $t = 1.6$ on the top and $t = 2.4$ s on the bottom.
Figure 17 presents the solutions at $t = 3.2$ on the top and at all times on the bottom. It is noted that the solutions at $t = 0$ s in the two applications coincide, as

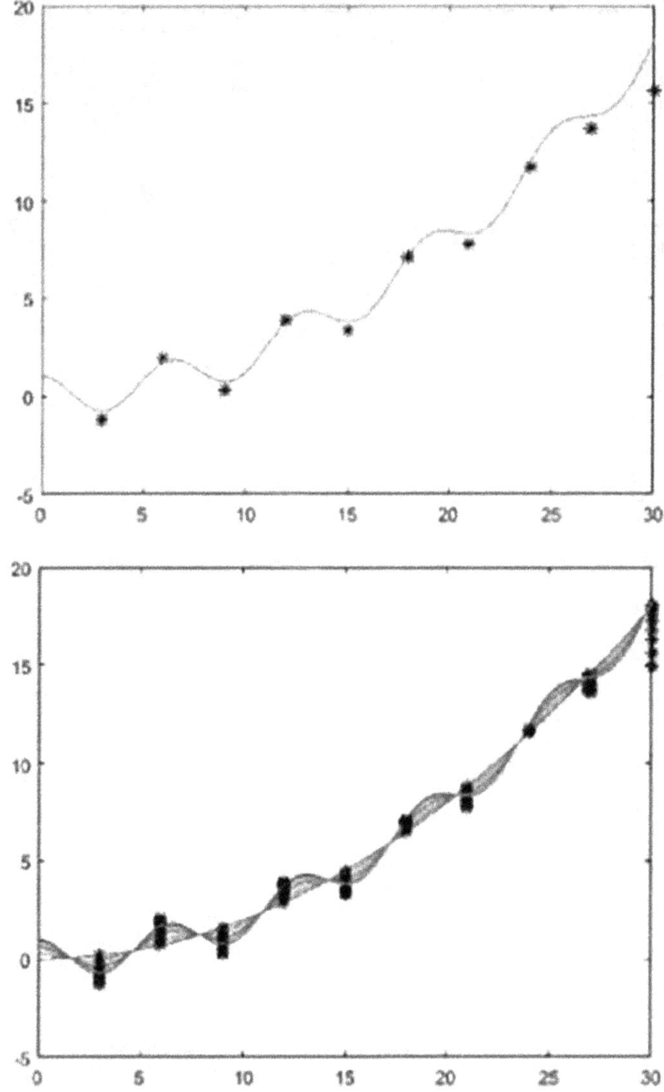

Fig. 14 Numerical solution (points) and analytical (line) of $u(x)$ as a function of x_i for $\alpha = \frac{1}{16}$ and $c = 1$ m/s. On the top at $t = 1.6$ and on the bottom at all times, [1]

these are related to the initial condition on displacement, which is the same for both cases. As expected, the numerical solutions better approximate the analytical solution as the number of nodes increases and the step of temporal discretisation decreases.

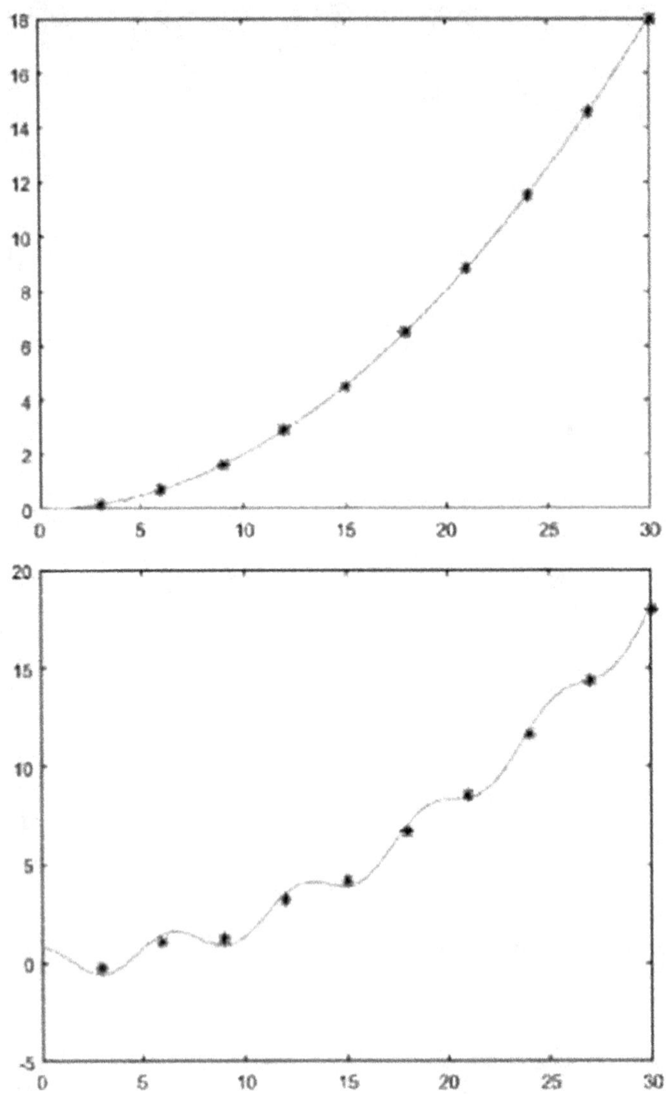

Fig. 15 Numerical solution (points) and analytical (line) of $u(x)$ as a function of x_i for $\alpha = \frac{1}{16}$ and $c = 1.2$ m/s. On the top at $t = 0$ and on the bottom at $t = 0.8$ s, [1]

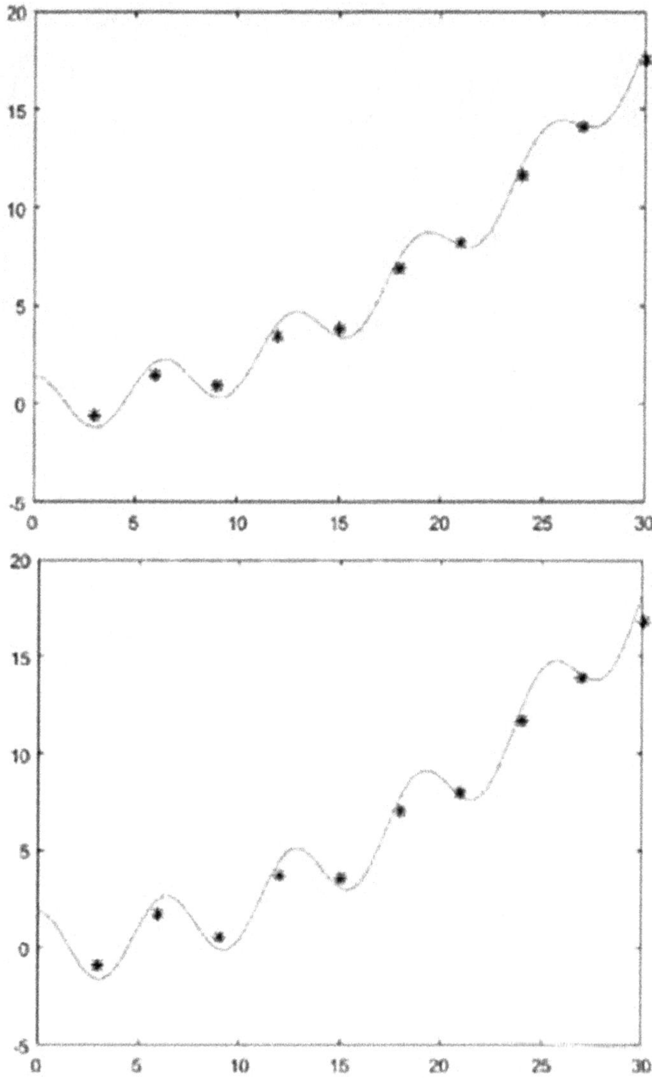

Fig. 16 Numerical solution (points) and analytical (line) of $u(x)$ as a function of x_i for $\alpha = \frac{1}{16}$ and $c = 1.2$ m/s. On the top at $t = 1.6$ and on the bottom at $t = 2.4$ s, [1]

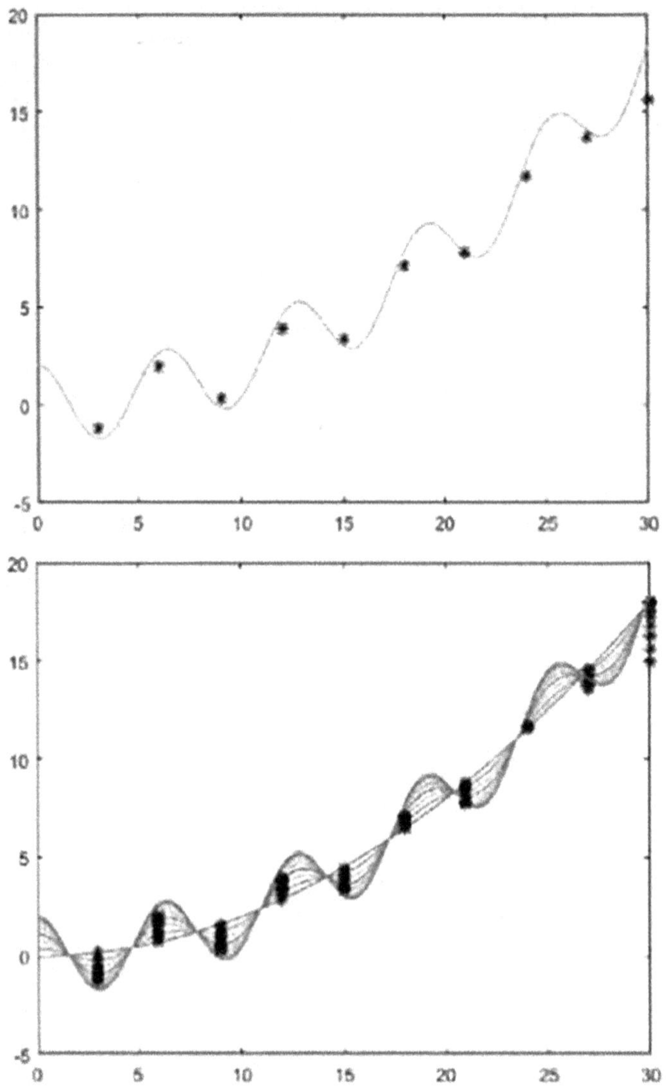

Fig. 17 Numerical solution (points) and analytical (line) of $u(x)$ as a function of x_i for $\alpha = \frac{1}{16}$ and $c = 1.2$ m/s. On the top at $t = 1.6$ and on the bottom at all times, [1]

5 Computational Thermo-Haemo-Dynamics (CTHD)

5.1 Geometry

The fluid dynamic simulations are carried out on a geometry that reproduces the model of a severely stenotic aortic valve (*VSS*) with *EOA* of 0.7 cm^2, in which we solve the *THD* (Thermo-Haemo-Dynamic equations) by applying a turbulence model of the type *RANS (Reynolds Averaged Navier-Stokes Equations)*. The fluid domain is modelled with a 2D geometry in which there is a reproduction of the valve flaps, assumed rigid, of a stenotic aortic valve. The assumption of rigid flaps is based on the pathophysiology of valve stenosis, as in the case of severe stenosis, where the flaps stiffen and can no longer achieve a tight closure in the time span of the cardiac cycle, which corresponds to diastole.

For the dimensions of the fluid domain, reproduced in Fig. 18, we refer to models used in other studies present in the literature, [14, 15]. As characteristic dimensions, we assume: in the inlet section, which corresponds to the section of the left ventricle, a diameter of 0.05 m, while in the outlet section, which corresponds to a section of the descending aortic arch, a diameter of 0.035 m. To size the area of the valve opening, *EOA*, according to the choice of the geometry and considering studies present in the literature on valve stenosis, [8], we model the stenotic valve with rigid flaps, with *EOA* of 0.7 cm^2. Figure 18 shows that part of the left ventricle, the intersection area of the valve flaps the Valsalva sinuses and the aortic arch are modelled. The ventricle and the aortic arch are included in the computational domain to ensure that the boundary conditions are sufficiently far from the valve so as not to influence the fluid dynamics.

In order to verify that the solutions obtained are independent from the grid used, and before carrying out the simulations in the transient regime, 3 simulations are performed in steady state on a less dense grid and on a denser grid compared to the

Fig. 18 Geometry of the fluid domain, [1]

standard one used subsequently. The main characteristics of the 3 grids are the following:

1. Coarse grid or mesh 1: number of nodes: 68,196; number of cells: 131,580
2. Standard grid or mesh 2: number of nodes: 116,032; number of cells: 227,260
3. Fine grid or mesh 3: number of nodes: 237,056; number of cells: 469,308

Figure 19 shows an enlargement of mesh 1 in the area of the upper valve flap, and the same area is shown in Figs. 20 and 21, respectively, for meshes 2 and 3.

The grid independence of the results is verified by comparing the velocity profile on a section downstream of the valve. The coarse grid or mesh 1 indeed provides slightly less approximate results at the point where the maximum speed is reached,

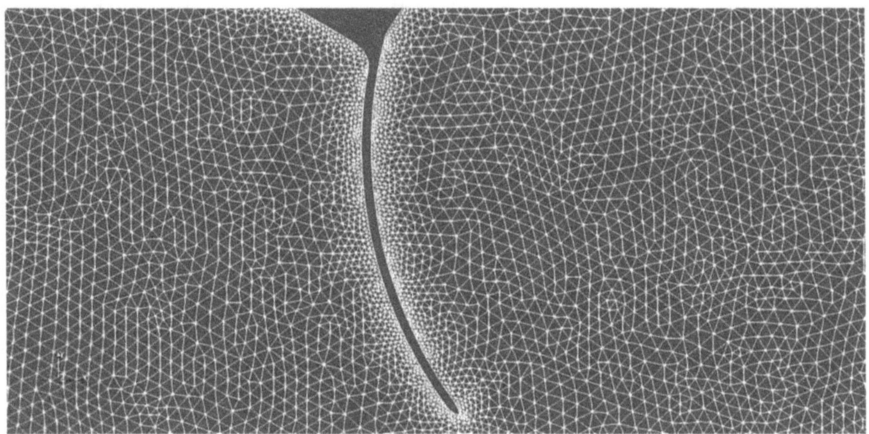

Fig. 19 Coarse grid, or mesh 1, around the upper valve flap, [1]

Fig. 20 Standard grid, or mesh 2, around the upper valve flap, [1]

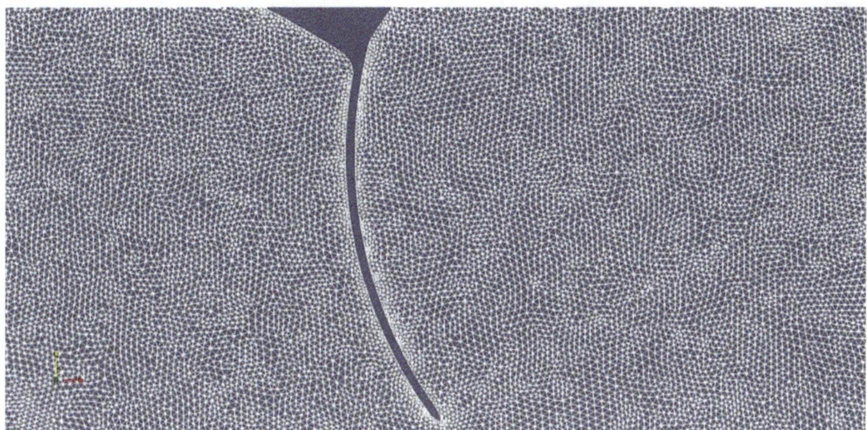

Fig. 21 Fine grid, or mesh 3, around the upper valve flap, [1]

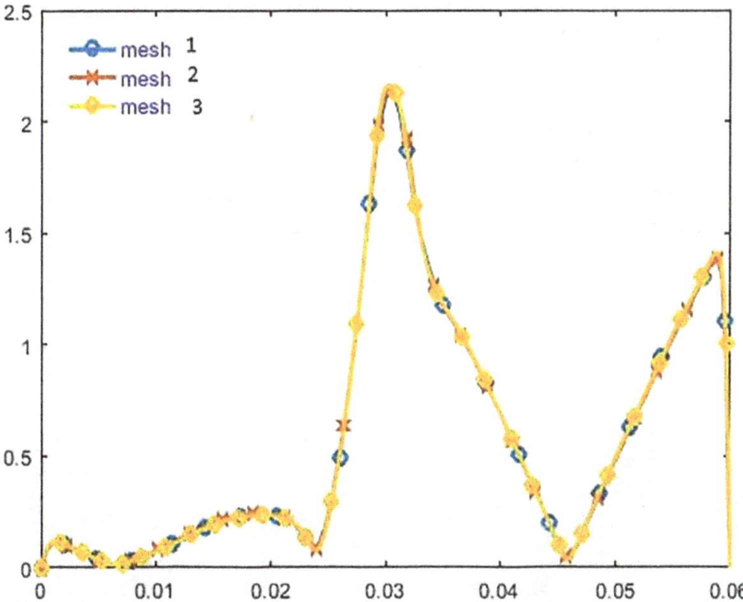

Fig. 22 Velocity profile (m/s) versus $x(m)$ for the three meshes, [1]

while increasing the number of cells of the grid, compared to the standard one or mesh 2, it is noted how the results of mesh 3 also coincide at the maximum point with those of mesh 2. Precisely, at the peak of speed, there is the maximum difference between mesh 1 and 2 and between mesh 2 and mesh 3, with percentage errors, respectively, of 0.6% and 0.3%. In Fig. 22, we report the trend of the velocity profile, in *m/s*, as a function of *x(m)*, on which the grid independence is verified.

5.1.1   The Fluid Domain

The standard grid or mesh 2, used for all simulations, is created with the software *WORKBENCH* 16.0 and is of an unstructured type. The domain is discretised with triangular elements because, given the complexity of the geometry, quadrangular elements would lead to having a mesh with a high *ratio* (maximum ratio between the sides of the cells) and therefore of lower quality from a computational point of view. As previously described, the standard grid or mesh 2 has a number of nodes equal to 116,032, while the fine grid or mesh 3 has a number of cells equal to 227,260. The grid is densified near the surface of the valve flaps, on the walls of the adjacent valve structure and on the remaining edges of the entire domain in order to evaluate more rigorously some characteristic parameters at the wall.

In Fig. 23, an enlargement of the grid around the valve is shown.

5.2   Steady State

In this paragraph, we report the setup and the results of the simulation carried out on the standard grid, or mesh 2, analysing some parameters taken into consideration in other studies on blood fluid dynamics [16, 17].

5.2.1   Simulation Setup

As far as the simulation setup is concerned, a *Pressure − Based* solver is chosen as the velocity considered is definitely subsonic. Being a steady-state problem, a spatial discretisation is necessary and not a temporal one, so in the *General* section of the software used, the *Steady* option is selected. As already mentioned at the beginning of this chapter, a turbulence model *RANS k − ε realisable* is chosen. Also in the solver settings, a *SIMPLE* scheme is chosen as the coupling algorithm between speed and pressure, and the method of resolving spatial variables is set to the second

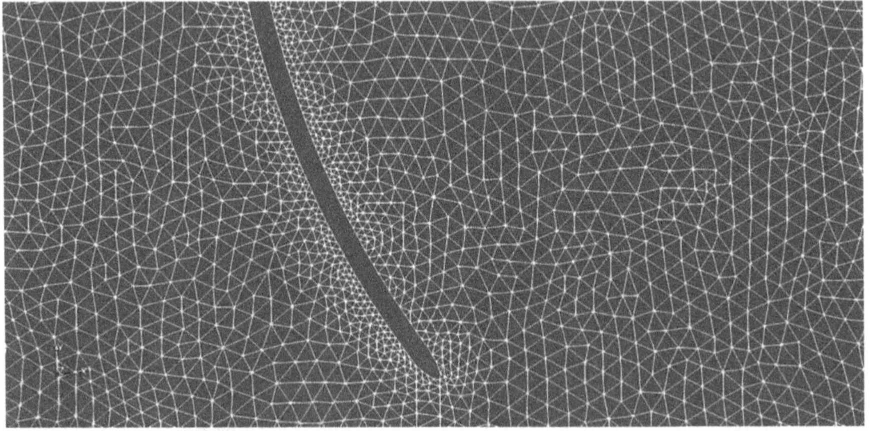

Fig. 23 Enlargement of the standard grid, or mesh 2, around the upper valve flap, [1]

order. The simulation is done by iterating until the residuals reach an order of magnitude equal to 10^{-6}. For the fluid characteristics, reference is made to the software library from which water liquid is taken as the base fluid, which is modified with the parameters of the blood. The density is set to 1060 kg/m^3; for the viscosity, referring to [16], a *User-Defined Function* is implemented with the *Casson* model, which is a function programmed by the user, written in *C* language, that is interpreted by the solver.

5.2.2 List of the Code of User-Defined Function for Casson Model

The list of the code in C language of the *User-Defined Function* used to implement the equations of the non-Newtonian fluid, modelled according to Casson, is reported below.

```
1 # include "udf.h"
2
3 DEFINE_PROPERTY ( cell_viscosity,c,t)
4
5 {
6
7 real mu_C;
8
9 real mu_inf;
10
11 real mu_0;
12
13 real S_0;
14
15 real tau_0;
16
17 real S= C_STRAIN_RATE_MAG (c,t);
18 User Defined Function of the pressure profile 185
19 tau_0 =0.01;
20
21 mu_inf =0.00333;
22
23 mu_0 =100* mu_inf;
24
25 S_0 = ( tau_0 / mu_inf ) /pow (( sqrt ( mu_0 / mu_inf ) -1.0),2.0);
26
27 if (S <=0.0)
28
29 mu_C = mu_inf *pow ((1.0+ sqrt ( tau_0 /( mu_inf *S_0) ) ),2.0);
30
31 else
32
```

```
33 mu_C = mu_inf *pow ((1.0+ sqrt ( tau_0 *(1.0 -exp ( -S/S_0) ) ) /( mu_inf
*S ) ) ),2.0);
34
35 return mu_C;
36
37 }
```

For the boundary conditions, a pressure equal to 2700 Pa is set on the inlet section, *Inlet*, a value slightly higher than the peak of the pressure gradient variable within the cardiac cycle that occurs later in the transient regime simulations. On the outlet section, *Outlet*, a null pressure is set. An adherence condition is imposed on the entire wall of the domain.

5.2.3 Results

Figure 24 shows the static pressure field over the entire fluid domain. It can be seen how the pressure gradient of the transvalvular pressure is very high due to the severe stenosis that obstructs the passage of blood from the ventricle to the aorta during ventricular systole.

From the velocity field shown in Fig. 25, it is noticeable how the fluid dynamics behave like that of a jet of fluid in the same stagnant fluid, also called a submerged jet, [18–34], with a maximum velocity value quite high compared to physiological fields, always as a consequence of the pathology itself.

From the streamlines, reported in Fig. 26, it is noticeable how recirculation zones are created around the sinuses of *Valsalva* and in the first part of the ascending aortic

Fig. 24 Static pressure (Pa) at steady state, [1]

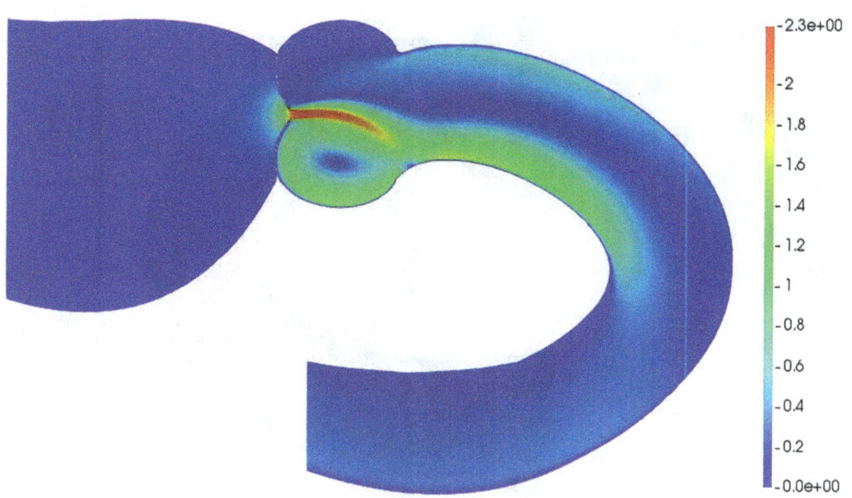

Fig. 25 Velocity field (m/s) at steady state, [1]

Fig. 26 Streamlines (m/s) at steady state, [1]

arch; zones are created where the blood stagnates. It is noteworthy that the jet and the vortices appear quite similar to Fig. 16 of the Introduction.

The apparent viscosity is derived from the equation of the non-Newtonian fluid constitutive model of *Casson*. The apparent viscosity is reported in Fig. 27, divided by the viscosity at infinity $\mu_\infty = 0.0033$ Pa $\cdot$ *s*, [35, 36], implemented in the *User-Defined Function*. In a Casson fluid, the viscosity is inversely proportional to the velocity gradient; therefore, the non-Newtonian behaviour can be significant only in slow flow conditions. In a rigid conduit, the *shear rate* is higher when the diameter of the tube is smaller and vice versa. Therefore, the apparent viscosity is smaller where

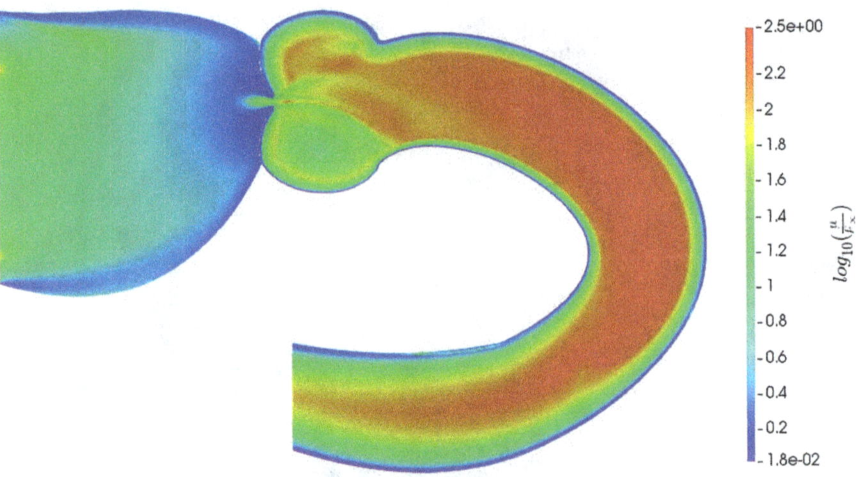

Fig. 27 Apparent viscosity, $\log_{10}(\mu/\mu_\infty)$, of the Casson model, [1]

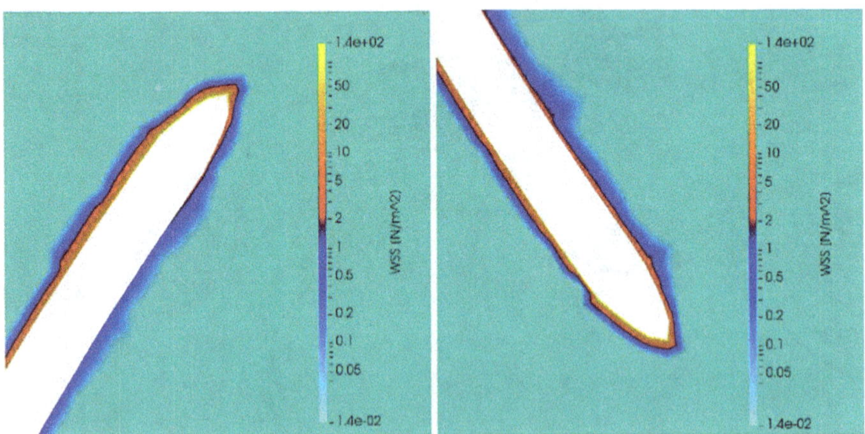

Fig. 28 WSS (N/m^2) on the two valve leaflets, [1]

there is a narrowing of the section due to stenosis and greater when the flow rate is lower.

Of significant importance in the field of blood fluid dynamics is the evaluation of the *wall shear stress* (*WSS*); in this regard, we report in Fig. 28 the values of *WSS*, evaluated at the most significant points, namely on the borders of the valve leaflets.

The wall shear stress is proportional to the velocity gradient; therefore, near the borders of the valve leaflets, affected by severe stenosis, the blood reaches high velocity, and a high velocity gradient is generated, which is going from the valve wall towards the valve lumen. Indeed, at these points, the *WSS* reaches values much higher than the physiological ones, which are the following. The physiological field

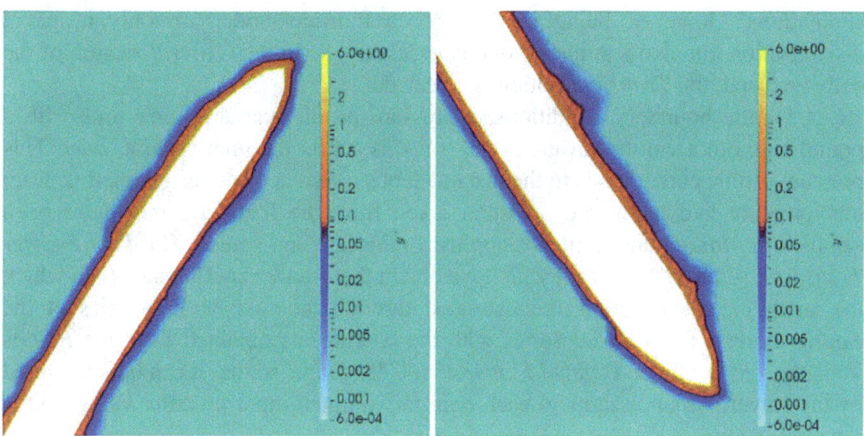

Fig. 29 Values of y+ on the two valve leaflets, [1]

in the arteries is 1–7 Pa; in the veins, 0.1–0.6 Pa [37]; in the ascending aorta, 0.4–0.7 Pa with an average of 0.5 Pa, a field that reaches up to the unit value in the case of aortic bicuspidia [38–40]. The physiological field obtained with 3D-CFD simulations in the ascending aorta, aortic arch and descending aorta is 0.012–2.5 Pa, with values that can also reach 6 Pa in the branches of the aortic arch [41].

To obtain an optimal wall resolution, it is necessary that the dimensionless distance from the wall, defined as

$$y^+ = u_\tau y/\nu \tag{70}$$

is smaller than 5. In Eq. 70, y is the height of the first wall cell, υ is the kinematic viscosity, and u_t is the wall shear stress velocity, defined as

$$u_\tau = \sqrt{\tau_0/\rho} \tag{71}$$

with τ_0 the wall shear stress. This limit increases the quality of the solution at the expense of an increased computational cost.

In Fig. 29, it is possible to note that the value of y^+ is almost everywhere smaller than 5, and only in a small area near the apex of the valve leaflet does it exceed this value.

5.3 Unsteady or Transient State

In this paragraph, we report the setting and the results of the simulations carried out with the standard grid or mesh 2 in the transient or unsteady state. As for the solver settings, the turbulence model and the implementation of the *Casson* constitutive model for blood, they remain those described previously in the steady-state simulations. However, being in this case a transient state, not only a spatial discretisation but also a temporal one is necessary, so in the *General* section of the software used, the *Transient* option is selected.

As for the boundary conditions, a pressure profile, variable over time with a period equal to a whole cardiac cycle, 0.8 s, is set on the inlet section, *Inlet*. This pressure profile corresponds to the instantaneous difference between the left ventricular pressure and the aortic pressure; taken from the literature, they have been sampled in *Matlab,* and a filter is applied to the sampled values. The *Fast Fourier Transform* is performed, from which we obtain the *Fourier* coefficients, with which the analytical waveform is reconstructed through the first 31 harmonics of the *Fourier* series. The *Fourier* series with the coefficients obtained from the *FFT* is implemented in a *User-Defined Function* in *C* language, so that it can be interpreted by the solver, which assigns to each *time step* a given input pressure value, and is reported below.

5.3.1 List of the Code of User-Defined Function for the Pressure Profile

The *User-Defined Function* code used to implement the pressure profile over time is reported below.

```
1 # include "udf.h"
2
3 # ifndef PI
4 # define PI 3.1415926535
5 # endif
6 double pressure ( double T, int narm , double *ccoef , double * scoef ,
real t);
7
8 DEFINE_PROFILE ( pressure_gradient , thread , position )
9 {
10 face_t f;
11 double ccoef [31] = { -11561.58071989 , -4950.48027878
,1940.61057406 , -70.10289338 ,
12 -216.89988345 , -115.48537585 , -190.97027968 ,372.50938443 ,
13 -110.46497433 , -69.22805996 ,18.39929200 ,9.13894232 , User Defined
Function of the pressure profile 186
14 -12.29188814 ,5.93207109 ,25.75520461 , -32.89408178 ,6.47395525 ,
15 -0.66816925 ,6.83284964 , -10.23868550 ,8.38522782 , -1.14154731 ,
16 -4.09749978 ,0.94419399 , 0.88760840 ,0.93914545 , -2.40004841 ,
17 2.53477858 , -2.36558272 , -0.00310458 , -0.99954701};
18
19 double scoef [31] = { -0.00000000 ,3145.21828087 , -1528.63531798
, -297.12470991 ,
20 1373.86415639 , -774.67384441 , -135.06504925 ,225.93776313 ,
21 -63.33398563 ,42.65425934 , -14.91676421 ,2.73799841 ,
22 -51.02192829 ,47.10487953 , -4.19045145 , -8.40009663 ,
```

```
23 5.49603128 ,0.43947157 , -7.54393248 , -1.87199993 ,12.53518335 ,
24 -8.34190342 ,2.58549258 , -0.14808927 ,1.24945476 , -5.28871006 ,
25 4.50171802 ,1.24315644 , -2.31608100 ,0.21507703 ,0.31430267};
26
27 begin_f_loop (f, thread )
28 {
29 real t = CURRENT_TIME;
30
31 F_PROFILE (f, thread , position ) = pressure (0.8 , 30 , ccoef , scoef ,
t) ;
32
33 }
34 end_f_loop (f, thread )
35
36 }
37
38
39 double pressure ( double T, int narm , double *ccoef , double * scoef ,
real t)
40 {
41 double p;
42 int k;
43 p=( ccoef [0]/2) ;
44
45 for (k=1;k <= narm;k++)
46 {
47 User Defined Function of the pressure profile 187
48 p += ( ccoef [k] * cos (2* PI * (k) * ((N -1) /T) *t/N) + scoef [k] * sin
(2* PI * (k) * ((N -1) /T) *t/N) );
49 }
50 return p *1.2248;
51 }
```

In Fig. 30, we report the waveform of the reconstructed pressure profile. The remaining boundary conditions on the outlet section, *Outlet*, and on the domain wall, *Wall*, remain set to the previous values.

A second-order implicit method is used for the solutions in the transient state, in which the simulations are iterated for two cardiac cycles, that is, 1.6 s, with a fixed *time step* of 10^{-5} s. Only the results of the second cycle are recorded, while the first cycle is used to initialise the solution. Despite the time discretisation scheme used being of the implicit type and therefore unconditionally stable, the simulation is still set up to satisfy the condition of *Courant Friedrichs Levy* (CFL), that is, the number of *Courant* must be ≤ 1. The Courant number in two dimensions is defined as

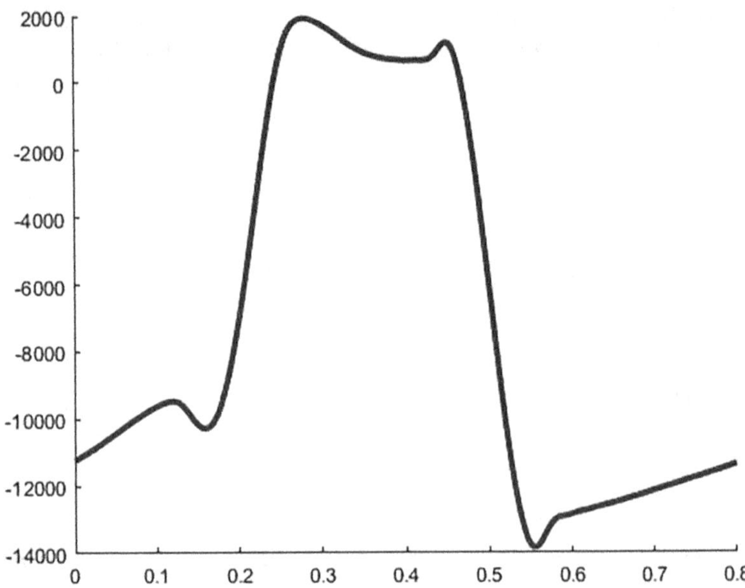

Fig. 30 Pressure profile (Pa) versus time (s) set at the Inlet, [1]

$$C = \frac{U_x \Delta t}{\Delta x} + \frac{U_y \Delta t}{\Delta y} \tag{72}$$

where U_x is the generic velocity component in the cell in the direction of the x, Δt the time step and Δx the size of the cell in the direction parallel to the velocity along x, while U_y is the generic velocity component in the cell in the direction of the y and Δy is the size of the cell in the direction parallel to the velocity along y. Satisfying this condition, even if a temporal discretisation is used, is an indication of the quality of the simulation, as it physically indicates that a fluid particle does not cross more than one grid cell for each *time step*. It is verified that the number of *Courant* respects this condition in all cells and for every time instant.

We report in Fig. 31 the values that this parameter assumes in all cells during the entire cardiac cycle.

We now want to make considerations on the number of *Courant* concerning the simulations just described. Compared to the simplified numerical approach used for the solution of the wave equation in the 2D simulations reported in this section and in the following ones, the discretisation scheme is of the type of implicit time discretisation and therefore unconditionally stable. The condition on the parameter is due to the criterion on the method of *Euler* explicit to ensure that the method is stable and valid for the partial differential equation. When solving the equations of *Navier-Stokes* at finite volumes, the number of *Courant* ≤ 1 applies to the equations of momentum and pressure interconnected. It is a limit on the time increment so that the solution, in addition to being stable, is also convergent; consequently, the wave

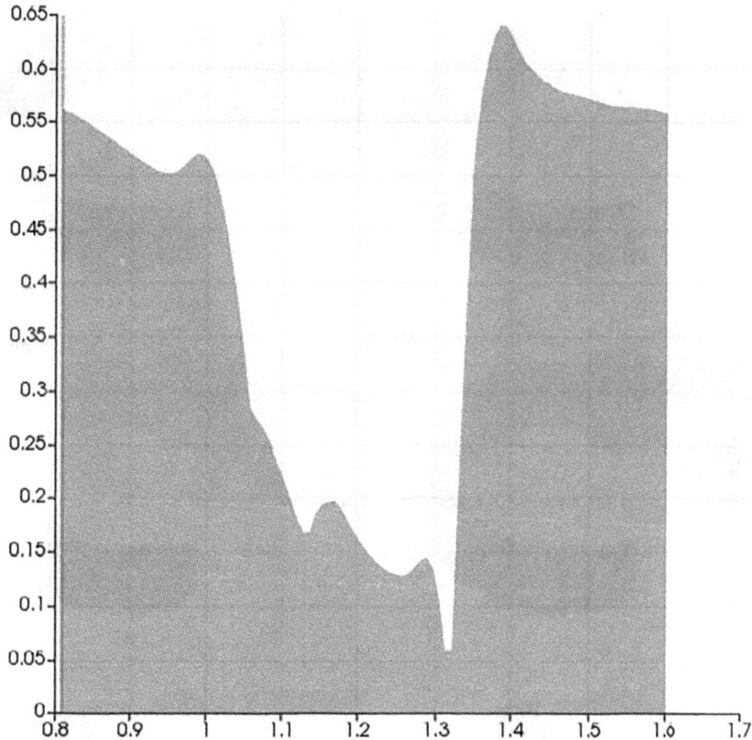

Fig. 31 Courant number on all the cells of the grid versus time (s), [1]

equation, which we solve in a further period of the cardiac cycle with the analogy of *FW-H*, will give rise to a stable and convergent numerical solution.

5.3.2 Results

In Fig. 32, we report the static pressure field over the entire fluid domain at the most significant instants of time. In addition to what was already said in the section on the results of the simulations in steady state, we note how, in most of the cardiac cycle, in the ventricular diastole phase, the transvalvular pressure gradient is very high between the downstream and upstream sections of the valve. From the point of view of a valvular stenosis, in the hypothesis of severe stenosis, the valve fails to close completely due to high rigidity and impairment of the valve flaps. Instead, from another point of view, the failure of the valve to seal in the ventricular diastole phase can be evaluated as the effect of valvular insufficiency, in which the closing valve presents a residual orifice that causes a reflux of blood into the ventricle.

These considerations can also be evidenced from the velocity field shown in Fig. 33, where the reflux of blood into the ventricle during diastole is more markedly noticeable.

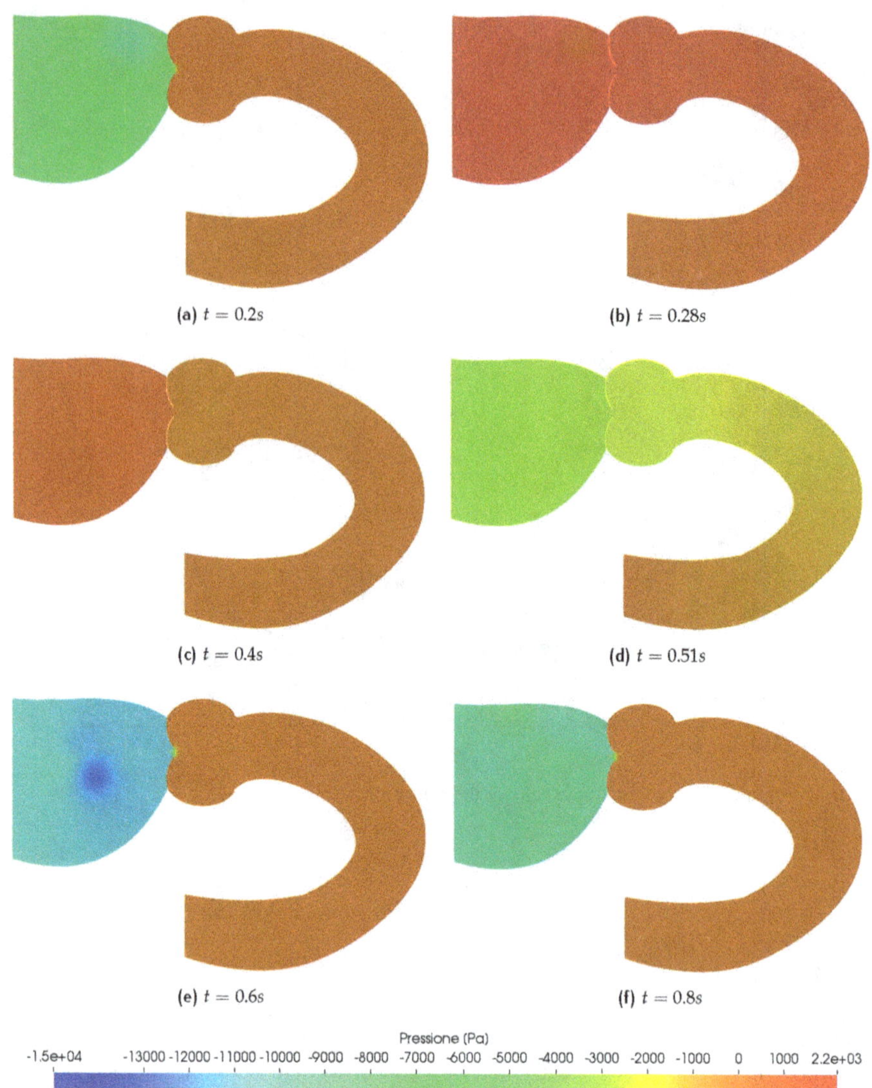

Fig. 32 Static pressure field (Pa) at some instants (s) of the cardiac cycle: from the top, $t = 0.2$ s, $t = 0.28$ s, $t = 0.4$ s, $t = 0.51$ s, $t = 0.6$ s, $t = 0.8$ s, [1]

For completeness, we report in Figs. 34 and 35, respectively, the static pressure and the velocity averaged over the entire cardiac cycle.

As a result of the obstruction to the fluid flow, Fig. 36 shows how, in the different phases of the cardiac cycle, recirculation zones are created near the valve flaps: both in the diastole phase, Fig. 36 *a*, *f*, and in the systole phase, Fig. 36 *d*, *e*, and in the phase of pressure gradient transition, Fig. 36 *c*.

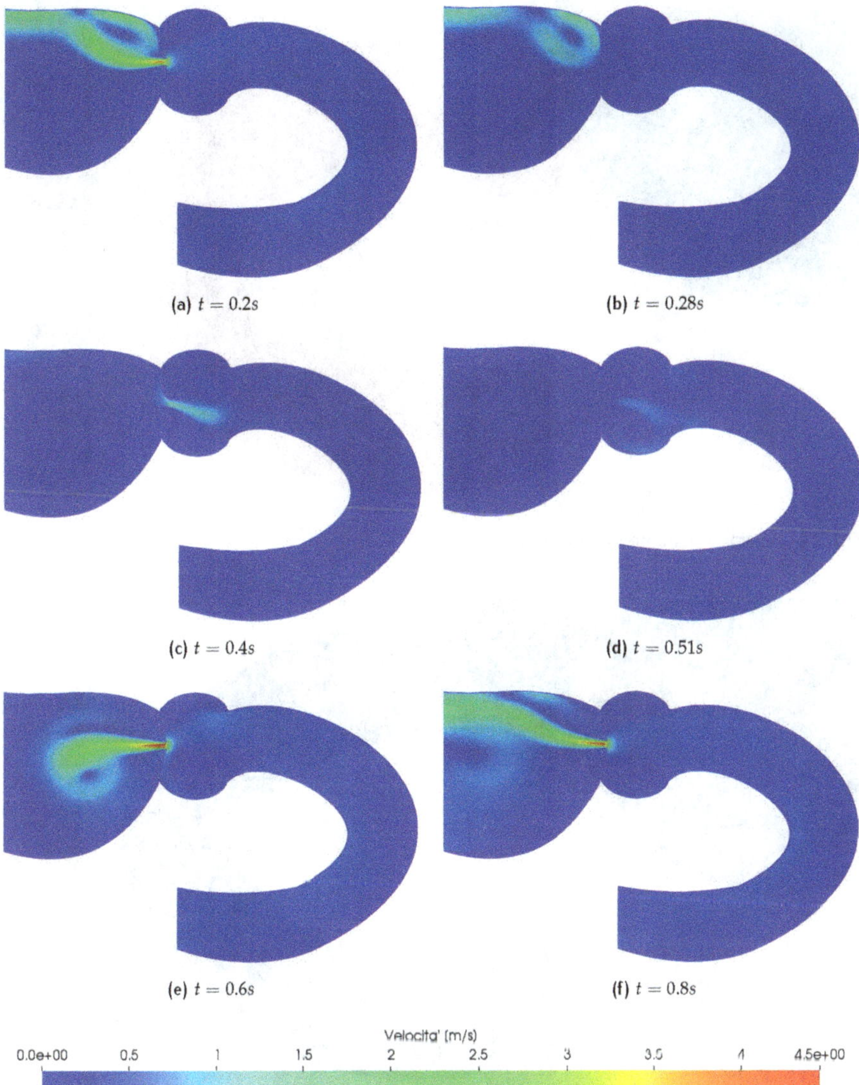

Fig. 33 Velocity field (m/s) at some instants of the cardiac cycle: from the top, $t = 0.2$ s, $t = 0.28$ s, $t = 0.4$ s, $t = 0.51$ s, $t = 0.6$ s, $t = 0.8$ s, [1]

In Fig. 37, we report the field of apparent viscosity, normalised with the viscosity at infinity, at all times of the entire cardiac cycle. It is noticeable how the viscosity is, in first approximation, inversely proportional to the velocity gradient; indeed, the non-Newtonian behaviour is significant only under conditions of slow flow. The apparent viscosity is smaller where the shear rate is greater, that is, on the wall and where there is a narrowing of the section, and it is greater towards the lumen of the duct, in which there is a low velocity gradient. This is in agreement with the

Fig. 34 Average static pressure (Pa), [1]

Fig. 35 Average velocity (m/s), [1]

physiological haemo-dynamics of blood vessels, in which all the corpuscles concentrate in the centre of the duct and the plasma on the walls. In all the figures that report the viscosity field, this characteristic is noticeable, especially in the tract of the aortic arch. These transient results are in agreement with what has already been said about the *Casson* model in the results of the simulations in steady state; there is also a behaviour of the blood fluid similar to that reported in other studies present in the literature, [17], in which this fluid is modelled with the same relationships.

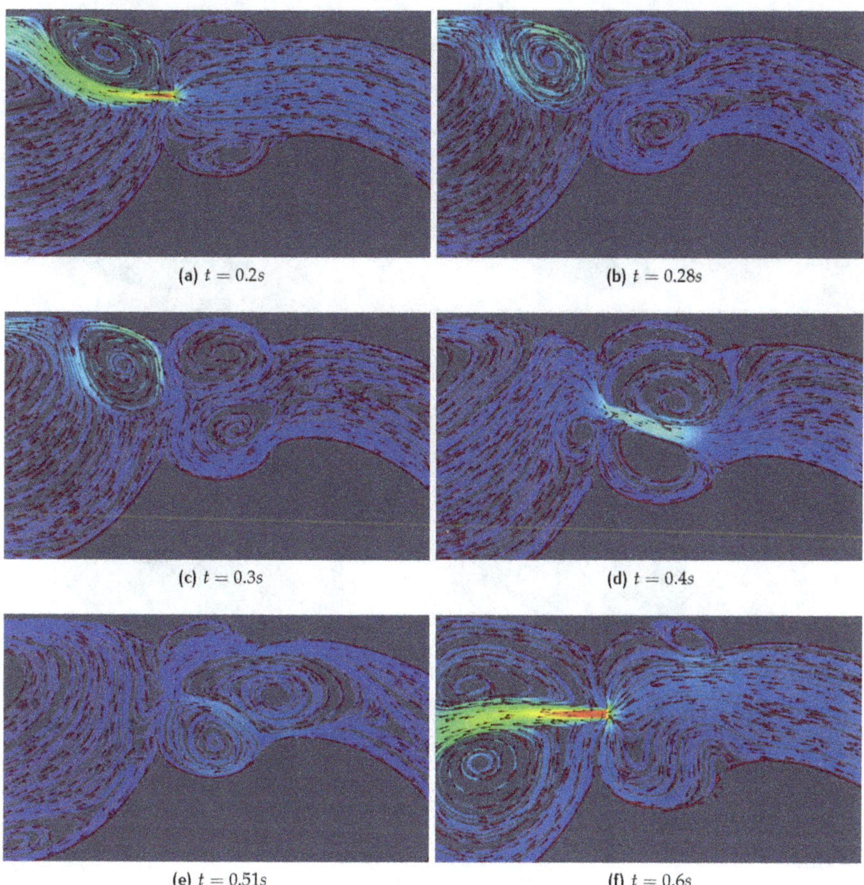

(a) $t = 0.2s$ (b) $t = 0.28s$

(c) $t = 0.3s$ (d) $t = 0.4s$

(e) $t = 0.51s$ (f) $t = 0.6s$

Fig. 36 Streamlines at some instants of the cardiac cycle: from the top, $t = 0.2$ s, $t = 0.28$ s, $t = 0.3$ s, $t = 0.4$ s, $t = 0.51$ s, $t = 0.6$ s, [1]

In Fig. 38, it is noticeable how in the phase of systole and of ventricular diastole, the vorticity is marked in the area where the jet of blood breaks into the stagnant fluid. This is justified by the fact that, around the jet, recirculation zones are created, as can also be seen from the streamlines of Fig. 36.

In Fig. 39, we report the *wall shear stress, WSS,* evaluated on the valve flaps at certain instants of time of the cardiac cycle. It is noted that during the entire cardiac cycle, the valve flaps are subject to WSS values higher than the physiological ones (0.4–0.7 Pa). The physiological field in the ascending aorta has an average of 0.5 Pa; a field that reaches up to the unit value in the case of aortic bicuspidia. The values of 0.012–2.5 Pa, as a physiological field, are obtained with 3D *CFD* simulations in the ascending aorta, aortic arch and descending aorta, with values that can also reach 6 Pa in the branches of the aortic arch. The valve thus modelled, in the systole phase, causes an obstruction to the flow that goes into the aortic arch, consequently,

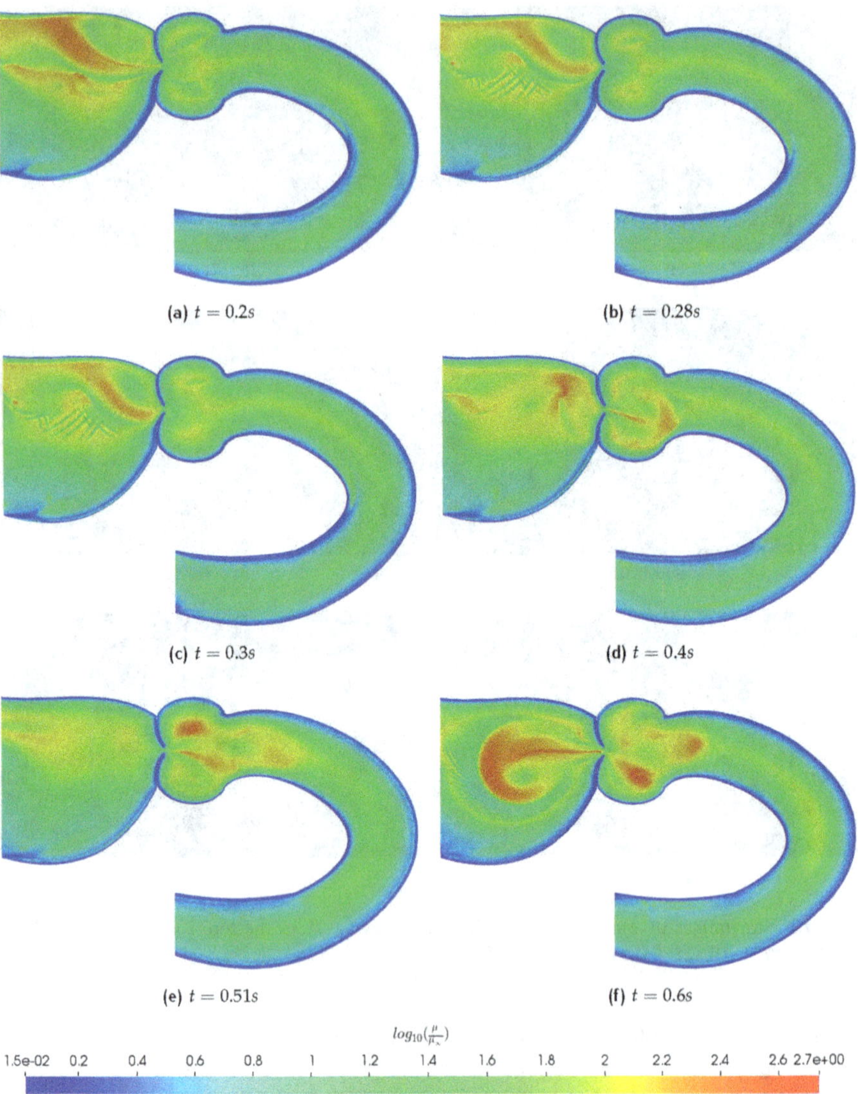

Fig. 37 Apparent viscosity, $\log_{10}(\mu/\mu_0)$, at some instants of the cardiac cycle: from the top, $t = 0.2$ s, $t = 0.28$ s, $t = 0.3$ s, $t = 0.4$ s, $t = 0.51$ s, $t = 0.6$ s, [1]

generating high velocity gradients at the ends of the valve flaps, which lead to high wall shear stress values. If we consider valvular insufficiency, during the diastole phase, a *backflow* of blood from the aortic arch towards the ventricle is generated with an analogous fluid dynamics and consequences.

Figure 40 underlines what has just been said by showing the average value over the entire cardiac cycle of the *wall shear stress*, WSS. As far as an optimal wall

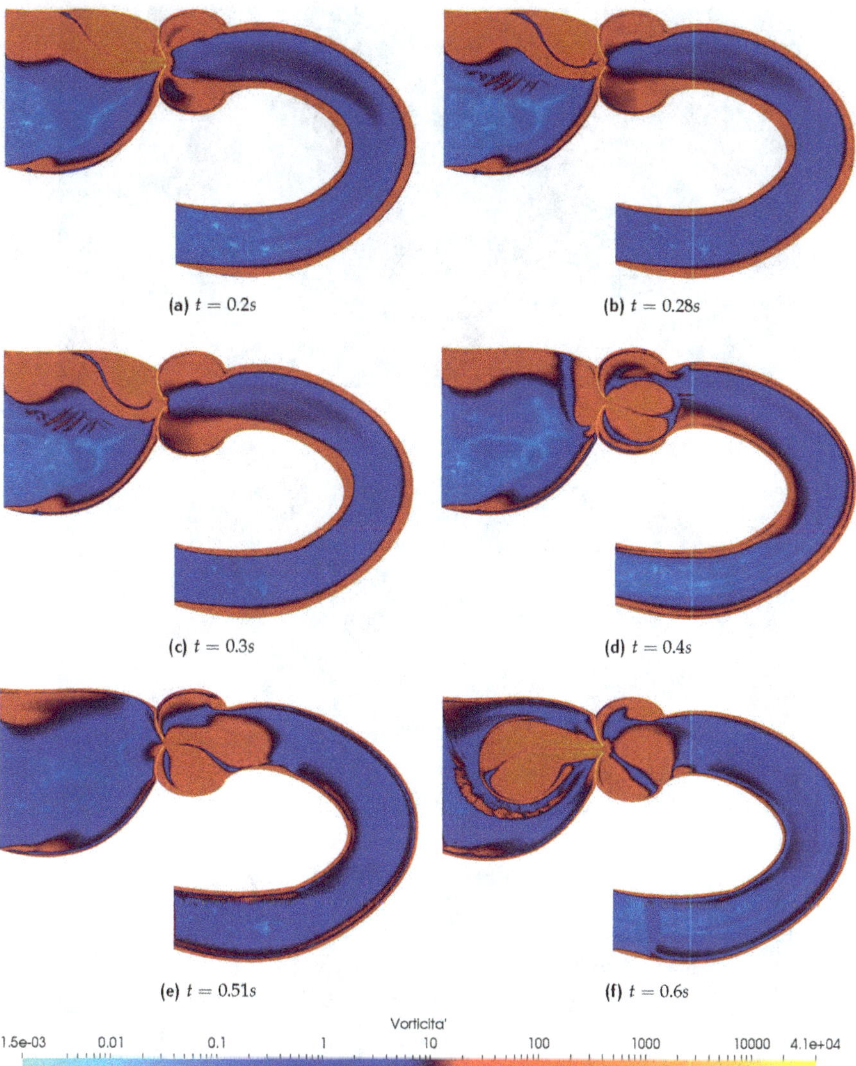

Fig. 38 Vorticity at some instants of the cardiac cycle: from the top, $t = 0.2$ s, $t = 0.28$ s, $t = 0.3$ s, $t = 0.4$ s, $t = 0.51$ s, $t = 0.6$ s, [1]

resolution is concerned, as already discussed in the analysis of the results in steady state, it is necessary that the dimensionless wall distance $y +$ is smaller than 5.

Figure 41 shows that the parameter y^+, evaluated on the wall of the domain, is always strictly smaller than this threshold. On the valve wall, instead, at a few points, there is a value of at most around 7. In order not to further overload the computational cost, already quite onerous, we have accepted the exceeding of this threshold in these points of the valve wall.

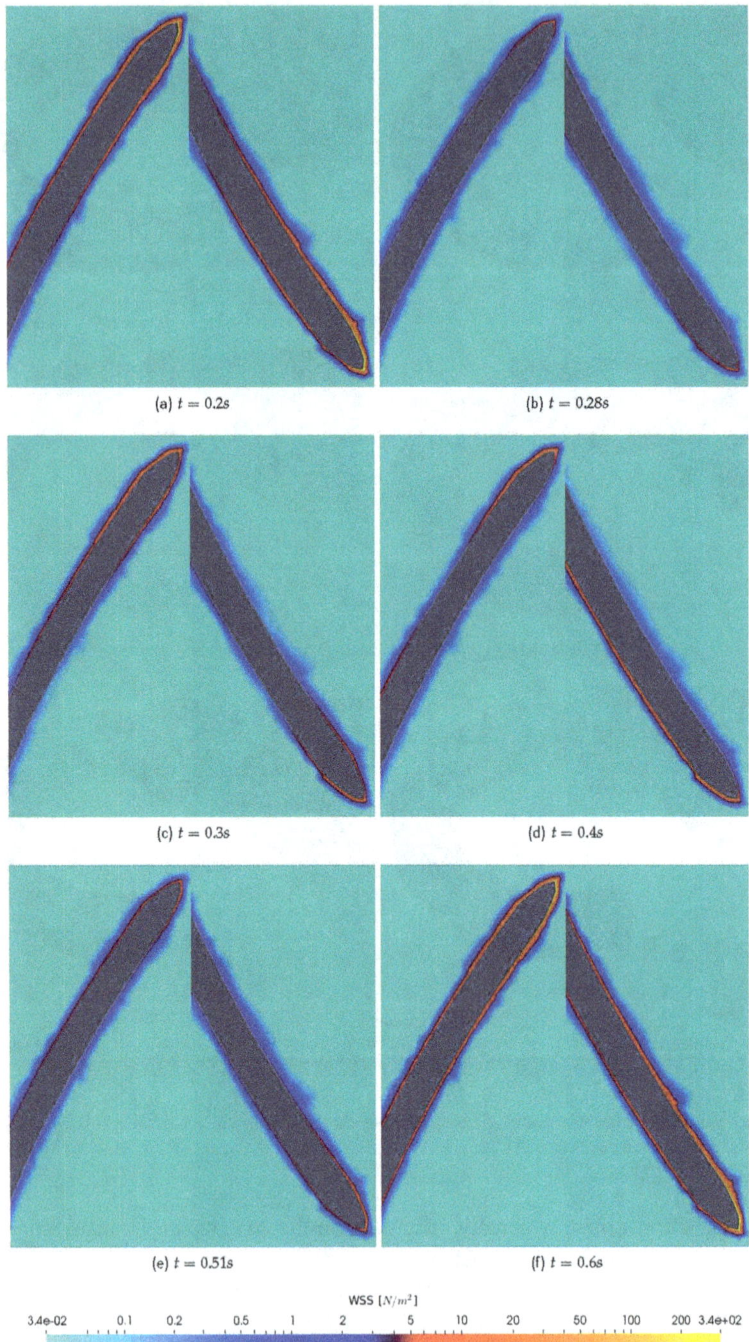

Fig. 39 WSS (N/m^2) on the valve flaps at some instants of the cardiac cycle: from the top, $t = 0.2$ s, $t = 0.28$ s, $t = 0.3$ s, $t = 0.4$ s, $t = 0.51$ s, $t = 0.6$ s, [1]

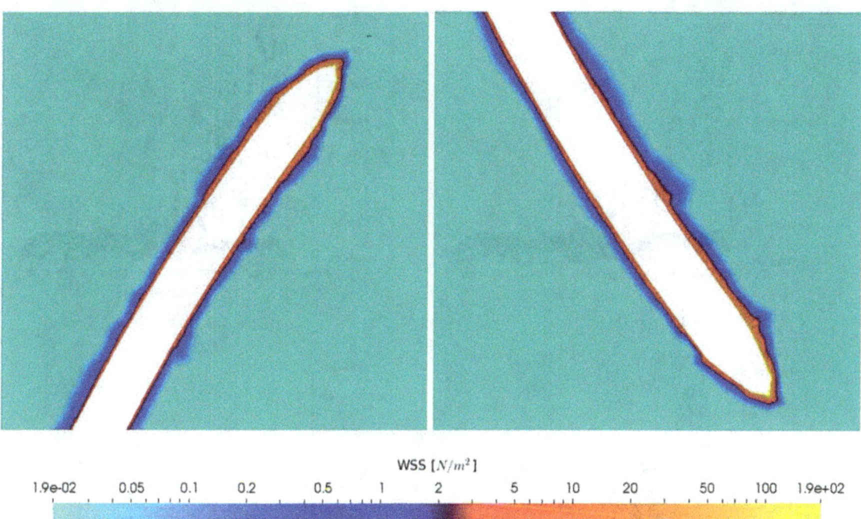

Fig. 40 Average WSS /N/m^2) on the ends of the two valve flaps: (left) inferior, (right) superior, [1]

In the final analysis, we evaluate two parameters related to the *WSS*. The first one is the *Oscillatory Shear Index (OSI)* defined as follows:

$$
\text{OSI} = 0.5 \left(1 - \frac{\left| \int_0^T \overrightarrow{\text{WSS}}\, dt \right|}{\left| \int_0^T \text{WSS}\, dt \right|} \right)
\tag{73}
$$

This parameter gives indications on the portions of the wall where, during the cardiac cycle, there is a reversal in the direction of the *WSS*. Its value can vary from 0, when there is no reversal, to 0.5, when there is a complete reversal. Studies present in the literature, in which this parameter is analysed on some blood vessels, [38], state that in areas where this parameter is high, there is a propensity for the development of wall lesions.

In Fig. 42, it can be seen how both valve flaps are subject to high reversal of direction of the *WSS* throughout the cardiac cycle; this confirms the fact that in these pathologies, the turbulent flow around the stiffened and stenotic valve flaps contributes to further self-increasing the lesions of the same valve flaps, with consequent and further reactions of inflammation and thickening of the ends.

The last parameter on which we focus is the *Relative Residence Time (RRT)*, defined as follows:

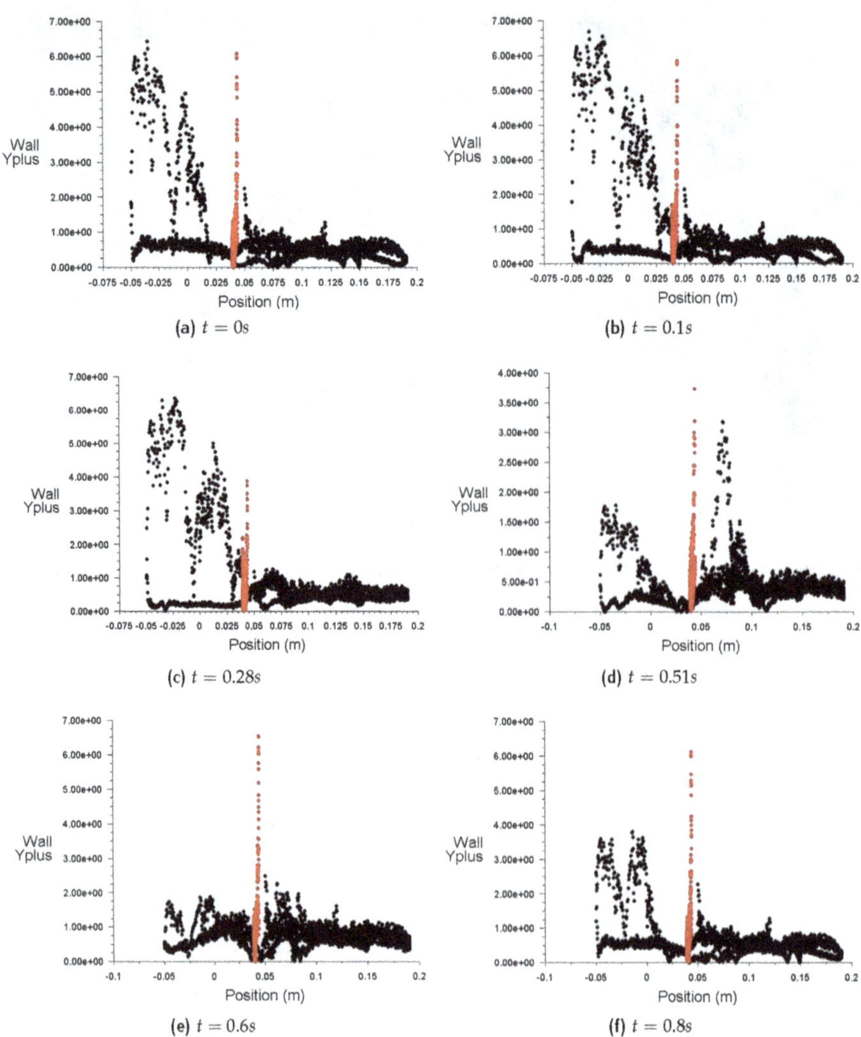

Fig. 41 Values of y^+ (black on wall, red on wall valve) at some instants when they assume maximum values. From the top, $t = 0.0$ s, $t = 0.1$ s, $t = 0.28$ s, $t = 0.51$ s, $t = 0.6$ s, $t = 0.8$ s, [1]

$$RRT = 1/[(1 - 2\, OSI)\, \text{TAWSS}] = 1/\left[\frac{1}{T}\left|\int_0^T \overrightarrow{WSS}\, dt\right|\right] \tag{74}$$

where TAWSS is the modulus of the *WSS* averaged over the entire cardiac cycle,

$$\text{TAWSS} = \frac{1}{T}\int_0^T WSS\, dt \tag{75}$$

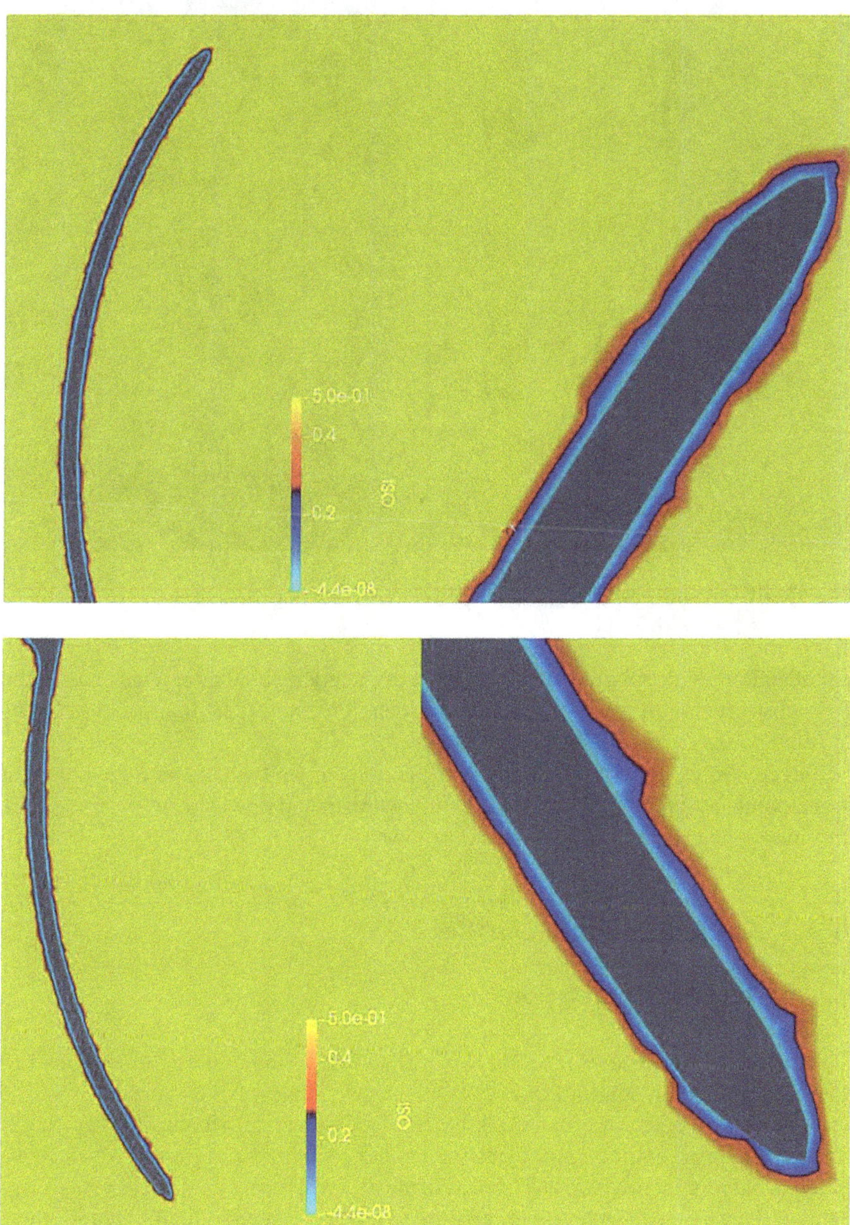

Fig. 42 OSI on the lower (top) and upper (bottom) valve flap, [1]

As reported in the literature, [42], the *RRT* identifies regions subject to stagnant flow. This parameter assumes higher values in the regions of the endothelium that are more prone to cell and macromolecule infiltrations, but it only provides a qualitative assessment of the fluid residence near the wall. In the case under

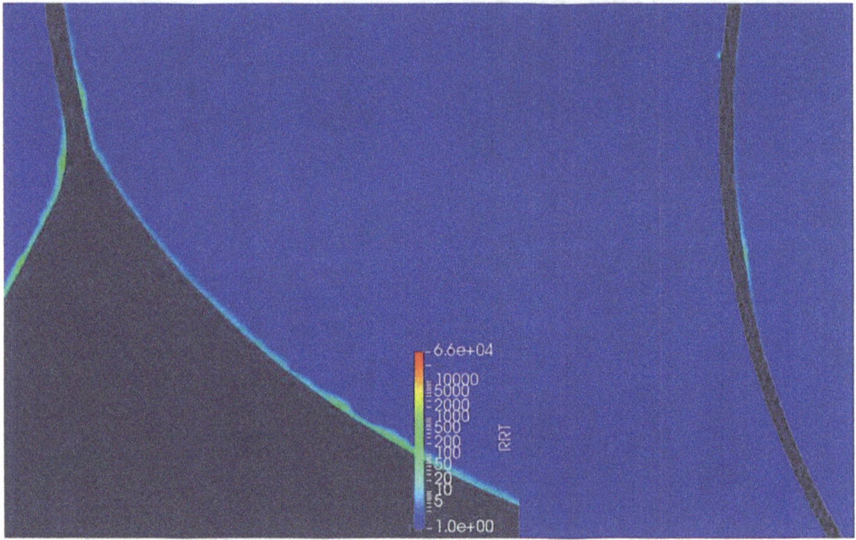

Fig. 43 RRT at some significant points [1]

examination, as already noted earlier, there are many recirculation zones due to the flow obstacle caused by the stenotic valve; therefore, we expect high values of *RRT* at some points on the wall near the valve.

In Fig. 43, we report two portions of the domain wall where we notice how, at some points at the base of the valve, both upstream and downstream of it, the *RRT* assumes quite considerable values.

6   Computational Acoustics

6.1   2D RANS Simulations

The numerical simulations are initialised with a first period of the cardiac cycle, and the fluid dynamics data are extracted at the beginning of the procedure for the evaluation of the acoustic model. After that, the numerical simulations are run in the second period of the cardiac cycle for another cycle, at the beginning of which we activate the formulation of *FW-H*, because the acoustic model uses a mean stationary motion. However, since the input pressure waveform is a periodic function in time, after the first period of numerical simulation, we can expect that the flow field is on average steady state. In the 2D model of the stenotic valve, the fluid dynamic field allows applying this formulation since the geometry with a degree of severe stenosis also falls into aortic valve insufficiency. Therefore, it is possible to apply the periodic pressure waveform without having to consider the opening and the closure of the valve, with the possibility of being able to run the simulations for more periods.

For the settings of the acoustic model, the source surface of *FW-H* is chosen as the surface of the valve alone, *wall valve*, considered separately from the wall of the entire domain, but with the boundary condition of adherence. For the extrapolation of the *acoustic source data,* we choose a frequency of writing the results equal to two *time steps*, and this also allows us not to overload the phase of *post-processing* of the acoustic signals. However, in this way, the highest frequency that the acoustic signals can generate is reduced to $f = \frac{1}{2}\ (2\Delta t)$, which, with a time step $\Delta t = 10^{-5}$ s, is equal to 25 KHz. This upper limit does not impose any restriction, as we know from the literature that the frequencies of pathological heart murmurs range between 50 and 100 Hz.

Finally, the *Far-Field Density* and the *Far-Field Sound Speed,* or the density of the fluid outside the computational domain and the speed of sound waves in the same fluid, which we have set equal to that in water ($1000\ Kg/m^3$), equal to 1500 m/s. The *Reference Acoustic Pressure* is set equal to $2\ 10^{-5}$ Pa. The software code employed assumes that the sound sources are perfectly correlated within the volume, with characteristic length equal to the correlation length and null outside of it. The *software* constructs a volume with characteristic dimensions equal to the specified correlation length, within which it considers the sound sources and neglects those outside.

6.2   Valve with Severe Stenosis (VSS)

We now evaluate the signals obtained from the simulations carried out on the first geometry taken into consideration, namely the model of severely stenotic aortic valve *VSS* with *EOA* equal to $0.7\ cm^2$. In Fig. 44, we report the radiation diagram, obtained by positioning 40 receivers along a circumference around the valve surface. As predicted by the theory of *FW-H* the radiation diagram presents a characteristic dipole shape, although not very pronounced, given the symmetry of the emitting surface. From the numerical simulations, carried out with the turbulence model *RANS*, we calculate the *OASPL* on the individual receivers, and we report the values on a polar diagram. This parameter represents the energy contained in the entire resolved frequency spectrum.

The positions of the receivers placed upstream and downstream of the valve leaflets are reported in Fig. 45.

The acoustic waves perceived by some receivers, over time, placed upstream and downstream of the valve leaflets, are reported in Fig. 46.

6.3   Systolic and Diastolic Acoustic Signals

As seen from the trends of the signals, evaluated on the different receivers, the acoustic waves are obtained by solving the wave equation along one direction rather than another. To take this into account, we have put together all the information obtained from the individual receivers with the aim of comparing our signals with

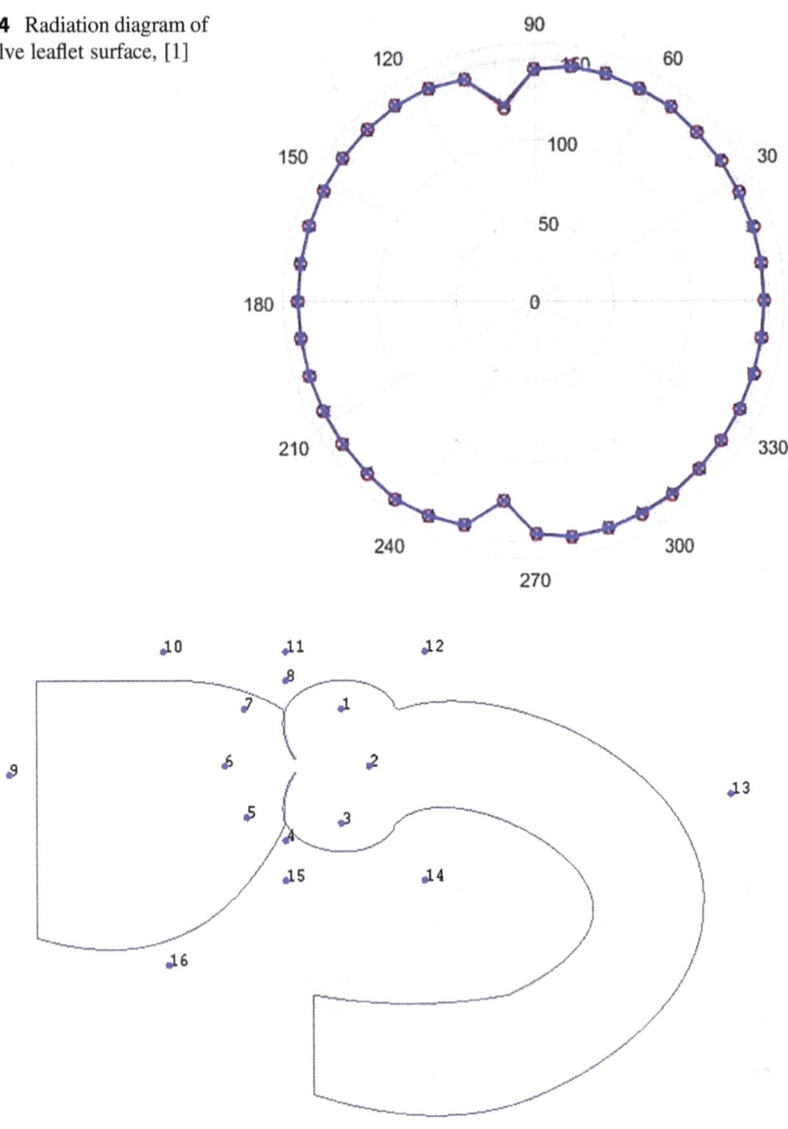

Fig. 44 Radiation diagram of the valve leaflet surface, [1]

Fig. 45 Positions of the receivers 1 ÷ 16, [1]

those reported on pathological clinical phonograms. We sum the acoustic signals, evaluated on the different receivers, and from the resulting signal, we remove the average trend to highlight the oscillations. All this is done for the acoustic signals obtained with the model *VSS*, with *EOA* equal to 0.7 cm^2. The signal obtained over the entire period of the cardiac cycle is divided into the systole and diastole phases to compare this with the heart murmurs present in aortic stenosis or in aortic insufficiency.

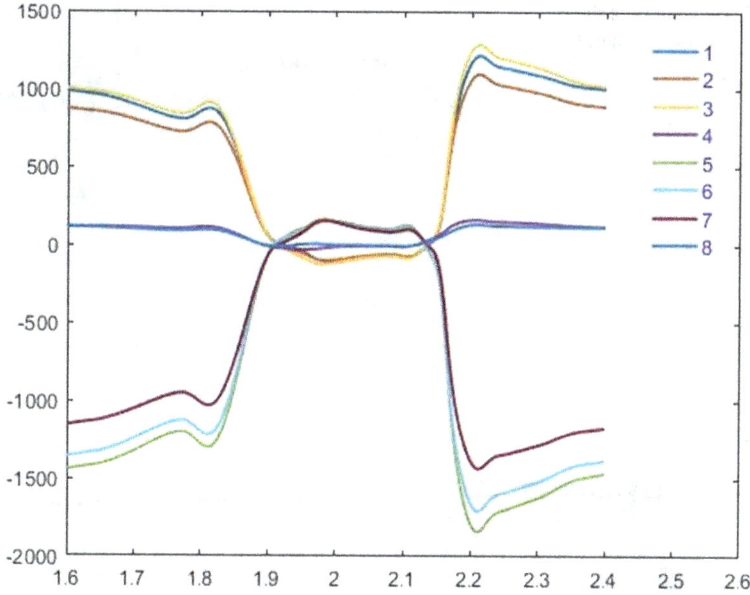

Fig. 46 Sound waves (Pa) versus time (s) on eight receivers around the valve, [1]

The mesosystolic ejection murmur, due to aortic stenosis, and the early diastolic murmur, due to blood regurgitation into the ventricle due to aortic valve insufficiency, are evaluated. In our signals, the physiological tones do not appear due to the opening and closing of the valves, as the model does not take into account such valve dynamics.

As for the systolic murmur, the blood ejected from the ventricle must face a resistance to pass into the aorta, for which an ejection murmur is identified that grows from the beginning of the systole to a mesosystolic peak and then decreases. This is due to the fact that at the beginning, less blood is ejected through the stenotic orifice, while, as the ventricular contraction progresses, the volume of blood that passes through the stenosis is greater and the flow more turbulent, with the accentuation of the murmur. As the ventricular contraction decreases, so does the blood flow and the murmur, as shown in Fig. 47, where a signal is seen between the first and second tone, which corresponds precisely to the systolic murmur.

Now, let us consider a pathological phonocardiogram with a systolic murmur from aortic stenosis, Fig. 48, and compare this with the signals obtained for the geometry *VSS* with *EOA* equal to 0.7 cm^2, Fig. 49.

In both models, a signal is obtained that increases and then decreases, with a typical diamond shape, characteristic of the systolic murmur from aortic stenosis. The murmur, obtained for the geometry with a slightly wider orifice, *VSL,* has more evident oscillations for the same reason discussed in the previous paragraph, in relation to the different evolution of the turbulent flow. We cannot highlight the quantitative characteristics in terms of intensity since we have subtracted the average

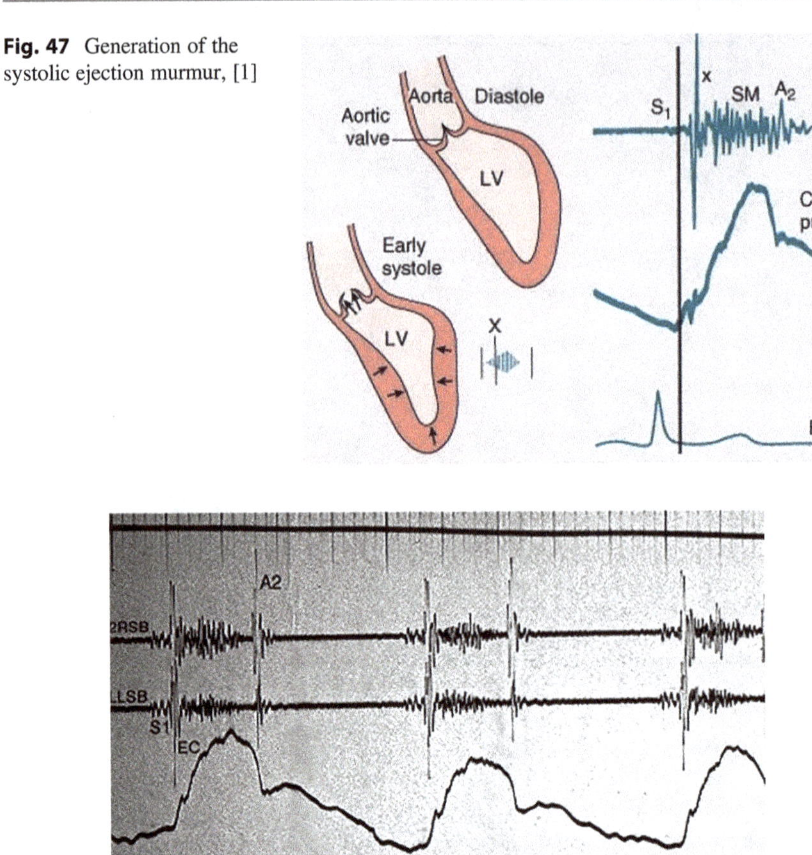

Fig. 47 Generation of the systolic ejection murmur, [1]

Fig. 48 Systolic ejection murmur in pathological phonocardiogram, [1]

trend from the signals, so we qualitatively evaluate the oscillations obtained from the sum of the signals, resolved in the direction of the individual receivers.

We now examine a pathological phonocardiogram, in which, after the second tone, a regurgitation murmur is noted due to aortic valve insufficiency, Fig. 50. We remember that the murmur, at the beginning of diastole, is particularly strong because the pressure difference between this and the aorta is maximum, causing a large regurgitation. As diastole progresses and ventricular filling occur, this pressure difference is minimised, and consequently so is the regurgitation.

In the signals obtained for the geometry *VSS*, with *EOA* equal to 0.7 cm^2 the trend of Fig. 51 is observed, where, at the beginning of diastole, the oscillations are of greater amplitude and then dampen. In this case, too, we can make a qualitative comparison with the pathological phonocardiogram.

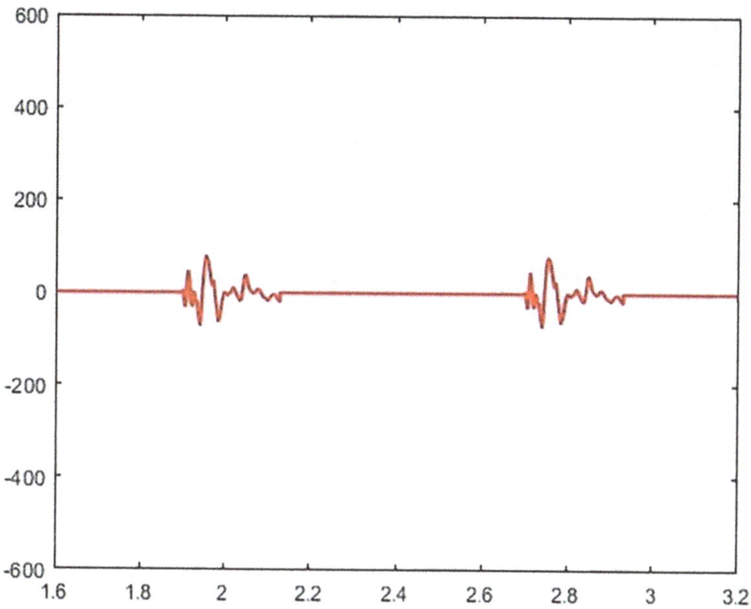

Fig. 49 Systolic ejection murmur: pressure (Pa) versus time (s) in VSS geometry with EOA $= 0.7$ cm^2, [1]

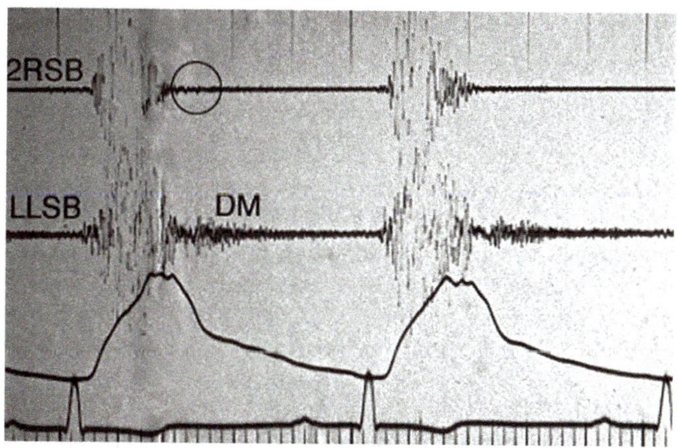

Fig. 50 Diastolic regurgitation murmur in pathological phonocardiogram, [1]

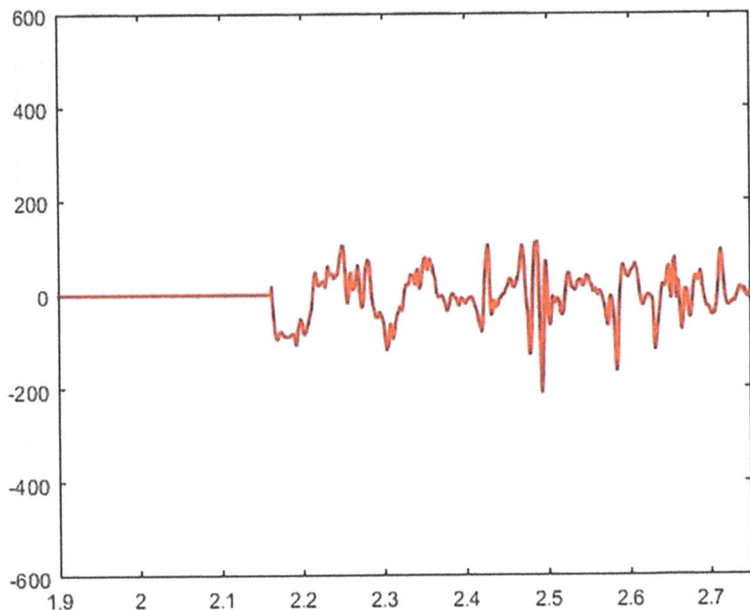

Fig. 51 Diastolic regurgitation murmur: pressure (Pa) versus time (s) in VSS geometry, with $EOA = 0.7 \text{ cm}^2$, [1]

7 Conclusions

An extensive literature search is conducted covering all the different fields that come into play in the physics of heart murmurs, from clinical aspects to computational acoustics, through fluid dynamics and rheological models. The acoustics of heart murmurs are studied using the methodology of *Computational Aero Acoustics (CAA)*, applied to turbulent cardiac *Computational Thermo-Haemo-Dynamics (CTHD)*. In the simulations presented, 2D aortic valve models are employed, developed exclusively for this study, according to appropriate assumptions, which allow us to study computational acoustics in the Haemo-Dynamics of heart valves.

To obtain the fluid flow fields, the turbulence model (*RANS*), implemented in numerical software, is used, while the noise is calculated using the formulation of *FW-H*, modified by *B-F*, as a generalisation of the acoustic analogy of *Lighthill*. The very long computation time of the simulations is one of the main problems encountered; then, simplifying assumptions are made in order to speed up the simulations, which are validated through comparison with results and models reported in other studies of the literature. Much attention is paid to the quality of the simulations, always keeping under control parameters such as the dimensionless coordinate $y+$ and the number of *Courant*.

From the simulations, it emerges that heart murmurs are acoustic signals with a significant low-frequency spectrum, in agreement with medical literature, while high-frequency components become significantly less meaningful. A dependence of the noise on the degree of stenosis emerges, and, in particular, the presence of peaks in the spectra of the signals obtained on the various receivers, with clear differences both between different degrees of stenosis and between signals obtained from the 2D modelling. This is believed to be due to the relationship between the dimensions of the jet and the vortices that develop because a jet with a larger section generates vortices that give rise to oscillations with different frequencies. Similarly, a valve orifice modelled in 3D gives rise to vortices responsible for oscillations with base harmonics, comparable to the 2D, and sub-harmonics of lower intensity, due to the 3D development of anisotropic scales.

Taking into account the signals analysed in the time domain on all the receivers, we reproduce the typical diamond and damped signals, which are evident from the pathological clinical phonograms, respectively, of stenosis and aortic valve insufficiency. These signals are more significant where more oscillations occur, in relation to the size of the jet and the refinement of the grid used to discretise the fluid domain, on which the solution of the perceptible fluctuations depends.

The application of *CAA* to *CTHD* therefore proves to be a useful and valid tool for the analysis of the physics of heart murmurs, but the results, while arousing particular interest, especially in relation to the fact that they represent the first application of computational acoustics to *CTHD*, require further investigation.

Acknowledgements for the Figures The Author thanks the publisher Elsevier for the kind permission to reproduce the images 1 of the Ref. [8]; 6 of the Ref. [10]; 2 of the Ref. [11]; 1, 4, and 5 of the Ref. [12]; and 1 of the Ref. [13].

References

1. Colucci A (2016/2017) Acoustics in cardiovascular thermofluidodynamics. Physiological, stenosis and aortic valve insufficiency cases, Master's Thesis in Medical Engineering (Thermofluidodynamics of Biological Systems), A.A
2. Sherwood L (2015) Human physiology: from cells to systems, 7th edn. Cengage learning
3. Koeppen BM, Stanton BA (2010) Berne & levy, physiology, 7th edn. Elsevier Health Sciences
4. Perrino AC Jr, Scott TR (2008) A practical approach to transesophageal echocardiography. Lippincott Williams & Wilkins
5. McCance K, Huether S (2016) Pathophysiology and elements of general pathology. Edra
6. Pontieri GM (2015) General pathology and general pathophysiology. Piccin
7. Perrino AC Jr, Popescu WM, Skubas NJ (2013) In: Barash PA (ed) Echocardiography, 28, Handbook of clinical anesthesia. Lippincott Williams & Wilkins
8. Garcia D (2006) Analytical modeling of the instantaneous maximal transvalvular pressure gradient in aortic stenosis. J Biomech 39(16):3036–3044
9. Gori F et al (2012) Lezioni di termofluidodinamica (Lectures in Thermo-Fluid-Dynamics). Texmat
10. Garcia D, Pibarot P, Durand LG (2005) Analytical modeling of the instantaneous pressure gradient across the aortic valve. J Biomech 38:1303–1311

11. Arsneault M, Masani N, Magni G, Yao J, Deras L, Pandian N (1998) Variation of anatomic valve area during ejection in patients with valvular aortic stenosis evaluated by two-dimensional echocardiographic planimetry: comparison with traditional doppler data. JACC 32(7):1931–1937

12. Handke M, Heinrichs G, Beyersdorf F, Olschewski M, Bode C, Geibel A (2003) In vivo analysis of aortic valve dynamics by transesophageal 3-dimensional echocardiography with high temporal resolution. J Thorac Cardiovasc Surg 125:1412–1419

13. Beauchesne LM, deKemp R, Chan KL, Burwash IG (2003 Sep) Temporal variation in effective orifice area during ejection in patients with valvular aortic stenosis. J Am Soc Echocardiogr 16(9):958–964

14. De Hart J et al (2000) A two-dimensional fluid–structure interaction model of the aortic valve. J Biomech 33(9):1079–1088

15. De Hart J et al (2003) A three-dimensional computational analysis of fluid–structure interaction in the aortic valve. J Biomech 36(1):103–112

16. Di Venuta I et al (2017) Three-dimensional numerical simulation of a failed coronary stent implant at different degrees of residual stenosis. Part I: Fluid dynamics and shear stress on the vascular wall. Numer Heat Transfer Part A 71(6):638–652

17. Boghi A et al (2017) Three-dimensional numerical simulation of a failed coronary stent implant at different degrees of residual stenosis. Part II: Apparent viscosity and wall permeability. Numer Heat Transfer Part A 71(6):653–665

18. Gori F et al (2002) Fluid dynamics measurements and optical visualization of the evolution of a submerged Slot Jet of Air. In: Proceedings of the 12th international heat transfer conference, Grenoble (France), Heat transfer 2002, vol 2. Editions Scientifique et Medicales, Elsevier SAS, pp 303–308

19. Gori F et al (2003) Fluid dynamics measurements and numerical simulations around a circular cylinder impinged by a submerged slot Jet of air. In: American Society of Mechanical Engineers, Fluids Engineering Division (Publication) FED, vol 259. ASME International Mechanical Engineering Congress, Washington, DC, pp 179–185; 15 November 2003 through 21 November 2003

20. Gori F et al (2003) Fluid dynamics measurements and flow visualizations of a free slot Jet of Air. In: American Society of Mechanical Engineers, Fluids Engineering Division (Publication) FED, vol 259. ASME International Mechanical Engineering Congress, Washington, DC, pp 187–192; 15 November 2003 through 21 November 2003; Code 62622

21. Gori F et al (2007) Shadowgraph visualizations of a submerged free slot jet of air. Int J Heat Technol 25(1):157–164. ISSN: 0392-8764

22. Gori F et al (2012) Preliminary numerical solutions of the evolution of free jets. In: ASME International Mechanical Engineering Congress and Exposition, Proceedings (IMECE), Volume 7, Issue PARTS A, B, C, D. ASME 2012 International Mechanical Engineering Congress and Exposition, IMECE 2012, Houston, pp 463–469; 9 November 2012 through 15 November 2012; Code 100737

23. Gori F et al (2013) Flow evolution of a turbulent submerged two-dimensional rectangular free jet of air. Average Particle Image Velocimetry (PIV) visualizations and measurements. Int J Heat Fluid Flow 44:764–775

24. Gori F et al (2014) Influence of the Reynolds number on the instant flow evolution of a turbulent rectangular free jet of air. Int J Heat Fluid Flow 50:386–401

25. Boghi A et al (2016) Numerical evidence of an undisturbed region of flow in a turbulent rectangular submerged free jet. Numer Heat Transf A Appl 70(1):14–29

26. Angelino M et al (2016) Numerical solution of three-dimensional rectangular submerged jets with evidence of the undisturbed region of flow. Numer Heat Transf A Appl 70(8):815–830

27. Boghi A, Di Venuta I, Gori F (2017) Passive scalar diffusion in the near field region of turbulent rectangular submerged free jets. Int J Heat Mass Transf 112:1017–1031

28. Di Venuta I et al (2018) Numerical simulation of mass transfer and fluid flow evolution of a rectangular free jet of air. Int J Heat Mass Transf 117:235–251

29. Di Venuta I et al (2018) Passive scalar diffusion in three-dimensional turbulent rectangular free jets with numerical evaluation of turbulent Prandtl/Schmidt number. Int Commun Heat Mass Transfer 95:106–115

30. Di Venuta I et al (2019) Flow evolution and mass transfer in a turbulent rectangular free jet of air with small laminar Schmidt number. Int Commun Heat Mass Transfer 107:44–54

31. Petracci I et al (2019) Experiments and numerical simulations of mass transfer and flow evolution in transient rectangular free jet of air. Int Commun Heat Mass Transfer 108:104290

32. Boghi A et al (2019) Large eddy simulation and self-similarity analysis of the momentum spreading in the near field region of turbulent submerged round jets. Int J Heat Fluid Flow 80: 108466

33. Di Venuta I et al (2021) Numerical simulation and self-similarity of the mean mass transfer in turbulent round jets. Int Commun Heat Mass Transfer 122:105146

34. Angelino M et al (2023) Further results on the mean mass transfer and fluid flow in a turbulent round jet. Int Commun Heat Mass Transfer 141:106568

35. Gori F et al (2010) Image-Based computational fluid dynamics in a carotid artery. In: ASME international mechanical engineering congress and exposition, proceedings, vol 2. ASME International Mechanical Engineering Congress and Exposition, IMECE2009, Lake Buena Vista, pp 123–128; 13 November 2009 through 19 November 2009; Code 80879

36. Gori F et al (2010) Three-dimensional numerical simulation of non-Newtonian blood in two coronary stents. In: 14th International Heat Transfer Conference, IHTC 14, vol 1. Washington, DC, pp 109–114; United States; 8 August 2010 through 13 August 2010; Code 89511

37. Papaioannou T (2005) Stefanadis, christodoulos, vascular wall shear stress: basic principles and methods. Hellenic J Cardiol 46:9–15

38. Meierhofer C, Schneider EP, Lyko C, Hutter A, Martinoff S, Markl M, Hager A, Hess J, Stern H, Fratz S (2013) Wall shear stress and flow patterns in the ascending aorta in patients with bicuspid aortic valves differ significantly from tricuspid aortic valves: a prospective study. Eur Heart J Cardiovasc Imaging 14:797–804

39. Barker AJ, Markl M, Bürk J, Lorenz R, Bock J, Bauer S, Schulz-Menger J, von Knobelsdorff-Brenkenhoff F (2012) Bicuspid aortic valve is associated with altered wall shear stress in the ascending aortaclinical perspective. Circ Cardiovasc Imaging 5:457–466

40. Hope MD, Hope TA, Crook SES, Ordovas KG, Urbania TH, Alley MT, Higgins CB (2011) 4d flow CMR in assessment of valve-related ascending aortic disease. JACC: Cardiovascular Imaging 4:781–787

41. Caballero AD, Laín S (2013) A review on computational fluid dynamics modelling in human thoracic aorta. Cardiovasc Eng Technol 4(2):103–130

42. Murray CD (1926) The physiological principle of minimum work. I. The vascular system and the cost of blood volume. Proc Nat Acad Sci 12(3):207–214

Alex Colucci was born in Velletri on 26 May 1991. He attended the University of Rome Tor Vergata, where he obtained both a Bachelor's and a Master's degree in Medical Engineering. During his Master's degree, he attended the course held by Prof. Fabio Gori on Thermo-fluid Dynamics of Biological Systems, studying and applying turbulence and computational aero-acoustics models to the blood flow in pathological heart valves in his Master's degree thesis, titled "Acoustics in cardiac thermofluid dynamics. Physiological cases, stenosis and aortic valve insufficiency." After graduating, he worked for a year as an industrial engineer, dealing with fluid structure interaction studies between air and large buildings, as well as heat transfer studies and thermo-fluid dynamic simulations in the field of industrial and civil constructions.

Since 2019, he has worked at AbbVie, a multinational company in the field of industrialisation of new technologies and production processes with high technological content in the pharmaceutical sector, dealing with pharmaceutical products and products combined with devices and medical devices. In 2024, Alex joined the pharmaceutical production supervision team for the packaging department, and by the end of the same year, he joined the global packaging engineering team, supporting both the Italian site and other sites in Europe and America in the introduction of new technologies and new production processes.

Fabio Gori Ammannati was born in Montale (Pistoia) on 5 August 1947 to Emilio Gori and Cesarina Ammannati. On 16 July 2021, he added his mother's surname, Ammannati. He graduated in Chemical Engineering on 11 November 1971 with honours. In December 1971, he won a scholarship for young graduates at the Faculty of Engineering in Bologna, which he undertook from January 1972. In 1974, he became assistant professor and, in 1982, associate professor of Technical Physics at the Faculty of Engineering of the University of Florence, where he taught, also as a contract professor, until 1990. In 1978, he won a scholarship from the *British Council,* which he carried out at the *Imperial College* in London, where he collaborated with *Prof. D. Brian Spalding* on the topic of numerical analysis in the turbulent flow of liquid metals. During his time at the University of Florence, he established international scientific collaborations with *Prof. R. Echigo, Tokyo Institute of Technology, Tokyo, Japan; Prof. D.R. Chaudhary, Department of Physics, University of Rajasthan, Jaipur, India;* International Centre for Theoretical Physics *(ICTP)* in Trieste; *Centralny Osrodek Techniki Medycznej, Warsaw, Poland and Prof. T. Aihara, Institute of Fluid Science, Sendai, Japan.* In 1986, he won a CNR scholarship, which he carried out at *Cornell University, Ithaca, New York,* where he collaborated with *Prof. R. Miller* on the topic of ground freezing. In the same year, he became professor in the Faculty of Engineering at the University of Reggio Calabria. In 1988, he was appointed as *professor* at the *University of New York at Stony Brook,* teaching the course, *Introduction to Fluid Dynamics,* in the autumn semester of the same year. During his stay, he collaborated with *Prof. T.F. Irvine Jr.* on the method of measuring thermal conductivity with a thermal probe and on the measurement of the isobaric thermal expansion coefficient of non-Newtonian fluids. In 1990, he moved to the Milano Polytechnic, where he taught Technical Physics and Systems until 1993. From 1991 until 1998, he was an adjunct professor at the Faculty of Engineering of the University of Siena. From 1992 to 2017, he was a professor of Technical Physics in the Faculty of Engineering at the University of Rome Tor Vergata, and from 2017, he is a contract professor. In 1994, he proposed the Research Doctorate (Italian *Ph.D.*) in Energy-Environment Engineering, of which he was the coordinator until 2011. In 1995, he proposed and directed the second-level Master's in Thermo-fluid-dynamics until 2016. From 1998 to 2001, he was the director of the Department of Mechanical Engineering. In 2001, he proposed and coordinated the programme, *Master of Science, Energy Engineering and Thermal and Fluid Dynamics,* in collaboration with the *Department of Mechanical Engineering, College of Engineering, University of Illinois at Chicago, USA,* which takes place at the University of Rome Tor Vergata but awards the title of *Master of Science* from the same American University. Since the late 90s, he has been a coordinator of the joint *Ph.D.* research programme with *Prof. R.J. Goldstein, Department of Mechanical Engineering, University of Minnesota, Minneapolis, USA,* and with *Prof. J.P. Hartnett, Prof. L. Kennedy, Prof. W. Minkowycz and Prof. W.M. Worek, Department of Mechanical Engineering, College of Engineering, University of Illinois at Chicago, Chicago, USA.* Since the mid-90s, he has proposed and directed the *Socrates–Erasmus* programme with *Prof. Mayinger, Technical University of Munich, Germany; Prof. van Steenhoven, Eindhoven Technical University, the Netherlands and Prof. C. Caro, Imperial College of Science, Technology and Medicine.* During his stay at Tor Vergata, he established scientific collaborations with the *Department of Mechanical Engineering, University of Minnesota, Minneapolis, USA,* collaborating with *Prof. R.J. Goldstein* on Thermo-fluid-dynamics and mass transport in gas turbine blades; the *Energy Resources Center, University of Illinois at Chicago,* collaborating with *Prof. J.P. Hartnett, Prof. W. Minkowycz* and *Prof. W.M. Worek*; the *Department of Mechanical Engineering, College of Engineering, University of Illinois at Chicago, collaborating with Prof. L. Kennedy and Prof. W.M. Worek; Duke University, collaborating with Prof. A. Bejan* and *S. Mary's University, Halifax, Nova Scotia, Canada, collaborating with Prof. W. R. Tarnawski.* The bibliographic review of documents and citations, titled *PlosBiology Career,* places him within the top 2% of researchers in the field of *Mechanical Engineering and Transports.* Since 1992, he has been a tutor for over 30 PhD theses, supervisor for over 40 second-level university Master's theses, supervisor for over 70 degree theses (five-year, Specialist and Master's), supervisor for over ten *Master's degree in Mechanical Engineering* from the *University of Illinois at Chicago* and supervisor for over 20 theses *Erasmus-Socrates.*

Since the early 2000s, he has been a reviewer of international research projects on behalf of *Portuguese Science and Technology Foundation (FCT); Czech Science Foundation (GACR); European Science Foundation, Strasbourg, France,* on behalf of *FCT* and the *Shota Rustaveli National Science Foundation, Georgia.* Since 2017, the year of his retirement due to age limits, he has been a contract professor at the University of Rome Tor Vergata, where in the academic year 2024–2025, he has taught Technical Physics for the Master's Degree Course in Medical Engineering. M.D. Salvatore Mangiafico (Florence), scientific head of the NeuroVascular Base Camp, 2021, invited him to give a scientific presentation titled *"The engineering approach to the study of the Circle of Willis"* on 23 September 2021 at the Congress Centre, University of Rome La Sapienza, thus opening up a possible future collaboration.

GPSR Compliance
The European Union's (EU) General Product Safety Regulation (GPSR) is a set
of rules that requires consumer products to be safe and our obligations to
ensure this.

If you have any concerns about our products, you can contact us on

ProductSafety@springernature.com

In case Publisher is established outside the EU, the EU authorized
representative is:

Springer Nature Customer Service Center GmbH
Europaplatz 3
69115 Heidelberg, Germany